Elliptic Equations in Polyhedral Domains

Mathematical
Surveys
and
Monographs

Volume 162

Elliptic Equations in Polyhedral Domains

Vladimir Maz′ya
Jürgen Rossmann

American Mathematical Society
Providence, Rhode Island

2010 *Mathematics Subject Classification.* Primary 35J57, 35J58, 35J25, 35J40, 35J08, 35J05, 35Q30.

For additional information and updates on this book, visit
www.ams.org/bookpages/surv-162

Library of Congress Cataloging-in-Publication Data

Maz'ia, V. G.
Elliptic equations in polyhedral domains / Vladimir Maz'ya, Jürgen Rossmann.
p. cm. — (Mathematical surveys and monographs ; v. 162)
Includes bibliographical references and index.
ISBN 978-0-8218-4983-5 (alk. paper)
1. Differential equations, Elliptic. 2. Polyhedra—Models. 3. Boundary value problems. I. Rossmann, J. (Jürgen), 1954– II. Title.

QA377.M296 2010
515'.3533—dc22

2009053203

∞ The paper used in this book is acid-free and falls within the guidelines
established to ensure permanence and durability.
Visit the AMS home page at http://www.ams.org/

10 9 8 7 6 5 4 3 2 1   15 14 13 12 11 10

Contents

Introduction

Motivation, subject and method. Stationary phenomena in structural mechanics and physics can be frequently modelled by solutions of elliptic equations in domains with edges and vertices on the boundary. It is known that standard facts of classical elliptic theory (normal solvability, smoothness of solutions etc.) fail for these domains, because the solutions acquire singularities even if the coefficients and other function data are regular. This stimulated the development of a theory of elliptic boundary value problems in polyhedral domains during the last 40 years. A variant of such a theory is presented in this book together with applications to particular boundary value problems of mathematical physics. The results obtained belong to the following four areas:

- Pointwise estimates for the Green and Poisson kernels
- Solvability in weighted and nonweighted Sobolev spaces
- Solvability in weighted and nonweighted Hölder spaces
- Miranda-Agmon type maximum principles

The inclusion of the L_p and Hölder scales is useful for applications to nonlinear problems and one of them, a mixed boundary value problem for the Navier-Stokes system in a polyhedral domain, is treated in detail in the third part of the book.

Dealing with polyhedral domains, one is able to study derivatives of solutions of arbitrarily high order. The weights in the norms of the above mentioned function spaces are products of powers of distances from a point to the vertices and edges. The use of such weights is natural because it enables one to control the singularities of solutions and their derivatives in a very efficient way.

In short, the approach to boundary value problems employed here is as follows. We pass subsequently from simpler geometrical configurations to more complicated ones, beginning with dihedral angles, then turning to polyhedral cones, and ending up with domains diffeomorphic to polyhedra. We systematically start with a variational statement of the problem and investigate the dependence of regularity of its weak solutions on the regularity of the data using only L_2 Sobolev type norms. This is attained by more or less standard techniques based on localization and the use of the Mellin and Fourier transforms. With sharp L_2 estimates at hand, we evaluate the kernels of the inverse integral operators of boundary value problems, so that the majorants depend explicitly on the position of the arguments of these kernels with respect to edges and vertices. This information about Green's and Poisson's kernels is a powerful tool leading to the aforementioned L_p and C^α estimates for derivatives of the solutions, both local and global.

Although the approach just described is applicable to quite general elliptic boundary value problems, we implement it exclusively for the Dirichlet, Neumann and certain mixed boundary operators. This restriction is made in order to state solvability conditions explicitly, without referring to requirements of triviality of kernels and cokernels of auxiliary operators (operator Fourier-Mellin symbols) arising inevitably in the treatment of arbitrary boundary value problems.

Illustrative examples. As a rule, we demonstrate applications of general results to special geometric configurations. To give an idea of the level of understanding which the analytic machinery developed in the book allows to achieve, we quote Section 8.3, where the Neumann problem for the equation

$$-\Delta u = f$$

with zero Neumann data on the boundary of a convex bounded three-dimensional polyhedron $\mathcal{G}$ is considered. We show that, for every $p \geq 2$, this problem has a unique (up to a constant term) solution in the Sobolev space $W^{1,p}(\mathcal{G})$ provided f is a distribution in the dual space $(W^{1,p'}(\mathcal{G}))^*$ orthogonal to 1.

Another typical result is the following assertion on the regularity of solutions to the above Neumann problem in a convex polyhedron. Let $\{\mathcal{O}\}$ be the collection of all vertices and let $\{U_{\mathcal{O}}\}$ be an open finite covering of $\overline{\mathcal{G}}$ such that $\mathcal{O}$ is the only vertex in $U_{\mathcal{O}}$. Let also $\{E\}$ be the collection of all edges and let α_E denote the opening of the dihedral angle with edge E, $0 < \alpha_E < \pi$. The notation $r_E(x)$ stands for the distance between $x \in U_{\mathcal{O}}$ and the edge E such that $\mathcal{O} \in \overline{E}$.

With every vertex $\mathcal{O}$ and edge E we associate real numbers $\beta_{\mathcal{O}}$ and δ_E, and we introduce the weighted L_p-norm

$$\|v\|_{L_p(\mathcal{G};\{\beta_{\mathcal{O}}\},\{\delta_E\})} := \Big(\sum_{\{\mathcal{O}\}} \int_{U_{\mathcal{O}}} |x-\mathcal{O}|^{p\beta_{\mathcal{O}}} \prod_{\{E:\mathcal{O}\in\overline{E}\}} \Big(\frac{r_E(x)}{|x-\mathcal{O}|}\Big)^{p\delta_E} |v(x)|^p dx\Big)^{1/p},$$

where $1 < p < \infty$. Under the conditions

$$\begin{aligned} 3/p' &> \beta_{\mathcal{O}} > -2 + 3/p', \\ 2/p' &> \delta_E > -\min\{2,\, \pi/\alpha_E\} + 2/p', \end{aligned}$$

the inclusion of the function f in $L_p(\mathcal{G};\{\beta_{\mathcal{O}}\},\{\delta_E\})$ implies the unique solvability of the Neumann problem in the class of functions with all derivatives of the second order belonging to $L_p(\Omega;\{\beta_{\mathcal{O}}\},\{\delta_E\})$. An important particular case when all $\beta_{\mathcal{O}}$ and δ_E vanish, i.e. when we deal with a standard Sobolev space $W^{2,p}(\Omega)$, is also included here. To be more precise, if

$$1 < p < \min\Big\{3,\, \frac{2\alpha_E}{(2\alpha_E - \pi)_+}\Big\} \tag{1}$$

for all edges E, then the inverse operator of the Neumann problem:

$$L_p(\mathcal{G}) \ni f \to u \in W^{2,p}(\mathcal{G})$$

is continuous whatever the convex polyhedron $\mathcal{G} \subset \mathbb{R}^3$ is. The bounds for p in (1) are sharp for the class of all convex polyhedra.

A deeper illustrative example can be found in Section 11.3, where the Navier-Stokes system

$$-\nu\,\Delta u + (u \cdot \nabla)\,u + \nabla p = f, \qquad \nabla \cdot u = 0$$

with the zero Dirichlet condition is considered in the complements of five regular Plato's polyhedra. We prove in particular that the variational solution (u, p) belongs to the Cartesian product $W^{2,s} \times W^{1,s}$ of Sobolev spaces in a neighborhood of all vertices and edges provided $f \in L_s$, $s > 1$, where the best possible values for s are as follows (with all digits shown correct):

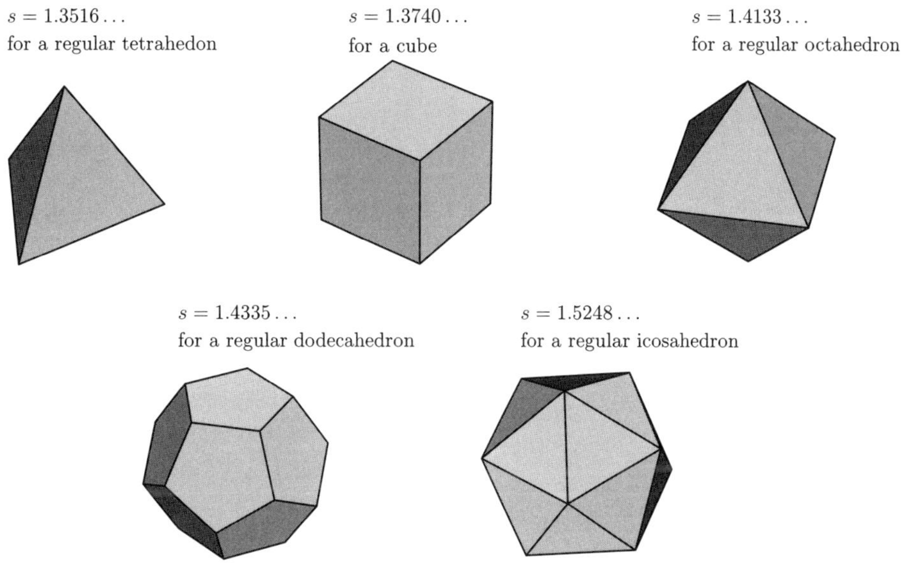

FIGURE 1. The $W^{2,s}$-regularity for solutions of the Navier-Stokes system outside of Plato's polyhedra

Monographic literature. The classical boundary value problems for harmonic functions in nonsmooth domains attracted attention of some of the best mathematicians such as POINCARÉ, PLEMELJ, RADON, CARLEMAN, LEBESGUE, WIENER at the end of the 19th and beginning of the 20th century. Afterwards, a gap followed, until the development of functional analytic methods led since the 1960's to a considerable progress in the study of special and general elliptic differential equations in domains with nonsmooth boundaries. Long lists of references related to domains with isolated point singularities can be found, for example, in [**84**] and [**85**]. Bibliographical information concerning piecewise smooth domains with boundary singularities of positive dimension and other classes of nonsmooth domains is collected at the end of the present volume.

Different aspects of the theory of elliptic boundary value problems in domains with vertices and edges were discussed in several monographs (GRISVARD [**59**, **60**], REMPEL, SCHULZE [**175**], DAUGE [**30**], MAZ'YA, NAZAROV, PLAMENEVSKIĬ [**110**], SCHULZE [**186**], NICAISE [**162**], NAZAROV, PLAMENEVSKIĬ [**160**], KOZLOV,

Maz'ya, Rossmann [**84, 85**], Borsuk, Kondrat'ev [**14**], Nazaikinski, Savin, Schulze, Sternin [**154**] *et al.*)

The present volume which is mostly based on our papers [**133**]–[**140**], has no essential intersections with the just mentioned books. None the less, there are close relations of this book with [**84**] and [**85**]. In fact, the monographs [**84**] and [**85**] can be considered as the first two volumes of a trilogy, of which the present book is the third volume.

In the first part [**84**] of the trilogy, a systematic exposition of the theory of general elliptic boundary value problems for domains with isolated singularities is given. This theory becomes satisfactory only being completed by information on the spectrum of model operator pencils (Mellin operator symbols) corresponding to the original problem. Obtaining this information is the goal of the second part [**85**] of the trilogy.

The theme of the present volume - analysis of boundary value problems for domains with boundary singularities of positive dimension - depends crucially on results in [**84**] concerning domains with isolated singularities. These results form the induction base when considering multi-dimensional geometric singularities. Furthermore, the systematic use in the present book of explicit estimates of eigenvalues of Mellin symbols obtained in [**85**] leads to the best known and sometimes optimal regularity results for solutions of boundary value problems in polyhedral domains. In the sequel, we only formulate auxiliary results proved in [**84**] and [**85**] but this does not prevent the text from being self-contained.

It is necessary to say that the large theme of asymptotic representations for solutions near edges and vertices is not in the scope of the present book. Its comprehensive exposition would require another monograph.

Structure of the book. The book consists of three parts. Part 1 is dedicated to the Dirichlet problem for strongly elliptic systems of arbitrary order. In Part 2 we are concerned with mixed boundary value problems for a class of second order elliptic systems which contains, for example, the system of linear elasticity. Mixed problems for the Stokes and Navier-Stokes systems are treated in Part 3.

There are five chapters in Part 1. Chapter 1 is auxiliary. It contains formulations of well-known properties of elliptic boundary value problems in domains with smooth boundaries and in infinite angles and cones. Chapters 2–4 are dedicated respectively to the Dirichlet problem in a dihedron, a cone with edges and a bounded polyhedral domain. Here we justify the solvability of the problem in weighted Sobolev and Hölder spaces and obtain regularity results for the variational solution. In Chapter 5 we obtain maximum modulus estimates for the solutions and their derivatives in a three-dimensional polyhedral domain.

In the second part of the book consisting of Chapters 6–8 we treat mixed problems for second order systems with Dirichlet and Neumann conditions prescribed on different faces of the boundary. The so-called homogeneous weighted norms employed in the case of the Dirichlet problem prove to be insufficient for the Neumann problem. This difficulty is overcome by using nonhomogeneous weighted norms.

The third part includes Chapters 9–11. Here we investigate mixed boundary value problems for the Stokes and Navier-Stokes systems with four boundary conditions given in arbitrary combinations on different faces of the polyhedron. It is again one of our main goals to study the regularity of the variational solutions. We conclude this part with maximum modulus estimates, first for solutions of the linear system and later of the nonlinear system of hydrodynamics in bounded polyhedral domains.

Readership. This volume is addressed to mathematicians who work in partial differential equations, spectral analysis, asymptotic methods and their applications. It will be also of interest for those who work in numerical analysis, mathematical elasticity and hydrodynamics. A prerequisite for this book is advanced calculus. The reader should be familiar with basic fasts of functional analysis and the theory of partial differential equations.

Acknowledgements. J. Rossmann thanks the University of Liverpool for the hospitality during various stages of the preparation of this book. V. Maz'ya was partially supported by the NSF grant DMS 0500029. Both authors were partially supported by the EPSRC grant EP/F005563/1.

Part 1

The Dirichlet problem for strongly elliptic systems in polyhedral domains

CHAPTER 1

Prerequisites on elliptic boundary value problems in domains with conical points

This auxiliary chapter provides background material on the theory of elliptic boundary value problems in domains with smooth boundaries and in domains with conical points. We restrict ourselves to a minimum of results of the classical theory which are used in the following chapters. All theorems are presented without proof. As in the whole book, we are interested in particular in solvability theorems. Here, the solutions are considered both in Sobolev and Hölder spaces – with weights if the boundary contains singular points and without weights if the boundary is smooth. Furthermore, this chapter contains a priori estimates for the solutions and regularity assertions. In the first section, we deal also with the Green's matrix and with parameter-dependent problems.

1.1. Elliptic boundary value problems in domains with smooth boundaries

The first section concerns general elliptic boundary value problems for systems of differential equations in smooth bounded domains. The a priori estimates and regularity assertions given in Theorems 1.1.5 and 1.1.7 were first proved by AGMON, DOUGLIS and NIRENBERG [**6, 7**]. Proofs of these results can be also found in a number of books. We mention here the books of MORREY [**152**], TRIEBEL [**195**], and for L_2 Sobolev spaces also the books of LIONS and MAGENES [**91**], WLOKA [**203**] KOZLOV, MAZYA and ROSSMANN [**84**]. In these books, also the normal solvability (Fredholm property of the operator) of the boundary value problem (see Theorem 1.1.4) is proved. For the estimates of Green's matrix in Theorem 1.1.8, we refer to SOLONNIKOV [**188**, Theorem 1.1]. The solvability result for the parameter-dependent boundary value problem in Section 1.1.7 and the estimate for its solutions (see Theorem 1.1.10) were obtained by AGMON and NIRENBERG [**8**], Agranovich and Vishik [**9**]. For the proof, we refer also to [**84**, Section 3.6].

1.1.1. Notation. Let $\mathbb{R}$ denote the set of real numbers and $\mathbb{C}$ the set of complex numbers. The N-dimensional Euclidean space is denote by $\mathbb{R}^N$. For partial derivatives we will use the following notation. If $x = (x_1, \dots, x_N) \in \mathbb{R}^N$ and $\alpha = (\alpha_1, \dots, \alpha_N)$ is a multi-index (i.e. α_j is a nonnegative integer for $j = 1, \dots, N$), then

$$\partial_x^\alpha = \partial_{x_1}^{\alpha_1} \cdots \partial_{x_n}^{\alpha_N} \quad \text{and} \quad D_x^\alpha = D_{x_1}^{\alpha_1} \cdots D_{x_N}^{\alpha_N} ,$$

where

$$\partial_{x_j} = \frac{\partial}{\partial x_j}, \qquad D_{x_j} = -i\,\partial_{x_j} ,$$

i denotes the imaginary unit. The length $\alpha_1 + \cdots + \alpha_N$ of a multi-index $\alpha = (\alpha_1, \dots, \alpha_N)$ is denoted by $|\alpha|$. Furthermore, we write

$$x^\alpha = x_1^{\alpha_1} \cdots x^{\alpha_N}, \qquad \alpha! = \alpha_1! \cdots \alpha_N!, \qquad \binom{\alpha}{\beta} = \binom{\alpha_1}{\beta_1} \cdots \binom{\alpha_N}{\beta_N}$$

if $x \in \mathbb{R}^N$ and α, β are multi-indices.

Let Ω be a domain (an open and connected set) in $\mathbb{R}^N$, and let l be a nonnegative integer. The space $C^l(\Omega)$ is the set of all (complex-valued) functions on Ω having bounded and continuous derivatives up to order l on Ω. The norm in $C^l(\Omega)$ is defined as

$$\|u\|_{C^l(\Omega)} = \sum_{|\alpha| \le l} \sup_{x \in \Omega} \big| \partial_x^\alpha u(x) \big|.$$

The subspace of all functions which are continuous together with the derivatives of order $\le l$ up to the boundary is denoted by $C^l(\overline{\Omega})$. Furthermore, let $C_0^\infty(\Omega)$ be the set of all infinitely differentiable functions u on Ω having compact support $\operatorname{supp} u \subset \Omega$. This means in particular that any function $u \in C_0^\infty(\Omega)$ vanishes in a neighborhood of the boundary. If S is a subset of the boundary $\partial\Omega$, then $C_0^\infty(\overline{\Omega} \backslash S)$ denotes the set of all functions u which are infinitely differentiable on $\overline{\Omega}$ and have compact support $\operatorname{supp} u \subset \overline{\Omega} \backslash S$.

Let σ be a real number, $0 < \sigma \le 1$. A function u is called Hölder-continuous with exponent σ (Lipschitz-continuous if $\sigma = 1$) if there exists a constant $C < \infty$ such that

$$\frac{|u(x) - u(y)|}{|x - y|^\sigma} \le C \quad \text{for all } x, y \in \Omega.$$

The space $C^{l,\sigma}(\Omega)$ is defined as the set of all functions u on Ω having bounded and continuous derivatives up to order l and Hölder-continuous (with exponent σ) derivatives of order l. The norm in the *Hölder space* $C^{l,\sigma}(\Omega)$ is defined as

$$\|u\|_{C^{l,\sigma}(\Omega)} = \|u\|_{C^l(\Omega)} + \sum_{|\alpha| = l} \sup_{x, y \in \Omega,\ x \ne y} \frac{|\partial_x^\alpha u(x) - \partial_y^\alpha u(y)|}{|x - y|^\sigma}.$$

We introduce the *Sobolev space* $W^{l,p}(\Omega)$. For $1 < p < \infty$, let $L_p(\Omega)$ be the set of all measurable functions u on Ω such that

$$\|u\|_{L_p(\Omega)} = \Big(\int_\Omega |u(x)|^p \, dx \Big)^p < \infty.$$

Then the space $W^{l,p}(\Omega)$ is defined for integer $l \ge 0$ as the set of all functions u on Ω such that $\partial_x^\alpha u \in L_p(\Omega)$ for every multi-index α, $|\alpha| \le l$. This space is provided with the norm

$$\|u\|_{W^{l,p}(\Omega)} = \Big(\int_\Omega \sum_{|\alpha| \le l} \big| \partial_x^\alpha u(x) \big|^p \, dx \Big)^{1/p}.$$

The subspace $\overset{\circ}{W}{}^{l,p}(\Omega)$ is defined as the closure of the set $C_0^\infty(\Omega)$ in $W^{l,p}(\Omega)$. Finally, we denote the trace space for $W^{l,p}(\Omega)$, $l \ge 1$, on the boundary $\partial\Omega$ by $W^{l-1/p,p}(\partial\Omega)$. The norm in this space is defined by

$$\|u\|_{W^{l-1/p,p}(\partial\Omega)} = \inf \big\{ \|v\|_{W^{l,p}(\Omega)} \ : \ v \in W^{l,p}(\Omega),\ v = u \ \text{ on } \partial\Omega \big\}.$$

1.1.2. Ellipticity of the boundary value problem. Let Ω be a bounded domain in $\mathbb{R}^N$ with smooth (of class C^∞) boundary $\partial\Omega$, and let

$$L(x, D_x) = \big(L_{i,j}(x, D_x)\big)_{1\le i,j\le \ell}$$

be a matrix of linear differential operators on Ω. We assume that the determinant of the matrix $\big(L_{i,j}(x,\xi)\big)_{1\le i,j\le\ell}$ is a polynomial in ξ of even order $2m \ge 2$, there exist integer numbers $s_i \le 0$ and $t_j \ge 0$ such that $\max(s_1, \dots, s_\ell) = 0$,

$$\text{(1.1.1)} \qquad \operatorname{ord} L_{i,j} \le s_i + t_j\,, \quad L_{i,j} = 0 \text{ if } s_i + t_j < 0,$$

$$\text{(1.1.2)} \qquad s_1 + s_2 + \cdots + s_\ell + t_1 + t_2 + \cdots + t_\ell = 2m\,,$$

and that the coefficients of the differential operators

$$L_{i,j}(x, D_x) = \sum_{|\alpha|\le s_i+t_j} a_\alpha^{i,j}(x)\, D_x^\alpha$$

are infinitely differentiable in $\overline{\Omega}$. Furthermore, let

$$B(x, D_x) = \Big(B_{k,j}(x, D_x)\Big)_{1\le k\le m,\ 1\le j\le \ell}$$

be a $m\times\ell$-matrix of linear differential operators with smooth coefficients satisfying the condition

$$\text{(1.1.3)} \qquad \operatorname{ord} B_{k,j} \le \mu_k + t_j\,, \quad B_{k,j} \equiv 0 \text{ if } \mu_k + t_j < 0,$$

where μ_k are given integer numbers.

We consider the boundary value problem

$$\text{(1.1.4)} \qquad L(x, D_x)\, u = f \quad \text{in } \Omega,$$

$$\text{(1.1.5)} \qquad B(x, D_x)\, u = g \quad \text{on } \partial\Omega.$$

Let $L_{i,j}^\circ$ denote the *principal part* of the operator $L_{i,j}$, i.e.

$$L_{i,j}^\circ(x, D_x) = \sum_{|\alpha|= s_i+t_j} a_\alpha^{i,j}(x)\, D_x^\alpha.$$

In the cases $s_i+t_j < 0$ and $\operatorname{ord} L_{i,j} < s_i+t_j$ we define $L_{i,j}^\circ(x, D_x) = 0$. Analogously, the principal parts $B_{k,j}^\circ$ of the operators $B_{k,j}$ are defined. The corresponding matrix operators are denoted by L° and B°.

DEFINITION 1.1.1. The matrix operator $L(x, D_x)$ is said to be *elliptic* in $\overline{\Omega}$ if

$$\det L^\circ(x,\xi) \ne 0 \qquad \text{for all } x \in \overline{\Omega},\ \xi \in \mathbb{R}^N\backslash\{0\}.$$

If moreover the polynomial $\tau \to \det L^\circ(x, \xi + \tau\zeta)$ has exactly m zeros (counting multiplicity) in the upper half-plane $\operatorname{Im}\tau > 0$ for arbitrary linearly independent vectors ξ, $\zeta \in \mathbb{R}^N$, then the operator $L(x, D_x)$ is said to be *properly elliptic.*

Note that every elliptic operator is properly elliptic if $N \ge 3$.

Suppose that the operator $L(x, D_x)$ is properly elliptic. For arbitrary $x^{(0)} \in \partial\Omega$ and arbitrary vectors ξ' tangential to $\partial\Omega$ in $x^{(0)}$, let $\mathcal{M}^+(\xi')$ be the linear m-dimensional space of the stable solutions of the equation

$$L^\circ\big(x^{(0)}, \xi' + n(x^{(0)})\, D_t\big)\, u(t) = 0, \qquad t > 0,$$

which tend to zero as $t \to \infty$. Here $n(x^{(0)})$ denotes the outer unit normal to $\partial\Omega$ at the point $x^{(0)}$.

DEFINITION 1.1.2. The system $B(x, D_x)$ of the boundary operators is said to satisfy the *complementing condition* on $\partial\Omega$ (with respect to the system (1.1.4)) if for every $x^{(0)} \in \partial\Omega$, every vector ξ' tangential to $\partial\Omega$ in $x^{(0)}$, and every $g \in \mathbb{C}^m$, there exists exactly one vector function $u \in \mathcal{M}^+(\xi')$ satisfying the equation

$$B^\circ\big(x^{(0)}, \xi' + n(x^{(0)})\, D_t\big)\, u(t)\big|_{t=0} = g.$$

The boundary value problem (1.1.4), (1.1.5) is said to be *elliptic* if

(i) the operator L is properly elliptic in $\overline{\Omega}$,
(ii) the system of the boundary operators satisfies the complementing condition on $\partial\Omega$.

1.1.3. Examples. 1) Let the operator $L(x, D_x)$ have the form

$$L(x, D_x) = \sum_{|\alpha|\le 2m} A_\alpha(x)\, D_x^\alpha\,, \tag{1.1.6}$$

where $A_\alpha(x)$ are $\ell \times \ell$-matrices. In this case, we have $s_1 = \cdots = s_\ell = 0$ and $t_1 = \cdots = t_\ell = 2m$. The operator (1.1.6) is called *strongly elliptic* if there exists a positive constant c such that

$$\mathrm{Re} \sum_{|\alpha|=2m} \big(A_\alpha(x)\, \xi^\alpha\, \eta, \eta\big)_{\mathbb{C}^\ell} \ge c\, |\xi|^{2m}\, |\eta|^2_{\mathbb{C}^\ell}$$

for all $x \in \overline{\Omega}$, $\xi \in \mathbb{R}^N\backslash\{0\}$, $\eta \in \mathbb{C}^\ell$. If $L(x, D_x)$ is strongly elliptic, then the *Dirichlet problem*

$$L(x, D_x)\, u = f \ \text{ in } \Omega, \qquad \frac{\partial^{k-1} u}{\partial n^{k-1}} = g_k \ \text{ on } \partial\Omega, \ k = 1, \ldots, m,$$

is elliptic (see [**7**] and [**152**, Theorem 6.5.5]).

2) Let

$$L(D_x)\, u = \Delta u + \gamma\, \mathrm{grad}\, \mathrm{div}\, u = f$$

be the *Lamé system* in $\mathbb{R}^3$. Then the Dirichlet problem

$$L(D_x)\, u = f \ \text{ in } \Omega, \qquad u = g \ \text{ on } \partial\Omega,$$

is elliptic for $\gamma \ne -1$, $\gamma \ne -2$, while the *Neumann problem*

$$L(D_x)\, u = f \ \text{ in } \Omega,$$
$$\sum_{j=1}^{3} \Big(\frac{\partial u_i}{\partial x_j} + \frac{\partial u_j}{\partial x_i}\Big)\, n_j + n_i\, (\gamma - 1)\, \mathrm{div}\, u = g_i \ \text{on } \partial\Omega, \ i = 1, 2, 3$$

is elliptic for $\gamma \ne -1$ (see e.g. [**84**, Section 4.2]).

3) The Dirichlet problem for the *Stokes system* of hydrodynamics

$$L(D_x)\begin{pmatrix} u \\ p \end{pmatrix} = \begin{pmatrix} -\Delta u + \mathrm{grad}\, p \\ \mathrm{div}\, u \end{pmatrix} = \begin{pmatrix} f \\ g \end{pmatrix} \ \text{ in } \Omega,$$
$$u = h \ \text{ on } \partial\Omega,$$

where $u = (u_1, u_2, u_3)$ denotes the velocity and p the pressure, is elliptic. The same is true for the Stokes system with the Neumann boundary condition

$$-pn_i + \sum_{j=1}^{3} \Big(\frac{\partial u_i}{\partial x_j} + \frac{\partial u_j}{\partial x_i} \Big) n_j = \phi_i, \quad i = 1, 2, 3.$$

1.1.4. Normal solvability of elliptic boundary value problems. Obviously the operator $\mathcal{A} = (L, B)$ of the boundary value problem (1.1.4), (1.1.5) realizes continuous mappings

$$\prod_{j=1}^{\ell} W^{l+t_j,p}(\Omega) \ni u \to \mathcal{A}u \in \prod_{i=1}^{\ell} W^{l-s_i,p}(\Omega) \times \prod_{k=1}^{m} W^{l-\mu_k-1/p,p}(\partial\Omega) \tag{1.1.7}$$

and

$$\prod_{j=1}^{\ell} C^{l+t_j,\sigma}(\Omega) \ni u \to \mathcal{A}u \in \prod_{i=1}^{\ell} C^{l-s_i,\sigma}(\Omega) \times \prod_{k=1}^{m} C^{l-\mu_k,\sigma}(\partial\Omega) \tag{1.1.8}$$

for $l \geq \max(0, \mu_1 + 1, \dots, \mu_m + 1)$ and $l \geq \max(0, \mu_1, \dots, \mu_m)$, respectively.

DEFINITION 1.1.3. Let $\mathcal{A}$ be a linear and continuous operator from the Banach space $\mathcal{X}$ into the Banach space $\mathcal{Y}$. Then $\mathcal{A}$ is said to be *Fredholm* if

(i) $\dim \ker \mathcal{A} < \infty$, $\mathcal{R}(\mathcal{A})$ is closed,
(ii) $\dim \operatorname{coker} \mathcal{A} = \dim(\mathcal{Y}/\mathcal{R}(\mathcal{A})) < \infty$.

Here $\ker \mathcal{A}$ denotes the kernel and $\mathcal{R}(\mathcal{A})$ the range of the operator $\mathcal{A}$.

THEOREM 1.1.4. *Suppose that the domain Ω is bounded and its boundary is smooth. If the boundary value problem* (1.1.4), (1.1.5) *is elliptic, then the operators* (1.1.7) *and* (1.1.8) *are Fredholm.*

1.1.5. A priori estimates for solutions of elliptic boundary value problems. If the boundary of the domain Ω is smooth, then the solutions of elliptic boundary value problems satisfy the following a priori estimates in Sobolev and Hölder spaces.

THEOREM 1.1.5. *Let u be a solution of the elliptic boundary value problem* (1.1.4), (1.1.5) *in the bounded domain Ω with smooth boundary $\partial\Omega$.*

1) *If $u_j \in W^{l+t_j,p}(\Omega)$ for $j = 1, \dots, \ell$, $l \geq \max(0, \mu_1 + 1, \dots, \mu_m + 1)$, then*

$$\sum_{j=1}^{\ell} \|u_j\|_{W^{l+t_j,p}(\Omega)} \leq c\Big(\sum_{i=1}^{\ell} \|f_i\|_{W^{l-s_i,p}(\Omega)} + \sum_{k=1}^{m} \|g_k\|_{W^{l-\mu_k-1/p,p}(\partial\Omega)} + \|u\|_{L_1(\Omega)^\ell}\Big).$$

2) *If $u_j \in C^{l+t_j,\sigma}(\Omega)$ for $j = 1, \dots, \ell$, $l \geq \max(0, \mu_1, \dots, \mu_m)$, then*

$$\sum_{j=1}^{\ell} \|u_j\|_{C^{l+t_j,\sigma}(\Omega)} \leq c\Big(\sum_{i=1}^{\ell} \|f_i\|_{C^{l-s_i,\sigma}(\Omega)} + \sum_{k=1}^{m} \|g_k\|_{C^{l-\mu_k,\sigma}(\partial\Omega)} + \|u\|_{L_1(\Omega)^\ell}\Big).$$

Here the constant c is independent of u.

We will also use the following local estimate which follows immediately from the last theorem.

COROLLARY 1.1.6. *Let ζ, η be infinitely differentiable functions on $\overline{\Omega}$ such that $\zeta\eta = \zeta$. If u is a solution of problem* (1.1.4), (1.1.5), $u_j \in W^{l+t_j,p}(\Omega)$ *for* $j = 1, \dots, \ell$, $l \ge \max(0, \mu_1 + 1, \dots, \mu_m + 1)$, *then*

$$\sum_{j=1}^{\ell} \|\zeta u_j\|_{W^{l+t_j,p}(\Omega)} \le c\Big(\sum_{i=1}^{\ell} \|\eta f_i\|_{W^{l-s_i,p}(\Omega)} + \sum_{k=1}^{m} \|\eta g_k\|_{W^{l-\mu_k-1/p,p}(\partial\Omega)} + \sum_{j=1}^{\ell} \|\eta u_j\|_{W^{l+t_j-1,p}(\Omega)^\ell}\Big)$$

with a constant c independent of u.

Note that the constant in the last estimate depends only on Ω, l, p, the support of ζ and the maximum of the derivatives of ζ up to order $l + \max t_j$.

1.1.6. Elliptic regularity. The smoothness of the solutions of elliptic problems in domains with smooth boundaries is determined by the smoothness of the right-hand sides f_j and g_k. In particular, the following regularity result is valid.

THEOREM 1.1.7. *Let $u = (u_1, \dots, u_\ell)$ be a solution of the elliptic boundary value problem* (1.1.4), (1.1.5) *in a bounded domain Ω with smooth boundary $\partial\Omega$. Suppose that $u_j \in W^{l+t_j,p}(\Omega)$ for $j = 1, \dots, \ell$, $l \ge \max(0, \mu_1 + 1, \dots, \mu_m + 1)$, or $u_j \in C^{l+t_j,\sigma}(\Omega)$ for $j = 1, \dots, \ell$, $l \ge \max(0, \mu_1, \dots, \mu_m)$. If*

$$f_j \in W^{l-s_j,p}(\Omega) \cap W^{l'-s_j,q}(\Omega), \quad g_k \in W^{l-\mu_k-1/p,p}(\partial\Omega) \cap W^{l'-\mu_k-1/q,q}(\partial\Omega)$$

for $j = 1, \dots, \ell$, $k = 1, \dots, m$, where $l' \ge \max(0, \mu_1 + 1, \dots, \mu_m + 1)$, then $u_j \in W^{l'+t_j,q}(\Omega)$ for $j = 1, \dots, \ell$. If

$$f_j \in W^{l-s_j,p}(\Omega) \cap C^{l'-s_j,\sigma}(\Omega), \quad g_k \in W^{l-\mu_k-1/p,p}(\partial\Omega) \cap C^{l'-\mu_k,\sigma}(\partial\Omega)$$

for $j = 1, \dots, \ell$, $k = 1, \dots, m$, where $l' \ge \max(0, \mu_1, \dots, \mu_m)$, then $u_j \in C^{l'+t_j,\sigma}(\Omega)$ for $j = 1, \dots, \ell$.

The regularity result in Sobolev spaces can be also extended to the case $l \le \max(0, \mu_1 + 1, \dots, \mu_m + 1)$ (see [**84**, Section 3.2] and [**176**])

1.1.7. Green's matrix of the boundary value problem. Let $\mathcal{A}$ be the operator of problem (1.1.4), (1.1.5) given in Theorem 1.1.4. If the boundary value problem is elliptic, then $\dim \ker \mathcal{A} < \infty$ and $\dim \mathcal{R}(\mathcal{A}) < \infty$. We denote by P the projection operator onto $\mathcal{R}(\mathcal{A})$. Then the problem

$$\mathcal{A}u = P\begin{pmatrix} f \\ 0 \end{pmatrix}, \quad u \perp \ker \mathcal{A} \tag{1.1.9}$$

is uniquely solvable.

THEOREM 1.1.8. *The solution of the problem* (1.1.9) *can be represented in the form*

$$u_i(x) = \sum_{j=1}^{\ell} \sum_{|\alpha^{(j)}| \le l-s_j} \int_\Omega G_{i,j,\alpha^{(j)}}(x,\xi)\, D_\xi^{\alpha^{(j)}} f_j(\xi)\, d\xi, \quad i = 1, \dots, \ell, \tag{1.1.10}$$

where $l \ge \max(0, \mu_1 + 1, \dots, \mu_m + 1)$ *and* $G_{i,j,\alpha^{(j)}}(x,\xi)$ *are functions satisfying the estimate*

$$\big|D_x^\beta D_\xi^\gamma G_{i,j,\alpha^{(j)}}(x,\xi)\big| \le c\,\big(1 + |x-\xi|^{-N+l+t_i-|\beta|-|\gamma|}\big)$$

if $-N + l + t_i - |\beta| - |\gamma| \ne 0$ *and*

$$\big|D_x^\beta D_\xi^\gamma G_{i,j,\alpha^{(j)}}(x,\xi)\big| \le c\,\Big(1 + \Big|\log \frac{1}{|x-\xi|}\Big|\Big)$$

if $-N + l + t_i - |\beta| - |\gamma| = 0$.

In the case $s_1 = \dots = s_\ell = 0$, $\mu_k < 0$, the representation (1.1.10) with $l = 0$ takes the form

$$u_i(x) = \sum_{j=1}^{\ell} \int_\Omega G_{i,j}(x,\xi)\, f_j(\xi)\, d\xi, \quad i = 1, \dots, \ell,$$

where $G_{i,j}$ satisfies the estimates in Theorem 1.1.8 with $l = 0$. The matrix with the elements $G_{i,j}(x,\xi)$ is called *Green's matrix.*

1.1.8. Elliptic problems with parameter. Suppose that $L_{i,j}$ and $B_{k,j}$ are differential operators depending polynomially on a complex parameter λ. More precisely, $L_{i,j}$ and $B_{k,j}$ are operators of the form

$$L_{i,j}(x, D_x, \lambda) = \sum_{\nu+|\alpha|\le s_i+t_j} a_\alpha^{i,j}(x)\,\lambda^\nu\, D_x^\alpha,$$
$$B_{k,j}(x, D_x, \lambda) = \sum_{\nu+|\alpha|\le \mu_k+t_j} b_\alpha^{k,j}(x)\,\lambda^\nu\, D_x^\alpha,$$

$i, j = 1, \dots, \ell$, $k = 1, \dots, m$, where s_i, t_j, μ_k are the same integer numbers as in the previous subsections. We consider the boundary value problem

$$L(x, D_x, \lambda)\, u = f \ \text{ in } \Omega, \quad B(x, D_x, \lambda)\, u = g \ \text{ on } \partial\Omega. \tag{1.1.11}$$

Here again L and B denote the matrices with the elements $L_{i,j}$ and $B_{k,j}$, respectively.

DEFINITION 1.1.9. The boundary value problem (1.1.11) is said to be *elliptic with parameter* if the boundary value problem

$$L(x, D_x, iD_t)\, u(x,t) = f(x,t) \quad \text{for } x \in \Omega,\ t \in \mathbb{R},$$
$$B(x, D_x, iD_t)\, u(x,t) = g(x,t) \quad \text{for } x \in \partial\Omega,\ t \in \mathbb{R}$$

is elliptic in the cylinder $\Omega \times \mathbb{R}$.

THEOREM 1.1.10. *Suppose that* Ω *is a bounded domain with smooth boundary* $\partial\Omega$ *and that the boundary value problem* (1.1.11) *is elliptic with parameter. Then there exist positive real constants* ρ *and* δ *such that for all* $\lambda \in \mathbb{C}$ *satisfying the conditions*

$$|\lambda| > \rho \quad \text{and} \quad \big|\mathrm{Re}\,\lambda\big| < \delta\,\big|\mathrm{Im}\,\lambda\big|,$$

the boundary value problem (1.1.11) *has a unique solution*

$$u = (u_1, \dots, u_\ell) \in \prod_{j=1}^{\ell} W^{l+t_j,2}(\Omega), \ l \ge \max(0, \mu_1 + 1, \dots, \mu_m + 1),$$

for arbitrary $f_j \in W^{l-s_j,2}(\Omega)$, $j = 1, \dots, \ell$, *and* $g_k \in W^{l-\mu_k-1/2,2}(\partial\Omega)$. *This solution satisfies the estimate*

$$\sum_{j=1}^{\ell} \sum_{\nu=0}^{l+t_j} |\lambda|^\nu \, \|u_j\|_{W^{l+t_j-\nu,2}(\Omega)} \le c \Big(\sum_{j=1}^{\ell} \sum_{\nu=0}^{l-s_j} |\lambda|^\nu \, \|u_j\|_{W^{l-s_j-\nu,2}(\Omega)} + \sum_{k=1}^{m} \big(\|g_k\|_{W^{l-\mu_k-1/2,2}(\partial\Omega)} + |\lambda|^{l-\mu_k-1/2} \, \|g_k\|_{L_2(\partial\Omega)} \big) \Big).$$

Here the constant c *is independent of* f, g *and* λ.

The operator $\mathcal{A}(\lambda) = \big(L(x, D_x, \lambda), B(x, D_x, \lambda)\big)$ of the boundary value problem (1.1.11) is called an *operator pencil*. From Theorem 1.1.10 it follows that the operator

$$\prod_{j=1}^{\ell} W^{l+t_j,2}(\Omega) \ni u \to \mathcal{A}(\lambda)\, u \in \prod_{j=1}^{\ell} W^{l-s_j,2}(\Omega) \times \prod_{k=1}^{m} W^{l-\mu_k-1/2,2}(\partial\Omega)$$

is an isomorphism for all $\lambda \in \mathbb{C}$ except for a countable set of isolated points, the eigenvalues of the pencil $\mathcal{A}(\lambda)$. All eigenvalues have finite geometric and algebraic multiplicities (see [**85**, Theorem 1.1.1]).

1.2. Elliptic boundary value problems in angles and cones

This section is dedicated to general elliptic boundary value problems in domains with angular and conical points. Here we restrict ourselves to "model problems" in a 2-dimensional angle or a N-dimensional cone $\mathcal{K}$, $N > 2$. This means, we assume that the differential operators $L_{i,j}$ are homogeneous ($L_{i,j} = L^{\circ}_{i,j}$) and have constant coefficients. In contrast to the following chapters, the boundary of the cone $\mathcal{K}$ is assumed to be smooth outside the vertex. Again without proof, we present some solvability and regularity results in weighted Sobolev spaces. Theorems 1.2.5, 1.2.6 and 1.2.8 below were proved by KONDRAT'EV [**74**] for the case $\ell = 1$ and $p = q = 2$ and by MAZYA and PLAMENEVSKIĬ [**117**] for arbitrary ℓ and $p, q > 1$. For the proofs, we refer also to the books [**84, 160**].

1.2.1. Weighted Sobolev spaces in a cone.

Let $\mathcal{K}$ be a cone (angle if $N = 2$) in the Euclidean space $\mathbb{R}^N$ with vertex at the origin. This means that $\mathcal{K}$ has the representation

$$\mathcal{K} = \{x \in \mathbb{R}^N : \ 0 < \rho < \infty, \ \omega = x/|x| \in \Omega\}, \tag{1.2.1}$$

where $\rho = |x|$, $\omega = x/|x|$, and Ω is a subdomain of the unit sphere S^{N-1}.

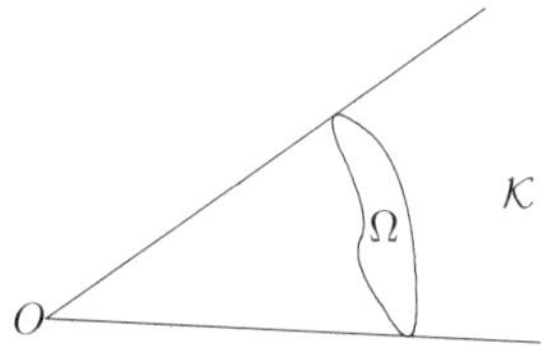

FIGURE 2. The cone $\mathcal{K}$

The boundary $\partial\Omega$ of Ω is assumed to be smooth (of class C^∞) for $N \ge 3$. In the case $N = 2$, the boundary $\partial\mathcal{K}$ of $\mathcal{K}$ consists of two half-lines γ^+ and γ^- and the vertex O.

Let l be a nonnegative integer, $\beta \in \mathbb{R}$ and $1 < p < \infty$. Then $V_\beta^{l,p}(\mathcal{K})$ is defined as the weighted Sobolev space of all functions u on $\mathcal{K}$ with finite norm

$$\|u\|_{V_\beta^{l,p}(\mathcal{K})} = \Big(\int_{\mathcal{K}} \sum_{|\alpha|\le l} \rho^{p(\beta-l+|\alpha|)} |\partial_x^\alpha u(x)|^p \, dx \Big)^{1/p}. \tag{1.2.2}$$

The space $V_\beta^{l,p}(\mathcal{K})$ can be also defined as the closure of the set $C_0^\infty(\overline{\mathcal{K}}\backslash\{0\})$ (the set of infinitely differential functions on $\overline{\mathcal{K}}$ with compact support vanishing in a neighborhood of the origin) with respect to the norm (1.2.2). We will frequently use another equivalent norm to (1.2.2) which is given in the following lemma (see [**84**, Lemma 6.1.1]). A proof of the analogous result in a cone with edges can be found in Section 3.1.

LEMMA 1.2.1. *Let ζ_ν be infinitely differentiable functions on $\overline{\mathcal{K}}$ with support in $\{x \in \overline{\mathcal{K}} :\ 2^{\nu-1} < |x| < 2^{\nu+1}\}$, $\nu = 0, \pm 1, \pm 2, \ldots$, such that*

$$\big|\partial_x^\alpha \zeta_\nu(x)\big| \le c_\alpha\, 2^{-\nu|\alpha|} \ \textit{for all } \alpha, \quad \textit{and} \quad \sum_{\nu=-\infty}^{+\infty} \zeta_\nu = 1, \tag{1.2.3}$$

where c_α are constants depending only on α. Then the norm (1.2.2) *and*

$$|||u|||_{V_\beta^{l,p}(\mathcal{K})} = \Big(\sum_{\nu=-\infty}^{+\infty} \|\zeta_\nu u\|_{V_\beta^{l,p}(\mathcal{K})} \Big)^{1/p}$$

are equivalent. This means that there exist positive constants c_1 and c_2 such that

$$c_1 \, \|u\|_{V_\beta^{l,p}(\mathcal{K})} \le |||u|||_{V_\beta^{l,p}(\mathcal{K})} \le c_2 \, \|u\|_{V_\beta^{l,p}(\mathcal{K})}$$

for all $u \in V_\beta^{l,p}(\mathcal{K})$.

1.2.2. Imbeddings. Obviously,

$$V_\beta^{l,p}(\mathcal{K}) \subset V_{\beta-l+k}^{k,p}(\mathcal{K})$$

for $l > k$. The last imbedding is a particular case of the following lemma.

LEMMA 1.2.2. *Suppose that*

$$l > k, \quad p \le q, \quad \textit{and} \quad l - \frac{N}{p} \ge k - \frac{N}{q}$$

Then $V_\beta^{l,p}(\mathcal{K})$ is continuously imbedded in $V_\gamma^{k,q}(\mathcal{K})$, where $\gamma = \beta - l + k + \frac{N}{p} - \frac{N}{q}$.

P r o o f. Let $u \in V_\beta^{l,p}(\mathcal{K})$, and let ζ_ν be the functions introduced in Lemma 1.2.1. We define $v_\nu(x) = u(2^\nu x)$ and $\eta_\nu(x) = \zeta_\nu(2^\nu x)$. Since $\eta_\nu(x) = 0$ for $|x| < 1/2$ and $|x| > 2$, it follows from the continuity of the imbedding $W^{l,p} \subset W^{k,q}$ that

$$\|\eta_\nu v_\nu\|_{V_\gamma^{k,q}(\mathcal{K})} \le c \, \|\eta_\nu v_\nu\|_{V_\beta^{l,p}(\mathcal{K})}$$

with a constant c independent of u and ν. This implies

$$\|\zeta_\nu u\|_{V_\gamma^{k,q}(\mathcal{K})} \le c \, \|\zeta_\nu u\|_{V_\beta^{l,p}(\mathcal{K})}$$

with the same constant c. Applying Lemma 1.2.1, we obtain

$$\begin{aligned}\|u\|_{V_\gamma^{k,q}(\mathcal{K})} &\le c\Big(\sum_{\nu=-\infty}^{+\infty}\|\zeta_\nu u\|^q_{V_\gamma^{k,q}(\mathcal{K})}\Big)^{1/q} \le c'\Big(\sum_{\nu=-\infty}^{+\infty}\|\zeta_\nu u\|^q_{V_\beta^{l,p}(\mathcal{K})}\Big)^{1/q}\\ &\le c'\Big(\sum_{\nu=-\infty}^{+\infty}\|\zeta_\nu u\|^p_{V_\beta^{l,p}(\mathcal{K})}\Big)^{1/p} \le c''\,\|u\|_{V_\beta^{l,p}(\mathcal{K})}.\end{aligned}$$

This proves the lemma. □

LEMMA 1.2.3. *Let $u \in V_\beta^{l,p}(\mathcal{K})$, $l > N/p$. Then $\rho^{\beta-l+N/p}\,u \in L_\infty(\mathcal{K})$ and*

$$\|\rho^{\beta-l+N/p}\,u\|_{L_\infty(\mathcal{K})} \le c\,\|u\|_{V_\beta^{l,p}(\mathcal{K})}$$

with a constant c independent of u.

P r o o f. Let $B(x)$ be the ball with radius $|x|/2$ centered about x. If $x \in \mathcal{K}$ and $|x| = 1$, then by Sobolev's imbedding theorem,

$$|u(x)| \le c\,\|u\|_{W^{l,p}(\mathcal{K}\cap B(x))}$$

for every $u \in V_\beta^{l,p}(\mathcal{K})$. Now let $|x| = t$, $y = t^{-1}x$ and $v(x) = u(tx)$. Then it follows from the last inequality that

$$|u(x)|^p = |v(y)|^p \le c\,\|v\|^p_{W^{l,p}(\mathcal{K}\cap B(y))} = c\sum_{|\alpha|\le l} t^{p|\alpha|-N}\,\|D^\alpha u\|^p_{L_p(\mathcal{K}\cap B(x))}.$$

Since $t/2 < |\xi| < 3t/2$ for $\xi \in B(x)$, we conclude that

$$|u(x)|^p \le c\,t^{p(l-\beta)-N}\,\|u\|^p_{V_\beta^{l,p}(\mathcal{K})}.$$

This proves the lemma. □

1.2.3. Trace spaces. Let $l \ge 1$. Then the space of the traces of functions from $V_\beta^{l,p}(\mathcal{K})$ on $\partial\mathcal{K}\backslash\{0\}$ is denoted by $V_\beta^{l-1/p,p}(\partial\mathcal{K})$. The norm in this space is defined as

$$\|u\|_{V_\beta^{l-1/p,p}(\partial\mathcal{K}\backslash\{0\})} = \inf\big\{\|v\|_{V_\beta^{l,p}(\mathcal{K})}\ :\ v \in V_\beta^{l,p}(\mathcal{K}),\ v = u \text{ on } \partial\mathcal{K}\backslash\{0\}\big\}.$$

In the case $N = 2$, where the boundary of $\mathcal{K}$ consists of two half-lines $\gamma^\pm$ and the vertex $x = 0$, we denote by $V_\beta^{l-1/p,p}(\gamma^\pm)$ the space of the traces of $V_\beta^{l,p}(\mathcal{K})$-functions on $\gamma^\pm$ and provide this space with the analogous norm. From Lemma 1.2.1, one can easily deduce the following lemma (see [**84**, Lemma 6.1.1]). The proofs of analogous results for weighted Sobolev spaces in a dihedron and in a cone with edges are given in Sections 2.1 and 3.1.

LEMMA 1.2.4. *Let ζ_ν be the same functions as in Lemma 1.2.1. Then there exist positive constants c_1 and c_2 such that*

$$c_1\,\|u\|_{V_\beta^{l-1/p,p}(\partial\mathcal{K}\backslash\{0\})} \le \Big(\sum_{\nu=-\infty}^{+\infty}\|\zeta_\nu u\|_{V_\beta^{l-1/p,p}(\partial\mathcal{K}\backslash\{0\})}\Big)^{1/p} \le c_2\,\|u\|_{V_\beta^{l-1/p,p}(\partial\mathcal{K}\backslash\{0\})}$$

for all $u \in V_\beta^{l-1/p,p}(\partial\mathcal{K}\backslash\{0\})$.

1.2.4. Solvability of elliptic boundary value problems in a cone. Let $\mathcal{K}$ be the cone (1.2.1), where Ω is a subdomain of the unit sphere S^{N-1} with smooth boundary $\partial\Omega$. Furthermore, let $L(D_x)$ and $B(D_x)$ be matrices of differential operators

$$L_{i,j}(D_x) = \sum_{|\alpha|=s_i+t_j} a_\alpha^{i,j}\, D_x^\alpha\,, \quad L_{i,j} = 0 \text{ if } s_i + t_j < 0,$$

and

$$B_{k,j}(D_x) = \sum_{|\alpha|=\mu_k+t_j} b_\alpha^{k,j}\, D_x^\alpha\,, \quad B_{k,j} = 0 \text{ if } \mu_k + t_j < 0,$$

respectively, with constant coefficients ($i,j = 1,\dots,\ell$, $k = 1,\dots,m$). Here again s_i, t_j, μ_k are integer numbers such that

$$\max_{1\le i\le \ell} s_i = 0, \quad t_j \ge 0, \quad \text{and} \quad \sum_{j=1}^{\ell} (s_j + t_j) = 2m.$$

We consider the boundary value problem

$$L(D_x)\,u = f \ \text{ in } \mathcal{K}, \quad B(D_x)\,u = g \ \text{ on } \partial\mathcal{K}\backslash\{0\}. \tag{1.2.4}$$

In the case $N = 2$, it is allowed that different boundary conditions

$$B^{\pm}(D_x)\,u = g^{\pm}$$

are prescribed on the sides γ^+ and γ^- of the angle $\mathcal{K}$. We will assume in the sequel that the boundary value problem (1.2.4) is elliptic. This means that the differential operator $L(D_x)$ is elliptic (see Definition 1.1.1) and the system $B(D_x)$ of the boundary operators satisfies the complementing condition on $\partial\mathcal{K}\backslash\{0\}$ (see Definition 1.1.2).

The differential operator $L_{i,j}(D_x)$ admits the representation

$$L_{i,j}(D_x) = \rho^{-(s_i+t_j)}\, \mathcal{L}_{i,j}(\omega, D_\omega, \rho\partial_\rho) = \rho^{-(s_i+t_j)} \sum_{\nu=0}^{s_i+t_j} p_{i,j,\nu}(\omega, D_\omega)\, (\rho\partial_\rho)^\nu$$

in the coordinates $\rho = |x|$, $\omega = x/|x|$, where $p_{i,j,\nu}(\omega, D_\omega)$ are differential operators of order $\le s_i + t_j - \nu$ on Ω. Analogously,

$$B_{k,j}(D_x) = \rho^{-(\mu_k+t_j)}\, \mathcal{B}_{k,j}(\omega, D_\omega, \rho\partial_\rho) = \rho^{-(\mu_k+t_j)} \sum_{\nu=0}^{\mu_k+t_j} q_{k,j,\nu}(\omega, D_\omega)\, (\rho\partial_\rho)^\nu$$

with differential operators $q_{k,j,\nu}(\omega, D_\omega)$ of order $\le \mu_k + t_j - \nu$. We denote by $\mathfrak{A}(\lambda)$ the operator of the parameter-dependent boundary value problem

$$\sum_{j=1}^{\ell} \mathcal{L}_{i,j}(\omega, D_\omega, \lambda + t_j)\, U_j(\omega) = F_i(\omega) \ \text{ for } \omega \in \Omega,\ i = 1,\dots,\ell, \tag{1.2.5}$$

$$\sum_{j=1}^{\ell} \mathcal{B}_{k,j}(\omega, D_\omega, \lambda + t_j)\, U_j(\omega) = G_k(\omega) \ \text{ for } \omega \in \partial\Omega,\ k = 1,\dots,m. \tag{1.2.6}$$

It realizes a continuous mapping

$$\prod_{j=1}^{\ell} W^{l+t_j,2}(\Omega) \to \prod_{j=1}^{\ell} W^{l-s_j,2}(\Omega) \times \prod_{k=1}^{m} W^{l-\mu_k-1/2,2}(\partial\Omega)$$

for arbitrary integer

$$l \geq l_0 \stackrel{def}{=} \max(0, \mu_1 + 1, \dots, \mu_m + 1).$$

The boundary value problem (1.2.5), (1.2.6) is elliptic with parameter (see Definition 1.1.9). Therefore, the spectrum of the operator pencil $\mathfrak{A}(\lambda)$ consists only of eigenvalues with finite geometric and algebraic multiplicities.

THEOREM 1.2.5. *Suppose that the boundary value problem* (1.2.4) *is elliptic and the line* $\operatorname{Re}\lambda = l - \beta - N/p$ *is free of eigenvalues of the pencil* $\mathfrak{A}(\lambda)$. *Then for all* $f_j \in V_\beta^{l-s_j,p}(\mathcal{K})$, $j = 1, \dots, \ell$, *and* $g_k \in V_\beta^{l-\mu_k-1/p,p}(\partial\mathcal{K}\backslash\{0\})$, $k = 1, \dots, m$, $l \geq l_0$, *there exists a unique solution*

$$u = (u_1, \dots, u_\ell) \in \prod_{j=1}^{\ell} V_\beta^{l+t_j,p}(\mathcal{K})$$

of the boundary value problem (1.2.4).

1.2.5. Asymptotics of solutions. A precise description of the behavior of solutions of the problem (1.2.4) near the vertex of the cone is given in the next theorem.

THEOREM 1.2.6. *Let*

$$u = (u_1, \dots, u_\ell) \in \prod_{j=1}^{\ell} V_\beta^{l+t_j,p}(\mathcal{K}), \quad l \geq l_0,$$

be a solution of the elliptic boundary value problem (1.2.4). *We suppose that*

$$f_j \in V_\beta^{l-s_j,p}(\mathcal{K}) \cap V_{\beta'}^{l'-s_j,q}(\mathcal{K}), \quad g_k \in V_\beta^{l-\mu_k-1/p,p}(\partial\mathcal{K}\backslash\{0\}) \cap V_{\beta'}^{l-\mu_k-1/q,q}(\partial\mathcal{K}\backslash\{0\})$$

for $j = 1, \dots, \ell$, $k = 1, \dots, m$, *and that the lines* $\operatorname{Re}\lambda = l - \beta - N/p$ *and* $\operatorname{Re}\lambda = l' - \beta' - N/q$ *do not contain eigenvalues of the pencil* $\mathfrak{A}(\lambda)$. *Then the components of the vector function* u *admit the decomposition*

$$u_j = \sum_{\nu=1}^{n} \sum_{k=1}^{I_\nu} \sum_{s=0}^{\kappa_{\nu,k}-1} c_{\nu,k,s}\, \rho^{\lambda_\nu + t_j} \sum_{\sigma=0}^{s} \frac{1}{\sigma!} (\log\rho)^\sigma\, U_j^{\nu,k,s-\sigma}(\omega) + v_j\,,$$

where $v_j \in V_{\beta'}^{l'+t_j,q}(\mathcal{K})$, $\lambda_1, \dots, \lambda_n$ *are the eigenvalues of the pencil* $\mathfrak{A}(\lambda)$ *in the strip between the lines* $\operatorname{Re}\lambda = l - \beta - N/p$ *and* $\operatorname{Re}\lambda = l' - \beta' - N/q$, *and* $(U_1^{\nu,k,s}, \dots, U_\ell^{\nu,k,s})$ *are eigenvectors (for* $s = 0$*) and generalized eigenvectors(*$s > 0$*) of the pencil* $\mathfrak{A}(\lambda)$ *corresponding to the eigenvalue* λ_ν. *Moreover,*

$$\sum_{j=1}^{\ell} \|v_j\|_{V_{\beta'}^{l'+t_j,q}(\mathcal{K})} \leq c \Big(\sum_{j=1}^{\ell} \|f_j\|_{V_{\beta'}^{l'-s_j,q}(\mathcal{K})} + \sum_{k=1}^{m} \|g_k\|_{V_{\beta'}^{l'-\mu_k-1/q,q}(\partial\mathcal{K}\backslash\{0\})} \Big),$$

where c *is a constant independent of* f_j *and* g_k.

Note that the vector-functions with the components

$$\rho^{\lambda_\nu + t_j} \sum_{\sigma=0}^{s} \frac{1}{\sigma!} (\log\rho)^\sigma\, U_j^{\nu,k,s-\sigma}(\omega), \quad j = 1, \dots, \ell$$

in Theorem 1.2.6 are solutions of the boundary value problem (1.2.4) with zero right-hand sides. As a consequence of the last theorem, the following regularity result holds.

COROLLARY 1.2.7. *Let*

$$u = (u_1, \dots, u_\ell) \in \prod_{j=1}^{\ell} V_\beta^{l+t_j,p}(\mathcal{K}), \quad l \ge l_0,$$

be a solution of the elliptic boundary value problem (1.2.4). *If the right-hand sides* f *and* g *of the boundary value problem are as in Theorem* 1.2.6 *and the closed strip between the lines* $\operatorname{Re}\lambda = l - \beta - N/p$ *and* $\operatorname{Re}\lambda = l' - \beta' - N/q$ *is free of eigenvalues of the pencil* $\mathfrak{A}(\lambda)$, *then* $u_j \in V_{\beta'}^{l'+t_j,q}(\mathcal{K})$ *for* $j = 1, \dots, \ell$.

The next regularity assertion is valid without any restrictions on β and follows essentially from the analogous result in nonweighted spaces for domains with smooth boundaries (cf. Theorems 1.1.5 and 1.1.7).

THEOREM 1.2.8. *Let* $u = (u_1, \dots, u_\ell)$ *be a solution of the elliptic boundary value problem* (1.2.4), *where*

$$f_j \in V_\beta^{l-s_j,p}(\mathcal{K}) \quad \text{and} \quad g_k \in V_\beta^{l-\mu_k-1/p,p}(\partial\mathcal{K}\backslash\{0\}),$$

$l \ge l_0$. *Suppose that* $u_j \in V_{\beta-1}^{l+t_j-1,p}(\mathcal{K})$ *and* $\chi u_j \in W^{l_0+t_j,p}(\mathcal{K})$ *for every* $\chi \in C_0^\infty(\overline{\mathcal{K}}\backslash\{0\})$, $j = 1, \dots, \ell$. *Then* $u_j \in V_\beta^{l+t_j,p}(\mathcal{K})$ *for* $j = 1, \dots, \ell$, *and*

$$\begin{aligned} \sum_{j=1}^{\ell} \|u_j\|_{V_\beta^{l+t_j,p}(\mathcal{K})} \le\ & c\Big(\sum_{j=1}^{\ell} \|f_j\|_{V_\beta^{l-s_j,p}(\mathcal{K})} + \sum_{k=1}^{m} \|g_k\|_{V_\beta^{l-\mu_k-1/p,p}(\partial\mathcal{K}\backslash\{0\})} \\ & + \sum_{j=1}^{\ell} \|u_j\|_{V_{\beta-1}^{l+t_j-1,p}(\mathcal{K})}\Big), \end{aligned}$$

where c *is a constant independent of* u.

CHAPTER 2

The Dirichlet problem for strongly elliptic systems in a dihedron

In this chapter, we deal with the Dirichlet problem for a strongly elliptic differential operator

$$L(D_x) = \sum_{|\alpha|=2m} A_\alpha\, D_x^\alpha$$

of order $2m$ in a dihedron $\mathcal{D}$, where the coefficients A_α are constant $\ell \times \ell$-matrices. The edge of $\mathcal{D}$ coincides with the x_3-axis. We consider solutions both in weighted Sobolev spaces $V_\delta^{l,p}(\mathcal{D})$ and in weighted Hölder classes $C_\delta^{l,\sigma}(\mathcal{D})$. Here, the space $V_\delta^{l,p}(\mathcal{D})$ is defined as the set of all functions u on $\mathcal{D}$ such that

$$r^{\delta-l+|\alpha|}\, \partial_x^\alpha u \in L_p(\mathcal{D}) \quad \text{for } |\alpha| \le l,$$

where $r = r(x)$ denotes the distance of the point x from the edge M of the dihedron. Similarly, the weighted Hölder space $N_\delta^{l,\sigma}(\mathcal{D})$ is defined. It is also possible to establish a theory for the Dirichlet problem in weighted Sobolev and Hölder spaces with "nonhomogeneous" norms. This is done in Chapter 5, where we consider mixed boundary value problems for second order elliptic equations which include the Dirichlet problem as a particular case.

The main results of this chapter are theorems on the unique solvability of the Dirichlet problem in the spaces $V_\delta^{l,p}(\mathcal{D})$ and $N_\delta^{l,\sigma}(\mathcal{D})$ and regularity assertions for these solutions. Here, l is an arbitrary integer, $l \ge m$, that is, we include variational solutions as well. In particular, it is shown in this chapter that a unique solution in $V_\delta^{l,p}(\mathcal{D})$ exists (for arbitrary data in the corresponding weighted Sobolev spaces) if and only if

$$-\delta_+ < \delta - l + m - 1 + 2/p < \delta_- ,$$

where δ_+ and δ_- are certain positive numbers depending on the operator L and on the dihedron $\mathcal{D}$.

We start with solutions in weighted L_2-Sobolev spaces $V_\delta^{l,2}(\mathcal{D})$ (Sections 2.2–2.4). The a priori estimates obtained for these solutions are used in Section 2.5 for the study of the Green's matrix $G(x,\xi)$. We obtain majorants for the elements of this matrix depending explicitly on the distances of x and ξ from the edge of the dihedron. For example, the inequality

$$\big|D_{x'}^\alpha D_{x_3}^j D_{\xi'}^\beta D_{\xi_3}^k G(x,\xi)\big| \le c\, \frac{|x'|^{m-1+\delta_+-|\alpha|-\varepsilon}\, |\xi'|^{m-1+\delta_--|\beta|-\varepsilon}}{|x-\xi|^{1+\delta_++\delta_-+j+k-2\varepsilon}}$$

holds for $|x-\xi| \ge \min(|x'|,|\xi'|)$. Here, we use the notation $x = (x', x_3)$, where $x' = (x_1, x_2)$. Then $|x'|$ and $|\xi'|$ are the distances of the points x and ξ from the the edge of the dihedron. The point estimates of the Green's matrix allow us to extend the results of Sections 2.2–2.4 to weighted L_p Sobolev and weighted Hölder spaces (Sections 2.6–2.8). We close the chapter with Section 2.9, where the Dirichlet problem for a strongly elliptic differential operator with variable coefficients is studied.

2.1. Weighted Sobolev spaces in a dihedron

In this section, we introduce the weighted Sobolev space $V_\delta^{l,p}(\mathcal{D})$ in a dihedron $\mathcal{D}$ both for positive and negative integer l. For $l \ge 1$ we consider also the corresponding trace spaces on the faces Γ^+ and Γ^- of the dihedron. We prove imbedding theorems and the equivalence of different norms.

2.1.1. The space $V_\delta^{l,p}(\mathcal{D})$. Let r, φ be the polar coordinates of the point $x' = (x_1, x_2)$, i.e. $r = (x_2^2 + x_2^2)^{1/2}$, $\tan \varphi = x_2/x_1$, and let

$$K = \{x' = (x_1, x_2) : \ 0 < r < \infty, -\theta/2 < \varphi < \theta/2\} \tag{2.1.1}$$

be a two-dimensional wedge with the sides $\gamma^\pm = \{x' : \ \varphi = \pm\theta/2\}$ and opening θ, $0 < \theta \le 2\pi$. Furthermore, let

$$\mathcal{D} = K \times \mathbb{R} = \{x = (x', x_3) : \ x' = (x_1, x_2) \in K, \ x_3 \in \mathbb{R}\} \tag{2.1.2}$$

be a *dihedron* in $\mathbb{R}^3$ with the faces $\Gamma^\pm = \gamma^\pm \times \mathbb{R}$ and the edge $M = \overline{\Gamma^+} \cap \overline{\Gamma^-}$.

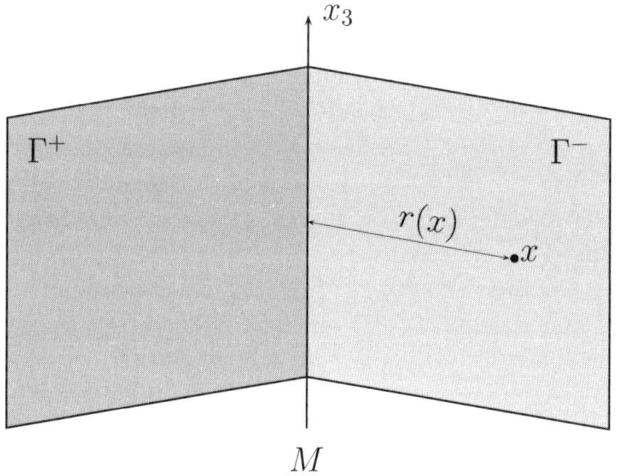

FIGURE 3. A dihedron in $\mathbb{R}^3$

For arbitrary integer l, real δ and real p, $1 < p < \infty$, let $V_\delta^{l,p}(K)$ and $V_\delta^{l,p}(\mathcal{D})$ be the weighted Sobolev spaces of functions on K and $\mathcal{D}$ with finite norms

$$\|u\|_{V_\delta^{l,p}(K)} = \Big(\int_K \sum_{|\alpha| \le l} r^{p(\delta - l + |\alpha|)} \, \big|\partial_{x'}^\alpha u(x')\big|^p \, dx' \Big)^{1/p} \tag{2.1.3}$$

and

$$\|u\|_{V_\delta^{l,p}(\mathcal{D})} = \Big(\int_{\mathcal{D}} \sum_{|\alpha| \le l} r^{p(\delta - l + |\alpha|)} \, \big|\partial_{x}^\alpha u(x)\big|^p \, dx \Big)^{1/p}, \tag{2.1.4}$$

respectively. Note that the set $C_0^\infty(\overline{\mathcal{D}}\backslash M)$ of all infinitely differentiable functions with compact support in $\overline{\mathcal{D}}\backslash M$ is dense in $V_\delta^{l,p}(\mathcal{D})$. Therefore, the spaces $V_\delta^{l,p}(\mathcal{D})$ can be also defined as the closure of the set $C_0^\infty(\overline{\mathcal{D}}\backslash M)$ with respect to the norm

(2.1.4). Analogously, $V_\delta^{l,p}(K)$ is the closure of the set $C_0^\infty(\overline{K}\backslash\{0\})$ with respect to the norm (2.1.3).

2.1.2. Imbeddings. We conclude immediately from the definition of the space $V_\delta^{l,p}(\mathcal{D})$ that

$$V_\delta^{l,p}(\mathcal{D}) \subset V_{\delta-1}^{l-1,p}(\mathcal{D}) \subset \cdots \subset V_{\delta-l}^{0,p}(\mathcal{D}).$$

Furthermore, the following more general assertion is true.

LEMMA 2.1.1. *Let $1 < p \le q < \infty$, $l-3/p > k-3/q$, and $\delta-l+3/p = \gamma-k+3/q$. Then the space $V_\delta^{l,p}(\mathcal{D})$ is continuously imbedded in $V_\gamma^{k,q}(\mathcal{D})$.*

P r o o f. Let $\mathcal{D}_j$ denote the set of all $x \in \mathcal{D}$ such that $2^{-j} < |x'| < 2^{-j+1}$. Since the space $W^{l,p}(\mathcal{D}_0)$ is continuously imbedded in $W^{k,q}(\mathcal{D}_0)$, the estimate

$$\sum_{|\alpha|\le k} \int_{\mathcal{D}_0} |\partial_x^\alpha v(x)|^q\, dx \le c\Big(\sum_{|\alpha|\le l} \int_{\mathcal{D}_0} |\partial_x^\alpha v(x)|^p\, dx\Big)^{q/p}$$

is satisfied for all $v \in W^{l,p}(\mathcal{D}_0)$. Multiplying this inequality by $2^{-qj(\delta-l+3/p)}$ and substituting $x = 2^j y$, we obtain

$$\sum_{|\alpha|\le k} \int_{\mathcal{D}_j} |y'|^{q(\gamma-k+|\alpha|)}\, |\partial_y^\alpha u(y)|^q\, dy \le c\Big(\sum_{|\alpha|\le l} \int_{\mathcal{D}_j} |y'|^{p(\delta-l+|\alpha|)} |\partial_y^\alpha u(y)|^p\, dy\Big)^{q/p},$$

where $u(y) = v(2^j y)$. Here the constant c is independent of u and j. Summing up over all integer j, we arrive at the inequality

$$\|u\|^q_{V_\gamma^{k,q}(\mathcal{D})} \le c\, \|u\|^q_{V_\delta^{l,p}(\mathcal{D})}\, .$$

The proof is complete. □

Similarly, we prove the following lemma .

LEMMA 2.1.2. *Let $1 < p < q < \infty$, $l - 3/p > -3/q$, and $\gamma + 2/q = \delta - l + 2/p$. If $\partial_{x_3}^k u \in V_{\delta-k}^{l-k,p}(\mathcal{D})$ for $k = 0, \dots, l$, then $u \in V_\gamma^{0,q}(\mathcal{D})$ and*

$$\|u\|_{V_\gamma^{0,q}(\mathcal{D})} \le c \sum_{k=0}^{l} \|\partial_{x_3}^k u\|_{V_{\delta-k}^{l-k,p}(\mathcal{D})}$$

with a constant c independent of u.

P r o o f. Let $\mathcal{D}_j$ be the same set as in the proof of Lemma 2.1.1. Then it follows from the continuity of the imbedding $W^{l,p}(\mathcal{D}_0) \subset L_q(\mathcal{D}_0)$ that

$$\int_{\mathcal{D}_0} |v(x)|^q\, dx \le c\Big(\sum_{|\alpha|+k\le l} \int_{\mathcal{D}_0} |\partial_{x'}^\alpha \partial_{x_3}^k v(x)|^p\, dx\Big)^{q/p}$$

for all $v \in W^{l,p}(\mathcal{D}_0)$. Multiplying the last inequality by $2^{-qj(\delta-l+2/p)}$ and substituting $x' = 2^j y'$, $x_3 = y_3$, we obtain

$$\int_{\mathcal{D}_j} |y'|^{q\gamma}\, |u(y)|^q\, dy \le c\Big(\sum_{|\alpha|+k\le 2} \int_{\mathcal{D}_j} |y'|^{p(\delta-l+|\alpha|)} |\partial_{y'}^\alpha \partial_{y_3}^k u(y)|^p\, dy\Big)^{q/p},$$

where $u(y) = v(2^j y', y_3)$. Summing up over all integer j, we arrive at the inequality

$$\|u\|^q_{V^{0,q}_\gamma(\mathcal{D})} \le c \sum_{k=0}^{l} \|\partial^k_{y_3} u\|^q_{V^{l-k,p}_{\delta-k}(\mathcal{D})} .$$

The result follows. □

Finally, we prove a weighted L_∞-estimate for functions of the spaces $V^{l,p}_\delta(\mathcal{D})$.

LEMMA 2.1.3. *If $u \in V^{l,p}_\delta(\mathcal{D})$ and $l > 3/p$, then $r^{\delta-l+3/p}\, u \in L_\infty(\mathcal{D})$ and*

$$\|r^{\delta-l+3/p}\, u\|_{L_\infty(\mathcal{D})} \le c\, \|u\|_{V^{l,p}_\delta(\mathcal{D})}$$

with a constant c independent of u.

P r o o f. Let $B(x)$ be the ball with radius $|x'|/2$ centered about x. If $x \in \mathcal{D}$ and $|x'| = 1$, then by Sobolev's imbedding theorem, the inequality

$$|u(x)| \le c\, \|u\|_{W^{l,p}(\mathcal{D}\cap B(x))}$$

is valid for every $u \in V^{l,p}_\delta(\mathcal{D})$. Now let $|x'| = t$. Substituting $y = t^{-1}x$ and $v(x) = u(tx)$, we obtain

$$|u(x)|^p = |v(y)|^p \le c\, \|v\|^p_{W^{l,p}(\mathcal{D}\cap B(y))} = c \sum_{|\alpha|\le l} t^{p|\alpha|-3}\, \|D^\alpha u\|^p_{L_p(\mathcal{D}\cap B(x))} .$$

Since $t/2 < r(\xi) < 3t/2$ for $\xi \in B(x)$, we conclude that

$$|u(x)|^p \le c\, t^{p(l-\delta)-3}\, \|u\|^p_{V^{l,p}_\delta(\mathcal{D})} .$$

This completes the proof. □

2.1.3. Equivalent norms. As in the case of a cone, we will frequently use another equivalent norm in $V^{l,p}_\delta(\mathcal{D})$. For this end, we introduce infinitely differentiable real-valued functions ζ_ν depending only on $r = |x'|$ such that

$$\text{(2.1.5)} \qquad \operatorname{supp}\zeta_\nu \subset \big(2^{\nu-1}, 2^{\nu+1}\big), \quad \big|\partial^j_r \zeta_\nu(r)\big| \le c_j\, 2^{-j\nu}, \quad \sum_{\nu=-\infty}^{+\infty} \zeta_\nu = 1.$$

Here c_j are constants independent of r and ν.

LEMMA 2.1.4. *There exist positive constants c_1 and c_2 such that*

$$c_1\, \|u\|_{V^{l,p}_\delta(\mathcal{D})} \le \Big(\sum_{\nu=-\infty}^{+\infty} \|\zeta_\nu u\|^p_{V^{l,p}_\delta(\mathcal{D})} \Big)^{1/p} \le c_2\, \|u\|_{V^{l,p}_\delta(\mathcal{D})}$$

for all $u \in V^{l,p}_\delta(\mathcal{D})$.

P r o o f. From (2.1.5) it follows that

$$\|\zeta_\nu u\|^p_{V^{l,p}_\delta(\mathcal{D})} \le c \int\limits_{\substack{\mathcal{D}\\ 2^{\nu-1}<|x'|<2^{\nu+1}}} \sum_{|\alpha|\le l} r^{p(\delta-l+|\alpha|)}\, |\partial^\alpha_x u(x)|^p\, dx$$

with a constant c independent of ν. Hence

$$\sum_{\nu=-\infty}^{+\infty} \|\zeta_\nu u\|^p_{V^{l,p}_\delta(\mathcal{D})} \le 2c\, \|u\|^p_{V^{l,p}_\delta(\mathcal{D})}.$$

Since $\zeta_{\nu-1}(x)+\zeta_\nu(x)+\zeta_{\nu+1}(x)=1$ for $2^{\nu-1}<|x'|<2^{\nu+1}$, we obtain

$$\begin{aligned}\|u\|^p_{V^{l,p}_\delta(\mathcal{D})} &= \frac12 \sum_{\nu=-\infty}^{+\infty} \int\limits_{\substack{\mathcal{D}\\ 2^{\nu-1}<|x'|<2^{\nu+1}}} \sum_{|\alpha|\le l} r^{p(\delta-l+|\alpha|)}\, |\partial_x^\alpha u(x)|^p\, dx \\ &\le \frac12 \sum_{\nu=-\infty}^{+\infty} \|(\zeta_{\nu-1}+\zeta_\nu+\zeta_{\nu+1})u\|^p_{V^{l,p}_\delta(\mathcal{D})} \le \frac{3^{p-1}}{2} \sum_{\nu=-\infty}^{+\infty} \|\zeta_\nu u\|^p_{V^{l,p}_\delta(\mathcal{D})}.\end{aligned}$$

This proves the lemma. □

By means of the last lemma, we prove the following weighted version of Ehrling's lemma.

LEMMA 2.1.5. *Let $l\ge 1$. Then for every positive ε, there exists a constant $c(\varepsilon)$ such that*

$$\|u\|_{V^{l-1,p}_{\delta-1}(\mathcal{D})} \le \varepsilon\, \|u\|_{V^{l,p}_\delta(\mathcal{D})} + c(\varepsilon)\, \|u\|_{V^{0,p}_{\delta-l}(\mathcal{D})}$$

for all $u\in V^{l,p}_\delta(\mathcal{D})$.

P r o o f. We introduce the subdomains

$$\mathcal{D}_0=\{x\in\mathcal{D}:\ 1/2<r<2\},\quad \mathcal{D}_{0,j}=\{x\in\mathcal{D}:\ 1/2<r<2,\ j-1<x_3<j\}$$

of $\mathcal{D}$. By Ehrling's lemma, the inequality

$$\|v\|^p_{W^{l-1,p}(\mathcal{D}_{0,j})} \le \varepsilon\, \|v\|^p_{W^{l,p}(\mathcal{D}_{0,j})} + c(\varepsilon)\, \|v\|^p_{L_p(\mathcal{D}_{0,j})}$$

is satisfied for all $v\in W^{l,p}(\mathcal{D}_0)$. Here, the constant $c(\varepsilon)$ is independent of v and j. Summing up over all integer j, we obtain

$$\|v\|^p_{W^{l-1,p}(\mathcal{D}_0)} \le \varepsilon\, \|v\|^p_{W^{l,p}(\mathcal{D}_0)} + c(\varepsilon)\, \|v\|^p_{L_p(\mathcal{D}_0)}. \tag{2.1.6}$$

Let $u\in V^{l,p}_\delta(\mathcal{D})$, and let ζ_ν be the same functions as in Lemma 2.1.4. Then the function $v(x)=\zeta_\nu(2^\nu x)\, u(2^\nu x)$ vanishes outside $\mathcal{D}_0$. Since on the set of functions with support in $\overline{\mathcal{D}}_0$, the $W^{l,p}$-norm is equivalent to the $V^{l,p}_\delta$-norm, the estimate (2.1.6) implies

$$\|v\|^p_{V^{l-1,p}_{\delta-1}(\mathcal{D})} \le \varepsilon\, \|v\|^p_{V^{l,p}_\delta(\mathcal{D})} + c_1(\varepsilon)\, \|v\|^p_{V^0_{\delta-l}(\mathcal{D})}.$$

Using the equality

$$\|v\|_{V^{l,p}_\delta(\mathcal{D})} = 2^{\nu(l-\delta-3/p)}\, \|\zeta_\nu u\|_{V^{l,p}_\delta(\mathcal{D})}$$

and analogous equalities for the norms of v in $V^{l-1,p}_{\delta-1}(\mathcal{D})$ and $V^{0,p}_{\delta-l}(\mathcal{D})$, we get

$$\|\zeta_\nu u\|^p_{V^{l-1,p}_{\delta-1}(\mathcal{D})} \le \varepsilon\, \|\zeta_\nu u\|^p_{V^{l,p}_\delta(\mathcal{D})} + c_1(\varepsilon)\, \|\zeta_\nu u\|^p_{V^0_{\delta-l}(\mathcal{D})}.$$

The reference to Lemma 2.1.4 completes the proof. □

By Lemma 2.1.5, the norm in $V^{l,p}_\delta(\mathcal{D})$ is equivalent to the norm

$$\Big(\int_{\mathcal{D}} \Big(r^{p(\delta-l)}\, |u(x)|^p + r^{p\delta} \sum_{|\alpha|=l} |\partial_x^\alpha u(x)|^p\Big)\, dx\Big)^{1/p}.$$

Moreover, the estimate given in the next lemma is valid in the space $V^{l,p}_\delta(\mathcal{D})$.

LEMMA 2.1.6. *There exists a constant c such that*

$$\|u\|^p_{V^{l,p}_\delta(\mathcal{D})} \le c \int_{\mathcal{D}} \Big(r^{p(\delta-l)}\, |u(x)|^p + r^{p\delta} \sum_{j=1}^3 |\partial^l_{x_j} u(x)|^p \Big)\, dx \tag{2.1.7}$$

for all $u \in V^{l,p}_\delta(\mathcal{D})$.

P r o o f. Let $\mathcal{D}_0$, $\mathcal{D}_{0,j}$ be the same subdomains of $\mathcal{D}$ as in the proof of Lemma 2.1.5. By [**195**, Theorem 4.2.4], there exists a constant c independent of v and j such that

$$\|v\|^p_{W^{l,p}(\mathcal{D}_{0,j})} \le c \int_{\mathcal{D}_{0,j}} \Big(|v(x)|^p + \sum_{j=1}^3 \big|\partial^l_{x_j} v(x)\big|^p \Big)\, dx$$

for all $v \in W^{l,p}(\mathcal{D}_0)$. Summing up over all integer j, we obtain the same estimate with $\mathcal{D}_0$ instead of $\mathcal{D}_{0,j}$. Arguing as in the proof of Lemma 2.1.5, we conclude that

$$\begin{aligned}
\|\zeta_\nu u\|^p_{V^{l,p}_\delta(\mathcal{D})} &\le c \int_{\mathcal{D}} \Big(r^{p(\delta-l)}\, |\zeta_\nu u|^p + r^{p\delta} \sum_{j=1}^3 \big|\partial^l_{x_j}(\zeta_\nu u)\big|^p \Big)\, dx \\
&\le c \int_{\mathcal{D}} \Big(r^{p(\delta-l)}\, |\zeta_\nu u|^p + r^{p\delta} \sum_{j=1}^3 \big|\zeta_\nu \partial^l_{x_j} u\big|^p \Big)\, dx + c'\, \|\eta_\nu u\|^p_{V^{l-1,p}_{\delta-1}(\mathcal{D})},
\end{aligned}$$

where ζ_ν is the same function as in Lemma 2.1.4 and $\eta_\nu = \zeta_{\nu-1} + \zeta_\nu + \zeta_{\nu+1}$. Summing up over all ν and using Lemmas 2.1.4 and 2.1.5, we arrive at (2.1.7). □

2.1.4. The space $V^{l,p}_\delta(\mathcal{D})$ with negative l. Let l be a nonnegative integer. Then the spaces $\overset{\circ}{V}{}^{l,p}_\delta(K)$ and $\overset{\circ}{V}{}^{l,p}_\delta(\mathcal{D})$ are defined as the closure of the sets $C_0^\infty(K)$ and $C_0^\infty(\mathcal{D})$ with respect to the norms (2.1.3) and (2.1.4), respectively. Obviously, $\overset{\circ}{V}{}^{0,p}_\delta(\mathcal{D}) = V^{0,p}_\delta(\mathcal{D})$, whereas

$$\overset{\circ}{V}{}^{l,p}_\delta(\mathcal{D}) = \big\{ u \in V^{l,p}_\delta(\mathcal{D}) :\ \partial^\alpha_x u = 0 \text{ on } \Gamma^\pm \text{ for } |\alpha| \le l-1 \big\}$$

if $l \ge 1$. The dual space of $V^{0,p}_\delta(\mathcal{D})$ coincides with $V^{0,p'}_{-\delta}(\mathcal{D})$, where $p' = p/(p-1)$. This means that every linear and continuous functional on $V^{0,p}_\delta(\mathcal{D})$ has the form

$$f(v) = (v, u)_{\mathcal{D}}\,,$$

where $u \in V^{0,p'}_{-\delta}(\mathcal{D})$ and $(\cdot,\cdot)_{\mathcal{D}}$ denotes the scalar product in $L_2(\mathcal{D})$. We define the space $V^{-l,p}_\delta(\mathcal{D})$ as the dual space of $\overset{\circ}{V}{}^{l,p'}_{-\delta}(\mathcal{D})$ with the norm

$$\|u\|_{V^{-l,p}_\delta(\mathcal{D})} = \sup \big\{ |(u,v)_{\mathcal{D}}| \ :\ v \in \overset{\circ}{V}{}^{l,p'}_{-\delta}(\mathcal{D}),\ \|v\|_{V^{l,p'}_{-\delta}(\mathcal{D})} = 1 \big\}. \tag{2.1.8}$$

Analogously, the space $V^{-l,p}_\delta(K)$ is defined as the dual space of $\overset{\circ}{V}{}^{l,p'}_{-\delta}(K)$. In the sequel, we will restrict ourselves to the spaces $V^{-l,p}_\delta(\mathcal{D})$. The same results are valid for the spaces $V^{-l,p}_\delta(K)$.

LEMMA 2.1.7. $V^{-l,p}_\delta(\mathcal{D})$ *is the set of all distributions of the form*

$$u = \sum_{|\alpha| \le l} \partial^\alpha_x u_\alpha, \quad \textit{where } u_\alpha \in V^{0,p}_{\delta+l-|\alpha|}(\mathcal{D}).$$

The norm

$$\|u\| = \inf\Big\{ \sum_{|\alpha|\le l} \|u_\alpha\|_{V^{0,p}_{\delta+l-|\alpha|}(\mathcal{D})} \ :\ u_\alpha \in V^{0,p}_{\delta+l-|\alpha|}(\mathcal{D}),\ \sum_{|\alpha|\le l} \partial_x^\alpha u_\alpha = u\Big\}$$

is equivalent to the norm (2.1.8).

P r o o f. Let B be the Banach space of all vector functions $\phi = (\phi_\alpha)_{|\alpha|\le l}$, $\phi_\alpha \in V^{0,p'}_{-\delta-l+|\alpha|}(\mathcal{D})$, with the norm

$$\|\phi\|_B = \sum_{|\alpha|\le l} \|\phi_\alpha\|_{V^{0,p'}_{-\delta-l+|\alpha|}(\mathcal{D})}.$$

By M, we denote the subspace of all $\phi = (\phi_\alpha)_{|\alpha|\le l}$ such that $\phi_\alpha = \partial_x^\alpha v$, where $v \in \overset{\circ}{V}{}^{l,p'}_{-\delta}(\mathcal{D})$. The subset M is closed since $\overset{\circ}{V}{}^{l,p'}_{-\delta}(\mathcal{D})$ is a Banach space. Suppose that f is a linear and continuous functional on $\overset{\circ}{V}{}^{l,p'}_{-\delta}(\mathcal{D})$. Then we define the functional F on M by

$$F(\phi) = f(v) \quad \text{for } \phi = (\partial_x^\alpha v)_{|\alpha|\le l},\ v \in \overset{\circ}{V}{}^{l,p'}_{-\delta}(\mathcal{D}).$$

By Hahn-Banach's theorem, there exists an extension of the functional F to B with the same norm. However, every linear and continuous functional F_1 on B has the form

$$F_1(\phi) = \int_{\mathcal{D}} \sum_{|\alpha|\le l} u_\alpha\, \phi_\alpha\, dx \quad \text{for all } \phi = (\phi_\alpha)_{|\alpha|\le l} \in \prod_{|\alpha|\le l} V^{0,p'}_{-\delta-l+|\alpha|}(\mathcal{D}),$$

where $u_\alpha \in V^{0,p}_{\delta+l-|\alpha|}(\mathcal{D})$. The result follows. □

Next, we prove the assertion of Lemma 2.1.4 for negative l.

LEMMA 2.1.8. *The norm* (2.1.8) *is equivalent to the norm*

$$\Big(\sum_{\nu=-\infty}^{+\infty} \|\zeta_\nu u\|^p_{V_\delta^{-l,p}(\mathcal{D})}\Big)^{1/p},$$

where ζ_ν are functions satisfying (2.1.5).

P r o o f. We show first that there exist a constant c such that

$$\|u\|_{V_\delta^{-l,p}(\mathcal{D})} \le c\Big(\sum_{\nu=-\infty}^{+\infty} \|\zeta_\nu u\|^p_{V_\delta^{-l,p}(\mathcal{D})}\Big)^{1/p} \tag{2.1.9}$$

for all $u \in V_\delta^{-l,p}(\mathcal{D})$. Let $v \in \overset{\circ}{V}{}^{l,p'}_{-\delta}(\mathcal{D})$, $p' = p/(p-1)$, and let $\eta_\nu = \zeta_{\nu-1} + \zeta_\nu + \zeta_{\nu+1}$ for $\nu = 0, \pm1, \pm2, \dots$. Then $\zeta_\nu\eta_\nu = \zeta_\nu$ and

$$\big|(u,v)_{\mathcal{D}}\big| = \Big|\sum_\nu (\zeta_\nu u, \eta_\nu v)_{\mathcal{D}}\Big| \le \Big(\sum_\nu \|\zeta_\nu u\|^p_{V_\delta^{-l,p}(\mathcal{D})}\Big)^{1/p} \Big(\sum_\nu \|\eta_\nu v\|^{p'}_{V^{l,p'}_{-\delta}(\mathcal{D})}\Big)^{1/p'}.$$

It follows from Lemma 2.1.4 that the norm in $V^{l,p'}_{-\delta}(\mathcal{D})$ is equivalent to

$$\Big(\sum_\nu \|\eta_\nu v\|^{p'}_{V^{l,p'}_{-\delta}(\mathcal{D})}\Big)^{1/p'}.$$

Consequently,

$$\big|(u,v)_{\mathcal{D}}\big| \le c \Big(\sum_\nu \|\zeta_\nu u\|^p_{V_\delta^{-l,p}(\mathcal{D})}\Big)^{1/p} \|v\|^p_{V_{-\delta}^{l,p'}(\mathcal{D})}$$

for all $u \in V_\delta^{-l,p}(\mathcal{D})$, $v \in \overset{\circ}{V}{}^{l,p'}_{-\delta}(\mathcal{D})$. This proves (2.1.9).

We prove the converse inequality. Let $u \in V_\delta^{-l,p}(\mathcal{D})$. Then there exist functions $u_\alpha \in V^{0,p}_{\delta+l-|\alpha|}(\mathcal{D})$ such that

$$u = \sum_{|\alpha|\le l} \partial_x^\alpha u_\alpha \quad \text{and} \quad \sum_{|\alpha|\le l} \|u_\alpha\|_{V^{0,p}_{\delta+l-|\alpha|}(\mathcal{D})} \le c\, \|u\|_{V_\delta^{-l,p}(\mathcal{D})}$$

with a constant c independent of u (see Lemma 2.1.7). Thus,

$$\begin{aligned} \big|(\zeta_\nu u, v)_{\mathcal{D}}\big| &\le \sum_{|\alpha|\le l} \big|\big(\eta_\nu u_\alpha, \partial_x^\alpha(\zeta_\nu v)\big)_{\mathcal{D}}\big| \\ &\le \sum_{|\alpha|\le l} \|\eta_\nu u_\alpha\|_{V^{0,p}_{\delta+l-|\alpha|}(\mathcal{D})} \, \|\partial_x^\alpha(\zeta_\nu v)\|_{V^{0,p'}_{-\delta-l+|\alpha|}(\mathcal{D})} \\ &\le c \sum_{|\alpha|\le l} \|\eta_\nu u_\alpha\|_{V^{0,p}_{\delta+l-|\alpha|}(\mathcal{D})} \, \|v\|_{V^{l,p'}_{-\delta}(\mathcal{D})} \end{aligned}$$

for all $v \in \overset{\circ}{V}{}^{l,p'}_{-\delta}(\mathcal{D})$. Consequently,

$$\begin{aligned} \sum_{\nu=-\infty}^{+\infty} \|\zeta_\nu u\|^p_{V_\delta^{-l,p}(\mathcal{D})} &\le c_1 \sum_{\nu=-\infty}^{+\infty} \sum_{|\alpha|\le l} \|\eta_\nu u_\alpha\|^p_{V^{0,p}_{\delta+l-|\alpha|}(\mathcal{D})} \\ &\le c_2 \sum_{|\alpha|\le l} \|u_\alpha\|^p_{V^{0,p}_{\delta+l-|\alpha|}(\mathcal{D})} \le c_3\, \|u\|^p_{V_\delta^{-l,p}(\mathcal{D})}. \end{aligned}$$

The lemma is proved. □

2.1.5. Trace spaces. We denote the trace space for $V_\delta^{l,p}(\mathcal{D})$, $l \ge 1$, on the face $\Gamma^\pm$ by $V_\delta^{l-1/p,p}(\Gamma^\pm)$. The norm in this space is given by

$$(2.1.10) \qquad \|u\|_{V_\delta^{l-1/p,p}(\Gamma^\pm)} = \inf\big\{ \|v\|_{V_\delta^{l,p}(\mathcal{D})} \,:\, v \in V_\delta^{l,p}(\mathcal{D}),\ v = u \text{ on } \Gamma^\pm \big\}.$$

We will often use an equivalent norm given in the following lemma.

LEMMA 2.1.9. *Let ζ_ν be infinitely differentiable functions depending only on $r = |x'|$ and satisfying the conditions* (2.1.5). *Then the norm* (2.1.10) *is equivalent to*

$$\Big(\sum_{\nu=-\infty}^{+\infty} \|\zeta_\nu u\|^p_{V_\delta^{l-1/p,p}(\Gamma^\pm)} \Big)^{1/p}.$$

P r o o f. Let u be an arbitrary function in $V_\delta^{l-1/p,p}(\Gamma^+)$. Then there exists an extension $v \in V_\delta^{l,p}(\mathcal{D})$ of u satisfying the inequality

$$\|v\|^p_{V_\delta^{l,p}(\mathcal{D})} \le 2\, \|u\|^p_{V_\delta^{l-1/p,p}(\Gamma^+)}.$$

Consequently by means of Lemma 2.1.4, we obtain

$$\sum_{\nu=-\infty}^{+\infty} \|\zeta_\nu u\|^p_{V_\delta^{l-1/p,p}(\Gamma^+)} \le \sum_{\nu=-\infty}^{+\infty} \|\zeta_\nu v\|^p_{V_\delta^{l,p}(\mathcal{D})} \le c\, \|v\|^p_{V_\delta^{l,p}(\mathcal{D})} \le 2c\, \|u\|^p_{V_\delta^{l-1/p,p}(\Gamma^+)}.$$

On the other hand, for every $\nu = 0, \pm 1, \pm 2, \ldots$, there exists an extension $v_\nu \in V_\delta^{l,p}(\mathcal{D})$ of $\zeta_\nu u$ satisfying the inequality

$$\|v_\nu\|^p_{V_\delta^{l,p}(\mathcal{D})} \le 2\, \|\zeta_\nu u\|^p_{V_\delta^{l-1/p,p}(\Gamma+)} .$$

Obviously, the function $w_\nu = (\zeta_{\nu-1} + \zeta_\nu + \zeta_{\nu+1}) v_\nu$ is also an extension of $\zeta_\nu u$ which satisfies the estimate

$$\|w_\nu\|^p_{V_\delta^{l,p}(\mathcal{D})} \le c\, \|\zeta_\nu u\|^p_{V_\delta^{l-1/p,p}(\Gamma+)}$$

with a constant c independent of u and ν. Since $w = \sum_{\nu=-\infty}^{+\infty} w_\nu$ is an extension of u and $w_\nu(x) = 0$ for $|x'| < 2^{\nu-2}$ and $|x'| > 2^{\nu+2}$, we obtain

$$\begin{aligned}
\|u\|^p_{V_\delta^{l-1/p,p}(\Gamma+)} &\le \Big\| \sum_{\nu=-\infty}^{+\infty} w_\nu \Big\|^p_{V_\delta^{l,p}(\mathcal{D})} \\
&= \frac{1}{2} \sum_{\mu=-\infty}^{+\infty} \sum_{|\alpha|\le l} \int\limits_{\substack{\mathcal{D} \\ 2^{\mu-1}<|x'|<2^{\mu+1}}} r^{p(\delta-l+|\alpha|)} \Big| \sum_{\nu=\mu-2}^{\mu+2} \partial_x^\alpha w_\nu \Big|^p \, dx \\
&\le c \sum_{\mu=-\infty}^{+\infty} \sum_{\nu=\mu-2}^{\mu+2} \|w_\nu\|^p_{V_\delta^{l,p}(\mathcal{D})} \le c' \sum_{\nu=-\infty}^{+\infty} \|\zeta_\nu u\|^p_{V_\delta^{l-1/p,p}(\Gamma+)} .
\end{aligned}$$

The proof is complete. □

In the next lemma, we give two explicit integral representations for the norm in the space $V_\delta^{l-1/p,p}(\Gamma^\pm)$.

LEMMA 2.1.10. *The norm (2.1.10) is equivalent to*

$$\begin{aligned}
\Big(\int\limits_{\Gamma^\pm} \sum_{|\alpha|\le l-1} & r^{p(\delta-l+|\alpha|)+1} \, |\partial^\alpha u(r,x_3)|^p \, dr\, dx_3 \\
&+ \sum_{|\alpha|=l-1} \int\limits_{\Gamma^\pm} \int\limits_{\Gamma^\pm} \big| r^\delta \, (\partial^\alpha) u(r,x_3) - \rho^\delta \, (\partial^\alpha u)(\rho, x_3') \big|^p \, \frac{dr\, dx_3\, d\rho\, dx_3'}{(|r-\rho| + |x_3 - x_3'|)^{p+1}} \Big)^{1/p} .
\end{aligned}$$

Another equivalent norm in $V_\delta^{l-1/p,p}(\Gamma^\pm)$ *is*

$$\begin{aligned}
\Big(\int_{\gamma^\pm} \int_{\mathbb{R}} \int_{\mathbb{R}} & r^{2\delta} \, \big| \partial_{x_3}^{l-1} u(r,x_3) - \partial_{y_3}^{l-1}(r,y_3) \big|^2 \, \frac{dx_3\, dy_3}{|x_3-y_3|^2} \, dr \\
&+ \int_{\mathbb{R}} \int_{\gamma^\pm} \int_{\gamma^\pm} \big| r_1^\delta (\partial_r^{l-1} u)(r_1,x_3) - r_2^\delta (\partial_r^{l-1} u)(r_2,x_3) \big|^2 \, \frac{dr_1\, dr_2}{|r_1-r_2|^2} \, dx_3 \\
&+ \int_{\Gamma^\pm} \sum_{j=0}^{l-1} r^{2(\delta-l+j)+1} \, |\partial_r^j u(r,x_3)|^p \, dr\, dx_3 \Big)^{1/2} .
\end{aligned}$$

P r o o f. We denote the first norm of the lemma by $\|\cdot\|_1$. Let $u \in V_\delta^{l-1/p,p}(\Gamma^\pm)$, and let ζ_ν be infinitely differentiable functions depending only on $r = |x'|$ and satisfying the conditions (2.1.5). Furthermore, let $\tilde{\zeta}_\nu(x) = \zeta_\nu(2^\nu x)$ and $v_\nu(x) = u(2^\nu x)$. Then $\tilde{\zeta}_\nu(x)\, v_\nu(x) = 0$ for $|x'| < 1/2$ and $|x'| > 2$. On the set of all functions vanishing for $|x'| < 1/2$ and $|x'| > 2$, the $V_\delta^{l-1/p,p}$ norm is equivalent to the nonweighted Sobolev-Slobodetskiĭ norm

$$\|u\|_{W^{l-1/p,p}(\Gamma^\pm)} = \Big(\int\limits_{\Gamma^\pm} \sum_{|\alpha|\le l-1} \big|\partial^\alpha u(r,x_3)\big|^p \, dr\, dx_3 + \sum_{|\alpha|=l-1} \int\limits_{\Gamma^\pm}\int\limits_{\Gamma^\pm} \frac{\big|(\partial^\alpha)u(r,x_3) - (\partial^\alpha u)(\rho,x_3')\big|^p}{(|r-\rho| + |x_3 - x_3'|)^{p+1}} \, dr\, dx_3\, d\rho\, dx_3' \Big)^{1/p}$$

and consequently also to the norm $\|\cdot\|_1$. This means that there exist positive constants c_1, c_2 independent of u and ν such that

$$c_1 \|\tilde{\zeta}_\nu v_\nu\|_{V_\delta^{l-1/p,p}(\Gamma^\pm)} \le \|\tilde{\zeta}_\nu v_\nu\|_1 \le c_2 \|\tilde{\zeta}_\nu v_\nu\|_{V_\delta^{l-1/p,p}(\Gamma^\pm)},$$

Multiplying this inequality by $2^{\nu(\delta-l+3/p)}$, we obtain

$$c_1 \|\zeta_\nu u\|_{V_\delta^{l-1/p,p}(\Gamma^\pm)} \le \|\zeta_\nu u\|_1 \le c_2 \|\zeta_\nu u\|_{V_\delta^{l-1/p,p}(\Gamma^\pm)}$$

with the same constants c_1 and c_2. Applying Lemma 2.1.9 and the analogous assertion for the norm $\|\cdot\|_1$, we obtain the equivalence of the norms (2.1.10) and $\|\cdot\|_1$. Analogously the equivalence of the second norm of the lemma holds. □

2.2. Variational solutions of the Dirichlet problem

We consider the Dirichlet problem

(2.2.1) $$L(D_x)\, u = f \ \text{ in } \mathcal{D}, \quad \frac{\partial^{k-1} u}{\partial n^{k-1}} = g_k^\pm \ \text{ on } \Gamma^\pm,\ k = 1,\dots,m,$$

for the differential operator

(2.2.2) $$L(D_x) = \sum_{|\alpha|=2m} A_\alpha\, D_x^\alpha .$$

Here A_α are constant $\ell\times\ell$-matrices, and n denotes the outer unit normal vector on Γ^+ and Γ^-, respectively. Throughout this chapter, we suppose that $L(D_x)$ is strongly elliptic. We prove in this section that the boundary value problem (2.2.1), (2.2.2) is uniquely solvable in the weighted Sobolev space $V_0^{m,2}(\mathcal{D})^\ell$ for arbitrary $f \in V_0^{-m,2}(\mathcal{D})^\ell$ and $g_k^\pm \in V_0^{m-k+1/2,2}(\Gamma^\pm)^\ell$, $k = 1,\dots,m$.

2.2.1. Reduction to zero boundary data.

When solving the boundary value problem (2.2.1), it is useful to apply the following lemma which allows one to restrict oneself to zero boundary data $g_k^\pm$.

LEMMA 2.2.1. *Let $g_k^\pm \in V_\delta^{l-k+1-1/p,p}(\Gamma^\pm)$, $k = 1,2,\dots,m$, $l \ge m$. Then there exists a function $u \in V_\delta^{l,p}(\mathcal{D})$ such that*

(2.2.3) $$\frac{\partial^{k-1} u}{\partial n^{k-1}} = g_k^\pm \quad \textit{on } \Gamma^\pm \quad \textit{for } k = 1,\dots,m,$$

and

(2.2.4) $$\|u\|_{V_\delta^{l,p}(\mathcal{D})} \le c \sum_\pm \sum_{k=1}^m \|g_k^\pm\|_{V_\delta^{l-k+1-1/p,p}(\Gamma^\pm)}$$

with a constant c independent of the functions $g_k^\pm$.

P r o o f. Let $\Gamma_1^\pm = \{x \in \Gamma^\pm : 1/2 < |x'| < 2\}$. As is known (see e.g. [**195**, Theorem 2.9.1]), there exists a linear mapping

$$\prod_\pm \prod_{k=1}^m \overset{\circ}{W}{}^{l-k+1-1/p,p}(\Gamma_1^\pm) \ni \{\phi_k^\pm\} \to T\{\phi_k^\pm\} = v \in W^{l,p}(\mathcal{D})$$

such that $v = 0$ outside the set $\{x : 1/4 < |x| < 4\}$,

$$\frac{\partial^{k-1} v}{\partial n^{k-1}} = \phi_k^\pm \quad \text{on } \Gamma^\pm,\ k = 1, \dots, m,$$

and

$$\|v\|_{W^{l,p}(\mathcal{D})} \le c \sum_\pm \sum_{k=1}^m \|\phi_k^\pm\|_{W^{l-k+1-1/p,p}(\Gamma^\pm)}$$

with a constant c independent of the functions $\phi_k^\pm$.

Let ζ_ν be infinitely differentiable functions depending only on $r = |x'|$ satisfying the conditions (2.1.5). Furthermore, let $\tilde\zeta_\nu(x) = \zeta_\nu(2^\nu x)$ and $g_{k,\nu}^\pm(x) = g_k^\pm(2^\nu x)$. Then $\tilde\zeta_\nu(x)\, g_{k,\nu}^\pm(x) = 0$ for $|x| > 2$ and $|x| < 1/2$. We define

$$v_\nu = T\{2^{(k-1)\nu}\, \tilde\zeta_\nu\, g_{k,\nu}^\pm\}, \quad u_\nu(x) = v_\nu(2^{-\nu} x) \quad \text{and} \quad u = \sum_{\nu=-\infty}^{+\infty} u_\nu\,.$$

Then

$$\frac{\partial^{k-1} u_\nu}{\partial n^{k-1}} = \zeta_\nu\, g_k^\pm \quad \text{on } \Gamma_\pm,\ k = 1, \dots, m,$$

which implies (2.2.3). Furthermore, from the estimate

$$\|v_\nu\|_{W^{l,p}(\mathcal{D})} \le c \sum_\pm \sum_{k=1}^m 2^{(k-1)\nu}\, \|\tilde\zeta_\nu\, g_{k,\nu}^\pm\|_{W^{l-k+1-1/p,p}(\Gamma^\pm)}$$

it follows that

$$\|u_\nu\|_{V_\delta^{l,p}(\mathcal{D})} \le c \sum_\pm \sum_{k=1}^m \|\zeta_\nu\, g_k^\pm\|_{V_\delta^{l-k+1-1/p,p}(\Gamma^\pm)} \tag{2.2.5}$$

with a constant c independent of ν. Since $\zeta_\mu u_\nu = 0$ for $|\mu - \nu| \ge 3$, we obtain

$$\|u\|^p_{V_\delta^{l,p}(\mathcal{D})} \le c \sum_{\mu=-\infty}^{+\infty} \|\zeta_\mu u\|^p_{V_\delta^{l,p}(\mathcal{D})} \le c' \sum_{\nu=-\infty}^{+\infty} \|u_\nu\|^p_{V_\delta^{l,p}(\mathcal{D})}\,.$$

This together with (2.2.5) and Lemma 2.1.9 implies (2.2.4). □

REMARK 2.2.2. The operator T used in the proof of the last lemma can be chosen independent of l and p. Thus, there exists an operator $\{g_k^\pm\} \to u$ which realizes a linear and continuous mapping

$$\prod_\pm \prod_{k=1}^m V_\delta^{l-k+1-1/p,p}(\Gamma^\pm) \to V_\delta^{l,p}(\mathcal{D})$$

for all $l \ge m$, $p \in (1, \infty)$, $\delta \in \mathbb{R}$, and satisfies (2.2.3).

The assertion of Lemma 2.2.1 can be easily extended to other Dirichlet systems. Here, a system of differential operators

$$B_k(x, D_x) = \sum_{|\alpha|\le k-1} b_{k,\alpha}(x)\, D_x^\alpha\,,$$

$k = 1, \dots, m$, is said to be a *Dirichlet system* of order m on Γ if

$$\sum_{|\alpha|=k-1} b_{k,\alpha}(x)\,\xi^\alpha \ne 0$$

for all $x \in \Gamma$ and all $\xi \in \mathbb{R}^3$ normal to Γ at the point x.

LEMMA 2.2.3. *Let $\{B_k^\pm\}$ be Dirichlet systems of order $m \le l$ on Γ^+ and Γ^-, respectively, with infinitely differentiable coefficients. Furthermore, let $g_k^\pm \in V_\delta^{l-k+1-1/p,p}(\Gamma^\pm)$, $k = 1, \dots, m$, be given functions with compact supports. Then there exists a function $u \in V_\delta^{l,p}(\mathcal{D})$ satisfying the equalities*

$$B_k^\pm u = g_k^\pm \quad \text{on } \Gamma^\pm, \ k = 1, \dots, m,$$

and the estimate (2.2.4). *In the case where the operators $B_k^\pm$ are homogeneous and have constant coefficients, the restriction on the support of the functions $g_k^\pm$ can be omitted.*

P r o o f. By [**91**, Chapter 2, Lemma 2.1], the operators $B_k^\pm$ have the form

$$B_k^\pm(x, D_x) = \sum_{j=1}^{k} \Lambda_{j,k}^\pm(x, D_x)\, \frac{\partial^{j-1}}{\partial n^{j-1}},$$

where $\Lambda_{k,k}$ are infinitely differentiable functions and $\Lambda_{j,k}$ ($j < k$) are differential operators of order $k-j$ containing only derivatives tangent to Γ^+ and Γ^-, respectively. Conversely, the representation

$$\frac{\partial^{k-1}}{\partial n^{k-1}} = \sum_{j=1}^{k} \Phi_{j,k}^\pm(x, D_x)\, B_j^\pm(x, D_x)$$

is valid for the normal derivatives on $\Gamma^\pm$, where $\Phi_{j,k}^\pm$ are also differential operators of order $k - j$ with infinitely differentiable coefficients containing only derivatives tangent to Γ^+ and Γ^-, respectively. Furthermore,

$$\sum_{j=s}^{k} \Lambda_{j,k}^\pm\, \Phi_{s,j}^\pm = \delta_{s,k} \quad \text{for } 1 \le s \le k \le m.$$

Since the supports of the functions $g_k^\pm$ are compact, we have

$$\|\Phi_{j,k}^\pm\, g_j^\pm\|_{V_\delta^{l-k+1-1/p,p}(\Gamma^\pm)} \le c\, \|g_j^\pm\|_{V_\delta^{l-j+1-1/p,p}(\Gamma^\pm)} \tag{2.2.6}$$

for $1 \le j \le k \le m$. If $B_k^\pm$ are homogeneous operators with constant coefficients, then the operators $\Lambda_{j,k}^\pm$ and $\Phi_{j,k}^\pm$ are also homogeneous and have constant coefficients. In this case, the estimate (2.2.6) holds without the restriction on the supports of the functions $g_k^\pm$. Hence by Lemma 2.2.1, there exists a function $u^+ \in V_\delta^{l,p}(\mathcal{D})$ satisfying the equalities

$$\frac{\partial^{k-1} u^+}{\partial n^{k-1}} = \psi_k^+ \ \text{on } \Gamma^+ \ \text{ for } k = 1, \dots, m, \quad \text{where } \psi_k^+ = \sum_{j=1}^{k} \Phi_{j,k}^+\, g_j^+\,.$$

Then

$$B_k^+ u^+ = \sum_{j=1}^{k} \Lambda_{j,k}^+ \psi_j^+ = \sum_{j=1}^{k} \Lambda_{j,k}^+ \sum_{s=1}^{j} \Phi_{s,j}^+ \, g_s^+ = \sum_{s=1}^{k} \Big(\sum_{j=s}^{k} \Lambda_{j,k}^+ \, \Phi_{s,j}^+ \Big) \, g_s^+ = g_k^+$$

on Γ^+, $k = 1, \dots, m$. Moreover, u^+ satisfies the estimate (2.2.4). Analogously, there exists a function $u^- \in V_\delta^{l,p}(\mathcal{D})$ satisfying the equalities $B_k^- u^- = g_k^-$ on Γ^- and the estimate (2.2.4). Let χ be an infinitely differentiable function on the interval $[-\theta/2, +\theta/2]$ equal to one near $-\theta/2$ and equal to zero near $+\theta/2$. We introduce the function

$$u(x) = \chi(\varphi) \, u^-(x) + (1 - \chi(\varphi)) \, u^+(x),$$

where φ is the angle in the polar coordinates of $x' = (x_1, x_2)$. Then $B_k^\pm u = g_k^\pm$ on $\Gamma^\pm$ for $k = 1, \dots, m$. Since $r^\alpha \partial_x^\alpha \chi(\varphi)$ is bounded for every α, the function u satisfies the estimate

$$\|u\|_{V_\delta^{l,p}(\mathcal{D})} \le c \left(\|u^+\|_{V_\delta^{l,p}(\mathcal{D})} + \|u^-\|_{V_\delta^{l,p}(\mathcal{D})} \right) \le c' \sum_{\pm} \sum_{k=1}^{m} \|g_k^\pm\|_{V_\delta^{l-k+1-1/p,p}(\Gamma^\pm)} .$$

The proof of the lemma is complete. □

2.2.2. Existence of solutions in $V_0^{m,2}(\mathcal{D})$. We consider the special case $\delta = 0$ and prove that the Dirichlet problem (2.2.1) is uniquely solvable in $V_0^{m,2}(\mathcal{D})^\ell$ for arbitrary $f \in V_0^{-m,2}(\mathcal{D})^\ell$ and $g_k \in V_0^{m-k+1/2,2}(\Gamma^\pm)^\ell$, $k = 1, \dots, m$. To this end, we write the differential operator (2.2.2) as

$$L(D_x) = \sum_{|\beta|=|\gamma|=m} B_{\beta,\gamma} \, D_x^{\beta+\gamma}$$

(this representation is not unique) and show that the form

$$b(u, v) = \int_{\mathcal{D}} \sum_{|\beta|=|\gamma|=m} B_{\beta,\gamma} \, D_x^\beta u(x) \cdot \overline{D_x^\gamma v(x)} \, dx \tag{2.2.7}$$

is $\overset{\circ}{V}{}_0^{m,2}(\mathcal{D})^\ell$*-elliptic*, i.e. there exists a positive constant c such that

$$\operatorname{Re} b(u, u) \ge c \, \|u\|^2_{V_0^{m,2}(\mathcal{D})^\ell} \quad \textit{for all } u \in \overset{\circ}{V}{}_0^{m,2}(\mathcal{D})^\ell.$$

We start with the following lemma.

LEMMA 2.2.4. *The $V_0^{m,p}(\mathcal{D})$-norm is equivalent to*

$$\|u\|_{L^{m,p}(\mathcal{D})} = \Big(\int_{\mathcal{D}} \sum_{|\alpha|=m} |\partial_x^\alpha u(x)|^p \, dx \Big)^{1/p}$$

in $\overset{\circ}{V}{}_0^{m,p}(\mathcal{D})$.

P r o o f. The inequality

$$\|u\|_{L^{m,p}(\mathcal{D})} \le \|u\|_{V_0^{m,p}(\mathcal{D})}$$

is obvious. For the the proof of the contrary estimate

$$\|u\|_{V_0^{m,p}(\mathcal{D})} \le c \, \|u\|_{L^{m,p}(\mathcal{D})} \tag{2.2.8}$$

we apply *Hardy's inequality*

$$\int_0^\infty r^{\delta-p} \, |f(r)|^p \, dr \le \Big(\frac{p}{|\delta - p + 1|} \Big)^p \int_0^\infty r^\delta \, |f'(r)|^p \, dr \tag{2.2.9}$$

which is valid for $\delta \neq p-1$ and arbitrary differentiable functions f on $[0,\infty)$ satisfying the condition $f(0)=0$ if $\delta < p-1$ and $f(\infty)=0$ if $\delta > p-1$. Let $u \in C_0^\infty(\mathcal{D})$. Using (2.2.9), we obtain

$$\int_{\mathcal{D}} r^{-pm}|u|^p\,dx \le c_1 \int_{\mathcal{D}} r^{p(1-m)}\,|\nabla u|^p\,dx \le \cdots \le c_{m-1} \int_{\mathcal{D}} r^{-p} \sum_{|\alpha|=m-1} \left|\partial_x^\alpha u\right|^p dx$$

with certain constants $c_1,\dots,c_{m-1}$ independent of u. Furthermore by Friedrichs' inequality, every $v \in C_0^\infty(\mathcal{D})$ satisfies the estimate

$$\int_{\mathcal{D}} r^{-p}|v|^p\,dx = \int_{\mathbb{R}} \int_0^\infty \int_{-\theta/2}^{+\theta/2} r^{1-p}|v|^p\,d\varphi\,dr\,dx_3 \le c \int_{\mathbb{R}} \int_0^\infty \int_{-\theta/2}^{+\theta/2} r^{1-p}|\partial_\varphi v|^p\,d\varphi\,dr\,dx_3$$

$$\le c \int_{\mathcal{D}} \big(|\partial_{x_1} v|^p + |\partial_{x_2} v|^p\big)\,dx.$$

Applying this inequality to $v = \partial_x^\alpha u$, we obtain

$$\sum_{|\alpha|=m-1} \int_{\mathcal{D}} r^{-p}|\partial_x^\alpha u|^p\,dx \le c \sum_{|\alpha|=m} \int_{\mathcal{D}} |\partial_x^\alpha u|^p\,dx.$$

This proves the inequality (2.2.8). □

LEMMA 2.2.5. *Suppose that L is strongly elliptic. Then the form (2.2.7) is $\overset{\circ}{V}{}_0^{m,2}(\mathcal{D})^\ell$-elliptic.*

P r o o f: Since L is strongly elliptic, the corresponding form b satisfies *Gårding's inequality*

$$\operatorname{Re} b(u,u) \ge c_1\, \|u\|_{L^{m,2}(\mathcal{D})^\ell} - c_2\, \|u\|^2_{L_2(\mathcal{D})^\ell}$$

for all $u \in C_0^\infty(\mathcal{D})^\ell$, $u(x)=0$ for $|x|\ge 1$ (see [**152**, Theorem 6.5.1]). Here c_1 and c_2 are positive constants independent of u. If $u(x)=0$ for $|x|>\varepsilon$, then by Hardy's inequality,

$$\|u\|^2_{L_2(\mathcal{D}} \le c \sum_{|\alpha|=m} \int_{\mathcal{D}} |x|^{2m}\,|\partial_x^\alpha u(x)|^2\,dx \le c\,\varepsilon^{2m}\|u\|^2_{L^{m,2}(\mathcal{D})^\ell}.$$

Consequently for sufficiently small ε, we have

$$\operatorname{Re} b(u,u) \ge \frac{c_1}{2}\,\|u\|^2_{L^{m,2}(\mathcal{D})^\ell}\,. \tag{2.2.10}$$

Now let u be an arbitrary infinitely differentiable vector function with support in $\{x : |x|\le N\}$. Then the function $v(x) = u(Nx/\varepsilon)$ vanishes for $|x|\ge\varepsilon$ and satisfies (2.2.10). From the equalities

$$b(v,v) = (N/\varepsilon)^{2m-3}\,b(u,u) \quad \text{and} \quad \|v\|^2_{L^{m,2}(\mathcal{D})^\ell} = (N/\varepsilon)^{2m-3}\,\|u\|^2_{L^{m,2}(\mathcal{D})^\ell}$$

it follows that u satisfies (2.2.10) with the same constant c_1. Consequently, the inequality (2.2.10) is satisfied for all $u \in C_0^\infty(\mathcal{D})^\ell$. This proves the theorem. □

The last lemma together with Lemma 2.2.1 imply the following result.

THEOREM 2.2.6. *Suppose that the operator $L(D_x)$ is strongly elliptic. Then the problem (2.2.1) has a unique solution $u \in V_0^{m,2}(\mathcal{D})^\ell$ for arbitrary $f \in V_0^{-m,2}(\mathcal{D})^\ell$ and $g_k^\pm \in V_0^{m-k+1/2,2}(\Gamma^\pm)^\ell$, $k=1,\dots,m$.*

P r o o f. By Lemma 2.2.1, there exists a vector function $U \in V_0^{m,2}(\mathcal{D})^\ell$ satisfying the boundary conditions

$$\frac{\partial^{k-1}U}{\partial n^{k-1}} = g_k^\pm \quad \text{on } \Gamma^\pm$$

and the estimate

$$\|U\|_{V_0^{m,2}(\mathcal{D})^\ell} \le c \sum_\pm \sum_{k=1}^m \|g_k\|_{V_0^{m-k+1/2,2}(\Gamma^\pm)^\ell}$$

with a constant c independent of $g_k^\pm$. By Lax-Milgram's lemma, the boundary value problem

$$b(w,v) = (f,v)_\mathcal{D} - b(U,v) \quad \text{for all } v \in \overset{\circ}{V}{}_0^{m,2}(\mathcal{D})^\ell$$

has a uniquely determined solution $w \in \overset{\circ}{V}{}_0^{m,2}(\mathcal{D})^\ell$. Obviously, the vector function $u = U + w$ is a solution of the problem (2.2.1). This proves the theorem. □

2.2.3. Solvability of the formally adjoint problem. Let $L^+(D_x)$ denote the *formally adjoint* differential operator to L, i.e.

$$L^+(D_x) = \sum_{|\alpha|=2m} A_\alpha^* D_x^\alpha \,, \tag{2.2.11}$$

where A_α^* is the adjoint matrix of A_α. It is evident, that the differential operators $L(D_x)$ and $L^+(D_x)$ are simultaneously strongly elliptic. Consequently, the Dirichlet problem

$$L^+(D_x)u = f \ \text{ in } \mathcal{D}, \quad \frac{\partial^{k-1}u}{\partial n^{k-1}} = g_k^\pm \ \text{ on } \Gamma^\pm, \ k=1,\dots,m, \tag{2.2.12}$$

is also uniquely solvable in $V_0^{m,2}(\mathcal{D})^\ell$ for arbitrary $f \in V_0^{-m,2}(\mathcal{D})^\ell$ and $g_k \in V_0^{m-k+1/2,2}(\Gamma^\pm)^\ell$.

2.2.4. An a priori estimate for the solution. Our goal is to obtain a regularity result for the variational solution analogous to that given in Theorem 1.2.8 for boundary value problems in a cone. For the proof of the following lemma, we refer to [**152**, Theorem 6.4.8] (in the case $l \ge 2m$ see also Theorem 1.1.7).

LEMMA 2.2.7. *Let Ω be a bounded domain in $\mathbb{R}^N$ with smooth (of class C^∞) boundary $\partial\Omega$. Furthermore, let $L(x,D_x)$ be a strongly elliptic differential operator of order $2m$ with infinitely differentiable coefficients on $\overline{\Omega}$. If $u \in W^{m,p}(\Omega)^\ell$ is a solution of the Dirichlet problem*

$$L(x,D_x)\, u = f \ \text{ in } \Omega, \quad \frac{\partial^{k-1}u}{\partial n^{k-1}} = g_k \ \text{ on } \partial\Omega, \ k=1,\dots,m,$$

where $f \in W^{l-2m,q}(\Omega)^\ell$, $g_k \in W^{l-k+1-1/q,q}(\partial\Omega)^\ell$, $l \ge m$, then $u \in W^{l,q}(\Omega)^\ell$ and the estimate

$$\|u\|_{W^{l,q}(\Omega)^\ell} \le c\Big(\|f\|_{W^{l-2m,q}(\Omega)^\ell} + \sum_{k=1}^m \|g_k\|_{W^{l-k+1-1/q,q}(\partial\Omega)^\ell} + \|u\|_{L_1(\Omega)}\Big)$$

is valid with a constant c independent of u.

Using this lemma, we can easily prove the following local regularity assertion for weak solutions.

LEMMA 2.2.8. *Let ζ,η be infinitely differentiable cut-off functions depending only on $x'=(x_1,x_2)$ vanishing for $|x'|<1/4$ and $|x'|>4$. Furthermore, let $\eta=1$ on the support of ζ. Suppose that*

$$\eta u\in W^{m-1,p}(\mathcal{D})^\ell\,,\quad \eta\, L(D_x)\,u\in W^{l-2m,p}(\mathcal{D})^\ell,\ l\ge m,$$

and $\chi u\in \overset{\circ}{W}{}^{m,p}(\mathcal{D})^\ell$ for every $\chi\in C_0^\infty(\overline{\mathcal{D}}\backslash M)$. Then $\zeta u\in W^{l,p}(\mathcal{D})^\ell$ and

$$\|\zeta u\|_{W^{l,p}(\mathcal{D})^\ell}\le c\Big(\|\eta\, L(D_x)\,u\|_{W^{l-2m,p}(\mathcal{D})^\ell}+\|\eta u\|_{W^{m-1,p}(\mathcal{D})^\ell}\Big) \tag{2.2.13}$$

with a constant c independent of u.

P r o o f. Let σ_ν be an infinitely differentiable functions depending only on x_3 such that

$$\sigma_\nu(x_3)=0 \text{ for } x_3\not\in(\nu-1,\nu+1),\quad \partial_{x_3}^j\sigma_\nu<c_j,\quad \sum_{\nu=-\infty}^{+\infty}\sigma_\nu=1.$$

Furthermore, let $\tau_\nu=\sigma_{\nu-1}+\sigma_\nu+\sigma_{\nu+1}$. It can be shown analogously to Lemmas 2.1.4 and 2.1.8 that the norm in $W^{l,p}(\mathcal{D})$ is equivalent to the norm

$$\Big(\sum_{\nu=-\infty}^{+\infty}\|\sigma_\nu u\|^p_{W^{l,p}(\mathcal{D})}\Big)^{1/p}.$$

Moreover, it follows from the assumptions of the lemma and from Lemma 2.2.7 that $\zeta\sigma_\nu u\in W^{l,p}(\mathcal{D})$ and

$$\begin{aligned}\|\zeta\sigma_\nu u\|^p_{W^{l,p}(\mathcal{D})^\ell} &\le c\Big(\|L(D_x)\,(\zeta\sigma_\nu u)\|_{W^{l-2m,p}(\mathcal{D})^\ell}+\|\zeta\sigma_\nu u\|_{L_1(\mathcal{D})^\ell}\Big)\\ &\le c\Big(\|\zeta\sigma_\nu L(D_x)u\|^p_{W^{l-2m,p}(\mathcal{D})^\ell}+\|\eta\tau_\nu u\|^p_{W^{l-1,p}(\mathcal{D})^\ell}\Big).\end{aligned}$$

Here, we used the fact that the commutator $[L(D_x),\zeta\sigma_\nu]=L(D_x)\zeta\sigma_\nu-\zeta\sigma_\nu L(D_x)$ is a differential operator of order $2m-1$ and that $\eta\tau_\nu=1$ on the support of $\zeta\sigma_\nu$. Obviously, the constant c is independent of ν. Summing up over all ν, we obtain (2.2.13) for $l=m$. For $l>m$, the assertion of the lemma holds by induction. □

In the proof of the next regularity result, we use the last lemma and the equivalent $V_\delta^{l,p}(\mathcal{D})$-norm given in Lemma 2.1.4.

THEOREM 2.2.9. *Let $u\in V^{m-1,p}_{\delta-l+m-1}(\mathcal{D})^\ell$ and $\chi u\in W^{m,p}(\mathcal{D})^\ell$ for every function $\chi\in C_0^\infty(\overline{\mathcal{D}}\backslash M)$. If u is a solution of the boundary value problem (2.2.1), where $f\in V_\delta^{l-2m}(\mathcal{D})^\ell$, $l\ge m$, and $g_k^\pm\in V_\delta^{l-k+1-1/p,p}(\Gamma^\pm)^\ell$ for $k=1,\dots,m$, then $u\in V_\delta^{l,p}(\mathcal{D})^\ell$ and*

$$\begin{aligned}\|u\|_{V_\delta^{l,p}(\mathcal{D})^\ell}\le c\Big(\|f\|_{V_\delta^{l-2m,p}(\mathcal{D})^\ell}+\sum_\pm\sum_{k=1}^m\|g_k^\pm\|_{V_\delta^{l-k+1-1/p,p}(\Gamma^\pm)^\ell}\\ +\|u\|_{V^{0,p}_{\delta-l}(\mathcal{D})^\ell}\Big).\end{aligned} \tag{2.2.14}$$

P r o o f. By Lemma 2.2.1, we may restrict ourselves to the case of zero boundary data $g_k^\pm$. Let ζ_ν be infinitely differentiable functions depending only on $r=|x'|$ and satisfying (2.1.5). Furthermore, let $\eta_\nu=\zeta_{\nu-1}+\zeta_\nu+\zeta_{\nu+1}$, $\tilde\zeta_\nu(x)=\zeta_\nu(2^\nu x)$ and $\tilde\eta_\nu(x)=\eta_\nu(2^\nu x)$, $\nu=0,\pm1,\pm2,\dots$. The functions $\tilde\zeta_\nu$ and $\tilde\eta_\nu$ vanish for $|x'|<1/4$ and $|x'|>4$, and their derivatives are bounded by constants independent of ν. Moreover, $\tilde\eta_\nu = 1$ on the support of $\tilde\zeta_\nu$. If $L(D_x)\,u = f$, then the function $\tilde u(x) = u(2^\nu x)$ satisfies the equation $L(D_x)\,\tilde u = \tilde f$, where $\tilde f(x) = 2^{2m\nu} f(2^\nu x)$. Therefore, it follows from Lemma 2.2.8 that $\tilde\zeta_\nu \tilde u \in V_\delta^{l,p}(\mathcal{D})^\ell$ and

$$\|\tilde\zeta_\nu \tilde u\|_{V_\delta^{l,p}(\mathcal{D})^\ell} \le c\,\Big(\|\tilde\eta_\nu \tilde f\|_{V_\delta^{l-2m,p}(\mathcal{D})^\ell} + \|\tilde\eta_\nu \tilde u\|_{V_{\delta-l+m-1}^{m-1,p}(\mathcal{D})^\ell}\Big),$$

where the constant c is independent of u and ν. It can be easily verified that

$$\|\tilde\zeta_\nu \tilde u\|_{V_\delta^{l,p}(\mathcal{D})^\ell} = 2^{\nu(l-\delta-3/p)}\, \|\zeta_\nu u\|_{V_\delta^{l,p}(\mathcal{D})^\ell}$$

and

$$\|\tilde\eta_\nu \tilde f\|_{V_\delta^{l-2m,p}(\mathcal{D})^\ell} = 2^{\nu(l-\delta-3/p)}\, \|\eta_\nu f\|_{V_\delta^{l-2m,p}(\mathcal{D})^\ell}\,.$$

An analogous equality holds for the norm of $\tilde\eta_\nu \tilde u$ in $V_{\delta-l+m-1}^{m-1,p}(\mathcal{D})^\ell$. Consequently, the estimate

$$\|\zeta_\nu u\|_{V_\delta^{l,p}(\mathcal{D})^\ell} \le c\,\Big(\|\eta_\nu f\|_{V_\delta^{l-2m,p}(\mathcal{D})^\ell} + \|\eta_\nu u\|_{V_{\delta-l+m-1}^{m-1,p}(\mathcal{D})^\ell}\Big) \tag{2.2.15}$$

holds, where the constant c does not depend on u and ν. From the last inequality together with Lemmas 2.1.4 and 2.1.8, we conclude that $u \in V_\delta^{l,p}(\mathcal{D})^\ell$ and

$$\|u\|_{V_\delta^{l,p}(\mathcal{D})^\ell} \le c\,\Big(\|f\|_{V_\delta^{l-2m,p}(\mathcal{D})^\ell} + \|u\|_{V_{\delta-l+m-1}^{m-1,p}(\mathcal{D})^\ell}\Big).$$

Applying Lemma 2.1.5, we obtain the estimate (2.2.14). □

2.3. The parameter-depending problem in a 2-dimensional wedge

Let

$$L(D_{x'},\eta) = \sum_{|\alpha|=2m} A_\alpha\, \eta^{\alpha_3}\, D_{x'}^{\alpha'}$$

be the parameter-depending differential operator which arises if one replaces the derivative D_{x_3} in $L(D_x)$ by the real number η. Here we used the notation $\alpha = (\alpha',\alpha_3)$ for the multi-index α. Our goal is to prove the solvability of the Dirichlet problem for the differential equation

$$L(D_{x'},\eta)\,u = f \quad \text{in } K \tag{2.3.1}$$

in weighted Sobolev spaces and to obtain regularity results for the solutions. For this, one has to consider the eigenvalues of the operator pencil $A(\lambda)$ generated by the problem (2.3.1). In the case $\eta = 0$, it follows from Theorem 1.2.5 that the equation (2.3.1) is uniquely solvable in the weighted Sobolev space $\overset{\circ}{V}{}_\delta^{m,2}(K)^\ell$ for arbitrary $f \in V_\delta^{-m,2}(K)^\ell$ if the line $\operatorname{Re}\lambda = m-1-\delta$ is free of eigenvalues of the operator pencil $A(\lambda)$. For $\eta \ne 0$, we employ the subspaces $E_\delta^{l,p}(K) \subset V_\delta^{l,p}(K)$ with the norm

$$\|u\|_{E_\delta^{l,p}(K)} = \Big(\int_K \sum_{|\alpha|\le l} r^{p\delta}\, \big(1 + r^{p(|\alpha|-l)}\big)\, \big|\partial_{x'}^\alpha u(x')\big|^p\, dx' \Big)^{1/p}. \tag{2.3.2}$$

Since $L(D_{x'},\eta) = |\eta|^{2m}\, L(D_{|\eta|x'}, |\eta|^{-1}\eta)$, it suffices to study the equation (2.3.1) for $\eta = \pm 1$. It is shown in this section that the operator

$$\overset{\circ}{E}{}_\delta^{m,2}(K)^\ell \ni u \to L(D_{x'},\pm 1)\,u \in E_\delta^{-m,2}(K)^\ell$$

is Fredholm if and only if the line $\operatorname{Re}\lambda = m-1-\delta$ does not contain eigenvalues of the operator pencil $A(\lambda)$. However, the unique solvability of the problem (2.2.1) in the dihedron $\mathcal{D}$ requires the bijectivity of this operator. For the bijectivity, we obtain the necessary and sufficient condition

$$-\delta_+ < \delta < \delta_- ,$$

where δ_+ and δ_- are certain positive numbers introduced below.

2.3.1. The operator pencil generated by the model problem. For an arbitrary complex parameter λ, we define the operator $A(\lambda)$ by the equality

$$A(\lambda)\, U(\varphi) = r^{2m-\lambda}\, L(D_{x'},0)\, \big(r^\lambda U(\varphi)\big), \tag{2.3.3}$$

where r, φ are the polar coordinates of the point $x' = (x_1, x_2)$. This operator realizes a linear and continuous mapping

$$W^{2m,2}((-\tfrac{\theta}{2}, +\tfrac{\theta}{2}))^\ell \cap \overset{\circ}{W}{}^{m,2}((-\tfrac{\theta}{2}, +\tfrac{\theta}{2}))^\ell \to L_2((-\tfrac{\theta}{2}, +\tfrac{\theta}{2}))^\ell.$$

Analogously, we define the operator $A^+(\lambda)$ by

$$A^+(\lambda)\, U(\varphi) = r^{2m-\lambda}\, L^+(D_{x'},0)\, \big(r^\lambda U(\varphi)\big).$$

By [**85**, Theorem 10.1.2], the number λ_0 is an eigenvalue of the pencil $A(\lambda)$ if and only if $2m-2-\overline{\lambda}_0$ is an eigenvalue of the pencil $A^+(\lambda)$.

It is known that the strip $\operatorname{Re}\lambda = m-1$ does not contain eigenvalues of the pencil $A(\lambda)$ (see [**85**, Theorem 10.1.3]). We denote by δ_+ and δ_- the greatest positive real numbers such that the strip

$$m-1-\delta_- < \operatorname{Re}\lambda < m-1+\delta_+ \tag{2.3.4}$$

is free of eigenvalues of $A(\lambda)$. Then the strip $m-1-\delta_+ < \operatorname{Re}\lambda < m-1+\delta_-$ is free of eigenvalues of the pencil $A^+(\lambda)$.

2.3.2. Bijectivity of the operator $\boldsymbol{L(D_{x'},0)}$. Applying Theorem 1.2.5 to the Dirichlet problem for the equations $L(D_{x'},0)\,u = f$ and $L^+(D_{x'},0)\,u = f$, we obtain the following assertions.

LEMMA 2.3.1. *If the line* $\operatorname{Re}\lambda = m-1-\delta$ *is free of eigenvalues of the pencil* $A(\lambda)$*, then the operators*

$$L(D_{x'},0) : \overset{\circ}{V}{}^{m,2}_{\delta}(K)^\ell \cap V^{2m}_{\delta+m}(K)^\ell \to V^{0,2}_{\delta+m}(K)^\ell$$

and

$$L^+(D_{x'},0) : \overset{\circ}{V}{}^{m,2}_{-\delta}(K)^\ell \cap V^{2m}_{-\delta+m}(K)^\ell \to V^{0,2}_{-\delta+m}(K)^\ell$$

are isomorphisms.

Obviously, the differential operator $L(D_{x'},0)$ realizes also a continuous mapping from $\overset{\circ}{V}{}^{m,2}_{\delta}(K)^\ell$ into $V_\delta^{-m,2}(K)^\ell$ for arbitrary real δ. We show that this mapping is bijective under the condition of Lemma 2.3.1. For this, we need the following result.

LEMMA 2.3.2. *Let* $u \in \overset{\circ}{V}{}^{m,2}_{\delta-l+m}(K)^\ell$ *be a solution of the equation* $L(D_{x'},0)\,u = f$*, where* $f \in V_\delta^{l-2m,2}(K)^\ell$*,* $l \ge m$*. Then* $u \in V_\delta^{l,2}(K)^\ell$ *and*

$$\|u\|_{V_\delta^{l,2}(K)^\ell} \le c\Big(\|f\|_{V_\delta^{l-2m,2}(K)^\ell} + \|u\|_{V^{0,2}_{\delta-l}(K)^\ell}\Big). \tag{2.3.5}$$

P r o o f. Let ζ_ν, η_ν be the same cut-off functions as in the proof of Theorem 2.2.9. Analogously to (2.2.15), we obtain the estimate

$$\|\zeta_\nu u\|_{V_\delta^{l,2}(K)^\ell} \le c\left(\|\eta_\nu f\|_{V_\delta^{l-2m,2}(K)^\ell} + \|\eta_\nu u\|_{V^{m-1,2}_{\delta-l+m-1}(K)^\ell}\right),$$

with a constant c independent of u and ν. Using the analogues to Lemmas 2.1.4, 2.1.5 and 2.1.8 for the spaces $V_\delta^{l,p}(K)$, we conclude from the last estimate that $u \in V_\delta^{l,2}(K)^\ell$ and that u satisfies the inequality (2.3.5). □

Now we are able to prove the existence and uniqueness of variational solutions in the weighted space $\overset{\circ}{V}{}_\delta^{m,2}(K)^\ell$.

THEOREM 2.3.3. *Suppose that the line* $\operatorname{Re}\lambda = m-1-\delta$ *is free of eigenvalues of the pencil* $A(\lambda)$. *Then the operator*

$$L(D_{x'},0) : \overset{\circ}{V}{}_\delta^{m,2}(K)^\ell \to V_\delta^{-m,2}(K)^\ell \tag{2.3.6}$$

is an isomorphism.

P r o o f. First we show that there exists a constant c such that

$$\|u\|_{V_\delta^{m,2}(K)^\ell} \le c\,\|L(D_{x'},0)\,u\|_{V_\delta^{-m,2}(K)^\ell} \quad \text{for all } u \in \overset{\circ}{V}{}_\delta^{m,2}(K)^\ell. \tag{2.3.7}$$

If $u \in \overset{\circ}{V}{}_\delta^{m,2}(K)^\ell$, then $r^{2(\delta-m)}u \in V^{0,2}_{-\delta+m}(K)^\ell$. Consequently by Lemma 2.3.1, there exists a vector function $v \in \overset{\circ}{V}{}_{-\delta}^{m,2}(K)^\ell \cap V^{2m,2}_{-\delta+m}(K)^\ell$ satisfying the equation $L^+(D_{x'},0)v = r^{2(\delta-m)}u$ and the estimate

$$\|v\|_{V_{-\delta}^{m,2}(K)^\ell} \le \|v\|_{V^{2m,2}_{-\delta+m}(K)^\ell} \le c\,\|r^{2(\delta-m)}u\|_{V^{0,2}_{-\delta+m}(K)^\ell} = c\,\|u\|_{V^{0,2}_{\delta-m}(K)^\ell}\,. \tag{2.3.8}$$

Thus,

$$\begin{aligned}\|u\|^2_{V^{0,2}_{\delta-m}(K)^\ell} &= \int_K u\cdot r^{2(\delta-m)}\,\overline{u}\,dx' = \int_K u\cdot \overline{L^+(D_{x'},0)v}\,dx' \\ &= \big(L(D_{x'},0)u, v\big)_K \le \|L(D_{x'},0)\,u\|_{V_\delta^{-m,2}(K)^\ell}\,\|v\|_{V_{-\delta}^{m,2}(K)^\ell}\end{aligned}$$

This together with (2.3.8) implies

$$\|u\|_{V^{0,2}_{\delta-m}(K)^\ell} \le c\,\|L(D_{x'},0)u\|_{V_\delta^{-m,2}(K)^\ell}.$$

Using this inequality and (2.3.5), we obtain (2.3.7). From (2.3.7) it follows that the kernel of the operator $L(D_{x'},0) : \overset{\circ}{V}{}_\delta^{m,2}(K)^\ell \to V_\delta^{-m,2}(K)^\ell$ is trivial and its range is closed. Since analogously the adjoint operator $L^+(D_{x'},0) : \overset{\circ}{V}{}_{-\delta}^{m,2}(K)^\ell \to V_{-\delta}^{-m,2}(K)^\ell$ has trivial kernel, we conclude that $L(D_{x'},0)$ is an isomorphism. □

Moreover, the regularity result of Corollary 1.2.7 can be extended to variational solutions of the Dirichlet problem for the equation (2.3.1). This is done in the next theorem.

THEOREM 2.3.4. *Let* $u \in \overset{\circ}{V}{}_\delta^{m,2}(K)^\ell$ *be a solution of the equation*

$$L(D_{x'},0)\,u = f \quad \text{in } K$$

where $f \in V_\delta^{-m,2}(K)^\ell \cap V_{\delta'}^{-m,2}(K)^\ell$. *Suppose that the closed strip between the lines* $\operatorname{Re}\lambda = m-1-\delta$ *and* $\operatorname{Re}\lambda = m-1-\delta'$ *is free of eigenvalues of the pencil* $A(\lambda)$. *Then* $u \in \overset{\circ}{V}{}_{\delta'}^{m,2}(K)^\ell$.

P r o o f. By Theorem 2.3.3, there exists a solution $v \in \overset{\circ}{V}{}^{m,2}_{\delta'}(K)^\ell$ of the equation $L(D_{x'},0)\, v = f$. We show that $u = v$. Without loss of generality, we may assume that $\delta > \delta'$. We prove by induction that $\zeta(u-v) \in V^{l,2}_{\delta+l-m}(K)$ for arbitrary $l \ge m$ and for each $\zeta \in C_0^\infty(\overline{K})$, $\zeta = 1$ near the origin. For $l = m$ this is obvious. Suppose that this is true for a certain $l \ge m$. Since $L(D_{x'},0)\,(u-v) = 0$, it follows that

$$L(D_{x'},0)\,(\zeta u - \zeta v) \in V^{l-2m+1,2}_{\delta+l-m+1}(K)^\ell.$$

This together with Lemma 2.3.2 implies $\zeta(u-v) \in V^{l+1,2}_{\delta+l-m+1}(K)^\ell$. Hence, $\zeta(u-v) \in V^{l,2}_{\delta+l-m}(K)^\ell$ for arbitrary $l \ge m$.

In particular, $\zeta(u-v) \in V^{2m,2}_{\delta+m}(K)^\ell$. Analogously, we obtain $(1-\zeta)(u-v) \in V^{2m,2}_{\delta'+m}(K)^\ell$. Since

$$L(D_{x'},0)\,\big((1-\zeta)(u-v)\big) = -L(D_{x'},0)\,\big(\zeta(u-v)\big) \in V^{0,2}_{\delta+m}(K)$$

it follows from Corollary 1.2.7 that $(1-\zeta)(u-v) \in V^{2m,2}_{\delta+m}(K)^\ell$. Consequently, $u - v \in V^{2m,2}_{\delta+m}(K)^\ell$. Using Lemma 2.3.1, we obtain $u - v = 0$. This proves the theorem. □

2.3.3. The space $E^{l,p}_\delta(K)$. Let l be a nonnegative integer, and let δ and p be real numbers, $1 < p < \infty$. We define the weighted Sobolev space $E^{l,p}_\delta(K)$ as the closure of the set $C_0^\infty(\overline{K}\backslash\{0\})$ with respect to the norm (2.3.2). The closure of the set $C_0^\infty(K)$ with respect to this norm is denoted by $\overset{\circ}{E}{}^{l,p}_\delta(K)$. Furthermore, let $E^{-l,p}_\delta(K)$ be the dual space of $\overset{\circ}{E}{}^{l,p'}_{-\delta}(K)$, $p' = p/(p-1)$, with respect to the L_2 scalar product $(\cdot,\cdot)_K$. The norm in this space is defined as

$$\|u\|_{E^{-l,p}_\delta(K)} = \sup\Big\{\big|(u,v)_K\big| \ :\ v \in \overset{\circ}{E}{}^{l,p'}_{-\delta}(K),\ \|v\|_{\overset{\circ}{E}{}^{l,p'}_{-\delta}(K)} = 1\Big\}.$$

We introduce the weighted space $E^{0,p}_{\beta,\delta}(K)$ with the norm

$$\|u\|_{E^{0,p}_{\beta,\delta}(K)} = \Big(\int\limits_{\substack{K\\ r<1}} r^{p\beta}|u(x')|^p\, dx' + \int\limits_{\substack{K\\ r>1}} r^{p\delta}|u(x')|^p\, dx'\Big)^{1/p}.$$

Obviously, $E^{0,p}_{\beta,\delta}(K) = E^{0,p}_\beta(K) \cap E^{0,p}_\delta(K)$ if $\beta \le \delta$ and $E^{0,p}_{\beta,\delta}(K) = E^{0,p}_\beta(K) + E^{0,p}_\delta(K)$ if $\beta \ge \delta$. The space $E^{0,p}_{\beta,\delta}(K)$ is used in the following description of the space $E^{-l,p}_\delta(K)$.

LEMMA 2.3.5. *$E^{-l,p}_\delta(K)$ is the set of all distributions of the form*

$$u = \sum_{|\alpha|\le l} \partial^\alpha_{x'} u_\alpha, \quad \textit{where } u_\alpha \in E^{0,p}_{\delta+l-|\alpha|,\delta}(K).$$

Furthermore,

$$|||u|||_{E^{-l,p}_\delta(K)} = \inf\Big\{\sum_{|\alpha|\le l} \|u_\alpha\|_{E^{0,p}_{\delta+l-|\alpha|,\delta}(K)} \ :\ u_\alpha \in E^{0,p}_{\delta+l-|\alpha|,\delta}(K),\ \sum_{|\alpha|\le l} \partial^\alpha_{x'} u_\alpha = u\Big\}$$

is an equivalent norm in $E^{-l,p}_\delta(K)$.

P r o o f. Let B be the Banach space of all vector functions $\phi = (\phi_\alpha)_{|\alpha|\le l}$, $\phi_\alpha \in E^{0,p'}_{-\delta-l+|\alpha|,-\delta}(K)$ with the norm

$$\|\phi\|_B = \sum_{|\alpha|\le l} \|\phi_\alpha\|_{E^{0,p'}_{-\delta-l+|\alpha|,-\delta}(K)}.$$

By M, we denote the subspace of all $\phi = (\phi_\alpha)_{|\alpha|\le l}$ such that $\phi_\alpha = \partial^\alpha_{x'} v$, where $v \in \overset{\circ}{E}{}^{l,p'}_{-\delta}(K)$. Suppose that f is a linear and continuous functional on $\overset{\circ}{E}{}^{l,p'}_{-\delta}(K)$. Then we define the functional F on M by

$$F(\phi) = f(v) \quad \text{for } \phi = (\partial^\alpha_{x'} v)_{|\alpha|\le l},\ v \in \overset{\circ}{E}{}^{l,p'}_{-\delta}(K).$$

By Hahn-Banach's theorem, there exists an extension of the functional F to B with the same norm. However, every linear and continuous functional F_1 on B has the form

$$F_1(\phi) = \int_K \sum_{|\alpha|\le l} u_\alpha\, \phi_\alpha\, dx' \quad \text{for all } \phi = (\phi_\alpha)_{|\alpha|\le l} \in \prod_{|\alpha|\le l} E^{0,p'}_{-\delta-l+|\alpha|,-\delta}(K),$$

where $u_\alpha \in E^{0,p}_{\delta+l-|\alpha|,\delta}(K)$. The result follows. □

The proof of the next lemma is a word-by-word repetition of the proof of the analogous result for the space $V^{l,p}_\delta(\mathcal{D})$ (cf. Lemmas 2.1.4 and 2.1.8).

LEMMA 2.3.6. *Let ζ_ν, $\nu = 0, \pm 1, \cdots$, be infinitely differentiable functions in K depending only on $r = |x'|$ and satisfying the conditions* (2.1.5). *Then the norm in $E^{l,p}_\delta(K)$ is equivalent to*

$$\Big(\sum_{\nu=-\infty}^{+\infty} \|\zeta_\nu u\|^2_{E^{l,p}_\delta(K)}\Big)^{1/2} \tag{2.3.9}$$

for each integer l.

2.3.4. An estimate for solutions of the equation $L(D_{x'}, \pm 1)u = f$. Our goal is to estimate the $E^{l,2}_\delta(K)^\ell$-norm of the solution to the equation

$$L(D_{x'}, \pm 1)\, u = f \ \text{ in } K. \tag{2.3.10}$$

To this end, we prove first a local estimate for the solutions. In the next lemma, let ζ_ν be infinitely differentiable functions in K depending only on $r = |x'|$ and satisfying the conditions (2.1.5). Furthermore, let $\eta_\nu = \zeta_{\nu-1} + \zeta_\nu + \zeta_{\nu+1}$. Then $\eta_\nu = 1$ on the support of ζ_ν.

LEMMA 2.3.7. *Suppose that u is a solution of the equation* (2.3.10) *and that $\psi u \in \overset{\circ}{W}{}^{m,2}(K)^\ell$ for every $\psi \in C_0^\infty(\overline{K}\backslash\{0\})$. Then the inequality*

$$\|\zeta_\nu u\|_{E^{l,2}_\delta(K)^\ell} \le c\Big(\|\eta_\nu f\|_{E^{l-2m,2}_\delta(K)^\ell} + \|\eta_\nu u\|_{E^{m-1,2}_{\delta-l+m-1}(K)^\ell}\Big)$$

is valid with a constant c independent of u and ν.

P r o o f. By (2.2.15), every $v \in V^{l,2}_\delta(\mathcal{D})^\ell \cap \overset{\circ}{V}{}^{m,2}_{\delta-l+m}(\mathcal{D})^\ell$ satisfies the estimate

$$\|\zeta_\nu v\|_{V^{l,2}_\delta(\mathcal{D})^\ell} \le c\Big(\|\eta_\nu L(D_x)\, v\|_{V^{l-2m,2}_\delta(\mathcal{D})^\ell} + \|\eta_\nu v\|_{V^{m-1,2}_{\delta-l+m-1}(\mathcal{D})^\ell}\Big). \tag{2.3.11}$$

We put $v(x', x_3) = \chi(x_3/N)\, e^{\pm i x_3}\, u(x')$, where $N \geq 1$ and χ is an arbitrary infinitely differentiable function such that $0 \leq \chi \leq 1$, $\chi(t) = 1$ for $|t| < 1$, $\chi(t) = 0$ for $|t| > 2$. Then obviously,

$$
\begin{aligned}
(2.3.12)\quad \|\zeta_\nu v\|^2_{V_\delta^{l,2}(\mathcal{D})^\ell} &\geq \int_{-N}^{+N} \int_K \sum_{|\alpha| \leq l} r^{2(\delta - l + |\alpha|)} \left|\partial_x^\alpha (e^{\pm i x_3} \zeta_\nu u(x'))\right|^2 dx'\, dx_3 \\
&\geq 2N\, \|\zeta_\nu u\|^2_{E_\delta^{l,2}(K)^\ell}
\end{aligned}
$$

and

$$
(2.3.13)\qquad \|\eta_\nu v\|^2_{V^{m-1,2}_{\delta-l+m-1}(\mathcal{D})^\ell} \leq 2Nm\, \|\eta_\nu u\|^2_{E^{m-1,2}_{\delta-l+m-1}(K)^\ell}\,.
$$

It remains to estimate the norm of $\eta_\nu L(D_x) v$ in $V_\delta^{l-2m,2}(\mathcal{D})^\ell$. Obviously,

$$
L(D_{x'}, D_{x_3})\, v = \chi(x_3/N)\, e^{\pm i x_3}\, L(D_{x'}, \pm 1)\, u + L_1 u,
$$

where L_1 is a differential operator of the form

$$
(2.3.14)\qquad L_1 u = \sum_{k=1}^{2m} \sum_{|\alpha| \leq 2m-k} N^{-k}\, \chi^{(k)}(x_3/N)\, e^{\pm i x_3}\, L_{k,\alpha}\, D_{x'}^\alpha u.
$$

Let $\psi_\nu = \eta_{\nu-1} + \eta_\nu + \eta_{\nu+1}$. Since

$$
\|\chi^{(k)}(x_3/N)\, e^{\pm i x_3} \psi_\nu u\|^2_{V_\delta^{l,2}(\mathcal{D})^\ell} \leq cN\, \|\psi_\nu u\|^2_{E_\delta^{l,2}(K)^\ell}\,,
$$

we obtain

$$
\|\eta_\nu \chi^{(k)}(x_3/N)\, e^{\pm i x_3}\, D_{x'}^\alpha u\|^2_{V_\delta^{l-|\alpha|,2}(\mathcal{D})^\ell} \leq c\, N\, \|\psi_\nu u\|^2_{E_\delta^{l,2}(K)^\ell}
$$

and consequently

$$
\|\eta_\nu \chi^{(k)}(x_3/N)\, e^{\pm i x_3}\, D_{x'}^\alpha u\|^2_{V_\delta^{l-2m,2}(\mathcal{D})^\ell} \leq c\, 2^{\nu(2m-|\alpha|)}\, N\, \|\psi_\nu u\|^2_{E_\delta^{l,2}(K)^\ell}
$$

for $|\alpha| \leq 2m$. Therefore,

$$
(2.3.15)\qquad \|\eta_\nu L_1 u\|^2_{V_\delta^{l-2m,2}(\mathcal{D})^\ell} \leq c\, N^{-1}\, (1 + 2^{2m\nu})\, \|\psi_\nu u\|^2_{E_\delta^{l,2}(K)^\ell}
$$

with a constant c independent of N, ν and u.

Finally, we consider the norm of the function $\eta_\nu \chi(x_3/N)\, e^{\pm i x_3} L(D_{x'}, \pm 1)\, u$ in $V_\delta^{l-2m,2}(\mathcal{D})^\ell$. Let $f = L(D_{x'}, \pm 1) u \in E_\delta^{l-2m,2}(K)^\ell$, $l < 2m$. Then there exist vector functions $f_\alpha \in E^{0,2}_{\delta+2m-l-|\alpha|}(K)^\ell$ and $g_\alpha \in E_\delta^{0,2}(K)^\ell$ such that

$$
\eta_\nu f = \sum_{|\alpha| \leq 2m-l} D_{x'}^\alpha (f_\alpha + g_\alpha)
$$

and

$$
\sum_{|\alpha| \leq 2m-l} \left(\|f_\alpha\|_{E^{0,2}_{\delta+2m-l-|\alpha|}(K)^\ell} + \|g_\alpha\|_{E_\delta^{0,2}(K)^\ell} \right) \leq c\, \|\eta_\nu f\|_{E_\delta^{l-2m,2}(K)^\ell}\,.
$$

Let $\chi_N(x) = \chi(x_3/N)$. Then we obtain

$$
\left(\eta_\nu \chi_N\, e^{\pm i x_3} f, v\right)_{\mathcal{D}} = \sum_{|\alpha| \leq 2m-l} \int_{\mathcal{D}} \chi_N\, e^{\pm i x_3} D_{x'}^\alpha (f_\alpha + g_\alpha) \cdot \overline{\psi_\nu v}\, dx
$$

for arbitrary $v \in \overset{\circ}{V}{}^{2m-l,2}_{-\delta}(\mathcal{D})^\ell$. Here,

$$\begin{aligned}\Big| \int_{\mathcal{D}} \chi_N\, e^{\pm ix_3} D^\alpha_{x'} f_\alpha \cdot \overline{\psi_\nu v}\, dx\Big| &= \Big| \int_{\mathcal{D}} \chi_N\, e^{\pm ix_3} f_\alpha \cdot \overline{D^\alpha_{x'}(\psi_\nu v)}\, dx\Big| \\ &\le 2N^{1/2} \|f_\alpha\|_{E^{0,2}_{\delta+2m-l-|\alpha|}(K)^\ell}\, \|D^\alpha_{x'}(\psi_\nu v)\|_{V^{0,2}_{-\delta-2m+l+|\alpha|}(\mathcal{D})^\ell} \\ &\le c\, N^{1/2}\, \|f_\alpha\|_{E^{0,2}_{\delta+2m-l-|\alpha|}(K)^\ell}\, \|v\|_{V^{2m-l,2}_{-\delta}(\mathcal{D})^\ell}\end{aligned}$$

and

$$\begin{aligned}\Big| \int_{\mathcal{D}} \chi_N\, e^{\pm ix_3} D^\alpha_{x'} g_\alpha \cdot \overline{\psi_\nu v}\, dx\Big| &= \Big| \int_{\mathcal{D}} \chi_N\, D^{2m-l-|\alpha|}_{x_3} e^{\pm ix_3}\, g_\alpha \cdot \overline{D^\alpha_{x'}(\psi_\nu v)}\, dx\Big| \\ = \Big| \int_{\mathcal{D}} e^{\pm ix_3} g_\alpha \cdot \overline{D^\alpha_{x'} D^{2m-l-|\alpha|}_{x_3}(\chi_N \psi_\nu v)}\, dx\Big| & \\ \le 2N^{1/2} \|g_\alpha\|_{E^{0,2}_\delta(K)^\ell}\, \|D^\alpha_{x'} D^{2m-l-|\alpha|}_{x_3}(\chi_N\psi_\nu v)\|_{V^{0,2}_{-\delta}(\mathcal{D})^\ell}. &\end{aligned}$$

Using the inequality $|D^k_{x_3}\chi_N| \le c\, N^{-k}$, we get

$$\begin{aligned}&\Big| \int_{\mathcal{D}} \chi_N\, e^{\pm ix_3} D^\alpha_{x'} g_\alpha \cdot \overline{\psi_\nu v}\, dx\Big| \\ &\quad \le cN^{1/2} \|g_\alpha\|_{E^{0,2}_\delta(K)^\ell} \sum_{k=0}^{2m-l-|\alpha|} N^{-k}\, \|\psi_\nu v\|_{V^{2m-l-k,2}_{-\delta}(\mathcal{D})^\ell} \\ &\quad \le cN^{1/2} \|g_\alpha\|_{E^{0,2}_\delta(K)^\ell} \sum_{k=0}^{2m-l-|\alpha|} N^{-k}\, 2^{\nu k} \|v\|_{V^{2m-l,2}_{-\delta}(\mathcal{D})^\ell}\,.\end{aligned}$$

Consequently,

$$\|\eta_\nu \chi_N\, e^{\pm ix_3} f\|_{V^{l-2m,2}_\delta(\mathcal{D})^\ell} \le c\, N^{1/2}\big(1 + (2^\nu/N)^{2m-l}\big) \|\eta_\nu f\|_{E^{l-2m,2}_\delta(K)^\ell} \tag{2.3.16}$$

if $l < 2m$. In the case $l \ge 2m$, it can be easily verified that

$$\|\eta_\nu \chi_N\, e^{\pm ix_3} f\|_{V^{l-2m,2}_\delta(\mathcal{D})^\ell} \le c\, N^{1/2} \|\eta_\nu f\|_{E^{l-2m,2}_\delta(K)^\ell}\,.$$

We conclude from the estimates (2.3.11)–(2.3.13), (2.3.15), (2.3.16) that

$$\begin{aligned}\|\zeta_\nu u\|_{E^{l,2}_\delta(K)^\ell} \;\le\; & c\Big(\big(1 + (2^\nu/N)^{\max(2m-l,0)}\big)\, \|\eta_\nu f\|_{E^{l-2m,2}_\delta(K)^\ell} \\ & + N^{-1}(1 + 2^{m\nu})\, \|\psi_\nu u\|_{E^{l,2}_\delta(K)^\ell} + \|\eta_\nu u\|_{E^{m-1,2}_{\delta-l+m-1}(K)^\ell}\Big).\end{aligned}$$

Since N is an arbitrarily large number and c is independent of N, we obtain the inequality given in the lemma. □

The next theorem is a corollary of the last lema and of Lemma 2.3.6.

THEOREM 2.3.8. *Let $u \in E^{m-1,2}_{\delta-l+m-1}(K)^\ell$ and $\psi u \in \overset{\circ}{W}{}^{m,2}(K)^\ell$ for all $\psi \in C_0^\infty(\overline{K}\backslash\{0\})$. If $L(D_{x'}, \pm 1)\, u \in E^{l-2m,2}_\delta(K)^\ell$, $l \ge m$, then $u \in E^{l,2}_\delta(K)^\ell$ and*

$$\|u\|_{E^{l,2}_\delta(K)^\ell} \le c\Big(\|L(D_{x'}, \pm 1)\, u\|_{E^{l-2m,2}_\delta(K)^\ell} + \|u\|_{E^{m-1,2}_{\delta-l+m-1}(K)^\ell}\Big).$$

2.3.5. Fredholm property of the operators $L(D_{x'},\pm 1)$. We consider the operator

$$L(D_{x'},\pm 1)\,:\, E_\delta^{l,2}(K)^\ell \cap \overset{\circ}{E}{}^{m,2}_{\delta-l+m}(K)^\ell \to E_\delta^{l-2m,2}(K)^\ell,\ l\ge m. \tag{2.3.17}$$

In order to prove that this operator is Fredholm for certain δ, we employ the following well-known lemma which can be found e.g. in [**203**, Theorem 12.12].

LEMMA 2.3.9. *Let $\mathcal{X},\mathcal{Y},\mathcal{Z}$ be Banach spaces, $\mathcal{X}\subset\mathcal{Z}$. Suppose that the last imbedding is compact and that $\mathcal{A}$ is a linear and continuous mapping from $\mathcal{X}$ into $\mathcal{Y}$. Then the following conditions are equivalent.*

(1) *The range of $\mathcal{A}$ is closed in $\mathcal{Y}$ and* $\dim\,\ker\mathcal{A}<\infty$.
(2) *There exists a positive constant c such that*

$$\|x\|_{\mathcal{X}}\le c\,\big(\|\mathcal{A}x\|_{\mathcal{Y}}+\|x\|_{\mathcal{Z}}\big)\quad \textit{for all } x\in\mathcal{X}.$$

We prove an estimate of the form (2) for the operator (2.3.17).

LEMMA 2.3.10. *Suppose that $l\ge m$ and that the line* $\operatorname{Re}\lambda=l-1-\delta$ *does not contain eigenvalues of the operator pencil $A(\lambda)$. Then every vector function $u\in E_\delta^{l,2}(K)^\ell\cap \overset{\circ}{E}{}^{m,2}_{\delta-l+m}(K)^\ell$ satisfies the estimate*

$$\|u\|_{E_\delta^{l,2}(K)^\ell}\le c\,\Big(\|L(D_{x'},\pm 1)u\|_{E_\delta^{l-2m,2}(K)^\ell}+\|u\|_{L_2(S)^\ell}\Big), \tag{2.3.18}$$

where c is a constant independent of u and S is a compact subset of K.

P r o o f. 1) First we consider the case $l=m$. Let again ζ_ν, $\nu=0,\pm1,\dots$, be infinitely differentiable functions in K depending only on $r=|x'|$ and satisfying the conditions (2.1.5). Furthermore, let $\eta_\nu=\zeta_{\nu-1}+\zeta_\nu+\zeta_{\nu+1}$,

$$\zeta=\sum_{\nu=-\infty}^{-N}\zeta_\nu,\quad\text{and}\quad \eta=\sum_{\nu=-\infty}^{-N+1}\zeta_\nu,$$

where N is a sufficiently large positive integer. Then ζ vanishes for $|x'|>2^{-N+1}$. In particular, $\zeta(x')=0$ for $|x'|>0$. Consequently by Theorem 2.3.3,

$$\begin{aligned}\|\zeta u\|_{E_\delta^{m,2}(K)^\ell} &\le c\,\|\zeta u\|_{V_\delta^{m,2}(K)^\ell}\le c\,\|L(D_{x'},0)\,(\zeta u)\|_{V_\delta^{-m,2}(K)^\ell}\\ &\le c\,\Big(\|\zeta L(D_{x'},\pm1)u\|_{V_\delta^{-m,2}(K)^\ell}+\|[L(D_{x'},0),\zeta]u\|_{V_\delta^{-m,2}(K)^\ell}\\ &\qquad+\|\zeta(L(D_{x'},0)-L(D_{x'},\pm1))u\|_{V_\delta^{-m,2}(K)^\ell}\Big).\end{aligned} \tag{2.3.19}$$

Here, $[L(D_{x'},0),\zeta]$ denotes the *commutator* of $L(D_{x'},0)$ and ζ, i.e.,

$$[L(D_{x'},0),\zeta]u=L(D_{x'},0)\,(\zeta u)-\zeta L(D_{x'},0)u.$$

We estimate the right-hand side of (2.3.19). Obviously,

$$\|\zeta L(D_{x'},\pm1)u\|_{V_\delta^{-m,2}(K)^\ell}\le c\,\|\zeta L(D_{x'},\pm1)u\|_{E_\delta^{-m,2}(K)^\ell}$$

and

$$\begin{aligned}\|\zeta(L(D_{x'},0)-L(D_{x'},\pm1))u\|_{V_\delta^{-m,2}(K)^\ell} &\le c\,\|\eta u\|_{V_\delta^{m-1,2}(K)^\ell}\\ &\le c'\,2^{-N}\,\|\eta u\|_{E_\delta^{m,2}(K)^\ell}.\end{aligned}$$

Since $[L(D_{x'},0),\zeta]$ is a linear differential operator of order $2m-1$ with coefficients vanishing outside the set $K_N = \{x' \in K : 2^{-N} < |x'| < 2^{-N+1}\}$, we furthermore obtain

$$\|[L(D_{x'},0),\zeta]u\|_{V_\delta^{-m,2}(K)^\ell} \le c_N \|u\|_{W^{m-1,2}(K_N)^\ell} .$$

This implies

$$\begin{aligned}\|\zeta u\|_{E_\delta^{m,2}(K)^\ell} &\le c\Big(\|\zeta L(D_{x'},\pm 1)u\|_{E_\delta^{-m,2}(K)^\ell} + 2^{-N}\|\eta u\|_{E_\delta^{m,2}(K)^\ell}\Big)\\ &\quad + c_N \|u\|_{W^{m-1,2}(K_N)^\ell}.\end{aligned}$$

Furthermore, according to Lemma 2.3.7,

$$\|\zeta_\nu u\|_{E_\delta^{m,2}(K)^\ell} \le c\Big(\|\eta_\nu L(D_{x'},\pm 1)u\|_{E_\delta^{-m,2}(K)^\ell} + \|\eta_\nu u\|_{E_{\delta-1}^{m-1,2}(K)^\ell}\Big)$$

for $\nu = -N+1, -N+2, \ldots$. We introduce the set

$$K_N' = \{x' \in K : 2^{-N-1} < |x'| < 2^{N+1}\}.$$

By Lemma 2.3.6, the $E_\delta^{l,2}(K)$-norm is equivalent to

$$\Big(\|\zeta u\|^2_{E_\delta^{l,2}(K)} + \sum_{\nu=-N+1}^{\infty} \|\zeta_\nu u\|^2_{E_\delta^{l,2}(K)}\Big)^{1/2}.$$

Using the inequalities

$$\|\eta_\nu u\|_{E_{\delta-1}^{m-1,2}(K)^\ell} \le c_N \|u\|_{W^{m-1,2}(K_N')^\ell} \quad \text{for } |\nu| \le N-1$$

and

$$\|\eta_\nu u\|_{E_{\delta-1}^{m-1,2}(K)^\ell} \le 2^{3-N} \|\eta_\nu u\|_{E_\delta^{m,2}(K)^\ell} \quad \text{for } \nu \ge N,$$

we obtain

$$\|u\|_{E_\delta^{m,2}(K)^\ell} \le c\,\|L(D_{x'},\pm 1)u\|_{E_\delta^{-m,2}(K)^\ell} + c_N \|u\|_{W^{m-1,2}(K_N')^\ell} .$$

The last estimate together with the inequality

$$\|u\|_{W^{m-1,2}(K_N')^\ell} \le \varepsilon \|u\|_{W^{m,2}(K_N')^\ell} + c(\varepsilon)\|u\|_{L_2(K_N')^\ell},$$

where ε is an arbitrarily small positive number, implies (2.3.18) for $l = m$.

2) If $l > m$, then it follows from Theorem 2.3.8 that

$$\|u\|_{E_\delta^{l,2}(K)^\ell} \le c\Big(\|f\|_{E_\delta^{l-2m,2}(K)^\ell} + \|u\|_{E_{\delta-l+m}^{m,2}(K)^\ell}\Big).$$

By the first part of the proof, the right-hand side can be estimated by the right-hand side of (2.3.18). The proof is complete. □

We show now that the condition on δ in Lemma 2.3.10 guarantees also the Fredholm property of the operator (2.3.17).

THEOREM 2.3.11. *The operator* (2.3.17) *is Fredholm if and only if the line* $\mathrm{Re}\,\lambda = l-1-\delta$ *does not contain eigenvalues of the pencil* $A(\lambda)$. *If this line if free of eigenvalues of the pencil* $A(\lambda)$, *then for the solvability of the equation* $L(D_{x'},\pm 1)\,u = f$ *in* $E_\delta^{l,2}(K)^\ell \cap \overset{\circ}{E}{}_{\delta-l+m}^{m,2}(K)^\ell$ *it is necessary and sufficient that* $f \in E_\delta^{l-2m,2}(K)^\ell$ *and*

$$(2.3.20)\qquad (f,v)_K = 0 \quad \textit{for all } v \in \overset{\circ}{E}{}_{-\delta+l-m}^{m,2}(K)^\ell \textit{ satisfying } L^+(D_{x'},\pm 1)v = 0.$$

P r o o f. 1) Suppose that the line $\operatorname{Re}\lambda = l-1-\delta$ is free of eigenvalues of the pencil $A(\lambda)$. Then by Lemmas 2.3.9 and 2.3.10, the dimension of the kernel of the operator (2.3.17) is finite and its range is closed. Since by the assumption of the theorem the line $\operatorname{Re}\lambda = 2m-2-(l-1-\delta)$ is free of eigenvalues of $A^+(\lambda)$, the same is true for the operator

$$L^+(D_{x'},\pm 1) : \overset{\circ}{E}{}^{m,2}_{-\delta+l-m}(K)^\ell \to E^{-m,2}_{-\delta+l-m}(K)^\ell \tag{2.3.21}$$

which is adjoint to the operator

$$L(D_{x'},\pm 1) : \overset{\circ}{E}{}^{m,2}_{\delta-l+m}(K)^\ell \to E^{-m,2}_{\delta-l+m}(K)^\ell. \tag{2.3.22}$$

It follows from the closed range theorem (see e.g. [**203**, Theorem 12.3]) that the range of the last operator consists of all $f \in E^{-m,2}_{\delta-l+m}(K)^\ell$ satisfying the condition (2.3.20). Since the range of the operator (2.3.17) is contained in the range of the operator (2.3.22), the condition (2.3.20) is also satisfied for all elements of the range of the operator (2.3.17). On the other hand for every $f \in E^{l-2m,2}_\delta(K)^\ell$ satisfying the condition (2.3.20), there exists a solution $u \in \overset{\circ}{E}{}^{m,2}_{\delta-l+m}(K)^\ell$ of the equation $L(D_{x'},\pm 1)u = f$. By Theorem 2.3.8, this solution belongs to $E^{l,2}_\delta(K)^\ell$. Hence, the condition (2.3.20) is necessary and sufficient for the solvability of the equation $L(D_{x'},\pm 1)u = f$ in $E^{l,2}_\delta(K)^\ell \cap \overset{\circ}{E}{}^{m,2}_{\delta-l+m}(K)^\ell$.

2) Let λ_0 be an eigenvalue of the pencil $A(\lambda)$ in the line $\operatorname{Re}\lambda = l-1-\delta$ and let U be an eigenfunction corresponding to this eigenvalue. Then we consider the function

$$u(x') = \zeta(x')\, r^{\lambda_0+\varepsilon}\, U(\varphi),$$

where $\varepsilon > 0$, $\zeta \in C_0^\infty(\overline{K})$, $\zeta(x') = 1$ for small $|x'|$, $\zeta = 0$ on the compact subset S of K. The norm of u in $E^{l,2}_\delta(K)$ tends to infinity as $\varepsilon \to 0$, while the norm of $L(D_{x'},\pm 1)\,u$ in $E^{l-2m,2}_\delta(K)$ is bounded. The last follows from the equation $L(D_{x'},0)\, r^{\lambda_0}\, U(\varphi) = 0$. Consequently, the estimate in Lemma 2.3.10 is not valid with a constant c independent of u. This means that the operator (2.3.17) is not Fredholm. The proof of the theorem is complete. □

2.3.6. Regularity results for solutions of the equation (2.3.10). Our goal is to prove regularity results for the solution $u \in \overset{\circ}{E}{}^{m,2}_\delta(K)^\ell$ of the equation (2.3.10) analogous to those in Theorem 2.3.4. We consider the cases $\delta < \delta'$ and $\delta > \delta'$ separately.

LEMMA 2.3.12. *Let $u \in \overset{\circ}{E}{}^{m,2}_\delta(K)^\ell$ be a solution of the equation* (2.3.10), *where $f \in E^{-m,2}_\delta(K)^\ell \cap E^{-m,2}_{\delta'}(K)^\ell$, $\delta' > \delta$. Then $u \in \overset{\circ}{E}{}^{m,2}_{\delta'}(K)^\ell$.*

P r o o f. Suppose first that $\delta < \delta' \le \delta + 1$. Then $E^{m,2}_\delta(K) \subset E^{m-1,2}_{\delta'-1}(K)$, and from Theorem 2.3.8 it follows that $u \in E^{m,2}_{\delta'}(K)^\ell$.

If $\delta' > \delta + 1$, then $f \in E^{-m,2}_\delta(K)^\ell \cap E^{-m,2}_{\delta+1}(K)^\ell$ and, according to the first part of the proof, we obtain $u \in E^{m,2}_{\delta+1}(K)^\ell$. Repeating this argument, we obtain $u \in E^{m,2}_{\delta'}(K)$ after a finite number of steps. □

LEMMA 2.3.13. *Let $u \in \overset{\circ}{E}{}^{m,2}_\delta(K)^\ell$ be a solution of the equation* (2.3.10), *where $f \in E^{-m,2}_\delta(K)^\ell \cap E^{-m,2}_{\delta'}(K)^\ell$, $\delta' < \delta$. We suppose that the strip $m-1-\delta \le \operatorname{Re}\lambda \le m-1-\delta'$ is free of eigenvalues of the pencil $A(\lambda)$. Then $u \in \overset{\circ}{E}{}^{m,2}_{\delta'}(K)^\ell$.*

P r o o f. Suppose first that $\delta - 1 \le \delta' < \delta$. We show that $L(D_{x'},0)u \in V_{\delta'}^{-m,2}(K)^\ell$. Let $v \in \overset{\circ}{V}{}^{m,2}_{-\delta'}(K)^\ell$, and let $\zeta \in C_0^\infty(\overline{K})$ be a cut-off function equal to one near the origin. Then obviously

$$\begin{aligned} \Big|\big(L(D_{x'},0)u,(1-\zeta)v\big)|_K\Big| &\le c\,\|u\|_{V_\delta^{m,2}(K)^\ell}\;\|(1-\zeta)v\|_{V_{-\delta}^{m,2}(K)^\ell} \\ &\le c\,\|u\|_{E_\delta^{m,2}(K)^\ell}\;\|v\|_{V_{-\delta'}^{m,2}(K)^\ell}\,. \end{aligned}$$

Since $\partial_x^\alpha u \in E_{\delta'}^{-m,2}(K)$ for $|\alpha| \le 2m-1$, we obtain

$$L(D_{x'},0)\,u = f - \big(L(D_{x'},\pm 1) - L(D_{x'},0)\big)\,u \in E_{\delta'}^{-m,2}(K)^\ell$$

and

$$\begin{aligned} \Big|\big(L(D_{x'},0)u,\zeta v\big)|_K\Big| &\le c\,\|L(D_{x'},0)u\|_{E_{\delta'}^{-m,2}(K)^\ell}\;\|\zeta v\|_{E_{-\delta'}^{m,2}(K)^\ell} \\ &\le c\,\|L(D_{x'},0)u\|_{E_{\delta'}^{-m,2}(K)^\ell}\;\|v\|_{V_{-\delta'}^{m,2}(K)^\ell}\,. \end{aligned}$$

This proves that $L(D_{x'},0)u \in V_{\delta'}^{-m,2}(K)^\ell$. From this and from Theorem 2.3.4 it follows that $u \in V_{\delta'}^{m,2}(K)^\ell$. Using the obvious imbedding $E_\delta^{m,2}(K) \cap V_{\delta'}^{m,2}(K) \subset E_{\delta'}^{m,2}(K)$, we obtain $u \in E_{\delta'}^{m,2}(K)^\ell$.

If $\delta' < \delta - 1$, then $f \in E_\delta^{-m,2}(K)^\ell \cap E_{\delta-1}^{-m,2}(K)^\ell$ and from the first part of the proof it follows that $u \in E_{\delta-1}^{m,2}(K)^\ell$. Repeating this argument, we obtain $u \in E_{\delta'}^{m,2}(K)^\ell$. □

As a corollary of the last two lemmas and of Theorem 2.3.8, we obtain the following higher order regularity result.

THEOREM 2.3.14. *Let $u \in \overset{\circ}{E}{}_\delta^{m,2}(K)^\ell$ be a solution of the equation* (2.3.10), *where $f \in E_\delta^{-m,2}(K) \cap E_{\delta'}^{l-2m,2}(K)^\ell$, $l \ge m$. Suppose that either $m - \delta \ge l - \delta'$ or the strip $m - 1 - \delta \le \operatorname{Re}\lambda \le l - 1 - \delta'$ is free of eigenvalues of the pencil $A(\lambda)$. Then $u \in E_{\delta'}^{l,2}(K)^\ell$.*

P r o o f. First note that $E_{\delta'}^{l-2m,2}(K) \subset E_{\delta'-l+m}^{-m,2}(K)$. Consequently, Lemmas 2.3.12 and 2.3.13 imply $u \in E_{\delta'-l+m}^{m,2}(K)$. Using Theorem 2.3.8, we obtain $u \in E_{\delta'}^{l,2}(K)^\ell$. □

2.3.7. Existence and uniqueness of solutions in $E_0^{m,2}(K)^\ell$.

We show that the operator (2.3.17) is an isomorhism for $l = m$ and $\delta = 0$.

THEOREM 2.3.15. *For arbitrary $f \in E_0^{-m,2}(K)^\ell$, there exists a unique solution $u \in \overset{\circ}{E}{}_0^{m,2}(K)^\ell$ of the equation* (2.3.10).

P r o o f. Let

$$b_K(u,v;\pm 1) = \int_K \sum_{|\beta|=|\gamma|=m} B_{\beta,\gamma}\,(\pm 1)^{\beta_3+\gamma_3}\, D_{x'}^{\beta'} u(x')\,\overline{D_{x'}^{\gamma'} v(x')}\,dx'$$

be the sesquilinear form corresponding to the differential operator

$$L(D_{x'},\pm 1) = \sum_{|\alpha|=2m} (\pm 1)^{\alpha_3} A_\alpha\, D_{x'}^{\alpha'} = \sum_{|\beta|=|\gamma|=m} (\pm 1)^{\beta_3+\gamma_3} B_{\beta,\gamma}\, D_{x'}^{\beta'+\gamma'}.$$

Here, we used the notation $\alpha = (\alpha', \alpha_3)$, $\beta = (\beta', \beta_3)$, $\gamma = (\gamma', \gamma_3)$. By Lemma 2.2.5, there exists a constant $c > 0$ such that

$$\operatorname{Re} b(v,v) \ge c\, \|v\|^2_{V_0^{m,2}(\mathcal{D})^\ell} \quad \textit{for all } v \in C_0^\infty(\mathcal{D}).$$

We set $v(x',x_3) = N^{-1/2}\, e^{\pm i x_3}\, \zeta(x_3/N)\, u(x')$, where $u \in \overset{\circ}{E}{}_0^{m,2}(K)^\ell$ and $\zeta \in C_0^\infty(\mathbb{R})$. Then

$$\operatorname{Re} b(v,v) \to \|\zeta\|^2_{L_2(\mathbb{R})} \operatorname{Re} b_K(u,u;\pm 1) \quad \text{as } N \to \infty$$

and

$$\|v\|_{V_0^{m,2}(\mathcal{D})^\ell} \to \|\zeta\|_{L_2(\mathbb{R})}\, \|u\|_{E_0^{m,2}(K)^\ell} \quad \text{as } N \to \infty.$$

Consequently, the inequality

$$\operatorname{Re} b_K(u,u;\pm 1) \ge c\, \|u\|^2_{E_0^{m,2}(K)^\ell} \tag{2.3.23}$$

holds for all $u \in \overset{\circ}{E}{}_0^{m,2}(K)$, where the constant c is independent of u. Thus, the assertion of the theorem follows from Lax-Milgram's lemma. □

2.3.8. Existence and uniqueness of solutions in $E_\delta^{m,2}(K)^\ell$. Let A_δ denote the operator

$$\overset{\circ}{E}{}_\delta^{m,2}(K)^\ell \ni u \to L(D_{x'}, \pm 1)\, u \in E_\delta^{-m,2}(K)^\ell.$$

Then the next lemma is an immediate consequence of Lemmas 2.3.12 and 2.3.13.

LEMMA 2.3.16. *Suppose that $\delta < \delta'$ or $\delta > \delta'$ and the strip $m-1-\delta \le \operatorname{Re}\lambda \le m-1-\delta'$ is free of eigenvalues of the pencil $A(\lambda)$. Then* $\ker A_\delta \subset \ker A_{\delta'}$.

An analogous result holds for the adjoint operator A_δ^*.

LEMMA 2.3.17. *Suppose that $\delta > \delta'$ or $\delta < \delta'$ and the strip $m-1-\delta' \le \operatorname{Re}\lambda \le m-1-\delta$ is free of eigenvalues of the pencil $A(\lambda)$. Then* $\ker A_\delta^* \subset \ker A_{\delta'}^*$.

P r o o f. Applying Lemma 2.3.16 to the adjoint operator

$$L^+(D_{x'}, \pm 1) \,:\, \overset{\circ}{E}{}_{-\delta}^{m,2}(K)^\ell \to E_{-\delta}^{-m,2}(K)^\ell,$$

we conclude that $\ker A_\delta^* \subset \ker A_{\delta'}^*$ if $-\delta < -\delta'$ or $-\delta > -\delta'$ and the strip $m-1+\delta \le \operatorname{Re}\lambda \le m-1+\delta'$ is free of eigenvalues of the pencil $A^+(\lambda)$. The last assumption is satisfied if the strip $m-1-\delta' \le \operatorname{Re}\lambda \le m-1-\delta$ does not contain eigenvalues of the pencil $A(\lambda)$. □

Let again δ_+ and δ_- be the greatest positive real numbers such that the strip (2.3.4) is free of eigenvalues of $A(\lambda)$. These are also the greatest real numbers such that the strip $m-1-\delta_+ < \operatorname{Re}\lambda < m-1+\delta_-$ is free of eigenvalues of the pencil $A^+(\lambda)$.

LEMMA 2.3.18. 1) *If $\delta < \delta_-$ then* $\ker A_\delta = \{0\}$, *if $\delta > -\delta_+$ then* $\ker A_\delta^* = \{0\}$.
2) *The kernel of A_δ is not trivial for $\delta > \delta_-$, the kernel of A_δ^* is not trivial for $\delta < -\delta_+$.*

P r o o f. 1) The first assertion follows immediately from Theorem 2.3.15 and Lemmas 2.3.16 and 2.3.17.

2) Suppose that $\ker A_{\delta'}^* = \{0\}$ for some $\delta' < -\delta_+$. Then by Lemma 2.3.17, $\ker A_\delta^* = \{0\}$ for all $\delta > \delta'$. Let β, δ be real numbers such that

$$\delta' < \delta < -\delta_+ < \beta < 0, \qquad \beta - \delta < 1,$$

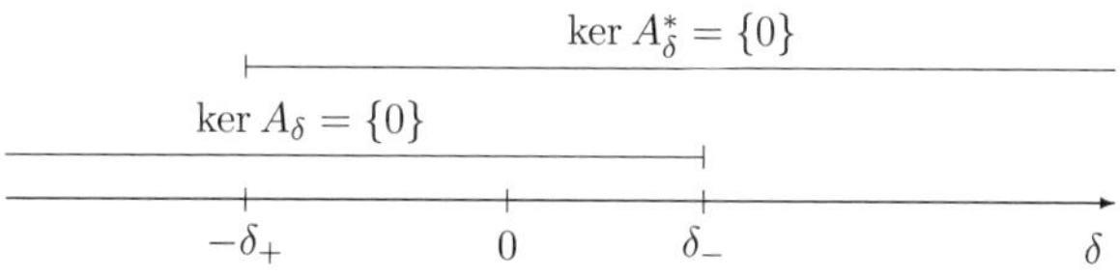

FIGURE 4. Bijectivity of the operator A_δ

and there are no eigenvalues of the pencil $A(\lambda)$ on the line $\operatorname{Re}\lambda = m-1-\delta$. Furthermore, let λ_0 be an eigenvalue of $A(\lambda)$ on the line $\operatorname{Re}\lambda = m-1+\delta_+$ and Φ an eigenvector corresponding to this eigenvalue. We consider the vector function

$$u(x') = \zeta(r)\, r^{\lambda_0}\, \Phi(\varphi),$$

where ζ is a smooth function on $\mathbb{R}_+ = (0,\infty)$, $\zeta(r) = 1$ for $r<1$, $\zeta(r)=0$ for $r>2$. Obviously, $u \in E^{2m,2}_{\beta+m}(K)^\ell$ and $u \notin E^{2m,2}_{\delta+m}(K)^\ell$. Furthermore, $L(D_{x'},0)u(x') = 0$ for $|x'|<1$ and for $|x'|>2$. Consequently,

$$\begin{aligned} L(D_{x'},\pm 1)u &= L(D_{x'},0)u + \big(L(D_{x'},\pm 1) - L(D_{x'},0)\big)\, u \\ &\in E^{1,2}_{\beta+m}(K)^\ell \subset E^{0}_{\beta+m}(K)^\ell \cap E^{0,2}_{\delta+m}(K)^\ell. \end{aligned}$$

In particular, we have $L(D_{x'},\pm 1)u \in E^{-m,2}_\delta(K)^\ell$. Since the operator A_δ is Fredholm (see Theorem 2.3.11) and $\ker A_\delta^* = \{0\}$, there exists a solution $w \in \mathring{E}^{m,2}_\delta(K)^\ell$ of the equation

$$L(D_{x'},\pm 1)w = L(D_{x'},\pm 1)u.$$

By Theorem 2.3.8, this solution belongs to $E^{2m,2}_{\delta+m}(K)^\ell$, and from Lemma 2.3.12 it follows that $w \in E^{2m,2}_{\beta+m}(K)^\ell$. Thus, $u-w \in \ker A_\beta$. However by the first part of the lemma, the kernel of A_β is trivial. This implies $u = w \in E^{2m,2}_{\delta+m}(K)^\ell$ which is not true for $\operatorname{Re}\lambda_0 = m-1+\delta_+$. Therefore, $\ker A_\delta^* \neq \{0\}$ for $\delta < -\delta_+$. Analogously, we obtain $\ker A_\delta \neq \{0\}$ for $\delta > \delta_-$. □

The last two lemmas together with Theorem 2.3.11 imply the following statement.

THEOREM 2.3.19. *The operator A_δ is an isomorphism if and only if δ satisfies the inequalities $-\delta_+ < \delta < \delta_-$.*

2.3.9. Existence and uniqueness of solutions in $E^{l,2}_\delta(K)^\ell$. Next we consider the operator (2.3.17), i.e. the restriction of the operator $A_{\delta-l+m}$ to the space $E^{l,2}_\delta(K)^\ell \cap \mathring{E}^{m,2}_{\delta-l+m}(K)^\ell$, $l \ge m$. Our goal is to find a condition on δ which guarantees that this operator is an isomorphism. We start with the proof of an a priori estimate.

LEMMA 2.3.20. *If $\delta - l + m < \delta_-$ and the line $\operatorname{Re}\lambda = l-1-\delta$ does not contain eigenvalues of the pencil $A(\lambda)$, then there exists a constant c such that*

$$\|u\|_{E^{l,2}_\delta(K)^\ell} \le c\, \|L(D_{x'},\pm 1)u\|_{E^{l-2m,2}_\delta(K)^\ell} \qquad (2.3.24)$$

for all $u \in E^{l,2}_\delta(K)^\ell \cap \mathring{E}^{m,2}_{\delta-l+m}(K)^\ell$.

Conversely, if (2.3.24) is satisfied for all $u \in E^{l,2}_\delta(K)^\ell \cap \mathring{E}^{m,2}_{\delta-l+m}(K)^\ell$, then $\delta - l + m < \delta_-$ and the line $\operatorname{Re}\lambda = l-1-\delta$ is free of eigenvalues of the pencil $A(\lambda)$.

P r o o f. 1) Suppose that $\delta - l + m < \delta_-$ and the $\mathrm{Re}\,\lambda = l - 1 - \delta$ does not contain eigenvalues of the pencil $A(\lambda)$. Then the operator (2.3.17) is Fredholm (cf. Theorem 2.3.11). By Theorem 2.3.8, the kernel of this operator coincides with the kernel of the operator $A_{\delta-l+m}$ and, by Lemma 2.3.18, the kernel is trivial. This implies (2.3.24).

2) If the line $\mathrm{Re}\,\lambda = l-1-\delta$ contains eigenvalues of the pencil $A(\lambda)$, then even the estimate (2.3.18) does not hold as was shown in the proof of Theorem 2.3.11. If $\delta - l + m > \delta_-$, then it follows from Lemma 2.3.18 that the kernel of $A_{\delta-l+m}$ and therefore also the kernel of the operator (2.3.17) are not trivial. This contradicts (2.3.24). □

COROLLARY 2.3.21. *If $l \ge 2m$, $\delta - l + m < \delta_-$ and the line $\mathrm{Re}\,\lambda = l - 1 - \delta$ does not contain eigenvalues of the pencil $A(\lambda)$, then*

$$\sum_{j=0}^{l} |\xi|^j \, \|u\|_{V_\delta^{l-j,2}(K)^\ell} \le c \sum_{j=0}^{l-2m} |\xi|^j \, \|L(D_{x'},\xi)\, u\|_{V_\delta^{l-2m-j,2}(K)^\ell}$$

for all $u \in E_\delta^{l,2}(K)^\ell \cap \overset{\circ}{E}{}^{m,2}_{\delta-l+m}(K)^\ell$, $\xi \in \mathbb{R}$, $\xi \ne 0$. Here, the constant c is independent of u and ξ.

P r o o f. Let $L(D_{x'},\xi)u = f$. We put $\tilde{u}(x') = u(|\xi|^{-1}x')$ and $\tilde{f}(x') = |\xi|^{-2m} f(|\xi|^{-1}x')$. Then

$$L(D_{x'}, |\xi|^{-1}\xi)\, \tilde{u}(x') = \tilde{f}(x')$$

and from Lemma 2.3.20 it follows that

$$\|\tilde{u}\|_{E_\delta^{l,2}(K)^\ell} \le c\, \|\tilde{f}\|_{E_\delta^{l-2m,2}(K)^\ell}\,.$$

Using the inequalities

$$\|\tilde{u}\|^2_{E_\delta^{l,2}(K)^\ell} \ge l^{-1} \sum_{j=0}^{l} \|\tilde{u}\|^2_{V_\delta^{l-j,2}(K)^\ell} = l^{-1}\, |\xi|^{2(\delta-l+1)} \sum_{j=0}^{l} |\xi|^{2j}\, \|u\|^2_{V_\delta^{l-j,2}(K)^\ell}$$

and

$$\|\tilde{f}\|^2_{E_\delta^{l-2m,2}(K)^\ell} \le |\xi|^{2(\delta-l+1)} \sum_{j=0}^{l-2m} |\xi|^{2j}\, \|f\|^2_{V_\delta^{l-2m-j,2}(K)^\ell}\,,$$

we obtain the desired estimate. □

Furthermore, we obtain the following generalization of Theorem 2.3.19 by means of Lemma 2.3.20.

THEOREM 2.3.22. *The operator* (2.3.17) *is an isomorphism if and only if*

$$-\delta_+ < \delta - l + m < \delta_-\,.$$

P r o o f. By Theorem 2.3.11, the operator (2.3.17) is Fredholm if and only if the line $\mathrm{Re}\,\lambda = l - 1 - \delta$ does not contain eigenvalues of the pencil $A(\lambda)$. Furthermore, the cokernel of this operator is trivial if and only if the equation $L^+(D_{x'}, \pm 1)\, v = 0$ has only the trivial solution in $\overset{\circ}{E}{}^{m,2}_{-\delta+l-m}(K)^\ell$, i.e. if $-\delta + l - m < \delta_+$ (see Lemma 2.3.18). According to Lemma 2.3.20, the kernel of the operator (2.3.17) is trivial if and only if $\delta - l + m < \delta_-$. This proves the theorem. □

2.4. Solvability of the Dirichlet problem in weighted L_2 Sobolev spaces

We return to the Dirichlet problem

$$L(D_x)u = f \ \text{ in } \mathcal{D}, \quad \frac{\partial^{k-1}u}{\partial n^{k-1}} = g_k^\pm \ \text{ on } \Gamma^\pm, \ k = 1, \dots, m,$$

for the strongly elliptic differential operator (2.2.2) in the dihedron $\mathcal{D}$. Applying the Fourier transform with respect to x_3 to this problem, we obtain a parameter-depending boundary value problem in the angle K. By means of the results of the preceding section, we prove that the Dirichlet problem in the dihedron $\mathcal{D}$ is uniquely solvable in $V_\delta^{l,2}(\mathcal{D})^\ell$ for arbitrary $f \in V_\delta^{l,2}(\mathcal{D})^\ell$, $g_k \in V_\delta^{l-k+1/2,2}(\Gamma^\pm)^\ell$ if and only if $-\delta_+ < \delta - l + m < \delta_-$. Furthermore, we obtain regularity assertions for the solutions and their x_3-derivatives.

2.4.1. A relation between the $V_\delta^{l,2}$-norm of a function and the $E_\delta^{l,2}$-norm of its Fourier transform.

Let $\tilde{u}(x',\eta)$ denote the Fourier transform of the function $u(x',x_3)$ with respect to the variable x_3. Using Parseval's equality, we obtain the following lemma.

LEMMA 2.4.1. *Let $f \in V_\delta^{l,2}(\mathcal{D})$, $l \ge 0$, and $F(x',\eta) = \tilde{f}(|\eta|^{-1}x',\eta)$. Then there exists a positive constant c independent of f such that*

$$c\,\|f\|^2_{V_\delta^{l,2}(\mathcal{D})} \le \int_{\mathbb{R}} |\eta|^{2(l-\delta-1)}\, \|F(\cdot,\eta)\|^2_{E_\delta^{l,2}(K)}\, d\eta \le \|f\|^2_{V_\delta^{l,2}(\mathcal{D})}\,.$$

We prove an analogous estimate for the case $l < 0$.

LEMMA 2.4.2. *Let $f \in V_\delta^{-l,2}(\mathcal{D})$, $l \ge 0$, and $F(x',\eta) = \tilde{f}(|\eta|^{-1}x',\eta)$. Then there exists a positive constant c independent of f such that*

$$\int_{\mathbb{R}} |\eta|^{2(-l-\delta-1)}\, \|F(\cdot,\eta)\|^2_{E_\delta^{-l,2}(K)}\, d\eta \le c\,\|f\|^2_{V_\delta^{-l,2}(\mathcal{D})}\,.$$

P r o o f. By Lemma 2.1.7, there exist functions $f_{j,\alpha} \in V^{0,2}_{\delta+l-j-|\alpha|}(\mathcal{D})$ such that

$$f(x) = \sum_{j+|\alpha|\le l} D^\alpha_{x'} D^j_{x_3} f_{j,\alpha}(x), \qquad \sum_{j+|\alpha|\le l} \|f_{j,\alpha}\|_{V^{0,2}_{\delta+l-j-|\alpha|}(\mathcal{D})} \le c\,\|f\|_{V_\delta^{-l,2}(\mathcal{D})}\,.$$

Applying the Fourier transform $x_3 \to \eta$ and substituting $x' = |\eta|^{-1}y'$, we obtain

$$F(y',\eta) = \sum_{j+|\alpha|\le l} \eta^j\, |\eta|^{|\alpha|}\, D^\alpha_{y'} F_{j,\alpha}(y',\eta),$$

where $F_{j,\alpha}(y',\eta) = \tilde{f}_{j,\alpha}(|\eta|^{-1}y',\eta)$. The last equation can be also written as

$$F(y',\eta) = \sum_{|\alpha|\le l} D^\alpha_{y'} G_\alpha(y',\eta), \quad \text{where } G_\alpha = \sum_{j=0}^{l-|\alpha|} \eta^j\, |\eta|^{|\alpha|}\, F_{j,\alpha}\,.$$

Obviously, $G_\alpha(\cdot,\eta) \in E^{0,2}_{\delta+l-|\alpha|,\delta}(K)$. Consequently,

$$\begin{aligned}
\|F(\cdot,\eta)\|^2_{E_\delta^{-l,2}(K)} &\le c_1 \sum_{|\alpha|\le l} \|G_\alpha(\cdot,\eta)\|^2_{E^{0,2}_{\delta+l-|\alpha|,\delta}(K)} \\
&\le c_2 \sum_{j+|\alpha|\le l} |\eta|^{2(j+|\alpha|)}\, \|F_{j,\alpha}(\cdot,\eta)\|^2_{E^{0,2}_{\delta+l-j-|\alpha|}(K)}
\end{aligned}$$

(cf. Lemma 2.3.5). Multiplying the last inequality by $|\eta|^{2(-l-\delta-1)}$ and integrating, we obtain

$$\begin{aligned}&\int_{\mathbb{R}} |\eta|^{2(-l-\delta-1)}\, \|F(\cdot,\eta)\|^2_{E_\delta^{-l,2}(K)}\, d\eta \\ &\le c \sum_{j+|\alpha|\le l} \int_{\mathbb{R}} |\eta|^{2(j+|\alpha|-l-\delta-1)}\, \|F_{j,\alpha}(\cdot,\eta)\|^2_{E^{0,2}_{\delta+l-j-|\alpha|}(K)}\, d\eta \\ &= c \sum_{j+|\alpha|\le l} \|f_{j,\alpha}\|^2_{V^{0,2}_{\delta+l-j-|\alpha|}(\mathcal{D})}\,.\end{aligned}$$

The result follows. □

2.4.2. An a priori estimate for the solution of the problem (2.2.1). Let again δ_+ and δ_- be the greatest numbers such that the strip $m-1-\delta_- < \operatorname{Re}\lambda < m-1+\delta_+$ is free of eigenvalues of the pencil $A(\lambda)$.

THEOREM 2.4.3. *If $\delta - l + m < \delta_-$ and the line $\operatorname{Re}\lambda = l-1-\delta$ does not contain eigenvalues of the pencil $A(\lambda)$, then every solution $u \in V_\delta^{l,2}(\mathcal{D})^\ell$, $l \ge m$, of the problem* (2.2.1) *satisfies the estimate*

$$\|u\|_{V_\delta^{l,2}(\mathcal{D})^\ell} \le c\Big(\|f\|_{V_\delta^{l-2m,2}(\mathcal{D})^\ell} + \sum_\pm \sum_{k=1}^m \|g_k^\pm\|_{V_\delta^{l-k+1/2,2}(\Gamma^\pm)^\ell}\Big) \tag{2.4.1}$$

with a constant c independent of u. The above conditions on δ are also necessary for the validity of the estimate (2.4.3).

P r o o f. By Lemma 2.2.1, we may restrict ourselves to the case of zero Dirichlet data $g_k^\pm$. If $u \in V_\delta^{l,2}(\mathcal{D})^\ell \cap \overset{\circ}{V}{}^{m,2}_{\delta-l+m}(\mathcal{D})^\ell$ is a solution of the equation $L(D_x)u = f$, then the vector function $U(x',\eta) = |\eta|^{2m}\, \tilde u(|\eta|^{-1}x',\eta)$ satisfies the equation

$$L(D_{x'}, |\eta|^{-1}\eta)\, U(x',\eta) = F(x',\eta) = \tilde f(|\eta|^{-1}x',\eta). \tag{2.4.2}$$

Suppose that $\delta-l+m < \delta_-$ and the line $\operatorname{Re}\lambda = l-1-\delta$ does not contain eigenvalues of the pencil $A(\lambda)$. Then by Lemma 2.3.20, the inequality

$$\|U(\cdot,\eta)\|^2_{E_\delta^{l,2}(K)^\ell} \le c\, \|F(\cdot,\eta)\|^2_{E_\delta^{l-2m,2}(K)^\ell}$$

holds with a constant c independent of u and η. Multiplying this inequality by $|\eta|^{2(l-2m-\delta-1)}$, integrating with respect to η, and applying Lemmas 2.4.1, 2.4.2, we obtain (2.4.1).

We prove the necessity of the conditions on δ. Suppose that every vector function $u \in V_\delta^{l,2}(\mathcal{D})^\ell \cap \overset{\circ}{V}{}^{m,2}_{\delta-l+m}(\mathcal{D})^\ell$ satisfies the estimate

$$\|u\|_{V_\delta^{l,2}(\mathcal{D})^\ell} \le c\, \|Lu\|_{V_\delta^{l-2m,2}(\mathcal{D})^\ell}\,. \tag{2.4.3}$$

If $l < 2m$, then it follows from (2.4.3) and Theorem 2.2.9 that

$$\|u\|_{V^{2m,2}_{\delta-l+2m}(\mathcal{D})^\ell} \le c\, \|Lu\|_{V^{0,2}_{\delta-l+2m}(\mathcal{D})^\ell}$$

for all $u \in V^{2m,2}_{\delta-l+2m}(\mathcal{D})^\ell \cap \overset{\circ}{V}{}^{m,2}_{\delta-l+m}(\mathcal{D})^\ell$. Therefore, we may assume without loss of generality that $l \ge 2m$ in (2.4.3). We put $u(x) = \chi(x_3/N)\, e^{\pm i x_3}\, v(x')$, where $v \in E_\delta^{l,2}(K)^\ell \cap \overset{\circ}{E}{}^{m,2}_{\delta-l+m}(K)$, $N \ge 1$, and χ is a smooth function such that $0 \le \chi \le 1$, $\chi(t) = 1$ for $|t| < 1$ and $\chi(t) = 0$ for $|t| > 2$. Then

$$\|u\|^2_{V_\delta^{l,2}(\mathcal{D})^\ell} \ge 2N\, \|v\|^2_{E_\delta^{l,2}(K)^\ell}$$

with a certain constant c independent of v and N (cf. (2.3.12)). Furthermore,

$$L(D_{x'}, D_{x_3})\, u = \chi(x_3/N)\, e^{\pm i x_3}\, L(D_{x'}, \pm 1)\, v + L_1 v,$$

where L_1 is a differential operator of the form (2.3.14). Here,

$$\|\chi(x_3/N)\, e^{\pm i x_3}\, L(D_{x'}, \pm 1)\, v\|^2_{V_\delta^{l-2m,2}(\mathcal{D})^\ell} \le cN\, \|L(D_{x'}, \pm 1)\, v\|^2_{E_\delta^{l-2m,2}(K)^\ell}$$

and

$$\|L_1 v\|^2_{V_\delta^{l-2m,2}(\mathcal{D})^\ell} \le cN^{-1}\, \|v\|^2_{E_\delta^{l,2}(K)}\,.$$

Thus, the inequality (2.4.3) implies

$$\|v\|_{E_\delta^{l,2}(K)^\ell} \le c\Big(\|L(D_{x'}, \pm 1)v\|_{E_\delta^{l-2m,2}(K)^\ell} + N^{-1}\, \|v\|_{E_\delta^{l,2}(K)^\ell}\Big)$$

for arbitrary $v \in E_\delta^{l,2}(K)^\ell \cap \overset{\circ}{E}{}^{m,2}_{\delta-l+m}(K)$. Since N can be chosen arbitrarily large and c is independent of N, we obtain

$$\|v\|_{E_\delta^{l,2}(K)^\ell} \le c\, \|L(D_{x'}, \pm 1)v\|_{E_\delta^{l-2m,2}(K)^\ell}$$

with a constant c independent of v. From Lemma 2.3.20 it follows that $\delta - l + m < \delta_-$ and that the line $\operatorname{Re}\lambda = l - 1 - \delta$ is free of eigenvalues of the pencil $A(\lambda)$. The proof of the theorem is complete. □

An analogous result holds for the solution of the adjoint problem.

COROLLARY 2.4.4. *The estimate*

$$\|u\|_{V_\delta^{m,2}(\mathcal{D})^\ell} \le c\, \|L^+(D_x)u\|_{V_\delta^{-m,2}(\mathcal{D})^\ell}$$

is valid for all $u \in \overset{\circ}{V}{}^{m,2}_\delta(\mathcal{D})^\ell$ *if and only if* $\delta < \delta_+$ *and the line* $\operatorname{Re}\lambda = m - 1 - \delta$ *is free of eigenvalues of the pencil* $A^+(\lambda)$.

Recall that the line $\operatorname{Re}\lambda = m - 1 - \delta$ does not contain eigenvalues of the pencil $A^+(\lambda)$ if and only if the line $\operatorname{Re}\lambda = m - 1 + \delta$ is free of eigenvalues of the pencil $A(\lambda)$.

2.4.3. Existence and uniqueness of solutions in $V_\delta^{l,2}(\mathcal{D})^\ell$. The a priori estimate for the solutions in Theorem 2.4.3 and the analogous result for the formally adjoint problem are used in the proof of the following existence and uniqueness theorem.

THEOREM 2.4.5. *The boundary value problem* (2.2.1) *is uniquely solvable in* $V_\delta^{l,2}(\mathcal{D})^\ell$, $l \ge m$, *for arbitrary* $f \in V_\delta^{l-2m,2}(\mathcal{D})^\ell$, $g_k^\pm \in V_\delta^{l-k+1/2,2}(\Gamma^\pm)^\ell$, $k = 1, \dots, m$, *if and only if* $-\delta_+ < \delta - l + m < \delta_-$.

P r o o f. By Lemma 2.2.1, we may restrict ourselves to the case of zero boundary data $g_k^\pm$. We consider the operator

$$L(D_x) : V_\delta^{l,2}(\mathcal{D})^\ell \cap \overset{\circ}{V}{}^{m,2}_{\delta-l+m}(\mathcal{D})^\ell \to V_\delta^{l-2m,2}(\mathcal{D})^\ell. \tag{2.4.4}$$

1) If $l = m$, then for the bijectivity of the operator (2.4.4) it is necessary and sufficient that both the operator (2.4.4) and the adjoint operator

$$L^+(D_x) : \overset{\circ}{V}{}^{m,2}_{-\delta}(\mathcal{D})^\ell \to V_{-\delta}^{-m,2}(\mathcal{D})^\ell$$

are injective and have closed ranges. By Theorem 2.4.3 and Corollary 2.4.4, this is true if and only if $-\delta_+ < \delta < \delta_-$.

2) Suppose that $l > m$ and $-\delta_+ < \delta - l + m < \delta_-$. Then by Theorem 2.4.3, the operator (2.4.4) is injective and has closed range. Let f be an arbitrary element of the space $V_\delta^{l-2m,2}(\mathcal{D})^\ell$. The last space is a subspace of $V_{\delta-l+m}^{-m,2}(\mathcal{D})^\ell$. Consequently by the first part of the proof, there exists a solution $u \in \overset{\circ}{V}{}^{m,2}_{\delta-l+m}(\mathcal{D})^\ell$ of the equation $L(D_x)u = f$. According to Theorem 2.2.9, this solution belongs to $V_\delta^{l,2}(\mathcal{D})^\ell$. Hence the operator (2.4.4) is surjective.

3) We assume now that the operator (2.4.4) is an isomorphism. Then it follows from Theorem 2.4.3 that $\delta - l + m < \delta_-$ and that the line $\operatorname{Re}\lambda = l - 1 - \delta$ does not contain eigenvalues of the pencil $A(\lambda)$. Consequently by Theorem 2.4.3, the estimate

$$\|u\|_{V_{\delta-l+m}^{m,2}(\mathcal{D})^\ell} \le c\, \|L(D_x)u\|_{V_{\delta-l+m}^{-m,2}(\mathcal{D})^\ell}$$

is satisfied for all $u \in \overset{\circ}{V}{}^{m,2}_{\delta-l+m}(\mathcal{D})^\ell$. Thus, the operator

$$L(D_x)\, :\, \overset{\circ}{V}{}^{m,2}_{\delta-l+m}(\mathcal{D})^\ell \to V_{\delta-l+m}^{-m,2}(\mathcal{D})^\ell \tag{2.4.5}$$

is injective and has closed range. Since the operator (2.4.5) is an extension of the operator (2.4.4), it follows that the range of the operator (2.4.5) contains the range of operator (2.4.4), i.e. the whole space $V_\delta^{l-2m,2}(\mathcal{D})^\ell$. The last space is dense in $V_{\delta-l+m}^{-m,2}(\mathcal{D})^\ell$. Therefore, the operator (2.4.5) is an isomorphism. By the first part of the proof, we have $-\delta_+ < \delta - l + m < \delta_-$. The proof is complete. □

2.4.4. A regularity assertion for the solution. Now we are interested in the smoothness of an arbitrary solution $u \in V_\delta^{l,2}(\mathcal{D})^\ell$, $l \ge m$, $-\delta_+ < \delta - l + m < \delta_-$, of the boundary value problem (2.2.1).

THEOREM 2.4.6. *Let $u \in V_\delta^{m,2}(\mathcal{D})^\ell$ be a solution of the problem* (2.2.1), *where*

$$f \in V_\delta^{-m,2}(\mathcal{D})^\ell \cap V_{\delta'}^{l-2m,2}(\mathcal{D})^\ell, \quad g_k^\pm \in V_\delta^{m-k+1/2,2}(\Gamma^\pm)^\ell \cap V_{\delta'}^{l-k+1/2,2}(\Gamma^\pm)^\ell,$$

$l \ge m$, $-\delta_+ < \delta < \delta_-$, $-\delta_+ < \delta' - l + m < \delta_-$. Then $u \in V_{\delta'}^{l,2}(\mathcal{D})^\ell$.

P r o o f. By Lemma 2.2.1 (see also Remark 2.2.2), it suffices to prove the lemma for zero boundary data $g_k^\pm$. As in the proof of Theorem 2.4.3, we define $U(x',\eta) = |\eta|^{2m}\, \tilde{u}(|\eta|^{-1}x',\eta)$. Then U is a solution of the equation (2.4.2), where $F(\cdot,\eta) \in E_\delta^{-m,2}(K)^\ell \cap E_{\delta'}^{l-2m,2}(K)^\ell$ for almost all η. This is a consequence of Lemma 2.4.2. Furthermore it follows from Lemma 2.4.1 that $U(\cdot,\eta) \in E_\delta^{m,2}(K)^\ell$ for almost all η. Applying Theorem 2.3.14 and Lemma 2.3.20, we obtain $U(\cdot,\eta) \in E_{\delta'}^{l,2}(K)^\ell$ and

$$\|U(\cdot,\eta)\|^2_{E_{\delta'}^{l,2}(K)^\ell} \le c\, \|F(\cdot,\eta)\|^2_{E_{\delta'}^{l-2m,2}(K)^\ell}\, .$$

Thus by Lemma 2.4.2,

$$\int_{\mathbb{R}} |\eta|^{2(l-2m-\delta'-1)}\, \|U(\cdot,\eta)\|^2_{E_{\delta'}^{l,2}(K)^\ell}\, d\eta < \infty.$$

This means that $u \in V_{\delta'}^{l,2}(\mathcal{D})^\ell$ (see Lemma 2.4.1). The theorem is proved. □

COROLLARY 2.4.7. *Let ζ, η be infinitely differentiable functions with compact supports such that $\eta = 1$ in a neighborhood of* $\operatorname{supp}\zeta$. *Suppose that $\eta u \in \overset{\circ}{V}{}^{m,2}_\delta(\mathcal{D})^\ell$*

and $\eta Lu \in V_{\delta'}^{l-2m,2}(\mathcal{D})^\ell$, *where* $l \ge m$, $-\delta_+ < \delta < \delta_-$, $-\delta_+ < \delta' - l + m < \delta_-$. *Then* $\zeta u \in V_{\delta'}^{l,2}(\mathcal{D})^\ell$ *and*

$$\|\zeta u\|_{V_{\delta'}^{l,2}(\mathcal{D})^\ell} \le c\left(\|\eta u\|_{V_\delta^{m,2}(\mathcal{D})^\ell} + \|\eta Lu\|_{V_{\delta'}^{l-2m,2}(\mathcal{D})^\ell}\right).$$

P r o o f. First we prove the corollary for the case $l = m$. Let $\chi_1, \chi_2, \ldots$ be a sequence of infinitely differentiable cut-off functions such that $\eta = 1$ in a neighborhood of $\operatorname{supp}\chi_1$ and $\chi_k = 1$ in a neighborhood of $\operatorname{supp}\chi_{k+1}$, $k = 1, 2, \ldots$. Then

$$[L, \chi_1]u \stackrel{def}{=} L(\chi_1 u) - \chi_1 Lu \in V_\delta^{-m+1,2}(\mathcal{D})^\ell \subset V_{\delta-1}^{-m,2}(\mathcal{D})^\ell$$

and, therefore, $L(\chi_1 u) \in V_{\max(\delta',\delta-1)}^{-m,2}(\mathcal{D})^\ell$. From Theorem 2.4.6 it follows that $\chi_1 u \in V_{\max(\delta',\delta-1)}^{m,2}(\mathcal{D})^\ell$. Using this result, we conclude in the same way that $\chi_2 u \in V_{\max(\delta',\delta-2)}^{m,2}(\mathcal{D})^\ell$. After a finite number of steps, we arrive at $\chi_k u \in V_{\delta'}^{m,2}(\mathcal{D})^\ell$. The sequence $\chi_1, \chi_2, \ldots$ can be chosen such that $\chi_k = \zeta$. This proves the corollary for $l = m$.

Now let $l > m$. Then it follows from the first part of the proof that $\chi_1 u \in V_{\delta'-l+m}^{m,2}(\mathcal{D})^\ell$ and consequently

$$L(\chi_2 u) = \chi_2 Lu + [L, \chi_2]u \in V_{\delta'-l+m+1}^{-m+1,2}(\mathcal{D})^\ell.$$

This together with Theorem 2.4.6 implies $\chi_2 u \in V_{\delta'-l+m+1}^{m+1,2}(\mathcal{D})^\ell$. Repeating this argument, we obtain $\chi_3 u \in V_{\delta'-l+m+2}^{m+2,2}(\mathcal{D})^\ell, \ldots, \chi_{l-m+1}u \in V_{\delta'}^{l,2}(\mathcal{D})^\ell$. We may assume that $\chi_{l-m+1} = \zeta$. This completes the proof. □

2.4.5. A regularity assertion for the x_3-derivatives of the solution. Suppose that $-\delta_+ < \delta < \delta_-$ and that $u \in \overset{\circ}{V}{}_\delta^{m,2}(\mathcal{D})^\ell$ is a solution of the equation $L(D_x)\, u = f$, where f is a sufficiently smooth vector function ($f \in V_{\delta'}^{l-2m+1,2}(\mathcal{D})^\ell$). Then it follows from Theorem 2.4.6 that $\partial_{x_j} u \in V_{\delta'}^{l,2}(\mathcal{D})^\ell$ provided δ' satisfies the inequalities $l - m + 1 - \delta_+ < \delta' < l - m + 1 + \delta_-$. For the x_3 derivatives of the solution, we obtain sharper regularity results. In particular, we conclude from the next lemma that $\partial_{x_3} u \in V_{\delta'}^{l,2}(\mathcal{D})^\ell$ if $f \in V_{\delta'}^{l-2m+1,2}(\mathcal{D})^\ell \cap V_{\delta'}^{l-2m,2}(\mathcal{D})^\ell$ and $l - m - \delta_+ < \delta' < l - m + 1 + \delta_-$.

LEMMA 2.4.8. *Let* $u \in \overset{\circ}{V}{}_\delta^{m,2}(\mathcal{D})^\ell$ *be a solution of the equation* $L(D_x)\, u = f$ *in* $\mathcal{D}$, *where* $\partial_{x_3}^j f \in V_{\delta'}^{l-2m,2}(\mathcal{D})^\ell$ *for* $j = 0, 1, \ldots, k$, $l \ge 2m$. *Suppose that* δ, δ' *satisfy the inequalities* $-\delta_+ < \delta < \delta_-$ *and* $-\delta_+ < \delta' - l + m < \delta_-$. *Then* $\partial_{x_3}^j u \in V_{\delta'}^{l,2}(\mathcal{D})^\ell$ *for* $j = 0, 1, \ldots, k$.

P r o o f. By Theorem 2.4.6, the solution u belongs to the space $V_{\delta'}^{l,2}(\mathcal{D})^\ell$ Consequently, the function

$$u_h(x', x_3) = h^{-1}\left(u(x', x_3 + h) - u(x', x_3)\right)$$

belongs to $V^{2m,2}_{\delta'-l+2m}(\mathcal{D})^\ell \cap \overset{\circ}{V}{}^{m,2}_{\delta'-l+m}(\mathcal{D})^\ell$ for every real $h \neq 0$. Since $L(D_x)\,u_h = f_h$ in $\mathcal{D}$, it follows from Theorem 2.4.5 that

$$\begin{aligned} \|u_h\|^2_{V^{2m,2}_{\delta'-l+2m}(\mathcal{D})^\ell} &\le c\,\|f_h\|_{V^{0,2}_{\delta'-l+2m}(\mathcal{D})^\ell} \\ &= c\int_{\mathcal{D}} r^{2(\delta'-l+2m)}\Big|\int_0^1 (\partial_{x_3} f)(x', x_3+th)\,dt\Big|^2\,dx \\ &\le c\int_{\mathcal{D}} r^{2(\delta'-l+2m)}\big|\partial_{x_3} f(x)\big|^2\,dx = \|\partial_{x_3} f\|^2_{V^{0,2}_{\delta'-l+2m}(\mathcal{D})^\ell} \end{aligned}$$

with a constant c independent of u and h. Consequently, $\partial_{x_3} u \in V^{2m,2}_{\delta'-l+2m}(\mathcal{D})^\ell$. This together with Theorem 2.2.9 implies $\partial_{x_3} u \in V^{l,2}_{\delta'}(\mathcal{D})^\ell$. Repeating this argument, we obtain $\partial^j_{x_3} u \in V^{l,2}_{\delta}(\mathcal{D})^\ell$ for $j = 2, \dots, k$. □

We furthermore need a local regularity result for the x_3-derivatives of the solution.

LEMMA 2.4.9. *Let ζ, η be infinitely differentiable functions with compact supports such that $\eta = 1$ in a neighborhood of* $\operatorname{supp}\zeta$. *Suppose that $\eta u \in \overset{\circ}{V}{}^{m,2}_{\delta}(\mathcal{D})^\ell$ and $\eta\partial^j_{x_3} L(D_x)u \in V^{l-2m,2}_{\delta'}(\mathcal{D})^\ell$ for $j = 0, \dots, k$, where $l \ge 2m$, $-\delta_+ < \delta < \delta_-$ and $-\delta_+ < \delta' - l + m < \delta_-$. Then $\zeta\,\partial^j_{x_3} u \in V^{l,2}_{\delta'}(\mathcal{D})^\ell$ for $j = 0, \dots, k$ and*

$$\sum_{j=0}^k \|\zeta\partial^j_{x_3} u\|_{V^{l,2}_{\delta'}(\mathcal{D})^\ell} \le c\Big(\sum_{j=0}^k \|\eta\partial^j_{x_3} L(D_x)u\|_{V^{l-2m,2}_{\delta'}(\mathcal{D})^\ell} + \|\eta u\|_{V^{m,2}_{\delta}(\mathcal{D})^\ell}\Big). \tag{2.4.6}$$

P r o o f. For $k = 0$, the assertion of the lemma is true according to Corollary 2.4.7. We assume that the assumptions of the lemma are satisfied for $k = 1$. Let χ be a smooth cut-off function such that $\eta = 1$ in a neighborhood of $\operatorname{supp}\chi$ and $\chi = 1$ in a neighborhood of $\operatorname{supp}\zeta$. From Corollary 2.4.7 it follows that $\chi u \in V^{l,2}_{\delta'}(\mathcal{D})^\ell$ and

$$\|\chi u\|_{V^{l,2}_{\delta'}(\mathcal{D})^\ell} \le c\Big(\|\eta L(D_x)u\|_{V^{l-2m,2}_{\delta'}(\mathcal{D})^\ell} + \|\eta u\|_{V^{m,2}_{\delta}(\mathcal{D})^\ell}\Big).$$

Consequently, $L(D_x)\,(\zeta u) \in V^{l-2m,2}_{\delta'}(\mathcal{D})^\ell$ and

$$\partial_{x_3} L(D_x)\,(\zeta u) = \partial_{x_3}\,[L(D_x),\zeta]u + \partial_{x_3}\big(\zeta L(D_x)u\big) \in V^{l-2m,2}_{\delta'}(\mathcal{D})^\ell$$

(here again $[L,\zeta]$ denotes the commutator of L and ζ). Using Lemma 2.4.8 and Theorem 2.4.5, we obtain $\partial_{x_3}(\zeta u) \in V^{l,2}_{\delta'}(\mathcal{D})^\ell$ and

$$\begin{aligned} \|\partial_{x_3}(\zeta u)\|_{V^{l,2}_{\delta'}(\mathcal{D})^\ell} &\le c\Big(\|\partial_{x_3}\,[L(D_x),\zeta]u\|_{V^{l-2m,2}_{\delta'}(\mathcal{D})^\ell} + \|\partial_{x_3}(\zeta f)\|_{V^{l-2m,2}_{\delta'}(\mathcal{D})^\ell}\Big) \\ &\le c\Big(\|\chi u\|_{V^{l,2}_{\delta'}(\mathcal{D})^\ell} + \|\partial_{x_3}(\zeta f)\|_{V^{l-2m,2}_{\delta'}(\mathcal{D})^\ell}\Big), \end{aligned}$$

where $f = L(D_x)u$. From this and from the inclusion $\chi u \in V^{l,2}_{\delta'}(\mathcal{D})^\ell$, we conclude that $\zeta\partial_{x_3} u \in V^{l,2}_{\delta'}(\mathcal{D})^\ell$. Moreover, the estimate (2.4.6) holds. This proves the lemma for $k = 1$. For $k > 1$, the assertion of the lemma holds by induction. □

2.5. Green's matrix of the Dirichlet problem in a dihedron

In this section, we study the Green's matrix of the Dirichlet problem for the strongly elliptic differential operator $L(D_x)$ in the dihedron $\mathcal{D}$. The matrix

$$G(x,\xi) = \big(G_{i,j}(x,\xi)\big)_{i,j=1}^{\ell}$$

is called *Green's matrix* of the problem (2.2.1) if

$$L(D_x)\, G(x,\xi) = \delta(x-\xi)\, I_\ell \quad \text{for } x,\xi \in \mathcal{D}, \tag{2.5.1}$$

$$\frac{\partial^k G(x,\xi)}{\partial n_x^k} = 0 \quad \text{for } x \in \Gamma^\pm,\ \xi \in \mathcal{D},\ k = 0,\dots,m-1. \tag{2.5.2}$$

Here I_ℓ denotes the $\ell \times \ell$ identity matrix. We are interested in solutions $G(x,\xi)$ of the problem (2.5.1), (2.5.2) which belong to the space $\overset{\circ}{W}{}^{m,2}(\mathcal{D})^{\ell\times\ell}$ outside an arbitrarily small neighborhood of the point $x = \xi$. The existence and uniqueness of such solutions can be easily deduced from Theorem 2.2.6. The main goal of this section is to prove point estimates for the matrix $G(x,\xi)$. We consider the cases $|x-\xi| < \min(|x'|,|\xi'|)$ and $|x-\xi| > \min(|x'|,|\xi'|)$ separately.

2.5.1. Existence of Green's matrix. We prove the existence and uniqueness of Green's matrix and some basis properties.

THEOREM 2.5.1. 1) *There exists a unique Green's matrix such that the function $x \to \zeta(x,\xi)\, G_{i,j}(x,\xi)$ belongs to the space $V_0^{m,2}(\mathcal{D})$ for $i,j = 1,\dots,\ell$, for each $\xi \in \mathcal{D}$ and for every smooth function $\zeta(\cdot,\xi)$ which is equal to zero in a neighborhood of the point $x = \xi$ and bounded together with all derivatives.*

2) *The equality*

$$G_{i,j}(tx, t\xi) = t^{2m-3}\, G_{i,j}(x,\xi) \tag{2.5.3}$$

is valid for all $x,\xi \in \mathcal{D}$, $t > 0$, $i,j = 1,\dots,\ell$.

3) *The adjoint matrix $G^*(x,\xi)$ is the unique solution of the problem*

$$L^+(D_\xi)\, G^*(x,\xi) = \delta(x-\xi)\, I_\ell \quad \textit{for } x,\xi \in \mathcal{D}, \tag{2.5.4}$$

$$\frac{\partial^k G^*(x,\xi)}{\partial n_\xi^k} = 0 \quad \textit{for } \xi \in \Gamma^\pm,\ x \in \mathcal{D},\ k = 0,\dots,m-1. \tag{2.5.5}$$

such that the function $\xi \to \zeta(x,\xi)\, G^*(x,\xi)$ belongs to the space $V_0^{m,2}(\mathcal{D})^{\ell\times\ell}$ for every $x \in \mathcal{D}$ and for every smooth function $\zeta(x,\cdot)$ which is equal to zero in a neighborhood of the point $\xi = x$ and bounded together with all derivatives.

4) *The solution $u \in \overset{\circ}{V}{}_0^{m,2}(\mathcal{D})^\ell$ of the equation $L(D_x)u = f$ in $\mathcal{D}$ admits the representation*

$$u(x) = \int_{\mathcal{D}} G(x,\xi)\, f(\xi)\, d\xi \tag{2.5.6}$$

for arbitrary $f \in V_0^{-m,2}(\mathcal{D})^\ell$.

P r o o f. 1) There exists a solution $\mathcal{G}(x,\xi)$ of the equation

$$L(D_x)\, \mathcal{G}(x,\xi) = \delta(x-\xi)\, I_\ell \quad \text{in } \mathbb{R}^3$$

having the form $\mathcal{G}(x,\xi) = h(x-\xi)\, |x-\xi|^{2m-3}$, where h is positively homogeneous of degree 0 (i.e. $h(\lambda x) = h(x)$ for $\lambda > 0$, see e.g. [**13**, Part 2, Section 5.1]). The function $\mathcal{G}(\cdot,\xi)$ is real-analytic for $x \neq \xi$. Let χ be a smooth function on $(0,\infty)$, $\chi(t) = 1$ for $t < 1/4$, $\chi(t) = 0$ for $t > 1/2$. Furthermore, let $\psi(x,\xi) = \chi(|x-\xi|/|\xi'|)$. For every fixed $\xi \in \mathcal{D}$, we define $R(x,\xi)$ as the unique solution (in the space

$V_0^{m,2}(\mathcal{D})^{\ell\times\ell}$) of the problem

$$(2.5.7)\qquad L(D_x)\,R(x,\xi) = \delta(x-\xi)\,I_\ell - L(D_x)\big(\psi(x,\xi)\,\mathcal{G}(x,\xi)\big) \quad \text{for } x \in \mathcal{D},$$

$$(2.5.8)\qquad \frac{\partial^k R(x,\xi)}{\partial n_x^k} = -\frac{\partial^k(\psi(x,\xi)\mathcal{G}(x,\xi))}{\partial n_x^k} \quad \text{for } x \in \Gamma^\pm,\ k = 0,\dots,m-1.$$

Obviously for every fixed $\xi \in \mathcal{D}$, the right-hand sides of (2.5.7), (2.5.8) are smooth functions on $\overline{\mathcal{D}}$ and $\Gamma^\pm$, respectively, with compact supports vanishing in a neighborhood of the edge M. Therefore by Theorem 2.2.6, the solution of (2.5.7), (2.5.8) exists for every $\xi \in \mathcal{D}$. The Green's function $G(x,\xi)$ is then defined as

$$G(x,\xi) = \psi(x,\xi)\,\mathcal{G}(x,\xi) + R(x,\xi).$$

We prove the uniqueness. Suppose that $G(x,\xi)$ and $\tilde{G}(x,\xi)$ are Green's matrices such that $\zeta(\cdot,\xi)G(\cdot,\xi)$ and $\zeta(\cdot,\xi)\tilde{G}(\cdot,\xi)$ are from $V_0^{1,2}(\mathcal{D})^{\ell\times\ell}$ for every smooth function ζ vanishing in a neighborhood of the point $x=\xi$. Then in particular,

$$\big(1-\psi(\cdot,\xi)\big)\,\big(G(\cdot,\xi)-\tilde{G}(\cdot,\xi)\big) \in V_0^{1,2}(\mathcal{D})^{\ell\times\ell}.$$

Since $G(\cdot,\xi)-\tilde{G}(\cdot,\xi)$ is a solution of the Dirichlet problem with zero right-hand sides, we further have

$$\psi(\cdot,\xi)\,\big(G(\cdot,\xi)-\tilde{G}(\cdot,\xi)\big) \in W^{1,2}(\mathcal{D})^{\ell\times\ell}.$$

Consequently, $G(\cdot,\xi)-\tilde{G}(\cdot,\xi) \in V_0^{1,2}(\mathcal{D})^{\ell\times\ell}$ for arbitrary ξ. Using Theorem 2.2.6, we obtain $G(x,\xi)-\tilde{G}(x,\xi)=0$.

2) By (2.5.1), we have $L(D_x)\,G(tx,t\xi) = t^{2m}\,\delta(tx-t\xi)\,I_\ell = t^{2m-3}\,\delta(x-\xi)\,I_\ell$ for $x,\xi\in\mathcal{D}$. This implies (2.5.3).

3) Analogously to the assertion 1), there exists a unique solution $H(x,\xi)$ of the problem

$$(2.5.9)\qquad L^+(D_x)H(x,\xi) = \delta(x-\xi)\,I_\ell \quad \text{for } x,\xi \in \mathcal{D},$$

$$(2.5.10)\qquad \frac{\partial^k H(x,\xi)}{\partial n_x^k} = 0 \quad \text{for } x\in\Gamma^\pm,\ \xi\in\mathcal{D},\ k=0,\dots,m-1.$$

such that the function $x \to \zeta(x,\xi)\,H(x,\xi)$ belongs to the space $V_0^{m,2}(\mathcal{D})^{\ell\times\ell}$ for every function ζ which is equal to zero in a neighborhood of the point $x=\xi$ and bounded together with all derivatives. We prove that $G(x,\xi) = H(\xi,x)^*$.

If $m\ge 2$, then the δ-function belongs to $V_0^{-m,2}(\mathcal{D})$. Consequently, the problems (2.5.1), (2.5.2) and (2.5.9), (2.5.10) have unique solutions $G(\cdot,\xi)$, $H(\cdot,\xi)$ in $V_0^{m,2}(\mathcal{D})^{\ell\times\ell}$ (see Theorem 2.2.6). Let $G^{(i)} = (G_{1,i},\dots,G_{\ell,i})^t$ denote the i-th column of the matrix G and $H^{(j)} = (H_{1,j},\dots,H_{\ell,j})^t$ the j-th column of H (here t means transposition). Then from the equation

$$(2.5.11)\qquad \big(L(D_x)\,G^{(i)}(\cdot,\xi),H^{(j)}(\cdot,\eta)\big)_{\mathcal{D}} = \big(G^{(i)}(\cdot,\xi),L^+(D_x)H^{(j)}(\cdot,\eta)\big)_{\mathcal{D}}$$

it follows that $\overline{H_{i,j}(\xi,\eta)} = G_{j,i}(\eta,\xi)$.

Let $m=1$, and let ε be positive and sufficiently small. Then the functions $u(x) = (1-\chi(\frac{|x-\xi|}{\varepsilon}))\,G^{(i)}(x,\xi)$ and $v(x) = (1-\chi(\frac{|x-\eta|}{\varepsilon}))\,H^{(j)}(x,\eta)$ belong to $\overset{\circ}{V}{}_0^{m,2}(\mathcal{D})^\ell$ and are infinitely differentiable in $\overline{\mathcal{D}}\backslash M$. Therefore, the equality

$$(2.5.12)\qquad \big(L(D_x)u,v\big)_{\mathcal{D}} = \big(u,L^+(D_x)v\big)_{\mathcal{D}}$$

is valid for every of the pairs

$$u(x) = \chi\Big(\frac{|x-\xi|}{\varepsilon}\Big)\, G^{(i)}(x,\xi), \quad v(x) = \Big(1-\chi\Big(\frac{|x-\eta|}{\varepsilon}\Big)\Big)\, H^{(j)}(x,\eta),$$
$$u(x) = \Big(1-\chi\Big(\frac{|x-\xi|}{\varepsilon}\Big)\Big)\, G^{(i)}(x,\xi), \quad v(x) = \chi\Big(\frac{|x-\eta|}{\varepsilon}\Big)\, H^{(j)}(x,\eta),$$
$$u(x) = \Big(1-\chi\Big(\frac{|x-\xi|}{\varepsilon}\Big)\Big)\, G^{(i)}(x,\xi), \quad v(x) = \Big(1-\chi\Big(\frac{|x-\eta|}{\varepsilon}\Big)\Big)\, H^{(j)}(x,\eta).$$

Furthermore, both sides of (2.5.12) are zero for $u(x) = \chi(\frac{|x-\xi|}{\varepsilon})\, G^{(i)}(x,\xi)$ and $v(x) = \chi(\frac{|x-\eta|}{\varepsilon})\, H^{(j)}(x,\eta)$ if ε is sufficiently small, since then u and v have disjoint supports. Thus, (2.5.11) holds also for $m = 1$, and we obtain $\overline{H_{i,j}(\xi,\eta)} = G_{j,i}(\eta,\xi)$. This proves the assertion 3).

4) Let $f \in C_0^\infty(\overline{\mathcal{D}}\backslash M)^\ell$, and let $u \in \overset{\circ}{V}{}_0^{m,2}(\mathcal{D})^\ell$ be the unique solution of the equation $L(D_x)u = f$ in $\mathcal{D}$ (see Theorem 2.2.6). Furthermore, let $H^{(j)}(\xi,x)$ be the j-th column of the matrix $H(\xi,x) = G^*(x,\xi)$. Since $u \in V_k^{m+k,2}(\mathcal{D})^\ell$ for arbitrary integer $k \geq 0$ (cf. Theorem 2.2.9), we have

$$\int_{\mathcal{D}} f(\xi)\cdot\overline{H^{(j)}(\xi,x)}\, d\xi = \int_{\mathcal{D}} L(D_\xi)u(\xi)\cdot\overline{H^{(j)}(\xi,x)}\, d\xi = \int_{\mathcal{D}} u(\xi)\cdot\overline{L^+(D_\xi)\, H^{(j)}(\xi,x)}\, d\xi$$

for each $x \in \mathcal{D}$. By (2.5.9), the right-hand side of the last equality is equal to $u_j(x)$. Hence, the formula (2.5.6) is true for $f \in C_0^\infty(\overline{\mathcal{D}}\backslash M)^\ell$. Since this set is dense in $V_0^{-m,2}(\mathcal{D})^\ell$, the representation (2.5.6) can be extended by continuity to all $f \in V_0^{-m,2}(\mathcal{D})^\ell$. The proof of the theorem is complete. □

2.5.2. Estimates of Green's matrix: the case $|x-\xi| < \min(|x'|,|\xi'|)$. Next, we prove point estimates of Green's matrix. The estimates in the case $|x-\xi| < \min(|x'|,|\xi'|)$ are essentially based on Theorem 1.1.8.

THEOREM 2.5.2. *Let $G(x,\xi)$ be the Green's matrix introduced in Theorem 2.5.2. Then the inequalities*

$$\big|D_x^\alpha\, D_\xi^\beta G_{i,j}(x,\xi)\big| \leq c_{\alpha,\beta}\, \big(|x-\xi|^{2m-3-|\alpha|-|\beta|} + |\xi'|^{2m-3-|\alpha|-|\beta|}\big) \quad \text{if } |\alpha|+|\beta| \neq 2m-3,$$
$$\big|D_x^\alpha\, D_\xi^\beta G_{i,j}(x,\xi)\big| \leq c_{\alpha,\beta}\, \Big(\Big|\log\frac{|x-\xi|}{|\xi'|}\Big| + 1\Big) \quad \text{if } |\alpha|+|\beta| = 2m-3$$

are satisfied for $|x-\xi| < \min(|x'|,|\xi'|)$.

P r o o f. We write the matrix $G(x,\xi)$ in the form

$$G(x,\xi) = \psi(x,\xi)\, \mathcal{G}(x,\xi) + R(x,\xi),$$

where $\mathcal{G}(x,\xi)$ is Green's matrix of the Dirichlet problem for the differential operator $L(D_x)$ in the half-space bounded by the plane $\Gamma^-\cup(-\Gamma^-)\cup M$, $R(x,\xi)$ is the solution of the problem (2.5.7), (2.5.8) in $V_0^{m,2}(\mathcal{D})^{\ell\times\ell}$, and ψ is the same cut-off function as in the proof of Theorem 2.5.1. The elements of the matrix $\mathcal{G}(x,\xi)$ satisfy the estimate (cf. Theorem 1.1.8)

$$\big|D_x^\alpha\, D_\xi^\beta \mathcal{G}_{i,j}(x,\xi)\big| \leq c_{\alpha,\beta}\, \big(|x-\xi|^{2m-3-|\alpha|-|\beta|} + 1\big) \quad \text{if } |\alpha|+|\beta| \neq 2m-3,$$
$$\big|D_x^\alpha\, D_\xi^\beta \mathcal{G}_{i,j}(x,\xi)\big| \leq c_{\alpha,\beta}\, \big(\big|\log|x-\xi|\big| + 1\big) \quad \text{if } |\alpha|+|\beta| = 2m-3.$$

The function $x \to D_\xi^\beta R(x,\xi)$ is a solution of the problem

$$L(D_x)\, D_\xi^\beta R(x,\xi) = F(x,\xi) \quad \text{for } x \in \mathcal{D},$$
$$\frac{\partial^k D_\xi^\beta R(x,\xi)}{\partial n_x^k} = \Phi_k^\pm(x,\xi) \quad \text{for } x \in \Gamma^\pm,\ k = 0,\dots,m-1,$$

where

$$F(x,\xi) = D_\xi^\beta \delta(x-\xi)\, I_\ell - L(D_x) D_\xi^\beta \big(\psi(x,\xi)\, \mathcal{G}(x,\xi)\big),$$
$$\Phi_k^-(x,\xi) = 0, \quad \Phi_k^+(x,\xi) = -\frac{\partial^k D_\xi^\beta (\psi(x,\xi)\mathcal{G}(x,\xi))}{\partial n_x^k}.$$

Suppose that $|\xi'| = 1$ and $\text{dist}(\xi,\Gamma^-) \le \text{dist}(\xi,\Gamma^+)$. Then the supports of $F(\cdot,\xi)$ and $\Phi_k^+(x,\xi)$ are contained in the ball $|x-\xi| < 1/2$. Furthermore, all derivatives $D_x^\alpha F(x,\xi)$ and $D_x^\alpha \Phi_k^+(x,\xi)$ are bounded by constants independent of x and ξ. Consequently, the norms of $F(\cdot,\xi)$ in $V_{l-m}^{l-2m,2}(\mathcal{D})^{\ell\times\ell}$ and $\Phi_k^+(\cdot,\xi)$ in $V_{l-m}^{l-k+1/2,2}(\Gamma^+)^{\ell\times\ell}$ are also bounded by constants independent of ξ. From Theorems 2.2.6 and 2.2.9 it follows that $D_\xi^\beta R(\cdot,\xi) \in V_{l-m}^{l,2}(\mathcal{D})^{\ell\times\ell}$ for arbitrary $l \ge m$ and

$$\|D_\xi^\beta R(\cdot,\xi)\|_{V_{l-m}^{l,2}(\mathcal{D})^{\ell\times\ell}} \le c_{l,\beta}\,,$$

where the constant $c_{l,\beta}$ is independent of ξ. Using the continuity of the imbedding $W^{l,2}(\Omega) \subset C^{l-2}(\Omega)$, we conclude that

$$\big|D_x^\alpha D_\xi^\beta R(x,\xi)\big| \le c_{\alpha,\beta}$$

for $|x-\xi| < \min(|x'|,1)$, $|\xi'| = 1$, $\text{dist}(\xi,\Gamma^-) \le \text{dist}(\xi,\Gamma^+)$. (Note that the relations $|x-\xi| < \min(|x'|,1)$ and $|\xi'| = 1$ imply $1/2 < |x'| < 2$.) Together with the estimates of $\mathcal{G}(x,\xi)$ this yields

$$\big|D_x^\alpha\, D_\xi^\beta G_{i,j}(x,\xi)\big| \le c_{\alpha,\beta}\,\big(|x-\xi|^{2m-3-|\alpha|-|\beta|} + 1\big) \quad \text{for } |\alpha|+|\beta| \ne 2m-3,$$
$$\big|D_x^\alpha\, D_\xi^\beta G_{i,j}(x,\xi)\big| \le c_{\alpha,\beta}\,\big(\big|\log|x-\xi|\big| + 1\big) \quad \text{for } |\alpha|+|\beta| = 2m-3$$

if $|x-\xi| < \min(|x'|,1)$, $|\xi'| = 1$, $\text{dist}(\xi,\Gamma^-) \le \text{dist}(\xi,\Gamma^+)$. Analogously, this estimates hold for $|x-\xi| < \min(|x'|,1)$, $|\xi'| = 1$, $\text{dist}(\xi,\Gamma^-) > \text{dist}(\xi,\Gamma^+)$. Using (2.5.3), we obtain the desired estimates for arbitrary ξ, $|x-\xi| < \min(|x'|,|\xi'|)$. □

2.5.3. Estimates of Green's matrix: the case $|x-\xi| > \min(|x'|,|\xi'|)$. As in the previous section, let δ_+ and δ_- be the greatest real numbers such that the strip

$$m-1-\delta_- < \text{Re}\,\lambda < m-1+\delta_+$$

is free of eigenvalues of the pencil $A(\lambda)$. These numbers will also appear in the estimates of Green's matrix for the case $|x-\xi| > \min(|x'|,|\xi'|)$. For the proof of these estimates, we need the following lemma.

LEMMA 2.5.3. *Let $\mathcal{B}$ be a ball with radius 1 and center at x_0, $\text{dist}(x_0,M) \le 4$. Furthermore, let ζ and η be smooth functions with support in $\mathcal{B}$ such that $\eta = 1$ in a neighborhood of* $\text{supp}\,\zeta$. *If $\eta u \in \overset{\circ}{V}{}_0^{m,2}(\mathcal{D})^\ell$ and $L(D_x)\,u = 0$ in $\mathcal{D}\cap\mathcal{B}$, then*

$$\sup_{x\in\mathcal{D}} |x'|^{|\alpha|-m+1-\delta_++\varepsilon}\,\big|\zeta(x)\,\partial_{x'}^\alpha \partial_{x_3}^j u(x)\big| \le c\,\|\eta u\|_{V_0^{m,2}(\mathcal{D})^\ell} \tag{2.5.13}$$

for all integer $l \ge 0$ and all multi-indices α. Here, ε is an arbitrarily small positive number.

P r o o f. Let χ, ψ be smooth cut-off functions such that $\eta = 1$ in a neighborhood of $\operatorname{supp}\psi$, $\psi = 1$ in a neighborhood of $\operatorname{supp}\chi$, and $\chi = 1$ in a neighborhood of $\operatorname{supp}\zeta$. Furthermore, let $\delta = l - m - \delta_+ + \varepsilon$, where l is an integer, $l \ge 2m$, $l \ge |\alpha| + 2$. Then by Lemma 2.4.9, we have $\chi\partial_{x_3}^j u \in V_\delta^{l,2}(\mathcal{D})^\ell$ for every j and

$$\|\psi\partial_{x_3}^j u\|_{V_\delta^{l,2}(\mathcal{D})^\ell} \le c\,\|\eta u\|_{V_0^{m,2}(\mathcal{D})^\ell}\,. \tag{2.5.14}$$

According to Lemma 1.2.3,

$$\sup_{x'\in K} |x'|^{\delta-l+|\alpha|+1}\,\big|\partial_{x'}^\alpha(\chi\partial_{x_3}^j u(x',x_3))\big| \le c\,\Phi(x_3),$$

where

$$\Phi(x_3) = \|\partial_{x'}^\alpha(\chi\partial_{x_3}^j u(\cdot,x_3))\|_{V_\delta^{l-|\alpha|,2}(K)^\ell}\,.$$

Using the continuity of the imbedding $W^{1,2}(\mathbb{R}) \subset C(\mathbb{R})$, we conclude that

$$\sup_{x_3\in\mathbb{R}}\ \sup_{x'\in K} |x'|^{\delta-l+|\alpha|+1}\,\big|\partial_{x'}^\alpha(\chi\partial_{x_3}^j u(x',x_3))\big| \le c\,\|\Phi\|_{W^{1,2}(\mathbb{R})^\ell}\,.$$

Since

$$\begin{aligned} \big|\Phi'(x_3)\big| &= \|\partial_{x'}^\alpha(\chi\partial_{x_3}^j u(\cdot,x_3))\|^{-1}_{V_\delta^{l-|\alpha|,2}(K)^\ell} \\ &\quad\times\Big|\operatorname{Re}\int_K \sum_{|\gamma|\le l-|\alpha|} r^{2(\delta-l+|\alpha|+|\gamma|)}\,\big(\partial_{x_3}\partial_{x'}^\alpha(\chi\partial_{x_3}^j u(x))\big)\cdot\overline{\partial_{x'}^\alpha(\chi\partial_{x_3}^j u(x))}\,dx'\Big| \\ &\le \|\partial_{x_3}\partial_{x'}^\alpha(\chi\partial_{x_3}^j u(\cdot,x_3))\|_{V_\delta^{l-|\alpha|,2}(K)^\ell}\,, \end{aligned}$$

we obtain

$$\begin{aligned} &\sup_{x\in\mathcal{D}} |x'|^{\delta-l+|\alpha|+1}\,\big|\partial_{x'}^\alpha(\chi\partial_{x_3}^j u(x))\big| \\ &\le c\Big(\|\partial_{x'}^\alpha(\chi\partial_{x_3}^j u)\|_{V_\delta^{l-|\alpha|,2}(\mathcal{D})^\ell} + \|\partial_{x_3}\partial_{x'}^\alpha(\chi\partial_{x_3}^j u)\|_{V_\delta^{l-|\alpha|,2}(\mathcal{D})^\ell}\Big) \\ &\le c\Big(\|\psi\partial_{x_3}^j u\|_{V_\delta^{l,2}(\mathcal{D})^\ell} + \|\psi\partial_{x_3}^{j+1} u\|_{V_\delta^{l,2}(\mathcal{D})^\ell}\Big). \end{aligned} \tag{2.5.15}$$

This together with (2.5.14) implies (2.5.13). □

The last lemma allows us to estimate the derivatives of Green's matrix in the case $|x-\xi| \ge \min(|x'|,|\xi'|)$.

THEOREM 2.5.4. *Let $G(x,\xi)$ be the Green's matrix introduced in Theorem 2.5.1. If $|x-\xi| \ge \min(|x'|,|\xi'|)$, then the estimate*

$$\big|D_{x'}^\alpha D_{x_3}^j D_{\xi'}^\beta D_{\xi_3}^k G(x,\xi)\big| \le c\,\frac{|x'|^{m-1+\delta_+-|\alpha|-\varepsilon}\,|\xi'|^{m-1+\delta_--|\beta|-\varepsilon}}{|x-\xi|^{1+\delta_++\delta_-+j+k-2\varepsilon}}$$

is valid for arbitrary α,β,j,k with an arbitrarily small positive ε.

P r o o f. By (2.5.3), it suffices to prove the estimate for $|x-\xi| = 2$. Then under the assumption $|x-\xi| \ge \min(|x'|,|\xi'|)$, we have $\max(|x'|,|\xi'|) \le 4$. Let $\mathcal{B}_x$ and $\mathcal{B}_\xi$ be balls with radius 1 and centers x and ξ, respectively. Furthermore, let ζ and η be infinitely differentiable functions with supports in $\mathcal{B}_x$ and $\mathcal{B}_\xi$ equal to one in neighborhoods of x and ξ, respectively. Applying Lemma 2.5.3 to the function

$z \to D^\alpha_{x'} D^j_{x_3} G(x,z)$ which satisfies the equation $L^+(D_z)\, D^\alpha_{x'} D^j_{x_3} G(x,z) = 0$ for $z \in \mathcal{D} \cap \mathcal{B}_\xi$, we obtain

$$(2.5.16) \qquad \begin{aligned} &|\xi'|^{|\beta|-m+1-\delta_-+\varepsilon} \left| D^\alpha_{x'} D^j_{x_3} D^\beta_{\xi'} D^k_{\xi_3} G(x,\xi) \right| \\ &\le c\, \|\eta(\cdot)\, D^\alpha_{x'} D^j_{x_3} G(x,\cdot)\|_{V_0^{m,2}(\mathcal{D})^\ell} \,. \end{aligned}$$

Let f be an arbitrary vector-function in $C_0^\infty(\overline{\mathcal{D}}\backslash M)^\ell$. Then the function

$$u(y) = \int_{\mathcal{D}} \eta(z)\, G(y,z)\, f(z)\, dz$$

satisfies the equation $L(D_y)u(y) = \eta(y)\, f(y)$ for $y \in \mathcal{D}$ and the homogeneous Dirichlet conditions on Γ^+ and Γ^-. Since $\eta(y) = 0$ for $y \in \mathcal{B}_x$, it follows from Lemma 2.5.3 that

$$|x'|^{|\alpha|-m+1-\delta_++\varepsilon} \left| D^\alpha_{x'} D^j_{x_3} u(x) \right| \le c\, \|\zeta u\|_{V_0^{m,2}(\mathcal{D})^\ell} \le c'\, \|f\|_{V_0^{-m,2}(\mathcal{D})^\ell} \,.$$

Consequently, the mappings

$$\begin{aligned} f \to\ & |x'|^{|\alpha|-m+1-\delta_++\varepsilon}\, D^\alpha_{x'} D^j_{x_3} u_i(x) \\ &= |x'|^{|\alpha|-m+1-\delta_++\varepsilon} \int_{\mathcal{D}} \eta(z) \sum_{k=1}^{\ell} D^\alpha_{x'} D^j_{x_3} G_{i,k}(x,z)\, f_k(z)\, dz, \end{aligned}$$

$i = 1, \dots, \ell$, can be extended to linear and continuous functionals on $V_0^{-m,2}(\mathcal{D})^\ell$. The norms of these functionals are bounded by a constant independent of x. Thus, the inequality

$$|x'|^{|\alpha|-m+1-\delta_++\varepsilon}\, \|\eta(\cdot)\, D^\alpha_{x'} D^j_{x_3} G(x,\cdot)\|_{V_0^{m,2}(\mathcal{D})^\ell} \le c$$

holds with a constant c independent of x. This together with (2.5.16) yields the estimate of the theorem. □

2.6. Solvability in weighted L_p Sobolev spaces

The estimates of Green's matrix in Theorems 2.5.2 and 2.5.4 enable us to extend the results of Section 2.4 to weighted L_p Sobolev spaces. By Lemma 2.2.1, we may restrict ourselves to the case of zero Dirichlet data. Then the solution of the problem (2.2.1) admits the representation

$$(2.6.1) \qquad u(x) = \int_{\mathcal{D}} G(x,\xi)\, f(\xi)\, d\xi,$$

where $G(x,\xi)$ is the Green's matrix introduced in the foregoing section. We prove that this function belongs to the space $V_\delta^{l,p}(\mathcal{D})^\ell$, $l \ge m$, if $f \in V_\delta^{l-2m,p}(\mathcal{D})^\ell$ and δ satisfies the inequalities

$$(2.6.2) \qquad -\delta_+ < \delta - l + m - 1 + 2/p < \delta_- \,.$$

As is shown in this section, the condition (2.6.2) is necessary and sufficient for the unique solvability of the boundary value problem (2.2.1) in $V_\delta^{l,p}(\mathcal{D})^\ell$.

2.6.1. Auxiliary estimates. Suppose that $f \in V_\delta^{-m,p}(\mathcal{D})^\ell$, i.e.

$$f(x) = \sum_{|\beta| \le m} D_x^\beta f^{(\beta)}(x), \quad \text{where } f^{(\beta)} \in V^{0,p}_{\delta+m-|\beta|}(\mathcal{D})^\ell \tag{2.6.3}$$

(cf. Lemma 2.1.7). Then the vector function (2.6.1) has the representation

$$u(x) = \sum_{|\beta| \le m} (-1)^{|\beta|} \int_{\mathcal{D}} D_\xi^\beta G(x,\xi)\, f^{(\beta)}(\xi)\, d\xi. \tag{2.6.4}$$

We estimate the integrals on the right-hand side of (2.6.4) over different subsets of $\mathcal{D}$. For the estimation of the integrals over the subset $\{\xi \in \mathcal{D} : |x-\xi| > \min(|x'|, |\xi'|)\}$, we need the following two lemmas.

LEMMA 2.6.1. *Let x be a point in $\mathcal{D}$ and $R > \delta\, r(x)$, where δ is a given positive number. Then*

$$\int\limits_{\substack{\mathcal{D} \\ |\xi-x|>R}} |\xi - x|^\alpha\, r(\xi)^\gamma\, d\xi \le c\, R^{3+\alpha+\gamma} \quad \text{if } \alpha+\gamma < -3,\ \gamma > -2 \tag{2.6.5}$$

and

$$\int\limits_{\substack{\mathcal{D} \\ \delta\, r(x)<|\xi-x|<R}} |\xi - x|^\alpha\, r(\xi)^\gamma\, d\xi \le c\, R^{3+\alpha+\gamma} \quad \text{if } \alpha+\gamma > -3,\ \gamma > -2. \tag{2.6.6}$$

Here the constant c is independent of x and R.

P r o o f. 1) We start with the case $\alpha + \gamma < -3$. Then in particular $\alpha < -1$. If $\gamma \ge 0$, then $\alpha < -3$,

$$r(\xi)^\gamma \le c\,\big(r(x)^\gamma + |x-\xi|^\gamma\big) \le c\,\big(\delta^{-\gamma} R^\gamma + |x-\xi|^\gamma\big)$$

and

$$\int\limits_{|x-\xi|>R} |x-\xi|^\alpha\, r(\xi)^\gamma\, d\xi \le c \int\limits_{|x-\xi|>R} \big(|x-\xi|^\alpha R^\gamma + |x-\xi|^{\alpha+\gamma}\big)\, d\xi \le c\, R^{3+\alpha+\gamma}.$$

Let $\gamma \le 0$. Then

$$\int\limits_{\substack{\mathcal{D} \\ r(\xi)>|x-\xi|>R}} |x-\xi|^\alpha\, r(\xi)^\gamma\, d\xi \le c \int\limits_{\substack{\mathcal{D} \\ |x-\xi|>R}} |x-\xi|^{\alpha+\gamma}\, d\xi \le c\, R^{3+\alpha+\gamma}.$$

We denote by $x^* = (0,0,x_3)$ and $\xi^* = (0,0,\xi_3)$ the nearest points to x and ξ on M, respectively. If $|x-\xi| > r(\xi) > R$, then $|\xi - x^*| \le |\xi-\xi^*| + |\xi^* - x^*| \le r(\xi) + |\xi - x| < 2|\xi - x|$ and $|\xi - x^*| \ge r(\xi) > R$. Consequently,

$$\int\limits_{\substack{\mathcal{D} \\ |x-\xi|>r(\xi)>R}} |x-\xi|^\alpha\, r(\xi)^\gamma\, d\xi \le c \int\limits_{\substack{\mathcal{D} \\ |\xi-x^*|>R}} |\xi-x^*|^\alpha\, r(\xi-x^*)^\gamma\, d\xi \le c\, R^{3+\alpha+\delta}.$$

Finally, since $2|x-\xi| > R + |x_3 - \xi_3|$ for $|x-\xi| > R$, we obtain

$$\int\limits_{\substack{\mathcal{D} \\ |x-\xi|>R>r(\xi)}} |x-\xi|^\alpha\, r(\xi)^\gamma\, d\xi \le c \int_{-\infty}^{+\infty} \big(R + |x_3-\xi_3|\big)^\alpha\, d\xi_3 \int_0^R r^{1+\gamma}\, dr = c\, R^{3+\alpha+\delta}.$$

This proves (2.6.5).

2) Let $\alpha+\gamma>-3$. We denote the left-hand side of (2.6.6) by A. Substituting $x/r(x)=y$ and $\xi/r(x)=\eta$, we obtain

$$A = r(x)^{3+\alpha+\gamma} \int\limits_{\substack{\mathcal{D}\\ \delta<|y-\eta|<R/r(x)}} |y-\eta|^\alpha\, r(\eta)^\gamma\, d\eta.$$

Since $r(y)=1$, the integral

$$\int\limits_{\substack{\mathcal{D}\\ \delta<|y-\eta|<2}} |y-\eta|^\alpha\, r(\eta)^\gamma\, d\eta$$

can be estimated by a finite constant independent of y. Let again $y^*=(0,0,y_3)$ denote the nearest point to y on M. If $2<|y-\eta|<R/r(x)$, then $1<|\eta-y^*|<1+R/r(x)$ and $2|\eta-y^*|/3<|y-\eta|<2|\eta-y^*|$. Therefore,

$$\begin{aligned}\int\limits_{2<|y-\eta|<R/r(x)} |y-\eta|^\alpha\, r(\eta)^\gamma\, d\eta &\le c \int\limits_{1<|\eta-y^*|<1+R/r(x)} |\eta-y^*|^\alpha\, r(\eta-y^*)^\gamma\, d\eta \\ &\le c\,\big(R/r(x)\big)^{3+\alpha+\gamma}.\end{aligned}$$

This proves (2.6.6). □

LEMMA 2.6.2. *Let*

$$v(x) = \int_{\mathcal{D}} K(x,\xi)\, f(\xi)\, d\xi, \tag{2.6.7}$$

where $f\in V^{0,p}_{\delta+m-k}(\mathcal{D})$, $K(x,\xi)$ *satisfies the estimate*

$$|K(x,\xi)| \le c\, \frac{|x'|^{m-1-j+\delta_+-\varepsilon}\, |\xi'|^{m-1-k+\delta_--\varepsilon}}{|x-\xi|^{1+\delta_++\delta_--2\varepsilon}},$$

and $K(x,\xi)=0$ *for* $|x-\xi|<\min(|x'|,|\xi'|)$. *If* $-\delta_++\varepsilon<\delta-1+2/p<\delta_--\varepsilon$, *then* $v\in V^{0,p}_{\delta-m+j}(\mathcal{D})$ *and*

$$\|v\|_{V^{0,p}_{\delta-m+j}(\mathcal{D})} \le c\, \|f\|_{V^{0,p}_{\delta+m-k}(\mathcal{D})}\,. \tag{2.6.8}$$

P r o o f. First note that $2|x-\xi|>\max(|x'|,|\xi'|)$ if $|x-\xi|>\min(|x'|,|\xi'|)$. By our assumptions on δ, there exist real numbers β_1 and γ_1 such that

$$\begin{aligned}&-2+2/p<\beta_1<\delta_--1-\delta-\varepsilon,\\ &-2-\delta-\delta_++\varepsilon+1/p<\beta_1+\gamma_1<-3+3/p\,.\end{aligned}$$

We put $\beta_2=m-1-k+\delta_--\varepsilon-\beta_1$ and $\gamma_2=-1-\delta_+-\delta_-+2\varepsilon-\gamma_1$. Then by Hölder's inequality,

$$\begin{aligned}|v(x)|^p &\le c\,|x'|^{p(m-1-j+\delta_+-\varepsilon)} \Big(\int\limits_{|\xi-x|>r(x)/2} |\xi'|^{p'\beta_1}\, |x-\xi|^{p'\gamma_1}\, d\xi \Big)^{p-1} \\ &\quad\times \int\limits_{\substack{\mathcal{D}\\ |\xi-x|>\min(r(x),r(\xi))}} |\xi'|^{p\beta_2}\, |x-\xi|^{p\gamma_2}\, |f(\xi)|^p\, d\xi,\end{aligned}$$

where $p' = p/(p-1)$. Using the first part of Lemma 2.6.1 (with $R = r(x)/2$), we obtain

$$|v(x)|^p \le c\,|x'|^{p(m+2-j+\delta_+ + \beta_1+\gamma_1-\varepsilon)-3} \int\limits_{\substack{\mathcal{D}\\ |\xi-x|>\min(r(x),r(\xi))}} |\xi'|^{p\beta_2}\,|x-\xi|^{p\gamma_2}\,|f(\xi)|^p\,d\xi.$$

Thus,

$$\int_{\mathcal{D}} |x'|^{p(\delta-m+j)}\,|v(x)|^p\,dx$$
$$\le c\int_{\mathcal{D}} |\xi'|^{p\beta_2}\,|f(\xi)|^p \Big(\int\limits_{|x-\xi|>r(\xi)/2} |x'|^{p(\delta+2+\delta_+ +\beta_1+\gamma_1-\varepsilon)-3}\,|x-\xi|^{p\gamma_2}\,dx\Big)\,d\xi.$$

Estimating the inner integral by means of Lemma 2.6.1, we arrive at (2.6.8). □

The next lemma is of importance for the estimation of the integral on the right-hand side of (2.6.4) over the subset $\{\xi\in\mathcal{D}: |x-\xi| < \min(|x'|,|\xi'|)\}$.

LEMMA 2.6.3. *Suppose that $j+k<2m$ and that $K(x,\xi)$ satisfies the estimate*

$$|K(x,\xi)| \le c\,|\xi'|^{2m-3-j-k} \quad \text{if } j+k<2m-3,$$
$$|K(x,\xi)| \le c\Big(\Big|\log\frac{|x-\xi|}{|\xi'|}\Big| + 1\Big) \quad \text{if } j+k=2m-3,$$
$$|K(x,\xi)| \le c\,|x-\xi|^{2m-3-j-k} \quad \text{if } 2m-3<j+k<2m,$$

and that $K(x,\xi)=0$ for $|x-\xi|>\min(|x'|,|\xi'|)$. Then the function (2.6.7) *satisfies the estimate* (2.6.8) *for arbitrary $f\in V^{0,p}_{\delta+m-k}(\mathcal{D})$.*

P r o o f. First note that $|x'|/2 < |\xi'| < 2|x'|$ for $|x-\xi|<\min(|x'|,|\xi'|)$.

1) Let $j+k<2m-3$. Then by Hölder's inequality,

$$\begin{aligned} |v(x)|^p &\le c \int\limits_{\substack{\mathcal{D}\\ |\xi-x|<\min(r(x),r(\xi))}} |\xi'|^{p(2m-3-j-k)}\,|f(\xi)|^p\,d\xi\,\Big(\int\limits_{|x-\xi|<r(x)} d\xi\Big)^{p-1} \\ &\le c\,r(x)^{3(p-1)} \int\limits_{\substack{\mathcal{D}\\ |\xi-x|<\min(r(x),r(\xi))}} r(\xi)^{p(2m-3-j-k)}\,|f(\xi)|^p\,d\xi \end{aligned}$$

and consequently

$$\int_{\mathcal{D}} r(x)^{p(\delta-m+j)}\,|v(x)|^p\,dx \le c\int_{\mathcal{D}} r(\xi)^{p(\delta+m-k)-3}\,|f(\xi)|^p\Big(\int_{|x-\xi|<r(\xi)} dx\Big)\,d\xi$$

which implies (2.6.8).

2) If $j+k=2m-3$, then by Hölder's inequality,

$$\begin{aligned} |v(x)|^p &\le c \int\limits_{\substack{\mathcal{D}\\ |\xi-x|<\min(|x'|,|\xi'|)}} |f(\xi)|^p\,d\xi\,\Big(\int\limits_{|x-\xi|<|x'|} \Big(\Big|\log\frac{|x-\xi|}{|x'|}\Big|+1\Big)^{p/(p-1)} d\xi\Big)^{p-1} \\ &= c\,|x'|^{3(p-1)} \int\limits_{\substack{\mathcal{D}\\ |\xi-x|<\min(|x'|,|\xi'|)}} |f(\xi)|^p\,d\xi. \end{aligned}$$

As in the first case, this implies (2.6.8).

3) In the case $j+k>2m-3$, we obtain

$$\begin{aligned} |v(x)|^p &\le c \int\limits_{\substack{\mathcal{D}\\ |\xi-x|<\min(r(x),r(\xi))}} |x-\xi|^{2m-3-j-k}\, |f(\xi)|^p\, d\xi \\ &\quad \times \Big(\int\limits_{|x-\xi|<r(x)} |x-\xi|^{2m-3-j-k} d\xi\Big)^{p-1} \\ &\le c\, r(x)^{(p-1)(2m-j-k)} \int\limits_{\substack{\mathcal{D}\\ |\xi-x|<\min(r(x),r(\xi))}} |x-\xi|^{2m-3-j-k}\, |f(\xi)|^p\, d\xi. \end{aligned}$$

Thus,

$$\begin{aligned} &\int_{\mathcal{D}} r(x)^{p(\delta-m+j)}\, |v(x)|^p\, dx \\ &\le c \int_{\mathcal{D}} r(\xi)^{p(\delta+m-k)-2m+j+k}\, |f(\xi)|^p \Big(\int\limits_{|x-\xi|<r(\xi)} |x-\xi|^{2m-3-j-k}\, dx\Big)\, d\xi \\ &\le c \int_{\mathcal{D}} r(\xi)^{p(\delta+m-k)}\, |f(\xi)|^p\, d\xi. \end{aligned}$$

This completes the proof. □

2.6.2. Weighted L_p estimates for the solution and its derivatives. Now we apply the last two lemmas in order to estimate the $V_{\delta-1}^{m-1,p}(\mathcal{D})^\ell$-norm of the function (2.6.1).

THEOREM 2.6.4. *Let $G(x,\xi)$ be the Green's matrix introduced in Theorem 2.5.1. Furthermore, let $f \in V_\delta^{-m,p}(\mathcal{D})^\ell$, where $1<p<\infty$ and $-\delta_+ < \delta - 1 + 2/p < \delta_-$. Then the vector function (2.6.1) belongs to $V_{\delta-1}^{m-1,p}(\mathcal{D})^\ell$ and satisfies the estimate*

$$\|u\|_{V_{\delta-1}^{m-1,p}(\mathcal{D})^\ell} \le c\, \|f\|_{V_\delta^{-m,p}(\mathcal{D})^\ell}$$

with a constant c independent of f.

Proof. The vector function u has the form (2.6.4), where $f^{(\beta)} \in V_{\delta+m-|\beta|}^{0,p}(\mathcal{D})^\ell$, $|\beta|\le m$. Here, the functions $f^{(\beta)}$ can be chosen such that

$$\sum_{|\beta|\le m} \|f^{(\beta)}\|_{V_{\delta+m-|\beta|}^{0,p}(\mathcal{D})^\ell} \le c\, \|f\|_{V_\delta^{-m,p}(\mathcal{D})^\ell} \tag{2.6.9}$$

with a constant c independent of f (see Lemma 2.1.7). Differentiating (2.6.4), we obtain

$$D_x^\alpha u(x) = \sum_{|\beta|\le m} \int_{\mathcal{D}} D_x^\alpha D_\xi^\beta G(x,\xi)\, f^{(\beta)}(\xi)\, d\xi.$$

Let

$$\chi_+(x,\xi) = \begin{cases} 1 & \text{for } |x-\xi| < \min(r(x),r(\xi)), \\ 0 & \text{else,} \end{cases}$$

and

$$\chi_-(x,\xi) = 1 - \chi_+(x,\xi).$$

We consider the functions

$$w_\pm^{(\alpha,\beta)}(x) = \int_{\mathcal{D}} \chi_\pm(x,\xi)\, D_x^\alpha D_\xi^\beta G(x,\xi)\, f^{(\beta)}(\xi)\, d\xi$$

for arbitrary multi-indices α, β, $|\alpha| \le m-1$, $|\beta| \le m$. By Theorems 2.5.2 and 2.5.4, the matrices $\chi_-(x,\xi)\, D_x^\alpha D_\xi^\beta G(x,\xi)$ and $\chi_+(x,\xi)\, D_x^\alpha D_\xi^\beta G(x,\xi)$ satisfy the conditions of Lemmas 2.6.2 and 2.6.3, respectively, where $j = |\alpha|$ and $k = |\beta|$. Consequently,

$$\|w_\pm^{(\alpha,\beta)}\|_{V^{0,p}_{\delta-m+|\alpha|}(\mathcal{D})^\ell} \le c\, \|f^{(\beta)}\|_{V^{0,p}_{\delta+m-|\beta|}(\mathcal{D})^\ell}\, .$$

This together with (2.6.9 implies

$$\|D_x^\alpha u\|_{V^{0,p}_{\delta-m+|\alpha|}(\mathcal{D})^\ell} \le c\, \|f\|_{V_\delta^{-m,p}(\mathcal{D})^\ell}$$

for $|\alpha| \le m-1$. This proves the theorem. □

2.6.3. Existence and uniqueness of solutions in $V_\delta^{l,p}(\mathcal{D})^\ell$. Now we are able to prove the main result of this section.

THEOREM 2.6.5. *Let $f \in V_\delta^{l-2m,p}(\mathcal{D})^\ell$ and $g_k^\pm \in V_\delta^{l-k+1-1/p,p}(\Gamma^\pm)^\ell$, $k = 1,\dots,m$, where $l \ge m$ and δ satisfies the condition* (2.6.2). *Then there exists a unique solution $u \in V_\delta^{l,p}(\mathcal{D})^\ell$ of the boundary value problem* (2.2.1).

P r o o f. 1) First let $l = m$. By Lemma 2.2.1, we may assume without loss of generality that $g_k^\pm = 0$ for $k = 1,\dots,m$. For each $f \in V_\delta^{-m,p}(\mathcal{D})^\ell$, there exists a sequence $\{f^{(\nu)}\} \subset C_0^\infty(\overline{\mathcal{D}}\backslash M)^\ell$ converging to f in $V_\delta^{-m,p}(\mathcal{D})^\ell$. We consider the corresponding sequence of solutions $u^{(\nu)} \in \overset{\circ}{V}{}_0^{m,2}(\mathcal{D})^\ell$ of the equation $L(D_x)u = f^{(\nu)}$. According to Theorem 2.5.1, these solutions have the representation

$$u^{(\nu)}(x) = \int_{\mathcal{D}} G(x,\xi)\, f^{(\nu)}(\xi)\, d\xi.$$

By Theorem 2.6.4, the sequence $\{u^{(\nu)}\}$ is a Cauchy sequence in $V_{\delta-1}^{m-1,p}(\mathcal{D})^\ell$. Applying Theorem 2.2.9, we conclude that $\{u^{(\nu)}\}$ is even a Cauchy sequence in the space $\overset{\circ}{V}{}_\delta^{m,p}(\mathcal{D})^\ell$. The limit u is a solution of the equation $L(D_x)u = f$ and satisfies the estimate

$$\|u\|_{V_\delta^{m,p}(\mathcal{D})^\ell} \le c\, \|f\|_{V_\delta^{-m,p}(\mathcal{D})^\ell}\, .$$

We prove the uniqueness of the solution. As was shown above, the operator $L(D_x)$ realizes a mapping from $\overset{\circ}{V}{}_\delta^{m,p}(\mathcal{D})^\ell$ onto $V_\delta^{-m,p}(\mathcal{D})^\ell$. Analogously, the adjoint operator

$$L^+(D_x) : \overset{\circ}{V}{}_{-\delta}^{m,p'}(\mathcal{D})^\ell \to V_{-\delta}^{-m,p'}(\mathcal{D})^\ell \tag{2.6.10}$$

($p' = p/(p-1)$) is surjective if $-\delta_- < -\delta - 1 + 2/p' < \delta_+$ or, what is the same, if $-\delta_+ < \delta - 1 + 2/p < \delta_-$. Let $u \in \overset{\circ}{V}{}_\delta^{m,p}(\mathcal{D})^\ell$ be a solution of the equation $L(D_x)\, u = 0$. Then

$$\big(u, L^+(D_x)v\big)_{\mathcal{D}} = \big(L(D_x)u, v\big)_{\mathcal{D}} = 0 \quad \text{for all} v \in \overset{\circ}{V}{}_{-\delta}^{m,p'}(\mathcal{D})^\ell.$$

Since the operator (2.6.10) is surjective, it follows that

$$(u, w)_{\mathcal{D}} = 0 \quad \text{for all } w \in V_{-\delta}^{-m,p'}(\mathcal{D})^\ell.$$

Consequently, $u = 0$. This proves the theorem for the case $l = m$.

2) Let $l > m$. Then by the first part of the proof, there exists a unique solution $u \in V^{m,p}_{\delta-l+m}(\mathcal{D})^\ell$ of the boundary value problem (2.2.1). Using Theorem 2.2.9, we obtain $u \in V^{l,p}_\delta(\mathcal{D})^\ell$. Since the solution is unique in $V^{m,p}_{\delta-l+m}(\mathcal{D})^\ell$, it is also unique in the subspace $V^{l,p}_\delta(\mathcal{D})^\ell$. The proof of the theorem is complete. □

2.6.4. Necessity of the condition on δ. By Theorem 2.4.5, the condition (2.6.2) is necessary for the unique solvability in $V^{l,p}_\delta(\mathcal{D})^\ell$ if $p = 2$. We prove the necessity of this condition for $p \neq 2$.

LEMMA 2.6.6. *Suppose that $l \geq m$ and that there exists a constant c such that*

$$\|u\|_{V^{l,p}_\delta(\mathcal{D})^\ell} \leq c\,\|L(D_x)\,u\|_{V^{l-2m,p}_\delta(\mathcal{D})^\ell} \tag{2.6.11}$$

for all $u \in V^{l,p}_\delta(\mathcal{D})^\ell \cap \overset{\circ}{V}{}^{m,p}_{\delta-l+m}(\mathcal{D})^\ell$. Then $\delta - l + m - 1 + 2/p < \delta_-$ and the line $\operatorname{Re}\lambda = l - \delta - 2/p$ does not contain eigenvalues of the pencil $A(\lambda)$.

P r o o f. From (2.6.11) it follows (cf. the proof of Theorem 2.4.3) that

$$\|u\|_{E^{l,p}_\delta(K)^\ell} \leq c\,\|L(D_{x'},\pm 1)\,u\|_{E^{l-2m,p}_\delta(K)^\ell} \tag{2.6.12}$$

for all $u \in E^{l,p}_\delta(K)^\ell \cap \overset{\circ}{E}{}^{m,p}_{\delta-l+m}(K)^\ell$. Arguing as in the second part of the proof of Theorem 2.3.11, we conclude that the line $\operatorname{Re}\lambda = l - \delta - 2/p$ is free of eigenvalues of the pencil $A(\lambda)$.

We show that every solution $u \in \overset{\circ}{E}{}^{m,2}_{\delta-l+m-1+2/p}(K)^\ell$ of the equation

$$L(D_{x'},\pm 1)\,u = 0 \ \text{ in } K$$

belongs to $E^{l,p}_\delta(K)^\ell$. If $p > 2$, then Theorem 2.3.8 and Lemma 1.2.2 imply $u \in E^{l+1,2}_{\delta+2/p}(K)^\ell \subset E^{l,p}_\delta(K)^\ell$. Let $p < 2$. According to Theorem 2.3.14, the vector function u belongs both to $E^{l,2}_{\delta-1+\varepsilon+2/p}(K)^\ell$ and $E^{l,2}_{\delta-1-\varepsilon+2/p}(K)^\ell$ if ε is a sufficiently small positive number. Using Hölder's inequality, we obtain

$$\begin{aligned}
&\int\limits_K r^{p\gamma}\,\big|\partial_x^\alpha u(x)\big|^p\,dx \\
&\leq \Big(\int\limits_{\substack{K\\ r<1}} r^{2(\gamma-1-\varepsilon+2/p)}\,\big|\partial_x^\alpha u(x)\big|^2\,dx\Big)^{p/2}\Big(\int\limits_{\substack{K\\ r<1}} r^{-2+2p\varepsilon/(2-p)}\,dx\Big)^{(2-p)/2} \\
&\quad + \Big(\int\limits_{\substack{K\\ r>1}} r^{2(\gamma-1+\varepsilon+2/p)}\,\big|\partial_x^\alpha u(x)\big|^2\,dx\Big)^{p/2}\Big(\int\limits_{\substack{K\\ r>1}} r^{-2-2p\varepsilon/(2-p)}\,dx\Big)^{(2-p)/2} \\
&\leq c\Big(\int_K \big(r^{2(\gamma-1-\varepsilon+2/p)} + r^{2(\gamma-1+\varepsilon+2/p)}\big)\,\big|\partial_x^\alpha u(x)\big|^2\,dx\Big)^{p/2}
\end{aligned}$$

for arbitrary $p < 2$ and arbitrary γ. Thus,

$$E^{l,2}_{\delta-1+\varepsilon+2/p}(K)^\ell \cap E^{l,2}_{\delta-1-\varepsilon+2/p}(K)^\ell \subset E^{l,p}_\delta(K).$$

Hence, both in the cases $p > 2$ and $p < 2$, we conclude that $u \in E^{l,p}_\delta(K)^\ell$.

By (2.6.12), the equation $L(D_{x'},\pm 1)\,u = 0$ has only the trivial solution in $E^{l,p}_\delta(K)^\ell \cap \overset{\circ}{E}{}^{m,p}_{\delta-l+m}(K)^\ell$. Thus, the kernel of the operator $L(D_{x'},\pm 1)$ is also trivial in $\overset{\circ}{E}{}^{m,2}_{\delta-l+m-1+2/p}(K)^\ell$. Applying Lemma 2.3.18, we get the inequality

$\delta - l + m - 1 + 2/p < \delta_-$. □

Using the analogous result for the formally adjoint boundary value problem, we conclude that the condition (2.6.2) is necessary for the unique solvability of the boundary value problem (2.2.1) in $V_\delta^{l,p}(\mathcal{D})^\ell$.

2.6.5. A regularity assertion for the solution. We close the section with a regularity result for the solution $u \in V_\delta^{l,p}(\mathcal{D})^\ell$.

THEOREM 2.6.7. *Let $u \in V_\delta^{l,p}(\mathcal{D})^\ell$ be a solution of the boundary value problem* (2.2.1), *where*

$$f \in V_\delta^{l-2m,p}(\mathcal{D})^\ell \cap V_{\delta'}^{l'-2m,q}(\mathcal{D})^\ell,$$
$$g_k^\pm \in V_\delta^{l-k+1-1/p,p}(\Gamma^\pm)^\ell \cap V_{\delta'}^{l'-k+1-1/q,q}(\Gamma^\pm)^\ell,$$

$l \geq m$, $l' \geq m$. Suppose that $-\delta_+ < \delta - l + m - 1 + 2/p < \delta_-$ and $-\delta_+ < \delta' - l' + m - 1 + 2/q < \delta_-$. Then $u \in V_{\delta'}^{l,q}(\mathcal{D})^\ell$.

P r o o f. By Lemma 2.2.1 (see also Remark 2.2.2), there exists a vector function $v \in V_\delta^{l,p}(\mathcal{D})^\ell \cap V_{\delta'}^{l',q}(\mathcal{D})^\ell$ such that

$$\frac{\partial^{k-1} v}{\partial n^{k-1}} = g_k^\pm \quad \text{on } \Gamma^\pm, \ k = 1, \dots, m.$$

Furthermore by Theorem 2.6.5, there exists a unique solution $w \in V_{\delta'}^{l,q}(\mathcal{D})^\ell$ of the Dirichlet problem

$$L(D_x)\, w = f - L(D_x)\, v \quad \text{in } \mathcal{D}, \qquad \frac{\partial^{k-1} w}{\partial n^{k-1}} = 0 \quad \text{on } \Gamma^\pm, \ k = 1, \dots, m,$$

which is given by the formula

$$w(x) = \int_{\mathcal{D}} G(x,\xi)\, \big(f(\xi) - L(D_\xi)\, v(\xi)\big)\, d\xi,$$

where $G(x,\xi)$ is the Green's function introduced in Theorem 2.5.1. The same representation is valid for the solution $u - v$ of the last problem. Consequently, $u - v = w$. This proves the theorem. □

2.6.6. Example. We consider the Dirichlet problem

$$-\Delta u = f \quad \text{in } \mathcal{D}, \qquad u = g^\pm \quad \text{on } \Gamma^\pm.$$

For this problem, the pencil $A(\lambda)$ introduced in Subsection 2.3.1 is

$$A(\lambda) = -\frac{\partial^2}{\partial \varphi^2} - \lambda^2.$$

The first positive eigenvalue is $\lambda = \pi/\theta$, where θ is the inner angle of the dihedron. Consequently the assertion of Theorem 2.6.5 holds for $\delta_\pm = \pi/\theta$. The result can be stated as follows.

COROLLARY 2.6.8. *The operator*

$$V_\delta^{l,p}(\mathcal{D}) \ni u \to \big(-\Delta u, u|_{\Gamma^+}, u|_{\Gamma^-}\big) \in V_\delta^{l-2,p}(\mathcal{D}) \times \prod_\pm V_\delta^{l-1/p,p}(\Gamma^\pm)$$

is an isomorphism if and only if $|\delta - l + 2/p| < \pi/\theta$.

2.7. Weighted Hölder spaces in a dihedron

Now, we are interested in solutions in weighted Hölder spaces, where the weights are powers of the distance from the edge of the dihedron. We introduce the weighted Hölder space $N_\delta^{l,\sigma}(\mathcal{D})$ and the corresponding trace spaces $N_\delta^{l,\sigma}(\Gamma^\pm)$ and prove imbeddings for these spaces. Furthermore, we show that any function $u \in N_\delta^{l,\sigma}(\mathcal{D})$ can be approximated by functions with compact supports in $\overline{\mathcal{D}}\backslash M$.

2.7.1. The space $N_\delta^{l,\sigma}$. Let K be the 2-dimensional angle (2.1.1). For arbitrary integer $l \ge 0$ and real δ, σ, $0 < \sigma < 1$, we define $N_\delta^{l,\sigma}(K)$ as the space of all functions with continuous derivatives up to order l on $\overline{K}\backslash\{0\}$ such that

$$\|u\|_{N_\delta^{l,\sigma}(K)} = \sum_{|\alpha|\le l} \sup_{x'\in K} |x'|^{\delta-l-\sigma+|\alpha|} |\partial_{x'}^\alpha u(x')| + \langle u\rangle_{l,\sigma,\delta;K} < \infty,$$

where

$$\langle u\rangle_{l,\sigma,\delta;K} = \sum_{|\alpha|=l} \sup_{\substack{x',y'\in K \\ |x'-y'|<|x'|/2}} |x'|^\delta \frac{|\partial_{x'}^\alpha u(x') - \partial_{y'}^\alpha u(y')|}{|x'-y'|^\sigma}.$$

Analogously, we define the weighted Hölder spaces in the dihedron $\mathcal{D} = K \times \mathbb{R}$. The space $N_\delta^{l,\sigma}(\mathcal{D})$ consists of all functions with continuous derivatives up to order l on $\overline{\mathcal{D}}\backslash M$ such that

$$\|u\|_{N_\delta^{l,\sigma}(\mathcal{D})} = \sum_{|\alpha|\le l} \sup_{x\in\mathcal{D}} |x'|^{\delta-l-\sigma+|\alpha|} |\partial_x^\alpha u(x)| + \langle u\rangle_{l,\sigma,\delta;\mathcal{D}} < \infty, \tag{2.7.1}$$

where

$$\langle u\rangle_{l,\sigma,\delta;\mathcal{D}} = \sum_{|\alpha|=l} \sup_{\substack{x,y\in\mathcal{D} \\ |x-y|<|x'|/2}} |x'|^\delta \frac{|\partial_x^\alpha u(x) - \partial_y^\alpha u(y)|}{|x-y|^\sigma}.$$

Here by $|x-y|$, we always mean the distance of x from y within $\mathcal{D}$. Furthermore, we define the space $N_\delta^{-l,\sigma}(\mathcal{D})$ for integer $l \ge 0$ as the set of all distributions of the form

$$u = \sum_{|\alpha|\le l} \partial_x^\alpha u_\alpha, \quad \text{where } u_\alpha \in N^{0,\sigma}_{\delta+l-|\alpha|}(\mathcal{D}). \tag{2.7.2}$$

The norm in $N_\delta^{-l,\sigma}(\mathcal{D})$ is defined as

$$\|u\|_{N_\delta^{-l,\sigma}(\mathcal{D})} = \inf \sum_{|\alpha|\le l} \|u_\alpha\|_{N^{0,\sigma}_{\delta+l-|\alpha|}(\mathcal{D})},$$

where the infimum is taken over all decompositions of the form (2.7.2).

On the faces Γ^+ and Γ^- of the dihedron $\mathcal{D}$, we introduce the coordinates $r = |x'|$ and $t = x_3$. Then the weighted Hölder space $N_\delta^{l,\sigma}(\Gamma^\pm)$ is defined for integer $l \ge 0$ as the set of all l times continuously differentiable functions on $\Gamma^\pm$ with finite norm

$$\|u\|_{N_\delta^{l,\sigma}(\Gamma^\pm)} = \sum_{j+k\le l} \sup_{r\in\mathbb{R}_+,\ t\in\mathbb{R}} r^{\delta-l-\sigma+j+k} |\partial_r^j \partial_t^k u(r,t)| + \langle u\rangle_{l,\sigma,\delta;\Gamma^\pm},$$

where

$$\langle u\rangle_{l,\sigma,\delta;\Gamma^\pm} = \sum_{j+k=l} \sup r^\delta \frac{|\partial_r^j\partial_t^k u(r,t) - \partial_\rho^j\partial_\tau^k u(\rho,\tau)|}{(|r-\rho|^2+|t-\tau|^2)^{\sigma/2}}.$$

Here the supremum is taken over the set of all $(r,t),(\rho,\tau) \in \mathbb{R}_+ \times \mathbb{R}$ such that $|(r,t)-(\rho,\tau)| < r/2$.

2.7.2. Imbeddings. Similar imbeddings as for the weighted Sobolev spaces $V_\delta^{l,p}$ are valid for the weighted Hölder spaces $N_\delta^{l,\sigma}$.

LEMMA 2.7.1. *Let $l+\sigma > l'+\sigma'$ and $l+\sigma-\delta = l'+\sigma'-\delta'$. Then*

$$N_\delta^{l,\sigma}(K) \subset N_{\delta'}^{l',\sigma'}(K), \quad N_\delta^{l,\sigma}(\mathcal{D}) \subset N_{\delta'}^{l',\sigma'}(\mathcal{D}) \quad and \quad N_\delta^{l,\sigma}(\Gamma^\pm) \subset N_{\delta'}^{l',\sigma'}(\Gamma^\pm).$$

The imbeddings are continuous.

P r o o f. It suffices to prove the lemma for real-valued functions. For $l=l'$ the assertion of the lemma is obvious. Let $|\alpha| = l' < l$ and $x,y \in \mathcal{D}$. By the mean value theorem, there exists a real number $t \in (0,1)$ such that

$$\partial_x^\alpha u(x) - \partial_y^\alpha u(y) = (\nabla \partial^\alpha u)\big(x+t(y-x)\big)\cdot(x-y).$$

Furthermore, the inequalities $r(x)/2 < r\big(x+t(y-x)\big) < 3r(x)/2$ and $|x-y| < r\big(x+t(y-x)\big)$ are satisfied for $|x-y| < r(x)/2$. Consequently,

$$r(x)^{\delta'}\,\frac{|\partial_x^\alpha u(x) - \partial_y^\alpha u(y)|}{|x-y|^{\sigma'}} \le c\, r\big(x+t(y-x)\big)^{\delta'-\sigma'+1}\,\big|(\nabla\partial^\alpha u)\big(x+t(y-x)\big)\big|.$$

Thus, $\langle u\rangle_{l',\sigma',\delta';\mathcal{D}}$ can be estimated by the norm of u in $N_\delta^{l,\sigma}(\mathcal{D})$. This proves the imbedding $N_\delta^{l,\sigma}(\mathcal{D}) \subset N_{\delta'}^{l',\sigma'}(\mathcal{D})$ for $l > l'$. Analogously, the imbeddings $N_\delta^{l,\sigma}(K) \subset N_{\delta'}^{l',\sigma'}(K)$ and $N_\delta^{l,\sigma}(\Gamma^\pm) \subset N_{\delta'}^{l',\sigma'}(\Gamma^\pm)$ can be proved. □

2.7.3. Approximation by functions with compact supports. We will furthermore use the following property of the weighted Hölder space $N_\delta^{l,\sigma}(\mathcal{D})$.

LEMMA 2.7.2. *Let u be an arbitrary function in $N_\delta^{l,\sigma}(\mathcal{D})$. Then there exists a sequence of functions $u_\nu \in N_\delta^{l,\sigma}(\mathcal{D})^\ell$ with compact supports in $\overline{\mathcal{D}}\backslash M$ such that*

$$\partial_x^\alpha u_\nu(x) \to \partial_x^\alpha u(x) \quad as\ \nu\to\infty \quad for\ x \in \overline{\mathcal{D}}\backslash M,\ |\alpha| \le l \tag{2.7.3}$$

and

$$\lim_{\nu\to\infty} \|u_\nu\|_{N_\delta^{l,\sigma}(\mathcal{D})} = \|u\|_{N_\delta^{l,\sigma}(\mathcal{D})}\,. \tag{2.7.4}$$

P r o o f. Let η be an infinitely differentiable function on $\mathbb{R}^2$ such that

$$0 \le \eta \le 1, \quad \eta(y) = 1 \text{ for } |y| \le 1, \quad \eta(y) = 0 \text{ for } |y| \ge 2.$$

We set

$$\zeta_\nu(x) = \zeta_\nu(x',x_3) = \eta\Big(\frac{\log r}{\log \nu}, \frac{x_3}{\nu^3}\Big) \tag{2.7.5}$$

for $\nu = 2,3,\ldots$, where $r = |x'|$. Since $\zeta_\nu(x)$ vanishes for $r > \nu^2$, we obtain

$$r^j\,\partial_{x_3}^j\,(r\partial_r)^k \zeta_\nu(x) = r^j\,\nu^{-3j}\,(\log\nu)^{-k}\,\partial_{y_1}^k\partial_{y_2}^j \eta(y)\Big|_{y_1=\log r/\log\nu\,,\ y_2 = x_3/\nu^3} \to 0$$

for $j+k > 0$. Consequently,

$$\lim_{\nu\to\infty}\ \sup_{x\in\mathcal{D}}\ |x'|^{|\alpha|}\,\partial_x^\alpha \zeta_\nu(x) = 0 \quad \text{for } |\alpha| > 0. \tag{2.7.6}$$

Obviously, the sequence $\{u_\nu\} = \{\zeta_\nu u\}$ satisfies (2.7.3). From the equality

$$(2.7.7)\quad |x'|^{\delta-l-\sigma+|\alpha|}\,\partial_x^\alpha(\zeta_\nu u) = \zeta_\nu(x)\,|x'|^{\delta-l-\sigma+|\alpha|}\,\partial_x^\alpha u(x) + \sum_{\substack{\gamma\le\alpha\\ \gamma\ne\alpha}} \binom{\alpha}{\gamma} |x'|^{\delta-l-\sigma+|\gamma|}\,\big|\partial_x^\gamma u(x)\big|\,|x'|^{|\alpha|-|\gamma|}\,\big|\partial_x^{\alpha-\gamma}\zeta_\nu(x)\big|$$

and (2.7.6) it follows that

$$(2.7.8)\quad |x'|^{\delta-l-\sigma+|\alpha|}\,\big|\partial_x^\alpha u_\nu(x)\big| \le \zeta_\nu(x)\,|x'|^{\delta-l-\sigma+|\alpha|}\,\big|\partial_x^\alpha u(x)\big| + c\varepsilon_\nu\,\|u\|_{N_\delta^{l,\sigma}(\mathcal{D})},$$

where

$$\varepsilon_\nu = \sum_{1\le|\alpha|\le l}\ \sup_{x\in\mathcal{D}} |x'|^{|\alpha|}\,\big|\partial_x^\alpha\zeta_\nu(x)\big| \to 0 \quad \text{as } \nu\to\infty.$$

It remains to estimate the expression $\langle u_\nu\rangle_{l,\sigma,\delta;\mathcal{D}}$. Let $|\alpha| = l$ and $|x-y| < |x'|/2$. Then

$$(2.7.9)\quad \Big|\,|x'|^\delta\,\frac{|\partial_x^\alpha u_\nu(x) - \partial_y^\alpha u_\nu(y)|}{|x-y|^\sigma} - \zeta_\nu(x)\,|x'|^\delta\,\frac{|\partial_x^\alpha u(x) - \partial_y^\alpha u(y)|}{|x-y|^\sigma}\Big|$$
$$\le |x'|^\delta\,\big|\zeta_\nu(x)-\zeta_\nu(y)\big|\,\frac{|\partial_y^\alpha u(y)|}{|x-y|^\sigma} + c\sum_{\substack{\gamma\le\alpha\\ \gamma\ne\alpha}} |x'|^\delta\,\frac{|\partial_x^\gamma u(x)\,\partial_x^{\alpha-\gamma}\zeta_\nu(x) - \partial_y^\gamma u(y)\,\partial_y^{\alpha-\gamma}\zeta_\nu(y)|}{|x-y|^\sigma}.$$

Here

$$\begin{aligned} |x'|^\delta\,\big|\zeta_\nu(x)-\zeta_\nu(y)\big|\,\frac{|\partial_y^\alpha u(y)|}{|x-y|^\sigma} &\le |x'|^\delta\,\big|\nabla\zeta_\nu(z)\big|\,|x-y|^{1-\sigma}\,\big|\partial_y^\alpha u(y)\big| \\ &\le c\,|z'|\,\big|\nabla\zeta_\nu(z)\big|\,|y'|^{\delta-\sigma}\,\big|\partial_y^\alpha u(y)\big|, \end{aligned}$$

where $z = x + t(y-x)$, $0<t<1$. For the last estimate, we used the inequalities $|y'|/2 < |x'| < 2|y'|$ and $|x'| < 2|z'|$. Thus,

$$|x'|^\delta\,\big|\zeta_\nu(x)-\zeta_\nu(y)\big|\,\frac{|\partial_y^\alpha u(y)|}{|x-y|^\sigma} \le c\,\varepsilon_\nu\,\|u\|_{N_\delta^{l,\sigma}(\mathcal{D})}.$$

Analogously, we obtain

$$|x'|^\delta\,\frac{|\partial_x^\gamma u(x)\,\partial_x^{\alpha-\gamma}\zeta_\nu(x) - \partial_y^\gamma u(y)\,\partial_y^{\alpha-\gamma}\zeta_\nu(y)|}{|x-y|^\sigma} \le c\,\varepsilon_\nu\,\|u\|_{N_\delta^{l,\sigma}(\mathcal{D})}.$$

Consequently,

$$\langle u_\nu\rangle_{l,\sigma,\delta;\mathcal{D}} \le \langle u\rangle_{l,\sigma,\delta;\mathcal{D}} + c\,\varepsilon_\nu\,\|u\|_{N_\delta^{l,\sigma}(\mathcal{D})}.$$

This together with (2.7.8) implies

$$\|u_\nu\|_{N_\delta^{l,\sigma}(\mathcal{D})} \le \|u\|_{N_\delta^{l,\sigma}(\mathcal{D})} + c\,\varepsilon_\nu\,\|u\|_{N_\delta^{l,\sigma}(\mathcal{D})}.$$

In the same way, it follows from (2.7.7) and (2.7.9) that

$$\|u_\nu\|_{N_\delta^{l,\sigma}(\mathcal{D})} \ge \|u\|_{N_\delta^{l,\sigma}(\mathcal{D})} - c\,\varepsilon_\nu\,\|u\|_{N_\delta^{l,\sigma}(\mathcal{D})}.$$

This proves (2.7.4). The proof is complete. $\square$

2.8. Solvability in weighted Hölder spaces

The estimates of Green's matrix in Theorems 2.5.2 and 2.5.4 can be also used for the proof of weighted Hölder estimates for solutions of the Dirichlet problem. We show in this section that the boundary value problem (2.2.1) is uniquely solvable in the weighted Hölder space $N_\delta^{l,\sigma}(\mathcal{D})^\ell$ for arbitrary $f \in N_\delta^{l-2m,\sigma}(\mathcal{D})^\ell$, $g_k^\pm \in N_\delta^{l-k+1,\sigma}(\Gamma^\pm)^\ell$, $l \ge m$, if δ satisfies the inequalities

$$-\delta_+ < \delta - l + m - 1 - \sigma < \delta_- . \tag{2.8.1}$$

Furthermore, we obtain regularity results for the solutions.

2.8.1. Reduction to zero boundary data. Similarly to Lemma 2.2.1, we prove the following lemma.

LEMMA 2.8.1. *Let arbitrary functions $g_k^\pm \in N_\delta^{l-k,\sigma}(\Gamma^\pm)$, $k = 0, \dots, l$, be given. Then there exists a function $u \in N_\delta^{l,\sigma}(\mathcal{D})$ such that*

$$\frac{\partial^k u}{\partial n^k} = g_k^\pm \quad on\ \Gamma^\pm \tag{2.8.2}$$

for $k = 0, 1, \dots, l$ and

$$\|u\|_{N_\delta^{l,\sigma}(\mathcal{D})} \le c \sum_\pm \sum_{k=0}^{l} \|g_k^\pm\|_{N_\delta^{l-k,\sigma}(\Gamma^\pm)} \tag{2.8.3}$$

with a constant c independent of the functions $g_k^\pm$.

P r o o f. Let ζ_ν be infinitely differentiable functions depending only on $r = |x'|$ and satisfying the conditions (2.1.5). Furthermore, let $\tilde\zeta_\nu(x) = \zeta_\nu(2^\nu x)$ and $g_{k,\nu}^\pm(x) = g_k^\pm(2^\nu x)$. Since $\tilde\zeta_\nu(x) = 0$ for $|x'| < 1/2$ and $|x'| > 2$, there exists a function $v_\nu \in C^{l,\sigma}(\mathcal{D})$ with support in $\{x : 1/4 < |x'| < 4\}$ such that

$$\frac{\partial^k v_\nu}{\partial n^k} = 2^{\nu k}\, \tilde\zeta_\nu\, g_{k,\nu}^\pm \quad \text{on } \Gamma^\pm \quad \text{for } k = 0, 1, \dots, l,$$

and

$$\begin{aligned}
\|v_\nu\|_{C^{l,\sigma}(\mathcal{D})} &\le c \sum_\pm \sum_{k=0}^{l} 2^{\nu k}\, \|\tilde\zeta_\nu g_{k,\nu}^\pm\|_{C^{l-k,\sigma}(\Gamma^\pm)} \\
&\le c'\, 2^{\nu(l+\sigma-\delta)} \sum_\pm \sum_{k=0}^{l} \|\zeta_\nu g_k^\pm\|_{N_\delta^{l-k,\sigma}(\Gamma^\pm)} .
\end{aligned}$$

We put

$$u_\nu(x) = v_\nu(2^{-\nu}x) \quad \text{and} \quad u(x) = \sum_{\nu=-\infty}^{+\infty} u_\nu(x)$$

(since $u_\nu(x) = 0$ for $|x| < 2^{\nu-2}$ and $|x| > 2^{\nu+2}$, the right sum consists only of finitely many nonzero terms for every fixed x). Then

$$\frac{\partial^k u_\nu}{\partial n^k} = \zeta_\nu\, g_k^\pm \quad \text{on } \Gamma^\pm \quad \text{for } k = 0, 1, \dots, l,$$

and

$$\|u_\nu\|_{N_\delta^{l,\sigma}(\mathcal{D})} \le c\, 2^{\nu(\delta-l-\sigma)}\, \|v_\nu\|_{C^{l,\sigma}(\mathcal{D})} \le c' \sum_\pm \sum_{k=0}^{l} \|\zeta_\nu g_k^\pm\|_{N_\delta^{l-k,\sigma}(\Gamma^\pm)} .$$

Since $u_\nu(x) = 0$ for $|x| < 2^{\nu-2}$ and $|x| > 2^{\nu+2}$, the last inequality implies

$$\|u\|_{N_\delta^{l,\sigma}(\mathcal{D})} \le c \sup_\nu \|u_\nu\|_{N_\delta^{l,\sigma}(\mathcal{D})} \le c' \sum_\pm \sum_{k=0}^{l} \|g_k^\pm\|_{N_\delta^{l-k,\sigma}(\Gamma^\pm)} .$$

Moreover, u satisfies (2.8.2). The lemma is proved. □

The result of Lemma 2.8.1 can be extended to other Dirichlet systems of boundary operators. Repeating the proof of Lemma 2.2.3 and replacing there the weighted Sobolev norms by the corresponding weighted Hölder norms, we obtain the following assertion.

LEMMA 2.8.2. *Let $\{B_k^\pm\}_{k=0}^l$ be Dirichlet systems of order $l+1$ on Γ^+ and Γ^-, respectively, with infinitely differentiable coefficients. Furthermore, let $g_k^\pm \in N_\delta^{l-k,\sigma}(\Gamma^\pm)$, $k = 1,\dots,l$, be given functions with compact supports. Then there exists a function $u \in N_\delta^{l,p}(\mathcal{D})$ satisfying the equalities*

$$B_k^\pm u = g_k^\pm \quad \text{on } \Gamma^\pm, \ k = 1,\dots,m,$$

and the estimate (2.8.3). In the case where the operators $B_k^\pm$ are homogeneous and have constant coefficients, the restriction on the support of the functions $g_k^\pm$ can be omitted.

2.8.2. An a priori estimate for the solution. For the next lemma, we refer to [**6**, Theorem 9.3] (see also [**152**, Theorem 6.4.8]).

LEMMA 2.8.3. *Let $\mathcal{G}_1$, $\mathcal{G}_2$ be bounded subdomains of $\mathbb{R}^3$ such that $\overline{\mathcal{G}}_1 \subset \mathcal{G}_2$, $\mathcal{G}_1 \cap \mathcal{D} \ne \emptyset$ and $\overline{\mathcal{G}}_1 \cap M = \emptyset$. Suppose that u is a solution of the boundary value problem (2.2.1) such that $u \in W^{m,p}(\mathcal{D}\cap\mathcal{G}_2)^\ell$. If $f \in C^{l-2m,\sigma}(\mathcal{D}\cap\mathcal{G}_2)^\ell$ and $g_k^\pm \in C^{l-k+1,\sigma}(\Gamma^\pm\cap\mathcal{G}_2)^\ell$, $l \ge m$, $0<\sigma<1$, then $u \in C^{l,\sigma}(\mathcal{D}\cap\mathcal{G}_1)^\ell$ and*

$$\begin{aligned}\|u\|_{C^{l,\sigma}(\mathcal{D}\cap\mathcal{G}_1)^\ell} \ \le\ & c\Big(\|f\|_{C^{l-2m,\sigma}(\mathcal{D}\cap\mathcal{G}_2)^\ell} + \sum_\pm\sum_{k=1}^m \|g_k\|_{C^{l-k+1,\sigma}(\Gamma^\pm\cap\mathcal{G}_2)^\ell} \\ & + \|u\|_{L_\infty(\mathcal{D}\cap\mathcal{G}_2)^\ell}\Big)\end{aligned}$$

with a constant c independent of u.

We introduce the weighted space $L_\delta^\infty(\mathcal{D})$ which is defined as the set of all functions on $\mathcal{D}$ such that

$$\|u\|_{L_\delta^\infty(\mathcal{D})} = \|r^\delta u\|_{L_\infty(\mathcal{D})} < \infty. \tag{2.8.4}$$

Using Lemma 2.8.3, we obtain the following estimate for the solution of the boundary value problem in the space $N_\delta^{l,\sigma}(\mathcal{D})$.

THEOREM 2.8.4. *Let $u \in L_{\delta-l-\sigma}^\infty(\mathcal{D})^\ell$ be a solution of the Dirichlet problem (2.2.1) such that $\chi u \in W^{l,p}(\mathcal{D})^\ell$ for every $\chi \in C_0^\infty(\overline{\mathcal{D}}\backslash M)$. If $f \in N_\delta^{l-2m,\sigma}(\mathcal{D})^\ell$, $l \ge m$, $0<\sigma<1$, and $g_k^\pm \in N_\delta^{l-k+1,\sigma}(\Gamma^\pm)^\ell$ for $k=1,\dots,m$, then $u \in N_\delta^{l,\sigma}(\mathcal{D})^\ell$ and*

$$\|u\|_{N_\delta^{l,\sigma}(\mathcal{D})^\ell} \le c\Big(\|f\|_{N_\delta^{l-2m,\sigma}(\mathcal{D})^\ell} + \sum_\pm\sum_{k=1}^m \|g_k^\pm\|_{N_\delta^{l-k+1,\sigma}(\Gamma^\pm)^\ell} + \|u\|_{L_{\delta-l-\sigma}^\infty(\mathcal{D})^\ell}\Big).$$

P r o o f. By Lemma 2.8.1, we may restrict ourselves to the case of zero boundary data $g_k^{\pm}$. For arbitrary $x \in \mathcal{D}$, let $\mathcal{B}_x$ denote the ball with radius $3|x'|/4$ and center x. Furthermore, let x be an arbitrary point in $\mathcal{D}$, $|x'| = r$. We introduce the vector function $U(\xi) = u(r\xi)$. Obviously, $L(D_\xi)\, U(\xi) = F(\xi)$, where $F(\xi) = r^{2m} f(r\xi)$. If $|\xi'| = 1$, then by Lemma 2.8.3,

$$|\partial_\xi^\alpha U(\xi)| \le c \Big(\|F\|_{C^{l-2m,\sigma}(\mathcal{D}\cap\mathcal{B}_\xi)^\ell} + \|U\|_{L_\infty(\mathcal{D}\cap\mathcal{B}_\xi)^\ell} \Big) \tag{2.8.5}$$

for $|\alpha| \le l$. Moreover for $|\xi - \eta| < |\xi'|/2 = 1/2$, we have

$$\frac{|\partial_\xi^\alpha U(\xi) - \partial_\eta^\alpha U(\eta)|}{|\xi - \eta|^\sigma} \le c \Big(\|F\|_{C^{l-2m,\sigma}(\mathcal{D}\cap\mathcal{B}_\xi)^\ell} + \|U\|_{L_\infty(\mathcal{D}\cap\mathcal{B}_\xi)^\ell} \Big) \tag{2.8.6}$$

for $|\alpha| = l$. Here,

$$\|F\|_{C^{l-2m,\sigma}(\mathcal{D}\cap\mathcal{B}_\xi)^\ell} \le c\, r^{l+\sigma-\delta}\, \|f\|_{N_\delta^{l-2m,\sigma}(\mathcal{D})^\ell}$$

and

$$\|U\|_{L_\infty(\mathcal{D}\cap\mathcal{B}_\xi)^\ell} \le c\, r^{l+\sigma-\delta}\, \|u\|_{L^\infty_{\delta-l-\sigma}(\mathcal{D})^\ell}\,.$$

Consequently, the substitution $x = r\xi$ and $y = r\eta$ in (2.8.5) and (2.8.6) yields

$$r^{\delta-l-\sigma+|\alpha|}\, |\partial_x^\alpha u(x)| \le c \Big(\|f\|_{N_\delta^{l-2m,\sigma}(\mathcal{D})^\ell} + \|u\|_{L^\infty_{\delta-l-\sigma}(\mathcal{D})^\ell} \Big)$$

for $|x'| = r$, $|\alpha| \le l$, and

$$r^\delta\, \frac{|\partial_x^\alpha u(x) - \partial_y^\alpha u(y)|}{|x-y|^\sigma} \le c \Big(\|f\|_{N_\delta^{l-2m,\sigma}(\mathcal{D})^\ell} + \|u\|_{L^\infty_{\delta-l-\sigma}(\mathcal{D})^\ell} \Big)$$

for $|x - y| < |x'|/2 = r/2$, $|\alpha| = l$. This proves the theorem. □

2.8.3. Existence of solutions in $N_\delta^{l,\sigma}(\mathcal{D})$, $l \ge 2m$. We consider again the vector function

$$u(x) = \int_{\mathcal{D}} G(x,\xi)\, f(\xi)\, d\xi, \tag{2.8.7}$$

where $G(x,\xi)$ is the Green's matrix introduced in Theorem 2.5.1. Our goal is to show that $u \in N_\delta^{l,\sigma}(\mathcal{D})^\ell$ if $f \in N_\delta^{l-2m,\sigma}(\mathcal{D})^\ell$. We start with a weighted L_∞-estimate for u.

LEMMA 2.8.5. *Let u be given by (2.8.7). If $f \in L^\infty_{\delta-l+2m-\sigma}(\mathcal{D})^\ell$ and δ satisfies the condition (2.8.1), then $u \in L^\infty_{\delta-l-\sigma}(\mathcal{D})^\ell$ and*

$$\sup_{x\in\mathcal{D}} |x'|^{\delta-l-\sigma}\, |u(x)| \le c\, \|f\|_{L^\infty_{\delta-l+2m-\sigma}(\mathcal{D})^\ell} \tag{2.8.8}$$

with a constant c independent of x.

P r o o f. Let

$$v(x) = \int\limits_{\substack{\mathcal{D}\\ |x-\xi|<\min(r(x),r(\xi))}} G(x,\xi)\, f(\xi)\, d\xi.$$

Since $|x'|/2 < |\xi'| < 2|x'|$ for $|x - \xi| < \min(r(x), r(\xi))$, we have

$$|v(x)| \le c\, |x'|^{l+\sigma-\delta-3}\, \|r^{\delta-l+2m-\sigma} f\|_{L_\infty(\mathcal{D})^\ell} \int\limits_{\substack{\mathcal{D}\\ |x-\xi|<\min(r(x),r(\xi))}} |x'|^{3-2m}\, |G(x,\xi)|\, d\xi.$$

Here $|G(x,\xi)| \le c\,(|x'|^{2m-3} + |x-\xi|^{2m-3})$ for $|x-\xi| < \min(r(x), r(\xi))$ and consequently

$$\int\limits_{\substack{\mathcal{D}\\ |x-\xi|<\min(r(x),r(\xi))}} |x'|^{3-2m}\,|G(x,\xi)|\,d\xi \le c \int\limits_{|\xi-x|<|x'|} \Big(1+\frac{|\xi-x|^{2m-3}}{|x'|^{2m-3}}\Big)\,d\xi = c'\,|x'|^3.$$

This proves the estimate (2.8.8) for the function v.

We consider the function $w = u - v$. Using the estimate in Theorem 2.5.4 for the matrix $G(x,\xi)$, we obtain

$$\begin{aligned} |w(x)| &\le c\,|x'|^{m-1+\delta_+-\varepsilon} \int\limits_{\substack{\mathcal{D}\\ |x-\xi|>|x'|/2}} \frac{|\xi'|^{m-1+\delta_--\varepsilon}}{|x-\xi|^{1+\delta_++\delta_--2\varepsilon}}\,|f(\xi)|\,d\xi \\ &\le c\,|x'|^{m-1+\delta_+-\varepsilon}\,\|r^{\delta-l+2m-\sigma}f\|_{L_\infty(\mathcal{D})^\ell} \\ &\quad\times \int\limits_{\substack{\mathcal{D}\\ |x-\xi|>|x'|/2}} \frac{|\xi'|^{l+\sigma-\delta-m-1+\delta_--\varepsilon}}{|x-\xi|^{1+\delta_++\delta_--2\varepsilon}}\,d\xi. \end{aligned}$$

Here we used the fact that $|x-\xi| > r(x)/2$ for $|x-\xi| > \min(r(x),r(\xi))$. By (2.6.5), the integral on the right-hand side of the last inequality does not exceed $c\,|x'|^{l+\sigma-\delta-m+1-\delta_++\varepsilon}$ if ε is such that $-\delta_+ + \varepsilon < \delta - l + m - 1 - \sigma < \delta_- - \varepsilon$. Thus, the estimate (2.8.8) is also valid for w. This completes the proof. □

The last lemma together with Theorem 2.8.4 lead to the following existence and uniqueness theorem.

THEOREM 2.8.6. *Let $f \in N_\delta^{l-2m,\sigma}(\mathcal{D})^\ell$, $g_k^\pm \in N_\delta^{l-k+1,\sigma}(\Gamma^\pm)^\ell$, where $l \ge 2m$ and δ satisfies the condition (2.8.1). Then there exists a unique solution $u \in N_\delta^{l,\sigma}(\mathcal{D})^\ell$ of the problem (2.2.1) satisfying the estimate*

$$(2.8.9)\qquad \|u\|_{N_\delta^{l,\sigma}(\mathcal{D})^\ell} \le c\Big(\|f\|_{N_\delta^{l-2m,\sigma}(\mathcal{D})^\ell} + \sum_\pm \sum_{k=1}^m \|g_k^\pm\|_{N_\delta^{l-k+1,\sigma}(\Gamma^\pm)^\ell}\Big).$$

P r o o f. The existence of solutions can be easily deduced from Theorem 2.8.4 and Lemma 2.8.5. We prove the uniqueness. Let $u \in N_\delta^{l,\sigma}(\mathcal{D})^\ell$ be a solution of the homogeneous problem (2.2.1), and let $u_\nu = \zeta_\nu u$, where ζ_ν is the function (2.7.5) introduced in the proof of Lemma 2.7.2. Since ζ_ν has compact support in $\overline{\mathcal{D}}\backslash M$, it follows that $u_\nu \in V_\kappa^{2m,2}(\mathcal{D})^\ell$, where κ is an arbitrary real number in the interval $(m-\delta_+, m-\delta_-)$. Consequently, u_ν admits the representation

$$u_\nu(x) = \int_{\mathcal{D}} G(x,\xi)\,f_\nu(\xi)\,d\xi$$

(see Section 2.6), where $f_\nu = L(D_x)\,(\zeta_\nu u) = [L,\zeta_\nu]\,u$ and $G(x,\xi)$ is the Green's matrix introduced in Theorem 2.5.1. Here $[L,\zeta_\nu]$ denotes the commutator of $L(D_x)$ and ζ_ν. Therefore by Lemma 2.8.5,

$$\|u_\nu\|_{L^\infty_{\delta-l-\sigma}(\mathcal{D})^\ell} \le c\,\|f_\nu\|_{L^\infty_{\delta-l+2m-\sigma}(\mathcal{D})^\ell}\,.$$

Using Theorem 2.8.4, we obtain the estimate

$$\|u_\nu\|_{N_\delta^{l,\sigma}(\mathcal{D})^\ell} \le c\,\|f_\nu\|_{N_\delta^{l-2m,\sigma}(\mathcal{D})^\ell}\,.$$

Since f_ν is a sum of terms of the form $A\,(\partial_x^\alpha \zeta_\nu)\,\partial_x^\gamma u$, where $\alpha \neq 0$ and $|\alpha|+|\gamma| = 2m$, we conclude from (2.7.6) that

$$\|f_\nu\|_{N_\delta^{l-2m,\sigma}(\mathcal{D})} \to 0 \quad \text{as } \nu \to \infty.$$

Applying Lemma 2.7.2, we get $u = 0$. This proves the uniqueness of the solution. □

2.8.4. Existence of solutions in $N_\delta^{l,\sigma}(\mathcal{D})$, $l < 2m$. Now we assume that $f \in N_\delta^{l-2m,\sigma}(\mathcal{D})^\ell$, where $l < 2m$. This means that f has the form

$$f = \sum_{|\alpha|\le 2m-l} \partial_x^\alpha f^{(\alpha)}, \quad \text{where } f^{(\alpha)} \in N^{0,\sigma}_{\delta-l+2m-|\alpha|}(\mathcal{D})^\ell.$$

Then the representation

$$u(x) = \sum_{|\alpha|\le 2m-l} (-1)^{|\alpha|} \int_{\mathcal{D}} \partial_\xi^\alpha G(x,\xi)\; f^{(\alpha)}(\xi)\, d\xi, \tag{2.8.10}$$

holds for the vector function (2.8.7), where $G(x,\xi)$ is the Green's matrix introduced in Theorem 2.5.1. In order to show that $u \in N_\delta^{l,\sigma}(\mathcal{D})^\ell$, we prove the following lemma.

LEMMA 2.8.7. *Let*

$$u^{(\alpha)}(x) = \int_{\mathcal{D}} \partial_\xi^\alpha G(x,\xi)\; f^{(\alpha)}(\xi)\, d\xi,$$

where $f^{(\alpha)} \in L^\infty_{\delta-l+2m-|\alpha|-\sigma}(\mathcal{D})^\ell$. *If* $|\alpha| < 2m$ *and* δ *satisfies the condition* (2.8.1), *then* $u^{(\alpha)} \in L^\infty_{\delta-l-\sigma}(\mathcal{D})^\ell$ *and*

$$\|u^{(\alpha)}\|_{L^\infty_{\delta-l-\sigma}(\mathcal{D})^\ell} \le c\,\|f^{(\alpha)}\|_{L^\infty_{\delta-l+2m-|\alpha|-\sigma}(\mathcal{D})^\ell} \tag{2.8.11}$$

Proof. We introduce the function

$$v^{(\alpha)}(x) = \int\limits_{\substack{\mathcal{D}\\ |x-\xi|<\min(|x'|,|\xi'|)}} \partial_\xi^\alpha G(x,\xi)\; f^{(\alpha)}(\xi)\, d\xi.$$

As in the proof of Lemma 2.8.5, we get

$$\begin{aligned} |v^{(\alpha)}(x)| \;&\le\; c\,|x'|^{l+\sigma-\delta-3}\, \|r^{\delta-l+2m-|\alpha|-\sigma} f^{(\alpha)}\|_{L_\infty(\mathcal{D})^\ell} \\ &\quad\times \int\limits_{\substack{\mathcal{D}\\ |x-\xi|<\min(r(x),r(\xi))}} |x'|^{|\alpha|-2m+3}\, |\partial_\xi^\alpha G(x,\xi)|\, d\xi. \end{aligned}$$

According to Theorem 2.5.2, the matrix $\partial_\xi^\alpha G(x,\xi)$ satisfies the estimate

$$|x'|^{|\alpha|-2m+3}\, |\partial_\xi^\alpha G(x,\xi)| \le c\, H\Big(\frac{|x-\xi|}{|x'|}\Big) \quad \text{for } |x-\xi| < \min(r(x), r(\xi)),$$

where

$$H(t) = \begin{cases} 1 + t^{2m-3-|\alpha|} & \text{for } 2m-3-|\alpha| \neq 0, \\ 1 + |\log t| & \text{for } 2m-3-|\alpha| = 0. \end{cases}$$

Consequently,

$$\int\limits_{|x-\xi|<|x'|} |x'|^{|\alpha|-2m+3}\, |\partial_\xi^\alpha G(x,\xi)|\, d\xi \le c\, |x'|^3 \int\limits_{|\eta|\le 1} H(|\eta|)\, d\eta$$

The integral on the right hand side exists for $|\alpha| < 2m$. Thus, $v^{(\alpha)}$ satisfies (2.8.11).

We consider the function $w^{(\alpha)} = u^{(\alpha)} - v^{(\alpha)}$. Using Theorem 2.5.4 and the inequality (2.6.5), we obtain (as in the proof of Lemma 2.8.5)

$$\begin{aligned} |w^{(\alpha)}(x)| &\leq c\,|x'|^{m-1+\delta_+-\varepsilon}\, \|r^{\delta-l+2m-|\alpha|-\sigma} f\|_{L_\infty(\mathcal{D})^\ell} \\ &\quad \times \int\limits_{\substack{\mathcal{D} \\ |x-\xi|>|x'|/2}} \frac{|\xi'|^{l+\sigma-\delta-m-1+\delta_--\varepsilon}}{|x-\xi|^{1+\delta_++\delta_--2\varepsilon}}\, d\xi \\ &\leq c\,|x'|^{l+\sigma-\delta}\, \|r^{\delta-l+2m-|\alpha|-\sigma} f\|_{L_\infty(\mathcal{D})^\ell} \end{aligned}$$

if δ satisfies (2.8.1). This proves the lemma. □

Using the last lemma, we extend the assertion Theorem 2.8.6 to integer $l \geq m$.

THEOREM 2.8.8. *Let $f \in N_\delta^{l-2m,\sigma}(\mathcal{D})^\ell$, $g_k^\pm \in N_\delta^{l-k+1,\sigma}(\Gamma^\pm)^\ell$, where $l \geq m$ and δ satisfies the condition* (2.8.1). *Then there exists a unique solution $u \in N_\delta^{l,\sigma}(\mathcal{D})^\ell$ of the problem* (2.2.1) *satisfying the estimate* (2.8.9).

P r o o f. For $l \geq 2m$ the theorem was proved in the foregoing subsection. Therefore, we may assume that $l < 2m$. Then the existence of solutions is guaranteed by Theorem 2.8.4 and Lemma 2.8.7. Let $u \in N_\delta^{l,\sigma}(\mathcal{D})^\ell$ be a solution of the problem (2.2.1) with zero right-hand sides. From Theorem 2.8.4 we conclude that $u \in N_{\delta-l+2m}^{2m,\sigma}(\mathcal{D})^\ell$. Hence, Theorem 2.8.6 yields $u = 0$. This proves the uniqueness of the solution in $N_\delta^{l,\sigma}(\mathcal{D})^\ell$. □

2.8.5. Regularity assertions for the solution. The representation (2.8.7) is valid for solutions of the problem (2.2.1) both in weighted Sobolev and Hölder spaces. This leads to the following results.

THEOREM 2.8.9. 1) *Let $u \in N_\delta^{l,\sigma}(\mathcal{D})^\ell$ be a solution of the boundary value problem* (2.2.1), *where*

$$f \in N_\delta^{l-2m,\sigma}(\mathcal{D})^\ell \cap N_{\delta'}^{l'-2m,\sigma'}(\mathcal{D})^\ell$$

and

$$g_k^\pm \in N_\delta^{l-k+1,\sigma}(\Gamma^\pm)^\ell \cap N_{\delta'}^{l'-k+1,\sigma'}(\Gamma^\pm)^\ell,$$

$l, l' \geq m$, $-\delta_+ < \delta - l + m - 1 - \sigma < \delta_-$, and $-\delta_+ < \delta' - l' + m - 1 - \sigma' < \delta_-$. Then $u \in N_{\delta'}^{l',\sigma'}(\mathcal{D})^\ell$.

2) *Let $u \in V_\delta^{l,p}(\mathcal{D})^\ell$ be a solution of the problem* (2.2.1), *where*

$$f \in V_\delta^{l-2m,p}(\mathcal{D})^\ell \cap N_{\delta'}^{l'-2m,\sigma}(\mathcal{D})^\ell$$

and

$$g_k^\pm \in V_\delta^{l-k+1-1/p,p}(\Gamma^\pm)^\ell \cap N_{\delta'}^{l'-k+1,\sigma}(\Gamma^\pm)^\ell,$$

$l, l' \geq m$, $-\delta_+ < \delta - l + m - 1 + 2/p < \delta_-$, and $-\delta_+ < \delta' - l' + m - 1 - \sigma < \delta_-$. Then $u \in N_{\delta'}^{l',\sigma}(\mathcal{D})^\ell$.

P r o o f. It suffices to prove the theorem for zero boundary data $g_k^\pm$. Then under the assumptions of item 1) there exist unique solutions in $N_\delta^{l,\sigma}(\mathcal{D})^\ell$ and

$N_{\delta'}^{l',\sigma'}(\mathcal{D})^\ell$. By what has been shown in the preceding subsections, both solutions are given by the formula

$$u(x) = \int_{\mathcal{D}} G(x,\xi)\, f(\xi)\, d\xi$$

with the same Green's matrix $G(x,\xi)$. Thus, these solutions coincide. Analogously, the second assertion holds. □

2.8.6. Example. We consider the Dirichlet problem

$$-\Delta u = f \ \text{ in } \mathcal{D}, \quad u = g^\pm \ \text{ on } \Gamma^\pm.$$

As was shown at the end of Section 2.6, the first positive eigenvalue of the operator pencil $A(\lambda)$ corresponding to this pencil is equal to π/θ, where θ is the inner angle of the dihedron. Consequently the assertion of Theorem 2.8.8 holds for $\delta_\pm = \pi/\theta$. This means that the following assertion is true.

COROLLARY 2.8.10. *Suppose that* $|\delta - l - \sigma| < \pi/\theta$. *Then the operator*

$$N_\delta^{l,\sigma}(\mathcal{D}) \ni u \to \big(-\Delta u, u|_{\Gamma^+}, u|_{\Gamma^-}\big) \in N_\delta^{l-2,\sigma}(\mathcal{D}) \times \prod_\pm N_\delta^{l,\sigma}(\Gamma^\pm)$$

is an isomorphism.

2.9. The problem with variable coefficients in a dihedron

In this section, we consider the Dirichlet problem for a strongly elliptic differential operator $L(x, D_x)$ of order $2m$ with variable coefficients A_α. The conditions imposed on the coefficients A_α are such that the operator norm of the difference $L(x,D_x) - L^\circ(0,D_x)$ is small. Here $L^\circ(0,D_x)$ denotes the principal part of $L(x,D_x)$ with coefficients frozen at the origin. Applying the results of the preceding sections, we obtain the same solvability and regularity assertions in weighted Sobolev and Hölder spaces as for differential operators with constant coefficients. Note that the conditions on the coefficients are only restrictive outside a neighborhood of the origin. Arbitrary infinitely differentiable functions A_α can be changed outside a sufficiently small neighborhood of the origin such that the new functions satisfy the conditions (2.9.3) and (2.9.4) below (see Remark 2.9.3). Thus, the results of this section can be applied, for example, when constructing a regularizer for the Dirichlet problem in a bounded domain with edges (see Section 4.1).

2.9.1. Solvability in weighted Sobolev and Hölder spaces. We consider the Dirichlet problem

$$L(x,D_x)\, u = f \ \text{ in } \mathcal{D}, \quad \frac{\partial^{k-1} u}{\partial n^{k-1}} = g_k^\pm \ \text{ on } \Gamma^\pm \tag{2.9.1}$$

for the differential operator

$$L(x,D_x) = \sum_{|\alpha|\le 2m} A_\alpha(x)\, D_x^\alpha. \tag{2.9.2}$$

Throughout this section, we assume that the coefficients A_α satisfy the following conditions for $|\beta| \le N$:

$$r^{|\beta|}\, \big|\partial_x^\beta (A_\alpha(x) - A_\alpha(0))\big| < \varepsilon_0 \quad \text{if } |\alpha| = 2m, \tag{2.9.3}$$

$$r^{2m-|\alpha|+|\beta|}\, \big|\partial_x^\beta A_\alpha(x)\big| < \varepsilon_0 \quad \text{if } |\alpha| < 2m. \tag{2.9.4}$$

Here ε_0 is a sufficiently small positive real number.

LEMMA 2.9.1. *Let l be a nonnegative integer and $\delta \in \mathbb{R}$. If the coefficients A_α satisfy the conditions* (2.9.3)–(2.9.4) *for $N = |l-2m|$, then*

$$\|L(x,D_x)u - L^\circ(0,D_x)u\|_{V_\delta^{l-2m,p}(\mathcal{D})^\ell} \le c\,\varepsilon_0\,\|u\|_{V_\delta^{l,p}(\mathcal{D})^\ell}$$

for all $u \in V_\delta^{l,p}(\mathcal{D})^\ell$. Here, c is a constant depending only on l and m. If the conditions (2.9.3)–(2.9.4) *are satisfied for $N = |l-2m|+1$, then*

$$\|L(x,D_x)u - L^\circ(0,D_x)u\|_{N_\delta^{l-2m,\sigma}(\mathcal{D})^\ell} \le c\,\varepsilon_0\,\|u\|_{N_\delta^{l,\sigma}(\mathcal{D})^\ell}$$

for all $u \in N_\delta^{l,\sigma}(\mathcal{D})^\ell$.

P r o o f. We define

$$B_\alpha(x) = A_\alpha(x) \text{ for } |\alpha| < 2m, \quad B_\alpha(x) = A_\alpha(x) - A_\alpha(0) \text{ for } |\alpha| = 2m.$$

For $l \ge 2m$, the inequality

$$\|B_\alpha D_x^\alpha u\|_{N_\delta^{l-2m,\sigma}(\mathcal{D})^\ell} \le c\,\varepsilon_0\,\|u\|_{N_\delta^{l,\sigma}(\mathcal{D})^\ell}\,, \tag{2.9.5}$$

follows directly from the definition of the norm in $N_\delta^{l,\sigma}(\mathcal{D})$ and from the assumptions on A_α. We consider the norm of $B_\alpha D_x^\alpha u$ in $N_\delta^{l-2m,\sigma}(\mathcal{D})^\ell$ for $l < 2m$. There exists a multi-index β such that $\beta \le \alpha$, $|\beta| \le 2m-l$, $|\alpha|-|\beta| \le l$. Obviously,

$$v \overset{def}{=} D_x^{\alpha-\beta}u \in N_\delta^{l-|\alpha|+|\beta|,\sigma}(\mathcal{D})^\ell \subset N^{0,\sigma}_{\delta-l+|\alpha|-|\beta|}(\mathcal{D})^\ell\,.$$

By (2.9.3)–(2.9.4), we have

$$\|(D_x^{\beta-\gamma}B_\alpha)v\|_{N^{0,\sigma}_{\delta-l+2m-|\gamma|}(\mathcal{D})^\ell} \le c\,\varepsilon_0\,\|v\|_{N^{0,\sigma}_{\delta-l+|\alpha|-|\beta|}(\mathcal{D})^\ell}$$

for all $\gamma \le \beta$. Since

$$B_\alpha D_x^\alpha u = B_\alpha D_x^\beta v = \sum_{\gamma\le\beta}(-1)^{|\beta|+|\gamma|}\begin{pmatrix}\beta\\ \gamma\end{pmatrix} D_x^\gamma\big((D_x^{\beta-\gamma}B_\alpha)v\big),$$

we get

$$\begin{aligned}\|B_\alpha D_x^\alpha u\|_{N_\delta^{l-2m,\sigma}(\mathcal{D})^\ell} &\le c\sum_{\gamma\le\beta}\|(D_x^{\beta-\gamma}B_\alpha)v\|_{N^{0,\sigma}_{\delta-l+2m-|\gamma|}(\mathcal{D})^\ell}\\ &\le c\,\varepsilon_0\,\|v\|_{N^{0,\sigma}_{\delta-l+|\alpha|-|\beta|}(\mathcal{D})^\ell}\,.\end{aligned}$$

This proves (2.9.5) for the case $l < 2m$. Analogously, the corresponding inequality in the spaces $V_\delta^{l,p}(\mathcal{D})$ can be proved. □

The estimates of the last lemma together with Theorems 2.6.5 and 2.8.8 imply the following existence and uniqueness theorem for the case of variable coefficients.

THEOREM 2.9.2. 1) *Let $f \in V_\delta^{l-2m,p}(\mathcal{D})^\ell$ and $g_k^\pm \in V_\delta^{l-k+1-1/p,p}(\Gamma^\pm)^\ell$, where $l \ge m$ and*

$$-\delta_+ < \delta - l + m - 1 + 2/p < \delta_-\,.$$

Furthermore, we assume that the coefficients A_α satisfy the conditions (2.9.3) and (2.9.4) *for $N = |l-2m|$, where ε_0 is sufficiently small. Then there exists a unique* *solution* $u \in V_\delta^{l,p}(\mathcal{D})^\ell$ *of the boundary value problem* (2.9.1). *This solution satisfies the estimate*

$$(2.9.6) \qquad \|u\|_{V_\delta^{l,p}(\mathcal{D})^\ell} \le c\Big(\|f\|_{V_\delta^{l-2m,p}(\mathcal{D})^\ell} + \sum_{\pm}\sum_{k=1}^{m} \|g_k^\pm\|_{V_\delta^{l-k+1-1/p,p}(\Gamma^\pm)^\ell}\Big).$$

2) *Let* $f \in N_\delta^{l-2m,\sigma}(\mathcal{D})^\ell$, $g_k^\pm \in N_\delta^{l-k+1,\sigma}(\Gamma^\pm)^\ell$, *where* $l \ge m$ *and*

$$-\delta_+ < \delta - l + m - 1 - \sigma < \delta_- .$$

Furthermore, let the coefficients A_α *satisfy the conditions* (2.9.3)–(2.9.4) *for* $N = |l-2m|+1$ *with a sufficiently small* ε_0. *Then there exists a unique solution* $u \in N_\delta^{l,\sigma}(\mathcal{D})^\ell$ *of the problem* (2.9.1)

2.9.2. Some remarks concerning the conditions on the coefficients. The conditions (2.9.3)–(2.9.4) seem to be very restrictive. However in the subsequent chapters, we need these conditions only for small $|x|$. Then we can make use of the following remarks.

REMARK 2.9.3. Let $L(x, D_x)$ be an operator of the form (2.9.2) with infinitely differentiable coefficients in $\overline{\mathcal{D}}$ and let χ be an infinitely differentiable function on $[0,\infty)$ such that $\chi = 1$ in $[0,1]$, $\chi = 0$ in $(2,\infty)$. Then the coefficients of the differential operator

$$L_\varepsilon(x, D_x) = \chi(|x|/\varepsilon)\, L(x, D_x) + \big(1 - \chi(|x|/\varepsilon)\big)\, L^\circ(0, D_x)$$

satisfy the conditions (2.9.3)–(2.9.4) if ε is sufficiently small.

REMARK 2.9.4. Suppose that the coefficients of the operator $L(x, D_x)$ satisfy the conditions (2.9.3)–(2.9.4) for $|\beta| \le N+2m$. Then the coefficients of the formally adjoint operator

$$L^+(x, D_x)\, u = \sum_{|\alpha|\le 2m} D_x^\alpha \big(A_\alpha^*(x)\, u\big)$$

satisfy the conditions (2.9.3)–(2.9.4) for $|\beta| \le N$.

REMARK 2.9.5. Suppose that the coefficients of the operator $L(x, D_x)$ satisfy the conditions (2.9.3)–(2.9.4). Then the coefficients of the operator

$$\rho^{2m}\, L(\rho x, \rho^{-1} D_x) = \sum_{|\alpha|\le 2m} \rho^{2m-|\alpha|}\, A_\alpha(\rho x)\, D_x^\alpha\,, \quad 0 < \rho < \infty,$$

satisfy these conditions with the same ε_0.

Note that the constant c in (2.9.6) depends only on ε_0 and on the norm of the inverse to the operator

$$V_\delta^{l,p}(\mathcal{D})^\ell \cap \overset{\circ}{V}{}^m_{\delta-l+m}(\mathcal{D})^\ell \ni u \to L^\circ(0, D_x) u \in V_\delta^{l-2m,p}(\mathcal{D})^\ell.$$

In particular, this estimate holds with the same constant c if we consider solutions of the Dirichlet problem for the operator $\rho^{2m} L(\rho x, \rho^{-1} D_x)$.

2.9.3. Regularity assertions for the solutions. First, we extend Theorem 2.2.9 to differential operators with variable coefficients.

THEOREM 2.9.6. *Let u be a solution of the boundary value problem* (2.9.1), $\psi u \in W^{m,p}(\mathcal{D})^\ell$ *for every* $\psi \in C_0^\infty(\overline{\mathcal{D}}\backslash M)$. *We assume that the coefficients* A_α *satisfy the conditions* (2.9.3)–(2.9.4) *for* $N = |l-2m|$ *with a sufficiently small* ε_0. *If* $u \in V_{\delta-1}^{l-1,p}(\mathcal{D})^\ell$, $l \ge m$ $f \in V_\delta^{l-2m,p}(\mathcal{D})^\ell$, *and* $g_k^\pm \in V_\delta^{l-k+1-1/p,p}(\Gamma^\pm)^\ell$, *then* $u \in V_\delta^{l,p}(\mathcal{D})^\ell$ *and*

$$\|u\|_{V_\delta^{l,p}(\mathcal{D})^\ell} \le c\Big(\|f\|_{V_\delta^{l-2m,p}(\mathcal{D})^\ell} + \sum_\pm \sum_{k=1}^m \|g_k^\pm\|_{V_\delta^{l-k+1-1/p,p}(\Gamma^\pm)^\ell} + \|u\|_{V_{\delta-l}^{0,p}(\mathcal{D})^\ell}\Big). \tag{2.9.7}$$

P r o o f. By Lemma 2.2.1, we may restrict ourselves to the case of zero boundary data $g_k^\pm = 0$. Let $\delta' = l-m+1-2/p$ and let ζ_ν be infinitely differentiable functions on $\mathcal{D}$ depending only on $r = |x'|$ which satisfy the conditions (2.1.5). Then $\zeta_\nu u \in V_{\delta'}^{l,p}(\mathcal{D})$ for arbitrary integer ν and, by Theorem 2.9.2,

$$\begin{aligned}\|\zeta_\nu u\|_{V_{\delta'}^{l,p}(\mathcal{D})} &\le c\,\|L(x,D_x)\,(\zeta_\nu u)\|_{V_{\delta'}^{l-2m,p}(\mathcal{D})^\ell} \\ &\le c\Big(\|\zeta_\nu f\|_{V_{\delta'}^{l-2m,p}(\mathcal{D})^\ell} + \|[L(x,D_x),\zeta_\nu]u\|_{V_{\delta'}^{l-2m,p}(\mathcal{D})^\ell}\Big)\end{aligned}$$

with a constant c independent of ν and u. Here, the commutator $[L,\zeta_\nu] = L\zeta_\nu - \zeta_\nu L$ is a differential operator of order $2m-1$. Multiplying the last inequality by $2^{\nu(\delta-\delta')}$, we obtain

$$\|\zeta_\nu u\|_{V_\delta^{l,p}(\mathcal{D})} \le c\Big(\|\zeta_\nu f\|_{V_\delta^{l-2m,p}(\mathcal{D})^\ell} + \|[L(x,D_x),\zeta_\nu]u\|_{V_\delta^{l-2m,p}(\mathcal{D})^\ell}\Big).$$

By the conditions on ζ_ν and on the coefficients A_α, we have

$$\|[L(x,D_x),\zeta_\nu]u\|_{V_\delta^{l-2m,p}(\mathcal{D})^\ell} \le c\,\|\eta_\nu u\|_{V_{\delta-1}^{l-1,p}(\mathcal{D})^\ell},$$

where $\eta_\nu = \zeta_{\nu-1} + \zeta_\nu + \zeta_{\nu+1}$. Consequently,

$$\|\zeta_\nu u\|_{V_\delta^{l,p}(\mathcal{D})^\ell} \le c\Big(\|\zeta_\nu f\|_{V_\delta^{l-2m,p}(\mathcal{D})^\ell} + \|\eta_\nu u\|_{V_{\delta-1}^{l-1,p}(\mathcal{D})^\ell}\Big).$$

This estimate and Lemmas 2.1.4, 2.1.8 imply $u \in V_\delta^{l,p}(\mathcal{D})^\ell$ and

$$\|u\|_{V_\delta^{l,p}(\mathcal{D})^\ell} \le c\Big(\|f\|_{V_\delta^{l-2m,p}(\mathcal{D})^\ell} + \|u\|_{V_{\delta-1}^{l-1,p}(\mathcal{D})^\ell}\Big).$$

Using Lemma 2.1.5, we obtain (2.9.7). □

The next theorem is a generalization of Theorems 2.6.7 and 2.8.9 to differential operators with variable coefficients.

THEOREM 2.9.7. *Let* $u \in V_\delta^{l,p}(\mathcal{D})^\ell$ *be a solution of the boundary value problem* (2.9.1), *where*

$$f \in V_\delta^{l-2m,p}(\mathcal{D})^\ell \cap V_{\delta'}^{l'-2m,q}(\mathcal{D})^\ell,\ l \ge m,\ l' \ge m,$$
$$g_k^\pm \in V_\delta^{l-k+1-1/p,p}(\Gamma^\pm)^\ell \cap V_{\delta'}^{l'-k+1-1/q,q}(\Gamma^\pm)^\ell,$$

$-\delta_+ < \delta-l+m-1+2/p < \delta_-$, *and* $-\delta_+ < \delta'-l'+m-1+2/q < \delta_-$. *Suppose that the coefficients* A_α *satisfy the conditions* (2.9.3)–(2.9.4) *for* $N = \max(|l-2m|, |l'-2m|)$ *with a sufficiently small* ε_0. *Then* $u \in V_{\delta'}^{l',q}(\mathcal{D})^\ell$.

Analogously, the assertions of Theorem 2.8.9 *are valid for the solution of the boundary value problem* (2.9.1) *if the coefficients of the differential operator* $L(x, D_x)$ *satisfy the conditions* (2.9.3)–(2.9.4) *with a sufficiently small* ε_0.

P r o o f. We prove the first assertion. By Theorems 2.6.5 and 2.6.7, the operator of the boundary value problem (2.2.1) is an isomorphism from $V_\delta^{l,p}(\mathcal{D})^\ell \cap V_{\delta'}^{l',q}(\mathcal{D})^\ell$ onto

$$(V_\delta^{l-2m,p}(\mathcal{D})^\ell \cap V_{\delta'}^{l'-2m,q}(\mathcal{D})^\ell) \times \prod_{\pm}\prod_{k=1}^{m}(V_\delta^{l-k+1-1/p,p}(\Gamma^\pm)^\ell \cap V_{\delta'}^{l'-k+1-1/q,q}(\Gamma^\pm)^\ell)$$

Since the the operator $L(x, D_x) - L^\circ(0, D_x)$ is small in the operator norm

$$V_\delta^{l,p}(\mathcal{D})^\ell \cap V_{\delta'}^{l',q}(\mathcal{D})^\ell \to V_\delta^{l-2m,p}(\mathcal{D})^\ell \times V_{\delta'}^{l'-2m,q}(\mathcal{D})^\ell,$$

the same is true for the operator of the problem (2.9.1). Together with Theorem 2.9.2 this implies the regularity result in the spaces $V_\delta^{l,p}$. The second assertion follows analogously from Theorems 2.8.6–2.8.9 and Theorem 2.9.2. □

2.9.4. Local estimates for solutions of the Dirichlet problem. In order to obtain analogous local regularity result, we need an estimate for the commutator

$$[L(x, D_x), \zeta] = L(x, D_x)\,\zeta - \zeta\, L(x, D_x)$$

of the differential operator $L(x, D_x)$ and a function ζ.

LEMMA 2.9.8. *Let* ζ, η *be infinitely differentiable functions with compact supports such that* $\eta = 1$ *in a neighborhood of* $\operatorname{supp}\zeta$. *If* $\eta u \in V_\delta^{l,p}(\mathcal{D})^\ell$, $l \ge m$, *then*

$$[L(x, D_x), \zeta]\, u \in V_\delta^{l-2m+1,p}(\mathcal{D})^\ell$$

and

$$\big\|[L(x, D_x), \zeta]\, u\big\|_{V_\delta^{l-2m+1,p}(\mathcal{D})^\ell} \le c\, \|\eta u\|_{V_\delta^{l,p}(\mathcal{D})^\ell}\,.$$

Here, the constant c *depends only on*

$$c_0 = \sum_{|\beta| \le |l-2m+1|} \sup_{x\in\mathcal{D}} r^{2m-|\alpha|+|\beta|}\, \big|D_x^\beta A_\alpha(x)\big|$$

and on ζ.

P r o o f. Let $l \ge 2m-1$ and $|\beta| \le l-2m+1$. Then $D_x^\beta[L(x, D_x), \zeta]\, u$ is a linear combination of the terms

$$\big(D_x^{\beta'} A_\alpha(x)\big)\ \big(D_x^{\beta''}\zeta\big)\ D_x^\gamma(\eta u),$$

where $|\beta'| + |\beta''| + |\gamma| = |\alpha| + |\beta|$, $|\gamma| \le l$, and $|\beta''| \ge 1$. One can easily check that

$$\int_{\mathcal{D}} r^{p(\delta-l+2m-1+|\beta|)}\, \big|D_x^{\beta'} A_\alpha(x)\big|^p\ |D_x^{\beta''}\zeta|^p\ \big|D_x^\gamma(\eta u)\big|^p\, dx$$
$$\le c_0\, \|\eta u\|_{V_\delta^{l,p}(\mathcal{D})^\ell}\ \max\big(r^{|\beta''|-1} D_x^{\beta''}\zeta\big)$$

Consequently,

$$\big\|[L(x, D_x), \zeta]\, u\big\|_{V_\delta^{l-2m+1,p}(\mathcal{D})^\ell} \le c\, c_0\, \|\eta u\|_{V_\delta^{l,p}(\mathcal{D})^\ell}\,,$$

where c depends only on ζ. This proves the lemma for $l \ge 2m-1$. Analogously, the assertion of the lemma holds in the case $l < 2m-1$. □

Applying Theorem 2.9.6 to the vector function ζu and using the last lemma, we obtain a local estimate for the solution of the problem (2.9.1).

LEMMA 2.9.9. *Let ζ, η be infinitely differentiable functions on $\overline{\mathcal{D}}$ with compact supports such that $\eta = 1$ in a neighborhood of* $\operatorname{supp}\zeta$. *Furthermore, let u be a solution of problem (2.9.1), $\psi u \in W^{m,p}(\mathcal{D})^\ell$ for every $\psi \in C_0^\infty(\overline{\mathcal{D}}\backslash M)$. Suppose that the coefficients A_α satisfy the conditions (2.9.3)–(2.9.4) for $N = |l-2m|$ with a sufficiently small ε_0. If*

$$\eta u \in V_{\delta-1}^{l-1,p}(\mathcal{D})^\ell, \quad \eta f \in V_\delta^{l-2m,p}(\mathcal{D})^\ell \quad \textit{and} \quad \eta g_k^\pm \in V_\delta^{l-k+1-1/p,p}(\Gamma^\pm)^\ell,$$

$l \geq m$, then $\zeta u \in V_\delta^{l,p}(\mathcal{D})^\ell$ and

$$\|\zeta u\|_{V_\delta^{l,p}(\mathcal{D})^\ell} \leq c\Big(\|\zeta f\|_{V_\delta^{l-2m,p}(\mathcal{D})^\ell} + \sum_\pm \sum_{k=1}^m \|\eta g_k^\pm\|_{V_\delta^{l-k+1-1/p,p}(\Gamma^\pm)^\ell} + \|\eta u\|_{V_{\delta-1}^{l-1,p}(\mathcal{D})^\ell}\Big).$$

Another local estimate is given in the following lemma.

LEMMA 2.9.10. *Let ζ, η be infinitely differentiable functions with compact supports such that $\eta = 1$ in a neighborhood of* $\operatorname{supp}\zeta$. *Suppose that the coefficients A_α satisfy the conditions (2.9.3)–(2.9.4) for $N = \max(m, |l'-2m|)$ with a sufficiently small ε_0. Furthermore, we assume that $\eta u \in V_\delta^{l,p}(\mathcal{D})^\ell \cap \overset{\circ}{V}{}_{\delta-l+m}^{m,2}(\mathcal{D})^\ell$ and $\eta L(x,D_x)u \in V_{\delta'}^{l'-2m,q}(\mathcal{D})^\ell$, where $l \geq m$, $l' \geq m$,*

$$-\delta_+ < \delta - l + m - 1 + 2/p < \delta_- \quad \textit{and} \quad -\delta_+ < \delta' - l' + m - 1 + 2/q < \delta_-\,.$$

Then $\zeta u \in V_{\delta'}^{l',q}(\mathcal{D})^\ell$ and

$$\|\zeta u\|_{V_{\delta'}^{l',q}(\mathcal{D})^\ell} \leq c\Big(\|\eta u\|_{V_\delta^{l,p}(\mathcal{D})^\ell} + \|\eta L(x,D_x)u\|_{V_{\delta'}^{l'-2m,q}(\mathcal{D})^\ell}\Big). \tag{2.9.8}$$

Here, the constant c depends only on ζ, ε_0 and on the norm of the inverse of the operator

$$L^\circ(0,D_x):\ V_\delta^{l,p}(\mathcal{D})^\ell \cap \overset{\circ}{V}{}_{\delta-l+m}^{m,p}(\mathcal{D})^\ell \cap V_{\delta'}^{l',q}(\mathcal{D})^\ell \to V_\delta^{l-2m,p}(\mathcal{D})^\ell \cap V_{\delta'}^{l'-2m,q}(\mathcal{D})^\ell.$$

P r o o f. By Lemma 2.9.9, it suffices to prove the theorem for $l = l'$. The space $V_\delta^{l-2m+1,p}(\mathcal{D})$ is imbedded in $V_\gamma^{l-2m,q}(\mathcal{D})$ if $3/q \leq 3/p < 1 + 3/q$ and $\gamma + 3/q = \delta - 1 + 3/p$ (cf. Lemma 2.1.1). Furthermore for $p > q$ and $\gamma + 2/q > \delta - 1 + 2/p$, one can easily deduce the estimate

$$\|f\|_{V_\gamma^{l-2m,q}(\mathcal{D})} \leq c\,\|f\|_{V_{\delta-1}^{l-2m,p}(\mathcal{D})} \leq c\,\|f\|_{V_\delta^{l-2m+1,p}(\mathcal{D})}$$

from Hölders inequality. Consequently, it follows from the assumptions of the theorem that

$$L(x,D_x)\,(\zeta u) = \zeta\, L(x,D_x)\,u + [L(x,D_x),\zeta]\,(\eta u) \in V_\gamma^{l-2m,q}(\mathcal{D})^\ell$$

if $3/p < 1+3/q$ and $\gamma + \frac{2}{q} > \delta + \frac{2}{p} - \frac{2}{3}$. In this case, the statement of the Theorem follows immediately from Theorem 2.9.7. If the above inequalities on p, q, β, δ are not satisfied, then one has to repeat the argument. □

In particular, the estimate

$$\|\zeta u\|_{V_{\delta'}^{l',q}(\mathcal{D})^\ell} \leq c\Big(\|\eta u\|_{V_\delta^{l,p}(\mathcal{D})^\ell} + \|\eta \rho^{2m} L(\rho x, \rho^{-1}D_x)u\|_{V_{\delta'}^{l'-2m,q}(\mathcal{D})^\ell}\Big)$$

holds with a constant c independent of ρ if the coefficients A_α satisfy the conditions (2.9.3)–(2.9.4) for $N = \max(m, |l' - 2m|)$ with a sufficiently small ε_0 (cf. Remark 2.9.5). This will be used in the proof of the following lemma.

LEMMA 2.9.11. *Let $\mathcal{B}_\rho(x_0)$ be the ball with radius $\rho < 1$ and center x_0, where $r(x_0) \le C\rho$ with a certain constant C. Furthermore, let η be an infinitely differentiable function with support in $\mathcal{B}_\rho(x_0)$ such that $\eta(x) = 1$ for $|x - x_0| < \rho/2$. We suppose that the coefficients of the operator $L(x, D_x)$ satisfy the conditions* (2.9.3) and (2.9.4) *for $N = \max(m, l - 2m)$ with a sufficiently small ε_0. If $\eta u \in \overset{\circ}{V}{}_0^{m,2}(\mathcal{D})^\ell$ and $\eta\, L(x, D_x)u = 0$, then*

$$r(x_0)^{|\alpha|-m+1-\delta_+ +\varepsilon}\, \big|(D_x^\alpha u)(x_0)\big| \le c\, \rho^{\varepsilon-\delta_+ -1/2}\, \|\eta u\|_{V_0^{m,2}(\mathcal{D})^\ell}$$

for $|\alpha| < l$. Here ε is an arbitrarily small positive number, and the constant c is independent of u, x_0 and ρ.

P r o o f. We put $y_0 = x_0/\rho$, $v(x) = u(\rho x)$ and $\eta_\rho(x) = \eta(\rho x)$. Then $r(y_0) \le C$, $\eta_\rho(x) = 1$ for $|x - y_0| < 1/2$, and $\eta_\rho(x) = 0$ for $|x - y_0| > 1$. From the equation $\eta\, L(x, D_x)u = 0$ it follows that

$$\eta_\rho(x)\, \rho^{2m}\, L(\rho x, \rho^{-1} D_x)\, v(x) = 0.$$

Let ζ be an infinitely differentiable function such that $\zeta = 1$ in a neighborhood of y_0, and $\eta_\rho = 1$ in a neighborhood of $\operatorname{supp}\zeta$. Furthermore, let $\delta = l - m - \delta_+ + \varepsilon$. Then by Lemma 2.9.10,

$$\|\zeta\, v\|_{V_\delta^{l,p}(\mathcal{D})^\ell} \le c\, \|\eta_\rho v\|_{V_0^{m,2}(\mathcal{D})^\ell}\,,$$

where $-\delta_+ < \delta - l + m - 1 + 2/p < \delta_-$. Here the constant c is independent of v, y_0 and ρ. Applying Lemma 2.1.3, we get

$$r(y_0)^{\delta-l+|\alpha|+3/p}\, \big|(D^\alpha v)(y_0)\big| \le c\, \|\eta_\rho v\|_{V_0^{m,2}(\mathcal{D})^\ell}$$

for $|\alpha| < l - 3/p$. We can choose l, p and δ such that $\delta - l + 3/p = 1 - m - \delta_+ + \varepsilon$. Using the equality

$$\|\eta_\rho v\|_{V_0^{m,2}(\mathcal{D})^\ell} = \rho^{m-3/2}\, \|\eta u\|_{V_0^{m,2}(\mathcal{D})^\ell}\,,$$

we obtain the desired estimate for u. □

CHAPTER 3

The Dirichlet problem for strongly elliptic systems in a cone with edges

In this chapter, we study the Dirichlet problem for the differential operator

$$L(D_x) = \sum_{|\alpha|=2m} A_\alpha \, D_x^\alpha$$

in a cone $\mathcal{K}$ whose boundary contains the edges $M_1, \dots, M_d$ meeting at the vertex of the cone. The coefficients A_α of the operator $L(D_x)$ are assumed to be constant $\ell \times \ell$-matrices. As in the foregoing chapter, we suppose that the operator $L(D_x)$ is strongly elliptic. We consider solutions of the Dirichlet problem in weighted Sobolev spaces $V_{\beta,\delta}^{l,p}(\mathcal{K})$ and weighted Hölder classes $N_{\beta,\delta}^{l,\sigma}(\mathcal{K})$ with arbitrary integer $l \ge m$. Here, the space $V_{\beta,\delta}^{l,p}(\mathcal{K})$ is defined as the set of all functions u on $\mathcal{K}$ such that

$$|x|^{\beta-l+|\alpha|} \prod_{k=1}^{d} \Big(\frac{r_k(x)}{|x|}\Big)^{\delta_k-l+|\alpha|} \partial_x^\alpha u \in L_p(\mathcal{K}) \quad \text{for } |\alpha| \le l,$$

where $r_k(x)$ denotes the distance of the point x from the edge M_k and $\delta_1, \dots, \delta_k$ are the components of δ. Similarly, the weighted Hölder space $N_{\beta,\delta}^{l,\sigma}(\mathcal{K})$ is defined.

The main results of this chapter concern the solvability of the boundary value problem in weighted Sobolev and Hölder spaces. We obtain necessary and sufficient conditions for the existence and uniqueness of the solutions. The conditions on the components δ_k of δ are the same as in Chapter 2, while on β one must impose a condition analogous to that for "smooth" cones (cf. Theorem 1.2.5). Furthermore, we prove regularity assertions for the solutions.

The method is analogous to the approach in Chapter 2. We start with solutions in weighted L_2-Sobolev spaces (Sections 3.2 and 3.3). The a priori estimates for the solutions obtained here are used in Section 3.4 for the evaluation of the Green's matrix $G(x,\xi)$. We obtain majorants for the elements of $G(x,\xi)$ depending explicitly on the distances of x and ξ from the edges and from the vertex of the cone. For example, in the case $|x| > 2|\xi|$, the estimate has the form

$$\big|D_x^\alpha D_\xi^\gamma G_{i,j}(x,\xi)\big| \le c_{\alpha,\gamma} |x|^{\Lambda_- -|\alpha|+\varepsilon} \, |\xi|^{2m-3-\Lambda_- -|\gamma|-\varepsilon}$$
$$\times \prod_{k=1}^{d} \Big(\frac{r_k(x)}{|x|}\Big)^{m-1+\delta_+^{(k)}-|\alpha|-\varepsilon} \prod_{k=1}^{d} \Big(\frac{r_k(\xi)}{|\xi|}\Big)^{m-1+\delta_-^{(k)}-|\gamma|-\varepsilon},$$

where Λ_- is a certain real number. Such point estimates are applied in Sections 3.5 and 3.6, where we extend the results of Sections 3.3 to weighted L_p Sobolev spaces $V_{\beta,\delta}^{l,p}(\mathcal{K})$ with arbitrary $p \in (1,\infty)$ and to weighted Hölder spaces $N_{\beta,\delta}^{l,\sigma}(\mathcal{K})$. The last Section 3.7 is concerned with unique solvability and regularity assertions for strongly elliptic differential equations with variable coefficients.

3.1. Weighted Sobolev spaces in a cone

Let

$$\mathcal{K} = \{x \in \mathbb{R}^3 : \ x/|x| = \omega \in \Omega\}$$

be a cone with vertex at the origin. Suppose that the boundary $\partial\mathcal{K}$ consists of the vertex $x = 0$, the edges (half-lines) $M_1, \dots, M_d$, and smooth (of class C^∞) faces $\Gamma_1, \dots, \Gamma_d$. This means that $\Omega = \mathcal{K} \cap S^2$ is a domain of polygonal type on the unit sphere S^2 with sides $\gamma_k = \Gamma_k \cap S^2$.

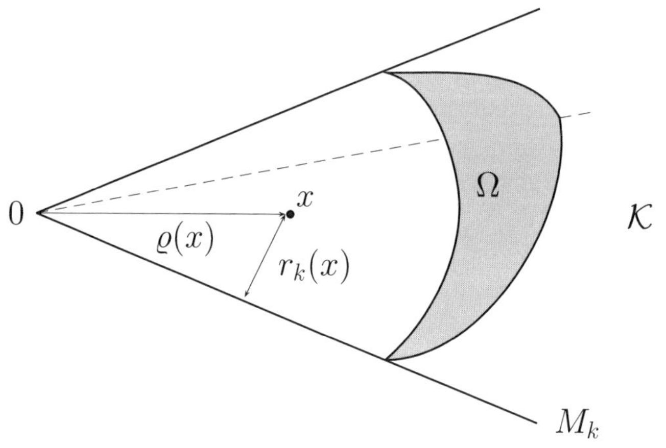

FIGURE 5. A cone $\mathcal{K}$ with edges $M_1, \dots, M_d$

We assume that for every edge point $x \in M_k$, there exist a neighborhood $\mathcal{U}$ and a diffeomorphism κ mapping $\mathcal{K} \cap \mathcal{U}$ onto the intersection of a dihedron $\mathcal{D}$ with the unit ball.

3.1.1. The space $V^{l,p}_{\beta,\delta}(\mathcal{K})$. Let l be a nonnegative integer, $\beta \in \mathbb{R}$, $\delta = (\delta_1, \dots, \delta_d) \in \mathbb{R}^d$, and $1 < p < \infty$. Furthermore, let $\mathcal{S} = \{0\} \cup M_1 \cup \cdots \cup M_d$ be the set of the singular boundary points. Then the space $V^{l,p}_{\beta,\delta}(\mathcal{K})$ is defined as the closure of the set $C_0^\infty(\overline{\mathcal{K}}\backslash\mathcal{S})$ with respect to the norm

$$\|u\|_{V^{l,p}_{\beta,\delta}(\mathcal{K})} = \Big(\int_{\mathcal{K}} \sum_{|\alpha| \le l} \rho^{p(\beta - l + |\alpha|)} \prod_{k=1}^{d} \Big(\frac{r_k}{\rho}\Big)^{p(\delta_k - l + |\alpha|)} \big|\partial_x^\alpha u(x)\big|^p \, dx \Big)^{1/p}. \tag{3.1.1}$$

Here, $\rho(x) = |x|$ is the distance of the point x from the vertex of the cone, while $r_k(x)$ denotes the distance from the edge M_k. The closure of the set $C_0^\infty(\mathcal{K})$ with respect to the norm (3.1.1) is denoted by $\overset{\circ}{V}{}^{l,p}_{\beta,\delta}(\mathcal{K})$. If $\delta = (\delta_1, \dots, \delta_d) \in \mathbb{R}^d$ and t is an arbitrary real number, then by $V^{l,p}_{\beta,\delta+t}(\mathcal{K})$ we mean the space $V^{l,p}_{\beta,\delta'}(\mathcal{K})$, where $\delta' = (\delta_1 + t, \dots, \delta_d + t)$. Let

$$r(x) = \min_{1 \le k \le d} r_k(x).$$

Obviously, there exist positive constants c_1 and c_2 independent of x such that

$$c_1\, \rho(x) \prod_{k=1}^{d} \frac{r_k(x)}{\rho(x)} \le r(x) \le c_2\, \rho(x) \prod_{k=1}^{d} \frac{r_k(x)}{\rho(x)} \quad \text{for all } x \in \mathcal{K}. \tag{3.1.2}$$

Hence, the norm in $V^{l,p}_{\beta,\beta}(\mathcal{K})$ is equivalent to

$$\Big(\int_{\mathcal{K}} \sum_{|\alpha|\le l} r^{p(\beta-l+|\alpha|)} \big|\partial_x^\alpha u(x)\big|^p \, dx \Big)^{1/p}.$$

Note that the dual space of $\overset{\circ}{V}{}^{0,p}_{\beta,\delta}(\mathcal{K}) = V^{0,p}_{\beta,\delta}(\mathcal{K})$ coincides with $V^{0,p'}_{-\beta,-\delta}(\mathcal{K})$, where $p' = p/(p-1)$. This means that every linear and continuous functional on $V^{0,p}_{\beta,\delta}(\mathcal{K})$ has the form

$$f(u) = (u,v)_{\mathcal{K}}\,,$$

where $v \in V^{0,p'}_{-\beta,-\delta}(\mathcal{K})$ and $(\cdot,\cdot)_{\mathcal{K}}$ denotes the scalar product in $L_2(\mathcal{K})$. We define the space $V^{-l,p}_{\beta,\delta}(\mathcal{K})$ for integer $l \ge 0$ as the dual space of $\overset{\circ}{V}{}^{l,p'}_{-\beta,-\delta}(\mathcal{K})$. This space is provided with the norm

$$\|u\|_{V^{-l,p}_{\beta,\delta}(\mathcal{K})} = \sup\big\{ |(u,v)_{\mathcal{K}}| \,:\, v \in \overset{\circ}{V}{}^{l,p'}_{-\beta,-\delta}(\mathcal{K}),\ \|v\|_{V^{l,p'}_{-\beta,-\delta}(\mathcal{K})} = 1 \big\}. \tag{3.1.3}$$

An equivalent norm to (3.1.3) is given in the following lemma which can be proved in the exact same manner as Lemma 2.1.7.

Lemma 3.1.1. *$V^{-l,p}_{\beta,\delta}(\mathcal{K})$ is the set of all distributions of the form*

$$u = \sum_{|\alpha|\le l} \partial_x^\alpha u_\alpha, \quad \text{where } u_\alpha \in V^{0,p}_{\beta+l-|\alpha|,\delta+l-|\alpha|}(\mathcal{K}).$$

The expression

$$\inf\Big\{ \sum_{|\alpha|\le l} \|u_\alpha\|_{V^{0,p}_{\beta+l-|\alpha|,\delta+l-|\alpha|}(\mathcal{K})} \,:\, u_\alpha \in V^{0,p}_{\beta+l-|\alpha|,\delta+l-|\alpha|}(\mathcal{K}),\ \sum_{|\alpha|\le l} \partial_x^\alpha u_\alpha = u \Big\}$$

defines an equivalent norm to (3.1.3).

3.1.2. An equivalent norm in $V^{l,p}_{\beta,\delta}(\mathcal{K})$. Let ζ_ν be infinitely differentiable functions depending only on $\rho = |x|$ such that

$$\operatorname{supp}\zeta_\nu \subset \big(2^{\nu-1}, 2^{\nu+1}\big), \quad \big|\partial_\rho^j \zeta_\nu(\rho)\big| \le c_j\, 2^{-j\nu}, \quad \sum_{\nu=-\infty}^{+\infty} \zeta_\nu = 1. \tag{3.1.4}$$

Here c_j are constants independent of r and ν.

Lemma 3.1.2. *Let l be an arbitrary integer, $\beta \in \mathbb{R}$, $\delta \in (\delta_1,\dots,\delta_d) \in \mathbb{R}^d$ and $1 < p < \infty$. Then there exist positive constants c_1 and c_2 such that*

$$c_1\, \|u\|_{V^{l,p}_{\beta,\delta}(\mathcal{K})} \le \Big(\sum_{\nu=-\infty}^{+\infty} \|\zeta_\nu u\|^p_{V^{l,p}_{\beta,\delta}(\mathcal{K})} \Big)^{1/p} \le c_2\, \|u\|_{V^{l,p}_{\beta,\delta}(\mathcal{K})}$$

for all $u \in V^{l,p}_{\beta,\delta}(\mathcal{K})$.

P r o o f. First let l be nonnegative. Then it follows from (3.1.4) that

$$\|\zeta_\nu u\|^p_{V^{l,p}_{\beta,\delta}(\mathcal{K})} \le c \int\limits_{\substack{\mathcal{K} \\ 2^{\nu-1}<|x|<2^{\nu+1}}} \sum_{|\alpha|\le l} \rho^{p(\delta-l+|\alpha|)} \prod_{k=1}^d \Big(\frac{r_k}{\rho}\Big)^{p(\delta_k-l+|\alpha|)} |\partial_x^\alpha u(x)|^p\, dx$$

with a constant c independent of ν. Hence,

$$\sum_{\nu=-\infty}^{+\infty} \|\zeta_\nu u\|^p_{V^{l,p}_{\beta,\delta}(\mathcal{K})} \le 2c\, \|u\|^p_{V^{l,p}_{\beta,\delta}(\mathcal{K})}.$$

Since $\zeta_{\nu-1}(x) + \zeta_\nu(x) + \zeta_{\nu+1}(x) = 1$ for $2^{\nu-1} < |x| < 2^{\nu+1}$, we obtain

$$\|u\|^p_{V^{l,p}_{\beta,\delta}(\mathcal{K})} = \frac12 \sum_{\nu=-\infty}^{+\infty} \int\limits_{\substack{\mathcal{K}\\ 2^{\nu-1}<|x|<2^{\nu+1}}} \sum_{|\alpha|\le l} \rho^{p(\beta-l+|\alpha|)} \prod_{k=1}^d \Big(\frac{r_k}{\rho}\Big)^{p(\delta_k-l+|\alpha|)} |\partial_x^\alpha u(x)|^p\, dx$$

$$\le \frac12 \sum_{\nu=-\infty}^{+\infty} \|(\zeta_{\nu-1}+\zeta_\nu+\zeta_{\nu+1})u\|^p_{V^{l,p}_{\beta,\delta}(\mathcal{K})} \le c \sum_{\nu=-\infty}^{+\infty} \|\zeta_\nu u\|^p_{V^{l,p}_{\beta,\delta}(\mathcal{K})}.$$

This proves the lemma for $l \ge 0$. If $l < 0$, then the estimate

$$\|u\|_{V^{l,p}_{\beta,\delta}(\mathcal{K})} \le c\Big(\sum_{\nu=-\infty}^{+\infty} \|\zeta_\nu u\|^p_{V^{l,p}_{\beta,\delta}(\mathcal{K})}\Big)^{1/p}$$

holds by the same arguments as in the proof of Lemma 2.1.8. Furthermore for an arbitrary function $u \in V^{l,p}_{\beta,\delta}(\mathcal{K})$, there exist functions $u_\alpha \in V^{0,p}_{\beta-l-|\alpha|,\delta-l-|\alpha|}(\mathcal{K})$ such that

$$u = \sum_{|\alpha|\le -l} \partial_x^\alpha u_\alpha \quad\text{and}\quad \sum_{|\alpha|\le -l} \|u_\alpha\|_{V^{0,p}_{\beta-l-|\alpha|,\delta-l-|\alpha|}(\mathcal{K})} \le c\, \|u\|_{V^{l,p}_{\beta,\delta}(\mathcal{K})}$$

with a constant c independent of u. We set $\eta_\nu = \zeta_{\nu-1}+\zeta_\nu+\zeta_{\nu+1}$. Then

$$\begin{aligned}
\big|(\zeta_\nu u, v)_{\mathcal{K}}\big| &\le \sum_{|\alpha|\le -l} \big|\big(\eta_\nu u_\alpha, \partial_x^\alpha(\zeta_\nu v)\big)_{\mathcal{K}}\big| \\
&\le \sum_{|\alpha|\le -l} \|\eta_\nu u_\alpha\|_{V^{0,p}_{\beta-l-|\alpha|,\delta-l-|\alpha|}(\mathcal{K})}\ \|\partial_x^\alpha(\zeta_\nu v)\|_{V^{0,p'}_{-\beta+l+|\alpha|,-\delta+l+|\alpha|}(\mathcal{K})} \\
&\le c \sum_{|\alpha|\le -l} \|\eta_\nu u_\alpha\|_{V^{0,p}_{\beta-l-|\alpha|,\delta-l-|\alpha|}(\mathcal{K})}\ \|v\|_{V^{-l,p'}_{-\beta,-\delta}(\mathcal{D})}
\end{aligned}$$

for all $v \in \overset{\circ}{V}{}^{-l,p'}_{-\beta,-\delta}(\mathcal{K})$. This implies

$$\begin{aligned}
\sum_{\nu=-\infty}^{+\infty} \|\zeta_\nu u\|^p_{V^{l,p}_{\beta,\delta}(\mathcal{K})} &\le c_1 \sum_{\nu=-\infty}^{+\infty} \sum_{|\alpha|\le -l} \|\eta_\nu u_\alpha\|^p_{V^{0,p}_{\beta-l-|\alpha|,\delta-l-|\alpha|}(\mathcal{K})} \\
&\le c_2 \sum_{|\alpha|\le -l} \|u_\alpha\|^p_{V^{0,p}_{\beta-l-|\alpha|,\delta-l-|\alpha|}(\mathcal{K})} \le c\, \|u\|^p_{V^{l,p}_{\beta,\delta}(\mathcal{K})}.
\end{aligned}$$

The proof of the lemma is complete. □

3.1.3. Imbeddings. Obviously,

$$V^{l,p}_{\beta,\delta}(\mathcal{K}) \subset V^{l-1,p}_{\beta-1,\delta-1}(\mathcal{K}) \subset \cdots \subset V^{0,p}_{\beta-l,\delta-l}(\mathcal{K}).$$

This follows immediately from the definition of the norm in $V^{l,p}_{\beta,\delta}(\mathcal{K})$. A more general result is given in the next lemma.

LEMMA 3.1.3. *Let* $1 < p \le q < \infty$, $l - 3/p > l' - 3/q$, $\beta - l + 3/p = \beta' - l' + 3/q$ *and* $\delta_k - l + 3/p \le \delta'_k - l' + 3/q$ *for* $k = 1, \dots, d$. *Then the space* $V^{l,p}_{\beta,\delta}(\mathcal{K})$ *is continuously imbedded in* $V^{l',q}_{\beta',\delta'}(\mathcal{K})$.

P r o o f. Let $u \in V^{l,p}_{\beta,\delta}(\mathcal{K})$ be given. Furthermore, let ζ_ν be the same functions as in Lemma 3.1.2. We define $\tilde{\zeta}_\nu(x) = \zeta_\nu(2^\nu x)$ and $v(x) = u(2^\nu x)$. Since $\tilde{\zeta}_\nu(x) = 0$ for $|x| < 1/2$ and $|x| > 2$, it follows from Lemma 2.1.1 that $\tilde{\zeta}_\nu v \in V^{l',q}_{\beta',\delta'}(\mathcal{K})$ and

$$\|\tilde{\zeta}_\nu v\|^q_{V^{l',q}_{\beta',\delta'}(\mathcal{K})} \le c\, \|\tilde{\zeta}_\nu v\|^q_{V^{l,p}_{\beta,\delta}(\mathcal{K})} .$$

Here the constant c is independent of u and ν. Multiplying the last inequality by $2^{\nu q(\beta - l + 3/p)}$ and substituting $x = 2^{-\nu} y$, we obtain

$$\|\zeta_\nu u\|^q_{V^{l',q}_{\beta',\delta'}(\mathcal{K})} \le c\, \|\zeta_\nu u\|^q_{V^{l,p}_{\beta,\delta}(\mathcal{K})} .$$

with the same constant c. Consequently, Lemma 3.1.2 yields

$$\|u\|^q_{V^{l',q}_{\beta',\delta'}(\mathcal{K})} \le c \sum_{\nu=-\infty}^{+\infty} \|\zeta_\nu u\|^q_{V^{l,p}_{\beta,\delta}(\mathcal{K})} \le c \Big(\sum_{\nu=-\infty}^{+\infty} \|\zeta_\nu u\|^p_{V^{l,p}_{\beta,\delta}(\mathcal{K})} \Big)^{q/p} \le c'\, \|u\|^q_{V^{l,p}_{\beta,\delta}(\mathcal{K})} .$$

This proves the lemma. □

Next, we prove a weighted L_∞-estimate for functions in $V^{l,p}_{\beta,\delta}(\mathcal{K})$.

LEMMA 3.1.4. *Let* $u \in V^{l,p}_{\beta,\delta}(\mathcal{K})$ *and* $l > 3/p$. *Then*

$$\Big\| \rho^{\beta - l + 3/p} \prod_{k=1}^{d} \Big(\frac{r_k}{\rho}\Big)^{\delta_k - l + 3/p} u \Big\|_{L_\infty(\mathcal{K})} \le c\, \|u\|_{V^{l,p}_{\beta,\delta}(\mathcal{K})} \tag{3.1.5}$$

Here the constant c *is independent of* u.

P r o o f. Let $x \in \mathcal{K}$, $\xi = x/r(x)$, and $v(y) = u(r(x)y)$. Since $r(\xi) = 1$, it follows from Sobolev's imbedding theorem that

$$|u(x)| = |v(\xi)| \le c\, \|v\|_{W^{l,p}(\mathcal{K} \cap B(\xi))} , \tag{3.1.6}$$

where $B(\xi)$ denotes the ball with radius $1/2$ centered about ξ. Here

$$\begin{aligned} \|v\|^p_{W^{l,p}(\mathcal{K} \cap B(\xi))} &= \sum_{|\alpha| \le l} \int\limits_{\substack{\mathcal{K} \\ |y-\xi| < 1/2}} |\partial_y^\alpha v(y)|^p\, dy \\ &= \sum_{|\alpha| \le l} r(x)^{p|\alpha| - 3} \int\limits_{\substack{\mathcal{K} \\ |y-x| < r(x)/2}} |\partial_y^\alpha u(y)|^p\, dy. \end{aligned}$$

Using the inequalities $\rho(x)/2 < \rho(y) < 3\rho(x)/2$ and $r_k(x)/2 < r_k(y) < 3r_k(x)/2$, which are valid for $|x - y| < r(x)/2$, and the inequality (3.1.2), we obtain

$$\|v\|_{W^{l,p}(\mathcal{K} \cap B(\xi))} \le c\, \rho(x)^{l - \beta - 3/p} \prod_{k=1}^{d} \Big(\frac{r_k(x)}{\rho(x)}\Big)^{l - \delta_k - 3/p} \|u\|_{V^{l,p}_{\beta,\delta}(\mathcal{K})} .$$

This together with (3.1.6) implies (3.1.5). □

3.1.4. Trace spaces. We denote the trace space for $V^{l,p}_{\beta,\delta}(\mathcal{K})$, $l \ge 1$, on the face Γ_j by $V^{l-1/p,p}_{\beta,\delta}(\Gamma_j)$. The norm in this space is defined as

$$\|u\|_{V^{l-1/p,p}_{\beta,\delta}(\Gamma_j)} = \inf\big\{\|v\|_{V^{l,p}_{\beta,\delta}(\mathcal{K})} : v \in V^{l,p}_{\beta,\delta}(\mathcal{K}),\ v = u \text{ on } \Gamma_j\big\}.$$

An equivalent norm is given in the following lemma.

LEMMA 3.1.5. *There exist positive constants c_1 and c_2 such that*

$$c_1\, \|u\|_{V^{l-1/p,p}_{\beta,\delta}(\Gamma_j)} \le \Big(\sum_{\nu=-\infty}^{+\infty} \|\zeta_\nu u\|^p_{V^{l-1/p,p}_{\beta,\delta}(\Gamma_j)}\Big)^{1/p} \le c_2\, \|u\|_{V^{l-1/p,p}_{\beta,\delta}(\Gamma_j)}$$

for all $u \in V^{l-1/p,p}_{\beta,\delta}(\Gamma_j)$, $l \ge 1$.

P r o o f. Let u be an arbitrary function in $V^{l-1/p,p}_{\beta,\delta}(\Gamma_j)$. Then there exists an extension $v \in V^{l,p}_{\beta,\delta}(\mathcal{K})$ of u satisfying the inequality

$$\|v\|^p_{V^{l,p}_{\beta,\delta}(\mathcal{K})} \le 2\, \|u\|^p_{V^{l-1/p,p}_{\beta,\delta}(\Gamma_j)}.$$

Consequently by Lemma 3.1.2,

$$\sum_{\nu=-\infty}^{+\infty} \|\zeta_\nu u\|^p_{V^{l-1/p,p}_{\beta,\delta}(\Gamma_j)} \le \sum_{\nu=-\infty}^{+\infty} \|\zeta_\nu v\|^p_{V^{l,p}_{\beta,\delta}(\mathcal{K})} \le c\, \|v\|^p_{V^{l,p}_{\beta,\delta}(\mathcal{K})} \le 2c\, \|u\|^p_{V^{l-1/p,p}_{\beta,\delta}(\Gamma_j)}.$$

On the other hand, for every $\nu = 0, \pm1, \pm2, \dots$ there exists an extension $v_\nu \in V^{l,p}_{\beta,\delta}(\mathcal{K})$ of $\zeta_\nu u$ such that

$$\|v_\nu\|^p_{V^{l,p}_{\beta,\delta}(\mathcal{K})} \le 2\, \|\zeta_\nu u\|^p_{V^{l-1/p,p}_{\beta,\delta}(\Gamma_j)}.$$

Obviously, the functions $w_\nu = (\zeta_{\nu-1} + \zeta_\nu + \zeta_{\nu+1}) v_\nu$ are also extensions of $\zeta_\nu u$. Furthermore,

$$\|w_\nu\|^p_{V^{l,p}_{\beta,\delta}(\mathcal{K})} \le c\, \|\zeta_\nu u\|^p_{V^{l-1/p,p}_{\beta,\delta}(\Gamma_j)}$$

with a constant c independent of u and ν. Consequently, the sum $w = \sum_{\nu=-\infty}^{+\infty} w_\nu$ is an extension of u. Since $w_\nu(x) = 0$ for $|x| < 2^{\nu-2}$ and $|x| > 2^{\nu+2}$, we obtain

$$\begin{aligned}
\|u\|^p_{V^{l-1/p,p}_{\beta,\delta}(\Gamma_j)} &\le \Big\| \sum_{\nu=-\infty}^{+\infty} w_\nu \Big\|^p_{V^{l,p}_{\beta,\delta}(\mathcal{K})} \\
&= \frac{1}{2} \sum_{\mu=-\infty}^{+\infty} \sum_{|\alpha|\le l} \int\limits_{\substack{\mathcal{K} \\ 2^{\mu-1}<|x|<2^{\mu+1}}} \rho^{p(\beta-l+|\alpha|)} \prod_{k=1}^{d} \Big(\frac{r}{\rho}\Big)^{p(\delta_k-l+|\alpha|)} \Big| \sum_{\nu=\mu-2}^{\mu+2} \partial_x^\alpha w_\nu \Big|^p dx \\
&\le c \sum_{\mu=-\infty}^{+\infty} \sum_{\nu=\mu-2}^{\mu+2} \|w_\nu\|^p_{V^{l,p}_{\beta,\delta}(\mathcal{K})} \le c' \sum_{\nu=-\infty}^{+\infty} \|\zeta_\nu u\|^p_{V^{l-1/p,p}_{\beta,\delta}(\Gamma_j)}.
\end{aligned}$$

This proves the lemma. □

3.1.5. A relation between the spaces $\overset{\circ}{W}{}^{l,p}(\mathcal{K})$ and $\overset{\circ}{V}{}^{l,p}_{0,0}(\mathcal{K})$. We denote the closure of the set $C_0^\infty(\mathcal{K})$ in the Sobolev space $W^{l,p}(\mathcal{K})$ by $\overset{\circ}{W}{}^{l,p}(\mathcal{K})$ and show that this space is a subspace of the weighted space $V^{l,p}_{\beta,\delta}(\mathcal{K})$ with $\beta = 0$ and $\delta = 0$.

LEMMA 3.1.6. *The space $\overset{\circ}{W}{}^{l,p}(\mathcal{K})$ is continuously imbedded in $\overset{\circ}{V}{}^{l,p}_{0,0}(\mathcal{K})$.*

P r o o f. First we prove that

$$\int_{\mathcal{K}} \rho^{p(\beta-1)} |u|^p\, dx \le c \int_{\mathcal{K}} \rho^{p\beta}\, |\nabla u|^p\, dx \tag{3.1.7}$$

for all $u \in C_0^\infty(\mathcal{K})$, where c is independent of u. Let the spherical coordinates ρ, φ, θ be defined by $x_1 = \rho \cos\varphi\, \sin\theta$, $x_2 = \rho \sin\varphi \sin\theta$, $x_3 = \rho\cos\theta$. Without loss of generality, we may assume that the negative x_3-axis belongs to the complement of $\mathcal{K}$. We extend u by zero outside $\mathcal{K}$. Then $u = 0$ for $\theta = \pi$ and consequently

$$\int_0^\pi |u|^p\, d\theta \le c \int_0^\pi |\partial_\theta u|^p\, d\theta.$$

Multiplying this inequality by $\rho^{p(\beta-1)+2}$ and integrating with respect to ρ and φ, we obtain

$$\int_{\mathbb{R}^3} \rho^{p(\beta-1)}\, |u|^p\, dx \le c \int_{\mathbb{R}^3} \rho^{p(\beta-1)}\, |\partial_\theta u|^p\, dx.$$

This implies (3.1.7).

Let $\overset{\circ}{V}{}^{l,p}_0(\mathcal{K})$ denote the closure of $C_0^\infty(\mathcal{K})$ with respect to the norm

$$\|u\|_{V_0^{l,p}(\mathcal{K})} = \Big(\int_{\mathcal{K}} \sum_{|\alpha|\le l} |x|^{p(|\alpha|-l)}\, \big|\partial_x^\alpha u(x)\big|^p\, dx\Big)^{1/p}.$$

By (3.1.7), the estimate

$$\|u\|^p_{V_0^{l,p}(\mathcal{K})} \le c \int_{\mathcal{K}} \sum_{|\alpha|=l} \big|\partial_x^\alpha u(x)\big|^p\, dx$$

holds for all $u \in C_0^\infty(\mathcal{K})$. This means that the space $\overset{\circ}{W}{}^{l,p}(\mathcal{K})$ is continuously imbedded in $\overset{\circ}{V}{}^{l,p}_0(\mathcal{K})$. We show that the spaces $\overset{\circ}{V}{}^{l,p}_0(\mathcal{K})$ and $\overset{\circ}{V}{}^{l,p}_{0,0}(\mathcal{K})$ coincide. By Lemmas 1.2.1 and 3.1.2, the norms in the space $\overset{\circ}{V}{}^{l,p}_0(\mathcal{K})$ and $\overset{\circ}{V}{}^{l,p}_{0,0}(\mathcal{K})$ are equivalent to the norms

$$\|u\|_1 = \Big(\sum_{\nu=-\infty}^{+\infty} \|\zeta_\nu u\|^p_{V_0^{l,p}(\mathcal{K})}\Big)^{1/p} \quad \text{and} \quad \|u\|_2 = \Big(\sum_{\nu=-\infty}^{+\infty} \|\zeta_\nu u\|^2_{V_{0,0}^{l,p}(\mathcal{K})}\Big)^{1/p},$$

respectively, where ζ_ν are infinitely differentiable functions depending only on $\rho = |x|$ and satisfying (3.1.4). Let u be an arbitrary infinitely differentiable function with compact support in $\mathcal{K}$. We set $\tilde{u}(x) = u(2^\nu x)$ and $\tilde{\zeta}_\nu(x) = \zeta(2^\nu x)$. Since $\tilde{\zeta}_\nu(x) = 0$ for $|x| < 1/2$ and $|x| > 2$, it follows from Lemma 2.2.4 that

$$c_1\, \|\tilde{\zeta}_\nu \tilde{u}\|_{V_0^{l,p}(\mathcal{K})} \le \|\tilde{\zeta}_\nu \tilde{u}\|_{V_{0,0}^{l,p}(\mathcal{K})} \le c_2\, \|\tilde{\zeta}_\nu \tilde{u}\|_{V_0^{l,p}(\mathcal{K})}$$

with certain positive constants c_1, c_2 independent of u and ν. Hence,

$$c_1\, \|\zeta_\nu u\|_{V_0^{l,p}(\mathcal{K})} \le \|\zeta_\nu u\|_{V_{0,0}^{l,p}(\mathcal{K})} \le c_2\, \|\zeta_\nu u\|_{V_0^{l,p}(\mathcal{K})}$$

with the same constants c_1 and c_2. Thus, the norms $\|\cdot\|_1$ and $\|\cdot\|_2$ are equivalent. The lemma is proved. □

3.2. Operator pencils generated by the Dirichlet problem

The boundary value problem

$$L(D_x)u = f \ \text{ in } \mathcal{K}, \quad \frac{\partial^{k-1}u}{\partial n^{k-1}} = 0 \ \text{ on } \Gamma_j, \ j=1,\dots,d, \ k=1,\dots,m, \tag{3.2.1}$$

generates two different types of operator pencils for the edges M_k and for the vertex of the cone. The pencils $A_k(\lambda)$ for the edges M_k are defined as the pencil $A(\lambda)$ for the boundary value problem in a dihedron (see Section 2.3). The operator pencil $\mathfrak{A}(\lambda)$ for the vertex arises if one sets $u(x) = |x|^\lambda U(\omega)$ in (3.2.1), where $\omega = x/|x|$. Then one obtains a boundary value problem for the function U on the subdomain Ω of the unit sphere with the complex parameter λ. This boundary value problem is uniquely solvable for all λ except for a denumerable set of eigenvalues of the pencil $\mathfrak{A}(\lambda)$. We prove a parameter dependent a priori estimate for the solution in weighted Sobolev spaces on Ω.

3.2.1. The operator pencil $A_k(\lambda)$. Let M_k be an edge of the cone $\mathcal{K}$, and let Γ_{k_+} and Γ_{k_-} be the faces adjacent to M_k. The half-planes tangent to Γ_{k_+}, Γ_{k_-} at M_k are denoted $\Gamma^\circ_{k_+}$ and $\Gamma^\circ_{k_-}$, respectively. They bound a dihedron $\mathcal{D}_k$ whose edge contains the points of M_k. We consider the Dirichlet problem

$$L(D_x)\, u = f \ \text{ in } \mathcal{D}_k, \quad \frac{\partial^{j-1}u}{\partial n^{j-1}} = 0 \text{ on } \Gamma^\circ_{k_\pm}, \ j=1,\dots,m.$$

The operator pencil for this problem (see the definition of the pencil $A(\lambda)$ in Section 2.3) is denoted by $A_k(\lambda)$. Let r and φ be polar coordinates in the plane perpendicular to M_k such that $\varphi = \pm\theta_k/2$ for $x \in \Gamma^\circ_{k_\pm}$ and $r = \operatorname{dist}(x, M_k)$. Then the operator $A_k(\lambda)$ is defined by

$$A_k(\lambda)\, U(\varphi) = r^{2m-\lambda} L(D_x)\, \big(r^\lambda U(\varphi)\big).$$

This differential operator realizes a continuous mapping

$$W^{2m,2}(-\theta_k/2, +\theta_k/2)^\ell \cap \overset{\circ}{W}{}^{m,2}(-\theta_k/2, +\theta_k/2)^\ell \to L_2(-\theta_k/2, +\theta_k/2)^\ell$$

for each $\lambda \in \mathbb{C}$. We denote by $\delta_+^{(k)}$ and $\delta_-^{(k)}$ the greatest positive real numbers such that the strip

$$m - 1 - \delta_-^{(k)} < \operatorname{Re}\lambda < m - 1 + \delta_+^{(k)}$$

is free of eigenvalues of the pencil $A_k(\lambda)$.

3.2.2. The operator pencil $\mathfrak{A}(\lambda)$. Let

$$\mathcal{K} = \{x \in \mathbb{R}^3 : \omega = x/|x| \in \Omega\}$$

be the cone introduced in Section 3.1. The weighted Sobolev space $V_\delta^{l,p}(\Omega)$ is defined for arbitrary integer $l \ge 0$, real $p > 1$ and $\delta = (\delta_1, \dots, \delta_d) \in \mathbb{R}^d$ as the set of all functions on Ω with finite norm

$$\|u\|_{V_\delta^{l,p}(\Omega)} = \Big(\int\limits_{\substack{\mathcal{K}\\ 1<|x|<2}} \sum_{|\alpha|\le l} \prod_{k=1}^{d} r_k^{p(\delta_k - l + |\alpha|)}\, |\partial_x^\alpha u(x)|^p\, dx \Big)^{1/p}, \tag{3.2.2}$$

where in the integral the function u is extended to $\mathcal{K}$ by $u(x) = u(x/|x|)$. The closure of the set $C_0^\infty(\Omega)$ with respect to the norm (3.2.2) is denoted by $\overset{\circ}{V}{}_\delta^{l,p}(\Omega)$.

We define the differential operator $\mathcal{L}(\lambda)$ by

$$\mathcal{L}(\lambda)\, U = \rho^{2m-\lambda}\, L(D_x)\, \big(\rho^\lambda\, U(\omega)\big). \tag{3.2.3}$$

Here λ is an arbitrary complex number, $\rho = |x|$, $\omega = x/|x|$, and U is a function on Ω. Obviously, $\mathcal{L}(\lambda)$ is a polynomial of degree $2m$ in λ. The operator

$$\overset{\circ}{W}{}^{m,2}(\Omega)^\ell \ni U \to \mathcal{L}(\lambda)\, U \in W^{-m,2}(\Omega)^\ell$$

is denoted by $\mathfrak{A}(\lambda)$. Here $W^{-m,2}(\Omega)$ is the dual space of $\overset{\circ}{W}{}^{m,2}(\Omega) = \overset{\circ}{V}{}_0^{m,2}(\Omega)$.

As is known, the operator $\mathfrak{A}(\lambda)$ is an isomorphism for all $\lambda \in \mathbb{C}$ with the possible exception of a denumerable set of isolated points. The mentioned denumerable set consists of eigenvalues with finite algebraic multiplicities which are situated, except for finitely many, outside a double sector $|\mathrm{Re}\,\lambda| < \epsilon\,|\mathrm{Im}\,\lambda|$ containing the imaginary axis (cf. [**85**, Theorem 10.1.1]). The vector function

$$u(x) = \rho^{\lambda_0} \sum_{k=0}^{s} \frac{1}{k!}\, (\log\rho)^k\, U_{s-k}(\omega),$$

where $U_k \in \overset{\circ}{W}{}^{m,2}(\Omega)^\ell$ for $k = 0, 1, \dots, s$, is a solution of the equation $L(D_x)u = 0$ in $\mathcal{K}$ if and only if λ_0 is an eigenvalue of the pencil $\mathfrak{A}(\lambda)$ and $U_0, \dots, U_s$ is a Jordan chain corresponding to this eigenvalue (cf. [**85**, Theorem 1.1.5]).

3.2.3. The adjoint operator pencil. Let $L^+(D_x)$ be the formally adjoint differential operator to $L(D_x)$. We define the operator pencil $\mathfrak{A}^+(\lambda)$ by

$$\mathfrak{A}^+(\lambda)\, U(\omega) = \rho^{2m-\lambda}\, L^+(D_x)\, \big(\rho^\lambda U(\omega)\big), \quad U \in \overset{\circ}{W}{}^{m,2}(\Omega)^\ell.$$

By [**85**, Lemma 10.1.2], we have

$$\mathfrak{A}^+(2m-3-\bar\lambda) = \big(\mathfrak{A}(\lambda)\big)^*, \tag{3.2.4}$$

where $\big(\mathfrak{A}(\lambda)\big)^*$ is the adjoint operator of $\mathfrak{A}(\lambda)$. Hence λ_0 is an eigenvalue of the pencil $\mathfrak{A}(\lambda)$ if and only if $2m-3-\bar\lambda_0$ is an eigenvalue of the pencil $\mathfrak{A}^+(\lambda)$ (cf. [**85**, Theorem 1.1.7]).

3.2.4. Solvability of the parameter-dependent problem on Ω. The parameter-dependent differential operator (3.2.3) realizes a continuous mapping

$$V_\delta^{l,2}(\Omega)^\ell \cap \overset{\circ}{V}{}^{m,2}_{\delta-l+m}(\Omega)^\ell \to V_\delta^{l-2m,2}(\Omega)^\ell$$

for $l \ge 2m$ and $\delta \in \mathbb{R}$. We denote this operator by $\mathfrak{A}_{l,\delta}(\lambda)$.

THEOREM 3.2.1. *Suppose that the components of δ satisfy the inequalities*

$$-\delta_+^{(k)} < \delta_k - l + m < \delta_-^{(k)}\,. \tag{3.2.5}$$

Then the following assertions are true.

1) *There exist positive real constants C and ϵ such that the operator $\mathfrak{A}_{l,\delta}(\lambda)$ is an isomorphism for all λ in the double angle*

$$\{\lambda \in \mathbb{C} :\ |\lambda| > C,\ |\mathrm{Re}\,\lambda| < \epsilon\,|\mathrm{Im}\,\lambda|\}. \tag{3.2.6}$$

Furthermore, every solution $u \in V_\delta^{l,2}(\Omega)^\ell \cap \overset{\circ}{V}{}^{m,2}_{\delta-l+m}(\Omega)^\ell$ *of the equation* $\mathcal{L}(\lambda)u = f$ *with* λ *in the set* (3.2.6) *satisfies the estimate*

$$(3.2.7)\qquad \sum_{j=0}^{l} |\lambda|^j \, \|u\|_{V_\delta^{l-j,2}(\Omega)^\ell} \le c \sum_{j=0}^{l-2m} |\lambda|^j \, \|f\|_{V_\delta^{l-2m-j,2}(\Omega)^\ell}\,,$$

where c *is a constant independent of* u *and* λ.

2) *The spectra of the pencils* $\mathfrak{A}$ *and* $\mathfrak{A}_{l,\delta}$ *coincide.*

P r o o f. 1) For $k = 1, \dots, d$ let ζ_k be a smooth cut-off function on Ω with sufficiently small support equal to one in a neighborhood of the corner $M_k \cap S^2$. We consider the parameter-dependent differential operator $\mathcal{L}(\lambda)$ in a neighborhood of the corner $M_1 \cap S^2$. Without loss of generality, we may assume that M_1 coincides with the x_3-axis. By $\omega' = (\omega_1, \omega_2) = \big(x_1/\rho, x_2/\rho\big)$ we denote local coordinates on the sphere S^2 in a neighborhood of the north pole $N = M_1 \cap S^2$. Since

$$D_{x_1} = \omega_1 D_\rho + \frac{1-\omega_1^2}{\rho} D_{\omega_1} - \frac{\omega_1\omega_2}{\rho}\, D_{\omega_2},$$
$$D_{x_2} = \omega_2 D_\rho - \frac{\omega_1\omega_2}{\rho} D_{\omega_1} + \frac{1-\omega_2^2}{\rho}\, D_{\omega_2},$$
$$D_{x_3} = (1-\omega_1^2-\omega_2^2)^{1/2}\,\Big(D_\rho - \frac{\omega_1}{\rho}\, D_{\omega_1} - \frac{\omega_2}{\rho}\, D_{\omega_2}\Big),$$

the differential equation $\mathcal{L}(\lambda)u = f$ in Ω can be written in a neighborhood of the north pole as

$$L(D_{\omega'}, -i\lambda)\, u(\omega') + \tilde{L}(\omega', D_{\omega'}, \lambda)\, u(\omega') = f(\omega'), \quad \omega' \in \tilde{K},$$

where $\tilde{K}$ is a plane domain diffeomorphic to an angle K in a neighborhood of the point $\omega' = 0$, $L(D_x) = L(D_{x'}, D_{x_3})$ is the given differential operator of the problem (3.2.1), and $\tilde{L}$ is a differential operator of the form

$$\tilde{L}(\omega', D_{\omega'}, \lambda) = \sum_{|\alpha|\le 2m} B_\alpha(\omega')\, \lambda^{\alpha_3}\, D_{\omega_1}^{\alpha_1}\, D_{\omega_2}^{\alpha_2}\,.$$

Here the coefficients B_α are smooth near the point $\omega' = 0$ and satisfy the condition $B_\alpha(0) = 0$ for $|\alpha| = 2m$. Let ε be an arbitrarily small positive number. If the support of ζ_1 is sufficiently small and $|\lambda| > C$, where C is sufficiently large, then

$$|\lambda|^\nu \, \|\tilde{L}(\omega', D_{\omega'}, \lambda)\, (\zeta_1 u)\|_{V_\delta^{l-2m-\nu,2}(\Omega)^\ell} \le \varepsilon \sum_{j=0}^{l} |\lambda|^j \, \|\zeta_1 u\|_{V_\delta^{l-j,2}(\Omega)^\ell}$$

for $\nu = 0, \dots, l-2m$. This together with Corollary 2.3.21 implies

$$(3.2.8)\qquad \sum_{j=0}^{l} |\lambda|^j \, \|\zeta_1 u\|_{V_\delta^{l-j,2}(\Omega)^\ell} \le c \sum_{j=0}^{l-2m} |\lambda|^j \, \|\mathcal{L}(\lambda)\, (\zeta_1 u)\|_{V_\delta^{l-2m-j,2}(\Omega)^\ell}$$

if λ is purely imaginary and $|\lambda| > C$. Suppose that $\lambda = i\xi\, e^{i\theta}$, where ξ is real $|\xi| > C$, and $|\theta|$ is sufficiently small. Then obviously

$$\sum_{j=0}^{l-2m} |\lambda|^j \, \big\|\big(\mathcal{L}(\lambda) - \mathcal{L}(i\xi)\big)\, (\zeta_1 u)\big\|_{V_\delta^{l-2m-j,2}(\Omega)^\ell} \le \varepsilon \sum_{j=0}^{l} |\lambda|^j \|\zeta_1 u\|_{V_\delta^{l-j,2}(\Omega)^\ell}\,.$$

Since ε can be chosen arbitrarily small, it follows that (3.2.8) is also valid for complex λ of the form $i\xi\, e^{i\theta}$, where ξ is real $|\xi| > C$, and $|\theta|$ is sufficiently small. Analogously, we obtain

$$(3.2.9) \qquad \sum_{j=0}^{l} |\lambda|^j \, \|\zeta_k u\|_{V_\delta^{l-j,2}(\Omega)^\ell} \le c \sum_{j=0}^{l-2m} |\lambda|^j \, \|\mathcal{L}(\lambda)\,(\zeta_k u)\|_{V_\delta^{l-2m-j,2}(\Omega)^\ell}$$

for such λ and $k = 2,\dots,d$. Let $\zeta_0 = 1-\zeta_1-\cdots-\zeta_d$. Since $\zeta_0 = 0$ in a neighborhood of the corners of Ω, it follows from Theorem 1.1.10 that the estimate (3.2.9) is also valid for $k = 0$. Using (3.2.9) for $k = 0,\dots,d$ and the inequality

$$\|\mathcal{L}(\lambda)\,(\zeta_k u)\|_{V_\delta^{l-2m-\nu,2}(\Omega)^\ell} \le \|\zeta_k \mathcal{L}(\lambda) u\|_{V_\delta^{l-2m-\nu,2}(\Omega)^\ell} + \varepsilon \sum_{j=0}^{l} |\lambda|^{j-\nu} \|\zeta_k u\|_{V_\delta^{l-j,2}(\Omega)^\ell}$$

which is valid for sufficiently large $|\lambda|$, we obtain (3.2.6).

2) In particular, by the first part of the proof, the operator $\mathfrak{A}_{l,\delta}(\lambda)$ is injective for sufficiently large purely imaginary λ, and its range is closed. The operator $\mathfrak{A}(\lambda)$ is even an isomorphism for sufficiently large purely imaginary λ (see [**85**, Theorem 10.1.1]). By Lemma 2.3.2 and Theorem 2.3.4, every solution $u \in \overset{\circ}{V}{}_0^{m,2}(\Omega)^\ell$ of the equation $\mathcal{L}(\lambda)u = f$ belongs also the space $V_\delta^{l,2}(\Omega)^\ell$ if $f \in V_\delta^{l-2m,2}(\Omega)^\ell \cap W^{-m,2}(\Omega)^\ell$. Consequently, the set $V_\delta^{l-2m,2}(\Omega)^\ell \cap W^{-m,2}(\Omega)^\ell$ is contained in the range of the operator $\mathfrak{A}_{l,\delta}(\lambda)$ if $|\lambda|$ is purely imaginary and sufficiently large. This implies that $\mathfrak{A}_{l,\delta}(\lambda)$ is an isomorphism for those λ. Hence, the spectrum of the operator pencil $\mathfrak{A}_{l,\delta}(\lambda)$ consists only of eigenvalues with finite algebraic multiplicities (see [**85**, Theorem 1.1.1]).

We prove that the eigenvalues and eigenvectors of the pencils $\mathfrak{A}(\lambda)$ and $\mathfrak{A}_{l,\delta}(\lambda)$ coincide. Suppose that $u \in \ker \mathfrak{A}(\lambda)$, i.e. $u \in \overset{\circ}{V}{}_0^{m,2}(\Omega)^\ell$ and $\mathcal{L}(\lambda)u = 0$. Using Lemma 2.3.2 and Theorem 2.3.4, we obtain $\zeta_k u \in V_\delta^{l,2}(\Omega)^\ell$ for $k = 1,\dots,d$. Consequently, $u \in \ker \mathfrak{A}_{l,\delta}(\lambda)$. In the same way, it follows that every $u \in \ker \mathfrak{A}_{l,\delta}(\lambda)$ belongs also to $\overset{\circ}{V}{}_0^{m,2}(\Omega)^\ell$. This means that $\ker \mathfrak{A}(\lambda) = \ker \mathfrak{A}_{l,\delta}(\lambda)$, i.e. the pencils $\mathfrak{A}$ and $\mathfrak{A}_{l,\delta}$ have the same eigenvalues and eigenvectors. Analogously, the generalized eigenvectors of the pencils $\mathfrak{A}$ and $\mathfrak{A}_{l,\delta}$ coincide. □

3.3. Solvability in weighted L_2 Sobolev spaces

In this section, we establish a solvability and regularity theory for the boundary value problem

$$(3.3.1) \quad L(D_x)\,u = f \ \text{ in } \mathcal{K}, \quad \frac{\partial^{k-1} u}{\partial n^{k-1}} = g_{j,k} \ \text{ on } \Gamma_j,\ j = 1,\dots,d,\ k = 1,\dots,m,$$

in the class of the spaces $V_{\beta,\delta}^{l,2}(\mathcal{K})$. In particular, we show that the problem (3.3.1) has a unique solution in the space $V_{\beta,\delta}^{l,2}(\mathcal{K})^\ell$ for corresponding data f and $g_{j,k}$ if the line $\operatorname{Re}\lambda = l - \beta - 3/2$ does not contain eigenvalues of the pencil $\mathfrak{A}(\lambda)$ and the components of δ satisfy the condition (3.2.5). Here l is an arbitrary integer, $l \ge m$. It will be shown in Section 3.5 that these conditions on β and δ are also necessary for the existence and uniqueness of solutions in $V_{\beta,\delta}^{l,2}(\mathcal{K})^\ell$. Furthermore, we prove regularity assertions for the solutions and their ρ-derivatives.

3.3.1. Reduction to zero boundary data. The following lemma allows us to restrict ourselves to the Dirichlet problem (3.3.1) with zero boundary data $g_{j,k}$.

LEMMA 3.3.1. *Let $g_{j,k} \in V_{\beta,\delta}^{l-k+1-1/p,p}(\Gamma_j)$, $j=1,\dots,d$, $k=1,\dots,m$, $m \le l$. Then there exists a function $u \in V_{\beta,\delta}^{l,p}(\mathcal{K})$ such that*

$$\frac{\partial^{k-1} u}{\partial n^{k-1}} = g_{j,k} \quad \text{on } \Gamma_j,\ j=1,\dots,d,\ k=1,\dots,m, \tag{3.3.2}$$

and

$$\|u\|_{V_{\beta,\delta}^{l,p}(\mathcal{K})} \le c \sum_{j=1}^{d} \sum_{k=1}^{m} \|g_{j,k}\|_{V_{\beta,\delta}^{l-k+1-1/p,p}(\Gamma_j)} \tag{3.3.3}$$

with a constant c independent of the functions $g_{j,k}$.

P r o o f. Let ζ_ν be infinitely differentiable functions depending only on $\rho = |x|$ satisfying the conditions (3.1.4). Furthermore, let $\tilde{\zeta}_\nu(x) = \zeta_\nu(2^\nu x)$ and $g_{j,k}^{(\nu)}(x) = g_{j,k}(2^\nu x)$. Then $\tilde{\zeta}_\nu(x)\, g_{j,k}^{(\nu)}(x) = 0$ for $|x| > 2$ and $|x| < 1/2$. By Lemma 2.2.3, there exists a function $v_\nu \in V_{\beta,\delta}^{l,p}(\mathcal{K})$ vanishing outside the set $\{x : 1/4 < |x| < 4\}$ such that

$$\frac{\partial^{k-1} v_\nu}{\partial n^{k-1}} = 2^{(k-1)\nu}\, \tilde{\zeta}_\nu\, g_{j,k}^{(\nu)} \quad \text{on } \Gamma_j\,, j=1,\dots,d,\ k=1,\dots,m,$$

and

$$\|v_\nu\|_{V_{\beta,\delta}^{l,p}(\mathcal{K})} \le c \sum_{j=1}^{d} \sum_{k=1}^{m} 2^{(k-1)\nu}\, \|\tilde{\zeta}_\nu\, g_{j,k}^{(\nu)}\|_{V_{\beta,\delta}^{l-k+1-1/p,p}(\Gamma_j)} \tag{3.3.4}$$

with a constant c independent of the functions $g_{j,k}$ and ν. We define

$$u_\nu(x) = v_\nu(2^{-\nu} x) \quad \text{and} \quad u = \sum_{\nu=-\infty}^{+\infty} u_\nu\,. \tag{3.3.5}$$

Then, as in the proof of Lemma 2.2.1, we obtain (3.3.2) and (3.3.3). □

REMARK 3.3.2. The mapping $\{g_{j,k}\} \to u$ in Lemma 3.3.1 can be chosen independent of l, p, β, δ (cf. Remark 2.2.2). This means that there exists an operator $\{g_{j,k}\} \to u$ which satisfies (3.3.2) and realizes a linear and continuous mapping

$$\prod_{j=1}^{d} \prod_{k=1}^{m} V_{\beta,\delta}^{l-k+1-1/p,p}(\Gamma_j) \to V_{\beta,\delta}^{l,p}(\mathcal{K})$$

for all l, p, β, δ.

In the same way as for the dihedron (see Lemma 2.2.3), the result of Lemma 3.3.1 can be extended to more general Dirichlet systems on the faces Γ_j.

LEMMA 3.3.3. *Let $\{B_{j,k}\}_{k=1}^m$ be Dirichlet systems of order $m \le l$ on Γ_j, $j=1,\dots,d$, with infinitely differentiable coefficients. Furthermore, let $g_{j,k} \in V_{\beta,\delta}^{l-k+1-1/p,p}(\Gamma_j)$, $j=1,\dots,d$, $k=1,\dots,m$, be given functions with compact supports. Then there exists a function $u \in V_{\beta,\delta}^{l,p}(\mathcal{K})$ satisfying the equalities*

$$B_{j,k} u = g_{j,k} \quad \text{on } \Gamma_j \quad \text{for } j=1,\dots,d,\ k=1,\dots,m, \tag{3.3.6}$$

and the estimate (3.3.3).

P r o o f. Analogously to the proof of Lemma 2.2.3, we can construct functions $u_j \in V^{l,p}_{\beta,\delta}(\mathcal{K})$ satisfying the equalities

$$B_{j,k} u_j = g_{j,k} \quad \text{on } \Gamma_j \quad \text{for } k = 1, \dots, m,$$

and the estimate (3.3.3). There exist positively homogeneous (of degree 0) functions χ_j such that $\chi_j = 1$ in a neighborhood of an arbitrary point $x \in \Gamma_j$, $\chi_j = 0$ in a neighborhood of an arbitrary point $x \in \Gamma_k$, $k \neq j$, and

$$r^{|\alpha|} |\partial_x^\alpha \chi_j(x)| \leq c_\alpha \tag{3.3.7}$$

for every multi-index α, where c_α are constants independent of x. Obviously, the function $u = \chi_1 u_1 + \dots + \chi_d u_d$ satisfies (3.3.2). Furthermore, it follows from (3.3.7) that $\chi_j u_j \in V^{l,p}_{\beta,\delta}(\mathcal{K})$ and

$$\|\chi_j u_j\|_{V^{l,p}_{\beta,\delta}(\mathcal{K})} \leq c\, \|u_j\|_{V^{l,p}_{\beta,\delta}(\mathcal{K})}$$

with a constant c independent of u_j. Thus, the function u satisfies the estimate (3.3.3). □

3.3.2. An a priori estimate for the solution. Our goal is to obtain an estimate analogous to (2.2.14) for the solutions of the boundary value problem (3.3.1). First we prove a local estimate for the solution.

LEMMA 3.3.4. *Let ζ, η be infinitely differentiable cut-off function vanishing for $|x| > 8$ and $|x| < 1/8$ such that $\eta = 1$ in a neighborhood of supp ζ. Furthermore, let u be a solution of the boundary value problem* (3.3.1), *$\psi u \in W^{m,p}(\mathcal{D})^\ell$ for every $\psi \in C_0^\infty(\overline{\mathcal{K}}\backslash\mathcal{S})$. If $\eta u \in V^{l-1,p}_{\beta-1,\delta-1}(\mathcal{K})^\ell$, $l \geq m$, $\eta f \in V^{l-2m,p}_{\beta,\delta}(\mathcal{K})^\ell$, $l \geq m$, and $\eta g_{j,k} \in V^{l-k+1-1/p,p}_{\beta,\delta}(\Gamma_j)^\ell$, then $\zeta u \in V^{l,p}_{\beta,\delta}(\mathcal{K})^\ell$ and*

$$\|\zeta u\|_{V^{l,p}_{\beta,\delta}(\mathcal{K})^\ell} \leq c \Big(\|\eta f\|_{V^{l-2m,p}_{\beta,\delta}(\mathcal{K})^\ell} + \sum_{j=1}^{d} \sum_{k=1}^{m} \|\eta g_{j,k}\|_{V^{l-k+1-1/p,p}_{\beta,\delta}(\Gamma_j)^\ell} + \|\eta u\|_{V^{l-1,p}_{\beta-1,\delta-1}(\mathcal{K})^\ell} \Big). \tag{3.3.8}$$

Here, the constant c depends only on $\mathcal{K}, L(D_x), l, p, \beta, \delta$, the support of ζ, and on the maximum of the derivatives of ζ up to order l.

P r o o f. By Lemma 3.3.1, we may restrict ourselves to the case where $g_{j,k} = 0$ for all j and k. If the support of η does not contain any edge point, then the assertion of the lemma follows from Lemma 2.2.7. Suppose the support of η contains the edge point $x^{(0)} \in M_1$. There exist a neighborhood $\mathcal{U}$ of $x^{(0)}$ and a diffeomorphism κ such that $\kappa(x^{(0)}) = 0$, $\kappa'(x^{(0)}) = I$, and $\kappa(\mathcal{K} \cap \mathcal{U}) = \mathcal{D} \cap B_\varepsilon$, where B_ε is the ball with radius ε centered about the origin and $\mathcal{D}$ is a dihedron of the form (2.1.2). We assume that the support of η is sufficiently small and that in particular $\eta = 0$ outside $\mathcal{U}$. In the new coordinates $y = \kappa(x)$, the equation $L(D_x)u = f$ takes the form

$$\tilde{L}(y, D_y)\, u(y) = f(y) \quad \text{for } x \in \mathcal{D} \cap B_\varepsilon, \tag{3.3.9}$$

where $\tilde{L}(y, D_y)$ is a strongly elliptic differential operator of order $2m$ with infinitely differentiable coefficients. We can assume that the differential operator $\tilde{L}$ in (3.3.9) satisfies the conditions (2.9.3)–(2.9.4), since the coefficients of this operator coincide with the coefficients of the operator $\tilde{L}_\varepsilon(y, D_y)$ (see Remark 2.9.3) on $\mathcal{D} \cap B_\varepsilon$. Then it follows from Lemma 2.9.9 that $\zeta u \in V^{l,p}_{\beta,\delta}(\mathcal{K})^\ell$ and that ζu satisfies the estimate (3.3.8). This proves the lemma for the case where the supports of ζ and η are small. Since ζ can be written as a finite sum of smooth functions with small supports, the assertion of the lemma is true for all functions ζ and η satisfying the conditions of the lemma. □

A global estimate analogous to (3.3.8) is given in the next theorem.

THEOREM 3.3.5. *Let u be a solution of problem (3.3.1), $\psi u \in W^{m,p}(\mathcal{K})^\ell$ for every $\psi \in C_0^\infty(\overline{\mathcal{K}}\backslash\mathcal{S})$. If $u \in V^{l-1,p}_{\beta-1,\delta-1}(\mathcal{K})^\ell$, $l \ge m$, $f \in V^{l-2m,p}_{\beta,\delta}(\mathcal{K})^\ell$, and $g_{j,k} \in V^{l-k+1-1/p,p}_{\beta,\delta}(\Gamma_j)^\ell$, then $u \in V^{l,p}_{\beta,\delta}(\mathcal{K})^\ell$ and*

$$\|u\|_{V^{l,p}_{\beta,\delta}(\mathcal{K})^\ell} \le c\Big(\|f\|_{V^{l-2m,p}_{\beta,\delta}(\mathcal{K})^\ell} + \sum_{j=1}^{d}\sum_{k=1}^{m} \|g_{j,k}\|_{V^{l-k+1-1/p,p}_{\beta,\delta}(\Gamma_j)^\ell} + \|u\|_{V^{0,p}_{\beta-l,\delta-l}(\mathcal{K})^\ell}\Big). \tag{3.3.10}$$

P r o o f. By Lemma 3.3.1, we may restrict ourselves to the case of zero boundary data $g_{j,k}$. Let ζ_ν be infinitely differentiable functions depending only on $\rho = |x|$ and satisfying the conditions (3.1.4). Furthermore, let

$$\eta_\nu = \sum_{j=\nu-2}^{\nu+2} \zeta_j, \quad \tilde{\zeta}_\nu(x) = \zeta_\nu(2^\nu x) \ \text{ and } \tilde{\eta}_\nu(x) = \eta_\nu(2^\nu x).$$

The functions $\tilde{\zeta}_\nu$ and $\tilde{\eta}_\nu$ vanish for $|x| < 1/8$ and $|x| > 8$, and their derivatives are bounded by constants independent of ν. Moreover $\tilde{\eta}_\nu = 1$ in a neighborhood of the support of $\tilde{\zeta}_\nu$. The equation $L(D_x)u = f$ implies $L(D_x)\tilde{u} = \tilde{f}$, where $\tilde{u}(x) = u(2^\nu x)$ and $\tilde{f}(x) = 2^{2m\nu} f(2^\nu x)$. Therefore, it follows from Lemma 3.3.4 that $\tilde{\zeta}_\nu \tilde{u} \in V^{l,p}_{\beta,\delta}(\mathcal{K})^\ell$ and

$$\|\tilde{\zeta}_\nu \tilde{u}\|_{V^{l,p}_{\beta,\delta}(\mathcal{K})^\ell} \le c\Big(\|\tilde{\eta}_\nu \tilde{f}\|_{V^{l-2m,p}_{\beta,\delta}(\mathcal{K})^\ell} + \|\tilde{\eta}_\nu \tilde{u}\|_{V^{l-1,p}_{\beta-1,\delta-1}(\mathcal{K})^\ell}\Big), \tag{3.3.11}$$

where the constant c is independent of u and ν. Using the equality

$$\|\tilde{\zeta}_\nu \tilde{u}\|_{V^{l,p}_{\beta,\delta}(\mathcal{K})^\ell} = 2^{\nu(l-\beta-3/p)} \|\zeta_\nu u\|_{V^{l,p}_{\beta,\delta}(\mathcal{K})^\ell}$$

and analogous equalities for the norms of $\tilde{\eta}_\nu \tilde{f}$ and $\tilde{\eta}_\nu \tilde{u}$ in the spaces $V^{l-2m,p}_{\beta,\delta}(\mathcal{K})^\ell$ and $V^{l-1,p}_{\beta-1,\delta-1}(\mathcal{K})^\ell$, respectively, we obtain the estimate

$$\|\zeta_\nu u\|_{V^{l,p}_{\beta,\delta}(\mathcal{K})^\ell} \le c\Big(\|\eta_\nu f\|_{V^{l-2m,p}_{\beta,\delta}(\mathcal{K})^\ell} + \|\eta_\nu u\|_{V^{l-1,p}_{\beta-1,\delta-1}(\mathcal{K})^\ell}\Big).$$

Applying Lemma 3.1.2, we conclude that $u \in V^{l,p}_{\beta,\delta}(\mathcal{K})^\ell$ and

$$\|u\|_{V^{l,p}_{\beta,\delta}(\mathcal{K})^\ell} \le c\Big(\|f\|_{V^{l-2m,p}_{\beta,\delta}(\mathcal{K})^\ell} + \|u\|_{V^{l-1,p}_{\beta-1,\delta-1}(\mathcal{K})^\ell}\Big).$$

By Lemma 2.1.5, the $V^{l-1,p}_{\beta-1,\delta-1}(\mathcal{K})^\ell$-norm of $\tilde{\eta}_\nu \tilde{u}$ in (3.3.11) can be estimated by

$$\varepsilon\,\|\tilde{\eta}_\nu \tilde{u}\|_{V^{l,p}_{\beta,\delta}(\mathcal{K})^\ell} + C(\varepsilon)\,\|\tilde{\eta}_\nu \tilde{u}\|_{V^{0,p}_{\beta-l,\delta-l}(\mathcal{K})^\ell}$$

with an arbitrarily small positive ε. Consequently,

$$\|\zeta_\nu u\|_{V^{l,p}_{\beta,\delta}(\mathcal{K})^\ell} \le c\Big(\|\eta_\nu f\|_{V^{l-2m,p}_{\beta,\delta}(\mathcal{K})^\ell} + \varepsilon\,\|\eta_\nu u\|_{V^{l,p}_{\beta,\delta}(\mathcal{K})^\ell} + C(\varepsilon)\,\|\eta_\nu u\|_{V^{0,p}_{\beta-l,\delta-l}(\mathcal{K})^\ell}\Big).$$

Using again Lemma 3.1.2, we get (3.3.10). □

3.3.3. Solvability in the space $V^{l,2}_{\beta,\delta}(\mathcal{K})^\ell$, $l \ge 2m$. For the proof of the existence and uniqueness of solutions, we will employ the *Mellin transform* $\mathcal{M}_{\rho\to\lambda}$ with respect to the variable ρ. This transform is defined for a function $u \in C_0^\infty((0,\infty))$ by

$$\tilde{u}(\lambda) = (\mathcal{M}_{\rho\to\lambda}u)(\lambda) = \int_0^\infty \rho^{-\lambda-1}\, u(\rho)\, d\lambda. \tag{3.3.12}$$

The Mellin transform can be also defined by the equality

$$(\mathcal{M}_{\rho\to\lambda}u)(\lambda) = \mathcal{L}_{t\to\lambda}u(e^t),$$

where $\mathcal{L}_{t\to\lambda}$ denotes the *Laplace transform*,

$$(\mathcal{L}_{t\to\lambda}u)(\lambda) = \int_{-\infty}^{+\infty} e^{-\lambda t}\, u(t)\, dt.$$

The basic properties of the Mellin transform given in the next lemma result directly from the analogous properties of the Laplace transform (see e.g. [**84**, Lemma 5.2.3]).

LEMMA 3.3.6. 1) *The transformation* (3.3.12) *realizes a linear and continuous mapping from $C_0^\infty(\mathbb{R}_+)$ into the space $\mathcal{A}(\mathbb{C})$ of analytic functions on $\mathbb{C}$.*

2) *Every $u \in C_0^\infty(\mathbb{R}_+)$ satisfies the equality*

$$\mathcal{M}_{\rho\to\lambda}(\rho\partial_\rho u) = \lambda\, \mathcal{M}_{\rho\to\lambda}u.$$

3) *The* inverse Mellin transform *is given by*

$$u(\rho) = \frac{1}{2\pi i} \int_{-\beta-i\infty}^{-\beta+i\infty} \rho^\lambda\, \tilde{u}(\lambda)\, d\lambda. \tag{3.3.13}$$

4) *The transformation* (3.3.12) *can be extended to an isomorhism*

$$\{u : \rho^{\beta-1/2}u \in L_2(\mathbb{R}_+)\} \to L_2(\ell_{-\beta}),$$

where $\ell_{-\beta}$ denotes the line $\mathrm{Re}\,\lambda = -\beta$ *in the complex plane. Furthermore, the* Parseval equality

$$\int_0^\infty \rho^{2\beta-1}\, u(\rho)\, \overline{v(\rho)}\, d\rho = \frac{1}{2\pi i} \int_{-\beta-i\infty}^{-\beta+i\infty} \tilde{u}(\lambda)\, \overline{\tilde{v}(\lambda)}\, d\lambda \tag{3.3.14}$$

is satisfied for all $u, v \in C_0^\infty(\mathbb{R}_+)$.

5) *If $\rho^{\beta-1/2}u \in L_2(\mathbb{R}_+)$ and $\rho^{\beta'-1/2}u \in L_2(\mathbb{R}_+)$, $\beta < \beta'$, then $\tilde{u}$ is holomorphic in the strip $-\beta' < \mathrm{Re}\,\lambda < -\beta$.*

Applying the Mellin transform with respect to $\rho = |x|$ to the equation $L(D_x)u = f$, we obtain the equation

$$\mathcal{L}(\lambda)\, \tilde{u} = \tilde{F},$$

where $\mathcal{L}(\lambda)$ is the parameter-depending differential operator (3.2.3) and $\tilde{F}$ is the Mellin transform (with respect to ρ) of $F = \rho^{2m} f$. Using Theorem 3.2.1, we prove the following statement.

THEOREM 3.3.7. *Suppose that the line* $\text{Re}\,\lambda = l - \beta - 3/2$ *does not contain eigenvalues of the pencil* $\mathfrak{A}(\lambda)$ *and that the components of* δ *satisfy the inequalities* (3.2.5). *Then the boundary value problem* (3.3.1) *is uniquely solvable in* $V^{l,2}_{\beta,\delta}(\mathcal{K})^\ell$ *for arbitrary* $f \in V^{l-2,2}_{\beta,\delta}(\mathcal{K})^\ell$ *and* $g_{j,k} \in V^{l-k+1/2,2}_{\beta,\delta}(\Gamma_j)^\ell$, $l \ge 2m$.

P r o o f. By Lemma 3.3.1, we may assume without loss of generality that $g_{j,k} = 0$ for $j = 1, \dots, d$, $k = 1, \dots, m$. An equivalent norm in $V^{l,2}_{\beta,\delta}(\mathcal{K})$ is

$$|||u|||_{V^{l,2}_{\beta,\delta}(\mathcal{K})} = \Big(\int_0^\infty \sum_{j=0}^{l} \rho^{2(\beta-l+1)} \, \|(\rho\partial_\rho)^j u(\rho,\cdot)\|^2_{V^{l-j,2}_\delta(\Omega)} \, d\rho\Big)^{1/2}.$$

By Parseval's equality, the last norm is equal to

$$\Big(\frac{1}{2\pi i} \int\limits_{l-\beta-3/2-i\infty}^{l-\beta-3/2+i\infty} \sum_{j=0}^{l} \lambda^{2j} \, \|\tilde{u}(\lambda,\cdot)\|^2_{V^{l-j,2}_\delta(\Omega)} \, d\lambda\Big)^{1/2}.$$

Applying the Mellin transform with respect to $\rho = |x|$ to the equation $L(D_x)u = f$, we obtain the parameter-dependent problem

$$\mathfrak{A}_{l,\delta}(\lambda)\, \tilde{u}(\lambda,\omega) = \tilde{F}(\lambda,\omega) \quad \text{for } \omega \in \Omega, \tag{3.3.15}$$

where $\tilde{F}(\lambda,\cdot) = \mathcal{M}_{\rho\to\lambda}(\rho^{2m} f) \in V^{l-2m,2}_\delta(\Omega)^\ell$ for almost all λ. If the line $\text{Re}\,\lambda = l - \beta - 3/2$ is free of eigenvalues of the pencil $\mathfrak{A}(\lambda)$, then $\mathfrak{A}_{l,\delta}(\lambda)$ is an isomorphism

$$V^{l,2}_\delta(\Omega)^\ell \to V^{l-2m,2}_\delta(\Omega)^\ell$$

for every λ on this line, and the solution of (3.3.15) satisfies the estimate

$$\sum_{j=0}^{l} |\lambda|^{2j} \, \|\tilde{u}(\lambda,\cdot)\|^2_{V^{l-j,2}_\delta(\Omega)^\ell} \le c \sum_{j=0}^{l-2m} |\lambda|^{2j} \, \|\tilde{F}(\cdot,\omega)\|^2_{V^{l-2m-j,2}_\delta(\Omega)^\ell} \tag{3.3.16}$$

(cf. Theorem 3.2.1). Then the function

$$u(\rho,\omega) = \frac{1}{2\pi i} \int\limits_{l-\beta-3/2-i\infty}^{l-\beta-3/2+i\infty} \rho^\lambda \, \tilde{u}(\lambda,\omega)\, d\lambda.$$

solves the equation $L(D_x)u = f$. Furthermore by (3.3.16),

$$\|u\|_{V^{l,2}_{\beta,\delta}(\mathcal{K})^\ell} \le c\, \|f\|_{V^{l-2m,2}_{\beta,\delta}(\mathcal{K})^\ell}\,.$$

This proves the theorem. □

COROLLARY 3.3.8. *Suppose that* $l \ge m$, *the line* $\text{Re}\,\lambda = 2m - l + \beta - 3/2$ *does not contain eigenvalues of the pencil* $\mathfrak{A}(\lambda)$, *and the components of* δ *satisfy the inequalities*

$$-\delta_-^{(k)} < \delta_k - l + m < \delta_+^{(k)}\,.$$

Then for arbitrary $f \in V^{l,2}_{\beta,\delta}(\mathcal{K})^\ell$ *and* $g_{j,k} \in V^{l-k+1/2,2}_{\beta,\delta}(\Gamma_j)^\ell$, *there exists a uniquely determined solution* $u \in V^{l,2}_{\beta,\delta}(\mathcal{K})^\ell$ *of the problem*

$$L^+(D_x)u = f \ \text{ in } \mathcal{K}, \quad \frac{\partial^{k-1}u}{\partial n^{k-1}} = g_{j,k} \ \text{ on } \Gamma_j,$$

$j = 1, \dots, d$, $k = 1, \dots, m$.

P r o o f. According to Theorem 3.3.7, the assertion of the corollary is true if the line $\operatorname{Re}\lambda = l - \beta - 3/2$ does not contain eigenvalues of the pencil $\mathfrak{A}^+(\lambda)$. By (3.2.4), the last condition is equivalent to the absence of eigenvalues of the pencil $\mathfrak{A}(\lambda)$ on the line $\operatorname{Re}\lambda = 2m - l + \beta - 3/2$. □

3.3.4. Solvability in the space $V^{l,2}_{\beta,\delta}(\mathcal{K})^\ell$, $l \ge m$. The assertion of Theorem 3.3.7 can be also proved for $l \ge m$. This is done in the next theorem.

THEOREM 3.3.9. *Suppose that $l \ge m$, the line $\operatorname{Re}\lambda = l-\beta-3/2$ does not contain eigenvalues of the pencil $\mathfrak{A}(\lambda)$, and the components of δ satisfy the inequalities (3.2.5). Then for arbitrary $f \in V^{l,2}_{\beta,\delta}(\mathcal{K})^\ell$ and $g_{j,k} \in V^{l-k+1/2,2}_{\beta,\delta}(\Gamma_j)^\ell$, there exists a unique solution $u \in V^{l,2}_{\beta,\delta}(\mathcal{K})^\ell$ of the problem* (3.3.1).

P r o o f. First we prove the theorem for $l = m$ and for zero boundary data $g_{j,k}$. We show that there exists a constant c such that

$$\|u\|_{V^{m,2}_{\beta,\delta}(\mathcal{K})^\ell} \le c\, \|L(D_{x'},0)u\|_{V^{-m,2}_{\beta,\delta}(\mathcal{K})^\ell} \quad \text{for all } u \in \overset{\circ}{V}{}^{m,2}_{\beta,\delta}(\mathcal{K})^\ell. \tag{3.3.17}$$

If $u \in \overset{\circ}{V}{}^{m,2}_{\beta,\delta}(\mathcal{K})^\ell$, then

$$F \overset{def}{=} \rho^{2(\beta-m)} \prod_k \Big(\frac{r_k}{\rho}\Big)^{2(\delta_k-m)} u \in V^{0,2}_{-\beta+m,-\delta+m}(\mathcal{K})^\ell.$$

Consequently by Corollary 3.3.8, there exists a vector function $v \in \overset{\circ}{V}{}^{m,2}_{-\beta,-\delta}(\mathcal{K})^\ell \cap V^{2m,2}_{-\beta+m,-\delta+m}(\mathcal{K})^\ell$ satisfying the equation $L^+(D_x)v = F$ and the estimate

$$\begin{aligned} \|v\|_{V^{m,2}_{-\beta,-\delta}(\mathcal{K})^\ell} &\le \|v\|_{V^{2m,2}_{-\beta+m,-\delta+m}(\mathcal{K})^\ell} \\ &\le c\,\|F\|_{V^{0,2}_{-\beta+m,-\delta+m}(\mathcal{K})^\ell} = c\,\|u\|_{V^{0,2}_{\beta-m,\delta-m}(\mathcal{K})^\ell}\,. \end{aligned} \tag{3.3.18}$$

Thus,

$$\begin{aligned} \|u\|^2_{V^{0,2}_{\beta-m,\delta-m}(K)^\ell} &= \int_K u\cdot \overline{F}\, dx' = \int_{\mathcal{K}} u\cdot \overline{L^+(D_x)v}\, dx' \\ &= \big(L(D_x)u, v\big)_{\mathcal{K}} \le \|L(D_x)u\|_{V^{-m,2}_{\beta,\delta}(\mathcal{K})^\ell}\, \|v\|_{V^{m,2}_{-\beta,-\delta}(\mathcal{K})^\ell}\,. \end{aligned}$$

This together with (3.3.18) implies

$$\|u\|_{V^{0,2}_{\beta-m,\delta-m}(\mathcal{K})^\ell} \le c\,\|L(D_x)u\|_{V^{-m,2}_{\beta,\delta}(\mathcal{K})^\ell}.$$

Using Theorem 3.3.5, we obtain (3.3.17). Consequently, the kernel of the operator

$$L(D_x) : \overset{\circ}{V}{}^{m,2}_{\beta,\delta}(\mathcal{K})^\ell \to V^{-m,2}_{\beta,\delta}(\mathcal{K})^\ell \tag{3.3.19}$$

is trivial, and its range is closed. Since analogously the adjoint operator $L^+(D_x)$ has a trivial kernel in $\overset{\circ}{V}{}^{m,2}_{-\beta,-\delta}(\mathcal{K})^\ell$, we conclude that (3.3.19) is an isomorphism. This proves the theorem for $l = m$ and $g_{j,k} = 0$, $j = 1,\dots,d$, $k = 1,\dots,l$.

If $l > m$, then we obtain the assertion by means of Theorem 3.3.5. In the case of nonzero boundary data, it suffices to apply Lemma 3.3.1. □

3.3.5. Asymptotics of the solution. In Theorem 1.2.6, an asymptotic formula for the solutions of elliptic boundary value problems in a neighborhood of the vertex of a "smooth" cone was given. We prove an analogous result for the solution of the Dirichlet problem in a cone with edges.

THEOREM 3.3.10. *Let* $u \in V^{l,2}_{\beta,\delta}(\mathcal{K})^\ell$ *be a solution of the boundary value problem* (3.3.1), *where*

$$f \in V^{l-2m,2}_{\beta,\delta}(\mathcal{K})^\ell \cap V^{l'-2m,2}_{\beta',\delta'}(\mathcal{K})^\ell, \quad g_{j,k} \in V^{l-k+1/2,2}_{\beta,\delta}(\Gamma_j)^\ell \cap V^{l'-k+1/2,2}_{\beta',\delta'}(\Gamma_j)^\ell,$$

$l \geq 2m$, $l' \geq 2m$. *Suppose that the lines* $\operatorname{Re}\lambda = l-\beta-3/2$ *and* $\operatorname{Re}\lambda = l'-\beta'-3/2$ *do not contain eigenvalues of the pencil* $\mathfrak{A}(\lambda)$ *and that the components of* δ *and* δ' *satisfy the inequalities*

$$-\delta^{(k)}_+ < \delta_k - l + m < \delta^{(k)}_-, \quad -\delta^{(k)}_+ < \delta'_k - l' + m < \delta^{(k)}_- .$$

Then u *admits the decomposition*

$$u = \sum_{\nu=1}^{N}\sum_{j=1}^{I_\nu}\sum_{s=0}^{\kappa_{\nu,j}-1} c_{\nu j s}\, \rho^{\lambda_\nu} \sum_{\sigma=0}^{s} \frac{1}{\sigma!} (\log\rho)^\sigma\, U^{\nu,j,s-\sigma}(\omega) + v, \tag{3.3.20}$$

where v *is the uniquely determined solution of the boundary value problem* (3.3.1) *in the space* $V^{l',2}_{\beta',\delta'}(\mathcal{K})^\ell$, λ_ν *are the eigenvalues of the pencil* $\mathfrak{A}(\lambda)$ *between the lines* $\operatorname{Re}\lambda = l-\beta-3/2$ *and* $\operatorname{Re}\lambda = l'-\beta'-3/2$ *and* $U^{\nu,j,s}$ *are eigenvectors* ($s=0$) *and generalized eigenvectors* ($s>0$) *corresponding to the eigenvalue* λ_ν.

P r o o f. Lemma 3.3.1 and Remark 3.3.2 allow us to restrict ourselves in the proof to the case of zero boundary data $g_{j,k}$. Then the uniquely determined solution u in $V^{l,2}_{\beta,\delta}(\mathcal{K})^\ell$ is given by the formula

$$u(\rho,\omega) = \frac{1}{2\pi i} \int\limits_{l-\beta-3/2-i\infty}^{l-\beta-3/2+i\infty} \rho^\lambda\, (\mathfrak{A}_{l,\delta}(\lambda))^{-1}\, \tilde{F}(\lambda,\omega)\, d\lambda,$$

where $\tilde{F}(\lambda,\omega)$ is the Mellin transform of $F = \rho^{2m} f$ and $\mathfrak{A}_{l,\delta}(\lambda)$ is the operator considered in Theorem 3.2.1. By our assumptions on the eigenvalues of the pencil $\mathfrak{A}(\lambda)$, both $\mathfrak{A}_{l,\delta}(\lambda)$ and $\mathfrak{A}_{l',\delta'}(\lambda)$ are isomorphisms for all λ in the closed strip between the lines $\operatorname{Re}\lambda = l-\beta-3/2$ and $\operatorname{Re}\lambda = l'-\beta'-3/2$. From the conditions on f it follows that $\tilde{F}(\lambda,\cdot) \in V^{l-2m,2}_{\delta}(\Omega)^\ell \cap V^{l'-2m,2}_{\delta'}(\Omega)^\ell$. Thus by Corollary 1.2.7, the vector function $(\mathfrak{A}_{l,\delta}(\lambda))^{-1}\, \tilde{F}(\lambda,\cdot)$ belongs to $V^{l',2}_{\delta'}(\Omega)^\ell \cap \overset{\circ}{V}{}^{m,2}_{\delta'-l'+m}(\Omega)^\ell$. This means that

$$(\mathfrak{A}_{l,\delta}(\lambda))^{-1}\, \tilde{F}(\lambda,\cdot) = (\mathfrak{A}_{l',\delta'}(\lambda))^{-1}\, \tilde{F}(\lambda,\cdot).$$

Furthermore, $\tilde{F}$ is holomorphic in the strip between the lines $\operatorname{Re}\lambda = l-\beta-3/2$ and $\operatorname{Re}\lambda = l'-\beta'-3/2$. Hence, the only singularities of $\rho^\lambda\, (\mathfrak{A}_{l,\delta}(\lambda))^{-1}\tilde{F}(\lambda,\omega)$ between the lines $\operatorname{Re}\lambda = l-\beta-3/2$ and $\operatorname{Re}\lambda = l'-\beta'-3/2$ are the poles of $(\mathfrak{A}_{l,\delta}(\lambda))^{-1}$, i.e. the eigenvalues $\lambda_1,\dots,\lambda_N$ of the pencil $\mathfrak{A}(\lambda)$ (see Theorem 3.2.1). Since the integral

$$\int\limits_{l-\beta-3/2}^{l'-\beta'-3/2} \rho^{t+iR}\, (\mathfrak{A}_{l,\delta}(t+iR))^{-1}\, \tilde{F}(t+iR,\omega)\, dt$$

tends to zero as $R \to \pm\infty$ (cf. [**84**, Lemma 5.4.1]), Cauchy's formula yields

$$u = \sum_{\nu=1}^{N} \operatorname{Res} \rho^{\lambda} \left(\mathfrak{A}_{l,\delta}(\lambda)\right)^{-1} \tilde{F}(\lambda,\omega)\big|_{\lambda=\lambda_\nu} + v,$$

where

$$v(\rho,\omega) = \frac{1}{2\pi i} \int\limits_{l'-\beta'-3/2-i\infty}^{l'-\beta'-3/2+i\infty} \rho^{\lambda} \left(\mathfrak{A}_{l',\delta'}(\lambda)\right)^{-1} \tilde{F}(\lambda,\omega)\, d\lambda$$

is the solution of problem (3.3.1) in the space $V^{l',2}_{\beta',\delta'}(\mathcal{K})^\ell$. Using the representation

$$\left(\mathfrak{A}_{l,\delta}(\lambda)\right)^{-1} = \sum_{j=1}^{I_\nu} \sum_{s=0}^{\kappa_{\nu,j}-1} \frac{P^{(\nu)}_{j,s}}{(\lambda-\lambda_\nu)^{\kappa_{\nu,j}-s}} + \Gamma_\nu(\lambda)$$

for the resolvent in a neighborhood of λ_ν, where Γ_ν is holomorphic in a neighborhood of λ_ν and $P^{(\nu)}_{j,s}$ are mappings into the space of eigenvectors and generalized eigenvectors (see [**81**, Theorem 1.2.1]), we obtain the desired decomposition for u. □

Next, we generalize the last theorem to arbitrary integer $l \geq m$.

THEOREM 3.3.11. *Let $u \in V^{m,2}_{\beta,\delta}(\mathcal{K})^\ell$ be a solution of the boundary value problem* (3.3.1), *where*

$$f \in V^{-m,2}_{\beta,\delta}(\mathcal{K})^\ell \cap V^{l-2m,2}_{\beta',\delta'}(\mathcal{K})^\ell, \quad g_{j,k} \in V^{m-k+1/2,2}_{\beta,\delta}(\Gamma_j)^\ell \cap V^{l-k+1/2,2}_{\beta',\delta'}(\Gamma_j)^\ell,$$

$l \geq m$. Suppose that the lines $\operatorname{Re}\lambda = m-\beta-3/2$ and $\operatorname{Re}\lambda = l-\beta'-3/2$ do not contain eigenvalues of the pencil $\mathfrak{A}(\lambda)$ and that the components of δ and δ' satisfy the inequalities

$$(3.3.21) \qquad -\delta_+^{(k)} < \delta_k < \delta_-^{(k)}, \quad -\delta_+^{(k)} < \delta'_k - l + m < \delta_-^{(k)}.$$

Then u admits the decomposition (3.3.20), *where $v \in V^{l,2}_{\beta',\delta'}(\mathcal{K})^\ell$, λ_ν are the eigenvalues of the pencil $\mathfrak{A}(\lambda)$ between the lines $\operatorname{Re}\lambda = m-\beta-3/2$ and $\operatorname{Re}\lambda = l-\beta'-3/2$ and $U^{\nu,j,s}$ are eigenvectors and generalized eigenvectors corresponding to the eigenvalue λ_ν.*

P r o o f. Suppose first that $l = m$. Then by Theorem 3.3.9, there exists a unique solution $v \in V^{m,2}_{\beta',\delta'}(\mathcal{K})^\ell$ of the Dirichlet problem (3.3.1). We have to show that

$$u - v = \Sigma,$$

where Σ denotes the sum on the right-hand side of (3.3.20). Without loss of generality, we may assume that $\beta \geq \beta'$. We put $\gamma_k = \max(\delta_k, \delta'_k)$ for $k = 1,\dots,d$. If χ is an arbitrary infinitely differentiable function on $\overline{\mathcal{K}}$ with compact support vanishing in a neighborhood of the origin, then

$$\chi(u-v) \in V^{2m,2}_{\kappa,\gamma+m}(\mathcal{K})^\ell$$

for each $\kappa \in \mathbb{R}$. Indeed, we have $L(D_x)\,(u-v) = 0$ and $\partial_n^k(u-v)|_{\Gamma_j} = 0$ for $k = 0,\dots,m-1$. Since the commutator $[L,\chi] = L\chi - \chi L$ is a differential operator of order $2m-1$ with coefficients vanishing outside $\operatorname{supp}\chi$, we conclude that

$$L(D_x)\big(\chi(u-v)\big) = [L(D_x),\chi]\,(u-v) \in V^{-m+1,2}_{\kappa,\gamma+1}(\mathcal{K})^\ell$$

and $\partial_n^k\big(\chi(u-v)\big) = 0$ on Γ_j for $k = 1, \dots, m$. This together with Theorem 3.3.5 implies $\chi(u-v) \in V^{m+1,2}_{\kappa,\gamma+1}(\mathcal{K})^\ell$. Repeating this argument, we obtain $\chi(u-v) \in V^{2m,2}_{\kappa,\gamma+m}(\mathcal{K})^\ell$ for arbitrary $\chi \in C_0^\infty(\overline{\mathcal{K}}\backslash\{0\})$, $\kappa \in \mathbb{R}$.

Now let ζ be an arbitrary infinitely differential function with compact support equal to one in a neighborhood of the origin. Furthermore, let χ be a function in $C_0^\infty(\overline{\mathcal{K}}\backslash\{0\})$ equal to one on the support of $\nabla\zeta$. Then

$$\zeta(u-v) \in V^{m,2}_{\beta,\gamma}(\mathcal{K})^\ell, \qquad (1-\zeta)\,(u-v) \in V^{m,2}_{\beta',\gamma}(\mathcal{K})^\ell$$

and

$$L(D_x)\,\big(\zeta(u-v)\big) = [L(D_x), \zeta]\,\big(\chi(u-v)\big) \in V^{0,2}_{\kappa,\gamma+m}(\mathcal{K})^\ell$$

with arbitrary $\kappa \in \mathbb{R}$. Consequently by Theorem 3.3.5, $\zeta(u-v) \in V^{2m,2}_{\beta+m,\gamma+m}(\mathcal{K})^\ell$. Using Theorem 3.3.10, we obtain

$$\zeta(u-v) = \Sigma + w,$$

where $w \in V^{2m,2}_{\beta'+m,\gamma+m}(\mathcal{K})^\ell$. This implies

$$u = \Sigma + w', \quad \text{where} \quad w' = v + (1-\zeta)\,(u-v) + w \in V^{m,2}_{\beta',\gamma}(\mathcal{K})^\ell.$$

Since

$$L(D_x)w' = L(D_x)u - L(D_x)\Sigma = f \ \text{ in } \mathcal{K}, \qquad \frac{\partial^{k-1}w'}{\partial n^{k-1}} = g_{j,k} \text{ on } \Gamma_j$$

for $j = 1, \dots, d$, $k = 1, \dots, m$, and the solution of the problem (3.3.1) is unique in $V^{m,2}_{\beta',\gamma}(\mathcal{K})^\ell$ (cf. Theorem 3.3.9), we conclude that $w' = v$. This proves the theorem for the case $l = m$.

If $l > m$, then it follows from the first part of the proof, that $u = \Sigma + v$, where $v \in V^{m,2}_{\beta'-l+m,\delta'-l+m}(\mathcal{K})^\ell$. Applying Theorem 3.3.5, we obtain $v \in V^{l,2}_{\beta',\delta'}(\mathcal{K})^\ell$. The proof of the theorem is complete. □

3.3.6. Regularity assertions for the solution. As a consequence of the last theorem, the following regularity result holds.

COROLLARY 3.3.12. *Let $u \in V^{m,2}_{\beta,\delta}(\mathcal{K})^\ell$ be a solution of the boundary value problem* (3.3.1), *where*

$$f \in V^{-m,2}_{\beta,\delta}(\mathcal{K})^\ell \cap V^{l-2m,2}_{\beta',\delta'}(\mathcal{K})^\ell, \quad g_{j,k} \in V^{m-k+1/2,2}_{\beta,\delta}(\Gamma_j)^\ell \cap V^{l-k+1/2,2}_{\beta',\delta'}(\Gamma_j)^\ell,$$

$l \ge m$. If the closed strip between the lines $\operatorname{Re}\lambda = m-\beta-3/2$ *and* $\operatorname{Re}\lambda = l-\beta'-3/2$ *is free of eigenvalues of the pencil $\mathfrak{A}(\lambda)$ and the components of δ and δ' satisfy the inequalities* (3.3.21), *then $u \in V^{l,2}_{\beta',\delta'}(\mathcal{K})^\ell$.*

An analogous result holds for the ρ-derivatives of the solutions.

THEOREM 3.3.13. *Let $u \in \overset{\circ}{V}{}^{m,2}_{\beta,\delta}(\mathcal{K})^\ell$ be a solution of the equation $L(D_x)u = f$ in $\mathcal{K}$, where $(\rho\partial_\rho)^j f \in V^{l-2m,2}_{\beta',\delta'}(\mathcal{K})^\ell$ for $j = 0, 1, \dots, k$. Suppose that $l \ge 2m$, the closed strip between the lines* $\operatorname{Re}\lambda = m-\beta-3/2$ *and* $\operatorname{Re}\lambda = l-\beta'-3/2$ *is free of eigenvalues of the pencil $\mathfrak{A}(\lambda)$, and the components of δ and δ' satisfy the inequalities* (3.3.21). *Then $(\rho\partial_\rho)^j u \in V^{l,2}_{\beta',\delta'}(\mathcal{K})^\ell$ for $j = 0, \dots, k$ and*

$$\sum_{j=0}^{k} \|(\rho\partial_\rho)^j u\|_{V^{l,2}_{\beta',\delta'}(\mathcal{K})^\ell} \le c \sum_{j=0}^{k} \|(\rho\partial_\rho)^j f\|_{V^{l-2m,2}_{\beta',\delta'}(\mathcal{K})^\ell}$$

with a constant c independent of u.

P r o o f. From Corollary 3.3.12 we conclude that $u \in V^{l,2}_{\beta',\delta'}(\mathcal{K})^\ell$. We assume that $\rho\partial\rho f \in V^{l-2m,2}_{\beta',\delta'}(\mathcal{K})^\ell$ and consider the function

$$u_t(x) = \frac{u(x) - u(tx)}{1-t}$$

for positive $t \neq 1$. Obviously, $u_t \in V^{l,2}_{\beta',\delta'}(\mathcal{K})^\ell$,

$$u_t(x) \to \sum_{j=1}^{3} x_j\, \partial_{x_j} u(x) = \rho\partial_\rho u(x) \quad \text{as } t \to 1,$$

and

$$L(D_x)\, u_t(x) = f_t(x) + \frac{1-t^{2m}}{1-t}\, f(tx) \quad \text{for } x \in \mathcal{K}.$$

Moreover, u_t satisfies the homogeneous Dirichlet conditions on the faces of $\mathcal{K}$. Therefore by Theorem 3.3.7, u_t satisfies the estimate

$$(3.3.22) \quad \|u_t\|_{V^{2m,2}_{\beta'-l+2m,\delta'-l+2m}(\mathcal{K})^\ell} \le c \Big\| f_t(\cdot) + \frac{1-t^{2m}}{1-t}\, f(t\cdot)\Big\|_{V^{0,2}_{\beta'-l+2m,\delta'-l+2m}(\mathcal{K})^\ell},$$

where c is independent of u and t. Here,

$$\|f(t\cdot)\|_{V^{0,2}_{\beta'-l+2m,\delta'-l+2m}(\mathcal{K})^\ell} = t^{l-2m-\beta'-3/2}\, \|f\|_{V^{0,2}_{\beta'-l+2m,\delta'-l+2m}(\mathcal{K})^\ell}.$$

Using the representation

$$\begin{aligned} f_t(x) &= \frac{1}{1-t}\int_0^1 \frac{\partial}{\partial\tau} f\big((t+\tau-t\tau)x\big)\, d\tau = \int_0^1 \sum_{j=1}^{3} x_j\, (\partial_{x_j} f)\big((t+\tau-t\tau)x\big)\, d\tau \\ &= \int_0^1 (\rho\partial_\rho f)\big((t+\tau-t\tau)x\big)\, d\tau, \end{aligned}$$

we obtain

$$\|f_t\|_{V^{0,2}_{\beta'-l+2m,\delta'-l+2m}(\mathcal{K})^\ell} \le \max(1, t^{l-2m-\beta'-3/2})\|\rho\partial_\rho f\|_{V^{0,2}_{\beta'-l+2m,\delta'-l+2m}(\mathcal{K})^\ell}.$$

Thus, it follows from (3.3.22) that $\rho\partial_\rho u \in V^{2m,2}_{\beta'-l+2m,\delta'-l+2m}(\mathcal{K})^\ell$ and

$$\|\rho\partial_\rho u\|_{V^{2m,2}_{\beta'-l+2m,\delta'-l+2m}(\mathcal{K})^\ell} \le c \sum_{j=0}^{1} \|(\rho\partial_\rho)^j f\|_{V^{0,2}_{\beta'-l+2m,\delta'-l+2m}(\mathcal{K})^\ell}.$$

Since $\rho\partial_\rho u$ is a solution of the equation

$$L(D_x)\, \rho\partial_\rho u = (\rho\partial_\rho + 2m) f,$$

Theorem 3.3.5 implies $\rho\partial_\rho u \in V^{l,2}_{\beta',\delta'}(\mathcal{K})^\ell$ and

$$\|\rho\partial_\rho u\|_{V^{l,2}_{\beta',\delta'}(\mathcal{K})^\ell} \le c\Big(\|f\|_{V^{l-2m,2}_{\beta',\delta'}(\mathcal{K})^\ell} + \|\rho\partial_\rho f\|_{V^{l-2m,2}_{\beta',\delta'}(\mathcal{K})^\ell}\Big)$$

This proves the theorem for $k = 1$. For $k > 1$, the assertion follows by induction. □

Using the last theorem, we prove the following local regularity result for the ρ-derivatives.

COROLLARY 3.3.14. *Let ζ, η be infinitely differentiable functions in $\overline{\mathcal{K}}$ with compact supports such that $\eta = 1$ in a neighborhood of* $\operatorname{supp}\zeta$. *Furthermore, let u be a solution of the equation $L(D_x)u = f$ in $\mathcal{K}$ such that $\eta u \in \overset{\circ}{V}{}^{m,2}_{\beta,\delta}(\mathcal{K})^\ell$ and $\eta(\rho\partial_\rho)^j f \in V^{l-2m,2}_{\beta',\delta'}(\mathcal{K})^\ell$ for $j = 0, 1, \dots, k$. Suppose that $l \ge 2m$, the closed strip between the lines $\operatorname{Re}\lambda = m - \beta - 3/2$ and $\operatorname{Re}\lambda = l - \beta' - 3/2$ is free of eigenvalues of the pencil $\mathfrak{A}(\lambda)$, and the components of δ and δ' satisfy the inequalities* (3.3.21). *Then $\zeta(\rho\partial_\rho)^j u \in V^{l,2}_{\beta',\delta'}(\mathcal{K})^\ell$ for $j = 0, \dots, k$ and*

$$\sum_{j=0}^{k} \|\zeta(\rho\partial_\rho)^j u\|_{V^{l,2}_{\beta',\delta'}(\mathcal{K})^\ell} \le c\Big(\sum_{j=0}^{k} \|\eta(\rho\partial_\rho)^j f\|_{V^{l-2m,2}_{\beta',\delta'}(\mathcal{K})^\ell} + \|\eta u\|_{V^{m,2}_{\beta,\delta}(\mathcal{K})^\ell}\Big).$$

with a constant c independent of u.

P r o o f. Let χ be an infinitely differentiable function such that $\chi = 1$ in a neighborhood of $\operatorname{supp}\zeta$ and $\eta = 1$ in a neighborhood of $\operatorname{supp}\chi$. By means of Corollary 3.3.12, it can be easily proved (see the proof of Corollary 2.4.7) that $\chi u \in V^{l,2}_{\beta',\delta'}(\mathcal{K})^\ell$ and

$$\|\chi u\|_{V^{l,2}_{\beta',\delta'}(\mathcal{K})^\ell} \le c\Big(\|\eta f\|_{V^{l-2m,2}_{\beta',\delta'}(\mathcal{K})^\ell} + \|\eta u\|_{V^{m,2}_{\beta,\delta}(\mathcal{K})}\Big). \tag{3.3.23}$$

Suppose that $\eta\rho\partial_\rho f \in V^{l-2m,2}_{\beta',\delta'}(\mathcal{K})^\ell$. Since the commutator $[L(D_x),\zeta] = L(D_x)\zeta - \zeta L(D_x)$ is a differential operator of order $2m-1$, we conclude that

$$\rho\partial_\rho L(D_x)\,(\zeta u) = \zeta\,\rho\partial_\rho f + \rho(\partial_\rho\zeta)\,f + \rho\partial_\rho[L(D_x),\zeta]u \in V^{l-2m,2}_{\beta',\delta'}(\mathcal{K})^\ell\,.$$

Thus by Theorem 3.3.11, $\rho\partial_\rho(\zeta u) \in V^{l,2}_{\beta',\delta'}(\mathcal{K})^\ell$ and

$$\|\rho\partial_\rho(\zeta u)\|_{V^{l,2}_{\beta',\delta'}(\mathcal{K})^\ell} \le c\Big(\|\eta f\|_{V^{l-2m,2}_{\beta',\delta'}(\mathcal{K})^\ell} + \|\eta\rho\partial_\rho f\|_{V^{l-2m,2}_{\beta',\delta'}(\mathcal{K})^\ell} + \|\chi u\|_{V^{l,2}_{\beta',\delta'}(\mathcal{K})^\ell}\Big).$$

Together with (3.3.23) this implies

$$\|\zeta\rho\partial_\rho u\|_{V^{l,2}_{\beta',\delta'}(\mathcal{K})^\ell} \le c\Big(\|\eta f\|_{V^{l-2m,2}_{\beta',\delta'}(\mathcal{K})^\ell} + \|\eta\rho\partial_\rho f\|_{V^{l-2m,2}_{\beta',\delta'}(\mathcal{K})^\ell} + \|\eta u\|_{V^{m,2}_{\beta,\delta}(\mathcal{K})^\ell}\Big).$$

This proves the corollary for $k = 1$. For $k > 1$ the assertion holds by induction. □

3.4. Green's matrix of the Dirichlet problem in a cone

In this section, we deal with the Green's matrix of the boundary value problem (3.3.1). The $\ell \times \ell$ matrix $G(x,\xi)$ is called *Green's matrix* for this problem if

$$L(D_x)\,G(x,\xi) = \delta(x-\xi)\,I_\ell \quad \text{for } x,\xi \in \mathcal{K}, \tag{3.4.1}$$

$$\frac{\partial^k G(x,\xi)}{\partial n_x^k} = 0 \quad \text{for } x \in \Gamma_j,\ \xi \in \mathcal{K},\ k = 0,\dots,m-1,\ j = 1,\dots,d. \tag{3.4.2}$$

Here I_ℓ denotes the $\ell \times \ell$ identity matrix. By Theorem 3.3.9, we can consider solutions $G(x,\xi)$ of the problem (3.4.1), (3.4.2) which belong to the space $\overset{\circ}{V}{}^{m,2}_{\kappa,0}(\mathcal{K})^{\ell\times\ell}$ outside an arbitrarily small neighborhood of the point $x = \xi$ provided the line $\operatorname{Re}\lambda = m - \kappa - 3/2$ does not contain eigenvalues of the pencil $\mathfrak{A}(\lambda)$. So we obtain different Green's matrices for different κ. The main goal of this section is to prove point estimates for the matrix $G(x,\xi)$. We consider the cases $|x|/2 < |\xi| < 2|x|$, $|x| > 2|\xi|$ and $|\xi| > 2|x|$ separately.

3.4.1. Existence of Green's matrix. First we prove the existence and some basic properties of Green's matrix.

THEOREM 3.4.1. *Let κ be an arbitrary real number such that the line* $\text{Re}\,\lambda = m-\kappa-3/2$ *is free of eigenvalues of the pencil $\mathfrak{A}(\lambda)$. Then the following assertions are true.*

1) *There exists a unique Green matrix such that the function $x \to \zeta(x,\xi)\, G(x,\xi)$ belongs to the space $V^{m,2}_{\kappa,0}(\mathcal{K})^{\ell\times\ell}$ for every $\xi \in \mathcal{K}$ and for every smooth function $\zeta(\cdot,\xi)$ which is equal to zero in a neighborhood of the point $x = \xi$ and bounded together with all derivatives.*

2) *The equality*

$$G(tx,t\xi) = t^{2m-3}\, G(x,\xi) \tag{3.4.3}$$

is valid for all $x,\xi \in \mathcal{D}$, $t > 0$.

3) *The adjoint matrix $G^*(x,\xi)$ is the unique solution of the problem*

$$L^+(D_\xi)\, G^*(x,\xi) = \delta(x-\xi)\, I_\ell \quad \text{for } x,\xi \in \mathcal{K}, \tag{3.4.4}$$

$$\frac{\partial^k G^*(x,\xi)}{\partial n_\xi^k} = 0 \quad \text{for } \xi \in \Gamma_j,\ x \in \mathcal{K},\ k = 0,\dots,m-1,\ j = 1,\dots,d, \tag{3.4.5}$$

such that the function $\xi \to \zeta(x,\xi)\, G^*(x,\xi)$ belongs to the space $V^{m,2}_{-\kappa,0}(\mathcal{K})^{\ell\times\ell}$ for every smooth function ζ which is equal to zero in a neighborhood of the point $\xi = x$ and bounded together with all derivatives.

4) *The solution $u \in \overset{\circ}{V}{}^{m,2}_{\kappa,0}(\mathcal{K})^\ell$ of the equation $L(D_x)\, u = f$ in $\mathcal{K}$ admits the representation*

$$u(x) = \int_{\mathcal{K}} G(x,\xi)\, f(\xi)\, d\xi \tag{3.4.6}$$

for arbitrary $f \in V^{-m,2}_{\kappa,0}(\mathcal{K})^\ell$.

P r o o f. 1) There exists a solution $\mathcal{G}(x,\xi)$ of the equation

$$L(D_x)\, \mathcal{G}(x,\xi) = \delta(x-\xi)\, I_\ell \quad \text{in } \mathbb{R}^3$$

which is real analytic for $x \neq \xi$ and has the form $\mathcal{G}(x,\xi) = h(x-\xi)\, |x-\xi|^{2m-3}$, where h is positively homogeneous of degree 0 (see [**13**, Part 2, Section 5.1]). Let χ be a smooth function on $(0,\infty)$, $\chi(t) = 1$ for $t < 1/4$, $\chi(t) = 0$ for $t > 1/2$. We put $\psi(x,\xi) = \chi(|x-\xi|/r(\xi))$ and define $R(x,\xi)$ for every fixed $\xi \in \mathcal{K}$ as the unique solution (in the space $V^{m,2}_{\kappa,0}(\mathcal{K})^{\ell\times\ell}$) of the problem

$$L(D_x)\, R(x,\xi) = \delta(x-\xi)\, I_\ell - L(D_x)\big(\psi(x,\xi)\, \mathcal{G}(x,\xi)\big) \quad \text{for } x \in \mathcal{K}, \tag{3.4.7}$$

$$\frac{\partial^k R(x,\xi)}{\partial n_x^k} = -\frac{\partial^k (\psi(x,\xi)\mathcal{G}(x,\xi))}{\partial n_x^k} \quad \text{for } x \in \Gamma_j,\ k = 0,\dots,m-1, \tag{3.4.8}$$

$j = 1,\dots,d$. Obviously for every fixed $\xi \in \mathcal{K}$, the right-hand sides of (3.4.7), (3.4.8) are smooth functions on $\overline{\mathcal{K}}$ and Γ_j, respectively, with compact supports vanishing in a neighborhood of the edges. Therefore by Theorem 3.3.9, there exists a unique solution $R(\cdot,\zeta)$ of the problem (3.4.7), (3.4.8) in $V^{m,2}_{\kappa,0}(\mathcal{K})^{\ell\times\ell}$. Obviously, the matrix

$$G(x,\xi) = \psi(x,\xi)\, \mathcal{G}(x,\xi) + R(x,\xi)$$

satisfies (3.4.1), (3.4.2). The uniqueness of Green's matrix follows from Theorem 3.3.9 (analogously to the first part of the proof of Theorem 2.5.1).

2) By (3.4.1), the equation $L(D_x)\,G(tx,t\xi) = t^{2m}\,\delta(tx-t\xi) = t^{2m-3}\,\delta(x-\xi)$ is satisfied for $x,\xi \in \mathcal{K}$. Moreover, $G(tx,t\xi)$ satisfies the homogeneous Dirichlet conditions for arbitrary $t>0$. This implies (3.4.3).

3) By the first assertion, there exists a unique solution $H(x,\xi)$ of the problem

$$L^+(D_x)\,H(x,\xi) = \delta(x-\xi)\,I_\ell \quad \text{for } x,\xi \in \mathcal{K}, \tag{3.4.9}$$

$$\frac{\partial^k H(x,\xi)}{\partial n_x^k} = 0 \quad \text{for } x \in \Gamma_j,\ \xi \in \mathcal{K},\ k=0,\dots,m-1,\ j=1,\dots,d, \tag{3.4.10}$$

such that the function $x \to \zeta(x,\xi)\,H(x,\xi)$ belongs to the space $V^{m,2}_{-\kappa,0}(\mathcal{K})^{\ell\times\ell}$ for every smooth function ζ which is equal to zero in a neighborhood of the point $x=\xi$ and bounded together with all derivatives. Let $G^{(i)} = (G_{1,i},\dots,G_{\ell,i})^t$ denote the i-th column of the matrix G and $H^{(j)} = (H_{1,j},\dots,H_{\ell,j})^t$ the j-th column of H (here t means transposition). Then analogously to the proof of Theorem 2.5.1, we obtain

$$\big(L(D_x)\,G^{(i)}(\cdot,\xi), H^{(j)}(\cdot,\eta)\big)_{\mathcal{K}} = \big(G^{(i)}(\cdot,\xi), L^+(D_x)H^{(j)}(\cdot,\eta)\big)_{\mathcal{K}}. \tag{3.4.11}$$

This implies the equality $\overline{H_{j,i}(\xi,\eta)} = G_{i,j}(\eta,\xi)$ for $i,j=1,\dots,\ell$.

4) Let $f \in C_0^\infty(\overline{\mathcal{K}}\backslash\mathcal{S})^\ell$, where $\mathcal{S} = M_1 \cup \dots \cup M_d \cup \{0\}$ denotes the set of the singular boundary points. By Theorem 3.3.9, there exists a unique solution $u \in \overset{\circ}{V}{}^{m,2}_{\kappa,0}(\mathcal{K})^\ell \cap V^{m+k,2}_{\kappa+k,k}(\mathcal{K})^\ell$ of the equation $L(D_x)u = f$ in $\mathcal{K}$. Here k is an arbitrary nonnegative integer. If $H^{(j)}(\xi,x)$ is the j-th column of the matrix $H(\xi,x) = (G(x,\xi))^*$, then

$$\int\limits_{\mathcal{K}} f(\xi)\cdot\overline{H^{(j)}(\xi,x)}\,d\xi = \int\limits_{\mathcal{K}} L(D_\xi)u(\xi)\cdot\overline{H^{(j)}(\xi,x)}\,d\xi = \int\limits_{\mathcal{K}} u(\xi)\cdot\overline{L^+(D_\xi)\,H^{(j)}(\xi,x)}\,d\xi$$

for each $x \in \mathcal{K}$. According to (3.4.9), the right-hand side of the last equality is equal to $u_j(x)$. Hence, the formula (3.4.6) is true for $f \in C_0^\infty(\overline{\mathcal{K}}\backslash\mathcal{S})^\ell$. Since this set is dense in $V^{-m,2}_{\kappa,0}(\mathcal{K})^\ell$, the representation (3.4.6) can be extended by continuity to functions $f \in V^{-m,2}_{\kappa,0}(\mathcal{K})^\ell$. The proof of the theorem is complete. □

3.4.2. Estimates of Green's matrix. The case $|x|/2 < |\xi| < 2|x|$. Let again $\delta_+^{(k)}$ and $\delta_-^{(k)}$ be the greatest real numbers such that the strip $m-1-\delta_-^{(k)} < \operatorname{Re}\lambda < m-1+\delta_+^{(k)}$ is free of eigenvalues of the pencil $A_k(\lambda)$. We denote by $k(x)$ the smallest integer k such that $r_k(x) = r(x)$ and define

$$\delta_+(x) = \delta_+^{(k(x))}, \quad \delta_-(x) = \delta_-^{(k(x))}.$$

THEOREM 3.4.2. *Suppose that the line* $\operatorname{Re}\lambda = m-\kappa-3/2$ *does not contain eigenvalues of the pencil* $\mathfrak{A}(\lambda)$. *Then the elements of the Green's matrix* $G(x,\xi)$ *introduced in Theorem* 3.4.1 *satisfy the following estimate for* $|x|/2 < |\xi| < 2|x|$, $|x-\xi| > \min(r(x),r(\xi))$:

$$\big|D_x^\alpha D_\xi^\gamma G_{i,j}(x,\xi)\big| \le c_{\alpha,\gamma}|x-\xi|^{2m-3-|\alpha|-|\gamma|}\Big(\frac{r(x)}{|x-\xi|}\Big)^{m-1+\delta_+(x)-|\alpha|-\varepsilon} \times\Big(\frac{r(\xi)}{|x-\xi|}\Big)^{m-1+\delta_-(\xi)-|\gamma|-\varepsilon}.$$

Here ε *is an arbitrarily small positive number, the constant* $c_{\alpha,\gamma}$ *is independent of* x *and* ξ.

P r o o f. First note that the inequality $|x-\xi| > \min(r(x), r(\xi))$ implies

$$\max(r(x), r(\xi)) < 2\,|x-\xi|.$$

Let $\mathcal{B}(x)$ and $\mathcal{B}(\xi)$ be balls with radius $C|x-\xi|$ centered about x and ξ, respectively, where C is a certain (sufficiently small) constant, $C < 1/2$. Furthermore, let ζ and η be smooth cut-off functions with supports in $\mathcal{B}(x)$ and $\mathcal{B}(\xi)$, respectively, which are equal to one in balls with radius $C|x-\xi|/2$ centered about x and ξ, respectively. Then

$$L^+(D_y)\, D_x^\alpha G^*(x,y) = 0 \quad \text{for } y \in \mathcal{K} \cap \mathcal{B}(\xi).$$

Suppose first that $r(\xi)/|\xi|$ is small, $r(\xi)/|\xi| < \varepsilon_0$. Furthermore, let C be sufficiently small. Then $\mathcal{K} \cap \mathcal{B}(\xi)$ is diffeomorphic to a subset of a dihedron. Consequently, it follows from Lemma 2.9.11 (see also Remark 2.9.3) that

$$\begin{aligned}(3.4.12)\qquad & r(\xi)^{|\gamma|-m+1-\delta_-(\xi)+\varepsilon}\, \big|D_x^\alpha D_\xi^\gamma G_{i,j}(x,\xi)\big| \\ & \le c\,|x-\xi|^{\varepsilon-\delta_-(\xi)-1/2}\, \|\eta(\cdot)\, D_x^\alpha G(x,\cdot)\|_{V_{0,0}^{m,2}(\mathcal{K})^{\ell\times\ell}} \\ & \le c\,|\xi|^\kappa\, |x-\xi|^{\varepsilon-\delta_-(\xi)-1/2}\, \|\eta(\cdot)\, D_x^\alpha G(x,\cdot)\|_{V_{-\kappa,0}^{m,2}(\mathcal{K})^{\ell\times\ell}}\,.\end{aligned}$$

If $r(\xi)/|\xi| > \varepsilon_0$, then

$$\varepsilon_0\,|\xi| < r(\xi) < |\xi| \quad \text{and} \quad \varepsilon_0\,|\xi|/2 < |x-\xi| < 3|\xi|$$

and, consequently,

$$\begin{aligned}\big|D_x^\alpha D_\xi^\gamma G_{i,j}(x,\xi)\big| &\le c\,|\xi|^{m-|\gamma|-3/2}\, \|\eta(\cdot)\, D_x^\alpha G(x,\cdot)\|_{V_{0,0}^{m,2}(\mathcal{K})^{\ell\times\ell}} \\ &\le c\,|\xi|^{\kappa+m-|\gamma|-3/2}\, \|\eta(\cdot)\, D_x^\alpha G(x,\cdot)\|_{V_{-\kappa,0}^{m,2}(\mathcal{K})^{\ell\times\ell}}\,.\end{aligned}$$

Thus, the estimate (3.4.12) is valid both for $r(\xi)/|\xi| < \varepsilon_0$ and $r(\xi)/|\xi| > \varepsilon_0$. Let $f \in V_{\kappa,0}^{-m,2}(\mathcal{K})^\ell$. We consider the solution

$$u(x) = \int_{\mathcal{K}} G(x,z)\,\eta(z)\,f(z)\,dz$$

of the equation $L(D_x)u = \eta f$ in the space $\overset{\circ}{V}{}_{\kappa,0}^{m,2}(\mathcal{K})^\ell$. Since $\eta f = 0$ in $\mathcal{B}(x)$, we obtain (again by Lemma 2.9.11) the estimate

$$\begin{aligned}r(x)^{|\alpha|-m+1-\delta_+(x)+\varepsilon}\, \big|\partial_x^\alpha u(x)\big| &\le c\,|x-\xi|^{\varepsilon-\delta_+(x)-1/2}\, \|\zeta u\|_{V_{0,0}^{m,2}(\mathcal{K})^\ell} \\ &\le c\,|x|^{-\kappa}\, |x-\xi|^{\varepsilon-\delta_+(x)-1/2}\, \|\zeta u\|_{V_{\kappa,0}^{m,2}(\mathcal{K})^\ell}\,.\end{aligned}$$

We may assume that $|D^\alpha\zeta| + |D^\alpha\eta| \le c_\alpha\,|x-\xi|^{-|\alpha|}$ for every multi-index α, where c_α is a constant depending only on α. Then there exists a constant c such that

$$\|\zeta u\|_{V_{\kappa,0}^{m,2}(\mathcal{K})^\ell} \le c\,\|u\|_{V_{\kappa,0}^{m,2}(\mathcal{K})^\ell} \le c'\,\|f\|_{V_{\kappa,0}^{-m,2}(\mathcal{K})^\ell}\,.$$

Therefore, the norm of the mapping

$$\begin{aligned}V_{\kappa,0}^{-m,2}(\mathcal{K})^\ell \ni f \to\ & r(x)^{|\alpha|-m+1-\delta_+(x)+\varepsilon}\partial_x^\alpha u(x) \\ &= r(x)^{|\alpha|-m+1-\delta_+(x)+\varepsilon} \int_{\mathcal{K}} \partial_x^\alpha G(x,z)\,\eta(z)\,f(z)\,dz \in \mathbb{C}^\ell\end{aligned}$$

is dominated by $c\,|x|^{-\kappa}\,|x-\xi|^{\varepsilon-\delta_+(x)-1/2}$. Thus, there exists a constant c independent of x and ξ such that

$$\|\eta(\cdot)\, D_x^\alpha G(x,\cdot)\|_{V_{-\kappa,0}^{m,2}(\mathcal{K})^{\ell\times\ell}} \le c\,|x|^{-\kappa}\,|x-\xi|^{\varepsilon-\delta_+(x)-1/2}\, r(x)^{m-1+\delta_+(x)-|\alpha|-\varepsilon}\,.$$

The last inequality together with (3.4.12) implies the desired estimate for the derivatives of Green's matrix $G(x,\xi)$. □

Next, we consider the case $|x-\xi| < \min(r(x), r(\xi))$. From the last inequality it follows in particular that $|x|/2 < |\xi| < 2|x|$.

THEOREM 3.4.3. *Let $G(x,\xi)$ be the Green's matrix introduced in Theorem 3.4.1. Then the following estimates are satisfied for $|x-\xi| < \min(r(x), r(\xi))$:*

$$\big|D_x^\alpha D_\xi^\gamma G_{i,j}(x,\xi)\big| \le c_{\alpha,\gamma}\,\big(|x-\xi|^{2m-3-|\alpha|-|\gamma|} + r(\xi)^{2m-3-|\alpha|-|\gamma|}\big) \quad \text{if } |\alpha|+|\gamma| \ne 2m-3,$$

$$\big|D_x^\alpha D_\xi^\gamma G_{i,j}(x,\xi)\big| \le c_{\alpha,\gamma}\,\Big(\Big|\log\frac{|x-\xi|}{r(\xi)}\Big| + 1\Big) \quad \text{if } |\alpha|+|\gamma| = 2m-3.$$

P r o o f. Suppose that $|x-\xi| < \min(r(x), r(\xi)) = 2$. Then x and ξ lie in a ball B with radius 1 and distance ≥ 1 from the set $\mathcal{S}$ of the edge points. Let B' be a ball concentric to B with radius 3/2. There exists a Green's matrix $g(x,\xi)$ satisfying the equations

$$L(D_x)\, g(x,\xi) = \delta(x-\xi)\, I_\ell \ \text{ for } x,\xi \in \mathcal{K}\cap B',$$
$$\partial_{n(x)}^{k-1} g(x,\xi) = 0 \ \text{ for } x \in \partial\mathcal{K}\cap B',\ \xi \in \mathcal{K}\cap B',\ k=1,\dots,m,$$

and the estimates

$$\text{(3.4.13)}\quad \big|D_x^\alpha D_\xi^\gamma g(x,\xi)\big| \le c_{\alpha,\gamma}\,\big(|x-\xi|^{2m-3-|\alpha|-|\gamma|} + 1\big) \text{ if } |\alpha|+|\gamma| \ne 2m-3,$$

$$\text{(3.4.14)}\quad \big|D_x^\alpha D_\xi^\gamma g(x,\xi)\big| \le c_{\alpha,\gamma}\,\Big(\log|x-\xi| + 1\Big) \quad \text{if } |\alpha|+|\gamma| = 2m-3.$$

We consider the function

$$u(x,\xi) = D_\xi^\gamma G(x,\xi) - \eta(x)\, D_\xi^\gamma g(x,\xi)$$

where $\xi \in \mathcal{K}\cap B$, $\eta \in C_0^\infty(B')$ and $\eta = 1$ in a certain neighborhood of $\bar{B}$. Obviously,

$$L(D_x)\, u(x,\xi) = f(x,\xi) \text{ for } x \in \mathcal{K},$$
$$\partial_{n(x)}^{k-1} u(x,\xi) = 0 \ \text{ for } x \in \partial\mathcal{K}\backslash\mathcal{S},\ k=1,\dots,m,$$

where $f(x,\xi) = 0$ for $x \in B$ and for $x \in \mathcal{K}\backslash B'$. Furthermore, the derivatives of f are bounded by constants independent of x and ξ. In particular, $f(\cdot,\xi)$ belongs to a weighted Sobolev space $V_{\beta,\delta}^{l-2m,2}(\mathcal{K})^{\ell\times\ell}$ with arbitrary l, β, δ. Using Corollary 3.3.14, we conclude that

$$\|u(\cdot,\xi)\|_{W^{l,2}(\mathcal{K}\cap B)^{\ell\times\ell}} \le c_l\,,$$

where c_l is a constant independent of $\xi \in B$. Hence, the exist constants $C_{\alpha,\gamma}$ independent of x and ξ such that

$$\big|D_x^\alpha u(x,\xi)\big| \le C_{\alpha,\gamma}$$

for $x \in \mathcal{K}$. Consequently, the matrix $G(x,\xi)$ satisfies the estimates (3.4.13), (3.4.14) for $x,\xi \in \mathcal{K}\cap B$. In particular, these estimates are valid for $|x-\xi| < \min(r(x), r(\xi)) = 2$. Applying (3.4.3), we obtain the estimates of the theorem for arbitrary $x,\xi \in \mathcal{K}$, $|x-\xi| < \min(r(x), r(\xi))$. □

3.4.3. Estimates of Green's matrix. The cases $|x| > 2|\xi|$ and $|\xi| > 2|x|$. For the proof of estimates in these cases, we need the following lemma.

LEMMA 3.4.4. *If $u \in V^{l,2}_{\beta,\delta}(\mathcal{K})$, $\rho\partial_\rho u \in V^{l,2}_{\beta,\delta}(\mathcal{K})$, $l \geq 2$, then there exists a constant c independent of u and x such that*

$$\rho^{\beta-l+3/2} \prod_{k=1}^{d} \Big(\frac{r_k}{\rho}\Big)^{\delta_k-l+1} |u(x)| \leq c\Big(\|u\|_{V^{l,2}_{\beta,\delta}(\mathcal{K})} + \|\rho\partial_\rho u\|_{V^{l,2}_{\beta,\delta}(\mathcal{K}))}\Big). \tag{3.4.15}$$

P r o o f. Applying the estimate

$$\sup_{0<\rho<\infty} |v(\rho)|^2 \leq c \int_0^\infty \big(|v(\rho)|^2 + |\rho v'(\rho)|^2\big)\, \frac{d\rho}{\rho}$$

(which follows immediately from Sobolev's imbedding theorem) to the function $\rho^{\beta-l+3/2}u(\rho,\omega)$, we obtain

$$\begin{aligned} &\sup_{0<\rho<\infty} \rho^{2(\beta-l)+3}\, \big|u(\rho,\omega)\big|^2 \\ &\quad\leq c \int_0^\infty \rho^{2(\beta-l+1)} \big(|u(\rho,\omega)|^2 + |\rho\partial_\rho u(\rho,\omega)|^2\big)\, d\rho. \end{aligned} \tag{3.4.16}$$

Furthermore by Lemma 1.2.3,

$$\sup_{\omega\in\Omega} \prod_{k=1}^{d} r_k(\omega)^{\delta_k-l+1}\, \big|(\rho\partial_\rho)^j u(\rho,\omega)\big| \leq c\, \big\|(\rho\partial_\rho)^j u(\rho,\cdot)\big\|_{V^{l,2}_\delta(\Omega)}$$

for $j=0$ and $j=1$. Since the norm in $V^{l,2}_{\beta,\delta}(\mathcal{K})$ is equivalent to

$$\|u\| = \Big(\int_0^\infty \rho^{2(\beta-l+1)} \sum_{j=0}^{l} \big\|(\rho\partial_\rho)^j u(\rho,\cdot)\big\|^2_{V^{l-j,2}_\delta(\Omega)}\, d\rho\Big)^{1/2},$$

the last inequality together with (3.4.16) implies (3.4.15). □

As in Theorem 3.4.1, we assume that the line $\operatorname{Re}\lambda = m-\kappa-3/2$ is free of eigenvalues of the pencil $\mathfrak{A}(\lambda)$. We denote by

$$\Lambda_- < \operatorname{Re}\lambda < \Lambda_+$$

the widest strip in the complex plane which contains the line $\operatorname{Re}\lambda = m-\kappa-3/2$ and no eigenvalues of the pencil $\mathfrak{A}(\lambda)$. Then the following statement holds.

THEOREM 3.4.5. *The elements of Green's matrix $G(x,\xi)$ introduced in Theorem 3.4.1 satisfy the estimate*

$$\begin{aligned} \big|D_x^\alpha D_\xi^\gamma G_{i,j}(x,\xi)\big| &\leq c_{\alpha,\gamma} |x|^{\Lambda_- -|\alpha|+\varepsilon}\, |\xi|^{2m-3-\Lambda_- -|\gamma|-\varepsilon} \\ &\quad\times \prod_{k=1}^{d} \Big(\frac{r_k(x)}{|x|}\Big)^{m-1+\delta_+^{(k)}-|\alpha|-\varepsilon} \prod_{k=1}^{d} \Big(\frac{r_k(\xi)}{|\xi|}\Big)^{m-1+\delta_-^{(k)}-|\gamma|-\varepsilon}. \end{aligned}$$

for $|x| > 2|\xi|$. If $|\xi| > 2|x|$, then

$$\begin{aligned} \big|D_x^\alpha D_\xi^\gamma G_{i,j}(x,\xi)\big| &\leq c_{\alpha,\gamma} |x|^{\Lambda_+ -|\alpha|-\varepsilon}\, |\xi|^{2m-3-\Lambda_+ -|\gamma|+\varepsilon} \\ &\quad\times \prod_{k=1}^{d} \Big(\frac{r_k(x)}{|x|}\Big)^{m-1+\delta_+^{(k)}-|\alpha|-\varepsilon} \prod_{k=1}^{d} \Big(\frac{r_k(\xi)}{|\xi|}\Big)^{m-1+\delta_-^{(k)}-|\gamma|-\varepsilon}. \end{aligned}$$

Here, ε is an arbitrarily small real positive number.

Proof. By item 3) of Theorem 3.4.1, it suffices to prove the first estimate. Suppose that $|x| = 1$. We denote by ζ and η smooth functions on $\overline{\mathcal{K}}$ such that $\zeta(\xi) = 1$ for $|\xi| < 1/2$, $\eta = 1$ in a neighborhood of $\operatorname{supp}\zeta$ and $\eta(\xi) = 0$ for $|\xi| > 3/4$. By Theorem 3.4.1, we have

$$\eta(\xi)\, L^+(D_\xi)\, D_x^\alpha G^*(x,\xi) = 0 \quad \text{for } \xi \in \mathcal{K}.$$

Since the function $\xi \to \eta(\xi)\, D_x^\alpha G(x,\xi)$ belongs to $\overset{\circ}{V}{}^{m,2}_{-\kappa,0}(\mathcal{K})^{\ell\times\ell}$, it follows from Corollary 3.3.14 that $\zeta(\cdot)\,(|\xi|\partial_{|\xi|})^j D_x^\alpha G(x,\cdot) \in V^{l,2}_{\beta,\delta}(\mathcal{K})^{\ell\times\ell}$ for every integer $j \ge 0$, where l is an arbitrary integer, $l \ge m$,

$$2m-3-\Lambda_+ < l-\beta-3/2 < 2m-3-\Lambda_-, \qquad -\delta_-^{(k)} < \delta_k - l + m < \delta_+^{(k)}.$$

Furthermore, the estimate

$$\|\zeta(\cdot)\,(|\xi|\partial_{|\xi|})^j D_x^\alpha D_\xi^\gamma G(x,\cdot)\|_{V^{l-|\gamma|,2}_{\beta,\delta}(\mathcal{K})^{\ell\times\ell}} \le c\,\|\eta(\cdot) D_x^\alpha G(x,\cdot)\|_{V^{m,2}_{-\kappa,0}(\mathcal{K})^{\ell\times\ell}}$$

holds. Thus by Lemma 3.4.4,

$$\begin{aligned} (3.4.17)\qquad & |\xi|^{\beta-l+|\gamma|+3/2} \prod_{k=1}^d \Big(\frac{r_k(\xi)}{|\xi|}\Big)^{\delta_k-l+|\gamma|+1} \big|D_x^\alpha D_\xi^\gamma G(x,\xi)\big| \\ & \le c\,\|\eta(\cdot)\, D_x^\alpha G(x,\cdot)\|_{V^{m,2}_{-\kappa,0}(\mathcal{K})^{\ell\times\ell}} \end{aligned}$$

for $|\xi| < 1/2$. Let $f \in V^{-m,2}_{\kappa,0}(\mathcal{K})^\ell$, and let

$$u(y) = \int_{\mathcal{K}} G(y,z)\,\eta(z)\,f(z)\,dz$$

be the uniquely determined solution of the equation $L(D_x)u = \eta f$ in $\overset{\circ}{V}{}^{m,2}_{\kappa,0}(\mathcal{K})^\ell$. We denote by χ and ψ smooth cut-off functions such that $\chi = 1$ in a neighborhood of the point x, $\psi = 1$ in a neighborhood of $\operatorname{supp}\chi$, $\psi(y) = 0$ for $|x-y| > 1/4$. Since ψ and η have disjoint supports, we have $\psi L(D_x)u = 0$ in $\mathcal{K}$. Corollary 3.3.14 implies $\chi(\rho\partial_\rho)^j u \in V^{l,2}_{\beta',\delta'}(\mathcal{K})^\ell$ for every integer $j \ge 0$, where l is an arbitrary integer, $l \ge m$,

$$\Lambda_- < l-\beta'-3/2 < \Lambda_+, \qquad -\delta_+^{(k)} < \delta_k' - l + m < \delta_-^{(k)}.$$

Furthermore,

$$\|\chi(\rho\partial_\rho)^j D_x^\alpha u\|_{V^{l-|\alpha|,2}_{\beta',\delta'}(\mathcal{K})^\ell} \le c\,\|\psi u\|_{V^{m,2}_{\kappa,0}(\mathcal{K})^\ell} \le c'\,\|f\|_{V^{-m,2}_{\kappa,0}(\mathcal{K})^\ell}.$$

Applying Lemma 3.4.4, we obtain

$$\prod_{k=1}^d \Big(\frac{r_k(x)}{|x|}\Big)^{\delta_k'-l+|\alpha|+1} \big|D_x^\alpha u(x)\big| \le c\,\|f\|_{V^{-m,2}_{\kappa,0}(\mathcal{K})^\ell}.$$

This means that the mapping

$$\begin{aligned} V^{-m,2}_{\kappa,0}(\mathcal{K})^\ell \ni f \to & \prod_{k=1}^d \Big(\frac{r_k(x)}{|x|}\Big)^{\delta_k'-l+|\alpha|+1} D_x^\alpha u(x) \\ & = \prod_{k=1}^d \Big(\frac{r_k(x)}{|x|}\Big)^{\delta_k'-l+|\alpha|+1} \int_{\mathcal{K}} D_x^\alpha G(x,z)\,\eta(z)\,f(z)\,dz \in \mathbb{C}^\ell \end{aligned}$$

is continuous and its norm is bounded by a constant independent of x. Consequently, there exists a constant c independent of x such that

$$\|\eta(\cdot)\, D_x^\alpha G(x,\cdot)\|_{V^{m,2}_{-\kappa,0}(\mathcal{K})^{\ell\times\ell}} \le c \prod_{k=1}^d \Big(\frac{r_k(x)}{|x|}\Big)^{l-\delta_k'-|\alpha|-1}.$$

Combining the last inequality with (3.4.17), we obtain

$$\big|D_x^\alpha D_\xi^\gamma G(x,\xi)\big| \le c\,|\xi|^{l-\beta-|\gamma|-3/2} \prod_{k=1}^d \Big(\frac{r_k(x)}{|x|}\Big)^{l-\delta_k'-|\alpha|-1} \prod_{k=1}^d \Big(\frac{r_k(\xi)}{|\xi|}\Big)^{l-\delta_k-|\gamma|-1}$$

for $|x|=1$, $|\xi|<1/2$. For $\beta = l-2m+\Lambda_- +\varepsilon+3/2$, $\delta_k = l-m-\delta_-^{(k)}+\varepsilon$ and $\delta_k' = l-m-\delta_+^{(k)}+\varepsilon$, the inequality

$$\big|D_x^\alpha D_\xi^\gamma G_{i,j}(x,\xi)\big| \le c\,|\xi|^{2m-3-\Lambda_- -|\gamma|-\varepsilon} \prod_{k=1}^d \Big(\frac{r_k(x)}{|x|}\Big)^{m-1+\delta_+^{(k)}-|\alpha|-\varepsilon} \times \prod_{k=1}^d \Big(\frac{r_k(\xi)}{|\xi|}\Big)^{m-1+\delta_-^{(k)}-|\gamma|-\varepsilon}$$

holds. This proves the theorem for $|x|=1$, $|\xi|<1/2$. If x is an arbitrary point of $\mathcal{K}$ and $|\xi|<|x|/2$, then we obtain the assertion of the theorem by means of the second item of Theorem 3.4.1. This completes the proof. □

3.5. Solvability in weighted L_p Sobolev spaces

Using the estimates of the Green's matrix in Theorems 3.4.2, 3.4.3 and 3.4.5, we are able to extend the results of Section 3.3 to the weighted spaces $V^{l,p}_{\beta,\delta}$ with arbitrary $p\in(1,\infty)$. By Lemma 3.3.1, we may restrict ourselves to the case of zero Dirichlet data $g_{j,k}$. Let κ be such that the line $\operatorname{Re}\lambda = m-\kappa-3/2$ is free of eigenvalues of the pencil $\mathfrak{A}(\lambda)$, and let $G(x,\xi)$ be the Green's matrix introduced in Theorem 3.4.1. Then the vector function

$$u(x) = \int_{\mathcal{D}} G(x,\xi)\, f(\xi)\, d\xi, \tag{3.5.1}$$

is a solution of the Dirichlet problem (3.3.1). We prove that this function belongs to the space $V^{l,p}_{\beta,\delta}(\mathcal{K})^\ell$, $l\ge m$, if $f\in V^{l-2m,p}_{\beta,\delta}(\mathcal{K})^\ell$, $l-\beta-3/p = m-\kappa-3/2$ and the components of δ satisfy the condition

$$-\delta_+^{(k)} < \delta_k - l + m - 1 + 2/p < \delta_-^{(k)} \quad \text{for } k=1,\dots,d. \tag{3.5.2}$$

In this way, we show that the absence of eigenvalues of the pencil $\mathfrak{A}(\lambda)$ on the line $\operatorname{Re}\lambda = l-\beta-3/p$ and the inequalities (3.5.2) are sufficient conditions for the existence and uniqueness of solutions in $V^{l,p}_{\beta,\delta}(\mathcal{K})^\ell$. Furthermore, we prove the necessity of these conditions and obtain regularity results for the solutions of the boundary value problem (3.3.1).

3.5.1. Auxiliary inequalities. Let the function χ_μ in $\mathcal{K}$ be defined as

$$\chi_\mu(x) = 1 \ \text{ for } 2^{\mu-1}\le |x| \le 2^{\mu+1}, \quad \chi_\mu(x) = 0 \ \text{ else.}$$

The subsequent lemmas enable us to estimate the vector function (3.5.1) in the case where the support of f is contained in the set of all x such that $2^{\nu-1}\le|x|\le 2^{\nu+1}$.

LEMMA 3.5.1. *Let* $f \in V^{0,p}_{\beta',\delta'}(\mathcal{K})^\ell$, $f(x) = 0$ *for* $|x| < 2^{\nu-1}$ *and* $|x| > 2^{\nu+1}$. *Furthermore, let* v *be defined as*

$$v(x) = \int_{\mathcal{K}} K(x,\xi)\, f(\xi)\, d\xi, \tag{3.5.3}$$

where $K(x,\xi)$ *satisfies the estimate*

$$\big|K(x,\xi)\big| \le c\,|x|^{\Lambda}\,|\xi|^{T-\Lambda-3} \prod_{k=1}^{d} \Big(\frac{r_k(x)}{|x|}\Big)^{\gamma_k} \prod_{k=1}^{d} \Big(\frac{r_k(\xi)}{|\xi|}\Big)^{\gamma'_k} \tag{3.5.4}$$

for $2^{\mu-1} < |x| < 2^{\mu+1}$, $2^{\nu-1} < |\xi| < 2^{\nu+1}$. *We suppose that the components of* δ' *and* δ'' *satisfy the inequalities*

$$\gamma_k + \delta''_k > -2/p \quad \text{and} \quad \gamma'_k - \delta'_k > -2 + 2/p$$

for $k = 1, \dots, d$. *Then*

$$\|\chi_\mu v\|_{V^{0,p}_{\beta'-T,\delta''}(\mathcal{K})} \le c\, 2^{(\mu-\nu)(\beta'-T+\Lambda+3/p)}\, \|f\|_{V^{0,p}_{\beta',\delta'}(\mathcal{K})} \tag{3.5.5}$$

with a constant c *independent of* f, μ *and* ν.

P r o o f. Since $2^{\mu-1} < |x| < 2^{\mu+1}$ and $2^{\nu-1} < |\xi| < 2^{\nu+1}$ for $x \in \operatorname{supp}\chi_\mu$, $\xi \in \operatorname{supp} f$, we have

$$\begin{aligned} \|\chi_\mu v\|^p_{V^{0,p}_{\beta'-T,\delta''}(\mathcal{K})} \le\ & c\, 2^{\mu p(\beta'-T+\Lambda)} \int\limits_{\substack{\mathcal{K} \\ 2^{\mu-1}<|x|<2^{\mu+1}}} \prod_{k=1}^{d} \Big(\frac{r_k(x)}{|x|}\Big)^{p(\gamma_k+\delta''_k)} dx \\ & \times 2^{\nu p(T-\Lambda-3)} \Big(\int_{\mathcal{K}} \prod_{k=1}^{d} \Big(\frac{r_k(\xi)}{|\xi|}\Big)^{\gamma'_k} |f(\xi)|\, d\xi \Big)^p. \end{aligned}$$

Here,

$$\int\limits_{\substack{\mathcal{K} \\ 2^{\mu-1}<|x|<2^{\mu+1}}} \prod_{k=1}^{d} \Big(\frac{r_k(x)}{|x|}\Big)^{p(\gamma_k+\delta''_k)} dx \le c\, 2^{3\mu}$$

if $\gamma_k + \delta''_k > -2/p$. Furthermore, Hölder's inequality yields

$$\begin{aligned} & \Big(\int_{\mathcal{K}} \prod_{k=1}^{d} \Big(\frac{r_k(\xi)}{|\xi|}\Big)^{\gamma'_k} |f(\xi)|\, d\xi \Big)^p \\ & \le 2^{-\nu p\beta'}\, \|f\|^p_{V^{0,p}_{\beta',\delta'}(\mathcal{K})} \Big(\int\limits_{\substack{\mathcal{K} \\ 2^{\nu-1}<|\xi|<2^{\nu+1}}} \prod_{k=1}^{d} \Big(\frac{r_k(\xi)}{|\xi|}\Big)^{p'(\gamma'_k-\delta'_k)} d\xi \Big)^{p-1} \\ & \le 2^{-\nu p\beta' + 3\nu(p-1)}\, \|f\|^p_{V^{0,p}_{\beta',\delta'}(\mathcal{K})} \end{aligned}$$

if $\gamma'_k - \delta'_k > -2/p'$, $p' = p/(p-1)$. This implies (3.5.5). □

Note that the Green's matrix $G(x,\xi)$ and its derivatives satisfy an estimate of the form (3.5.4) if $2^{\mu-1} < |x| < 2^{\mu+1}$, $2^{\nu-1} < |\xi| < 2^{\nu+1}$, and $|\mu-\nu| > 2$. In Lemmas 3.5.3 and 3.5.4, we consider the situation when $2^{\mu-1} < |x| < 2^{\mu+1}$, $2^{\nu-1} < |\xi| < 2^{\nu+1}$, and $|\mu-\nu| \le 2$. For the proof of Lemma 3.5.3, we will employ the following lemma.

LEMMA 3.5.2. *Let $\mathcal{K}_x$ denote the set*

$$\{\xi \in \mathcal{K} : c_1|x| < |\xi| < c_2|x|,\ |\xi - x| > \delta\, r(x)\},$$

where c_1, c_2 and δ are given positive numbers. Furthermore, let $\alpha, \beta_k, \gamma_k$ and γ'_k be real numbers satisfying the inequalities

$$\beta_k + \gamma'_k > -2, \qquad \alpha + \beta_k - \gamma_k \neq -3$$

for $k = 1, \dots, d$. By $k(x)$, we denote the smallest index k such that $r_k(x) = r(x)$. Then the estimate

$$\int\limits_{\mathcal{K}_x} |x-\xi|^\alpha \Big(\frac{r(x)}{|x-\xi|}\Big)^{\gamma_{k(x)}} \Big(\frac{r(\xi)}{|x-\xi|}\Big)^{\gamma'_{k(\xi)}} \prod_{k=1}^d \Big(\frac{r_k(\xi)}{|\xi|}\Big)^{\beta_k} d\xi \tag{3.5.6}$$
$$\leq c\,|x|^{3+\alpha} \prod_{k=1}^d \Big(\frac{r_k(x)}{|x|}\Big)^{\min(\gamma_k, 3+\alpha+\beta_k)}$$

is valid with a constant c independent of x.

P r o o f. Without loss of generality, we may assume that $r(x) = r_1(x)$, i.e. $k(x) = 1$. Then the left-hand side of (3.5.6) is equal to

$$A \stackrel{def}{=} r_1(x)^{\gamma_1} \int\limits_{\mathcal{K}_x} |x-\xi|^{\alpha-\gamma_1} \Big(\frac{r(\xi)}{|x-\xi|}\Big)^{\gamma'_{k(\xi)}} \prod_{k=1}^d \Big(\frac{r_k(\xi)}{|\xi|}\Big)^{\beta_k} d\xi.$$

Suppose first that $r_1(x) > |x|/2$. Then the inequality $\delta|x|/2 < |x-\xi| < (c_2+1)\,|x|$ holds for $\xi \in \mathcal{K}_x$. Consequently,

$$A \leq c\,|x|^{\alpha-\gamma_1}\, r_1(x)^{\gamma_1} \int_{\mathcal{K}_x} \prod_{k=1}^d \Big(\frac{r_k(\xi)}{|\xi|}\Big)^{\beta_k+\gamma'_k} d\xi \leq c\,|x|^{3+\alpha-\gamma_1}\, r_1(x)^{\gamma_1},$$

which proves (3.5.6). In the case $r(x) < |x|/2$ we consider the expressions

$$A_j = r_1(x)^{\gamma_1} \int\limits_{\mathcal{K}_x^{(j)}} |x-\xi|^{\alpha-\gamma_1} \Big(\frac{r(\xi)}{|x-\xi|}\Big)^{\gamma'_{k(\xi)}} \prod_{k=1}^d \Big(\frac{r_k(\xi)}{|\xi|}\Big)^{\beta_k} d\xi$$

for $j = 1$ and $j = 2$, where $\mathcal{K}_x^{(1)} = \{\xi \in \mathcal{K}_x : r_1(\xi) > 2r(\xi)\}$ and $\mathcal{K}_x^{(2)} = \mathcal{K}_x \backslash \mathcal{K}_x^{(1)}$. If $\xi \in \mathcal{K}_x^{(1)}$, then the inequality $c_0\,|x| < |x-\xi| < (c_2+1)\,|x|$ is satisfied with a certain positive constant c_0 independent of x and ξ, and we obtain

$$A_1 \leq c\,|x|^{\alpha-\gamma_1}\, r_1(x)^{\gamma_1} \int_{\mathcal{K}_x^{(1)}} \prod_{k=1}^d \Big(\frac{r_k(\xi)}{|\xi|}\Big)^{\beta_k+\gamma'_k} d\xi \leq c\,|x|^{3+\alpha-\gamma_1}\, r_1(x)^{\gamma_1}.$$

If $\xi \in \mathcal{K}_x^{(2)}$, then

$$\Big(\frac{r(\xi)}{|x-\xi|}\Big)^{\gamma'_{k(\xi)}} \leq c \Big(\frac{r_1(\xi)}{|x-\xi|}\Big)^{\gamma'_1} \quad \text{and} \quad \prod_{k=1}^d \Big(\frac{r_k(\xi)}{|\xi|}\Big)^{\beta_k} \leq c \Big(\frac{r_1(\xi)}{|\xi|}\Big)^{\beta_1}$$

with a certain constant c independent of x and ξ. Thus, we obtain

$$A_2 \leq c\,|x|^{-\beta_1}\, r_1(x)^{\gamma_1} \int_{\mathcal{K}_x^{(2)}} |x-\xi|^{\alpha-\gamma_1-\gamma'_1}\, r_1(\xi)^{\beta_1+\gamma'_1}\, d\xi.$$

If $\alpha+\beta_1-\gamma_1<-3$, then the inequality (2.6.5) with $R=\delta r_1(x)$ yields

$$A_2 \le c\,|x|^{-\beta_1}\, r_1(x)^{\gamma_1}\, r_1(x)^{3+\alpha+\beta_1-\gamma_1} = c\,|x|^{3+\alpha}\Big(\frac{r_1(x)}{|x|}\Big)^{3+\alpha+\beta_1}.$$

In the case $\alpha+\beta_1-\gamma_1>-3$ we apply the inequality (2.6.6) with $R=3|x|$ and obtain

$$A_2 \le c\,|x|^{3+\alpha-\gamma_1}\, r_1(x)^{\gamma_1} = c\,|x|^{3+\alpha}\Big(\frac{r_1(x)}{|x|}\Big)^{\gamma_1}.$$

This proves the lemma. □

We apply the last lemma in order to estimate the norm of $\chi_\mu v$, where v is the vector function (3.5.3) and $K(x,\xi)$ satisfies the same estimates as the derivatives of Green's matrix in the case $|x|/2<|\xi|<2|x|$, $|x-\xi|>\min(r(x),r(\xi))$.

LEMMA 3.5.3. *Let T, γ_k, γ'_k, δ'_k and δ''_k be real numbers satisfying the inequalities*

$$\gamma_k+\gamma'_k>T-2,\quad \delta'_k+2/p<2+\gamma'_k,\quad \delta''_k+2/p>-\gamma_k,\quad \delta''_k\ge\delta'_k-T$$

for $k=1,\dots,d$. Furthermore, let

$$\big|K(x,\xi)\big| \le c\,|x-\xi|^{T-3}\Big(\frac{r(x)}{|x-\xi|}\Big)^{\gamma_{k(x)}}\Big(\frac{r(\xi)}{|x-\xi|}\Big)^{\gamma'_{k(\xi)}}$$

and $K(x,\xi)=0$ for $|x-\xi|<r(x)/4$. If $f\in V^{0,p}_{\beta',\delta'}(\mathcal{K})$, $f(x)=0$ for $|x|\notin(2^{\mu-3},2^{\mu+3})$, then the function (3.5.3) satisfies the inequality

$$\|\chi_\mu v\|_{V^{0,p}_{\beta'-T,\delta''}(\mathcal{K})} \le c\,\|f\|_{V^{0,p}_{\beta',\delta'}(\mathcal{K})} \tag{3.5.7}$$

with a constant c independent of f, μ and ν.

P r o o f. By the assumptions of the lemma, there exist real numbers $t_k\ge0$ such that

$$t_k-\gamma_k<\delta''_k+2/p<t_k+2+\gamma'_k-T \quad\text{and}\quad \delta''_k\ge\delta'_k-T+t_k\,.$$

Let s_k be real numbers satisfying the inequalities

$$\frac{T-\gamma_k}{p'}<s_k<\frac{\gamma'_k+2}{p'}$$

where $p'=p/(p-1)$, and

$$\delta''_k-t_k+T-\frac{\gamma'_k}{p}<s_k<\delta''_k-t_k+T+\frac{\gamma_k-T+2}{p}\,.$$

By the conditions on t_k, the intersection of these intervals for s_k is not empty. By Hölder's inequality, we have

$$\begin{aligned}|v(x)|^p \le c\int_{\mathcal{K}_x}|x-\xi|^{T-3}\Big(\frac{r(x)}{|x-\xi|}\Big)^{\gamma_{k(x)}}\Big(\frac{r(\xi)}{|x-\xi|}\Big)^{\gamma'_{k(\xi)}}\prod_k\Big(\frac{r_k(\xi)}{|\xi|}\Big)^{ps_k}|f(\xi)|^p\,d\xi\\ \times\Big(\int_{\mathcal{K}_x}|x-\xi|^{T-3}\Big(\frac{r(x)}{|x-\xi|}\Big)^{\gamma_{k(x)}}\Big(\frac{r(\xi)}{|x-\xi|}\Big)^{\gamma'_{k(\xi)}}\prod_k\Big(\frac{r_k(\xi)}{|\xi|}\Big)^{-p's_k}d\xi\Big)^{p-1}\end{aligned}$$

for $2^{\mu-1} < |x| < 2^{\mu+1}$, where $\mathcal{K}_x = \{\xi \in \mathcal{K} : |x|/16 < |\xi| < 16|x|, |x-\xi| > r(x)/4\}$. Applying Lemma 3.5.2, we obtain

$$|v(x)|^p \le c\,|x|^{T(p-1)} \prod_{k=1}^d \Big(\frac{r_k(x)}{|x|}\Big)^{p(T-s_k)-T} \times \int_{\mathcal{K}_x} |x-\xi|^{T-3} \Big(\frac{r(x)}{|x-\xi|}\Big)^{\gamma_{k(x)}} \Big(\frac{r(\xi)}{|x-\xi|}\Big)^{\gamma'_{k(\xi)}} \prod_k \Big(\frac{r_k(\xi)}{|\xi|}\Big)^{ps_k} |f(\xi)|^p\, d\xi.$$

Since $|x|/16 < |\xi| < 16|x|$ for $\xi \in \mathcal{K}_x$, it follows that

$$\begin{aligned}
&\int_{\mathcal{K}} |x|^{p(\beta'-T)} \prod_{k=1}^d \Big(\frac{r_k(x)}{|x|}\Big)^{p\delta''_k} \big|\chi_\mu(x)\, v(x)\big|^p\, dx \\
&\le c \int_{\mathcal{K}} |x|^{p(\beta'-T)} \prod_{k=1}^d \Big(\frac{r_k(x)}{|x|}\Big)^{p(\delta''_k - t_k)} \big|\chi_\mu(x)\, v(x)\big|^p\, dx \\
&\le c \int_{\mathcal{K}} |\xi|^{p\beta'-T} \prod_k \Big(\frac{r_k(\xi)}{|\xi|}\Big)^{ps_k} |f(\xi)|^p \Big(\int_{\mathcal{K}_\xi} |x-\xi|^{T-3} \Big(\frac{r(x)}{|x-\xi|}\Big)^{\gamma_{k(x)}} \\
&\qquad\qquad \times \Big(\frac{r(\xi)}{|x-\xi|}\Big)^{\gamma'_{k(\xi)}} \prod_{k=1}^d \Big(\frac{r_k(x)}{|x|}\Big)^{p(\delta''_k - t_k - s_k + T) - T} dx\Big)\, d\xi,
\end{aligned}$$

where $\mathcal{K}_\xi = \{x \in \mathcal{K} : |\xi|/16 < |x| < 16|\xi|, |x-\xi| > r(\xi)/5\}$. Estimating the inner integral on the right-hand side by means of Lemma 3.5.2, we obtain (3.5.7). □

In the next lemma, we assume that $K(x,\xi)$ satisfies the same estimates as the derivatives of Green's matrix in the case $|x-\xi| < \min(r(x), r(\xi))$.

LEMMA 3.5.4. *Let T be a positive real number, and let $K(x,\xi)$ be such that*

$$\big|K(x,\xi)\big| \le c\,\big(|x-\xi|^{T-3} + r(\xi)^{T-3}\big) \quad \text{for } T \ne 3,$$
$$\big|K(x,\xi)\big| \le c\,\Big(\Big|\log\frac{|x-\xi|}{r(\xi)}\Big| + 1\Big) \quad \text{for } T = 3.$$

Furthermore, let $K(x,\xi) = 0$ for $|x-\xi| > 3r(x)/4$. Then the function (3.5.3) *satisfies the inequality*

$$\|v\|_{V^{0,p}_{\beta'-T,\delta'-T}(\mathcal{K})} \le c\,\|f\|_{V^{0,p}_{\beta',\delta'}(\mathcal{K})} \tag{3.5.8}$$

for arbitrary $f \in V^{0,p}_{\beta',\delta'}(\mathcal{K})$.

P r o o f. First note that $|x-\xi| < 3r(\xi)$,

$$\frac{|x|}{4} < |\xi| < \frac{7|x|}{4}, \quad \frac{r_k(x)}{4} < r_k(\xi) < \frac{7r_k(\xi)}{4}, \quad \text{and} \quad \frac{r(x)}{4} < r(\xi) < \frac{7r(x)}{4}$$

if $|x-\xi| < 3r(x)/4$. Let $\mathcal{K}_x$ denote the set $\{\xi \in \mathcal{K} : |x-\xi| < 3r(x)/4\}$. Then by Hölder's inequality,

$$\begin{aligned} |v(x)|^p &\le c \int_{\mathcal{K}_x} \big(|x-\xi|^{T-3} + r(\xi)^{T-3}\big)\, |f(\xi)|^p \, d\xi \\ &\qquad \times \Big(\int\limits_{|x-\xi|<3r(x)/4} \big(|x-\xi|^{T-3} + r(x)^{T-3}\big)\, d\xi \Big)^{p-1} \\ &\le c\, r(x)^{T(p-1)} \int_{\mathcal{K}_x} \big(|x-\xi|^{T-3} + r(\xi)^{T-3}\big)\, |f(\xi)|^p \, d\xi \end{aligned}$$

for $T \neq 3$. This implies

$$\begin{aligned} &\int_{\mathcal{K}} |x|^{p(\beta'-T)} \prod_{k=1}^{d} \Big(\frac{r_k(x)}{|x|}\Big)^{p(\delta'_k - T)} \big|v(x)\big|^p \, dx \\ &\le c \int_{\mathcal{K}} |\xi|^{p(\beta'-T)} \prod_{k=1}^{d} \Big(\frac{r_k(\xi)}{|\xi|}\Big)^{p(\delta'_k - T)} r(\xi)^{T(p-1)}\, |f(\xi)|^p \\ &\qquad \times \Big(\int\limits_{|x-\xi|<3r(\xi)} \big(|x-\xi|^{T-3} + r(\xi)^{T-3}\big)\, dx \Big)\, d\xi \\ &\le c \int_{\mathcal{K}} |\xi|^{p(\beta'-T)} \prod_{k=1}^{d} \Big(\frac{r_k(\xi)}{|\xi|}\Big)^{p(\delta'_k - T)} r(\xi)^{Tp}\, |f(\xi)|^p \, d\xi \le c\, \|f\|^p_{V^{0,p}_{\beta',\delta'}(\mathcal{K})} . \end{aligned}$$

This proves (3.5.8) for $T \neq 3$. The proof in the case $T = 3$ proceeds analogously. □

For the cases $T = 1$ and $T = 2$, we also need the following lemma.

LEMMA 3.5.5. *Let $T = 1$ or $T = 2$, and let K and v be the same functions as in Lemma 3.5.4. Then there exists a constant c such that*

$$\int_{\Gamma_j} |x|^{p(\beta'-T)+1} \prod_{k=1}^{d} \Big(\frac{r_k(x)}{|x|}\Big)^{p(\delta'_k - T)+1} \big|v(x)\big|^p \, dx \le c\, \|f\|^p_{V^{0,p}_{\beta',\delta'}(\mathcal{K})}$$

for arbitrary $f \in V^{0,p}_{\beta',\delta'}(\mathcal{K})$, $j = 1, \dots, d$.

P r o o f. Let $\mathcal{K}_x$ denote the set $\{\xi \in \mathcal{K} : |x-\xi| < 3r(x)/4\}$. Furthermore, let α be a real number such that $-T + 3/p < \alpha < 2/p$. Then

$$\begin{aligned} |v(x)|^p &\le c \Big(\int_{\mathcal{K}_x} |x-\xi|^{T-3} |f(\xi)|\, d\xi \Big)^p \\ &\le c \int_{\mathcal{K}_x} |x-\xi|^{-p\alpha}\, |f(\xi)|^p \, d\xi \; \Big(\int\limits_{|x-\xi|<3r(x)/4} |x-\xi|^{p'(T-3+\alpha)}\, d\xi \Big)^{p-1} \\ &\le c\, r(x)^{p(T+\alpha)-3} \int_{\mathcal{K}_x} |x-\xi|^{-p\alpha}\, |f(\xi)|^p \, d\xi. \end{aligned}$$

Consequently,

$$\int_{\Gamma_j} |x|^{p(\beta'-T)+1} \prod_{k=1}^d \Big(\frac{r_k(x)}{|x|}\Big)^{p(\delta'_k-T)+1} |v(x)|^p\, dx$$

$$\le c \int_{\Gamma_j} |x|^{p\beta'} \prod_{k=1}^d \Big(\frac{r_k(x)}{|x|}\Big)^{p\delta'_k} r(x)^{p\alpha-2}\Big(\int_{\mathcal{K}_x} |x-\xi|^{-p\alpha}\, |f(\xi)|^p\, d\xi\Big)\, dx$$

$$\le c \int_{\Gamma_j} |\xi|^{p\beta'} \prod_{k=1}^d \Big(\frac{r_k(\xi)}{|\xi|}\Big)^{p\delta'_k} r(\xi)^{p\alpha-2}\, |f(\xi)|^p \Big(\int\limits_{\substack{\Gamma_j\\ |x-\xi|<3r(\xi)}} |x-\xi|^{-p\alpha}\, dx\Big)\, d\xi$$

$$\le c\, \|f\|^p_{V^{0,p}_{\beta',\delta'}(\mathcal{K})}\, .$$

This proves the lemma. □

3.5.2. A local estimate for the solution. We suppose that the line $\operatorname{Re}\lambda = l-\beta-3/p$ does not contain eigenvalues of the pencil $\mathfrak{A}(\lambda)$ and denote by

$$\Lambda_- < \operatorname{Re}\lambda < \Lambda_+$$

the widest strip in the complex plane which contains this line and which is free of eigenvalues of the pencil $\mathfrak{A}(\lambda)$. Furthermore, let ζ_ν be infinitely differentiable functions depending only on $\rho = |x|$ satisfying the conditions (3.1.4). In the subsequent two lemmas, we consider the vector function

$$v(x) = \int_{\mathcal{K}} G(x,\xi)\, \zeta_\nu(\xi)\, f(\xi)\, d\xi, \tag{3.5.9}$$

where $G(x,\xi)$ is the Green's matrix introduced in Theorem 3.4.1 for an arbitrary κ in the interval

$$m-\Lambda_+-3/2 < \kappa < m-\Lambda_- -3/2\, . \tag{3.5.10}$$

Using Lemmas 3.5.1–3.5.4 together with the estimates for Green's matrix given in the preceding section, we are able show that the function (3.5.9) satisfies the estimate

$$\|\zeta_\mu v\|_{V^{m-1,p}_{\beta-l+m-1,\delta-l+m-1}(\mathcal{K})^\ell} \le c\, 2^{-|\mu-\nu|\varepsilon_0}\, \|\zeta_\nu f\|_{V^{l-2m,p}_{\beta,\delta}(\mathcal{K})^\ell} \tag{3.5.11}$$

with a certain positive ε_0 for all μ and ν. We start with the case $|\mu-\nu|>2$.

LEMMA 3.5.6. *Let v be defined by* (3.5.9), *where $f\in V^{l-2m,p}_{\beta,\delta}(\mathcal{K})^\ell$ and $G(x,\xi)$ is the Green's matrix introduced in Theorem* 3.4.1 *for an arbitrary κ satisfying* (3.5.10). *We assume that the components of δ satisfy the inequalities* (3.5.2). *Then the estimate* (3.5.11) *holds for $|\mu-\nu|>2$, where c and ε_0 are positive constants independent of f, μ and ν.*

P r o o f. First let $l\ge 2m$ and $\mu>\nu+2$. Then $|x|>2|\xi|$ for $2^{\mu-1}<|x|<2^{\mu+1}$, $2^{\nu-1}<|\xi|<2^{\nu+1}$, and $D_x^\alpha G(x,\xi)$ satisfies the first estimate of Theorem 3.4.5. Applying Lemma 3.5.1 with

$$\beta' = \beta-l+2m,\quad \delta'_k=\delta_k-l+2m,\quad \delta''_k=\delta_k-l+|\alpha|,\quad \Lambda=\Lambda_- -|\alpha|+\varepsilon,$$

$$T=2m-|\alpha|,\quad \gamma_k = m-1+\delta_+^{(k)}-|\alpha|-\varepsilon,\quad \gamma'_k = m-1+\delta_-^{(k)}-\varepsilon\, ,$$

we obtain

$$\|\chi_\mu D_x^\alpha v\|_{V^{0,p}_{\beta-l+|\alpha|,\delta-l+|\alpha|}(\mathcal{K})^\ell} \le c\, 2^{(\mu-\nu)(\Lambda_- - l+\beta+\varepsilon+3/p)}\, \|\zeta_\nu f\|_{V^{l-2m,p}_{\beta,\delta}(\mathcal{K})^\ell}$$

for an arbitrary multi-index α. Here, $\Lambda_- - l+\beta+\varepsilon+3/p$ is negative for sufficiently small ε. Since

$$\|\zeta_\mu v\|_{V^{m-1,p}_{\beta-l+m-1,\delta-l+m-1}(\mathcal{K})^\ell} \le c \sum_{|\alpha|\le m-1} \|\chi_\mu D_x^\alpha v\|_{V^{0,p}_{\beta-l+|\alpha|,\delta-l+|\alpha|}(\mathcal{K})^\ell}, \tag{3.5.12}$$

we obtain (3.5.11) for $\mu > \nu+2$. If $\mu < \nu-2$, then we can apply Lemma 3.5.1 with the same $\beta', \delta_k', \delta_k'', \gamma_k, \gamma_k', T$ as above and $\Lambda = \Lambda_+ - |\alpha| - \varepsilon$.

Now let $l < 2m$. Then $\zeta_\nu f$ has the form

$$\zeta_\nu f = \sum_{|\gamma|\le 2m-l} D_x^\gamma f^{(\gamma)}, \tag{3.5.13}$$

where $f^{(\gamma)} \in V^{0,p}_{\beta-l+2m-|\gamma|,\delta-l+2m-|\gamma|}(\mathcal{K})^\ell$, $f^{(\gamma)}(x) = 0$ for $|x| \not\in (2^{\nu-1}, 2^{\nu+1})$, and

$$\sum_{|\gamma|\le 2m-l} \|f^{(\gamma)}\|_{V^{0,p}_{\beta-l+2m-|\gamma|,\delta-l+2m-|\gamma|}(\mathcal{K})^\ell} \le c\, \|\zeta_\nu f\|_{V^{l-2m}_{\beta,\delta}(\mathcal{K})^\ell} \tag{3.5.14}$$

(cf. Lemma 3.1.1). Thus, (3.5.9) can be written as

$$v = \sum_{|\gamma|\le 2m-l} v^{(\gamma)}, \quad \text{where} \quad v^{(\gamma)}(x) = \int_{\mathcal{K}} \big(D_\xi^\gamma G(x,\xi)\big)\, f^{(\gamma)}(\xi)\, d\xi.$$

By Theorem 3.4.5, the elements of the matrix $D_x^\alpha D_\xi^\alpha G(x,\xi)$ satisfy the estimate

$$\big|D_x^\alpha D_\xi^\alpha G(x,\xi)\big| \le c|x|^{\Lambda_- -|\alpha|+\varepsilon}\, |\xi|^{2m-3-\Lambda_- -|\gamma|-\varepsilon} \times \prod_{k=1}^d \Big(\frac{r_k(x)}{|x|}\Big)^{m-1+\delta_+^{(k)}-|\alpha|-\varepsilon} \prod_{k=1}^d \Big(\frac{r_k(\xi)}{|\xi|}\Big)^{m-1+\delta_-^{(k)}-|\gamma|-\varepsilon}$$

for $|x| > 2|\xi|$. Consequently, Lemma 3.5.1 with

$$\beta' = \beta - l + 2m - |\gamma|, \quad \delta_k' = \delta_k - l + 2m - |\gamma|, \quad \delta_k'' = \delta_k - l + |\alpha|,$$
$$\Lambda = \Lambda_- - |\alpha| + \varepsilon, \quad T = 2m - |\alpha| - |\gamma|,$$
$$\gamma_k = m - 1 + \delta_+^{(k)} - |\alpha| - \varepsilon, \quad \gamma_k' = m - 1 + \delta_-^{(k)} - |\gamma| - \varepsilon,$$

yields

$$\|\chi_\mu D_x^\alpha v^{(\gamma)}\|_{V^{0,p}_{\beta-l+|\alpha|,\delta-l+|\alpha|}(\mathcal{K})^\ell} \le c\, 2^{(\mu-\nu)(\Lambda_- - l+\beta+\varepsilon+3/p)}\, \|f^{(\gamma)}\|_{V^{0,p}_{\beta-l+2m-|\gamma|,\delta-l+2m-|\gamma|}(\mathcal{K})^\ell}$$

for $\mu > \nu + 2$. Analogously, the estimate

$$\|\chi_\mu D_x^\alpha v^{(\gamma)}\|_{V^{0,p}_{\beta-l+|\alpha|,\delta-l+|\alpha|}(\mathcal{K})^\ell} \le c\, 2^{(\mu-\nu)(\Lambda_+ - l+\beta-\varepsilon+3/p)}\, \|f^{(\gamma)}\|_{V^{0,p}_{\beta-l+2m-|\gamma|,\delta-l+2m-|\gamma|}(\mathcal{K})^\ell}$$

holds for $\mu < \nu - 2$. This proves the inequality (3.5.11). □

Next, we prove (3.5.11) for the case $|\mu - \nu| \le 2$.

LEMMA 3.5.7. *Let the same assumptions on β and δ as in Lemma 3.5.6 be satisfied. Then the function (3.5.9) satisfies the estimate*

$$\|\zeta_\mu v\|_{V^{m-1,p}_{\beta-l+m-1,\delta-l+m-1}(\mathcal{K})^\ell} \le c\,\|\zeta_\nu f\|_{V^{l-2m,p}_{\beta,\delta}(\mathcal{K})^\ell} \tag{3.5.15}$$

for arbitrary μ,ν, $|\mu-\nu|\le 2$, where c is a constant independent of f, μ and ν.

P r o o f. Suppose first that $l\ge 2m$. We consider the function $w=D^\alpha_x v$ for an arbitrary multi-index α, $|\alpha|\le m-1$. Let the functions $\chi_\pm(x,\xi)$ be defined by

$$\chi_+(x,\xi)=\begin{cases} 1 & \text{for } |x-\xi|\le r(x)/2,\\ 0 & \text{for } |x-\xi|> r(x)/2,\end{cases} \qquad \chi_-(x,\xi)=1-\chi_+(x,\xi).$$

Then

$$w=w_+ + w_-\,, \quad \text{where} \quad w_\pm(x)=\int_{\mathcal{K}} \chi_\pm(x,\xi)\, D^\alpha_x G(x,\xi)\,\zeta_\nu(\xi)\,f(\xi)\,d\xi.$$

By Theorems 3.4.2 and 3.4.3, the kernels $\chi_\pm(x,\xi)\,D^\alpha_x G(x,\xi)$ satisfy the conditions of Lemmas 3.5.3 and 3.5.4, respectively, with

$$T=2m-|\alpha|,\quad \gamma_k=m-1+\delta_+^{(k)}-|\alpha|-\varepsilon,\quad \gamma'_k=m-1+\delta_-^{(k)}-\varepsilon.$$

We put $\beta'=\beta-l+2m$ and $\delta'_k=\delta_k-l+2m$ for $k=1,\dots,d$. Then by Lemma 3.5.3, we have

$$\|\chi_\mu w_-\|_{V^{0,p}_{\beta-l+|\alpha|,\delta-l+|\alpha|}(\mathcal{K})^\ell} \le c\,\|\zeta_\nu f\|_{V^{l-2m,p}_{\beta-l+2m,\delta-l+2m}(\mathcal{K})^\ell}\,.$$

Lemma 3.5.4 implies the same estimate for $\chi_\mu w_+$. Using (3.5.12), we obtain (3.5.15)

In the case $l<2m$, the function $\zeta_\nu f$ admits the representation (3.5.13), where $f^{(\gamma)}$ are functions in $V^{0,p}_{\beta-l+2m-|\gamma|,\delta-l+2m-|\gamma|}(\mathcal{K})^\ell$ which are zero outside the set $\{x:\, 2^{\nu-1}<|x|<2^{\nu+1}\}$ and satisfy the inequality (3.5.14). Then

$$v=\sum_{|\gamma|\le 2m-l} v^{(\gamma)},\quad \text{where} \quad v^{(\gamma)}(x)=\int_{\mathcal{K}} \big(D^\alpha_\xi G(x,\xi)\big)\,f^{(\gamma)}(\xi)\,d\xi.$$

We write the function $D^\alpha_x v^{(\gamma)}$ in the form

$$D^\alpha_x v^{(\gamma)}=w_+ + w_-\,,\quad \text{where} \quad w_\pm(x)=\int_{\mathcal{K}} \chi_\pm(x,\xi)\,\big(D^\alpha_x D^\alpha_\xi G(x,\xi)\big)\,f^{(\gamma)}(\xi)\,d\xi.$$

Applying Lemma 3.5.3 with $T=2m-|\alpha|-|\gamma|$,

$$\beta'=\beta-l+2m-|\gamma|,\quad \delta'_k=\delta_k-l+2m-|\gamma|,\quad \delta''_k=\delta_k-l+|\alpha|,$$
$$\gamma_k=m-1+\delta_+^{(k)}-|\alpha|-\varepsilon,\quad \gamma'_k=m-1+\delta_-^{(k)}-|\gamma|-\varepsilon\,,$$

we obtain the estimate

$$\|\chi_\mu w_-\|_{V^{0,p}_{\beta-l+|\alpha|,\delta-l+|\alpha|}(\mathcal{K})^\ell} \le c\,\|f^{(\gamma)}\|_{V^{0,p}_{\beta-l+2m-|\gamma|,\delta-l+2m-|\gamma|}(\mathcal{K})^\ell}\,.$$

The same estimate for $\chi_\mu w_+$ results from Lemma 3.5.4. This together with (3.5.12) and (3.5.14) implies (3.5.15). The proof of the lemma is complete. □

3.5.3. Existence of solutions in $V^{l,p}_{\beta,\delta}(\mathcal{K})^\ell$. We consider the operator

$$V^{l-2m,p}_{\beta,\delta}(\mathcal{K})^\ell \ni f \to \mathcal{O}f = u$$

defined by (3.5.1). Our goal is to show that this operator realizes a continuous mapping from $V^{l-2m,p}_{\beta,\delta}(\mathcal{K})^\ell$ into $V^{l,p}_{\beta,\delta}(\mathcal{K})^\ell$ if the line $\operatorname{Re}\lambda = l - \beta - 3/p$ is free of eigenvalues of the pencil $\mathfrak{A}(\lambda)$ and the inequalities (3.5.2) are satisfied for the components of δ. For this, we need the following lemma.

LEMMA 3.5.8. *Let $\mathcal{X}$, $\mathcal{Y}$ be Banach spaces of functions on $\mathcal{K}$ in each of them the multiplication with $C_0^\infty(\overline{\mathcal{K}}\backslash\{0\})$-functions is defined. We suppose that the inequalities*

$$(3.5.16)\qquad \|f\|_{\mathcal{X}} \ge c\Big(\sum_{\nu=-\infty}^{+\infty} \|\zeta_\nu f\|^p_{\mathcal{X}}\Big)^{1/p}, \qquad \|u\|_{\mathcal{Y}} \le c\Big(\sum_{\nu=-\infty}^{+\infty} \|\zeta_\nu u\|^p_{\mathcal{Y}}\Big)^{1/p}$$

are satisfied for all $f \in \mathcal{X}$, $u \in \mathcal{Y}$. Here, ζ_ν are infinitely differentiable functions depending only on $\rho = |x|$ and satisfying the conditions (3.1.4). *Furthermore, let $\mathcal{O}$ be a linear operator from $\mathcal{X}$ into $\mathcal{Y}$ defined on functions with compact support in $\overline{\mathcal{K}}\backslash\{0\}$ such that*

$$(3.5.17)\qquad \|\zeta_\mu \mathcal{O}\zeta_\nu f\|_{\mathcal{Y}} \le c\, 2^{-|\mu-\nu|\varepsilon_0}\, \|\zeta_\nu f\|_{\mathcal{X}}$$

with positive constants c, ε_0 independent of μ, ν and f. Then $\|\mathcal{O}f\|_{\mathcal{Y}} \le c\,\|f\|_{\mathcal{X}}$ for all $f \in \mathcal{X}$ with compact support in $\overline{\mathcal{K}}\backslash\{0\}$.

P r o o f. According to (3.5.16),

$$\begin{aligned}\|\mathcal{O}f\|_{\mathcal{Y}} &= \Big\|\mathcal{O}\sum_{\nu=-\infty}^{+\infty} \zeta_\nu f\Big\|_{\mathcal{Y}} \le c\Big(\sum_{\mu=-\infty}^{+\infty}\Big\|\sum_{\nu=-\infty}^{+\infty} \zeta_\mu \mathcal{O}\zeta_\nu f\Big\|^p_{\mathcal{Y}}\Big)^{1/p} \\ &\le c\Big[\sum_{\mu=-\infty}^{+\infty}\Big(\sum_{\nu=-\infty}^{+\infty} \|\zeta_\mu \mathcal{O}\zeta_\nu f\|_{\mathcal{Y}}\Big)^p\Big]^{1/p}.\end{aligned}$$

This together with (3.5.17) implies

$$\|\mathcal{O}f\|_{\mathcal{Y}} \le c\Big[\sum_{\mu=-\infty}^{+\infty}\Big(\sum_{\nu=-\infty}^{+\infty} 2^{-|\mu-\nu|\varepsilon_0}\, \|\zeta_\nu f\|_{\mathcal{X}}\Big)^p\Big]^{1/p}.$$

Since the operator of discrete convolution with kernel $\{2^{-|k|\varepsilon_0}\}_{k=-\infty}^{+\infty}$ acts continuously in l_p, it follows that

$$\|\mathcal{O}f\|_{\mathcal{Y}} \le c\Big(\sum_{\nu=-\infty}^{+\infty} \|\zeta_\nu f\|^p_{\mathcal{X}}\Big)^{1/p}.$$

The last inequality together with (3.5.16) leads to the desired estimate for the norm of $\mathcal{O}f$. □

As a consequence of the last lemma, we obtain the following assertion for the operator (3.5.1).

COROLLARY 3.5.9. *We assume that the line $\operatorname{Re}\lambda = l-\beta-3/p$ does not contain eigenvalues of the pencil $\mathfrak{A}(\lambda)$ and that the components of δ satisfy the inequalities*

(3.5.2). *If* $G(x,\xi)$ *is the Green's matrix introduced in Theorem* 3.4.1 *for an arbitrary* κ *satisfying* (3.5.10), *then the operator*

$$V_{\beta,\delta}^{l-2m,p}(\mathcal{K})^\ell \ni f \to \mathcal{O}f = u$$

defined by (3.5.1) *realizes a continuous mapping from* $V_{\beta,\delta}^{l-2m,p}(\mathcal{K})^\ell$ *into the space* $V_{\beta-l+m-1,\delta-l+m-1}^{m-1,p}(\mathcal{K})^\ell$.

P r o o f. By Lemmas 3.5.6 and 3.5.7, the inequality

$$\|\zeta_\mu \mathcal{O} \zeta_\nu f\|_{V_{\beta-l+m-1,\delta-l+m-1}^{m-1,p}(\mathcal{K})^\ell} \le c\, 2^{-|\mu-\nu|\varepsilon_0}\, \|\zeta_\nu f\|_{V_{\beta,\delta}^{l-2m,p}(\mathcal{K})^\ell}$$

is satisfied for all μ and ν. Since the subset of functions with compact support in $\overline{\mathcal{K}}\backslash\{0\}$ is dense in $V_{\beta,\delta}^{l-2m,p}(\mathcal{K})$, it follows from Lemma 3.5.8 that $u \in V_{\beta-l+m-1,\delta-l+m-1}^{m-1,p}(\mathcal{K})^\ell$ and

$$\|u\|_{V_{\beta-l+m-1,\delta-l+m-1}^{m-1,p}(\mathcal{K})^\ell} \le c\, \|f\|_{V_{\beta,\delta}^{l-2m,p}(\mathcal{K})^\ell}\,.$$

This proves the corollary. □

Now we are able to prove the main result of this section.

THEOREM 3.5.10. *Let* $f \in V_{\beta,\delta}^{l-2m,p}(\mathcal{K})^\ell$, $l \ge m$, *and* $g_{j,k} \in V_{\beta,\delta}^{l-k+1-1/p}(\Gamma_j)^\ell$ *for* $j = 1,\dots,d$, $k = 1,\dots,m$. *We assume that the line* $\operatorname{Re}\lambda = l - \beta - 3/p$ *does not contain eigenvalues of the pencil* $\mathfrak{A}(\lambda)$ *and that the components of* δ *satisfy the inequalities* (3.5.2). *Then there exists a unique solution* $u \in V_{\beta,\delta}^{l,p}(\mathcal{K})^\ell$ *of the problem* (3.3.1) *satisfying the estimate*

$$\|u\|_{V_{\beta,\delta}^{l,p}(\mathcal{K})^\ell} \le c \Big(\|f\|_{V_{\beta,\delta}^{l-2m,p}(\mathcal{K})^\ell} + \sum_{j=1}^{d} \sum_{k=1}^{m} \|g_{j,k}\|_{V_{\beta,\delta}^{l-k+1-1/p}(\Gamma_j)^\ell} \Big)$$

with a constant c *independent of* f *and* $g_{j,k}$.

P r o o f. By Lemma 3.3.1, we may restrict ourselves in the proof to the case of zero boundary data $g_{j,k}$. Let $f \in V_{\beta,\delta}^{l-2m,p}(\mathcal{K})^\ell$ and let $\{f^{(\nu)}\} \subset C_0^\infty(\overline{\mathcal{K}}\backslash\mathcal{S})^\ell$ be a sequence converging to f in $V_{\beta,\delta}^{l-2m,p}(\mathcal{D})^\ell$. We consider the corresponding sequence of solutions $u^{(\nu)} \in \overset{\circ}{V}{}_{\kappa,0}^{m,2}(\mathcal{K})^\ell$ of the equation $L(D_x)u = f^{(\nu)}$ (see Theorem 3.3.9). According to Theorem 3.4.1, these solutions have the representation

$$u^{(\nu)}(x) = \int_{\mathcal{K}} G(x,\xi)\, f^{(\nu)}(\xi)\, d\xi.$$

It follows from Corollary 3.5.9 that $\{u^{(\nu)}\}$ is a Cauchy sequence in the space $V_{\beta-l+m-1,\delta-l+m-1}^{m-1,p}(\mathcal{K})^\ell$. By Theorem 3.3.5, it is also a Cauchy sequence in $V_{\beta,\delta}^{l,p}(\mathcal{K})^\ell$. The limit u is a solution of the equation $L(D_x)u = f$ in $\mathcal{K}$ satisfying the Dirichlet conditions $\partial_n^{k-1}u = 0$ on the faces Γ_j for $k = 1,\dots,m$ and the estimate

$$\|u\|_{V_{\beta,\delta}^{l,p}(\mathcal{K})^\ell} \le c\, \|f\|_{V_{\beta,\delta}^{l-2m,p}(\mathcal{K})^\ell}\,.$$

We prove the uniqueness of the solution. Let $u \in V_{\beta,\delta}^{l,p}(\mathcal{K})^\ell \cap \overset{\circ}{V}{}_{\beta-l+m,\delta-l+m}^{m,p}(\mathcal{K})^\ell$ be a solution of the equation $L(D_x)u = 0$ in $\mathcal{K}$. Then

$$\big(u, L^+(D_x)v\big)_{\mathcal{K}} = \big(L(D_x)u, v\big)_{\mathcal{K}} = 0 \quad \text{for all } v \in \overset{\circ}{V}{}_{-\beta+l-m,-\delta+l-m}^{m,p'}(\mathcal{K})^\ell,$$

$p' = p/(p-1)$. Applying the solvability result obtained in the first part of the proof to the Dirichlet problem for the formally adjoint operator $L^+(D_x)$, we conclude that for every $w \in V^{-m,p'}_{-\beta+l-m,-\delta+l-m}(\mathcal{K})^\ell$ there exists a solution $v \in \overset{\circ}{V}{}^{m,p'}_{-\beta+l-m,-\delta+l-m}(\mathcal{K})^\ell$ of the equation $L^+(D_x)v = w$. Consequently, we have

$$(u, w)_{\mathcal{K}} = 0 \quad \text{for all } w \in V^{-m,p'}_{-\beta+l-m,-\delta+l-m}(\mathcal{K})^\ell.$$

This means that $u = 0$. The proof of the theorem is complete. □

3.5.4. Necessity of the conditions on β and δ. We prove that the conditions in Theorem 3.5.10 are necessary. We start with the condition on δ.

LEMMA 3.5.11. *Suppose that $l \ge m$ and there exists a constant c such that*

$$\|u\|_{V^{l,p}_{\beta,\delta}(\mathcal{K})^\ell} \le c\|L(D_x)\,u\|_{V^{l-2m,p}_{\beta,\delta}(\mathcal{K})^\ell} \tag{3.5.18}$$

for all $u \in V^{l,p}_{\beta,\delta}(\mathcal{K})^\ell \cap \overset{\circ}{V}{}^{m,p}_{\beta-l+m,\delta-l+m}(\mathcal{K})^\ell$. Then $\delta_k - l + m - 1 + 2/p < \delta_-^{(k)}$ for $k = 1, \dots, m$.

P r o o f. From Theorem 3.3.5 and (3.5.18) and from the assumptions of the lemma it follows that

$$\|u\|_{V^{l+k,p}_{\beta+k,\delta+k}(\mathcal{K})^\ell} \le c\|L(D_x)\,u\|_{V^{l+k,p}_{\beta+k,\delta+k}(\mathcal{K})^\ell}$$

for all $u \in V^{l+k,p}_{\beta+k,\delta+k}(\mathcal{K})^\ell \cap \overset{\circ}{V}{}^{m,p}_{\beta-l+m,\delta-l+m}(\mathcal{K})^\ell$. Therefore, we may assume without loss of generality that $l \ge 2m$. Let ξ be a point on the edge M_k, $|\xi| = 1$. There exist a neighborhood $\mathcal{U}$ of ξ and a diffeomorphism κ mapping ξ onto the origin and $\mathcal{K} \cap \mathcal{U}$ onto a subset of a dihedron $\mathcal{D}$. Furthermore, we may assume that the Jacobian matrix $\kappa'(\xi)$ coincides with the identity matrix. Suppose that $\operatorname{supp} u \subset \mathcal{U}$. In the new coordinates $y = \kappa(x)$, the equation $L(D_x)\,u = f$ takes the form $\tilde{L}(y, D_y)\,\tilde{u}(y) = \tilde{f}(y)$, where $\tilde{u} = u \circ \kappa^{-1}$, $\tilde{f} = f \circ \kappa^{-1}$, and $\tilde{L}(y, D_y)$ is a differential operator whose principal part coincides with $L(D_y)$ at the origin. By (3.5.18), the inequality

$$\|u\|_{V^{l,p}_{\delta_k}(\mathcal{D})^\ell} \le c\,\|\tilde{L}(x, D_x)\,u\|_{V^{l-2m,p}_{\delta_k}(\mathcal{D})^\ell}$$

holds for all $u \in V^{l,p}_{\delta_k}(\mathcal{D})^\ell \cap \overset{\circ}{V}{}^{m,p}_{\delta_k-l+m}(\mathcal{D})^\ell$, $\operatorname{supp} u \in \kappa(\mathcal{U})$. If $u(x) = 0$ for $|x| > \varepsilon$ and ε is sufficiently small, then it follows that

$$\|u\|_{V^{l,p}_{\delta_k}(\mathcal{D})^\ell} \le 2c\,\|L(D_x)\,u\|_{V^{l-2m,p}_{\delta_k}(\mathcal{D})^\ell}\,.$$

Using the mapping $y = Nx/\varepsilon$, we obtain this estimate with the same constant $2c$ for functions on $\mathcal{D}$ with support in a ball with radius N. Hence, we conclude from Lemma 2.6.6 that $\delta_k - l + m - 1 + 2/p < \delta_-^{(k)}$. □

THEOREM 3.5.12. *Suppose that the equation $L(D_x)\,u = f$ has a uniquely determined solution $u \in V^{l,p}_{\beta,\delta}(\mathcal{K})^\ell \cap \overset{\circ}{V}{}^{m,p}_{\beta-l+m,\delta-l+m}(\mathcal{K})^\ell$ for arbitrary $f \in V^{l-2m,p}_{\beta,\delta}(\mathcal{K})^\ell$. Then the line $\operatorname{Re}\lambda = l - \beta - 3/p$ does not contain eigenvalues of the pencil $\mathfrak{A}(\lambda)$ and the components of δ satisfy the inequalities* (3.5.2).

P r o o f. The conditions on δ follow from Lemma 3.5.11 and from the analogous result for the formally adjoint problem. Suppose that λ_0 is an eigenvalue of the pencil $\mathfrak{A}(\lambda)$ on the line $\operatorname{Re}\lambda = l - \beta - 3/p$ and that U is an eigenfunction corresponding to this eigenvalue, $U \in \overset{\circ}{W}{}^{m,2}(\Omega)^\ell$. From the regularity results for solutions of elliptic boundary value problems in domains with angular points (cf. Corollary 1.2.7) it follows that $U \in W_\delta^{l,p}(\Omega)^\ell$. We consider the vector functions

$$u^{(N)}(x) = \chi_N(\rho)\, \rho^{\lambda_0}\, U(\omega),$$

where χ_N are infinitely differentiable functions of the variable $\rho = |x|$ satisfying the conditions $\chi_N(\rho) = 1$ for $2^{-N} < \rho < 2^N$, $\chi_N(\rho) = 0$ for $\rho < 2^{-N-1}$ and $\rho > 2^{N+1}$, and $|\partial_\rho^j \chi_N(\rho)| \le c_j\, \rho^{-j}$ for all integer $j \ge 0$. Here, c_j are constants independent of N. Obviously, the $V^{0,p}_{\beta-l,\delta-l}$-norm and therefore also the $V^{l,p}_{\beta,\delta}$-norm of $u^{(N)}$ tend to infinity as $N \to \infty$. On the other hand, $L(D_x)\, u^{(N)}$ is a sum of terms of the form

$$c\,(\partial_x^\alpha \chi_N)\, \partial_x^\gamma (\rho^{\lambda_0}\, U(\omega))$$

where $|\alpha| + |\gamma| = 2m$, $|\alpha| \ne 0$. In particular, $L(D_x)\, u^{(N)}$ vanishes outside the sets $2^{-N-1} < |x| < 2^{-N}$ and $2^N < |x| < 2^{N+1}$. Consequently, the norm of $L(D_x)\, u^{(N)}$ in $V^{l-2m,p}_{\beta,\delta}(\mathcal{K})^\ell$ can be estimated by a constant independent of N. Thus, the inequality (3.5.18) cannot be valid for all $u \in V^{l,p}_{\beta,\delta}(\mathcal{K})^\ell \cap \overset{\circ}{V}{}^{m,p}_{\beta-l+m,\delta-l+m}(\mathcal{K})^\ell$. This contradicts the assumptions of the theorem. □

3.5.5. A regularity assertion for the solution. Since the solutions in different spaces $V^{l,p}_{\beta,\delta}(\mathcal{K})^\ell$ have an integral representation with the same Green's matrix, we obtain the following regularity result.

THEOREM 3.5.13. *Let $u \in V^{l,p}_{\beta,\delta}(\mathcal{K})^\ell$ be a solution of the boundary value problem* (3.3.1), *where*

$$f \in V^{l-2m,p}_{\beta,\delta}(\mathcal{K})^\ell \cap V^{l'-2m,q}_{\beta',\delta'}(\mathcal{K})^\ell,\ l \ge m,\ l' \ge m,$$

$$g_{j,k} \in V^{l-k+1-1/p,p}_{\beta,\delta}(\Gamma_j)^\ell \cap V^{l'-k+1-1/q,q}_{\beta',\delta'}(\Gamma_j)^\ell.$$

Suppose that the closed strip between the lines $\operatorname{Re}\lambda = l - \beta - 3/p$ *and* $\operatorname{Re}\lambda = l' - \beta' - 3/q$ *is free of eigenvalues of the pencil* $\mathfrak{A}_j(\lambda)$ *and that the components of* δ *and* δ' *satisfy the inequalities*

$$-\delta_+^{(k)} < \delta_k - l + m - 1 + 2/p < \delta_-^{(k)}, \quad -\delta_+^{(k)} < \delta'_k - l' + m - 1 + 2/q < \delta_-^{(k)}.$$

Then $u \in V^{l,q}_{\beta',\delta'}(\mathcal{K})^\ell$.

P r o o f. Lemma 3.3.1 allows us to restrict ourselves in the proof to the case of zero boundary data $g_{j,k}$. By Theorem 3.5.10, there exist solutions $u \in V^{l,p}_{\beta,\delta}(\mathcal{K})^\ell$ and $v \in V^{l',q}_{\beta',\delta'}(\mathcal{K})^\ell$ of problem (3.3.1). Both solutions are given by the formula (3.5.1), where $G(x,\xi)$ is the Green's matrix introduced in Theorem 3.4.1 for $\kappa = \beta + m - l + (6-3p)/(2p)$. This proves the theorem. □

3.6. Solvability in weighted Hölder spaces

This section is dedicated to solutions of the Dirichlet problem in weighted Hölder spaces. The norm in the weighted Hölder space $N^{l,\sigma}_{\beta,\delta}(\mathcal{K})$ has a similar structure as in the weighted Sobolev space $V^{l,p}_{\beta,\delta}(\mathcal{K})$. Therefore, we obtain solvability and regularity assertions in the spaces $N^{l,\sigma}_{\beta,\delta}(\mathcal{K})$ which are analogous to the results of the preceding section. In particular, we prove that the problem (3.3.1) is uniquely solvable in $N^{l,\sigma}_{\beta,\delta}(\mathcal{K})^\ell$ for arbitrary $f \in N^{l-2m,\sigma}_{\beta,\delta}(\mathcal{K})^\ell$, $g_{j,k} \in N^{l-k+1,\sigma}_{\beta,\delta}(\Gamma_j)^\ell$ if the line

$\mathrm{Re}\,\lambda = l+\sigma-\beta$ is free of eigenvalues of the pencil $\mathfrak{A}(\lambda)$ and the components of δ satisfy the condition

$$-\delta_+^{(k)} < \delta_k - l + m - 1 - \sigma < \delta_-^{(k)} \quad \text{for } k=1,\dots,d. \tag{3.6.1}$$

For the proof, we apply again the point estimates of the Green's matrix obtained in Section 3.4.

3.6.1. Weighted Hölder spaces in a cone. Let l be a nonnegative integer, $\beta \in \mathbb{R}$, $\delta = (\delta_1,\dots,\delta_d) \in \mathbb{R}^d$, and $0<\sigma<1$. Then the weighted Hölder space $N_{\beta,\delta}^{l,\sigma}(\mathcal{K})$ is defined as the set of all l times continuously differentiable functions on $\overline{\mathcal{K}}\backslash\mathcal{S}$ with finite norm

$$\|u\|_{N_{\beta,\delta}^{l,\sigma}(\mathcal{K})} = \sum_{|\alpha|\le l} \sup_{x\in\mathcal{K}} |x|^{\beta-l-\sigma+|\alpha|} \prod_{k=1}^d \Big(\frac{r_k(x)}{|x|}\Big)^{\delta_k-l-\sigma+|\alpha|} \big|\partial_x^\alpha u(x)\big| + \langle u\rangle_{l,\sigma,\beta,\delta;\mathcal{K}}\,, \tag{3.6.2}$$

where

$$\langle u\rangle_{l,\sigma,\beta,\delta;\mathcal{K}} = \sum_{|\alpha|=l} \sup_{|x-y|<r(x)/2} |x|^\beta \prod_{k=1}^d \Big(\frac{r_k(x)}{|x|}\Big)^{\delta_k} \frac{|\partial_x^\alpha u(x) - \partial_y^\alpha u(y)|}{|x-y|^\sigma}\,.$$

The corresponding trace space on the face Γ_j is denoted by $N_{\beta,\delta}^{l,\sigma}(\Gamma_j)$. If $\delta = (\delta_1,\dots,\delta_d) \in \mathbb{R}^d$ and t is an arbitrary real number, then by $N_{\beta,\delta+t}^{l,\sigma}(\mathcal{K})$ we mean the weighted space $N_{\beta,\delta'}^{l,\sigma}(\mathcal{K})$, where $\delta' = (\delta_1+t,\dots,\delta_d+t)$.

Analogously to Lemma 2.7.1, the following assertions hold.

LEMMA 3.6.1. *Let $l+\sigma \ge l'+\sigma'$, $l+\sigma-\beta = l'+\sigma'-\beta'$, and $l+\sigma-\delta_k \ge l'+\sigma'-\delta_k'$ for $k=1,\dots,d$. Then*

$$N_{\beta,\delta}^{l,\sigma}(\mathcal{K}) \subset N_{\beta',\delta'}^{l',\sigma'}(\mathcal{K}) \quad \textit{and} \quad N_{\beta,\delta}^{l,\sigma}(\Gamma_j) \subset N_{\beta',\delta'}^{l',\sigma'}(\Gamma_j).$$

These imbeddings are continuous.

P r o o f. It is sufficient, to prove the lemma for real-valued functions. The assertion is obvious for $l=l'$ and $\sigma \ge \sigma'$. Suppose that $l > |\alpha| = l'$. For arbitrary $x\in\mathcal{K}$, let B_x denote the set $\{z\in\mathcal{K}:\ |x-z|<r(x)/2\}$. Obviously

$$|x|/2 < |z| < 3|x|/2 \ \text{ and } \ r_k(x)/2 < r_k(z) < 3r_k(x)/2 \ \text{ for } z\in B_x\,.$$

Furthermore for every $y\in B_x$ there exist a constant c independent of x and y and a point $z\in B_x$ such that

$$\partial_x^\alpha u(x) - \partial_y^\alpha u(y) \le c\,|x-y|\, \big|\nabla_z \partial_z^\alpha u(z)\big|.$$

Consequently,

$$\begin{aligned}\frac{|\partial_x^\alpha u(x) - \partial_y^\alpha u(y)|}{|x-y|^{\sigma'}} &\le c\,|x-y|^{1-\sigma'}\, \big|\nabla_z\partial_z^\alpha u(z)\big| \le c\,r(x)^{1-\sigma'}\, \big|\nabla_z\partial_z^\alpha u(z)\big| \\ &\le c\,2^{1-\sigma'}\, r(z)^{1-\sigma'}\, \big|\nabla_z\partial_z^\alpha u(z)\big|\end{aligned}$$

and

$$(3.6.3)\quad \langle u\rangle_{l',\sigma',\beta',\delta';\mathcal{K}} \le c \sum_{|\alpha|=l'+1} \sup_{x\in\mathcal{K}} |x|^{\beta'+1-\sigma'} \prod_{k=1}^{d} \Big(\frac{r_k(x)}{|x|}\Big)^{\delta'_k+1-\sigma'} \big|\partial_x^\alpha u(x)\big| \le c\,\|u\|_{N^{l,\sigma}_{\beta,\delta}(\mathcal{K})}\,.$$

This proves the imbedding $N^{l,\sigma}_{\beta,\delta}(\mathcal{K}) \subset N^{l',\sigma'}_{\beta',\delta'}(\mathcal{K})$. Analogously the imbedding $N^{l,\sigma}_{\beta,\delta}(\Gamma_j) \subset N^{l',\sigma'}_{\beta',\delta'}(\Gamma_j)$ holds. □

Next, we prove a relation between the spaces $V^{l,s}_{\beta,\delta}$ and $N^{l,\sigma}_{\beta,\delta}$.

LEMMA 3.6.2. *Suppose that $l-3/s > l'+\sigma$, $\beta - l + 3/s = \beta' - l' - \sigma$ and $\delta_k - l + 3/s \le \delta'_k - l' - \sigma$ for $k=1,\dots,d$. Then $V^{l,s}_{\beta,\delta}(\mathcal{K})$ is continuously imbedded in $N^{l',\sigma}_{\beta',\delta'}(\mathcal{K})$.*

P r o o f. It suffices to prove the lemma for the case where $\delta_k - l+3/s = \delta'_k - l' - \sigma$, $k=1,\dots,d$. Let $u \in V^{l,s}_{\beta,\delta}(\mathcal{K})$ be given, and let B_x be the same set as in the proof of Lemma 3.6.1. Obviously,

$$(3.6.4)\quad \frac{|x|}{2} \le |y| < \frac{3|x|}{2},\quad \frac{r_k(x)}{2} \le r_k(y) < \frac{3r_k(x)}{2},\quad \frac{r(x)}{2} \le r(y) \le \frac{3r(x)}{2}$$

for $y \in B_x$. First let $r(x)=1$. From the continuity of the imbedding $W^{l,s}(B_x) \subset C^{l',\sigma}(B_x)$ it follows that there exists a constant c independent of u and x such that

$$\big|(\partial^\alpha u)(x)\big| \le c\,\|u\|_{W^{l,s}(B_x)} \quad \text{for } |\alpha| \le l'$$

and

$$\frac{|(\partial^\alpha u)(x) - (\partial^\alpha u)(y)|}{|x-y|^\sigma} \le c\,\|u\|_{W^{l,s}(B_x)} \quad \text{for } |\alpha| = l',\ y \in B_x\,.$$

By (3.1.2) and (3.6.4), this implies

$$|x|^\beta \prod_k \Big(\frac{r_k(x)}{|x|}\Big)^{\delta_k} |(\partial^\alpha u)(x)| \le c \sum_{|\gamma|\le l} \Big\| r^{|\gamma|-l} \rho^\beta \prod_k \Big(\frac{r_k}{\rho}\Big)^{\delta_k} \partial^\gamma u \Big\|_{L_s(B_x)} \le c\,\|u\|_{V^{l,s}_{\beta,\delta}(\mathcal{K})}$$

for $|\alpha| \le l'$ and analogously

$$|x|^\beta \prod_k \Big(\frac{r_k(x)}{|x|}\Big)^{\delta_k} \frac{|(\partial^\alpha u)(x) - (\partial^\alpha u)(y)|}{|x-y|^\sigma} \le c\,\|u\|_{V^{l,s}_{\beta,\delta}(\mathcal{K})}$$

for $|\alpha| = l'$, $y \in B_x$. Now let x be an arbitrary point in $\mathcal{K}$ and $y \in B_x$. We set $\xi = x/r(x)$, $\eta = y/r(x)$. Then $r(\xi) = 1$ and $\eta \in B_\xi$. Consequently, the function $v(z) = u\big(r(x)\,z\big)$ satisfies the inequalities

$$|\xi|^\beta \prod_k \Big(\frac{r_k(\xi)}{|\xi|}\Big)^{\delta_k} |(\partial^\alpha v)(\xi)| \le c\,\|v\|_{V^{l,s}_{\beta,\delta}(\mathcal{K})} \le c\,r(x)^{l-\beta-3/s}\,\|u\|_{V^{l,s}_{\beta,\delta}(\mathcal{K})}$$

for $|\alpha| \le l'$ and

$$|y|^\beta \prod_k \Big(\frac{r_k(\xi)}{|\xi|}\Big)^{\delta_k} \frac{|(\partial^\alpha v)(\xi) - (\partial^\alpha v)(\eta)|}{|\xi-\eta|^\sigma} \le c\,r(x)^{l-\beta-3/s}\,\|u\|_{V^{l,s}_{\beta,\delta}(\mathcal{K})}$$

for $|\alpha| = l'$. Here the constant c is independent of u, x and y. Using (3.1.2), we obtain the inequalities

$$|x|^{\beta-l+|\alpha|+3/s} \prod_k \Big(\frac{r_k(x)}{|x|}\Big)^{\delta_k-l+|\alpha|+3/s} \big|(\partial^\alpha u)(x)\big| \le c\, \|u\|_{V^{l,s}_{\beta,\delta}(\mathcal{K})} \quad \text{for } |\alpha| \le l'$$

and

$$|x|^{\beta'} \prod_k \Big(\frac{r_k(x)}{|x|}\Big)^{\delta'_k} \frac{|(\partial^\alpha u)(x) - (\partial^\alpha u)(y)|}{|x-y|^\sigma} \le c\, \|u\|_{V^{l,s}_{\beta,\delta}(\mathcal{K})} \quad \text{for } |\alpha| = l'.$$

This proves the lemma. □

Finally, we introduce weighted Hölder spaces of negative order. If l is a nonnegative integer, then $N^{l,\sigma}_{\beta,\delta}(\mathcal{K})$ denotes the set of all distributions of the form

$$u = \sum_{|\alpha|\le l} \partial_x^\alpha\, u_\alpha\,, \quad \text{where } u_\alpha \in N^{0,\sigma}_{\beta+l-|\alpha|,\delta+l-|\alpha|}(\mathcal{K}). \tag{3.6.5}$$

The norm in this space is defined as

$$\|u\|_{N^{l,\sigma}_{\beta,\delta}(\mathcal{K})} = \inf \sum_{|\alpha|\le l} \|u_\alpha\|_{N^{0,\sigma}_{\beta+l-|\alpha|,\delta+l-|\alpha|}(\mathcal{K})}\,,$$

where the infimum is taken over all decompositions of the form (3.6.5).

3.6.2. Reduction to zero boundary data. The following lemma allows us to restrict ourselves to zero boundary data.

LEMMA 3.6.3. *Let $g_{j,k} \in N^{l-k,\sigma}_{\beta,\delta}(\Gamma_j)$, $j = 1,\dots,d$, $k = 0,1,\dots,l$. Then there exists a function $u \in N^{l,\sigma}_{\beta,\delta}(\mathcal{K})$ such that*

$$\frac{\partial^k u}{\partial n^k} = g_{j,k} \quad \text{on } \Gamma_j,\ j = 1,\dots,d,\ k = 0,\dots,l, \tag{3.6.6}$$

and

$$\|u\|_{N^{l,\sigma}_{\beta,\delta}(\mathcal{K})} \le c \sum_{j=1}^{d}\sum_{k=0}^{l} \|g_{j,k}\|_{N^{l-k,\sigma}_{\beta,\delta}(\Gamma_j)}\,, \tag{3.6.7}$$

where c is a constant independent of the functions $g_{j,k}$.

P r o o f. The proof is essentially the same as for Lemma 3.3.1. Let ζ_ν be infinitely differentiable functions depending only on $\rho = |x|$ satisfying the conditions (3.1.4). Furthermore, let $\tilde\zeta_\nu(x) = \zeta_\nu(2^\nu x)$ and $g^{(\nu)}_{j,k}(x) = g_{j,k}(2^\nu x)$. Then $\tilde\zeta_\nu(x)\, g^{(\nu)}_{j,k}(x) = 0$ for $|x| > 2$ and $|x| < 1/2$. By Lemma 2.8.2, there exists a function $v_\nu \in N^{l,\sigma}_{\beta,\delta}(\mathcal{K})$ vanishing outside the set $\{x : 1/4 < |x| < 4\}$ such that

$$\frac{\partial^k v_\nu}{\partial n^k} = 2^{k\nu}\, \tilde\zeta_\nu\, g^{(\nu)}_{j,k} \quad \text{on } \Gamma_j\,, j = 1,\dots,d,\ k = 1,\dots,l,$$

and

$$\|v_\nu\|_{N^{l,\sigma}_{\beta,\delta}(\mathcal{K})} \le c \sum_{j=1}^{d} \sum_{k=0}^{l} 2^{k\nu} \|\tilde{\zeta}_\nu\, g^{(\nu)}_{j,k}\|_{N^{l-k,\sigma}_{\beta,\delta}(\Gamma_j)} \le c'\, 2^{\nu(l+\sigma-\beta)} \sum_{j=1}^{d} \sum_{k=0}^{l} \|\zeta_\nu\, g_{j,k}\|_{N^{l-k,\sigma}_{\beta,\delta}(\Gamma_j)}$$

with constants c and c' independent of the functions $g_{j,k}$ and ν. We define

$$u_\nu(x) = v_\nu(2^{-\nu}x) \quad \text{and} \quad u = \sum_{\nu=-\infty}^{+\infty} u_\nu\,.$$

Then we obtain (3.6.6) and (3.6.7). □

Analogously to Lemma 3.3.3, the result of the last lemma can be extended to other Dirichlet systems of boundary operators on the sides Γ_j.

LEMMA 3.6.4. *Let $\{B_{j,k}\}_{k=0}^{l}$ be Dirichlet systems of order $l+1$ on Γ_j, $j = 1,\dots,d$, with infinitely differentiable coefficients. Furthermore, let $g_{j,k}$ be given functions in $N^{l-k,\sigma}_{\beta,\delta}(\Gamma_j)$ with compact supports, $j = 1,\dots,d$, $k = 0,\dots,l$. Then there exists a function $u \in N^{l,\sigma}_{\beta,\delta}(\mathcal{K})$ satisfying the equalities*

$$B_{j,k}u = g_{j,k} \quad \text{on } \Gamma_j \quad \text{for } j = 1,\dots,d,\ k = 0,\dots,l,$$

and the estimate (3.6.7).

3.6.3. An a priori estimate for the solution. We denote by $L^\infty_{\beta,\delta}(\mathcal{K})$ the space of all functions on $\mathcal{K}$ such that

$$\|u\|_{L^\infty_{\beta,\delta}(\mathcal{K})} = \Big\|\rho^\beta \prod_k \Big(\frac{r_k}{\rho}\Big)^{\delta_k} u\Big\|_{L_\infty(\mathcal{K})} < \infty\,.$$

THEOREM 3.6.5. *Let $u \in L^\infty_{\beta-l-\sigma,\delta-l-\sigma}(\mathcal{K})^\ell$ be a solution of the Dirichlet problem (3.3.1) such that $\chi u \in W^{l,p}(\mathcal{K})^\ell$ for every $\chi \in C_0^\infty(\overline{\mathcal{K}}\backslash\mathcal{S})$. If $f \in N^{l-2m,\sigma}_{\beta,\delta}(\mathcal{K})^\ell$, $l \ge 2m$, $0 < \sigma < 1$, and $g_{j,k} \in N^{l-k+1,\sigma}_{\beta,\delta}(\Gamma_j)^\ell$ for $j = 1,\dots,d$, $k = 1,\dots,m$, then $u \in N^{l,\sigma}_{\beta,\delta}(\mathcal{K})^\ell$ and*

$$\|u\|_{N^{l,\sigma}_{\beta,\delta}(\mathcal{K})^\ell} \le c\Big(\|f\|_{N^{l-2m,\sigma}_{\beta,\delta}(\mathcal{K})^\ell} + \sum_{j,k} \|g_{j,k}\|_{N^{l-k+1,\sigma}_{\beta,\delta}(\Gamma_j)^\ell} + \|u\|_{L^\infty_{\beta-l-\sigma,\delta-l-\sigma}(\mathcal{K})^\ell}\Big).$$

P r o o f. By Lemma 3.6.3, we may assume without loss of generality $g_{j,k} = 0$ for $j = 1,\dots,d$, $k = 1,\dots,m$. From Lemma 6.8.6 it follows that $\zeta u \in N^{l,\sigma}_{\beta,\delta}(\mathcal{K})^\ell$ for every smooth function ζ with compact support vanishing in a neighborhood of the origin. Let ρ be a positive integer, $\mathcal{K}_\rho = \{x \in \mathcal{K} : \rho/2 < |x| < 2\rho\}$, and $\mathcal{K}'_\rho = \{x \subset \mathcal{K} : \rho/4 < |x| < 4\rho\}$. Furthermore, let $\tilde{u}(x) = u(\rho x)$ and $\tilde{f}(x) = \rho^{2m} f(\rho x)$. Obviously, the function $\tilde{u}$ satisfies the differential equation $L(D_x)\tilde{u} = \tilde{f}$ in $\mathcal{K}$ and the homogeneous Dirichlet boundary conditions. Consequently by Lemma 2.8.3,

$$\|\tilde{u}\|_{N^{l,\sigma}_{\beta,\delta}(\mathcal{K}_1)^\ell} \le c\Big(\|\tilde{f}\|_{N^{l-2m,\sigma}_{\beta,\delta}(\mathcal{K}'_1)^\ell} + \|\tilde{u}\|_{L^\infty_{\beta-l-\sigma,\delta-l-\sigma}(\mathcal{K}'_1)^3}\Big) \tag{3.6.8}$$

with a constant c independent of u and ρ. Here the norm in $N^{l,\sigma}_{\beta,\delta}(\mathcal{K}_\rho)$ is defined by the right-hand side of (3.6.2), where $\mathcal{K}$ has to be replaced by $\mathcal{K}_\rho$. Since

$$\|\tilde{u}\|_{N^{l,\sigma}_{\beta,\delta}(\mathcal{K}_1)^\ell} = \rho^{l+\sigma-\beta}\|u\|_{N^{l,\sigma}_{\beta,\delta}(\mathcal{K}_\rho)^\ell},$$

we obtain an analogous estimate for the norm of u in $N^{l,\sigma}_{\beta,\delta}(\mathcal{K}_\rho)^\ell$. Here, the constant c is the same as in (3.6.8). The result follows. □

3.6.4. Existence of solutions in $N^{l,\sigma}_{\beta,\delta}(\mathcal{K})$, $l \ge 2m$. We suppose that the line $\operatorname{Re}\lambda = l+\sigma-\beta$ does not contain eigenvalues of the pencil $\mathfrak{A}(\lambda)$ and denote by $\Lambda_- < \operatorname{Re}\lambda < \Lambda_+$ the widest strip in the complex plane which contains this line and which is free of eigenvalues of the pencil $\mathfrak{A}(\lambda)$. Furthermore, let $G(x,\xi)$ be the Green's matrix introduced in Theorem 3.4.1, where κ is an arbitrary real number in the interval

$$m - \Lambda_+ - 3/2 < \kappa < m - \Lambda_- - 3/2.$$

Again, we consider the vector function

$$u(x) = \int_{\mathcal{K}} G(x,\xi)\, f(\xi)\, d\xi \tag{3.6.9}$$

which is a solution of the Dirichlet problem (3.3.1) with zero boundary data.

LEMMA 3.6.6. *Suppose that the line* $\operatorname{Re}\lambda = l+\sigma-\beta$ *does not contain eigenvalues of the pencil* $\mathfrak{A}(\lambda)$ *and the components of* δ *satisfy the inequalities* (3.6.1). *If* $f \in L^\infty_{\beta-l+2m-\sigma,\delta-l+2m-\sigma}(\mathcal{K})^\ell$, *then the function* (3.6.9) *satisfies the estimate*

$$\|u\|_{L^\infty_{\beta-l-\sigma,\delta-l-\sigma}(\mathcal{K})^\ell} \le c\,\|f\|_{L^\infty_{\beta-l+2m-\sigma,\delta-l+2m-\sigma}(\mathcal{K})^\ell} \tag{3.6.10}$$

with a constant c independent of f.

P r o o f. We define the vector functions $v^{(j)}$, $j=1,2,3,4$, by

$$v^{(j)}(x) = \int_{\mathcal{K}^{(j)}_x} G(x,\xi)\, f(\xi)\, d\xi,$$

where

$$\begin{aligned}
\mathcal{K}^{(1)}_x &= \{\xi\in\mathcal{K} : |x-\xi| < \min(r(x),r(\xi))\},\\
\mathcal{K}^{(2)}_x &= \{\xi\in\mathcal{K} : |x-\xi| > \min(r(x),r(\xi)),\ |x|/2<|\xi|<2|x|\},\\
\mathcal{K}^{(3)}_x &= \{\xi\in\mathcal{K} : |x-\xi| > \min(r(x),r(\xi)),\ |\xi|<|x|/2\},\\
\mathcal{K}^{(4)}_x &= \{\xi\in\mathcal{K} : |x-\xi| > \min(r(x),r(\xi)),\ |\xi|>2|x|\}.
\end{aligned}$$

Obviously, $|x|/2 < |\xi| < 2|x|$ and $r_k(x)/2 < r_k(\xi) < 2r_k(x)$ for $\xi\in\mathcal{K}^{(1)}_x$. Consequently by Theorem 3.4.3 and (3.1.2), we have

$$\begin{aligned}
|v^{(1)}(x)| &\le c\,|x|^{l-2m+\sigma-\beta}\prod_{k=1}^d\Big(\frac{r_k(x)}{|x|}\Big)^{l-2m+\sigma-\delta_k}\|f\|_{L^\infty_{\beta-l+2m-\sigma,\delta-l+2m-\sigma}(\mathcal{K})^\ell}\\
&\quad\times\int\limits_{|\xi-x|<r(x)}\big(|x-\xi|^{2m-3}+r(x)^{2m-3}\big)\,d\xi\\
&\le c\,|x|^{l+\sigma-\beta}\prod_{k=1}^d\Big(\frac{r_k(x)}{|x|}\Big)^{l+\sigma-\delta_k}\|f\|_{L^\infty_{\beta-l+2m-\sigma,\delta-l+2m-\sigma}(\mathcal{K})^\ell}.
\end{aligned}$$

Analogously, it follows from Theorem 3.4.2 that

$$\begin{aligned}|v^{(2)}(x)| &\leq c\,|x|^{l-2m+\sigma-\beta}\,\|f\|_{L^\infty_{\beta-l+2m-\sigma,\delta-l+2m-\sigma}(\mathcal{K})^\ell} \\ &\quad\times \int_{\mathcal{K}_x^{(2)}} |x-\xi|^{2m-3} \Big(\frac{r(x)}{|x-\xi|}\Big)^{m-1+\delta_+(x)-\varepsilon} \Big(\frac{r(\xi)}{|x-\xi|}\Big)^{m-1+\delta_-(x)-\varepsilon} \\ &\quad\times \prod_{k=1}^d \Big(\frac{r_k(\xi)}{|\xi|}\Big)^{l-2m+\sigma-\delta_k} d\xi.\end{aligned}$$

Applying Lemma 3.5.2, we obtain

$$|v^{(2)}(x)| \leq c\,|x|^{l+\sigma-\beta} \prod_{k=1}^d \Big(\frac{r_k(x)}{|x|}\Big)^{l+\sigma-\delta_k} \|f\|_{L^\infty_{\beta-l+2m-\sigma,\delta-l+2m-\sigma}(\mathcal{K})^\ell}.$$

Furthermore by the first part of Theorem 3.4.5,

$$\begin{aligned}|v^{(3)}(x)| &\leq c\,|x|^{\Lambda_-+\varepsilon} \prod_{k=1}^d \Big(\frac{r_k(x)}{|x|}\Big)^{m-1+\delta_+^{(k)}-\varepsilon} \|f\|_{L^\infty_{\beta-l+2m-\sigma,\delta-l+2m-\sigma}(\mathcal{K})^\ell} \\ &\quad\times \int_{\mathcal{K}_x^{(3)}} |\xi|^{l-3+\sigma-\beta-\Lambda_--\varepsilon} \prod_{k=1}^d \Big(\frac{r_k(\xi)}{|\xi|}\Big)^{l-m-1+\sigma-\delta_k+\delta_-^{(k)}-\varepsilon} d\xi.\end{aligned}$$

Since $l-m-1+\sigma-\delta_k+\delta_-^{(k)}-\varepsilon > -2$ for sufficiently small ε, the integral on the right does not exceed $c\,|x|^{l+\sigma-\beta-\Lambda_--\varepsilon}$. Since moreover $m-1+\delta_+^{(k)} > l+\sigma-\delta_k$, we obtain

$$|v^{(3)}(x)| \leq c\,|x|^{l+\sigma-\beta} \prod_{k=1}^d \Big(\frac{r_k(x)}{|x|}\Big)^{l+\sigma-\delta_k} \|f\|_{L^\infty_{\beta-l+2m-\sigma,\delta-l+2m-\sigma}(\mathcal{K})^\ell}.$$

The analogous estimate for $v^{(4)}(x)$ follows in the same way from the second part of Theorem 3.4.5. This proves the lemma. □

We can easily deduce the following existence and uniqueness theorem from the last lemma and from Theorem 3.6.5.

THEOREM 3.6.7. *Suppose that* $l \geq 2m$, *the line* $\operatorname{Re}\lambda = l+\sigma-\beta$ *does not contain eigenvalues of the pencil* $\mathfrak{A}(\lambda)$, *and the components of* δ *satisfy the inequalities* (3.6.1). *Then for arbitrary* $f \in N_{\beta,\delta}^{l-2m,\sigma}(\mathcal{K})^\ell$, $g_{j,k} \in N_{\beta,\delta}^{l-k+1,\sigma}(\Gamma_j)^\ell$, $j=1,\dots,d$, $k=1,\dots,m$, *there exists a unique solution* $u \in N_{\beta,\delta}^{l,\sigma}(\mathcal{K})^\ell$ *of the boundary value problem* (3.3.1).

P r o o f. By Lemma 3.6.3, we may restrict ourselves to the case of zero boundary data $g_{j,k}$. Then the vector function (3.6.9) is a solution of the problem (3.3.1). From Theorem 3.6.5 and Lemma 3.6.6 it follows that $u \in N_{\beta,\delta}^{l,\sigma}(\mathcal{K})^\ell$. We prove the uniqueness of the solution. Let $u \in N_{\beta,\delta}^{l,\sigma}(\mathcal{K})^\ell$ be a solution of the problem

$$L(D_x)u = 0 \ \text{ in } \mathcal{K}, \qquad \frac{\partial^{k-1}u}{\partial n^{k-1}} = 0 \ \text{ on } \Gamma_j,\ j=1,\dots,d,\ k=1,\dots,m,$$

and let χ be an infinitely differentiable function on $\overline{\mathcal{K}}$ with compact support equal to one in a neighborhood of the origin. Then $\chi u \in V_{\beta'+\varepsilon,\delta'+\varepsilon}^{l,2}(\mathcal{K})^\ell$ and $(1-\chi)u \in$

$V^{l,2}_{\beta'-\varepsilon,\delta'+\varepsilon}(\mathcal{K})^\ell$, where $\beta' = \beta - \sigma - 3/2$, $\delta'_k = \delta_k - \sigma - 1$ for $k = 1, \dots, d$, and ε is an arbitrarily small positive number. From the equations

$$L(D_x)(\chi u) = -L(D_x)\big((1-\chi)u\big) \subset V^{l-2m,2}_{\beta'-\varepsilon,\delta'+\varepsilon}(\mathcal{K})^\ell,$$
$$\frac{\partial^{k-1}(\chi u)}{\partial n^{k-1}}\Big|_{\Gamma_j} = 0,\ j = 1, \dots, d,\ k = 1, \dots, m,$$

and from Corollary 3.3.12 we conclude that $\chi u \in V^{l,2}_{\beta'-\varepsilon,\delta'+\varepsilon}(\mathcal{K})^\ell$ and therefore also $u \in V^{l,2}_{\beta'-\varepsilon,\delta'+\varepsilon}(\mathcal{K})^\ell$. Applying Theorem 3.3.9, we obtain $u = 0$. The proof is complete. □

3.6.5. Existence of solutions in $N^{l,\sigma}_{\beta,\delta}(\mathcal{K})$, $l < 2m$. Now let $m \le l < 2m$. Then the distribution $f \in N^{l-2m,\sigma}_{\beta,\delta}(\mathcal{K})^\ell$ has the form

$$f = \sum_{|\alpha|\le 2m-l} D_x^\alpha f^{(\alpha)}, \quad \text{where} \quad f^{(\alpha)} \in N^{0,\sigma}_{\beta-l+2m-|\alpha|,\delta-l+2m-|\alpha|}(\mathcal{K})^\ell$$

and

$$\sum_{|\alpha|\le 2m-l} \|f^{(\alpha)}\|_{N^{0,\sigma}_{\beta-l+2m-|\alpha|,\delta-l+2m-|\alpha|}(\mathcal{K})^\ell} \le 2\,\|f\|_{N^{l-2m,\sigma}_{\beta,\delta}(\mathcal{K})^\ell}\,.$$

As in the preceding subsection, we suppose that the line $\operatorname{Re}\lambda = l+\sigma-\beta$ does not contain eigenvalues of the pencil $\mathfrak{A}(\lambda)$ and that κ is an arbitrary real number in the interval

$$m - \Lambda_+ - 3/2 < \kappa < m - \Lambda_- - 3/2.$$

Here again $\Lambda_- < \operatorname{Re}\lambda < \Lambda_+$ is the widest strip in the complex plane which contains the line $\operatorname{Re}\lambda = l+\sigma-\beta$ and which is free of eigenvalues of the pencil $\mathfrak{A}(\lambda)$. Then we consider the vector function

$$u(x) = \sum_{|\alpha|\le 2m-l} \int_{\mathcal{K}} \big(D_\xi^\alpha G(x,\xi)\big)\, f^{(\alpha)}(\xi)\, d\xi, \tag{3.6.11}$$

where $G(x,\xi)$ is the Green's matrix introduced in Theorem 3.4.1. Using the estimates of $G(x,\xi)$ given in Theorems 3.4.2–3.4.5, we can prove the following lemma analogously to Lemma 3.6.6.

LEMMA 3.6.8. *Suppose that the line* $\operatorname{Re}\lambda = l+\sigma-\beta$ *does not contain eigenvalues of the pencil* $\mathfrak{A}(\lambda)$ *and that the components of* δ *satisfy the inequalities* (3.6.1). *If* $f \in N^{l-2m,\sigma}_{\beta,\delta}(\mathcal{K})^\ell$, $m \le l < 2m$, *then the function* (3.6.11) *satisfies the estimate*

$$\|u\|_{L^\infty_{\beta-l-\sigma,\delta-l-\sigma}(\mathcal{K})^\ell} \le c\,\|f\|_{L^\infty_{\beta-l+2m-\sigma,\delta-l+2m-\sigma}(\mathcal{K})^\ell}$$

with a constant c *independent of* f.

P r o o f. Let α be a multi-index, $|\alpha| \le 2m-l$. We have to show that the vector function

$$v(x) = \int_{\mathcal{K}} \big(D_\xi^\alpha G(x,\xi)\big)\, f^{(\alpha)}(\xi)\, d\xi$$

satisfies the estimate

$$\|v\|_{L^\infty_{\beta-l-\sigma,\delta-l-\sigma}(\mathcal{K})^\ell} \le c\,\|f^{(\alpha)}\|_{L^\infty_{\beta-l+2m-\sigma-|\alpha|,\delta-l+2m-\sigma-|\alpha|}(\mathcal{K})^\ell}\,. \tag{3.6.12}$$

Let $\mathcal{K}_x^{(1)}, \dots, \mathcal{K}_x^{(4)}$ be the same subsets as in the proof of Lemma 3.6.6 and

$$v^{(j)}(x) = \int_{\mathcal{K}_x^{(j)}} \big(D_\xi^\alpha G(x,\xi)\big)\, f^{(\alpha)}(\xi)\, d\xi, \quad j = 1,2,3,4.$$

Analogously to the proof of Lemma 3.6.6, we obtain the estimate (3.6.12) for the vector functions $v^{(1)}, \dots, v^{(4)}$. Thus, the vector function v satisfies the same estimate. □

Using Lemma 3.6.8 instead of Lemma 3.6.6, one can prove the following theorem in the same way as Theorem 3.6.7.

THEOREM 3.6.9. *The assertion of Theorem* 3.6.7 *is valid for arbitrary* $l \ge m$.

3.6.6. Regularity assertions for the solution. We close the section with the following two regularity results.

THEOREM 3.6.10. *Let* $u \in N^{l,\sigma}_{\beta,\delta}(\mathcal{K})^\ell$ *be a solution of the problem* (3.3.1).

1) *Suppose that the closed strip between the lines* $\text{Re}\,\lambda = l+\sigma-\beta$ *and* $\text{Re}\,\lambda = l'+\sigma'-\beta'$ *is free of eigenvalues of the pencil* $\mathfrak{A}(\lambda)$, *and that the components of* δ *and* δ' *satisfy the inequalities*

$$-\delta_+^{(k)} < \delta_k - l + m - 1 - \sigma < \delta_-^{(k)}, \quad -\delta_+^{(k)} < \delta'_k - l' + m - 1 - \sigma' < \delta_-^{(k)}.$$

If $f \in N^{l-2m,\sigma}_{\beta,\delta}(\mathcal{K})^\ell \cap N^{l'-2m,\sigma'}_{\beta',\delta'}(\mathcal{K})^\ell$, $g_{j,k} \in N^{l-k+1,\sigma}_{\beta,\delta}(\Gamma_j)^\ell \cap N^{l'-k+1,\sigma'}_{\beta',\delta'}(\Gamma_j)^\ell$, *then* $u \in N^{l',\sigma'}_{\beta',\delta'}(\mathcal{K})^\ell$.

2) *Suppose that the closed strip between the lines* $\text{Re}\,\lambda = l+\sigma-\beta$ *and* $\text{Re}\,\lambda = l'-\beta'-3/p$ *is free of eigenvalues of the pencil* $\mathfrak{A}(\lambda)$, *and that the components of* δ *and* δ' *satisfy the inequalities*

$$-\delta_+^{(k)} < \delta_k - l + m - 1 - \sigma < \delta_-^{(k)}, \quad -\delta_+^{(k)} < \delta'_k - l' + m - 1 + 2/p < \delta_-^{(k)}.$$

If $f \in N^{l-2m,\sigma}_{\beta,\delta}(\mathcal{K})^\ell \cap V^{l'-2m,p}_{\beta',\delta'}(\mathcal{K})^\ell$, $g_{j,k} \in N^{l-k+1,\sigma}_{\beta,\delta}(\Gamma_j)^\ell \cap V^{l'-k+1-1/p,p}_{\beta',\delta'}(\Gamma_j)^\ell$, *then* $u \in V^{l',p}_{\beta',\delta'}(\mathcal{K})^\ell$.

P r o o f. It suffices to prove the theorem for the case of zero boundary data. Under the conditions of the first part of the theorem, there exist unique solutions $u \in N^{l,\sigma}_{\beta,\delta}(\mathcal{K})^\ell$ and $v \in N^{l',\sigma'}_{\beta',\delta'}(\mathcal{K})^\ell$ of problem (3.3.1). Both solutions have the representation (3.6.9) with the same Green's matrix $G(x,\xi)$. Hence, $u = v$. Analogously, the second part of the theorem holds. □

3.7. The boundary value problem with variable coefficients in a cone

Here, we extend the results of the last two sections to the Dirichlet problem for a differential operator $L(x, D_x)$ with variable coefficients. We make the same assumptions on the coefficients as in Section 2.9. These conditions, which are only restrictive outside a neighborhood of the vertex of the cone, ensure that the operator norm of the difference $L(x, D_x) - L^\circ(0, D_x)$ is small. Here $L^\circ(0, D_x)$ denotes the principal part of $L(x, D_x)$ with coefficients frozen at the origin.

3.7.1. Solvability in weighted Sobolev and Hölder spaces. We consider the boundary value problem

$$L(x, D_x)\, u = f \ \text{ in } \mathcal{K}, \quad \frac{\partial^{k-1} u}{\partial n^{k-1}} = g_{j,k} \ \text{ on } \Gamma_j, \tag{3.7.1}$$

$j = 1, \dots, d,\ k = 1, \dots, m$, where

$$L(x, D_x) = \sum_{|\alpha| \le 2m} A_\alpha(x)\, D_x^\alpha \tag{3.7.2}$$

is a strongly elliptic differential operator of order $2m$. Suppose that the coefficients A_α satisfy the following conditions:

$$r^{|\beta|}\,\big|\partial_x^\beta(A_\alpha(x)-A_\alpha(0))\big| < \varepsilon_0 \quad \text{for } |\alpha|=2m,\ |\beta|\le N, \tag{3.7.3}$$

$$r^{2m-|\alpha|+|\beta|}\,\big|\partial_x^\beta A_\alpha(x)\big| < \varepsilon_0 \quad \text{for } |\alpha|<2m,\ |\beta|\le N, \tag{3.7.4}$$

where $r(x)=\text{dist}(x,\mathcal{S})$ and ε_0 is a sufficiently small positive real number.

REMARK 3.7.1. Let $L(x,D_x)$ be an operator of the form (3.7.2) with infinitely differentiable coefficients in $\overline{\mathcal{K}}$, and let ε be a sufficiently small positive number. Then there exists an operator $L_\varepsilon(x,D_x)$ with coefficients satisfying (3.7.3) and (3.7.4) such that

$$L(x,D_x)\,u(x)=L_\varepsilon(x,D_x)\,u(x)$$

for all u and all x, $|x|<\varepsilon$. This operator can be defined in the exact same manner as in Remark 2.9.3.

The next lemma follows immediately from the definition of the norms in $V_{\beta,\delta}^{l,p}(\mathcal{K})$ and $N_{\beta,\delta}^{l,\sigma}(\mathcal{K})$ and from the conditions (3.7.3) and (3.7.4) (cf. Lemma 2.9.1).

LEMMA 3.7.2. *Let l be a nonnegative integer and $\delta\in\mathbb{R}$. If the coefficients A_α satisfy the conditions* (3.7.3)–(3.7.4) *for $N=|l-2m|$, then*

$$\|L(x,D_x)u-L^\circ(0,D_x)u\|_{V_{\beta,\delta}^{l-2m,p}(\mathcal{K})^\ell}\le c\,\varepsilon_0\,\|u\|_{V_{\beta,\delta}^{l,p}(\mathcal{K})^\ell}$$

for all $u\in V_{\beta,\delta}^{l,p}(\mathcal{K})^\ell$. Here, c is a constant depending only on l and m. If the conditions (3.7.3)–(3.7.4) *are satisfied for $N=|l-2m|+1$, then*

$$\|L(x,D_x)u-L^\circ(0,D_x)u\|_{N_{\beta,\delta}^{l-2m,\sigma}(\mathcal{K})^\ell}\le c\,\varepsilon_0\,\|u\|_{N_{\beta,\delta}^{l,\sigma}(\mathcal{K})^\ell}$$

for all $u\in N_{\beta,\delta}^{l,\sigma}(\mathcal{K})^\ell$.

We denote by $\mathfrak{A}(\lambda)$ the operator pencil generated by the Dirichlet problem for the equation $L^\circ(0,D_x)\,u=f$ in the cone $\mathcal{K}$ (see Subsection 3.2.2). Using the last lemma together with Theorems 3.5.10 and 3.6.9, we obtain an existence and uniqueness theorem for the boundary value problem (3.7.1).

THEOREM 3.7.3. 1) *Suppose that the line $\text{Re}\,\lambda=l-\beta-3/p$ is free of eigenvalues of the pencil $\mathfrak{A}(\lambda)$ and that*

$$-\delta_+^{(k)}<\delta_k-l+m-1+2/p<\delta_-^{(k)}\quad \textit{for } k=1,\dots,d.$$

Furthermore, we assume that the coefficients A_α satisfy the conditions (3.7.3)–(3.7.4) *for $N=|l-2m|$, where ε_0 is sufficiently small. Then for arbitrary $f\in V_{\beta,\delta}^{l-2m,p}(\mathcal{K})^\ell$ and $g_{j,k}\in V_{\beta,\delta}^{l-k+1-1/p,p}(\Gamma_j)^\ell$, $l\ge m$, there exists a unique solution $u\in V_{\beta,\delta}^{l,p}(\mathcal{K})^\ell$ of the boundary value problem* (3.7.1).

2) *Suppose that the line $\text{Re}\,\lambda=l+\sigma-\beta$ is free of eigenvalues of the pencil $\mathfrak{A}(\lambda)$ and that*

$$-\delta_+^{(k)}<\delta_k-l-\sigma+m-1<\delta_-^{(k)}\quad \textit{for } k=1,\dots,d.$$

Furthermore, we assume that the coefficients A_α satisfy the conditions (3.7.3)–(3.7.4) *for $N=|l-2m|+1$, where ε_0 is sufficiently small. Then for arbitrary $f\in N_{\beta,\delta}^{l-2m,\sigma}(\mathcal{K})^\ell$ and $g_{j,k}\in N_{\beta,\delta}^{l-k+1,\sigma}(\Gamma_j)^\ell$, $l\ge m$, there exists a unique solution $u\in N_{\beta,\delta}^{l,\sigma}(\mathcal{K})^\ell$ of the boundary value problem* (3.7.1).

3.7.2. Regularity assertions for the solutions. Furthermore, we can extend the assertions of Theorem 3.5.13 and 3.6.10 to differential operators with variable coefficients provided the coefficients satisfy the conditions (3.7.3) and (3.7.4) with sufficiently small ε_0.

THEOREM 3.7.4. *Let* $u \in V^{l,p}_{\beta,\delta}(\mathcal{K})^\ell$ *be a solution of the boundary value problem* (3.7.1)*, where*

$$f \in V^{l-2m,p}_{\beta,\delta}(\mathcal{K})^\ell \cap V^{l'-2m,q}_{\beta',\delta'}(\mathcal{K})^\ell, \; l \ge m, \; l' \ge m,$$

$$g_{j,k} \in V^{l-k+1-1/p,p}_{\beta,\delta}(\Gamma_j)^\ell \cap V^{l'-k+1-1/q,q}_{\beta',\delta'}(\Gamma_j)^\ell$$

for $j = 1, \dots, d$, $k = 1, \dots, m$. *Suppose that the closed strip between the lines* $\operatorname{Re}\lambda = l - \beta - 3/p$ *and* $\operatorname{Re}\lambda = l' - \beta' - 3/q$ *is free of eigenvalues of the pencil* $\mathfrak{A}(\lambda)$ *and that*

$$-\delta_+^{(k)} < \delta_k - l + m - 1 + 2/p < \delta_-, \quad -\delta_+^{(k)} < \delta'_k - l' + m - 1 + 2/q < \delta_-^{(k)}$$

for $k = 1, \dots, d$. *Furthermore, we assume that the coefficients* A_α *satisfy the conditions* (3.7.3)–(3.7.4) *for* $N = \max(|l - 2m|, |l' - 2m|)$ *with a sufficiently small* ε_0. *Then* $u \in V^{l',q}_{\beta'\delta'}(\mathcal{K})^\ell$.

Analogously, the assertions of Theorem 3.6.10 *are valid for the solution of the boundary value problem* (3.7.1) *if the coefficients of the differential operator* $L(x, D_x)$ *satisfy the conditions* (3.7.3)–(3.7.4) *with a sufficiently small* ε_0.

P r o o f. We prove the first assertion. By Theorems 3.5.10 and 3.5.13, the operator of the boundary value problem (3.3.1) is an isomorphism from $V^{l,p}_{\beta,\delta}(\mathcal{K})^\ell \cap V^{l',q}_{\beta',\delta'}(\mathcal{K})^\ell$ onto

$$\big(V^{l-2m,p}_{\beta,\delta}(\mathcal{K})^\ell \cap V^{l'-2m,q}_{\beta',\delta'}(\mathcal{K})^\ell\big) \times \prod_{j,k} \big(V^{l-k+1-1/p,p}_{\beta,\delta}(\Gamma_j)^\ell \cap V^{l'-k+1-1/q,q}_{\beta',\delta'}(\Gamma_j)^\ell\big).$$

Since the operator $L(x, D_x) - L^\circ(0, D_x)$ is small in the operator norm

$$V^{l,p}_{\beta,\delta}(\mathcal{K})^\ell \cap V^{l',q}_{\beta',\delta'}(\mathcal{K})^\ell \to V^{l-2m,p}_{\beta,\delta}(\mathcal{K})^\ell \times V^{l'-2m,q}_{\beta',\delta'}(\mathcal{K})^\ell,$$

the same is true for the operator of the problem (3.7.1). Together with Theorem 3.5.10 this implies the regularity result in the spaces $V^{l,p}_{\beta,\delta}$. The second assertion follows analogously from Theorems 3.6.9 and 3.6.10. □

As a consequence of the last theorem, the following local regularity assertion holds analogously to Lemma 2.9.10.

COROLLARY 3.7.5. *Let* ζ, η *be infinitely differentiable functions with compact supports such that* $\eta = 1$ *in a neighborhood of* $\operatorname{supp}\zeta$, *and let* u *be a solution of the boundary value problem* (3.7.1). *Suppose that* $l \ge m$, $l' \ge m$, $\eta u \in V^{l,p}_{\beta,\delta}(\mathcal{K})^\ell$,

$$\eta f \in V^{l-2m,p}_{\beta,\delta}(\mathcal{K})^\ell \cap V^{l'-2m,q}_{\beta',\delta'}(\mathcal{K})^\ell,$$

$$\eta g_{j,k} \in V^{l-k+1-1/p,p}_{\beta,\delta}(\Gamma_j)^\ell \cap V^{l'-k+1-1/q,q}_{\beta',\delta'}(\Gamma_j)^\ell$$

for $j = 1, \dots, d$, $k = 1, \dots, m$, *the closed strip between the lines* $\operatorname{Re}\lambda = l - \beta - 3/p$ *and* $\operatorname{Re}\lambda = l' - \beta' - 3/q$ *is free of eigenvalues of the pencil* $\mathfrak{A}(\lambda)$ *and that*

$$-\delta_+^{(k)} < \delta_k - l + m - 1 + 2/p < \delta_-, \quad -\delta_+^{(k)} < \delta'_k - l' + m - 1 + 2/q < \delta_-^{(k)}$$

for $k = 1, \dots, d$. *Furthermore, we assume that the coefficients* A_α *satisfy the conditions* (3.7.3)–(3.7.4) *for* $N = \max(|l-2m|, |l'-2m|)$ *with a sufficiently small* ε_0. *Then* $\zeta u \in V^{l',q}_{\beta',\delta'}(\mathcal{K})^\ell$ *and*

$$\|\zeta u\|_{V^{l',q}_{\beta',\delta'}(\mathcal{K})^\ell} \le c\Big(\|\eta f\|_{V^{l'-2m,q}_{\beta',\delta'}(\mathcal{K})^\ell} + \sum_{j,k} \|\eta g_{j,k}\|_{V^{l'-k+1-1/q,q}_{\beta',\delta'}(\Gamma_j)^\ell} + \|\eta u\|_{V^{l,p}_{\beta,\delta}(\mathcal{K})^\ell}\Big)$$

with a constant c *independent of* u.

CHAPTER 4

The Dirichlet problem in a bounded domain of polyhedral type

Now we study the Dirichlet problem for a strongly elliptic differential operator of order $2m$ with variable coefficients in a bounded domain $\mathcal{G}$ of polyhedral type. Using the existence and uniqueness results for the Dirichlet problem in dihedra and cones, we prove the normal solvability (Fredholm property of the operator of the boundary value problem) both in weighted Sobolev and weighted Hölder spaces. Furthermore, we deal with regularity assertions for the solutions. As an example, let us consider the variational solution $u \in \overset{\circ}{W}{}^{1,2}(\mathcal{G})^3$ of the Lamé system

$$\Delta u + \frac{1}{1-2\nu}\,\mathrm{grad}\,\mathrm{div}\,u = f$$

in an arbitrary bounded convex domain of polyhedral type. Then the following typical assertions are deduced from the regularity results in Sections 4.1, 4.2 and from estimates for the eigenvalues of the operator pencils generated by the Lamé system in [**85**, Chapter 3]:

$$\text{If } f \in W^{-1,p}(\mathcal{G})^3,\ 1 < p < \infty, \text{ then } u \in W^{1,p}(\mathcal{G})^3,$$
$$\text{if } f \in L_p(\mathcal{G}),\ p < 2+\varepsilon, \text{ then } u \in W^{2,p}(\mathcal{G})^3.$$

Here ε is a positive number depending on the domain and on ν. Furthermore, we conclude that $u \in C^{1,\sigma}(\mathcal{G})^3$ with sufficiently small $\sigma > 0$ if $f \in C^{-1,\sigma}(\mathcal{G})^3$. Other results of this kind are discussed in Section 4.3.

4.1. Solvability of the boundary value problem in weighted Sobolev spaces

In this section, we establish a solvability and regularity theory for the Dirichlet problem

$$(4.1.1) \qquad L(x,D_x)\,u = f \ \text{ in } \mathcal{G}, \qquad \frac{\partial^{k-1}u}{\partial n^{k-1}} = g_{j,k} \ \text{ on } \Gamma_j\,,$$

$j = 1,\dots,N,\ k = 1,\dots,m$, in the weighted Sobolev spaces $V^{l,p}_{\beta,\delta}(\mathcal{G})$. Here

$$L(x,D_x) = \sum_{|\alpha|\le 2m} A_\alpha(x)\, D_x^\alpha$$

is a strongly elliptic differential operator of order $2m$, the coefficients A_α are infinitely differentiable $\ell\times\ell$-matrices on $\overline{\mathcal{G}}$. As for the problem in a cone, the solvability results hold only under certain conditions on the weight parameters β and δ. In particular, we show that the operator

$$V_{\beta,\delta}^{l,p}(\mathcal{G}) \ni u \to (f, \{g_{j,k}\}) \in V_{\beta,\delta}^{l-2m,p}(\mathcal{G})^\ell \times \prod_{j=1}^{N} \prod_{k=1}^{m} V_{\beta,\delta}^{l-k+1-1/p,p}(\Gamma_j)^\ell$$

of the problem (4.1.1) is Fredholm if the line $\operatorname{Re}\lambda = l - \beta_j - 3/p$ does not contain eigenvalues of the pencil $\mathfrak{A}_j(\lambda)$ for each j and the components of δ satisfy the condition $-\delta_+^{(k)} < \delta_k - l + m - 1 + 2/p < \delta_-^{(k)}$. Here $\mathfrak{A}_j(\lambda)$ are the operator pencils generated by the boundary value problem for the vertices of $\mathcal{G}$ and $\delta_+^{(k)}, \delta_-^{(k)}$ are certain positive numbers introduced below. The proof of the Fredholm property is based on the construction of a regularizer. Here, the existence of local regularizers follows from the solvability theorems in Chapters 2 and 3. The results of the last chapters are also used in the proof of a regularity assertion for the solutions.

4.1.1. The domain. The bounded domain $\mathcal{G} \subset \mathbb{R}^3$ is said to be a *domain of polyhedral type* if

(i) the boundary $\partial\mathcal{G}$ consists of smooth (of class C^∞) open two-dimensional manifolds Γ_j (the faces of $\mathcal{G}$), $j = 1, \dots, N$, smooth curves M_k (the edges), $k = 1, \dots, d$, and vertices $x^{(1)}, \dots, x^{(d')}$,

(ii) for every $\xi \in M_k$ there exist a neighborhood $\mathcal{U}_\xi$ and a diffeomorphism (a C^∞ mapping) κ_ξ which maps $\mathcal{G} \cap \mathcal{U}_\xi$ onto $\mathcal{D}_\xi \cap B_1$, where $\mathcal{D}_\xi$ is a dihedron and B_1 is the unit ball,

(iii) for every vertex $x^{(j)}$ there exist a neighborhood $\mathcal{U}_j$ and a diffeomorphism κ_j mapping $\mathcal{G} \cap \mathcal{U}_j$ onto $\mathcal{K}_j \cap B_1$, where $\mathcal{K}_j$ is a cone with edges and vertex at the origin.

The set $M_1 \cup \dots \cup M_d \cup \{x^{(1)}, \dots, x^{(d')}\}$ of the singular boundary points is denoted by $\mathcal{S}$. We do not exclude the cases $d = 0$ and $d' = 0$. In the last case, the set $\mathcal{S}$ consists only of smooth non-intersecting edges.

4.1.2. Weighted Sobolev spaces. We denote the distance of x from the edge M_k by $r_k(x)$, the distance from the vertex $x^{(j)}$ by $\rho_j(x)$, and the distance from $\mathcal{S}$ (the set of all edge points and vertices) by $r(x)$. Furthermore, we denote by X_j the set of all indices k such that the vertex $x^{(j)}$ is an end point of the edge M_k. Let $\mathcal{U}_1, \dots, \mathcal{U}_{d'}$ be domains in $\mathbb{R}^3$ such that

$$\mathcal{U}_1 \cup \dots \cup \mathcal{U}_{d'} \supset \overline{\mathcal{G}}, \quad x^{(i)} \notin \overline{\mathcal{U}}_j \text{ if } i \neq j, \quad \text{and} \quad \overline{\mathcal{U}}_j \cap \overline{M}_k = \emptyset \text{ if } k \notin X_j.$$

Then for arbitrary integer $l \geq 0$, real $p > 1$ and real tuples $\beta = (\beta_1, \dots, \beta_{d'})$, $\delta = (\delta_1, \dots, \delta_d)$, we define $V_{\beta,\delta}^{l,p}(\mathcal{G})$ as the weighted Sobolev space with the norm

$$\|u\|_{V_{\beta,\delta}^{l,p}(\mathcal{G})} = \Big(\sum_{j=1}^{d'} \int_{\mathcal{G}\cap\mathcal{U}_j} \sum_{|\alpha|\leq l} \rho_j^{p(\beta_j - l + |\alpha|)} \prod_{k\in X_j} \Big(\frac{r_k}{\rho_j}\Big)^{p(\delta_k - l + |\alpha|)} |\partial_x^\alpha u|^p \, dx \Big)^{1/p}.$$

Obviously, the space $V_{\beta,\delta}^{l,p}(\mathcal{G})$ does not depend on the choice of the domains $\mathcal{U}_j$. Note that the set $C_0^\infty(\overline{\mathcal{G}}\backslash\mathcal{S})$ is dense in $V_{\beta,\delta}^{l,p}(\mathcal{G})$. If $\beta \in \mathbb{R}^{d'}$, $\delta \in \mathbb{R}^d$ and $s, t \in \mathbb{R}$, then by $V_{\beta+s,\delta+t}^{l,p}(\mathcal{G})$, we mean the space $V_{\beta',\delta'}^{l,p}(\mathcal{G})$ with the weight parameters $\beta' = (\beta_1 + s, \dots, \beta_{d'} + s)$ and $\delta' = (\delta_1 + t, \dots, \delta_d + t)$.

The closure of the set $C_0^\infty(\mathcal{G})$ in $V_{\beta,\delta}^{l,p}(\mathcal{G})$ is denoted by $\overset{\circ}{V}{}_{\beta,\delta}^{l,p}(\mathcal{G})$. Furthermore, let $V_{\beta,\delta}^{-l,p}(\mathcal{G})$ denote the dual space of $\overset{\circ}{V}{}_{-\beta,-\delta}^{l,p'}(\mathcal{G})$, $p' = p/(p-1)$, with respect to the L_2 scalar product $(\cdot,\cdot)_{\mathcal{G}}$. This space is provided with the norm

$$\|u\|_{V_{\beta,\delta}^{-l,p}(\mathcal{G})} = \sup\big\{|(u,v)_{\mathcal{G}}| \, : \, v \in \overset{\circ}{V}{}_{-\beta,-\delta}^{l,p'}(\mathcal{G}), \ \|v\|_{V_{-\beta,-\delta}^{l,p'}(\mathcal{G})} = 1\big\}. \tag{4.1.2}$$

An equivalent definition of the space $V_{\beta,\delta}^{-l,p}(\mathcal{G})$ is given in the following lemma (cf. Lemmas 2.1.7 and 3.1.1).

LEMMA 4.1.1. *The space $V_{\beta,\delta}^{-l,p}(\mathcal{G})$ consists of all distributions of the form*

$$u = \sum_{|\alpha|\le l} \partial_x^\alpha u_\alpha, \quad \text{where } u_\alpha \in V^{0,p}_{\beta+l-|\alpha|,\delta+l-|\alpha|}(\mathcal{G}).$$

The norm

$$\|u\| = \inf\Big\{ \sum_{|\alpha|\le l} \|u_\alpha\|_{V^{0,p}_{\beta+l-|\alpha|,\delta+l-|\alpha|}(\mathcal{G})} : u_\alpha \in V^{0,p}_{\beta+l-|\alpha|,\delta+l-|\alpha|}(\mathcal{G}), \ \sum_{|\alpha|\le l} \partial_x^\alpha u_\alpha = u\Big\}$$

is equivalent to the norm (4.1.2).

Note that

$$\overset{\circ}{W}{}^{l,p}(\mathcal{G}) = \overset{\circ}{V}{}^{l,p}_{0,0}(\mathcal{G})$$

for arbitrary integer $l \ge 0$. This can be easily deduced from Lemmas 2.2.4 and 3.1.6.

4.1.3. Imbeddings. In the subsequent two lemmas, we state conditions on $l, l', \beta, \beta', \delta, \delta'$ under which the imbedding

$$V_{\beta,\delta}^{l,p}(\mathcal{G}) \subset V_{\beta',\delta'}^{l',q}(\mathcal{G}).$$

is valid.

LEMMA 4.1.2. *Let $1 < q < p < \infty$, $\beta_j + 3/p < \beta_j' + 3/q$ for $j = 1,\dots,d'$, and $\delta_k + 2/p < \delta_k' + 2/q$ for $k = 1,\dots,d$. Then $V_{\beta,\delta}^{l,p}(\mathcal{G})$ is continuously imbedded in $V_{\beta',\delta'}^{l,q}(\mathcal{G})$.*

P r o o f. Let $\mathcal{G}_j = \mathcal{G} \cap \mathcal{U}_j$ and $s = pq/(p-q)$. By Hölder's inequality,

$$\Big\|\rho_j^{\beta_j'-l+|\alpha|} \prod_{k\in X_j} \Big(\frac{r_k}{\rho_j}\Big)^{\delta_k'-l+|\alpha|} \partial_x^\alpha u\Big\|_{L_q(\mathcal{G}_j)} \le c \,\Big\|\rho_j^{\beta_j-l+|\alpha|} \prod_{k\in X_j} \Big(\frac{r_k}{\rho_j}\Big)^{\delta_k-l+|\alpha} \partial_x^\alpha u\Big\|_{L_p(\mathcal{G}_j)},$$

where

$$c = \Big\|\rho_j^{\beta_j'-\beta_j} \prod_{k\in X_j} \Big(\frac{r_k}{\rho_j}\Big)^{\delta_k'-\delta_k}\Big\|_{L_s(\mathcal{G}_j)} < \infty$$

if $\beta_j' - \beta_j > -3/s$ and $\delta_k' - \delta_k > -2/s$ for $k \in X_j$. This proves the lemma. □

The next result can be easily obtained by means of Lemma 3.1.3.

LEMMA 4.1.3. *Let l, l' be nonnegative integers, $1 < p \le q < \infty$, $l - 3/p \ge l' - 3/q$, $\beta_j - l + 3/p \le \beta_j' - l' + 3/q$ for $j = 1,\dots,d'$, and $\delta_k - l + 3/p \le \delta_k' - l' + 3/q$ for $k = 1,\dots,d$. Then $V_{\beta,\delta}^{l,p}(\mathcal{G})$ is continuously imbedded in $V_{\beta',\delta'}^{l',q}(\mathcal{G})$.*

4.1.4. Compact imbeddings. By Lemma 4.1.3, the space $V^{l,p}_{\beta,\delta}(\mathcal{G})$ is continuously imbedded in $V^{l',p}_{\beta',\delta'}(\mathcal{G})$ if $l \ge l'$, $\beta_j - l \le \beta'_j - l'$ and $\delta_k - l \le \delta'_k - l'$. If even the strong inequalities $l > l'$, $\beta_j - l < \beta'_j - l'$ and $\delta_k - l < \delta'_k - l'$ are satisfied, then the following assertion holds.

LEMMA 4.1.4. *Suppose that* $\beta_j - l < \beta'_j - l'$ *for* $j = 1, \dots, d'$, *and* $\delta_k - l < \delta'_k - l'$. *Then the imbeddings* $V^{l,p}_{\beta,\delta}(\mathcal{G}) \subset V^{l',p}_{\beta',\delta'}(\mathcal{G})$ *for arbitrary integer* $l > l'$ *and* $\overset{\circ}{V}{}^{l,p}_{\beta,\delta}(\mathcal{G}) \subset \overset{\circ}{V}{}^{l',p}_{\beta',\delta'}(\mathcal{G})$ *for arbitrary integer* $l > l' \ge 0$ *are compact.*

P r o o f. First let $l > l' \ge 0$. We denote by $\mathfrak{M}$ the set of all $u \in V^{l,p}_{\beta,\delta}(\mathcal{G})$ with norm $\le c_0$ and show that $\mathfrak{M}$ is precompact in $V^{l',p}_{\beta',\delta'}(\mathcal{G})$, i.e. that for arbitrary positive ε there exists a finite ε-net $\{u_1, \dots, u_n\}$ in $\mathfrak{M}$ such that

$$\min_{1\le j\le n} \|u - u_j\|_{V^{l',p}_{\beta',\delta'}(\mathcal{G})} \le \varepsilon \tag{4.1.3}$$

for all $u \in \mathfrak{M}$. Let $\mathcal{G}_\sigma = \{x \in \mathcal{G} : r(x) < \sigma\}$, $\sigma > 0$, and let the $V^{l,p}_{\beta,\delta}(\mathcal{G}_\sigma)$-norm be defined by the same integral as the $V^{l,p}_{\beta,\delta}(\mathcal{G})$-norm, where the domain of integration is replaced by $\mathcal{G}_\sigma$. Obviously,

$$\|u\|_{V^{l',p}_{\beta',\delta'}(\mathcal{G}_\sigma)} \le c(\sigma)\, \|u\|_{V^{l',p}_{\beta-l+l',\delta-l+l'}(\mathcal{G})} \le c(\sigma)\, \|u\|_{V^{l,p}_{\beta,\delta}(\mathcal{G})}$$

for all $u \in V^{l,p}_{\beta,\delta}(\mathcal{G})$, where $c(\sigma) \to 0$ as $\sigma \to 0$. Consequently, δ can be chosen such that

$$\|u\|_{V^{l',p}_{\beta',\delta'}(\mathcal{G}_\sigma)} \le \varepsilon/3 \tag{4.1.4}$$

for all $u \in \mathfrak{M}$. Since the imbedding $W^{l,p}(\mathcal{G}\backslash\overline{\mathcal{G}}_\sigma) \subset W^{l',p}(\mathcal{G}\backslash\overline{\mathcal{G}}_\sigma)$ is compact and the $W^{l,p}(\mathcal{G}\backslash\overline{\mathcal{G}}_\sigma)$-norm is equivalent to the $V^{l,p}_{\beta,\delta}(\mathcal{G}\backslash\overline{\mathcal{G}}_\sigma)$-norm, there exist finitely many functions $u_1, \dots, u_n \in \mathfrak{M}$ such that

$$\min_{1\le j\le n} \|u - u_j\|_{V^{l',p}_{\beta',\delta'}(\mathcal{G}\backslash\overline{\mathcal{G}}_\sigma)} \le \varepsilon/3 .$$

The last inequality together with (4.1.4) implies (4.1.3). This proves the compactness of the imbedding $V^{l,p}_{\beta,\delta}(\mathcal{G}) \subset V^{l',p}_{\beta',\delta'}(\mathcal{G})$ for case $l > l' \ge 0$. Analogously the compactness of the imbedding $\overset{\circ}{V}{}^{l,p}_{\beta,\delta}(\mathcal{G}) \subset \overset{\circ}{V}{}^{l',p}_{\beta',\delta'}(\mathcal{G})$ holds.

If $l' < l \le 0$, then the compactness of the imbedding $V^{l,p}_{\beta,\delta}(\mathcal{G}) \subset V^{l',p}_{\beta',\delta'}(\mathcal{G})$ follows from the compactness of the imbedding $\overset{\circ}{V}{}^{-l',p'}_{-\beta',-\delta'}(\mathcal{G}) \subset \overset{\circ}{V}{}^{-l,p'}_{-\beta,-\delta}(\mathcal{G})$, $p' = p/(p-1)$. If $l' < 0 < l$, then

$$V^{l,p}_{\beta,\delta}(\mathcal{G}) \subset V^{0,p}_{\beta'-l',\delta'-l'}(\mathcal{G}) \subset V^{l',p}_{\beta',\delta'}(\mathcal{G}),$$

where the first imbedding is compact under the assumptions of the lemma and the second one is continuous. Consequently, the imbedding $V^{l,p}_{\beta,\delta}(\mathcal{G}) \subset V^{l',p}_{\beta',\delta'}(\mathcal{G})$ is compact. The proof of the lemma is complete. □

4.1.5. Trace spaces. For $l \ge 1$, let $V^{l-1/p,p}_{\beta,\delta}(\Gamma_j)$ be the space of traces of $V^{l,p}_{\beta,\delta}(\mathcal{G})$-functions on Γ_j. The norm in $V^{l-1/p,p}_{\beta,\delta}(\Gamma_j)$ is defined as

$$\|u\|_{V^{l-1/p,p}_{\beta,\delta}(\Gamma_j)} = \inf\{\|v\|_{V^{l,p}_{\beta,\delta}(\mathcal{G})} \, : \, v \in V^{l,p}_{\beta,\delta}(\mathcal{G}),\ v = u \text{ on } \Gamma_j\}.$$

The next lemma can be easily deduced from Lemma 4.1.3.

LEMMA 4.1.5. *Let $1 < p \le q < \infty$, $l \ge l' \ge 1$, $l - 3/p \ge l' - 3/q$, $\beta_j - l + 3/p \le \beta'_j - l' + 3/q$ for $j = 1, \dots, d'$, and $\delta_k - l + 3/p \le \delta'_k - l' + 3/q$ for $k = 1, \dots, d$. Then $V^{l-1/p,p}_{\beta,\delta}(\Gamma_j)$ is continuously imbedded in $V^{l'-1/q,q}_{\beta',\delta'}(\Gamma_j)$.*

Furthermore using Lemma 4.1.4, we can prove the compactness of the last imbedding.

LEMMA 4.1.6. *Suppose that $l > l' \ge 1$, $\beta_j - l < \beta'_j - l'$ for $j = 1, \dots, d'$, and $\delta_k - l < \delta'_k - l'$. Then the imbedding $V^{l-1/p,p}_{\beta,\delta}(\Gamma_j) \subset V^{l'-1/p,p}_{\beta',\delta'}(\Gamma_j)$ is compact.*

P r o o f. We show that the set

$$\mathfrak{M} = \{u \in V^{l-1/p,p}_{\beta,\delta}(\Gamma_j) : \|u\|_{V^{l-1/p,p}_{\beta,\delta}(\Gamma_j)} \le c_0\}$$

is precompact in $V^{l'-1/p,p}_{\beta',\delta'}(\Gamma_j)$ for arbitrary $c_0 > 0$. Let $\mathfrak{M}'$ be the set of all $v \in V^{l,p}_{\beta,\delta}(\mathcal{G})$ with norm $\le 2c_0$. It follows from Lemma 4.1.4 that $\mathfrak{M}'$ is precompact in $V^{l',p}_{\beta',\delta'}(\mathcal{G})$, i.e. for arbitrary $\varepsilon > 0$ there exists a finite ε-net $\{u_1, \dots, u_l\}$ in $\mathfrak{M}'$ such that

$$\min_{1\le k\le l} \|v - v_k\|_{V^{l'-1/p,p}_{\beta',\delta'}(\Gamma_j)} < \varepsilon \quad \text{for all } v \in \mathfrak{M}'.$$

We denote the trace of v_k on Γ_j by u_k. Then for an arbitrary function $u \in \mathfrak{M}$, there exist a function $v \in \mathfrak{M}'$ and an index k such that $v|_{\Gamma_j} = u$ and

$$\|u - u_k\|_{V^{l'-1/p,p}_{\beta',\delta'}(\Gamma_j)} \le \|v - v_k\|_{V^{l',p}_{\beta',\delta'}(\mathcal{G})} < \varepsilon.$$

Thus, the set $\mathfrak{M}$ is precompact in $V^{l'-1/p,p}_{\beta',\delta'}(\Gamma_j)$. The lemma is proved. □

The following lemma is an immediate consequence of Lemmas 2.2.3 and 3.3.1.

LEMMA 4.1.7. *Let the functions $g_{j,k} \in V^{l-k+1-1/p,p}_{\beta,\delta}(\Gamma_j)$, $j = 1, \dots, N$, $k = 1, \dots, l$, be given. Then there exists a function $u \in V^{l,p}_{\beta,\delta}(\mathcal{G})$ such that*

$$\frac{\partial^{k-1} u}{\partial n^{k-1}} = g_{j,k} \quad \text{on } \Gamma_j,\ j = 1, \dots, N,\ k = 1, \dots, l,$$

and

$$\|u\|_{V^{l,p}_{\beta,\delta}(\mathcal{G})} \le c \sum_{j=1}^{N} \sum_{k=1}^{l} \|g_{j,k}\|_{V^{l-k+1-1/p,p}_{\beta,\delta}(\Gamma_j)},$$

where c is a constant independent of the functions $g_{j,k}$.

4.1.6. Operator pencils generated by the boundary value problem. Analogously to the problem in a cone, the boundary value problem (4.1.1) generates two types of operator pencils $A_\xi(\lambda)$ and $\mathfrak{A}_j(\lambda)$ for the edge points and vertices, respectively.

1) Let ξ be a point on an edge M_k, and let $\Gamma_{k_+}, \Gamma_{k_-}$ be the faces of $\mathcal{G}$ adjacent to ξ. Then by $\mathcal{D}_\xi$ we denote the dihedron which is bounded by the half-planes $\Gamma^\circ_{k_\pm}$ tangent to $\Gamma_{k_\pm}$ at ξ and the edge $M^\circ_\xi = \bar{\Gamma}^\circ_{k_+} \cap \bar{\Gamma}^\circ_{k_-}$. Furthermore, let r, φ be polar coordinates in the plane perpendicular to M°_ξ such that

$$\Gamma^\circ_{k_\pm} = \{x \in \mathbb{R}^3 : r > 0,\ \varphi = \pm\theta_\xi/2\}.$$

Then we define the operator $A_\xi(\lambda)$ as follows:

$$A_\xi(\lambda)\,U(\varphi) = r^{2m-\lambda}\,L^\circ(\xi, D_x)\,u$$

where $u(x) = r^\lambda U(\varphi)$. Here,

$$L^\circ(\xi, D_x) = \sum_{|\alpha|=2m} A_\alpha(\xi)\,D_x^\alpha$$

denotes the *principal part* of the differential operator $L(x, D_x)$ with coefficients frozen at ξ. The operator $A_\xi(\lambda)$ realizes a continuous mapping

$$W^{2m,2}(I_\xi)^\ell \to L_2(I_\xi)^\ell$$

for every $\lambda \in \mathbb{C}$, where I_ξ denotes the interval $(-\theta_\xi/2, +\theta_\xi/2)$. We denote by $\delta_+(\xi)$ and $\delta_-(\xi)$ the greatest positive real numbers such that the strip

$$m - 1 - \delta_-(\xi) < \operatorname{Re}\lambda < m - 1 + \delta_+(\xi)$$

is free of eigenvalues of the pencil $A_\xi(\lambda)$. Furthermore, we define

$$\delta_\pm^{(k)} = \inf_{\xi \in M_k} \delta_\pm(\xi)$$

for $k = 1, \dots, d$.

2) Let $x^{(i)}$ be a vertex of $\mathcal{G}$, and let J_i be the set of all indices j such that $x^{(i)} \in \overline{\Gamma}_j$. By our assumptions, there exist a neighborhood $\mathcal{U}$ of $x^{(i)}$ and a diffeomorphism κ mapping $\mathcal{G} \cap \mathcal{U}$ onto $\mathcal{K}_i \cap B_1$ and $\Gamma_k \cap \mathcal{U}$ onto $\Gamma_k^\circ \cap B_1$ for $k \in J_i$, where

$$\mathcal{K}_i = \{x : x/|x| \in \Omega_i\}$$

is a cone with vertex at the origin, and $\Gamma_k^\circ = \{x : x/|x| \in \gamma_k\}$ are the faces of this cone. Without loss of generality, we may assume that the Jacobian matrix $\kappa'(x)$ is equal to the identity matrix I at the point $x^{(i)}$. We introduce spherical coordinates $\rho = |x|$, $\omega = x/|x|$ in $\mathcal{K}_i$ and define

$$\mathfrak{A}_i(\lambda)\,U(\omega) = \rho^{2m-\lambda}\,L^\circ(x^{(i)}, D_x)u,$$

where $u(x) = \rho^\lambda U(\omega)$, $U \in \overset{\circ}{W}{}^{m,2}(\Omega_i)^\ell$. The operator $\mathfrak{A}_i(\lambda)$ realizes a continuous mapping

$$\overset{\circ}{W}{}^{m,2}(\Omega_i)^\ell \to W^{-m,2}(\Omega_i)^\ell$$

for arbitrary $\lambda \in \mathbb{C}$.

4.1.7. Normal solvability of the boundary value problem in weighted Sobolev spaces. We denote the operator

$$(4.1.5) \qquad V_{\beta,\delta}^{l,p}(\mathcal{G})^\ell \ni u \to \big(f, \{g_{j,k}\}\big) \in V_{\beta,\delta}^{l-2m,p}(\mathcal{G})^\ell \times \prod_{j=1}^{N}\prod_{k=1}^{m} V_{\beta,\delta}^{l-k+1-1/p,p}(\Gamma_j)^\ell$$

of the boundary value problem (4.1.1) by $\mathcal{A}_{l,p,\beta,\delta}$. The restriction of $\mathcal{A}_{l,p,\beta,\delta}$ to the space $V_{\beta,\delta}^{l,p}(\mathcal{G})^\ell \cap \overset{\circ}{V}{}^{m,p}_{\beta-l+m,\delta-l+m}(\mathcal{G})^\ell$ is denoted by $\mathcal{A}^\circ_{l,p,\beta,\delta}$. Our goal is to show that $\mathcal{A}_{l,p,\beta,\delta}$ and $\mathcal{A}^\circ_{l,p,\beta,\delta}$ are *Fredholm operators* under certain conditions on β and δ. To this end, we construct a regularizer for the operator $\mathcal{A}_{l,p,\beta,\delta}$. We start with the construction of local regularizers.

LEMMA 4.1.8. *Let $\mathcal{U}$ be a sufficiently small open subset of $\mathcal{G}$ and let φ be a smooth function with support in $\overline{\mathcal{U}}$. Suppose that $l \ge m$ and that the following conditions are satisfied.*

(i) *The line* $\operatorname{Re}\lambda = l - \beta_j - 3/p$ *does not contain eigenvalues of the pencil* $\mathfrak{A}_j(\lambda)$ *for* $j = 1, \dots, d'$,

(ii) $-\delta_+^{(k)} < \delta_k - l + m - 1 + 2/p < \delta_-^{(k)}$ *for* $k = 1, \dots, d$.

Then there exists an operator $\mathcal{R}$ continuously mapping the space of all

$$f \in V_{\beta,\delta}^{l-2m,p}(\mathcal{G})^\ell, \quad \operatorname{supp} f \subset \overline{\mathcal{U}},$$

into the space $V_{\beta,\delta}^{l,p}(\mathcal{G})^3 \cap \overset{\circ}{V}{}^{m,p}_{\beta-l+m,\delta-l+m}(\mathcal{G})^\ell$ *such that*

$$\varphi \mathcal{A}^\circ_{l,p,\beta,\delta} \mathcal{R} f = \varphi f$$

for all $f \in V_{\beta,\delta}^{l-2m,p}(\mathcal{G})^\ell$, $\operatorname{supp} f \subset \overline{\mathcal{U}}$, *and*

$$\varphi \mathcal{R} \mathcal{A}^\circ_{l,p,\beta,\delta} u = \varphi u$$

for all $u \in V_{\beta,\delta}^{l,p}(\mathcal{G})^\ell \cap \overset{\circ}{V}{}^{m,p}_{\beta-l+m,\delta-l+m}(\mathcal{G})^\ell$, $\operatorname{supp} u \in \overline{\mathcal{U}}$.

Suppose that $\beta'_j < \beta_j + 1$ for $j = 1, \dots, d'$, the strip $l - \beta_j - 3/p \le \operatorname{Re}\lambda \le l + 1 - \beta'_j - 3/p$ does not contain eigenvalues of the pencil $\mathfrak{A}_j(\lambda)$ for $j = 1, \dots, d'$, and that the components of δ' satisfy the inequalities $-\delta_+^{(k)} < \delta'_k - l + m - 2 + 2/p < \delta_k - l + m - 1 + 2/p$. Then the operator $\mathcal{R}$ realizes also a continuous mapping from the subspace

$$\{f \in V_{\beta'\delta'}^{l-2m+1,p}(\mathcal{G})^\ell : \operatorname{supp} f \subset \overline{\mathcal{U}}\} \tag{4.1.6}$$

into $V_{\beta',\delta'}^{l+1,p}(\mathcal{G})^\ell$.

P r o o f. Suppose first that $\overline{\mathcal{U}}$ contains the vertex $x^{(1)}$ of $\mathcal{G}$. Then there exists a diffeomorphism κ mapping $\mathcal{U}$ onto a subset $\mathcal{V}$ of a cone $\mathcal{K}$ with vertex at the origin such that $\kappa(x^{(1)}) = 0$ and the Jacobian matrix κ' coincides with the identity matrix I at $x^{(1)}$. We assume that the support of u is contained in $\overline{\mathcal{U}}$. Then the coordinate change $y = \kappa(x)$ transforms (4.1.1) into the equation

$$\tilde{L}(y, D_y)\, \tilde{u}(y) = \tilde{f}(y), \tag{4.1.7}$$

where $\tilde{u} = u \circ \kappa^{-1}$, $\tilde{f} = f \circ \kappa^{-1}$, $\tilde{L}(y, D_y)$ is a differential operator of order $2m$ such that $\tilde{L}^\circ(0, D_y) = L(x^{(1)}, D_y)$.

Let ζ be an infinitely differentiable cut-off function on $[0, \infty)$ equal to 1 in $[0, 1)$ and to zero in $(2, \infty)$. For arbitrary positive ϵ, we define $\zeta_\epsilon(y) = \zeta(|y|/\epsilon)$ and consider the operator

$$V_{\beta_1,\delta}^{l,p}(\mathcal{K})^\ell \cap \overset{\circ}{V}{}^{m,p}_{\beta_1-l+m,\delta-l+m}(\mathcal{K})^\ell \ni \tilde{u} \to \tilde{L}_\epsilon(y, D_y)\, u \in V_{\beta_1,\delta}^{l-2m,p}(\mathcal{K})^\ell \tag{4.1.8}$$

defined by

$$\tilde{L}_\epsilon(y, D_y) = \zeta_\epsilon\, \tilde{L}(y, D_y) + (1 - \zeta_\epsilon)\, \tilde{L}^\circ(0, D_y)$$

If ϵ is sufficiently small, then the coefficients of this operator satisfy the conditions (3.7.3) and (3.7.4). Hence by Theorems 3.7.3 and 3.7.4, the operator $\tilde{L}_\epsilon(0, D_y)$ realizes isomorphisms

$$V_{\beta_1,\delta}^{l,p}(\mathcal{K})^\ell \cap \overset{\circ}{V}{}^{m,p}_{\beta_1-l+m,\delta-l+m}(\mathcal{K})^\ell \to V_{\beta_1,\delta}^{l-2m,p}(\mathcal{K})^\ell$$

and

$$V^{l,p}_{\beta_1,\delta}(\mathcal{K})^\ell \cap \overset{\circ}{V}{}^{m,p}_{\beta_1-l+m,\delta-l+m}(\mathcal{K})^\ell \cap V^{l+1,p}_{\beta_1',\delta'}(\mathcal{K})^\ell \to V^{l-2m,p}_{\beta_1,\delta}(\mathcal{K})^\ell \cap V^{l-2m+1,p}_{\beta_1',\delta'}(\mathcal{K})^\ell .$$

We may assume that $\zeta_\epsilon = 1$ on $\mathcal{V}$ for $\epsilon = \epsilon_0$. Then the equation (4.1.7) can be written as

$$\tilde{L}_{\epsilon_0}(y, D_y)\, \tilde{u}(y) = \tilde{f}(y)$$

if the support of $\tilde{u}$ is contained in $\overline{\mathcal{V}}$. Let

$$(4.1.9) \qquad u(x) = \tilde{u}(\kappa(x)) \ \text{ for } x \in \mathcal{U}, \quad \text{where } \tilde{u} = \big(\tilde{L}_{\epsilon_0}(y, D_y)\big)^{-1} \tilde{f}.$$

Outside $\mathcal{U}$, let u be continuously extended to a vector function in $V^{l,p}_{\beta,\delta}(\mathcal{G})^\ell \cap \overset{\circ}{V}{}^{m,p}_{\beta-l+m,\delta-l+m}(\mathcal{G})^\ell$. The so defined mapping $f \to u$ is denoted by $\mathcal{R}$ and has the desired properties. In particular, the continuity of the mapping $\mathcal{R}$ from the subspace (4.1.6) into $V^{l+1,p}_{\beta',\delta'}(\mathcal{G})^\ell$ is a consequence of Theorem 3.5.13.

Analogously, the operator $\mathcal{R}$ can be constructed in the case where $\overline{\mathcal{U}}$ contains an edge point $\xi \in M_1$ but no points of other edges and no vertices of $\mathcal{G}$. Then instead of Theorem 3.5.10, one can use Theorem 2.9.2.

If $\bar{\mathcal{U}}$ contains no edge points of $\mathcal{G}$, then there exists a diffeomorphism κ mapping $\mathcal{U}$ onto a subset $\mathcal{V}$ of the half-space $\mathbb{R}^3_+$. The half-space can be also considered as a dihedron with edge angle π. It can be assumed that the "edge" of this dihedron has a positive distance from $\mathcal{V}$. Then for functions with support in $\mathcal{V}$, the norm in the weighted space $V^{l,p}_\delta$ is equivalent to the usual $W^{l,p}$-norm. Therefore by means Theorem 2.9.2 (with suitable δ), we can construct an operator $\mathcal{R}$ continuously mapping the space of all vector functions $f \in W^{l-2m,p}(\mathcal{G})^\ell$ with support in $\overline{\mathcal{U}}$ into the space $W^{l,p}(\mathcal{G})^\ell \cap \overset{\circ}{W}{}^{m,p}(\mathcal{G})^\ell$ such that

$$\varphi \mathcal{A}^\circ_{l,p,\beta,\delta} \mathcal{R} f = \varphi f$$

for all $f \in W^{l-2m,p}(\mathcal{G})^\ell$, $\operatorname{supp} f \subset \overline{\mathcal{U}}$, and

$$\mathcal{R}\mathcal{A}^\circ_{l,p,\beta,\delta} u = u$$

for all $u \in W^{l,p}(\mathcal{G})^\ell \cap \overset{\circ}{W}{}^{m,p}(\mathcal{G})^\ell$, *supp* $u \in \overline{\mathcal{U}}$. The proof of the lemma is complete. □

THEOREM 4.1.9. *Suppose that $l \ge m$ and the conditions* (i) *and* (ii) *of Lemma* 4.1.8 *are satisfied. Then the operators $\mathcal{A}_{l,p,\beta,\delta}$ and $\mathcal{A}^\circ_{l,p,\beta,\delta}$ are Fredholm.*

P r o o f. In order to prove the Fredholm property for the operator $\mathcal{A}^\circ_{l,p,\beta,\delta}$ it suffices to show that there exists a linear and continuous operator

$$\mathcal{R}^\circ \,:\, V^{l-2m,p}_{\beta,\delta}(\mathcal{G})^\ell \to \overset{\circ}{V}{}^{m,p}_{\beta-l+m,\delta-l+m}(\mathcal{G})^\ell$$

such that

$$\mathcal{R}^\circ\, \mathcal{A}^\circ_{l,p,\beta,\delta} - I \qquad \text{and} \qquad \mathcal{A}^\circ_{l,p,\beta,\delta}\, \mathcal{R}^\circ - I$$

are compact operators in $V^{l,p}_{\beta,\delta}(\mathcal{G})^\ell$ and $V^{l-2m,p}_{\beta,\delta}(\mathcal{G})^\ell$, respectively (see e.g. [**88**, Theorem 5.6]). An operator with these properties is called (left and right) *regularizer.*

For the sake of brevity, we write $\mathcal{A}^\circ$ instead of $\mathcal{A}^\circ_{l,p,\beta,\delta}$. Let $\{\mathcal{U}_j\}$ be a sufficiently fine open covering of $\mathcal{G}$, and let φ_j, ψ_j be infinitely differentiable functions such that

$$\operatorname{supp} \varphi_j \subset \operatorname{supp} \psi_j \subset \mathcal{U}_j, \quad \varphi_j\, \psi_j = \varphi_j, \quad \text{and} \quad \sum_j \varphi_j = 1.$$

For every j, there exists an operator $\mathcal{R}_j$ having the properties of Lemma 4.1.8 for $\mathcal{U} = \mathcal{U}_j \cap \mathcal{G}$. We consider the operator $\mathcal{R}^\circ$ defined by

$$\mathcal{R}^\circ f = \sum_j \varphi_j \mathcal{R}_j (\psi_j f).$$

Obviously,

$$\mathcal{R}^\circ \mathcal{A}^\circ u = \sum_j \varphi_j \mathcal{R}_j \left(\mathcal{A}^\circ \psi_j u - [\mathcal{A}^\circ, \psi_j]\, u\right) = u - \sum_j \varphi_j \mathcal{R}_j^\circ [\mathcal{A}^\circ, \psi_j]\, u,$$

where $[\mathcal{A}^\circ, \psi_j] = \mathcal{A}^\circ \psi_j - \psi_j \mathcal{A}^\circ$ is the commutator of $\mathcal{A}^\circ$ and ψ_j. The mapping

$$u \to [\mathcal{A}^\circ, \psi_j]\, u$$

is continuous from $\overset{\circ}{V}{}^{l,p}_{\beta,\delta}(\mathcal{G})^\ell$ into $V^{l-2m+1,p}_{\beta,\delta}(\mathcal{G})$. Suppose that $\beta_j \le \beta'_j < \beta_j + 1$ for $j = 1, \dots, d'$, $\delta_k \le \delta'_k < \delta_k + 1$ for $k = 1, \dots, d$ and that in addition the conditions of Lemma 4.1.8 are satisfied for β' and δ'. Then the mapping

$$u \to \mathcal{R}_j\, [\mathcal{A}^\circ, \psi_j]\, u$$

is continuous from $\overset{\circ}{V}{}^{l,p}_{\beta,\delta}(\mathcal{G})^\ell$ into $V^{l+1,p}_{\beta',\delta'}(\mathcal{G})^\ell$. Hence by Lemma 4.1.4, the operator

$$\mathcal{R}^\circ \mathcal{A}^\circ - I = -\sum_j \varphi_j \mathcal{R}_j^\circ [\mathcal{A}^\circ, \psi_j]$$

is compact on $\overset{\circ}{V}{}^{l,p}_{\beta,\delta}(\mathcal{G})^\ell$. Furthermore,

$$\mathcal{A}^\circ \mathcal{R}^\circ f = \sum_j \varphi_j \mathcal{A}^\circ \mathcal{R}_j(\psi_j f) + \sum_j [\mathcal{A}^\circ, \varphi_j]\, \mathcal{R}_j(\psi_j f) = f + \sum_j [\mathcal{A}^\circ, \varphi_j]\, \mathcal{R}_j(\psi_j f).$$

Since the commutator $[\mathcal{A}^\circ, \varphi_j]$ is a differential operator of order $2m-1$, the mapping

$$f \to \sum_j [\mathcal{A}^\circ, \varphi_j]\, \mathcal{R}_j(\psi_j f)$$

is continuous from $V^{l-2m,p}_{\beta,\delta}(\mathcal{G})$ into $V^{l-2m+1,p}_{\beta,\delta}(\mathcal{G})$. This means that it is compact on $V^{l-2m,p}_{\beta,\delta}(\mathcal{G})$. Thus, the operator $\mathcal{R}^\circ$ is a regularizer for the operator $\mathcal{A}^\circ$. By means of this regularizer, we can easily construct a regularizer for the operator $\mathcal{A}_{l,p,\beta,\delta}$. By Lemma 4.1.7, there exists a linear and continuous mapping

$$\prod_{j=1}^{N} \prod_{k=1}^{m} V^{l-k+1-1/p,p}_{\beta,\delta}(\Gamma_j)^\ell \ni \{g_{j,k}\} = g \to Tg \in V^{l,p}_{\beta,\delta}(\mathcal{G})^\ell$$

such that

$$\frac{\partial^{k-1} Tg}{\partial n^{k-1}} = g_{j,k} \quad \text{on } \Gamma_j \quad \text{for } j = 1, \dots, N,\ k = 1, \dots, m.$$

Then the operator $\mathcal{R}$ defined by

$$\mathcal{R}(f, g) = Tg + \mathcal{R}^0 \big(f - L(x, D_x) Tg\big)$$

is a regularizer for the operator $\mathcal{A}_{l,p,\beta,\delta}$. The proof of the theorem is complete. □

In particular, it follows from the last theorem that the solutions of the boundary value problem (4.1.1) satisfy the following a priori estimate (cf. Lemma 2.3.9).

COROLLARY 4.1.10. *Suppose that the conditions of Theorem* 4.1.9 *are satisfied. Then every solution* $u \in V^{l,p}_{\beta,\delta}(\mathcal{G})^\ell$ *of the boundary value problem* (4.1.1) *satisfies the estimate*

$$\|u\|_{V^{l,p}_{\beta,\delta}(\mathcal{G})^\ell} \le c\Big(\|f\|_{V^{l-2m,p}_{\beta,\delta}(\mathcal{G})^\ell} + \sum_{j,k} \|g_{j,k}\|_{V^{l-k+1-1/p,p}_{\beta,\delta}(\Gamma_j)^\ell} + \|u\|_{V^{0,p}_{\beta,\delta}(\mathcal{G})}\Big)$$

with a constant c independent of u.

4.1.8. Regularity assertions in weighted Sobolev spaces. Applying the regularity results for the Dirichlet problems in a dihedron and a cone obtained in the preceding chapters, we can easily prove the following regularity result.

THEOREM 4.1.11. *Let* $u \in V^{l,p}_{\beta,\delta}(\mathcal{G})^\ell$ *be a solution of the boundary value problem* (4.1.1), *where*

$$f \in V^{l-2m,p}_{\beta,\delta}(\mathcal{G})^\ell \cap V^{l'-2m,q}_{\beta',\delta'}(\mathcal{G})^\ell,$$
$$g_{j,k} \in V^{l-k+1-1/p,p}_{\beta,\delta}(\Gamma_j)^\ell \cap V^{l'-k+1-1/q,q}_{\beta',\delta'}(\Gamma_j)^\ell,$$

$l, l' \ge m$. *Suppose that the closed strip between the lines* $\operatorname{Re}\lambda = l - \beta_j - 3/p$ *and* $\operatorname{Re}\lambda = l' - \beta'_j - 3/q$ *is free of eigenvalues of the pencil* $\mathfrak{A}_j(\lambda)$, $j = 1, \dots, d'$, *and that*

$$-\delta_+^{(k)} < \delta_k - l + m - 1 + 2/p < \delta_-^{(k)}, \quad -\delta_+^{(k)} < \delta'_k - l' + m - 1 + 2/q < \delta_-^{(k)}$$

for $k = 1, \dots, d$. *Then* $u \in V^{l',q}_{\beta',\delta'}(\mathcal{G})^\ell$.

P r o o f. Suppose first that the support of u is sufficiently small. If this support contains an edge point or a vertex $x^{(j)}$, then the assertion of the theorem follows from Theorem 2.9.7 and Theorem 3.7.4 (see Remark 3.7.1). If $\operatorname{supp} u \cap \mathcal{S} = \emptyset$, then we can apply Lemma 2.2.7.

Suppose now that the support of u is an arbitrary subset of $\overline{\mathcal{G}}$. Then we consider the vector functions $\zeta_j u$, $j = 1, \dots, k$, where ζ_j are infinitely differentiable functions with small supports such that $\zeta_1 + \dots + \zeta_k = 1$ in $\mathcal{G}$. Using local regularity results (cf. Lemma 2.9.10 and Corollary 3.7.5), we obtain $\zeta_j u \in V^{l',q}_{\beta',\delta'}(\mathcal{G})^\ell$ for $j = 1, \dots, k$. This proves the theorem. □

4.2. Solvability of the boundary value problem in weighted Hölder spaces

Next, we extend the results of Section 4.1 to the class of the weighted Hölder spaces $N^{l,\sigma}_{\beta,\delta}(\mathcal{G})$ introduced below. In order to prove the normal solvability, we construct again a regularizer for the boundary value problem, while the regularity assertions can be directly deduced from the results in Chapters 2 and 3.

4.2.1. Weighted Hölder spaces. Let $\mathcal{U}_1, \dots, \mathcal{U}_{d'}$ be domains in $\mathbb{R}^3$ such that

$$\mathcal{U}_1 \cup \dots \cup \mathcal{U}_{d'} \supset \overline{\mathcal{G}}, \quad x^{(i)} \notin \overline{\mathcal{U}}_j \text{ if } i \ne j, \quad \text{and} \quad \overline{\mathcal{U}}_j \cap \overline{M}_k = \emptyset \text{ if } k \notin X_j.$$

Here again X_j denotes the set of the indices k such that $x^{(j)}$ is an end point of the edge M_k. Furthermore, let l be a nonnegative integer, σ a real number, $0 < \sigma < 1$, and let $\beta = (\beta_1, \dots, \beta_{d'})$, $\delta = (\delta_1, \dots, \delta_d)$ be arbitrary real tuples. We define the weighted Hölder space $N^{l,\sigma}_{\beta,\delta}(\mathcal{G})$ as the space of all l times continuously differentiable functions on $\bar{\mathcal{G}}\backslash\mathcal{S}$ with finite norm

$$\|u\|_{N^{l,\sigma}_{\beta,\delta}(\mathcal{G})} = \sum_{j=1}^{d'}\sum_{|\alpha|\le l} \sup_{x\in\mathcal{G}\cap\mathcal{U}_j} \rho_j(x)^{\beta_j-l-\sigma+|\alpha|} \prod_{k\in X_j}\Big(\frac{r_k(x)}{\rho_j(x)}\Big)^{\delta_k-l-\sigma+|\alpha|} \big|\partial_x^\alpha u(x)\big|$$
$$+\sum_{j=1}^{d'}\sum_{|\alpha|=l} \sup_{\substack{x,y\in\mathcal{G}\cap\mathcal{U}_j\\ |x-y|<r(x)/2}} \rho_j(x)^{\beta_j} \prod_{k\in X_j}\Big(\frac{r_k(x)}{\rho_j(x)}\Big)^{\delta_k} \frac{\big|\partial_x^\alpha u(x)-\partial_y^\alpha u(y)\big|}{|x-y|^\sigma}.$$

Furthermore, we introduce the following notation. If $\beta\in\mathbb{R}^{d'}$, $\delta\in\mathbb{R}^d$ and $s,t\in\mathbb{R}$, then by $N^{l,\sigma}_{\beta+s,\delta+t}(\mathcal{G})$, we mean the space $N^{l,\sigma}_{\beta',\delta'}(\mathcal{G})$ with $\beta'=(\beta_1+s,\dots,\beta_{d'}+s)$, $\delta'=(\delta_1+t,\dots,\delta_d+t)$.

For arbitrary integer $l\ge 0$, we define $N^{-l,\sigma}_{\beta,\delta}(\mathcal{G})$ as the space of all distributions of the form

$$u=\sum_{|\alpha|\le l}\partial_x^\alpha u_\alpha, \quad \text{where } u_\alpha\in N^{0,\sigma}_{\beta+l-|\alpha|,\delta+l-|\alpha|}(\mathcal{G}).$$

The norm in this space is defined as

$$\|f\|_{N^{-l,\sigma}_{\beta,\delta}(\mathcal{G})}=\inf\Big\{\sum_{|\alpha|\le l}\|u_\alpha\|_{N^{0,\sigma}_{\beta+l-|\alpha|,\delta+l-|\alpha|}(\mathcal{G})}\Big\},$$

where the infimum is taken over the set of all $u_\alpha\in N^{0,\sigma}_{\beta+l-|\alpha|,\delta+l-|\alpha|}(\mathcal{G})$, $|\alpha|\le l$, such that

$$\sum_{|\alpha|\le l}\partial_x^\alpha u_\alpha = u.$$

The trace space on Γ_j for $N^{l,\sigma}_{\beta,\delta}(\mathcal{G})$, $l\ge 0$, is denoted by $N^{l,\sigma}_{\beta,\delta}(\Gamma_j)$. Obviously

$$N^{l,\sigma}_{\beta,\delta}(\mathcal{G})\subset N^{l',\sigma'}_{\beta',\delta'}(\mathcal{G}) \quad \text{if} \quad l+\sigma\ge l'+\sigma',\ \beta_j-l-\sigma\le\beta_j'-l'-\sigma',$$
$$\delta_k-l-\sigma\le\delta_k'-l'-\sigma'$$

for $j=1,\dots,d'$, $k=1,\dots,d$. The same imbedding holds for the space $N^{l,\sigma}_{\beta,\delta}(\Gamma_j)$. Furthermore, it can be proved analogously to Lemma 4.1.4 that the imbeddings

$$N^{l,\sigma}_{\beta,\delta}(\mathcal{G})\subset N^{l',\sigma'}_{\beta',\delta'}(\mathcal{G}) \quad\text{and}\quad N^{l,\sigma}_{\beta,\delta}(\Gamma_j)\subset N^{l',\sigma'}_{\beta',\delta'}(\Gamma_j)$$

are compact if

$$l+\sigma>l'+\sigma',\quad \beta_j-l-\sigma<\beta_j'-l'-\sigma' \quad\text{and}\quad \delta_k-l-\sigma<\delta_k'-l'-\sigma'$$

for all j and k.

4.2.2. Normal solvability of the Dirichlet problem in weighted Hölder spaces. Let the operator

$$N^{l,\sigma}_{\beta,\delta}(\mathcal{G})^\ell \ni u\to\big(f,\{g_{j,k}\}\big)\in N^{l-2m,\sigma}_{\beta,\delta}(\mathcal{G})^\ell\times\prod_{j=1}^{d}\prod_{k=1}^{m}N^{l-k+1,\sigma}_{\beta,\delta}(\Gamma_j)^\ell$$

of the boundary value problem (4.1.1) be denoted by $\mathcal{A}$. We will prove that this operator is Fredholm if the line $\operatorname{Re}\lambda=l+\sigma-\beta_j$ is free of eigenvalues of the pencil $\mathfrak{A}_j(\lambda)$ for $j=1,\dots,d'$ and the components of δ satisfy the inequalities

$$-\delta_+^{(k)}<\delta_k-l-\sigma+m-1<\delta_-^{(k)} \quad \text{for } k=1,\dots,d. \tag{4.2.1}$$

As in the foregoing section, we construct a regularizer for $\mathcal{A}$. We start with the construction of a local regularizer.

LEMMA 4.2.1. *Let $\mathcal{U}$ be a sufficiently small subdomain of $\mathcal{G}$ and let φ be a smooth function with support in $\overline{\mathcal{U}}$. Suppose that there are no eigenvalues of the pencil $\mathfrak{A}_j(\lambda)$ on the line $\operatorname{Re}\lambda = l+\sigma-\beta_j$ for $j=1,\dots,d'$ and that the components of δ satisfy the condition* (4.2.1). *Then there exists an operator $\mathcal{R}$ continuously mapping the space of all*

$$(f,g) \in N^{l-2,\sigma}_{\beta,\delta}(\mathcal{G})^\ell \times \prod_{j=1}^{N}\prod_{k=1}^{m} N^{l-k+1,\sigma}_{\beta,\delta}(\Gamma_j)^\ell \tag{4.2.2}$$

with support in $\overline{\mathcal{U}}$ into the space $N^{l,\sigma}_{\beta,\delta}(\mathcal{G})^\ell$ such that

$$\varphi\mathcal{A}\mathcal{R}(f,g) = \varphi(f,g)$$

for all vector functions (4.2.2) *with support in $\overline{\mathcal{U}}$, and*

$$\varphi\mathcal{R}\mathcal{A}u = \varphi u$$

for all $u \in N^{l,\sigma}_{\beta,\delta}(\mathcal{G})^\ell$, $\operatorname{supp} u \subset \overline{\mathcal{U}}$.

Suppose that $\beta'_j < \beta_j + 1$ for $j=1,\dots,d'$, the strip $l+\sigma-\beta_j \le \operatorname{Re}\lambda \le l+1+\sigma-\beta'_j$ does not contain eigenvalues of the pencil $\mathfrak{A}_j(\lambda)$ for $j=1,\dots,d'$, and that the components of δ' satisfy the inequalities $l-m+2-\delta_+^{(k)} < \delta'_k-\sigma < \delta_k+1-\sigma$ for $k=1,\dots,d$. Then the operator $\mathcal{R}$ realizes also a continuous mapping from the subspace of all

$$(f,g) \in N^{l-1,\sigma}_{\beta',\delta'}(\mathcal{G})^\ell \times \prod_{j=1}^{N}\prod_{k=1}^{m} N^{l-k+2,\sigma}_{\beta',\delta'}(\Gamma_j)^\ell$$

with support in $\overline{\mathcal{U}}$ into the space $N^{l+1,\sigma}_{\beta',\delta'}(\mathcal{G})^\ell$.

Proof. This lemma can be proved analogously to Lemma 4.1.8. We restrict ourselves to the case where $\overline{\mathcal{U}}$ contains the vertex $x^{(1)}$ of $\mathcal{G}$. Then there exists a diffeomorphism κ mapping $x^{(1)}$ onto the origin and $\mathcal{U}$ onto a subset $\mathcal{V}$ of a cone $\mathcal{K}$ with vertex at the origin and faces Γ_j°, $j \in J_1$. Applying the coordinate transformation $y=\kappa(x)$ to the problem (4.1.1), we obtain the equations

$$\tilde{L}\tilde{u} = \tilde{f} \;\text{ in } \mathcal{V}, \quad \frac{\partial^{k-1}\tilde{u}}{\partial n^{k-1}} = \tilde{g}_{j,k} \;\text{ on } \kappa(\Gamma_j)\cap\overline{\mathcal{V}}, \;\; j=1,\dots,N,\; k=1,\dots,m,$$

where $\tilde{u}(x) = u(\kappa^{-1}x)$ for $x\in\mathcal{V}$. Let ε be a sufficiently small positive real number, and let $\tilde{L}_\varepsilon$ be the same differential operators as in the proof of Lemma 4.1.8. By Theorem 3.7.3, the operator

$$\tilde{u} \to \tilde{\mathcal{A}}_\varepsilon\tilde{u} = \big(\tilde{L}_\varepsilon\tilde{u}, \{\partial_n^{k-1}\tilde{u}|_{\Gamma_j^\circ}\}_{j\in J_1,\,1\le k\le m}\big)$$

realizes an isomorphism

$$N^{l,\sigma}_{\beta_1,\delta}(\mathcal{K})^\ell \to N^{l-2,\sigma}_{\beta_1,\delta}(\mathcal{K})^\ell \times \prod_{j\in J_1}\prod_{k=1}^{m} N^{l-k+1,\sigma}_{\beta_1,\delta}(\Gamma_j^\circ)^\ell.$$

We assume that $\mathcal{V}$ is contained in the ball $|x|<\varepsilon$. Then the coefficients of $\tilde{L}_\varepsilon$ and $\tilde{L}$ coincide on the support of $\tilde{f}$ and $\tilde{g}$. Let $\tilde{f}$ and $\tilde{g}$ be extended by zero outside $\mathcal{U}$. Then we define

$$u(x) = \tilde{u}(\kappa(x)) \;\text{ for } x\in\mathcal{U}, \quad \text{where } \tilde{u} = \tilde{\mathcal{A}}_\varepsilon^{-1}(\tilde{f},\tilde{g}).$$

This function can be continuously extended to a function in $u \in N^{l,\sigma}_{\beta,\delta}(\mathcal{G})^\ell$. The so defined mapping $(f,g) \to u$ is denoted by $\mathcal{R}$ and has the desired properties. □

THEOREM 4.2.2. *Suppose that the line* $\operatorname{Re}\lambda = l+\sigma-\beta_j$ *does not contain eigenvalues of the pencil* $\mathfrak{A}_j(\lambda)$ *for* $j=1,\dots,d'$ *and that the components* δ_k *of* δ *satisfy the inequalities* (4.2.1). *Then the operator* $\mathcal{A}$ *is Fredholm.*

P r o o f. Let the operator $\mathcal{R}$ be defined by

$$\mathcal{R}(f,g) = \sum_\nu \phi_\nu \mathcal{R}_\nu\, \psi_\nu(f,g),$$

where ϕ_ν, ψ_ν are the same cut-off functions as in the proof of Theorem 8.1.5 and $\mathcal{R}_\nu$ is the local regularizer of Lemma 4.2.1 for the subdomain $\mathcal{U} = \mathcal{U}_\nu \cap \mathcal{G}$. Repeating the proof of Theorem 4.1.9, we conclude from Lemma 4.2.1 that $\mathcal{R}\mathcal{A}-I$ is compact on $N^{l,\sigma}_{\beta,\delta}(\mathcal{G})^\ell$ and $\mathcal{A}\mathcal{R}-I$ is compact on the space (4.2.2). This proves the theorem. □

4.2.3. A regularity result for the solution. Using the regularity results for solutions of the Dirichlet problem in a dihedron and a cone, we prove the following theorem.

THEOREM 4.2.3. *Let* $u \in V^{l,p}_{\beta,\delta}(\mathcal{G})^\ell$ *be a solution of the boundary value problem* (4.1.1), *where*

$$f \in V^{l-2m,p}_{\beta,\delta}(\mathcal{G})^\ell \cap N^{l'-2m,\sigma}_{\beta',\delta'}(\mathcal{G})^\ell,$$
$$g_{j,k} \in V^{l-k+1-1/p,p}_{\beta,\delta}(\Gamma_j)^\ell \cap N^{l'-k+1,\sigma}_{\beta',\delta'}(\Gamma_j)^\ell,$$

$l, l' \ge m$. *Suppose that the closed strip between the lines* $\operatorname{Re}\lambda = l-\beta_j-3/p$ *and* $\operatorname{Re}\lambda = l'+\sigma-\beta'_j$ *is free of eigenvalues of the pencil* $\mathfrak{A}_j(\lambda)$, $j=1,\dots,d'$, *and that*

$$-\delta_+^{(k)} < \delta_k - l + m - 1 + 2/p < \delta_-^{(k)}, \quad -\delta_+^{(k)} < \delta'_k - l' - \sigma + m - 1 < \delta_-^{(k)}$$

for $k=1,\dots,d$. *Then* $u \in N^{l',\sigma}_{\beta',\delta'}(\mathcal{G})^\ell$.

P r o o f. For the case where $\operatorname{supp} u \cap \mathcal{S} = \emptyset$, we refer to Lemma 2.8.3. If the support of u is sufficiently small and contains an edge point or a vertex $x^{(j)}$, then the assertion of the theorem follows from the second parts of Theorem 2.9.7 and 3.7.4. Arguing as in the proof of Theorem 4.1.11, we obtain the assertion for solutions with arbitrary support. □

4.3. Examples

In this section, we consider the Dirichlet problem for some special strongly elliptic differential equations in a bounded domain of polyhedral type. In particular, we are concerned with regularity assertions for the variational solutions. In Sections 4.1 and 4.2, we obtained inequalities for the weight parameter β and δ which guarantee that the solution belongs to the weighted spaces $V^{l,p}_{\beta,\delta}$ and $N^{l,\sigma}_{\beta,\delta}$. The bounds in these inequalities depend on the eigenvalues of the pencils $A_k(\lambda)$ and $\mathfrak{A}_j(\lambda)$ introduced in Section 4.1. In general, the eigenvalues of these pencils can be calculated only numerically. However, for a number of regularity results it is sufficient to have estimates for the eigenvalues with smallest positive real parts greater than $m-1$ nd $m-3/2$, respectively. Such estimates can be found e.g. in the book [**85**] both for general and for particular elliptic boundary value problems.

We are interested here in results in nonweighted Sobolev and Hölder spaces. Then the weight parameters β and δ are zero.

4.3.1. The Dirichlet problem for the Laplace equation. We consider the Dirichlet problem

$$-\Delta u = f \ \text{ in } \mathcal{G}, \quad u = 0 \ \text{ on } \partial\mathcal{G}. \tag{4.3.1}$$

For the following information concerning the eigenvalues of the pencils $A_\xi(\lambda)$ and $\mathfrak{A}_j(\lambda)$ introduced in Section 4.1, we refer to [**85**, Chapter 2]. The eigenvalues of the operator pencil $A_\xi(\lambda)$ are

$$\lambda_k = k\pi/\theta_\xi, \quad k = \pm 1, \pm 2, \dots,$$

where θ_ξ is the inner angle at the edge point ξ. Let $\tilde{\lambda}_k$ be the eigenvalues of the Laplace-Beltrami operator $-\delta$ (with Dirichlet boundary condition) on the subdomain Ω_j of the unit sphere (for the definition of Ω_j see Subsection 4.1.6). Then the eigenvalues of the pencils $\mathfrak{A}_j(\lambda)$ are given by

$$\Lambda_{\pm k} = -\frac{1}{2} \pm \sqrt{\tilde{\lambda}_k + 1/4}.$$

Clearly, the interval $[-1, 0]$ is free of eigenvalues of the pencil $\mathfrak{A}_j(\lambda)$. If Ω_j is a proper subset of a half-sphere, then even the interval $[-2, +1]$ does not contain eigenvalues of the pencil $\mathfrak{A}_j(\lambda)$ (cf. [**85**, Cororollary 2.2.1 and Theorem 2.2.4]). We denote the smallest positive eigenvalue of the pencil $\mathfrak{A}_j(\lambda)$ by Λ_j^+ and set

$$\theta_k = \inf_{\xi \in M_k} \theta_\xi\,.$$

Applying Theorems 4.1.11 and 4.2.3, we obtain the following regularity results for the variational solution of the problem (4.3.1).

LEMMA 4.3.1. *Let $u \in \overset{\circ}{W}{}^{1,2}(\mathcal{G})$ be the variational solution of the boundary value problem* (4.3.1).

1) *Suppose that $f \in W^{-1,2}(\mathcal{G}) \cap V_{\beta,\delta}^{l-2,p}(\mathcal{G})$, $l \ge 1$, and that*

$$-1 - \Lambda_j^+ < l - \beta_j - 3/p < \Lambda_j^+ \quad \text{for } j = 1, \dots, d',$$
$$|l - \delta_k - 2/p| < \pi/\theta_k \quad \text{for } k = 1, \dots, d.$$

Then $u \in V_{\beta,\delta}^{l,p}(\mathcal{G})^\ell$.

2) *If $f \in W^{-1,2}(\mathcal{G}) \cap N_{\beta,\delta}^{l-2,\sigma}(\mathcal{G})$, $l \ge 1$,*

$$-1 - \Lambda_j^+ < l + \sigma - \beta_j < \Lambda_j^+ \quad \text{for } j = 1, \dots, d',$$
$$|l + \sigma - \delta_k| < \pi/\theta_k \quad \text{for } k = 1, \dots, d,$$

then $u \in N_{\beta,\delta}^{l,\sigma}(\mathcal{G})^\ell$.

In the special cases $\beta = 0$ and $\delta = 0$, we obtain regularity results in weighted Sobolev spaces withought weight.

THEOREM 4.3.2. *Let $u \in \overset{\circ}{W}{}^{1,2}(\mathcal{G})$ be the variational solution of the boundary value problem* (4.3.1).

1) *Suppose that $f \in W^{-1,p}(\mathcal{G})$, $p \ge 2$, $3/p > 1 - \Lambda_j^+$ for $j = 1, \dots, d'$ and $2/p > 1 - \pi/\theta_k$ for $k = 1, \dots, d$. Then $u \in W^{1,p}(\mathcal{G})$.*

2) *Suppose that $f \in L_p(\mathcal{G}) \cap W^{-1,2}(\mathcal{G})$, $3/p > 2 - \Lambda_j^+$ for $j = 1, \dots, d'$ and $2/p > 2 - \pi/\theta_k$ for $k = 1, \dots, d$. Then $u \in W^{2,p}(\mathcal{G})$.*

P r o o f. Using the equalities

$$W^{-1,p}(\mathcal{G}) = V_{0,0}^{-1,p}(\mathcal{G}), \quad L_p(\mathcal{G}) = V_{0,0}^{0,p}(\mathcal{G})$$

and the imbedding $V_{0,0}^{l,p}(\mathcal{G}) \subset W^{l,p}(\mathcal{G})$, one can deduce the theorem from the assertions of Lemma 4.3.1 for the case $\beta = 0$, $\delta = 0$. □

Since the eigenvalues of the pencil $\mathfrak{A}_j(\lambda)$ lie outside the interval $[-1, 0]$, the $W^{1,p}(\mathcal{G})$ regularity result in Theorem 4.3.2 holds for arbitrary p, $2 \le p < 3 + \varepsilon$, where ε is a positive number depending on the domain $\mathcal{G}$. If $\theta_k < \pi$ for all k and every Ω_j is contained in a half-sphere, then the $W^{1,p}(\mathcal{G})$ regularity results holds for arbitrary p.

The $W^{2,p}(\mathcal{G})$ regularity result in Theorem 4.3.2 is always true for $1 < p < \frac{3}{2} + \varepsilon$, where again ε is a positive number depending on the domain $\mathcal{G}$. For a convex domain $\mathcal{G}$ of polyhedral type, the $W^{2,p}(\mathcal{G})$ regularity result is true for $1 < p < 2 + \varepsilon$. Under additional assumptions on the angles θ_k, the condition on p can be weakened once again. For example in the case of a cube $\mathcal{G}$ (then $\Lambda_j^+ = 3$), the $W^{2,p}(\mathcal{G})$ regularity result holds for all p.

4.3.2. The Dirichlet problem for the Lamé system. We consider the Dirichlet problem for the Lamé system

$$\Delta u + \frac{1}{1-2\nu} \operatorname{grad} \operatorname{div} u = f \ \text{ in } \mathcal{G}, \qquad u = 0 \ \text{ on } \partial\mathcal{G}. \tag{4.3.2}$$

Here u is the displacement vector and ν denotes the Poisson ratio, $\nu < 1/2$. We are again interested in the eigenvalue of the pencil $A_\xi(\lambda)$ with smallest positive real part. By [**85**, Theorem 3.1.2], this eigenvalue is

$$\lambda_1(\xi) = \frac{z(\theta_\xi)}{\theta_\xi},$$

where $z(\theta)$ is the smallest positive solution of the equation

$$\frac{\sin z}{z} - \frac{\sin\theta}{(3-4\nu)\theta} = 0 \quad \text{if } \theta < \pi,$$
$$\frac{\sin z}{z} + \frac{\sin\theta}{(3-4\nu)\theta} = 0 \quad \text{if } \theta > \pi.$$

Note that $\lambda_1(\xi) > 1/2$. If $\theta_\xi < \pi$, then even $\lambda_1(\xi) > 1$.

For the operator pencils $\mathfrak{A}_j(\lambda)$ it is known, that the strip $-1 \le \operatorname{Re}\lambda \le 0$ is free of eigenvalues of this pencil (cf. [**85**, Theorem 3.6.1]). We denote the eigenvalue with smallest positive real part by Λ_j^+. Note that the eigenvalues with positive real parts ≤ 1 are real. If Ω_j is a proper subset of a half-sphere, then $\operatorname{Re}\Lambda_j^+ > 1$ (cf. [**85**, Theorem 3.5.3]).

The following theorem holds as an immediate consequence of Theorem 4.1.11.

THEOREM 4.3.3. *Let $u \in \overset{\circ}{W}{}^{1,2}(\mathcal{G})^3$ be the variational solution of the boundary value problem* (4.3.2).

1) *Suppose that $f \in W^{-1,p}(\mathcal{G})^3$, $p \ge 2$, $3/p > 1 - \operatorname{Re}\Lambda_j^+$ for $j = 1, \dots, d'$ and and $2/p > 1 - \delta_+^{(k)}$ for $k = 1, \dots, d$, where*

$$\delta_+^{(k)} = \inf_{\xi \in M_k} \lambda_1(\xi).$$

Then $u \in W^{1,p}(\mathcal{G})^3$.

2) *Suppose that* $f \in L_p(\mathcal{G})^3 \cap W^{-1,2}(\mathcal{G})^3$, $3/p > 2 - \operatorname{Re}\Lambda_j^+$ *for* $j = 1, \dots, d'$ *and* $2/p > 2 - \delta_+^{(k)}$ *for* $k = 1, \dots, d$. *Then* $u \in W^{2,p}(\mathcal{G})^3$.

As in the case of the Laplace equation, the $W^{1,p}(\mathcal{G})$ regularity result in Theorem 4.3.3 holds for arbitrary p, $2 \le p \le 3 + \varepsilon$, where ε is a positive number depending on the domain $\mathcal{G}$. If $\theta_k < \pi$ for all k and every Ω_j is contained in a half-sphere, then the $W^{1,p}(\mathcal{G})$ regularity results holds for arbitrary p.

The $W^{2,p}(\mathcal{G})$ regularity result in Theorem 4.3.3 is always true for $1 < p < \frac{3}{2} + \varepsilon$, where again ε is a positive number depending on the domain $\mathcal{G}$. For convex domains $\mathcal{G}$ of polyhedral type, the $W^{2,p}(\mathcal{G})$ regularity result is true for $1 < p < 2 + \varepsilon$.

4.3.3. The Dirichlet problem for elliptic self-adjoint differential operators. We consider the Dirichlet problem

$$L(x, D_x)\, u = f \quad \text{in } \mathcal{G}, \qquad \frac{\partial^{k-1} u}{\partial n^{k-1}} = 0 \quad \text{on } \partial\mathcal{G}\backslash\mathcal{S},\ k = 1, \dots, m, \tag{4.3.3}$$

where

$$L(x, D_x) = \sum_{|\alpha| \le 2m} A_\alpha(x)\, D_x^\alpha.$$

with infinitely differentiable coefficients A_α on $\overline{\mathcal{G}}$. We assume that the operator $L(x, D_x)$ is elliptic in $\overline{\mathcal{G}}$ and that

$$A_\alpha(x) = (A_\alpha(x))^* \quad \text{for } |\alpha| = 2m.$$

In addition, we assume that for every vertex $x^{(j)}$ there exist a neighborhood $\mathcal{U}_j$ and a diffeomorphism κ_j mapping $\mathcal{G} \cap \mathcal{U}_j$ onto $\mathcal{K}_j \cap B_1$, where B_1 is the unit ball and $\mathcal{K}_j$ is a Lipschitz graph cone with edges, i.e. $\mathcal{K}_j$ has the representation $x_3 > \phi(x_1, x_2)$ in Cartesian coordinates. Note that not every polyhedron is Lipschitz (cf. Figure 7 in Section 11.3).

Then the following results hold for the spectra of the operator pencils $A_\xi(\lambda)$ and $\mathfrak{A}_j(\lambda)$ introduced in Section 4.1 (see [**85**, Theorems 8.5.1 and 11.1.1])

LEMMA 4.3.4. 1) *The eigenvalues of the pencils* $A_\xi(\lambda)$ *lie outside the strip* $m - 3/2 \le \operatorname{Re}\lambda \le m - 1/2$. *This means that* $\delta_+^{(k)} = \delta_-^{(k)} > 1/2$.

2) *The strip* $m - 2 \le \operatorname{Re}\lambda \le m - 1$ *is free of eigenvalues of the pencil* $\mathfrak{A}_j(\lambda)$.

Using the last lemma together with the results of Section 4.1, we obtain the following assertions.

THEOREM 4.3.5. *Suppose that for every vertex* $x^{(j)}$ *there exist a neighborhood* $\mathcal{U}_j$ *and a diffeomorphism* κ_j *mapping* $\mathcal{G} \cap \mathcal{U}_j$ *onto the intersection of a Lipschitz graph cone with the unit ball. Then the following assertion are true.*

1) *The operator*

$$\overset{\circ}{W}{}^{m,p}(\mathcal{G}) \ni u \to f \in W^{-m,p}(\mathcal{G})$$

of the problem (4.3.3) *is Fredholm for* $3/2 - \varepsilon < p < 3 + \varepsilon$. *Here* ε *is a certain positive number depending on the domain* $\mathcal{G}$.

2) *If* $u \in \overset{\circ}{W}{}^{m,2}(\mathcal{G})^\ell$ *is a solution of the boundary value problem* (4.3.3), *where* $f \in W^{-m,p}(\mathcal{G})^\ell$ *and* $p < 3 + \varepsilon$, *then* $u \in \overset{\circ}{W}{}^{m,p}(\mathcal{G})^\ell$.

P r o o f. The assertions of the theorem follow immediately from Theorems 4.1.9 and 4.1.11 and from the equalities $\overset{\circ}{W}{}^{m,p}(\mathcal{G}) = \overset{\circ}{V}{}^{m,p}_{0,0}(\mathcal{G})$, $W^{-m,p}(\mathcal{G}) = V^{-m,p}_{0,0}(\mathcal{G})$. □

4.3.4. The Dirichlet problem for second order systems in convex polyhedral domains. We consider the Dirichlet problem

$$L(x, D_x)u = f \ \text{ in } \mathcal{G}, \quad u = 0 \quad \text{on } \partial\mathcal{G}, \tag{4.3.4}$$

where

$$L(D_x) = \sum_{|\alpha|\le 2} A_\alpha(x)\, D_x^\alpha$$

is a second order elliptic differential operator with infinitely differentiable coefficients A_α satisfying the condition

$$A_\alpha(x) = \big(A_\alpha(x)\big)^* \quad \text{for } |\alpha| = 2.$$

Then the results of Lemma 4.3.4 with $m = 1$ are valid for the spectra of the operator pencils $A_\xi(\lambda)$ and $\mathfrak{A}_j(\lambda)$. Under the additional assumption that the edge angles are less than π and the cones $\mathcal{K}_j$ are contained in half-spaces (this means for example if $\mathcal{G}$ is convex), the following sharper result holds (cf. [**85**, Theorems 8.4.1 and 11.4.1]).

LEMMA 4.3.6. *Suppose that $\theta_\xi < \pi$ for every $\xi \in M_k$, $k = 1, \dots, d$. Then the following assertions are true.*

1) The eigenvalues of the pencil $A_\xi(\lambda)$ lie outside the strip $-1 \le \operatorname{Re}\lambda \le +1$ for every ξ. This means that $\delta_+^{(k)} = \delta_-^{(k)} > 1$ for $k = 1, \dots, d$.

2) If the cone $\mathcal{K}_j$ is a subset of a half-space, then the strip $-3/2 \le \operatorname{Re}\lambda \le 1/2$ is free of eigenvalues of the pencil $\mathfrak{A}_j(\lambda)$.

Applying Theorems 4.1.9 and 4.1.11, we obtain the following theorem.

THEOREM 4.3.7. *Suppose that $\theta_\xi < \pi$ for every $\xi \in M_k$, $k = 1, \dots, d$, and that every of the cones $\mathcal{K}_j$ is a subset of a half-space. Then the following assertions are true.*

1) The operator

$$\overset{\circ}{W}{}^{m,p}(\mathcal{G}) \ni u \to f \in W^{-m,p}(\mathcal{G})$$

of the problem (4.3.4) is Fredholm for $6/5 - \varepsilon < p < 6 + \varepsilon$. Here ε is a certain positive number depending on the domain $\mathcal{G}$.

2) If $u \in \overset{\circ}{W}{}^{m,2}(\mathcal{G})^\ell$ is a solution of the problem (4.3.4), where $f \in W^{-m,p}(\mathcal{G})^\ell$ and $p < 6 + \varepsilon$, then $u \in \overset{\circ}{W}{}^{m,p}(\mathcal{G})^\ell$.

Note that both assertions of Theorem 4.3.7 are valid for arbitrary $p \in (1, \infty)$ if the domain $\mathcal{G}$ has no vertices and all edge angles are less than π.

4.3.5. The Dirichlet problem for the biharmonic operator. Now we deal with the Dirichlet problem for the biharmonic equation

$$\Delta^2 u = f \ \text{ in } \mathcal{G}, \qquad u = \frac{\partial u}{\partial n} = 0 \ \text{ on } \partial\mathcal{G}\backslash\mathcal{S}. \tag{4.3.5}$$

We consider the operator pencils generated by this problem at the edge points and vertices. Let ξ is a point on the edge M_k and let θ_ξ denote the inner angle at this edge point. Then the eigenvalues of the pencil $A_\xi(\lambda)$ are the solutions of the equation

$$\sin^2\big((\lambda-1)\theta_\xi\big) = (\lambda-1)^2 \sin^2\theta_\xi\,,$$
$$\lambda \neq 1 \text{ for all } \theta_\xi, \quad \lambda \neq 2 \text{ if } \theta_\xi \neq \pi, 2\pi, \theta_*,$$

where θ_* denotes the unique solution of the equation $\tan\theta = \theta$ in the interval $(0, 2\pi)$, $\theta_* \approx 1.4303\,\pi$ (see e.g. [**85**, Section 7.1]). The eigenvalue with smallest real part > 1 is

$$\lambda = 1 + \frac{z(\theta_\xi)}{\theta_\xi}\,,$$

where $z(\theta_\xi)$ denotes the solution with smallest positive real part of the equation

$$\frac{\sin z}{z} + \frac{\sin\theta_\xi}{\theta_\xi} = 0. \tag{4.3.6}$$

Note that $z(\theta_\xi)$ is real if $\theta_\xi \geq \theta_0$, where θ_0 is the smallest positive solution of the equation $\theta^{-1}\sin\theta = -\cos\theta_*$, $\theta_0 \approx 0.8128\,\pi$. If $\theta_\xi < \theta_0$, then (4.3.6) has no real solutions, and the real part of the solutions $z = x + iy$ of (4.3.6) satisfies the equation

$$\cos x \sqrt{\Big(\frac{x\sin\theta_\xi}{\theta_\xi \sin x}\Big)^2 - 1} + \frac{\sin\theta_\xi}{\theta_\xi}\, \operatorname{arcosh}\Big(-\frac{x\sin\theta_\xi}{\theta_\xi\sin x}\Big) = 0. \tag{4.3.7}$$

Consequently, the numbers $\delta_+(\xi)$ and $\delta_-(\xi)$ are given for the problem (4.3.5) by the equality

$$\delta_\pm(\xi) = \frac{x(\theta_\xi)}{\theta_\xi}\,,$$

where $x(\theta_\xi)$ is the smallest positive solution of (4.3.7) if $\theta_\xi < \theta_0$ and the smallest positive solution of the equation (4.3.6) if $\theta_\xi \geq \theta_0$. Note that $\delta_+(\xi)$ is a decreasing function with respect to θ_ξ, $\delta_+(\xi) = 1$ for $\theta_\xi = \pi$, $\delta_+(\xi) = 1/2$ for $\theta_\xi = 2\pi$.

Let $x^{(j)}$ be a vertex of $\mathcal{G}$. The operator pencil $\mathfrak{A}_j(\lambda)$ is given by the equality

$$\mathfrak{A}_j(\lambda) = \big(\delta + (\lambda-2)(\lambda-1)\big)\,\big(\delta + \lambda(\lambda+1)\big).$$

Here δ is the Beltrami operator on the subdomain Ω_j of the sphere S^2.

We denote by m_j the positive number such that $m_j(m_j+1)$ is the first eigenvalue of the operator $-\delta$ with Dirichlet conditions on $\partial\Omega_j$. For following lemma, we refer to [**85**, Theorems 7.2.5 and 7.2.6].

LEMMA 4.3.8. 1) *The strip*

$$1 - \max(m_j, 1) \leq \operatorname{Re}\lambda \leq \max(m_j, 1)$$

does not contain eigenvalues of the pencil $\mathfrak{A}_j(\lambda)$.

2) *If* Ω_j *is a subset of a hemisphere* S^2_+ *and* $S^2_+\backslash\Omega_j$ *contains a nonempty open set, then the strip* $-1 \leq \operatorname{Re}\lambda \leq 2$ *is free of eigenvalues of the pencil* $\mathfrak{A}_j(\lambda)$.

Using this information on the spectra of the pencils $A_\xi(\lambda)$ and $\mathfrak{A}_j(\lambda)$, we can deduce the following regularity results from Theorem 4.1.11.

THEOREM 4.3.9. *Let $\mathcal{G}$ be an arbitrary bounded domain of polyhedral type, and let $u \in \overset{\circ}{W}{}^{2,2}(\mathcal{G})$ be the variational solution of the problem* (4.3.5).

1) *If $f \in W^{-2,p}(\mathcal{G})$ and $p < 3+\varepsilon$, then $u \in W^{2,p}(\mathcal{G})$.*

2) *If $f \in W^{-1,p}(\mathcal{G})$ and $p < \varepsilon + 4/3$, then $u \in W^{3,p}(\mathcal{G})$.*

Here ε is a positive constant depending on $\mathcal{G}$.

Proof. If $f \in W^{-2,p}(\mathcal{G}) = V_{0,0}^{-2,p}(\mathcal{G})$, $p \ge 2$, then we conclude from Theorem 4.1.11 that $u \in V_{0,0}^{2,p}(\mathcal{G}) \subset W^{2,2}(\mathcal{G})$ provided that $1 - 2/p < \delta_+^{(k)}$ and the strip $1/2 \le \mathrm{Re}\,\lambda \le 2 - 3/p$ is free of eigenvalues of the pencils $\mathfrak{A}_j(\lambda)$. Since $\delta_+^{(k)} > 1/2$ for all k and the eigenvalues of the pencils $\mathfrak{A}_j(\lambda)$ lie outside the strip $0 \le \mathrm{Re}\,\lambda \le 1$, the first assertion of the theorem holds for $p < 3 + \varepsilon$. Analogously, the $W^{3,p}(\mathcal{G})$-regularity result holds if $2 - 2/p < \delta_+^{(k)}$ and the closed strip between the lines $\mathrm{Re}\,\lambda = 1/2$ and $\mathrm{Re}\,\lambda = 3 - 3/p$ does not contain eigenvalues of the pencils $\mathfrak{A}_j(\lambda)$. This is the case for $p < \varepsilon + 4/3$. □

Note that the bounds for p in Theorem 4.3.9 are sharp. This means that the numbers 3 and 4/3 in this theorem cannot be replaced by greater numbers without additional conditions on $\mathcal{G}$. A $W^{4,p}$-regularity result does not hold for general polyhedral domains.

For special domains, it is possible to obtain more precise regularity results. Let for example $\mathcal{G}$ be a polyhedron which arises if one cuts a small cube from a bigger one (see Figure 6 below).

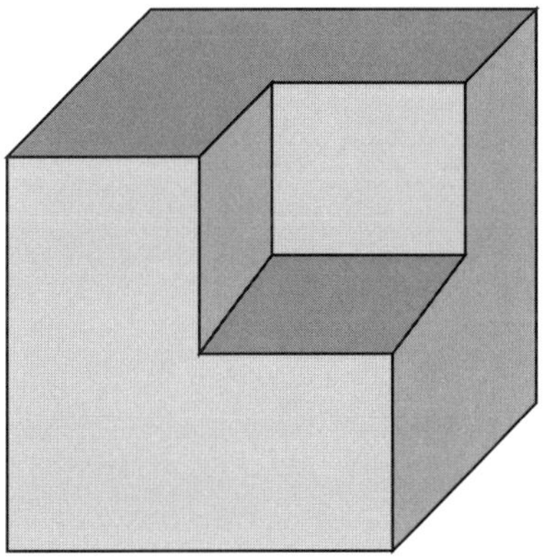

FIGURE 6. A polyhedral domain for which a $W^{3,p}$-regularity result with $p < 1.374$ holds

For the edge angle $\theta_\xi = 3\pi/2$, one obtains $\delta_+(\xi) = 0.54448$. Thus, the $W^{3,p}$-regularity result of Theorem 4.3.9 holds for $p < 1.37408$.

If the domain $\mathcal{G}$ is convex, then $\delta_\pm(\xi) > 1$ for every edge point. Furthermore, we can apply the second part of Lemma 4.3.8. Thus, the regularity results of Theorem 4.3.9 hold under weaker conditions on p.

THEOREM 4.3.10. *Let $\mathcal{G}$ be an arbitrary bounded convex domain of polyhedral type, and let $u \in \overset{\circ}{W}{}^{2,2}(\mathcal{G})$ be the variational solution of the problem* (4.3.5).

1) *If $f \in W^{-2,p}(\mathcal{G})$, $1 < p < \infty$, then $u \in W^{2,p}(\mathcal{G})$.*
2) *If $f \in W^{-1,p}(\mathcal{G})$ and $p < 2 + \varepsilon$, then $u \in W^{3,p}(\mathcal{G})$.*
3) *If $f \in L_p(\mathcal{G})$, $p < 1 + \varepsilon$, then $u \in W^{4,p}(\mathcal{G})$.*

Here ε is a positive constant depending on $\mathcal{G}$.

Note again, that the conditions on p in the last theorem are sharp. However, it is possible to improve the result for special domains. As an example, we consider the boundary value problem (4.3.5) in a cube. Here, we obtain $\delta_+(\xi) = 2.73959$ for every edge point ξ, since the smallest positive solution of (4.3.7) is equal to 4.303343 for $\theta_\xi = \pi/2$. Furthermore, we have $m_j = 3$. This follows from the obvious fact that $u = x_1x_2x_3$ is a positive solution of the Laplace equation in the cone $x_1 > 0$, $x_2 > 0$, $x_3 > 0$ with zero Dirichlet data on the planes $x_1 = 0$, $x_2 = 0$ and $x_3 = 0$. Hence, by Lemma 4.3.8, the strip $-2 \le \operatorname{Re}\lambda \le 3$ does not contain eigenvalues of the pencils $\mathfrak{A}_j(\lambda)$. Therefore, the following assertions are true for the solution $u \in \overset{\circ}{W}{}^{2,2}(\mathcal{G})$ of the boundary value problem (4.3.5) in a cube:

- If $f \in W^{-1,p}(\mathcal{G})$, $1 < p < \infty$, then $u \in W^{3,p}(\mathcal{G})$.
- If $f \in L_p(\mathcal{G})$, $p \le 3$, then $u \in W^{4,p}(\mathcal{G})$.

CHAPTER 5

The Miranda-Agmon maximum principle

It is well-known that harmonic functions in $C(\bar{\mathcal{G}})$ satisfy the maximum principle

$$\max_{x\in\bar{\mathcal{G}}} |u(x)| \le \max_{x\in\partial\mathcal{G}} |u(x)|\,.$$

As was proved by MIRANDA [**144, 145**] for two dimensions and by AGMON [**5**] for the higher-dimensional case, the solutions of an arbitrary strongly elliptic equation

$$L(x, D_x)\, u = 0$$

in a bounded domain $\mathcal{G}$ with smooth boundary $\partial\mathcal{G}$ satisfy the estimate

$$\|u\|_{C^{m-1}(\bar{\mathcal{G}})} \le c\Big(\sum_{k=1}^{m} \|\partial_n^{k-1}u\|_{C^{m-k}(\partial\mathcal{G})} + \|u\|_{L(\mathcal{G})}\Big)$$

with a constant c independent of u. Our goal is to extend this result to solutions of strongly elliptic equations in domains with nonsmooth boundaries. Throughout this chapter, we assume that the differential equation $L(x, D_x)\, u = f$ is uniquely solvable in $\overset{\circ}{W}{}^{m,2}(\mathcal{G})^\ell$ for arbitrary given $f \in W^{-m,2}(\mathcal{G})^\ell$.

The first two sections of this chapter deal with the Dirichlet problem in three-dimensional domains of polyhedral type. We consider the Green's matrix of this problem and obtain point estimates for its elements. These inequalities are applied in Section 5.2, where we prove weighted and nonweighted L_∞ estimates for generalized solutions and their derivatives. In particular, it follows from the main result of Section 5.2 that the Miranda-Agmon maximum principle is valid for strongly elliptic self-adjoint systems in three-dimensional Lipschitz graph domains of polyhedral type.

The last two sections of this chapter are concerned with the Dirichlet problem for strongly elliptic systems in a N-dimensional domain with conical points. In Section 5.3, we state conditions which ensure the validity of weighted and non-weighted $W^{m-1,\infty}$-estimates for generalized solutions. In particular, we show that the generalized solution of the problem

$$Lu = 0 \ \text{ in } \mathcal{G}, \quad \frac{\partial^{k-1}u}{\partial n^{k-1}} = g_k \ \text{ on } \partial\mathcal{G}\backslash\mathcal{S},\ k = 1, \dots, m,$$

satisfies the classical Miranda-Agmon estimate

$$\|u\|_{W^{m-1,\infty}(\mathcal{G})} \le c \sum_{k=1}^{m} \|\partial_n^{k-1}u\|_{W^{m-k,\infty}(\partial\mathcal{G})}$$

if the strip $m - N/2 < \operatorname{Re}\lambda \le m-1$ does not contain eigenvalues of the pencils $\mathfrak{A}_j(\lambda)$ generated by the boundary value problem at the vertices of the domain $\mathcal{G}$. In Section 5.4, we consider only smooth (i.e., C^∞) solutions of the equation $Lu = 0$. The condition for the validity of a $W^{m-1,\infty}$-estimate for such solutions is weaker than the above condition given for generalized solutions. Here, it is sufficient and necessary that only the line $\mathrm{Re}\,\lambda = m-1$ is free of eigenvalues of the pencils $\mathfrak{A}_j(\lambda)$. As is shown at the end of the chapter, the last requirement may fail for any dimension $N \geq 4$ and for any strongly elliptic differential operator of order $2m \geq 4$. A counterexample for the biharmonic equation which shows this effect is given at the end of the chapter.

5.1. Green's matrix for the Dirichlet problem in a bounded domain

We consider the Dirichlet problem (4.1.1) for the same differential operator $L(x,D_x)$ and the same domain $\mathcal{G}$ as in Chapter 4. For the sake of simplicity, we assume that the differential equation $L(x,D_x)\,u = f$ is uniquely solvable in $\overset{\circ}{W}{}^{m,2}(\mathcal{G})^\ell$ for arbitrary given $f \in W^{-m,2}(\mathcal{G})^\ell$. Then there exists a uniquely determined solution $G(\cdot,\xi)$ of the problem

$$L(x,D_x)\,G(x,\xi) = \delta(x-\xi)\,I_\ell \quad \text{for } x,\xi \in \mathcal{G}, \tag{5.1.1}$$

$$\frac{\partial^k G(x,\xi)}{\partial n_x^k} = 0 \quad \text{for } x \in \Gamma_j,\ \xi \in \mathcal{G},\ j = 1,\dots,N,\ k = 0,\dots,m-1, \tag{5.1.2}$$

which belongs to the Sobolev space $W^{m,2}$ outside an arbitrarily small neighborhood of the point $\xi \in \mathcal{G}$. The $\ell \times \ell$-matrix $G(x,\xi)$ is called *Green's matrix* for the boundary value problem (4.1.1). Using the same methods as for the Dirichlet problem in a dihedron and in a cone, we can establish point estimates for this matrix.

5.1.1. Existence of Green's matrix.

We prove the existence and uniqueness of the Green's matrix analogously to Theorems 2.5.1 and 3.4.1. Again we denote by $L^+(x,D_x)$ the formally adjoint operator to $L(x,D_x)$, i.e.,

$$L^+(x,D_x)\,v = \sum_{|\alpha|\leq 2m} D_x^\alpha\big(A_\alpha^*(x)\,v\big).$$

Here, $A_\alpha^*(x)$ is the adjoint matrix to $A_\alpha(x)$.

THEOREM 5.1.1. *Suppose that the equation $L(x,D_x)\,u = f$ has a unique solution $u \in \overset{\circ}{W}{}^{m,2}(\mathcal{G})^\ell$ for arbitrary given $f \in W^{-m,2}(\mathcal{G})^\ell$. Then the following assertions are true.*

1) *There exists a uniquely determined solution $G(x,\xi)$ of the problem* (5.1.1), (5.1.2) *such that the function $x \to \zeta(x,\xi)\,G(x,\xi)$ belongs to $\overset{\circ}{W}{}^{m,2}(\mathcal{G})^{\ell\times\ell}$ for every $\xi \in \mathcal{G}$ and for every infinitely differentiable function $\zeta(\cdot,\xi)$ on $\overline{\mathcal{G}}$ equal to zero in a neighborhood of the point $x = \xi$.*

2) *The adjoint matrix $G^*(x,\xi)$ is the unique solution of the problem*

$$L^+(\xi,D_\xi)\,G^*(x,\xi) = \delta(x-\xi)\,I_\ell \quad \textit{for } x,\xi \in \mathcal{G}, \tag{5.1.3}$$

$$\frac{\partial^k G^*(x,\xi)}{\partial n_\xi^k} = 0 \quad \textit{for } \xi \in \Gamma^\pm,\ x \in \mathcal{G},\ k = 0,\dots,m-1. \tag{5.1.4}$$

such that the function $\xi \to \zeta(x,\xi)\,G^*(x,\xi)$ belongs to the space $\overset{\circ}{W}{}^{m,2}(\mathcal{G})^{\ell\times\ell}$ for every infinitely differentiable function $\zeta(x,\cdot)$ on $\overline{\mathcal{G}}$ which is equal to zero in a neighborhood of the point $\xi = x$.

3) *The solution* $u \in \overset{\circ}{W}{}^{m,2}(\mathcal{G})^\ell$ *of the equation* $L(x,D_x)u = f$ *in* $\mathcal{G}$ *admits the representation*

$$u(x) = \int_{\mathcal{G}} G(x,\xi)\, f(\xi)\, d\xi \tag{5.1.5}$$

for arbitrary $f \in W^{-m,2}(\mathcal{G})^\ell$.

P r o o f. Let ξ be an arbitrary point in $\mathcal{G}$ and $\mathcal{U}(\xi) = \{x : |x-\xi| < r(\xi)/2\}$. Then there exists a function $g(x,\xi)$ such that

$$L(x,D_x)g(x,\xi) = \delta(x-\xi)\, I_\ell \quad \text{for } x \in \mathcal{G}\cap\mathcal{U}(\xi),$$
$$\frac{\partial^k g(x,\xi)}{\partial n_x^k} = 0 \in \Gamma_j \cap \mathcal{U}(\xi),\ j=1,\dots,N,\ k=0,\dots,m-1.$$

Let η be an infinitely differentiable function on the interval $(0,\infty)$ equal to one in $(0,1/4)$ and to zero in $(1/2,\infty)$. Then the function

$$h(x,\xi) = \eta\Big(\frac{|x-\xi|}{r(\xi)}\Big)\, g(x,\xi) \ \text{ for } x\in\mathcal{U}(\xi), \quad h(x,\xi)=0 \ \text{ for } x\notin\mathcal{U}(\xi)$$

satisfies the equations

$$L(x,D_x)\, h(x,\xi) = \delta(x-\xi)\, I_\ell + \phi(x,\xi) \ \text{ for } x\in\mathcal{G}; \qquad \frac{\partial^k h(x,\xi)}{\partial n_x^k} = 0 \ \text{ for } x \in \Gamma_j\,,$$

$j=1,\dots,N$, $k=0,\dots,m-1$, where $\phi(\cdot,\xi)$ is an infinitely differentiable matrix-valued function vanishing in a neighborhood of $\mathcal{S}$. By the assumption of the theorem, there exists a solution $R(\cdot,\xi) \in \overset{\circ}{W}{}^{m,2}(\mathcal{G})^{\ell\times\ell}$ of the problem

$$L(x,D_x)\, R(\cdot,\xi) = -\phi(\cdot,\xi) \ \text{ in } \mathcal{G}, \qquad \frac{\partial^k R(,\xi)}{\partial n^k} = 0 \ \text{ on } \Gamma_j\,,$$

$j=1,\dots,N$, $k=0,\dots,m-1$. The function $G(x,\xi) = h(x,\xi) + R(x,\xi)$ is a solution of the problem (5.1.1), (5.1.2). This proves the first assertion. The proof of the second and third parts proceeds analogously to items 3) and 4) of Theorem 3.4.1. □

5.1.2. Estimates of Green's matrix: the case where x and ξ lie in neighborhoods of different vertices. For $j=1,\dots,d'$, let

$$\Lambda_j^- < \operatorname{Re}\lambda < \Lambda_j^+$$

be the widest strip in the complex plane containing the line $\operatorname{Re}\lambda = m-3/2$ which is free of eigenvalues of the pencil $\mathfrak{A}_j(\lambda)$.

THEOREM 5.1.2. *Let* $\mathcal{U}_i$ *and* $\mathcal{U}_j$ *be sufficiently small neighborhoods of the vertices* $x^{(i)}$ *and* $x^{(j)}$, *respectively. If* $x\in\mathcal{G}\cap\mathcal{U}_i$ *and* $\xi\in\mathcal{G}\cap\mathcal{U}_j$, $i\neq j$, *then*

$$\begin{aligned}\big|D_x^\alpha D_\xi^\gamma G(x,\xi)\big| \ &\le\ c_{\alpha,\gamma}\, \rho_i(x)^{\Lambda_i^+ - |\alpha| - \varepsilon}\, \rho_j(\xi)^{2m-3-\Lambda_j^- - |\gamma| - \varepsilon}\\ &\quad \times \prod_{k\in X_i} \Big(\frac{r_k(x)}{\rho_i(x)}\Big)^{m-1+\delta_+^{(k)} - |\alpha| - \varepsilon} \prod_{k\in X_j} \Big(\frac{r_k(\xi)}{\rho_j(\xi)}\Big)^{m-1+\delta_-^{(k)} - |\gamma| - \varepsilon},\end{aligned}$$

where ε *is an arbitrarily small positive number. The constant* $c_{\alpha,\gamma}$ *is independent of* x *and* ξ.

P r o o f. Let ζ, η be smooth cut-off functions with sufficiently small supports such that $\zeta = 1$ on $\mathcal{U}_j$ and $\eta = 1$ in a neighborhood of $\operatorname{supp}\zeta$. Since $\eta(\xi)\, L^+(\xi, D_\xi)\, G^*(x,\xi) = 0$, it follows from Corollary 3.7.5 that $\zeta(\cdot)\, D_x^\alpha G(x,\cdot) \in V_{\beta,\delta}^{l,p}(\mathcal{G})^{\ell\times\ell}$ and

$$\|\zeta(\cdot)\, D_x^\alpha G(x,\cdot)\|_{V_{\beta,\delta}^{l,p}(\mathcal{G})^{\ell\times\ell}} \le c\, \|\eta(\cdot) D_x^\alpha G(x,\cdot)\|_{W^{m,2}(\mathcal{G})^{\ell\times\ell}}$$

for $x \in \mathcal{G} \cap \mathcal{U}_i$, where $2m-3-\Lambda_j^+ < l - \beta_j - 3/p < 2m-3-\Lambda_j^-$ and $-\delta_-^{(k)} < \delta_k - l + m - 1 + 2/p < \delta_+^{(k)}$. Hence, Lemma 3.1.4 yields

$$\rho_j(\xi)^{\beta_j - l + |\gamma| + 3/p} \prod_{k\in X_j} \Big(\frac{r_k(\xi)}{\rho_j(\xi)}\Big)^{\delta_k - l + |\gamma| + 3/p} \big|D_x^\alpha D_\xi^\gamma G(x,\xi)\big| \le c\, \|\eta(\cdot) D_x^\alpha G(x,\cdot)\|_{W^{m,2}(\mathcal{G})^{\ell\times\ell}} \tag{5.1.6}$$

for $\xi \in \mathcal{U}_j$. Let $f \in W^{-m,2}(\mathcal{G})^\ell$, and let

$$u(y) = \int_{\mathcal{G}} G(y,z)\, \eta(z)\, f(z)\, dz$$

be the unique solution of the equation $L(x, D_x)\, u = \eta f$ in $\overset{\circ}{W}{}^{m,2}(\mathcal{G})^\ell$. We denote by χ and ψ smooth cut-off functions such that $\chi = 1$ in $\mathcal{U}_i$ and $\psi = 1$ in a neighborhood of $\operatorname{supp}\chi$. Since $\psi L(x, D_x)\, u = \psi\eta f = 0$, we obtain (again by Corollary 3.7.5) the estimate

$$\|\chi u\|_{V_{\beta',\delta'}^{l,q}(\mathcal{G})^\ell} \le c\, \|\psi u\|_{W^{m,2}(\mathcal{G})^\ell} \le c\, \|f\|_{W^{-m,2}(\mathcal{G})^\ell}\,,$$

where $\Lambda_i^- < l - \beta_j' - 3/q < \Lambda_i^+$ and $-\delta_+^{(k)} < \delta_k' - l + m - 1 + 2/q < \delta_-^{(k)}$. Applying Lemma 3.1.4, we get

$$\rho_i(x)^{\beta_i' - l + |\alpha| + 3/q} \prod_{k\in X_i} \Big(\frac{r_k(x)}{\rho_i(x)}\Big)^{\delta_k' - l + |\alpha| + 3/q} \big|D_x^\alpha u(x)\big| \le c\, \|f\|_{W^{-m,2}(\mathcal{G})^\ell}$$

with a constant c independent of x. Thus, the mapping

$$\begin{aligned} W^{-m,2}(\mathcal{G})^\ell \ni f \to \rho_i(x)^{\beta_i' - l + |\alpha| + 3/q} \prod_{k\in X_i} \Big(\frac{r_k(x)}{\rho_i(x)}\Big)^{\delta_k' - l + |\alpha| + 3/q} D_x^\alpha u(x) \\ = \rho_i(x)^{\beta_i' - l + |\alpha| + 3/q} \prod_{k\in X_i} \Big(\frac{r_k(x)}{\rho_i(x)}\Big)^{\delta_k' - l + |\alpha| + 3/q} \int_{\mathcal{G}} D_x^\alpha G(x,z)\, \eta(z)\, f(z)\, dz \in \mathbb{C}^\ell \end{aligned}$$

is continuous, and its norm is bounded by a constant independent of x. Therefore,

$$\|\eta(\cdot)\, D_x^\alpha G(x,\cdot)\|_{W^{m,2}(\mathcal{G})^{\ell\times\ell}} \le c\, \rho_i(x)^{l - |\alpha| - \beta_i' - 3/q} \prod_{k\in X_i} \Big(\frac{r_k(x)}{\rho_i(x)}\Big)^{l - |\alpha| - \delta_k' - 3/q}.$$

This together with (5.1.6) implies

$$\begin{aligned} \big|D_x^\alpha D_\xi^\gamma G(x,\xi)\big| \ \le\ & c\, \rho_i(x)^{l - |\alpha| - \beta_i' - 3/q}\, \rho_j(\xi)^{l - |\gamma| - \beta_j - 3/p} \\ & \times \prod_{k\in X_i} \Big(\frac{r_k(x)}{\rho_i(x)}\Big)^{l - |\alpha| - \delta_k' - 3/q} \prod_{k\in X_j} \Big(\frac{r_k(\xi)}{\rho_j(\xi)}\Big)^{l - |\gamma| - \delta_k - 3/p} \end{aligned}$$

for $x \in \mathcal{U}_i$, $\xi \in \mathcal{U}_j$. We can choose $\beta_i', \beta_j, \delta_k, \delta_k', p$ and q such that $l - \beta_i' - 3/q = \Lambda_i^+ - \varepsilon$, $l - \beta_j - 3/p = 2m - 3 - \Lambda_j^- - \varepsilon$, $\delta_k - l + m - 1 + 3/p = -\delta_-^{(k)} + \varepsilon$ and $\delta_k' - l + m - 1 + 3/q = -\delta_+^{(k)} + \varepsilon$. Thus, we obtain the desired inequality. □

5.1.3. Estimates of Green's matrix: the case where x and ξ lie in a neighborhood of the same vertex. Now we consider the case where x and ξ lie in a sufficiently small neighborhood $\mathcal{U}$ of the same vertex $x^{(j)}$. As in the case of a cone, we obtain different estimates for the cases $\rho_j(x)/2 < \rho_j(\xi) < 2\rho_j(x)$, $\rho_j(\xi) < \rho_j(x)/2$ and $\rho_j(\xi) > 2\rho_j(x)$. We start with the case $\rho_j(x)/2 < \rho_j(\xi) < 2\rho_j(x)$.

THEOREM 5.1.3. 1) *Let $x, \xi \in \mathcal{G} \cap \mathcal{U}$, $\rho_j(x)/2 < \rho_j(\xi) < 2\rho_j(x)$ and $|x-\xi| > \min(r(x), r(\xi))$. Then*

$$\big|D_x^\alpha D_\xi^\gamma G(x,\xi)\big| \le c_{\alpha,\gamma} |x-\xi|^{2m-3-|\alpha|-|\gamma|} \Big(\frac{r(x)}{|x-\xi|}\Big)^{m-1+\delta_+^{(k(x))}-|\alpha|-\varepsilon} \times \Big(\frac{r(\xi)}{|x-\xi|}\Big)^{m-1+\delta_-^{(k(\xi))}-|\gamma|-\varepsilon},$$

where $k(x)$ denotes the smallest integer such that $r_k(x) = r(x)$ and ε is an arbitrarily small positive number.

2) *Let $x, \xi \in \mathcal{G} \cap \mathcal{U}$, $\rho_j(x)/2 < \rho_j(\xi) < 2\rho_j(x)$ and $|x-\xi| < \min(r(x), r(\xi))$. Then*

$$\big|D_x^\alpha D_\xi^\gamma G(x,\xi)\big| \le c_{\alpha,\gamma} \big(|x-\xi|^{2m-3-|\alpha|-|\gamma|} + r(\xi)^{2m-3-|\alpha|-|\gamma|}\big) \quad \text{if } |\alpha|+|\gamma| \ne 2m-3,$$

$$\big|D_x^\alpha D_\xi^\gamma G(x,\xi)\big| \le c_{\alpha,\gamma} \Big(\Big|\log\frac{|x-\xi|}{r(\xi)}\Big| + 1\Big) \quad \text{if } |\alpha|+|\gamma| = 2m-3.$$

The constant $c_{\alpha,\gamma}$ is independent of x and ξ.

P r o o f. The proof of the first part is a word-by-word repetition of the proof of Theorem 3.4.2. It suffices to prove the second part for small $\min(r(x), r(\xi))$. Suppose that $|x-\xi| < \min(r(x), r(\xi)) = 2\delta$. Then x and ξ lie in a ball B_δ with radius δ and distance $> 3\delta/4$ from the set $\mathcal{S}$ of the edge points. Let B'_δ be a ball concentric to B_δ with radius $3\delta/2$. We put $\mathcal{G}_\delta = \mathcal{G} \cap B'_\delta$, $\Gamma_\delta = \partial\mathcal{G} \cap B'_\delta$, $\tilde{\mathcal{G}}_\delta = \delta^{-1}\mathcal{G}_\delta = \{y = x/\delta : x \in \mathcal{G}_\delta\}$, and $\tilde\Gamma_\delta = \delta^{-1}\Gamma_\delta$. The distance of the sets $\tilde{\mathcal{G}}_\delta$ and $\tilde\Gamma_\delta$ from the edges of $\delta^{-1}\mathcal{G}$ is greater than $1/4$. Thus, there exists a Green's matrix $g(y,\eta)$ satisfying the equations

$$\delta^{2m}\, L(\delta y, \delta^{-1}D_y)\, g(y,\eta) = \delta(y-\eta)\, I_\ell \quad \text{for } y,\eta \in \tilde{\mathcal{G}}_\delta\,,$$
$$\partial_{n(y)}^{k-1} g(y,\eta) = 0 \quad \text{for } y \in \tilde\Gamma_\delta\,,\ \eta \in \tilde{\mathcal{G}}_\delta\,,\ k = 1,\dots,m,$$

and the estimates

$$\big|D_y^\alpha D_\eta^\gamma g(y,\eta)\big| \le c_{\alpha,\gamma} \big(|y-\eta|^{2m-3-|\alpha|-|\gamma|} + 1\big) \text{ if } |\alpha|+|\gamma| \ne 2m-3,$$
$$\big|D_y^\alpha D_\eta^\gamma g(y,\eta)\big| \le c_{\alpha,\gamma} \big(\log|y-\eta| + 1\big) \quad \text{if } |\alpha|+|\gamma| = 2m-3.$$

Since the difference of the operators $\delta^{2m}\, L(\delta y, \delta^{-1}D_y)$ and $L^\circ(0, D_y)$ is small for small δ, we can assume that the constants $c_{\alpha,\gamma}$ are independent of δ. Obviously, the matrix $\tilde G(x,\xi) = \delta^{2m-3}\, g(\delta^{-1}x, \delta^{-1}\xi)$ satisfies the equations

$$L(x, D_x)\, \tilde G(x,\xi) = \delta(x-\xi)\, I_\ell \quad \text{for } x,\xi \in \mathcal{G}_\delta\,,$$
$$\partial_{n(x)}^{k-1} \tilde G(x,\xi) = 0 \quad \text{for } x \in \Gamma_\delta\,,\ \xi \in \mathcal{G}_\delta\,,\ k = 1,\dots,m,$$

and the inequalities

$$\big|D_x^\alpha D_\xi^\gamma \tilde{G}(x,\xi)\big| \le c_{\alpha,\gamma}\left(|x-\xi|^{2m-3-|\alpha|-|\gamma|} + \delta^{2m-3-|\alpha-|\gamma|}\right) \quad \text{if } |\alpha|+|\gamma| \ne 2m-3, \tag{5.1.7}$$

$$\big|D_x^\alpha D_\xi^\gamma \tilde{G}(x,\xi)\big| \le c_{\alpha,\gamma}\left(\log\Big|\frac{x-\xi}{\delta}\Big| + 1\right) \quad \text{if } |\alpha|+|\gamma| = 2m-3. \tag{5.1.8}$$

Let $\xi \in \mathcal{G}\cap B_\delta$, and let $\zeta \in C_0^\infty(B'_\delta)$ be a cut-off function equal to one in ball concentric to B_δ with radius $5\delta/4$. Then the matrix-valued function

$$U(x,\xi) = G(x,\xi) - \zeta(x)\,\tilde{G}(x,\xi)$$

satisfies the equations

$$L(D_x)\,U(x,\xi) = F(x,\xi) \text{ for } x \in \mathcal{K},$$
$$\partial_{n(x)}^{k-1} U(x,\xi) = 0 \quad \text{for } x \in \partial\mathcal{K}\backslash\mathcal{S},\ k=1,\dots,m,$$

where $F(x,\xi)=0$ for $x \in \mathcal{G}\cap B_\delta$ and for $x \in \mathcal{G}\backslash B'_\delta$. Furthermore,

$$\big|D_x^\alpha D_\xi^\gamma F(x,\xi)\big| \le c\,\delta^{-3-|\alpha|-|\gamma|}.$$

Using Corollary 4.1.10 with $\beta_j = \delta_k = l-m$, $p=2$, and the uniqueness of the solutions of the Dirichlet problem (4.1.1), we obtain the estimate

$$\sum_{|\alpha|\le l} \delta^{|\alpha|-m} \|D_x^\alpha D_\xi^\gamma U(\cdot,\xi)\|_{L_2(\mathcal{G}\cap B_\delta)} \le c\,\|D_\xi^\gamma U(\cdot,\xi)\|_{V^{l,2}_{\beta,\delta}(\mathcal{G})}$$
$$\le c\,\|D_\xi^\gamma F(\cdot,\xi)\|_{V^{l-2m,2}_{\beta,\delta}(\mathcal{G})} \le c\,\delta^{m-|\gamma|-3/2},$$

i.e.

$$\|D_x^\alpha D_\xi^\gamma U(\cdot,\xi)\|_{L_2(\mathcal{G}\cap B_\delta)} \le c\,\delta^{2m-|\alpha|-|\gamma|-3/2}.$$

Thus, it follows from the imbedding $W^{2,2} \subset L_\infty$ that

$$\begin{aligned}\|D_x^\alpha D_\xi^\gamma U(\cdot,\xi)\|_{L_\infty(\mathcal{G}\cap B_\delta)} &\le c \sum_{|\alpha'|\le 2} \delta^{|\alpha'|-3/2}\, \|D_x^{\alpha+\alpha'} D_\xi^\gamma U(\cdot,\xi)\|_{L_2(\mathcal{G}\cap B_\delta)} \\ &\le c\,\delta^{2m-|\alpha|-|\gamma|-3}.\end{aligned}$$

Consequently, the matrix $G(x,\xi)$ satisfies the estimates (5.1.7), (5.1.8) for $x,\xi \in \mathcal{G}\cap B_\delta$. This proves the theorem. □

COROLLARY 5.1.4. *Let $x,\xi \in \mathcal{G}\cap\mathcal{U}$, $\rho_j(x)/2 < \rho_j(\xi) < 2\rho_j(x)$, and $|x-\xi| < \min(r(x),r(\xi))$. If $|\alpha| \le m-1$ and $|\alpha|+|\gamma| > 2m-3$, then*

$$\big|D_x^\alpha D_\xi^\gamma G(x,\xi)\big| \le c_{\alpha,\gamma}\, d(x)\, |x-\xi|^{2m-4-|\alpha|-|\gamma|} \tag{5.1.9}$$

where $d(x)$ is the distance of x from the boundary $\partial\mathcal{G}$.

P r o o f. If $|x-\xi| < 4d(x)$, then (5.1.9) follows immediately from Theorem 5.1.3. Suppose that $|x-\xi| > 4d(x)$. Then let y be the nearest point to x on $\partial\mathcal{G}$. Since the function $x \to G_{i,j}^{\alpha,\gamma}(x,\xi) = \partial_x^\alpha \partial_\xi^\gamma G_{i,j}(x,\xi)$ vanishes on $\partial\mathcal{G}$ for $|\alpha| \le m-1$, we obtain

$$\begin{aligned}\big|\mathrm{Re}\,\partial_x^\alpha \partial_\xi^\gamma G_{i,j}(x,\xi)\big| &= \big|\mathrm{Re}\,\big(G_{i,j}^{\alpha,\gamma}(x,\xi) - G_{i,j}^{\alpha,\gamma}(y,\xi)\big)\big| \\ &= \Big|\mathrm{Re}\sum_{j=1}^3 \partial_{z_j} G_{i,j}^{\alpha,\gamma}(z,\xi)\,(x_j-y_j)\Big| \le d(x) \sum_{j=1}^3 \big|\partial_{z_j} G_{i,j}^{\alpha,\gamma}(z,\xi)\big|,\end{aligned}$$

where $z = \theta x+(1-\theta)y$, $0 < \theta < 1$. The same estimate (with another z) holds for the imaginary part of $\partial_x^\alpha \partial_\xi^\gamma G_{i,j}(x,\xi)$. By the conditions on x and ξ, we have $\rho_j(\xi)/8 < \rho_j(z) < 3\rho_j(\xi)$ and $3|y-\xi| < 5\min(r(y), r(\xi))$. Consequently by Theorem 5.1.3,

$$\left|\partial_{z_j} G_{i,j}^{\alpha,\beta}(z,\xi)\right| \le c\,|z-\xi|^{2m-4-|\alpha|-|\gamma|}$$

Using the inequality $3|x-\xi| < 4|z-\xi| < 5|x-\xi|$, we get (5.1.9). □

For the proof of estimates of Green's matrix in the cases $\rho_j(\xi) < \rho_j(x)/2$ and $\rho_j(\xi) > 2\rho_j(x)$, we need the following lemma.

LEMMA 5.1.5. *Let η_ε be an infinitely differentiable function such that $\eta_\varepsilon(x) = 1$ for $c_1\varepsilon < \rho_j(x) < c_2\varepsilon$, $\eta_\varepsilon(x) = 0$ for $|x| > c_3\varepsilon$, where ε is a sufficiently small positive number. Suppose that $\Lambda_j^- < l - \beta_j - 3/p < \Lambda_j^+$ and that the components of δ satisfy the inequalities $-\delta_+^{(k)} < \delta_k - l + m - 1 + 2/p < \delta_-^{(k)}$. If $\eta_\varepsilon u \in \overset{\circ}{W}{}^{m,2}(\mathcal{G})^\ell$ and $\eta_\varepsilon L(x, D_x)\, u = 0$, then*

$$\|\zeta_\varepsilon u\|_{V^{l,p}_{\beta,\delta}(\mathcal{G})^\ell} \le c\,\varepsilon^{m-l+\beta_j+3/p-3/2}\,\|\eta_\varepsilon u\|_{V^{m,2}_{0,0}(\mathcal{G})^\ell}$$

with a constant c independent of u and ε. Here ζ_ε is an infinitely differentiable function such that $\zeta_\varepsilon(x) = 1$ for $c_1'\varepsilon < \rho_j(x) < c_2'\varepsilon$, $0 \le c_1 \le c_1' < c_2' < c_2$, and $\zeta_\varepsilon(x) = 0$ for $\rho_j(x) < c_1\varepsilon$ and $\rho_j(x) > c_2\varepsilon$.

P r o o f. Without loss of generality, we may assume that $x^{(j)}$ is the origin and $\mathcal{G}$ coincides with a cone $\mathcal{K}$ in a neighborhood of this vertex. We introduce the functions $v(x) = u(\varepsilon x)$ and $\tilde\eta_\varepsilon(x) = \eta_\varepsilon(\varepsilon x)$. Obviously, $\tilde\eta_\varepsilon(x) = 1$ for $c_1 < |x| < c_2$. Furthermore, let $\tilde\zeta$ be an infinitely differentiable function such that $\tilde\zeta(x) = 1$ for $c_1' < |x| < c_2'$, $\tilde\zeta(x) = 0$ for $|x| < c_1$ and $|x| > c_2$. It follows from the assumptions of the lemma that

$$\tilde\eta_\varepsilon v \in \overset{\circ}{W}{}^{m,2}(\mathcal{K})^\ell \quad \text{and} \quad \tilde\eta_\varepsilon(x)\, L(\varepsilon x, \varepsilon^{-1}D_x)\, v(x) = 0.$$

Thus by Corollary 3.7.5, $\tilde\zeta v \in V^{l,p}_{\beta_j,\delta}(\mathcal{K})^\ell$ and

$$\|\tilde\zeta v\|_{V^{l,p}_{\beta_j,\delta}(\mathcal{K})^\ell} \le c\,\|\tilde\eta_\varepsilon v\|_{V^{m,2}_{0,0}(\mathcal{K})^\ell}\,. \tag{5.1.10}$$

Since the operator $\tilde\eta_\varepsilon\big(\varepsilon^{2m}L(\varepsilon x, \varepsilon^{-1}D_x) - L^\circ(0, D_x)\big)$ is small in the operator norm $V^{l,p}_{\beta_j,\delta}(\mathcal{K})^\ell \to V^{l-2m,p}_{\beta_j,\delta}(\mathcal{K})^\ell$ for small ε, the constant c in (5.1.10) can be chosen independent of ε. Using the equalities

$$\|\tilde\zeta v\|_{V^{l,p}_{\beta_j,\delta}(\mathcal{K})^\ell} = \varepsilon^{l-\beta_j-3/p}\,\|\zeta_\varepsilon u\|_{V^{l,p}_{\beta_j,\delta}(\mathcal{K})^\ell},$$

where $\zeta_\varepsilon(x) = \tilde\zeta(x/\varepsilon)$, and

$$\|\tilde\eta_\varepsilon v\|_{V^{m,2}_{0,0}(\mathcal{K})^\ell} = \varepsilon^{m-3/2}\,\|\eta_\varepsilon u\|_{V^{m,2}_{0,0}(\mathcal{K})^\ell}\,,$$

we obtain the desired inequality. □

The last lemma enables us to estimate $G(x,\xi)$ in the cases $\rho_j(\xi) < \rho_j(x)/2$ and $\rho_j(\xi) > 2\rho_j(x)$.

THEOREM 5.1.6. *If $x, \xi \in \mathcal{G} \cap \mathcal{U}$ and $\rho_j(\xi) < \rho_j(x)/2$, then*

$$\begin{aligned}\left|D_x^\alpha D_\xi^\gamma G(x,\xi)\right| &\le c\,\rho_j(x)^{\Lambda_j^- - |\alpha| + \varepsilon}\,\rho_j(\xi)^{2m-3-\Lambda_j^- - |\gamma| - \varepsilon}\\ &\quad\times \prod_{k\in X_j}\left(\frac{r_k(x)}{\rho_j(x)}\right)^{m-1+\delta_+^{(k)}-|\alpha|-\varepsilon}\prod_{k\in X_j}\left(\frac{r_k(\xi)}{\rho_j(\xi)}\right)^{m-1+\delta_-^{(k)}-|\gamma|-\varepsilon},\end{aligned}$$

where ε is an arbitrarily small positive number. Analogously, the estimate

$$\begin{aligned}\left|D_x^\alpha D_\xi^\gamma G(x,\xi)\right| &\le c\,\rho_j(x)^{\Lambda_j^+ - |\alpha| - \varepsilon}\,\rho_j(\xi)^{2m-3-\Lambda_j^+ - |\gamma| + \varepsilon}\\ &\quad\times \prod_{k\in X_j}\left(\frac{r_k(x)}{\rho_j(x)}\right)^{m-1+\delta_+^{(k)}-|\alpha|-\varepsilon}\prod_{k\in X_j}\left(\frac{r_k(\xi)}{\rho_j(\xi)}\right)^{m-1+\delta_-^{(k)}-|\gamma|-\varepsilon}\end{aligned}$$

holds for $x, \xi \in \mathcal{U}$, $\rho_j(\xi) > 2\rho_j(x)$.

P r o o f. Let $x \in \mathcal{G} \cap \mathcal{U}$ and let ζ, η be smooth functions such that $\zeta(\xi) = 1$ for $\rho_j(\xi) < \frac{1}{2}\rho_j(x)$, $\eta = 1$ in a neighborhood of $\operatorname{supp}\zeta$, and $\eta(\xi) = 0$ for $\rho_j(\xi) > \frac{3}{4}\rho_j(x)$. The function $\xi \to \eta(\xi) D_x^\alpha G(x,\xi)$ belongs to the space $\overset{\circ}{W}{}^{m,2}(\mathcal{G})^{\ell\times\ell}$ and

$$\eta(\xi)\, L^+(\xi, D_\xi)\, G^*(x,\xi) = 0.$$

Applying Lemma 5.1.5 with $\varepsilon = \rho_j(x)$, we get

$$\|\zeta(\cdot)\, D_x^\alpha G(x,\cdot)\|_{V^{l,p}_{\beta,\delta}(\mathcal{G})^{\ell\times\ell}} \le c\,\rho_j(x)^{m-l+\beta_j+3/p-3/2}\,\|\eta(\cdot) D_x^\alpha G(x,\cdot)\|_{W^{m,2}(\mathcal{G})^{\ell\times\ell}},$$

where $2m-3-\Lambda_j^+ < l-\beta_j-3/p < 2m-3-\Lambda_j^-$ and $-\delta_-^{(k)} < \delta_k - l + m - 1 + 2/p < \delta_+^{(k)}$. Thus by Lemma 3.1.4,

$$\begin{aligned}\text{(5.1.11)}\qquad \rho_j(\xi)^{\beta_j-l+|\gamma|+3/p}\prod_{k\in X_j}\left(\frac{r_k(\xi)}{\rho_j(\xi)}\right)^{\delta_k-l+|\gamma|+3/p}&\left|D_x^\alpha D_\xi^\gamma G(x,\xi)\right|\\ &\le c\,\rho_j(x)^{m-l+\beta_j+3/p-3/2}\,\|\eta(\cdot)D_x^\alpha G(x,\cdot)\|_{W^{m,2}(\mathcal{G})^{\ell\times\ell}}\end{aligned}$$

for $\rho_j(\xi) < \rho_j(x)/2$. Let $f \in W^{-m,2}(\mathcal{G})^\ell$, and let

$$u(y) = \int_{\mathcal{G}} G(y,z)\,\eta(z)\,f(z)\,dz$$

be the unique solution of the equation $L(x, D_x)\,u = \eta f$ in $\overset{\circ}{W}{}^{m,2}(\mathcal{G})^\ell$. We denote by ψ a smooth cut-off function such that $\psi(y) = 1$ for $|\rho_j(x) - \rho_j(y)| < \rho_j(x)/8$ and $\psi(y) = 0$ for $|\rho_j(x) - \rho_j(y)| > \rho_j(x)/4$. Since $\psi L(x, D_x)\,u = \psi\eta f = 0$, we obtain (again by Lemma 5.1.5) the estimate

$$\begin{aligned}\|\chi u\|_{V^{l,q}_{\beta',\delta'}(\mathcal{G})^\ell} &\le c\,\rho_j(x)^{m-l+\beta_j'+3/q-3/2}\|\psi u\|_{W^{m,2}(\mathcal{G})^\ell}\\ &\le c\,\rho_j(x)^{m-l+\beta_j'+3/q-3/2}\,\|f\|_{W^{-m,2}(\mathcal{G})^\ell},\end{aligned}$$

where $\chi = 1$ in a neighborhood of the point x, $\Lambda_j^- < l - \beta_j' - 3/q < \Lambda_j^+$ and $-\delta_+^{(k)} < \delta_k' - l + m - 1 + 2/q < \delta_-^{(k)}$. Lemma 3.1.4 yields

$$\rho_j(x)^{|\alpha|-m+3/2}\prod_{k\in X_j}\left(\frac{r_k(x)}{\rho_j(x)}\right)^{\delta_k'-l+|\alpha|+3/q}\left|D_x^\alpha u(x)\right| \le c\,\|f\|_{W^{-m,2}(\mathcal{G})^\ell}.$$

Thus, the mapping

$$W^{-m,2}(\mathcal{G})^\ell \ni f \to \rho_j(x)^{|\alpha|-m+3/2} \prod_{k\in X_j} \Big(\frac{r_k(x)}{\rho_j(x)}\Big)^{\delta'_k-l+|\alpha|+3/q} D_x^\alpha u(x)$$

$$= \rho_j(x)^{|\alpha|-m+3/2} \prod_{k\in X_j} \Big(\frac{r_k(x)}{\rho_j(x)}\Big)^{\delta'_k-l+|\alpha|+3/q} \int_{\mathcal{G}} D_x^\alpha G(x,z)\,\eta(z)\,f(z)\,dz \in \mathbb{C}^\ell$$

is continuous, and its norm is bounded by a constant independent of x. Therefore,

$$\|\eta(\cdot)\,D_x^\alpha G(x,\cdot)\|_{W^{m,2}(\mathcal{G})^{\ell\times\ell}} \le c\,\rho_j(x)^{m-|\alpha|-3/2} \prod_{k\in X_j} \Big(\frac{r_k(x)}{\rho_j(x)}\Big)^{l-|\alpha|-\delta'_k-3/q}.$$

This together with (5.1.11) implies

$$\begin{aligned}\big|D_x^\alpha D_\xi^\gamma G(x,\xi)\big| \;&\le\; c\,\rho_j(x)^{2m-3-|\alpha|-l+\beta_j+3/p}\,\rho_j(\xi)^{l-|\gamma|-\beta_j-3/p}\\ &\quad\times \prod_{k\in X_j} \Big(\frac{r_k(x)}{\rho_j(x)}\Big)^{l-|\alpha|-\delta'_k-3/q} \prod_{k\in X_j} \Big(\frac{r_k(\xi)}{\rho_j(\xi)}\Big)^{l-|\gamma|-\delta_k-3/p}\end{aligned}$$

for $\rho_j(\xi) < \rho_j(x)/2$. Here we can choose $\beta_j, \delta_k, \delta'_k, p$ and q such that $l-\beta_j-3/p = 2m-3-\Lambda_j^- -\varepsilon$, $\delta_k-l+m-1+3/p = -\delta_-^{(k)}+\varepsilon$ and $\delta'_k-l+m-1+3/q = -\delta_+^{(k)}+\varepsilon$. This proves the theorem for the case $\rho_j(\xi) < \rho_j(x)/2$. The proof for the case $\rho_j(\xi) > 2\rho_j(x)$ proceeds analogously. □

5.1.4. Further estimates of Green's matrix. Finally, we consider the cases where x lies in neighborhood of a vertex, ξ lies in neighborhood of an edge point or when both x and ξ lie in small neighborhoods of different edge points.

THEOREM 5.1.7. 1) *Let z be a point on the edge M_j, $|z-x^{(k)}| > c > 0$ for $k=1,\dots,d'$. Furthermore, let $\mathcal{U}_i$ and $\mathcal{U}_z$ be sufficiently small neighborhoods of $x^{(i)}$ and z, respectively. Then the estimate*

$$\begin{aligned}\big|D_x^\alpha D_\xi^\gamma G(x,\xi)\big| \;&\le\; c_{\alpha,\gamma}\,\rho_i(x)^{\Lambda_i^+-|\alpha|-\varepsilon}\, r_j(\xi)^{m-1+\delta_-^{(k)}-|\gamma|-\varepsilon}\\ &\quad\times \prod_{k\in X_i} \Big(\frac{r_k(x)}{\rho_i(x)}\Big)^{m-1+\delta_+^{(k)}-|\alpha|-\varepsilon}\end{aligned}$$

is valid for $x \in \mathcal{G}\cap\mathcal{U}_i$ and $\xi \in \mathcal{G}\cap\mathcal{U}_z$. Here ε is an arbitrarily small positive number.

2) *Let $y \in M_i$ and $z \in M_j$ be given edge points such that $|y-z| > c > 0$, $|y-x^{(k)}| > c$ and $|z-x^{(k)}| > c$ for $k=1,\dots,d'$. Furthermore, let $\mathcal{U}_y$ and $\mathcal{U}_z$ be sufficiently small neighborhoods of y and z, respectively. Then*

$$\big|D_x^\alpha D_\xi^\gamma G(x,\xi)\big| \le c_{\alpha,\gamma}\, r_i(x)^{m-1+\delta_+^{(i)}-|\alpha|-\varepsilon}\; r_j(\xi)^{m-1+\delta_-^{(k)}-|\gamma|-\varepsilon}$$

is valid for $x \in \mathcal{G}\cap\mathcal{U}_y$ and $\xi \in \mathcal{G}\cap\mathcal{U}_z$.

P r o o f. Items 1) and 2) can be proved in the same manner as Theorem 5.1.2. Using Lemmas 2.9.10 and 2.1.3, we obtain the estimate

$$r_j(\xi)^{\delta_j-l+|\gamma|+3/p}\,\big|D_x^\alpha D_\xi^\gamma G(x,\xi)\big| \le c\,\|\eta(\cdot) D_x^\alpha G(x,\cdot)\|_{\overset{\circ}{W}{}^{m,2}(\mathcal{G})^{\ell\times\ell}}$$

instead of (5.1.6), where η is a smooth cut-off function with sufficiently small support, $\eta = 1$ in a neighborhood of $\mathcal{U}_z$, and δ_j satisfies the inequalities $-\delta_-^{(j)} < \delta_j - l + m - 1 + 2/p < \delta_+^{(j)}$. Furthermore in the case 1), the estimate

$$\|\eta(\cdot)\, D_x^\alpha G(x,\cdot)\|_{\overset{\circ}{W}{}^{m,2}(\mathcal{G})^{\ell\times\ell}} \le c\, \rho_i(x)^{l-|\alpha|-\beta_i'-3/q} \prod_{k\in X_i} \Big(\frac{r_k(x)}{\rho_i(x)}\Big)^{l-|\alpha|-\delta_k'-3/q}.$$

holds by the same arguments as in the proof of Theorem 3.4.5, where $\Lambda_i^- < l - \beta_j' - 3/q < \Lambda_i^+$ and $-\delta_+^{(k)} < \delta_k' - l + m - 1 + 2/q < \delta_-^{(k)}$. In the case 2), the estimate

$$\|\eta(\cdot)\, D_x^\alpha G(x,\cdot)\|_{\overset{\circ}{W}{}^{m,2}(\mathcal{G})^{\ell\times\ell}} \le c\, r_i(x)^{l-|\alpha|-\delta_i'-3/q}$$

can be deduced from Lemmas 2.9.10 and 2.1.3, where $-\delta_+^{(i)} < \delta_i' - l + m - 1 + 2/q < \delta_-^{(i)}$. The result follows. □

5.1.5. Estimates for the Green's function of some special second order equations in convex polyhedral domains. In this subsection, we consider the Green's matrix $G(x,\xi)$ of the Dirichlet problem for a system of second order equations, i.e. $m = 1$. If the domain $\mathcal{G}$ is convex, then it is known for the Laplace equation that

$$\Lambda_j^+ > 1, \quad \Lambda_j^- < -2 \quad \text{for } j = 1, \dots, d', \tag{5.1.12}$$

$$\delta_\pm^{(k)} > 1 \quad \text{for } k = 1, \dots, d \tag{5.1.13}$$

(see e.g. [**85**, Corollary 2.2.1, Theorem 2.2.4]). The same estimates hold in the case of the Dirichlet problem for the Lamé system if the polyhedral domain $\mathcal{G}$ is convex. Theorems 5.1.2–5.1.7 together with the estimates (5.1.12) and (5.1.13) lead to the following result.

THEOREM 5.1.8. *Suppose that $m = 1$ and that the inequalities (5.1.12) and (5.1.13) are satisfied. Then*

$$\big|\partial_x^\alpha \partial_\xi^\gamma G(x,\xi)\big| \le c\, |x-\xi|^{-1-|\alpha|-|\gamma|} \tag{5.1.14}$$

for $|\alpha| \le 1$ and $|\gamma| \le 1$, where c is a constant independent of x and ξ, $x \ne \xi$.

Note that the estimate $\big|\partial_x^\alpha \partial_\xi^\gamma G(x,\xi)\big| \le c\, |x-\xi|^{2-n-|\alpha|-|\gamma|}, |\alpha| \le 1$, $|\gamma| \le 1$, is true for the Green's function of the Laplace equation in convex domains of arbitrary dimension n (see the paper by GRÜTER and WIDMAN [**63**]). KOZLOV [**78**] proved the estimate

$$\sum_{|\alpha|=|\gamma|=m} \big|\partial_x^\alpha \partial_\xi^\gamma G(x,\xi)\big| \le c\, |x-\xi|^{-n}$$

for the Green's matrix of a class of strongly elliptic systems of order $2m$ in convex domains provided the outward unit normal has no big jumps on the boundary. The class of differential equations in [**78**] includes, e.g., the polyharmonic equation and the Lamé system. In the case $n = 2$, the last estimate was proved for arbitrary symmetric second order systems in convex domains without the assumption on the outward normal.

Next, we show that the derivatives $\partial_x^\alpha \partial_\xi^\gamma G(x,\xi)$ satisfy a Hölder estimate for $|\alpha| \le 1$ and $|\gamma| \le 1$. To this end, we prove the following two lemmas for $|x - \xi| > a|x - y|$, where a is a sufficiently large positive constant.

LEMMA 5.1.9. *Suppose that* $m = 1$ *and that the inequalities* (5.1.12) *and* (5.1.13) *are satisfied. Then*

$$\frac{\left|\partial_{x_i}\partial_\xi^\gamma G(x,\xi) - \partial_{y_i}\partial_\xi^\gamma G(y,\xi)\right|}{|x-y|^\sigma} \le c\,|x-\xi|^{-2-\sigma-|\gamma|} \tag{5.1.15}$$

for $i = 1,2,3$, $|\gamma| \le 1$, $|x-\xi| > a|x-y| > r(x)$, *where* σ *is an arbitrary positive real number such that* $\sigma < 1$, $\sigma < \Lambda_j^+ - 1$, $\sigma < -2 - \Lambda_j^-$ *for* $j = 1,\dots,d'$, *and* $\sigma < \delta_\pm^{(k)}$ *for* $k = 1,\dots,d$.

P r o o f. Since $|x-\xi| > a\,|x-y|$ with a sufficiently large m, we may assume that x and y lie in a neighborhood $\mathcal{V}_j$ of the same vertex $x^{(j)}$. We suppose that $\mathcal{V}_j$ has a positive distance from the edges M_k, $k \notin X_j$. From the condition $r(x) < a\,|x-y|$ it follows that

$$r(y) < (a+1)\,|x-y|.$$

Furthermore, the condition $a|x-y| < |x-\xi|$ implies

$$\Big(1-\frac{1}{a}\Big)\,|x-\xi| < |y-\xi| < \Big(1+\frac{1}{a}\Big)\,|x-\xi|.$$

We consider the following cases

1) $\xi \in \mathcal{V}_j$ and $\rho_j(x) < \rho_j(\xi)/2$
2) $\xi \in \mathcal{V}_j$ and $\rho_j(x) > 2\rho_j(\xi)$
3) $\xi \in \mathcal{V}_j$ and $\rho_j(\xi)/2 < \rho_j(x) < 2\rho_j(\xi)$
4) ξ lies in a neighborhood of another vertex $x^{(\nu)}$ and $|x-\xi| > \delta$, where δ is a fixed positive number.

We start with Case 1). Then obviously $|x-\xi| < \rho_j(x) + \rho_j(\xi) < 3\rho_j(\xi)/2$ and

$$\rho_j(y) < \rho_j(x) + |x-y| < \rho_j(x) + \frac{1}{a}\,|x-\xi| < \Big(\frac{1}{2}+\frac{3}{2a}\Big)\,\rho_j(\xi).$$

Consequently, Theorems 5.1.3 and 5.1.6 yield

$$\begin{aligned}
&\left|\partial_{x_i}\partial_\xi^\gamma G(x,\xi) - \partial_{y_i}\partial_\xi^\gamma G(y,\xi)\right| \le \left|\partial_{x_i}\partial_\xi^\gamma G(x,\xi)\right| + \left|\partial_{y_i}\partial_\xi^\gamma G(y,\xi)\right| \\
&\quad \le c\,\rho_j(\xi)^{-1-\Lambda_j^+-|\gamma|+\varepsilon}\left(\rho_j(x)^{\Lambda_j^+-1-\varepsilon}\Big(\frac{r(x)}{\rho_j(x)}\Big)^\sigma + \rho_j(y)^{\Lambda_j-1-\varepsilon}\Big(\frac{r(y)}{\rho_j(y)}\Big)^\sigma\right)
\end{aligned}$$

for $|\gamma| \le 1$. Here, ε can be chosen such that $\Lambda_j^+ - 1 - \varepsilon - \sigma \ge 0$. Thus,

$$\begin{aligned}
\left|\partial_{x_i}\partial_\xi^\gamma G(x,\xi) - \partial_{y_i}\partial_\xi^\gamma G(y,\xi)\right| &\le c\,\rho_j(\xi)^{-2-\sigma-|\gamma|}\left(r(x)^\sigma + r(y)^\sigma\right) \\
&\le c'\,|x-\xi|^{-2-\sigma-|\gamma|}\,|x-y|^\sigma.
\end{aligned}$$

Case 2): In this case, $|x-\xi| < \rho_j(x) + \rho_j(\xi) < 3\rho_j(x)/2$ and

$$\begin{aligned}
\rho_j(y) &> \rho_j(x) - |x-y| > \rho_j(x) - \frac{1}{a}\,|x-\xi| > \rho_j(x) - \frac{1}{a}\left(\rho_j(x)+\rho_j(\xi)\right) \\
&> \Big(1-\frac{3}{2a}\Big)\,\rho_j(x) > \Big(2-\frac{3}{a}\Big)\,\rho_j(\xi).
\end{aligned}$$

Therefore by 5.1.3 and 5.1.6, the inequalities

$$\begin{aligned}
\left|\partial_{x_i}\partial_\xi^\gamma G(x,\xi) - \partial_{y_i}\partial_\xi^\gamma G(y,\xi)\right| &\le \left|\partial_{x_i} G(x,\xi)\right| + \left|\partial_{y_i} G(y,\xi)\right| \\
&\le c\,\rho_j(\xi)^{-1-\Lambda_j^- -|\gamma|-\varepsilon}\left(\rho_j(x)^{\Lambda_j^- -1+\varepsilon}\Big(\frac{r(x)}{\rho_j(x)}\Big)^\sigma + \rho_j(y)^{\Lambda_j^- -1+\varepsilon}\Big(\frac{r(y)}{\rho_j(y)}\Big)^\sigma\right) \\
&\le c'\left(\rho_j(x)^{-2-\sigma-|\gamma|} r(x)^\sigma + \rho_j(y)^{-2-\sigma-|\gamma|} r(y)^\sigma\right) \\
&\le c''\,|x-\xi|^{-2-\sigma-|\gamma|}\,|x-y|^\sigma
\end{aligned}$$

hold for $|\gamma| \le 1$.

Case 3): Then $|x-\xi| < 3\rho_j(\xi)$ and

$$\Big(\frac{1}{2} - \frac{3}{a}\Big)\rho_j(\xi) < \rho_j(y) < \Big(2 + \frac{3}{a}\Big)\rho_j(\xi).$$

Since $r(x) < |x-\xi|$ and

$$r(y) < (a+1)\,|x-y| < \frac{a+1}{a}\,|x-\xi| < \frac{a+1}{a-1}\,|y-\xi|,$$

we can apply Theorem 5.1.3 and obtain (for $|\gamma| \le 1$)

$$\begin{aligned}
&\left|\partial_{x_i}\partial_\xi^\gamma G(x,\xi) - \partial_{y_i}\partial_\xi^\gamma G(y,\xi)\right| \\
&\quad\le c\left(|x-\xi|^{-2-|\gamma|}\Big(\frac{r(x)}{|x-\xi|}\Big)^\sigma + |y-\xi|^{-2-|\gamma|}\Big(\frac{r(y)}{|y-\xi|}\Big)^\sigma\right) \\
&\quad\le c'\,|x-\xi|^{-2-\sigma-|\gamma|}\,|x-y|^\sigma.
\end{aligned}$$

Case 4): Finally, we consider the case where x and y lie in the neighborhood $\mathcal{V}_j$ of the vertex $x^{(j)}$ and ξ lies in a neighborhood of another vertex $x^{(\nu)}$ such that $|x-\xi| > \delta$, where δ is a fixed positive number. Then Theorem 5.1.2 yields

$$\begin{aligned}
&\left|\partial_{x_i}\partial_\xi^\gamma G(x,\xi) - \partial_{y_i}\partial_\xi^\gamma G(y,\xi)\right| \\
&\quad\le c\,\rho_\nu(\xi)^{-1-\Lambda_\nu^- -|\gamma|-\varepsilon}\left(\rho_j(x)^{\Lambda_j^+ -1-\varepsilon}\Big(\frac{r(x)}{\rho_j(x)}\Big)^\sigma + \rho_j(y)^{\Lambda_j^+ -1-\varepsilon}\Big(\frac{r(y)}{\rho_j(y)}\Big)^\sigma\right) \\
&\quad\le c'\left(r(x)^\sigma + r(y)^\sigma\right) \le 2c'\,(a+1)^\sigma\,|x-y|^\sigma
\end{aligned}$$

if $|\gamma| \le 1$. This completes the proof. □

In the next lemma, we consider the case $\min(|x-\xi|, r(x)) > a|x-y|$.

LEMMA 5.1.10. *Suppose that $m = 1$, the domain $\mathcal{G}$ is convex and that the inequalities (5.1.12) and (5.1.13) are satisfied. Then the matrix $G(x,\xi)$ satisfies the estimate (5.1.15) for $i = 1,2,3$, $|\gamma| \le 1$, $|x-\xi| > a|x-y|$, $r(x) > a|x-y|$, where σ is an arbitrary positive real number such that $\sigma < 1$, $\sigma < \Lambda_j^+ - 1$, $\sigma < -2 - \Lambda_j^-$ for $j = 1,\dots,d'$, and $\sigma < \delta_\pm^{(k)}$ for $k = 1,\dots,d$.*

Proof. From the mean value theorem it follows that

$$\left|\partial_{x_i}\partial_\xi^\gamma G(x,\xi) - \partial_{y_i}\partial_\xi^\gamma G(y,\xi)\right| \le |x-y|\,\left|\nabla_z \partial_{z_i}\partial_\xi^\gamma G(z,\xi)\right|, \tag{5.1.16}$$

where $z = x + t(y-x)$, $0 < t < 1$. We assume again that x and y lie in the neighborhood $\mathcal{V}_j$ of the vertex $x^{(j)}$ and consider the same cases 1)–4) as in the proof of Lemma 5.1.9.

Case 1): Since

$$\begin{aligned}\rho_j(z) &< \rho_j(x) + |x-y| < \rho_j(x) + \frac{1}{a}\,|x-\xi| < \rho_j(x) + \frac{1}{a}\,(\rho_j(x)+\rho_j(\xi)) \\ &< \Big(\frac{1}{2}+\frac{3}{2a}\Big)\,\rho_j(\xi),\end{aligned}$$

the derivatives of G at the point (z,ξ) satisfy the estimates (cf. Theorems 5.1.3 and 5.1.6)

$$\big|\nabla_z\partial_{z_i}\partial_\xi^\gamma G(z,\xi)\big| \le c\,\rho_j(z)^{\Lambda_j^+-2-\varepsilon}\,\rho_j(\xi)^{-1-\Lambda_j^+-|\gamma|+\varepsilon}\Big(\frac{r(z)}{\rho_j(z)}\Big)^{\sigma-1}$$

for $|\gamma|\le 1$. The number ε can be chosen such that $\Lambda_j^+-1-\varepsilon-\sigma\ge 0$. Consequently,

$$\big|\partial_{x_i}\partial_\xi^\gamma G(x,\xi) - \partial_{y_i}\partial_\xi^\gamma G(y,\xi)\big| \le |x-y|\,\rho_j(\xi)^{-2-\sigma-|\gamma|}\,r(z)^{\sigma-1}.$$

Using the inequalities $r(z)>(a-1)\,|x-y|$ and $\rho_j(\xi)>2|x-\xi|/3$, we get (5.1.15).

In Case 2) we have

$$\begin{aligned}\rho_j(z) &> \rho_j(x) - |x-y| > \rho_j(x) - \frac{1}{a}\,|x-\xi| > \rho_j(x) - \frac{1}{a}\,\big(\rho_j(x)+\rho_j(\xi)\big) \\ &> \Big(1-\frac{3}{2a}\Big)\,\rho_j(x) > \Big(2-\frac{3}{a}\Big)\,\rho_j(\xi).\end{aligned}$$

Therefore,

$$\begin{aligned}\big|\nabla_z\partial_{z_i}\partial_\xi^\gamma G(z,\xi)\big| &\le c\,\rho_j(z)^{\Lambda_j^--2+\varepsilon}\,\rho_j(\xi)^{-1-\Lambda_j^--|\gamma|-\varepsilon}\Big(\frac{r(z)}{\rho_j(z)}\Big)^{\sigma-1} \\ &\le c'\,\rho_j(z)^{-2-\sigma-|\gamma|}\,r(z)^{\sigma-1},\end{aligned}$$

This together with the inequalities $r(z)>(a-1)\,|x-y|$ and

$$\rho_j(z) > \rho_j(x) - |x-y| > \frac{2}{3}\,|x-\xi| - \frac{1}{a}\,|x-\xi|,$$

implies (5.1.15).

Case 3): From the inequalities $\rho_j(\xi)/2<\rho_j(x)<2\rho_j(\xi)$ and $|x-\xi|>a|x-z|$ it follows that

$$\Big(\frac{1}{2}-\frac{3}{a}\Big)\,\rho_j(\xi) < \rho_j(z) < \Big(2+\frac{3}{a}\Big)\,\rho_j(\xi)\,.$$

Furthermore, the inequalities $|x-\xi|>a\,|x-z|$ and $r(x)>a\,|x-z|$ yield

$$\Big(1-\frac{1}{a}\Big)\,|x-\xi| < |z-\xi| < \Big(1+\frac{1}{a}\Big)\,|x-\xi| \tag{5.1.17}$$

and

$$\Big(1-\frac{1}{a}\Big)\,r(x) < r(z) < \Big(1+\frac{1}{a}\Big)\,r(x). \tag{5.1.18}$$

If $|z-\xi|>\min(r(z),r(\xi))$, then Theorem 5.1.3 and the inequalities (5.1.17), (5.1.18) imply

$$\begin{aligned}&\big|\nabla_z\partial_{z_i}\partial_\xi^\gamma G(z,\xi)\big| \le c\,|z-\xi|^{-3-|\gamma|}\Big(\frac{r(z)}{|z-\xi|}\Big)^{\sigma-1} \\ &\quad\le c'\,|x-\xi|^{-2-\sigma-|\gamma|}\,r(x)^{\sigma-1} \le c'\,a^{\sigma-1}\,|x-\xi|^{-2-\sigma-|\gamma|}\,|x-y|^{\sigma-1}.\end{aligned}$$

for $|\gamma| \le 1$. In the case $|z-\xi| < \min(r(z), r(\xi))$, we obtain

$$\begin{aligned}\left|\nabla_z \partial_{z_i} \partial_\xi^\gamma G(z,\xi)\right| &\le c\,|z-\xi|^{-3-|\gamma|} \le c'\,|x-\xi|^{-3-|\gamma|} \\ &\le c'\,a^{\sigma-1}\,|x-\xi|^{-2-\sigma-|\gamma|}\,|x-y|^{\sigma-1}.\end{aligned}$$

This together with (5.1.16) implies (5.1.15).

Case 4): Suppose that ξ lies in a neighborhood $\mathcal{V}_\nu$ of the vertex $x^{(\nu)}$ and that $|x-\xi| > \delta$, where δ is a fixed positive number. Then

$$\left|\nabla_z \partial_{z_i} \partial_\xi^\gamma G(z,\xi)\right| \le c\,\rho_j(z)^{\Lambda_j^+ -2-\varepsilon}\,\rho_\nu(\xi)^{-1-\Lambda_\nu^- -|\gamma|-\varepsilon}\left(\frac{r(z)}{\rho_j(z)}\right)^{\sigma-1} \le c'\,r(z)^{\sigma-1}$$

for $|\gamma| \le 1$. Using the inequalities (5.1.16) and $r(z) > (a-1)\,|x-y|$, we obtain (5.1.15). The proof of the lemma is complete. □

Now it is easy to prove the following theorem by means of the last two lemmas and Theorem 5.1.8.

THEOREM 5.1.11. *Suppose that $m = 1$, the domain $\mathcal{G}$ is convex and that the inequalities (5.1.12) and (5.1.13) are satisfied. Furthermore, let σ be a positive real number such that $\sigma < 1$, $\sigma < \Lambda_j^+ - 1$, $\sigma < -2 - \Lambda_j^-$ for $j = 1, \dots, d'$, and $\sigma < \delta_\pm^{(k)}$ for $k = 1, \dots, d$. Then the matrix $G(x,\xi)$ satisfies the estimates*

$$\frac{\left|\partial_{x_i}\partial_\xi^\gamma G(x,\xi) - \partial_{y_i}\partial_\xi^\gamma G(y,\xi)\right|}{|x-y|^\sigma} \le c\left(|x-\xi|^{-2-\sigma-|\gamma|} + |y-\xi|^{-2-\sigma-|\gamma|}\right) \tag{5.1.19}$$

for $|\gamma| \le 1$, $x \ne y \ne \xi \ne x$ and

$$\frac{\left|\partial_x^\alpha \partial_{\xi_i} G(x,\xi) - \partial_x^\alpha \partial_{\xi_i} G(x,\eta)\right|}{|\xi-\eta|^\sigma} \le c\left(|x-\xi|^{-2-\sigma-|\alpha|} + |x-\eta|^{-2-\sigma-|\alpha|}\right) \tag{5.1.20}$$

for $|\alpha| \le 1$, $x \ne \xi \ne \eta \ne x$.

P r o o f. For $|x-\xi| > a|x-y|$, the inequality (5.1.19) is already proved (see Lemmas 5.1.9 and 5.1.10. If $|x-\xi| < a\,|x-y|$, then $|y-\xi| < (a+1)\,|x-y|$ and (5.1.14) implies

$$\begin{aligned}\left|\partial_{x_i}\partial_\xi^\gamma G(x,\xi) - \partial_{y_i}\partial_\xi^\gamma G(y,\xi)\right| &\le c\left(|x-\xi|^{-2-|\gamma|} + |y-\xi|^{-2-|\gamma|}\right) \\ &\le c\,(a+1)^\sigma\,|x-y|^\sigma\left(|x-\xi|^{-2-\sigma} + |y-\xi|^{-2-\sigma}\right).\end{aligned}$$

This proves (5.1.19). The inequality (5.1.20) holds analogously. □

Note that the estimate (5.1.14) was proved by FROMM [**52**] for the Green's function of the Dirichlet problem for the Laplace equation in convex, not necessarily polyhedral, domains. However, the counterexample given in [**52**] indicates that (5.1.19) and (5.1.20) cannot be extended to general convex domains.

5.2. The Miranda-Agmon maximum principle in domains of polyhedral type

The goal of this section is to prove the Miranda-Agmon maximum principle. For this, we apply the estimates of the Green's matrix obtained Section 5.1. As in the preceding section, we suppose that the equation $L(x, D_x)\,u = f$ is uniquely solvable in $\overset{\circ}{W}{}^{m,2}(\mathcal{G})^\ell$ for arbitrary given $f \in W^{-m,2}(\mathcal{G})^\ell$. The proof of the maximum principle is organized as follows. First, we get a weighted L_∞ estimate for the generalized solution of the boundary value problem

$$(5.2.1)\quad L(x,D_x)u = 0 \ \text{ in } \mathcal{G}, \quad \frac{\partial^{k-1}u}{\partial n^{k-1}} = g_{j,k} \ \text{ on } \Gamma_j,\ j=1,\dots,N,\ k=1,\dots,m,$$

if the boundary data $g_{j,k}$ belong to a certain weighted class $\mathcal{V}^{m-1,\infty}_{\beta,\delta}(\partial\mathcal{G})$ introduced below. Using a local maximum principle, we obtain weighted L_∞-estimates for the derivatives of the generalized solution up to order $m-1$. In the last step, we prove unweighted L_∞ estimates for boundary data $g_{j,k} \in C^{m-k}(\overline{\Gamma}_j)$.

5.2.1. The space $V^{l,\infty}_{\beta,\delta}(\mathcal{G})$. Let $\mathcal{U}_1,\dots,\mathcal{U}_{d'}$ be domains in $\mathbb{R}^3$ such that

$$\mathcal{U}_1 \cup \cdots \cup \mathcal{U}_{d'} \supset \overline{\mathcal{G}}, \quad x^{(i)} \notin \overline{\mathcal{U}}_j \ \text{ if } i \neq j, \quad \text{and} \quad \overline{\mathcal{U}}_j \cap \overline{M}_k = \emptyset \text{ if } k \notin X_j\,.$$

Then for arbitrary $\beta = (\beta_1,\dots,\beta_{d'}) \in \mathbb{R}^{d'}$ and $\delta = (\delta_1,\dots,\delta_d) \in \mathbb{R}^d$, we define $L^\infty_{\beta,\delta}(\mathcal{G})$ as the set of all functions on $\mathcal{G}$ such that

$$\|u\|_{L^\infty_{\beta,\delta}(\mathcal{G})} = \max_{1\le j\le d'} \Big\| \rho_j^{\beta_j} \prod_{k\in X_j} \Big(\frac{r_k}{\rho_j}\Big)^{\delta_k} u \Big\|_{L_\infty(\mathcal{G}\cap\mathcal{U}_j)} < \infty.$$

Here again $\rho_j(x) = |x - x^{(j)}|$, $r_k(x) = \text{dist}(x, M_k)$, and X_j is the set of all indices k such that the vertex $x^{(j)}$ is an end point of the edge M_k. Obviously, the set $L^\infty_{\beta,\delta}(\mathcal{G})$ does not depend on the choice of the domains $\mathcal{U}_j$. The space $V^{l,\infty}_{\beta,\delta}(\mathcal{G})$ is defined as the set of all functions on $\mathcal{G}$ such that

$$\|u\|_{V^{l,\infty}_{\beta,\delta}(\mathcal{G})} = \sum_{|\alpha|\le l} \|r^{|\alpha|-l}\partial_x^\alpha u\|_{L^\infty_{\beta,\delta}(\mathcal{G})} < \infty,$$

where $r(x) = \min(r_1(x),\dots,r_d(x))$.

Analogously, we define the spaces $L^\infty_{\beta,\delta}(\Gamma_j)$ and $V^{l,\infty}_{\beta,\delta}(\Gamma_j)$.

5.2.2. Generalized solutions of the boundary value problem. Let $\mathcal{V}^{l,\infty}_{\beta,\delta}(\mathcal{G})$ be the set of all $u \in V^{l,\infty}_{\beta,\delta}(\mathcal{G})$ for which there exists a sequence of functions $u^{(\nu)} \in C_0^\infty(\overline{\mathcal{G}}\backslash\mathcal{S})$, $\nu = 1,2,\dots$, such that

$$\|u^{(\nu)}\|_{V^{l,\infty}_{\beta,\delta}(\mathcal{G})} \le c < \infty \quad \text{for all } \nu,$$

$$u^{(\nu)} \to u \ \text{ a.e. on } \mathcal{G}.$$

The sequence $\{u^{(\nu)}\}$ is called *approximating* for u in $\mathcal{V}^{l,\infty}_{\beta,\delta}(\mathcal{G})$. Analogously, we denote by $\mathcal{V}^{m-1,\infty}_{\beta,\delta}(\partial\mathcal{G})$ the set of all tuples

$$g = \{g_{j,k} :\ j=1,\dots,N,\ k=1,\dots,m\} \in \prod_{j=1}^N \prod_{k=1}^m V^{m-k,\infty}_{\beta,\delta}(\Gamma_j)$$

for which there exist sequences $\{g^{(\nu)}_{j,k}\}_{\nu=1}^\infty \subset C_0^\infty(\Gamma_j)$ such that

$$\|g^{(\nu)}_{j,k}\|_{V^{m-k,\infty}_{\beta,\delta}(\Gamma_j)^\ell} \le c < \infty \quad \text{for all } \nu,$$

$$g^{(\nu)}_{j,k} \to g_{j,k} \ \text{ a.e. on } \Gamma_j\,.$$

The sequence $\{g^{(\nu)}\}_{\nu=1}^\infty$, where

$$g^{(\nu)} = \{g^{(\nu)}_{j,k} :\ j=1,\dots,N,\ k=1,\dots,m\},$$

is said to be *approximating* for g in $\mathcal{V}^{m-1,\infty}_{\beta,\delta}(\partial\mathcal{G})$.

As in Section 5.1, let

$$\Lambda_j^- < \operatorname{Re}\lambda < \Lambda_j^+$$

be the widest strip in the complex plane containing the line $\operatorname{Re}\lambda = m - 3/2$ which is free of eigenvalues of the pencil $\mathfrak{A}_j(\lambda)$. We assume that $g \in \mathcal{V}^{m-1,\infty}_{\beta,\delta}(\partial\mathcal{G})^\ell$, $f \in \mathcal{V}^{0,\infty}_{\beta+m+1,\delta+m+1}(\mathcal{G})^\ell$, i.e. $r^{m+1}f \in \mathcal{V}^{0,\infty}_{\beta,\delta}(\mathcal{G})^\ell$, and that the components of β and δ satisfy the inequalities

$$(5.2.2)\qquad m-1-\Lambda_j^+ < \beta_j < m-1-\Lambda_j^-, \qquad -\delta_+^{(k)} < \delta_k < \delta_-^{(k)}$$

for $j = 1, \dots, d'$, $k = 1, \dots, d$. By $f^{(\nu)}$ and $\psi^{(\nu)}$, we denote approximating sequences for f and g.

There exist differential operators $S_{j,k}(x, D_x)$ and $T_{j,k}(x, D_x)$ of order $2m - k$ such that the *Green's formula*

$$\int_{\mathcal{G}} \big(L(x,D_x)\,u\cdot\bar{v} - u\cdot\overline{L^+(x,D_x)\,v}\big)\,dx$$
$$= \sum_{j=1}^{N}\sum_{k=1}^{m}\int_{\Gamma_j}\Big(S_{j,k}(x,D_x)\,u\cdot\partial_n^{k-1}\bar{v} - \partial_n^{k-1}u\cdot\overline{T_{j,k}(x,D_x)\,v}\Big)\,dx$$

is valid for all $u, v \in C_0^\infty(\overline{\mathcal{G}}\backslash\mathcal{S})^\ell$.

If the differential equation $L(x, D_x)\,u = f$ is uniquely solvable in $\overset{\circ}{W}{}^{m,2}(\mathcal{G})^\ell$ for arbitrary given $f \in W^{-m,2}(\mathcal{G})^\ell$, then there exists a uniquely determined Green's matrix $G(x,\xi)$ satisfying the estimates in Theorems 5.1.2–5.1.7. Let

$$H^{(i)}(x,\xi) = \begin{pmatrix} \overline{G_{i,1}(x,\xi)} \\ \vdots \\ \overline{G_{i,\ell}(x,\xi)} \end{pmatrix}$$

be the i-th column of the adjoint matrix $G^*(x,\xi)$. Then the vector function $u^{(\nu)}$ with the components

$$(5.2.3)\qquad u_i^{(\nu)}(x) = \int_{\mathcal{G}} f^{(\nu)}(\xi)\cdot\overline{H^{(i)}(x,\xi)}\,d\xi + \sum_{j=1}^{N}\sum_{k=1}^{m}\int_{\Gamma_j} g_{j,k}^{(\nu)}(\xi)\cdot\overline{T_{j,k}(\xi,D_\xi)\,H^{(i)}(x,\xi)}\,d\xi,$$

$i = 1, \dots, \ell$, is a solution of the boundary value problem

$$L(x,D_x)\,u^{(\nu)} = f^{(\nu)} \text{ in } \mathcal{G}, \quad \frac{\partial^{k-1}u}{\partial n^{k-1}} = g_{j,k}^{(\nu)} \text{ on } \Gamma_j,\ j = 1,\dots,N,\ k = 1,\dots,m.$$

If $f^{(\nu)}$ and $\{g^{(\nu)}\}_{\nu=1}^\infty$ are approximating sequences for f and g, respectively, then the sequence $\{u^{(\nu)}(x)\}$ converges for every $x \in \mathcal{G}$. The limit $u(x)$ does not depend on the choice of the approximating sequences and its components admit the representation

$$(5.2.4)\quad u_i(x) = \int_{\mathcal{G}} f(\xi)\cdot\overline{H^{(i)}(x,\xi)}\,d\xi + \sum_{j=1}^{N}\sum_{k=1}^{m}\int_{\Gamma_j} g_{j,k}(\xi)\cdot\overline{T_{j,k}(\xi,D_\xi)\,H^{(i)}(x,\xi)}\,d\xi.$$

We call the vector function u *generalized solution* of the boundary value problem (4.1.1).

5.2.3. Weighted L_∞ estimates of generalized solutions. Using the estimates of the Green's matrix $G(x,\xi)$ given in the preceding section, we can estimate the $L^\infty_{\beta-m+1,\delta-m+1}$-norm of the generalized solution. This is done in the next two lemmas.

LEMMA 5.2.1. *Let the components of β and δ satisfy the inequalities* (5.2.2). *Then the function*

$$u_i(x) = \sum_{j=1}^N \sum_{k=1}^m \int_{\Gamma_j} g_{j,k}(\xi) \cdot \overline{T_{j,k}(\xi, D_\xi)\, H^{(i)}(x,\xi)}\, d\xi,$$

satisfies the estimate

$$\|r^{1-m} u_i\|_{L^\infty_{\beta,\delta}(\mathcal{G})} \le c \sum_{j=1}^N \sum_{k=1}^m \|r^{k-m} g_{j,k}\|_{L^\infty_{\beta,\delta}(\Gamma_j)^\ell} \tag{5.2.5}$$

for arbitrary $g_{j,k} \in L^\infty_{\beta-m+k,\delta-m+k}(\Gamma_j)^\ell$, $i = 1,\dots,\ell$.

P r o o f. For $j = 1,\dots,d'$, let $\mathcal{U}_j$ and $\mathcal{V}_j$ be neighborhoods of the vertex $x^{(j)}$ such that $\mathcal{U}_1 \cup \dots \cup \mathcal{U}_{d'} \supset \mathcal{G}$, $\overline{\mathcal{U}}_j \subset \mathcal{V}_j$,

$$r(x) = \min_{k \in X_j} r_k(x) \quad \text{and} \quad \rho_j(x) = \min_{1 \le k \le d'} \rho_k(x) \quad \text{for } x \in \mathcal{V}_j\,.$$

Suppose that $x \in \mathcal{U}^{(1)}$ and that M_1 is the nearest edge to x. We introduce the subsets

$$\begin{aligned}
A_1 &= \{\xi \in \Gamma_j \cap \mathcal{V}_1 :\ \rho_1(\xi) > 2\rho_1(x)\},\\
A_2 &= \{\xi \in \Gamma_j \cap \mathcal{V}_1 :\ \rho_1(\xi) < \rho_1(x)/2\},\\
A_3 &= \{\xi \in \Gamma_j \cap \mathcal{V}_1 :\ \rho_1(x)/2 < \rho_1(\xi) < 2\rho_1(x),\ |x-\xi| > \min(r(x), r(\xi))\},\\
A_4 &= \{\xi \in \Gamma_j \cap \mathcal{V}_1 :\ \rho_1(x)/2 < \rho_1(\xi) < 2\rho_1(x),\ |x-\xi| < \min(r(x), r(\xi))\},\\
A_5 &= \Gamma_j \backslash \mathcal{V}_1
\end{aligned}$$

of the face Γ_j and show that the function

$$v(x) = \int_{A_\nu} g_{j,k}(\xi) \cdot \overline{T_{j,k}(\xi, D_\xi)\, H^{(i)}(x,\xi)}\, d\xi \tag{5.2.6}$$

satisfies the estimate

$$|v(x)| \le c\, \rho_1(x)^{m-1-\beta_1} \prod_{\mu \in X_1} \Big(\frac{r_\mu(x)}{\rho_1(x)}\Big)^{m-1-\delta_\mu} \|r^{k-m} g_{j,k}\|_{L^\infty_{\beta,\delta}(\Gamma_j)^\ell} \tag{5.2.7}$$

for $\nu = 1,\dots,5$, $j = 1,\dots,N$, $k = 1,\dots,m$.

We start with the case $\nu = 1$. Then it follows from the second estimate of Theorem 5.1.6 that

$$\begin{aligned}|v(x)| &\le c\,\rho_1(x)^{\Lambda_1^+-\varepsilon} \prod_{\mu\in X_1} \Big(\frac{r_\mu(x)}{\rho_1(x)}\Big)^{m-1+\delta_+^{(\mu)}-\varepsilon} \\ &\quad \times \int_{A_1} \rho_1(\xi)^{k-3-\Lambda_1^++\varepsilon} \prod_{\mu\in X_1} \Big(\frac{r_\mu(\xi)}{\rho_1(\xi)}\Big)^{k-m-1+\delta_-^{(\mu)}-\varepsilon} |g_{j,k}(\xi)|\, d\xi \\ &\le c\,\rho_1(x)^{\Lambda_1^+-\varepsilon} \prod_{\mu\in X_1} \Big(\frac{r_\mu(x)}{\rho_1(x)}\Big)^{m-1+\delta_+^{(\mu)}-\varepsilon} \|r^{k-m} g_{j,k}\|_{L^\infty_{\beta,\delta}(\Gamma_j)^\ell} \\ &\quad \times \int_{A_1} \rho_1(\xi)^{m-3-\beta_1-\Lambda_1^++\varepsilon} \prod_{\mu\in X_1} \Big(\frac{r_\mu(\xi)}{\rho_1(\xi)}\Big)^{-\delta_\mu-1+\delta_-^{(\mu)}-\varepsilon} d\xi \\ &\le c\,\rho_1(x)^{m-1-\beta_1} \prod_{\mu\in X_1} \Big(\frac{r_\mu(x)}{\rho_1(x)}\Big)^{m-1-\delta_\mu} \|r^{k-m} g_{j,k}\|_{L^\infty_{\beta,\delta}(\Gamma_j)^\ell}\,.\end{aligned}$$

For $\nu = 2$ this estimate holds analogously by the first estimate of Theorem 5.1.6.

If $\xi \in A_3$, then by Theorem 5.1.3

$$\begin{aligned}\big|T_{j,k}(\xi, D_\xi)\, H^{(i)}(x,\xi)\big| &\le c\,|x-\xi|^{k-3} \Big(\frac{r(x)}{|x-\xi|}\Big)^{m-1+\delta_+^{(1)}-\varepsilon} \\ &\quad \times \Big(\frac{r(\xi)}{|x-\xi|}\Big)^{k-m-1+\delta_-^{(k(\xi))}-\varepsilon}.\end{aligned}$$

Let A_3', A_3'' be the following subsets of A_3:

$$A_3' = \{\xi \in A_3 :\ r_1(\xi) < \min_{2\le k\le d} r_k(\xi)\}, \qquad A_3'' = A_3\backslash A_3'\,.$$

Since $|x-\xi| > c\,\rho_1(x)$ for $\xi \in A_3''$, we obtain

$$\begin{aligned}&\Big|\int_{A_3''} g_{j,k}(\xi)\cdot \overline{T_{j,k}(\xi, D_\xi)\, H^{(i)}(x,\xi)}\, d\xi\Big| \\ &\le c\,\rho_1(x)^{-\beta_1-1-\delta_+^{(1)}+2\varepsilon}\, r(x)^{m-1+\delta_+^{(1)}-\varepsilon}\, \|r^{k-m} g_{j,k}\|_{L^\infty_{\beta,\delta}(\Gamma_j)^\ell} \\ &\quad \times \int_{A_3''} \rho_1(\xi)^{-\delta_-^{(k(\xi))}}\, r(\xi)^{-1+\delta_-^{(k(\xi))}-\varepsilon} \prod_{k\in X_1} \Big(\frac{r_k(\xi)}{\rho_1(\xi)}\Big)^{-\delta_k} d\xi.\end{aligned}$$

The integral on the right-hand side is dominated by $c\,\rho_1(x)^{1-\varepsilon}$. Consequently,

$$\begin{aligned}&\Big|\int_{A_3''} g_{j,k}(\xi)\cdot \overline{T_{j,k}(\xi, D_\xi)\, H^{(i)}(x,\xi)}\, d\xi\Big| \\ &\le c\,\rho_1(x)^{m-1-\beta_1} \Big(\frac{r(x)}{\rho_1(x)}\Big)^{m-1+\delta_+^{(1)}-\varepsilon} \|r^{k-m} g_{j,k}\|_{L^\infty_{\beta,\delta}(\Gamma_j)^\ell} \\ &\le c\,\rho_1(x)^{m-1-\beta_1} \Big(\frac{r(x)}{\rho_1(x)}\Big)^{m-1-\delta_1} \|r^{k-m} g_{j,k}\|_{L^\infty_{\beta,\delta}(\Gamma_j)^\ell}\end{aligned}$$

if $\delta_1 > -\delta_+^{(1)} + \varepsilon$. Furthermore, it follows from the above estimate of $T_{j,k}(\xi, D_\xi)\, H^{(i)}$ in A_3 that

$$\Big| \int_{A_3'} g_{j,k}(\xi) \cdot \overline{T_{j,k}(\xi, D_\xi)\, H^{(i)}(x,\xi)}\, d\xi \Big|$$
$$\le c\, \rho_1(x)^{\delta_1-\beta_1}\, r(x)^{m-1+\delta_+^{(1)}-\varepsilon}\, \|r^{k-m} g_{j,k}\|_{L^\infty_{\beta,\delta}(\Gamma_j)^\ell}$$
$$\times \int_{A_3'} |x-\xi|^{-1-\delta_+^{(1)}-\delta_-^{(1)}+2\varepsilon}\, r_1(\xi)^{-1+\delta_-^{(1)}-\delta_1-\varepsilon}\, d\xi.$$

Let x' and ξ' be the nearest points on M_1 to x and ξ, respectively. Suppose that A_3' is a subset of a half-plane. Using the inequality

$$4\,|x-\xi| > r(x) + r(\xi) + |x'-\xi'| \quad \text{for } \xi \in A_3',$$

we obtain

$$\int_{A_3'} |x-\xi|^{-1-\delta_+^{(1)}-\delta_-^{(1)}+2\varepsilon}\, r_1(\xi)^{-1+\delta_-^{(1)}-\delta_1-\varepsilon}\, d\xi$$
$$\le c \int_0^\infty \int_0^\infty (r + r_1(x) + |x'-\xi'|)^{-1-\delta_+^{(1)}-\delta_-^{(1)}+2\varepsilon}\, r^{-1+\delta_-^{(1)}-\delta_1-\varepsilon}\, d\xi'\, dr$$
$$\le c\, r_1(x)^{-\delta_1-\delta_+^{(1)}+\varepsilon}.$$

Passing to suitable local coordinates, we obtain the same estimate in the case where A_3' is not contained in a half-plane. Thus,

$$\Big| \int_{A_3'} g_{j,k}(\xi) \cdot \overline{T_{j,k}(\xi, D_\xi)\, H^{(i)}(x,\xi)}\, d\xi \Big|$$
$$\le c\, \rho_1(x)^{m-1-\beta_1} \Big(\frac{r_1(x)}{\rho_1(x)} \Big)^{m-1-\delta_1} \|r^{k-m} g_{j,k}\|_{L^\infty_{\beta,\delta}(\Gamma_j)^\ell}.$$

This proves (5.2.7) for $\nu = 3$.

We consider the case $\nu = 4$. If $\xi \in A_4$, then

$$\big|T_{j,k}(\xi, D_\xi)\, H^{(i)}(x,\xi)\big| \le c\, \big(|x-\xi|^{k-3} + r(x)^{k-3} \big) \quad \text{for } k \ne 3,$$
$$\big|T_{j,k}(\xi, D_\xi)\, H^{(i)}(x,\xi)\big| \le c\, \Big(\Big|\log \frac{|x-\xi|}{r(x)}\Big| + 1 \Big) \quad \text{for } k = 3.$$

Using the inequalities $\rho_1(x)/2 < \rho_1(\xi) < 2\rho_1(x)$, $r_k(x)/2 < r_k(\xi) < 2r_k(x)$,

$$\int_{A_4} \Big(\Big|\log \frac{|x-\xi|}{r(x)}\Big| + 1 \Big)\, d\xi \le c\, r(x)^2$$

and

$$\int_{A_4} \big(|x-\xi|^{k-3} + r(x)^{k-3} \big)\, d\xi \le c\, r(x)^{k-1} \quad \text{for } k > 1,$$

we obtain

$$\begin{aligned} |v(x)| &\le c\, r(x)^{m-k}\, \rho_1(x)^{-\beta_1} \prod_{\mu \in X_1} \Big(\frac{r_\mu(x)}{\rho_1(x)} \Big)^{-\delta_\mu} \|r^{k-m} g_{j,k}\|_{L^\infty_{\beta,\delta}(\Gamma_j)^\ell} \\ &\quad \times \int_{A_4} \big|T_{j,k}(\xi, D_\xi)\, H^{(i)}(x,\xi)\big|\, d\xi \\ &\le c\, r(x)^{m-1}\, \rho_1(x)^{-\beta_1} \prod_{\mu \in X_1} \Big(\frac{r_\mu(x)}{\rho_1(x)} \Big)^{-\delta_\mu} \|r^{k-m} g_{j,k}\|_{L^\infty_{\beta,\delta}(\Gamma_j)^\ell} \end{aligned}$$

for $k > 1$. If $k = 1$, then

$$\big|T_{j,1}(\xi, D_\xi)\, H^{(i)}(x,\xi)\big| \le c\, d(x)\, |x-\xi|^{-3}.$$

(cf. Corollary 5.1.4). This together with the inequality

$$\int_{\Gamma_j} |x-\xi|^{-3}\, d\xi \le c\, d(x)^{-1}$$

yields

$$\begin{aligned} \big|v(x)\big| &\le c\, r(x)^{m-1}\, \rho_1(x)^{-\beta_1} \prod_{k\in X_1} \Big(\frac{r_k(x)}{\rho_1(x)}\Big)^{-\delta_k} \|r^{1-m} g_{j,1}\|_{L^\infty_{\beta,\delta}(\Gamma_j)^\ell} \\ &\quad \times \int_{A_4} \big|T_{j,1}(\xi, D_\xi)\, H^{(i)}(x,\xi)\big|\, d\xi \\ &\le c\, r(x)^{m-1}\, \rho_1(x)^{-\beta_1} \prod_{k\in X_1} \Big(\frac{r_k(x)}{\rho_1(x)}\Big)^{-\delta_k} \|r^{1-m} g_{j,1}\|_{L^\infty_{\beta,\delta}(\Gamma_j)^\ell} \,. \end{aligned}$$

Thus, (5.2.7) is shown for $\nu = 4$.

It remains to consider the case $\nu = 5$. If $\xi \in A_5 \cap \mathcal{U}_s$, $s \ge 2$, then

$$\begin{aligned} \big|T_{j,k}(\xi, D_\xi)\, H^{(i)}(x,\xi)\big| &\le c\, \rho_1(x)^{\Lambda_1^+ - \varepsilon} \prod_{\mu\in X_1} \Big(\frac{r_\mu(x)}{\rho_1(x)}\Big)^{m-1+\delta_+^{(\mu)}-\varepsilon} \\ &\quad \times \rho_s(\xi)^{k-3-\Lambda_s^- -\varepsilon} \prod_{\mu\in X_s} \Big(\frac{r_\mu(\xi)}{\rho_s(\xi)}\Big)^{k-m-1+\delta_+^{(\mu)}-\varepsilon} . \end{aligned}$$

Since $\Lambda_1^+ > m-1-\beta_1$ and $\delta_+^{(\mu)} > -\delta_\mu$, we obtain

$$\begin{aligned} |v(x)| &\le c\, \rho_1(x)^{m-1-\beta_1} \prod_{\mu\in X_1} \Big(\frac{r_\mu(x)}{\rho_1(x)}\Big)^{m-1-\delta_\mu} \|r^{k-m} g_{j,k}\|_{L^\infty_{\beta,\delta}(\Gamma_j)^\ell} \\ &\quad \times \sum_{s=2}^{d'} \int_{A_5} \rho_s(\xi)^{-\beta_s+m-3-\Lambda_s^- -\varepsilon} \sum_{\mu\in X_s} \Big(\frac{r_\mu(\xi)}{\rho_s(\xi)}\Big)^{\delta_-^{(\mu)}-\delta_\mu-1-\varepsilon} d\xi. \end{aligned}$$

The integral on the right-hand side of the last inequality is finite for sufficiently small ε since $-\beta_s + m - 3 - \Lambda_s^- > -2$ and $\delta_-^{(\mu)} - \delta_\mu > 0$. Therefore, the estimate (5.2.7) holds for $\nu = 5$. The proof is complete. □

Next we consider the solution

$$u(x) = \int_{\mathcal{G}} G(x,\xi)\, f(\xi)\, d\xi \tag{5.2.8}$$

of the problem

$$L(x, D_x)\, u = f \ \text{ in } \mathcal{G}, \quad \frac{\partial^{k-1} u}{\partial n^{k-1}} = 0 \ \text{ on } \Gamma_j,\ j = 1, \dots, N,\ k = 1, \dots, m,$$

where $f \in L^\infty_{\beta+m+1,\delta+m+1}(\mathcal{G})^\ell$. The proof of the next lemma proceeds analogously to Lemma 3.6.6, where this result was obtained for the solution of the problem in a cone.

LEMMA 5.2.2. *Let the components of* β *and* δ *satisfy the inequalities* (5.2.2). *Then the vector function* (5.2.8) *satisfies the estimate*

$$\|r^{1-m}u\|_{L^\infty_{\beta,\delta}(\mathcal{G})^\ell} \le c\,\|r^{1+m}f\|_{L^\infty_{\beta,\delta}(\mathcal{G})^\ell}$$

for arbitrary $f \in L^\infty_{\beta+m+1,\delta+m+1}(\mathcal{G})^\ell$.

As an immediate consequence of the last two lemmas, the following statement holds.

COROLLARY 5.2.3. *Suppose that* $r^{m+1}f \in L^\infty_{\beta,\delta}(\mathcal{G})^\ell$ *and* $r^{k-m}g_{j,k} \in L^\infty_{\beta,\delta}(\Gamma_j)^\ell$ *for* $j = 1, \dots, N$, $k = 1, \dots, m$. *If the components of* β *and* δ *satisfy the inequalities* (5.2.2), *then the function* (5.2.4) *satisfies the estimate*

$$\|r^{1-m}u_i\|_{L^\infty_{\beta,\delta}(\mathcal{G})} \le c\Big(\|r^{1+m}f\|_{L^\infty_{\beta,\delta}(\mathcal{G})^\ell} + \sum_{j=1}^N\sum_{k=1}^m \|r^{k-m}g_{j,k}\|_{L^\infty_{\beta,\delta}(\Gamma_j)^\ell}\Big)$$

with a constant c *independent of* f *and* $g_{j,k}$.

5.2.4. A local maximum principle. Let $L(x, D_x)$ be a strongly elliptic differential operator with infinitely differentiable coefficients on $\overline{\Omega}$, where Ω is a bounded domain in $\mathbb{R}^N$ with smooth (of class C^∞) boundary $\partial\Omega$. Then the *Miranda-Agmon maximum principle* is valid for the solutions of the problem

$$L(x, D_x)\,u = 0 \ \text{ in } \Omega, \quad \frac{\partial^{k-1}u}{\partial n^{k-1}} = g_k \ \text{ on } \partial\Omega,\ k = 1, \dots, m.$$

This means that there exists a constant c such that

$$\|u\|_{C^{m-1}(\bar{\Omega})^\ell} \le c\Big(\sum_{k=1}^m \|g_k\|_{C^{m-k}(\partial\Omega)^\ell} + \|u\|_{L_1(\Omega)^\ell}\Big) \tag{5.2.9}$$

(cf. [**5, 182, 183**]). In the case of uniqueness, the L_1 norm of u can be omitted. As a consequence of (5.2.9), we obtain the following result.

LEMMA 5.2.4. *Let* $p > N$. *Then there exists a constant* c *such that the inequality*

$$\|u\|_{C^{m-1}(\bar{\Omega})^\ell} \le c\Big(\|L(x, D_x)\,u\|_{W^{-m,p}(\Omega)^\ell} + \sum_{k=1}^m \|\partial_n^{k-1}u\|_{C^{m-k}(\partial\Omega)^\ell} + \|u\|_{L_1(\Omega)^\ell}\Big)$$

is satisfied for all $u \in C^\infty(\overline{\Omega})^\ell$.

P r o o f. There exists a vector function $v \in W^{m,p}(\Omega)^\ell$ such that $L(x, D_x)v = L(x, D_x)u$ and

$$\|v\|_{W^{m,p}(\Omega)^\ell} \le c\,\|L(x, D_x)u\|_{W^{-m,p}(\Omega)^\ell}$$

with a constant c independent of u. Since the space $W^{m,p}(\Omega)$ is continuously imbedded in $C^{m-1}(\overline{\Omega})$ for $p > N$, it follows that

$$\|v\|_{C^{m-1}(\bar{\Omega})^\ell} \le c\,\|L(x, D_x)u\|_{W^{-m,p}(\Omega)^\ell}\,. \tag{5.2.10}$$

Applying (5.2.9) to $u - v$, we obtain

$$\begin{aligned}\|u\|_{C^{m-1}(\bar{\Omega})^\ell} &\le \|v\|_{C^{m-1}(\bar{\Omega})^\ell} + \|u - v\|_{C^{m-1}(\bar{\Omega})^\ell} \\ &\le c\Big(\|v\|_{C^{m-1}(\bar{\Omega})^\ell} + \sum_{k=1}^m \|\partial_n^{k-1}u\|_{C^{m-k}(\partial\Omega)^\ell} + \|u\|_{L_1(\Omega)^\ell}\Big).\end{aligned}$$

This together with (5.2.10) leads to the desired inequality. □

The last lemma is used in the proof of a local C^{m-1} estimate.

LEMMA 5.2.5. *Let ζ, η be infinitely differentiable functions such that $\eta = 1$ in a neighborhood of* $\operatorname{supp}\zeta$. *Then the inequality*

$$\|\zeta u\|_{C^{m-1}(\bar{\Omega})^\ell} \leq c\Big(\|\eta L(x, D_x)\, u\|_{L_\infty(\Omega)^\ell} + \sum_{k=1}^{m} \|\eta \partial_n^{k-1} u\|_{C^{m-k}(\partial\Omega)^\ell} + \|\eta u\|_{L_\infty(\Omega)^\ell}\Big)$$

is satisfied for all $u \in C^\infty(\bar{\Omega})^\ell$, where c is a constant independent of u.

Proof. Let ψ, χ be infinitely differentiable functions such that $\psi = 1$ in a neighborhood of $\operatorname{supp}\zeta$, $\chi = 1$ in a neighborhood of $\operatorname{supp}\psi$ and $\eta = 1$ in a neighborhood of $\operatorname{supp}\chi$. Since the commutator of $L(x, D_x)$ and ζ is a $2m-1$ order differential operator, we obtain

$$\|L(x, D_x)\,(\zeta u)\|_{W^{-m,p}(\Omega)^\ell} \leq c\Big(\|\zeta\, L(x, D_x)\, u\|_{W^{-m,p}(\Omega)^\ell} + \|\psi u\|_{W^{m-1,p}(\Omega)^\ell}\Big). \tag{5.2.11}$$

We denote by $\tilde{W}^{l,p}(\Omega)$ the closure of the set $C^\infty(\bar{\Omega})$ with respect to the norm

$$\|u\|_{\tilde{W}^{l,p}(\Omega)} = \|u\|_{W^{l,p}(\Omega)} + \sum_{k=0}^{2m-1} \|\partial_n^k u\|_{W^{l-k-1/p,p}(\partial\Omega)}\,.$$

By [**176**, Theorem 7.2.1] (for $p = 2$ this estimate is also proved in [**84**, Lemma 3.2.4]), the vector function u satisfies the estimate

$$\|\psi u\|_{W^{m-1,p}(\Omega)^\ell} \leq c\Big(\|\chi L(x, D_x)\, u\|_{(W^{m+1,p'}(\Omega)^*)^\ell} + \sum_{k=1}^{m} \|\chi \partial_n^{k-1} u\|_{W^{m-k-1/p,p}(\partial\Omega)^\ell} + \|\chi u\|_{\tilde{W}^{0,p}(\Omega)^\ell}\Big),$$

where $p' = p/(p-1)$. Furthermore by [**84**, Lemma 3.2.3], the inequality

$$\|\chi u\|_{\tilde{W}^{0,p}(\Omega)^\ell} \leq c\Big(\|\eta u\|_{L_p(\Omega)^\ell} + \|\eta L(x, D_x) u\|_{(W^{2m,p'}(\Omega)^*)^\ell}\Big)$$

is valid. Consequently,

$$\|\psi u\|_{W^{m-1,p}(\Omega)^\ell} \leq c\Big(\|\eta u\|_{L_p(\Omega)^\ell} + \|\eta L(x, D_x)\, u\|_{(W^{m+1,p'}(\Omega)^*)^\ell} + \sum_{k=1}^{m} \|\eta \partial_n^{k-1} u\|_{W^{m-k-1/p,p}(\partial\Omega)^\ell}\Big).$$

This together with (5.2.11) yields

$$\|L(x, D_x)\,(\zeta u)\|_{W^{-m,p}(\Omega)^\ell}$$
$$\leq c\Big(\|\eta L(x, D_x)\, u\|_{L_p(\Omega)^\ell} + \sum_{k=1}^{m} \|\eta \partial_n^{k-1} u\|_{W^{m-k-1/p,p}(\partial\Omega)^\ell} + \|\eta u\|_{L_p(\Omega)^\ell}\Big).$$

Applying Lemma 5.2.4 to the vector function ζu and using the last inequality, we obtain the assertion of the lemma. □

5.2.5. A weighted maximum principle. First we prove the following assertion by means of Lemma 5.2.4.

LEMMA 5.2.6. *Let $u \in C^\infty(\overline{\mathcal{G}}\backslash\mathcal{S})^\ell \cap L^\infty_{\beta-m+1,\delta-m+1}(\mathcal{G})^\ell$ be such that*

$$L(x,D_x)u \in L^\infty_{\beta+m+1,\delta+m+1}(\mathcal{G})^\ell \quad \text{and} \quad \frac{\partial^{k-1}u}{\partial n^{k-1}}\Big|_{\Gamma_j} \in V^{m-k,\infty}_{\beta,\delta}(\Gamma_j)^\ell$$

for $j=1,\dots,N$, $k=1,\dots,m$. Then $u \in V^{m-1,\infty}_{\beta,\delta}(\mathcal{G})^\ell$ and

$$\begin{aligned}\|u\|_{V^{m-1,\infty}_{\beta,\delta}(\mathcal{G})^\ell} \;\le\; & c\Big(\|r^{m+1}\,L(x,D_x)u\|_{L^\infty_{\beta,\delta}(\mathcal{G})^\ell} + \sum_{j=1}^N\sum_{k=1}^m \|\partial_n^{k-1}u\|_{V^{m-k,\infty}_{\beta,\delta}(\Gamma_j)^\ell} \\ & + \|r^{1-m}\,u\|_{L^\infty_{\beta,\delta}(\mathcal{G})^\ell}\Big).\end{aligned}$$

P r o o f. Let y be a point in a sufficiently small neighborhood of the vertex $x^{(1)}$. Without loss of generality, we may assume that $x^{(1)}$ is the origin and that $\mathcal{G}$ coincides with a cone $\mathcal{K}$ in a neighborhood of $x^{(1)}$. We set $\theta = r(y)$ and $z = y/\theta$. Furthermore, let η be an infinitely differentiable function equal to one near z such that $\eta(x) = 0$ for $|x-z| > 1/2$ and $|\partial_x^\alpha \eta(x)| < c_\alpha$ for all α, where c_α are constants which do not depend on x and y. Then the function $\zeta(x) = \eta(x/\theta)$ is equal to one near y and vanishes for $|x-y| > \theta/2$. Let

$$L_\theta(x,D_x) = \theta^{2m} L(\theta x, \theta^{-1}D_x).$$

By Lemma 5.2.5, the vector function $v(x) = u(\theta x)$ satisfies the estimate

$$\begin{aligned}(5.2.12)\qquad |\partial^\alpha v(z)| \le c\Big(&\|\eta L_\theta(x,D_x)\,v\|_{L_\infty(\mathcal{K})^\ell} + \sum_j\sum_{k=1}^m \|\eta\partial_n^{k-1}v\|_{C^{m-k}(\Gamma_j)^\ell} \\ & + \|\eta v\|_{L_\infty(\mathcal{K})^\ell}\Big),\end{aligned}$$

where c is independent of u and y. The constant c can be also assumed to be independent of θ for small θ, because the coefficients of the operator

$$\theta^{2m}L(\theta x, \theta^{-1}D_x) - L^\circ(0,D_x)$$

and their derivatives are small for small θ. (Here $L^\circ(0,D_x)$ is the principal part of $L(x,D_x)$ with coefficients frozen at the origin.) Obviously,

$$\partial^\alpha v(z) = \theta^{|\alpha|}\,\partial^\alpha u(y) \quad \text{and} \quad \|\eta L_\theta(x,D_x)\,v\|_{L_\infty(\mathcal{K})^\ell} = \theta^{2m}\,\|\zeta L(x,D_x)\,u\|_{L_\infty(\mathcal{K})^\ell}\,.$$

Furthermore, it follows from the inequality $\theta/2 < r(x) < 2\theta$ for $x \in \operatorname{supp}\zeta$ that

$$\|\eta\partial_n^{k-1}v\|_{C^{m-k}(\Gamma_j)^\ell} \le c\,\theta^{m-1}\|\zeta\partial_n^{k-1}u\|_{V^{m-k,\infty}_{0,0}(\Gamma_j)^\ell}\,.$$

Multiplying (5.2.12) by

$$\theta^{1-m}\rho_1(y)^{\beta_1}\prod_{k\in X_1}\Big(\frac{r_k(y)}{\rho_1(y)}\Big)^{\delta_k}$$

and using the fact that $\rho_1(y)/2 < \rho_1(x) < 2\rho_1(y)/2$ and $r_k(y)/2 < r_k(x) < 2r_k(y)$ for $x \in \operatorname{supp}\zeta$, we consequently obtain

$$r(y)^{|\alpha|-m+1}\,\rho_1(y)^{\beta_1}\prod_{k\in X_1}\Big(\frac{r_k(y)}{\rho_1(y)}\Big)^{\delta_k}\,|\partial^\alpha u(y)|$$
$$\le c\Big(\|\zeta\, r^{m+1}\,L(x,D_x)\,u\|_{L^\infty_{\beta,\delta}(\mathcal{G})^\ell} + \sum_j\sum_{k=1}^m \|\zeta\partial_n^{k-1}u\|_{V^{m-k,\infty}_{\beta,\delta}(\Gamma_j)^\ell}$$
$$+\ \|\zeta\, r^{1-m}\,u\|_{L^\infty_{\beta,\delta}(\mathcal{G})^\ell}\Big).$$

Analogously, this estimate holds if y lies in a small neighborhood of an edge point. In the case where y lies outside a neighborhood of the vertices and edges, this estimate follows immediately from Lemma 5.2.5. This proves the lemma. □

Combining the last lemma with Corollary 5.2.3, we obtain weighted L_∞ estimates for the derivatives of generalized solutions.

THEOREM 5.2.7. *Suppose that the components of β and δ satisfy the inequalities* (5.2.2). *Then the generalized solution of problem* (4.1.1) *satisfies the estimate*

$$\|u\|_{V^{m-1,\infty}_{\beta,\delta}(\mathcal{G})^\ell} \le c\Big(\|r^{m+1}f\|_{L^\infty_{\beta,\delta}(\mathcal{G})^\ell} + \sum_{j=1}^m\sum_{k=1}^m \|g_{j,k}\|_{V^{m-k,\infty}_{\beta,\delta}(\Gamma_j)^\ell}\Big) \tag{5.2.13}$$

for arbitrary $f \in \mathcal{V}^{0,\infty}_{\beta+m+1,\delta+m+1}(\mathcal{G})^\ell$, $g \in \mathcal{V}^{m-1,\infty}_{\beta,\delta}(\partial\mathcal{G})^\ell$.

P r o o f. Let u be the vector function with the components (5.2.4). Since $L(x,D_x)\,u = f \in L^\infty_{\beta+m+1,\delta+m+1}(\mathcal{G})$, it follows that $u \in W^{2m,p}_{loc}(\mathcal{G})^\ell$ for arbitrary $p<\infty$. In particular, u has continuous derivatives up to order $m-1$ at every point $x\in\mathcal{G}$. By $\{f^{(\nu)}\}$ and $\{\psi^{(\nu)}\}$, we denote approximating sequences for f and g in $\mathcal{V}^{0,\infty}_{\beta+m+1,\delta+m+1}(\mathcal{G})^\ell$ and $\mathcal{V}^{m-1,\infty}_{\beta,\delta}(\partial\mathcal{G})$, respectively. The vector functions $u^{(\nu)}$ with the components (5.2.3) are infinitely differentiable on $\overline{\mathcal{G}}\backslash\mathcal{S}$ and satisfy the estimate

$$\|r^{1-m}\,u^{(\nu)}\|_{L^\infty_{\beta,\delta}(\mathcal{G})^\ell} \le c\Big(\|r^{m+1}f^{(\nu)}\|_{L^\infty_{\beta,\delta}(\mathcal{G})^\ell} + \sum_{j=1}^N\sum_{k=1}^m \|r^{k-m}\psi^{(\nu)}_{j,k}\|_{L^\infty_{\beta,\delta}(\Gamma_j)^\ell}\Big)$$

(see Corollary 5.2.3). Using Lemma 5.2.6, we obtain the estimate

$$\|u^{(\nu)}\|_{V^{m-1,\infty}_{\beta,\delta}(\mathcal{G})^\ell} \le c\Big(\|r^{m+1}\,f^{(\nu)}\|_{L^\infty_{\beta,\delta}(\mathcal{G})^\ell} + \sum_{j=1}^N\sum_{k=1}^m \|\psi^{(\nu)}_{j,k}\|_{V^{m-k,\infty}_{\beta,\delta}(\Gamma_j)^\ell}\Big)$$

with a constant c independent of f, g and ν. Since $u^{(\nu)}(x)\to u(x)$ for every $x\in\mathcal{G}$ as $\nu\to\infty$, it follows that u satisfies (5.2.13). □

5.2.6. The Miranda-Agmon maximum principle. Obviously, the condition (5.2.2) allows us to choose $\delta_k = 0$ in Theorem 5.2.7. Moreover, if $\Lambda_j^+ > m-1$, i.e. the strip $m-3/2 < \operatorname{Re}\lambda \le m-1$ does not contain eigenvalues of the pencil $\mathfrak{A}(\lambda)$, it is possible to set $\beta_j = 0$. In this case, Theorem 5.2.7 yields a C^{m-1}-estimate for the solution. However, the assumption $g_{j,k}\in V^{m-k,\infty}_{0,0}(\Gamma_j)$ means that the derivatives of $g_{j,k}$ up to order $m-k-1$ vanish on the edges bounding the face Γ_j. Our goal is to avoid this restriction on the functions $g_{j,k}$.

Let $g_{j,k}$ be arbitrary functions in $C^{m-k}(\Gamma_j)$, $j = 1,\dots,N$, $k = 1,\dots,m$. We say that

$$g = \{g_{j,k} : j = 1,\dots,N, \ k = 1,\dots,m\}$$

belongs to $\mathfrak{C}^{m-1}(\partial\mathcal{G})$ if there exists a function $v \in C^{m-1}(\overline{\mathcal{G}})$ such that

$$\frac{\partial^{k-1} v}{\partial n^{k-1}} = g_{j,k} \quad \text{on } \Gamma_j, \ j = 1,\dots,N, \ k = 1,\dots,m. \tag{5.2.14}$$

LEMMA 5.2.8. *For arbitrary $g \in \mathfrak{C}^{m-1}(\partial\mathcal{G})$, there exists a vector function $v \in C^{m-1}(\overline{\mathcal{G}})$ satisfying the inequalities*

$$\|r^{\max(0,|\alpha|-m+1)}\, \partial_x^\alpha v\|_{L_\infty(\mathcal{G})} \le c_\alpha \sum_{j=1}^{N}\sum_{k=1}^{m} \|g_{j,k}\|_{C^{m-k}(\overline{\Gamma}_j)} \quad \text{for all } \alpha, \tag{5.2.15}$$

$$\|\partial_n^{\mu-1} v - g_{\nu,\mu}\|_{V_{0,0}^{m-\mu,\infty}(\Gamma_\nu)} \le c \sum_{j=1}^{N}\sum_{k=1}^{m} \|g_{j,k}\|_{C^{m-k}(\overline{\Gamma}_j)} \tag{5.2.16}$$

for $\mu = 1,\dots,m$, $\nu = 1,\dots,N$, where c_α and c are constants independent of $g_{j,k}$.

P r o o f. We prove the assertion of the lemma for a cone $\mathcal{K}$ and for functions $g_{j,k}$ with compact supports. For a bounded polyhedral domain $\mathcal{G}$, the assertion can be obtained then by means of a partition of unity and suitable diffeomorphisms.

Let $\mathcal{K}$ be a cone with vertex at the origin and faces $\Gamma_1,\dots,\Gamma_d$. Furthermore, let $g_{j,k} \in C^{m-k}(\overline{\Gamma}_j)$ be functions vanishing outside the unit ball. We assume that there exists a function $u \in C^{m-1}(\overline{\mathcal{K}})$ satisfying the conditions

$$\partial_n^{k-1} u = g_{j,k} \quad \text{on } \Gamma_j, \ j = 1,\dots,d, \ k = 1,\dots,m.$$

First we consider the case where $(\partial_x^\alpha u)(0) = 0$ for $|\alpha| \le m-2$, i.e. the derivatives of the functions $g_{j,k}$ up to order $m-k-1$ are zero at the origin. On the set $\{u \in C^l(\overline{\mathcal{K}}) : (\partial_x^\alpha u)(0) = 0 \text{ for } |\alpha| \le m-2\}$, we introduce the norm

$$\|u\|_{V_0^{l,\infty}(\mathcal{K})} = \sum_{|\alpha|\le l} \sup_{x\in\mathcal{K}} |x|^{|\alpha|-l}\, \big|\partial_x^\alpha u(x)\big|.$$

Analogously, we define the $V_0^{l,\infty}(\Gamma_j)$-norm on the set of all functions $u \in C^l(\overline{\Gamma}_j)$ vanishing at the origin together with the derivatives up to order $l-1$. By the mean value theorem, there exists a real $t \in (0,1)$ such that

$$|x|^{-l}\, \big|\mathrm{Re}\, u(x)\big| = |x|^{1-l}\, \frac{|\mathrm{Re}\, u(x) - \mathrm{Re}\, u(0)|}{|x|} \le |x|^{1-l}\, \big|(\nabla u)(tx)\big|,$$

if $u(0) = 0$. The same estimate (with another t) holds for the imaginary part of u. Consequently,

$$|x|^{-l}\, |u(x)| \le \sup_{y\in\mathcal{K}} |y|^{1-l} \big|(\nabla u)(y)\big|$$

if $u(0) = 0$. Repeating this argument, we obtain the inequality

$$\|u\|_{V_0^{l,\infty}(\mathcal{K})} \le c\, \|u\|_{C^l(\overline{\mathcal{K}})}$$

for all $u \in C^l(\overline{\mathcal{K}})$ such that $(\partial_x^\alpha u)(0) = 0$ for $|\alpha| \le l-1$. Analogously, the inequality

$$\|u\|_{V_0^{l,\infty}(\Gamma_j)} \le c\, \|u\|_{C^l(\overline{\Gamma}_j)} \tag{5.2.17}$$

holds for all functions $u \in C^l(\overline{\Gamma}_j)$ which are zero at the origin together with all derivatives up to order $l-1$.

Let ζ_ν be infinitely differentiable functions depending only on $\rho = |x|$ and satisfying the conditions (3.1.4). Then the functions

$$\psi_{j,k,\nu}(x) = 2^{\nu(k-1)}\,\zeta_\nu(2^\nu x)\,g_{j,k}(2^\nu x)$$

are zero for $|x| < 1/2$ and $|x| > 2$. By Corollary 6.7.14, there exist functions w_ν satisfying the estimates

$$\|r^{\max(0,|\alpha|-m+1)}\,\partial_x^\alpha w_\nu\|_{L_\infty(\mathcal{K})} \le c_\alpha \sum_{j=1}^{d}\sum_{k=1}^{m} \|\psi_{j,k,\nu}\|_{V_0^{m-k,\infty}(\Gamma_j)} \tag{5.2.18}$$

for all α and

$$\|\partial_n^{t-1} w_\nu - \psi_{s,t,\nu}\|_{V_{0,0}^{m-t,\infty}(\Gamma_s)} \le c \sum_{j=1}^{d}\sum_{k=1}^{m} \|\psi_{j,k,\nu}\|_{V_0^{m-k,\infty}(\Gamma_j)} \tag{5.2.19}$$

for $s = 1,\dots,d$, $k = 1,\dots,m$. Here the constants c and c_α are independent of ν. By the construction of the function in Corollary 6.7.14, the functions w_ν can be chosen such that $w_\nu(x) = 0$ for $|x| < 1/4$ and $|x| > 4$. We define

$$v_\nu(x) = w_\nu(2^{-\nu}x) \quad\text{and}\quad v(x) = \sum_{\nu=-\infty}^{0} v_\nu(x)\,.$$

Since v_ν vanishes outside the set $\{x : 2^{\nu-2} < |x| < 2^{\nu+2}\}$, the last sum consists only of four terms for every x. Consequently according to (5.2.18),

$$\begin{aligned}
\|r^{\max(0,|\alpha|-m+1)}\,\partial_x^\alpha v\|_{L_\infty(\mathcal{G})} &\le 4\sup_\nu \|r^{\max(0,|\alpha|-m+1)}\,\partial_x^\alpha v_\nu\|_{L_\infty(\mathcal{G})}\\
&= 4\sup_\nu 2^{-\nu(m-1)}\,2^{\nu\max(0,m-1-|\alpha|)}\,\|r^{\max(0,|\alpha|-m+1)}\,\partial_x^\alpha w_\nu\|_{L_\infty(\mathcal{G})}\\
&\le c_\alpha \sup_\nu 2^{-\nu(m-1)} \sum_{j=1}^{d}\sum_{k=1}^{m} \|\psi_{j,k,\nu}\|_{V_0^{m-k,\infty}(\Gamma_j)}\\
&\le c\sup_\nu \sum_{j=1}^{d}\sum_{k=1}^{m} \|\zeta_\nu g_{j,k}\|_{V_0^{m-k,\infty}(\Gamma_j)} \le c\sum_{j=1}^{d}\sum_{k=1}^{m} \|g_{j,k}\|_{V_0^{m-k,\infty}(\Gamma_j)}\,.
\end{aligned}$$

Using the equality

$$\partial_n^{t-1} v|_{\Gamma_s} - g_{s,t} = \sum_{\nu=-\infty}^{0}\Big(\partial_n^{t-1} v_\nu|_{\Gamma_s} - \zeta_\nu g_{s,t}\Big)$$

and (5.2.19), we obtain

$$\begin{aligned}
\|\partial_n^{t-1} v - g_{s,t}\|_{V_{0,0}^{m-t,\infty}(\Gamma_s)} &\le c\sup_\nu \|\partial_n^{t-1} v_\nu - \zeta_\nu g_{s,t}\|_{V_{0,0}^{m-t,\infty}(\Gamma_s)}\\
&\le c\sup_\nu 2^{-\nu(m-1)}\,\|\partial_n^{t-1} w_\nu - \psi_{s,t,\nu}\|_{V_{0,0}^{m-t,\infty}(\Gamma_s)}\\
&\le c\sup_\nu 2^{-\nu(m-1)} \sum_{j=1}^{d}\sum_{k=1}^{m} \|\psi_{j,k,\nu}\|_{V_0^{m-k,\infty}(\Gamma_j)} \le c\sum_{j=1}^{d}\sum_{k=1}^{m} \|g_{j,k}\|_{V_0^{m-k,\infty}(\Gamma_j)}\,.
\end{aligned}$$

Applying (5.2.17), we arrive at (5.2.15) and (5.2.16).

Now we consider the case where $\partial_x^\alpha u(x) \neq 0$ for at least one multi-index α, $|\alpha| \le m-2$. We introduce the function

$$u_0(x) = \chi(x) \sum_{|\alpha|\le m-2} (\partial_x^\alpha u)(0)\, \frac{x^\alpha}{\alpha!}\, ,$$

where χ is an infinitely differentiable function with compact support equal to one near the origin. Then by the first part of the proof, there exists a function w satisfying the inequalities

$$\| r^{\max(0,|\alpha|-m+1)}\, \partial_x^\alpha w \|_{L_\infty(\mathcal{K})} \le c_\alpha \sum_{j=1}^{d} \sum_{k=1}^{m} \| g_{j,k} - \partial_n^{k-1} u_0 \|_{C^{m-k}(\overline{\Gamma}_j)}\, ,$$

$$\| \partial_n^{t-1} w + \partial_n^{t-1} u_0 - g_{s,t} \|_{V_{0,0}^{m-t,\infty}(\Gamma_s)} \le c \sum_{j=1}^{d} \sum_{k=1}^{m} \| g_{j,k} - \partial_n^{k-1} u_0 \|_{C^{m-k}(\overline{\Gamma}_j)}\, .$$

Since the C^{m-1}-norm of u_0 can be estimated by the C^{m-k}-norms of the functions $g_{j,k}$, we obtain the desired inequalities for the function $v = u_0 + w$. The proof is complete. □

Suppose that $g \in \mathfrak{C}^{m-1}(\partial\mathcal{G})^\ell$ and that $\Lambda_j^+ > m-1$ for $j = 1, \dots, d'$. Then we can construct the solution of the problem (5.2.1) by means of Theorem 5.2.7 as follows. There exists a vector function $v \in C^{m-1}(\overline{\mathcal{G}})^\ell$ satisfying (5.2.15) and (5.2.16). This means in particular that

$$r^{m+1} L v \in V_{0,0}^{0,\infty}(\mathcal{G})^\ell \quad \text{and} \quad \{\partial_n^{k-1} v - g_{j,k}\} \in V_{0,0}^{m-k,\infty}(\partial\mathcal{G})^\ell.$$

By Theorem 5.2.7, there exists a generalized solution w of the problem

$$L(x, D_x)\, w = -L(x, D_x) v \ \text{ in } \mathcal{G},$$
$$\partial_n^{k-1} w = g_{j,k} - \partial_n^{k-1} v \text{ on } \Gamma_j,\ j = 1, \dots, N,\ k = 1, \dots, m,$$

satisfying the estimate

$$\text{(5.2.20)} \quad \|w\|_{V_{0,0}^{m-1,\infty}(\mathcal{G})^\ell} \le c \Big(\| r^{m+1} L(x, D_x)\, v \|_{L_\infty(\mathcal{G})^\ell} + \sum_{j=1}^{m} \sum_{k=1}^{m} \| g_{j,k} - \partial_n^{k-1} v \|_{V_{0,0}^{m-k,\infty}(\Gamma_j)^\ell} \Big).$$

Then $u = v + w$ is a solution of problem (5.2.1), and the following statement holds.

THEOREM 5.2.9. *Suppose that the differential operator* $L(x, D_x)$ *is strongly elliptic and that*

$$A_\alpha(x^{(j)}) = \big(A_\alpha(x^{(j)})\big)^* \quad \textit{for } |\alpha| = 2m,\ j = 1, \dots, d'.$$

Furthermore, we assume that the domain $\mathcal{G}$ *is diffeomorphic to a Lipschitz graph cone* $\mathcal{K}_j$ *in a neighborhood of the vertex* $x^{(j)}$ *for* $j = 1, \dots, d'$, *i.e.* $\mathcal{K}_j$ *has the representation* $x_3 > \phi_j(x_1, x_2)$ *in Cartesian coordinates, where* ϕ_j *is positively homogeneous of degree 1. Then the solution of the problem* (5.2.1) *satisfies the estimate*

$$\|u\|_{W^{m-1,\infty}(\mathcal{G})^\ell} \le c \sum_{j=1}^{N} \sum_{k=1}^{m} \|g_{j,k}\|_{W^{m-k,\infty}(\Gamma_j)^\ell}$$

for arbitrary $g = \{g_{j,k}\} \in \mathfrak{C}^{m-1}(\partial\mathcal{G})^\ell$.

P r o o f. Under the assumptions of the theorem on the operator $L(x, D_x)$ and on the domain $\mathcal{G}$, the strip $m-2 \le \operatorname{Re}\lambda \le m-1$ does not contain eigenvalues of the pencil $\mathfrak{A}(\lambda)$ (cf. [**85**, Theorem 11.1.1]). Let $u = v + w$ be the above constructed solution of problem (5.2.1). Using (5.2.20) and Lemma 5.2.8, we obtain

$$\begin{aligned} \|u\|_{W^{m-1,\infty}(\mathcal{G})^\ell} &\le c\left(\|w\|_{V_{0,0}^{m-1,\infty}(\mathcal{G})^\ell} + \|v\|_{W^{m-1,\infty}(\mathcal{G})^\ell}\right) \\ &\le c \sum_{j=1}^{m} \sum_{k=1}^{m} \|g_{j,k}\|_{C^{m-k,\infty}(\Gamma_j)^\ell} . \end{aligned}$$

This proves the theorem. □

REMARK 5.2.10. The Lipschitz graph condition on the domain $\mathcal{G}$ in Theorem 5.2.9 is sufficient but not necessary for the validity of the Miranda-Agmon maximum principle. Needless to say, the maximum principle for harmonic functions is valid without this assumption on the boundary. In fact, we showed above that the inequality $\Lambda_j^+ > m-1$ guarantees the validity of the Miranda-Agmon maximum principle for solutions of strongly elliptic equations in arbitrary three-dimensional domains of polyhedral type. In [**84**, Theorems 3.6.1, 7.2.5, 7.3.1], it is shown that this inequality is satisfied for the biharmonic and polyharmonic equations and for the Lamé system in domains which are not necessarily in the class $C^{0,1}$. Thus for example, the estimate

$$\|u\|_{L_\infty(\mathcal{G})^3} \le c \sum_{j=1}^{N} \|g_j\|_{L_\infty(\Gamma_j)^3}$$

holds for solutions of the Dirichlet problem for the Lamé system in an arbitrary three-dimensional domain of polyhedral type (cf. [**124**]).

5.3. The Miranda-Agmon maximum principle for generalized solutions in domains with conical points

Now let $\mathcal{G}$ be a domain in $\mathbb{R}^N$ with angular ($N = 2$) or conical ($N > 2$) points $x^{(1)}, \dots, x^{(d')}$. This means that $\mathcal{G}$ is diffeomorphic to a cone

$$\mathcal{K}_j = \{x \in \mathbb{R}^N : x/|x| \in \Omega_j\}$$

in a neighborhood of the vertex $x^{(j)}$ for $j = 1, \dots, d$, where Ω_j is a subdomain of the unit sphere. Outside the set $\mathcal{S} = \{x^{(1)}, \dots, x^{(d')}\}$, the boundary $\partial\mathcal{G}$ is assumed to be smooth (of class C^∞). The distance of the point x from the vertex $x^{(j)}$ is denoted by $\rho_j(x)$.

The goal of this section is to prove the estimate

$$\|u\|_{W^{m-1,\infty}(\mathcal{G})^\ell} \le c \sum_{k=1}^{m} \|\partial_n^{k-1} u\|_{W^{m-k,\infty}(\partial\mathcal{G}\backslash\mathcal{S})^\ell} \tag{5.3.1}$$

for generalized solutions of the differential equation $L(x, D_x)\, u = 0$. As in the last two sections, it is assumed that the equation $L(x, D_x)\, u = f$ is uniquely solvable in $\overset{\circ}{W}{}^{m,2}(\mathcal{G})^\ell$ for arbitrary $f \in W^{-m,2}(\mathcal{G})^\ell$. Furthermore, we suppose that the strip $m - N/2 < \operatorname{Re}\lambda \le m-1$ is free of eigenvalues of the pencils $\mathfrak{A}_j(\lambda)$. In the case $N = 3$, this is the same condition as in the main result of Section 5.2.

5.3.1. Estimates of Green's matrix. The following theorem was proved for $N = 3$ in Section 5.1. Without any changes in the proof, the same results hold for $N \neq 3$.

THEOREM 5.3.1. *Suppose that the equation $L(x, D_x)\, u = f$ has a unique solution $u \in \overset{\circ}{W}{}^{m,2}(\mathcal{G})^\ell$ for arbitrary given $f \in W^{-m,2}(\mathcal{G})^\ell$. Then the following assertions are true.*

1) *There exists a uniquely determined solution $G(x, \xi)$ of the problem*

$$L(x, D_x)\, G(x, \xi) = \delta(x - \xi)\, I_\ell \quad \text{for } x, \xi \in \mathcal{G},$$
$$\frac{\partial^{k-1} u}{\partial n_x^{k-1}} = 0 \quad \text{for } x \in \partial\mathcal{G}\backslash\mathcal{S},\ \xi \in \mathcal{G},\ k = 1, \dots, m,$$

such that the function $x \to \zeta(x, \xi)\, G(x, \xi)$ belongs to $\overset{\circ}{W}{}^{m,2}(\mathcal{G})^{\ell\times\ell}$ for every $\xi \in \mathcal{G}$ and for every infinitely differentiable function $\zeta(\cdot, \xi)$ equal to zero in a neighborhood of the point $x = \xi$.

2) *The adjoint matrix $G^*(x, \xi)$ is the unique solution of the problem*

$$L^+(\xi, D_\xi)\, G^*(x, \xi) = \delta(x - \xi)\, I_\ell \quad \text{for } x, \xi \in \mathcal{G},$$
$$\frac{\partial^k G^*(x, \xi)}{\partial n_\xi^k} = 0 \quad \text{for } \xi \in \partial\mathcal{G}\backslash\mathcal{S},\ x \in \mathcal{G},\ k = 0, \dots, m - 1,$$

such that the function $\xi \to \zeta(x, \xi)\, G^*(x, \xi)$ belongs to the space $\overset{\circ}{W}{}^{m,2}(\mathcal{G})^{\ell\times\ell}$ for every infinitely differentiable function $\zeta(x, \cdot)$ in $\overline{\mathcal{G}}$ which is equal to zero in a neighborhood of the point $\xi = x$.

Let the operator pencil $\mathfrak{A}_j(\lambda)$ be the same as in Subsection 4.1.6. As is known (cf. [**85**, Theorem 10.1.3]), the line $\operatorname{Re}\lambda = m - N/2$ does not contain eigenvalues of the pencils $\mathfrak{A}_j(\lambda)$, $j = 1, \dots, d'$. We denote by

$$\Lambda_j^- < \operatorname{Re}\lambda < \Lambda_j^+$$

the widest strip in the complex plane which contains this line and which is free of eigenvalues of the pencil $\mathfrak{A}_j(\lambda)$. Furthermore, let $\mathcal{U}_1, \dots, \mathcal{U}_{d'}$ denote neighborhoods of the vertices $x^{(1)}, \dots, x^{(d')}$, respectively, such that

$$\mathcal{U}_1 \cup \dots \cup \mathcal{U}_{d'} \supset \overline{\mathcal{G}} \quad \text{and} \quad x^{(i)} \notin \overline{\mathcal{U}}_j \text{ for } i \neq j.$$

As for the case $N = 3$ (cf. Theorems 5.1.2, 5.1.3, 5.1.6), the following estimates of the Green's matrix hold.

THEOREM 5.3.2. 1) *If $x \in \mathcal{G} \cap \mathcal{U}_i$ and $\xi \in \mathcal{G} \cap \mathcal{U}_j$, $i \neq j$, then*

$$\left|D_x^\alpha D_\xi^\gamma G(x, \xi)\right| \le c_{\alpha,\gamma}\, \rho_i(x)^{\Lambda_i^+ - |\alpha| - \varepsilon}\, \rho_j(\xi)^{2m - N - \Lambda_j^- - |\gamma| - \varepsilon},$$

where ε is an arbitrarily small positive number. The constant $c_{\alpha,\gamma}$ is independent of x and ξ.

2) *Suppose that $x, \xi \in \mathcal{G} \cap \mathcal{U}_j$ and $\rho_j(x)/2 < \rho_j(\xi) < 2\rho_j(x)$. Then*

$$\left|D_x^\alpha D_\xi^\gamma G(x, \xi)\right| \le c_{\alpha,\gamma} \left(|x - \xi|^{2m - N - |\alpha| - |\gamma|} + \rho_j(x)^{2m - N - |\alpha| - |\gamma|}\right)$$
$$\text{if } |\alpha| + |\gamma| \neq 2m - N,$$

$$\left|D_x^\alpha D_\xi^\gamma G(x, \xi)\right| \le c_{\alpha,\gamma} \left(\left|\log \frac{|x - \xi|}{\rho_j(x)}\right| + 1\right) \quad \text{if } |\alpha| + |\gamma| = 2m - N.$$

3) *If* $x, \xi \in \mathcal{G} \cap \mathcal{U}_j$ *and* $\rho_j(\xi) < \rho_j(x)/2$, *then*

$$\left|D_x^\alpha D_\xi^\gamma G(x,\xi)\right| \le c\, \rho_j(x)^{\Lambda_j^- - |\alpha| + \varepsilon}\, \rho_j(\xi)^{2m-N-\Lambda_j^- - |\gamma| - \varepsilon},$$

where ε *is an arbitrarily small positive number. Analogously, the estimate*

$$\left|D_x^\alpha D_\xi^\gamma G(x,\xi)\right| \le c\, \rho_j(x)^{\Lambda_j^+ - |\alpha| - \varepsilon}\, \rho_j(\xi)^{2m-N-\Lambda_j^+ - |\gamma| + \varepsilon}$$

is valid for $x, \xi \in \mathcal{U}_j$, $\rho_j(\xi) > 2\rho_j(x)$.

COROLLARY 5.3.3. *Let* $d(x)$ *denote the distance of the point* x *from* $\partial\mathcal{G}$. *If* $|\alpha| \le m-1$, $|\alpha| + |\gamma| \ne 2m - N$ *and* $|\alpha| + |\gamma| \ne 2m - N - 1$, *then*

$$\left|D_x^\alpha D_\xi^\gamma G(x,\xi)\right| \le c_{\alpha,\gamma} \left(d(x)\, |x-\xi|^{2m-N-|\alpha|-|\gamma|-1} + \rho_j(x)^{2m-N-|\alpha|-|\gamma|}\right)$$

for $x, \xi \in \mathcal{U}_j$, $\rho_j(x)/2 < \rho_j(\xi) < 2\rho_j(x)$.

P r o o f. If $|x-\xi| < 4d(x)$, then the estimate of the corollary follows immediately from item 2) of the last theorem. Suppose that $|x-\xi| > 4d(x)$. We denote by y the nearest point to x on $\partial\mathcal{G}$. Since the function $x \to G_{i,j}^{\alpha,\gamma}(x,\xi) = \partial_x^\alpha \partial_\xi^\gamma G_{i,j}(x,\xi)$ vanishes on $\partial\mathcal{G}$ for $|\alpha| \le m-1$, we obtain

$$\begin{aligned} \left|\operatorname{Re} \partial_x^\alpha \partial_\xi^\gamma G_{i,j}(x,\xi)\right| &= \left|\operatorname{Re}\left(G_{i,j}^{\alpha,\gamma}(x,\xi) - G_{i,j}^{\alpha,\gamma}(y,\xi)\right)\right| \\ &= \left|\operatorname{Re} \sum_{j=1}^N \partial_{z_j} G_{i,j}^{\alpha,\gamma}(z,\xi)\,(x_j - y_j)\right| \le d(x) \sum_{j=1}^N \left|\partial_{z_j} G_{i,j}^{\alpha,\gamma}(z,\xi)\right|, \end{aligned}$$

where $z = \theta x + (1-\theta) y$, $0 < \theta < 1$. The same estimate (with another z) holds for the imaginary part of $\partial_x^\alpha \partial_\xi^\gamma G_{i,j}(x,\xi)$. Here, the estimate $\rho_j(\xi)/8 < \rho_j(z) < 3\rho_j(\xi)$ holds for z. Consequently by Theorem 5.3.2,

$$\left|\partial_{z_j} G_{i,j}^{\alpha,\gamma}(z,\xi)\right| \le c\left(|z-\xi|^{2m-N-|\alpha|-|\gamma|-1} + \rho_j(x)^{2m-N-|\alpha|-|\gamma|-1}\right).$$

Using the inequality $3|x-\xi| < 4|z-\xi| < 5|x-\xi|$, we get the desired estimate for $D_x^\alpha D_\xi^\gamma G(x,\xi)$. □

5.3.2. Generalized solutions of the boundary value problem. For arbitrary $\beta = (\beta_1, \dots, \beta_{d'}) \in \mathbb{R}^{d'}$, let $V_\beta^{l,\infty}(\mathcal{G})$ be defined as the set of all functions on $\mathcal{G}$ such that

$$\|u\|_{V_\beta^{l,\infty}(\mathcal{G})} = \max_{1 \le j \le d'} \max_{|\alpha| \le l} \left\|\rho_j^{\beta_j - l + |\alpha|}\, \partial_x^\alpha u\right\|_{L_\infty(\mathcal{G} \cap \mathcal{U}_j)}.$$

Here $\mathcal{U}_1, \dots, \mathcal{U}_{d'}$ are domains in $\mathbb{R}^N$ such that

$$\mathcal{U}_1 \cap \cdots \cap \mathcal{U}_{d'} \supset \mathcal{G} \quad \text{and} \quad x^{(i)} \notin \mathcal{U}_j \ \text{ if } i \ne j. \tag{5.3.2}$$

Analogously, the space $V_\beta^{l,\infty}(\partial\mathcal{G})$ is defined.

Furthermore, let $\mathcal{V}_\beta^{l,\infty}(\mathcal{G})$ be the set of all functions $u \in V_\beta^{l,\infty}(\mathcal{G})$ for which there exists a sequence $\{u^{(\nu)}\} \subset C_0^\infty(\overline{\mathcal{G}} \backslash \mathcal{S})$ such that

$$\|u^{(\nu)}\|_{V_\beta^{l,\infty}(\mathcal{G})} \le c < \infty \quad \text{for all } \nu,$$

$$u^{(\nu)} \to u \quad \text{a.e. on } \mathcal{G}.$$

The space $\mathcal{V}_\beta^{m-1,\infty}(\partial\mathcal{G})$ is defined as the set of all tuples

$$g = \{g_k : k = 1, \dots, m\} \in \prod_{k=1}^m V_\beta^{m-k,\infty}(\partial\mathcal{G})^\ell$$

for which there exist sequences $\{g_k^{(\nu)}\}_{\nu=1}^\infty \subset C_0^\infty(\partial\mathcal{G}\backslash\mathcal{S})^\ell$ such that

$$\|g_k^{(\nu)}\|_{V_\beta^{m-k,\infty}(\partial\mathcal{G})^\ell} \le c < \infty,$$
$$g_k^{(\nu)} \to g_k \quad \text{a.e. on } \partial\mathcal{G}\,.$$

The sequence $\{g^{(\nu)}\}_{\nu=1}^\infty = \{(g_1^{(\nu)},\dots,g_m^{(\nu)})\}_{\nu=1}^\infty$ is called *approximating* for g in $\mathcal{V}_\beta^{m-1,\infty}(\partial\mathcal{G})$.

We assume that $g \in \mathcal{V}_\beta^{m-1,\infty}(\partial\mathcal{G})$, $f \in \mathcal{V}_{\beta+m+1}^{0,\infty}(\mathcal{G})^\ell$, and that the components of β satisfy the inequalities

$$m-1-\Lambda_j^+ < \beta_j < m-1-\Lambda_j^- \tag{5.3.3}$$

for $j=1,\dots,d'$. By $f^{(\nu)}$ and $\psi^{(\nu)}$, we denote approximating sequences for f and g. There exist differential operators $S_k(x,D_x)$ and $T_k(x,D_x)$ of order $2m-k$ such that the *Green's formula*

$$\begin{aligned} &\int_{\mathcal{G}} \Big(L(x,D_x)\,u\cdot\bar v - u\cdot\overline{L^+(x,D_x)\,v}\Big)\,dx \\ &\quad = \sum_{k=1}^m \int_{\partial\mathcal{G}\backslash\mathcal{S}} \Big(S_k(x,D_x)\,u\cdot\partial_n^{k-1}\bar v - \partial_n^{k-1}u\cdot\overline{T_k(x,D_x)\,v}\Big)\,dx \end{aligned} \tag{5.3.4}$$

is valid for all $u,v \in C_0^\infty(\overline{\mathcal{G}}\backslash\mathcal{S})^\ell$. If the differential equation $L(x,D_x)\,u = f$ is uniquely solvable in $\overset{\circ}{W}{}^{m,2}(\mathcal{G})^\ell$ for arbitrary given $f \in W^{-m,2}(\mathcal{G})^\ell$, then there exists a uniquely determined Green's matrix $G(x,\xi)$ satisfying the estimates in Theorem 5.3.2. Let

$$H^{(i)}(x,\xi) = \begin{pmatrix} \overline{G_{i,1}(x,\xi)} \\ \vdots \\ \overline{G_{i,\ell}(x,\xi)} \end{pmatrix}$$

be the i-th column of the adjoint matrix $G^*(x,\xi)$. The vector function $u^{(\nu)}$ with the components

$$\begin{aligned} u_i^{(\nu)}(x) = &\int_{\mathcal{G}} f^{(\nu)}(\xi)\cdot\overline{H^{(i)}(x,\xi)}\,d\xi \\ &+ \sum_{k=1}^m \int_{\partial\mathcal{G}\backslash\mathcal{S}} g_k^{(\nu)}(\xi)\cdot\overline{T_k(\xi,D_\xi)\,H^{(i)}(x,\xi)}\,d\xi, \end{aligned} \tag{5.3.5}$$

$i=1,\dots,\ell$, is a solution of the boundary value problem

$$L(x,D_x)\,u^{(\nu)} = f^{(\nu)} \text{ in } \mathcal{G}, \quad \frac{\partial^{k-1}u}{\partial n^{k-1}} = g_k^{(\nu)} \text{ on } \partial\mathcal{G}\backslash\mathcal{S},\ k=1,\dots,m.$$

If $f^{(\nu)}$ and $\{g^{(\nu)}\}_{\nu=1}^\infty$ are approximating sequences for f and g, respectively, then the sequence $\{u^{(\nu)}(x)\}$ converges for every $x \in \mathcal{G}$. The limit $u(x)$ is the vector function with the components

$$u_i(x) = \int_{\mathcal{G}} f(\xi)\cdot\overline{H^{(i)}(x,\xi)}\,d\xi + \sum_{k=1}^m \int_{\partial\mathcal{G}\backslash\mathcal{S}} g_k(\xi)\cdot\overline{T_k(\xi,D_\xi)\,H^{(i)}(x,\xi)}\,d\xi. \tag{5.3.6}$$

The vector function u is called *generalized solution* of the boundary value problem

$$L(x, D_x)\, u = f \quad \text{in } \mathcal{G}, \tag{5.3.7}$$

$$\frac{\partial^{k-1} u}{\partial n^{k-1}} = g_k \quad \text{on } \partial\mathcal{G}\backslash\mathcal{S},\ k = 1, \dots, m. \tag{5.3.8}$$

5.3.3. Weighted L_∞ estimates of generalized solutions. Our goal is to estimate the $V^{0,\infty}_{\beta-m+1}$ norm of the generalized solution u. We start with the first integral on the right-hand side of (5.3.6).

LEMMA 5.3.4. *Suppose that $f \in \mathcal{V}^{0,\infty}_{\beta+m+1}(\mathcal{G})$ and that the components of β satisfy the inequalities* (5.3.3). *Then the function*

$$v(x) = \int_{\mathcal{G}} f(\xi) \cdot \overline{H^{(i)}(x, \xi)}\, d\xi$$

satisfies the estimate

$$\|v\|_{V^{0,\infty}_{\beta-m+1}(\mathcal{G})} \le c\, \|f\|_{V^{0,\infty}_{\beta+m+1}(\mathcal{G})}$$

with a constant c independent of f.

Proof. Let x be a point in $\mathcal{G}$, and let $x^{(1)}$ be the nearest vertex to x. Furthermore, let $\mathcal{U}_1, \dots, \mathcal{U}_{d'}$ be neighborhoods of the vertices $x^{(1)}, \dots, x^{(d')}$, respectively, satisfying the conditions (5.3.2) and

$$\operatorname{dist}(x, \mathcal{U}_j) \ge \rho_1(x)/2 \quad \text{for } j = 2, \dots, d'.$$

Then

$$\big|v(x)\big| \le \sum_{j=1}^{d'} \int_{\mathcal{G}\cap\mathcal{U}_j} \big|f(\xi) \cdot H^{(i)}(x, \xi)\big|\, d\xi.$$

Using item 1) of Theorem 5.3.2, we obtain

$$\int\limits_{\mathcal{G}\cap\mathcal{U}_j} \big|f(\xi) \cdot H^{(i)}(x, \xi)\big|\, d\xi \le c\, \|f\|_{V^{0,\infty}_{\beta+m+1}(\mathcal{G})} \int\limits_{\mathcal{G}\cap\mathcal{U}_j} \rho_j(\xi)^{-\beta_j - m - 1} \big|H^{(i)}(x, \xi)\big|\, d\xi$$

$$\le c\, \|f\|_{V^{0,\infty}_{\beta+m+1}(\mathcal{G})}\, \rho_1(x)^{\Lambda_1^+ - \varepsilon} \int\limits_{\mathcal{G}\cap\mathcal{U}_j} \rho_j(\xi)^{m-1-\Lambda_j^- - \beta_j - N - \varepsilon}\, d\xi$$

for $j = 2, \dots, d'$. Since

$$m - 1 - \Lambda_j^+ - \beta_j < 0 < m - 1 - \Lambda_j^- - \beta_j \quad \text{for all } j,$$

it follows that

$$\int_{\mathcal{G}\cap\mathcal{U}_j} \big|f(\xi) \cdot H^{(i)}(x, \xi)\big|\, d\xi \le c\, \rho_1(x)^{m-1-\beta_1}\, \|f\|_{V^{0,\infty}_{\beta+m+1}(\mathcal{G})} \tag{5.3.9}$$

for $j \neq 1$. It remains to prove the same estimate for $j = 1$. To this end, we introduce the subsets

$$\begin{aligned}
\mathcal{G}_x^{(1)} &= \big\{\xi \in \mathcal{G} \cap \mathcal{U}_1 : \rho_1(\xi) < \rho_1(x)/2\big\},\\
\mathcal{G}_x^{(2)} &= \big\{\xi \in \mathcal{G} \cap \mathcal{U}_1 : \rho_1(\xi) > 2\rho_1(x)\big\},\\
\mathcal{G}_x^{(3)} &= \big\{\xi \in \mathcal{G} \cap \mathcal{U}_1 : \rho_1(x)/2 < \rho_1(\xi) < 2\rho_1(x)\big\}
\end{aligned}$$

and consider the integrals

$$I_\nu(x) = \int_{\mathcal{G}_x^{(\nu)}} \big|f(\xi) \cdot H^{(i)}(x, \xi)\big|\, d\xi\,, \quad \nu = 1, 2, 3.$$

Obviously,

$$I_\nu(x) \le c\,\|f\|_{V^{0,\infty}_{\beta+m+1}(\mathcal{G})} \int_{\mathcal{G}_x^{(\nu)}} \rho_1(\xi)^{-\beta_1-m-1} \left|H^{(i)}(x,\xi)\right| d\xi$$

for $\nu = 1,2,3$. Using item 3) of Theorem 5.3.2, we obtain

$$\begin{aligned} I_1(x) &\le c\,\|f\|_{V^{0,\infty}_{\beta+m+1}(\mathcal{G})}\, \rho_1(x)^{\Lambda_1^-+\varepsilon} \int_{\mathcal{G}_x^{(1)}} \rho_1(\xi)^{m-1-\Lambda_1^--\beta_1-N-\varepsilon}\, d\xi \\ &\le c\,\rho_1(x)^{m-1-\beta_1}\, \|f\|_{V^{0,\infty}_{\beta+m+1}(\mathcal{G})} \end{aligned}$$

and analogously

$$I_2(x) \le c\,\rho_1(x)^{m-1-\beta_1}\, \|f\|_{V^{0,\infty}_{\beta+m+1}(\mathcal{G})}\,.$$

Furthermore, by item 2) of Theorem 5.3.2,

$$I_3(x) \le c\,\rho_1(x)^{-\beta_1-m-1}\, \|f\|_{V^{0,\infty}_{\beta+m+1}(\mathcal{G})} \int_{\mathcal{G}_x^{(3)}} \left(|x-\xi|^{2m-N} + \rho_1(x)^{2m-N}\right) d\xi$$

if $2m \ne N$ and

$$I_3(x) \le c\,\rho_1(x)^{-\beta_1-m-1}\, \|f\|_{V^{0,\infty}_{\beta+m+1}(\mathcal{G})} \int_{\mathcal{G}_x^{(3)}} \left(\Big|\log\frac{|x-\xi|}{\rho_j(x)}\Big| + 1\right) d\xi$$

if $2m = N$. Obviously,

$$\int_{\mathcal{G}_x^{(1)}} d\xi \le c\,\rho_1(x)^N.$$

Since $|x-\xi| < \rho_1(x) + \rho_1(\xi) < 3\rho_1(x)$ in $G_x^{(3)}$, we furthermore obtain

$$\int_{\mathcal{G}_x^{(3)}} |x-\xi|^{2m-N}\, d\xi \le \int\limits_{|x-\xi|<3\rho_1(x)} |x-\xi|^{2m-N}\, d\xi = c\,\rho_1(x)^{2m}$$

and

$$\int_{\mathcal{G}_x^{(3)}} \Big|\log\frac{|x-\xi|}{\rho_1(x)}\Big|\, d\xi \le \int\limits_{|x-\xi|<3\rho_1(x)} \Big|\log\frac{|x-\xi|}{\rho_1(x)}\Big| = \rho_1(x)^N \int\limits_{|\eta|<3} \big|\log|\eta|\,\big|\, d\eta.$$

Consequently, the inequality

$$I_3(x) \le c\,\rho_1(x)^{m-1-\beta_1}\, \|f\|_{V^{0,\infty}_{\beta+m+1}(\mathcal{G})}$$

holds both for $2m \ne N$ and $2m = N$. Thus, (5.3.9) is valid for all j. This proves the lemma. □

We prove an analogous estimate for the second integral on the right-hand side of (5.3.6).

LEMMA 5.3.5. *Suppose that $g_k \in V^{0,\infty}_{\beta-m+k}(\partial\mathcal{G})^\ell$ and that the components of β satisfy the inequalities* (5.3.3). *Then the function*

$$v(x) = \int_{\partial\mathcal{G}\backslash\mathcal{S}} g_k(\xi)\cdot \overline{T_k(\xi,D_\xi)\, H^{(i)}(x,\xi)}\, d\xi,$$

satisfies the estimate

(5.3.10) $$\|v\|_{V^{0,\infty}_{\beta-m+1}(\mathcal{G})^\ell} \le c\,\|g_k\|_{V^{0,\infty}_{\beta-m+k}(\partial\mathcal{G})^\ell}$$

with a constant c independent of g_k.

P r o o f. Let x be a point in $\mathcal{G}$, and let $x^{(1)}$ be the nearest vertex to x. Furthermore, let $\mathcal{U}_1, \dots, \mathcal{U}_{d'}$ be the same neighborhoods of the vertices $x^{(1)}, \dots, x^{(d')}$, respectively, as in the proof of Lemma 5.3.4. Then

$$|v(x)| \le \sum_{j=1}^{d'} \int_{\partial\mathcal{G}\cap\mathcal{U}_j} \big|g_k(\xi)\cdot T_k(\xi, D_\xi)\, H^{(i)}(x,\xi)\big|\, d\xi.$$

If $j \neq 1$, then it follows from item 1) of Theorem 5.3.2 that

$$\begin{aligned}\int_{\partial\mathcal{G}\cap\mathcal{U}_j} & \big|g_k(\xi)\cdot T_k(\xi, D_\xi)\, H^{(i)}(x,\xi)\big|\, d\xi \\ &\le c\, \|g_k\|_{V^{0,\infty}_{\beta-m+k}(\mathcal{G})^\ell} \int\limits_{\partial\mathcal{G}\cap\mathcal{U}_j} \rho_j(\xi)^{-\beta_j+m-k}\, \big|H^{(i)}(x,\xi)\big|\, d\xi \\ &\le c\, \|g_k\|_{V^{0,\infty}_{\beta-m+k}(\partial\mathcal{G})^\ell}\, \rho_1(x)^{\Lambda_1^+-\varepsilon} \int\limits_{\partial\mathcal{G}\cap\mathcal{U}_j} \rho_j(\xi)^{m-\Lambda_j^- -\beta_j - N-\varepsilon}\, d\xi.\end{aligned}$$

Thus, by the conditions of the lemma on β,

$$\int\limits_{\partial\mathcal{G}\cap\mathcal{U}_j} \big|g_k(\xi)\cdot T_k(\xi, D_\xi)\, H^{(i)}(x,\xi)\big|\, d\xi \le c\, \rho_1(x)^{m-1-\beta_1}\, \|g_k\|_{V^{0,\infty}_{\beta-m+k}(\partial\mathcal{G})^\ell} \tag{5.3.11}$$

for $j \neq 1$. In order to prove this estimate for $j = 1$, we introduce the subsets

$$\begin{aligned} A_1 &= \{\xi \in \partial\mathcal{G}\cap\mathcal{U}_1 :\ \rho_1(\xi) < \rho_1(x)/2\}, \\ A_2 &= \{\xi \in \partial\mathcal{G}\cap\mathcal{U}_1 :\ \rho_1(\xi) > 2\rho_1(x)\}, \\ A_3 &= \{\xi \in \partial\mathcal{G}\cap\mathcal{U}_1 :\ \rho_1(x)/2 < \rho_1(\xi) < 2\rho_1(x)\} \end{aligned}$$

and consider the integrals

$$I_\nu(x) = \int_{A_\nu} \big|g_k(\xi)\cdot T_k(\xi, D_\xi)\, H^{(i)}(x,\xi)\big|\, d\xi.$$

For $\nu = 1$ and $\nu = 2$ the estimate

$$I_\nu(x) \le c\, \rho_1(x)^{m-1-\beta_1}\, \|g_k\|_{V^{0,\infty}_{\beta-m+k}(\partial\mathcal{G})^\ell} \tag{5.3.12}$$

holds analogously to the estimate of $I_\nu(x)$ in the proof of Lemma 5.3.4. Obviously,

$$I_3(x) \le c\, \rho_1(x)^{m-k-\beta}\, \|g_k\|_{V^{0,\infty}_{\beta-m+k}(\partial\mathcal{G})^\ell} \int_{A_3} \big|T_k(\xi, D_\xi)\, H^{(i)}(x,\xi)\big|\, d\xi.$$

By item 2) of Theorem 5.3.2, the integrand on the right-hand side satisfies the inequalities

$$\big|T_k(\xi, D_\xi)\, H^{(i)}(x,\xi)\big| \le c\,\big(|x-\xi|^{k-N} + \rho_1(x)^{k-N}\big) \quad \text{for } \xi \in A_3,\ k \neq N$$

and

$$\big|T_k(\xi, D_\xi)\, H^{(i)}(x,\xi)\big| \le c\,\Big(\Big|\log\frac{|x-\xi|}{\rho_1(x)}\Big| + 1\Big) \quad \text{for } \xi \in A_3,\ k = N.$$

Using the inequalities

$$\int_{A_3} \big(|x-\xi|^{k-N} + \rho_1(x)^{k-N}\big)\, d\xi \le c\, \rho_1(x)^{k-1} \quad \text{for } k > 1$$

and

$$\int_{A_3} \Big(\Big|\log\frac{|x-\xi|}{\rho_1(x)}\Big| + 1\Big)\, d\xi \le c\, \rho_1(x)^{N-1},$$

we obtain

$$I_3(x) \le c\, \rho_1(x)^{m-1-\beta_1} \, \|g_k\|_{V^{0,\infty}_{\beta-m+k}(\partial\mathcal{G})^\ell}$$

for $k > 1$. In the case $k = 1$, the estimates

$$\big|T_1(\xi, D_\xi)\, H^{(i)}(x,\xi)\big| \le c\,\big(d(x)\,|x-\xi|^{-N} + \rho_1(x)^{1-N}\big) \quad \text{for } \xi \in A_3$$

(cf. Corollary 5.3.3) and

$$\int_{\partial\mathcal{G}} |x-\xi|^{-N}\, d\xi \le c\, d(x)^{-1}$$

lead to the same result. Thus, the inequality (5.3.12) is satisfied for $\nu = 3$ and all k. This proves the lemma. □

As a direct consequence of Lemmas 5.3.4 and 5.3.5, we obtain the following result.

COROLLARY 5.3.6. *Suppose that* $f \in \mathcal{V}^{0,\infty}_{\beta+m+1}(\mathcal{G})$, $g \in \mathcal{V}^{m-1,\infty}_{\beta}(\partial\mathcal{G})$, *and that the components of* β *satisfy the inequalities* (5.3.3). *Then the generalized solution of the boundary value problem* (5.3.7), (5.3.8) *satisfies the estimate*

$$\|u\|_{V^{0,\infty}_{\beta-m+1}(\mathcal{G})} \le c\,\Big(\|f\|_{V^{0,\infty}_{\beta+m+1}(\mathcal{G})} + \sum_{k=1}^{m} \|g_k\|_{V^{m-k,\infty}_{\beta}(\partial\mathcal{G})}\Big).$$

5.3.4. Weighted L_∞ estimates for the derivatives of the solutions. Using the local maximum principle in Subsection 5.2.4, we obtain the following lemma.

LEMMA 5.3.7. *Let* $u \in C^\infty(\overline{\mathcal{G}}\backslash\mathcal{S})^\ell \cap V^{0,\infty}_{\beta-m+1}(\mathcal{G})^\ell$ *be such that*

$$L(x, D_x)\, u \in V^{0,\infty}_{\beta+m+1}(\mathcal{G})^\ell \quad \textit{and} \quad \frac{\partial^{k-1} u}{\partial n^{k-1}}\Big|_{\partial\mathcal{G}\backslash\mathcal{S}} \in V^{m-k,\infty}_{\beta}(\partial\mathcal{G})^\ell$$

for $k = 1, \ldots, m$. *Then* $u \in V^{m-1,\infty}_{\beta}(\mathcal{G})^\ell$ *and*

$$\begin{aligned} \|u\|_{V^{m-1,\infty}_{\beta}(\mathcal{G})^\ell} \ &\le\ c\,\Big(\|L(x,D_x)u\|_{V^{0,\infty}_{\beta+m+1}(\mathcal{G})^\ell} + \sum_{k=1}^{m} \|\partial_n^{k-1} u\|_{V^{m-k,\infty}_{\beta,\delta}(\partial\mathcal{G})^\ell} \\ &\qquad + \|u\|_{V^{0,\infty}_{\beta-m+1}(\mathcal{G})^\ell}\Big). \end{aligned}$$

P r o o f. Let y be a point in a sufficiently small neighborhood of the vertex $x^{(1)}$. Without loss of generality, we may assume that $x^{(1)}$ is the origin and that $\mathcal{G}$ coincides with a cone $\mathcal{K}$ in a neighborhood of $x^{(1)}$. We set $z = y/|y|$. Furthermore, let η be an infinitely differentiable function equal to one near z such that $\eta(x) = 0$ for $|x - z| > 1/2$ and $|\partial_x^\alpha \eta(x)| < c_\alpha$ for all α, where c_α are constants which do not depend on x and y. Then the function $\zeta(x) = \eta(x/|y|)$ is equal to one near y and vanishes for $|x - y| > |y|/2$. We introduce the operator

$$L_t(x, D_x) = t^{2m} L(tx, t^{-1}D_x) = \sum_{|\alpha|\le 2m} A_\alpha(tx)\, t^{2m-|\alpha|}\, D_x^\alpha,$$

where $t = |y|$. Obviously, the coefficients of the operator $L_t(x, D_x) - L^\circ(0, D_x)$ and their derivatives are small for small $t = |y|$ (here $L^\circ(0, D_x)$ denotes the principal part of $L(x,D_x)$ with coefficients frozen at the origin). By Lemma 5.2.5, the vector function $v(x)=u(tx)$ satisfies the estimate

$$|\partial^\alpha v(z)| \le c\Big(\|\eta L_t(x,D_x)\,v\|_{L_\infty(\mathcal{K})^\ell} + \sum_{k=1}^{m} \|\eta\partial_n^{k-1} v\|_{C^{m-k}(\partial\mathcal{K})^\ell} + \|\eta v\|_{L_\infty(\mathcal{K})^\ell}\Big), \tag{5.3.13}$$

where c is independent of u and y. Here,

$$\partial^\alpha v(z) = t^{|\alpha|}\,\partial^\alpha u(y) \quad\text{and}\quad \|\eta L_t(x,D_x)\,v\|_{L_\infty(\mathcal{K})^\ell} = t^{2m}\,\|\zeta L(x,D_x)\,u\|_{L_\infty(\mathcal{K})^\ell}\,.$$

Since $\zeta(x)=0$ for $|x|<t/2$ and $|x|>2t$, we conclude that

$$\|\eta\partial_n^{k-1} v\|_{C^{m-k}(\partial\mathcal{K})^\ell} \le c\,t^{m-1-\beta_1}\|\zeta\partial_n^{k-1} u\|_{V_{\beta_1}^{m-k,\infty}(\partial\mathcal{K})^\ell}\,.$$

Multiplying (5.3.13) by $|y|^{\beta_1-m+1}$ and using the fact that $|y|/2<|x|<2|y|$ for $x\in\operatorname{supp}\zeta$, we consequently obtain

$$\rho_1(y)^{\beta_1-m+1+|\alpha|}\,|\partial^\alpha u(y)| \le c\Big(\|\zeta\,L(x,D_x)\,u\|_{V^{0,\infty}_{\beta+m+1}(\mathcal{G})^\ell} + \sum_{k=1}^{m}\|\zeta\partial_n^{k-1}u\|_{V_\beta^{m-k,\infty}(\partial\mathcal{G})^\ell} + \|\zeta u\|_{V^{0,\infty}_{\beta-m+1}(\mathcal{G})^\ell}\Big).$$

In the case where y lies outside a neighborhood of a vertex, this estimate follows immediately from Lemma 5.2.5. This proves the lemma. □

Using the last lemma and Corollary 5.3.6, we get a weighted maximum principle for generalized solutions.

THEOREM 5.3.8. *Suppose that $f\in\mathcal{V}^{0,\infty}_{\beta+m+1}(\mathcal{G})$, $g\in\mathcal{V}_\beta^{m-1,\infty}(\partial\mathcal{G})$, and that the components of β satisfy the inequalities* (5.3.3). *Then the generalized solution of the boundary value problem* (5.3.7), (5.3.8) *satisfies the estimate*

$$\|u\|_{V_\beta^{m-1,\infty}(\mathcal{G})^\ell} \le c\Big(\|f\|_{V^{0,\infty}_{\beta+m+1}(\mathcal{G})^\ell} + \sum_{k=1}^{m}\|g_k\|_{V_\beta^{m-k,\infty}(\partial\mathcal{G})^\ell}\Big). \tag{5.3.14}$$

P r o o f. Let $\{f^{(\nu)}\}$ and $\{g^{(\nu)}\}=\{(g_1^{(\nu)},\dots,g_m^{(\nu)}\}$ be approximating sequences for f and g in $\mathcal{V}^{0,\infty}_{\beta+m+1}(\mathcal{G})^\ell$ and $\mathcal{V}_\beta^{m-1,\infty}(\partial\mathcal{G})$, respectively. Furthermore, let $u^{(\nu)}$ be the vector function with the components (5.3.5). This vector function is infinitely differentiable on $\overline{\mathcal{G}}\backslash\mathcal{S}$. Moreover, according to Corollary 5.3.6 and Lemma 5.3.7, it satisfies the estimate

$$\|u^{(\nu)}\|_{V_\beta^{m-1,\infty}(\mathcal{G})^\ell} \le c\Big(\|f^{(\nu)}\|_{V^{0,\infty}_{\beta+m+1}(\mathcal{G})^\ell} + \sum_{k=1}^{m}\|g_k^{(\nu)}\|_{V_{\beta,\delta}^{m-k,\infty}(\partial\mathcal{G})^\ell}\Big)$$

with a constant c independent of f, g and ν. We denote by $u(x)$ the limit of the sequence $\{u^{(\nu)}(x)\}$, $x\in\mathcal{G}$. Then u satisfies (5.3.14). □

5.3.5. The Miranda-Agmon maximum principle. If $\Lambda_j^+>m-1$, i.e. if the strip $m-N/2<\operatorname{Re}\lambda\le m-1$ does not contain eigenvalues of the pencil $\mathfrak{A}_j(\lambda)$, then the condition (5.3.3) in Theorem 5.3.8 allows us to put $\beta_j=0$, and we obtain a $W^{m-1,\infty}$ estimate for the generalized solutions. However, the assumption $g\in\mathcal{V}_0^{m-1,\infty}(\partial\mathcal{G})$ means that the derivatives of g_k up to order $m-k-1$ vanish at the vertices.

Let g_k be arbitrary functions in $C^{m-k}(\partial\mathcal{G}\backslash\mathcal{S})$, $k = 1, \dots, m$. We say that $g = (g_1, \dots, g_m)$ belongs to $\mathfrak{C}^{m-1}(\partial\mathcal{G})$ if there exists a function $v \in C^{m-1}(\overline{\mathcal{G}})$ such that

$$\frac{\partial^{k-1} v}{\partial n^{k-1}} = g_k \text{ on } \partial\mathcal{G}\backslash\mathcal{S}$$

for $k = 1, \dots, m$.

THEOREM 5.3.9. *Suppose that* $\Lambda_j^+ > m-1$ *for* $j = 1, \dots, d'$ *and that* $g = (g_1, \dots, g_m) \in \mathfrak{C}^{m-1}(\partial\mathcal{G})^\ell$. *Then there exists a solution of the problem*

$$(5.3.15) \qquad L(x, D_x)\, u = 0 \ \text{ in } \mathcal{G}, \quad \frac{\partial^{k-1} u}{\partial n^{k-1}} = g_k \ \text{ on } \partial\mathcal{G}\backslash\mathcal{S}, \ k = 1, \dots, m,$$

satisfying the estimate

$$(5.3.16) \qquad \|u\|_{W^{m-1,\infty}(\mathcal{G})^\ell} \le c \sum_{k=1}^m \|g_k\|_{W^{m-k,\infty}(\partial\mathcal{G})^\ell}$$

with a constant c *independent of* g.

P r o o f. Since $g \in \mathfrak{C}^{m-1}(\partial\mathcal{G})^\ell$, there exists a vector function $W \in C^{m-1}(\overline{\mathcal{G}})^\ell$ such that $\partial_n^{k-1} W = g_k$ on $\partial\mathcal{G}\backslash\mathcal{S}$ for $k = 1, \dots, m$. We introduce the vector function

$$w(x) = \sum_{j=1}^{d'} \chi_j(x) \sum_{|\alpha| \le m-2} \frac{1}{\alpha!}\, (\partial_x^\alpha W)(x^{(j)})\, (x - x^{(j)})^\alpha,$$

where χ_j are infinitely differentiable cut-off functions equal to one near $x^{(j)}$ and to zero near the other vertices of $\mathcal{G}$. Obviously, w is infinitely differentiable on $\overline{\mathcal{G}}$, and the derivatives of w can be estimated by the right-hand side of (5.3.16). In particular,

$$(5.3.17) \qquad \|w\|_{W^{m-1,\infty}(\mathcal{G})^\ell} + \|L(x, D_x)\, w\|_{V^{0,\infty}_{\beta+m+1}(\mathcal{G})^\ell} \le c \sum_{k=1}^m \|g_k\|_{W^{m-k,\infty}(\partial\mathcal{G})^\ell}\,.$$

Let ψ_k be defined as

$$\psi_k = g_k - \partial_n^{k-1} w\big|_{\partial\mathcal{G}\backslash\mathcal{S}} \quad \text{for } k = 1, \dots, m.$$

Then $\psi = (\psi_1, \dots, \psi_m) \in V_0^{m-1,\infty}(\partial\mathcal{G})$. By Theorem 5.3.8, there exists a generalized solution v of the problem

$$L(x, D_x)\, v = -L(x, D_x)\, w \ \text{ in } \mathcal{G}, \quad \frac{\partial^{k-1} v}{\partial n^{k-1}} = \psi_k \ \text{ on } \partial\mathcal{G}\backslash\mathcal{S}, \ k = 1, \dots, m,$$

satisfying the estimate

$$\|v\|_{V_0^{m-1,\infty}(\mathcal{G})^\ell} \le c \Big(\|L(x, D_x)\, w\|_{V^{0,\infty}_{\beta+m+1}(\mathcal{G})^\ell} + \sum_{k=1}^m \|\psi_k\|_{V_0^{m-k,\infty}(\partial\mathcal{G})^\ell} \Big).$$

Then $u = v + w$ is a solution of the problem (5.3.15). Since ψ_k is zero at the vertices together with the derivatives up to order $m-k-1$, there exist constants c and c' such that

$$\|\psi_k\|_{V_0^{m-k,\infty}(\partial\mathcal{G})^\ell} \le c\, \|\psi_k\|_{W^{m-k,\infty}(\partial\mathcal{G})^\ell} \le c'\, \|g_k\|_{W^{m-k,\infty}(\partial\mathcal{G})^\ell}\,.$$

This together with the estimates

$$\|u\|_{W^{m-1,\infty}(\mathcal{G})^\ell} \le c \left(\|v\|_{V_0^{m-1,\infty}(\mathcal{G})^\ell} + \|w\|_{W^{m-1,\infty}(\mathcal{G})^\ell} \right)$$

and (5.3.17) implies (5.3.16). The theorem is proved. □

5.4. The Miranda-Agmon maximum principle for smooth solutions in a domain with a conical point

In the previous section, we established a condition on the eigenvalues of the pencils $\mathfrak{A}_j(\lambda)$ which guarantees the validity of the estimate (5.3.16) for generalized solutions of the problem (5.3.15). Now, we consider only smooth solutions of the differential equation $Lu = 0$ in $\mathcal{G}$. We are interested in conditions which ensure the estimate (5.3.1) for C^∞-solutions of the equation $Lu = 0$.

Again we suppose that the equation $L(x, D_x)\, u = f$ is uniquely solvable in $\overset{\circ}{W}{}^{m,2}(\mathcal{G})^\ell$ for arbitrary $f \in W^{-m,2}(\mathcal{G})^\ell$. For simplicity, we assume that the boundary $\partial\mathcal{G}$ contains only one conical point at the origin and that $\partial\mathcal{G}\backslash\{0\}$ is smooth. Furthermore, we suppose that

$$L(D_x) = \sum_{|\alpha|=2m} a_\alpha\, D_x^\alpha$$

is a (scalar) homogenous strongly elliptic differential operator with constant coefficients a_α and that the domain $\mathcal{G}$ coincides with the cone

$$\mathcal{K} = \{x \in \mathbb{R}^N : x/|x| \in \Omega\}$$

in a neighborhood of the origin, where Ω is a subdomain of the unit sphere S^{N-1} with smooth boundary $\partial\Omega$. By $\mathfrak{A}(\lambda)$, we denote the operator

$$\overset{\circ}{W}{}^{m,2}(\Omega)^\ell \ni U \to \mathfrak{A}(\lambda)\, U = \rho^{2m-\lambda}\, L(D_x)\, \big(\rho^\lambda U(\omega)\big) \in W^{-m,2}(\Omega)^\ell$$

where $\rho = |x|$, $\omega = x/|x|$. We prove in this section that the estimate (5.3.1) is valid for all smooth solutions of the equation

$$L(D_x)\, u = 0 \quad \text{in } \mathcal{G} \tag{5.4.1}$$

if and only if the line $\operatorname{Re}\lambda = m - 1$ is free of eigenvalues of the pencil $\mathfrak{A}(\lambda)$. In Subsection 5.4.1, we prove that the last condition on the eigenvalues is sufficient for the estimate (5.3.1), the necessity of this condition is proved in Subsection 5.4.2.

5.4.1. A C^{m-1} estimate for smooth solutions of strongly elliptic equations. Let $\lambda_1, \lambda_2, \dots$ be all eigenvalues of the operator pencil $\mathfrak{A}(\lambda)$ with real part $> m - N/2$ enumerated in the order of increasing real parts. The first eigenvalue with real part $> m-1$ is denoted by $\lambda_{\mu+1}$. This means that

$$m - N/2 < \operatorname{Re}\lambda_1 \le \dots \le \operatorname{Re}\lambda_\mu \le m - 1 < \operatorname{Re}\lambda_{\mu+1}\,.$$

Furthermore, let $\mathfrak{A}^+(\lambda)$ be the operator pencil generated by the Dirichlet problem for the formally adjoint equation

$$L^+(D_x)\, u = \sum_{|\alpha|=2m} \overline{a}_\alpha\, D_x^\alpha = f \quad \text{in } \mathcal{G},$$

i.e. $\mathfrak{A}^+(\lambda)\, U(\omega) = \rho^{2m-\lambda}\, L^+(D_x)\, \big(\rho^\lambda U(\omega)\big)$. By [**85**, Lemma 10.1.2], the operator $\mathfrak{A}^+(\lambda)$ coincides with the adjoint operator of $\mathfrak{A}(2m - N - \overline{\lambda})$. Consequently, the numbers $2m - N - \overline{\lambda}_j$ are eigenvalues of the pencil $\mathfrak{A}^+(\lambda)$ for $j = 1, 2, \dots$ (cf. [**85**, Theorem 1.1.7]). In the sequel, let

$$\phi_j^{(0,\sigma)}, \dots, \phi_j^{(\kappa_{\sigma j}-1,\sigma)}, \quad \sigma = 1, \dots, \chi_j,$$

be a canonical system of Jordan chains of the pencil $\mathfrak{A}(\lambda)$ corresponding to the eigenvalue λ_j, and let

$$\psi_j^{(0,\sigma)}, \dots, \psi_j^{(\kappa_{\sigma j}-1,\sigma)}, \quad \sigma = 1, \dots, \chi_j,$$

be Jordan chains of the pencil $\mathfrak{A}^+(\lambda)$ corresponding to the eigenvalue $2m - N - \overline{\lambda}_j$ satisfying the biorthogonality condition (cf. [**81**, Theorem 1.2.1])

$$\sum_{s=0}^{\kappa_{\sigma j}-1} \sum_{t=1}^{q+1} \frac{1}{(s+t)!} \Big(\frac{\partial^{s+t}\mathfrak{A}}{\partial \lambda^{s+t}}(\lambda_j)\, \phi_j^{(\kappa_{\sigma j}-1,\sigma)}\, ,\, \psi_j^{(q+1-t,\zeta)} \Big)_\Omega = \delta_{\sigma,\zeta}\, \delta_{0,q} \tag{5.4.2}$$

for $q = 0, 1, \dots, \kappa_{\sigma j} - 1$. For the following lemma, we refer to [**121**, Theorem 2.2].

LEMMA 5.4.1. *Let $G(x, \xi)$ be the Green's function of the boundary value problem* (5.3.7), (5.3.8) *(cf. Theorem* 5.3.1*). Then $G(x, \xi)$ admits the representation*

$$G(x, \xi) = A(x, \xi) + R(x, \xi)$$

for $|x| < |\xi|/2$, where

$$\begin{aligned} A(x, \xi) \quad = \quad & -\sum_{j=1}^{\mu} |x|^{\lambda_j}\, |\xi|^{2m-N-\lambda_j} \sum_{\sigma=1}^{\chi_j} \sum_{s=0}^{\kappa_{\sigma j}-1} \frac{1}{(\kappa_{\sigma j} - s - 1)!} \\ & \times \Big(\log \frac{|x|}{|\xi|} \Big)^{\kappa_{\sigma j} - s - 1} \sum_{t=0}^{s} \phi_j^{(t,\sigma)}(\omega_x)\, \overline{\psi_j^{(s-t,\sigma)}(\omega_\xi)} \end{aligned} \tag{5.4.3}$$

and

$$\big| \partial_x^\alpha \partial_\xi^\gamma R(x, \xi) \big| \le c\, |x|^{\mathrm{Re}\lambda_{\mu+1} - |\alpha| - \varepsilon}\, |\xi|^{2m-N-\mathrm{Re}\lambda_{\mu+1} - |\gamma| + \varepsilon}. \tag{5.4.4}$$

Here, ε is an arbitrarily small positive number.

The asymptotic decomposition of Green's function is used in the proof of the next theorem.

THEOREM 5.4.2. *Let $u \in C^\infty(\overline{\mathcal{G}})$ be a solution of the equation* (5.4.1). *If the line $\mathrm{Re}\,\lambda = m - 1$ is free of eigenvalues of the pencil $\mathfrak{A}(\lambda)$, then u satisfies the estimate* (5.3.1). *Here, the constant c is independent of u.*

P r o o f. We can assume that $(\partial_x^\alpha u)(0) = 0$ for every multi-index α, $|\alpha| = m-2$. Otherwise, we consider the function

$$v = u - \sum_{|\alpha| \le m-2} \frac{1}{\alpha!}\, (\partial_x^\alpha u)(0)\, x^\alpha$$

instead of u. If $(\partial_x^\alpha u)(0) = 0$ for $|\alpha| = m - 2$, then

$$\|\rho^{-m+1} u\|_{L_\infty(\mathcal{G})} < \infty, \qquad \|\rho^{-m+k} g_k\|_{L_\infty(\partial\mathcal{G})} \le c\, \|g_k\|_{C^{m-k}(\partial\mathcal{G}\setminus\{0\})}$$

and u admits the representation

$$u(x) = \sum_{k=1}^{m} \int_{\partial\mathcal{G}\setminus\{0\}} g_k(\xi)\, \overline{T_k(\xi, D_\xi)\, \overline{G(x, \xi)}}\, d\xi,$$

where T_k are the differential operators in the Green's formula (5.3.4). We introduce the subsets

$$A_1 = \{\xi \in \partial\mathcal{G}\backslash\{0\} : \ |\xi| < |x|/2\},$$
$$A_2 = \{\xi \in \partial\mathcal{G} : \ |\xi| > 2|x|\},$$
$$A_3 = \{\xi \in \partial\mathcal{G} : \ |x|/2 < |\xi| < 2|x|\}$$

and consider the integrals

$$I_\nu(x) = \int_{A_\nu} g_k(\xi)\, \overline{T_k(\xi, D_\xi)\, \overline{G(x,\xi)}}\, d\xi$$

for $\nu = 1, 2, 3$. Using item 3) of Theorem 5.3.2, we obtain

$$\big|I_1(x)\big| \le c\,\|\rho^{-m+k} g_k\|_{L_\infty(\partial\mathcal{G})}\ |x|^{\Lambda_- +\varepsilon} \int_{A_1} |\xi|^{m-N-\Lambda_- -\varepsilon}\, d\xi,$$

where $\Lambda_- < m - N/2$. This implies

$$\big|I_1(x)\big| \le c\,|x|^{m-1}\, \|\rho^{-m+k} g_k\|_{L_\infty(\partial\mathcal{G})}\,.$$

The estimate

$$\big|I_3(x)\big| \le c\,|x|^{m-1}\, \|\rho^{-m+k} g_k\|_{L_\infty(\partial\mathcal{G})}$$

can be obtained in the same way as in the proof of Lemma 5.3.5 for $\beta = 0$.

We prove the same estimate for $I_2(x)$. Obviously,

$$\begin{aligned} I_2(x) \ &= \ \int_{A_2} g_k(\xi)\, \overline{T_k(\xi, D_\xi)\, \overline{R(x,\xi)}}\, d\xi - \int_{\partial\mathcal{G}\backslash A_2} g_k(\xi)\, \overline{T_k(\xi, D_\xi)\, \overline{A(x,\xi)}}\, d\xi \\ &\quad + \int_{\partial\mathcal{G}} g_k(\xi)\, \overline{T_k(\xi, D_\xi)\, \overline{A(x,\xi)}}\, d\xi, \end{aligned}$$

where $A(x,\xi)$ and $R(x,\xi)$ are the functions introduced in Lemma 5.4.1. Here

$$\begin{aligned} \Big|\int_{A_2} g_k(\xi)\, \overline{T_k(\xi, D_\xi)\, \overline{R(x,\xi)}}\, d\xi\Big| \ &\le \ c\,\|\rho^{-m+k} g_k\|_{L_\infty(\partial\mathcal{G})}\ |x|^{\mathrm{Re}\lambda_{\mu+1}-\varepsilon} \\ &\qquad \times \int_{A_2} |\xi|^{m-N-\mathrm{Re}\lambda_{\mu+1}-\varepsilon}\, d\xi, \end{aligned}$$

where $m - N - \mathrm{Re}\,\lambda_{\mu+1} < N - 1$. Thus,

$$\Big|\int_{A_2} g_k(\xi)\, \overline{T_k(\xi, D_\xi)\, \overline{R(x,\xi)}}\, d\xi\Big| \le c\,|x|^{m-1}\, \|\rho^{-m+k} g_k\|_{L_\infty(\partial\mathcal{G})}\,.$$

Since the integral of $g_k(\xi)\, \overline{T_k(\xi, D_\xi)\, \overline{A(x,\xi)}}$ over the set $\partial\mathcal{G}\backslash A_2$ is a finite sum of terms of the form

$$|x|^{\lambda_j} \phi(\omega_x) \int_{\partial\mathcal{G}\backslash A_2} g_k(\xi)\, |\xi|^{k-N-\lambda_j} \Big(\log \frac{|x|}{|\xi|}\Big)^s\, \psi(\omega_\xi)\, d\xi,$$

where $m - N/2 < \mathrm{Re}\lambda_j < m - 1$, we obtain analogously

$$\Big|\int_{\partial\mathcal{G}\backslash A_2} g_k(\xi)\, \overline{T_k(\xi, D_\xi)\, \overline{A(x,\xi)}}\, d\xi\Big| \le c\,|x|^{m-1}\, \|\rho^{-m+k} g_k\|_{L_\infty(\partial\mathcal{G})}\,.$$

The expression $A(x,\xi)$ can be written as

$$A(x,\xi) = \sum_{j=1}^{k} \sum_{\sigma=1}^{\chi_j} \sum_{s=0}^{\kappa_{\sigma,j}-1} u_{j,\sigma,s}(x)\, \overline{v_{j,\sigma,\kappa_{\sigma,j}-1-s}(\xi)}\,,$$

where

$$u_{j,\sigma,s}(x) = |x|^{\lambda_j} \sum_{t=0}^{s} \frac{1}{t!} (\log|x|)^t \, \phi_j^{s-t,\sigma}(\omega_x)$$

and

$$v_{j,\sigma,s}(\xi) = -|\xi|^{2m-N-\bar{\lambda}_j} \sum_{t=0}^{s} \frac{1}{t!} (-\log|\xi|)^t \, \psi_j^{s-t,\sigma}(\omega_\xi).$$

Hence,

$$\int_{\partial\mathcal{G}} g_k(\xi)\, \overline{T_k(\xi, D_\xi)\, \overline{A(x,\xi)}}\, d\xi = \sum_{j=1}^{\mu} \sum_{\sigma=1}^{\chi_j} \sum_{s=0}^{\kappa_{\sigma,j}-1} c_{j,\sigma,s}\, u_{j,\sigma,s}(x),$$

where $c_{j,\sigma,s}$ are constants. Consequently, the solution u admits the representation

$$u = \sum_{j=1}^{\mu} \sum_{\sigma=1}^{\chi_j} \sum_{s=0}^{\kappa_{\sigma,j}-1} c_{j,\sigma,s}\, u_{j,\sigma,s}(x) + v(x), \tag{5.4.5}$$

where

$$|v(x)| \le c\, |x|^{m-1} \sum_{k=1}^{m} \|\rho^{-m+k} g_k\|_{L_\infty(\partial\mathcal{G})}.$$

Since $\rho^{-m+1} u \in L_\infty(\mathcal{G})$ and $\operatorname{Re}\lambda_j < m-1$ for $j = 1, \dots, \mu$, the coefficients $c_{j,\sigma,s}$ in (5.4.5) must be equal to zero. Applying Lemma 5.3.7 with $\beta = 0$, we get (5.3.1). □

5.4.2. Necessity of the condition on the eigenvalues. Our goal is to show that the absence of eigenvalues of the pencil $\mathfrak{A}(\lambda)$ on the line $\operatorname{Re}\lambda = m-1$ is a necessary condition for the validity of the estimate (5.3.1). For this purpose, we prove the subsequent three lemmas.

LEMMA 5.4.3. *Suppose that every solution $u \in C^\infty(\overline{\mathcal{G}})$ of the equation* (5.4.1) *satisfies the inequality* (5.3.1). *If* $\operatorname{supp} f \subset \{x : \delta < |x| < 2\delta\}$ *and $v \in C^\infty(\overline{\mathcal{G}})$ is a solution of the boundary value problem*

$$L(D_x)\, v = f \quad \text{in } \mathcal{G}, \qquad \frac{\partial^{k-1} v}{\partial n^{k-1}} = g_k \quad \text{on } \partial\mathcal{G}\backslash\{0\},\ k = 1, \dots, m,$$

satisfying the condition $(\partial_x^\alpha v)(0) = 0$ for $|\alpha| \le m-2$, then

$$\|\rho^{-m+1} v\|_{L_\infty(\mathcal{G})} \le c \Big(\sum_{k=1}^{m} \|g_k\|_{W^{m-k,\infty}(\partial\mathcal{G})} + \delta^{m+1}\, \|f\|_{L_\infty(\mathcal{G})} \Big) \tag{5.4.6}$$

with a constant c independent of f and δ.

Proof. Since the function $x \to f(\delta x)$ vanishes for $|x| < 1$ and $|x| > 2$, there exists a function $u \in C_0^\infty(\mathbb{R}^N)$, $u = 0$ near the origin, such that

$$L(D_x)\, u(x) = \delta^{2m} f(\delta x) \quad \text{and} \quad \|u\|_{V_0^{m-1,\infty}(\mathbb{R}^N)} \le c\, \delta^{2m}\, \|f\|_{L_\infty(\mathcal{G})},$$

where

$$\|u\|_{V_0^{m-1,\infty}(\mathbb{R}^N)} = \sum_{|\alpha| \le m-1} \sup_x |x|^{|\alpha|-m+1} \big|\partial_x^\alpha u(x)\big|.$$

Then the function $w(x) = u(\delta^{-1}x)$ satisfies the equation $L(D_x)\, w = f$ and the estimate

$$\|w\|_{V_0^{m-1,\infty}(\mathbb{R}^N)} \le c\, \delta^{m+1}\, \|f\|_{L_\infty(\mathcal{G})}.$$

Furthermore, by the assumption of the lemma, the inequality

$$\|v - w\|_{W^{m-1,\infty}(\mathcal{G})} \le c \sum_{k=1}^{m} \|g_k - \partial_n^{k-1} w\|_{W^{m-k,\infty}(\partial\mathcal{G})}$$

holds. Using the equivalence of the $W^{m-1,\infty}$- and $V_0^{m-1,\infty}$-norms on the set of all functions v such that $(\partial_x^\alpha v)(0) = 0$ for $|\alpha| \le m-2$, we obtain (5.4.6). □

LEMMA 5.4.4. *Let z_j and $a_{j,s}$ be given complex numbers, $j = 1, \dots, \mu$, $s = 0, 1, \dots, \kappa_j - 1$. Then there exists a polynomial $p(t)$ of degree $k = \kappa_1 + \dots + \kappa_\mu - 1$ such that*

$$\int_{-\infty}^{+\infty} t^s\, p(t)\, e^{-t^2/2}\, e^{-iz_j t}\, dt = a_{j,s}$$

for $j = 1, \dots, \mu$, $s = 0, 1, \dots, \kappa_j - 1$. The coefficients of $p(t)$ depend linearly on the numbers $a_{j,s}$.

P r o o f. The polynomial

$$p(t) = \sum_{\nu=0}^{k} c_\nu\, t^\nu$$

satisfies the conditions of the lemma if its coefficients c_ν are solutions of the algebraic system

$$(5.4.7)\quad \int_{-\infty}^{+\infty} t^s\, p(t)\, e^{-t^2/2}\, e^{-iz_j t}\, dt = \frac{d^s}{d\lambda^s}\Big(\sum_{\nu=0}^{k} c_\nu\, \mathrm{He}_\nu(\lambda)\, e^{-\lambda^2/2}\Big)\Big|_{\lambda=z_j} = a_{j,s}\,,$$

where $\mathrm{He}_\nu(\lambda)$ are the Hermite polynomials. Let the polynomial $q(\lambda)$ be defined as

$$q(\lambda) = \sum_{\nu=0}^{k} d_\nu\, \lambda^\nu = \sum_{\nu=0}^{k} c_\nu\, \mathrm{He}_\nu(\lambda).$$

Then the system (5.4.7) is equivalent to the algebraic system

$$\frac{d^s}{d\lambda^s}\, q(\lambda)\big|_{\lambda=z_j} = b_{j,s}$$

for the coefficients $d_0, \dots, d_k$. Here $b_{j,s}$ are linear combinations of the numbers $a_{j,s}$. Since the last system is always solvable and the Hermite polynomials are linearly independent, there exist coefficients c_ν satisfying (5.4.7). This proves the lemma. □

LEMMA 5.4.5. *Let κ be an integer, $\kappa \ge 1$. Then there exists a set of functions $\chi_\delta \in C^\infty(\mathbb{R}_+)$, $\delta \in (0, e^{-1})$, with support in $[0,1]$ such that*

$$\Big|\int_0^{\delta^2} \rho^{i\beta-1}\, \chi_\delta(\rho)\, |\log\rho|^k\, d\rho\Big| \le c\, |\log\delta|^{k-\kappa+1} \quad \textit{for } k = 0, \dots, \kappa-1,\ \beta \in \mathbb{R},$$

$$\Big|\int_{\delta^2}^{\delta} \rho^{i\beta-1}\, \chi_\delta(\rho)\, |\log\rho|^k\, d\rho\Big| \le c\, |\log\delta|^{k-\kappa+1} \quad \textit{for } k = 0, \dots, \kappa-1,\ \beta \in \mathbb{R}\backslash\{0\},$$

$$\int_{\delta^2}^{\delta} \rho^{-1}\, \chi_\delta(\rho)\, |\log\rho|^k\, d\rho = C_k\, |\log\delta|^{k-\kappa+2} \quad \textit{for } k = 0, 1, \dots, \kappa-1,$$

where $C_k = (k-\kappa+2)^{-1}(2^{k-\kappa+2}-1)$ *if* $k \neq \kappa-2$, $C_k = \log 2$ *if* $k=\kappa-2$, *and*

$$\Big| \int_0^1 \rho^{i\beta-1}\,\chi_\delta(\rho)\,|\log\rho|^k\,d\rho\Big| \le c_\beta < \infty \quad \textit{if}\ \ \mathrm{Im}\,\beta = 0,\ k\le \kappa-1 \quad \textit{or}\ \mathrm{Im}\,\beta<0.$$

Here, the constants c *and* c_β *are independent of* δ.

P r o o f. We define the functions ψ_δ for $0<\delta<1/e$ as follows.

$$\begin{aligned} \psi_\delta(\rho) &= \delta^{-2}\,\rho\,|\log\rho|^{1-\kappa} \quad \text{if } 0<\rho<\delta^2,\\ \psi_\delta(\rho) &= |\log\rho|^{1-\kappa} \qquad\quad \text{if } \delta^2<\rho<\delta,\\ \psi_\delta(\rho) &= -\kappa\,|\log\rho|^{1-\kappa} \quad \text{if } \delta<\rho<1. \end{aligned}$$

Furthermore, we set $\psi_\delta(\rho)=0$ for $\rho>1$. Then

$$\Big|\int_0^{\delta^2} \rho^{i\beta-1}\,\psi_\delta(\rho)\,|\log\rho|^k\,d\rho\Big| \le \sum_{j=1}^{\infty}\delta^{-2}\int_{\delta^2/(j+1)}^{\delta^2/j} |\log\rho|^{k-\kappa+1}\,d\rho$$

$$\le \sum_{j=1}^{\infty}\delta^{-2}\Big(\frac{\delta^2}{j}-\frac{\delta^2}{j+1}\Big)\,\big|\log(\delta^2/j)\big|^{k-\kappa+1} \le c\,|\log\delta|^{k-\kappa+1}$$

if $k\le\kappa-1$. For real $\beta\neq 0$, we obtain

$$\Big|\int_{\delta^2}^{\delta} \rho^{i\beta-1}\,\psi_\delta(\rho)\,|\log\rho|^k\,d\rho\Big| = \Big|\int_{2\log\delta}^{\log\delta} |s|^{k-\kappa+1}\,e^{i\beta s}\,ds\Big|$$

$$= |\log\delta|^{k-\kappa+2}\Big|\int_1^2 t^{k-\kappa+1}e^{i\beta t\log\delta}\,dt\Big| \le c\,|\log\delta|^{k-\kappa+1}$$

and, analogously,

$$\Big|\int_{\delta}^{1} \rho^{i\beta-1}\,\psi_\delta(\rho)\,|\log\rho|^k\,d\rho\Big| = \kappa\,|\log\delta|^{k-\kappa+2}\Big|\int_0^1 t^k e^{i\beta t\log\delta}\,dt\Big| \le c\,|\log\delta|^{k-\kappa+1}.$$

Furthermore,

$$\int_{\delta^2}^{\delta} \rho^{-1}\,\psi_\delta(\rho)\,|\log\rho|^k\,d\rho = |\log\delta|^{k-\kappa+2}\int_1^2 t^{k-\kappa+1}\,dt = C_k\,|\log\delta|^{k-\kappa+2}$$

and

$$\int_{\delta}^{1} \rho^{-1}\,\psi_\delta(\rho)\,|\log\rho|^k\,d\rho = -\frac{\kappa}{k+1}\,|\log\delta|^{k-\kappa+2}$$

In particular, ψ_δ satisfies the inequality

$$\Big|\int_0^1 \rho^{i\beta-1}\,\chi_\delta(\rho)\,|\log\rho|^k\,d\rho\Big| \le c_\beta < \infty$$

for $k=0,1,\dots,\kappa-1$ and real β. Obviously, the last inequality holds also for complex β with negative imaginary part. Thus, all inequalities of the lemma are satisfied for the functions ψ_δ. It is possible to change the functions ψ_δ in sufficiently small neighborhoods of the points δ^2, δ and 1 such that the so arising functions are smooth and satisfy also the inequalities of the lemma. The proof is complete. □

Now we are able to prove the main theorem of this subsection.

THEOREM 5.4.6. *Suppose that the line* $\operatorname{Re}\lambda = m-1$ *contains eigenvalues of the pencil* $\mathfrak{A}(\lambda)$. *Then there does not exist a constant* c *such that the estimate* (5.3.1) *is valid for all solutions* $u \in C^\infty(\overline{\mathcal{G}})$ *of the equation* $L(D_x)\,u = 0$ *in* $\mathcal{G}$.

P r o o f. Let λ_μ be an eigenvalue of the pencil $\mathfrak{A}(\lambda)$ on the line $\operatorname{Re}\lambda = m-1$, and let

$$\kappa = \max_{1\le\sigma\le\chi_\mu} \kappa_{\sigma,\mu}$$

be the maximal length of the Jordan chains corresponding to this eigenvalue. We assume that κ is also the maximal length of Jordan chains corresponding to all eigenvalues on the line $\operatorname{Re}\lambda = m-1$. Furthermore, let $\phi_\mu^{(0,1)},\dots,\phi_\mu^{(\kappa-1,1)}$ be a Jordan chain of length κ corresponding to the eigenvalue λ_μ. By $\psi_\mu^{(0,1)},\dots,\psi_\mu^{(\kappa-1,1)}$, we denote a Jordan chain of the operator pencil $\mathfrak{A}^+(\lambda)$ corresponding to the eigenvalue $2m-N-\overline{\lambda}_\mu$ satisfying the biorthogonality condition (5.4.2). Then the function

$$w(x) = \rho^{2m-N-\bar\lambda_\mu}\,\psi_\mu^{(0,1)}(\omega)$$

satisfies the equations

$$L^+(D_x)\,w = 0 \ \text{ in } \mathcal{K}, \quad \partial_n^{k-1}w = 0 \ \text{ on } \partial\mathcal{K}\backslash\{0\},\ k=1,\dots,m.$$

Let $T_k(x,D_x)$ be the differential operators in Green's formula (5.3.4), and let $T_k^\circ(0,D_x)$ be the principal part of $T_k(x,D_x)$ with coefficients frozen at the origin. From the uniqueness of solutions of the Cauchy problem it follows that $T_k^\circ(0,D_x)\,w \not\equiv 0$ on $\partial\mathcal{K}\backslash\{0\}$ for at least one of the indices $k=1,\dots,m$. Suppose that

$$T_m^\circ(0,D_x)\,w \ne 0 \ \text{ on } \partial\mathcal{K}\backslash\{0\}$$

(for another index k, the proof proceeds analogously). The last equation implies

$$\mathcal{T}_m(\omega,\partial_\omega,2m-N-\overline{\lambda}_\mu)\,\psi_\mu^{(0,1)} \ne 0 \ \text{ on } \partial\Omega,$$

where $\mathcal{T}_m(\omega,\partial_\omega,\lambda)$ is defined by the equality

$$\mathcal{T}_m(\omega,\partial_\omega,\lambda)\,U(\omega) = r^{m-\lambda}\,T_m^\circ(0,D_x)\,r^\lambda\,U(\omega).$$

Consequently, there exists a smooth function ψ on $\partial\Omega$ such that

$$\int_{\partial\Omega}\psi(\omega)\,\overline{\mathcal{T}_m(\omega,\partial_\omega,2m-N-\overline{\lambda}_\mu)\,\psi_\mu^{(0,1)}(\omega)}\,d\omega = 1.$$

Let χ_δ be the functions introduced in Lemma 5.4.5, $0<\delta<1/e$. We set

$$\eta_\delta(\rho) = \chi_\delta(\rho)\,\rho^{\lambda_\mu-m+1} + p_\delta(\log\rho)\,e^{-(\log\rho)^2/2},$$

where p_δ are polynomials such that

$$\int_0^\infty \rho^{m-2-\lambda_j}\,\eta_\delta(\rho)\,(\log\rho)^l\,d\rho = 0 \tag{5.4.8}$$

for $j=1,\dots,\mu$, $l=0,1,\dots,\max_\sigma(\kappa_{\sigma,j}-1)$. By Lemmas 5.4.4 and 5.4.5, such polynomials exist, and their coefficients are bounded with respect to $\delta\in(0,e^{-1})$. We consider the solutions v_δ of the boundary value problem

$$L(D_x)\,v_\delta = 0 \ \text{ in } \mathcal{K},$$
$$v_\delta = \partial_n v_\delta = \cdots = \partial_n^{m-2}v_\delta = 0, \quad \partial_n^{m-1}v_\delta = g_\delta = \eta_\delta(\rho)\,\psi(\omega) \ \text{ on } \partial\mathcal{K}\backslash\{0\}.$$

Obviously,

$$\|g_\delta\|_{L_\infty(\partial\mathcal{K})} \le c$$

with a constant c independent of δ. The solution v_δ has the representation

$$\begin{aligned} v_\delta(x) &= \int_{\partial\mathcal{K}} g_\delta(\xi)\, \overline{T_m^\circ(0,D_\xi)\, \overline{G(x,\xi)}}\, d\xi \\ &= \int\limits_{\substack{\partial\mathcal{K}\\ |\xi|<2|x|}} g_\delta(\xi)\, \overline{T_m^\circ(0,D_\xi)\, \overline{G(x,\xi)}}\, d\xi \\ &\quad + \int\limits_{\substack{\partial\mathcal{K}\\ |\xi|>2|x|}} g_\delta(\xi)\, \overline{T_m^\circ(0,D_\xi)\, (\overline{A(x,\xi)} + \overline{R(x,\xi)})}\, d\xi, \end{aligned}$$

where $A(x,\xi)$ is defined by (5.4.3) and $R(x,\xi)$ satisfies the inequality (5.4.4). As in the proof of Theorem 5.4.2, we get

$$(5.4.9)\qquad |x|^{-m+1} \Big| \int\limits_{\substack{\partial\mathcal{K}\\ |\xi|<2|x|}} g_\delta(\xi)\, \overline{T_m^\circ(0,D_\xi)\, \overline{G(x,\xi)}}\, d\xi \Big| \le c\, \|g_\delta\|_{L_\infty(\partial\mathcal{K})}$$

and

$$(5.4.10)\qquad |x|^{-m+1} \Big| \int\limits_{\substack{\partial\mathcal{K}\\ |\xi|>2|x|}} g_\delta(\xi)\, \overline{T_m^\circ(0,D_\xi)\, \overline{R(x,\xi)}}\, d\xi \Big| \le c\, \|g_\delta\|_{L_\infty(\partial\mathcal{K})},$$

where c is a constant independent of δ. Furthermore, (5.4.8) yields

$$\begin{aligned} |x|^{-m+1} & \int\limits_{\substack{\partial\mathcal{K}\\ |\xi|>2|x|}} g_\delta(\xi)\, \overline{T_m^\circ(0,D_\xi)\, \overline{A(x,\xi)}}\, d\xi \\ &= -\sum_{j=1}^{\mu} \sum_{\sigma=1}^{\chi_j} \sum_{t=0}^{\kappa_{\sigma j}-1} |x|^{\lambda_j - m+1}\, \phi_j^{(t,\sigma)}(\omega_x) \sum_{s=0}^{\kappa_{\sigma j}-1-t} \frac{1}{s!}\, (\log|x|)^s \\ &\quad \times \sum_{l=0}^{\kappa_{\sigma j}-t-s-1} \frac{1}{l!} \int_0^{2|x|} |\xi|^{m-2-\lambda_j}\, \eta_\delta(|\xi|)\, (-\log|\xi|)^l\, d|\xi| \\ &\quad \times \Big(\psi\, , \sum_{\nu=0}^{\kappa_{\sigma j}-t-s-l-1} \frac{1}{\nu!}\, \frac{\partial^\nu \mathcal{T}_m}{\partial \lambda^\nu} (2m-N-\overline{\lambda}_j)\, \psi_j^{(\kappa_{\sigma j}-t-s-l-1,\sigma)} \Big)_{\partial\Omega}. \end{aligned}$$

For $|x| = \delta/2$, the expression

$$|x|^{\lambda_j-m+1}\, (\log|x|)^s \int_0^{2|x|} |\xi|^{m-2-\lambda_j}\, \eta_\delta(|\xi|)\, (-\log|\xi|)^l\, d|\xi|$$

is bounded with respect to δ for all j, s, l, except for $j = \mu$, $s + l = \kappa - 1$ (cf. Lemma 5.4.5). We assume that $\kappa_{\sigma,\mu} = \kappa$ for $\sigma = 1, \dots, k$ and $\kappa_{\sigma,\mu} < \kappa$ for $\sigma > k$. Then, by means of Lemma 5.4.5, we obtain the following representation for $|x| = \delta/2$:

$$|x|^{-m+1} \int\limits_{\substack{\partial\mathcal{K} \\ |\xi|>2|x|}} g_\delta(\xi)\, \overline{T_m^\circ(0, D_\xi)\, \overline{A(x,\xi)}}\, d\xi$$

$$= -\sum_{\sigma=1}^{\chi_\mu} |x|^{\lambda_\mu - m+1}\, \phi_\mu^{(0,\sigma)}(\omega_x) \sum_{s=0}^{\kappa-1} \frac{(\log|x|)^s}{s!\,(\kappa-1-s)!} \int_0^\delta \frac{\eta_\delta(\rho)\,(-\log\rho)^{\kappa-1-s}}{\rho^{\lambda_\mu - m+2}}\, d\rho$$
$$\times \big(\psi,\, \mathcal{T}_m(2m - N - \overline{\lambda}_\mu)\, \psi_\mu^{(0,\sigma)}\big)_{\partial\Omega} + M_{\delta,1}(x)$$
$$= -\sum_{\sigma=1}^{k} \delta^{\lambda_\mu - m+1}\, \phi_\mu^{(0,\sigma)}(\omega_x) \sum_{s=0}^{\kappa-1} \frac{(\log|x|)^s}{s!\,(\kappa-1-s)!} \int_{\delta^2}^{\delta} \rho^{-1}\, \chi_\delta(\rho)\,(-\log\rho)^{\kappa-1-s}\, d\rho$$
$$\times \big(\psi,\, \mathcal{T}_m(2m - N - \overline{\lambda}_\mu)\, \psi_\mu^{(0,\sigma)}\big)_{\partial\Omega} + M_{\delta,2}(x)$$
$$= \delta^{\lambda_\mu - m+1} \sum_{\sigma=1}^{k} c_\sigma \phi_\mu^{(0,\sigma)}(\omega_x)\, \log\delta + M_{\delta,3}(x),$$

where

$$\big|M_{\delta,j}(x)\big| \le c \quad \text{for } j = 1,2,3$$

and c_σ are certain constants, $c_1 \neq 0$. Since the eigenfunctions $\phi_\mu^{(0,1)}, \dots, \phi_\mu^{(0,k)}$ are linearly independent, there exists al lest one ω_x such that

$$\sum_{\sigma=1}^{k} c_\sigma \phi_\mu^{(0,\sigma)}(\omega_x) \neq 0.$$

Consequently,

$$\sup_{|x|=\delta/2} |x|^{-m+1}\, \big|v_\delta(x)\big| \ge c\,|\log\delta|.$$

Let $\zeta \in C^\infty((0,\infty))$ be a cut-off function equal to one in $(0,1)$ and to zero in $(2,\infty)$. Then the function

$$u_\delta(x) = \big(\zeta(\rho) - \zeta(4\delta^{-2}\rho)\big)\, v_\delta(x)$$

belongs to $C_0^\infty(\overline{\mathcal{G}}\backslash\{0\})$ and satisfies the estimate

$$\|\rho^{-m+1} u_\delta\|_{C(\overline{\mathcal{G}})} \ge \sup_{|x|=\delta/2} |x|^{-m+1}\, \big|v_\delta(x)\big| \ge c\,|\log\delta| \tag{5.4.11}$$

for $\delta < 1/e$. Furthermore,

$$u = \partial_n u_\delta = \cdots = \partial_n^{m-2} u_\delta = 0 \quad \text{on } \partial\mathcal{G}\backslash\{0\}$$

and

$$\partial_n^{m-1} u_\delta = \big(\zeta(\rho) - \zeta(4\delta^{-2}\rho)\big)\, \partial_n^{m-1} v_\delta = \big(\zeta(\rho) - \zeta(4\delta^{-2}\rho)\big)\, \eta_\delta(\rho)\, \psi(\omega)$$

on $\partial\mathcal{G}\backslash\{0\}$. In particular,

$$\|\partial_n^{m-1} u_\delta\|_{L_\infty(\partial\mathcal{G})} \le c, \tag{5.4.12}$$

where c is a constant independent of δ. We show that

$$\|\rho^{m+1}\, L(D_x)\, u_\delta\|_{L_\infty(\partial\mathcal{G})} \le c \tag{5.4.13}$$

with a constant c independent of δ. Since $L(D_x)\, v_\delta = 0$ in $\mathcal{K}$, the function $L(D_x)\, u_\delta$ vanishes outside the set $\{x : |x| \in (\delta^2/4, \delta^2/2) \cup (1,2)\}$. For $x \in \mathcal{G}$, $\delta^2/4 < |x| < \delta^2/2$, we obtain

$$\begin{aligned}
|x|^{m+1} \big|L(D_x)\, v_\delta(x)\big| &\le c \sum_{|\alpha|\le 2m} |x|^{|\alpha|-m+1} \big|\partial_x^\alpha v_\delta(x)\big| \\
&= c \sum_{|\alpha|\le 2m} |x|^{|\alpha|-m+1} \Big| \int_{\partial\mathcal{K}} g_\delta(\xi)\, \overline{\partial_x^\alpha\, T_m^\circ(0,D_\xi)\, \overline{G(x,\xi)}}\, d\xi \Big| \\
&\le c \sum_{|\alpha|\le 2m} |x|^{|\alpha|-m+1} \Big| \int\limits_{\substack{\partial\mathcal{K}\\ |\xi|<2|x|}} g_\delta(\xi)\, \overline{\partial_x^\alpha\, T_m^\circ(0,D_\xi)\, \overline{G(x,\xi)}}\, d\xi \Big| \\
&\quad + c \sum_{|\alpha|\le 2m} |x|^{|\alpha|-m+1} \Big| \int\limits_{\substack{\partial\mathcal{K}\\ |\xi|>2|x|}} g_\delta(\xi)\, \overline{\partial_x^\alpha\, T_m^\circ(0,D_\xi)\, \big(\overline{A(x,\xi)} + \overline{R(x,\xi)}\big)}\, d\xi \Big|.
\end{aligned}$$

Analogously to (5.4.9) and (5.4.10), we get

$$|x|^{|\alpha|-m+1} \Big| \int_{\partial\mathcal{K}} g_\delta(\xi)\, \overline{\partial_x^\alpha\, T_m^\circ(0,D_\xi)\, \overline{G(x,\xi)}}\, d\xi \Big| \le c\, \|g_\delta\|_{L_\infty(\partial\mathcal{K})}$$

and

$$|x|^{|\alpha|-m+1} \Big| \int\limits_{\substack{\partial\mathcal{K}\\ |\xi|>2|x|}} g_\delta(\xi)\, \overline{\partial_x^\alpha\, T_m^\circ(0,D_\xi)\, \overline{R(x,\xi)}}\, d\xi \Big| \le c\, \|g_\delta\|_{L_\infty(\partial\mathcal{K})}$$

for $|\alpha| \le 2m$. Using the first inequality in Lemma 5.4.5, we furthermore obtain

$$\begin{aligned}
&|x|^{|\alpha|-m+1} \Big| \int\limits_{\substack{\partial\mathcal{K}\\ |\xi|>2|x|}} g_\delta(\xi)\, \overline{\partial_x^\alpha\, T_m^\circ(0,D_\xi)\, \overline{A(x,\xi)}}\, d\xi \Big| \\
&\quad = |x|^{|\alpha|-m+1} \Big| \partial_x^\alpha \sum_{j=1}^{\mu} \sum_{\sigma=1}^{\chi_j} \sum_{t=0}^{\kappa_{\sigma j}-1} |x|^{\lambda_j}\, \phi_j^{(t,\sigma)}(\omega_x) \sum_{s=0}^{\kappa_{\sigma j}-1-t} \frac{1}{s!} (\log|x|)^s \\
&\qquad \times \sum_{l=0}^{\kappa_{\sigma j}-t-s-1} \frac{1}{l!} \int_0^{2|x|} |\xi|^{m-2-\lambda_j}\, \eta_\delta(|\xi|)\, (-\log|\xi|)^l\, d|\xi| \\
&\qquad \times \Big(\psi, \sum_{\nu=0}^{\kappa_{\sigma j}-t-s-l-1} \frac{1}{\nu!} \frac{\partial^\nu \mathcal{T}_m}{\partial\lambda^\nu} (2m-N-\overline{\lambda}_j)\, \psi_j^{(\kappa_{\sigma j}-t-s-l-1,\sigma)} \Big)_{\partial\Omega} \Big| \le c
\end{aligned}$$

for $\delta^2/4 < |x| < \delta^2/2$, where c is independent of δ. This proves (5.4.13).

Suppose now that (5.3.1) is valid for all solutions $u \in C^\infty(\overline{\mathcal{G}})$ of the equation $L(D_x)\, u = 0$ in $\mathcal{G}$. Then by Lemma 5.4.3, the inequality

$$\|\rho^{-m+1} u_\delta\|_{L_\infty(\mathcal{G})} \le c \Big(\sum_{k=1}^{m} \|\partial_n^{k-1} u_\delta\|_{W^{m-k,\infty}(\partial\mathcal{G})} + \|\rho^{m+1}\, L(D_x)\, u_\delta\|_{L_\infty(\mathcal{G})} \Big)$$

holds with a constant c independent of δ. This contradicts (5.4.11)–(5.4.13). The theorem is proved. □

5.4.3. The failure of the condition on the eigenvalues. In Theorems 5.4.2 and 5.4.6 it was shown that the inequality (5.3.1) is valid for all infinitely differentiable solutions of the equation $L(D_x)\, u = 0$ if and only if the line $\operatorname{Re}\lambda = m-1$ is free of eigenvalues of the pencil $\mathfrak{A}(\lambda)$. We prove now that for $N \ge 4$ one can always find a domain with a conical point for which the last condition fails.

THEOREM 5.4.7. *Suppose that $\mathcal{G}$ is a domain in $\mathbb{R}^N$, $N \ge 4$, which coincides in a neighborhood of the origin with a circular cone*

$$\mathcal{K} = \mathcal{K}_\alpha = \{x \in \mathbb{R}^N :\ x_N > |x| \cos\alpha\}$$

with opening 2α. Then for every strongly elliptic differential operator of order $2m > 2$, there exists an angle $\alpha \in [\pi/2, \pi)$ such that the line $\operatorname{Re}\lambda = m-1$ contains at least one eigenvalue of the pencil $\mathfrak{A}(\lambda)$.

P r o o f. By [**85**, Theorem 10.1.3], the line $\operatorname{Re}\lambda = m - N/2$ is free of eigenvalues of the pencil $\mathfrak{A}(\lambda)$. Furthermore, the eigenvalues of the pencil $\mathfrak{A}(\lambda) = \mathfrak{A}_\alpha(\lambda)$ depend continuously on the angle α. We assume that no eigenvalues of the pencil $\mathfrak{A}_\alpha(\lambda)$ lie on the line $\operatorname{Re}\lambda = m-1$ for $\pi/2 \le \alpha < \pi$. Then it follows from the operator generalization of Rouché's theorem (cf. [**56**]) that the sum of the algebraic multiplicities of the eigenvalues of the pencil $\mathfrak{A}_\alpha(\lambda)$ in the strip $m - N/2 < \operatorname{Re}\lambda < m-1$ does not depend on $\alpha \in [\pi/2, \pi)$. However by [**85**, Theorem 10.3.3], the strip $m - N/2 < \operatorname{Re}\lambda < m-1$ does not contain eigenvalues of the pencil $\mathfrak{A}_\alpha(\lambda)$ if $\alpha = \pi/2$. For $\alpha = \pi - \varepsilon$ with small positive ε, there exists an eigenvalue near the line $\operatorname{Re}\lambda = m - (N-1)/2$ if $2m \ge N$ and near the line $\operatorname{Re}\lambda = 0$ if $2m < N$ (see [**85**, Corollary 11.5.1 and Theorem 11.8.1]). This contradicts the independence of the sum of the algebraic multiplicities in the strip $m - N/2 < \operatorname{Re}\lambda < m-1$ on α. The theorem is proved. □

As an example, we consider the Dirichlet problem for the biharmonic equation

$$\Delta^2 u = 0 \ \text{ in } \mathcal{G}, \quad u = g_1, \ \frac{\partial u}{\partial n} = g_2 \ \text{ on } \partial\mathcal{G}\backslash\{0\}, \tag{5.4.14}$$

where $\mathcal{G}$ is a domain in $\mathbb{R}^4$ which coincides with the circular cone

$$\mathcal{K}_\alpha = \{x \in \mathbb{R}^N :\ x_4 > |x| \cos\alpha\} \tag{5.4.15}$$

in a neighborhood of the origin. Then $\mathfrak{A}(\lambda) = \mathfrak{A}_\alpha(\lambda)$ is the operator of the boundary value problem

$$\big(\delta + \lambda(\lambda-2)\big)\ \big(\delta + \lambda(\lambda+2)\big)v = f \ \text{ for } \vartheta < \alpha \tag{5.4.16}$$

$$v = \frac{\partial v}{\partial \vartheta} = 0 \ \text{ for } \vartheta = \alpha, \tag{5.4.17}$$

where $\cos\vartheta = x_4/|x|$ and δ denotes the Laplace-Beltrami operator on the sphere S^3. We show that there exists an angle α such that $\lambda = 1$ is an eigenvalue of the pencil $\mathfrak{A}(\lambda)$ with an eigenfunction depending only on ϑ. If $v = v(\vartheta)$, then

$$\delta v(\vartheta) = (\sin^2\vartheta)^{-1} \frac{d}{d\vartheta}\Big(\sin^2\vartheta\, \frac{dv}{d\vartheta}\Big)$$

For $\lambda = 1$ and $f = 0$, the equation (5.4.16) takes the form

$$(\delta - 1)(\delta + 3)\, v = 0. \tag{5.4.18}$$

It can be easily checked that

$$v(\vartheta) = c_1 \cos\vartheta + c_2 \frac{\vartheta}{\sin\vartheta}$$

is a smooth solution of (5.4.18) for arbitrary constants c_1 and c_2. The boundary conditions (5.4.17) imply

$$c_1 \cos\alpha + c_2 \frac{\alpha}{\sin\alpha} = 0,$$

$$-c_1 \sin\alpha + c_2 \frac{\sin\alpha - \alpha\cos\alpha}{\sin^2\alpha} = 0.$$

The coefficient determinant of this system is

$$(2\sin^2\alpha)^{-1}\big(2\alpha\cos(2\alpha) - \sin 2\alpha\big).$$

Consequently, there exists an eigenfunction $v = v(\vartheta)$ corresponding to the eigenvalue $\lambda = 1$ if $2\alpha = \tan 2\alpha$, i.e. $\mathcal{K}_\alpha$ is a cone with opening $2\alpha = 4.49341...$ $(= 257.45...^\circ)$. In this case, even the smooth solutions of the problem (5.4.14) do not satisfy the Miranda-Agmon maximum principle.

Part 2

Neumann and mixed boundary value problems for second order systems in polyhedral domains

CHAPTER 6

Boundary value problems for second order systems in a dihedron

In this chapter, we consider the Dirichlet, Neumann and mixed boundary value problems for elliptic systems of second order in a three-dimensional dihedron

$$\mathcal{D} = \{x = (x', x_3) : \ x' = (x_1, x_2) \in K, \ x_3 \in \mathbb{R}\},$$

where K is the two-dimensional angle (2.1.1). For the mixed problem, Dirichlet and Neumann data are prescribed on different sides of the dihedron. In contrast to the preceding chapters, we are interested in solutions in weighted Sobolev and Hölder spaces with "nonhomogeneous" norms. The Dirichlet and the mixed boundary value problems can be also handled in the classes of the weighted spaces $V_\delta^{l,p}(\mathcal{D})$ and $N_\delta^{l,p}(\mathcal{D})$ introduced in Sections 2.1 and 2.7, respectively. For the Dirichlet problem this was done in Chapter 2. We prove in Section 6.1 that also the mixed boundary value problem is uniquely solvable in the weighted Sobolev space $V_\delta^{l,2}(\mathcal{D})$ for a certain set of weight parameters δ. However, such an existence and uniqueness result is not valid for the Neumann problem (see Theorem 6.1.12). This boundary value problem requires the use of other weighted spaces. We deal in this chapter with solutions in the function spaces $W_\delta^{l,p}(\mathcal{D})$ and $C_\delta^{l,\sigma}(\mathcal{D})$. Here, the weighted Sobolev space $W_\delta^{l,p}(\mathcal{D})$ is the set of all functions u on $\mathcal{D}$ such that

$$r^\delta \, \partial_x^\alpha u \in L_p(\mathcal{D}) \quad \text{for } |\alpha| \le l,$$

where again $r = r(x)$ denotes the distance of the point x from the edge M of the dihedron. The weighted Hölder space $C_\delta^{l,\sigma}(\mathcal{D})$ is defined similarly. The scales of the spaces $W_\delta^{l,p}$ and $C_\delta^{l,\sigma}$ include the standard Sobolev and Hölder spaces for $\delta = 0$. An advantage of a theory involving the spaces $W_\delta^{l,p}$ and $C_\delta^{l,\sigma}$ is that it is applicable not only to the Dirichlet and mixed boundary value problems but also to the Neumann problem. In fact, this theory can be also extended to other boundary value problems in the variational form. On the other hand, we obtain results for a larger class of right-hand sides f and $g^\pm$ in the differential equation and in the boundary conditions.

In Sections 6.2 and 6.7, we study the properties of the spaces $W_\delta^{l,p}(\mathcal{D})$ and $C_\delta^{l,\sigma}(\mathcal{D})$. In particular, we are interested in the traces on the edge M and in relations between the spaces with homogeneous and nonhomogeneous norms. In order to prove the solvability of the boundary value problem in a dihedron, we must consider first the parameter-dependent problem in a plane angle K which arises after the application of the Fourier transform with respect to the variable x_3 (the variable on the edge). We show in Section 6.3 that this problem is uniquely solvable in the weighted Sobolev space $W_\delta^{2,2}(K)$ for certain δ and obtain an a priori estimate for this solution. This result for the parameter-dependent problem is used in Section 6.4, where we prove that the second order derivatives of the variational solution are square integrable with the weight $r^{2\delta}$ under corresponding conditions on the right hand sides. In Section 6.5, we obtain regularity results and a priori estimates for the variational solutions in the class of the spaces $W_\delta^{l,2}$ with arbitrary integer $l \ge 2$. Let δ_+ be the greatest real number such that the strip $0 < \operatorname{Re}\lambda < \delta_+$ is free of eigenvalues of the pencil $A(\lambda)$ generated by the boundary value problem. The just mentioned $W_\delta^{l,2}$-regularity result is proven for all δ satisfying the condition

$$\max(0, l-1-\delta_+) < \delta < l-1.$$

For some important boundary value problems (e.g., the Neumann problem for the Lamé system), the condition on δ can be weakened, i.e., the number δ_+ can be replaced by μ_+ which is the greatest real number such that the strip $0 < \operatorname{Re}\lambda < \mu_+$ contains at most the eigenvalue $\lambda = 1$.

In Section 6.6, we establish point estimates of Green's matrix $G(x,\xi)$. As for the Dirichlet problem in Chapter 2, we have to consider the cases $|x-\xi| < \min(|x'|,|\xi'|)$ and $|x-\xi| > \min(|x'|,|\xi'|)$ separately. For example, the estimate

$$\begin{aligned} &\big|\partial_{x'}^\alpha \partial_{x_3}^j \partial_{\xi'}^\beta \partial_{\xi_3}^k G(x,\xi)\big| \\ &\le c\,|x-\xi|^{-1-|\alpha|-|\beta|-j-k} \Big(\frac{|x'|}{|x-\xi|}\Big)^{\min(0,\delta_+-|\alpha|-\varepsilon)} \Big(\frac{|\xi'|}{|x-\xi|}\Big)^{\min(0,\delta_--|\beta|-\varepsilon)} \end{aligned}$$

with an arbitrarily small positive real ε holds for $|x-\xi| > \min(|x'|,|\xi'|)$. This estimate is improved in the following two cases. For the Dirichlet and mixed problems, the sharper estimate obtained in Section 2.5 is valid. In the case of the Neumann problem for some special systems (e.g. the Lamé system), one can replace again δ_+ by μ_+.

6.1. Solvability in weighted Sobolev spaces with homogeneous norms

This section is concerned with the solvability of the Dirichlet, Neumann and mixed boundary value problems for a second order elliptic differential operator

$$L(D_x) = \sum_{j,k=1}^{3} A_{j,k}\, D_{x_j}\, D_{x_k} \tag{6.1.1}$$

in the class of the weighted Sobolev spaces $V_\delta^{l,2}(\mathcal{D})$. Here $A_{j,k}$ are constant $\ell \times \ell$-matrices. Throughout this chapter, we suppose that the corresponding sesquilinear form

$$b_{\mathcal{D}}(u,v) = \int_{\mathcal{D}} \sum_{j,k=1}^{3} A_{j,k} D_{x_j} u \cdot \overline{D_{x_k} v}\, dx. \tag{6.1.2}$$

is $\mathcal{H}_{\mathcal{D}}$-elliptic (see the definition below). We show that both the Dirichlet and the mixed boundary value problem have uniquely determined solutions in $V_\delta^{l,2}(\mathcal{D})^\ell$, $l \ge 1$, if the data are from corresponding weighted spaces and δ satisfies the inequalities

$$-\delta_+ < \delta - l + 1 < \delta_-\,,$$

where δ_+ and δ_- are certain positive real numbers. For the Dirichlet problem this was already proved in Chapter 2. As it is shown at the end of this section, an analogous result is not valid for the Neumann problem.

6.1.1. Formulation of the problem. Let $\mathcal{D} = K \times \mathbb{R}$ be the same dihedron as in Chapter 2. Furthermore, let $L(D_x)$ be the differential operator (6.1.1) and let

$$N^\pm(D_x) = \sum_{j,k=1}^{3} A_{j,k}\, n_k^\pm\, \partial_{x_j}$$

be the conormal derivative associated to $L(D_x)$ on the face $\Gamma^\pm$. Here $n^\pm = (n_1^\pm, n_2^\pm, 0)$ denotes the outer normal to the face $\Gamma^\pm$ of $\mathcal{D}$. We consider the boundary value problem

$$L(D_x)\, u = f \ \text{ in } \mathcal{D}, \quad B^\pm(D_x)\, u = g^\pm \text{ on } \Gamma^\pm, \tag{6.1.3}$$

where

$$B^\pm(D_x)\, u = (1 - d^\pm)\, u + d^\pm N^\pm(D_x)\, u$$

and $d^\pm \in \{0, 1\}$. In the case $d^+ = d^- = 0$ we are concerned with the Dirichlet problem, for $d^+ = d^- = 1$ with the Neumann problem, and for $d^+ \neq d^-$ with the mixed boundary value problem. The numbers $d^\pm$ are the orders of the differential operators $B^\pm$.

Using the sesquilinear form (6.1.2), we define variational solutions of the problem (6.1.3). Let $L^{1,2}(\mathcal{D})$ be the closure of the set $C_0^\infty(\overline{\mathcal{D}})$ with respect to the norm

$$\|u\|_{L^{1,2}(\mathcal{D})} = \|\nabla u\|_{L_2(\mathcal{D})^3} \tag{6.1.4}$$

and let $L^{1/2,2}(\Gamma^\pm)$ be the corresponding trace space on $\Gamma^\pm$. Furthermore, let the subspace $\mathcal{H}_\mathcal{D}$ of $L^{1,2}(\mathcal{D})^\ell$ be defined as

$$\mathcal{H}_\mathcal{D} = \{u \in L^{1,2}(\mathcal{D})^\ell : \ (1 - d^\pm)u = 0 \text{ on } \Gamma^\pm\}.$$

The vector function $u \in L^{1,2}(\mathcal{D})^\ell$ is called a *variational solution* of the boundary value problem (6.1.3) if

$$b_\mathcal{D}(u, v) = (F, v)_\mathcal{D} \text{ for all } v \in \mathcal{H}_\mathcal{D}, \tag{6.1.5}$$

$$u = g^\pm \ \text{ on } \Gamma^\pm \ \text{ for } d^\pm = 0. \tag{6.1.6}$$

Here F is a given linear and continuous functional on $\mathcal{H}_\mathcal{D}$, $g^\pm$ are given vector-valued functions in $L^{1/2,2}(\Gamma^\pm)^\ell$, and $(\cdot, \cdot)_\mathcal{D}$ denotes the extension of the $L_2(\mathcal{D})^\ell$ inner product to $\mathcal{H}_\mathcal{D}^* \times \mathcal{H}_\mathcal{D}$.

Note that every solution $u \in W^{2,2}(\mathcal{D})^\ell$ of the boundary value problem (6.1.3) is a solution of problem (6.1.5), (6.1.6), where

$$(F, v)_\mathcal{D} = \int_\mathcal{D} f \cdot \bar{v}\, dx + \sum_{d^\pm = 1} \int_{\Gamma^\pm} g^\pm \cdot \bar{v}\, dx \ \text{ for all } v \in \mathcal{H}_\mathcal{D}\,. \tag{6.1.7}$$

We will always assume that the form (6.1.2) is $\mathcal{H}_\mathcal{D}$-*elliptic*, i.e. there exists a positive constant c such that

$$\operatorname{Re} b_\mathcal{D}(u, u) \geq c\, \|u\|^2_{\mathcal{H}_\mathcal{D}} \quad \text{for all } u \in \mathcal{H}_\mathcal{D}\,. \tag{6.1.8}$$

By the Lax-Milgram lemma, this condition guarantees the existence and uniqueness of a variational solution $u \in \mathcal{H}_\mathcal{D}$ in the case of zero Dirichlet data $g^\pm$.

The following lemma is only valid for the Dirichlet and mixed problems.

LEMMA 6.1.1. *If $d^+ + d^- \leq 1$, then $\mathcal{H}_\mathcal{D} \subset V_0^{1,2}(\mathcal{D})^\ell$ and the $L^{1,2}(\mathcal{D})^\ell$- and $V_0^{1,2}(\mathcal{D})^\ell$-norms are equivalent in $\mathcal{H}_\mathcal{D}$.*

P r o o f. For the Dirichlet problem, we refer to Lemma 2.2.4, for the mixed problem ($d^+ \neq d^-$) the lemma can be proved in the exact same manner. □

COROLLARY 6.1.2. *If* $d^+ + d^- \le 1$, $u \in L^{1,2}(\mathcal{D})$ *and* $(1-d^\pm)u|_{\Gamma^\pm} \in V_0^{1/2,2}(\Gamma^\pm)^\ell$, *then* $u \in V_0^{1,2}(\mathcal{D})^\ell$.

P r o o f. Let $g^\pm = (1-d^\pm)u|_{\Gamma^\pm} \in V_0^{1/2,2}(\Gamma^\pm)^\ell$. By Lemma 2.2.1, there exists a vector function $w \in V_0^{1,2}(\mathcal{D})^\ell$ such that $w = g^\pm$ on $\Gamma^\pm$ for $d^\pm = 0$. Then $u - w \in \mathcal{H}_\mathcal{D}$ and consequently $u - w \in V_0^{1,2}(\mathcal{D})^\ell$. The result follows. □

Furthermore, it follows from Lemma 6.1.1 and (6.1.8) that

$$\text{Re}\, b_\mathcal{D}(u,u) \ge c\, \|u\|^2_{V_0^{1,2}(\mathcal{D})^\ell} \quad \text{for all } u \in \mathcal{H}_\mathcal{D} \tag{6.1.9}$$

if $d^+ + d^- \le 1$, where c is a positive constant independent of u.

6.1.2. An a priori estimate for the solution. We will frequently use the following assertion which was proved for the Dirichlet problem in Section 2.2. For mixed and Neumann problems the proof is a word by word repetition of the proof of Theorem 2.2.9.

THEOREM 6.1.3. *Suppose that* $u \in V_\delta^{0,p}(\mathcal{D})^\ell$, $\chi u \in W^{1,p}(\mathcal{D})^\ell$ *for every* $\chi \in C_0^\infty(\overline{\mathcal{D}}\backslash M)^\ell$, *and that* u *is a solution of the problem*

$$b_\mathcal{D}(u,v) = (F,v)_\mathcal{D} \text{ for all } v \in C_0^\infty(\overline{\mathcal{D}}\backslash M)^\ell,\ (1-d^\pm)v|_{\Gamma^\pm} = 0,$$
$$u|_{\Gamma^\pm} = g^\pm \quad \text{for } d^\pm = 0.$$

1) *If* $F \in (V^{1,p'}_{-\delta-1}(\mathcal{D})^*)^\ell$, $p' = p/(p-1)$, *and* $g^\pm \in V_{\delta+1}^{1-1/p,p}(\Gamma^\pm)^\ell$, *then* $u \in V_{\delta+1}^{1,p}(\mathcal{D})^\ell$ *and*

$$\|u\|_{V^{1,p}_{\delta+1}(\mathcal{D})^\ell} \le c\Big(\|F\|_{(V^{1,p'}_{-\delta-1}(\mathcal{D})^*)^\ell} + \sum_{d^\pm=0} \|g^\pm\|_{V^{1-1/p,p}_{\delta+1}(\Gamma^\pm)^\ell} + \|u\|_{V^{0,p}_\delta(\mathcal{D})^\ell}\Big). \tag{6.1.10}$$

2) *If* $g^\pm \in V_{\delta+l}^{l-1/p,p}(\Gamma^\pm)^\ell$ *for* $d^\pm = 0$, $l \ge 2$, *and* F *admits the representation* (6.1.7), *where* $f \in V_{\delta+l}^{l-2,p}(\mathcal{D})^\ell$ *and* $g^\pm \in V_{\delta+l}^{l-1-1/p,p}(\Gamma^\pm)^\ell$ *for* $d^\pm = 1$, *then* $u \in V_{\delta+l}^{l,p}(\mathcal{D})^\ell$ *and*

$$\|u\|_{V^{l,p}_{\delta+l}(\mathcal{D})^\ell} \le c\Big(\|f\|_{V^{l-2,p}_{\delta+l}(\mathcal{D})^\ell} + \sum_\pm \|g^\pm\|_{V^{l-d^\pm-1/p,p}_{\delta+l}(\Gamma^\pm)^\ell} + \|u\|_{V^{0,p}_\delta(\mathcal{D})^\ell}\Big). \tag{6.1.11}$$

The constant c *is independent of* u.

6.1.3. The operator pencil generated by the boundary value problem. Let

$$L(D_{x'},0) = \sum_{j,k=1}^{2} A_{j,k}\, D_{x_j}\, D_{x_k}\,, \quad N^\pm(D_{x'},0) = \sum_{j,k=1}^{2} A_{j,k}\, n_k^\pm\, \partial_{x_j}\,. \tag{6.1.12}$$

Then we define the parameter-depending ordinary differential operators $\mathcal{L}_0(\lambda)$ and $\mathcal{B}_0^\pm(\lambda)$ by

$$\mathcal{L}_0(\lambda)\, U(\varphi) = r^{2-\lambda}\, L(D_{x'},0)\, \big(r^\lambda U(\varphi)\big),$$
$$\mathcal{B}_0^\pm(\lambda)\, U(\varphi) = (1-d^\pm)\, U(\varphi) + d^\pm\, r^{1-\lambda}\, N^\pm(D_{x'},0)\, \big(r^\lambda U(\varphi)\big).$$

Here again r, φ denote the polar coordinates of the point $x' = (x_1, x_2)$. The operator

$$U \to \big(\mathcal{L}_0(\lambda)\, U\, ,\, \mathcal{B}_0^+(\lambda)U|_{\varphi=\theta/2}\, ,\, \mathcal{B}_0^-(\lambda)U|_{\varphi=-\theta/2}\big)$$

realizes a continuous mapping $W^{2,2}((-\theta/2, +\theta/2))^\ell \to L_2((-\theta/2, +\theta/2))^\ell \times \mathbb{C}^\ell \times \mathbb{C}^\ell$ for arbitrary $\lambda \in \mathbb{C}$. We denote this operator by $A(\lambda)$. Obviously, it depends quadratically on the parameter λ. As is known, the spectrum of the pencil $A(\lambda)$ consists of isolated points, the eigenvalues of this pencil. The line $\operatorname{Re} \lambda = 0$ contains no eigenvalues if $d^+ + d^- \le 1$ (see [**85**, Th.10.1.3]). In the case $d^+ = d^- = 1$ (the case of the Neumann problem), the line $\operatorname{Re} \lambda = 0$ contains the single eigenvalue $\lambda = 0$. Let δ_+ and δ_- be the greatest positive real numbers such that the strip

$$-\delta_- < \operatorname{Re} \lambda < \delta_+$$

contains at most the eigenvalue $\lambda = 0$ of the pencil $A(\lambda)$.

For the next lemma we refer to [**85**, Th.1.1.5].

LEMMA 6.1.4. *The vector function* $u = r^{\lambda_0} \sum_{k=0}^{s} \frac{1}{k!} (\log r)^k\, v_{s-k}(\varphi)$ *is a solution of the problem*

$$L(D_{x'}, 0)\, u = 0 \text{ in } K, \quad (1 - d^\pm)\, u + d^\pm\, N^\pm(D_{x'}, 0)\, u = 0 \text{ on } \gamma^\pm$$

if and only if λ_0 *is an eigenvalue of the pencil* $A(\lambda)$ *and* $v_0, v_1, \dots, v_s$ *is a Jordan chain corresponding to this eigenvalue.*

Analogously to the operator pencil $A(\lambda)$, we define the operator pencil $A^+(\lambda)$ generated by the *formally adjoint boundary value problem*

$$L^+(D_x)\, u = f \ \text{ in } \mathcal{D}, \quad C^\pm(D_x)\, u = g^\pm \text{ on } \Gamma^\pm. \tag{6.1.13}$$

Here,

$$L^+(D_x) = \sum_{j,k=1}^{3} A_{k,j}^*\, D_{x_j}\, D_{x_k}$$

($A_{k,j}^*$ denotes the adjoint matrix of $A_{k,j}$) is the formally adjoint differential operator to $L(D_x)$ and the operators $C^\pm(D_x)$ are defined by

$$C^\pm(D_x)\, u = (1 - d^\pm)\, u + d^\pm \sum_{j,k=1}^{3} A_{k,j}^*\, n_k^\pm\, \partial_{x_j} u.$$

We introduce the parameter-depending ordinary differential operators

$$\mathcal{L}_0^+(\lambda)\, U(\varphi) = r^{2-\lambda}\, L(D_x)\, \big(r^\lambda U(\varphi)\big),$$
$$\mathcal{C}_0^\pm(\lambda)\, U(\varphi) = r^{d^\pm - \lambda}\, C^\pm(D_{x'}, 0)\, \big(r^\lambda U(\varphi)\big)$$

and put

$$A^+(\lambda)\, U = \big(\mathcal{L}_0^+(\lambda)\, U\, ,\, \mathcal{C}_0^+(\lambda)U|_{\varphi=\theta/2}\, ,\, \mathcal{C}_0^-(\lambda)U|_{\varphi=-\theta/2}\big).$$

The spectrum of the pencil $A^+(\lambda)$ consists also of isolated points. Furthermore, the complex number λ_0 is an eigenvalue of the pencil $A(\lambda)$ if and only if $-\overline{\lambda}_0$ is an eigenvalue of the pencil $A^+(\lambda)$ (see [**85**, Theorems 10.1.2, 12.1.1]). Consequently, the strip

$$-\delta_+ < \operatorname{Re} \lambda < \delta_-$$

contains at most the eigenvalue $\lambda = 0$ of the pencil $A^+(\lambda)$.

6.1.4. The model problem in the angle K. Let K be the same two-dimensional angle as in Chapter 2, and let $\gamma^\pm$ be the sides of K. We define the differential operators $L(D_{x'},\xi)$ and $N^\pm(D_{x'},\xi)$ for arbitrary real ξ by the equalities

$$L(D_{x'},\xi)\,u(x') = e^{-i\xi x_3}\,L(D_x)\,\big(e^{i\xi x_3}u(x')\big),$$
$$N^\pm(D_{x'},\xi)\,u(x') = e^{-i\xi x_3}\,N^\pm(D_x)\,\big(e^{i\xi x_3}u(x')\big)$$

and consider the boundary value problem

$$L(D_{x'},\pm 1)\,u(x') = f \quad \text{in } K, \tag{6.1.14}$$

$$B^+(D_{x'},\pm 1)\,u(x') = 0 \quad \text{on } \gamma^+, \qquad B^-(D_{x'},\pm 1)\,u(x') = 0 \quad \text{on } \gamma^-, \tag{6.1.15}$$

where

$$B^\pm(D_{x'},\xi)\,u = (1-d^\pm)\,u + d^\pm\,N^\pm(D_{x'},\xi)\,u.$$

We are also interested in *variational solutions* $u \in E^{1,2}_\delta(K)^\ell$ which satisfy the equations

$$b_K(u,v;\xi) = (F,v)_K \quad \text{for all } v\in E^{1,2}_{-\delta}(K),\ (1-d^\pm)v|_{\gamma^\pm}=0, \tag{6.1.16}$$

$$(1-d^\pm)u|_{\gamma^\pm} = 0 \tag{6.1.17}$$

for $\xi=\pm 1$. Here the sesquilinear form $b_K(u,v;\xi)$ is defined as

$$b_K(u,v;\xi) = \int_K \Big(\sum_{j,k=1}^{2} A_{j,k}D_{x_j}u\cdot\overline{D_{x_k}v} + \xi\sum_{j=1}^{2} A_{j,3}D_{x_j}u\cdot\bar v + \xi\sum_{k=1}^{2} A_{3,k}u\cdot\overline{D_{x_k}v} + \xi^2\,A_{3,3}u\cdot\bar v\Big)\,dx', \tag{6.1.18}$$

and $(\cdot,\cdot)_K$ denotes the scalar product in $L_2(K)^\ell$.

The following lemma guarantees the existence and uniqueness of solutions of the problem (6.1.16), (6.1.17) in the nonweighted Sobolev space $W^{1,2}(K)^\ell$ if $\xi\neq 0$.

LEMMA 6.1.5. *Suppose that the form $b_{\mathcal D}$ satisfies the condition* (6.1.8). *Then*

$$\operatorname{Re} b_K(v,v,\xi) \ge c\int_K\Big(\sum_{j=1}^{2}|\partial_{x_j}v|^2 + |\xi|^2\,|v|^2\Big)\,dx'$$

for all $v\in W^{1,2}(K)^\ell$, $(1-d^\pm)v=0$ on $\gamma^\pm$. Here the constant c is independent of ξ and v.

P r o o f. Let $v\in W^{1,2}(K)^\ell$, $v=0$ on $\gamma^\pm$ for $d^\pm=0$, and let ζ be an infinitely differentiable function on $\mathbb{R}$ with compact support. We put $u(x',x_3) = N^{-1/2}\,e^{\pm ix_3}\,\zeta(x_3/N)\,v(x')$. Then

$$b_{\mathcal D}(u,u)\to\|\zeta\|^2_{L_2(\mathbb{R})}\,b_K(v,v;\xi) \quad \text{as } N\to\infty$$

and

$$\|u\|_{L^{1,2}(\mathcal D)^\ell}\to\|\zeta\|_{L_2(\mathbb{R})}\int_K\Big(\sum_{j=1}^{2}|\partial_{x_j}v|^2+|\xi|^2\,|v|^2\Big)\,dx' \quad \text{as } N\to\infty.$$

This proves the lemma. □

The assertion in the next lemma was proved in Section 2.3 for the Dirichlet problem (see Theorem 2.3.8). Using Theorem 6.1.3, the analogous assertion for the Neumann and mixed boundary value problems holds by the same arguments.

LEMMA 6.1.6. *Suppose that* $u \in E^{1,2}_{\delta-l+1}(K)^\ell$ *is a solution of the problem*

$$b_K(u,v;\pm 1) = (F,v)_K \text{ for all } v \in C_0^\infty(\overline{K}\backslash\{0\})^\ell, \quad (1-d^\pm)v|_{\gamma^\pm} = 0,$$
$$u|_{\gamma^\pm} = g^\pm \text{ for } d^\pm = 0,$$

where $F \in (E^{1,2}_{l-1-\delta}(K)^\ell)^*$, $g^\pm \in E^{1/2,2}_{\delta-l+1}(\gamma^\pm)^\ell$. *Then*

$$\|u\|_{E^{1,2}_{\delta-l+1}(K)^\ell} \le c\Big(\|F\|_{(E^{1,2}_{l-1-\delta}(K)^\ell)^*} + \sum_{d^\pm=0} \|g^\pm\|_{E^{1/2,2}_{\delta-l+1}(\gamma^\pm)^\ell} + \|u\|_{E^{0,2}_{\delta-l}(K)^\ell}\Big)$$

with a constant independent of u. *If* $g^\pm \in E^{l-1/2,2}_\delta(\gamma^\pm)^\ell$ *for* $d^\pm = 0$, $l \ge 2$, *and* F *is a functional of the form*

$$(F,v)_K = \int_K f\cdot\bar v\,dx' + \sum_{d^\pm=1}\int_{\gamma^\pm} g^\pm\cdot\bar v\,dx',$$

where $f \in E^{l-2,2}_\delta(K)^\ell$ *and* $g^\pm \in E^{l-3/2,2}_\delta(\gamma^\pm)^\ell$ *for* $d^\pm = 1$, *then* $u \in E^{l,2}_\delta(K)^\ell$ *and*

$$\|u\|_{E^{l,2}_\delta(K)^\ell} \le c\Big(\|f\|_{E^{l-2,2}_\delta(K)^\ell} + \sum_\pm \|g^\pm\|_{E^{l-d^\pm-1/2,2}_\delta(\gamma^\pm)^\ell} + \|u\|_{E^{0,2}_{\delta-l}(K)^\ell}\Big)$$

with a constant c *independent of* u.

Analogously to Lemma 6.1.1, the imbedding

$$\{v \in W^{1,2}(K)^\ell : (1-d^\pm)v|_{\gamma^\pm} = 0\} \subset E^{1,2}_0(K)^\ell$$

holds in the cases of the Dirichlet and mixed problems. Moreover, the $W^{1,2}(K)^\ell$- and $E^{1,2}_0(K)^\ell$-norms are equivalent on this subspace. Thus, the problem (6.1.16), (6.1.17) for $\xi = \pm 1$ and $\delta = 0$ is uniquely solvable in $E^{1,2}_0(K)^\ell$ if $d^+ + d^- \le 1$. Furthermore, the following theorem holds. For the Dirichlet problem we refer to Theorem 2.3.22. The proof for the mixed problem proceeds analogously.

THEOREM 6.1.7. *Suppose that* $d^+ + d^- \le 1$. *Then the problem* (6.1.16), (6.1.17) *with* $\xi = \pm 1$ *is uniquely solvable in* $E^{1,2}_\delta(K)^\ell$ *for every linear and continuous functional* F *on the set* $\{v \in E^{1,2}_{-\delta}(K)^\ell : (1-d^\pm)v|_{\gamma^\pm} = 0\}$ *if and only if* $-\delta_+ < \delta < \delta_-$.

The last theorem together with Lemma 6.1.6 imply

COROLLARY 6.1.8. *Suppose that* $d^+ + d^- \le 1$, $l \ge 2$, *and* $l - 1 - \delta_+ < \delta < l - 1 + \delta_-$. *Then the boundary value problem* (6.1.14), (6.1.15) *is uniquely solvable in* $E^{l,2}_\delta(K)^\ell$ *for all* $f \in E^{l-2,2}_\delta(K)^\ell$, $g^\pm \in E^{l-d^\pm-1/2,2}_\delta(\gamma^\pm)^\ell$.

As the next lemma shows, the result of Theorem 6.1.7 is not true for the Neumann problem

$$L(D_{x'},\xi)\,u = f \text{ in } K, \quad N^\pm(D_{x'},\xi)\,u = g^\pm \text{ on } \gamma^\pm.$$

We denote by A_δ the operator

$$E^{1,2}_\delta(K)^\ell \ni u \to F \in (E^{1,2}_{-\delta}(K)^*)^\ell,$$

where

$$(F,v) = b_k(u,v;\pm 1) \quad \text{for all } v \in E^{1,2}_{-\delta}(K)^\ell.$$

LEMMA 6.1.9. *Let $d^+ = d^- = 1$ and $\xi = \pm 1$. Suppose that $A_{j,k} = A^*_{k,j}$ for all $j,k = 1,2,3$ and that the form $b_{\mathcal{D}}$ is $L^{1,2}(\mathcal{D})^\ell$-elliptic. Then there does not exist a real δ such that $\mathcal{A}_\delta$ is an isomorphism.*

P r o o f. We prove that the kernel of $\mathcal{A}_\delta$ is not trivial for $\delta > 0$. By [**85**, Theorem 12.2.1], the spectrum of the pencil $A(\lambda)$ contains the eigenvalue $\lambda = 0$. The eigenvectors corresponding to this eigenvalue are the constant vectors, and for every eigenvector c, there exists exactly one generalized eigenvector ψ. Thus, there exists a solution u of the problem

$$L(D_{x'},0)\,u = 0 \text{ in } K, \quad N^\pm(D_{x'},0)\,u = 0 \text{ on } \gamma^\pm$$

which has the form

$$u(x') = c\,\log r + \psi(\varphi),$$

where c is a nonzero constant vector and ψ is an infinitely differentiable vector function. Let ζ be an infinitely differentiable function, $\zeta(r) = 1$ for $r < 1$, $\zeta(r) = 0$ for $r > 2$ and let $v(x') = \zeta(r)\,u(x')$. Obviously, $v \in E^{1,2}_\delta(K)^\ell$ for each $\delta > 0$. Furthermore, let

$$f = L(D_{x'},\pm 1)\,v = [L(D_{x'},0),\zeta]\,u + \big(L(D_{x'},\pm 1) - L(D_{x'},0)\big)\,v$$

and $g^\pm = N^\pm(D_{x'},\pm 1)\,v$. Since the commutator $[L(D_{x'},0),\zeta] = L(D_{x'},0)\zeta - \zeta L(D_{x'},0)$ is a first order differential operator with coefficients vanishing for $r < 1$ and $r > 2$, we have $f \in E^{0,2}_\delta(K)^\ell$ and analogously $g^\pm \in E^{1/2,2}_\delta(\gamma^\pm)^\ell$. Both f and $g^\pm$ vanish for $r > 2$. Therefore, the mapping

$$v \to \int_K \bar{f}\cdot v\,dx + \sum_\pm \int_{\gamma^\pm} \overline{g^\pm}\cdot v\,dx$$

defines a linear and continuous functional on $W^{1,2}(K)^\ell$. Let $w \in W^{1,2}(K)^\ell$ be the uniquely determined variational solution of the problem

$$L(D_{x'},\pm 1)\,w = f \text{ in } K, \quad N^\pm(D_{x'},\pm 1)\,w = g^\pm \text{ on } \gamma^\pm. \tag{6.1.19}$$

This solution exists by Lemma 6.1.5. Since $v \notin W^{1,2}(K)^\ell$, the vector function $v - w$ is a nontrivial solution of the homogeneous problem (6.1.19). We show that $w \in E^{1,2}_\delta(K)^\ell$ for arbitrary $\delta > 0$. Let η be an infinitely differentiable function, $\eta(x') = 0$ for $|x'| < 2$, $\eta(x') = 1$ for $|x'| > 3$. Obviously, $(1-\eta)w \in E^{1,2}_\delta(K)^\ell$ and $\eta w \in E^{1,2}_{-\delta}(K)^\ell$ for arbitrary $\delta > 0$. Since

$$\eta\,L(D_{x'},\pm 1)\,w = \eta\,L(D_{x'},\pm 1)\,v = 0,$$

it follows that $L(D_{x'},\pm 1)\,(\eta w) = [L(D_{x'},\pm 1),\eta]\,w \in E^{0,2}_\kappa(K)^\ell$ for all real κ. Analogously, $N^\pm(D_{x'},\pm 1)\,(\eta w) \in E^{1/2,2}_\kappa(K)^\ell$. Applying Lemma 6.1.6, we conclude that $\eta w \in E^{2,2}_{1-\delta}(K)^\ell \subset E^{1,2}_{1-\delta}(K)^\ell$ for arbitrary $\delta > 0$. Repeating this argument, we obtain $\eta w \in E^{1,2}_\delta(K)^\ell$ for arbitrary real δ. Thus, it is shown that $w \in E^{1,2}_\delta(K)^\ell$ for arbitrary $\delta > 0$. Consequently, the kernel of the operator $\mathcal{A}_\delta$ is not trivial for $\delta > 0$.

Since the adjoint operator of $\mathcal{A}_\delta$ coincides with $\mathcal{A}_{-\delta}$, it follows that the cokernel of $\mathcal{A}_\delta$ is not trivial for $\delta < 0$. Moreover, it can be shown analogously to Theorem 2.3.11 that the operator $\mathcal{A}_\delta$ is not Fredholm if the line $\operatorname{Re}\lambda = -\delta$ contains eigenvalues of the pencil $A(\lambda)$. In particular, the operator $\mathcal{A}_\delta$ is not Fredholm for $\delta = 0$. The proof of the lemma is complete. □

6.1.5. Solvability of the Dirichlet and mixed problems. The assertions in the next two theorems were proved in Section 2.4 for the Dirichlet problem. Theorem 6.1.10 below is based on the unique solvability of the model problem (6.1.14), (6.1.15) (see Theorem 6.1.7 and Corollary 6.1.8) and can be proved analogously to Theorem 2.4.5.

THEOREM 6.1.10. *Let $d^+ + d^- \le 1$ and let δ_+, δ_- be the same numbers as in Theorem 6.1.7. Then the problem*

$$b_{\mathcal{D}}(u,v) = (F,v)_{\mathcal{D}} \quad \textit{for all } v \in V^{1,2}_{-\delta}(\mathcal{D})^\ell, \ (1-d^\pm)v = 0 \ \textit{on } \Gamma^\pm,$$

$$u = g^\pm \quad \textit{on } \Gamma^\pm \ \textit{if } d^\pm = 0$$

is uniquely solvable in $V^{1,2}_\delta(\mathcal{D})^\ell$ for every linear and continuous functional F on the set $\{v \in V^{1,2}_{-\delta}(\mathcal{D})^\ell : \ (1-d^\pm)v|_{\Gamma^\pm} = 0\}$ and all $g^\pm \in V^{1/2,2}_\delta(\Gamma^\pm)^\ell$ if and only if $-\delta_+ < \delta < \delta_-$.

Analogously, the boundary value problem (6.1.3) is uniquely solvable in $V^{l,2}_\delta(\mathcal{D})^\ell$, $l \ge 2$, for arbitrary $f \in V^{l-2,2}_\delta(\mathcal{D})^\ell$, $g^\pm \in V^{l-d^\pm-1/2,2}_\delta(\Gamma^\pm)^\ell$ if and only if $-\delta_+ < \delta - l + 1 < \delta_-$.

Furthermore, we obtain a regularity assertion for the variational solution $u \in L^{1,2}(\mathcal{D})^\ell$ in the class of the spaces $V^{l,2}_\delta$.

THEOREM 6.1.11. *Suppose that $d^+ + d^- \le 1$,*

$$f \in V^{l-2,2}_\delta(\mathcal{D})^\ell, \quad g^\pm \in V^{l-d^\pm-1/2}_\delta(\Gamma^\pm)^\ell,$$

$l \ge 2$, $-\delta_+ < \delta - l + 1 < \delta_-$, and that $u \in L^{1,2}(\mathcal{D})^\ell$ is a solution of the problem (6.1.5), (6.1.6), where the functional $F \in \mathcal{H}^*_{\mathcal{D}}$ has the form (6.1.7). Then $u \in V^{l,2}_\delta(\mathcal{D})^\ell$ and

$$\|u\|_{V^{l,2}_\delta(\mathcal{D})^\ell} \le c\Big(\|f\|_{V^{l-2,2}_\delta(\mathcal{D})^\ell} + \sum_\pm \|g^\pm\|_{V^{l-d^\pm-1/2}_\delta(\Gamma^\pm)^\ell}\Big).$$

P r o o f. By Corollary 6.1.2, the vector function u belongs to the space $V^{1,2}_0(\mathcal{D})^\ell$. Thus for the Dirichlet problem the result follows from Theorem 2.4.6. The proof for the mixed boundary value problem proceeds analogously. □

6.1.6. The failure of analogous results for the Neumann problem. As the following assertion shows, analogous results do not hold for the Neumann problem.

THEOREM 6.1.12. *Suppose that $A_{j,k} = A^*_{k,j}$ for all $j,k = 1,2,3$ and that the form $b_{\mathcal{D}}(\cdot,\cdot)$ is $L^{1,2}(\mathcal{D})^\ell$-elliptic. Then there does not exist a real δ such that the problem*

$$b_{\mathcal{D}}(u,v) = (F,v)_{\mathcal{D}} \quad \textit{for all } v \in V^{1,2}_{-\delta}(\mathcal{D})^\ell \tag{6.1.20}$$

is uniquely solvable in $V^{1,2}_\delta(\mathcal{D})^\ell$ for arbitrary $F \in (V^{1,2}_{-\delta}(\mathcal{D})^*)^\ell$.

P r o o f. Suppose that the problem (6.1.20) is uniquely solvable in $V^{1,2}_\delta(\mathcal{D})^\ell$ for arbitrary $F \in (V^{1,2}_{-\delta}(\mathcal{D})^*)^\ell$. Then, by the self-adjointness of the operator $L(D_x)$, this problem is also uniquely solvable in $V^{1,2}_{-\delta}(\mathcal{D})^\ell$ for arbitrary $F \in (V^{1,2}_\delta(\mathcal{D})^*)^\ell$. From our assumption and from Theorem 6.1.3 it follows that the Neumann problem

$$L(D_x)\,u = f \in \mathcal{D}, \quad N^\pm(D_x)u|_{\Gamma^\pm} = g^\pm$$

is uniquely solvable in $V^{2,2}_{\delta+1}(\mathcal{D})^\ell$ for arbitrary $f \in V^{0,2}_{\delta+1}(\mathcal{D})^\ell$ and $g^\pm \in V^{1/2,2}_{\delta+1}(\Gamma^\pm)^\ell$. In particular, every $u \in V^{2,2}_{\delta+1}(\mathcal{D})^\ell$ satisfies the inequality

$$(6.1.21) \qquad \|u\|_{V^{2,2}_{\delta+1}(\mathcal{D})^\ell} \le c\Big(\|L(D_x)u\|_{V^{0,2}_{\delta+1}(\mathcal{D})^\ell} + \sum_\pm \|N^\pm(D_x)u\|_{V^{1/2,2}_{\delta+1}(\Gamma^\pm)^\ell}\Big).$$

Let $v \in E^{2,2}_{\delta+1}(K)^\ell$, and let χ be a smooth function on $(-\infty,+\infty)$ such that $0 \le \chi \le 1$, $\chi(t) = 1$ for $|t| < 1$ and $\chi(t) = 0$ for $|t| > 2$. We put $u(x) = \chi(x_3/N)\, e^{\pm i x_3}\, v(x')$, where N is sufficiently large. Analogously to the proof of Theorem 2.4.3, it follows from (6.1.21) that

$$\|v\|_{E^{2,2}_{\delta+1}(K)^\ell} \le c\Big(\|L(D_{x'},\pm 1)v\|_{E^{0,2}_{\delta+1}(K)^\ell} + \sum_\pm \|N^\pm(D_{x'},\pm 1)v\|_{E^{1/2,2}_{\delta+1}(\gamma^\pm)^\ell}\Big)$$

for all $v \in E^{2,2}_{\delta+1}(K)^\ell$. Therefore, the kernel of the operator

$$E^{2,2}_{\delta+1}(K)^\ell \ni v \to \big(L(D_{x'},\pm 1)v, N^\pm(D_{x'},\pm 1)v\big) \in E^{0,2}_{\delta+1}(K)^\ell \times \prod_\pm E^{1/2,2}_{\delta+1}(\gamma^\pm)^\ell$$

is trivial and the range of this operator is closed. The same is true for the operator

$$E^{2,2}_{1-\delta}(K)^\ell \ni v \to \big(L(D_{x'},\pm 1)v, N^\pm(D_{x'},\pm 1)v\big) \in E^{0,2}_{1-\delta}(K)^\ell \times \prod_\pm E^{1/2,2}_{1-\delta}(\gamma^\pm)^\ell .$$

By the self-adjointness, both operators are isomorphisms. Now let u be an arbitrary vector function in $E^{1,2}_\delta(K)^\ell$. Then $r^{2\delta-2}u \in E^{0,2}_{1-\delta}(K)^\ell$. Consequently, there exist a unique solution $v \in E^{2,2}_{1-\delta}(K)^\ell$ of the boundary value problem

$$L(D_{x'},\pm 1)v = r^{2\delta-2}\, u \ \text{ in } K, \quad N^\pm(D_{x'},\pm 1)v = 0 \ \text{ on } \gamma^\pm$$

satisfying the estimate

$$\|v\|_{E^{2,2}_{1-\delta}(K)^\ell} \le c\, \|r^{2\delta-2}\, u\|_{E^{0,2}_{1-\delta}(K)^\ell} = c\, \|u\|_{E^{0,2}_{\delta-1}(K)^\ell} .$$

This implies

$$\begin{aligned} \|u\|^2_{E^{0,2}_{\delta-1}(K)^\ell} &= \int_K u \cdot r^{2\delta-2}\, \bar u\, dx' = \int_K u \cdot \overline{L(D_{x'},\pm 1)\, v}\, dx' = b_K(u,v;\pm 1) \\ &\le \|A_\delta u\|_{(E^{1,2}_{-\delta}(K)^\ell)^*}\, \|v\|_{E^{1,2}_{-\delta}(K)^\ell} \le c\, \|A_\delta u\|_{(E^{1,2}_{-\delta}(K)^\ell)^*}\, \|u\|_{E^{0,2}_{\delta-1}(K)^\ell} , \end{aligned}$$

where A_δ denotes the operator in Lemma 6.1.9. Using Lemma 6.1.6, we obtain the estimate

$$\|u\|_{E^{1,2}_\delta(K)^\ell} \le c\, \|A_\delta u\|_{(E^{1,2}_{-\delta}(K)^\ell)^*} .$$

The same estimate holds for the operator $A_{-\delta}$. Since $A_{-\delta}$ is the adjoint operator of A_δ, it follows that A_δ is an isomorphism. However, this contradicts Lemma 6.1.9. The theorem is proved. □

6.2. Weighted Sobolev spaces with nonhomogeneous norms

In this section, we deal with the weighted spaces $W^{l,p}_\delta(\mathcal{D})$ and the corresponding trace spaces. In contrast to the weighted spaces $V^{l,p}_\delta(\mathcal{D})$, here also traces on the edge of the dihedron must be considered. We show that the trace space for $W^{l,p}_\delta(\mathcal{D})$ is the Sobolev-Slobodetskiĭ space $W^{l-\delta-2/p,p}(M)$ if $-2/p < \delta < l-2/p$ and $\delta+2/p$ is not integer. Furthermore, we study the relations between the spaces $V^{l,p}_\delta(\mathcal{D})$ and $W^{l,p}_\delta(\mathcal{D})$. As is shown in this section, any function $u \in W^{l,p}_\delta(\mathcal{D})$ is a sum of a "quasipolynomial" and a $V_\delta^{l,p}(\mathcal{D})$-function. In particular, it follows from this result in the case of noninteger $\delta+2/p<l$ that $u\in V_\delta^{l,p}(\mathcal{D})$ if and only if the traces of the derivatives $\partial_{x_1}^i\partial_{x_2}^j u$ on the edge M are zero for $i+j<l-\delta-2/p$. An analogous condition holds for integer $\delta+2/p$.

6.2.1. The spaces $L_\delta^{l,p}(\mathcal{D})$ and $W_\delta^{l,p}(\mathcal{D})$. Let $1<p<\infty$ and $\delta>-2/p$. For arbitrary nonnegative integer l, we define the space $L_\delta^{l,p}(\mathcal{D})$ as the closure of the set $C_0^\infty(\overline{\mathcal{D}})$ with respect to the norm

$$\|u\|_{L_\delta^{l,p}(\mathcal{D})}=\Big(\int_{\mathcal{D}} r^{p\delta}\sum_{|\alpha|=l}|\partial_x^\alpha u(x)|^p\,dx\Big)^{1/p},$$

where again $r=|x'|=(x_1^2+x_2^2)^{1/2}$ is the distance of the point x from the edge M of $\mathcal{D}$. Furthermore, let $W_\delta^{l,p}(\mathcal{D})=L_\delta^{0,p}(\mathcal{D})\cap\cdots\cap L_\delta^{l,p}(\mathcal{D})$ be the closure of the set $C_0^\infty(\overline{\mathcal{D}})$ with respect to the norm

$$\|u\|_{W_\delta^{l,p}(\mathcal{D})}=\Big(\int_{\mathcal{D}} r^{p\delta}\sum_{|\alpha|\le l}|\partial_x^\alpha u(x)|^p\,dx\Big)^{1/p}.$$

6.2.2. Imbeddings. By Hardy's inequality, every function $u\in C_0^\infty(\overline{\mathcal{D}})$ satisfies the inequality

$$\int_{\mathcal{D}} r^{p(\delta-1)}|u|^p\,dx\le c\int_{\mathcal{D}} r^{p\delta}\,|\nabla u|^p\,dx \tag{6.2.1}$$

for $\delta>1-2/p$ with a constant c depending only on p and δ . Consequently, the following lemma holds.

LEMMA 6.2.1. *Let $l>l'$ and $\delta>l-l'-2/p$. Then*

$$L_\delta^{l,p}(\mathcal{D})\subset L_{\delta-l+l'}^{l',p}(\mathcal{D})\quad\text{and}\quad W_\delta^{l,p}(\mathcal{D})\subset W_{\delta-l+l'}^{l',p}(\mathcal{D}).$$

Both imbeddings are continuous.

In particular,

$$W_\delta^{l,p}(\mathcal{D})\subset L_\delta^{l,p}(\mathcal{D})=V_\delta^{l,p}(\mathcal{D})\quad\text{if } \delta>l-2/p.$$

Furthermore, we note that (as a consequence of Hardy's inequality) every function $u\in L_\delta^{l,p}(\mathcal{D})$ with compact support belongs to the space $W_\delta^{l,p}(\mathcal{D})$.

6.2.3. Traces on the edge M. Let s be a positive noninteger number and $k=[s]$ its integral part. Then $W^{s,p}(\mathbb{R})$ denotes the Sobolev-Slobodetskiĭ space with the norm

$$\|f\|_{W^{s,p}(\mathbb{R})}=\Big(\|f\|^p_{L_p(\mathbb{R})}+\int\limits_{\mathbb{R}}\int\limits_{\mathbb{R}}\frac{|f^{(k)}(\zeta)-f^{(k)}(z)|^p}{|z|^{1+p(s-k)}}\,d\zeta\,dz\Big)^{1/p}.$$

LEMMA 6.2.2. *Suppose that $l\ge1$, $-2/p<\delta<l-2/p$, and that $\delta+2/p$ is not integer. Then the trace of an arbitrary function $u\in W_\delta^{l,p}(\mathcal{D})$ on M exists. Furthermore, $u|_M\in W^{l-\delta-2/p,p}(M)$ and*

$$\|u|_M\|_{W^{l-\delta-2/p,p}(M)}\le c\,\|u\|_{W_\delta^{l,p}(\mathcal{D})}, \tag{6.2.2}$$

where c is a constant independent of u.

P r o o f. Suppose first that $l-1-2/p<\delta<l-2/p$. Then the space $W^{l,p}_\delta(\mathcal{D})$ is continuously imbedded in $W^{1,p}_{\delta-l+1}(\mathcal{D})$. Let

$$\overset{\circ}{u}(r,x_3)=\frac{1}{\theta}\int\limits_{-\theta/2}^{+\theta/2} u(r\cos\varphi, r\sin\varphi, x_3)\, d\varphi$$

be the average of u with respect to the angle φ. We estimate the L_p norm of $\overset{\circ}{u}(0,\cdot)$. Obviously,

$$\begin{aligned}\|\overset{\circ}{u}(0,\cdot)\|^p_{L_p(\mathbb{R})} &= \big(p(\delta-l+1)+2\big)\int\limits_0^1\int\limits_{\mathbb{R}} \rho^{p(\delta-l+1)+1}\,|\overset{\circ}{u}(0,x_3)|^p\, dx_3\, d\rho \\ &\le c\int\limits_0^1\int\limits_{\mathbb{R}} \rho^{p(\delta-l+1)+1}\Big(|\overset{\circ}{u}(\rho,x_3)-\overset{\circ}{u}(0,x_3)|^p+|\overset{\circ}{u}(\rho,x_3)|^p\Big)\, dx_3\, d\rho.\end{aligned}$$

Using Hölder's inequality, we obtain

$$\begin{aligned}&\int\limits_0^1\int\limits_{\mathbb{R}} \rho^{p(\delta-l+1)+1}\,|\overset{\circ}{u}(\rho,x_3)-\overset{\circ}{u}(0,x_3)|^p\, dx_3\, d\rho \\ &\quad= \int\limits_0^1\int\limits_{\mathbb{R}} \rho^{p(\delta-l+1)+1}\Big|\frac{1}{\theta}\int\limits_0^\rho\int\limits_{-\theta/2}^{\theta/2} \partial_r u(r\cos\varphi, r\sin\varphi, x_3)\, d\varphi\, dr\Big|^p dx_3\, d\rho \\ &\quad\le c\int\limits_0^1\int\limits_{\mathbb{R}} \rho^{p-1}\Big(\int\limits_0^\rho\int\limits_{-\theta/2}^{\theta/2} r^{p(\delta-l+1)+1}|\partial_r u(r\cos\varphi, r\sin\varphi, x_3)|^p\, d\varphi\, dr\Big)\, dx_3\, d\rho \\ &\quad\le c\int\limits_{\mathbb{R}}\int\limits_0^1\int\limits_{-\theta/2}^{\theta/2} r^{p(\delta-l+1)+1}\,|\partial_r u|^p\, d\varphi\, dr\, dx_3 \le c\,\|u\|^p_{W^{1,p}_{\delta-l+1}(\mathcal{D})}.\end{aligned}$$

Since furthermore

$$\int\limits_0^1\int\limits_{\mathbb{R}} \rho^{p(\delta-l+1)+1}\,|\overset{\circ}{u}(\rho,x_3)|^p\, dx_3\, d\rho \le c\,\|u\|^p_{W^{1,p}_{\delta-l+1}(\mathcal{D})},$$

we obtain

$$\|\overset{\circ}{u}(0,\cdot)\|_{L_p(\mathbb{R})} \le c\,\|u\|_{W^{1,p}_{\delta-l+1}(\mathcal{D})}.$$

Next, we prove that

$$\int\limits_{\mathbb{R}}\int\limits_{\mathbb{R}} |z|^{1-p(l-\delta)}\,|\overset{\circ}{u}(0,x_3+z)-\overset{\circ}{u}(0,x_3)|^p\, dx_3\, dz \le c\,\|u\|^p_{W^{1,p}_{\delta-l+1}(\mathcal{D})}. \tag{6.2.3}$$

For this we use the decomposition

$$\begin{aligned}\overset{\circ}{u}(0,x_3+z)-\overset{\circ}{u}(0,x_3) &= \overset{\circ}{u}(|z|,x_3+z)-\overset{\circ}{u}(|z|,x_3)+\overset{\circ}{u}(|z|,x_3)-\overset{\circ}{u}(0,x_3) \\ &\quad - \overset{\circ}{u}(|z|,x_3+z)+\overset{\circ}{u}(0,x_3+z).\end{aligned}$$

Here

$$\overset{\circ}{u}(|z|, x_3 + z) - \overset{\circ}{u}(|z|, x_3) = z \int_0^1 \partial_{x_3} \overset{\circ}{u}(|z|, x_3 + tz)\, dt$$

and therefore

$$\int_{\mathbb{R}}\int_{\mathbb{R}} |z|^{p(\delta-l)+1} \left| \overset{\circ}{u}(|z|, x_3 + z) - \overset{\circ}{u}(|z|, x_3)\right|^p dx_3\, dz$$

$$\le \int_0^1 \int_{\mathbb{R}}\int_{\mathbb{R}} |z|^{p(\delta-l+1)+1} \left|\partial_{x_3} \overset{\circ}{u}(|z|, x_3 + tz)\right|^p dx_3\, dz\, dt$$

$$= 2 \int_0^\infty \int_{\mathbb{R}} r^{p(\delta-l+1)+1} \left|\partial_{x_3} \overset{\circ}{u}(r, x_3)\right|^p dx_3\, dr.$$

Furthermore,

$$\int_{\mathbb{R}}\int_{\mathbb{R}} |z|^{p(\delta-l)+1} \left| \overset{\circ}{u}(|z|, x_3) - \overset{\circ}{u}(0, x_3)\right|^p dx_3\, dz$$

$$= \int_{\mathbb{R}}\int_{\mathbb{R}} |z|^{p(\delta-l)+1} \left| \int_0^{|z|} \partial_\rho \overset{\circ}{u}(\rho, x_3)\, d\rho\right|^p dx_3\, dz$$

$$= 2 \int_{\mathbb{R}}\int_0^\infty r^{p(\delta-l)+1} \left| \int_0^{r} \partial_\rho \overset{\circ}{u}(\rho, x_3)\, d\rho\right|^p dr\, dx_3\, .$$

Estimating the integral on the right-hand side by means of Hardy's inequality, we get

$$\int_{\mathbb{R}}\int_{\mathbb{R}} |z|^{p(\delta-l)+1} \left| \overset{\circ}{u}(|z|, x_3) - \overset{\circ}{u}(0, x_3)\right|^p dx_3\, dz$$

$$\le c \int_{\mathbb{R}}\int_0^\infty r^{p(\delta-l+1)+1} \left|\partial_r \overset{\circ}{u}(r, x_3)\right|^p dr\, dx_3\, .$$

The same inequality holds for the expression $\overset{\circ}{u}(|z|, x_3 + z) - \overset{\circ}{u}(0, x_3 + z)$. The last estimates together with the obvious inequality

$$\int_{\mathbb{R}}\int_0^\infty r^{p(\delta-l+1)+1} \Big(\left|\partial_r \overset{\circ}{u}(r, x_3)\right|^p + \left|\partial_{x_3} \overset{\circ}{u}(r, x_3)\right|^p\Big)\, dr\, dx_3 \le c\, \|u\|^p_{W^{1,p}_{\delta-l+1}(\mathcal{D})}$$

imply (6.2.3). Thus, the lemma is proved for the case $l - 1 - 2/p < \delta < l - 2/p$.

Now let $l-k-1-2/p < \delta < l-k-2/p$, where k is a positive integer, $k \le l-1$. Since $\partial^j_{x_3} u \in W^{l-k,p}_\delta(\mathcal{D})$ for $j = 0, \dots, k$, it follows from the first part of the proof that $\partial^j_{x_3} u|_M \in W^{l-k-\delta-2/p,p}(M)$ for $j \le k$. Consequently, $u|_M \in W^{l-\delta-2/p,p}(M)$. Furthermore, the estimate (6.2.2) holds. □

Note that the trace of an arbitrary function $u \in V^{l,p}_\delta(\mathcal{D})$ on M is equal to zero if $\delta < l - 2/p$. If $\delta \ge l - 2/p$, then in general the traces of functions of the spaces $V^{l,p}_\delta(\mathcal{D})$ and $W^{l,p}_\delta(\mathcal{D})$ do not exist on M.

6.2.4. Extension from M to $\mathcal{D}$. Next, we prove that any function $f \in W^{l-\delta-2/p,p}(M)$ can be extended to a function $u \in W^{l,p}_\delta(\mathcal{D})$ if the conditions on l, p, δ in Lemma 6.2.2 are satisfied. To this end, we introduce the following extension operator E:

$$(Ef)(x', x_3) = \chi(r) \int_{\mathbb{R}} f(x_3 + tr)\, \psi(t)\, dt. \tag{6.2.4}$$

Here χ is an infinitely differentiable function on the interval $(0, +\infty)$ such that $\chi(r) = 1$ for $r < 1/2$, $\chi(r) = 0$ for $r > 1$, and ψ is an infinitely differentiable function on $(-\infty, +\infty)$ vanishing outside the interval $[-1, +1]$ such that

$$\int_{\mathbb{R}} \psi(t)\, dt = 1, \quad \int_{\mathbb{R}} t^j\, \psi(t)\, dt = 0 \ \text{ for } j = 1, \dots, [l - \delta - 2/p], \tag{6.2.5}$$

where $[l - \delta - 2/p]$ is the integral part of $l - \delta - 2/p$.

LEMMA 6.2.3. *Let $f \in W^{l-\delta-2/p,p}(M)$, where l is an integer, $l \ge 1$, $-2/p < \delta < l - 2/p$, and $\delta + 2/p$ is not integer. Then $Ef \in W^{l,p}_\delta(\mathcal{D})$, $\partial_{x_j} Ef \in V^{l-1,p}_\delta(\mathcal{D})$ for $j = 1, 2$, and $Ef = f$ on M. Furthermore, the estimate*

$$\|Ef\|_{W^{l,p}_\delta(\mathcal{D})} + \sum_{j=1}^{2} \|\partial_{x_j} Ef\|_{V^{l-1,p}_\delta(\mathcal{D})} \le c\, \|f\|_{W^{l-\delta-2/p,p}(M)} \tag{6.2.6}$$

is satisfied with a constant c independent of f.

P r o o f. Let v be the function

$$v(r, x_3) = \int_{\mathbb{R}} f(x_3 + tr)\, \psi(t)\, dt,$$

and let k be the integral part of $l - \delta - 2/p$. Then

$$\partial^j_{x_3} v(r, x_3) = \int_{\mathbb{R}} f^{(j)}(x_3 + tr)\, \psi(t)\, dt$$

for $j \le k$. Since $f^{(j)} \in L_p(M)$ for $j \le k$, it follows that

$$\int_0^1 \int_{\mathbb{R}} r^{-1+\varepsilon} \left|\partial^j_{x_3} v(r, x_3)\right|^p dx_3\, dr \le c\, \|f^{(j)}\|^p_{L_p(M)} \le c\, \|f\|^p_{W^{l-\delta-2/p,p}(M)} \tag{6.2.7}$$

for $j \le k$, where ε is an arbitrarily small positive number. We show that

$$\int_0^1 \int_{\mathbb{R}} r^{p(\delta-l+i+j)+1} \left|\partial^i_r \partial^j_{x_3} v(r, x_3)\right|^p dx_3\, dr \le c\, \|f\|^p_{W^{l-\delta-2/p,p}(M)} \tag{6.2.8}$$

for $i \ge 1$ or $j \ge k+1$. Let $j \ge k+1$. Then

$$\begin{aligned}
\partial^j_{x_3} v(r, x_3) &= \partial^{j-k}_{x_3} \int_{\mathbb{R}} f^{(k)}(x_3 + tr)\, \psi(t)\, dt = \frac{1}{r}\, \partial^{j-k}_{x_3} \int_{\mathbb{R}} f^{(k)}(\tau)\, \psi\Big(\frac{\tau - x_3}{r}\Big)\, d\tau \\
&= \frac{(-1)^{j-k}}{r^{j-k+1}} \int_{\mathbb{R}} f^{(k)}(\tau)\, \psi^{(j-k)}\Big(\frac{\tau - x_3}{r}\Big)\, d\tau \\
&= \Big(-\frac{1}{r}\Big)^{j-k} \int_{\mathbb{R}} \big(f^{(k)}(x_3 + tr) - f^{(k)}(x_3)\big)\, \psi^{(j-k)}(t)\, dt.
\end{aligned}$$

This implies

$$\int_0^1 \int_{\mathbb{R}} r^{p(\delta-l+j)+1} \left|\partial_{x_3}^j v(r,x_3)\right|^p dx_3\, dr$$
$$\le c \int_0^1 \int_{\mathbb{R}} \int_{-1}^{+1} |t|^{p(l-k-\delta)-1} \frac{\left|f^{(k)}(x_3+tr) - f^{(k)}(x_3)\right|^p}{|tr|^{p(l-k-\delta)-1}}\, dt\, dx_3\, dr.$$

Thus, the inequality (6.2.8) is valid for $i = 0$, $j \ge k+1$. Now let $i \ge 1$ and $i+j \ge k+1$. In this case, there exist nonnegative integers μ and ν such that $\mu < i$, $\nu \le j$, $\mu + \nu = k$. Then

$$\begin{aligned}\partial_r^i \partial_{x_3}^j v(r,x_3) &= \partial_r^{i-\mu}\partial_{x_3}^{j-\nu} \int_{\mathbb{R}} \big(f^{(k)}(x_3+tr) - f^{(k)}(x_3)\big)\, t^\mu\, \psi(t)\, dt \\ &= \int_{\mathbb{R}} \big(f^{(k)}(\tau) - f^{(k)}(x_3)\big)\, \partial_r^{i-\mu}\partial_{x_3}^{j-\nu} \frac{1}{r}\Big(\frac{\tau - x_3}{r}\Big)^\mu \psi\Big(\frac{\tau-x_3}{r}\Big)\, d\tau \\ &= r^{k-i-j} \int_{\mathbb{R}} \big(f^{(k)}(x_3+tr) - f^{(k)}(x_3)\big)\, \Psi(t)\, dt,\end{aligned}$$

where Ψ is an infinitely differentiable function equal to zero outside the interval $[-1,+1]$. Analogously to the case $i = 0$, $j \ge k+1$, the last inequality implies (6.2.8). Finally, let $i \ge 1$ and $i + j \le k$. Then

$$\partial_r^i \partial_{x_3}^j v(r,x_3) = \int_{\mathbb{R}} f^{(i+j)}(x_3+tr)\, t^i\, \psi(t)\, dt.$$

We introduce the functions

$$\psi_0(t) = t^i \psi(t), \quad \psi_1(t) = \int_{-1}^t \psi_0(\tau)\, d\tau, \ldots,$$
$$\psi_k(t) = \int_{-1}^t \psi_{k-1}(\tau)\, d\tau = \int_{-1}^t \tau^i \psi(\tau)\, \frac{(t-\tau)^{k-1}}{(k-1)!}\, d\tau.$$

According to the conditions on ψ, the functions $\psi_1, \ldots, \psi_{k+1-i}$ vanish outside the interval $[-1,+1]$. Furthermore, the integrals of the functions $\psi_1, \ldots, \psi_{k-i}$ on the interval $[-1,+1]$ are zero. Integrating by parts, we obtain

$$\begin{aligned}\partial_r^i \partial_{x_3}^j v(r,x_3) &= (-r)^{k-i-j} \int_{\mathbb{R}} f^{(k)}(x_3+tr)\, \psi_{k-i-j}(t)\, dt \\ &= (-r)^{k-i-j} \int_{\mathbb{R}} \big(f^{(k)}(x_3+tr) - f^{(k)}(x_3)\big)\, \psi_{k-i-j}(t)\, dt.\end{aligned}$$

Thus, (6.2.8) holds also in the case $i \ge 1$, $i + j \le k$. The estimate (6.2.6) follows immediately from (6.2.7), (6.2.8) and from the estimate

$$|\partial_{x'}^\alpha \partial_{x_3}^j v(r,x_3)| \le c \sum_{i=1}^{|\alpha|} r^{i-|\alpha|} \left|\partial_r^i \partial_{x_3}^j v(r,x_3)\right| \quad \text{for } |\alpha| \ge 1.$$

It remains to prove that the trace of Ef on M is equal to f. Obviously, it suffices to show this for $0 < l - \delta - 2/p < 1$, i.e. $k = 0$. Then

$$\int_0^1 \int_{\mathbb{R}} r^{p(\delta-l)+1} \big|v(r,x_3) - f(x_3)\big|^p \, dx_3 \, dr$$
$$\le \int_0^1 \int_{\mathbb{R}} r^{p(\delta-l)+1} \Big| \int_{\mathbb{R}} \big(f(x_3+tr) - f(x_3)\big)\, \psi(t)\, dt \Big|^p \, dx_3\, dr$$
$$\le c \int_0^1 \int_{\mathbb{R}} \int_{-1}^{+1} |t|^{p(l-\delta)-1} \frac{\big|f(x_3+tr) - f(x_3)\big|^p}{|tr|^{p(l-\delta)-1}} \, dt\, dx_3\, dr \le c\, \|f\|^p_{W^{l-\delta-2/p,p}(M)}\,.$$

Since $p(\delta - l) + 1 < -1$, it follows that $v|_M = f$. The proof is complete. □

The last lemma allows us also to answer the question of the existence of a function $u \in W_\delta^{l,p}(\mathcal{D})$ satisfying the trace conditions

$$\partial_{x_1}^i \partial_{x_2}^j v = f_{i,j} \quad \text{on } M \quad \text{for } i + j < l - \delta - 2/p \tag{6.2.9}$$

if $\delta + 2/p$ is not integer.

COROLLARY 6.2.4. *Let l be an integer, $l \ge 1$, and let δ be a real number such that $-2/p < \delta < l - 2/p$ and $\delta + 2/p$ is not integer. Furthermore, let $f_{i,j} \in W^{l-\delta-i-j-2/p,p}(M)$ for $i + j < l - \delta - 2/p$. Then the function*

$$v(x) = \sum_{i+j<l-\delta-2/p} (Ef_{i,j})(x)\, \frac{x_1^i\, x_2^j}{i!\, j!}\,,$$

where E denotes the extension operator (6.2.4), belongs to the space $W_\delta^{l,p}(\mathcal{D})$ and satisfies the equalities (6.2.4). Moreover, the estimate

$$\|v\|_{W_\delta^{l,p}(\mathcal{D})} \le c \sum_{i+j<l-\delta-2/p} \|f_{i,j}\|_{W^{l-\delta-i-j-2/p,p}(M)}$$

holds with a constant c independent of the functions $f_{i,j}$.

P r o o f. By Lemma 6.2.3, we have

$$Ef_{i,j} \in W^{l,p}_{\delta+i+j}(\mathcal{D}) \quad \text{and} \quad \partial_{x_k} Ef_{i,j} \in V^{l-1,p}_{\delta+i+j}(\mathcal{D}) \quad \text{for } k = 1, 2.$$

Consequently, $v \in W_\delta^{l,p}(\mathcal{D})$ and the traces of $\partial_{x_1}^i \partial_{x_2}^j v$ exist for $i + j < l - \delta - 2/p$. We consider the term

$$\partial_{x_1}^\mu \partial_{x_2}^\nu \Big(Ef_{i,j}\, \frac{x_1^i\, x_2^j}{i!\, j!} \Big) = \sum_{s=0}^{\min(i,\mu)} \sum_{t=0}^{\min(j,\nu)} \binom{\mu}{s} \binom{\nu}{t} \big(\partial_{x_1}^{\mu-s} \partial_{x_2}^{\nu-t} Ef_{i,j}\big)\, \frac{x_1^{i-s}\, x_2^{j-t}}{(i-s)!\,(j-t)!}$$

for $i + j < l - \delta - 2/p$ and $\mu + \nu < l - \delta - 2/p$. By Lemma 6.2.3,

$$\big(\partial_{x_1}^{\mu-s} \partial_{x_2}^{\nu-t} Ef_{i,j}\big)\, \frac{x_1^{i-s}\, x_2^{j-t}}{(i-s)!\,(j-t)!} \in V_\delta^{l-\mu-\nu,p}(\mathcal{D}) \quad \text{for } s + t < \mu + \nu.$$

Therefore, the trace of $\big(\partial_{x_1}^{\mu-s} \partial_{x_2}^{\nu-t} Ef_{i,j}\big)\, x_1^{i-s}\, x_2^{j-t}$ on M is zero for $s + t < \mu + \nu$. For $s = i = \mu$ and $t = j = \nu$, the trace of this expression on M is equal to $f_{\mu,\nu}$. Consequently,

$$\partial_{x_1}^\mu \partial_{x_2}^\nu v = f_{\mu,\nu} \quad \text{on } M \quad \text{for } \mu + \nu < l - \delta - 2/p.$$

The corollary is proved. □

6.2.5. A property of the space $V_\delta^{l,p}(\mathcal{D})$. As was mentioned above, the space $W_\delta^{l,p}(\mathcal{D})$ is continuously imbedded in $V_\delta^{l,p}(\mathcal{D})$ if $\delta > l-2/p$. We are interested here in relations between the spaces $V_\delta^{l,p}(\mathcal{D})$ and $W_\delta^{l,p}(\mathcal{D})$ for $\delta \le l-2/p$. First we consider the case of noninteger $\delta+2/p$.

LEMMA 6.2.5. *Suppose that $u \in W_\delta^{l,p}(\mathcal{D})$, where $-2/p < \delta < l-2/p$ and $\delta+2/p$ is not an integer. Then $u \in V_\delta^{l,p}(\mathcal{D})$ if and only if $\partial_{x_1}^i \partial_{x_2}^j u = 0$ on M for $i+j \le [l-\delta-2/p]$. Furthermore, the inequality*

$$\|u\|_{V_\delta^{l,p}(\mathcal{D})} \le c\,\|u\|_{W_\delta^{l,p}(\mathcal{D})} \tag{6.2.10}$$

holds in this case, where c is a constant independent of u.

P r o o f. First note that $W_\delta^{l,p}(\mathcal{D}) \subset W_{\delta-l+k+1}^{k+1,p}(\mathcal{D})$, where $k = [l-\delta-2/p]$. Furthermore by Lemma 6.2.2, the traces of $\partial_{x_1}^i \partial_{x_2}^j u$ on M exist for $i+j \le k$. Suppose that these traces are equal to zero. Then by Hardy's inequality,

$$\int_{\mathcal{D}} r^{p(\delta-l)}\,|u(x)|^p\,dx \le c_1 \int_{\mathcal{D}} r^{p(\delta-l+1)}\big(|\partial_{x_1} u|^p + |\partial_{x_2} u|^p\big)\,dx$$
$$\le \cdots \le c_k \int_{\mathcal{D}} \sum_{i+j=k} r^{p(\delta-l+k)}\,\big|\partial_{x_1}^i \partial_{x_2}^j u(x)\big|^p\,dx \le c\,\|u\|^p_{L^{k+1,p}_{\delta-l+k+1}(\mathcal{D})}\,.$$

Consequently, the function u belongs to the space $V_\delta^{l,p}(\mathcal{D})$ and satisfies (6.2.10). Conversely, all derivatives up to order k of an arbitrary function $u \in V_\delta^{l,p}(\mathcal{D})$ have traces equal to zero on M. The lemma is proved. □

In the case of integer $\delta+2/p$, the following modification of Lemma 6.2.5 holds.

LEMMA 6.2.6. *Suppose that $u \in W_\delta^{l,p}(\mathcal{D})$, where $l \ge 1$, $\delta+2/p$ is an integer, $1 \le \delta+2/p \le l$. Then $u \in V_\delta^{l,p}(\mathcal{D})$ if and only if the conditions*

$$\partial_{x_1}^i \partial_{x_2}^j u = 0 \quad \text{on } M \quad \text{for } i+j < l-\delta-2/p, \tag{6.2.11}$$

$$\int_{\mathcal{D}} r^{-2}\,\big|\partial_{x_1}^i \partial_{x_2}^j u\big|^p\,dx < \infty \text{ for } i+j = l-\delta-2/p \tag{6.2.12}$$

are satisfied.

P r o o f. The necessity of (6.2.11) and (6.2.12) is obvious. Let $u \in W_\delta^{l,p}(\mathcal{D})$ be a function satisfying (6.2.11) and (6.2.12). We show by induction in $k = l-\delta-2/p$ that $u \in V_\delta^{l,p}(\mathcal{D})$.

1) If $k=0$, i.e. $\delta+2/p = l$, then

$$W_\delta^{l,p}(\mathcal{D}) \subset W_{\delta-1}^{l-1,p}(\mathcal{D}) \subset \cdots \subset W_{\delta-l+1}^{1,p}(\mathcal{D}).$$

This together with (6.2.12) implies $u \in V_\delta^{l,p}(\mathcal{D})$.

2) Let $k=1$. Then

$$W_\delta^{l,p}(\mathcal{D}) \subset W_{\delta-1}^{l-1,p}(\mathcal{D}) \subset \cdots \subset W_{\delta-l+2}^{2,p}(\mathcal{D}). \tag{6.2.13}$$

Since $u|_M = 0$, it follows from Hardy's inequality and condition (6.2.12) that

$$\int_{\mathcal{D}} r^{-2-p}|u(x)|^p\,dx \le c \int_{\mathcal{D}} r^{-2}\,|\partial_{x_1} u|^p + |\partial_{x_2} u|^p)\,dx < \infty. \tag{6.2.14}$$

It remains to show that

$$\int_{\mathcal{D}} r^{-2}\, |\partial_{x_3} u(x)|^p\, dx < \infty. \tag{6.2.15}$$

We introduce the norm

$$\|u\|^* = \Big(\int_{\mathcal{D}} \Big(r^{-2-p}\, |u(x)|^p + r^{-2+p} \sum_{j=1}^{3} |\partial_{x_j}^2 u(x)|^p \Big)\, dx \Big)^{1/p}.$$

By (6.2.13) and (6.2.14), we have

$$\|u\|^* \le C < \infty.$$

Let ζ be an infinitely differentiable function on the interval $(0,\infty)$, $\zeta(r) = 0$ for $r < 1$, $\zeta(r) = 1$ for $r > 2$. For an arbitrary positive ε, we define $\zeta_\varepsilon(r) = \zeta(r/\varepsilon)$. We show that $\zeta_\varepsilon u \to u$ in the norm $\|\cdot\|^*$. Obviously,

$$\begin{aligned} \|(1-\zeta_\varepsilon)\, u\|^* \ &\le\ c\Big(\int_{\mathcal{D}} \big(r^{-2-p}\, |(1-\zeta_\varepsilon)\, u|^p + r^{-2+p} \sum_{j=1}^{3} |(1-\zeta_\varepsilon)\, \partial_{x_j}^2 u|^p \big)\, dx \\ &\quad + \int_{\mathcal{D}} r^{-2+p} \sum_{j=1}^{2} \big(|\partial_{x_j}(1-\zeta_\varepsilon)\, \partial_{x_j} u|^p + |u\, \partial_{x_j}^2 (1-\zeta_\varepsilon)|^p \big)\, dx \Big) \end{aligned}$$

The first integral on the right-hand side of the last inequality tends to zero as $\varepsilon \to 0$, since $\|u\|^* \le C$. Furthermore,

$$\begin{aligned} &\int_{\mathcal{D}} r^{-2+p} \big(|\partial_{x_j}(1-\zeta_\varepsilon)\, \partial_{x_j} u|^p + |u\, \partial_{x_j}^2 (1-\zeta_\varepsilon)|^p \big)\, dx \Big) \\ &\qquad \le c \int\limits_{\substack{\mathcal{D} \\ \varepsilon<r<2\varepsilon}} \big(r^{-2}\, |\partial_{x_j} u|^p + r^{-2-p}\, |u|^p \big)\, dx \end{aligned}$$

for $j = 1$ and $j = 2$, where c is independent of ε. Since the right-hand side of the last inequality also tends to zero as $\varepsilon \to 0$, it follows that $\zeta_\varepsilon u \to u$ in the norm $\|\cdot\|^*$. However, $\zeta_\varepsilon u \in V_\delta^{l,p}(\mathcal{D}) \subset V_{1-2/p}^{2,p}(\mathcal{D})$, and on the last space the $V_{1-2/p}^{2,p}(\mathcal{D})$-norm and the norm $\|\cdot\|^*$ are equivalent (cf. Lemma 2.1.6). Consequently, there exists a constant c independent of ε such that

$$\|\zeta_\varepsilon u\|^p_{V_{1-2/p}^{2,p}(\mathcal{D})} \le c$$

for all $\varepsilon < 1$. In particular,

$$\int_{\mathcal{D}} r^{-2}\, |\partial_{x_3}(\zeta_\varepsilon u)|^p\, dx \le c$$

which implies (6.2.15). This together with (6.2.13) and (6.2.14) implies $u \in V_\delta^{l,p}(\mathcal{D})$. Thus, the lemma is true for $k = 1$.

3) Suppose that $k = l - \delta - 2/p \ge 2$ and the assertion of the lemma is true for $l - \delta - 2/p = k - 1$. Then we conclude from the conditions (6.2.11) and (6.2.12) that $\partial_{x_j} u \in V_\delta^{l-1,p}(\mathcal{D})$ for $j = 1$ and $j = 2$. In particular, it follows that

$$\int_{\mathcal{D}} r^{-2}\, |\partial_{x_1}^i \partial_{x_2}^j \partial_{x_3} u|^p\, dx < \infty \text{ for } i + j = k - 1.$$

Moreover by (6.2.11), the traces of $\partial_{x_1}^i \partial_{x_2}^j \partial_{x_3} u$ on M are zero for $i+j<k-1$. Therefore by the induction hypothesis, we have $\partial_{x_3} u \in V_\delta^{l-1,p}(\mathcal{D})$. Furthermore by Hardy's inequality,

$$\int_{\mathcal{D}} r^{p(\delta-l)} \, |u(x)|^p \, dx \le c \int_{\mathcal{D}} \sum_{j=1}^{2} r^{p(\delta-l+1)} \, \big|\partial_{x_j} u\big|^p \, dx < \infty.$$

Thus, $u \in V_\delta^{l,p}(\mathcal{D})$ and the lemma is proved for all k. □

6.2.6. A relation between the spaces $V_\delta^{l,p}(\mathcal{D})$ and $W_\delta^{l,p}(\mathcal{D})$ in the case of noninteger $\delta + 2/p$. As a consequence of Corollary 6.2.4 and Lemma 6.2.5, we obtain the following theorem.

THEOREM 6.2.7. *Let $u \in W_\delta^{l,p}(\mathcal{D})$, where $-2/p < \delta < l - 2/p$ and $\delta + 2/p$ is not an integer, and let $f_{i,j}$ be the traces of $\partial_{x_1}^i \partial_{x_2}^j u$ on M for $i + j \le k = [l - \delta - 2/p]$. Furthermore, let v be the quasi-polynomial*

$$v(x) = \sum_{i+j\le k} (E f_{i,j})(x) \, \frac{x_1^i \, x_2^j}{i! \, j!} \, ,$$

where E denotes the extension operator (6.2.4). *Then $v \in W_{\delta+s}^{l+s,p}(\mathcal{D})$ for arbitrary integer $s \ge 0$ and $u - v \in V_\delta^{l,p}(\mathcal{D})$.*

P r o o f. The inclusion $v \in W_{\delta+s}^{l+s,p}(\mathcal{D})$ follows from Corollary 6.2.4. Furthermore by Corollary 6.2.4, the traces of $\partial_{x_1}^i \partial_{x_2}^j (u - v)$ on M are equal to zero for $i + j \le k$. Thus, Lemma 6.2.5 implies $u - v \in V_\delta^{l,p}(\mathcal{D})$. □

6.2.7. Weighted Sobolev spaces on a half-plane. Our goal is to obtain a relation between the spaces $V_\delta^{l,p}(\mathcal{D})$ and $W_\delta^{l,p}(\mathcal{D})$ analogous to that in Theorem 6.2.7 for the case of integer $\delta + 2/p$. To this end, we consider functions on the half-plane $\mathbb{R}_+^2 = (0,\infty) \times \mathbb{R}$ instead of the traces of $\partial_{x_1}^i \partial_{x_2}^j u$ on M.

For nonnegative integer l and real p, δ, $p > 1$, let $V_\delta^{l,p}(\mathbb{R}_+^2)$ denote the closure of the set $C_0^\infty(\mathbb{R}_+^2)$ with respect to the norm

$$\|u\|_{V_\delta^{l,p}(\mathbb{R}_+^2)} = \Big(\sum_{|\alpha|\le l} \int_{\mathbb{R}_+^2} r^{p(\delta-l+i+j)} \sum_{i+j\le l} \big|\partial_r^i \partial_z^j u(r,z)\big|^p \, dr \, dz \Big)^{1/p}.$$

Furthermore, we define the space $W_\delta^{l,p}(\mathbb{R}_+^2)$ for $\delta > -1/p$ as the closure of the set $C_0^\infty(\overline{\mathbb{R}_+^2})$ with respect to the norm

$$\|u\|_{W_\delta^{l,p}(\mathbb{R}_+^2)} = \Big(\sum_{|\alpha|\le l} \int_{\mathbb{R}_+^2} r^{p\delta} \sum_{i+j\le l} \big|\partial_r^i \partial_z^j u(r,z)\big|^p \, dr \, dz \Big)^{1/p}.$$

It follows immediately from Hardy's inequality that

$$W_\delta^{l,p}(\mathbb{R}_+^2) \subset W_{\delta-1}^{l-1,p}(\mathbb{R}_+^2) \subset \cdots \subset W_{\delta-k}^{l-k,p}(\mathbb{R}_+^2)$$

for $l \ge k$, $\delta > k - 1/p$.

Let u be a function in $\mathcal{D}$, and let r, φ denote the polar coordinates of the point $x' = (x_1, x_2)$. We consider the average $\overset{\circ}{u}(r, x_3)$ of $u(x)$ with respect to the variable

φ, i.e.

$$\overset{\circ}{u}(r,x_3) = \frac{1}{\theta} \int\limits_{-\theta/2}^{+\theta/2} u(r\cos\varphi, r\sin\varphi, x_3)\, d\varphi.$$

We show that $\overset{\circ}{u}$ belongs to the same weighted Sobolev space $W_\delta^{l,p}(\mathcal{D})$ as u.

LEMMA 6.2.8. *If $u \in W_\delta^{l,p}(\mathcal{D})$, then $\overset{\circ}{u} \in W_{\delta+1/p}^{l,p}(\mathbb{R}_+^2)$. If we consider $\overset{\circ}{u}$ as a function in $\mathcal{D}$ (which depends only on the coordinates $r = |x'|$ and $z = x_3$), then $\overset{\circ}{u} \in W_\delta^{l,p}(\mathcal{D})$. Furthermore,*

$$\int_{\mathcal{D}} r^{p(\delta-k-1)} \big|u(x) - \overset{\circ}{u}(r,x_3)\big|^p\, dx \le c\, \|u\|^p_{W_\delta^{l,p}(\mathcal{D})} \tag{6.2.16}$$

if $\delta > k - 2/p$, k is an integer, $0 \le k \le l-1$.

P r o o f. The first assertion is obvious. We prove the inequality (6.2.16). By Hardy's inequality,

$$\|u\|_{W_{\delta-k}^{1,p}(\mathcal{D})} \le c\, \|u\|_{W_\delta^{k+1,p}(\mathcal{D})} \le c\, \|u\|_{W_\delta^{l,p}(\mathcal{D})}\,. \tag{6.2.17}$$

Furthermore,

$$\int_{\mathcal{D}} r^{p(\delta-k-1)} \big|u(x) - \overset{\circ}{u}(r,x_3)\big|^p\, dx$$
$$= \frac{1}{\theta^p} \int_{\mathcal{D}} r^{p(\delta-k-1} \Big| \int\limits_{-\theta/2}^{+\theta/2} \big(u(r\cos\varphi, r\sin\varphi, x_3) - u(r\cos\psi, r\sin\psi, x_3)\big)\, d\psi\Big|^p\, dx$$
$$\le c \int_{\mathcal{D}} r^{p(\delta-k-1)}\, |\partial_\varphi u(x)|^p\, dx \le c\, 2^{p-1} \int_{\mathcal{D}} r^{p(\delta-k)} \big(|\partial_{x_1} u|^p + |\partial_{x_2} u|^p\big)\, dx.$$

This together with (6.2.17) implies (6.2.16). □

We can easily derive the following properties of the function $\overset{\circ}{u}$ from the last lemma.

COROLLARY 6.2.9. *1) If $u \in W_\delta^{l,p}(\mathcal{D})$, $\delta > l - 1 - 2/p$, then $u - \overset{\circ}{u} \in V_\delta^{l,p}(\mathcal{D})$.*
2) If $u \in W_\delta^{l,p}(\mathcal{D})$, $\delta < l - 2/p$, then $u|_M = \overset{\circ}{u}|_M$.

P r o o f. 1) If $\delta > l-1-2/p$, then the space $W_\delta^{l,p}(\mathcal{D})$ is continuously imbedded in $W_{\delta-l+1}^{1,p}(\mathcal{D})$. Hence,

$$\int_{\mathcal{D}} r^{p(\delta-l+|\alpha|)} \big|\partial_x^\alpha (u(x) - \overset{\circ}{u}(r,x_3))\big|^p\, dx \le c\, \|u\|^p_{W_\delta^{l,p}(\mathcal{D})}$$

for $1 \le |\alpha| \le l$. According to Lemma 6.2.8, the last estimate is also valid for $\alpha = 0$. Consequently, $u - \overset{\circ}{u} \in V_\delta^{l,p}(\mathcal{D})$.

2) If $\delta < l - 2/p$, then the traces of u and $\overset{\circ}{u}$ on M exist (cf. Lemma 6.2.2). By Lemma 6.2.8, the inequality (6.2.16) is valid, where k is the greatest integer less than $\delta + 2/p$. Since $p(\delta - k - 1) \le -2$, it follows that the trace of $u - \overset{\circ}{u}$ is zero on M. □

6.2.8. A relation between the spaces $V_\delta^{l,p}(\mathcal{D})$ and $W_\delta^{l,p}(\mathcal{D})$ in the case of integer $\delta + 2/p$. Let χ be an infinitely differentiable function on $(0,\infty)$, $\chi(r) = 1$ for $r < 1/2$, $\chi(r) = 0$ for $r > 1$. Furthermore, let ψ be an infinitely differentiable function vanishing outside the interval $[1/2, 1]$ such that

$$\int_{\mathbb{R}} \psi(t)\, dt = 1, \qquad \int_{\mathbb{R}} t^j\, \psi(t)\, dt = 0 \quad \text{for } j = 1, \dots, l. \tag{6.2.18}$$

We introduce the following operator $\mathcal{E}$ on the space $W_\delta^{l,p}(\mathbb{R}^2_+)$:

$$(\mathcal{E}g)(r,z) = \chi(r) \int_{\mathbb{R}^2} g(sr, z+tr)\, \psi(s)\, \psi(t)\, ds\, dt\,. \tag{6.2.19}$$

In the following lemma, some basic properties of the operator $\mathcal{E}$ are given.

LEMMA 6.2.10. *Let $g \in W_\delta^{l,p}(\mathbb{R}^2_+)$, $l \ge 1$, $\delta > -1/p$, and let k be the smallest integer less than $\delta + 1/p$.*

1) *Suppose that $-1/p < \delta \le l - 1/p$. Then*

$$\int_{\mathbb{R}^2_+} r^{p(\delta-l+i+j)} \left|\partial_r^i \partial_z^j (\mathcal{E}g)(r,z)\right|^p dr\, dz \le c\, \|g\|^p_{W_\delta^{l,p}(\mathbb{R}^2_+)} \tag{6.2.20}$$

if $i \ge 1$ or $j \ge l-k$. Furthermore, the estimates

$$\int_{\mathbb{R}^2_+} r^{p(\delta-k)} \left|\partial_z^j (\mathcal{E}g)(r,z)\right|^p dr\, dz \le c\, \|g\|^p_{W_\delta^{l,p}(\mathbb{R}^2_+)}\,, \tag{6.2.21}$$

$$\int_{\mathbb{R}^2_+} r^{p(\delta-k-1)} \left|\partial_z^j (g - \mathcal{E}g)\right|^p dr\, dz \le c\, \|g\|^p_{W_\delta^{l,p}(\mathbb{R}^2_+)}\,. \tag{6.2.22}$$

are valid for $j = 0, \dots, l-k-1$. The constant c in (6.2.20)–(6.2.22) *depends only on i and j.*

2) *If $\delta > l - 1/p$, then* (6.2.20) *is valid for all i and j.*

P r o o f. Suppose that $\delta \le l - 1/p$, i.e. $0 \le k \le l-1$. Then $W_\delta^{l,p}(\mathbb{R}^2_+) \subset W_{\delta-k}^{l-k,p}(\mathbb{R}^2_+)$. We consider the function

$$v(r,z) = \int_{\mathbb{R}^2_+} g(sr, z+tr)\, \psi(s)\, \psi(t)\, ds\, dt\,.$$

For an arbitrary multi-index $\alpha = (\alpha_1, \alpha_2)$, we denote the derivative $\partial_r^{\alpha_1} \partial_z^{\alpha_2} g(r,z)$ by $g^{(\alpha)}(r,z)$. If $i + j \ge l - k$, then $\partial_r^i \partial_z^j v(r,z)$ is a finite sum of terms of the form

$$\begin{aligned} A(r,z) &= \partial_r^\mu \partial_z^\nu \int_{\mathbb{R}^2} g^{(\alpha)}(sr, z+tr)\, s^{k_1}\, t^{k_2}\, \psi(s)\, \psi(t)\, ds\, dt \\ &= \int_{\mathbb{R}^2} g^{(\alpha)}(\sigma, \tau)\, \partial_r^\mu \partial_z^\nu\, r^{-2} \left(\frac{\sigma}{r}\right)^{k_1} \left(\frac{\tau - z}{r}\right)^{k_2} \psi\Big(\frac{\sigma}{r}\Big)\, \psi\Big(\frac{\tau - z}{r}\Big)\, d\sigma\, d\tau, \end{aligned}$$

where $|\alpha| = l - k$, $\mu + \nu = i + j - l + k$, $k_1 + k_2 = i - \mu$. Obviously, the expression A has the form

$$A(r,z) = r^{-\mu-\nu} \int_{\mathbb{R}^2} g^{(\alpha)}(sr, z+tr)\, \Psi(s,t)\, ds\, dt,$$

where Ψ is an infinitely differentiable function vanishing outside the square $[1/2,1]^2$. Consequently,

$$\int_0^1\int_{\mathbb{R}} r^{p(\delta-l+i+j)} \left|\partial_r^i\partial_z^j v(r,z)\right|^p dz\,dr \le c\,\|g\|^p_{W^{l-k,p}_{\delta-k}(\mathbb{R}^2_+)}$$

for $i+j\ge l-k$. This proves (6.2.20) for the case $i+j\ge l-k$. Analogously, the estimate (6.2.20) holds for all i,j if $\delta>l-1/p$.

Now let $i+j\le l-k-1$, $k\le l-1$ and $i\ge 1$. Then $\partial_r^i\partial_z^j v(r,z)$ is a finite sum of terms of the form

$$B(r,z)=\int_{\mathbb{R}^2} g^{(\alpha)}(sr,z+tr)\,s^{k_1}\,t^{k_2}\,\psi(s)\,\psi(t)\,ds\,dt,$$

where $|\alpha|=i+j$, $k_1+k_2=i\ge 1$. Suppose that $k_1\ge 1$. Then we introduce the functions

$$\psi_0(s)=s^{k_1}\psi(s),\quad \psi_1(s)=\int_0^s\psi_0(\tau)\,d\tau,\dots,$$
$$\psi_{l-k-i-j}(s)=\int_0^s\psi_{k-1}(\tau)\,d\tau=\int_0^s\tau^{k_1}\psi(\tau)\,\frac{(t-\tau)^{l-k-i-j-1}}{(l-k-i-j-1)!}\,d\tau.$$

According to the conditions on ψ, the functions $\psi_1,\dots,\psi_{l-k-i-j}$ vanish outside the interval $[1/2,+1]$. Therefore, integration by parts yields

$$B(r,z)=(-r)^{l-k-i-j}\int_{\mathbb{R}^2}(\partial_r^{l-k-i-j+\alpha_1}\partial_z^{\alpha_2}g)(sr,z+tr)\,t^{k_2}\,\psi_{l-k-i-j}(s)\,\psi(t)\,ds\,dt.$$

Consequently, we obtain

$$\int_0^1\int_{\mathbb{R}} r^{p(\delta-l+i+j)}\,|B(r,z)|^p\,dz\,dr\le c\,\|g\|^p_{W^{l-k,p}_{\delta-k}(\mathbb{R}^2_+)}.$$

Analogously, this estimate holds in the case $k_2\ge 1$. Thus, the estimate (6.2.20) is valid if $i\ge 1$.

It remains to prove the estimates (6.2.21) and (6.2.22). The inequality (6.2.21) is obvious. Let $\delta\le l-1/p$ and $j\le l-k-1$. Then

$$\partial_z^j v(r,z)-\partial_z^j g(r,z)=\int_{\mathbb{R}^2}\big(\partial_z^j g(sr,z+tr)-\partial_z^j g(r,z)\big)\,\psi(s)\,\psi(t)\,dt.$$

Using the inequality

$$\begin{aligned}&\left|\partial_z^j g(sr,z+tr)-\partial_z^j g(r,z)\right|\\ &\quad\le\left|\partial_z^j g(sr,z+tr)-\partial_z^j g(sr,z)\right|+\left|\partial_z^j g(sr,z)-\partial_z^j g(r,z)\right|\\ &\quad=\left|r\int_0^1 t\,g^{(0,j+1)}(sr,z+\tau tr)\,d\tau\right|+\left|r\int_s^1 g^{(1,j)}(\sigma r,z)\,d\sigma\right|\end{aligned}$$

and the inequalities (6.2.20), (6.2.21), we obtain

$$\int_0^1 \int_{\mathbb{R}} r^{p(\delta-k-1)} \left|\partial_z^j v(r,z) - \partial_z^j g(r,z)\right|^p dr\, dz$$

$$\leq c \sum_{|\alpha|=j+1} \int_0^1 \int_{\mathbb{R}} r^{p(\delta-k)} \left|g^{(\alpha)}(r,z)\right|^p dr\, dz \leq c\, \|g\|^p_{W_\delta^{l,p}(\mathbb{R}^2_+)} .$$

This proves (6.2.22). □

In particular, it follows from Lemma 6.2.10 that $\mathcal{E}$ realizes a continuous mapping from $W_\delta^{l,p}(\mathbb{R}^2_+)$ into $W_{\delta+s}^{l+s,p}(\mathbb{R}^2_+)$ for arbitrary integer $s \geq 0$. If $g \in W_\delta^{l,p}(\mathbb{R}^2_+)$, $\delta < l-1/p$, then the traces of $\partial_z^j g$ and $\partial_z^j \mathcal{E} g$ on the line $r=0$ exist for $j < l-\delta-1/p$ (cf. Lemma 6.2.2). Furthermore, according to (6.2.22),

$$(\partial_z^j g)(0,z) = (\partial_z^j \mathcal{E} g)(0,z) \quad \text{for } j < l-\delta-1/p$$

and for almost all z.

Next we prove a relation between the spaces $V_\delta^{l,p}(\mathcal{D})$ and $W_\delta^{l,p}(\mathcal{D})$ for the case of integer $\delta + 2/p$. Let $u \in W_\delta^{l,p}(\mathcal{D})$, where $\delta + 2/p$ is an integer, $1 \leq \delta + 2/p \leq l$. We put $u_{i,j} = \partial_{x_1}^i \partial_{x_2}^j u$ for $i + j \leq l - k$ and denote by

$$(6.2.23) \qquad \overset{\circ}{u}_{i,j}(r, x_3) = \frac{1}{\theta} \int_{-\theta/2}^{+\theta/2} u_{i,j}(r\cos\varphi, r\sin\varphi, x_3)\, d\varphi$$

the average of $u_{i,j}$ with respect to the variable φ.

THEOREM 6.2.11. *Let $u \in W_\delta^{l,p}(\mathcal{D})$, where $\delta+2/p$ is an integer, $1 \leq \delta+2/p \leq l$. Furthermore, let v be the quasi-polynomial*

$$v(x) = \sum_{i+j \leq l-\delta-2/p} (\mathcal{E}\, \overset{\circ}{u}_{i,j})(r, x_3)\, \frac{x_1^i\, x_2^j}{i!\, j!} .$$

Then $v \in W_{\delta+s}^{l+s,p}(\mathcal{D})$ for arbitrary integer $s \geq 0$ and $u - v \in V_\delta^{l,p}(\mathcal{D})$.

P r o o f. From Lemma 6.2.8 it follows that $\overset{\circ}{u}_{i,j} \in W_{\delta+1/p}^{l-i-j,p}(\mathbb{R}^2_+)$. Using Lemma 6.2.10, it can be easily verified that every of the terms

$$(\mathcal{E}\, \overset{\circ}{u}_{i,j})(r, x_3)\, \frac{x_1^i\, x_2^j}{i!\, j!}$$

belongs to the space $W_{\delta+s}^{l+s,p}(\mathcal{D})$ for $i + j \leq l - \delta - 2/p$. Hence, $v \in W_{\delta+s}^{l+s,p}(\mathcal{D})$ for every integer $s \geq 0$. Furthermore, the traces of $\partial_{x_1}^i \partial_{x_2}^j u$ and $\partial_{x_1}^i \partial_{x_2}^j v$ on M exist for $i+j < l-\delta-2/p$ (cf. Lemma 6.2.2). Obviously, the trace of $\partial_{x_1}^i \partial_{x_2}^j v$ coincides with the trace of the function

$$\sum_{\substack{\mu+\nu \leq l-\delta-2/p \\ \mu \leq i, \nu \leq j}} \partial_{x_1}^{i-\mu}\, \partial_{x_2}^{j-\nu}\, \mathcal{E}\, \overset{\circ}{u}_{\mu,\nu}$$

on M. However for $\mu \leq i$, $\nu \leq j$, $\mu + \nu \neq i + j$, the inequality

$$\int_{\mathcal{D}} r^{p(\delta-l+i+j)} \left|\partial_{x_1}^{i-\mu}\, \partial_{x_2}^{j-\nu}\, \mathcal{E}\, \overset{\circ}{u}_{\mu,\nu}\right|^p dx \leq c\, \|u\|^p_{W_\delta^{l,p}(\mathcal{D})}$$

is satisfied by (6.2.20). Consequently, the trace of $\partial_{x_1}^{i-\mu}\,\partial_{x_2}^{j-\nu}\,\mathcal{E}\,\overset{\circ}{u}_{\mu,\nu}$ is zero for $\mu \le i$, $\nu \le j$, $\mu+\nu < i+j < l-\delta-2/p$. Thus,

$$\partial_{x_1}^i \partial_{x_2}^j v = \mathcal{E}\,\overset{\circ}{u}_{i,j} = \overset{\circ}{u}_{i,j} = \partial_{x_1}^i \partial_{x_2}^j u \quad \text{on } M \quad \text{for } i+j < l-\delta-2/p .$$

If $i+j = \delta - l - 2/p$, then $\partial_{x_1}^i \partial_{x_2}^j v$ admits the representation

$$\partial_{x_1}^i \partial_{x_2}^j v = \mathcal{E}\,\overset{\circ}{u}_{i,j} + R,$$

where R is a sum of terms of the form

$$\big(\partial_{x_1}^{\alpha_1}\partial_{x_2}^{\alpha_2}\,\mathcal{E}\,\overset{\circ}{u}_{\mu,\nu}\big)\,\frac{x_1^{\beta_1}\,x_2^{\beta_2}}{\beta_1!\,\beta_2!}\,,$$

$\mu+\nu \le l-\delta-2/p$, $\alpha_1+\alpha_2 \ge 1$, $\mu+\nu+\alpha_1+\alpha_2-\beta_1-\beta_2 = l-\delta-2/p$. Using Lemma 6.2.10, we obtain the inequality

$$\int_{\mathcal{D}} r^{-2}\,\big|\partial_{x_1}^i \partial_{x_2}^j v - \mathcal{E}\,\overset{\circ}{u}_{i,j}\,\big|^p\,dx = \int_{\mathcal{D}} r^{-2}\,|R|^p\,dx < \infty.$$

Furthermore, it follows from Lemmas 6.2.8 and 6.2.10 that

$$\int_{\mathcal{D}} r^{-2}\,\big|\partial_{x_1}^i \partial_{x_2}^j u - \overset{\circ}{u}_{i,j}\,\big|^p\,dx < \infty \quad \text{and} \quad \int_{\mathcal{D}} r^{-2}\,\big|\mathcal{E}\,\overset{\circ}{u}_{i,j} - \overset{\circ}{u}_{i,j}\,\big|^p\,dx < \infty.$$

This implies

$$\int_{\mathcal{D}} r^{-2}\,\big|\partial_{x_1}^i \partial_{x_2}^j (u-v)\big|^p\,dx < \infty \text{ for } i+j = l-\delta-2/p .$$

Applying Lemma 6.2.6, we conclude that $u-v \in V_\delta^{l,p}(\mathcal{D})$. The proof of the theorem is complete. □

6.2.9. Weighted Sobolev spaces on the angle K. Let l be a nonnegative integer, $1 < p < \infty$, and $\delta > -2/p$. The spaces $L_\delta^{l,p}(K)$ and $W_\delta^{l,p}(K)$ are defined as the closure of the set $C_0^\infty(\overline{K})$ with respect to the norms

$$\|u\|_{L_\delta^{l,p}(K)} = \Big(\int_K |x'|^{p\delta} \sum_{|\alpha|=l} \big|\partial_{x'}^\alpha u(x')\big|^p\,dx'\Big)^{1/p} \quad \text{and}$$

$$\|u\|_{W_\delta^{l,p}(K)} = \Big(\int_K |x'|^{p\delta} \sum_{|\alpha|\le l} \big|\partial_{x'}^\alpha u(x')\big|^p\,dx'\Big)^{1/p},$$

respectively. By Hardy's inequality,

$$W_\delta^{l,p}(K) \subset W_{\delta-1}^{l-1,p}(K) \quad \text{if } l \ge 1,\ \delta > 1-2/p.$$

Thus, the imbedding $W_\delta^{l,p}(K) \subset V_\delta^{l,p}(K)$ holds for $\delta > l-2/p$. If $u \in W_\delta^{l,p}(K)$ and $\delta < k-2/p$, $k = 1,\dots,l$, then the derivatives of u up to order $l-k$ are continuous at the origin (cf. [**84**, Lemma 7.1.3]). We denote by

$$p_{l-k}(u) = \sum_{i+j\le l-k} \big(\partial_{x_1}^i \partial_{x_2}^j u\big)(0)\,\frac{x_1^i\,x_2^j}{i!\,j!}$$

the Taylor polynomial of degree $l-k$ of the function u. For the following lemma, we refer to [**84**, Theorem 7.1.1].

LEMMA 6.2.12. *Let $u \in W_\delta^{l,p}(K)$, $k-1-2/p < \delta < k-2/p$, where k is an integer, $1 \le k \le l$. Then $u - p_{l-k}(u) \in V_\delta^{l,p}(K)$ and*

$$\|u - p_{l-k}(u)\|_{V_\delta^{l,p}(K)} + \sum_{i+j\le l-k} \big|(\partial_{x_1}^i \partial_{x_2}^j u)(0)\big| \le c\, \|u\|_{W_\delta^{l,p}(K)}$$

with a constant c independent of u. In particular, $u \in V_\delta^{l,p}(K)$ if and only if $(\partial_{x_1}^i \partial_{x_2}^j u)(0) = 0$ for $i+j \le l-k$.

In the case of integer $\delta + 2/p$, we obtain a result analogous to Theorem 6.2.11. Let $u \in W_\delta^{l,p}(K)$, where $\delta + 2/p = k$ is an integer, $1 \le k \le l$, and let $u_{i,j} = \partial_{x_1}^i \partial_{x_2}^j u$ for $i+j = l-k$. Then the functions $\overset{\circ}{u}_{i,j}$ are defined by

$$\overset{\circ}{u}_{i,j}(r) = \frac{1}{\theta} \int_{-\theta/2}^{+\theta/2} u_{i,j}(r\cos\varphi, r\sin\varphi)\, d\varphi.$$

Analogously to Lemma 6.2.8, we have $\overset{\circ}{u}_{i,j} \in W_{\delta+1/p}^{k,p}(\mathbb{R}_+)$, where $W_\delta^{l,p}(\mathbb{R}_+)$ is the weighted Sobolev space with the norm

$$\|u\|_{W_\delta^{l,p}(\mathbb{R}_+)} = \Big(\int_0^\infty r^{p\delta} \sum_{j=0}^{l} \big|\partial_r^j u(r)\big|^p \, dr \Big)^{1/p}.$$

We introduce the following operator $\mathfrak{E}$ on $W_\delta^{l,p}(\mathbb{R}_+)$:

$$(\mathfrak{E} g)(r) = \chi(r) \int_0^\infty g(tr)\, \psi(t)\, dt,$$

where χ is an infinitely differentiable function with compact support, $\chi(r) = 1$ for $r < 1/2$, and ψ is an infinitely differentiable function vanishing outside the interval $[1/2, 1]$ and satisfying the conditions (6.2.18). The operator $\mathfrak{E}$ continuously maps the space $W_\delta^{l,p}(\mathbb{R}_+)$ into $W_{\delta+s}^{l+s,p}(\mathbb{R}_+)$ with arbitrary integer $s \ge 0$. Furthermore,

$$\int_0^\infty r^{p(\delta-l+j)} \big|\partial_r^j (\mathfrak{E} u)(r)\big|^p \, dr \le c\, \|u\|^p_{W_\delta^{l,p}(\mathbb{R}_+)}$$

for arbitrary $u \in W_\delta^{l,p}(\mathbb{R}_+)$ and $j \ge 1$, where c is a constant independent of u.

The following theorem is proved [**84**, Theorem 7.3.2].

THEOREM 6.2.13. *Let $u \in W_\delta^{l,p}(K)$, where $\delta + 2/p = k$ is an integer, $1 \le k \le l$, and let*

$$v(x) = (p_{l-k-1}(u))(x) + \sum_{i+j=l-k} (\mathfrak{E}\, \overset{\circ}{u}_{i,j})(r)\, \frac{x_1^i\, x_2^j}{i!\, j!}\,.$$

Then $u - v \in V_\delta^{l,p}(K)$.

6.2.10. Traces on $\Gamma^\pm$ and $\gamma^\pm$. The trace spaces for $L_\delta^{l,p}(\mathcal{D})$ and $W_\delta^{l,p}(\mathcal{D})$, $l \ge 1$, on $\Gamma^\pm$ are denoted by $L_\delta^{l-1/p,p}(\Gamma^\pm)$ and $W_\delta^{l-1/p,p}(\Gamma^\pm)$, respectively. The norms in $L_\delta^{l-1/p,p}(\Gamma^\pm)$ and $W_\delta^{l-1/p,p}(\Gamma^\pm)$ are defined as

$$\|u\|_{L_\delta^{l-1/p,p}(\Gamma^\pm)} = \inf\big\{ \|v\|_{L_\delta^{l,p}(\mathcal{D})} \,:\, v \in L_\delta^{l,p}(\mathcal{D}),\ v|_{\Gamma^\pm} = u \big\},$$

$$\|u\|_{W_\delta^{l-1/p,p}(\Gamma^\pm)} = \inf\big\{ \|v\|_{W_\delta^{l,p}(\mathcal{D})} \,:\, v \in W_\delta^{l,p}(\mathcal{D}),\ v|_{\Gamma^\pm} = u \big\}.$$

Analogously, the trace spaces $L_\delta^{l-1/p,p}(\gamma^\pm)$ and $W_\delta^{l-1/p,p}(\gamma^\pm)$ for $L_\delta^{l,p}(K)$ and $W_\delta^{l,p}(K)$, $l \ge 1$, are defined.

LEMMA 6.2.14. *Let $u \in W_\delta^{l-1/p,p}(\Gamma^\pm)$. Then*

$$\int_{\mathbb{R}} \|u(\cdot,x_3)\|^p_{W_\delta^{l-1/p,p}(\gamma^\pm)}\, dx_3 \le \|u\|^p_{W_\delta^{l-1/p,p}(\Gamma^\pm)}.$$

P r o o f. Let $v \in W_\delta^{l,p}(\mathcal{D})$ be an extension of u such that

$$\|v\|^p_{W_\delta^{l,p}(\mathcal{D})} \le (1+\varepsilon)\, \|u\|^p_{W_\delta^{l-1/p,p}(\Gamma^\pm)}.$$

Here the positive number ε can be chosen arbitrarily small. Then $v(\cdot,x_3)$ is an extension of $u(\cdot,x_3)$ for all x_3 and

$$\begin{aligned}\int_{\mathbb{R}} \|u(\cdot,x_3)\|^p_{W_\delta^{l-1/p,p}(\gamma^\pm)}\, dx_3 &\le \int_{\mathbb{R}} \|v(\cdot,x_3)\|^p_{W_\delta^{l,p}(K)}\, dx_3 \le \|v\|^p_{W_\delta^{l,p}(\mathcal{D})} \\ &\le (1+\varepsilon)\, \|u\|^p_{W_\delta^{l-1/p,p}(\Gamma^\pm)}.\end{aligned}$$

This proves the lemma. □

The following relation between the spaces $V_\delta^{l-1/p,p}(\Gamma^\pm)$ and $W_\delta^{l-1/p,p}(\Gamma^\pm)$ can be easily deduced from Theorem 6.2.7 and from the properties of the extension operator (6.2.4).

THEOREM 6.2.15. *Suppose that $v \in W_\delta^{l-1/p,p}(\Gamma^\pm)$, where $l-k-1-2/p < \delta < l-k-2/p$, k is an integer, $0 \le k \le l-1$. Furthermore, let $f_j = (\partial_r^j v)|_M$ for $j = 0,1,\dots,k$. Then*

$$v - \sum_{j=0}^{k} (Ef_j)\, \frac{r^j}{j!} \in V_\delta^{l-1/p,p}(\Gamma^\pm).$$

Here E denotes the extension operator (6.2.4). *In particular, $v \in V_\delta^{l-1/p,p}(\Gamma^\pm)$ if and only if $\partial_r^j v = 0$ on M for $j = 0,1,\dots,k$.*

We prove an analogous assertion for integer $\delta + 2/p$.

THEOREM 6.2.16. *Let $v \in W_\delta^{l-1/p,p}(\Gamma^\pm)$, where $\delta + 2/p$ is an integer, $1 \le \delta + 2/p \le l$. Then $v \in V_\delta^{l-1/p,p}(\Gamma^\pm)$ if and only if the conditions*

$$\partial_r^j v = 0 \quad \text{on } M \quad \text{for } j < l - \delta - 2/p, \tag{6.2.24}$$

$$\int_{-\infty}^{+\infty}\int_0^\infty r^{-1}\, \big|\partial_r^j v(r,x_3)\big|^p\, dr\, dx_3 < \infty \quad \text{for } j = l-\delta-2/p \tag{6.2.25}$$

are satisfied

P r o o f. The necessity of the conditions (6.2.24) and (6.2.25) follows immediately from Lemma 2.1.10. Suppose that the function $v \in W_\delta^{l-1/p,p}(\Gamma^\pm)$ satisfies the conditions (6.2.24) and (6.2.25). There exists an extension $u \in W_\delta^{l,p}(\mathcal{D})$ of v. Let

$$w(x) = \sum_{i+j \le l-\delta-2/p} (\mathcal{E}\, \overset{\circ}{u}_{i,j})(r,x_3)\, \frac{x_1^i\, x_2^j}{i!\, j!},$$

where $\overset{\circ}{u}_{i,j}$ is defined by (6.2.23) and $\mathcal{E}$ is defined by (6.2.19). Furthermore, let $W = w|_{\Gamma^\pm}$. Since $u - w \in V_\delta^{l,p}(\mathcal{D})$ (see Theorem 6.2.11), it follows that $v - W \in V_\delta^{l-1/p,p}(\Gamma^\pm)$. By Lemma 2.1.10, the function $v-W$ satisfies the conditions (6.2.24) and (6.2.25) of the theorem. Consequently, $\partial_r^j W = 0$ on M for $j < l - \delta - 2/p$ and

$$\int\limits_{-\infty}^{+\infty}\int\limits_0^{\infty} r^{-1}\,\big|\partial_r^j W(r,x_3)\big|^p\,dr\,dx_3 < \infty \quad \text{for } j = l-\delta-2/p$$

Moreover, the function W belongs to the space $W_{\delta+1/p}^{l,p}(\mathbb{R}_+^2)$ (cf. Theorem 6.2.11). Analogously to Lemma 6.2.6, the last conditions imply $W \in V_{\delta+1/p}^{l,p}(\mathbb{R}_+^2)$. Consequently, the $V_\delta^{l-1/p,p}(\Gamma^\pm)$-norm of W given in Lemma 2.1.10 is finite. This proves the theorem. □

6.3. Parameter-dependent problems in an angle

Let $L(D_{x'},\xi)$, $N^\pm(D_{x'},\xi)$ be the parameter-dependent differential operators introduced in Section 6.1. We consider the boundary value problem

(6.3.1) $\quad L(D_{x'},\xi)\,u = f \;$ in K,

(6.3.2) $\quad N^\pm(D_{x'},\xi)\,u = g^\pm$ on $\gamma^\pm$ for $d^\pm = 1$, $\quad u = 0$ on $\gamma^\pm$ for $d^\pm = 0$,

where ξ is a real parameter. As in Section 6.1, we assume that the form $b_{\mathcal{D}}$ is $\mathcal{H}_{\mathcal{D}}$-elliptic. Then by Lemma 6.1.5, the parameter-dependent form (6.1.18) is $W^{1,2}(K)$-elliptic for every real $\xi \neq 0$ which guarantees the existence and uniqueness of a variational solution of the problem (6.3.1), (6.3.2). We show that this solution belongs to the weighted space $W_\delta^{2,2}(K)^\ell$ if $f \in W_\delta^{0,2}(K)^\ell$, $g^\pm \in W_\delta^{1/2,2}(\gamma^\pm)^\ell$, $\delta < 1$, and $1-\delta$ is sufficiently small. Moreover, we obtain a parameter-dependent estimate for the solution. At the end of this section, we prove a $W_\delta^{l,p}$-regularity result for the solution in the case $\xi = 0$.

6.3.1. Existence of variational solutions. Let $\xi \neq 0$, $0 < \delta < 1$, and let $f \in W_\delta^{0,2}(K)^\ell$, $g^\pm \in W_\delta^{1/2,2}(\gamma^\pm)^\ell$ be given vector functions. We consider the variational solution $u \in W^{1,2}(K)^\ell$ of the boundary value problem (6.3.1), (6.3.2). This means that u is the solution of the problem

(6.3.3) $\quad b_K(u,v,\xi) = (F,v)_K \;$ for all $v \in W^{1,2}(K)^\ell$, $(1-d^\pm)v|_{\gamma^\pm} = 0$,

(6.3.4) $\quad u = 0 \;$ on $\gamma^\pm$ for $d^\pm = 0$,

where b_K is the sesquilinear form (6.1.18) and the functional F is given by

$$(F,v)_K = \int_K f\cdot\bar v\,dx' + \sum_{d^\pm=1}\int_{\gamma^\pm} g^\pm\cdot\bar v\,dx'. \tag{6.3.5}$$

LEMMA 6.3.1. *The functional* (6.3.5) *is continuous on* $W^{1,2}(K)^\ell$ *for arbitrary* $f \in W_\delta^{0,2}(K)^\ell$ *and* $g^\pm \in W_\delta^{1/2,2}(\gamma^\pm)^\ell$ *if* $0 < \delta < 1$.

P r o o f. Let $v \in W^{1,2}(K)^\ell$, and let ζ be a continuously differentiable function on K equal to 1 for $|x'| < 1$ and to zero for $|x'| > 2$. Then by Hardy's inequality,

$$\|\zeta v\|_{V_{1-\delta}^{1,2}(K)^\ell} \le c\,\|v\|_{W^{1,2}(K)^\ell}$$

with a constant c independent of v. Consequently,

$$\Big| \int_K f \cdot \bar{v}\, dx' \Big| \leq \|f\|_{V_\delta^{0,2}(K)^\ell}\, \|\zeta v\|_{V_{-\delta}^{0,2}(K)^\ell} + \|(1-\zeta)f\|_{L_2(K)^\ell}\, \|v\|_{L_2(K)^\ell}$$
$$\leq c\, \|f\|_{W_\delta^{0,2}(K)^\ell}\, \|v\|_{W^{1,2}(K)^\ell}\,.$$

Since the space $W_\delta^{1/2,2}(\gamma^\pm)$ is continuously imbedded in $V_\delta^{1/2,2}(\gamma^\pm)$, we obtain

$$\Big| \int_{\gamma^\pm} g^\pm \cdot \bar{v}\, dx' \Big|$$
$$\leq \|r^{\delta-1/2} g^\pm\|_{L_2(\gamma^\pm)^\ell}\, \|r^{-\delta+1/2}\zeta v\|_{L_2(\gamma^\pm)^\ell} + \|(1-\zeta)g^\pm\|_{L_2(\gamma^\pm)^\ell}\, \|v\|_{L_2(\gamma^\pm)^\ell}$$
$$\leq \|g^\pm\|_{V_\delta^{1/2,2}(\gamma^\pm)^\ell}\, \|\zeta v\|_{V_{1-\delta}^{1/2,2}(\gamma^\pm)^\ell} + \|(1-\zeta)g^\pm\|_{L_2(\gamma^\pm)^\ell}\, \|v\|_{L_2(\gamma^\pm)^\ell}$$
$$\leq c\, \|g^\pm\|_{W_\delta^{1/2,2}(\gamma^\pm)^\ell} \|v\|_{W^{1,2}(K)^\ell}\,.$$

This proves the Lemma. □

Using the last lemma together with Lemma 6.1.5, we obtain the following statement.

LEMMA 6.3.2. *Suppose that the form $b_{\mathcal{D}}$ is $\mathcal{H}_{\mathcal{D}}$-elliptic and the functional F has the form* (6.3.5), *where $f \in W_\delta^{0,2}(K)^\ell$, $g^\pm \in W_\delta^{1/2,2}(\gamma^\pm)^\ell$, $0<\delta<1$. Then the problem* (6.3.3), (6.3.4) *has a unique solution $u \in W^{1,2}(K)^\ell$ for real $\xi \neq 0$.*

6.3.2. Solvability in $W_\delta^{2,2}(K)^\ell$ for $\xi = \pm 1$. We show that the variational solution belongs to $W_\delta^{2,2}(K)^\ell$ if δ satisfies the inequalities

$$\max(0, 1-\delta_+) < \delta < 1, \tag{6.3.6}$$

where δ_+ is the greatest real number such that the strip $0 < \operatorname{Re}\lambda < \delta_+$ does not contain eigenvalues of the pencil $A(\lambda)$. First we consider the case $\xi = \pm 1$.

LEMMA 6.3.3. *Let $\xi = \pm 1$ and let the functional F have the form* (6.3.5), *where $f \in W_\delta^{0,2}(K)^\ell$, and $g^\pm \in W_\delta^{1/2,2}(\gamma^\pm)^\ell$. If δ satisfies* (6.3.6), *then the uniquely determined solution $u \in W^{1,2}(K)^\ell$ of the problem* (6.3.3), (6.3.4) *belongs to $W_\delta^{2,2}(K)^\ell$. Furthermore,*

$$\|u\|_{W_\delta^{2,2}(K)^\ell} \leq c \Big(\|f\|_{W_\delta^{0,2}(K)^\ell} + \sum_{d^\pm = 1} \|g^\pm\|_{W_\delta^{1/2,2}(\gamma^\pm)^\ell} \Big) \tag{6.3.7}$$

with a constant c independent of f and $g^\pm$.

P r o o f. It follows from Lemmas 6.1.5 and 6.3.1 that

$$\|u\|_{W^{1,2}(K)^\ell} \leq c \Big(\|f\|_{W_\delta^{0,2}(K)^\ell} + \sum_{d^\pm = 1} \|g^\pm\|_{W_\delta^{1/2,2}(\gamma^\pm)^\ell} \Big)$$

with a constant c independent of f and $g^\pm$. Let ζ be an infinitely differentiable function with compact support equal to one near the vertex of K. Then $\zeta u \in V_\varepsilon^{1,2}(K)^\ell$ with an arbitrary positive ε. Since the commutator $[L(D_{x'},0),\zeta] = L(D_{x'},0)\zeta - \zeta L(D'_x,0)$ and the difference $L(D_{x'},0) - L(D_{x'},\pm 1)$ are first order differential operators, it follows that

$$L(D_{x'},0)\,(\zeta u) = \zeta f + \zeta\,\big(L(D_{x'},0) - L(D_{x'},\pm 1)\big)\, u + [L(D_{x'},0),\zeta] u \in V_\kappa^{0,2}(K)^\ell$$

with an arbitrary $\kappa \ge \delta$. Analogously $N^\pm(D_{x'},0)\,(\zeta u)|_{\gamma^\pm} \in V_\kappa^{1/2,2}(\gamma^\pm)^\ell$ for $d^\pm = 1$. Thus, ζu belongs to the space $V_{\varepsilon+1}^{2,2}(K)^\ell$ and admits the decomposition

$$\zeta u = \sum_{j,k} c_{j,k}\, r^{\lambda_j}\, \log^k r\, U_{j,k}(\varphi) + w \tag{6.3.8}$$

where $w \in V_\delta^{2,2}(K)^\ell$, $c_{j,k}$ are constants, λ_j are the eigenvalues of the pencil $A(\lambda)$ with real part equal to zero, and $U_{j,k}$ are infinitely differentiable vector functions (cf. Theorems 1.2.6 and 1.2.8). Furthermore,

$$\|w\|_{V_\delta^{2,2}(K)^\ell} \le c\Big(\|f\|_{W_\delta^{0,2}(K)^\ell} + \sum_{d^\pm=1} \|g^\pm\|_{W_\delta^{1/2,2}(\gamma^\pm)^\ell}\Big),$$

where c is independent of f and $g^\pm$. Since $\zeta u \in W^{1,2}(K)^\ell$, all terms in the sum of (6.3.8) with the exception of a constant vector (which corresponds to the eigenvalue $\lambda = 0$ of $A(\lambda)$ in the case of the Neumann problem) vanish. This means that

$$\zeta u = c + w,$$

where $c \in \mathbb{C}^\ell$. Consequently,

$$\|\zeta u\|_{L_\delta^{2,2}(K)^\ell} = \|w\|_{L_\delta^{2,2}(K)^\ell} \le c\Big(\|f\|_{W_\delta^{0,2}(K)^\ell} + \sum_{d^\pm=1} \|g^\pm\|_{W_\delta^{1/2,2}(\gamma^\pm)^\ell}\Big).$$

We consider the vector function $(1-\zeta)u$. Obviously,

$$(1-\zeta)u \in E_{\delta-1}^{1,2}(K)^\ell, \quad L(D_{x'},\pm 1)\,((1-\zeta)u) \in E_\delta^{0,2}(K)^\ell$$

and

$$N^\pm(D_{x'},\pm 1)\,((1-\zeta)u)\big|_{\gamma^\pm} \in E_\delta^{1/2,2}(\gamma^\pm)^\ell$$

for $d^\pm = 1$. Applying Lemma 6.1.6, we obtain $(1-\zeta)\,u \in E_\delta^{2,2}(K)^\ell$ and

$$\|(1-\zeta)\,u\|_{E_\delta^{2,2}(K)^\ell} \le c\Big(\|f\|_{W_\delta^{0,2}(K)^\ell} + \sum_{d^\pm=1} \|g^\pm\|_{W_\delta^{1/2,2}(\gamma^\pm)^\ell}\Big).$$

The result follows. $\square$

6.3.3. Solvability in $W_\delta^{2,2}(K)^\ell$ for $\xi \ne 0$. An analogous result holds in the case $\xi \ne \pm 1$, $\xi \ne 0$. Then the norms in (6.3.7) must be replaced by parameter-depending norms. For the proof of this result, we will employ an equivalent norm in the space $W_\delta^{1/2,2}(\gamma^\pm)$ which is given in the following lemma.

LEMMA 6.3.4. *Let $0 < \delta < 1$. Then the norm in $W_\delta^{1/2,2}(\gamma^\pm)$ is equivalent to the expression*

$$\Big(\|g\|^2_{V_\delta^{1/2,2}(\gamma^\pm)} + \|r^\delta g\|^2_{L_2(\gamma^\pm)}\Big)^{1/2}. \tag{6.3.9}$$

P r o o f. The norm in $V_\delta^{1/2,2}(\gamma^\pm)$ is equivalent to the norm

$$\|g\| = \Big(\int_0^\infty r^{2\delta-1}\,|g(r)|^2\,dr + \int_0^\infty\!\!\int_0^\infty \frac{|r^\delta g(r) - \rho^\delta g(\rho)|^2}{|r-\rho|^2}\,dr\,d\rho\Big)^{1/2}$$

(cf. Lemma 2.1.10). From the continuity of the imbedding $W_\delta^{1/2,2}(\gamma^\pm) \subset V_\delta^{1/2,2}(\gamma^\pm)$ it follows that the $V_\delta^{1/2,2}(\gamma^\pm)$- and $W_\delta^{1/2,2}(\gamma^\pm)$-norms are equivalent on the set of all functions $g = g(r)$ vanishing for $r > 1$. Thus, the assertion of the lemma is true on the set of of all $g \in W_\delta^{1/2,2}(\gamma^\pm)$ vanishing for $r > 1$.

Let $g \in W_\delta^{1/2,2}(\gamma^\pm)$, $g(r) = 0$ for $r < 1/2$. Then there exists a function $u \in W_\delta^{1,2}(K)$, $u(x') = 0$ for $|x'| < 1/2$, such that

$$u = g \text{ on } \gamma^\pm \quad \text{and} \quad \|u\|_{W_\delta^{1,2}(K)} \le 2\,\|g\|_{W_\delta^{1/2,2}(\gamma^\pm)}.$$

Consequently,

$$\|r^\delta g\|_{W^{1/2,2}(\gamma^\pm)} \le \|r^\delta u\|_{W^{1,2}(K)} \le c\,\|u\|_{W_\delta^{1,2}(K)} \le 2c\,\|g\|_{W_\delta^{1/2,2}(\gamma^\pm)}.$$

On the other hand, there exists a function $v \in W^{1,2}(K)$, $v(x') = 0$ for $|x'| < 1/2$, such that

$$v = r^\delta g \text{ on } \gamma^\pm \quad \text{and} \quad \|v\|_{W^{1,2}(K)} \le 2\,\|r^\delta g\|_{W^{1/2,2}(\gamma^\pm)}.$$

This implies

$$\|g\|_{W_\delta^{1/2,2}(\gamma^\pm)} \le \|r^{-\delta} v\|_{W_\delta^{1,2}(K)} \le c\,\|v\|_{W^{1,2}(K)} \le 2c\,\|r^\delta g\|_{W^{1/2,2}(\gamma^\pm)}.$$

Thus on the set of all functions vanishing for $r < 1/2$, the $W_\delta^{1/2,2}(\gamma^\pm)$-norm of g is equivalent to the $W^{1/2,2}(\gamma^\pm)$-norm of $r^\delta g$ and, therefore, to the expression

$$\Big(\int_0^\infty r^{2\delta}\,|g(r)|^2\,dr + \int_0^\infty\!\!\int_0^\infty \frac{|r^\delta g(r) - \rho^\delta g(\rho)|^2}{|r-\rho|^2}\,dr\,d\rho\Big)^{1/2}.$$

This proves the lemma. □

Now we are in position to prove a parameter-dependent estimate for solutions of the problem (6.3.1), (6.3.2).

THEOREM 6.3.5. *Suppose that the form $b_{\mathcal{D}}$ is $\mathcal{H}_{\mathcal{D}}$-elliptic and δ satisfies the inequalities* (6.3.6). *Then the problem* (6.3.1), (6.3.2) *is uniquely solvable in $W_\delta^{2,2}(K)^\ell$ for arbitrary real $\xi \ne 0$, $f \in W_\delta^{0,2}(K)^\ell$, and $g^\pm \in W_\delta^{1/2,2}(\gamma^\pm)^\ell$. The solution u satisfies the estimate*

$$\text{(6.3.10)}\quad \sum_{k=0}^{2} |\xi|^{2-k}\,\|u\|_{L_\delta^{k,2}(K)^\ell} \le c\Big(\|f\|_{L_\delta^{0,2}(K)^\ell} + \sum_{d^\pm=1} \|g^\pm\|_{V_\delta^{1/2,2}(\gamma^\pm)^\ell} + \sum_{d^\pm=1} |\xi|^{1/2}\,\|r^\delta g^\pm\|_{L_2(\gamma^\pm)^\ell}\Big)$$

with a constant c independent of f, $g^\pm$ and ξ.

P r o o f. By Lemma 6.3.3, the problem (6.3.1), (6.3.2) is uniquely solvable in $W_\delta^{2,2}(K)^\ell$ for $\xi = \pm 1$. Let $v \in W_\delta^{2,2}(K)^\ell$ be the uniquely determined solution of the problem

$$L(D_{x'}, \xi/|\xi|)\, v = \phi \text{ in } K,$$
$$N^\pm(D_{x'}, \xi/|\xi|)\, v = \psi^\pm \text{ on } \gamma^\pm \text{ for } d^\pm = 1, \quad v = 0 \text{ on } \gamma^\pm \text{ for } d^\pm = 0,$$

where $\phi(x') = |\xi|^{-2} f(x'/|\xi|)$, $\psi^\pm(x') = |\xi|^{-1} g^\pm(x'/|\xi|)$. Then the vector function $u(x') = v(|\xi|x')$ is a solution of the boundary value problem (6.3.1), (6.3.2). Using the estimate

$$\|v\|_{W_\delta^{2,2}(K)^\ell} \le c\Big(\|\phi\|_{W_\delta^{0,2}(K)^\ell} + \sum_{d^\pm=1} \|\psi^\pm\|_{W_\delta^{1/2,2}(\gamma^\pm)^\ell}\Big)$$

and the relations

$$\|v\|^2_{W_\delta^{2,2}(K)^\ell} = |\xi|^{2\delta-2} \sum_{k=0}^{2} |\xi|^{4-2k} \|u\|^2_{L_\delta^{k,2}(K)^\ell}\,,$$

$$\|\phi\|^2_{W_\delta^{0,2}(K)^\ell} = |\xi|^{2\delta-2}\, \|f\|^2_{L_\delta^{0,2}(K)^\ell}\,,$$

$$\|\psi^\pm\|^2_{W_\delta^{1/2,2}(\gamma^\pm)^\ell} \le c\,|\xi|^{2\delta-2}\,\big(\|g^\pm\|^2_{V_\delta^{1/2,2}(\gamma^\pm)^\ell} + |\xi|\,\|r^\delta g^\pm\|^2_{L_2(\gamma^\pm)^\ell}\big),$$

we obtain (6.3.10). □

6.3.4. A regularity result for the solution in the case $\xi = 0$. We are interested in a regularity result for the solution $u \in W_\delta^{l,p}(K)^\ell$ of the boundary value problem

$$L(D_{x'},0)\,u = f \text{ in } K, \quad B^\pm(D_{x'},0)\,u = g^\pm \text{ on } \gamma^\pm, \tag{6.3.11}$$

where

$$B^\pm(D_{x'},0)u = (1-d^\pm)u + d^\pm\, N^\pm(D_{x'},0)\,u.$$

To this end, we prove the following lemma.

LEMMA 6.3.6. *Let the integer $k > 0$ be not an eigenvalue of the pencil $A(\lambda)$. Then for an arbitrary homogeneous polynomial p° of degree $k-2$ ($p^\circ = 0$ if $k = 1$) and for an arbitrary vector function $q^\pm = c^\pm r^{k-d^\pm}$, where $c^\pm$ is a constant vector, there exists a homogeneous polynomial P of degree k such that*

$$L(D_{x'},0)\,P = p^\circ \quad in\ K, \quad B^\pm(D_{x'},0)\,P = q^\pm \quad on\ \gamma^\pm. \tag{6.3.12}$$

P r o o f. Let

$$p^\circ = \sum_{j=0}^{k-2} b_j\, x_1^j\, x_2^{k-2-j} \quad \text{and} \quad q^\pm = c^\pm\, r^{k-d^\pm}$$

be given. Inserting

$$P = \sum_{j=0}^{k} a_j\, x_1^j\, x_2^{k-j} \tag{6.3.13}$$

into (6.3.12) and comparing the coefficients of $x_1^j x_2^{k-2-j}$ and r^{k-1}, respectively, we get a linear system of $k+1$ equations with $k+1$ unknowns $a_0, a_1, \dots, a_k$. Since k is not an eigenvalue of the pencil $A(\lambda)$, the corresponding homogeneous system has only the trivial solution (see Lemma 6.1.4). Therefore, there exists a unique polynomial (6.3.13) satisfying (6.3.12). □

Using the last lemma and regularity results in the class of the spaces $V_\delta^{l,p}(K)$ (cf. Corollary 1.2.7), we obtain analogous regularity assertions in the class of the spaces $W_\delta^{l,p}(K)$.

LEMMA 6.3.7. *Let $u \in W_\delta^{l,p}(K)^\ell$ be a solution of the problem (6.3.11), where $f \in W_\delta^{l-1,p}(K)^\ell$, $g^\pm \in W_\delta^{l+1-d^\pm-1/p,p}(\gamma^\pm)^\ell$, $l \ge 2$, $\delta > -2/p$. Suppose that the strip*

$$l - \delta - 2/p \le \operatorname{Re}\lambda \le l + 1 - \delta - 2/p$$

does not contain eigenvalues of the pencil $A(\lambda)$. Then $u \in W_\delta^{l+1,p}(K)^\ell$ and

$$\|u\|_{W_\delta^{l+1,p}(K)^\ell} \le c\Big(\|u\|_{W_\delta^{l,p}(K)^\ell} + \|f\|_{W_\delta^{l-1,p}(K)^\ell} + \sum_\pm \|g^\pm\|_{W_\delta^{l+1-d^\pm-1/p,p}(\gamma^\pm)^\ell}\Big)$$

with a constant c independent of u.

P r o o f. If $\delta + 2/p > l$, then $W_\delta^{l,p}(K) \subset V_\delta^{l,p}(K)$, $W_\delta^{l-1,p}(K) \subset V_\delta^{l-1,p}(K)$, and $W_\delta^{l-1/p,p}(\gamma^\pm) \subset V_\delta^{l-1/p,p}(\gamma^\pm)$. Therefore, it follows from Corollary 1.2.7 that $u \in V_\delta^{l+1,p}(K)^\ell \cap W_\delta^{l,p}(K)^\ell \subset W_\delta^{l+1,p}(K)^\ell$.

We suppose that $\delta+2/p < l$ and $\delta+2/p$ is not an integer. Then u has continuous derivatives up to order $k = [l-\delta-2/p]$ at $x = 0$, where $[s]$ denotes the integral part of s. Let ζ be a smooth cut-off function on $\overline{K}$ equal to one near the vertex $x = 0$. We denote by $p_k(x')$ the Taylor polynomial of degree k of u and set $v = u - \zeta p_k$. By Lemma 6.2.12, $v \in V_\delta^{l,p}(K)^\ell$. Furthermore,

$$L(D_{x'}, 0)\, v = f - L(D_{x'}, 0)\,(\zeta p_k) \in W_\delta^{l-1,p}(K)^\ell \cap V_\delta^{l-2,p}(K)^\ell.$$

Analogously, $B^\pm(D_{x'}, 0)\, v|_{\gamma^\pm} \in W_\delta^{l-1/p,p}(\gamma^\pm)^\ell \cap V_\delta^{l-1-1/p,p}(\gamma^\pm)^\ell$. Consequently, there are the representations

$$L(D_{x'}, 0)\, v = \zeta\, \overset{\circ}{p}_{k-1} + F, \quad B^\pm(D_{x'}, 0)\, v = \zeta\, q^\pm + G^\pm,$$

where $\overset{\circ}{p}_{k-1}$, $q^\pm$ are homogeneous polynomials of degrees $k-1$ and $k+1-d^\pm$, respectively, $F \in V_\delta^{l-1,p}(K)^\ell$, and $G^\pm|_{\gamma^\pm} \in V_\delta^{l-1/p,p}(\gamma^\pm)^\ell$. Since $\lambda = k+1$ is not an eigenvalue of the pencil $A(\lambda)$, there exists a homogeneous polynomial $\overset{\circ}{p}_{k+1}$ of degree $k+1$ such that

$$L(D_{x'}, 0)\, \overset{\circ}{p}_{k+1} = \overset{\circ}{p}_{k-1} \ \text{ in } K, \quad B^\pm(D_{x'}, 0)\, \overset{\circ}{p}_{k+1} = q^\pm \ \text{ on } \gamma^\pm$$

(see Lemma 6.3.6). Then the vector function $w = v - \zeta \overset{\circ}{p}_{k+1}$ belongs to $V_\delta^{l,p}(K)^\ell$. Furthermore,

$$L(D_{x'}, 0)\, w \in V_\delta^{l-1,p}(K)^\ell, \quad B(D_{x'}, 0)\, w \in V_\delta^{l+1-d^\pm-1/p,p}(\gamma^\pm)^\ell.$$

Applying Corollary 1.2.7, we obtain

$$w = v - \zeta \overset{\circ}{p}_{k+1} = u - \zeta(p_k + \overset{\circ}{p}_{k+1}) \in V_\delta^{l+1,p}(K)^\ell$$

and therefore $u \in W_\delta^{l+1,p}(K)^\ell$. Furthermore, the desired estimate holds.

We consider the case where $\delta + 2/p$ is an integer, $\delta + 2/p \le l$. Since $\zeta u \in W_{\delta+\varepsilon}^{l,p}(K)^\ell$, $L(D_{x'}, 0)\,(\zeta u) \in W_{\delta+\varepsilon}^{l-1,p}(K)^\ell$, $B^\pm(D_{x'}, 0)\,(\zeta u) \in W_{\delta+\varepsilon}^{l+1-d^\pm-1/p,p}(\gamma^\pm)^\ell$ with arbitrary positive ε, it follows from the first part of the proof that $\zeta u \in W_{\delta+\varepsilon}^{l+1,p}(K)^\ell$. Consequently, the vector function $v = \zeta(u - p_k)$, where $k = l-\delta-2/p$, belongs to $V_{\delta+\varepsilon}^{l+1,p}(K)^\ell$. Analogously to the first part of the proof,

$$L(D_{x'}, 0)\, v \in W_\delta^{l-1,p}(K)^\ell \cap V_{\delta+\varepsilon}^{l-1,p}(K)^\ell,$$

$$B^\pm(D_{x'}, 0)\, v|_{\gamma^\pm} \in W_\delta^{l+1-d^\pm-1/p,p}(\gamma^\pm)^\ell \cap V_{\delta+\varepsilon}^{l+1-d^\pm-1/p,p}(K)^\ell.$$

Consequently by Theorem 6.2.13, the representations

$$L(D_{x'},0)\,v = \sum_{i+j=k-1} f_{i,j}(r)\,x_1^i\,x_2^j + F \tag{6.3.14}$$

and

$$B^\pm(D_{x'},0)\,v|_{\gamma^\pm} = \psi^\pm(r)\,r^{k+1-d^\pm} + G^\pm, \tag{6.3.15}$$

hold, where $F \in V_\delta^{l-1,p}(K)^\ell$, $G^\pm \in W_\delta^{l+1-d^\pm-1/p,p}(\gamma^\pm)^\ell$, $f_{i,j}$, $\psi^\pm$ are functions in $W^{l,p}_{l-1/p}(\mathbb{R}_+)^\ell$ with support in $[0,1)$ such that

$$\int_0^1 r^{ps-1}|\partial_r^s f_{i,j}(r)|^p\,dr \le c_s\,\|L(\partial_{x'},0)\,v\|^p_{W_\delta^{l-1,p}(K)^\ell}, \tag{6.3.16}$$

$$\int_0^1 r^{ps-1}|\partial_r^s g_m^\pm(r)|^p\,dr \le c_s\,\|B^\pm(\partial_{x'},0)\,v\|^p_{W_\delta^{l+1-d^\pm-1/p,p}(\gamma^\pm)^\ell} \tag{6.3.17}$$

for s=1,2,.... Since $\lambda = k+1$ is not an eigenvalue of the pencil $A(\lambda)$, there exist homogeneous matrix-valued polynomials $p_{i,j}$ of degree $k+1$ such that

$$L(D_{x'},0)\,p_{i,j} = x_1^i x_2^j\,I_\ell \text{ in } K, \quad B^\pm(D_{x'},0)\,p_{i,j} = 0 \text{ on } \gamma^\pm$$

for $i+j=k-1$, where I_ℓ denotes the $\ell\times\ell$ identity matrix (see Lemma 6.3.6). Furthermore, there exist homogeneous matrix-valued polynomials q^+ and q^- of degree $k+1$ such that

$$\begin{aligned} &L(D_{x'},0)\,q^\pm = 0 \text{ in } K,\\ &B^+(D_{x'},0)\,q^+ = r^k\,I_\ell \text{ on } \gamma^+, \quad B^-(D_{x'},0)\,q^+ = 0 \text{ on } \gamma^-,\\ &B^+(D_{x'},0)\,q^- = 0 \text{ on } \gamma^+, \quad B^-(D_{x'},0)\,q^- = r^k\,I_\ell \text{ on } \gamma^-. \end{aligned}$$

We introduce the vector function

$$w = \sum_{i+j=k-1} p_{i,j}(x')\,f_{i,j}(r) + \sum_\pm q^\pm(x')\,\psi^\pm(r).$$

It follows from (6.3.16), (6.3.17) that $w \in W_\delta^{l+1,p}(K)^\ell \cap V_{\delta+\varepsilon}^{l+1,p}(K)^\ell$. Furthermore, according to (6.3.14)–(6.3.17), we have

$$L(D_{x'},0)\,(v-w) \in V_\delta^{l-1,p}(K)^\ell, \quad B^\pm(D_{x'},0)\,(v-w)\big|_{\gamma^\pm} \in V_\delta^{l+1-d^\pm-1/p,p}(\gamma^\pm)^\ell.$$

Applying Corollary 1.2.7, we obtain $v - w \in V_\delta^{l+1,p}(K)^\ell$ and consequently $u \in W_\delta^{l+1,p}(K)^\ell$. The proof is complete. □

6.4. Solvability of the boundary value problem in the dihedron in weighted L_2 Sobolev spaces

In this section, we prove the existence and uniqueness of solutions of the boundary value problem (6.1.3) in the weighted space $L_\delta^{2,2}(\mathcal{D})$. First we obtain an a priori estimate analogous to Theorem 6.1.3 in the class of the spaces $W_\delta^{l,p}(\mathcal{D})$. Then we establish conditions on the boundary data ensuring the existence of a function $u \in W_\delta^{l,p}(\mathcal{D})^\ell$ satisfying the boundary conditions $B^\pm(D_x)u = g^\pm$ on the faces $\Gamma^\pm$. In the last subsection, we consider the variational solution $u \in L_\delta^{1,2}(\mathcal{D})^\ell$ of the boundary value problem (6.1.3). We prove that $u \in L_\delta^{2,2}(\mathcal{D})^\ell$ if f and $g^\pm$ are functions in $W_\delta^{0,2}(\mathcal{D})^\ell$ and $W_\delta^{1-d^\pm-1/2,2}(\Gamma^\pm)^\ell$, respectively, with compact supports.

For the proof of this assertion, we employ the analogous result for the parameter-depending problem in the angle K (cf. Theorem 6.3.5).

6.4.1. An a priori estimate for the solution. In Section 6.1 it was proved that any solution $u \in V_{\delta-1}^{l-1,p}(\mathcal{D})^\ell$ of the boundary value problem (6.1.3) belongs to $V_\delta^{l,p}(\mathcal{D})^\ell$ if $f \in V_\delta^{l-2,p}(\mathcal{D})^\ell$ and $g^\pm \in V_\delta^{l-d^\pm-1/p,p}(\Gamma^\pm)^\ell$, $l \geq 2$. We extend this result to the spaces $W_\delta^{l,p}(\mathcal{D})$.

THEOREM 6.4.1. *Suppose that* $u \in W_{\delta-1}^{l-1,p}(\mathcal{D})^\ell$ *is a solution of the boundary value problem* (6.1.3), *where* $f \in W_\delta^{l-2,p}(\mathcal{D})^\ell$, $g^\pm \in W_\delta^{l-d^\pm-1/p,p}(\Gamma^\pm)^\ell$, $l \geq 2$, $\delta > 1-2/p$. *Then* $u \in L_\delta^{l,p}(\mathcal{D})^\ell$ *and*

$$\|u\|_{L_\delta^{l,p}(\mathcal{D})^\ell} \leq c\Big(\|f\|_{W_\delta^{l-2,p}(\mathcal{D})^\ell} + \sum_\pm \|g^\pm\|_{W_\delta^{l-d^\pm-1/p,p}(\Gamma^\pm)^\ell} + \|u\|_{W_{\delta-1}^{l-1,p}(\mathcal{D})^\ell}\Big) \tag{6.4.1}$$

with a constant c independent of u.

Proof. If $\delta > l-2/p$, then $W_{\delta-1}^{l-1,p}(\mathcal{D})^\ell \subset V_{\delta-1}^{l-1,p}(\mathcal{D})^\ell$ and the assertion follows immediately from Theorem 6.1.3. Let $1-2/p < \delta \leq l-2/p$. Then according to Theorems 6.2.7 and 6.2.11, there exists a vector function $v \in W_\delta^{l,p}(\mathcal{D})^\ell$ such that $u - v \in V_{\delta-1}^{l-1,p}(\mathcal{D})^\ell$ and

$$\|v\|_{W_\delta^{l,p}(\mathcal{D})^\ell} + \|u-v\|_{V_{\delta-1}^{l-1,p}(\mathcal{D})^\ell} \leq c\,\|u\|_{W_{\delta-1}^{l-1,p}(\mathcal{D})^\ell}\,.$$

Consequently,

$$L(D_x)\,(u-v) = f - L(D_x)\,v \in W_\delta^{l-2,p}(\mathcal{D})^\ell.$$

In addition, $L(D_x)\,(u-v) \in V_{\delta-1}^{l-3,p}(\mathcal{D})^\ell$ for $l \geq 3$ and therefore $L(D_x)\,(u-v) \in V_\delta^{l-2,p}(\mathcal{D})^\ell$. Analogously,

$$B^\pm(D_x)\,(u-v) = g^\pm - B^\pm(D_x)\,v \in V_\delta^{l-d^\pm-1/p,p}(\Gamma^\pm)^\ell.$$

Applying Theorem 6.1.3, we obtain $u - v \in V_\delta^{l,p}(\mathcal{D})^\ell$ and, consequently, $u \in L_\delta^{l,p}(\mathcal{D})^\ell$. Moreover, the estimate (6.4.1) can be deduced by means of Theorem 6.1.3 and Lemma 6.2.5. □

We will also employ the following local regularity result in the class of the spaces $W_\delta^{l,p}$.

COROLLARY 6.4.2. *Let u be a solution of the problem* (6.1.3), *and let* ζ, η *be infinitely differentiable functions with compact supports such that* $\eta = 1$ *in a neighborhood of* $\operatorname{supp}\zeta$. *If* $\eta u \in W_\delta^{l,p}(\mathcal{D})^\ell$, $\eta f \in W_{\delta+k}^{l+k-2,p}(\mathcal{D})^\ell$, $\eta g^\pm \in W_{\delta+k}^{l+k-d^\pm-1/p,p}(\Gamma^\pm)^\ell$, $k, l \geq 1$, $\delta > -2/p$, *and* $\delta + 2/p$ *is not integer, then* $\zeta u \in W_{\delta+k}^{l+k,p}(\mathcal{D})^\ell$ *and*

$$\|\zeta u\|_{W_{\delta+k}^{l+k,p}(\mathcal{D})^\ell} \leq c\Big(\|\eta f\|_{W_{\delta+k}^{l+k-2,p}(\mathcal{D})^\ell} + \sum_\pm \|\eta g^\pm\|_{W_{\delta+k}^{l+k-d^\pm-1/p,p}(\Gamma^\pm)^\ell} + \|\eta u\|_{W_\delta^{l,p}(\mathcal{D})^\ell}\Big).$$

Proof. For $k = 1$ the assertion of the corollary can be easily deduced from Theorem 6.4.1, for $k > 1$ the assertion holds by induction. □

6.4.2. Reduction to zero boundary data. If $g^\pm \in V_\delta^{l-d^\pm-1/p,p}(\Gamma^\pm)^\ell$, then there always exists a vector function $u \in V_\delta^{l,p}(\mathcal{D})^\ell$ satisfying the boundary conditions

$$B^\pm(D_x)\, u = g^\pm \quad \text{on } \Gamma^\pm \tag{6.4.2}$$

(cf Lemma 2.2.3). As the next lemma shows, an analogous result in the weighted spaces $W_\delta^{l,p}(\mathcal{D})$ holds only under additional compatibility conditions on the boundary data $g^\pm$.

LEMMA 6.4.3. *Let $g^\pm \in W_\delta^{l-d^\pm-1/p,p}(\Gamma^\pm)^\ell$ be given vector functions, $l \ge 2$, $\delta + 2/p > l - 1$. We assume that $g^\pm(x) = 0$ for $|x'| > 1$. In the case $d^+ = d^- = 0$ we assume in addition that*

$$g^+|_M = g^-|_M \quad \text{if } \delta + 2/p < l,$$

$$\int_{-\infty}^{+\infty}\int_0^1 r^{-1}\, |g^+(r,x_3) - g^-(r,x_3)|^p\, dr\, dx_3 < \infty \quad \text{if } \delta + 2/p = l.$$

Then there exists a vector function $u \in W_\delta^{l,p}(\mathcal{D})^\ell$ vanishing for $|x'| > 1$ and satisfying the boundary conditions (6.4.2).

P r o o f. 1) If $d^+ = d^- = 1$, then $g^\pm \in W_\delta^{l-1-1/p,p}(\Gamma^\pm)^\ell \subset V_\delta^{l-1-1/p,p}(\Gamma^\pm)^\ell$. Hence by Lemma 2.2.3, there exists a vector function $u \in V_\delta^{l,p}(\mathcal{D})^\ell$ satisfying (6.4.2).

2) Let $d^+ = 0$ and $d^- = 1$. Then there exists a vector function $v \in W_\delta^{l,p}(\mathcal{D})^\ell$ such that $B^+ v = v = g^+$ on Γ^+. Since $g^- - B^- v|_{\Gamma^-} \in W_\delta^{l-1-1/p,p}(\Gamma^-)^\ell = V_\delta^{l-1-1/p,p}(\Gamma^-)^\ell$, there exists a vector function $w \in V_\delta^{l,p}(\mathcal{D})^\ell$ such that

$$B^+ w = 0 \text{ on } \Gamma^+, \qquad B^- w = g^- - B^- v \quad \text{on } \Gamma^-.$$

Both v and w can be chosen such that $v(x') = w(x') = 0$ for $|x'| \ge 1$. Thus, the sum $u = v + w$ has the desired properties.

3) Let $d^+ = d^- = 0$. If $\delta > l - 2/p$, then $W_\delta^{l-1/p,p}(\Gamma^\pm)^\ell \subset V_\delta^{l-1/p,p}(\Gamma^\pm)^\ell$ and the assertion of the lemma follows from Lemma 2.2.3. If $\delta < l - 2/p$, then the traces of $g^\pm$ exist on the edge M. Let $v \in W_\delta^{l,p}(\mathcal{D})^\ell$ be a vector function such that $v = g^+$ on Γ^+. If $g^+|_M = g^-|_M$, then the trace of $g^- - v|_{\Gamma^-}$ is zero on M and consequently $g^- - v|_{\Gamma^-} \in V_\delta^{l-1/p,p}(\Gamma^-)^\ell$. Thus by Lemma 2.2.3, there exists a vector function $w \in V_\delta^{l,p}(\mathcal{D})^\ell$ satisfying the conditions

$$w = 0 \text{ on } \Gamma^+, \qquad w = g^- - v \quad \text{on } \Gamma^-. \tag{6.4.3}$$

As in the case 2), we set $u = v + w$.

Finally, we consider the case $d^+ = d^- = 0$, $\delta = l - 2/p$. Let again $v \in W_\delta^{l,p}(\mathcal{D})^\ell$ be a vector function satisfying the equation $v|_{\Gamma^+} = g^+$, i.e. $v(r, \theta/2, x_3) = g^+(r, x_3)$.

Using (6.2.1), we obtain

$$\int_{-\infty}^{+\infty}\int_0^1 r^{-1}\big|v(r,\theta/2,x_3)-v(r,-\theta/2,x_3)\big|^p\,dr\,dx_3$$

$$=\int_{-\infty}^{+\infty}\int_0^1 r^{-1}\Big|\int_{-\theta/2}^{+\theta/2}\partial_\varphi u(r,\varphi,x_3)\,d\varphi\Big|^p\,dr\,dx_3$$

$$\le c\sum_{j=1}^{2}\int_{\mathcal{D}} r^{p-2}|\partial_{x_j}v|^p\,dx\le c'\,\|v\|^p_{W^{l,p}_\delta(\mathcal{D})^\ell}$$

and consequently

$$\int_{-\infty}^{+\infty}\int_0^1 r^{-1}\big|v(r,-\theta/2,x_3)-g^-(r,x_3)\big|^p\,dr\,dx_3<\infty.$$

The last inequality together with Theorem 6.2.16 yields $g^- - v|_{\Gamma^-}\in V_\delta^{l-1/p,p}(\Gamma^-)^\ell$. Thus by Lemma 2.2.3, there exists a vector function $w\in V_\delta^{l,p}(\mathcal{D})^\ell$ satisfying (6.4.3). Then $u=v+w$ has the desired properties. The proof of the lemma is complete. □

We prove an analogous result for the case $l-2<\delta+2/p\le l-1$.

LEMMA 6.4.4. *Let $g^\pm\in W_\delta^{l-d^\pm-1/p,p}(\Gamma^+)^\ell$, $l\ge 2$, $l-2<\delta+2/p\le l-1$, and let $g^\pm(x)=0$ for $|x'|>1$. We suppose that $\lambda=1$ is not an eigenvalue of the pencil $A(\lambda)$. In the case $d^+=d^-=0$ we assume in addition that $g^+|_M=g^-|_M$. Then there exists a vector function $u\in W_\delta^{l,p}(\mathcal{D})^\ell$, $u(x)=0$ for $|x'|>1$, satisfying the boundary conditions* (6.4.2) *and the estimate*

$$\|u\|_{W_\delta^{l,p}(\mathcal{D})^\ell}\le c\sum_\pm\|g^\pm\|_{W_\delta^{l-d^\pm-1/p,p}(\Gamma^\pm)^\ell},\tag{6.4.4}$$

where the constant c is independent of g^+ and g^-.

Proof. 1) We start with the case $d^+=d^-=1$. Then the boundary conditions (6.4.2) take the form

$$B^\pm(D_x)\,u=\sum_{j=1}^{3}(A_{j,1}n_1^\pm+A_{j,2}n_2^\pm)\,\partial_{x_j}u=g^\pm\quad\text{on }\Gamma^\pm.$$

If $\delta+2/p<l-1$, then the traces $\psi^\pm=g^\pm|_M$ exist. By Lemma 6.3.6, there exists a function $w(x)=c^{(1)}(x_3)\,x_1+c^{(2)}(x_3)\,x_2$ such that

$$B^\pm(D_{x'},0)\,w(x)=\sum_{j=1}^{2}(A_{j,1}n_1^\pm+A_{j,2}n_2^\pm)\,c^{(j)}(x_3)=\psi^\pm(x_3)$$

for all x_3 and

$$\sum_{j=1}^{2}\|c^{(j)}\|_{W^{l-1-\delta-2/p,p}(\mathbb{R})^\ell}\le c\sum_\pm\|\psi^\pm\|_{W^{l-1-\delta-2/p,p}(\mathbb{R})^\ell}.$$

We put $v(x)=(Ec^{(1)})(x)\,x_1+(Ec^{(2)})(x)\,x_2$, where E is the extension operator (6.2.4). Then by Lemma 6.2.3, $B^\pm(D_x)\,v-g^\pm=0$ on M and consequently

$B^\pm(D_x)\, v - g^\pm \in V_\delta^{l-1-1/p,p}(\Gamma^\pm)^\ell$ (see Theorem 6.2.15). Now the existence of a function u with the desired properties follows from Lemma 2.2.3.

Let $\delta + 2/p = l-1$. By Theorem 6.2.11, g^+ and g^- admit the decompositions

$$g^\pm(r,x_3) = G^\pm(r,x_3) + v^\pm(r,x_3),$$

where $v^\pm \in V_\delta^{l-1-1/p,p}(\Gamma^\pm)^\ell$ and $G^\pm \in W^{l,p}_{\delta+1+1/p}(\mathbb{R}^2_+)^\ell$. Let $(c^{(1)}, c^{(2)})$ be the uniquely determined solution of the linear system

$$\sum_{j=1}^{2} (A_{j,1} n_1^\pm + A_{j,2} n_2^\pm)\, c^{(j)}(r,x_3) = G^\pm(r,x_3),$$

$c^{(j)} \in W^{l,p}_{\delta+1+1/p}(\mathbb{R}^2_+)^\ell$ for $j=1,2$. If we consider $c^{(j)}$ as a function in $\mathcal{D}$, then $c^{(j)} \in W^{l,p}_{\delta+1}(\mathcal{D})^\ell$. Furthermore, $\partial_{x_i} c^{(j)} \in W^{l-1,p}_{\delta+1}(\mathcal{D})^\ell \subset V^{l-1,p}_{\delta+1}(\mathcal{D})^\ell$ for $i=1,2,3$. We define $v(x) = c^{(1)}(r,x_3)\, x_1 + c^{(2)}(r,x_3)\, x_2$. Then $B^\pm(D_x)\, v - g^\pm \in V_\delta^{l-1-1/p,p}(\Gamma^\pm)^\ell$ and it remains to apply Lemma 2.2.3.

2) Let $d^+ = 0$ and $d^- = 1$. Since there exists a vector function $u \in W_\delta^{l,p}(\mathcal{D})^\ell$ satisfying the condition $B^+(D_x)u = u = g^+$ on Γ^+, we may assume without loss of generality that $g^+ = 0$. If $\delta + 2/p < l-1$, then $g^-|_M \in W^{l-1-\delta-2/p,p}(M)^\ell$. By Lemma 6.3.6, there exists a function $w(x) = c^{(1)}(x_3)\, x_1 + c^{(2)}(x_3)\, x_2$ such that $w = 0$ on Γ^+ and

$$B^-(D_{x'},0)\, w(x) = \sum_{j=1}^{2} (A_{j,1} n_1^- + A_{j,2} n_2^-)\, c^{(j)}(x_3) = g^-(0,x_3)$$

We define $v(x) = (Ec^{(1)})(x)\, x_1 + (Ec^{(2)})(x)\, x_2$. Then

$$v|_{\Gamma^+} \in V_\delta^{l-1/p,p}(\Gamma^+)^\ell \quad \text{and} \quad B^-(D_x) v|_{\Gamma^-} - g^- \in V_\delta^{l-1-1/p,p}(\Gamma^-)^\ell$$

and it remains to apply Lemma 2.2.3. In the case $\delta + 2/p = l-1$, one obtains the result analogously to part 1) by means of Theorem 6.2.11.

3) Let $d^+ = d^- = 0$. Then we may assume again without loss of generality that $g^+ = 0$. If $\delta + 2/p < l-1$, then the trace ψ^- of $\partial_r g^-$ on M exists, $\psi^- \in W^{l-1-\delta-2/p,p}(M)^\ell$. Since $\theta \neq \pi$, $\theta \neq 2\pi$ (otherwise, $\lambda = 1$ is an eigenvalue of the pencil $A(\lambda)$), the linear system

$$c^{(1)}(x_3) \cos\frac{\theta}{2} + c^{(2)}(x_3) \sin\frac{\theta}{2} = 0, \quad c^{(1)}(x_3) \cos\frac{\theta}{2} - c^{(2)}(x_3) \sin\frac{\theta}{2} = \psi^-(x_3)$$

has a uniquely determined solution $(c^{(1)}, c^{(2)})$. We define again

$$v(x) = (Ec^{(1)})(x)\, x_1 + (Ec^{(2)})(x)\, x_2.$$

Then $v \in W_\delta^{l,p}(\mathcal{D})^\ell$, and $v|_{\Gamma^\pm} - g^\pm \in V_\delta^{l-1/p,p}(\Gamma^\pm)^\ell$. The last follows from the fact that both the traces of $v|_{\Gamma^\pm} - g^\pm$ and of the first order derivatives of $v|_{\Gamma^\pm} - g^\pm$ on M are zero. Applying Lemma 2.2.3, we obtain the assertion of the lemma for $\delta + 2/p < l-1$. In the case $\delta + 2/p = l-1$, the assertion can be proved analogously by means of Theorem 6.2.11. □

6.4.3. Solvability of the boundary value problem in $L_\delta^{2,2}(\mathcal{D})^\ell$. We consider the boundary value problem (6.1.3) with the right-hand sides $f \in W_\delta^{0,2}(\mathcal{D})^\ell$ and $g^\pm \in W_\delta^{2-d^\pm-1/2,2}(\Gamma^\pm)^\ell$, $0 < \delta < 1$. First we show that the corresponding functional (6.1.7) is continuous on $\mathcal{H}_{\mathcal{D}}$.

LEMMA 6.4.5. *Let the functional F be given by (6.1.7), where $f \in W_\delta^{0,2}(\mathcal{D})^\ell$, $g^\pm \in W_\delta^{1/2,2}(\Gamma^\pm)^\ell$ are vector functions with compact supports and $0 < \delta < 1$. Then*

$$\big|(F, v)_{\mathcal{D}}\big| \le c\,\big(\|f\|_{W_\delta^{0,2}(\mathcal{D})^\ell} + \|g^\pm\|_{W_\delta^{1/2,2}(\Gamma^\pm)^\ell}\big)\, \|v\|_{\mathcal{H}_{\mathcal{D}}}$$

for all $v \in \mathcal{H}_{\mathcal{D}}$, where c is a constant depending only on the support of f and $g^\pm$.

Proof. First note that $W_\delta^{1/2,2}(\Gamma^\pm) \subset V_\delta^{1/2,2}(\Gamma^\pm)$ for $\delta > 0$. Let $v \in L^{1,2}(\mathcal{D})^\ell$, and let ζ be an infinitely differentiable function with compact support equal to one on the supports of f and $g^\pm$. Then $\zeta v \in L^{1,2}(\mathcal{D})^\ell$. This results from the estimate

$$\int_{\mathcal{D}} |x|^{-2}\, |v(x)|^2\, dx \le c \int_{\mathcal{D}} \sum_{j=1}^{3} \big|\partial_{x_j} v(x)\big|^2\, dx \tag{6.4.5}$$

which can be easily deduced from Hardy's inequality. Since the support of ζv is compact, it follows that $\zeta v \in L_{1-\delta}^{1,2}(\mathcal{D})^\ell = V_{1-\delta}^{1,2}(\mathcal{D})^\ell$ and

$$\|\zeta v\|_{V_{1-\delta}^{1,2}(\mathcal{D})^\ell} \le c\, \|v\|_{L^{1,2}(\mathcal{D})^\ell}\,.$$

Consequently,

$$\Big| \int_{\mathcal{D}} f \cdot \bar{v}\, dx\Big| \le \|f\|_{W_\delta^{0,2}(\mathcal{D})^\ell}\ \|\zeta v\|_{W_{-\delta}^{0,2}(\mathcal{D})^\ell} \le \|f\|_{W_\delta^{0,2}(\mathcal{D})^\ell}\ \|v\|_{L^{1,2}(\mathcal{D})^\ell}$$

and

$$\begin{aligned}\Big| \int_{\Gamma^\pm} g^\pm \cdot \bar{v}\, dx\Big|^2 \ &\le\ \int_{\Gamma^\pm} r^{2\delta-1} |g^\pm|^2\, dx \int_{\Gamma^\pm} r^{1-2\delta} |\zeta v|^2\, dx \\ &\le\ \|g^\pm\|^2_{V_\delta^{1/2,2}(\Gamma^\pm)}\ \|\zeta v\|^2_{V_{1-\delta}^{1/2,2}(\Gamma^\pm)} \le c\, \|g^\pm\|^2_{V_\delta^{1/2,2}(\Gamma^\pm)}\ \|v\|_{L^{1,2}(\mathcal{D})^\ell}\,.\end{aligned}$$

This proves the lemma. □

According to the last lemma, the problem (6.1.5), (6.1.6) with the functional (6.1.7) is uniquely solvable in $L^{1,2}(\mathcal{D})^\ell$ if the form $b_{\mathcal{D}}$ is $\mathcal{H}_{\mathcal{D}}$-elliptic and $f \in W_\delta^{0,2}(\mathcal{D})^\ell$, $g^\pm \in W_\delta^{2-d^\pm-1/2}(\Gamma^\pm)^\ell$ are vector functions with compact supports, $0 < \delta < 1$ ($g^+|_M = g^-|_M$ if $d^+ = d^- = 0$).

THEOREM 6.4.6. *Let $f \in W_\delta^{0,2}(\mathcal{D})^\ell$, $g^\pm \in W_\delta^{2-d^\pm-1/2,2}(\Gamma^\pm)^\ell$ be vector functions with compact supports, and let δ satisfy the condition (6.3.6). In the case $d^+ = d^- = 0$, we assume that $g^+|_M = g^-|_M$. Suppose that the form $b_{\mathcal{D}}$ is $\mathcal{H}_{\mathcal{D}}$-elliptic and that $u \in L^{1,2}(\mathcal{D})^\ell$ is the uniquely determined solution of the problem (6.1.5), (6.1.6), where F is given by (6.1.7). Then $u \in L_\delta^{2,2}(\mathcal{D})^\ell$ and*

$$\|u\|_{L_\delta^{2,2}(\mathcal{D})^\ell} \le c\,\Big(\|f\|_{W_\delta^{0,2}(\mathcal{D})^\ell} + \sum_\pm \|g^\pm\|_{W_\delta^{2-d^\pm-1/2,2}(\Gamma^\pm)^\ell}\Big) \tag{6.4.6}$$

with a constant c independent of f and $g^\pm$.

Proof. By Lemma 6.4.3, we may restrict ourselves to the case $g^\pm = 0$. Let $\tilde{u}(x',\xi)$, $\tilde{f}(x',\xi)$ denote the Fourier transforms with respect to the variable x_3 of $u(x',x_3)$ and $f(x',x_3)$, respectively. Then $\tilde{u}(\cdot,\xi)$ is the (variational) solution of the problem

$$L(D_{x'},\xi)\,\tilde{u}(\cdot,\xi) = \tilde{f}(\cdot,\xi) \ \text{ in } K, \quad B^\pm(D_{x'},\xi)\,\tilde{u}(\cdot,\xi) = 0 \ \text{ on } \gamma^\pm.$$

By Theorem 6.3.5, the function $\tilde{u}(\cdot,\xi)$ belongs to $W_\delta^{2,2}(K)^\ell$ and satisfies the estimate

$$\sum_{k=0}^{2} |\xi|^{4-2k}\, \|\tilde{u}(\cdot,\xi)\|^2_{L_\delta^{k,2}(K)^\ell} \le c\, \|\tilde{f}(\cdot,\xi)\|^2_{L_\delta^{0,2}(K)^\ell}$$

for $\xi \neq 0$. Here c is independent of f and ξ. Integrating the last inequality with respect to ξ and using Parseval's theorem, we obtain (6.4.6). □

6.5. Regularity results for solutions of the boundary value problem

The goal of this section is to obtain a $W_\delta^{l,2}$-regularity result for the variational solution $u \in L^{1,2}(\mathcal{D})^\ell$ of the boundary value problem (6.1.3), where l is an arbitrary integer, $l \ge 2$. For $l = 2$ this was done in the last section. In order to prove the analogous result for $l > 2$, we need a regularity assertion for the x_3-derivatives of the solution. For this reason, the first two subsections deal with the x_3-derivatives of the variational solution of first and higher order. These results together with a regularity assertion for the problem in the angle K are used in the third subsection, where we show that $\zeta u \in W_\delta^{l,2}(\mathcal{D})^\ell$ under certain conditions on the right-hand sides f, $g^\pm$ and on δ. In particular, δ is assumed to be a positive real number such that the strip $0 < \operatorname{Re}\lambda < l-1-\delta$ is free of eigenvalue of the pencil $A(\lambda)$. The last condition can be weakened for some special Neumann problems, e.g. the Neumann problem for the Lamé system. For these problems, the spectrum of the pencil $A(\lambda)$ contains always (i.e., for every edge angle θ) the eigenvalue $\lambda = 1$. Then we may allow that this eigenvalue lies in the strip $0 < \operatorname{Re}\lambda < l-1-\delta$.

6.5.1. Regularity results for the x_3-derivatives of the solution. By Theorem 6.4.6, the derivatives of the solution $u \in L^{1,2}(\mathcal{D})^\ell$ of the problem (6.1.5), (6.1.6) belong to the space $L_\delta^{1,2}(\mathcal{D})^\ell$ if the conditions of this theorem on f, $g^\pm$ and δ are satisfied. As the following theorem shows, this result holds for the x_3-derivative under weaker conditions on δ.

THEOREM 6.5.1. *Let $u \in \mathcal{H}_\mathcal{D}$ be the solution of the problem*

$$b_\mathcal{D}(u,v) = (f,v)_\mathcal{D} \quad \text{for all } v \in \mathcal{H}_\mathcal{D}\,,$$

where $f \in \mathcal{H}^*_\mathcal{D} \cap V_\delta^{0,2}(\mathcal{D})^\ell$, $0 < \delta < 1$. Then $\partial_{x_3} u \in V_\delta^{1,2}(\mathcal{D})^3$ and

$$\|\partial_{x_3} u\|_{V_\delta^{1,2}(\mathcal{D})^3} \le c\, \|f\|_{V_\delta^{0,2}(\mathcal{D})^3}$$

with a constant c independent of f.

P r o o f. For arbitrary real $h \neq 0$ let $u_h(x) = h^{-1}\big(u(x',x_3+h) - u(x',x_3)\big)$. Obviously,

$$b_\mathcal{D}(u_h,v) = \int_\mathcal{D} f_h \cdot \bar{v}\, dx$$

for all $v \in \mathcal{H}_\mathcal{D}$. Consequently, there exists a constant c independent of u and h such that

$$\|u_h\|^2_{\mathcal{H}_\mathcal{D}} \le c\, \|f_h\|^2_{\mathcal{H}^*_\mathcal{D}}\,. \tag{6.5.1}$$

We prove that

$$\int_0^\infty h^{2\delta-1} \|f_h\|^2_{\mathcal{H}^*_\mathcal{D}}\, dh \le c\, \|f\|^2_{V_\delta^{0,2}(\mathcal{D})^3}\,. \tag{6.5.2}$$

Indeed, let χ be a smooth function, $0 \le \chi \le 1$, $\chi = 1$ for $r > h$, $\chi = 0$ for $r < h/2$, $|\nabla \chi| \le c\,h^{-1}$, and ε is an arbitrary positive number. Then

$$\Big| \int_{\mathcal{D}} f_h \cdot v\, dx \Big| = \Big| \int_{\mathcal{D}} f \cdot v_{-h}\, dx \Big|$$

$$\le \|\chi f\|_{L_2(\mathcal{D})^3}\, \|v_{-h}\|_{L_2(\mathcal{D})^3} + \|(r/h)^{1-\varepsilon} f\|_{L_2(\mathcal{D}_h)^3}\, \|(r/h)^{\varepsilon-1}(1-\chi) v_{-h}\|_{L_2(\mathcal{D}_h)^3},$$

where $\mathcal{D}_h = \{x \in \mathcal{D} :\, r(x) < h\}$. Here

$$\|v_{-h}\|^2_{L_2(\mathcal{D})^3} = \int_{\mathcal{D}} \Big| \int_0^1 \partial_{x_3} v(x', x_3 - th)\, dt \Big|^2 dx \le \|\partial_{x_3} v\|^2_{L_2(\mathcal{D})^3} \le \|v\|^2_{\mathcal{H}_{\mathcal{D}}}.$$

By Hardy's inequality,

$$\begin{aligned} \|(r/h)^{\varepsilon-1}(1-\chi) v_{-h}\|^2_{L_2(\mathcal{D}_h)^3} &\le c \int\limits_{\mathcal{D}} r^{2\varepsilon} h^{2-2\varepsilon} \big|\partial_r (1-\chi) v_{-h}\big|^2 dx \\ &\le c \int\limits_{\substack{\mathcal{D} \\ r<h}} \big(|v_{-h}|^2 + |\partial_r v_{-h}|^2\big)\, dx \le c\, \|v\|^2_{\mathcal{H}_{\mathcal{D}}}. \end{aligned}$$

Consequently,

$$\int_0^\infty h^{2\delta-1}\, \|f_h\|^2_{\mathcal{H}^*_{\mathcal{D}}}\, dh \le c \int_0^\infty h^{2\delta-1} \big(\|\chi f\|^2_{L_2(\mathcal{D})^3} + \|(r/h)^{1-\varepsilon} f\|^2_{L_2(\mathcal{D}_h)^3} \big)\, dh$$

$$\le \int_{\mathcal{D}} |f(x)|^2 \Big(\int_0^{2r} h^{2\delta-1}\, dh + r^{2-2\varepsilon} \int_r^\infty h^{2\delta-3+2\varepsilon}\, dh \Big)\, dx \le c\, \|f\|^2_{V^{0,2}_\delta(\mathcal{D})^3}.$$

for $0 < \varepsilon < 1 - \delta$. Next we prove that

$$\|\partial_{x_3} u\|^2_{V^{0,2}_{\delta-1}(\mathcal{D})^3} \le c \int_0^\infty h^{2\delta-1}\, \|u_h\|^2_{\mathcal{H}_{\mathcal{D}}}\, dh. \tag{6.5.3}$$

Let $\tilde{u}(x', \xi)$ denote the Fourier transform of $u(x', x_3)$ with respect to the variable x_3. Obviously,

$$\begin{aligned} \|\partial_{x_3} u\|^2_{V^{0,2}_{\delta-1}(\mathcal{D})^3} &= \int_{\mathbb{R}} \int_K |\xi|^2\, |x'|^{2\delta-2}\, |\tilde{u}(x', \xi)|^2\, dx'\, d\xi \\ &= \int\limits_{|\xi|>1/|x'|} \int\limits_K |\xi|^2\, |x'|^{2\delta-2}\, |\tilde{u}|^2\, dx'\, d\xi + \int\limits_{|\xi|<1/|x'|} \int\limits_K |\xi|^2\, |x'|^{2\delta-2}\, |\tilde{u}|^2\, dx'\, d\xi. \end{aligned}$$

Applying again Hardy's inequality, we obtain

$$\begin{aligned} &\|\partial_{x_3} u\|^2_{V^{0,2}_{\delta-1}(\mathcal{D})^3} \\ &\le \int\limits_{\mathbb{R}} \int\limits_K |\xi|^{4-2\delta}\, |\tilde{u}|^2\, dx'\, d\xi + c \int\limits_{|\xi|<1/|x'|} \int\limits_K |\xi|^2\, |x'|^{2\delta} \sum_{j=1}^2 \big|\partial_{x_j} \tilde{u}\big|^2\, dx'\, d\xi \\ &\le c \int_{\mathbb{R}} \int_K |\xi|^{2-2\delta} \Big(|\xi|^2\, |\tilde{u}(x', \xi)|^2 + \sum_{j=1}^2 |\partial_{x_j} \tilde{u}(x', \xi)|^2 \Big)\, dx'\, d\xi \\ &= c \int\limits_{\mathbb{R}} \int\limits_K \Big(\int\limits_0^\infty h^{2\delta-3} |e^{i\xi h} - 1|^2\, dh \Big) \Big(|\xi|^2\, |\tilde{u}(x', \xi)|^2 + \sum_{j=1}^2 |\partial_{x_j} \tilde{u}(x', \xi)|^2 \Big)\, dx'\, d\xi. \end{aligned}$$

Thus, the inequality (6.5.3) holds. From (6.5.1)–(6.5.3) it follows that

$$\|\partial_{x_3} u\|_{V^{0,2}_{\delta-1}(\mathcal{D})^3} \le c\, \|f\|_{V^{0,2}_{\delta}(\mathcal{D})^3}\,.$$

Now the assertion of the theorem results directly from Theorem 6.1.3. □

6.5.2. A regularity result for the x_3-derivatives of higher order. In the next lemma, let ζ and η be real-valued infinitely differentiable functions on $\overline{\mathcal{D}}$ with compact supports such that $\eta = 1$ in a neighborhood of $\operatorname{supp}\zeta$.

THEOREM 6.5.2. *Suppose that*

$$\eta\partial^j_{x_3} f \in W^{0,2}_\delta(\mathcal{D})^\ell \quad \textit{and} \quad \eta\partial^j_{x_3} g^\pm \in W^{2-d^\pm-1/2,2}_\delta(\Gamma^\pm)^\ell$$

for $j = 0,\dots,k$, where δ satisfies the condition (6.3.6). If $u \in L^{1,2}(\mathcal{D})^\ell$ is a solution of the problem (6.1.5), (6.1.6), and the functional F is given by (6.1.7), then $\zeta\partial^j_{x_3} u \in L^{2,2}_\delta(\mathcal{D})^\ell$ for $j = 0,\dots,k$, and

$$(6.5.4) \qquad \sum_{j=0}^{k} \|\zeta\partial^j_{x_3} u\|_{L^{2,2}_\delta(\mathcal{D})^\ell} \le c\Big(\sum_{j=0}^{k} \|\eta\partial^j_{x_3} f\|_{L^{0,2}_\delta(\mathcal{D})^\ell} + \sum_{j=0}^{k}\sum_{\pm} \|\eta\partial^j_{x_3} g^\pm\|_{L^{2-d^\pm-1/2,2}_\delta(\Gamma^\pm)^\ell} + \|\eta u\|_{L^{1,2}_\delta(\mathcal{D})^\ell}\Big)$$

with a constant c independent of u.

P r o o f. We prove the theorem by induction in k. First let $k = 0$. From (6.4.5) it follows that $\zeta u \in L^{1,2}(\mathcal{D})^\ell$. Furthermore by (6.1.5),

$$b_{\mathcal{D}}(\zeta u, v) = (\zeta F, v)_{\mathcal{D}} + \sum_{d^\pm=1} \int_{\Gamma^\pm} A_{i,j}\,(\partial_{x_i}\zeta)\, n_j\, u\cdot v\, dx + \int_{\mathcal{D}} \sum_{i,j=1}^{3} A_{i,j}\Big((D_{x_i}\zeta)\, D_{x_j} u + D_{x_i}(u\, D_{x_j}\zeta)\Big)\cdot v\, dx.$$

By the assumptions of the theorem, the right-hand side of the last equation ist a functional of the form (6.1.7), where $f \in L^{0,2}_\delta(\mathcal{D})^\ell$ and $g^\pm \in L^{1/2,2}_\delta(\Gamma^\pm)^\ell$. Thus, we can apply Theorem 6.4.6 and obtain (6.5.4).

Suppose the theorem is proved for $k-1$. Then, under our assumptions on F, we have $\zeta\partial^j_{x_3} u \in L^{2,2}_\delta(\mathcal{D})^\ell$ for $j = 0,\dots,k-1$. Let $v = \partial^{k-1}_{x_3} u$, $\Phi = \partial^{k-1}_{x_3} f$ and $\Psi^\pm = \partial^{k-1}_{x_3} g^\pm$. We assume that $\eta\partial^j_{x_3} g^\pm \in V^{2-d^\pm-1/2,2}_\delta(\Gamma^\pm)^\ell$ for $j = 0,\dots,k$ and show that

$$(6.5.5) \quad \|\zeta\partial_{x_3} v\|_{L^{2,2}_\delta(\mathcal{D})^\ell} \le c\sum_{j=0}^{1}\Big(\|\eta\partial^j_{x_3}\Phi\|_{V^{0,2}_\delta(\mathcal{D})^\ell} + \sum_{\pm}\|\eta\partial^j_{x_3}\Psi^\pm\|_{V^{2-d^\pm-1/2,2}_\delta(\Gamma)^\ell} + \|\chi\partial^j_{x_3} v\|_{L^{1,2}_\delta(\mathcal{D})^\ell}\Big),$$

where χ is a real-valued infinitely differentiable function such that $\eta = 1$ in a neighborhood of $\operatorname{supp}\chi$ and $\chi = 1$ in a neighborhood of $\operatorname{supp}\zeta$. To this end, we consider the vector function

$$v_h(x) = h^{-1}\big(v(x', x_3+h) - v(x', x_3)\big),$$

where h is a real number not equal to zero. Obviously, v_h is a solution of the problem

$$L(D_x)\, v_h = \Phi_h \ \text{ in } \mathcal{D}, \quad B^\pm(D_x) v_h = \Psi_h^\pm \ \text{ on } \Gamma^\pm.$$

Let ψ be a real-valued infinitely differentiable function such that $\chi = 1$ in a neighborhood of $\operatorname{supp}\psi$ and $\psi = 1$ in a neighborhood of $\operatorname{supp}\zeta$. By the induction hypothesis,

$$\text{(6.5.6)} \quad \|\zeta v_h\|_{L_\delta^{2,2}(\mathcal{D})^\ell} \le c\Big(\|\psi\Phi_h\|_{V_\delta^{0,2}(\mathcal{D})^\ell} + \sum_\pm \|\psi\Psi_h^\pm\|_{V_\delta^{2-d^\pm-1/2,2}(\Gamma^\pm)^\ell} + \|\psi v_h\|_{L_\delta^{1,2}(\mathcal{D})}\Big)$$

with a constant c independent of h. For sufficiently small $|h|$, we get

$$\|(\psi\Phi)_h\|^2_{V_\delta^{0,2}(\mathcal{D})^\ell} = \int_{\mathcal{D}} r^{2\delta}\, h^{-2}\, \big|(\psi\Phi)(x',x_3+h) - (\psi\Phi)(x',x_3)\big|^2\, dx$$

$$= \int_{\mathcal{D}} r^{2\delta}\Big|\int_0^1 \frac{\partial(\psi\Phi)}{\partial x_3}(x',x_3+th)\, dt\Big|^2 dx \le \int_{\mathcal{D}} r^{2\delta}\, \big|\partial_{x_3}(\psi(x)\Phi(x))\big|^2\, dx$$

$$\le c\Big(\|\eta\Phi\|^2_{V_\delta^{0,2}(\mathcal{D})^\ell} + \|\eta\partial_{x_3}\Phi\|^2_{V_\delta^{0,2}(\mathcal{D})^\ell}\Big).$$

Since $\psi\Phi_h = (\psi\Phi)_h - \psi_h\Phi$, the last inequality implies

$$\|\psi\Phi_h\|_{V_\delta^{0,2}(\mathcal{D})^\ell} \le c\Big(\|\eta\Phi\|_{V_\delta^{0,2}(\mathcal{D})^\ell} + \|\eta\partial_{x_3}\Phi\|_{V_\delta^{0,2}(\mathcal{D})^\ell}\Big).$$

Analogously, we obtain the estimate

$$\|\psi\Psi_h^\pm\|_{V_\delta^{2-d^\pm-1/2,2}(\Gamma^\pm)^\ell} \le c\sum_{j=0}^{1} \|\eta\partial_{x_3}^j\Psi^\pm\|^2_{V_\delta^{2-d^\pm-1/2,2}(\Gamma^\pm)^\ell}$$

by means of Lemma 2.1.10. Furthermore,

$$\text{(6.5.7)} \qquad \|\psi v_h\|_{L_\delta^{1,2}(\mathcal{D})^\ell} \le c\Big(\|\chi v\|_{L_\delta^{1,2}(\mathcal{D})^\ell} + \|\chi\partial_{x_3} v\|_{L_\delta^{1,2}(\mathcal{D})^\ell}\Big).$$

Taking the limit $h \to 0$ in (6.5.6) and using the last estimates for $\psi\Phi_h$, $\psi\Psi_h^\pm$ and ψv_h, we obtain (6.5.5). This proves the theorem for the case where $\eta\partial_{x_3}^j g^\pm \in V_\delta^{2-d^\pm-1/2,2}(\Gamma^\pm)^\ell$ for $j = 0,\dots,k$. Since $W_\delta^{1/2,2}(\Gamma^\pm) \subset V_\delta^{1/2,2}(\Gamma^\pm)$ the theorem is in particular proved for $d^+ = d^- = 1$. Suppose that $d^+ = 0$. Then we consider the vector function

$$w = u - E(g^+|_M),$$

where E denotes the extension operator (6.2.4). Since $(\eta g^+)|_M \in W^{k+1-\delta,2}(M)^\ell$, it follows that $\eta E(g^+|_M) \in W_\delta^{k+2,2}(\mathcal{D})^\ell$ (cf. Lemmas 6.2.2 and 6.2.3). Consequently,

$$\eta\partial_{x_3}^j L(D_x) w \in W_\delta^{0,2}(\mathcal{D})^\ell \quad \text{and} \quad \eta\partial_{x_3}^j B^\pm(D_x) w \in W_\delta^{2-d^\pm-1/2,2}(\Gamma)^\ell$$

for $j = 0,1,\dots,k$. Since the trace of $\partial_{x_3}^j B^\pm(D_x)w$ on M is zero for $d^\pm = 0$, $j = 0,1,\dots,k$, we even have $\eta\partial_{x_3}^j B^\pm(D_x)w \in V_\delta^{2-d^\pm-1/2,2}(\Gamma)^\ell$. Thus, by what has been shown above, we obtain $\zeta\partial_{x_3}^j w \in L_\delta^{2,2}(\mathcal{D})^\ell$ for $j = 0,1,\dots,k$. Since $\eta E(g^+|_M) \in W_\delta^{k+2,2}(\mathcal{D})^\ell$, the same is true for u. Furthermore, the estimate (6.5.4) holds. This completes the proof of the theorem. $\square$

6.5.3. A $W_\delta^{l,2}$-regularity result for the solution. The proof of the following lemma is essentially based on Lemma 6.3.7.

LEMMA 6.5.3. *Let u be a solution of the boundary value problem* (6.1.3), *and let ζ, η be infinitely differentiable functions with compact supports such that $\eta = 1$ in a neighborhood of* $\operatorname{supp}\zeta$. *Furthermore, let $\eta u \in W_\delta^{l,p}(\mathcal{D})^\ell$, $\eta\partial_{x_3}u \in W_\delta^{l,p}(\mathcal{D})^\ell$, $\eta f \in W_\delta^{l-1,p}(\mathcal{D})^\ell$, $\eta g^\pm \in W_\delta^{l+1-d^\pm-1/p,p}(\Gamma^\pm)^\ell$, $l \ge 2$, $\delta > -2/p$. If the strip $l-\delta-2/p \le \operatorname{Re}\lambda \le l+1-\delta-2/p$ does not contain eigenvalues of the pencil $A(\lambda)$, then $\zeta u \in W_\delta^{l+1,p}(\mathcal{D})^\ell$ and*

$$\|\zeta u\|_{W_\delta^{l+1,p}(\mathcal{D})^\ell} \le c\Big(\sum_{j=0}^{1}\|\eta\partial_{x_3}^j u\|_{W_\delta^{l,p}(\mathcal{D})^\ell} + \|\eta f\|_{W_\delta^{l-1,p}(\mathcal{D})^\ell} + \sum_\pm \|\eta g^\pm\|_{W_\delta^{l+1-d^\pm-1/p,p}(\Gamma^\pm)^\ell}\Big). \tag{6.5.8}$$

Here the constant c depends only on the C^{l+1} norm of ζ.

Proof. From the equation $L(D_x)u = f$ it follows that

$$L(D_{x'},0)\,(\zeta u) = F, \quad \text{where } F = \zeta f + \zeta L_1 D_{x_3}u + [L(D_{x'},0),\zeta]\,u.$$

Here $[L(D_{x'},0),\zeta] = L(D_{x'},0)\,\zeta - \zeta L(D_{x'},0)$ is the commutator of $L(D_{x'},0)$ and ζ, and L_1 is a first order differential operator with constant coefficients,

$$L_1 D_{x_3}u = \big(L(D_{x'},0) - L(D_{x'},D_{x_3})\big)u.$$

An analogous representation holds for $G^\pm = B^\pm(D_{x'},0)\,(\zeta u)\big|_{\Gamma^\pm}$. By the conditions of the lemma,

$$\zeta(\cdot,x_3)\,u(\cdot,x_3) \in W_\delta^{l,p}(K)^\ell, \quad F(\cdot,x_3) \in W_\delta^{l-1,p}(K)^\ell, \quad \text{and}$$

$$G^\pm(\cdot,x_3) \in W_\delta^{l+1-d^\pm-1/p,p}(\gamma^\pm)^\ell$$

for almost all x_3. Using Lemma 6.3.7, we obtain $\zeta(\cdot,x_3)\,u(\cdot,x_3) \in W_\delta^{l+1,p}(K)^\ell$ and

$$\begin{aligned}
\|(\zeta u)(\cdot,x_3)\|^p_{W_\delta^{l+1,p}(K)^\ell} &\le c\Big(\|(\zeta u)(\cdot,x_3)\|^p_{W_\delta^{l,p}(K)^\ell} + \|F(\cdot,x_3)\|^p_{W_\delta^{l-1,p}(K)^\ell} \\
&\qquad + \sum_\pm \|G^\pm(\cdot,x_3)\|^p_{W_\delta^{l+1-d^\pm-1/p,p}(\gamma^\pm)^\ell}\Big) \\
&\le c\Big(\|(\eta u)(\cdot,x_3)\|^p_{W_\delta^{l,p}(K)^\ell} + \|(\eta\partial_{x_3}u)(\cdot,x_3)\|^p_{W_\delta^{l,p}(K)^\ell} \\
&\qquad + \|(\zeta f)(\cdot,x_3)\|^p_{W_\delta^{l-1,p}(K)^\ell} + \sum_\pm \|(\zeta g^\pm)(\cdot,x_3)\|^p_{W_\delta^{l+1-d^\pm-1/p,p}(\gamma^\pm)^\ell}\Big).
\end{aligned}$$

Integrating the last inequality with respect to x_3 and using Lemma 6.2.14, we obtain (6.5.8). □

Using the last lemma, we can generalize the regularity result in Theorem 6.4.6.

THEOREM 6.5.4. *Let $u \in L^{1,2}(\mathcal{D})^\ell$ be a solution of the boundary valu problem* (6.1.3), *and let ζ, η be the same functions as in Lemma* 6.5.3. *Furthermore, let $\eta f \in W_\delta^{l-2,2}(\mathcal{D})^\ell$ and $\eta g^\pm \in W_\delta^{l-d^\pm-1/2,2}(\Gamma^\pm)$, where $l \ge 2$, δ is not an integer, and*

$$\max(0, l-1-\delta_+) < \delta < l-1. \tag{6.5.9}$$

Then $\zeta u \in W_\delta^{l,2}(\mathcal{D})^\ell$ *and*

$$\|\zeta u\|_{W_\delta^{l,2}(\mathcal{D})^\ell} \le c\Big(\|\eta f\|_{W_\delta^{l-2,2}(\mathcal{D})^\ell} + \sum_{\pm} \|\eta g^\pm\|_{W_\delta^{l-d^\pm-1/2,2}(\Gamma^\pm)^\ell} + \|\eta u\|_{L_\delta^{1,2}(\mathcal{D})^\ell}\Big).$$

P r o o f. We prove the theorem by induction in $k = [l-\delta]$, where $[l-\delta]$ denotes the integral part of $l-\delta$, i.e. $l-k-1 < \delta < l-k$, $1 \le k \le l-1$. Let ψ be a smooth function equal to one near $\operatorname{supp}\zeta$ such that $\eta = 1$ near $\operatorname{supp}\psi$.

1) If $k=1$, then $0 < \delta - l + 2 < 1$, $\eta f \in W^{0,2}_{\delta-l+2}(\mathcal{D})^\ell$, $\eta g^\pm \in W^{2-d^\pm-1/2,2}_{\delta-l+2}(\Gamma^\pm)^\ell$. Consequently according to Theorem 6.5.2, we have $\psi u \in W^{2,2}_{\delta-l+2}(\mathcal{D})^\ell$. Applying Corollary 6.4.2, we obtain $\zeta u \in W_\delta^{l,2}(\mathcal{D})^\ell$.

2) If $k=2$, then $0 < \delta - l + 3 < 1$. Using Theorem 6.5.2, we obtain $\psi\partial^j_{x_3} u \in W^{2,2}_{\delta-l+3}(\mathcal{D})^\ell$ for $j = 0, 1$. Consequently, it follows from Lemma 6.5.3 and Corollary 6.4.2 that $\zeta u \in W_\delta^{l,2}(\mathcal{D})^\ell$.

3) Suppose that $k \ge 3$ and the theorem is proved for $l-\delta < k$. Then by the induction hypothesis, $\psi u \in W_\delta^{l-1,2}(\mathcal{D})^\ell$, $l \ge 4$, and $\psi\partial_{x_3} u \in W_\delta^{l-2,2}(\mathcal{D})^\ell \subset L^{1,2}(\mathcal{D})^\ell$. Since $\partial_{x_3}(\eta f) \in W_\delta^{l-3,2}(\mathcal{D})^\ell$ and $\partial_{x_3}(\eta g^\pm) \in W_\delta^{l-d^\pm-3/2,2}(\Gamma^\pm)^\ell$, it follows from the induction hypothesis that $\psi\partial_{x_3} u \in W_\delta^{l-1,2}(\mathcal{D})^\ell$. Thus, Lemma 6.5.3 implies $\zeta u \in W_\delta^{l,2}(\mathcal{D})^\ell$. Furthermore, the desired estimate for ζu holds. □

COROLLARY 6.5.5. *Let* $u \in L^{1,2}(\mathcal{D})^3$ *be a solution of the boundary value problem* (6.1.3), *and let* ζ, η *be the same functions as in Lemma* 6.5.3. *Suppose that* $\eta\partial^j_{x_3} f \in W_\delta^{l-2,2}(\mathcal{D})^\ell$ *and* $\eta\partial^j_{x_3} g^\pm \in W_\delta^{l-d^\pm-1/2,2}(\Gamma^\pm)^\ell$ *for* $j = 0,\dots,k$, *where* $l \ge 2$, $0 < \delta < l-1$ *and* δ *is not integer. If the strip* $0 < \operatorname{Re}\lambda \le l-1-\delta$ *is free of eigenvalues of the pencil* $A(\lambda)$, *then* $\zeta\partial^j_{x_3} u \in W_\delta^{l,2}(\mathcal{D})^\ell$ *and*

$$\begin{aligned}\sum_{j=0}^k \|\zeta\partial^j_{x_3} u\|_{W_\delta^{l,2}(\mathcal{D})^\ell} \;\le\;& c\Big(\|\eta u\|_{L_\delta^{1,2}(\mathcal{D})^\ell} + \sum_{j=0}^k \|\eta\partial^j_{x_3} f\|_{W_\delta^{l-2,2}(\mathcal{D})^\ell} \\ &+ \sum_{j=0}^k \sum_{\pm} \|\eta\partial^j_{x_3} g^\pm\|_{W_\delta^{l-d^\pm-1/2,2}(\Gamma^\pm)^\ell}\Big).\end{aligned}$$

P r o o f. First let $\delta > l-2$. Then $0 < \delta - l + 2 < 1$, $W_\delta^{l-2,2}(\mathcal{D}) \subset W^{0,2}_{\delta-l+2}(\mathcal{D})$ and $W_\delta^{l-d^\pm-1/2,2}(\Gamma^\pm) \subset W^{3/2-d^\pm}_{\delta-l+2}(\Gamma^\pm)$. Consequently by Theorem 6.5.2, we have $\psi\partial^j_{x_3} u \in W^{2,2}_{\delta-l+2}(\mathcal{D})^\ell$, where ψ is a smooth function such that $\eta = 1$ in a neighborhood of $\operatorname{supp}\psi$ and $\psi = 1$ in a neighborhood of $\operatorname{supp}\zeta$. Applying Corollary 6.4.2, we obtain $\zeta\partial^j_{x_3} u \in W_\delta^{l,2}(\mathcal{D})^\ell$ for $j = 0,\dots,k$.

Now let $\delta < l-2$. Then $l > 2$ and we conclude from Theorem 6.5.4 that $\psi u \in W_\delta^{l,2}(\mathcal{D})^\ell$, $\psi\partial_{x_3} u \in W_\delta^{l-1,2}(\mathcal{D})^\ell \subset \mathcal{H}_{\mathcal{D}}$. Since $\eta\partial_{x_3} f \in W_\delta^{l-2,2}(\mathcal{D})^\ell$ and $\eta\partial_{x_3} g^\pm \in W_\delta^{l-d^\pm-1/2,2}(\Gamma^\pm)^\ell$, it follows from Theorem 6.5.4 that $\zeta\partial_{x_3} u \in W_\delta^{l,2}(\mathcal{D})^\ell$. Repeating this argument, we get $\zeta\partial^j_{x_3} u \in W_\delta^{l,2}(\mathcal{D})^\ell$ for $j = 2,\dots,k$. Furthermore, the desired estimate holds. □

6.5.4. The Neumann problem for a special class of systems. We consider a special case where $\lambda = 1$ is the eigenvalue of $A(\lambda)$ with smallest positive real part. Then the conditions in Theorem 6.5.4 imply that $\delta > l-2$. However in some cases (for example, the Neumann problem for the Lamé system), it is possible to weaken this condition on δ. In the case considered here, the spectrum of the pencil $A(\lambda)$ contains the eigenvalue $\lambda = 1$ for every opening θ of the dihedron $\mathcal{D}$ and the eigenfunctions corresponding to this eigenvalue are of the form $U(\varphi) = c\cos\varphi + d\sin\varphi$, where c, d are constant vectors. This means that the homogeneous problem

$$L(D_{x'},0)\,u = 0 \quad \text{in } K, \qquad B^\pm(D_{x'},0)\,u = 0 \quad \text{on } \gamma^\pm$$

has nontrivial solutions of the form $u = c\,x_1 + d\,x_2$ for every θ. Clearly, this is only possible for the Neumann problem

$$L(D_x)\,u = f \quad \text{in } \mathcal{D}, \qquad N^\pm(D_x)\,u = g^\pm \quad \text{on } \Gamma^\pm. \tag{6.5.10}$$

LEMMA 6.5.6. *Let $\theta \neq \pi$ and $\theta \neq 2\pi$. Then the homogeneous boundary value problem*

$$L(D_{x'},0)\,u = 0 \quad \textit{in } K, \qquad N^\pm(D_{x'},0)\,u = 0 \quad \textit{on } \gamma^\pm. \tag{6.5.11}$$

has a nonzero solution of the form $u = c\,x_1 + d\,x_2$, where c and d are constant vectors, if and only if the $2\ell \times 2\ell$ matrix

$$A' = \begin{pmatrix} A_{1,1} & A_{2,1} \\ A_{1,2} & A_{2,2} \end{pmatrix} \tag{6.5.12}$$

is not invertible.

P r o o f. The function $u = c\,x_1 + d\,x_2$ satisfies the homogeneous boundary conditions $N^\pm(D_{x'},0)\,u = 0$ on $\gamma^\pm$ if and only if

$$\begin{pmatrix} n_1^+ I_\ell & n_2^+ I_\ell \\ n_1^- I_\ell & n_2^- I_\ell \end{pmatrix} \begin{pmatrix} A_{1,1} & A_{2,1} \\ A_{1,2} & A_{2,2} \end{pmatrix} \begin{pmatrix} c \\ d \end{pmatrix} = 0.$$

Here I_ℓ denotes the $\ell \times \ell$ identity matrix. The first matrix is invertible for $\theta \neq \pi$, $\theta \neq 2\pi$. This proves the lemma. □

In the sequel, we exclude the cases $\theta = \pi$ and $\theta = 2\pi$. The case $\theta = \pi$ is only interesting for mixed problems, and in the case $\theta = 2\pi$, the eigenvalue $\lambda = 1$ is in general not the eigenvalue with smallest positive real part.

Let r' denote the rank of the matrix A'. By the proof of the last lemma, there exist $2\ell - r'$ linearly independent eigenvectors of the form $c\cos\varphi + d\sin\varphi$ corresponding to the eigenvalue $\lambda = 1$. Furthermore, the inhomogeneous boundary conditions $N^\pm(D_{x'},0)u = g^\pm$ on $\gamma^\pm$ can be satisfied for a vector function $u \in W_\delta^{l,p}(K)^\ell$, $\delta \le l - 1 - 2/p$, only if g^+ and g^- satisfy $2\ell - r'$ compatibility conditions at $x = 0$.

Such compatibility conditions must be also satisfied for the boundary data of the Neumann problem in the dihedron $\mathcal{D}$. Let $u \in W_\delta^{l,p}(\mathcal{D})^\ell$, $\delta < l - 1 - 2/p$, and let b, c, d denote the trace on M of u, $\partial_{x_1} u$ and $\partial_{x_2} u$, respectively. By Lemma 6.2.2, $b \in W^{l-\delta-2/p,p}(M)^\ell$ and $c, d \in W^{l-1-\delta-2/p,p}(M)^\ell$. From the boundary conditions $N^\pm(D_x)\,u = g^\pm$ on $\Gamma^\pm$ it follows that

$$(A_{1,1}n_1^\pm + A_{1,2}n_2^\pm)\,c + (A_{2,1}n_1^\pm + A_{2,2}n_2^\pm)\,d + (A_{3,1}n_1^\pm + A_{3,2}n_2^\pm)\,\partial_{x_3} b = g^\pm|_M.$$

The last system can be written in the form

$$\begin{pmatrix} n_1^+ I_\ell & n_2^+ I_\ell \\ n_1^- I_\ell & n_2^- I_\ell \end{pmatrix} \begin{pmatrix} A_{1,1} & A_{2,1} & A_{3,1} \\ A_{1,2} & A_{2,2} & A_{3,2} \end{pmatrix} \begin{pmatrix} c(x_3) \\ d(x_3) \\ b'(x_3) \end{pmatrix} = \begin{pmatrix} g^+(0,x_3) \\ g^-(0,x_3) \end{pmatrix}. \tag{6.5.13}$$

For the existence of a vector function $u \in W_\delta^{3,2}(\mathcal{D})^\ell$, $0 < \delta < 1$, satisfying the boundary conditions $N^\pm(D_x)\,u = g^\pm$ on $\Gamma^\pm$ it is necessary (and sufficient) that the system (6.5.13) with the unknowns b, c, d is solvable. In order to guarantee the solvability of this system and to prove the regularity result in Lemma 6.5.8 below, we assume that the matrices A' and

$$(6.5.14) \qquad A'' = \begin{pmatrix} A_{1,1} & A_{2,1} & A_{3,1} \\ A_{1,2} & A_{2,2} & A_{3,2} \end{pmatrix}$$

have the same rank r'. Then for the solvability of the system (6.5.13) it is necessary and sufficient that

$$(6.5.15) \qquad w^{(k)} \cdot \big(g^+|_M, g^-|_M\big) = 0 \quad \text{for } k = 1, \dots, 2\ell - r',$$

where $w^{(1)}, \dots, w^{(2\ell-r')}$ are linearly independent solutions of the algebraic system

$$(6.5.16) \qquad \begin{pmatrix} A^t_{1,1} & A^t_{1,2} \\ A^t_{2,1} & A^t_{2,2} \end{pmatrix} \begin{pmatrix} n_1^+ I_\ell & n_1^- I_\ell \\ n_2^+ I_\ell & n_2^- I_\ell \end{pmatrix} w = 0.$$

Here A^t denotes the transposed matrix of A. Note that the condition (6.5.15) guarantees even the existence of solutions (b, c, d) of the system (6.5.13), where $b = 0$.

LEMMA 6.5.7. *Let $g^\pm \in W_\delta^{l-1-1/p,p}(\Gamma^\pm)^\ell$, $0 \le l - 2 < \delta + 2/p \le l - 1$, be given vector functions on $\Gamma^\pm$ such that $g^\pm(x) = 0$ for $|x'| > 1$. We suppose that the matrices A' and A'' have the same rank $r' < 2\ell$ and that g^+ and g^- satisfy the compatibility condition* (6.5.15) *if $\delta + 2/p < l - 1$ and the generalized trace conditions*

$$(6.5.17) \quad \int\limits_{-\infty}^{+\infty} \int\limits_0^\infty r^{-1} \left| w^{(k)} \cdot \big(g^+(r,x_3), g^-(r,x_3)\big)\right|^p dr\, dx_3 < \infty, \qquad k = 1, \dots, 2\ell - r',$$

if $\delta + 2/p = l - 1$. Then there exists a vector function $u \in W_\delta^{l,p}(\mathcal{D})^\ell$ satisfying the boundary conditions $N^\pm(D_x) = g^\pm$ on $\Gamma^\pm$.

P r o o f. 1) Let $l - 2 < \delta + 2/p < l - 1$. By (6.5.15), there exists a solution $(c, d) \in W^{l-1-\delta-2/p,p}(M)^\ell \times W^{l-1-\delta-2/p,p}(M)^\ell$ of the system

$$(A_{1,1} n_1^\pm + A_{1,2} n_2^\pm)\, c + (A_{2,1} n_1^\pm + A_{2,2} n_2^\pm)\, d = g^\pm|_M\,.$$

We put $v(x) = (Ec)(x)\, x_1 + (Ed)(x)\, x_2$, where E is the extension operator (6.2.4). By Lemma 6.2.3, $v \in W_\delta^{l,p}(\mathcal{D})^\ell$. Moreover, the traces of $g^\pm - N^\pm(D_x) v|_{\Gamma^\pm}$ on the edge M are zero. Consequently, it follows from Theorem 6.2.15 that

$$g^\pm - N^\pm(D_x) v|_{\Gamma^\pm} \in V_\delta^{l-1-1/p,p}(\Gamma^\pm)^\ell.$$

Applying Lemma 2.2.3, we obtain the assertion of the lemma.

2) Let $\delta + 2/p = l - 1$. Then by Theorem 6.2.11, the vector functions $g^\pm$ admit the decompositions

$$g^\pm = \psi^\pm + G^\pm,$$

where $\psi^\pm \in W^{l,p}_{l-1/p}(\mathbb{R}^2_+)^\ell$ and $G^\pm \in V_\delta^{l-1-1/p,p}(\Gamma^\pm)^\ell$. Here, the vector functions $\psi^\pm$ satisfy also the condition (6.5.17). Thus, there exist vector functions $c, d \in W^{l,p}_{l-1/p}(\mathbb{R}^2_+)^\ell$ such that

$$(A_{1,1} n_1^\pm + A_{1,2} n_2^\pm)\, c + (A_{2,1} n_1^\pm + A_{2,2} n_2^\pm)\, d - \psi^\pm \in V^{l,p}_{l-1/p}(\mathbb{R}^2_+)^\ell.$$

If we consider c and d as functions in $\mathcal{D}$ (which depend only on the variables $r = |x'|$ and x_3), then $c, d \in W^{l,p}_{l-2/p}(\mathcal{D})^\ell$ and

$$\partial_{x_j} c,\ \partial_{x_j} d \in W^{l-1,p}_{l-2/p}(\mathcal{D})^\ell \subset V^{l-1,p}_{l-2/p}(\mathcal{D})^\ell \quad \text{for } j = 1, 2, 3.$$

Hence for the vector function $v = cx_1 + dx_2$ we obtain $g^\pm - N^\pm(D_x)v|_{\Gamma^\pm} \in V_\delta^{l-1-1/p,p}(\Gamma^\pm)^\ell$, and it remains to apply Lemma 2.2.3. □

Note that the assertion of the last lemma is valid in the case $\delta + 2/p > l - 1$ without compatibility conditions on $g^\pm$ (see Lemma 6.4.3).

LEMMA 6.5.8. *Let $u \in L^{1,2}(\mathcal{D})^3$ be a solution of the boundary value problem* (6.5.10), *and let ζ, η be the same functions as in Lemma* 6.5.3. *Suppose that $\eta f \in W^{1,2}_\delta(\mathcal{D})^\ell$, $\eta g^\pm \in W_\delta^{3/2,2}(\Gamma^\pm)^\ell$, $0 < \delta < 1$, the ranks of the matrices A' and A'' coincide, and that the strip $0 < \operatorname{Re}\lambda < 2 - \delta$ contains the single eigenvalue $\lambda = 1$ which has geometric and algebraic multiplicity $2\ell - r'$, where r' denotes the rank of the matrices A' and A''. Furthermore, we assume that g^+ and g^- satisfy the compatibility condition* (6.5.15) *on M. Then $\zeta u \in W^{3,2}_\delta(\mathcal{D})^\ell$ and*

$$(6.5.18)\quad \|\zeta u\|_{W^{3,2}_\delta(\mathcal{D})^\ell} \le c\Big(\|\eta f\|_{W^{1,2}_\delta(\mathcal{D})^\ell} + \sum_\pm \|\eta g^\pm\|_{W_\delta^{3/2,2}(\Gamma^\pm)^\ell} + \|\eta u\|_{L^{1,2}_\delta(\mathcal{D})^\ell}\Big)$$

with a constant c independent of u.

P r o o f. Let ψ be an infinitely differentiable function on $\overline{\mathcal{D}}$ such that $\psi\zeta = \zeta$ and $\eta\psi = \psi$. From Corollary 6.5.5 it follows that $\psi u \in W^{2,2}_\delta(\mathcal{D})^\ell$ and $\psi \partial_{x_3} u \in W^{2,2}_\delta(\mathcal{D})^\ell$. Consequently,

$$L(D_{x'}, 0)\, u(\cdot, x_3) = f(\cdot, x_3) - \big(L(D_{x'}, D_{x_3}) - L(D_{x'}, 0)\big)\, u(\cdot, x_3) = F(\cdot, x_3),$$

$$N^\pm(D_{x'}, 0)\, u(\cdot, x_3) = g^\pm(\cdot, x_3) - \sum_{k=1}^{2} A_{3,k} n_k^\pm\, \partial_{x_3} u(\cdot, x_3) = G^\pm(\cdot, x_3),$$

where $\zeta(\cdot, x_3)F(\cdot, x_3) \in W^{1,2}_\delta(K)^\ell$, $\zeta(\cdot, x_3)G^\pm(\cdot, x_3) \in W_\delta^{3/2,2}(\gamma^\pm)^\ell$ for almost all x_3 and

$$\begin{aligned}
&\|\zeta(\cdot, x_3)F(\cdot, x_3)\|_{W^{1,2}_\delta(K)^\ell} + \|\zeta(\cdot, x_3)G^\pm(\cdot, x_3)\|_{W_\delta^{3/2,2}(\gamma^\pm)^\ell} \\
&\le c\Big(\|\eta(\cdot, x_3) f(\cdot, x_3)\|_{W^{1,2}_\delta(K)^\ell} + \|\eta(\cdot, x_3) g^\pm(\cdot, x_3)\|_{W_\delta^{3/2,2}(\gamma^\pm)^\ell} \\
&\qquad + \|\psi(\cdot, x_3)\, \partial_{x_3} u(\cdot, x_3)\|_{W^{2,2}_\delta(\mathcal{D})^\ell}\Big).
\end{aligned}$$

Here the constant c is independent of x_3. Let b denote the trace of the vector function u on M. Since the ranks of the matrices A' and A'' coincide and g^+, g^- satisfy the compatibility condition (6.5.15), there exist vectors $c(x_3), d(x_3) \in \mathbb{C}^\ell$ satisfying (6.5.13) with $G^+(0, x_3)$ and $G^-(0, x_3)$ on the right-hand side and the estimate

$$|c(x_3)| + |d(x_3)| \le c \sum_\pm |G^\pm(0, x_3)| \le c' \sum_\pm \|\zeta(\cdot, x_3)G^\pm(\cdot, x_3)\|_{W_\delta^{3/2,2}(\gamma^\pm)^\ell}$$

for almost all x_3. Consequently, the vector function $p(x) = c(x_3)\, x_1 + d(x_3)\, x_2$ satisfies the equations

$$N^\pm(D_x', 0)\, p(\cdot, x_3) = G^\pm(0, x_3) \text{ on } \gamma^\pm$$

for all x_3. We put $v(x) = u(x) - p(x) - b(x_3)$. Then $\zeta(\cdot,x_3)\,v(\cdot,x_3) \in V_\delta^{2,2}(K)^\ell$ (see Lemma 6.2.12),

$$L(D_{x'},0)\,v(\cdot,x_3) = F(\cdot,x_3) \ \text{ in } K,$$
$$N^\pm(D_{x'},0)\,v(\cdot,x_3) = G^\pm(\cdot,x_3) - G^\pm(0,x_3) \text{ on } \gamma^\pm.$$

Here $\eta(\cdot,x_3)\,F(\cdot,x_3) \in W_\delta^{1,2}(\mathcal{D})^\ell \subset V_\delta^{1,2}(\mathcal{D})^\ell$ and $\eta(\cdot,x_3)\,\big(G^\pm(\cdot,x_3) - G^\pm(0,x_3)\big) \in V_\delta^{3/2,2}(\gamma^\pm)^\ell$. By the assumptions of the lemma, all eigenfunctions of the operator pencil $A(\lambda)$ corresponding to the eigenvalue $\lambda = 1$ are restrictions of linear functions to the unit circle and generalized eigenfunctions corresponding to this eigenvalue do not exist. Thus by Theorem 1.2.6, we obtain

$$\zeta(x)v(x) = c^{(1)}(x_3)\,x_1 + c^{(2)}(x_3)\,x_2 + w(x),$$

where $w(\cdot,x_3) \in V_\delta^{3,2}(K)^\ell$ and

$$\begin{aligned}
&\|w(\cdot,x_3)\|^2_{V_\delta^{3,2}(K)^\ell} \\
&\le c\,\Big(\|\zeta(\cdot,x_3)\,F(\cdot,x_3)\|^2_{W_\delta^{1,2}(K)^\ell} + \sum_\pm \|\zeta(\cdot,x_3)\,G^\pm(\cdot,x_3)\|^2_{W_\delta^{3/2,2}(\gamma^\pm)^\ell} \\
&\qquad + \|\psi(\cdot,x_3)\,u(\cdot,x_3)\|^2_{W_\delta^{2,2}(K)^\ell} \Big) \\
&\le c\,\Big(\|\eta(\cdot,x_3)\,f(\cdot,x_3)\|^2_{W_\delta^{1,2}(K)^\ell} + \sum_\pm \|\eta(\cdot,x_3)\,g^\pm(\cdot,x_3)\|^2_{W_\delta^{3/2,2}(\gamma^\pm)^\ell} \\
&\qquad + \|\psi(\cdot,x_3)\,u(\cdot,x_3)\|^2_{W_\delta^{2,2}(K)^\ell} + \|\psi(\cdot,x_3)\,\partial_{x_3}u(\cdot,x_3)\|^2_{W_\delta^{2,2}(K)^\ell} \Big)
\end{aligned}$$

with a constant c independent of x_3. Since $\partial^\alpha_{x'}(\zeta u) = \partial^\alpha_{x'}(w + \zeta p + \zeta b)$ for $|\alpha| = 3$, the last estimate implies

$$\begin{aligned}
&\|\zeta(\cdot,x_3)u(\cdot,x_3)\|^2_{L_\delta^{3,2}(K)^\ell} \\
&\le c\,\Big(\|\eta(\cdot,x_3)\,f(\cdot,x_3)\|^2_{W_\delta^{1,2}(K)^\ell} + \sum_\pm \|\eta(\cdot,x_3)\,g^\pm(\cdot,x_3)\|^2_{W_\delta^{3/2,2}(\gamma^\pm)^\ell} \\
&\qquad + \|\psi(\cdot,x_3)\,u(\cdot,x_3)\|^2_{W_\delta^{2,2}(K)^\ell} + \|\psi(\cdot,x_3)\,\partial_{x_3}u(\cdot,x_3)\|^2_{W_\delta^{2,2}(K)^\ell} \Big).
\end{aligned}$$

Integrating this inequality with respect to x_3 and using Lemma 6.2.14 and Corollary 6.5.5, we obtain (6.5.18). The lemma is proved. □

Next we prove a generalization of Theorem 6.5.4.

THEOREM 6.5.9. *Let $u \in L^{1,2}(\mathcal{D})^3$ be a solution of the problem (6.5.10), and let ζ, η be the same functions as in Lemma 6.5.3. Suppose that $\eta f \in W_\delta^{l-2,2}(\mathcal{D})^\ell$, $\eta g^\pm \in W_\delta^{l-3/2,2}(\Gamma^\pm)^\ell$, where $0 < \delta < l-1$, δ is not integer. Furthermore, we suppose that the strip $0 < \operatorname{Re}\lambda < l-1-\delta$ contains either no eigenvalues of the pencil $A(\lambda)$ or the single eigenvalue $\lambda = 1$. In the last case, we assume in addition that the ranks of the matrices A' and A'' coincide, the eigenvalue $\lambda = 1$ has the geometric and algebraic multiplicity $2\ell - r'$, where r' denotes the rank of the matrices A' and A'', and that the boundary data g^+ and g^- satisfy the compatibility condition (6.5.15) on M. Then $\zeta u \in W_\delta^{l,2}(\mathcal{D})^\ell$ and*

$$(6.5.19)\quad \|\zeta u\|_{W_\delta^{l,2}(\mathcal{D})^\ell} \le c\,\Big(\|\eta f\|_{W_\delta^{l-2,2}(\mathcal{D})^\ell} + \sum_\pm \|\eta g^\pm\|_{W_\delta^{l-3/2,2}(\Gamma^\pm)^\ell} + \|\eta u\|_{L_\delta^{1,2}(\mathcal{D})^\ell} \Big)$$

with a constant c independent of u.

P r o o f. For the case where the strip $0 < \operatorname{Re}\lambda < l-1-\delta$ is free of eigenvalues of the pencil $A(\lambda)$, we refer to Theorem 6.5.4. Suppose that $\delta < l-2$ and the strip $0 < \operatorname{Re}\lambda < l-1-\delta$ contains the eigenvalue $\lambda = 1$. If $\delta > l-3$, then $0 < \delta - l + 3 < 1$ and Lemma 6.5.8 implies $\psi u \in W^{3,2}_{\delta-l+3}(\mathcal{D})^\ell$, where ψ is a smooth function such that $\psi = 1$ in a neighborhood of $\operatorname{supp}\zeta$ and $\eta = 1$ in a neighborhood of $\operatorname{supp}\psi$. Applying Theorem 6.4.1, we obtain $\zeta u \in W^{l,2}_\delta(\mathcal{D})^\ell$.

For $\delta < l-3$, we prove the theorem by induction. Suppose that $l-k-1 < \delta < l-k$ and the assertion of the theorem is proved for $\delta > l-k$, where k is an integer, $3 \le k \le l-1$. Then by the induction hypothesis, $\psi u \in W^{l-1,2}_\delta(\mathcal{D})^\ell$ and $\psi \partial_{x_3} u \in W^{l-2,2}_\delta(\mathcal{D})^\ell \subset L^{1,2}(\mathcal{D})^\ell$. Since $\partial_{x_3}(\eta f) \in W^{l-3,2}_\delta(\mathcal{D})$ and $\partial_{x_3}(\eta g^\pm) \in W^{l-1-d^\pm-1/2,2}_\delta(\Gamma^\pm)^\ell$, the induction hypothesis implies $\psi\partial_{x_3} u \in W^{l-1,2}_\delta(\mathcal{D})^\ell$. Applying Lemma 6.5.3, we obtain $\zeta u \in W^{l,2}_\delta(\mathcal{D})^\ell$. Furthermore, the desired estimate for ζu holds. □

Finally, we mention that an analogous assertion holds for the x_3-derivatives of the solution. This means that the assertion of Corollary 6.5.5 remains true if the strip $0 < \operatorname{Re}\lambda < l-1-\delta$ contains the single eigenvalue $\lambda = 1$ and the conditions of Theorem 6.5.9 on this eigenvalue and on the ranks of the matrices A' and A'' are satisfied.

6.5.5. The Neumann problem for the Lamé system. We apply the results of the preceding subsection to the boundary value problem

$$-\mu\Big(\Delta u + \frac{1}{1-2\nu}\nabla\nabla\cdot u\Big) = f \ \text{ in } \mathcal{D}, \quad \sigma(u)\, n^\pm = g^\pm \ \text{ on } \Gamma^\pm. \tag{6.5.20}$$

Here $\sigma(u) = \{\sigma_{i,j}(u)\}$ is the *stress tensor* which is connected with the *strain tensor*

$$\Big\{\varepsilon_{i,j}(u)\Big\} = \Big\{\frac{1}{2}\,(\partial_{x_j}u_i + \partial_{x_i}u_j)\Big\}$$

by the Hooke law

$$\sigma_{i,j} = 2\mu\,\Big(\varepsilon_{i,j} + \frac{\nu}{1-2\nu}(\varepsilon_{1,1}+\varepsilon_{2,2}+\varepsilon_{3,3})\Big),$$

μ is the shear modulus, ν is the Poisson ration, $\nu < 1/2$, and $\delta_{i,j}$ denotes the Kronecker symbol. Let I_3 denote the 3×3 identity matrix and let

$$E_{k,l} = \big\{\delta_{i-k,j-l}\big\}_{i,j=1}^3$$

(i.e. all elements of the matrix $E_{k,l}$ are zero with the exception of the element in the kth row and lth column which is one). Then the problem (6.5.20) can be written as

$$\sum_{j,k=1}^3 A_{j,k}\, D_{x_j} D_{x_k} u = f \ \text{ in } \mathcal{D}, \quad \sum_{j,k=1}^3 A_{j,k}\, n_k^\pm\, \partial_{x_j} u = g^\pm \ \text{ on } \Gamma^\pm,$$

where

$$A_{j,j} = \mu\,\Big(I_3 + \frac{1}{1-2\nu}\,E_{j,j}\Big), \quad A_{j,k} = \mu\,\Big(E_{j,k} + \frac{2\nu}{1-2\nu}\,E_{k,j}\Big) \ \text{ for } j \ne k.$$

If the opening θ of the angle K is greater than π, then the eigenvalue with smallest positive real part of the pencil $A(\lambda)$ is $\xi_+(\theta)/\theta$, where $\xi_+(\theta)$ is the smallest positive solution of the equation

$$\frac{\sin\xi}{\xi} + \frac{\sin\theta}{\theta} = 0 \tag{6.5.21}$$

(see e.g. [**85**, Section 4.2]). Note that $\xi_+(\theta) < \pi$ for $\pi < \theta < 2\pi$. Thus, the smallest positive eigenvalue of the pencil $A(\lambda)$ is less than 1 and the result of Theorem 6.5.4 cannot be improved.

If $\theta < \pi$, then the eigenvalues with smallest positive real parts are $\lambda_1 = 1$ and $\lambda_2 = \pi/\theta$. The eigenvector corresponding to the eigenvalue $\lambda_1 = 1$ is $U(\varphi) = (\sin\varphi, -\cos\varphi)$. Generalized eigenvectors corresponding to this eigenvalue do not exist if $\theta < \pi$. Furthermore, both the matrices A' and A'' have rank 5 (the second and fourth rows coincide). The algebraic system (6.5.16) has the nontrivial solution $w = (n_1^-, n_2^-, 0, -n_1^+, -n_2^+, 0)^t$. Thus, the system (6.5.13) is solvable if and only if

$$n^- \cdot g^+ = n^+ \cdot g^- \quad \text{on } M. \tag{6.5.22}$$

Note that this compatibility condition follows immediately from the equations $\sigma(u)|_M\, n^\pm = g^\pm|_M$ and from the symmetry of the matrix $\sigma(u)$. Applying Theorem 6.5.9, we obtain the following statement.

THEOREM 6.5.10. *Let $u \in L^{1,2}(\mathcal{D})^3$ be a solution of the problem* (6.5.20), *and let ζ, η be the same functions as in Lemma* 6.5.3.

1) *If $\eta f \in W_\delta^{0,2}(\mathcal{D})^3$ and $\eta g^\pm \in W_\delta^{1/2,2}(\Gamma^\pm)^3$, where $0 < \delta < 1$ for $\theta < \pi$ and $1 - \xi_+(\theta)/\theta < \delta < 1$ for $\theta > \pi$, then $\zeta u \in W_\delta^{2,2}(\mathcal{D})^3$.*

2) *Suppose that $\eta f \in W_\delta^{l-2,2}(\mathcal{D})^3$, $\eta g^\pm \in W_\delta^{l-3/2,2}(\Gamma^\pm)^3$, $l \ge 3$, $\theta < \pi$, δ is not integer, and $\max(0, l-1-\pi/\theta) < \delta < l-1$. In the case $\delta < l-2$, we assume in addition that g^+ and g^- satisfy the compatibility condition* (6.5.22) *on M. Then $\zeta u \in W_\delta^{l,2}(\mathcal{D})^3$.*

6.6. Green's matrix for the problem in the dihedron

This section is dedicated to the *Green's matrix* of the boundary value problem (6.1.3). By a Green's matrix for this problem, we mean a $\ell \times \ell$-matrix $G(x,\xi)$ satisfying

$$L(D_x)\, G(x,\xi) = \delta(x-\xi)\, I_\ell \quad \text{for } x,\xi \in \mathcal{D}, \tag{6.6.1}$$

$$B^\pm(D_x)\, G(x,\xi) = 0 \quad \text{for } x \in \Gamma^\pm,\ \xi \in \mathcal{D}. \tag{6.6.2}$$

Here I_ℓ denotes the $\ell\times\ell$ identity matrix. The existence and uniqueness of a solution of the problem (6.6.1), (6.6.2) follows from the $\mathcal{H}_\mathcal{D}$-ellipticity of the form $b_\mathcal{D}$. We are interested here in point estimates for the elements of the matrix $G(x,\xi)$. As in the case of the Dirichlet problem in Section 2.5, we have to consider the cases $|x-\xi| < \min(|x'|,|\xi'|)$ and $|x-\xi| > \min(|x'|,|\xi'|)$ separately.

6.6.1. Existence of Green's matrix.

First, we prove the existence, uniqueness and some basic properties of the Green's matrix.

THEOREM 6.6.1. *Suppose that the form $b_\mathcal{D}(\cdot,\cdot)$ satisfies the condition* (6.1.8). *Then the following assertions hold.*

1) *There exists a unique solution $G(x,\xi)$ of the boundary value problem* (6.6.1), (6.6.2) *such the function $x \to \zeta(x,\xi)\, G(x,\xi)$ belongs to $\mathcal{H}_\mathcal{D}^\ell$ for every $\xi \in \mathcal{D}$ and for*

every infinitely differentiable function $\zeta(\cdot,\xi)$ *which is equal to zero in a neighborhood of the point* $x=\xi$ *and bounded together with all derivatives.*

2) *The equality*

$$G_{i,j}(tx,t\xi)=t^{-1}\,G_{i,j}(x,\xi) \tag{6.6.3}$$

is valid for all $x,\xi\in\mathcal{D}$, $t>0$, $i,j=1,\dots,\ell$.

3) *The adjoint matrix* $G^*(x,\xi)$ *is the unique solution of the formally adjoint problem (see* (6.1.13)*)*

$$L^+(D_\xi)\,G^*(x,\xi)=\delta(x-\xi)\,I_\ell \quad \textit{for } x,\xi\in\mathcal{D},$$

$$C^\pm(D_\xi)\,G^*(x,\xi)=0 \quad \textit{for } x\in\mathcal{D},\ \xi\in\Gamma^\pm$$

such that the function $\xi\to\zeta(x,\xi)\,G(x,\xi)$ *belongs to* $\mathcal{H}_\mathcal{D}^\ell$ *for every* $x\in\mathcal{D}$ *and for every infinitely differentiable function* $\zeta(x,\cdot)$ *which is equal to zero in a neighborhood of the point* $\xi=x$ *and bounded together with all derivatives.*

4) *The solution* $u\in\mathcal{H}_\mathcal{D}$ *of the problem* (6.1.5), (6.1.6) *admits the representation*

$$u(x)=\int_\mathcal{D} G(x,\xi)\,F(\xi)\,d\xi \tag{6.6.4}$$

for arbitrary $F\in\mathcal{H}_\mathcal{D}^*$ *if* $g^\pm=0$ *for* $d^\pm=0$.

P r o o f. 1) Let $\mathcal{G}(x,\xi)$ be the same solution of the equation

$$L(D_x)\,\mathcal{G}(x,\xi)=\delta(x-\xi)\,I_\ell \quad \text{in } \mathbb{R}^3$$

as in the proof of Theorem 2.5.1. Furthermore, let χ be a smooth function on $(0,\infty)$, $\chi(t)=1$ for $t<1/4$, $\chi(t)=0$ for $t>1/2$. We put $\psi(x,\xi)=\chi(|x-\xi|/|\xi'|)$ and define $R(x,\xi)$ as the unique variational solution (in $\mathcal{H}_\mathcal{D}^\ell$) of the problem

$$L(D_x)\,R(x,\xi)=\delta(x-\xi)\,I_\ell-L(D_x)\big(\psi(x,\xi)\,\mathcal{G}(x,\xi)\big) \quad \text{for } x\in\mathcal{D}, \tag{6.6.5}$$

$$B^\pm(D_x)\,R(x,\xi)=-B^\pm(D_x)\,\big(\psi(x,\xi)\mathcal{G}(x,\xi)\big) \quad \text{for } x\in\Gamma^\pm. \tag{6.6.6}$$

Since the right-hand sides of (6.6.5), (6.6.6) are smooth functions with compact supports vanishing in a neighborhood of the edge M, the existence of the solution $R(x,\xi)$ is guaranteed by the condition (6.1.8). Obviously, the matrix

$$G(x,\xi)=\psi(x,\xi)\,\mathcal{G}(x,\xi)+R(x,\xi)$$

satisfies (6.6.1) and (6.6.2). We prove the uniqueness. Suppose that $G(x,\xi)$ and $H(x,\xi)$ are Green's matrices such that $\zeta(\cdot,\xi)\,G(\cdot,\xi)$ and $\zeta(\cdot,\xi)\,H(\cdot,\xi)$ belong to $\mathcal{H}_\mathcal{D}^\ell$ for every smooth function ζ which vanishes in a neighborhood of the point $x=\xi$ and which is bounded together with all derivatives. Then in particular,

$$\big(1-\psi(\cdot,\xi)\big)\,\big(G(\cdot,\xi)-H(\cdot,\xi)\big)\in\mathcal{H}_\mathcal{D}^\ell\,.$$

Since $G(\cdot,\xi)-H(\cdot,\xi)$ is a solution of problem (6.1.3) with zero right-hand sides and $\psi(\cdot,\xi)$ vanishes in a neighborhood of the edge M, we furthermore have

$$\psi(\cdot,\xi)\,\big(G(\cdot,\xi)-H(\cdot,\xi)\big)\in W^{1,2}(\mathcal{D})^{\ell\times\ell}.$$

Consequently, $G(\cdot,\xi)-H(\cdot,\xi)\in\mathcal{H}_\mathcal{D}^\ell$ for arbitrary $\xi\in\mathcal{D}$. By the uniqueness of variational solutions in $\mathcal{H}_\mathcal{D}$, we obtain $G(x,\xi)=H(x,\xi)$.

2) From (6.6.1), (6.6.2) it follows that

$$L(D_x)\,G(tx,t\xi)=t^2\,\delta(tx-t\xi)\,I_\ell=t^{-1}\,\delta(x-\xi)\,I_\ell \quad \text{for } x,\xi\in\mathcal{D}$$

and $B^\pm(D_x)\,G(tx,t\xi)=0$ for $x\in\Gamma^\pm$, $\xi\in\mathcal{D}$. This implies (6.6.3).

3) Analogously to the first assertion, there exists a unique solution $H(x,\xi)$ of the problem

$$L^+(D_x)H(x,\xi) = \delta(x-\xi)\, I_\ell \quad \text{for } x,\xi \in \mathcal{D}, \tag{6.6.7}$$

$$C^\pm(D_x)\, H(x,\xi) = 0 \quad \text{for } x \in \Gamma^\pm,\ \xi \in \mathcal{D} \tag{6.6.8}$$

such that the function $x \to \zeta(x,\xi)\, H(x,\xi)$ belongs to the space $\mathcal{H}^\ell_{\mathcal{D}}$ for every infinitely differentiable function ζ which is equal to zero in a neighborhood of the point $x=\xi$ and bounded together with all derivatives. We prove that $G(x,\xi) = (H(\xi,x))^*$. Let $G^{(i)} = (G_{1,i},\dots,G_{\ell,i})^t$ denote the i-th column of the matrix G and $H^{(j)} = (H_{1,j},\dots,H_{\ell,j})^t$ the j-th column of H (here t means transposition). Furthermore, let $\xi,\eta \in \mathcal{D}$, $\xi \neq \eta$, and let ε be a sufficiently small positive number, $\varepsilon < |\xi-\eta|$. Then the functions

$$u(x) = \Big(\chi\big(\frac{|x-\xi|}{N}\big) - \chi\big(\frac{|x-\xi|}{\varepsilon}\big)\Big)\, G^{(i)}(x,\xi), \quad v(x) = \Big(1-\chi\big(\frac{|x-\eta|}{\varepsilon}\big)\Big)\, H^{(j)}(x,\eta)$$

belong to $\mathcal{H}_{\mathcal{D}}$ and are infinitely differentiable in $\overline{\mathcal{D}}\backslash M$. Moreover, the functions u and v are from the space $W^{2,2}_\delta$ in a neighborhood of the edge with certain positive $\delta < 1$ (see Theorem 6.4.6), and the support of u is compact. Therefore, the formula

$$\begin{aligned} &\int_{\mathcal{D}} L(D_x)u \cdot \bar v\, dx + \sum_{d^\pm=1} \int_{\Gamma^\pm} B^\pm(D_x)\, u \cdot \bar v\, dx \\ &= \int_{\mathcal{D}} u \cdot \overline{L^+(D_x)v}\, dx + \sum_{d^\pm=1} \int_{\Gamma^\pm} u \cdot \overline{C^\pm(D_x)v}\, dx \end{aligned} \tag{6.6.9}$$

holds for every of the pairs

$$u = \Big(\chi\big(\frac{|x-\xi|}{N}\big) - \chi\big(\frac{|x-\xi|}{\varepsilon}\big)\Big)\, G^{(i)}(x,\xi),\ v = \Big(1-\chi\big(\frac{|x-\eta|}{\varepsilon}\big)\Big)\, H^{(j)}(x,\eta),$$

$$u = \chi\big(\frac{|x-\xi|}{\varepsilon}\big)\, G^{(i)}(x,\xi), \quad v = \Big(1-\chi\big(\frac{|x-\eta|}{\varepsilon}\big)\Big)\, H^{(j)}(x,\eta),$$

$$u = \Big(\chi\big(\frac{|x-\xi|}{N}\big) - \chi\big(\frac{|x-\xi|}{\varepsilon}\big)\Big)\, G^{(i)}(x,\xi), \quad v = \chi\big(\frac{|x-\eta|}{\varepsilon}\big)\, H^{(j)}(x,\eta),$$

$$u = \chi\big(\frac{|x-\xi|}{\varepsilon}\big)\, G^{(i)}(x,\xi), \quad v = \chi\big(\frac{|x-\eta|}{\varepsilon}\big)\, H^{(j)}(x,\eta)$$

(the functions of the last pair have disjoint supports). Thus, formula (6.6.9) is valid for

$$u(x) = \chi\big(\frac{|x-\xi|}{N}\big)\, G^{(i)}(x,\xi) \quad \text{and} \quad v(x) = H^{(j)}(x,\eta). \tag{6.6.10}$$

Then the right-hand side of (6.6.9) is equal to $G_{j,i}(\eta,\xi)$. Suppose that $N > 8\max(|\xi|,|\xi-\eta|)$. For the sake of brevity, we put $\chi_N(x) = \chi(|x-\xi|/N)$. Since the commutator $[L,\chi_N] = L\chi_N - \chi_N L$ is a first order differential operator with coefficients vanishing for $|x-\xi| < N/4$ and $|x-\xi| > N/2$, we obtain

$$\Big| \int_{\mathcal{D}} [L(D_x), \chi_N]\, G^{(i)}(x,\xi) \cdot \overline{H^{(j)}(x,\eta)}\, dx \Big|$$

$$\le c \int\limits_{\substack{\mathcal{D} \\ N/4<|x-\xi|<N/2}} \Big(N^{-2}\, \big|G^{(i)}(x,\xi)\big|\, \big|H^{(j)}(x,\eta)\big| + N^{-1} \sum_{k=1}^{3} \big|\partial_{x_k} G^{(i)}(x,\xi)\big|\, \big|H^{(j)}(x,\eta)\big| \Big)\, dx$$

$$\le c' \int\limits_{\substack{\mathcal{D} \\ N/8<|x|<N}} \Big(|x|^{-2}\, \big|G^{(i)}(x,\xi)\big|\, \big|H^{(j)}(x,\eta)\big| + |x|^{-1} \sum_{k=1}^{3} \big|\partial_{x_k} G^{(i)}(x,\xi)\big|\, \big|H^{(j)}(x,\eta)\big| \Big)\, dx.$$

Since the functions $\zeta(\cdot,\xi)\, G(\cdot,\xi)$ and $\zeta(\cdot,\xi)\, H(\cdot,\xi)$ belong to $\mathcal{H}^\ell_{\mathcal{D}}$ for every $\xi \in \mathcal{D}$ and for every smooth function $\zeta(\cdot,\xi)$ equal to zero in a neighborhood of ξ, it follows from (6.4.5) that

$$\Big| \int_{\mathcal{D}} [L(D_x), \chi_N]\, G^{(i)}(x,\xi) \cdot \overline{H^{(j)}(x,\eta)}\, dx \Big| \to 0 \quad \text{as } N \to \infty.$$

Analogously,

$$\sum_{d^\pm=1} \int_{\Gamma^\pm} [B^\pm(D_x), \chi_N]\, G^{(i)}(x,\xi) \cdot \overline{H^{(j)}(x,\eta)}\, dx$$

$$= \sum_{\pm} \int_{\Gamma^\pm} \sum_{l,k=1}^{3} A_{l,k}\, n_k\, (\partial_{x_l} \chi_N)\, G^{(i)}(x,\xi) \cdot \overline{H^{(j)}(x,\eta)}\, dx$$

$$= \int_{\mathcal{D}} \sum_{l,k=1}^{3} \partial_{x_k} \Big(A_{l,k}\, (\partial_{x_l} \chi_N) G^{(i)}(x,\xi) \cdot \overline{H^{(j)}(x,\eta)} \Big)\, dx \to 0 \quad \text{as } N \to \infty.$$

Hence, the left-hand side of (6.6.9) tends to $\overline{H_{i,j}(\xi,\eta)}$ for the pair (6.6.10) as $N \to \infty$. This proves the assertion 3).

4) Let $F \in C_0^\infty(\overline{\mathcal{D}})^\ell$, and let $u \in \mathcal{H}_{\mathcal{D}}$ be the unique solution of the problem

$$b_{\mathcal{D}}(u,v) = (F,v)_{\mathcal{D}} \quad \text{for all } v \in \mathcal{H}_{\mathcal{D}}.$$

We denote again by $H^{(j)}(\xi,x)$ the j-th column of the matrix $H(\xi,x) = (G(x,\xi))^*$. Then

$$\int_{\mathcal{D}} F(\xi) \cdot \overline{H^{(j)}(\xi,x)}\, d\xi = b_{\mathcal{D}}\big(u, H^{(j)}(\cdot,x)\big) = \int_{\mathcal{D}} u(\xi) \cdot \overline{L^+(D_\xi)\, H^{(j)}(\xi,x)}\, d\xi$$

for each $x \in \mathcal{D}$. By (6.6.7), the right-hand side of the last equality is equal to $u_j(x)$. Hence, the formula (6.6.4) is true for $F \in C^\infty(\overline{\mathcal{D}})^\ell$. Since this set is dense in $\mathcal{H}^*_{\mathcal{D}}$, the representation (6.6.4) can be extended by continuity to all $F \in \mathcal{H}^*_{\mathcal{D}}$. The proof of the theorem is complete. □

6.6.2. Estimates of Green's matrix: the case $|x-\xi| < \min(|x'|,|\xi'|)$. Now we are interested in point estimates for Green's matrix. In the first case, we obtain the same estimate as for the Dirichlet problem (cf. Theorem 2.5.2).

THEOREM 6.6.2. *Let $G(x,\xi)$ be the Green's matrix introduced in Theorem 6.6.1. Then its elements satisfy the inequality*

$$|\partial_x^\alpha \partial_\xi^\beta G_{i,j}(x,\xi)| \le c_{\alpha,\beta}\, |x-\xi|^{-1-|\alpha|-|\beta|}$$

for all multi-indices α, β and $|x-\xi| < \min(|x'|,|\xi'|)$. The constants $c_{\alpha,\beta}$ are independent of x and ξ.

P r o o f. We write the matrix $G(x,\xi)$ in the form

$$G(x,\xi) = \psi(x,\xi)\, \mathcal{G}(x,\xi) + R(x,\xi),$$

where $\mathcal{G}(x,\xi)$ is the Green's matrix of the Dirichlet (if $d^- = 0$) or Neumann problem (if $d^- = 1$) for the differential operator $L(D_x)$ in the half-space bounded by the plane $\Gamma^- \cup (-\Gamma^-) \cup M$, $R(x,\xi)$ is the solution of the problem (6.6.5), (6.6.6) in $\mathcal{H}^\ell_{\mathcal{D}}$, and ψ is the same cut-off function as in the proof of Theorem 6.6.1. By Theorem 1.1.8, the elements of the matrix $\mathcal{G}(x,\xi)$ satisfy the estimate

$$\big|\partial_x^\alpha\, \partial_\xi^\beta \mathcal{G}_{i,j}(x,\xi)\big| \le c_{\alpha,\beta}\, \big(|x-\xi|^{-1-|\alpha|-|\beta|} + 1\big).$$

The function $x \to \partial_\xi^\beta R(x,\xi)$ is a solution of the problem

$$L(D_x)\, \partial_\xi^\beta R(x,\xi) = F(x,\xi) \quad \text{for } x \in \mathcal{D},$$

$$B^\pm(D_x)\, \partial_\xi^\beta R(x,\xi) = \Phi^\pm(x,\xi) \quad \text{for } x \in \Gamma^\pm,$$

where

$$F(x,\xi) = \partial_\xi^\beta \delta(x-\xi)\, I_\ell - L(D_x)\partial_\xi^\beta \big(\psi(x,\xi)\, \mathcal{G}(x,\xi)\big),$$

$$\Phi^\pm(x,\xi) = -B^\pm(D_x)\, \partial_\xi^\beta \big(\psi(x,\xi)\mathcal{G}(x,\xi)\big).$$

Suppose that $|\xi'| = 1$ and $\text{dist}(\xi,\Gamma^-) \le \text{dist}(\xi,\Gamma^+)$. Then $F(\cdot,\xi)$ and $\Phi^\pm(\cdot,\xi)$ vanish outside the set $\{x : 1/4 < |x-\xi| < 1/2\}$, and all derivatives $\partial_x^\alpha F(x,\xi)$ and $\partial_x^\alpha \Phi^\pm(x,\xi)$ are bounded by constants independent of x and ξ. Consequently, the norms of $F(\cdot,\xi)$ in $W_\delta^{l-2,2}(\mathcal{D})^{\ell\times\ell}$ and $\Phi^\pm(\cdot,\xi)$ in $W_\delta^{l-d^\pm-1/2,2}(\Gamma^+)^{\ell\times\ell}$ are also bounded by constants independent of ξ for arbitrary $\delta > -1$. Suppose that δ satisfies the inequality (6.5.9). Then it follows from Theorem 6.5.4 that $\eta\partial_\xi^\beta R(\cdot,\xi) \in W_\delta^{l,2}(\mathcal{D})^{\ell\times\ell}$ for an arbitrary smooth cut-off function η and

$$\|\eta\partial_\xi^\beta R(\cdot,\xi)\|_{W_\delta^{l,2}(\mathcal{D})^{\ell\times\ell}} \le c_{l,\beta}\,,$$

where the constant $c_{l,\beta}$ is independent of ξ. Using the continuity of the imbedding $W^{l,2}(\Omega) \subset C^{l-2}(\Omega)$, we conclude that

$$\big|\partial_x^\alpha \partial_\xi^\beta R(x,\xi)\big| \le c_{\alpha,\beta}$$

for $|x-\xi| < \min(|x'|,1)$, $|\xi'| = 1$, $\text{dist}(\xi,\Gamma^-) \le \text{dist}(\xi,\Gamma^+)$. Together with the estimates of $\mathcal{G}(x,\xi)$ this implies

$$\big|\partial_x^\alpha\, \partial_\xi^\beta G_{i,j}(x,\xi)\big| \le c_{\alpha,\beta}\, |x-\xi|^{-1-|\alpha|-|\beta|}$$

if $|x-\xi| < \min(|x'|,1)$, $|\xi'| = 1$, $\operatorname{dist}(\xi,\Gamma^-) \le \operatorname{dist}(\xi,\Gamma^+)$. Analogously, this estimates hold for $|x-\xi| < \min(|x'|,1)$, $|\xi'| = 1$, $\operatorname{dist}(\xi,\Gamma^-) > \operatorname{dist}(\xi,\Gamma^+)$. Using (6.6.3), we obtain the desired estimates for arbitrary ξ, $|x-\xi| < \min(|x'|,|\xi'|)$. □

6.6.3. Estimates of Green's matrix for the Dirichlet and mixed problems in the case $|x-\xi| > \min(|x'|,|\xi'|)$. The following estimate was proved in Section 2.5 for the Dirichlet problem. It is essentially based on the regularity assertion for the variational solution given in Theorem 2.4.6. Since the same regularity assertion is valid for the mixed problem (cf. Theorem 6.1.11), we obtain the following result analogously to Theorem 2.5.4.

THEOREM 6.6.3. *Let $d^+ + d^- \le 1$, and let δ_+,δ_- be the greatest positive real numbers such that the strip*

$$-\delta_- < \operatorname{Re}\lambda < \delta_+$$

is free of eigenvalues of the pencil $A(\lambda)$. Then the Green's matrix $G(x,\xi)$ introduced in Theorem 6.6.1 satisfies the estimate

$$\left|\partial_{x'}^{\alpha}\partial_{x_3}^{j}\partial_{\xi'}^{\beta}\partial_{\xi_3}^{k} G(x,\xi)\right| \le c_{j,k,\alpha,\beta}\, \frac{|x'|^{\delta_+-|\alpha|-\varepsilon}\, |\xi'|^{\delta_--|\beta|-\varepsilon}}{|x-\xi|^{1+\delta_++\delta_-+j+k-2\varepsilon}}$$

for $|x-\xi| > \min(|x'|,|\xi'|)$ and for arbitrary α,β,j,k. Here ε is an arbitrarily small positive number.

6.6.4. Estimates of Green's matrix for the Neumann problem in the case $|x-\xi| > \min(|x'|,|\xi'|)$. In order to obtain an analogous result for the Neumann problem ($d^+ = d^- = 1$), we need the following lemma. Here δ_+ and δ_- are the greatest positive real numbers such that the strip

$$-\delta_- < \operatorname{Re}\lambda < \delta_+$$

contains only the eigenvalue $\lambda = 0$ of the pencil $A(\lambda)$.

LEMMA 6.6.4. *Let $x_0 \in \mathcal{D}$, $\operatorname{dist}(x_0,M) \le 4$, and let $\mathcal{B}$ be a ball with radius 1 centered about x_0. Furthermore, let ζ, η be infinitely differentiable functions with supports in $\mathcal{B}$ such that $\eta = 1$ in a neighborhood of $\operatorname{supp}\zeta$. If $\eta u \in \mathcal{H}_{\mathcal{D}}$, $L(D_x)\,u = 0$ in $\mathcal{D}\cap\mathcal{B}$ and $B^{\pm}(D_x)u = 0$ on $\Gamma^{\pm}\cap\mathcal{B}$, then*

$$\sup_{x\in\mathcal{D}} |x'|^{\max(0,|\alpha|-\delta_++\varepsilon)} \left|\zeta(x)\,\partial_{x'}^{\alpha}\partial_{x_3}^{j} u(x)\right| \le c\,\|\eta u\|_{\mathcal{H}_{\mathcal{D}}}\,, \tag{6.6.11}$$

where ε is an arbitrarily small positive number. The constant c in (6.6.11) is independent of u and x_0.

P r o o f. Let ε be such that $\delta_+ - \varepsilon \in (k,k+1)$. Then $\delta = k+1-\delta_++\varepsilon \in (0,1)$. Furthermore, let χ be a function in $C_0^\infty(\mathcal{B})$ such that $\zeta\chi = \zeta$ and $\chi\eta = \chi$. From Corollary 6.5.5 it follows that $\partial_{x_3}^j(\chi u) \in W_\delta^{k+2,2}(\mathcal{D})^\ell$ for $j = 0,1,\dots$ and

$$\|\partial_{x_3}^j(\chi u)\|_{W_\delta^{k+2,2}(\mathcal{D})^\ell} \le c\,\|\eta u\|_{\mathcal{H}_{\mathcal{D}}}\,.$$

Hence, $\partial_{x'}^{\alpha}\partial_{x_3}^{j}(\chi u) \in W_\delta^{2,2}(\mathcal{D})^\ell$ for $|\alpha| \le k$. Since the space $W_\delta^{2,2}(K)$ is continuously imbedded in $C(\overline{K})$ (cf. [**84**, Lemma 7.1.3]), we get

$$\sup_{x'\in K,\, x_3\in\mathbb{R}} \left|\partial_{x'}^{\alpha}\partial_{x_3}^{j}(\chi u)\right| \le c \sup_{x_3\in\mathbb{R}} \|\partial_{x'}^{\alpha}\partial_{x_3}^{j}(\chi u)(\cdot,x_3)\|_{W_\delta^{2,2}(K)^\ell}\,.$$

Using the imbedding $W^{1,2}(M) \subset C(M)$, we obtain

$$\sup_{x_3\in\mathbb{R}} \|\partial_{x'}^\alpha \partial_{x_3}^j (\chi u)(\cdot,x_3)\|_{W_\delta^{2,2}(K)^\ell}$$
$$\le c\Big(\|\partial_{x'}^\alpha \partial_{x_3}^j (\chi u)\|_{W_\delta^{2,2}(\mathcal{D})^\ell} + \|\partial_{x'}^\alpha \partial_{x_3}^{j+1} (\chi u)\|_{W_\delta^{2,2}(\mathcal{D})^\ell}\Big) \le c\,\|\eta u\|_{\mathcal{H}_\mathcal{D}}\,.$$

This proves (6.6.11) for $|\alpha|\le k$. Now let $|\alpha|\ge k+1$. It follows from Corollary 6.5.5 that $\partial_{x_3}^j(\chi u) \in W_{\delta-k+|\alpha|}^{|\alpha|+2,2}(\mathcal{D})^\ell$ and, consequently, $\partial_{x'}^\alpha\partial_{x_3}^j(\chi u) \in W_{\delta-k+|\alpha|}^{2,2}(\mathcal{D})^\ell \subset V_{\delta-k+|\alpha|}^{2,2}(\mathcal{D})^\ell$. According to Lemma 1.2.3,

$$\sup_{x'\in K} |x'|^{\delta-k+|\alpha|-1}\,|v(x')| \le c\,\|v\|_{V_{\delta-k+|\alpha|}^{2,2}(K)} \quad \text{for all } v\in V_{\delta-k+|\alpha|}^{2,2}(K). \tag{6.6.12}$$

Consequently

$$\sup_{x'\in K,\,x_3\in\mathbb{R}} |x'|^{\delta-k+|\alpha|-1}\,\big|\partial_{x'}^\alpha \partial_{x_3}^j(\chi u)\big| \le c\,\sup_{x_3\in\mathbb{R}} \|\partial_{x'}^\alpha\partial_{x_3}^j(\chi u)(\cdot,x_3)\|_{W_{\delta-k+|\alpha|}^{2,2}(K)^\ell}.$$

Using again the imbedding $W^{1,2}(M)\subset C(M)$, we arrive at (6.6.11). □

THEOREM 6.6.5. *Let $d^+=d^-=1$. Then the Green's matrix $G(x,\xi)$ introduced in Theorem 6.6.1 satisfies the estimate*

$$\big|\partial_{x'}^\alpha\partial_{x_3}^j\partial_{\xi'}^\beta\partial_{\xi_3}^k G(x,\xi)\big|$$
$$\le c\,|x-\xi|^{-1-|\alpha|-|\beta|-j-k}\Big(\frac{|x'|}{|x-\xi|}\Big)^{\min(0,\delta_+-|\alpha|-\varepsilon)}\Big(\frac{|\xi'|}{|x-\xi|}\Big)^{\min(0,\delta_--|\beta|-\varepsilon)}$$

for $|x-\xi|\ge\min(|x'|,|\xi'|)$, where ε is an arbitrarily small positive number.

P r o o f. Since $G(Tx,T\xi) = T^{-1}G(x,\xi)$, we may assume without loss of generality that $|x-\xi|=2$. Then $\max(|x'|,|\xi'|)\le 4$. Let $\mathcal{B}_x$ and $\mathcal{B}_\xi$ be balls with centers x and ξ, respectively, and radius 1. Furthermore, let ζ and η be infinitely differentiable functions with supports in $\mathcal{B}_x$ and $\mathcal{B}_\xi$, respectively. Since the function $y\to\partial_{x'}^\alpha\partial_{x_3}^j G^*(x,y)$ satisfies the equations

$$L^+(D_y)\partial_{x'}^\alpha\partial_{x_3}^j G^*(x,y) = 0 \text{ in } \mathcal{D}\cap\mathcal{B}_\xi,\quad C^\pm(D_y)\partial_{x'}^\alpha\partial_{x_3}^j G^*(x,y)=0 \text{ on } \Gamma^\pm\cap\mathcal{B}_\xi\,,$$

it follows from Lemma 6.6.4 that

$$|\xi'|^{\max(0,|\beta|-\delta_-+\varepsilon)}\,\big|\partial_{x'}^\alpha\partial_{x_3}^j\partial_{\xi'}^\beta D_{\xi_3}^k G(x,\xi)\big| \le c\,\|\eta(\cdot)\partial_{x'}^\alpha\partial_{x_3}^j G(x,\cdot)\|_{\mathcal{H}_\mathcal{D}}\,. \tag{6.6.13}$$

We consider the solution

$$u(x) = \int_\mathcal{D} G(x,y)\,\eta(y)\,F(y)\,dy$$

of the problem

$$b_\mathcal{D}(u,v) = (\eta F,v)_\mathcal{D} \quad \text{for all } v\in\mathcal{H}_\mathcal{D}\,,$$

where $F\in\mathcal{H}_\mathcal{D}^*$. Since ηF vanishes in the ball $\mathcal{B}_x$, we conclude from Lemma 6.6.4 that

$$|x'|^{\max(0,|\alpha|-\delta_++\varepsilon)}\,|\partial_{x'}^\alpha\partial_{x_3}^j u(x)| \le c\,\|\zeta u\|_{\mathcal{H}_\mathcal{D}} \le c'\,\|F\|_{\mathcal{H}_\mathcal{D}^*}\,. \tag{6.6.14}$$

Consequently, the mapping

$$F \;\to\; |x'|^{\max(0,|\alpha|-\delta_++\varepsilon)}\,\partial_{x'}^\alpha\,\partial_{x_3}^j u(x)$$
$$= |x'|^{\max(0,|\alpha|-\delta_++\varepsilon)}\int_\mathcal{D}\partial_{x'}^\alpha\,\partial_{x_3}^j G(x,y)\,\eta(y)\,F(y)\,dy$$

is linear and continuous from $\mathcal{H}_{\mathcal{D}}^*$ into $\mathbb{C}^\ell$ for arbitrary $x \in \mathcal{D}$. The norm of this mapping is bounded by a constant independent of x. This implies

$$|x'|^{\max(0,|\alpha|-\delta_+ + \varepsilon)} \, \|\eta(\cdot)\, \partial_{x'}^\alpha \partial_{x_3}^j G(x,\cdot)\|_{\mathcal{H}_{\mathcal{D}}} \le c$$

which together with (6.6.13) yields the desired estimate. □

Note that the estimate in Theorem 6.6.3 is stronger than that in Theorem 6.6.5. Thus, the assertion of the last theorem holds for the Dirichlet and mixed problems as well.

6.6.5. Estimates of Green's matrix for special Neumann problems. We show that, under the conditions of Theorem 6.5.9, the estimate in Theorem 6.6.5 can be improved. Suppose that the matrix

$$A' = \begin{pmatrix} A_{1,1} & A_{2,1} \\ A_{1,2} & A_{2,2} \end{pmatrix}$$

is not regular, i.e. the rank r' of this matrix is less than 2ℓ. Then by Lemma 6.5.6, the number $\lambda = 1$ is an eigenvalue of the pencil $A(\lambda)$. We assume that the line $\operatorname{Re}\lambda = 1$ does not contain other eigenvalues of the pencil $A(\lambda)$. Furthermore, we assume that the matrix (6.5.14) has also the rank r' and that the eigenvalue $\lambda = 1$ has the geometric and algebraic multiplicity $2\ell - r'$. Then we denote by μ_+ the greatest real number such that the strip

$$0 < \operatorname{Re}\lambda < \mu_+ \tag{6.6.15}$$

contains at most the eigenvalue $\lambda = 1$.

THEOREM 6.6.6. *Under the above assumptions, the Green's matrix of the Neumann problem satisfies the estimate*

$$\begin{aligned} &\big|D_{x'}^\alpha D_{x_3}^j D_{\xi'}^\beta D_{\xi_3}^k G(x,\xi)\big| \\ &\le c\,|x-\xi|^{-1-|\alpha|-|\beta|-j-k} \Big(\frac{|x'|}{|x-\xi|}\Big)^{\min(0,\mu_+-|\alpha|-\varepsilon)} \Big(\frac{|\xi'|}{|x-\xi|}\Big)^{\min(0,\delta_--|\beta|-\varepsilon)}, \end{aligned}$$

for $|x-\xi| \ge \min(|x'|,|\xi'|)$, *where* ε *is an arbitrarily small positive number. If the above conditions are also valid for the formally adjoint problem, then the number* δ_- *can be replaced by the greatest real number* μ_- *such that the strip* $0 < \operatorname{Re}\lambda < \mu_-$ *contains at most the eigenvalue* $\lambda = 1$ *of the pencil* $A^+(\lambda)$.

P r o o f. Using Theorem 6.5.9, one can show that the assertion of Lemma 6.6.4 is valid with μ_+ instead of δ_+. Thus, the estimate (6.6.14) in the proof of Theorem 6.6.5 can be replaced by

$$|x'|^{\max(0,|\alpha|-\mu_+ + \varepsilon)} \, |D_{x'}^\alpha D_{x_3}^j u(x)| \le c\, \|F\|_{\mathcal{H}_{\mathcal{D}}^*}.$$

If the conditions of the theorem for the formally adjoint problem are satisfied, then analogously the estimate (6.6.13) in the proof of Theorem 6.6.5 can be replaced by

$$|\xi'|^{\max(0,|\beta|-\mu_- + \varepsilon)} \, \big|D_{x'}^\alpha D_{x_3}^j D_{\xi'}^\beta D_{\xi_3}^k G(x,\xi)\big| \le c\, \|\eta(\cdot) D_{x'}^\alpha D_{x_3}^j G(x,\cdot)\|_{\mathcal{H}_{\mathcal{D}}}.$$

This leads to the desired estimates of Green's matrix. □

As an example, we consider the Neumann problem (6.5.20) for the Lamé system. According to Theorem 6.6.6, the Green's matrix for this problem satisfies the estimate

$$\begin{aligned}&\big|D^\alpha_{x'} D^j_{x_3} D^\beta_{\xi'} D^k_{\xi_3} G(x,\xi)\big| \\ &\le c\,|x-\xi|^{-1-|\alpha|-|\beta|-j-k}\Big(\frac{|x'|}{|x-\xi|}\Big)^{\min(0,\mu_+-|\alpha|-\varepsilon)}\Big(\frac{|\xi'|}{|x-\xi|}\Big)^{\min(0,\mu_+-|\beta|-\varepsilon)}.\end{aligned}$$

Here $\mu_+ = \pi/\theta$ for $\theta < \pi$ and $\mu_+ = \xi_+(\theta)/\theta$ for $\theta > \pi$, where $\xi_+(\theta)$ is the smallest positive solution of the equation (6.5.21).

6.7. Weighted Hölder spaces with nonhomogeneous norms

In Section 6.2, we studied the weighted Sobolev spaces $W^{l,p}_\delta(\mathcal{D})$. Now we consider the weighted Hölder spaces with analogous nonhomogeneous norms. The space $C^{l,\sigma}_\delta(\mathcal{D})$ introduced in this section consists of all l times continuously differentiable functions on $\overline{\mathcal{D}}\backslash M$ such that the derivatives of u satisfy the inequality

$$\sup_{x\in\mathcal{D}} |x'|^{\max(0,\delta-l-\sigma+|\alpha|)}|\partial^\alpha_x u(x)| < \infty$$

for $|\alpha| \le l$ and a weighted Hölder condition. Again we are interested in imbeddings, traces and in relations between the spaces with homogeneous and nonhomogeneous norm. By our definition, the spaces $C^{l,\sigma}_\delta(\mathcal{D})$ and $N^{l,\sigma}_\delta(\mathcal{D})$ coincide if $\delta \ge l+\sigma$. For the case $0 \le \delta < l+\sigma$ we show that every function $u \in C^{l,\sigma}_\delta(\mathcal{D})$ is a sum of a quasipolynomial and a $N^{l,\sigma}_\delta(\mathcal{D})$-function.

6.7.1. Weighted Hölder spaces in an angle.

Let K be the angle

$$K = \{x' = (x_1,x_2) :\ 0 < r < \infty,\ -\theta/2 < \varphi < \theta/2\},$$

where r,φ are the polar coordinates of the point x'. We introduce the weighted Hölder space $C^{l,\sigma}_\delta(K)$. For integer $l \ge 0$ and real δ,σ, $\delta \ge 0$, $0 < \sigma < 1$, we define this space as the set of all l times continuously differentiable functions on $\overline{K}\backslash\{0\}$ with finite norm

$$\|u\|_{C^{l,\sigma}_\delta(K)} = \sum_{|\alpha|\le l} \sup_{x'\in K} |x'|^{\max(0,\delta-l-\sigma+|\alpha|)}|\partial^\alpha_{x'} u(x')| + \langle u\rangle_{l,\sigma,\delta;K}\,,$$

where

$$\langle u\rangle_{l,\sigma,\delta;K} = \sum_{|\alpha|=l} \sup_{\substack{x',y'\in K\\ |x'-y'|\le|x'|/2}} |x'|^\delta \frac{|\partial^\alpha_{x'}u(x') - \partial^\alpha_{y'}u(y')|}{|x'-y'|^\sigma}\,.$$

Obviously, the space $C^{l,\sigma}_\delta(K)$ coincides with $N^{l,\sigma}_\delta(K)$ for $\delta \ge l+\sigma$, where $N^{l,\sigma}_\delta(K)$ is the weighted space introduced in Section 2.7.

LEMMA 6.7.1. *Let $l+\sigma > l'+\sigma'$, $l+\sigma-\delta = l'+\sigma'-\delta'$, and $\delta' \ge 0$. Then the space $C^{l,\sigma}_\delta(K)$ is continuously imbedded in $C^{l',\sigma'}_{\delta'}(K)$*

P r o o f. For $l = l'$ the assertion is obvious. Suppose that $l > l'$. Then as in the proof of Lemma 2.7.1,

$$\langle u\rangle_{l',\sigma',\delta';K} \le c \sum_{|\alpha|=l'+1} \sup_{x'\in K} |x'|^{\delta'-\sigma'+1}\,\big|\partial^\alpha_{x'}u(x')\big| \le c\,\|u\|_{C^{l,\sigma}_\delta(K)}$$

and consequently

$$\|u\|_{C_{\delta'}^{l',\sigma'}(K)} \le c\,\|u\|_{C_\delta^{l,\sigma}(K)}\,.$$

This proves the lemma. □

Let $C^{l,\sigma}(K)$ be the nonweighted Hölder space on K, i.e. the set of all l times continuously differentiable functions on $\overline{K}$ with finite norm

$$\|u\|_{C^{l,\sigma}(K)} = \sum_{|\alpha|\le l} \sup_{x'\in K} |\partial_{x'}^\alpha u(x')| + \sum_{|\alpha|=l} \sup_{x',y'\in K} \frac{|\partial_{x'}^\alpha u(x') - \partial_{y'}^\alpha u(y')|}{|x'-y'|^\sigma}\,.$$

We show that this space concides with the space $C_\delta^{l,\sigma}(K)$ for $\delta = 0$.

LEMMA 6.7.2. *The spaces $C_0^{l,\sigma}(K)$ and $C^{l,\sigma}(K)$ coincide, and the norms in these spaces are equivalent.*

Proof. Obviously $C^{l,\sigma}(K)$ is continuously imbedded in $C_0^{l,\sigma}(K)$. We prove the imbedding $C_0^{l,\sigma}(K) \subset C^{l,\sigma}(K)$ for $l = 0$. First we show that any function $u \in C_0^{0,\sigma}(K)$ is continuous at $x = 0$. We consider the sequence of the points $x_n = 2^{-n}x_0$, where x_0 is an arbitrary point in K. Since $|x_n - x_{n+1}| = |x_n|/2$, we have

$$|u(x_n) - u(x_{n+1})| \le c_0\,|x_n - x_{n+1}|^\sigma = c_0\,|x_0|^\sigma\,2^{-(n+1)\sigma},$$

where $c_0 = \langle u\rangle_{0,\sigma,0;K}$. Consequently,

$$|u(x_n) - u(x_m)| \le c_0\,|x_0|^\sigma \left(2^{-(n+1)\sigma} + 2^{-(n+2)\sigma} + \cdots + 2^{-m\sigma}\right) < c_0\,|x_0|^\sigma\,\frac{1}{2^\sigma - 1}\,2^{-n\sigma}$$

for $m > n$. Thus, $\{u(x_n)\}$ is a Cauchy sequence. The limit of this sequence is denoted by a. It can be easily shown that the sequence $\{u(x_n)\}$ has the same limit if $\{x_n\}$ is an arbitrary sequence on the ray from 0 to x_0 such that $x_n \to 0$ as $n \to \infty$. Now let $\{x_n\}$ be an arbitrary sequence in the angle $\{x' \in K : |\varphi - \varphi_0| < \arctan 1/2\}$, where r, φ are the polar coordinates of x' and r_0, φ_0 are the polar coordinates of x_0. We denote by y_n the orthogonal projection of x_n onto the line through 0 and x_0. Then $|x_n - y_n| \le |y_n|/2$ and, therefore, $|u(x_n) - u(y_n)| \le c_0\,|x_n - y_n|^\sigma < c_0\,|y_n|^\sigma$. If x_n tends to 0 as $n \to \infty$, then $\lim |y_n| = 0$ and $\lim u(y_n) = a$. This implies $\lim u(x_n) = a$. Repeating this argument, we conclude that $u(x_n) \to a$ for an arbitrary sequence $\{x_n\}$ in K converging to 0. Thus, the function u is continuous at the vertex of K. Next we show that

$$\frac{|u(x') - u(y')|}{|x'-y'|^\sigma} \le c\,\langle u\rangle_{0,\sigma,0;K} \quad \text{for } u \in C_0^{0,\sigma}(K). \tag{6.7.1}$$

For $|x'-y'| < |x'|/2$ this follows from the definition of $\langle u\rangle_{0,\sigma,0;K}$. If $|x'-y'| > |x'|/2$, then $|x'-y'| > |y'|/3$ and therefore

$$\frac{|u(x') - u(y')|}{|x'-y'|^\sigma} \le 2^\sigma\,\frac{|u(x') - u(0)|}{|x'|^\sigma} + 3^\sigma\,\frac{|u(y') - u(0)|}{|y'|^\sigma}\,.$$

Using the inequalities

$$\big|u(x') - u(0)\big| \le \sum_{n=0}^\infty \big|u(2^{-n}x') - u(2^{-n-1}x')\big| \le c_0 \sum_{n=0}^\infty |2^{-n-1}x'|^\sigma = \frac{c_0}{2^\sigma - 1}\,|x'|^\sigma,$$

we obtain (6.7.1). Thus, it is proved that $C_0^{0,\sigma}(K) = C^{0,\sigma}(K)$. From this we can easily deduce the equality $C_0^{l,\sigma}(K) = C^{l,\sigma}(K)$ for $l \ge 1$. □

COROLLARY 6.7.3. *If* $k-1 \le \delta - \sigma < k$, $k \in \{0,1,\dots,l\}$, *then* $C^{l,\sigma}_\delta(K)$ *is continuously imbedded in* $C^{l-k,\sigma-\delta+k}(K)$.

6.7.2. Relations between the spaces $C^{l,\sigma}_\delta(K)$ and $N^{l,\sigma}_\delta(K)$. We show that a $C^{l,\sigma}_\delta(K)$-function belongs to the space $N^{l,\sigma}_\delta(K)$ if and only if it is zero at the origin together with its derivatives up to a certain order. For this, we employ the following version of Hardy's inequality.

LEMMA 6.7.4. *Let u be a differentiable function on $\overline{K}\backslash\{0\}$ and let $r^\delta|\nabla u(x)| < c < \infty$ for $x \in K$. Furthermore, we assume that*

(i) $\delta > 1$ *and* $u(x) \to 0$ *as* $|x| \to \infty$ *or* (ii) $\delta < 1$ *and* $u(x) \to 0$ *as* $|x| \to 0$.

Then $\displaystyle \sup_{x\in K} r^{\delta-1}\,|u(x)| \le \frac{1}{|\delta-1|}\, \sup_{x\in K} r^\delta\, |\partial_r u(x)|.$

P r o o f. If the condition (i) is satisfied, then the assertion follows from the inequality

$$|u(r,\varphi)| \le \int_r^\infty |\partial_t u(t,\varphi)|\, dt \le \sup_{0\le t\le r} \big|t^\delta \partial_t u(t,\varphi)\big| \cdot \int_r^\infty t^{-\delta}\, dt.$$

Replacing the integration interval (r,∞) in the above inequality by $(0,r)$, we can estimate $|u(r,\varphi)|$ in the same way for the case (ii). □

Let $k-1 \le \delta-\sigma < k$, where k is an integer, $0 \le k \le l$. Then, by Corollary 6.7.3, the derivatives of an arbitrary function $u \in C^{l,\sigma}_\delta(K)$ up to order $l-k$ are continuous at the origin. We denote by

$$p_{l-k}(u) = \sum_{|\alpha|\le l-k} \frac{1}{\alpha!} (\partial^\alpha u)(0)\, x^\alpha$$

the Taylor polynomial of degree $l-k$ of u.

LEMMA 6.7.5. *Let $u \in C^{l,\sigma}_\delta(K)$, where $k-1 \le \delta-\sigma < k \le l$. Then $u \in N^{l,\sigma}_\delta(K)$ if and only if $(\partial^\alpha u)(0) = 0$ for $|\alpha| \le l-k$. Moreover, there exists a constant c such that*

$$\|u\|_{N^{l,\sigma}_\delta(K)} \le c\, \|u\|_{C^{l,\sigma}_\delta(K)}$$

for all $u \in C^{l,\sigma}_\delta(K)$, $(\partial^\alpha u)(0) = 0$ for $|\alpha| \le l-k$.

P r o o f. If $u \in N^{l,\sigma}_\delta(K)$, then $\partial^\alpha u(x') = O(|x'|^{l+\sigma-\delta-|\alpha|})$ and consequently $(\partial^\alpha u)(0) = 0$ for $|\alpha| \le l-k$.

Suppose that $u \in C^{l,\sigma}_\delta(K)$ and $(\partial^\alpha u)(0) = 0$ for $|\alpha| \le l-k$. We show that

$$\sup |x'|^{\delta-l-\sigma+|\alpha|}\, \big|\partial^\alpha u(x')\big| \le c\, \|u\|_{C^{l,\sigma}_\delta(K)} \tag{6.7.2}$$

for $|\alpha| \le l$. If $k > 0$, then (6.7.2) is obviously satisfied for $|\alpha| \ge l-k+1$. Furthermore, we conclude from Lemma 6.7.4 that

$$\begin{aligned} \sup |x'|^{\delta-\sigma-l} |u(x')| &\le c_1 \sup |x'|^{\delta-\sigma-l+1} |\nabla u(x')| \le \cdots \\ &\le c_{l-k+1} \sum_{|\alpha|=l-k+1} \sup |x'|^{\delta-\sigma-k+1} |\partial^\alpha u(x')| \end{aligned}$$

which proves (6.7.2) for $|\alpha| \le l$. If $k = 0$, then the space $C_\delta^{l,\sigma}(K)$ is continuously imbedded in $C^{l,\sigma-\delta}(K)$. Consequently,

$$\sup |x'|^{\delta-\sigma} |\partial^\alpha u(x')| = \sup |x'|^{\delta-\sigma} \big|(\partial^\alpha u)(x') - (\partial^\alpha u)(0)\big| \le c\, \|u\|_{C_\delta^{l,\sigma}(K)}$$

for $|\alpha| = l$. Using Lemma 6.7.4, we obtain (6.7.2) for $|\alpha| \le l$. Hence, $u \in N_\delta^{l,\sigma}(K)$. □

As an immediate consequence of the last lemma, we obtain the following statement.

THEOREM 6.7.6. *Let $u \in C_\delta^{l,\sigma}(K)$, where $k - 1 \le \delta - \sigma < k \le l$, k is a nonnegative integer, $k \le l$. Furthermore, let ζ be an infinitely differentiable function on $\overline{K}$ with compact support which is equal to one in a neighborhood of the origin. Then $u - \zeta\, p_{l-k}(u) \in N_\delta^{l,\sigma}(K)$ and*

$$\|u - \zeta\, p_{l-k}(u)\|_{N_\delta^{l,\sigma}(K)} \le c\, \|u\|_{C_\delta^{l,\sigma}(K)}$$

with a constant c independent of u.

P r o o f. Obviously, $\zeta p_{l-k}(u) \in C_\delta^{l,\sigma}(K)$ and

$$\|\zeta\, p_{l-k}(u)\|_{C_\delta^{l,\sigma}(K)} \le c\, \|u\|_{C_\delta^{l,\sigma}(K)}\,.$$

Since the derivatives of $u - \zeta p_{l-k}(u)$ up to order $l-k$ are zero at the origin, Lemma 6.7.5 implies

$$\|u - \zeta\, p_{l-k}(u)\|_{N_\delta^{l,\sigma}(K)} \le c\, \|u - \zeta\, p_{l-k}(u)\|_{C_\delta^{l,\sigma}(K)} \le c'\, \|u\|_{C_\delta^{l,\sigma}(K)}\,.$$

This proves the theorem. □

6.7.3. Weighted Hölder spaces in a dihedron. Let $\mathcal{D}$ be the dihedron $\{x = (x', x_3) : x' = (x_1, x_2) \in K,\ x_3 \in \mathbb{R}\}$, where K is an angle in the (x_1, x_2)-plane. For arbitrary integer $l \ge 0$ and real $\sigma \in (0,1)$, we denote by $C^{l,\sigma}(\mathcal{D})$ the Hölder space with the norm

$$\|u\|_{C^{l,\sigma}(\mathcal{D})} = \sum_{|\alpha| \le l} \sup_{x \in \mathcal{D}} |\partial^\alpha u(x)| + \sum_{|\alpha| = l} \sup_{\substack{x,y \in \mathcal{D} \\ |x-y|<1}} \frac{\big|\partial^\alpha u(x) - \partial^\alpha u(y)\big|}{|x-y|^\sigma}$$

An equivalent norm is given in the following lemma.

LEMMA 6.7.7. *The norm in $C^{l,\sigma}(\mathcal{D})$ is equivalent to*

$$\begin{aligned} \|u\| \ &= \ \sum_{|\alpha| \le l} \sup_{x \in \mathcal{D}} |\partial^\alpha u(x)| + \sum_{|\alpha| = l} \Big(\sup_{x'} \sup_{|x_3 - y_3| < 1} \frac{\big|\partial^\alpha u(x', x_3) - \partial^\alpha u(x', y_3)\big|}{|x_3 - y_3|^\sigma} \\ &\qquad + \sup_{x_3} \sup_{|x'-y'| < |x'|/2} \frac{\big|\partial^\alpha u(x', x_3) - \partial^\alpha u(y', x_3)\big|}{|x' - y'|^\sigma} \Big). \end{aligned}$$

P r o o f. It suffices to prove the lemma for $l = 0$. Obviously, the norm in $C^{0,\sigma}(\mathcal{D})$ is equivalent to

$$\begin{aligned} \|u\| \ &= \ \sup_{x \in \mathcal{D}} |u(x)| + \sup_{x'} \sup_{|x_3 - y_3| < 1} \frac{\big|u(x', x_3) - u(x', y_3)\big|}{|x_3 - y_3|^\sigma} \\ &\qquad + \sup_{x_3} \sup_{|x'-y'| < 1} \frac{\big|u(x', x_3) - u(y', x_3)\big|}{|x' - y'|^\sigma} \Big). \end{aligned}$$

We show, that there exists a constant c independent of u such that

$$\sup_{x_3} \sup_{x',y' \in K} \frac{|u(x',x_3) - u(y',x_3)|}{|x'-y'|^\sigma} \le c \sup_{x_3} \sup_{\substack{x',y' \in K \\ |x'-y'|<|x'|/2}} \frac{|u(x',x_3) - u(y',x_3)|}{|x'-y'|^\sigma} . \tag{6.7.3}$$

We denote the right-hand side of (6.7.3) by $A_\sigma(u)$. Let x', y' be arbitrary points in K such that $|x'-y'| > |x'|/2$, and let $x_3 \in \mathbb{R}$. We put $\xi_n = 2^{-n}x'$. Then

$$\big|u(\xi_n,x_3) - u(\xi_{n+1},x_3)\big| \le c_0\, |\xi_n - \xi_{n+1}|^\sigma = c_0\, |x'|^\sigma\, 2^{-(n+1)\sigma},$$

where $c_0 = A_\sigma(u)$. Consequently,

$$\begin{aligned} \big|u(x',x_3) - u(0,x_3)\big| &\le \sum_{n=0}^{\infty} \big|u(\xi_n,x_3) - u(\xi_{n+1},x_3)\big| \\ &\le c_0\, |x'|^\sigma \sum_{n=0}^{\infty} 2^{-(n+1)\sigma} = \frac{c_0}{2^\sigma - 1}\, |x'|^\sigma \end{aligned}$$

and, analogously,

$$\big|u(y',x_3) - u(0,x_3)\big| \le \frac{c_0}{2^\sigma - 1}\, |y'|^\sigma .$$

Since $|x'| < 2|x'-y'|$ and $|y'| < 3|x'-y'|$, it follows that

$$\big|u(x',x_3) - u(y',x_3)\big| \le \frac{c_0}{2^\sigma - 1}\, \big(|x'|^\sigma + |y'|^\sigma\big) \le \frac{2^\sigma + 3^\sigma}{2^\sigma - 1}\, A_\sigma(u)\, |x'-y'|^\sigma$$

which proves (6.7.3). The result follows. □

Next we introduce the weighted Hölder space $C_\delta^{l,\sigma}(\mathcal{D})$, where δ is a nonnegative real number. For $0 \le \delta < l + \sigma$ this space is defined as the set of all l times continuously differentiable functions on $\overline{\mathcal{D}} \backslash M$ with finite norm

$$\|u\|_{C_\delta^{l,\sigma}(\mathcal{D})} = \|u\|_{C^{l-k,k-\delta+\sigma}(\mathcal{D})} + \sum_{|\alpha|=l-k+1}^{l} \sup_{x\in\mathcal{D}} |x'|^{\delta-l-\sigma+|\alpha|} |\partial_x^\alpha u(x)| + \langle u \rangle_{l,\sigma,\delta;\mathcal{D}} ,$$

where $k = [\delta - \sigma] + 1$, $[s]$ denotes the integral part of s, and

$$\langle u \rangle_{l,\sigma,\delta;\mathcal{D}} = \sum_{|\alpha|=l} \sup_{\substack{x,y\in\mathcal{D} \\ |x-y|<|x'|/2}} |x'|^\delta \frac{|\partial_x^\alpha u(x) - \partial_y^\alpha u(y)|}{|x-y|^\sigma} .$$

In the case $\delta \ge l + \sigma$ we set $C_\delta^{l,\sigma}(\mathcal{D}) = N_\delta^{l,\sigma}(\mathcal{D})$. Analogously, the Hölder spaces $C_\delta^{l,\sigma}(\Gamma^\pm)$ on the faces of the dihedron $\mathcal{D}$ are defined.

LEMMA 6.7.8. *The space $C_\delta^{l,\sigma}(\mathcal{D})$ is continuously imbedded in $C_{\delta'}^{l',\sigma'}(\mathcal{D})$ if $l+\sigma > l'+\sigma'$, $\delta > \delta' \ge 0$, and $l+\sigma-\delta = l'+\sigma'-\delta'$.*

P r o o f. Suppose that $u \in C_\delta^{l,\sigma}(\mathcal{D})$, $l+\sigma > l'+\sigma'$ and $l+\sigma-\delta = l'+\sigma'-\delta'$. If $k' = [\delta'-\sigma']+1$, then $l'-k' = l-k$ and $k'-\delta'+\sigma' = k-\delta+\sigma$. Consequently

$$\|u\|_{C^{l'-k',k'-\delta'+\sigma'}(\mathcal{D})} + \sum_{|\alpha|=l'-k'+1}^{l'} \sup_{x\in\mathcal{D}} |x'|^{\delta'-l'-\sigma'+|\alpha|} |\partial_x^\alpha u(x)| \le \|u\|_{C_\delta^{l,\sigma}(\mathcal{D})}$$

Furthermore, it is evident that

$$\langle u \rangle_{l',\sigma',\delta';\mathcal{D}} \le \langle u \rangle_{l,\sigma,\delta;\mathcal{D}}$$

if $l = l'$. If $l > l'$, then as in the proof of Lemma 2.7.1

$$\langle u \rangle_{l',\sigma',\delta';\mathcal{D}} \le c \sum_{|\alpha|=l'+1} \sup_{x\in\mathcal{D}} r(x)^{\delta'-\sigma'+1} \left|\partial^\alpha u(x)\right| \le c \, \|u\|_{C_\delta^{l,\sigma}(\mathcal{D})}.$$

Thus, there exists a constant c such that

$$\|u\|_{C_{\delta'}^{l',\sigma'}(\mathcal{D})} \le c \, \|u\|_{C_\delta^{l,\sigma}(\mathcal{D})}$$

for all $u \in C_\delta^{l,\sigma}(\mathcal{D})$. This proves the lemma. □

6.7.4. Traces on the edge. Obviously, the trace of an arbitrary function $u \in C_\delta^{l,\sigma}(\mathcal{D})$ on the edge M lies in the space $C^{l-k,k-\delta+\sigma}(M)$ if $k-1 \le \delta - \sigma < k$, $k \in \{0,1,\dots,l\}$. The next lemma shows that every function $f \in C^{l-k,k-\delta+\sigma}(M)$ can be extended to a function $u \in C_\delta^{l,\sigma}(\mathcal{D})$.

LEMMA 6.7.9. *Let $f \in C^{l-k,k-\delta+\sigma}(M)$, where k, l are integers, $0 \le k \le l$, and σ, δ are real numbers, $0 < \sigma \le 1$, $k-1 \le \delta-\sigma < k$. Then there exists a function $u \in C_\delta^{l,\sigma}(\mathcal{D})$ satisfying the condition $\partial_{x_3}^j u|_M = \partial_{x_3}^j f$ for $j = 0,1,\dots,l-k$ and the estimates*

$$\|u\|_{C_\delta^{l,\sigma}(\mathcal{D})} \le c \, \|f\|_{C^{l-k,k-\delta+\sigma}(M)}, \tag{6.7.4}$$

$$\sup_{x\in\mathcal{D}} |x'|^{\delta-l-\sigma+|\alpha'|+j} \left|\partial_{x'}^{\alpha'} \partial_{x_3}^j u(x)\right| \le c_{j,\alpha'} \, \|f\|_{C^{l-k,k-\delta+\sigma}(M)} \quad \text{for } \alpha' \ne 0 \text{ or } j > l-k, \tag{6.7.5}$$

$$\sup_{x\in\mathcal{D}} |x'|^{\delta-l-\sigma+j} \left|\partial_{x_3}^j \big(u(x) - f(x_3)\big)\right| \le c \, \|f\|_{C^{l-k,k-\delta+\sigma}(M)} \quad \text{for } j \le l-k. \tag{6.7.6}$$

Here c and $c_{j,\alpha'}$ are constants independent of f.

P r o o f. Let us consider the function

$$u(x',x_3) = \int_{\mathbb{R}} f(x_3 + tr)\, \psi(t)\, dt, \tag{6.7.7}$$

where $r = |x'|$ and ψ is an infinitely differentiable function with support in the interval $[-1,1]$ satisfying the conditions

$$\int_{\mathbb{R}} \psi(t)\, dt = 1, \quad \int_{\mathbb{R}} t^j\, \psi(t)\, dt = 0 \quad \text{for } j = 1,\dots,l-k.$$

Since the derivatives of f up to order $l-k$ are continuous, it follows from the definition of u that $\partial_{x_3}^j u(0,x_3) = f^{(j)}(x_3)$ for $j = 0,1\dots,l-k$.

We prove that

$$\left|\partial_{x_3}^j \partial_r^\nu u(x)\right| \le c\, r^{l+\sigma-\delta-j-\nu} \, \|f\|_{C^{l-k,k-\delta+\sigma}(M)} \tag{6.7.8}$$

if $\nu \ge 1$ or $j > l-k$. First let $j+\nu \le l-k$, $\nu \ge 1$. We set $\psi_{j,\nu}(t) = t^\nu\, \psi(t)$ if $j+\nu = l-k$ and

$$\psi_{j,\nu}(t) = \int_{-1}^t \frac{(t-\tau)^{l-k-j-\nu-1}}{(l-k-j-\nu-1)!}\, \tau^\nu\, \psi(\tau)\, d\tau \quad \text{if } j+\nu < l-k.$$

Obviously,

$$\operatorname{supp}\psi_{j,\nu} \subset [-1,+1], \qquad \int_{\mathbb{R}} \psi_{j,\nu}(t)\,dt = 0 \quad \text{and} \quad \psi_{j,\nu}^{(l-k-j-\nu)}(t) = t^\nu\,\psi(t).$$

Hence, integration by parts yields

$$\begin{aligned}
\partial_{x_3}^j \partial_r^\nu u &= \int_{\mathbb{R}} t^\nu\, f^{(j+\nu)}(x_3+tr)\,\psi(t)\,dt \\
&= (-1)^{l-k-j-\nu} \int_{\mathbb{R}} r^{l-k-j-\nu} f^{(l-k)}(x_3+tr)\,\psi_{j,\nu}(t)\,dt \\
&= (-1)^{l-k-j-\nu} \int_{\mathbb{R}} r^{l-k-j-\nu} \big(f^{(l-k)}(x_3+tr) - f^{(l-k)}(x_3)\big)\,\psi_{j,\nu}(t)\,dt
\end{aligned}$$

which implies (6.7.8). Now let $j+\nu > l-k$. For $j = j'+j''$, $\nu = \nu'+\nu''$, and $j'+\nu' = l-k$, we obtain

$$\begin{aligned}
\partial_{x_3}^j \partial_r^\nu u &= \partial_{x_3}^{j''}\partial_r^{\nu''} \int_{\mathbb{R}} t^{\nu'}\, f^{(l-k)}(x_3+tr)\,\psi(t)\,dt \\
&= \int_{\mathbb{R}} f^{(l-k)}(\tau)\,\partial_{x_3}^{j''}\partial_r^{\nu''} \frac{1}{r}\Big(\frac{\tau-x_3}{r}\Big)^{\nu'} \psi\Big(\frac{\tau-x_3}{r}\Big)\,d\tau \\
&= \int_{\mathbb{R}} \big(f^{(l-k)}(\tau) - f^{(l-k)}(x_3)\big)\,\partial_{x_3}^{j''}\partial_r^{\nu''} \frac{1}{r}\Big(\frac{\tau-x_3}{r}\Big)^{\nu'} \psi\Big(\frac{\tau-x_3}{r}\Big)\,d\tau.
\end{aligned}$$

Consequently,

$$\begin{aligned}
\big|\partial_{x_3}^j\partial_r^\nu u\big| &\le c\,r^{-1-j''-\nu''} \int_{x_3-r}^{x_3+r} \big|f^{(l-k)}(\tau) - f^{(l-k)}(x_3)\big|\,d\tau \\
&\le c\,r^{-1-j''-\nu''} \sup_{|\tau-x_3|\le r} \frac{\big|f^{(l-k)}(\tau) - f^{(l-k)}(x_3)\big|}{|\tau-x_3|^{k+\sigma-\delta}} \int_{x_3-r}^{x_3+r} |\tau-x_3|^{k+\sigma-\delta}\,d\tau.
\end{aligned}$$

This implies (6.7.8) for $j+\nu > l-k$. Since

$$\big|\partial_x^\alpha u(x)\big| \le c \sum_{\nu=1}^{\alpha_1+\alpha_2} r^{\nu-\alpha_1-\alpha_2}\,\big|\partial_{x_3}^{\alpha_3}\partial_r^\nu u(x)\big| \tag{6.7.9}$$

for $\alpha = (\alpha_1,\alpha_2,\alpha_3)$, $\alpha_1+\alpha_2 > 0$, we obtain (6.7.5).

Next we prove the estimate (6.7.4). Obviously,

$$\big|\partial_{x_3}^j\partial_r^\nu u(x)\big| \le c\,\|f\|_{C^{l-k,k-\delta+\sigma}(M)}$$

for $j+\nu \le l-k$. If $\alpha = (\alpha_1,\alpha_2,\alpha_3)$ is a multi-index such that $|\alpha| \le l-k$ and $\alpha_1+\alpha_2 \ne 0$, then the last inequality together with (6.7.9) implies

$$\big|\partial_x^\alpha u(x)\big| \le c\,\big(1 + r^{1-\alpha_1-\alpha_2}\big)\,\|f\|_{C^{l-k,k-\delta+\sigma}(M)}$$

On the other hand by (6.7.5),

$$\big|\partial_x^\alpha u(x)\big| \le c\,r^{l+\sigma-\delta-|\alpha|}\,\|f\|_{C^{l-k,k-\delta+\sigma}(M)}$$

for $\alpha_1+\alpha_2 \ne 0$. Since the exponent $l+\sigma-\delta-|\alpha|$ is positive for $|\alpha| \le l-k$, we obtain

$$\big|\partial_x^\alpha u(x)\big| \le c\,\|f\|_{C^{l-k,k-\delta+\sigma}(M)} \quad \text{for } |\alpha| \le l-k. \tag{6.7.10}$$

Let α be a multi-index with length $|\alpha| = l$. Without loss of generality, we may assume that u is real-valued. Then by the mean value theorem, there exists a real number $t \in (0,1)$ such that

$$\partial_x^\alpha u(x) - \partial_y^\alpha u(y) = (\nabla \partial^\alpha u)\big(x + t(x-y)\big) \cdot (x-y).$$

If $|x-y| < |x'|/2$, then $|x'| < 2\,\big|x' + t(x'-y')\big|$ and therefore

$$|x'|^\delta \frac{\big|\partial_x^\alpha u(x) - \partial_y^\alpha u(y)\big|}{|x-y|^\sigma} \le c\,\big|x' + t(y'-x')\big|^{\delta-\sigma+1}\,\big|(\nabla\partial^\alpha u)\big(x+t(y-x)\big)\big|.$$

According to (6.7.5), the right-hand side of the last inequality can be estimated by the norm of f in $C^{l-k,k-\delta+\sigma}(M)$ for $|\alpha| = l$. Hence,

$$\langle u\rangle_{l,\sigma,\delta;\mathcal{D}} \le c\,\|f\|_{C^{l-k,k-\delta+\sigma}(M)}\,. \tag{6.7.11}$$

Together with (6.7.5) this implies that $\partial_x^\alpha u \in N_\delta^{k,\sigma}(\mathcal{D}) \subset N^{0,k-\delta+\sigma}(\mathcal{D})$ for $|\alpha| = l-k$, $\alpha_1+\alpha_2 \ne 0$. From this and from (6.7.10) we conclude that $\partial_x^\alpha u \in C^{0,k-\delta+\sigma}(\mathcal{D})$ and

$$\|\partial_x^\alpha u\|_{C^{0,k-\delta+\sigma}(\mathcal{D})} \le c\,\|f\|_{C^{l-k,k-\delta+\sigma}(M)} \quad \text{for } |\alpha| = l-k,\ \ \alpha_1+\alpha_2 \ne 0.$$

Furthermore, it follows from the definition of u that

$$\frac{\big|\partial_{x_3}^{l-k} u(x) - \partial_{y_3}^{l-k} u(y)\big|}{|x-y|^{k-\delta+\sigma}} \le c\,\|f^{(l-k)}\|_{C^{0,k-\delta+\sigma}(M)} \quad \text{for } x,y \in \mathcal{D}.$$

Consequently, $u \in C^{l-k,k-\delta+\sigma}(\mathcal{D})$ and

$$\|u\|_{C^{l-k,k-\delta+\sigma}(\mathcal{D})} \le c\,\|f\|_{C^{l-k,k-\delta+\sigma}(M)}\,.$$

Thus, (6.7.4) holds.

It remains to prove the estimate (6.7.6). Since $\partial_{x_3}^j u(0,x_3) = f^{(j)}(x_3)$ and $\delta - \sigma < k$, we have (see Lemma 6.7.4)

$$r^{\delta-l-\sigma+j}\,\big|\partial_{x_3}^j\big(u(x) - f(x_3)\big)\big| \le \sup_{x\in\mathcal{D}} r^{\delta-l-\sigma+j+1}\,\big|\partial_{x_3}^j \partial_r u\big| \le c\,\|f\|_{C^{l-k,k+\sigma-\delta}(M)}$$

for $j = 0,1,\dots,l-k$. The proof is complete. □

6.7.5. A relation between the spaces $C_\delta^{l,\sigma}(\mathcal{D})$ and $N_\delta^{l,\sigma}(\mathcal{D})$. The proof of the next lemma is a word-by-word repetition of the proof of Lemma 6.7.5.

LEMMA 6.7.10. *Let $u \in C_\delta^{l,\sigma}(\mathcal{D})$, where $k-1 \le \delta - \sigma < k$, $k \in \{0,1,\dots,l\}$. Then $u \in N_\delta^{l,\sigma}(\mathcal{D})$ if and only if $\partial^\alpha u = 0$ on M for $|\alpha| \le l-k$. Moreover,*

$$\|u\|_{N_\delta^{l,\sigma}(\mathcal{D})} \le c\,\|u\|_{C_\delta^{l,\sigma}(\mathcal{D})}$$

for all $u \in C_\delta^{l,\sigma}(\mathcal{D})$, $\partial^\alpha u = 0$ on M for $|\alpha| \le l-k$.

In the following, let E be the same extension operator as in Section 6.2, i.e.

$$(Ef)(x',x_3) = \chi(|x'|)\,u(x',x_3),$$

where u is the function (6.7.7) and χ is an infinitely differentiable function on the interval $(0,+\infty)$, $\chi(r) = 1$ for $r < 1/2$, $\chi(r) = 0$ for $r > 1$.

As an analog to Theorem 6.7.6, we obtain the following result.

THEOREM 6.7.11. *Let $u \in C_\delta^{l,\sigma}(\mathcal{D})$, where $\delta \ge 0$, $0 < \sigma \le 1$, $k-1 \le \delta - \sigma < k$, k is an integer, $0 \le k \le l$. Furthermore, let*

$$v(x) = \sum_{i+j\le l-k} \frac{1}{i!\,j!}\,(Ef_{i,j})(x)\,x_1^i x_2^j$$

where $f_{i,j} = \partial_{x_1}^i \partial_{x_2}^j u|_M$ for $i+j \le l-k$. Then $v \in C_{\delta+\nu}^{l+\nu,\sigma}(\mathcal{D})$ for every integer $\nu \ge 0$ and $u - v \in N_\delta^{l,\sigma}(\mathcal{D})$. Furthermore, the inequality

$$\|v\|_{C_{\delta+\nu}^{l+\nu,\sigma}(\mathcal{D})} + \|u-v\|_{N_\delta^{l,\sigma}(\mathcal{D})} \le c\,\|u\|_{C_\delta^{l,\sigma}(\mathcal{D})}$$

holds with a constant c independent of u.

P r o o f. We show first that $v \in C_{\delta+\nu}^{l+\nu,\sigma}(\mathcal{D})$. Since $f_{i,j} \in C^{l-k-i-j,k+\sigma-\delta}(M)$, it follows from Lemma 6.7.9 that

$$|x'|^{\max(0,\delta-l-\sigma+|\alpha|)}\,\big|\partial_x^\alpha v(x)\big| \le c \sum_{i+j\le l-k} \|f_{i,j}\|_{C^{l-k-i-j,k+\sigma-\delta}(M)} \le c\,\|u\|_{C_\delta^{l,\sigma}(\mathcal{D})}$$

for all α. Furthermore,

$$\langle v\rangle_{l+\nu,\sigma,\delta+\nu;\mathcal{D}} \le c \sum_{|\alpha|=l+\nu+1} \sup_{x\in\mathcal{D}} |x'|^{\delta-\sigma+j+1}\,\big|\partial_x^\alpha v(x)\big| \le c\,\|u\|_{C_\delta^{l,\sigma}(\mathcal{D})}\,.$$

In order to show that $v \in C^{l-k,k-\delta+\sigma}(\mathcal{D})$, we consider the function $\partial_x^\alpha\,(Ef_{i,j}\,x_1^i x_2^j)$ for $|\alpha| = l-k$. Obviously, this is a linear combination of functions

$$(6.7.12) \qquad (\partial_x^\gamma Ef_{i,j})\,x_1^{i-\mu}\,x_2^{j-\nu}, \quad \text{where } \mu \le i,\ \nu \le j,\ |\gamma| = l-k-\mu-\nu.$$

Since $Ef_{i,j} \in C^{l-k-i-j,k+\sigma-\delta}(\mathcal{D})$, the term (6.7.12) with $\mu = i$, $\nu = j$ belongs to $C^{0,k-\delta+\sigma}(\mathcal{D})$. If $\mu+\nu < i+j$, then it follows from Lemma 6.7.12 that $\partial_x^\gamma Ef_{i,j} \in N_{\delta+i+j+|\gamma|}^{l,\sigma}(\mathcal{D})$. Consequently, $(\partial_x^\gamma Ef_{i,j})\,x_1^{i-\mu}x_2^{j-\nu} \in N_{\delta-k+l}^{l,\sigma}(\mathcal{D}) \subset N_0^{0,k-\delta+\sigma}(\mathcal{D})$ if $\mu+\nu < i+j$. This implies that $\partial_x^\alpha v \in C^{0,k-\delta+\sigma}(\mathcal{D})$ for $|\alpha| = l-k$.

Thus, it is shown that $v \in C_{\delta+\nu}^{l+\nu,\sigma}(\mathcal{D})$. Furthermore, the norm of v in $C_{\delta+\nu}^{l+\nu,\sigma}(\mathcal{D})$ can be estimated by the norm of u in $C_\delta^{l,\sigma}(\mathcal{D})$.

By means of Lemma 6.7.12, it can be easily shown that $\partial_{x_1}^i \partial_{x_2}^j v = f_{i,j}$ on M for $i+j \le l-k$. From this we conclude that $\partial^\alpha(u-v) = 0$ on M for $|\alpha| \le l-k$. Applying Lemma 6.7.10, we obtain $u - v \in N_\delta^{l,\sigma}(\mathcal{D})$ and an estimate for the $N_\delta^{l,\sigma}(\mathcal{D})$-norm of $u - v$. This proves the theorem. □

6.7.6. An analogous result for the C^l-spaces. Let $C^l(M)$ be the space of all l times continuously differentiable functions on M with finite norm

$$\|u\|_{C^l(M)} = \sum_{j=0}^{l} \sup_{x_3\in M} \big|\partial_{x_3}^j u(x_3)\big|.$$

Analogously to Lemma 6.7.9, the following lemma holds.

LEMMA 6.7.12. *Let $f \in C^l(M)$, where l is a nonnegative integer. Then there exists a function $u \in C^l(\overline{\mathcal{D}})$ such that $u|_M = f$,*

$$\|u\|_{C^l(\overline{\mathcal{D}})} \le c\,\|f\|_{C^l(M)}\,, \tag{6.7.13}$$

$$\sup_{x\in\mathcal{D}} |x'|^{|\alpha'|+j-l}\,\big|\partial_{x'}^{\alpha'}\partial_{x_3}^j u(x',x_3)\big| \le c_\alpha\,\|f\|_{C^l(M)} \text{ for } \alpha' \ne 0 \text{ or } j \ge l, \tag{6.7.14}$$

$$\sup_{x\in\mathcal{D}} |x'|^{j-l}\,\big|\partial_{x_3}^j\big(u(x)-f(x_3)\big)\big| \le c\,\|f\|_{C^l(M)} \text{ for } j=0,1,\dots,l-1. \tag{6.7.15}$$

Here c and c_α are constants independent of f.

Proof. Let u be the same function as in the proof of Lemma 6.7.9. Here, we assume that the condition on ψ is satisfied for $j=0,1,\dots,l$. Then

$$\big|\partial_{x_3}^j\partial_r^\nu u(x)\big| = \Big|\int_{\mathbb{R}} t^\nu\, f^{(j+\nu)}(x_r+tr)\,\psi(t)\,dt\Big| \le \|f\|_{C^l(M)}$$

for $j+\nu \le l$ which implies (6.7.13). Moreover, as in the proof of Lemma 6.7.9, we obtain

$$\partial_{x_3}^j\partial_r^\nu u = (-r)^{l-j-\nu}\int_{\mathbb{R}} f^{(l)}(x_3+tr)\,\psi_{j,\nu}(t)\,dt$$

for $j+\nu \le l$, $\nu \ge 1$ and

$$\begin{aligned}\partial_{x_3}^j\partial_r^\nu u &= \partial_{x_3}^{j''}\partial_r^{\nu''}\int_{\mathbb{R}} f^{(l)}(x_3+tr)\,t^{\nu'}\psi(t)\,dt\\ &= \int_{\mathbb{R}} f^{(l)}(\tau)\,\partial_{x_3}^{j''}\partial_r^{\nu''}\frac{1}{r}\Big(\frac{\tau-x_3}{r}\Big)^{\nu'}\psi\Big(\frac{\tau-x_3}{r}\Big)\,d\tau\end{aligned}$$

for $j=j'+j''$, $\nu=\nu'+\nu''$, $j'+\nu'=l$. From these two equalities, one can easily deduce (6.7.14). Then the inequality (6.7.15) holds with the help of Lemma 6.7.4. This proves the lemma. □

THEOREM 6.7.13. *Let $u \in C^l(\overline{\mathcal{D}})$ and $f_{i,j} = \partial_{x_1}^i\partial_{x_2}^j u|_M$ for $i+j \le l-1$. Then the function*

$$v(x) = \sum_{i+j\le l-1} \frac{1}{i!\,j!}\,(Ef_{i,j})(x)\,x_1^i x_2^j, \tag{6.7.16}$$

where E is the same extension operator as in Theorem 6.7.11, has continuous and bounded derivatives up to order l and satisfies the inequalities

$$\sup_{x\in\mathcal{D}} r^{\max(0,|\alpha|-l)}\,\big|\partial_x^\alpha v(x)\big| \le c_\alpha\,\|u\|_{C^l(\overline{\mathcal{D}})} \quad \text{for all } \alpha, \tag{6.7.17}$$

$$\sup_{x\in\mathcal{D}} r^{|\alpha|-l}\,\big|\partial_x^\alpha(v(x)-u(x))\big| \le c\,\|u\|_{C^l(\overline{\mathcal{D}})} \quad \text{for } |\alpha| \le l, \tag{6.7.18}$$

where c and c_α are independent of u.

Proof. Using Lemma 6.7.12, we obtain

$$\sup_{x\in\mathcal{D}} r^{\max(0,|\alpha|-l)}\,\big|\partial_x^\alpha v(x)\big| \le c\sum_{i+j\le l}\|f_{i,j}\|_{C^{l-i-j}(M)} \tag{6.7.19}$$

which implies (6.7.17). We prove the inequality (6.7.18) by induction. Suppose this inequality is valid for $|\alpha| \ge k$, where k is an integer $0 < k \le l$. Let now α be a multi-index with length $|\alpha| = k-1$. Since $\partial_x^\alpha(u-v) = 0$ on M, it follows from Lemma 6.7.4 that

$$\begin{aligned}\sup_{x\in\mathcal{D}} r^{|\alpha|-l}\,\big|\partial_x^\alpha(v(x)-u(x))\big| &\le \sum_{j=1}^{2}\sup_{x\in\mathcal{D}} r^{|\alpha|+1-l}\,\big|\partial_x^\alpha\partial_{x_j}(v(x)-u(x))\big| \\ &\le c\,\|u\|_{C^l(\overline{\mathcal{D}})}\,.\end{aligned}$$

Thus, (6.7.17) is valid for all α. □

We use the last result for the proof of the following extension theorem.

COROLLARY 6.7.14. *Let $g_k^\pm \in C^{l-k}(\overline{\Gamma^\pm})$, $k=0,\dots,l$, be given functions on Γ^+ and Γ^-. Suppose that there exists a function $u\in C^l(\overline{\mathcal{D}})$ such that $\partial_n^k u = g_k^\pm$ on $\Gamma^\pm$ for $k=0,\dots,l$. Then there exists a function $v\in C^l(\overline{\mathcal{D}})$ which is infinitely differentiable in $\mathcal{D}$ and satisfies the inequalities*

$$\sup_{x\in\mathcal{D}} r^{\max(0,|\alpha|-l)}\,\big|\partial_x^\alpha v(x)\big| \le c_\alpha \sum_\pm\sum_{k=0}^{l} \|g_k^\pm\|_{C^{l-k}(\overline{\Gamma^\pm})} \quad \textit{for all } \alpha, \tag{6.7.20}$$

$$\begin{aligned}&\sup_{x\in\Gamma^\pm} r^{\mu+\nu-l+k}\,\big|\partial_r^\mu\partial_{x_3}^\nu(\partial_n^k v-g_k^\pm)(x)\big| \\ &\qquad\le c\sum_\pm\sum_{j=0}^{l}\|g_j^\pm\|_{C^{l-j}(\overline{\Gamma^\pm})} \quad \textit{for } \mu+\nu\le l-k\end{aligned} \tag{6.7.21}$$

with constants c, c_α independent of $g_k^\pm$.

P r o o f. Let $f_{i,j} = \partial_{x_1}^i\partial_{x_2}^j u|_M$ for $i+j\le l-1$, and let v be the function (6.7.16). Obviously,

$$\|f_{i,j}\|_{C^{l-i-j}(M)} \le \sum_\pm\sum_{k=0}^{l}\|g_k^\pm\|_{C^{l-k}(\overline{\Gamma^\pm})}$$

for $i+j\le l-1$. Thus, the inequality (6.7.20) can be deduced directly from (6.7.19). In particular,

$$\sup_{x\in\Gamma^\pm}\big|\partial_r^\mu\partial_{x_3}^\nu(\partial_n^k v(x)-g_k^\pm(x))\big| \le c\sum_\pm\sum_{j=0}^{l}\|g_j^\pm\|_{C^{l-j}(\overline{\Gamma^\pm})}$$

for $\mu+\nu = l-k$. Since $\partial_r^\mu\partial_{x_3}^\nu(\partial_n^k v - g_k^\pm) = 0$ on M for $\mu+\nu\le l-k-1$, it follows from Lemma 6.7.4 that

$$\sup_{x\in\Gamma^\pm} r^{\mu+\nu-l+k}\,\big|\partial_r^\mu\partial_{x_3}^\nu(\partial_n^k v-g_k^\pm)(x)\big| \le c\sup_{x\in\Gamma^\pm} r^{\mu+\nu-l+k+1}\big|\partial_r^{\mu+1}\partial_{x_3}^\nu(\partial_n^k v-g_k^\pm)(x)\big|$$

for $\mu+\nu\le l-k-1$. Consequently by induction in $\mu+\nu$, the estimate (6.7.21) holds. □

6.8. Some estimates of the solutions in weighted Hölder spaces

In the last section of this chapter, we prove some auxiliary lemmas which are needed in the next chapter, when we turn to the boundary value problem with Dirichlet and Neumann boundary conditions in a polyhedral cone. First we describe the conditions on the data g^+ and g^- which ensure the existence of a function $u\in C_\delta^{l,\sigma}(\mathcal{D})^\ell$ satisfying the boundary conditions $B^\pm(D_x)\,u = g^\pm$ on $\Gamma^\pm$. Then we derive some local regularity results for the solutions in the classes of the spaces $N_\delta^{l,\sigma}$ and $C_\delta^{l,\sigma}$.

6.8.1. Reduction to zero boundary data. In Section 2.8 (cf. Lemma 2.8.2) it was shown that for arbitrary given vector functions $g^\pm \in N_\delta^{l-d^\pm,\sigma}(\Gamma^\pm)^\ell$ there exists a vector function $u \in N_\delta^{l,\sigma}(\mathcal{D})^\ell$ satisfying the boundary conditions

$$B^\pm(D_x)\,u = (1-d^\pm)\,u + d^\pm\,N^\pm(D_x)\,u = g^\pm \quad \text{on } \Gamma^\pm. \tag{6.8.1}$$

As the subsequent four lemmas show, an analogous result is also true under certain additional conditions in the class of the weighted Hölder spaces $C_\delta^{l,\sigma}$.

LEMMA 6.8.1. *Let $l \ge 1$, $\delta > l+\sigma-1$, and let $g^\pm \in C_\delta^{l-d^\pm,\sigma}(\Gamma^\pm)^\ell$ be given vector functions such that $g^\pm(x) = 0$ for $|x'| > 1$. In the case $d^+ = d^- = 0$, $\delta < l+\sigma$, we assume that $g^+ = g^-$ on M. Then there exists a vector function $u \in C_\delta^{l,\sigma}(\mathcal{D})^\ell$ satisfying the boundary conditions* (6.8.1) *and the estimate*

$$\|u\|_{C_\delta^{l,\sigma}(\mathcal{D})^\ell} \le c \sum_\pm \|g^\pm\|_{C_\delta^{l-d^\pm,\sigma}(\Gamma^\pm)^\ell}, \tag{6.8.2}$$

where the constant c is independent of g^+ and g^-.

Proof. If $\delta \ge l+\sigma$ or $d^+ = d^- = 1$ and $\delta > l+\sigma-1$, then $C_\delta^{l-d^\pm,\sigma}(\Gamma^\pm) = N_\delta^{l-d^\pm,\sigma}(\Gamma^\pm)$. In these cases, the assertion of the lemma follows immediately from Lemma 2.8.2.

We consider the case where $d^+ + d^- \le 1$ and $l+\sigma-1 < \delta < l+\sigma$. Without loss of generality, we may assume that $d^+ = 0$. Then there exists a vector function $v \in C_\delta^{l,\sigma}(\mathcal{D})^\ell$ such that $v(x) = 0$ for $|x'| > 1$ and $B^+v = v = g^+$ on Γ^+. Moreover, v satisfies the estimate (6.8.2). If $d^- = 0$, then $g^- - v|_{\Gamma^-} \in C_\delta^{l,\sigma}(\Gamma^-)^\ell$. Since

$$\big(g^- - v|_{\Gamma^-}\big)\big|_M = g^-|_M - g^+|_M = 0,$$

it follows that $g^- - v|_{\Gamma^-} \in N_\delta^{l,\sigma}(\Gamma^-)^\ell$ (see Lemma 6.7.10). If $d^- = 1$, then

$$g^- - B^-v|_{\Gamma^-} \in C_\delta^{l-1,\sigma}(\Gamma^-)^\ell = N_\delta^{l-1,\sigma}(\Gamma^-)^\ell.$$

Thus both in the cases $d^- = 0$ and $d^- = 1$, there exists a vector function $w \in N_\delta^{l,\sigma}(\mathcal{D})^\ell$ such that

$$w = 0 \text{ on } \Gamma^+, \qquad B^-(D_x)\,w = g^- - B^-(D_x)\,v \ \text{ on } \Gamma^-,$$

and $w(x) = 0$ for $|x'| > 1$ (see Lemma 2.8.2). The sum $u = v + w$ has the desired properties. □

In the case $l+\sigma-2 < \delta < l+\sigma-1$, we obtain the following analog to Lemma 6.4.4.

LEMMA 6.8.2. *Let $g^\pm \in C_\delta^{l-d^\pm,\sigma}(\Gamma^\pm)^\ell$ be given vector functions, $l \ge 1$, $\delta \ge 0$, $l+\sigma-2 < \delta < l+\sigma-1$. We suppose that $g^\pm(x) = 0$ for $|x'| > 1$ and that $\lambda = 1$ is not an eigenvalue of the pencil $A(\lambda)$. In the case $d^+ = d^- = 0$, we assume in addition that $g^+|_M = g^-|_M$. Then there exists a vector function $u \in C_\delta^{l,\sigma}(\mathcal{D})^\ell$ satisfying the boundary conditions* (6.8.1) *and the estimate* (6.8.2).

P r o o f. Let $g_\pm^{(j)}$ be the trace of $\partial_r^j g^\pm$ on M for $j \le 1-d^\pm$. Obviously, $g_\pm^{(j)} \in C^{1-d^\pm-j,l+\sigma-1-\delta}(M)^\ell$. We assume that $d^- \le d^+$ and set

$$v^{(0)} = g^-|_M \ \text{ if } d^- = 0, \quad v^{(0)} = 0 \text{ else.}$$

Since $\lambda = 1$ is not an eigenvalue of the pencil $A(\lambda)$, there exist uniquely determined functions $v^{(1)}, v^{(2)} \in C^{0,l+\sigma-1-\delta}(M)^\ell$ satisfying the equations

$$B^\pm(D_{x'},0)\,\big(v^{(1)}(x_3)\,x_1 + v^{(2)}(x_3)\,x_2\big)$$
$$= \Big(g_\pm^{(1-d^\pm)}(x_3) + \big(B^\pm(D_{x'},0) - B^\pm(D_x)\big)\,v^{(0)}(x_3)\Big)\,r^{1-d^\pm}$$

for $x' \in \gamma^\pm$ and the estimate

$$\|v^{(0)}\|_{C^{1,l+\sigma-1-\delta}(M)^\ell} + \sum_{j=1}^{2} \|v^{(j)}\|_{C^{0,l+\sigma-1-\delta}(M)^\ell} \le c \sum_\pm \|g^\pm\|_{C_\delta^{l-d^\pm,\sigma}(\Gamma^+)^\ell}$$

(see Lemma 6.3.6). We introduce the vector function

$$v = Ev^{(0)} + x_1\,Ev^{(1)} + x_2\,Ev^{(2)}\,,$$

where E denotes the extension operator (6.2.4). If $d^\pm = 0$, then

$$v|_{\Gamma^\pm} = Ev^{(0)} + r\,Eg_\pm^{(1)} = Eg_\pm^{(0)} + r\,Eg_\pm^{(1)}.$$

Thus by Theorem 6.7.11, $v|_{\Gamma^\pm} - g^\pm \in N_\delta^{l,\sigma}(\Gamma^\pm)^\ell$. From the properties of the extension operator E (cf. Lemma 6.7.9) it follows that the trace of $N^\pm(D_{x'},0)\,Ev^{(0)}$ on M is zero. Consequently,

$$B^\pm(D_x)v|_M = g^\pm|_M$$

if $d^\pm = 1$. This together with Lemma 6.7.10 implies that $B^\pm(D_x)v|_{\Gamma^\pm} - g^\pm \in N_\delta^{l-1,\sigma}(\Gamma^\pm)^\ell$. By Lemma 2.8.2, there exists a vector function $w \in N_\delta^{l,\sigma}(\mathcal{D})^\ell$ vanishing for $|x'| > 1$ such that $B^\pm(D_x)\,w|_{\Gamma^\pm} = g^\pm - B^\pm(D_x)v|_{\Gamma^\pm}$. Then $u = v + w$ satisfies the boundary conditions (6.8.1). The proof is complete. □

The assertion of the last lemma is not true in general if $\lambda = 1$ is an eigenvalue of the pencil $A(\lambda)$. This is the case if one considers the Neumann problem for some special systems (see Subsection 6.5.4). However under the assumptions of Lemma 6.5.7, an additional compatibility condition ensures the existence of a vector function $u \in C_\delta^{l,\sigma}(\mathcal{D})^\ell$ satisfying the boundary conditions

(6.8.3) $$N^\pm(D_x)\,u = g^\pm \quad \text{on } \Gamma^\pm.$$

LEMMA 6.8.3. *Let $g^\pm \in C_\delta^{l-1,\sigma}(\Gamma^\pm)^\ell$ be given vector functions on $\Gamma^\pm$, $l \ge 1$, $\delta \ge 0$, $l+\sigma-2 < \delta < l+\sigma-1$. We suppose that the matrices* (6.5.12) *and* (6.5.14) *have the same rank $r' < 2\ell$, that $g^\pm(x) = 0$ for $|x'| > 1$ and that g^+ and g^- satisfy the compatibility condition* (6.5.15). *Then there exists a vector function $u \in C_\delta^{l,\sigma}(\mathcal{D})^\ell$ satisfying the boundary conditions* (6.8.3) *and the estimate* (6.8.2).

P r o o f. By (6.5.15), there exist vector functions $v^{(1)}, v^{(2)} \in C^{0,l+\sigma-1-\delta}(M)^\ell$ such that

$$(A_{1,1}n_1^\mp + A_{1,2}n_2^\mp)\,v^{(1)} + (A_{2,1}n_1^\mp + A_{2,2}n_2^\mp)\,v^{(2)} = g^\pm|_M\,.$$

We put $v(x) = (Ev^{(1)})(x)\, x_1 + (Ev^{(2)})(x)\, x_2$, where E is the extension operator (6.2.4). It follows from Lemma 6.7.9 that $v \in C_\delta^{l,\sigma}(\mathcal{D})^\ell$. Moreover, the traces of $N^\pm(D_x)v$ and $g^\pm$ on M coincide. Consequently,

$$g^\pm - N^\pm(D_x)v|_{\Gamma^\pm} \in N_\delta^{l-1,\sigma}(\Gamma^\pm)^\ell$$

(see Lemma 6.7.10). Applying Lemma 2.8.2, we obtain the assertion of the lemma. □

Finally, we prove a similar result for the case $l+\sigma-3 < \delta < l+\sigma-2$.

LEMMA 6.8.4. *Let $f \in C_\delta^{l-2,\sigma}(\mathcal{D})^\ell$ and $g^\pm \in C_\delta^{l-d^\pm,\sigma}(\Gamma^\pm)^\ell$ be given vector functions, $l \ge 2$, $\delta \ge 0$, $l+\sigma-3 < \delta < l+\sigma-2$. We suppose that $g^\pm(x) = 0$ for $|x'| > 1$ and that $\lambda = 1$ and $\lambda = 2$ are not eigenvalues of the pencil $A(\lambda)$. In the case $d^+ = d^- = 0$, we assume in addition that $g^+|_M = g^-|_M$. Then there exists a vector function $u \in C_\delta^{l,\sigma}(\mathcal{D})^\ell$ such that*

$$L(D_x)u - f \in N_\delta^{l-2,\sigma}(\mathcal{D})^\ell, \quad B^\pm(D_x)\, u = g^\pm \quad \text{on } \Gamma^\pm \tag{6.8.4}$$

and

$$\|u\|_{C_\delta^{l,\sigma}(\mathcal{D})^\ell} \le c\Big(\|f\|_{C_\delta^{l-2,\sigma}(\mathcal{D})^\ell} + \sum_\pm \|g^\pm\|_{C_\delta^{l-d^\pm,\sigma}(\Gamma^\pm)^\ell}\Big) \tag{6.8.5}$$

with a constant c independent of f and $g^\pm$.

P r o o f. Let $v^{(0)}, v^{(1)}, v^{(2)}$ and v be the same vector functions as in the proof of Lemma 6.8.2. If $l+\sigma-3 < \delta < l+\sigma-2$, then

$$v^{(0)} \in C^{2,l+\sigma-2-\delta}(M)^\ell, \quad v^{(1)}, v^{(2)} \in C^{1,l+\sigma-2-\delta}(M)^\ell$$

and $v \in C_\delta^{l,\sigma}(\mathcal{D})^\ell$. Furthermore,

$$\partial_r^j\big(B^\pm(D_x)v|_{\Gamma^\pm} - g^\pm\big) = 0 \quad \text{on } M \text{ for } j \le 1 - d^\pm.$$

Since $\lambda = 2$ is not an eigenvalue of the pencil $A(\lambda)$, there exists a vector function

$$W(x) = \sum_{i+j=2} w^{(i,j)}(x_3)\, \frac{x_1^i\, x_2^j}{i!\, j!}$$

with coefficients $w^{(i,j)} \in C^{0,l+\sigma-2-\delta}(M)^\ell$ such that

$$L(D_{x'}, 0)\, W = \big(f - L(D_x)v\big)\big|_M$$

and

$$\partial_r^{2-d^\pm} B^\pm(D_{x'}, 0)\, W = \partial_r^{2-d^\pm}\, \big(g^\pm - B^\pm(D_x)\, v\big) \quad \text{on } M$$

(see Lemma 6.3.6). We set

$$w(x) = \sum_{i+j=2} (Ew^{(i,j)})(x)\, \frac{x_1^i\, x_2^j}{i!\, j!}\,,$$

where E denotes the extension operator (6.2.4). Then

$$L(D_x)\, w = f - L(D_x)\, v \quad \text{on } M$$

and

$$\partial_r^{2-d^\pm}\, B^\pm(D_x)\, w = \partial_r^{2-d^\pm}\, \big(g^\pm - B^\pm(D_x)\, v\big) \quad \text{on } M$$

Furthermore, $\partial_r^j B^\pm(D_x)\, w = 0$ on M for $j \le 1 - d^\pm$. Consequently,

$$L(D_x)\,(v+w) - f \in N_\delta^{l-2,\sigma}(\mathcal{D})^\ell \quad\text{and}\quad B^\pm(D_x)\,(v+w) - g^\pm \in N_\delta^{l-d^\pm,\sigma}(\Gamma^\pm)^\ell$$

(cf. Lemma 6.7.10). It remains to apply Lemma 2.8.2. □

6.8.2. Local regularity results in the spaces $N_\delta^{l,\sigma}$. Our goal is to prove regularity assertions analogous to Lemmas 6.4.1 and 6.5.3 for the solution of the problem (6.1.3) in the class of the weighted Hölder spaces $C_\delta^{l,\sigma}$. First we prove such results for the weighted Hölder spaces $N_\delta^{l,\sigma}$.

LEMMA 6.8.5. *Let $u \in L^\infty_{\delta-l-\sigma}(\mathcal{D})^\ell$ be a solution of the boundary value problem (6.1.3) such that $\chi u \in W^{2,p}(\mathcal{D})^\ell$ for every $\chi \in C_0^\infty(\overline{\mathcal{D}}\backslash M)$. If $f \in N_\delta^{l-2,\sigma}(\mathcal{D})^\ell$, $l \ge 1$, $0<\sigma<1$, and $g^\pm \in N_\delta^{l-d^\pm,\sigma}(\Gamma^\pm)^\ell$, then $u \in N_\delta^{l,\sigma}(\mathcal{D})^\ell$ and*

$$\|u\|_{N_\delta^{l,\sigma}(\mathcal{D})^\ell} \le c\Big(\|f\|_{N_\delta^{l-2,\sigma}(\mathcal{D})^\ell} + \sum_\pm \|g^\pm\|_{N_\delta^{l-d^\pm,\sigma}(\Gamma^\pm)^\ell} + \|u\|_{L^\infty_{\delta-l-\sigma}(\mathcal{D})^\ell}\Big).$$

P r o o f. For the Dirichlet problem, we refer to Theorem 2.8.4. The proof of this theorem is based on the local estimate given in Lemma 2.7.1. Since such an estimate is valid for general boundary value problem, the assertion of Lemma 6.8.5 is also true for the problem (6.1.3). □

We furthermore need a local estimate for the solutions in the norm of the weighted Hölder space $N_\delta^{l,\sigma}$.

LEMMA 6.8.6. *Let ζ, η be smooth functions with compact supports, $\eta = 1$ in a neighborhood of* $\operatorname{supp}\zeta$. *Suppose that u is a solution of the boundary value problem (6.1.3) such that $\eta u \in L^\infty_{\delta-l-\sigma}(\mathcal{D})^\ell$ and $\chi u \in W^{2,p}(\mathcal{D})^\ell$ for every $\chi \in C_0^\infty(\overline{\mathcal{D}}\backslash M)$. If $\eta f \in N_\delta^{l-2,\sigma}(\mathcal{D})^\ell$ and $\eta g^\pm \in N_\delta^{l-d^\pm,\sigma}(\Gamma^\pm)^\ell$, $l \ge 2$, $0<\sigma<1$, then $\zeta u \in N_\delta^{l,\sigma}(\mathcal{D})^\ell$ and*

$$\|\zeta u\|_{N_\delta^{l,\sigma}(\mathcal{D})^\ell} \le c\Big(\|\eta f\|_{N_\delta^{l-2,\sigma}(\mathcal{D})^\ell} + \sum_\pm \|\eta g^\pm\|_{N_\delta^{l-d^\pm,\sigma}(\Gamma^\pm)^\ell} + \|\eta u\|_{L^\infty_{\delta-l-\sigma}(\mathcal{D})^\ell}\Big). \tag{6.8.6}$$

P r o o f. By Lemma 2.8.2, we may restrict ourselves to the case of zero boundary data $g^\pm$. Let $\mathcal{U}$ be a neighborhood of $\operatorname{supp}\zeta$ such that $\eta = 1$ in a neighborhood of $\mathcal{U}$. Obviously, we obtain an equivalent norm in $N_\delta^{l,\sigma}(\mathcal{D})$ if we replace the expression $\langle u\rangle_{l,\sigma,\delta;\mathcal{D}}$ in the norm (2.7.1) by

$$\langle u\rangle'_{l,\sigma,\delta;\mathcal{D}} = \sum_{|\alpha|=l} \sup_{\substack{x,y\in\mathcal{D}\\ |x-y|<\varepsilon|x'|}} |x'|^\delta \frac{|\partial_x^\alpha u(x) - \partial_y^\alpha u(y)|}{|x-y|^\sigma},$$

where ε is an arbitrarily small positive number, $\varepsilon < 1$. Let the $N_\delta^{l,\sigma}(\mathcal{U}\cap\mathcal{D})$-norm be defined for functions with support in $\overline{\mathcal{U}}\cap\overline{\mathcal{D}}$ as the expression (2.7.1), where the supremum is taken over all $x, y \in \mathcal{U}\cap\mathcal{D}$. Then obviously,

$$\|\zeta u\|_{N_\delta^{l,\sigma}(\mathcal{D})^3} \le c\,\|u\|_{N_\delta^{l,\sigma}(\mathcal{U}\cap\mathcal{D})^3}. \tag{6.8.7}$$

Furthermore, as in the proof of Theorem 2.8.4, we obtain the estimates

$$r^{\delta-l-\sigma+|\alpha|}\,|\partial_x^\alpha u(x)| \le c\Big(\|f\|_{N_\delta^{l-2,\sigma}(\mathcal{D}\cap\mathcal{B}_x)^\ell} + \|u\|_{L^\infty_{\delta-l-\sigma}(\mathcal{D}\cap\mathcal{B}_x)^\ell}\Big)$$

for $|x'| = r$, $|\alpha| \le l$, and

$$r^\delta \, \frac{|\partial_x^\alpha u(x) - \partial_y^\alpha u(y)|}{|x-y|^\sigma} \le c\Big(\|f\|_{N_\delta^{l-2,\sigma}(\mathcal{D}\cap\mathcal{B}_x)^\ell} + \|u\|_{L^\infty_{\delta-l-\sigma}(\mathcal{D}\cap\mathcal{B}_x)^\ell}\Big)$$

for $|x-y| < \varepsilon|x'| = \varepsilon r$, $|\alpha| = l$, where $\mathcal{B}_x$ is a ball with radius $2\varepsilon|x'|$ centered at the point x. Here the constant c depends on ε but not on x. The last two estimates together with (6.8.7) imply (6.8.6). □

Next we prove a local regularity assertion for the solutions of the boundary value problem in the class of the spaces $N_\delta^{l,\sigma}$.

LEMMA 6.8.7. *Let ζ, η be the same cut-off functions as in Lemma 6.8.6, and let u be a solution of the boundary value problem (6.1.3) such that $\eta\partial_{x_3}^j u \in N_\delta^{l,\sigma}(\mathcal{D})^\ell$ for $j=0$ and $j=1$, where $l \ge 2$, $0<\sigma<1$. If*

$$\eta f \in N_\delta^{l-1,\sigma}(\mathcal{D})^\ell, \quad \eta g^\pm \in N_\delta^{l+1-d^\pm,\sigma}(\Gamma^\pm)^\ell$$

and the strip $l+\sigma-\delta \le \operatorname{Re}\lambda \le l+1+\sigma-\delta$ does not contain eigenvalues of the pencil $A(\lambda)$, then $\zeta u \in N_\delta^{l+1,\sigma}(\mathcal{D})^\ell$.

Proof. Let χ be a smooth cut-off function such that $\chi = 1$ in a neighborhood of $\operatorname{supp}\zeta$ and $\eta = 1$ in a neighborhood of $\operatorname{supp}\chi$. Then $L(D_{x'},0)(\chi u) = F$, where

$$F = \chi f + \chi\big(L(D_{x'},0) - L(D_{x'},D_{x_3})\big)\, u + [L(D_{x'},0),\chi]\, u \in N_\delta^{l-1,\sigma}(\mathcal{D})^\ell.$$

Here $[L(D_{x'},0),\chi] = L(D_{x'},0)\chi - \chi\, L(D_{x'},0)$ denotes the commutator of $L(D_{x'},0)$ and χ. Analogously, $B^\pm(D_{x'},0)\,(\chi u)|_{\Gamma_\pm} = G^\pm \in N_\delta^{l+1-d^\pm,\sigma}(\Gamma^\pm)^\ell$. Consequently, we obtain $(\chi u)(\cdot,x_3) \in N_\delta^{l+1,\sigma}(K)^\ell$ and

$$\|(\chi u)(\cdot,x_3)\|_{N_\delta^{l+1,\sigma}(K)^\ell} \le c\Big(\|F(\cdot,x_3)\|_{N_\delta^{l-1,\sigma}(K)^3} + \sum_\pm \|G^\pm\|_{N_\delta^{l+1,\sigma}(\gamma^\pm)^\ell}\Big)$$

with a constant c independent of x_3 (cf. [**117**, Theorem 8.4]). In particular, $\chi u \in L^\infty_{\delta-l-1-\sigma}(\mathcal{D})^\ell$ which implies $\zeta u \in N_\delta^{l+1,\sigma}(\mathcal{D})^\ell$ (cf. Lemma 6.8.6). □

6.8.3. Local regularity results in the spaces $C_\delta^{l,\sigma}$.

In the next lemma, we extend the result of Lemma 6.8.6 to weighted Hölder spaces with nonhomogeneous norms.

LEMMA 6.8.8. *Let ζ, η be smooth functions with compact supports, $\eta = 1$ in a neighborhood of $\operatorname{supp}\zeta$, and let u be a solution of the boundary value problem (6.1.3) such that $\chi u \in W^{2,p}(\mathcal{D})^\ell$ for every $\chi \in C_0^\infty(\overline{\mathcal{D}}\backslash M)$. Suppose that $\eta u \in C_{\delta-1}^{l-1,\sigma}(\mathcal{D})^\ell$, $\eta f \in C_\delta^{l-2,\sigma}(\mathcal{D})^\ell$, and $\eta g^\pm \in C_\delta^{l-d^\pm,\sigma}(\Gamma^\pm)^\ell$, where $l\ge 2$, $\delta \ge 1$, $0<\sigma<1$. Then $\zeta u \in C_\delta^{l,\sigma}(\mathcal{D})^\ell$ and*

$$\|\zeta u\|_{C_\delta^{l,\sigma}(\mathcal{D})^\ell} \le c\Big(\|\eta f\|_{C_\delta^{l-2,\sigma}(\mathcal{D})^\ell} + \sum_\pm \|\eta g^\pm\|_{C_\delta^{l-d^\pm,\sigma}(\Gamma^\pm)^\ell} + \|\eta u\|_{C_{\delta-1}^{l-1,\sigma}(\mathcal{D})^\ell}\Big). \tag{6.8.8}$$

Proof. By Theorem 6.7.11, the vector function ηu admits the decomposition

$$\eta u = v + w, \quad \text{where } v \in N_{\delta-1}^{l-1,\sigma}(\mathcal{D})^\ell,\ w \in C_\delta^{l,\sigma}(\mathcal{D})^\ell.$$

Let χ be a smooth cut-off function equal to one in a neighborhood of $\operatorname{supp}\zeta$ such that $\eta = 1$ in a neighborhood of $\operatorname{supp}\chi$. Then

$$\chi\, L(D_x)\, v = \chi f - \chi\, L(D_x)\, w \in C_\delta^{l-2,\sigma}(\mathcal{D})^\ell.$$

If $\delta \ge l-2+\sigma$, then $C_\delta^{l-2,\sigma}(\mathcal{D}) = N_\delta^{l-2,\sigma}(\mathcal{D})$. In the case $1 \le \delta < l-2+\sigma$ it follows that $l \ge 3$ and consequently

$$\chi\, L(D_x)\, v \in N_{\delta-1}^{l-3,\sigma}(\mathcal{D})^\ell \cap C_\delta^{l-2,\sigma}(\mathcal{D})^\ell.$$

By Lemma 6.7.10, the space $N_{\delta-1}^{l-3,\sigma}(\mathcal{D}) \cap C_\delta^{l-2,\sigma}(\mathcal{D})$ is a subspace of $N_\delta^{l-2,\sigma}(\mathcal{D})$. Therefore $\chi\, L(D_x)\, v \in N_\delta^{l-2,\sigma}(\mathcal{D})^3$ for $\delta \ge 1$. Analogously, we conclude that $\chi B^\pm(D_x) v|_{\Gamma^\pm} \in N_\delta^{l-d^\pm,\sigma}(\Gamma^\pm)^\ell$. This together with Lemma 6.8.6 implies $\zeta v \in N_\delta^{l,\sigma}(\mathcal{D})^\ell$. Furthermore, the estimate (6.8.8) holds. □

As the next lemma shows, the assertion of Lemma 6.8.7 is also valid in the class of the weighted Hölder spaces $C_\delta^{l,\sigma}$.

LEMMA 6.8.9. *Let ζ, η be the same functions as in Lemma 6.8.8, and let u be a solution of the boundary value problem (6.1.3) such that $\eta \partial_{x_3}^j u \in C_\delta^{l,\sigma}(\mathcal{D})^\ell$ for $j = 0$ and $j = 1$, where $l \ge 2$, $0 < \sigma < 1$, $\delta - \sigma$ is not integer. If*

$$\eta f \in C_\delta^{l-1,\sigma}(\mathcal{D})^\ell, \quad \eta g^\pm \in C_\delta^{l+1-d^\pm,\sigma}(\Gamma^\pm)^\ell,$$

and the strip $l+\sigma-\delta \le \operatorname{Re}\lambda \le l+1+\sigma-\delta$ does not contain eigenvalues of the pencil $A(\lambda)$, then $\zeta u \in C_\delta^{l+1,\sigma}(\mathcal{D})^\ell$

P r o o f. Let k be the integral part of $l+\sigma-\delta$. Then both ηu and $\partial_{x_3}(\eta u)$ belong to $C^{k,l+\sigma-\delta-k}(\mathcal{D})^\ell$. Consequently, the traces $u^{(i,j)}$ of $\partial_{x_1}^i \partial_{x_2}^j (\eta u)$ on M are from $C^{k-i-j+1,l+\sigma-\delta-k}(M)^\ell$ for $i+j \le k$. By Theorem 6.7.11, the function ζu admits the decomposition

$$\zeta u = \zeta \sum_{i+j\le k} \frac{E u^{(i,j)}}{i!\, j!}\, x_1^i\, x_2^j + v,$$

where E is the extension operator (6.2.4) and $\partial_{x_3}^j v \in N_\delta^{l,\sigma}(\mathcal{D})^\ell$ for $j = 0, 1$. From the properties of the extension operator E (cf. Lemma 6.7.9) it follows that $\zeta u - v \in C_\delta^{l+1,\sigma}(\mathcal{D})^\ell$. Therefore,

$$L(D_x)\, v \in C_\delta^{l-1,\sigma}(\mathcal{D})^\ell, \quad B^\pm(D_x)\, v \in C_\delta^{l+1-d^\pm,\sigma}(\Gamma^\pm)^\ell.$$

Since $v \in N_\delta^{l,\sigma}(\mathcal{D})^\ell$, we conclude that $\partial_x^\alpha L(D_x)\, v = 0$ on M for $|\alpha| \le k-2$ and $\partial_r^j (B^\pm(D_x) v|_{\Gamma^\pm}) = 0$ on M for $j \le k - d^\pm$. We put

$$F^{(i,j)} = \partial_{x_1}^i \partial_{x_2}^j L(D_x)\, v\big|_M \quad \text{for } i+j = k-1$$

and $G^\pm = \partial_r^{k-d^\pm+1} B^\pm(D_x) v|_M$. Then by Theorem 6.7.11

$$\eta \Big(L(D_x)\, v - \sum_{i+j=k-1} \frac{E F^{(i,j)}}{i!\, j!}\, x_1^i\, x_2^j \Big) \in N_\delta^{l-1,\sigma}(\mathcal{D})^\ell$$

and

$$\eta \Big(B^\pm(D_x)\, v - \frac{E G^\pm}{(k-d^\pm+1)!}\, r^{k-d^\pm+1} \Big) \in N_\delta^{l+1-d^\pm,\sigma}(\Gamma^\pm)^\ell.$$

Since $\lambda = k+1$ is not an eigenvalue of the pencil $A(\lambda)$, there exist homogeneous matrix-valued polynomials $P^{(i,j)}(x_1,x_2)$ of degree $k+1$ satisfying the equations

$$L(D_{x'},0)\,P^{(i,j)} = \frac{x_1^i\,x_2^j}{i!\,j!}\,I_\ell \;\text{ in } K,\quad B^\pm(D_{x'},0)\,P^{i,j} = 0 \;\text{ on } \gamma^\pm \;\text{ for } i+j=k-1$$

(cf. Lemma 6.3.6). Let the vector function w be defined by

$$w(x) = \sum_{i+j=k-1} \frac{1}{i!\,j!}\,P^{(i,j)}(x')\,(EF^{(i,j)})(x).$$

Then $\eta w \in C_\delta^{l+1,\sigma}(\mathcal{D})^\ell$ and $\eta\,\partial_{x_3}^j w \in N_\delta^{l,\sigma}(\mathcal{D})^\ell$ for $j=0,1$. Moreover, it follows from the properties of the extension operator E that

$$\eta\Big(L(D_x)w - \sum_{i+j=k-1} \frac{EF^{(i,j)}}{i!\,j!}\,x_1^i\,x_2^j\Big) \in N_\delta^{l-1,\sigma}(\mathcal{D})^\ell,$$

$$\eta\,B^\pm(D_x)\,w \in N_\delta^{l+1-d^\pm,\sigma}(\mathcal{D})^\ell.$$

Analogously, there exists a vector function W such that $\eta W \in C_\delta^{l+1,\sigma}(\mathcal{D})^\ell$, $\eta\partial_{x_3}^j W \in N_\delta^{l,\sigma}(\mathcal{D})^\ell$ for $j=0,1$,

$$\eta\,L(D_x)\,W \in N_\delta^{l-1,\sigma}(\mathcal{D})^\ell,$$

$$\eta\Big(B^\pm(D_x)\,W - \frac{EG^\pm}{(k-d^\pm+1)!}\,r^{k-d^\pm+1}\Big) \in N_\delta^{l+1-d^\pm,\sigma}(\Gamma^\pm)^\ell.$$

Thus, we get $\eta\,L(D_x)\,(v-w-W) \in N_\delta^{l-1,\sigma}(\mathcal{D})^\ell$ and $\eta\,B^\pm(D_x)\,(v-w-W) \in N_\delta^{l+1-d^\pm,\sigma}(\Gamma^\pm)^\ell$. Applying Lemma 6.8.7 to the vector function $v-w-W$, we conclude that $\chi(v-w-W) \in N_\delta^{l+1,\sigma}(\mathcal{D})^\ell$ where χ is the same cut-off function as in the proof of Lemma 6.8.8. This proves the lemma. □

CHAPTER 7

Boundary value problems for second order systems in a polyhedral cone

In this chapter, we consider boundary value problems for elliptic systems of second order differential equations in a three-dimensional polyhedral cone $\mathcal{K}$, whose boundary contains the faces $\Gamma_1, \dots, \Gamma_d$ and the edges $M_1, \dots, M_d$ meeting at the origin. Here the Dirichlet and Neumann boundary conditions are arbitrarily combined on $\Gamma_1, \dots, \Gamma_d$. In contrast to Chapter 3, we consider solutions in weighted spaces $W^{l,p}_{\beta,\delta}(\mathcal{K})$ and $C^{l,\sigma}_{\beta,\delta}(\mathcal{K})$ with nonhomogeneous norms. Here, the weighted Sobolev space $W^{l,p}_{\beta,\delta}(\mathcal{K})$ is the set of all functions u in $\mathcal{K}$ such that

$$|x|^{p(\beta-l+|\alpha|)} \prod_{k=1}^{d} \Big(\frac{r_k}{\rho}\Big)^{p\delta_k} \partial_x^\alpha u \in L_p(\mathcal{K}) \quad \text{for } |\alpha| \le l,$$

where $\rho(x) = |x|$, $r_k(x) = \text{dist}(x, M_k)$, and $\delta_1, \dots, \delta_k$ are the components of δ. Similarly, the weighted Hölder space $N^{l,\sigma}_{\beta,\delta}(\mathcal{K})$ is defined.

If for every edge the Dirichlet condition is prescribed on at least one of the adjoining faces, then it is also possible to establish a solvability and regularity theory in the weighted spaces $V^{l,p}_{\beta,\delta}(\mathcal{K})$ and $N^{l,\sigma}_{\beta,\delta}(\mathcal{K})$. For the Dirichlet problem this was done in Chapter 3. The use of the weighted spaces $W^{l,p}_{\beta,\delta}(\mathcal{K})$ and $C^{l,\sigma}_{\beta,\delta}(\mathcal{K})$ allows us to include the Dirichlet and Neumann problems as well as any mixed problem with Dirichlet and Neumann conditions.

The main results of this chapter concern the existence and uniqueness of solutions in the weighted spaces $W^{l,p}_{\beta,\delta}(\mathcal{K})$ and $C^{l,\sigma}_{\beta,\delta}(\mathcal{K})$ and regularity assertions for the solutions. For example, we prove that the boundary value problem is uniquely solvable in the space $W^{l,p}_{\beta,\delta}(\mathcal{K})^\ell$ (for data in appropriate weighted spaces) if the line $\text{Re}\,\lambda = l-\beta-3/p$ is free of eigenvalues of a certain operator pencil $\mathfrak{A}(\lambda)$ introduced in Section 7.1 and the components of δ satisfy the inequalities

$$\max(0, l-\delta_+^{(k)}) < \delta_k + 2/p < l,$$

where $\delta_+^{(k)}$ are certain positive numbers. Here also variational solutions in the space $W^{1,p}_{\beta,\delta}(\mathcal{K})^\ell$ are included.

We start again with the case $p = 2$ (Sections 7.3 and 7.4). The a priori estimates for the solutions in the weighted spaces $W^{l,2}_{\beta,\delta}(\mathcal{K})^\ell$ are used in Section 7.5, where the Green's matrix for the boundary value problem is studied. In particular, we obtain point estimates for the elements of this matrix. As in Chapter 3, where we dealt with the Dirichlet problem, we consider the cases $|x| > 2|\xi|$, $|\xi| > 2|x|$ and

$|x|/2 < |\xi| < 2|x|$ separately. For example, the estimate

$$\begin{aligned} \big|\partial_x^\alpha \partial_\xi^\gamma G(x,\xi)\big| &\le c\,|x|^{\Lambda_- - |\alpha| + \varepsilon} \prod_{k=1}^{d} \Big(\frac{r_k(x)}{|x|}\Big)^{\min(0,\delta_+^{(k)} - |\alpha| - \varepsilon)} \\ &\quad \times |\xi|^{-1-\Lambda_- - |\gamma| - \varepsilon} \prod_{k=1}^{d} \Big(\frac{r_k(\xi)}{|\xi|}\Big)^{\min(0,\delta_-^{(k)} - |\gamma| - \varepsilon)} \end{aligned}$$

holds if $|x| > 2|\xi|$, where ε is an arbitrarily small positive number. The exponents $\min(0, \delta_+^{(k)} - |\alpha| - \varepsilon)$ and $\min(0, \delta_-^{(k)} - |\gamma| - \varepsilon)$ can be replaced by $\delta_+^{(k)} - |\alpha| - \varepsilon$ and $\delta_-^{(k)} - |\gamma| - \varepsilon$ if the Dirichlet condition is prescribed on at least one of the adjoining faces of the edge M_k. We also improve the above given estimate in the case of the Neumann problem for a special class of elliptic second order systems which includes the Lamé system.

The estimates of Green's matrix obtained in Section 7.5 enable us to extend the results of Sections 7.3 and 7.4 to weighted L_p Sobolev spaces (Sections 7.6 and 7.7) and weighted Hölder spaces (Section 7.8).

7.1. The boundary value problem and corresponding operator pencils

As was shown in Chapter 3 for the Dirichlet problem, the smoothness of the solutions depends not only on the smoothness of the data but also on the eigenvalues of certain operator pencils $A_k(\lambda)$ and $\mathfrak{A}(\lambda)$ which are generated by the boundary value problem at the edges M_k and at the vertex, respectively. Here, we introduce these pencils for the mixed boundary value problem. Furthermore, we prove some basic spectral properties of the pencil $\mathfrak{A}(\lambda)$.

7.1.1. Formulation of the problem. Let

$$\mathcal{K} = \{x \in \mathbb{R}^3 : \ x/|x| = \omega \in \Omega\}$$

be a polyhedral cone with vertex at the origin. Suppose that the boundary $\partial\mathcal{K}$ consists of the vertex $x = 0$, the edges (half-lines) $M_1, \dots, M_d$, and plane faces $\Gamma_1, \dots, \Gamma_d$. We consider the boundary value

$$L(D_x)\, u = \sum_{i,j=1}^{3} A_{i,j}\, D_{x_i} D_{x_j} u = f \ \text{ in } \mathcal{K}, \tag{7.1.1}$$

$$u = g_k \text{ on } \Gamma_k \text{ for } k \in I_0, \tag{7.1.2}$$

$$N(D_x)\, u = \sum_{i,j=1}^{3} A_{i,j}\, n_j\, \partial_{x_i} u = g_k \ \text{ on } \Gamma_k \text{ for } k \in I_1. \tag{7.1.3}$$

where $A_{i,j}$ are constant $\ell \times \ell$ matrices, $I_0 \cup I_1 = \{1, \dots, d\}$, $I_0 \cap I_1 = \emptyset$, and (n_1, n_2, n_3) denotes the exterior normal to $\partial\mathcal{K}\backslash\mathcal{S}$. Here $\mathcal{S} = \{0\} \cup M_1 \cup \dots \cup M_d$ denotes the set of the singular boundary points of the cone $\mathcal{K}$. In the following, let d_k be the order of the differential operators in the boundary conditions (7.1.2), (7.1.3), i.e. $d_k = 0$ for $k \in I_0$, $d_k = 1$ for $k \in I_1$.

We define the space $L^{1,2}(\mathcal{K})$ as the closure of the set $C_0^\infty(\overline{\mathcal{K}})$ with respect to the norm

$$\|u\|_{L^{1,2}(\mathcal{K})} = \Big(\int_{\mathcal{K}} |\nabla u(x)|^2\, dx\Big)^{1/2}.$$

The corresponding trace space on the face Γ_j is denoted by $L^{1/2,2}(\Gamma_j)$. Obviously, the sesquilinear form

$$b_{\mathcal{K}}(u,v) = \int_{\mathcal{K}} \sum_{i,j=1}^{3} A_{i,j} D_{x_i} u \cdot \overline{D_{x_j} v}\, dx \tag{7.1.4}$$

is continuous on $L^{1,2}(\mathcal{K})^\ell \times L^{1,2}(\mathcal{K})^\ell$. Let

$$\mathcal{H}_{\mathcal{K}} = \{u \in L^{1,2}(\mathcal{K})^\ell : \ u = 0 \text{ on } \Gamma_k \text{ for } k \in I_0\}.$$

A vector function $u \in L^{1,2}(\mathcal{K})^\ell$ is called *variational solution* of the problem (7.1.1)–(7.1.3) if

$$b_{\mathcal{K}}(u,v) = (F,v)_{\mathcal{D}} \text{ for all } v \in \mathcal{H}_{\mathcal{K}}, \tag{7.1.5}$$

$$u = g_k \quad \text{on } \Gamma_k \quad \text{for } k \in I_0. \tag{7.1.6}$$

Here F is a given linear and continuous functional on $\mathcal{H}_{\mathcal{K}}$, g_k are given vector functions in $L^{1/2,2}(\Gamma_k)^\ell$, and $(\cdot,\cdot)_{\mathcal{K}}$ denotes the extension of the $L_2(\mathcal{K})^\ell$ scalar product to $\mathcal{H}_{\mathcal{K}}^* \times \mathcal{H}_{\mathcal{K}}$.

Clearly, every strong solution $u \in W^{2,2}(\mathcal{K})^\ell$ of the boundary value problem (7.1.1)–(7.1.3) is a solution of problem (7.1.5), (7.1.6), where

$$(F,v)_{\mathcal{K}} = \int_{\mathcal{K}} f \cdot \bar{v}\, dx + \sum_{k\in I_1} \int_{\Gamma_k} g_k \cdot \bar{v}\, dx \quad \text{for all } v \in \mathcal{H}_{\mathcal{K}}\,. \tag{7.1.7}$$

Throughout this chapter, we assume that the form (7.1.4) is $\mathcal{H}_{\mathcal{K}}$-*elliptic*, i.e.

$$\operatorname{Re} b_{\mathcal{K}}(u,u) \ge c\, \|u\|^2_{\mathcal{H}_{\mathcal{K}}} \quad \text{for all } u \in \mathcal{H}_{\mathcal{K}} \tag{7.1.8}$$

with a positive constant c independent of u. Then the problem (7.1.5), (7.1.6) has a unique solution $u \in \mathcal{H}_{\mathcal{K}}$ for arbitrary $F \in \mathcal{H}_{\mathcal{K}}^*$ and zero Dirichlet data g_k, $k \in I_0$.

7.1.2. The operator pencil $A_k(\lambda)$. Let M_k be an edge of the cone $\mathcal{K}$, and let Γ_{k_+} and Γ_{k_-} be the faces adjacent to M_k. Furthermore, let $\Gamma^\circ_{k_+} \supset \Gamma_{k_+}$ and $\Gamma^\circ_{k_-} \supset \Gamma_{k_-}$ be the half-planes bounded by the line which contains the edge M_k. The dihedron with the boundary $\overline{\Gamma^\circ_{k_+}} \cup \overline{\Gamma^\circ_{k_-}}$ is denoted by $\mathcal{D}_k$, the angle at the edge M_k by θ_k. We consider the boundary value problem

$$L(D_x)u = f \ \text{ in } \mathcal{D}_k, \quad B^\pm(D_x)u = g^\pm \text{ on } \Gamma^\circ_{k_\pm}, \tag{7.1.9}$$

where

$$B^\pm(D_x)u = u \text{ if } k_\pm \in I_0, \quad B^\pm(D_x)u = N(D_x)u \text{ if } k_\pm \in I_1.$$

The operator pencil for this problem (see the definition of the pencil $A(\lambda)$ in Section 6.1) is denoted by $A_k(\lambda)$. Let r and φ be polar coordinates in the plane perpendicular to M_k such that $\varphi = \pm\theta_k/2$ for $x \in \Gamma^\circ_{k_\pm}$ and $r = \operatorname{dist}(x, M_k)$. Furthermore, let

$$\mathcal{L}_k(\lambda)\, U(\varphi) = r^{2-\lambda} L(D_x)\, \big(r^\lambda U(\varphi)\big),$$

$$\mathcal{B}_k^\pm(\lambda)\, U(\varphi) = r^{d^\pm-\lambda} B^\pm(D_x)\, \big(r^\lambda U(\varphi)\big),$$

where $d^\pm$ is the order of the differential operator $B^\pm(D_x)$. Then

$$A_k(\lambda)\, U(\varphi) = \big(\mathcal{L}_k(\lambda)\, U\,,\, \mathcal{B}_k^+(\lambda) U|_{\varphi=\theta_k/2}\,,\, \mathcal{B}_k^-(\lambda) U|_{\varphi=-\theta_k/2}\big).$$

This operator realizes a continuous mapping

$$W^{2,2}(-\theta_k/2, +\theta_k/2)^\ell \to L_2(-\theta_k/2, +\theta_k/2)^\ell \times \mathbb{C}^\ell \times \mathbb{C}^\ell$$

for each $\lambda \in \mathbb{C}$. We denote by $\delta_+^{(k)}, \delta_-^{(k)}$ the greatest positive real numbers such that the strip

$$-\delta_-^{(k)} < \operatorname{Re}\lambda < \delta_+^{(k)}$$

does not contain eigenvalues of the pencil $A_k(\lambda)$ with the exception of the eigenvalue $\lambda = 0$ in the case of the Neumann problem.

7.1.3. The operator pencil $\mathfrak{A}(\lambda)$. Let

$$\mathcal{H}_\Omega = \{u \in W^{1,2}(\Omega)^\ell \, : \, u = 0 \text{ on } \gamma_j \text{ for } j \in I_0\}.$$

Here $\gamma_j = \Gamma_j \cap S^2$, $j = 1, \dots, d$, are the sides of Ω. We introduce the parameter-dependent sesquilinear form

$$a(U, V; \lambda) = \frac{1}{\log 2} \int\limits_{\substack{\mathcal{K} \\ 1<|x|<2}} \sum_{i,j=1}^{3} A_{i,j}\, D_{x_i} u \cdot \overline{D_{x_j} v}\, dx \tag{7.1.10}$$

on $\mathcal{H}_\Omega \times \mathcal{H}_\Omega$, where $u(x) = \rho^\lambda U(\omega)$, $v(x) = \rho^{-1-\bar\lambda} V(\omega)$, $\rho = |x|$, $\omega = x/|x|$, and define the operator $\mathfrak{A}(\lambda) : \mathcal{H}_\Omega \to \mathcal{H}_\Omega^*$ by

$$\big(\mathfrak{A}(\lambda)U, V\big)_\Omega = a(U, V; \lambda), \quad U, V \in \mathcal{H}_\Omega\,.$$

Analogously, the operator pencil $\mathfrak{A}^+(\lambda)$ corresponding to the *formally adjoint problem*

$$L^+(D_x)\, u = \sum_{i,j=1}^{3} A_{j,i}^*\, D_{x_i}\, D_{x_j}\, u = f \ \text{ in } \mathcal{K}, \tag{7.1.11}$$

$$u = g_k \ \text{ on } \Gamma_k \text{ for } k \in I_0, \tag{7.1.12}$$

$$N^+(D_x)\, u = \sum_{i,j=1}^{3} A_{j,i}^*\, n_j\, \partial_{x_i} u = g_k \ \text{ on } \Gamma_k \text{ for } k \in I_1 \tag{7.1.13}$$

is defined. Let

$$a^+(U, V; \lambda) = \frac{1}{\log 2} \int\limits_{\substack{\mathcal{K} \\ 1<|x|<2}} \sum_{i,j=1}^{3} A_{j,i}^*\, D_{x_i} u \cdot \overline{D_{x_j} v}\, dx,$$

where u, v are the same vector functions as in (7.1.10). Then the operator $\mathfrak{A}^+(\lambda)$: $\mathcal{H}_\Omega \to \mathcal{H}_\Omega^*$ is defined by the equality

$$\big(\mathfrak{A}^+(\lambda)U, V\big)_\Omega = a^+(U, V; \lambda), \quad U, V \in \mathcal{H}_\Omega\,.$$

Both $\mathfrak{A}(\lambda)$ and $\mathfrak{A}^+(\lambda)$ depend quadratically on the complex parameter λ. The following lemma is proved for the Neumann problem in [**85**, Theorem 12.2.1]. The proof for the mixed boundary value problem proceeds analogously.

LEMMA 7.1.1. *Suppose that the form $b_\mathcal{K}$ is $\mathcal{H}_\mathcal{K}$-elliptic. Then the operators $\mathfrak{A}(\lambda)$ and $\mathfrak{A}^+(\lambda)$ are isomorphisms for* $\operatorname{Re}\lambda = -1/2$.

It follows from the last lemma that the spectra of the operator pencils $\mathfrak{A}(\lambda)$ and $\mathfrak{A}^+(\lambda)$ consist of isolated points, the eigenvalues of these pencils (see e.g. [**85**, Theorem 1.1.1]). Obviously,

$$a(U, V; \lambda) = \overline{a^+(V, U; -1-\bar\lambda)} \quad \text{for all } U, V \in \mathcal{H}_\Omega.$$

Thus according to [**85**, Theorem 1.1.7], the number λ_0 is an eigenvalue of the pencil $\mathfrak{A}(\lambda)$ if and only if $-1-\overline{\lambda}_0$ is an eigenvalue of the pencil $\mathfrak{A}^+(\lambda)$. Both eigenvalues have the same geometric and algebraic multiplicities.

7.1.4. Solvability of the parameter-depending boundary value problem on the sphere. Let l be an integer, $l \ge 0$, $1<p<\infty$, and $\delta=(\delta_1,\dots,\delta_d) \in \mathbb{R}^d$, $\delta_k > -2/p$ for $k=1,\dots,d$. Then we define the norm in the weighted Sobolev space $W_\delta^{l,p}(\Omega)$ by

$$\|u\|_{W_\delta^{l,p}(\Omega)} = \Big(\int\limits_{\substack{\mathcal{K}\\ 1<|x|<2}} \sum_{|\alpha|\le l} \prod_{k=1}^{d} r_k^{p\delta_k}\, |\partial_x^\alpha u(x)|^p\, dx \Big)^{1/p},$$

where u is extended by the equality $u(x)=u(x/|x|)$ to the cone $\mathcal{K}$ and $r_k(x) = \text{dist}\,(x, M_k)$. Obviously (as a consequence of Hardy's inequality), the space $W_\delta^{2,2}(\Omega)$ is continuously imbedded in $W^{1,2}(\Omega)$ if $\delta_k \le 1$ for $k=1,\dots,d$. We denote by $\mathfrak{A}_\delta(\lambda)$ the restriction of the operator $\mathfrak{A}(\lambda)$ to the space $W_\delta^{2,2}(\Omega)^\ell \cap \mathcal{H}_\Omega$. Let the parameter-depending operators $\mathcal{L}(\lambda)$ and $\mathcal{N}(\lambda)$ be defined by

$$\mathcal{L}(\lambda)\, u(\omega) = \rho^{2-\lambda} L(D_x)\, \big(\rho^\lambda u(\omega)\big), \tag{7.1.14}$$

$$\mathcal{N}(\lambda)\, u(\omega) = \rho^{1-\lambda} N(D_x)\, \big(\rho^\lambda u(\omega)\big), \tag{7.1.15}$$

where $\rho=|x|$, $\omega = x/|x|$. Then $\mathfrak{A}_\delta(\lambda)$ is the operator of the parameter-depending boundary value problem

$$\mathcal{L}(\lambda)\, u = f \quad \text{in } \Omega, \tag{7.1.16}$$

$$u=0 \quad \text{on } \gamma_k \text{ for } k\in I_0, \qquad \mathcal{N}(\lambda) u = g_k \quad \text{on } \gamma_k \text{ for } k \in I_1. \tag{7.1.17}$$

Here $\gamma_k = \Gamma_k \cap S^2$ are the sides of the domain Ω on the sphere S^2. More precisely, the operator

$$W_\delta^{2,2}(\Omega)^\ell \cap \mathcal{H}_\Omega \ni u \to \mathfrak{A}_\delta(\lambda)\, u \in \mathcal{H}_\Omega^*$$

is defined by

$$\big(\mathfrak{A}_\delta(\lambda) u, v\big)_\Omega = \int_\Omega f\cdot \bar{v}\, d\omega + \sum_{k\in I_1} \int_{\gamma_k} g_k \cdot \bar{v}\, d\omega,$$

where $f=\mathcal{L}(\lambda)u$ and $g_k = \mathcal{N}(\lambda)u|_{\gamma_k}$.

THEOREM 7.1.2. *Let* $\max(0, 1-\delta_+^{(k)}) < \delta_k < 1$ *for* $k=1,\dots,d$. *Then the following assertions are true.*

1) *The spectra of the pencils* $\mathfrak{A}$ *and* $\mathfrak{A}_\delta$ *coincide.*

2) *There exist positive real constants* N *and* ε *such that the operator* $\mathfrak{A}_\delta(\lambda)$ *is an isomorphism for all* λ *in the set*

$$\{\lambda \in \mathbb{C} :\ |\lambda| > N,\ |\text{Re}\,\lambda| < \varepsilon\, |\text{Im}\,\lambda|\}. \tag{7.1.18}$$

Furthermore, every solution $u \in W_\delta^{2,2}(\Omega)^\ell$ *of the boundary value problem* (7.1.16), (7.1.17) *satisfies the inequality*

$$\begin{aligned} &\sum_{j=0}^{2} |\lambda|^{2-j}\, \|u\|_{W_\delta^{j,2}(\Omega)^\ell} \\ &\le c \Big(\|f\|_{W_\delta^{0,2}(\Omega)^\ell} + \sum_{k\in I_1} \|g_k\|_{W_\delta^{1/2,2}(\gamma_k)^\ell} + \sum_{k\in I_1} |\lambda|^{1/2}\, \|g_k\|_{W_\delta^{0,2}(\gamma_k)^\ell} \Big) \end{aligned} \tag{7.1.19}$$

if λ lies the set (7.1.18). *Here the constant c is independent of f, g_k and λ.*

P r o o f. 1) Since $\mathfrak{A}_\delta(\lambda)$ is the restriction of $\mathfrak{A}(\lambda)$, every eigenvalue of the pencil $\mathfrak{A}_\delta(\lambda)$ is also an eigenvalue of the pencil $\mathfrak{A}(\lambda)$. The same is true for the eigenfunctions and generalized eigenfunctions. In order to prove the converse, we consider the differential operators $\mathcal{L}(\lambda)$ and $\mathcal{N}(\lambda)$ in a neighborhood of $M_1 \cap S^2$. Without loss of generality, we may assume that M_1 is the positive x_3-axis and that $\mathcal{K}$ coincides with the dihedron $\mathcal{D}_1 = K \times \mathbb{R}$ in a neighborhood of an arbitrary edge point $x \in M_1$, where K is an infinite angle in the (x_1, x_2)-plane. By $\omega_1 = x_1/\rho$ and $\omega_2 = x_2/\rho$, we denote local coordinates on the unit sphere near the north pole $N = M_1 \cap S^2$. Since

$$D_{x_1} = \omega_1 \partial_\rho + \frac{1-\omega_1^2}{\rho} D_{\omega_1} - \frac{\omega_1\omega_2}{\rho} D_{\omega_2},$$

$$D_{x_2} = \omega_2 D_\rho - \frac{\omega_1\omega_2}{\rho} D_{\omega_1} + \frac{1-\omega_2^2}{\rho} D_{\omega_2},$$

$$D_{x_3} = (1-\omega_1^2-\omega_2^2)^{1/2} \big(D_\rho - \frac{\omega_1}{\rho} D_{\omega_1} - \frac{\omega_2}{\rho} D_{\omega_2}\big),$$

the operator $\mathcal{L}(\lambda)$ has the form

$$\begin{aligned}\mathcal{L}(\lambda) \;=\; & \sum_{j,k=1}^{2} A_{j,k} D_{\omega_j} D_{\omega_k} - i(\lambda-1) \sum_{j=1}^{2} \big(A_{j,3} + A_{3,j}\big) D_{\omega_j} \\ & -\lambda\,(A_{1,1}+A_{2,2}) - \lambda(\lambda-1)\,A_{3,3} + \lambda^2 \mathcal{P}_0(\omega) + \lambda\,\mathcal{P}_1(\omega,\partial_\omega) + \mathcal{P}_2(\omega,\partial_\omega),\end{aligned}$$

where $\mathcal{P}_j$ are differential operators of order j with coefficients vanishing at the point $(\omega_1,\omega_2) = 0$. Analogously,

$$\mathcal{N}(\lambda) = \sum_{j,k=1}^{2} A_{j,k}\, n_k\, \partial_{\omega_i} + \lambda \sum_{k=1}^{2} A_{3,k} n_k + \lambda\, \mathcal{Q}_0(\omega) + \mathcal{Q}_1(\omega, \partial_\omega)$$

near N, where $\mathcal{Q}_j$ are differential operators of order j with coefficients vanishing at $(\omega_1,\omega_2) = 0$. Furthermore, Ω coincides with the angle K in the coordinate system ω_1, ω_2 near $M_1 \cap S^2$. The principal parts of the operators $\mathcal{L}(\lambda)$ and $\mathcal{N}(\lambda)$ with coefficients frozen at the point $\omega_1 = \omega_2 = 0$ coincide with the operators $L(D_{\omega_1}, D_{\omega_2}, 0)$ and $N(D_{\omega_1}, D_{\omega_2}, 0)$, respectively. Therefore, we can conclude analogously to the proof of Lemma 6.3.3 that every variational solution $u \in W^{1,2}(\Omega)^\ell$ of the problem (7.1.16), (7.1.17) with support near N belongs to the space $W_\delta^{2,2}(\Omega)^\ell$ if $f \in W_\delta^{0,2}(\Omega)^\ell$, $g_k \in W_\delta^{1/2,2}(\gamma_k)^\ell$. By means of a partition of unity on Ω, we obtain this result for variational solutions with arbitrary support. In particular, this means that every eigenfunction of the pencil $\mathfrak{A}(\lambda)$ is also an eigenfunction of $\mathfrak{A}_\delta(\lambda)$ corresponding to the same eigenvalue. The same is true for generalized eigenfunctions.

2) We prove the second assertion first for purely imaginary $\lambda = i\xi$. Let $\psi_0, \psi_1, \dots, \psi_d$ be a partition of unity on Ω such that $\psi_k = 1$ near $M_k \cap S^2$ and $\operatorname{supp}\psi_k$ is sufficiently small for $k = 1, \dots, d$. We consider the vector-function $\psi_1 u$ and assume as above that the edge M_1 coincides with the x_3-axis. The difference of the operator $\mathcal{L}(i\xi)$ (in the coordinates ω_1, ω_2 introduced above) and the operator $L(D_{\omega_1}, D_{\omega_2}, \xi)$ is small for large $|\lambda|$ and small $\omega_1^2 + \omega_2^2$. This means, there is the

inequality

$$\left\|\left(\mathcal{L}(i\xi) - L(D_{\omega_1}, D_{\omega_2}, i\xi)\right)(\psi_1 u)\right\|_{L^{0,2}_{\delta_1}(K)^\ell} \le \varepsilon \sum_{k=0}^{2} |\xi|^{2-k} \|\psi_1 u\|_{L^{k,2}_{\delta_1}(K)^\ell},$$

where ε is small if $\operatorname{supp}\psi_1$ is small and $|\lambda|$ is large. An analogous estimates holds for $(\mathcal{N}(i\xi) - N(D_{\omega_1}, D_{\omega_2}, \xi))(\psi_1 u)$. Hence by Theorem 6.3.5,

$$\begin{aligned}\sum_{j=0}^{2-j} |\lambda|^{2-j} \|\psi_1 u\|_{L^{j,2}_\delta(\Omega)^\ell} \le\ & c\Big(\|\mathcal{L}(\lambda)u\|_{W^{0,2}_\delta(\Omega)^\ell} + \sum_{k\in I_1} \|\mathcal{N}(\lambda)u\|_{V^{1/2,2}_\delta(\gamma_k)^\ell} \\ & + \sum_{k\in I_1} |\lambda|^{1/2}\, \|\mathcal{N}(\lambda)u\|_{V^{0,2}_\delta(\gamma_k)^\ell}\Big)\end{aligned}$$

for sufficiently large $|\lambda|$. The same inequality is true for the vector-functions $\psi_k u$, $k = 2, \dots, d$. The validity of this inequality for $\psi_0 u$ follows from Theorem 1.1.10. Thus, the estimate (7.1.19) is valid for purely imaginary λ, $|\lambda| > N$. Analogously to the proof of Theorem 3.2.1, this estimates holds for λ in the set (7.1.18). □

7.2. Weighted Sobolev spaces in a cone

In this section, we introduce the weighted Sobolev spaces $W^{l,p}_{\beta,\delta}(\mathcal{K})$ and the corresponding trace spaces $W^{l-1/p,p}_{\beta,\delta}(\Gamma_j)$. Furthermore, we prove some imbeddings for these spaces.

7.2.1. The spaces $W^{l,p}_{\beta,\delta}(\mathcal{K})$. We use the same notation as for the definition of the spaces $V^{l,p}_{\beta,\delta}(\mathcal{K})$ in Section 3.1. For an arbitrary point $x \in \overline{\mathcal{K}}$, let $\rho(x) = |x|$ be the distance from the vertex of the cone $\mathcal{K}$, $r_k(x)$ the distance from the edge M_k, and $r(x) = \min_k r_k(x)$ the distance from the set $\mathcal{S} = \overline{M}_1 \cup \dots \cup \overline{M}_d$.

Let l be a nonnegative integer, $1 < p < \infty$, $\beta \in \mathbb{R}$, and $\delta = (\delta_1, \dots, \delta_d) \in \mathbb{R}^d$, $\delta_k > -2/p$ for $k = 1, \dots, d$. We define the space $W^{l,p}_{\beta,\delta}(\mathcal{K})$ as the closure of the set $C_0^\infty(\overline{\mathcal{K}}\backslash\{0\})$ with respect to the norm

$$\text{(7.2.1)} \qquad \|u\|_{W^{l,p}_{\beta,\delta}(\mathcal{K})} = \Big(\int_{\mathcal{K}} \sum_{|\alpha|\le l} \rho^{p(\delta-l+|\alpha|)} \prod_{k=1}^{d} \Big(\frac{r_k}{\rho}\Big)^{p\delta_k} \left|\partial_x^\alpha u(x)\right|^p dx\Big)^{1/p}.$$

By $W^{l-1/p,p}_{\beta,\delta}(\Gamma_j)$ we denote the trace space for $W^{l,p}_{\beta,\delta}(\mathcal{K})$, $l \ge 1$, on the face Γ_j. The norm in this space is defined as

$$\|u\|_{W^{l-1/p,p}_{\beta,\delta}(\Gamma_j)} = \inf\Big\{\|v\|_{W^{l,p}_{\beta,\delta}(\mathcal{K})} \ :\ v \in W^{l,p}_{\beta,\delta}(\mathcal{K}),\ v|_{\Gamma_j} = u\Big\}.$$

Furthermore, we introduce the following notation. If $\delta = (\delta_1, \dots, \delta_d) \in \mathbb{R}^d$ and $t \in \mathbb{R}$, then $W^{l,p}_{\beta,\delta+t}(\mathcal{K})$ denotes the space $W^{l,p}_{\beta,\delta'}(\mathcal{K})$, where $\delta'_k = \delta_k + t$ for $k = 1, \dots, d$. The same notation is used for the trace spaces.

7.2.2. Equivalent norms. Passing to spherical coordinates ρ, ω, one obtains the following equivalent norm in $W^{l,p}_{\beta,\delta}(\mathcal{K})$:

$$\text{(7.2.2)} \qquad \|u\| = \Big(\int_0^\infty \rho^{p(\beta-l+1)} \sum_{k=0}^{l} \|(\rho\partial_\rho)^k u(\rho,\cdot)\|^2_{W^{l-k,p}_\delta(\Omega)}\, d\rho\Big)^{1/p}.$$

Let ζ_k be smooth functions depending only on $\rho = |x|$ such that

$$\text{(7.2.3)} \qquad \operatorname{supp}\zeta_k \subset (2^{k-1}, 2^{k+1}), \quad \sum_{k=-\infty}^{+\infty} \zeta_k = 1, \quad |\partial_\rho^j \zeta_k(\rho)| \le c_j\, \rho^{-j}$$

with constants c_j independent of k and ρ. Then the following assertion holds analogously to Lemmas 3.1.2 and 3.1.5.

LEMMA 7.2.1. *There exist positive constants c_1 and c_2 such that*

$$c_1 \|u\|_{W^{l,p}_{\beta,\delta}(\mathcal{K})} \le \Big(\sum_{k=-\infty}^{+\infty} \|\zeta_k u\|^p_{W^{l,p}_{\beta,\delta}(\mathcal{K})} \Big)^{1/p} \le c_2 \|u\|_{W^{l,p}_{\beta,\delta}(\mathcal{K})}$$

for all $u \in W^{l,p}_{\beta,\delta}(\mathcal{K})$, $l \ge 0$, and

$$c_1 \|u\|_{W^{l-1/p,p}_{\beta,\delta}(\Gamma_j)} \le \Big(\sum_{k=-\infty}^{+\infty} \|\zeta_k u\|^p_{W^{l-1/p,p}_{\beta,\delta}(\Gamma_j)} \Big)^{1/p} \le c_2 \|u\|_{W^{l-1/p,p}_{\beta,\delta}(\Gamma_j)}$$

for all $u \in W^{l-1/p,p}_{\beta,\delta}(\Gamma_j)$, $l \ge 1$.

7.2.3. Imbeddings. According to the definition of the space $W^{l,p}_{\beta,\delta}(\mathcal{K})$, the imbedding $W^{l+1,p}_{\beta+1,\delta}(\mathcal{K}) \subset W^{l,p}_{\beta,\delta}(\mathcal{K})$ is valid for arbitrary β, δ, l, p. Moreover, the following lemma holds.

LEMMA 7.2.2. *Let $\delta = (\delta_1, \dots, \delta_d)$, $\delta' = (\delta'_1, \dots, \delta'_d)$ be such that $\delta_k > -2/p$, $\delta'_k > -2/p$ and $\delta'_k - \delta_k \le 1$ for $k = 1, \dots, d$. Then $W^{l+1,p}_{\beta+1,\delta'}(\mathcal{K})$ is continuously imbedded in $W^{l,p}_{\beta,\delta}(\mathcal{K})$.*

Proof. The assertion of the lemma follows immediately from the continuity of the imbedding $W^{l+1-k,p}_{\delta'}(\Omega) \subset W^{l-k,p}_{\delta}(\Omega)$, $k = 0, \dots, l$. □

Obviously, $V^{l,p}_{\beta,\delta}(\mathcal{K}) \subset W^{l,p}_{\beta,\delta}(\mathcal{K})$. If $\delta_k > l - 2/p$ for all k, then according to Lemma 7.2.2, the spaces $V^{l,p}_{\beta,\delta}(\mathcal{K})$ and $W^{l,p}_{\beta,\delta}(\mathcal{K})$ coincide.

We are also interested in imbeddings for the spaces $W^{l,p}_{\beta,\delta}(\mathcal{K})$ with different p. Here we prove the following lemma.

LEMMA 7.2.3. *Let $1 < p \le q < \infty$ and $l - 3/p \ge \max(\delta_k, 0) - 3/q$. Then $W^{l,p}_{\beta,\delta}(\mathcal{K}) \subset W^{0,q}_{\beta-l+3/p-3/q,0}(\mathcal{K})$ and*

$$\text{(7.2.4)} \qquad \|\rho^{\beta-l+3/p-3/q} u\|_{L_q(\mathcal{K})} \le c\, \|u\|_{W^{l,p}_{\beta,\delta}(\mathcal{K})}$$

for all $u \in W^{l,p}_{\beta,\delta}(\mathcal{K})$ with a constant c independent of u. Furthermore, the inequality

$$\text{(7.2.5)} \qquad \|\rho^{\beta-l+3/p} u\|_{L_\infty(\mathcal{K})} \le c\, \|u\|_{W^{l,p}_{\beta,\delta}(\mathcal{K})}$$

is valid for arbitrary $u \in W^{l,p}_{\beta,\delta}(\mathcal{K})$ if $l - 3/p > \max(\delta_k, 0)$.

Proof. Let $u \in W^{l,p}_{\beta,\delta}(\mathcal{K})$, $1 < p \le q < \infty$ and $l - 3/p \ge \max(\delta_k, 0) - 3/q$. Furthermore, let ζ_k be smooth functions depending only on $\rho = |x|$ and satisfying the condition (7.2.3). We set $v(x) = u(2^k x)$ and $\eta_k(x) = \zeta_k(2^k x)$. Then $\eta_k(x)$ vanishes for $|x| < 1/2$ and $|x| > 2$. Using [**178**, Theorem 3], we obtain the estimate

$$\|\eta_k v\|_{W^{s,p}(\mathcal{K})} \le c\, \|\eta_k v\|_{W^{l,p}_{\beta,\delta}(\mathcal{K})},$$

where $s = l - \max(\delta_k, 0)$. This inequality together with the continuity of the imbedding $W^{s,p} \subset L_q$ implies

$$\|\eta_k v\|_{L_q(\mathcal{K})} \le c\, \|\eta_k v\|_{W^{l,p}_{\beta,\delta}(\mathcal{K})},$$

where c is independent of u and k. It can be easily verified that

$$\|\eta_k v\|_{L_q(\mathcal{K})} = 2^{-3k/q}\, \|\zeta_k u\|_{L_q(\mathcal{K})} \ \text{ and } \ \|\eta_k v\|_{W^{l,p}_{\beta,\delta}(\mathcal{K})} = 2^{-k(\beta-l)-3k/p}\, \|\zeta_k u\|_{W^{l,p}_{\beta,\delta}(\mathcal{K})}.$$

Consequently

$$\|\rho^{\beta-l+3/p-3/q}\zeta_k u\|_{L_q(\mathcal{K})} \le c\, \|\zeta_k u\|_{W^{l,p}_{\beta,\delta}(\mathcal{K})}$$

and

$$\|\rho^{\beta-l+3/p-3/q} u\|_{L_q(\mathcal{K})} \le c \Big(\sum_{k=-\infty}^{+\infty} \|\rho^{\beta-l+3/p-3/q}\zeta_k u\|^q_{L_q(\mathcal{K})} \Big)^{1/q}$$

$$\le c \Big(\sum_{k=-\infty}^{+\infty} \|\zeta_k u\|^q_{W^{l,p}_{\beta,\delta}(\mathcal{K})} \Big)^{1/q} \le c \Big(\sum_{k=-\infty}^{+\infty} \|\zeta_k u\|^p_{W^{l,p}_{\beta,\delta}(\mathcal{K})} \Big)^{1/p} \le c\, \|u\|_{W^{l,p}_{\beta,\delta}(\mathcal{K})}.$$

This proves (7.2.4). The proof of (7.2.5) proceeds analogously. □

We will also employ the following generalization of the second part of Lemma 7.2.3.

LEMMA 7.2.4. *If $u \in W^{l,p}_{\beta,\delta}(\mathcal{K})$, $l > 3/p$, then*

$$\rho^{\beta-l+3/p} \prod_{k=1}^{d} \Big(\frac{r_k}{\rho}\Big)^{\sigma_k} u \in L_\infty(\mathcal{K}), \tag{7.2.6}$$

where $\sigma_k = 0$ for $\delta_k < l - 3/p$, $\sigma_k = 1/p + \varepsilon$ for $l - 3/p \le \delta_k \le l - 2/p$, and $\sigma_k = \delta_k - l + 3/p$ for $\delta_k > l - 2/p$ (ε is an arbitrarily small positive number).

Proof. Let ψ_k be a smooth function on the unit sphere S^2 such that $\psi_k = 1$ in a neighborhood of the corner $S^2 \cap M_k$ and $\psi_k = 0$ in a neighborhood of the the other corners $S^2 \cap M_j$ of Ω. We extend ψ_k to $\mathbb{R}^3\backslash\{0\}$ by $\psi_k(x) = \psi_k(x/|x|)$. Then $\psi_k u \in W^{l,p}_{\beta,\delta_k}(\mathcal{K})$. Obviously, it suffices to prove (7.2.6) for the function $\psi_k u$. From Lemma 7.2.3 it follows that $\rho^{\beta-l+3/p}\psi_k u \in L_\infty(\mathcal{K})$ if $\delta_k < l - 3/p$. In the case $\delta_k > l - 2/p$, the space $W^{l,p}_{\beta,\delta_k}(\mathcal{K})$ coincides with $V^{l,p}_{\beta,\delta_k}(\mathcal{K})$. According to Lemma 3.6.2, the last space is imbedded in $N^{0,\sigma}_{\beta-l+\sigma+3/p,\delta_k-l+\sigma+3/p}(\mathcal{K})$ for arbitrary $\sigma < l - 3/p$. Therefore in particular,

$$\rho^{\beta-l+3/p} \prod_{k=1}^{d} \Big(\frac{r_k}{\rho}\Big)^{\delta_k-l+3/p} \psi_k u \in L_\infty(\mathcal{K}).$$

For $l - 3/p \le \delta_k < l - 2/p$, the assertion follows from the imbedding $W^{l,s}_{\beta,\delta_k}(\mathcal{K}) \subset W^{l,p}_{\beta,l-2/p+\varepsilon}(\mathcal{K}) = V^{l,p}_{\beta,l-2/p+\varepsilon}(\mathcal{K})$ and from Lemma 3.6.2. □

Finally, we prove the following assertion by means of Theorems 6.2.7 and 6.2.11.

LEMMA 7.2.5. *Let $u \in W^{l,p}_{\beta,\delta}(\mathcal{K})$, and let j be an integer, $j \ge 1$. Then u admits the decomposition $u = v + w$, where $v \in V^{l,p}_{\beta,\delta}(\mathcal{K})$, $w \in W^{l+j,p}_{\beta+j,\delta+j}(\mathcal{K})$ and*

$$\|v\|_{V^{l,p}_{\beta,\delta}(\mathcal{K})} + \|w\|_{W^{l+j,p}_{\beta+j,\delta+j}(\mathcal{K})} \le c\, \|u\|_{W^{l,p}_{\beta,\delta}(\mathcal{K})}.$$

P r o o f. Let ζ_k be smooth functions depending only on $\rho = |x|$ and satisfying the conditions (7.2.3). We set

$$\tilde{u}_k(x) = \zeta_k(2^k x)\, u(2^k x).$$

Obviously, $\tilde{u}_k \in W^{l,s}_{\beta,\delta}(\mathcal{K})$ and $\tilde{u}_k(x) = 0$ for $|x| < 1/2$ and for $|x| > 2$. Consequently by Theorems 6.2.7 and 6.2.11, there exist functions $\tilde{v}_k \in V^{l,p}_{\beta,\delta}(\mathcal{K})$ and $\tilde{w}_k \in W^{l+j,p}_{\beta+j,\delta+j}(\mathcal{K})$ with supports in $\{x : 1/4 < |x| < 4\}$ such that $\tilde{u}_k = \tilde{v}_k + \tilde{w}_k$ and

$$\|\tilde{v}_k\|_{V^{l,p}_{\beta,\delta}(\mathcal{K})} + \|\tilde{w}_k\|_{W^{l+j,p}_{\beta+j,\delta+j}(\mathcal{K})} \le c\, \|\tilde{u}_k\|_{W^{l,p}_{\beta,\delta}(\mathcal{K})},$$

where c is independent of u and k. Let $v_k(x) = \tilde{v}_k(2^{-k}x)$ and $w_k(x) = \tilde{w}_k(2^{-k}x)$. Then the supports of v_k and w_k are contained in $\{x : 2^{k-2} < |x| < 2^{k+2}\}$. Furthermore, $\zeta_k u = v_k + w_k$ for all k and

$$\begin{aligned}\|v_k\|_{V^{l,p}_{\beta,\delta}(\mathcal{K})} + \|w_k\|_{W^{l+j,p}_{\beta+j,\delta+j}(\mathcal{K})} &= 2^{k(\beta-l+3/p)}\Big(\|\tilde{v}_k\|_{V^{l,p}_{\beta,\delta}(\mathcal{K})} + \|\tilde{w}_k\|_{W^{l+j,p}_{\beta+j,\delta+j}(\mathcal{K})}\Big)\\ &\le c\, 2^{k(\beta-l+3/p)}\, \|\tilde{u}_k\|_{W^{l,p}_{\beta,\delta}(\mathcal{K})} = c\, \|\zeta_k u\|_{W^{l,p}_{\beta,\delta}(\mathcal{K})}\,.\end{aligned}$$

Thus $u = v + w$, where

$$v = \sum_{k=-\infty}^{+\infty} v_k \quad \text{and} \quad w = \sum_{k=-\infty}^{+\infty} w_k.$$

Since v_k and v_m have disjoint supports for $|k-m| \ge 4$, it follows that $v \in V^{l,p}_{\beta,\delta}(\mathcal{K})$ and

$$\|v\|^p_{V^{l,p}_{\beta,\delta}(\mathcal{K})} \le 7^{p-1} \sum_k \|v_k\|^p_{V^{l,p}_{\beta,\delta}(\mathcal{K})} \le c \sum_k \|\zeta_k u\|^p_{W^{l,p}_{\beta,\delta}(\mathcal{K})} \le c'\, \|u\|^p_{W^{l,p}_{\beta,\delta}(\mathcal{K})}.$$

Analogously, the norm of w in $W^{l+j,p}_{\beta+j,\delta+j}(\mathcal{K})$ can be estimated by the norm of u in $W^{l,p}_{\beta,\delta}(\mathcal{K})$. This proves the lemma. □

7.3. Solvability of the boundary value problem in weighted L_2 Sobolev spaces

Using the estimate for solutions of the parameter depending problem (7.1.16), (7.1.17) given in Theorem 7.1.2, we show in this section (cf. Theorem 7.3.2) that the boundary value problem (7.1.1)–(7.1.3) is uniquely solvable in the space $W^{2,2}_{\beta,\delta}(\mathcal{K})^\ell$ if the line $\operatorname{Re}\lambda = -\beta+1/2$ is free of eigenvalues of the pencil $\mathfrak{A}(\lambda)$ and the components δ_k of δ satisfy the inequalities

$$\max(0, 1-\delta_+^{(k)}) < \delta_k < 1 \quad \text{for } k = 1, \dots, d. \tag{7.3.1}$$

Furthermore, we prove the existence and uniqueness of variational solutions in the weighted space $V^{1,2}_\beta(\mathcal{K})^\ell = W^{1,2}_{\beta,0}(\mathcal{K})^\ell$ if the line $\operatorname{Re}\lambda = -\beta-1/2$ does not contain eigenvalues of the pencil $\mathfrak{A}(\lambda)$.

7.3.1. Reduction to zero boundary data. If $g_j \in V^{l-d_j-1/p,p}_{\beta,\delta}(\Gamma_j)^\ell$, then by Lemma 3.3.3, there exists a vector function $u \in V^{l,p}_{\beta,\delta}(\mathcal{K})^\ell$ satisfying the boundary conditions (7.1.2), (7.1.3). In the class of the spaces $W^{l,p}_{\beta,\delta}$, the same result is only valid under additional assumptions on the boundary data g_j. Using Lemmas 6.4.3 and 6.4.4, we obtain the following assertion.

LEMMA 7.3.1. *Let $g_j \in W_{\beta,\delta}^{l-d_j-1/p,p}(\Gamma_j)^\ell$ for $j = 1, \dots, d$, where $l \ge 2$ and the components δ_k of δ satisfy the inequality $\delta_k + 2/p > l - 2$. We suppose that $\lambda = 1$ is not an eigenvalue of the pencil $A_k(\lambda)$ if $\delta_k + 2/p \le l - 1$. If the edge M_k is adjacent to faces Γ_{k_+} and Γ_{k_-} with indices $k_\pm \in I_0$, then we assume in addition that $\delta_k + 2/p \ne l$ and*

$$g_{k_+}|_{M_k} = g_{k_-}|_{M_k} \tag{7.3.2}$$

for $\delta_k + 2/p < l$. Then there exists a vector function $u \in W_{\beta,\delta}^{l,p}(\mathcal{K})^\ell$ satisfying the boundary conditions (7.1.2), (7.1.3) and the estimate

$$\|u\|_{W_{\beta,\delta}^{l,p}(\mathcal{K})^\ell} \le c \sum_{j=1}^d \|g_j\|_{W_{\beta,\delta}^{l-d_j-1/p,p}(\Gamma_j)^\ell}, \tag{7.3.3}$$

where the constant c is independent of g_j.

P r o o f. Let ζ_k be smooth functions depending only on $\rho = |x|$ and satisfying the conditions (7.2.3). We put

$$h_{j,k}(x) = 2^{kd_j}\, \zeta_k(2^k x)\, g_j(2^k x).$$

Obviously, $h_{j,k}(x) = 0$ for $|x| < \frac{1}{2}$ and $|x| > 2$. Consequently by Lemmas 6.4.3 and 6.4.4, there exist vector functions $v_k \in W_{\beta,\delta}^{l,p}(\mathcal{K})^\ell$ such that

$$v_k = h_{j,k} \text{ on } \Gamma_j \text{ for } j \in I_0, \quad N(D_x)\, v_k = h_{j,k} \text{ on } \Gamma_j \text{ for } j \in I_1.$$

Furthermore, $v_k(x) = 0$ for $|x| < 1/4$ and $|x| > 4$, and there exists a constant c such that

$$\|v_k\|_{W_{\beta,\delta}^{l,p}(\mathcal{K})^\ell} \le c \sum_{j=1}^d \|h_{j,k}\|_{W_{\beta,\delta}^{l-d_j-1/p,p}(\Gamma_j)^\ell}. \tag{7.3.4}$$

We define $u_k(x) = v_k(2^{-k}x)$. Then $u_k = \zeta_k g_j$ on Γ_j for $j \in I_0$, $N(D_x)u_k = \zeta_k g_j$ on Γ_j for $j \in I_1$, $u_k(x) = 0$ for $|x| < 2^{k-2}$ and for $|x| > 2^{k+2}$. Moreover, u_k satisfies (7.3.4) with $\zeta_k g_j$ instead of $h_{k,j}$ and a constant c independent of k and g_j. Consequently, the vector function $u = \sum u_k$ satisfies the boundary conditions (7.1.2) and (7.1.3). Inequality (7.3.3) results from Lemma 7.2.1. □

7.3.2. Existence and uniqueness of solutions. Let again $\delta_+^{(k)}$ be the greatest real number such that the strip $0 < \operatorname{Re}\lambda < \delta_+^{(k)}$ is free of eigenvalues of the pencil $A_k(\lambda)$. In the same way as for the Dirichlet problem (cf. Theorem 3.3.7), we apply the Mellin transform (3.3.12) in order to prove the following existence and uniqueness theorem.

THEOREM 7.3.2. *Let $f \in W_{\beta,\delta}^{0,2}(\mathcal{K})^\ell$ and $g_j \in W_{\beta,\delta}^{2-d_j-1/2,2}(\Gamma_j)^\ell$ for $j = 1, \dots, d$. We suppose that the line $\operatorname{Re}\lambda = -\beta + 1/2$ does not contain eigenvalues of the pencil $\mathfrak{A}(\lambda)$, the components of δ satisfy the inequalities (7.3.1), and that the boundary data g_j satisfy the compatibility condition*

$$g_{k_+}\big|_{M_k} = g_{k_-}\big|_{M_k} \quad \text{if } d_{k_+} = d_{k_-} = 0. \tag{7.3.5}$$

(Here Γ_{k_+} and Γ_{k_-} are the faces adjacent to the edge M_k.) Then the boundary value problem (7.1.1)–(7.1.3) has a uniquely determined solution $u \in W_{\beta,\delta}^{2,2}(\mathcal{K})^\ell$.

P r o o f. By Lemma 7.3.1, we may assume without loss of generality that $g_j = 0$ for $j = 1, \dots, d$. Applying the Mellin transform $\mathcal{M}_{\rho\to\lambda}$ with respect to $\rho = |x|$ to the equations (7.1.1)–(7.1.3), we obtain the parameter-dependent problem

$$\mathcal{L}(\lambda)\,\tilde{u}(\lambda,\omega) = \tilde{F}(\lambda,\omega) \quad \text{for } \omega \in \Omega,$$
$$\tilde{u}(\lambda,\cdot) = 0 \text{ on } \gamma_j \text{ for } j \in I_0, \quad \mathcal{N}(\lambda)\,\tilde{u}(\lambda,\cdot) = 0 \text{ on } \gamma_j \text{ for } j \in I_1,$$

where $\tilde{F}(\lambda,\cdot) = \mathcal{M}_{\rho\to\lambda}(\rho^2 f) \in W_\delta^{0,2}(\Omega)^\ell$ for almost all λ. By Theorem 7.1.2, the last problem is uniquely solvable in $W_\delta^{2,2}(\Omega)^\ell$ for every λ on the line $\operatorname{Re}\lambda = -\beta + 1/2$, and the solution satisfies the estimate

$$\sum_{j=0}^{2} |\lambda|^{4-2j}\, \|\tilde{u}(\lambda,\cdot)\|^2_{W_\delta^{j,2}(\Omega)^\ell} \le c\, \|\tilde{F}(\cdot,\omega)\|^2_{W_\delta^{0,2}(\Omega)^\ell}\,. \tag{7.3.6}$$

Consequently, the function

$$u(\rho,\omega) = \frac{1}{2\pi i} \int\limits_{-\beta+1/2-i\infty}^{-\beta+1/2+i\infty} \rho^\lambda\, \tilde{u}(\lambda,\omega)\, d\lambda$$

solves the problem (7.1.1)–(7.1.3). By Parseval's equality (see also (7.2.2)), the norm of u in $W_{\beta,\delta}^{l,2}(\mathcal{K})$ is equivalent to

$$\Big(\frac{1}{2\pi i} \int\limits_{l-\beta-3/2-i\infty}^{l-\beta-3/2+i\infty} \sum_{j=0}^{l} |\lambda|^{2l-2j}\, \|\tilde{u}(\lambda,\cdot)\|^2_{W_\delta^{j,2}(\Omega)}\, d\lambda\Big)^{1/2}. \tag{7.3.7}$$

Consequently, we conclude from (7.3.6) that $u \in W_{\beta,\delta}^{2,2}(\mathcal{K})^\ell$ and

$$\|u\|_{W_{\beta,\delta}^{2,2}(\mathcal{K})^\ell} \le c\, \|f\|_{W_{\beta,\delta}^{0,2}(\mathcal{K})^\ell}\,.$$

This proves the theorem. □

7.3.3. Some estimates for solutions of the boundary value problem. The subsequent lemma was proved for the Dirichlet problem in Section 3.3. Using Theorem 6.1.3, we can prove this result in the same way for the Neumann and mixed problems.

LEMMA 7.3.3. *Let $u \in V_{\beta-l,\delta-l}^{0,p}(\mathcal{K})^\ell$ be a solution of the boundary value problem (7.1.1)–(7.1.3), where $f \in V_{\beta,\delta}^{l-2,p}(\mathcal{K})^\ell$ and $g_j \in V_{\beta,\delta}^{l-d_j-1/p,p}(\Gamma_j)^\ell$, $l \ge 2$. Then $u \in V_{\beta,\delta}^{l,p}(\mathcal{K})^\ell$ and*

$$\|u\|_{V_{\beta,\delta}^{l,p}(\mathcal{K})^\ell} \le c\Big(\|u\|_{V_{\beta-l,\delta-l}^{0,p}(\mathcal{K})^\ell} + \|f\|_{V_{\beta,\delta}^{l-2,p}(\mathcal{K})^\ell} + \sum_{j=1}^{d} \|g_j\|_{V_{\beta,\delta}^{l-d_j-1/p,p}(\Gamma_j)^\ell}\Big).$$

P r o o f. Let ζ_k be smooth functions depending only on $\rho = |x|$ and satisfying the conditions (7.2.3). Furthermore, let

$$\eta_k = \sum_{j=k-2}^{k+2} \zeta_j\,.$$

By Theorem 6.1.3, $\zeta_k u \in V^{l,p}_{\beta,\delta}(\mathcal{K})^\ell$ and

$$\|\zeta_k u\|^p_{V^{l,p}_{\beta,\delta}(\mathcal{K})^\ell} \le c\Big(\|\eta_k u\|^p_{V^{0,p}_{\beta-l,\delta-l}(\mathcal{K})^\ell} + \|\eta_k f\|^p_{V^{l-2,p}_{\beta,\delta}(\mathcal{K})^\ell} + \sum_{j=1}^d \|\eta_k g_j\|_{V^{l-d_j-1/p,p}_{\beta,\delta}(\Gamma_j)^\ell}\Big). \tag{7.3.8}$$

Here by the same argument as in the proof of Theorem 3.3.5, the constant c is independent of u and k. Using Lemma 7.2.1, we obtain the assertion of the lemma. □

A similar result in the spaces $W^{l,p}_{\beta,\delta}$ can be proved analogously by means of Corollary 6.4.2.

LEMMA 7.3.4. *Let $u \in W^{l,p}_{\beta,\delta}(\mathcal{K})^\ell$ be a solution of the boundary value problem (7.1.1)–(7.1.3), where $f \in W^{l+k-2,p}_{\beta+k,\delta+k}(\mathcal{K})^\ell$, $g_j \in W^{l+k-d_j-1/p,p}_{\beta+k,\delta+k}(\Gamma_j)^\ell$, $k \ge 0$. Then $u \in W^{l+k,p}_{\beta+k,\delta+k}(\mathcal{K})^\ell$ and*

$$\|u\|_{W^{l+k,p}_{\beta+k,\delta+k}(\mathcal{K})^\ell} \le c\Big(\|f\|_{W^{l+k-2,p}_{\beta+k,\delta+k}(\mathcal{K})^\ell} + \sum_{j=1}^d \|g_j\|_{W^{l+k-d_j-1/p,p}_{\beta+k,\delta+k}(\Gamma_j)^\ell} + \|u\|_{W^{l,p}_{\beta,\delta}(\mathcal{K})^\ell}\Big).$$

Finally, we prove the following regularity assertion which is an analog to Lemma 6.5.3.

LEMMA 7.3.5. *Let $u \in W^{l,p}_{\beta,\delta}(\mathcal{K})^\ell$ be a solution of the problem (7.1.1)–(7.1.3) such that $\rho\partial_\rho u \in W^{l,p}_{\beta,\delta}(\mathcal{K})^\ell$. If $f \in W^{l-1,p}_{\beta+1,\delta}(\mathcal{K})^\ell$, $g_j \in W^{l+1-d_j-1/p,p}_{\beta+1,\delta}(\Gamma_j)^\ell$, and the strip $l-\delta_k-2/p \le \operatorname{Re}\lambda \le l+1-\delta_k-2/p$ is free of eigenvalues of the pencil $A_k(\lambda)$ for $k=1,\dots,d$, then $u \in W^{l+1,p}_{\beta+1,\delta}(\mathcal{K})^\ell$ and*

$$\|u\|_{W^{l+1,p}_{\beta+1,\delta}(\mathcal{K})^\ell} \le c\Big(\sum_{j=0}^1 \|(\rho\partial_\rho)^j u\|_{W^{l,p}_{\beta,\delta}(\mathcal{K})^\ell} + \|f\|_{W^{l-1,p}_{\beta+1,\delta}(\mathcal{K})^\ell} + \sum_{j=1}^d \|g_j\|_{W^{l+1-d_j-1/p,p}_{\beta+1,\delta}(\Gamma_j)^\ell}\Big).$$

P r o o f. By Lemma 7.3.4, the solution u belongs to $W^{l+1,p}_{\beta+1,\delta+1}(\mathcal{K})^\ell$. We denote by ζ_k, η_k the same functions as in the proof of Lemma 7.3.3. Furthermore, we set $\tilde\zeta_k(x) = \zeta_k(2^k x)$, $\tilde\eta_k(x) = \eta_k(2^k x)$ and $v(x) = u(2^k x)$. Since $\tilde\zeta_k(x) = 0$ for $|x| < 1/2$ and $|x| > 2$, we conclude from Lemma 6.5.3 and from the assumptions on u, f and g_j that $\tilde\zeta_k v \in W^{l+1,p}_{\beta+1,\delta}(\mathcal{K})^\ell$ and

$$\|\tilde\zeta_k v\|^p_{W^{l+1,p}_{\beta+1,\delta}(\mathcal{K})^\ell} \le c\Big(\sum_{j=0}^1 \|\tilde\eta_k(\rho\partial_\rho)^j v\|^p_{W^{l,p}_{\beta,\delta}(\mathcal{K})^\ell} + \|\tilde\eta_k L(D_x)\, v\|^p_{W^{l-1,p}_{\beta+1,\delta}(\mathcal{K})^\ell} + \sum_{j\in I_0} \|\tilde\eta_k v\|^p_{W^{l+1-1/p,p}_{\beta+1,\delta}(\Gamma_j)^\ell} + \sum_{j\in I_1} \|\tilde\eta_k N(D_x)\, v\|^p_{W^{l-1/p,p}_{\beta+1,\delta}(\Gamma_j)^\ell}\Big),$$

where c is independent of k. Multiplying the last inequality by $2^{kp(\beta-l)+3k}$ and substituting $2^k = y$, we obtain the same inequality with ζ_k, η_k instead of $\tilde\zeta_k, \tilde\eta_k$ for the vector function u. Now the desired estimate for u results from Lemma 7.2.1. □

7.3.4. Existence of variational solutions. Let $V_\beta^{l,2}(\mathcal{K}) = W_{\beta,0}^{l,2}(\mathcal{K})$ be the weighted Sobolev space with the norm

$$\|u\|_{V_\beta^{l,2}(\mathcal{K})} = \Big(\int_{\mathcal{K}} \sum_{|\alpha|\le l} |x|^{2(\beta-l+|\alpha|)} \big|\partial_x^\alpha u(x)\big|^2 \, dx \Big)^{1/2}.$$

It follows from Hardy's inequality that $V_0^{1,2}(\mathcal{K})$ coincides with $L^{1,2}(\mathcal{K})$ and that the $V_0^{1,2}(\mathcal{K})$- and $L^{1,2}(\mathcal{K})$-norms are equivalent. Let

$$\mathcal{H}_\beta = \{ v \in V_\beta^{1,2}(\mathcal{K})^\ell,\ u|_{\Gamma_j} = 0 \text{ for } j \in I_0 \}.$$

Obviously, the sesquilinear form $b_{\mathcal{K}}(\cdot,\cdot)$ is continuous on $\mathcal{H}_\beta \times \mathcal{H}_{-\beta}$. We define the linear and continuous operator $\mathcal{A}_\beta : \mathcal{H}_\beta \to \mathcal{H}_{-\beta}^*$ by

$$\big(\mathcal{A}_\beta u, v\big)_{\mathcal{K}} = b_{\mathcal{K}}(u,v) \quad \text{for all } u \in \mathcal{H}_\beta\,,\ v \in \mathcal{H}_{-\beta}$$

and prove the following assertion.

LEMMA 7.3.6. *There exists a constant c such that*

$$\|u\|_{V_\beta^{1,2}(\mathcal{K})^\ell} \le c \Big(\|\mathcal{A}_\beta u\|_{\mathcal{H}_{-\beta}^*} + \|u\|_{V_{\beta-1}^{0,2}(\mathcal{K})^\ell} \Big) \tag{7.3.9}$$

for all $u \in \mathcal{H}_\beta$.

P r o o f. Let $\zeta_k = \zeta_k(\rho)$ be smooth real-valued functions satisfying the conditions (7.2.3), and let $\eta_k = \zeta_{k-1} + \zeta_k + \zeta_{k+1}$. By Lemma 7.2.1, the norm in $V_\beta^{1,2}(\mathcal{K})$ is equivalent to the norm

$$\|u\| = \Big(\sum_{k=-\infty}^{+\infty} \|\zeta_k u\|^2_{V_\beta^{1,2}(\mathcal{K})} \Big)^{1/2}.$$

Analogously, the norm in $\mathcal{H}_{-\beta}^*$ is equivalent to

$$\Big(\sum_{k=-\infty}^{+\infty} \|\zeta_k f\|^2_{\mathcal{H}_{-\beta}^*} \Big)^{1/2}$$

(cf. Lemma 3.1.2). We show that

$$\|\zeta_k u\|^2_{V_\beta^{1,2}(\mathcal{K})^\ell} \le c\, \|\zeta_k \mathcal{A}_\beta u\|^2_{\mathcal{H}_{-\beta}^*} + \varepsilon\, \|\eta_k u\|^2_{V_\beta^{1,2}(\mathcal{K})^\ell} + C(\varepsilon)\, \|\eta_k u\|^2_{V_{\beta-1}^{0,2}(\mathcal{K})^\ell}, \tag{7.3.10}$$

where c, ε and $C(\varepsilon)$ are independent of u and k and ε can be chosen arbitrarily small.

Let first $k = 0$. Integrating by parts, we get

$$b_{\mathcal{K}}(\zeta_0 u, v) = \big(\zeta_0 \mathcal{A}_\beta u, v\big)_{\mathcal{K}} + c_1(u,v) + c_2(u,v), \tag{7.3.11}$$

where

$$c_1(u,v) = \int_{\mathcal{K}} \sum_{i,j=1}^{3} \Big(A_{i,j}\, (\partial_{x_i}\zeta_0)\, u \cdot \partial_{x_j}\bar v \, dx - A_{i,j} u \cdot \partial_{x_i}(\bar v \partial_{x_j}\zeta_0) \Big)\, dx$$

and

$$c_2(u,v) = -\sum_{k\in I_1} \int_{\Gamma_k} \sum_{i,j=1}^{3} A_{i,j} (\partial_{x_j}\zeta_0)\, n_i\, u \cdot \bar v \, dx.$$

Obviously

$$|c_1(u,v)| \le c\, \|\eta_0 u\|_{L_2(\mathcal{K})^\ell}\, \|v\|_{V_0^{1,2}(\mathcal{K})^\ell}$$

with a constant c independent of u and v. Furthermore,

$$\begin{aligned} |c_2(u,v)| &\le c \sum_{k\in I_1} \|\eta_0 u\|_{L_2(\Gamma_k)^\ell}\, \|v\|_{L_2(\Gamma_k)^\ell} \\ &\le \Big(\varepsilon\, \|\eta_0 u\|_{V_0^{1,2}(\mathcal{K})^\ell} + C_0(\varepsilon)\, \|\eta_0 u\|_{L_2(\mathcal{K})^\ell}\Big)\, \|v\|_{V_0^{1,2}(\mathcal{K})^\ell}\,, \end{aligned}$$

where ε can be chosen arbitrarily small. Since the form $b_{\mathcal{K}}$ is $\mathcal{H}_{\mathcal{K}}$-elliptic, it follows from (7.3.11) that

$$\|\zeta_0 u\|^2_{V_0^{1,2}(\mathcal{K})^\ell} \le c\, \|\zeta_0 \mathcal{A}_\beta u\|^2_{\mathcal{H}^*_{\mathcal{K}}} + \varepsilon\, \|\eta_0 u\|^2_{V_0^{1,2}(\mathcal{K})^\ell} + C(\varepsilon)\, \|\eta_0 u\|^2_{L_2(\mathcal{K})^\ell}\,.$$

This proves (7.3.10) for $k=0$. By means of the transformation $x = 2^k y$, we obtain (7.3.10) with the same constants c, ε and $C(\varepsilon)$ for $k \neq 0$. Summing up in (7.3.10), we arrive at the estimate

$$\|u\|^2_{V_\beta^{1,2}(\mathcal{K})^\ell} \le c\, \|\mathcal{A}_\beta u\|^2_{\mathcal{H}^*_{-\beta}} + \varepsilon\, \|u\|^2_{V_\beta^{1,2}(\mathcal{K})^\ell} + C(\varepsilon)\, \|u\|^2_{V_{\beta-1}^{0,2}(\mathcal{K})^\ell}\,,$$

where ε can be chosen arbitrarily small. This proves the lemma. □

We turn to the main result of this subsection.

THEOREM 7.3.7. *Suppose that there are no eigenvalues of the pencil $\mathfrak{A}(\lambda)$ on the line* $\operatorname{Re}\lambda = -\beta - 1/2$. *Then the operator $\mathcal{A}_\beta$ is an isomorphism.*

P r o o f. Let $u \in \mathcal{H}_\beta$. Since $V_\beta^{1,2}(\mathcal{K}) \subset W^{0,2}_{\beta-1,\varepsilon-1}(\mathcal{K})$, where ε is an arbitrarily small positive number (see Lemma 7.2.2), the vector function

$$w \stackrel{def}{=} \rho^{2(\beta-1)} \prod_{k=1}^{d} \Big(\frac{r_k}{\rho}\Big)^{2(\varepsilon-1)} u$$

belongs to $W^{0,2}_{1-\beta,1-\varepsilon}(\mathcal{K})^\ell$. From the absence of eigenvalues of the pencil $\mathfrak{A}(\lambda)$ on the line $\operatorname{Re}\lambda = -\beta - 1/2$ it follows that the line $\operatorname{Re}\lambda = \beta - 1/2$ is free of eigenvalues of the operator pencil $\mathfrak{A}^+(\lambda)$. Consequently by Theorem 7.3.2, there exists a solution $v \in W^{2,2}_{1-\beta,1-\varepsilon}(\mathcal{K})^\ell$ of the formally adjoint problem (see (7.1.11)–(7.1.13))

$$\begin{aligned} &L^+(D_x)v = w \quad \text{in } \mathcal{K}, \\ &v = 0 \quad \text{on } \Gamma_j \text{ for } j \in I_0, \quad N^+(D_x)\, v = 0 \quad \text{on } \Gamma_j \text{ for } j \in I_1\,. \end{aligned}$$

Here ε is a sufficiently small positive real number. This solution satisfies the inequality

$$\|v\|_{W^{2,2}_{1-\beta,1-\varepsilon}(\mathcal{K})^\ell} \le c\, \|w\|_{W^{0,2}_{1-\beta,1-\varepsilon}(\mathcal{K})^\ell} = c\, \|u\|_{W^{0,2}_{\beta-1,\varepsilon-1}(\mathcal{K})^\ell} \tag{7.3.12}$$

with a constant c independent of u. This implies

$$\begin{aligned} \|u\|^2_{W^{0,2}_{\beta-1,\varepsilon-1}(\mathcal{K})^\ell} &= \int_{\mathcal{K}} u \cdot \overline{w}\, dx = \int_{\mathcal{K}} u \cdot \overline{L^+(D_x)v}\, dx = b_{\mathcal{K}}(u,v) = \big(\mathcal{A}_\beta u, v\big)_{\mathcal{K}} \\ &\le c\, \|\mathcal{A}_\beta u\|_{\mathcal{H}^*_{-\beta}}\, \|v\|_{V^{1,2}_{-\beta}(\mathcal{K})^\ell} \le c\, \|\mathcal{A}_\beta u\|_{\mathcal{H}^*_{-\beta}}\, \|v\|_{W^{2,2}_{1-\beta,1-\varepsilon}(\mathcal{K})^\ell} \\ &\le c\, \|\mathcal{A}_\beta u\|_{\mathcal{H}^*_{-\beta}}\, \|u\|_{W^{0,2}_{\beta-1,\varepsilon-1}(\mathcal{K})^\ell}. \end{aligned}$$

Consequently,

$$\|u\|_{V^{0,2}_{\beta-1}(\mathcal{K})^\ell} \le c\, \|u\|_{W^{0,2}_{\beta-1,\varepsilon-1}(\mathcal{K})^\ell} \le c\, \|\mathcal{A}_\beta u\|_{\mathcal{H}^*_{-\beta}}.$$

This estimate together with Lemma 7.3.6 yields

(7.3.13) $$\|u\|_{V_\beta^{1,2}(\mathcal{K})^\ell} \le c\, \|\mathcal{A}_\beta u\|_{\mathcal{H}^*_{-\beta}}.$$

Therefore, the kernel of $\mathcal{A}_\beta$ is trivial and its image is closed.

We prove that for every $F \in \mathcal{H}^*_{-\beta}$ there exists a solution of the equation $\mathcal{A}_\beta u = F$. Let $\{f_k\}_{k\ge 0} \subset C_0^\infty(\overline{\mathcal{K}})^\ell$ be a sequence which converges to F in $\mathcal{H}^*_{-\beta}$. For every k, there exists a solution $u_k \in W^{2,2}_{\beta+1,1-\varepsilon}(\mathcal{K})^\ell \subset V_\beta^{1,2}(\mathcal{K})^\ell$ of the problem

$$L(D_x)\, u_k = f_k \text{ in } \mathcal{K}, \quad u_k|_{\Gamma_j} = 0 \text{ for } j\in I_0, \quad N(D_x)u_k|_{\Gamma_j} = 0 \text{ for } j \in I_1$$

(cf. By Theorem 7.3.2). Since according to (7.3.13)

$$\|u_k - u_l\|_{V_\beta^{1,2}(\mathcal{K})^\ell} \le c\, \|f_k - f_l\|_{\mathcal{H}^*_{-\beta}}$$

with a constant c independent of k and l, the functions u_k form a Cauchy sequence in $V_\beta^{1,2}(\mathcal{K})^\ell$. Its limit u is the solution of the equation $\mathcal{A}_\beta u = F$. The proof is complete. □

7.4. Regularity results for variational solutions

We consider the variational solutions of the boundary value problem (7.1.1)–(7.1.3). The existence and uniqueness of these solutions in weighted Sobolev spaces were proved in the preceding section. Our goal is to establish conditions which ensure that the variational solution $u \in V^{1,2}_{\beta-l+1}(\mathcal{K})^\ell$ belongs to the weighted space $W^{l,2}_{\beta,\delta}(\mathcal{K})^\ell$. The answer to this question is given in Theorem 7.4.1 below. For the Dirichlet and mixed problems we obtain an analogous result in the subspaces $W^{l,2}_{\beta,\delta}(\mathcal{K};J)$ of $W^{l,2}_{\beta,\delta}(\mathcal{K})$. Furthermore, we improve the result of Theorem 7.4.1 for the Neumann problem for a special class of elliptic systems which includes the Lamé system. Other topics considered in this section are the asymptotics near the vertex and the regularity of the ρ-derivatives.

7.4.1. Solvability in $W^{l,2}_{\beta,\delta}(\mathcal{K})^\ell$.

We start with a regularity assertion which is based on Theorem 6.5.4.

THEOREM 7.4.1. *Let $u \in V^{1,2}_{\beta-l+1}(\mathcal{K})^\ell$ be a solution of the problem*

(7.4.1) $$b_{\mathcal{K}}(u,v) = \int_{\mathcal{K}} f\cdot \bar{v}\, dx + \sum_{j\in I_1} \int_{\Gamma_j} g_j \cdot \bar{v}\, dx \quad \textit{for all } v \in \mathcal{H}_{l-1-\beta}\,,$$

(7.4.2) $$u = g_j \quad \textit{on } \Gamma_j \textit{ for } j \in I_0\,,$$

where $f \in W^{l-2,2}_{\beta,\delta}(\mathcal{K})^\ell$ and $g_j \in W^{l-d_j-1/2,2}_{\beta,\delta}(\Gamma_j)^\ell$, $l\ge 2$. We suppose that the components δ_k of δ are not integer and satisfy the inequalities

(7.4.3) $$\max(0, l-1-\delta_+^{(k)}) < \delta_k < l-1 \quad \textit{for } k = 1,\dots,d.$$

Then $u \in W^{l,2}_{\beta,\delta}(\mathcal{K})^\ell$ and

$$\|u\|_{W^{l,2}_{\beta,\delta}(\mathcal{K})^\ell} \le c\Big(\|f\|_{W^{l-2,2}_{\beta,\delta}(\mathcal{K})^\ell} + \sum_{j=1}^d \|g_j\|_{W^{l-d_j-1/2,2}_{\beta,\delta}(\Gamma_j)^\ell} + \|u\|_{V^{1,2}_{\beta-l+1}(\mathcal{K})^\ell}\Big).$$

P r o o f. First note that the expression on the right-hand side of (7.4.1) defines a continuous functional on $V^{1,2}_{l-1-\beta}(\mathcal{K})^\ell$. This follows from the continuity of the imbeddings

$$W^{l-2,2}_{\beta,\delta}(\mathcal{K}) \subset V^{l-2,2}_{\beta,l-1-\varepsilon}(\mathcal{K}), \quad W^{l-3/2,2}_{\beta,\delta}(\Gamma_j) \subset V^{l-3/2,2}_{\beta,l-1-\varepsilon}(\Gamma_j),$$

and

$$V^{1,2}_{\beta-l+1}(\mathcal{K}) \subset V^{1,2}_{\beta-l+1,\varepsilon}(\mathcal{K}),$$

where ε is a sufficiently small positive number.

Let ζ_k be smooth functions depending only on $\rho = |x|$ and satisfying the conditions (7.2.3). Furthermore, let $\eta_k = \zeta_{k-1} + \zeta_k + \zeta_{k+1}$, $\tilde{\zeta}_k(x) = \zeta_k(2^k x)$, $\tilde{\eta}_k(x) = \eta_k(2^k x)$, and $v(x) = u(2^k x)$. Then the supports of the functions $\tilde{\zeta}_k$ and $\tilde{\eta}_k(x)$ are contained in the set of all x such that $1/4 \le |x| \le 4$. By Theorem 6.5.4, the vector functions $\zeta_k u$ and $\tilde{\zeta}_k v$ belong to $W^{l,2}_{\beta,\delta}(\mathcal{K})^\ell$ for every integer k. Furthermore,

$$\|\tilde{\zeta}_k v\|^2_{W^{l,2}_{\beta,\delta}(\mathcal{K})^\ell} \le c\Big(\|\tilde{\eta}_k L(D_x)\, v\|^2_{W^{l-2,2}_{\beta,\delta}(\mathcal{K})^\ell} + \sum_{j\in I_0} \|\tilde{\eta}_k v\|^2_{W^{l-1/2,2}_{\beta,\delta}(\Gamma_j)^\ell} + \sum_{j\in I_1} \|\tilde{\eta}_k N(D_x)\, v\|^2_{W^{l-3/2,2}_{\beta,\delta}(\Gamma_j)^\ell} + \|\tilde{\eta}_k v\|^2_{V^{1,2}_{\beta-l+1}(\mathcal{K})^\ell}\Big).$$

By (7.2.3), the constant c is independent of k. Multiplying the last estimate by $2^{2k(\beta-l)+3k}$ and substituting $2^k x = y$, we arrive at the inequality

$$\|\zeta_k u\|^2_{W^{l,2}_{\beta,\delta}(\mathcal{K})^\ell} \le c\Big(\|\eta_k L(D_x) u\|^2_{W^{l-2}_{\beta,\delta}(\mathcal{K})^\ell} + \sum_{j\in I_0} \|\eta_k u\|^2_{W^{l-1/2,2}_{\beta,\delta}(\Gamma_j)^\ell} + \sum_{j\in I_1} \|\eta_k N(D_x) u\|^2_{W^{l-3/2,2}_{\beta,\delta}(\Gamma_j)^\ell} + \|\eta_k u\|^2_{V^{1,2}_{\beta-l+1}(\mathcal{K})^\ell}\Big).$$

Summing up over all k and using Lemma 7.2.1, we obtain the assertion of the theorem. □

As a consequence of the last theorem, we obtain an existence and uniqueness result in the space $W^{l,2}_{\beta,\delta}(\mathcal{K})^\ell$.

COROLLARY 7.4.2. *Let $f \in W^{l-2,2}_{\beta,\delta}(\mathcal{K})^\ell$, $l \ge 2$, and $g_j \in W^{l-d_j-1/2,2}_{\beta,\delta}(\Gamma_j)^\ell$ for $j = 1,\dots,d$. Suppose that the line $\operatorname{Re}\lambda = l - \beta - 3/2$ does not contain eigenvalues of the pencil $\mathfrak{A}(\lambda)$ and that the components of δ are not integer and satisfy the inequalities* (7.4.3). *Furthermore, we assume that the boundary data g_j satisfy the compatibility condition* (7.3.5). *Then the boundary value problem* (7.1.1)–(7.1.3) *has a uniquely determined solution $u \in W^{l,2}_{\beta,\delta}(\mathcal{K})^\ell$ satisfying the estimate*

$$\|u\|_{W^{l,2}_{\beta,\delta}(\mathcal{K})^\ell} \le c\Big(\|f\|_{W^{l-2,2}_{\beta,\delta}(\mathcal{K})^\ell} + \sum_{j=1}^{d} \|g_j\|_{W^{l-d_j-1/2,2}_{\beta,\delta}(\Gamma_j)^\ell}\Big), \tag{7.4.4}$$

where c is independent of f and g_j.

P r o o f. By Lemma 7.3.1, there exists a vector function $u \in W^{l,2}_{\beta,l-1-\varepsilon}(\mathcal{K})^\ell \subset V^{1,2}_{\beta-l+1}(\mathcal{D})^\ell$ satisfying (7.1.2), (7.1.3). Let $U \in \mathcal{H}_{\beta-l+1}$ be the unique solution of the problem

$$b_{\mathcal{K}}(U,v) = -b_{\mathcal{K}}(u,v) + \int_{\mathcal{K}} f\cdot\bar{v}\,dx + \sum_{j\in I_1}\int_{\Gamma_j} g_j\cdot\bar{v}\,dx \quad \text{for all } v\in\mathcal{H}_{l-1-\beta}$$

(see Theorem 7.3.7). Then $w = u + U$ is a solution of the problem (7.4.1), (7.4.2). According to Theorem 7.4.1, this solution belongs to the space $W^{l,2}_{\beta,\delta}(\mathcal{K})^\ell$. The uniqueness of the solution and the estimate (7.4.4) follow immediately from Theorem 7.3.7 and from the imbedding $W^{l,2}_{\beta,\delta}(\mathcal{K}) \subset V^{1,2}_{\beta-l+1}(\mathcal{K})$. □

REMARK 7.4.3. Note that the assertions of Theorem 7.4.1 and Corollary 7.4.2 are even true if the components of δ are arbitrary real numbers satisfying the condition

$$\max(0, l-\delta_+^{(k)}) < \delta_k + 1 < l \quad \text{for } k = 1,\dots,d.$$

This will be proved in Sections 7.6 and 7.7 by means of estimates for the Green's matrix.

7.4.2. Asymptotics of variational solutions. Repeating the proof of Theorem 3.3.10, we obtain an asymptotic decomposition of the solution of the boundary value problem (7.1.1)–(7.1.3) near the vertex of the cone.

THEOREM 7.4.4. *Let $u \in W^{2,2}_{\beta,\delta}(\mathcal{K})^\ell$ be a solution of the boundary value problem* (7.1.1)–(7.1.3), *where*

$$f \in W^{0,2}_{\beta,\delta}(\mathcal{K})^\ell \cap W^{0,2}_{\beta',\delta'}(\mathcal{K})^\ell, \quad g_j \in W^{2-d_j-1/2,2}_{\beta,\delta}(\Gamma_j)^\ell \cap W^{2-d_j-1/2,2}_{\beta',\delta'}(\Gamma_j)^\ell.$$

Suppose that the lines $\operatorname{Re}\lambda = -\beta + 1/2$ *and* $\operatorname{Re}\lambda = -\beta' + 1/2$ *do not contain eigenvalues of the pencil* $\mathfrak{A}(\lambda)$ *and that the components of* δ *and* δ' *satisfy the inequalities* (7.3.1). *Then u admits the decomposition*

$$u = \sum_{\nu=1}^{N}\sum_{j=1}^{I_\nu}\sum_{s=0}^{\kappa_{\nu,j}-1} c_{\nu j s}\,\rho^{\lambda_\nu}\sum_{\sigma=0}^{s}\frac{1}{\sigma!}(\log\rho)^\sigma\, U^{\nu,j,s-\sigma}(\omega) + w, \tag{7.4.5}$$

where w is the uniquely determined solution of the boundary value problem (7.1.1)–(7.1.3) *in the space* $W^{2,2}_{\beta',\delta'}(\mathcal{K})^\ell$, λ_ν *are the eigenvalues of the pencil* $\mathfrak{A}(\lambda)$ *between the lines* $\operatorname{Re}\lambda = -\beta+1/2$ *and* $\operatorname{Re}\lambda = -\beta'+1/2$ *and* $U^{\nu,j,s}$ *are eigenvectors* ($s=0$) *and generalized eigenvectors* ($s>0$) *corresponding to the eigenvalue* λ_ν.

We prove an analogous result for the variational solutions of the boundary value problem.

THEOREM 7.4.5. *Let $u \in \mathcal{H}_\beta$ be a solution of the equation $\mathcal{A}_\beta u = F$, where $F \in \mathcal{H}^*_{-\beta} \cap \mathcal{H}^*_{-\beta'}$. If the lines* $\operatorname{Re}\lambda = -\beta - 1/2$ *and* $\operatorname{Re}\lambda = -\beta' - 1/2$ *are free of eigenvalues of the pencil* $\mathfrak{A}(\lambda)$, *then u admits the decomposition* (7.4.5), *where $w \in \mathcal{H}_{\beta'}$ is the uniquely determined solution of the equation $\mathcal{A}_{\beta'}w = F$ and λ_ν are the eigenvalues of the pencil $\mathfrak{A}(\lambda)$ between the lines* $\operatorname{Re}\lambda = -\beta - 1/2$ *and* $\operatorname{Re}\lambda = -\beta' - 1/2$.

P r o o f. Let χ be an infinitely differentiable function with compact support equal to one near the vertex of the cone $\mathcal{K}$. We assume without loss of generality that $\beta' < \beta$. Then $\chi(u-w) \in V_\beta^{1,2}(\mathcal{K})^\ell$ and $\chi v \in V_{-\beta}^{1,2}(\mathcal{K})^\ell \cap V_{-\beta'}^{1,2}(\mathcal{K})^\ell$ for arbitrary $v \in V_{-\beta}^{1,2}(\mathcal{K})^\ell$. Since $b_{\mathcal{K}}(u-w, \chi v) = 0$, we obtain

$$b_{\mathcal{K}}\big(\chi(u-w), v\big) = \int_{\mathcal{K}} f \cdot \bar{v}\, dx + \sum_{k \in I_1} \int_{\Gamma_k} g_k \cdot \bar{v}\, dx$$

for arbitrary $v \in \mathcal{H}_{-\beta}$, where

$$f = -\sum_{i,j=1}^{3} A_{i,j} \Big((\partial_{x_j}\chi)\, \partial_{x_i}(u-w) + \partial_{x_j}\big((\partial_{x_i}\chi)(u-w)\big)\Big),$$

and

$$g_k = \sum_{i,j=1}^{3} A_{i,j}\, (\partial_{x_i}\chi)\, n_j\, (u-w).$$

Obviously,

$$f \in W_{\beta+1,0}^{0,2}(\mathcal{K})^\ell \cap W_{\beta'+1,0}^{0,2}(\mathcal{K})^\ell \quad \text{and} \quad g_k \in W_{\beta+1,0}^{1/2,2}(\Gamma_j)^\ell \cap W_{\beta'+1,0}^{1/2,2}(\Gamma_j)^\ell.$$

Applying Theorem 7.4.1, we conclude that $\chi(u-w) \in W_{\beta+1,\delta}^{2,2}(\mathcal{K})^\ell$, where the components of δ are arbitrary real numbers satisfying the inequalities (7.3.1). Furthermore, it follows from Theorem 7.4.4 that

$$\chi(u-w) = \Sigma + W, \tag{7.4.6}$$

where Σ is the sum in the decomposition (7.4.5) and $W \in W_{\beta'+1,\delta}^{2,2}(\mathcal{K})^\ell \subset V_{\beta'}^{1,2}(\mathcal{K})^\ell$ is a solution of the boundary value problem

$$\begin{aligned} &L(D_x)\, W = L(D_x)\, (\chi u - \chi w) \quad \text{in } \mathcal{K}, \\ &W|_{\Gamma_j} = 0 \quad \text{for } j \in I_0, \quad N(D_x) W|_{\Gamma_j} = N(D_x)\, (\chi u - \chi w)|_{\Gamma_j} \quad \text{for } j \in I_1\,. \end{aligned}$$

Thus, we obtain

$$u = \Sigma + W + \chi w + (1-\chi) u.$$

Here, the vector function W as well as χw and $(1-\chi)u$ are elements of the space $V_{\beta'}^{1,2}(\mathcal{K})^\ell$. Furthermore, $\mathcal{A}_{\beta'}\big(W + \chi w + (1-\chi)u\big) = F$. Consequently, $W + \chi w + (1-\chi)u = w$. This proves the theorem. □

7.4.3. A regularity result for the ρ-derivatives of the solution. We extend the regularity result of Theorem 7.4.1 to ρ-derivatives of the solution.

THEOREM 7.4.6. *Let $u \in V_{\beta-l+1}^{1,2}(\mathcal{K})^\ell$ be a solution of the problem* (7.4.1),(7.4.2), *where $(\rho\partial_\rho)^\nu f \in W_{\beta,\delta}^{l-2,2}(\mathcal{K})^\ell$ for $\nu = 0,\dots,k$, $g_j = 0$ for $j = 1,\dots,d$. Suppose that the line $\operatorname{Re}\lambda = l - \beta - 3/2$ is free of eigenvalues of the pencil $\mathfrak{A}(\lambda)$ and that the components of δ are not integer and satisfy the inequalities* (7.4.3). *Then $(\rho\partial_\rho)^\nu u \in W_{\beta,\delta}^{l,2}(\mathcal{K})^\ell$ for $\nu = 0, 1, \dots, k$ and*

$$\sum_{\nu=0}^{k} \|(\rho\partial_\rho)^\nu u\|_{W_{\beta,\delta}^{l,2}(\mathcal{K})^\ell} \le c \sum_{\nu=0}^{k} \|(\rho\partial_\rho)^\nu f\|_{W_{\beta,\delta}^{l-2,2}(\mathcal{K})^\ell}\,.$$

P r o o f. Let t be an arbitrary real number, $1/2 < t < 1$. For an arbitrary function ϕ, we define $\phi_t(x) = \dfrac{\phi(x) - \phi(tx)}{1-t}$. Obviously,

$$u_t(x) \to \sum_{j=1}^{3} x_j \partial_{x_j} u(x) = \rho\partial_\rho u(x) \quad \text{as } t \to 1.$$

Furthermore,

$$L(D_x)\, u_t(x) = f_t(x) + (1+t) f(tx) \quad \text{in } \mathcal{K},$$
$$u_t|_{\Gamma_j} = 0 \quad \text{for } j \in I_0, \qquad N(D_x)\, u_t|_{\Gamma_j} = 0 \quad \text{for } j \in I_1.$$

Consequently by Theorem 7.4.1 and Corollary 7.4.2,

$$(7.4.7) \quad \|u_t\|_{W^{l,2}_{\beta,\delta}(\mathcal{K})^\ell} \le c \Big(\|f_t\|_{W^{l-2,2}_{\beta,\delta}(\mathcal{K})^\ell} + (1+t)\, \|f(t\cdot)\|_{W^{l-2,2}_{\beta,\delta}(\mathcal{K})^\ell} \Big)$$

with a constant c independent of u and t. Using the equality

$$(7.4.8) \quad f_t(x) = \int_0^1 \sum_{j=1}^{3} x_j (\partial_{x_j} f)\big((t + \tau - t\tau)x\big)\, d\tau = \int_0^1 (\rho\partial_\rho f)\big((t+\tau-t\tau)x\big)\, d\tau,$$

it can be easily shown that

$$\|f_t\|_{W^{l-2,2}_{\beta,\delta}(\mathcal{K})^\ell} \le c\, \|\rho\partial_\rho f\|_{W^{l-2,2}_{\beta,\delta}(\mathcal{K})^\ell}$$

with c independent of t. Consequently, it follows from (7.4.7) that $\rho\partial_\rho u \in W^{l,2}_{\beta,\delta}(\mathcal{K})^\ell$. Repeating this procedure, we obtain $(\rho\partial_\rho)^\nu u \in W^{l,2}_{\beta,\delta}(\mathcal{K})^\ell$ for $\nu = 2, \dots, k$ together with the desired estimate. □

7.4.4. A modification of the regularity result for Dirichlet and mixed problems. As was proved in Chapter 3 (see Theorem 3.3.13), an analogous result to Theorem 7.4.6 is valid for the Dirichlet problem in the class of the spaces $V^{l,2}_{\beta,\delta}(\mathcal{K})$. For the mixed problem it makes sense to consider solutions in another subspace of $W^{l,2}_{\beta,\delta}(\mathcal{K})^\ell$. Let J be a subset of $\{1, 2, \dots, d\}$. Then we define $W^{l,p}_{\beta,\delta}(\mathcal{K}; J)$ as the weighted Sobolev space with the norm

$$\|u\|_{W^{l,p}_{\beta,\delta}(\mathcal{K};J)} = \Big(\int_{\mathcal{K}} \sum_{|\alpha| \le l} \rho^{p(\beta - l + |\alpha|)} \prod_{k \in J} \Big(\frac{r_k}{\rho}\Big)^{p(\delta_k - l + |\alpha|)} \prod_{k \notin J} \Big(\frac{r_k}{\rho}\Big)^{p\delta_k} |\partial_x^\alpha u|^p \, dx \Big)^{1/p}.$$

Obviously, $V^{l,p}_{\beta,\delta}(\mathcal{K}) \subset W^{l,p}_{\beta,\delta}(\mathcal{K}; J) \subset W^{l,p}_{\beta,\delta}(\mathcal{K})$.

In the sequel, J is the set of the indices k such that the Dirichlet condition in problem (7.1.1)–(7.1.3) is given on at least one of the adjoining faces Γ_{k_+} and Γ_{k_-} to the edge M_k (i.e., at least one of the numbers k_+, k_- belongs to the set I_0).

THEOREM 7.4.7. *Let $u \in V^{1,2}_{\beta-l+1}(\mathcal{K})^\ell$ be a solution of the problem* (7.4.1),(7.4.2), *where $(\rho\partial_\rho)^\nu f \in W^{l-2,2}_{\beta,\delta}(\mathcal{K}; J)^\ell$ for $\nu = 0, \dots, k$, $g_j = 0$ for $j = 1, \dots, d$. Suppose that the line* $\operatorname{Re}\lambda = l - \beta - 3/2$ *is free of eigenvalues of the pencil $\mathfrak{A}(\lambda)$ and that the components of δ are not integer and satisfy the inequalities*

$$l - 1 - \delta_+^{(j)} < \delta_j < l - 1 + \delta_-^{(j)} \quad \textit{for } j \in J$$

and the inequalities (7.4.3) *for* $j \notin J$. *Then* $(\rho\partial_\rho)^\nu u \in W^{l,2}_{\beta,\delta}(\mathcal{K};J)^\ell$ *for* $\nu = 0,1,\dots,k$ *and*

$$\sum_{\nu=0}^{k} \|(\rho\partial_\rho)^\nu u\|_{W^{l,2}_{\beta,\delta}(\mathcal{K};J)^\ell} \le c \sum_{\nu=0}^{k} \|(\rho\partial_\rho)^\nu f\|_{W^{l-2,2}_{\beta,\delta}(\mathcal{K};J)^\ell}\,.$$

P r o o f. For $k = 0$, the assertion holds analogously to Theorem 7.4.1. It suffices to replace the space $W^{l,2}_{\beta,\delta}(\mathcal{K})$ by $W^{l,2}_{\beta,\delta}(\mathcal{K};J)$. The estimates for the norms of the functions $\tilde{\zeta}_k v$ and $\zeta_k u$ in these spaces hold by means of Theorems 6.1.11 and 6.5.4. Repeating the proof of Theorem 7.4.6, we obtain the assertion for $k > 0$. □

7.4.5. A regularity result for special Neumann problems. We consider the Neumann problem

$$(7.4.9) \qquad L(D_x)\,u = f \ \text{ in } \mathcal{K}, \quad N(D_x)\,u = g_j \ \text{ on } \Gamma_j,\ j = 1,\dots,d,$$

and assume that the following conditions are satisfied for $k = 1,\dots,d$ (cf. Subsection 6.5.4).

(i) $\lambda = 1$ is the only eigenvalue of the pencil $A_k(\lambda)$ on the line $\operatorname{Re}\lambda = 1$.
(ii) The eigenvectors of the pencils $A_k(\lambda)$ corresponding to the eigenvalue $\lambda = 1$ are restrictions of linear vector functions to the unit circle, while generalized eigenvectors corresponding to this eigenvalue do not exist.
(iii) The ranks of the matrices $\mathcal{N}\mathcal{A}$ and $\mathcal{N}\mathcal{A}\mathcal{N}^T$, where

$$\mathcal{A} = \begin{pmatrix} A_{1,1} & A_{2,1} & A_{3,1} \\ A_{1,2} & A_{2,2} & A_{3,2} \\ A_{1,3} & A_{2,3} & A_{3,3} \end{pmatrix} \quad \text{and} \quad \mathcal{N} = \begin{pmatrix} n_1^+ I_\ell & n_2^+ I_\ell & n_3^+ I_\ell \\ n_1^- I_\ell & n_2^- I_\ell & n_3^- I_\ell \end{pmatrix}$$

(n^+, n^- are the normal vectors to the faces Γ_{k_+} and Γ_{k_-} adjacent to the edge M_k, I_ℓ denotes the $\ell\times\ell$ identity matrix, and $\mathcal{N}^T$ denotes the transposed matrix of $\mathcal{N}$), coincide.

If M_k is the positive x_3-axis and θ_k/π is not an integer, then the conditions (iii) is the same as in Lemma 6.5.7 and Theorem 6.5.9.

Note that the conditions (i)–(iii) are satisfied e.g. for the Neumann problem for the Lamé system. We denote by $\mu_+^{(k)}$ the greatest positive real number such that the strip

$$0 < \operatorname{Re}\lambda < \mu_+^{(k)}$$

contains at most the eigenvalue $\lambda = 1$ of the operator pencil $A_k(\lambda)$.

Using Theorem 6.5.9, we can improve the assertion of Theorem 7.4.6 provided the conditions (i)–(iii) are satisfied.

THEOREM 7.4.8. *Let* $u \in V^{1,2}_{\beta-l+1}(\mathcal{K})^\ell$ *be a solution of the problem* (7.4.1),(7.4.2), *where* $I_0 = \emptyset$, $(\rho\partial_\rho)^\nu f \in W^{l-2,2}_{\beta,\delta}(\mathcal{K})^\ell$ *for* $\nu = 0,\dots,k$, $g_j = 0$ *for* $j = 1,\dots,d$. *Suppose that the conditions* (i)–(iii) *are satisfied. Furthermore, we assume that the line* $\operatorname{Re}\lambda = l-\beta-3/2$ *is free of eigenvalues of the pencil* $\mathfrak{A}(\lambda)$ *and that the components of* δ *are not integer and satisfy the inequalities*

$$\max(0, l-1-\mu_+^{(k)}) < \delta_k < l-1$$

for $k = 1,\dots,d$. *Then* $(\rho\partial_\rho)^\nu u \in W^{l,2}_{\beta,\delta}(\mathcal{K})^\ell$ *for* $\nu = 0,1,\dots,k$ *and*

$$\sum_{\nu=0}^{k} \|(\rho\partial_\rho)^\nu u\|_{W^{l,2}_{\beta,\delta}(\mathcal{K})^\ell} \le c \sum_{\nu=0}^{k} \|(\rho\partial_\rho)^\nu f\|_{W^{l-2,2}_{\beta,\delta}(\mathcal{K})^\ell}\,.$$

As an example, we consider the Neumann problem

$$(7.4.10)\quad -\mu\Big(\Delta u + \frac{1}{1-2\nu}\,\nabla\nabla\cdot u\Big) = f \ \text{ in } \mathcal{K}, \quad \sigma(u)\,n = 0 \ \text{ on } \Gamma_j\,,\ j = 1,\dots,d.$$

for the Lamé system. Let θ_k be the inner angle at the edge M_k. Then

$$(7.4.11)\qquad \mu_+^{(k)} = \frac{\pi}{\theta_k} \ \text{ if } \theta_k < \pi, \qquad \mu_+^{(k)} = \frac{\xi_+(\theta_k)}{\theta_k} \ \text{ if } \theta_k > \pi,$$

where $\xi_+(\theta)$ is the smallest positive solution of the equation

$$\frac{\sin\xi}{\xi} + \frac{\sin\theta}{\theta} = 0$$

(see Subsection 6.5.5).

7.5. Green's matrix of the boundary value problem in a polyhedral cone

By a *Green's matrix* for the boundary value problem (7.1.1)–(7.1.3), we mean a $\ell\times\ell$-matrix $G(x,\xi)$ satisfying the equations

$$(7.5.1)\qquad L(D_x)\,G(x,\xi) = \delta(x-\xi)\,I_\ell \ \text{ for } x,\xi \in \mathcal{K},$$

$$(7.5.2)\qquad G(x,\xi) = 0 \ \text{ for } x \in \Gamma_j,\ \xi \in \mathcal{K},\ j \in I_0,$$

$$(7.5.3)\qquad N(D_x)\,G(x,\xi) = 0 \ \text{ for } x \in \Gamma_j,\ \xi \in \mathcal{K},\ j \in I_1.$$

Here I_ℓ denotes the $\ell\times\ell$ identity matrix. Since variational solutions exist in different weighted spaces $V^{1,2}_\kappa(\mathcal{K})^\ell$ (the only condition is that the line $\text{Re}\,\lambda = -\kappa-1/2$ is free of eigenvalues of the pencil $\mathfrak{A}(\lambda)$), we can construct Green's matrices for different values of κ. The goal of this section is to prove point estimates for the elements of the matrix $G(x,\xi)$ and their derivatives.

7.5.1. Existence of Green's matrices. First we prove the existence and uniqueness of the Green's matrix for special κ and some basic properties.

THEOREM 7.5.1. *Suppose that the form* $b_{\mathcal{K}}(\cdot,\cdot)$ *is* $\mathcal{H}_{\mathcal{K}}$*-elliptic and that the line* $\text{Re}\,\lambda = -\kappa - 1/2$ *is free of eigenvalues of the pencil* $\mathfrak{A}(\lambda)$. *Then the following assertions are true.*

1) *There exists a unique solution* $G(x,\xi)$ *of the boundary value problem* (7.5.1)–(7.5.3) *such that the function* $x \to \zeta(x,\xi)\,G(x,\xi)$ *belongs to* $V^{1,2}_\kappa(\mathcal{K})^\ell$ *for every* $\xi\in\mathcal{K}$ *and for every infinitely differentiable function* $\zeta(\cdot,\xi)$ *which is equal to zero in a neighborhood of the point* $x=\xi$ *and bounded together with all derivatives.*

2) *The equality*

$$(7.5.4)\qquad G_{i,j}(tx,t\xi) = t^{-1}\,G_{i,j}(x,\xi)$$

is valid for all $x,\xi\in\mathcal{K}$, $t>0$, $i,j = 1,\dots,\ell$.

3) *The adjoint matrix* $G^*(x,\xi)$ *is the unique solution of the formally adjoint problem (cf.* (7.1.11)–(7.1.13)*)*

$$L^+(D_\xi)\, G^*(x,\xi) = \delta(x-\xi)\, I_\ell \quad \textit{for } x,\xi \in \mathcal{K},$$
$$G^*(x,\xi) = 0 \quad \textit{for } x \in \mathcal{K},\ \xi \in \Gamma_j\,,\ j \in I_0,$$
$$N^+(D_\xi)\, G^*(x,\xi) = 0 \ \textit{for } x \in \mathcal{K},\ \xi \in \Gamma_j\,,\ j \in I_1,$$

such that the function $\xi \to \zeta(x,\xi)\, G^*(x,\xi)$ *belongs to* $V^{1,2}_{-\kappa}(\mathcal{K})^\ell$ *for every* $x \in \mathcal{D}$ *and for every infinitely differentiable function* $\zeta(x,\cdot)$ *which is equal to zero in a neighborhood of the point* $\xi = x$ *and bounded together with all derivatives.*

4) *The solution* $u \in \mathcal{H}_\kappa$ *of the problem*

$$b_{\mathcal{K}}(u,v) = (F,v)_{\mathcal{K}} \quad \textit{for all } v \in \mathcal{H}_{-\beta}$$

admits the representation

$$u(x) = \int_{\mathcal{K}} G(x,\xi)\, F(\xi)\, d\xi \tag{7.5.5}$$

for arbitrary $F \in \mathcal{H}^*_{-\kappa}$.

P r o o f. 1) There exists a solution $\mathcal{G}(x,\xi)$ of the equation

$$L(D_x)\, \mathcal{G}(x,\xi) = \delta(x-\xi)\, I_\ell \quad \text{in } \mathbb{R}^3$$

which is real analytic for $x \neq \xi$ and has the form $\mathcal{G}(x,\xi) = h(x-\xi)\, |x-\xi|^{-1}$, where h is positively homogeneous of degree 0. Let χ be a smooth function on $(0,\infty)$, $\chi(t) = 1$ for $t < 1/4$, $\chi(t) = 0$ for $t > 1/2$. We put $\psi(x,\xi) = \chi(|x-\xi|/r(\xi))$ and define $R(x,\xi)$ for every fixed $\xi \in \mathcal{K}$ as the unique solution (in the space $V^{1,2}_\kappa(\mathcal{K})^{\ell\times\ell}$) of the problem

$$L(D_x)\, R(x,\xi) = \delta(x-\xi)\, I_\ell - L(D_x)\big(\psi(x,\xi)\, \mathcal{G}(x,\xi)\big) \quad \text{for } x \in \mathcal{K}, \tag{7.5.6}$$
$$R(x,\xi) = -\psi(x,\xi)\, \mathcal{G}(x,\xi) \quad \text{for } x \in \Gamma_j\,,\ j \in I_0, \tag{7.5.7}$$
$$N(D_x)\, R(x,\xi) = -N(D_x)\, (\psi(x,\xi)\mathcal{G}(x,\xi)) \quad \text{for } x \in \Gamma_j,\ j \in I_1\,. \tag{7.5.8}$$

For every fixed $\xi \in \mathcal{K}$, the right-hand sides of (7.5.6)–(7.5.8) are smooth functions with compact supports vanishing in a neighborhood of the edges. Consequently, there exists a unique solution $R(\cdot,\xi)$ of the problem (7.5.6)–(7.5.8) in $V^{1,2}_\kappa(\mathcal{K})^{\ell\times\ell}$ (cf. Theorem 7.3.7). This prove item 1).

2)–4) hold analogously to Theorem 6.6.1. □

7.5.2. Estimates of Green's matrix. The case $|x|/2 < |\xi| < 2|x|$. We denote again by $\delta_+^{(k)}$ and $\delta_-^{(k)}$ the greatest positive real numbers such that the strip

$$-\delta_-^{(k)} < \operatorname{Re}\lambda < \delta_+^{(k)}$$

contains at most the eigenvalue $\lambda = 0$ of the pencil $A_k(\lambda)$. Furthermore, let $k(x)$ denote the smallest integer k such that $r_k(x) = r(x)$.

THEOREM 7.5.2. *Let* $G(x,\xi)$ *be the Green's matrix introduced in Theorem* 7.5.1. *Furthermore, let*

$$\delta_\alpha^\pm(x) = \delta_\pm^{(k(x))} - |\alpha| - \varepsilon \quad \textit{for } k(x) \in J,$$
$$\delta_\alpha^\pm(x) = \min(0, \delta_\pm^{(k(x))} - |\alpha| - \varepsilon) \quad \textit{for } k(x) \notin J,$$

where ε is an arbitrarily small positive number. Then the estimate

$$\left|\partial_x^\alpha \partial_\xi^\gamma G(x,\xi)\right| \le c\,|x-\xi|^{-1-|\alpha|-|\gamma|} \Big(\frac{r(x)}{|x-\xi|}\Big)^{\delta_\alpha^+(x)} \Big(\frac{r(\xi)}{|x-\xi|}\Big)^{\delta_\gamma^-(\xi)}$$

is valid for $|x|/2 < |\xi| < 2|x|$, $|x-\xi| > \min(r(x), r(\xi))$. The constant c is independent of x and ξ.

P r o o f. Suppose that $|x|/2 < |\xi| < 2|x|$ and $2 = |x-\xi| > \min(r(x), r(\xi))$. Then $\max(r(x), r(\xi)) < 4$, $|x| > 2/3$ and $|\xi| > 2/3$. We denote by $\mathcal{B}(x)$ and $\mathcal{B}(\xi)$ the balls with radius $1/2$ centered at x and ξ, respectively. Furthermore, let ζ and η be smooth cut-off functions with supports in $\mathcal{B}(x)$ and $\mathcal{B}(\xi)$, respectively, which are equal to one in balls with radius $1/4$ concentric to $\mathcal{B}(x)$ and $\mathcal{B}(\xi)$, respectively. Then

$$L^+(D_y)\,\partial_x^\alpha G^*(x,y) = 0 \quad \text{for } y \in \mathcal{K} \cap \mathcal{B}(\xi)$$

and

$$(1-d_j)\,\partial_x^\alpha G^*(x,y) + d_j\,N^+(D_y)\,\partial_x^\alpha G^*(x,y) = 0$$

for $y \in \Gamma_j \cap \mathcal{B}(\xi)$, $j = 1,\dots,d$. Therefore, Lemmas 2.5.3 and 6.6.4 yield

$$\text{(7.5.9)} \quad \begin{aligned} r(\xi)^{-\delta_\gamma^-(\xi)} \left|\partial_x^\alpha \partial_\xi^\gamma G_{i,j}(x,\xi)\right| &\le c\,\|\eta(\cdot)\,\partial_x^\alpha G(x,\cdot)\|_{V_0^{1,2}(\mathcal{K})^{\ell\times\ell}} \\ &\le c\,|\xi|^\kappa\,\|\eta(\cdot)\,\partial_x^\alpha G(x,\cdot)\|_{V_{-\kappa}^{1,2}(\mathcal{K})^{\ell\times\ell}}\,. \end{aligned}$$

Let $F \in \mathcal{H}_{-\kappa}^*$. We consider the solution

$$u(x) = \big(\eta(\cdot)\,F(\cdot), \overline{G(x,\cdot)}\big)_{\mathcal{K}}$$

of the problem

$$b_{\mathcal{K}}(u,v) = \big(\eta F, v\big)_{\mathcal{K}} \quad \text{for all } v \in \mathcal{H}_\kappa\,.$$

Since $\eta F = 0$ in $\mathcal{B}(x)$, we conclude from Lemmas 2.5.3 and 6.6.4 that

$$r(x)^{-\delta_\alpha^+(x)}\left|\partial_x^\alpha u(x)\right| \le c\,|x|^{-\kappa}\,\|\zeta u\|_{V_\kappa^{1,2}(\mathcal{K})^\ell}\,.$$

Consequently, the mapping

$$F \to |x|^\kappa\,r(x)^{-\delta_\alpha^+(x)}\,\partial_x^\alpha u(x) \;= |x|^\kappa\,r(x)^{-\delta_\alpha^+(x)}\,\big(F(\cdot)\,,\,\overline{\eta(\cdot)\partial_x^\alpha\,G(x,\cdot)}\big)_{\mathcal{K}}$$

is linear and continuous from $\mathcal{H}_{-\kappa}^*$ into $\mathbb{C}^\ell$ for arbitrary $x \in \mathcal{K}$. The norm of this mapping is bounded by a constant independent of x. Thus,

$$|x|^\kappa\,r(x)^{-\delta_\alpha^+(x)}\,\|\eta(\cdot)\,\partial_x^\alpha\,G(x,\cdot)\|_{V_{-\kappa}^{1,2}(\mathcal{K})^{\ell\times\ell}} \le c.$$

This together with (7.5.9) proves the theorem for the case $|x-\xi| = 2$. If $|x-\xi| \ne 2$, then we apply (7.5.4). □

Next, we estimate the matrix $G(x,\xi)$ for $|x-\xi| < \min(r(x), r(\xi))$. In this case, in particular, the inequalities $|x|/2 < |\xi| < 2|x|$ are valid.

THEOREM 7.5.3. *Let $G(x,\xi)$ be the Green's matrix introduced in Theorem 7.5.1. Then the estimate*

$$\left|\partial_x^\alpha \partial_\xi^\gamma G(x,\xi)\right| \le c\,|x-\xi|^{-1-|\alpha|-|\gamma|}$$

is valid for $|x-\xi| < \min(r(x), r(\xi))$.

P r o o f. Suppose that $|x-\xi| < \min(r(x), r(\xi)) = 2$. Then x and ξ lie in a ball B with radius 1 and distance ≥ 1 from the set $\mathcal{S}$ of the edge points. Let B' be a ball concentric to B with radius 3/2. There exists a Green's matrix $g(x,\xi)$ satisfying the equations

$$L(D_x)\, g(x,\xi) = \delta(x-\xi)\, I_\ell \quad \text{for } x,\xi \in \mathcal{K}\cap B',$$
$$(1-d_j)\, g(x,\xi) + d_j\, N(D_x)\, g(x,\xi) = 0 \quad \text{for } x \in \Gamma_j \cap B',\ \xi \in \mathcal{K}\cap B',$$

$j = 1,\dots,d$, and the estimates

$$\big|\partial_x^\alpha \partial_\xi^\gamma g(x,\xi)\big| \le c_{\alpha,\gamma}\, |x-\xi|^{-1-|\alpha|-|\gamma|}. \tag{7.5.10}$$

We consider the matrix

$$u(x,\xi) = \partial_\xi^\gamma G(x,\xi) - \eta(x)\, \partial_\xi^\gamma g(x,\xi)$$

where $\xi \in \mathcal{K}\cap B$, $\eta \in C_0^\infty(B')$ and $\eta = 1$ in a certain neighborhood of $\bar{B}$. Then, by the same arguments as in the proof of Theorem 3.4.3, the norm of $u(\cdot,\xi)$ in the Sobolev space $W^{l,2}(\mathcal{K}\cap B)$ is bounded by a constant c_l independent of $\xi \in B$. Thus, we obtain the estimate

$$\big|\partial_x^\alpha u(x,\xi)\big| \le C_{\alpha,\gamma}$$

with a constant $C_{\alpha,\gamma}$ independent of x and ξ. This together with (7.5.10) proves the theorem for the case $\min(r(x), r(\xi)) = 2$. Applying (7.5.4), we obtain the assertion of the theorem for arbitrary $\min(r(x), r(\xi))$. □

7.5.3. Estimates of Green's matrix. The cases $|\xi| < |x|/2$ and $|\xi| > 2|x|$. For the proof of point estimates for $G(x,\xi)$ in the cases $|x| < |\xi|/2$ and $|x| > 2\,|\xi|$ we need the following modification of Lemma 3.4.4.

LEMMA 7.5.4. *Let $u \in W^{l,2}_{\beta,\delta}(\mathcal{K};J)$ and $\rho\partial_\rho u \in W^{l,2}_{\beta,\delta}(\mathcal{K};J)$, where $l \ge 2$. We assume that $\delta_k > -1$ and $\delta_k \ne l-1$ for $k \notin J$. Then*

$$\rho^{\beta-l+3/2} \prod_{k\in J} \Big(\frac{r_k}{\rho}\Big)^{\delta_k - l+1} \prod_{k\notin J} \Big(\frac{r_k}{\rho}\Big)^{\max(0,\delta_k-l+1)} |u(x)| \le c \sum_{j=0}^{1} \|(\rho\partial_\rho)^j u\|_{W^{l,2}_{\beta,\delta}(\mathcal{K})}$$

with a constant c independent of u and x.

P r o o f. Applying the inequality

$$\sup_{0<\rho<\infty} |v(\rho)|^2 \le c \int_0^\infty \big(|v(\rho)|^2 + |\rho v'(\rho)|^2\big)\, \frac{d\rho}{\rho}$$

(which follows immediately from Sobolev's imbedding theorem) to the function $\rho^{\beta-l+3/2} u(\rho,\omega)$, we obtain

$$\begin{aligned} &\rho^{2(\beta-l)+3}\, |u(\rho,\omega)|^2 \\ &\le c \int_0^\infty \rho^{2(\beta-l+1)} \big(|u(\rho,\omega)|^2 + |\rho\partial_\rho u(\rho,\omega)|^2\big)\, d\rho. \end{aligned} \tag{7.5.11}$$

Let $\psi_1,\dots,\psi_n$ be smooth functions on $\overline{\Omega}$ such that $\psi_k = 1$ near $M_k \cap S^2$, $\psi_k \ge 0$, and $\sum \psi_k = 1$. Furthermore, let v be an arbitrary function in $W^{l,2}_\delta(\Omega)$. If $k \notin J$ and $-1 < \delta_k < l-1$, then $\psi_k v$ is continuous on $\overline{\Omega}$, and the supremum of $\psi_k v$ can be estimated by its norm in $W_\delta^{l,2}(\Omega)$ (cf. Lemma 6.2.12). If $k \in J$ or $\delta_k > l-1$, then $\psi_k v$ belongs to $V_\delta^{l,2}(\Omega)$. Therefore in the last case, Lemma 1.2.3 implies

$$\Big(\frac{r_k}{\rho}\Big)^{\delta_k - l+1} |\psi_k(\omega)\, v(\omega)| \le c\, \|\psi_k v\|_{W_\delta^{l,2}(\Omega)}\,.$$

Thus,

$$\prod_{k\in J} \Big(\frac{r_k}{\rho}\Big)^{\delta_k-l+1} \prod_{k\notin J} \Big(\frac{r_k}{\rho}\Big)^{\max(0,\delta_k-l+1)} |v(\omega)| \le c\, \|v\|_{W_\delta^{l,2}(\Omega)}\,.$$

The last inequality and (7.5.11) yield

$$\rho^{2(\beta-l)+3} \prod_{k\in J} \Big(\frac{r_k}{\rho}\Big)^{2(\delta_k-l+1)} \prod_{k\notin J} \Big(\frac{r_k}{\rho}\Big)^{2\max(0,\delta_k-l+1)} |u(\rho,\omega)|^2$$

$$\le c \int_0^\infty \rho^{2(\beta-l+1)} \Big(\|u(\rho,\cdot)\|^2_{W_\delta^{l,2}(\Omega)} + \|\rho\partial_\rho u(\rho,\cdot)\|^2_{W_\delta^{l,2}(\Omega)}\Big)\, d\rho.$$

The result follows. □

In the sequel, let κ be a fixed number such that no eigenvalues of the pencil $\mathfrak{A}(\lambda)$ lie on the line $\operatorname{Re}\lambda = -\kappa - 1/2$. Furthermore let

$$\Lambda_- < \operatorname{Re}\lambda < \Lambda_+$$

be the widest strip in the complex plane which is free of eigenvalues of this pencil and contains the line $\operatorname{Re}\lambda = -\kappa - 1/2$. This means that

$$-1-\Lambda_+ < \operatorname{Re}\lambda < -1-\Lambda_-$$

is the widest strip which is free of eigenvalues of the pencil $\mathfrak{A}^+(\lambda)$ and contains the line $\operatorname{Re}\lambda = \kappa - 1/2$.

THEOREM 7.5.5. *Let $G(x,\xi)$ be Green's matrix introduced in Theorem 7.5.1. Furthermore, let*

$$\delta^\pm_{k,\alpha} = \delta_\pm^{(k)} - |\alpha| - \varepsilon \text{ for } k\in J, \quad \delta^\pm_{k,\alpha} = \min(0, \delta_\pm^{(k)} - |\alpha| - \varepsilon) \text{ for } k \notin J,$$

where ε is an arbitrarily small positive real number. If $|x| < |\xi|/2$, then

$$\big|\partial_x^\alpha \partial_\xi^\gamma G(x,\xi)\big| \le c\, |x|^{\Lambda_+-|\alpha|-\varepsilon}\, |\xi|^{-1-\Lambda_+-|\gamma|+\varepsilon} \prod_{k=1}^d \Big(\frac{r_k(x)}{|x|}\Big)^{\delta^+_{k,\alpha}} \prod_{k=1}^d \Big(\frac{r_j(\xi)}{|\xi|}\Big)^{\delta^-_{k,\gamma}}.$$

Analogously, the estimate

$$\big|\partial_x^\alpha \partial_\xi^\gamma G(x,\xi)\big| \le c\, |x|^{\Lambda_--|\alpha|+\varepsilon}\, |\xi|^{-1-\Lambda_--|\gamma|-\varepsilon} \prod_{k=1}^d \Big(\frac{r_k(x)}{|x|}\Big)^{\delta^+_{k,\alpha}} \prod_{k=1}^d \Big(\frac{r_k(\xi)}{|\xi|}\Big)^{\delta^-_{k,\gamma}}$$

is valid for $|x| > 2|\xi|$.

P r o o f. By item 3) of Theorem 7.5.1, it suffices to prove the theorem for $|\xi| < |x|/2$. Suppose that $|x| = 1$. We denote by ζ, η smooth functions on $\overline{\mathcal{K}}$ such that $\zeta(y) = 1$ for $|y| < 1/2$, $\eta = 1$ in a neighborhood of $\operatorname{supp}\zeta$, and $\eta(y) = 0$ for $|y| > 3/4$. Furthermore, let l be an integer, $l > \max \delta_\pm^{(k)} + 1$. By Theorem 7.5.1, the function $y \to \partial_x^\alpha G^*(x,y)$ satisfies the equation

$$\eta(y)\, L^+(D_y)\, \partial_x^\alpha G^*(x,y) = 0 \text{ for } y \in \mathcal{K}$$

and the boundary conditions

$$(1-d_j)\, \partial_x^\alpha G^*(x,y) + d_j\, N^+(D_y)\, \partial_x^\alpha G^*(x,y) = 0 \text{ for } y \in \Gamma_j,$$

$j = 1, \dots, d$. Since the function $y \to \eta(y)\, \partial_x^\alpha G^*(x,y)$ belongs to $V_{-\kappa}^{1,2}(\mathcal{K})^{\ell\times\ell}$, it follows from Theorems 7.4.5 and 7.4.7 that the function $y \to \zeta(y)\, (|y|\partial_{|y|})^j \partial_x^\alpha G^*(x,y)$ belongs to $W_{\beta,\delta}^{l,2}(\mathcal{K};J)^{\ell\times\ell}$ for $j = 0,1,\dots$, where l is an arbitrary integer, $l \ge 2$,

$$\begin{aligned} &\Lambda_- + l - 1/2 < \beta < \Lambda_+ + l - 1/2\,, \\ &l - 1 - \delta_-^{(k)} < \delta_k < l - 1 + \delta_+^{(k)} \quad \text{for } k \in J, \\ &\max(0, l-1-\delta_-^{(k)}) < \delta_k < l - 1 \quad \text{for } k \notin J. \end{aligned}$$

Furthermore, the estimate

$$\|\zeta(\cdot)\, (|y|\partial_{|y|})^j \partial_x^\alpha \partial_y^\gamma G(x,\cdot)\|_{W_{\beta,\delta}^{l-|\gamma|,2}(\mathcal{K};J)^{\ell\times\ell}} \le c\, \|\eta(\cdot)\partial_x^\alpha G(x,\cdot)\|_{V_{-\kappa}^{1,2}(\mathcal{K})^{\ell\times\ell}}$$

holds for $|\gamma| \le l$. Hence by Lemma 7.5.4, the inequality

$$(7.5.12)\quad \prod_{k\in J} \Big(\frac{r_k(\xi)}{|\xi|}\Big)^{\delta_k - l + |\gamma| + 1} \prod_{k\notin J} \Big(\frac{r_k(\xi)}{|\xi|}\Big)^{\max(0,\delta_k - l + |\gamma| + 1)} \big|\partial_x^\alpha \partial_\xi^\gamma G(x,\xi)\big| \\ \le c\, |\xi|^{l-\beta-|\gamma|-3/2}\, \|\eta(\cdot)\, \partial_x^\alpha G(x,\cdot)\|_{V_{-\kappa}^{1,2}(\mathcal{K})^{\ell\times\ell}}$$

holds for $|\xi| < 1/2$, where c is independent of x and ξ.

By Theorem 7.3.7, the problem

$$b_{\mathcal{K}}(u,v) = (\eta F, v)_{\mathcal{K}}\,, \quad v \in V_{-\kappa}^1(\mathcal{K})^\ell,$$

has a unique solution $u \in \mathcal{H}_\kappa$ for arbitrary $F \in \mathcal{H}_{-\kappa}^*$. According to Theorem 7.5.1, the components of this solution have the representation

$$u_j(y) = \big(\eta(\cdot)\, F(\cdot),\, H^{(j)}(\xi,x)\big)_{\mathcal{K}} \quad \text{for } j = 1,2,3,$$

where $H^{(j)}(\xi,x)$ denotes the jth column of the matrix $H(\xi,x) = (G(x,\xi))^*$. Let χ and ψ be a smooth cut-off functions such that, $\chi = 1$ near x, $\psi = 1$ in a neighborhood of $\operatorname{supp}\chi$, $\psi(y) = 0$ for $|x-y| > 1/4$. Then

$$\psi\, L(D_x)\, u = 0 \ \text{ in } \mathcal{K} \ \text{ and } \psi\, N(D_x)\, u = 0 \ \text{ on } \Gamma_j \text{ for } j \in I_1,$$

since ψ and η have disjunct supports. Thus, it follows from Theorems 7.4.5 and 7.4.7 that $\chi(\rho\partial_\rho)^j u \in W_{\beta',\delta'}^{l,2}(\mathcal{K};J)^\ell$ for $j = 0,1,\dots$, where l is an arbitrary integer, $l \ge 2$,

$$\begin{aligned} &\Lambda_- < l - \beta' - 3/2 < \Lambda_+\,, \\ &l - 1 - \delta_+^{(k)} < \delta_k' < l - 1 + \delta_-^{(k)} \quad \text{for } k \in J, \\ &\max(0, l-1-\delta_+^{(k)}) < \delta_k' < l - 1 \quad \text{for } k \notin J. \end{aligned}$$

Furthermore,

$$\|\chi(\rho\partial_\rho)^j \partial_x^\alpha u\|_{W_{\beta',\delta'}^{l-|\alpha|,2}(\mathcal{K};J)^\ell} \le c\, \|\psi u\|_{V_\kappa^{1,2}(\mathcal{K})^\ell} \le c'\, \|F\|_{\mathcal{H}_{-\kappa}^*}$$

for $|\alpha| \le l$. Applying Lemma 7.5.4, we obtain the inequality

$$\prod_{k\in J} \big(r_k(x)\big)^{\delta_k' - l + |\alpha| + 1} \prod_{k\notin J} \big(r_k(x)\big)^{\max(0,\delta_k' - l + |\alpha| + 1)} \big|\partial_x^\alpha u(x)\big| \\ \le c\, \|\psi u\|_{V_\kappa^{1,2}(\mathcal{K})^\ell} \le c'\, \|F\|_{\mathcal{H}_{-\kappa}^*}\,.$$

Thus, the mappings

$$\begin{aligned}\mathcal{H}^*_{-\kappa} \ni F \quad \to \quad & \prod_{k\in J} \big(r_k(x)\big)^{\delta'_k-l+|\alpha|+1} \prod_{k\notin J} \big(r_k(x)\big)^{\max(0,\delta'_k-l+|\alpha|+1)}\, \partial_x^\alpha u_j(x) \\ &= \prod_{k\in J} r_k^{\delta'_k-l+|\alpha|+1} \prod_{k\notin J} r_k^{\max(0,\delta'_k-l+|\alpha|+1)} \big(\eta F\,,\, \partial_x^\alpha H^{(j)}(\cdot,x)\big)_{\mathcal{K}},\end{aligned}$$

$j=1,2,3$, represent linear and continuous functionals on $\mathcal{H}^*_{-\kappa}$ with norms bounded by a constant independent of x. Consequently, there exists a constant c independent of x such that

$$\|\eta(\cdot)\,\partial_x^\alpha H^{(j)}(\cdot,x)\|_{V^{1,2}_{-\kappa}(\mathcal{K})^\ell} \le c \prod_{k\in J}\Big(\frac{r_k(x)}{|x|}\Big)^{l-\delta'_k-|\alpha|-1} \prod_{k\notin J}\Big(\frac{r_k(x)}{|x|}\Big)^{\min(0,l-\delta'_k-|\alpha|-1)}.$$

Combining the last inequality with (7.5.12), we obtain

$$\begin{aligned}&\big|\partial_x^\alpha\partial_\xi^\gamma G(x,\xi)\big| \\ &\le c\,|\xi|^{l-\beta-|\gamma|-3/2} \prod_{k\in J}\Big(\frac{r_k(\xi)}{|\xi|}\Big)^{l-\delta_k-|\gamma|-1} \prod_{k\notin J}\Big(\frac{r_k(\xi)}{|\xi|}\Big)^{\min(0,l-\delta_k-|\gamma|-1)} \\ &\quad\times \prod_{k\in J}\Big(\frac{r_k(x)}{|x|}\Big)^{l-\delta'_k-|\alpha|-1} \prod_{k\notin J}\Big(\frac{r_k(x)}{|x|}\Big)^{\min(0,l-\delta'_k-|\alpha|-1)}\end{aligned}$$

for $|x|=1$, $|\xi|<1/2$. We can choose l such that $l\ge \delta_\pm^{(k)}+1$ for $k=1,\dots,d$. If we put $\beta=\Lambda_-+l+\varepsilon-1/2$, $\delta_k=l-1-\delta_-^{(k)}+\varepsilon$ and $\delta'_k=l-1-\delta_+^{(k)}+\varepsilon$, we arrive at the inequality

$$\big|\partial_x^\alpha\partial_\xi^\gamma G_{i,j}(x,\xi)\big| \le c\,|\xi|^{-1-\Lambda_--|\gamma|-\varepsilon} \prod_{k=1}^d \Big(\frac{r_k(x)}{|x|}\Big)^{\delta^+_{k,\alpha}} \prod_{k=1}^d \Big(\frac{r_k(\xi)}{|\xi|}\Big)^{\delta^-_{k,\gamma}}.$$

This proves the theorem for $|x|=1$, $|\xi|<1/2$. If x is an arbitrary point of $\mathcal{K}$ and $|\xi|<|x|/2$, then we obtain the desired estimate by means of the second assertion of Theorem 7.5.1. The proof is complete. □

REMARK 7.5.6. The estimates in Theorems 7.5.2–7.5.5 for the derivatives of Green's matrix function can be improved if the direction of the derivatives is tangential to the edges. In particular, the estimate

$$\big|\partial_\rho\partial_x^\alpha\partial_\xi^\gamma G(x,\xi)\big| \le c\,|x-\xi|^{-2-|\alpha|-|\gamma|}\Big(\frac{r(x)}{|x-\xi|}\Big)^{\delta^+_\alpha(x)}\Big(\frac{r(\xi)}{|x-\xi|}\Big)^{\delta^-_\gamma(\xi)}$$

is valid for $\rho=|x|$, $|x|/2<|\xi|<2|x|$, $|x-\xi|>\min(r(x),r(\xi))$. Furthermore, the estimate

$$\big|\partial_\rho\partial_x^\alpha\partial_\xi^\gamma G(x,\xi)\big| \le c\,|x|^{\Lambda_+-|\alpha|-1-\varepsilon}\,|\xi|^{-1-\Lambda_+-|\gamma|+\varepsilon} \prod_{k=1}^d\Big(\frac{r_k(x)}{|x|}\Big)^{\delta^+_{k,\alpha}} \prod_{k=1}^d\Big(\frac{r_j(\xi)}{|\xi|}\Big)^{\delta^-_{k,\gamma}}.$$

holds for $|x|<|\xi|/2$, while the estimate

$$\big|\partial_\rho\partial_x^\alpha\partial_\xi^\gamma G(x,\xi)\big| \le c\,|x|^{\Lambda_--|\alpha|-1+\varepsilon}\,|\xi|^{-1-\Lambda_--|\gamma|-\varepsilon} \prod_{k=1}^d\Big(\frac{r_k(x)}{|x|}\Big)^{\delta^+_{k,\alpha}} \prod_{k=1}^d\Big(\frac{r_k(\xi)}{|\xi|}\Big)^{\delta^-_{k,\gamma}}$$

is valid for $|x|>2|\xi|$.

7.5.4. Estimates for Λ_+ and Λ_-. If the sesquilinear form $b_{\mathcal{K}}$ is $\mathcal{H}_{\mathcal{K}}$-elliptic, then in particular the line $\operatorname{Re}\lambda = -1/2$ is free of eigenvalues of the pencil $\mathfrak{A}(\lambda)$ (see Lemma 7.1.1). We consider the Green's matrix $G(x,\xi)$ introduced in Theorem 7.5.1 for $\kappa = 0$. Then

$$\Lambda_- < \operatorname{Re}\lambda < \Lambda_+$$

is the widest strip in the complex plane which contains the line $\operatorname{Re}\lambda = -1/2$ and which is free of eigenvalues of the pencil $\mathfrak{A}(\lambda)$. Here we give some estimates for Λ_+ and Λ_- which can be found in the book [**85**].

The Dirichlet problem ($I_1 = \emptyset$). If $A_{i,j} = A_{j,i}^*$ for $i,j = 1,2,3$, and the cone $\mathcal{K}$ is Lipschitz graph, then $\Lambda_+ > 0$, $\Lambda_- < -1$ (cf. [**85**, Theorem 11.1.1]). If all interior angles, with the possible exception of one, are less than π, $A_{i,j} = A_{j,i}^*$ for $i,j = 1,2,3$, and there exists a positive constant c such that

$$\sum_{i,j=1}^{3} A_{i,j} f^{(i)} \cdot \overline{f^{(j)}} \ge c \sum_{j=1}^{3} \big|f^{(j)}\big|^2 \quad \text{for all } f^{(1)}, f^{(2)}, f^{(3)} \in \mathbb{C}^\ell, \tag{7.5.13}$$

then even the estimates $\Lambda_+ > 1/2$, $\Lambda_- < -3/2$ (cf. [**85**, Theorem 11.4.1]).

The Neumann problem ($I_0 = \emptyset$). If the cone $\mathcal{K}$ is Lipschitz graph, $A_{i,j} = A_{j,i}^*$ for $i,j = 1,2,3$, and the condition (7.5.13) is satisfied, then $\Lambda_+ = 0$ and $\Lambda_- = -1$. In this case, the strip $-1 \le \operatorname{Re}\lambda \le 0$ contains only the eigenvalues $\lambda = 0$ and $\lambda = -1$. The eigenvectors corresponding to the eigenvalue $\lambda = 0$ are the constant vectors in $\mathbb{C}^\ell$, generalized eigenvectors corresponding to this eigenvalue do not exist. This means that the eigenvalues $\lambda = 0$ and $\lambda = -1$ have geometric and algebraic multiplicity ℓ (cf. [**85**, Theorem 12.3.3]). By [**85**, Theorem 4.3.1], the same is also true for the Neumann problem for the Lamé system.

7.5.5. Green's matrix for the Neumann problem. We consider the Green matrix for the Neumann problem (7.4.9) in the case $\kappa = 0$. For this problem, the numbers 0 and -1 belong always to the spectrum of the corresponding pencil $\mathfrak{A}(\lambda)$. We suppose in this subsection that these are the only eigenvalues in the strip $-1 \le \operatorname{Re}\lambda \le 0$, that the eigenvectors corresponding to the eigenvalue $\lambda = 0$ are constant, and that generalized eigenvectors corresponding to this eigenvalue do not exist. Then let

$$\Lambda'_- < \operatorname{Re}\lambda < \Lambda'_+$$

be the widest strip in the complex plane which contains $\lambda = 0$ and $\lambda = -1$ but no other eigenvectors of the pencil $\mathfrak{A}(\lambda)$.

LEMMA 7.5.7. *Let ζ, η be smooth functions on $\overline{\mathcal{K}}$ with compact supports such that $\zeta = 1$ in a neighborhood of the origin and $\eta = 1$ in a neighborhood of* $\operatorname{supp}\zeta$. *Furthermore, let $\eta u \in V_0^{1,2}(\mathcal{K})^\ell$,*

$$\eta\, L(D_x)\, u = 0 \ \text{ in } \mathcal{K}, \quad \eta\, N(D_x)\, u = 0 \ \text{ on } \Gamma_j,\ j = 1,\dots,d.$$

Suppose that the strip $-1/2 \le \operatorname{Re}\lambda \le 0$ contains only the eigenvalue $\lambda = 0$, that the eigenvectors corresponding to this eigenvalue are constant vectors in $\mathbb{C}^\ell$ and that there are no generalized eigenvectors corresponding to this eigenvalue. Then there exists a constant vector c_0 such that

$$\zeta u - c_0 \in W_{\beta,\delta}^{l,2}(\mathcal{K})^\ell, \quad \text{and} \quad \zeta \rho \partial_\rho u \in W_{\beta,\delta}^{l,2}(\mathcal{K})^\ell,$$

where l is an arbitrary integer, $l \ge 2$, $0 < l-\beta-3/2 < \Lambda'_+$, and $\max(0, l-1-\delta_+^{(k)}) < \delta_k < l-1$ *for* $k = 1, \dots, d$. *Furthermore,*

$$\|\zeta u - c_0\|_{W^{l,2}_{\beta,\delta}(\mathcal{K})^\ell} + \|\zeta \rho \partial_\rho u\|_{W^{l,2}_{\beta,\delta}(\mathcal{K})^\ell} \le c\, \|\eta u\|_{V_0^{1,2}(\mathcal{K})^\ell}\,.$$

P r o o f. Let χ_1, χ_2, χ_3 be smooth functions such that $\eta = 1$ in a neighborhood of $\operatorname{supp}\chi_1$, $\chi_1 = 1$ in a neighborhood of $\operatorname{supp}\chi_2$, $\chi_2 = 1$ in a neighborhood of $\operatorname{supp}\chi_3$, and $\chi_3 = 1$ in a neighborhood of $\operatorname{supp}\zeta$. Since the derivatives of χ_1 vanish in a neighborhood of the origin and of infinity, it follows that

$$L(D_x)\,(\chi_1 u) \in W^{0,2}_{\beta-l+2,0}(\mathcal{K})^\ell \quad \text{and} \quad N(D_x)\,(\chi_1 u)|_{\Gamma_j} \in W^{1/2,2}_{\beta-l+2,0}(\Gamma_j)^\ell.$$

Thus by Theorems 7.4.5 and 7.4.6, the decomposition $\chi_1 u = c_0 + w$ holds, where $w \in W^{2,2}_{\beta-l+2,\delta'}(\mathcal{K})^\ell$, $\max(0, 1-\delta_+^{(k)}) < \delta'_k < 1$ for $k = 1, \dots, d$, and

$$|c_0| + \|w\|_{W^{2,2}_{\beta,\delta'}(\mathcal{K})^\ell} \le c\, \|\psi u\|_{V_0^{1,2}(\mathcal{K})^\ell}\,.$$

This implies

$$L(D_x)\,(\chi_2 w) \in W^{1,2}_{\beta-l+3,\delta'}(\mathcal{K})^\ell \quad \text{and} \quad N(D_x)(\zeta u) \in W^{3/2,2}_{\beta-l+3,\delta'}(\Gamma_j)^\ell.$$

Using Theorem 7.4.6, we obtain $\chi_2 w \in W^{3,2}_{\beta-l+3,\delta''}(\mathcal{K})^\ell$, where $\max(0, 2-\delta_+^{(k)}) < \delta''_k < 2$ for $k = 1, \dots, d$. Repeating this argument, we get $\chi_3 w \in W^{l,2}_{\beta,\delta}(\mathcal{K})^\ell$. Consequently,

$$\zeta u = c_0 + v, \quad \text{where } v = \zeta w - (1-\zeta)\, c_0 \in W^{l,2}_{\beta,\delta}(\mathcal{K})^\ell.$$

Furthermore, we conclude that $\zeta\, \rho \partial_\rho u \in W^{l-1,2}_{\beta-1,\delta}(\mathcal{K})^\ell \subset W^{0,2}_{\beta-l,0}(\mathcal{K})^\ell$. From the inclusion $\eta u \in V_0^{1,2}(\mathcal{K})^\ell$ and from the equalities

$$\eta L(D_x)\,(\rho \partial_\rho u) = \eta\big(\rho \partial_\rho L(D_x) u + 2\, L(D_x)\, u\big) = 0$$

and $\eta N(D_x)\,(\rho\partial_\rho u)|_{\Gamma_j} = 0$ it follows analogously to Theorem 7.4.6 that $\chi_1 \rho \partial_\rho u \in V_0^{1,2}(\mathcal{K})^\ell$. Thus, by the first part of the proof, the function $\zeta \rho \partial_\rho u$ admits the decomposition $\zeta \rho \partial_\rho u = c_1 + V$, where c_1 is a constant vector and $V \in W^{l,2}_{\beta,\delta}(\mathcal{K})^\ell$. Since $\zeta\, \rho \partial_\rho u \in W^{0,2}_{\beta-l,0}(\mathcal{K})^\ell$ and $W^{l,2}_{\beta,\delta}(\mathcal{K}) \subset W^{0,2}_{\beta-l,0}(\mathcal{K})$, we conclude that $c_1 \in W^{0,2}_{\beta-l,0}(\mathcal{K})^\ell$, i.e. $c_1 = 0$. This proves the lemma. □

We apply the last lemma in order to improve the estimates of Green's matrix given in Theorem 7.5.5 for the cases $|\xi| < |x|/2$ and $|x| < |\xi|/2$.

THEOREM 7.5.8. *Let $G(x,\xi)$ be the Green's matrix introduced in Theorem* 7.5.1 *for the Neumann problem and $\kappa = 0$.*

1) *If $\lambda = 0$ is the only eigenvalue of the pencil $\mathfrak{A}(\lambda)$ in the strip $-1/2 < \operatorname{Re}\lambda \le 0$ and the algebraic multiplicity of this eigenvalue is equal to ℓ, then the estimate*

$$\big|\partial_x^\alpha \partial_\xi^\gamma G(x,\xi)\big| \le c\, \frac{|x|^{\Lambda'_+ - |\alpha| - \varepsilon}}{|\xi|^{1+\Lambda'_+ + |\gamma| - \varepsilon}} \prod_{k=1}^d \Big(\frac{r_k(x)}{|x|}\Big)^{\delta^+_{k,\alpha}} \prod_{k=1}^d \Big(\frac{r_k(\xi)}{|\xi|}\Big)^{\delta^-_{k,\gamma}}$$

holds for $|\alpha| \ne 0$, $|x| < |\xi|/2$, where $\delta^\pm_{k,\alpha} = \min(0, \delta_\pm^{(k)} - |\alpha| - \varepsilon)$, ε is an arbitrarily small positive number.

2) *Suppose that $\lambda = -1$ is the only eigenvalue of the pencil $\mathfrak{A}(\lambda)$ in the strip $-1 \le \operatorname{Re}\lambda < -1/2$ and that the algebraic multiplicity of this eigenvalue is ℓ. Then*

the estimate

$$\big|\partial_x^\alpha \partial_\xi^\gamma G(x,\xi)\big| \le c\,|x|^{\Lambda'_- -|\alpha|+\varepsilon}\,|\xi|^{-1-\Lambda'_- -|\gamma|-\varepsilon}\prod_{k=1}^d \Big(\frac{r_k(x)}{|x|}\Big)^{\delta^+_{k,\alpha}}\prod_{k=1}^d \Big(\frac{r_k(\xi)}{|\xi|}\Big)^{\delta^-_{k,\gamma}}$$

is satisfied for $|\gamma|\neq 0$, $|\xi|<|x|/2$.

P r o o f. By item 3) of Theorem 7.5.1, it suffices to prove the theorem for $|\xi|<|x|/2$. If $\lambda=-1$ is the only eigenvalue of the pencil $\mathfrak{A}(\lambda)$ in the strip $\Lambda'_- < \operatorname{Re}\lambda < -1/2$, then $\lambda=0$ is the only eigenvalue of the pencil $\mathfrak{A}^+(\lambda)$ in the strip $-1/2<\operatorname{Re}\lambda<-1-\Lambda'_-$ with the same algebraic multiplicity ℓ. Suppose that $|x|=1$. As in the proof of Theorem 7.5.5, we denote by ζ, η smooth functions on $\overline{\mathcal{K}}$ such that $\zeta(y)=1$ for $|y|<1/2$, $\eta=1$ in a neighborhood of $\operatorname{supp}\zeta$, and $\eta(y)=0$ for $|y|>3/4$. By Theorem 7.5.1, the function $y\to\partial_x^\alpha G^*(x,y)$ satisfies the equation

$$\eta(y)\,L^+(D_y)\,\partial_x^\alpha G^*(x,y)=0 \quad \text{for } y\in\mathcal{K}$$

and the boundary conditions $N^+(D_y)\,\partial_x^\alpha G^*(x,y)=0$ for $y\in\Gamma_j$, $j=1,\dots,d$. Furthermore, the function $y\to\eta(y)\,\partial_x^\alpha G^*(x,y)$ belongs to $V_0^{1,2}(\mathcal{K})^{\ell\times\ell}$. Thus by Lemma 7.5.7, the functions $y\to\zeta(y)\,\partial_y^\gamma\partial_x^\alpha G^*(x,y)$ and $y\to\zeta(y)\,|y|\partial_{|y|}\,\partial_y^\gamma\partial_x^\alpha G^*(x,y)$ belong to the space $W_{\beta,\delta}^{l-|\gamma|,2}(\mathcal{K})^{\ell\times\ell}$ for $|\gamma|\neq0$, where l is an arbitrary integer, $l\ge|\gamma|$, $0<l-\beta-3/2<-1-\Lambda'_-$, and $\max(0,l-1-\delta_-^{(k)})<\delta_k<l-1$ for $k=1,\dots,d$. The norms of the last functions can be estimated by the $V_0^{1,2}(\mathcal{K})$-norm of $\eta(\cdot)\,\partial_x^\alpha G(x,\cdot)$. Hence by Lemma 7.5.4, the inequality

$$\begin{aligned}|\xi|^{\beta-l+|\gamma|+3/2}\prod_{k=1}^d\Big(\frac{r_k(\xi)}{|\xi|}\Big)^{\max(0,\delta_k-l+|\gamma|+1)}&\big|\partial_x^\alpha\partial_\xi^\gamma G(x,\xi)\big|\\ &\le c\,\|\eta(\cdot)\,\partial_x^\alpha G(x,\cdot)\|_{V_0^{1,2}(\mathcal{K})^{\ell\times\ell}}\end{aligned}$$

holds for $|\xi|<1/2$, $|\gamma|\neq0$. Here c is independent of x and ξ. As in the proof of Theorem 7.5.5, we obtain the estimate

$$\|\eta(\cdot)\,\partial_x^\alpha G(x,\cdot)\|_{V_0^{1,2}(\mathcal{K})^{\ell\times\ell}}\le c\prod_{k=1}^d\Big(\frac{r_k(x)}{|x|}\Big)^{\min(0,l-\delta'_k-|\alpha|-1)},$$

where $\max(0,l-1-\delta_+^{(k)})<\delta'_k<l-1$ for $k=1,\dots,d$. If we put $\beta=l+\Lambda'_-+\varepsilon-1/2$, $\delta_k=l-1-\delta_-^{(k)}+\varepsilon$ and $\delta'_k=l-1-\delta_+^{(k)}+\varepsilon$, where $l\ge1+\max\delta_\pm^{(k)}$, we arrive at the inequality

$$\big|\partial_x^\alpha\partial_\xi^\gamma G_{i,j}(x,\xi)\big|\le c\,|\xi|^{-1-\Lambda'_- -|\gamma|-\varepsilon}\prod_{k=1}^d\Big(\frac{r_k(x)}{|x|}\Big)^{\delta^+_{k,\alpha}}\prod_{k=1}^d\Big(\frac{r_k(\xi)}{|\xi|}\Big)^{\delta^-_{k,\gamma}}$$

for $|\gamma|\neq0$. This proves the theorem for $2|\xi|<|x|=1$. Using item 2) of Theorem 7.5.1, we obtain the assertion of the theorem for arbitrary x, $|x|>2|\xi|$. □

Now we assume that the conditions (i)–(iii) in Subsection 7.4.5 are satisfied for $k=1,\dots,d$. Furthermore, we suppose that the same conditions are satisfied for the formally adjoint boundary value problem. Then we denote by $\mu_+^{(k)},\mu_-^{(k)}$ the greatest positive real numbers such that the strip

$$-\mu_-^{(k)}<\operatorname{Re}\lambda<\mu_+^{(k)}$$

contains at most the eigenvalues $\lambda = 0$ and $\lambda = \pm 1$ of the operator pencil $A_k(\lambda)$. For example, in the case of the Neumann problem (7.4.10) for the Lamé system, the numbers $\mu_+^{(k)} = \mu_-^{(k)}$ are given by (7.4.11). Then it is possible to improve the estimates in Theorems 7.5.2, 7.5.5 and 7.5.8.

THEOREM 7.5.9. *Suppose that the conditions* (i)–(iii) *of Subsection* 7.4.5 *are satisfied both for the Neumann problem* (7.4.9) *and for the formally adjoint problem. Then the Green's matrix introduced in Theorem* 7.5.1 *for* $\kappa = 0$ *satisfies the estimates of Theorems* 7.5.2 *and* 7.5.5 *with the exponents*

$$\delta_\alpha^+(x) = \min(0, \mu_+^{(k(x))} - |\alpha| - \varepsilon), \quad \delta_\gamma^-(\xi) = \min(0, \mu_-^{(k(\xi))} - |\gamma| - \varepsilon)$$

and

$$\delta_{k,\alpha}^\pm = \min(0, \mu_\pm^{(k)} - |\alpha| - \varepsilon). \tag{7.5.14}$$

Here ε *is an arbitrarily small positive real number. If in addition the assumptions of Theorem* 7.5.8 *concerning the eigenvalues of the pencil* $\mathfrak{A}(\lambda)$ *are satisfied, then* $G(x, \xi)$ *satisfies the estimates of Theorem* 7.5.8 *with the exponents* (7.5.14).

The proof is a word-by-word repetition of the proofs of Theorems 7.5.2, 7.5.5 and 7.5.8 with the only difference that we have to apply Theorems 6.6.6 and 7.4.8 instead of Theorems 6.6.5 and 7.4.6, respectively.

7.6. Solvability in weighted L_p Sobolev spaces

The goal of this section is to prove the existence and uniqueness of solutions of the boundary value problem (7.1.1)–(7.1.3) in the space $W_{\beta,\delta}^{2,p}(\mathcal{K})^\ell$. For this, we employ the estimates of the Green's matrix and its derivatives obtained in the preceding section. The existence and uniqueness result holds under the assumption that the line $\operatorname{Re}\lambda = 2 - \beta - 3/p$ is free of eigenvalues of the operator pencil $\mathfrak{A}(\lambda)$ and the components δ_k satisfy the inequalities

$$\max(0, 2 - \delta_+^{(k)}) < \delta_k + 2/p < 2 \quad \text{for } k = 1, \dots, d. \tag{7.6.1}$$

The method to obtain this result is essentially the same as for the Dirichlet problem in Section 3.5. We consider the solution of the boundary value problem

$$L(D_x)v = \zeta_\nu f \ \text{ in } \mathcal{K}, \quad (1 - d_j)\, u + d_j\, N(D_x)\, u = 0 \ \text{ on } \Gamma_j\,,$$

$j = 1, \dots, d$, where ζ_ν are smooth cut-off functions depending only on $\rho = |x|$ and satisfying the conditions (7.2.3), and estimate the norm of $\zeta_\mu v$. Then it suffices to apply Lemma 3.5.8.

7.6.1. A local estimate for the solution. Let $f \in W_{\beta,\delta}^{0,p}(\mathcal{K})^\ell$ and $g_j \in W_{\beta,\delta}^{2-d_j-1/p,p}(\Gamma_j)^\ell$, $j = 1, \dots, d$, be given vector functions. We assume that the following conditions are satisfied.

(i) The line $\operatorname{Re}\lambda = 2 - \beta - 3/p$ is free of eigenvalues of the pencil $\mathfrak{A}(\lambda)$.

(ii) The components of δ satisfy the inequalities (7.6.1).

(iii) The Dirichlet data satisfy the compatibility condition (7.3.5).

Note that the last condition is necessary and sufficient for the existence of a vector function $u \in W_{\beta,\delta}^{2,p}(\mathcal{K})^\ell$ satisfying the boundary conditions (7.1.2). Lemma 7.3.1

allows us to restrict ourselves to zero boundary data g_j. Then the solution of the boundary value problem (7.1.1)–(7.1.3) admits the representation

$$u(x) = \int_{\mathcal{K}} G(x,\xi)\, f(\xi)\, d\xi, \tag{7.6.2}$$

where $G(x,\xi)$ is the Green's matrix of the boundary value problem. Let

$$\Lambda_- < \operatorname{Re}\lambda < \Lambda_+$$

be the widest strip in the complex plane containing the line $\operatorname{Re}\lambda = 2-\beta-3/p$ which is free of eigenvalues of the pencil $\mathfrak{A}(\lambda)$. Furthermore, let κ be an arbitrary real number in the interval

$$-\Lambda_+ - 1/2 < \kappa < -\Lambda_- - 1/2\,.$$

This means that the closed strip between the lines $\operatorname{Re}\lambda = 2-\beta-3/p$ and $\operatorname{Re}\lambda = -\kappa - 1/2$ does not contain eigenvalues. In the sequel, let $G(x,\xi)$ be the Green's matrix introduced in Theorem 7.5.1 for our fixed κ. Furthermore, let ζ_ν be infinitely differentiable functions depending only on $\rho = |x|$ and satisfying the conditions (7.2.3). In order to estimate the integral (7.6.2), we consider the function

$$v(x) = \int_{\mathcal{K}} G(x,\xi)\, \zeta_\nu(\xi)\, f(\xi)\, d\xi \tag{7.6.3}$$

for $f \in W^{0,p}_{\beta,\delta}(\mathcal{K})^\ell$, $\nu = 0, \pm 1, \pm 2, \ldots$.

LEMMA 7.6.1. *Suppose that the conditions* (i) *and* (ii) *are satisfied. Then the function* (7.6.3) *satisfies the inequality*

$$\|\zeta_\mu v\|_{W^{2,p}_{\beta,\delta}(\mathcal{K})^\ell} \le c\, 2^{-|\mu-\nu|\varepsilon_0}\, \|\zeta_\nu f\|_{W^{0,p}_{\beta,\delta}(\mathcal{K})^\ell} \tag{7.6.4}$$

with positive constants c and ε_0 independent of f, μ and ν.

P r o o f. Let the function χ_μ in $\mathcal{K}$ be defined as

$$\chi_\mu(x) = 1 \ \text{ for } 2^{\mu-1} \le |x| \le 2^{\mu+1}, \quad \chi_\mu(x) = 0 \ \text{ else.}$$

Then there exists a constant c independent of v and μ such that

$$\|\zeta_\mu v\|_{W^{2,p}_{\beta,\delta}(\mathcal{K})^\ell} \le c \sum_{|\alpha|\le 2} \|\chi_\mu \partial_x^\alpha v\|_{W^{0,p}_{\beta-2+|\alpha|,\delta}(\mathcal{K})^\ell}\,.$$

First let $\mu > \nu + 2$. Then by Theorem 7.5.5,

$$\big|\partial_x^\alpha G(x,\xi)\big| \le c\, |x|^{\Lambda_- - |\alpha| + \varepsilon}\, |\xi|^{-1-\Lambda_- - \varepsilon} \prod_{k=1}^{d} \Big(\frac{r_k(x)}{|x|}\Big)^{\delta^+_{k,\alpha}}$$

for $2^{\mu-1} \le |x| \le 2^{\mu+1}$, $2^{\nu-1} \le |\xi| \le 2^{\nu+1}$, where $\delta^+_{k,\alpha} = \min(0, \delta_+^{(k)} - |\alpha| - \varepsilon)$ and ε is an arbitrarily small positive number. Applying Lemma 3.5.1, we obtain

$$\|\chi_\mu \partial_x^\alpha v\|_{W^{0,p}_{\beta-2+|\alpha|,\delta}(\mathcal{K})^\ell} \le c\, 2^{(\mu-\nu)(\beta-2+\Lambda_-+\varepsilon+3/p)}\, \|\zeta_\nu f\|_{W^{0,p}_{\beta,\delta}(\mathcal{K})^\ell}$$

for $|\alpha| \le 2$. Here the factor $\beta - 2 + \Lambda_- + \varepsilon + 3/p$ is negative for sufficiently small ε. This proves the lemma for the case $\mu > \nu + 2$. In the case $\nu > \mu + 2$ the matrix $\partial_x^\alpha G(x,\xi)$ satisfies the same estimate with Λ_+ instead of Λ_- and $-\varepsilon$ instead of ε. Thus, the desired estimate holds analogously by means of Lemma 3.5.1.

We consider the case $|\mu - \nu| \le 2$. Let the functions $\chi^\pm$ be defined as

$$\chi^+(x,\xi) = \chi\Big(\frac{|x-\xi|}{r(x)}\Big), \qquad \chi^-(x,\xi) = 1 - \chi^+(x,\xi),$$

where χ is a smooth function on $[0,\infty)$ such that $\chi(t) = 1$ for $0 \le t \le 1/2$ and $\chi(t) = 0$ for $t \ge 3/4$. Furthermore, let

$$v^\pm(x) = \int_{\mathcal{K}} \chi^\pm(x,\xi)\, G(x,\xi)\, \zeta_\nu(\xi)\, f(\xi)\, d\xi.$$

Then $v = v^+ + v^-$. By Theorem 7.5.2,

$$\big|\partial_x^\alpha\big(\chi^-(x,\xi)\, G(x,\xi)\big)\big| \le c\,|x-\xi|^{-1-|\alpha|} \Big(\frac{r(x)}{|x-\xi|}\Big)^{\delta_\alpha^+(x)}$$

for $2^{\mu-1} \le |x| \le 2^{\mu+1}$, $2^{\nu-1} \le |\xi| \le 2^{\nu+1}$, where $\delta_\alpha^+(x) = \min(0, \delta_+^{(k(x))} - |\alpha| - \varepsilon)$, $k(x)$ denotes the smallest integer k such that $r_k(x) = r(x)$. We prove that

$$\|\chi_\mu \partial_x^\alpha v^-\|_{W^{0,p}_{\beta-2+|\alpha|,\delta}(\mathcal{K})^\ell} \le c\, \|\zeta_\nu f\|_{W^{0,p}_{\beta,\delta}(\mathcal{K})^\ell} \tag{7.6.5}$$

for $|\alpha| \le 2$ with a constant c independent of f, μ, ν. For $|\alpha| = 0$ the estimate

$$\begin{aligned} |v^-(x)|^p &\le c\Big(\int_{\mathcal{K}} |\chi^-(x,\xi)\, G(x,\xi)\, \zeta_\nu(\xi)\, f(\xi)|\, d\xi\Big)^p \\ &\le c \int_{\mathcal{K}} |x-\xi|^{-1} \prod_k \Big(\frac{r_k(\xi)}{|\xi|}\Big)^{p\delta_k} |\zeta_\nu(\xi)\, f(\xi)|^p\, d\xi \\ &\quad \times \Big(\int\limits_{\substack{\mathcal{K} \\ 2^{\nu-1}<|\xi|<2^{\nu+1}}} |x-\xi|^{-1} \prod_k \Big(\frac{r_k(\xi)}{|\xi|}\Big)^{-p'\delta_k} d\xi\Big)^{p-1} \end{aligned}$$

holds, where $p' = p/(p-1)$. Substituting $x/|x| = y$, $\xi/|x| = \eta$, we obtain

$$\begin{aligned} &\int\limits_{\substack{\mathcal{K} \\ 2^{\nu-1}<|\xi|<2^{\nu+1}}} |x-\xi|^{-1} \prod_k \Big(\frac{r_k(\xi)}{|\xi|}\Big)^{-p'\delta_k} d\xi \\ &\qquad \le |x|^2 \int\limits_{\substack{\mathcal{K} \\ 2^{-4}<|\eta|<2^4}} |y-\eta|^{-1} \prod_k \Big(\frac{r_k(\eta)}{|\eta|}\Big)^{-p'\delta_k} d\eta \le c\,|x|^2. \end{aligned}$$

Therefore,

$$\|\chi_\mu v^-\|^p_{W^{0,p}_{\beta-2,\delta}(\mathcal{K})^\ell} \le c \int\limits_{\substack{\mathcal{K}\\ 2^{\mu-1}<|x|<2^{\mu+1}}} \Big(\int_{\mathcal{K}} |x-\xi|^{-1} \prod_j \Big(\frac{r_j(\xi)}{|\xi|}\Big)^{p\delta_j} |\zeta_\nu(\xi)\, f(\xi)|^p\, d\xi\Big) \times |x|^{p\beta-2} \prod_k \Big(\frac{r_k(x)}{|x|}\Big)^{p\delta_k} dx$$

$$\le c\, 2^{-2\mu} \int_{\mathcal{K}} |\xi|^{p\beta} \prod_k \Big(\frac{r_k(\xi}{|\xi|}\Big)^{p\delta_k} |\zeta_\nu(\xi)\, f(\xi)|^p \times \Big(\int\limits_{\substack{\mathcal{K}\\ 2^{\mu-1}<|x|<2^{\mu+1}}} |x-\xi|^{-1} \prod_k \Big(\frac{r_k(x)}{|x|}\Big)^{p\delta_k} dx\Big)\, d\xi$$

$$\le c\, \|\zeta_\nu f\|^p_{W^{0,p}_{\beta,\delta}(\mathcal{K})^\ell}\, .$$

Thus, the estimate (7.6.5) is valid for $|\alpha| = 0$. Applying Lemma 3.5.3 with $T = 2-|\alpha|$, $\gamma_k = \min(0, \delta_+^{(k)} - |\alpha| - \varepsilon)$, $\gamma'_k = 0$, $\delta'_k = \delta''_k = \delta_k$, we obtain the estimate (7.6.5) for $1 \le |\alpha| \le 2$. Consequently,

(7.6.6) $$\|\zeta_\mu v^-\|_{W^{2,p}_{\beta,\delta}(\mathcal{K})^\ell} \le c \sum_{|\alpha|\le 2} \|\chi_\mu \partial_x^\alpha v^-\|_{W^{0,p}_{\beta-2+|\alpha|,\delta}(\mathcal{K})^\ell} \le c'\, \|\zeta_\nu f\|_{W^{0,p}_{\beta,\delta}(\mathcal{K})^\ell}\, .$$

It remains to estimate the norm of $\zeta_\mu v^+$. By Theorem 7.5.3,

$$\big|\partial_x^\alpha \big(\chi^+(x,\xi)\, G(x,\xi)\big)\big| \le c\, |x-\xi|^{-1-|\alpha|}$$

for $2^{\mu-1} \le |x| \le 2^{\mu+1}$, $2^{\nu-1} \le |\xi| \le 2^{\nu+1}$. Using Lemmas 3.5.4 and 3.5.5, we obtain

(7.6.7) $$\|\zeta_\mu v^+\|_{V^{0,p}_{\beta-2,\delta-2}(\mathcal{K})^\ell} \le c\, \|\zeta_\nu f\|_{W^{0,p}_{\beta,\delta}(\mathcal{K})^\ell}$$

and

(7.6.8) $$\int_{\Gamma_j} |x|^{p(\beta-2+|\alpha|)+1} \prod_{k=1}^d \Big(\frac{r_k(x)}{|x|}\Big)^{p(\delta_k-2+|\alpha|)+1} \big|\zeta_\mu(x)\partial_x^\alpha v^+(x)\big|^p\, dx \le c\|\zeta_\nu f\|^p_{W^{0,p}_{\beta,\delta}(\mathcal{K})^\ell}$$

for $j = 1,\dots,d$, $|\alpha| \le 1$. Note that the estimates (7.6.6)–(7.6.8) hold also for $|\mu-\nu| \le 4$. Let $\eta_\mu = \zeta_{\mu-2} + \cdots + \zeta_{\mu+2}$. Then

$$\begin{aligned}\|\eta_\mu L(D_x)\, v^+\|_{W^{0,p}_{\beta,\delta}(\mathcal{K})^\ell} &\le \|\eta_\mu \zeta_\nu f\|_{W^{0,p}_{\beta,\delta}(\mathcal{K})^\ell} + \|\eta_\mu L(D_x)\, v^-\|_{W^{0,p}_{\beta,\delta}(\mathcal{K})^\ell}\\ &\le c\, \|\zeta_\nu f\|_{W^{0,p}_{\beta,\delta}(\mathcal{K})^\ell}\end{aligned}$$

and

$$\|\eta_\mu v^+\|_{W^{2-1/p,p}_{\beta,\delta}(\Gamma_j)^\ell} = \|\eta_\mu v^-\|_{W^{2-1/p,p}_{\beta,\delta}(\Gamma_j)^\ell} \le c\, \|\zeta_\nu f\|_{W^{0,p}_{\beta,\delta}(\mathcal{K})^\ell} \quad \text{for } j \in I_0\, .$$

This together with (7.6.8) implies

$$\|\eta_\mu v^+\|_{V^{2-1/p,p}_{\beta,\delta}(\Gamma_j)^\ell} \le c\, \|\zeta_\nu f\|_{W^{0,p}_{\beta,\delta}(\mathcal{K})^\ell}$$

for $j \in I_0$. Analogously,

$$\|\eta_\mu\, N(D_x)\, v^+\|_{V^{1-1/p,p}_{\beta,\delta}(\Gamma_j)^\ell} \le c\, \|\zeta_\nu f\|_{W^{0,p}_{\beta,\delta}(\mathcal{K})^\ell}$$

for $j \in I_1$. Thus according to (7.3.8),

$$\begin{aligned}\|\zeta_\mu v^+\|^p_{V^{2,p}_{\beta,\delta}(\mathcal{K})^\ell} &\le c\Big(\|\eta_\mu L(D_x)v^+\|^p_{V^{0,p}_{\beta,\delta}(\mathcal{K})^\ell} + \sum_{j\in I_0} \|\eta_\mu v^+\|^p_{V^{2-1/p,p}_{\beta,\delta}(\Gamma_j)^\ell} \\ &\quad + \sum_{j\in I_1} \|\eta_\mu N(D_x)\, v^+\|^p_{V^{1-1/p,p}_{\beta,\delta}(\Gamma_j)^\ell} + \|\eta_\mu v^+\|^p_{V^{0,p}_{\beta-2,\delta-2}(\mathcal{K})^\ell}\Big) \\ &\le c\,\|\zeta_\nu f\|^p_{W^{0,p}_{\beta,\delta}(\mathcal{K})^\ell}.\end{aligned}$$

The last estimate and (7.6.6) yield (7.6.4). □

7.6.2. Existence of solutions in $W^{2,p}_{\beta,\delta}(\mathcal{K})^\ell$. Using the last lemma and Theorem 3.5.8, it is easy to prove the following statement.

THEOREM 7.6.2. *Let $f \in W^{0,p}_{\beta,\delta}(\mathcal{K})^\ell$ and $g_j \in W^{2-d_j-1/p,p}_{\beta,\delta}(\Gamma_j)^\ell$ for $j = 1, \dots, d$. We assume that the line $\operatorname{Re}\lambda = 2-\beta-3/p$ is free of eigenvalues of the pencil $\mathfrak{A}(\lambda)$, the components of δ satisfy the inequalities* (7.6.1), *and the Dirichlet data satisfy the compatibility condition* (7.3.5). *Then there exists a solution $u \in W^{2,p}_{\beta,\delta}(\mathcal{K})^\ell$ of the boundary value problem* (7.1.1)–(7.1.3) *satisfying the estimate*

$$\|u\|_{W^{2,p}_{\beta,\delta}(\mathcal{K})^\ell} \le c\Big(\|f\|_{W^{0,p}_{\beta,\delta}(\mathcal{K})^\ell} + \sum_{j=1}^{d} \|g_j\|_{W^{2-d_j-1/p,p}_{\beta,\delta}(\Gamma_j)^\ell}\Big)$$

with a constant c independent of f and g_j.

P r o o f. By Lemma 7.3.1, there exists a vector function $w \in W^{2,p}_{\beta,\delta}(\mathcal{K})^\ell$ satisfying the boundary conditions (7.1.2) and (7.1.3). Thus, we may assume without loss of generality that $g_j = 0$ for $j = 1, \dots, d$. Let $G(x,\xi)$ be the Green's matrix introduced in Theorem 7.5.1 for an arbitrary κ in the interval $-\Lambda_+ - 1/2 < \kappa < -\Lambda_- - 1/2$, where $\Lambda_- < \operatorname{Re}\lambda < \Lambda_+$ is the widest strip in the complex plane which is free of eigenvalues of the pencil $\mathfrak{A}(\lambda)$ and contains the line $\operatorname{Re}\lambda = 2-\beta-3/p$. Then the vector function (7.6.2) is a solution of the problem (7.1.1)–(7.1.3). According to Lemma 7.6.1, the operator

$$W^{0,p}_{\beta,\delta}(\mathcal{K})^\ell \ni f \to \mathcal{O}f = u$$

satisfies the conditions of Lemma 3.5.8 for the spaces $\mathcal{X} = W^{0,p}_{\beta,\delta}(\mathcal{K})^\ell$ and $\mathcal{Y} = W^{2,p}_{\beta,\delta}(\mathcal{K})^\ell$. Consequently, this operator realizes a continuous mapping from $\mathcal{X}$ into $\mathcal{Y}$. This proves the theorem. □

7.6.3. Uniqueness of the solution. In the case $p \le 2$ we will prove the uniqueness of the solution in Theorem 7.6.2 by means of Lemma 2.1.2. For this end, we pass to the coordinates t, ω, where $t = \log\rho = \log|x|$ and $\omega = x/|x|$. We denote by $W^{l,p}_\delta(\mathbb{R}\times\Omega)$ the weighted Sobolev space with the norm

$$\|u\|_{W^{l,p}_\delta(\mathbb{R}\times\Omega)} = \Big(\int_{\mathbb{R}} \sum_{j=0}^{l} \|\partial_t^j u(t,\cdot)\|^p_{W^{l-j,p}_\delta(\Omega)}\, dt\Big)^{1/p}. \tag{7.6.9}$$

For an arbitrary function $v \in W^{l,p}_\delta(\mathbb{R}\times\Omega)$ we define by v_ε the mollification with respect to the variable t of v, i.e.,

$$v_\varepsilon(t,\omega) = \int_{\mathbb{R}} v(\tau,\omega)\, h_\varepsilon(t-\tau)\, d\tau,$$

where $h_\varepsilon(t) = \varepsilon^{-1}h(t/\varepsilon)$ and h is a smooth function with compact support such that $\int h(t)\,dt = 1$. Since

$$\partial_\omega^\alpha \partial_t^{j+k} v_\varepsilon(\omega,t) = \int_{\mathbb{R}} (\partial_\omega^\alpha \partial_t^k v)(\omega,\tau)\, h_\varepsilon^{(j)}(t-\tau)\,d\tau,$$

it follows that $\partial_t^j v_\varepsilon \in W_\delta^{l,p}(\mathbb{R}\times\Omega)$ for $v \in W_\delta^{3,p}(\mathbb{R}\times\Omega)$, $\varepsilon > 0$, $j = 0,1,\dots$.

LEMMA 7.6.3. *Let $1 < p \le 2$, and let the conditions* (i) *and* (ii) *of Subsection* 7.6.1 *be satisfied. Then the homogeneous boundary value problem* (7.1.1)–(7.1.3) *has only the trivial solution $u = 0$ in $W_{\beta,\delta}^{2,p}(\mathcal{K})^\ell$.*

Proof. Since $W_{\beta,\delta}^{2,p}(\mathcal{K}) \subset W_{\beta,\delta'}^{2,p}(\mathcal{K})$ if $\delta_k \le \delta_k'$ for $k = 1,\dots,d$, it suffices to prove the lemma for the case where $\max(1, 2-\delta_+^{(k)}) < \delta_k + 2/p < 2$.

Let $u \in W_{\beta,\delta}^{2,p}(\mathcal{K})^\ell$ be a solution of the homogeneous problem (7.1.1)–(7.1.3). It follows from Lemma 7.3.4 that $u \in W_{\beta+1,\delta+1}^{3,p}(\mathcal{K})^\ell$. We set $v = \rho^{\beta-2+3/p}\,u$. Then $v \in W_{\delta+1}^{3,p}(\mathbb{R}\times\Omega)^\ell$ in the coordinates (t,ω), where $t = \log\rho$, $\rho = |x|$, $\omega = x/|x|$, and therefore, $\partial_t^j v_\varepsilon \in W_{\delta+1}^{3,p}(\mathbb{R}\times\Omega)^\ell$ for $j = 0,1,2,\dots$. Furthermore, both v and v_ε are solutions of the problem

$$\begin{aligned}
&\mathcal{L}(\partial_t - \beta + 2 - 3/p)\,v = 0 \quad \text{for } t\in\mathbb{R},\ \omega\in\Omega,\\
&v = 0 \quad \text{for } t\in\mathbb{R},\ \omega\in\gamma_j,\ j\in I_0,\\
&\mathcal{N}(\partial_t - \beta + 2 - 3/p)\,v = 0 \quad \text{for } t\in\mathbb{R},\ \omega\in\gamma_j,\ j\in I_1,
\end{aligned}$$

where $\mathcal{L}(\lambda)$ and $\mathcal{N}(\lambda)$ are the parameter-dependent differential operators defined by (7.1.14) and (7.1.15), respectively. Since

$$\partial_t^j v_\varepsilon \in W_{\delta+1}^{2,p}(\mathbb{R}\times\Omega)^\ell = V_{\delta+1}^{2,p}(\mathbb{R}\times\Omega)^\ell$$

and $\partial_t^j \partial_\omega v_\varepsilon \in W_{\delta+1}^{2,p}(\mathbb{R}\times\Omega)^\ell = V_{\delta+1}^{2,p}(\mathbb{R}\times\Omega)^\ell$ for $j = 0,1,2$, it follows from Lemma 2.1.2 that $v_\varepsilon \in V_{\delta-2+2/p}^{0,2}(\mathbb{R}\times\Omega)^\ell$ and $\partial_\omega v_\varepsilon \in V_{\delta-2+2/p}^{0,2}(\mathbb{R}\times\Omega)^\ell$. Therefore, $v_\varepsilon \in W_{\delta-2+2/p}^{1,2}(\mathbb{R}\times\Omega)^\ell$, where $\delta_k - 2 + 2/p < 0$ for $k = 1,\dots,d$. Consequently,

$$\rho^{-\beta+2-3/p} v_\varepsilon \in W_{\beta+3/p-5/2,0}^{1,2}(\mathcal{K})^\ell = V_{\beta+3/p-5/2}^{1,2}(\mathcal{K})^\ell$$

(in Cartesian coordinates). Since the function $u_\varepsilon = \rho^{-\beta+2-3/p}v_\varepsilon$ is also a solution of the homogeneous problem (7.1.1)–(7.1.3), we conclude from Theorem 7.3.7 that $u_\varepsilon = 0$ for all $\varepsilon > 0$. This implies $u = 0$. □

THEOREM 7.6.4. *Let $f \in W_{\beta,\delta}^{0,p}(\mathcal{K})^\ell$ and $g_j \in W_{\beta,\delta}^{2-d_j-1/p,p}(\Gamma_j)^\ell$. If the conditions of Theorem* 7.6.2 *are satisfied, then the problem* (7.1.1)–(7.1.3) *is uniquely solvable in $W_{\beta,\delta}^{2,p}(\mathcal{K})^\ell$.*

Proof. The existence of solutions and the uniqueness in the case $1 < p \le 2$ are already shown (see Theorem 7.6.2 and Lemma 7.6.3). Let $p > 2$ and let $u \in W_{\beta,\delta}^{2,p}(\mathcal{K})^\ell$ be a solution of the homogeneous problem (7.1.1)–(7.1.3). By ζ we denote an infinitely differentiable function on $\overline{\mathcal{K}}$ such that $\zeta(x) = 1$ for $|x| < 1$ and $\zeta(x) = 0$ for $|x| > 2$. Furthermore, we set $\beta' = \beta - (3p-6)/(2p)$ and $\delta_k' = \delta_k - 1 + 2/p$ for

$k = 1, \dots, d$. Then by Hölder's inequality,

$$\int_{\mathcal{K}} \rho^{2(\beta'+\varepsilon-2+|\alpha|)} \prod_k \Big(\frac{r_k}{\rho}\Big)^{2(\delta'_k+\varepsilon)} |\partial_x^\alpha(\zeta u)|^2\, dx$$

$$\le \Big(\int_{\mathcal{K}} \rho^{p(\beta-2+|\alpha|)} \prod_k \Big(\frac{r_k}{\rho}\Big)^{p\delta_k} |\partial_x^\alpha(\zeta u)|^p\, dx\Big)^{2/p} \Big(\int\limits_{\substack{\mathcal{K}\\ |x|\le 2}} \prod \Big(\frac{r_k}{\rho}\Big)^{-2+q\varepsilon} \frac{dx}{\rho^{3-q\varepsilon}}\Big)^{2/q},$$

where $q = 2p/(p-2)$. The second integral on the right-hand side is finite for $|\alpha| \le 2$ if $\varepsilon > 0$. Consequently, $\zeta u \in W^{2,2}_{\beta'+\varepsilon,\delta'+\varepsilon}(\mathcal{K})^\ell$. In the same way, we obtain $(1-\zeta)u \in W^{2,2}_{\beta'-\varepsilon,\delta'+\varepsilon}(\mathcal{K})^\ell$. This implies

$$L(D_x)\,(\zeta u) = -L(D_x)\,\big((1-\zeta)u\big) \in W^{0,2}_{\beta'-\varepsilon,\delta'+\varepsilon}(\mathcal{K})^\ell$$

Analogously, $\zeta u|_{\Gamma_j} \in W^{3/2,2}_{\beta'-\varepsilon,\delta'+\varepsilon}(\Gamma_j)^\ell$ for $j \in I_0$ and $N(D_x)\,(\zeta u) \in W^{1/2,2}_{\beta'-\varepsilon,\delta'+\varepsilon}(\Gamma_j)^\ell$ for $j \in I_1$. From this and from Theorem 7.4.4 it follows that ζu and therefore also u belong to the space $W^{2,2}_{\beta'-\varepsilon,\delta'+\varepsilon}(\mathcal{K})^\ell$. Hence, $u = 0$ by Theorem 7.3.2. The proof of the theorem is complete. □

The solution in Theorem 7.6.2 was constructed by means of the Green's function introduced in Theorem 7.5.1 with an arbitrary κ such that the closed strip between the lines $\operatorname{Re}\lambda = 2-\beta-3/p$ and $\operatorname{Re}\lambda = -\kappa - 1/2$ is free of eigenvalues of the pencil of the pencil $\mathfrak{A}(\lambda)$. Thus, the solutions in the spaces $W^{2,p}_{\beta,\delta}(\mathcal{K})^\ell$ and $W^{2,q}_{\beta',\delta'}(\mathcal{K})^\ell$ can be represented by the same Green's matrix provided the closed strip between the lines $\operatorname{Re}\lambda = 2-\beta-3/p$ and $\operatorname{Re}\lambda = 2-\beta'-3/q$ does not contain eigenvalues of the pencil $\mathfrak{A}(\lambda)$. In particular, the following statement holds.

THEOREM 7.6.5. *Suppose that the closed strip between the lines* $\operatorname{Re}\lambda = 2-\beta-3/p$ *and* $\operatorname{Re}\lambda = 2-\beta'-3/q$ *does not contain eigenvalues of the pencil* $\mathfrak{A}$ *and that the components of* δ, δ' *satisfy the inequalities* $\max(0, 2-\delta_+^{(k)}) < \delta_k + 2/p < 2$ *and* $\max(0, 2-\delta_+^{(k)}) < \delta'_k + 2/q < 2$ *for* $k = 1, \dots, d$. *If*

$$f \in W^{0,p}_{\beta,\delta}(\mathcal{K})^\ell \cap W^{0,q}_{\beta',\delta'}(\mathcal{K})^\ell \quad \textit{and} \quad g_j \in W^{2-d_j-1/p,p}_{\beta,\delta}(\Gamma_j)^\ell \cap W^{2-d_j-1/q,q}_{\beta',\delta'}(\Gamma_j)^\ell$$

for $j = 1, \dots, d$ *and the Dirichlet data satisfy the compatibility condition* (7.3.5), *then the solution* $u \in W^{2,p}_{\beta,\delta}(\mathcal{K})^\ell$ *of the boundary value problem* (7.1.1)–(7.1.3) *belongs to the space* $W^{2,q}_{\beta',\delta'}(\mathcal{K})^\ell$.

7.7. Weak solutions in weighted L_p Sobolev spaces

Now we are interested in variational solutions of the boundary value problem (7.1.1)–(7.1.3) in the weighted Sobolev space $W^{1,p}_{\beta,\delta}(\mathcal{K})^\ell$. By a *variational solution* of this problem, we mean a vector function $u \in W^{1,p}_{\beta,\delta}(\mathcal{K})^\ell$ satisfying the equation

$$b_{\mathcal{K}}(u,v) = (F,v)_{\mathcal{K}} \quad \text{for all } v \in W^{1,p'}_{-\beta,-\delta}(\mathcal{K})^\ell,\ v|_{\Gamma_j} = 0 \text{ if } j \in I_0, \tag{7.7.1}$$

and the boundary condition

$$u = g_j \quad \text{on } \Gamma_j \text{ for } j \in I_0. \tag{7.7.2}$$

Here $g_j \in W^{1-1/p,p}_{\beta,\delta}(\Gamma_j)^\ell$ for $j \in I_0$, and F is a given linear and continuous functional on the subspace of all $v \in W^{1,p'}_{-\beta,-\delta}(\mathcal{K})^\ell$, $p' = p/(p-1)$, such that $v = 0$ on

Γ_j for $j \in I_0$. We assume that the line $\operatorname{Re}\lambda = 1-\beta-3/p$ is free of eigenvalues of the pencil $\mathfrak{A}(\lambda)$ and that the components of δ satisfy the inequalities

$$\max(0, 1-\delta_+^{(k)}) < \delta_k + 2/p < 1 \quad \text{for } k = 1,\dots,d. \tag{7.7.3}$$

In the case $p = 2$ it is allowed that $\delta_k = 0$. We prove in this section that these conditions on β and δ guarantee the existence and uniqueness of weak solutions in the weighted space $W^{1,p}_{\beta,\delta}(\mathcal{K})^\ell$. Furthermore, we obtain regularity assertions for the variational solutions. This means, we show that the solution $u \in W^{1,p}_{\beta,\delta}(\mathcal{K})^\ell$ belongs to the space $W^{l,q}_{\beta',\delta'}(\mathcal{K})^\ell$ under certain conditions on the data F, g_j and on the weight parameters β' and δ'.

7.7.1. Representation of variational solutions by Green's matrix. We introduce the subspace

$$\mathcal{H}_{p',-\beta,-\delta} = \{u \in W^{1,p'}_{-\beta,-\delta}(\mathcal{K})^\ell :\ u|_{\Gamma_j} = 0 \text{ for } j \in I_0\}.$$

Note that $\mathcal{H}_{p',-\beta,-\delta}$ is a subspace of $V^{1,p'}_{-\beta,-\delta}(\mathcal{K})^\ell$ if $-\delta_k > 1-2/p'$, i.e. $\delta_k + 2/p < 1$ for $k = 1,\dots,d$. In this case, the following characterization of the dual space of $\mathcal{H}_{p',-\beta,-\delta}$ holds analogously to Lemma 3.1.1.

LEMMA 7.7.1. *Let $F \in \mathcal{H}^*_{p',-\beta,-\delta}$, where $\delta_k + 2/p < 1$ for $k = 1,\dots,d$. Then there exist vector functions $f^{(0)} \in V^{0,p}_{\beta+1,\delta+1}(\mathcal{K})^\ell$ and $f^{(j)} \in V^{0,p}_{\beta,\delta}(\mathcal{K})^\ell$, $j = 1,2,3$, such that*

$$F(v) = \int_{\mathcal{K}} \Big(f^{(0)} \cdot v + \sum_{j=1}^{3} f^{(j)} \cdot \partial_{x_j} v \Big)\, dx \quad \textit{for all } v \in \mathcal{H}_{p',-\beta,-\delta} \tag{7.7.4}$$

and

$$\|f^{(0)}\|_{V^{0,p}_{\beta+1,\delta+1}(\mathcal{K})^\ell} + \sum_{j=1}^{3} \|f^{(j)}\|_{V^{0,p}_{\beta,\delta}(\mathcal{K})^\ell} \le c\, \|F\|_{\mathcal{H}^*_{p',-\beta,-\delta}}.$$

Here c is a constant independent of F.

P r o o f. Let B be the Banach space of all

$$\phi = (\phi^{(0)}, \phi^{(1)}, \phi^{(2)}, \phi^{(3)}) \in V^{0,p'}_{-\beta-1,-\delta-1}(\mathcal{K})^\ell \times V^{0,p'}_{-\beta,-\delta}(\mathcal{K})^{3\ell}.$$

Furthermore, let M be the subspace of all

$$\phi = \big(v, \partial_{x_1} v, \partial_{x_2} v, \partial_{x_3} v\big), \quad \text{where } v \in \mathcal{H}_{p',-\beta,-\delta}\,.$$

Suppose that F is a linear and continuous functional on $\mathcal{H}_{p',-\beta,-\delta}$. Then we define the functional G on M by

$$G(\phi) = F(v) \quad \text{for } \phi = \big(v, \partial_{x_1} v, \partial_{x_2} v, \partial_{x_3} v\big) \in M.$$

In view of the Hahn-Banach theorem, G can be extended to a functional G_1 on B with the same norm. However, every functional G_1 on B has the form

$$G_1(\phi) = \int_{\mathcal{K}} \sum_{j=0}^{3} f^{(j)} \cdot \phi^{(j)}\, dx \quad \text{for } \phi \in B,$$

where $f^{(0)} \in V^{0,p}_{\beta+1,\delta+1}(\mathcal{K})^\ell$ and $f^{(j)} \in V^{0,p}_{\beta,\delta}(\mathcal{K})^\ell$, $j = 1,2,3$. This proves the lemma.
□

We consider the problem (7.7.1), (7.7.2) for given

$$F \in \mathcal{H}^*_{p',-\beta,-\delta}, \quad g_j \in W^{1-1/p,p}_{\beta,\delta}(\Gamma_j)^\ell$$

and assume that the components of δ satisfy the inequalities (7.7.3). For the existence of a function $u \in W^{1,p}_{\beta,\delta}(\mathcal{K})^\ell$ satisfying the Dirichlet condition (7.7.2) it is necessary and sufficient that the vector functions g_j satisfy the compatibility condition (7.3.5). By Lemma 7.3.1, we may restrict ourselves to the case of zero Dirichlet data g_j. Then the vector function

$$u(x) = \int_{\mathcal{K}} G(x,\xi)\, F(\xi)\, d\xi$$

is a solution of the boundary value problem (7.7.1), (7.7.2). Using the representation (7.7.4) for the functional F, we obtain

$$(7.7.5) \qquad u(x) = \int_{\mathcal{K}} G(x,\xi)\, f^{(0)}(\xi)\, d\xi + \sum_{j=1}^{3} \int_{\mathcal{K}} \partial_{\xi_j} G(x,\xi)\, f^{(j)}(\xi)\, d\xi,$$

where $f^{(0)} \in W^{0,p}_{\beta+1,\delta+1}(\mathcal{K})^\ell$ and $f^{(j)} \in W^{0,p}_{\beta,\delta}(\mathcal{K})^\ell$ for $j = 1,2,3$.

7.7.2. A local estimate. Suppose that the line $\mathrm{Re}\,\lambda = 1 - \beta - 3/p$ does not contain eigenvalues of the pencil $\mathfrak{A}(\lambda)$. By

$$\Lambda_- < \mathrm{Re}\,\lambda < \Lambda_+ \,,$$

we denote the widest strip in the complex plane containing the line $\mathrm{Re}\,\lambda = 1-\beta-3/p$ which is free of eigenvalues of the pencil $\mathfrak{A}(\lambda)$. Furthermore, let κ be an arbitrary real number in the interval

$$-\Lambda_+ - 1/2 < \kappa < -\Lambda_- - 1/2 \,.$$

In the sequel, let $G(x,\xi)$ be the Green matrix introduced in Theorem 7.5.1 for our fixed κ. Furthermore, let ζ_ν be infinitely differentiable functions depending only on $\rho = |x|$ and satisfying the conditions (7.2.3). We consider the function

$$(7.7.6) \quad v(x) = \int_{\mathcal{K}} G(x,\xi)\, \zeta_\nu(\xi)\, f^{(0)}(\xi)\, d\xi + \sum_{j=1}^{3} \int_{\mathcal{K}} \partial_{\xi_j} G(x,\xi)\, \zeta_\nu(\xi)\, f^{(j)}(\xi)\, d\xi.$$

LEMMA 7.7.2. *Suppose that the line* $\mathrm{Re}\,\lambda = 1 - \beta - 3/p$ *does not contain eigenvalues of the operator pencil* $\mathfrak{A}(\lambda)$ *and that the components of* δ *satisfy the inequalities* (7.7.3). *Then the vector function* (7.7.5) *satisfies the inequality*

$$(7.7.7) \quad \|\zeta_\mu v\|_{W^{1,p}_{\beta,\delta}(\mathcal{K})^\ell} \le c\, 2^{-|\mu-\nu|\varepsilon_0} \Big(\|\zeta_\nu f^{(0)}\|_{W^{0,p}_{\beta+1,\delta+1}(\mathcal{K})^\ell} + \sum_{j=1}^{3} \|\zeta_\nu f^{(j)}\|_{W^{0,p}_{\beta,\delta}(\mathcal{K})^\ell} \Big)$$

with positive constants c *and* ε_0 *independent of* $f^{(j)}$, μ *and* ν.

P r o o f. For the term

$$\int_{\mathcal{K}} G(x,\xi)\, \zeta_\nu(\xi)\, f^{(0)}(\xi)\, d\xi,$$

the assertion of the lemma follows immediately from Lemma 7.6.1. Therefore, it suffices to consider the vector function

$$w(x) = \int_{\mathcal{K}} \partial_{\xi_j} G(x,\xi)\, \zeta_\nu(\xi)\, f^{(j)}(\xi)\, d\xi$$

for $j = 1,2,3$. Let the function χ_μ in $\mathcal{K}$ be defined as

$$\chi_\mu(x) = 1 \ \text{ for } 2^{\mu-1} \le |x| \le 2^{\mu+1}, \quad \chi_\mu(x) = 0 \ \text{ else.}$$

Then there exists a constant c independent of v and μ such that

$$\|\zeta_\mu w\|_{W^{1,p}_{\beta,\delta}(\mathcal{K})^\ell} \le c \sum_{|\alpha|\le 1} \|\chi_\mu \partial_x^\alpha w\|_{W^{0,p}_{\beta-1+|\alpha|,\delta}(\mathcal{K})^\ell} .$$

First let $\mu > \nu + 2$. Then by Theorem 7.5.5,

$$\big|\partial_x^\alpha \partial_{\xi_j} G(x,\xi)\big| \le c\,|x|^{\Lambda_- - |\alpha| + \varepsilon}\, |\xi|^{-2-\Lambda_- -\varepsilon} \prod_{k=1}^d \Big(\frac{r_k(x)}{|x|}\Big)^{\delta^+_{k,\alpha}} \prod_{k=1}^d \Big(\frac{r_k(\xi)}{|x|}\Big)^{\delta^-_{k,1}}$$

for $2^{\mu-1} \le |x| \le 2^{\mu+1}$, $2^{\nu-1} \le |\xi| \le 2^{\nu+1}$, where $\delta^+_{k,\alpha} = \min(0, \delta_+^{(k)} - |\alpha| - \varepsilon)$, $\delta^-_{k,1} = \min(0, \delta_-^{(k)} - 1 - \varepsilon)$ and ε is an arbitrarily small positive number. Applying Lemma 3.5.1, we obtain

$$\|\chi_\mu \partial_x^\alpha w\|_{W^{0,p}_{\beta-1+|\alpha|,\delta}(\mathcal{K})^\ell} \le c\, 2^{(\mu-\nu)(\beta-1+\Lambda_-+\varepsilon+3/p)}\, \|\zeta_\nu f^{(j)}\|_{W^{0,p}_{\beta,\delta}(\mathcal{K})^\ell}$$

for $|\alpha| \le 1$. Here the factor $\beta - 1 + \Lambda_- + \varepsilon + 3/p$ is negative for sufficiently small ε. This proves the lemma for the case $\mu > \nu + 2$. Analogously, the desired estimate holds for $\nu > \mu + 2$.

We consider the case $|\mu - \nu| \le 2$. Let $\chi^\pm$ be the same functions as in the proof of Lemma 7.6.1. Furthermore, let

$$w^\pm(x) = \int_{\mathcal{K}} \chi^\pm(x,\xi)\, \partial_{\xi_j} G(x,\xi)\, \zeta_\nu(\xi)\, f^{(j)}(\xi)\, d\xi.$$

By Theorem 7.5.2,

$$\big|\partial_x^\alpha \big(\chi^-(x,\xi)\, \partial_{\xi_j} G(x,\xi)\big)\big| \le c\,|x-\xi|^{-2-|\alpha|} \Big(\frac{r(x)}{|x-\xi|}\Big)^{\delta^+_\alpha(x)} \Big(\frac{r(\xi)}{|x-\xi|}\Big)^{\delta^-_1(\xi)}$$

for $2^{\mu-1} \le |x| \le 2^{\mu+1}$, $2^{\nu-1} \le |\xi| \le 2^{\nu+1}$, where

$$\delta^+_\alpha(x) = \min(0, \delta_+^{(k(x))} - |\alpha| - \varepsilon), \quad \delta^-_1(\xi) = \min(0, \delta_-^{(k(\xi))} - 1 - \varepsilon),$$

$k(x)$ denotes the smallest integer k such that $r_k(x) = r(x)$. Applying Lemma 3.5.3 with $T = 1 - |\alpha|$, $\gamma_k = \min(0, \delta_+^{(k)} - |\alpha| - \varepsilon)$, $\gamma'_k = \min(0, \delta_+^{(k)} - 1 - \varepsilon)$, $\delta'_k = \delta''_k = \delta_k$, we obtain the estimate

$$\|\chi_\mu \partial_x^\alpha w^-\|_{W^{0,p}_{\beta-1+|\alpha|,\delta}(\mathcal{K})^\ell} \le c\,\|\zeta_\nu f^{(j)}\|_{W^{0,p}_{\beta,\delta}(\mathcal{K})^\ell} \tag{7.7.8}$$

for $|\alpha| \le 1$. Consequently,

$$\|\zeta_\mu w^-\|_{W^{1,p}_{\beta,\delta}(\mathcal{K})^\ell} \le c \sum_{|\alpha|\le 1} \|\chi_\mu \partial_x^\alpha w^-\|_{W^{0,p}_{\beta-1+|\alpha|,\delta}(\mathcal{K})^\ell} \le c'\, \|\zeta_\nu f^{(j)}\|_{W^{0,p}_{\beta,\delta}(\mathcal{K})^\ell} . \tag{7.7.9}$$

It remains to estimate the norm of $\zeta_\mu w^+$. By Theorem 7.5.3,

$$\big|\chi^+(x,\xi)\, \partial_{\xi_j} G(x,\xi)\big| \le c\,|x-\xi|^{-2}$$

for $2^{\mu-1} \le |x| \le 2^{\mu+1}$, $2^{\nu-1} \le |\xi| \le 2^{\nu+1}$. Using Lemmas 3.5.4 and 3.5.5, we obtain

$$\|\zeta_\mu w^+\|_{V^{0,p}_{\beta-1,\delta-1}(\mathcal{K})^\ell} \le c\,\|\zeta_\nu f^{(j)}\|_{W^{0,p}_{\beta,\delta}(\mathcal{K})^\ell} . \tag{7.7.10}$$

and

$$(7.7.11)\quad \int_{\Gamma_j} |x|^{p(\beta-1)+1} \prod_{k=1}^{d} \Big(\frac{r_k(x)}{|x|}\Big)^{p(\delta_k-1)+1} \big|\zeta_\mu(x) w^+(x)\big|^p\, dx \le c \|\zeta_\nu f\|^p_{W^{0,p}_{\beta,\delta}(\mathcal{K})^\ell}$$

for $j = 1, \dots, d$. Note that the estimates (7.7.9)–(7.7.11) hold also for $|\mu - \nu| \le 4$. Let $\eta_\mu = \zeta_{\mu-2} + \dots + \zeta_{\mu+2}$. Since $w = w^+ + w^-$ is a solution of the problem

$$b_{\mathcal{K}}(w, v) = \int_{\mathcal{K}} f^{(j)}(x)\, \partial_{x_j} v\, dx \quad \text{for all } v \in C_0^\infty(\overline{\mathcal{K}}\backslash\mathcal{S})^\ell,\ v|_{\Gamma_k} = 0 \text{ for } k \in I_0,$$
$$w = 0 \quad \text{on } \Gamma_k \text{ for } k \in I_0,$$

we obtain the estimate

$$\|\eta_\mu w^+\|_{W^{1-1/p,p}_{\beta,\delta}(\Gamma_k)^\ell} = \|\eta_\mu w^-\|_{W^{1-1/p,p}_{\beta,\delta}(\Gamma_k)^\ell} \le c \, \|\zeta_\nu f^{(j)}\|_{W^{0,p}_{\beta,\delta}(\mathcal{K})^\ell} \quad \text{for } k \in I_0.$$

This together with (7.7.11) implies

$$(7.7.12)\qquad \|\eta_\mu w^+\|_{V^{1-1/p,p}_{\beta,\delta}(\Gamma_k)^\ell} \le c \, \|\zeta_\nu f^{(j)}\|_{W^{0,p}_{\beta,\delta}(\mathcal{K})^\ell}$$

for $k \in I_0$. Let the functional Φ be defined by

$$\Phi(v) = \int_{\mathcal{K}} f^{(j)}(x)\, \partial_{x_j} v\, dx - b_{\mathcal{K}}(w^-, v)\,.$$

Then analogously to (7.3.8), the function $\zeta_\mu w^+$ satisfies the estimate

$$\begin{aligned} \|\zeta_\mu w^+\|^p_{V^{1,p}_{\beta,\delta}(\mathcal{K})^\ell} \ &\le \ c\Big(\|\eta_\mu \Phi\|^p_{\mathcal{H}^*_{p',-\beta-\delta}} + \sum_{k\in I_0} \|\eta_\mu w^+\|^p_{V^{1-1/p,p}_{\beta,\delta}(\Gamma_k)^\ell} \\ &\qquad + \|\eta_\mu w^+\|^p_{V^{0,p}_{\beta-1,\delta-1}(\mathcal{K})^\ell}\Big). \end{aligned}$$

Using (7.7.9), (7.7.10) and (7.7.12), we get

$$\|\zeta_\mu w^+\|^p_{V^{1,p}_{\beta,\delta}(\mathcal{K})^\ell} \le c \, \|\zeta_\nu f^{(j)}\|^p_{W^{0,p}_{\beta,\delta}(\mathcal{K})^\ell}\,.$$

The proof is complete. □

7.7.3. Existence and uniqueness of variational solutions. Using the last lemma, we can prove the following statement analogously to Theorem 7.6.2.

THEOREM 7.7.3. *Let $F \in \mathcal{H}^*_{p',-\beta-\delta}$ and $g_j \in W^{1-1/p,p}_{\beta,\delta}(\Gamma_j)^\ell$ for $j \in I_0$. We suppose that the line* $\text{Re}\,\lambda = 1 - \beta - 3/p$ *is free of eigenvalues of the pencil $\mathfrak{A}(\lambda)$, the components of δ satisfy the inequalities* (7.7.3) *and the Dirichlet data satisfy the compatibility condition* (7.3.5). *Then the problem* (7.7.1), (7.7.2) *has a uniquely determined solution $u \in W^{1,p}_{\beta,\delta}(\mathcal{K})^\ell$.*

P r o o f. Lemma 7.3.1 allows us to restrict ourselves to the case of zero Dirichlet data g_j. By Lemma 7.7.1, the functional F has the form (7.7.4). We consider the operator

$$\big(f^{(0)}, f^{(1)}, f^{(2)}, f^{(3)}\big) \to u,$$

where u is given by (7.7.5) and $G(x,\xi)$ is the Green's matrix introduced in Theorem 7.5.1 for an arbitrary κ in the interval $-\Lambda_+ - 1/2 < \kappa < -\Lambda_- - 1/2$. According to Lemma 7.7.2, this operator satisfies the conditions of Lemma 3.5.8 for the spaces

$$\mathcal{X} = W^{0,p}_{\beta+1,\delta+1}(\mathcal{K})^\ell \times W^{0,p}_{\beta,\delta}(\mathcal{K})^{3\ell} \quad \text{and} \quad \mathcal{Y} = W^{1,p}_{\beta,\delta}(\mathcal{K})^\ell.$$

Hence, the vector function (7.7.5) is a solution of the problem (7.7.1), (7.7.2) in $W^{1,p}_{\beta,\delta}(\mathcal{K})^\ell$.

We prove the uniqueness of the solution. Let $u \in W^{1,p}_{\beta,\delta}(\mathcal{K})^\ell$ be a solution of the problem (7.7.1), (7.7.2), where $f = 0$ and $g_j = 0$ for all $j \in I_0$. Then it follows from Lemma 7.3.4 that $u \in W^{2,p}_{\beta+1,\delta+1}(\mathcal{K})^\ell$. Applying Theorem 7.6.4, we obtain $u = 0$. This proves the theorem. □

We consider the special case $p = 2$, $\delta = 0$. By the condition (7.7.3), this case is excluded in Theorem 7.7.3. However by Theorem 7.3.7, the problem

$$b_{\mathcal{K}}(u, v) = (F, v)_{\mathcal{K}} \quad \text{for all } v \in \mathcal{H}_{-\beta} \tag{7.7.13}$$

has a uniquely determined solution $u \in \mathcal{H}_\beta = \mathcal{H}_{2,\beta,0}$ for arbitrary $F \in \mathcal{H}^*_{-\beta}$ if the line $\operatorname{Re}\lambda = -\beta - 1/2$ does not contain eigenvalues of the pencil $\mathfrak{A}(\lambda)$. Note that the space $\mathcal{H}_{-\beta} = \mathcal{H}_{2,-\beta,0}$ is a subspace of $V^{1,2}_{-\beta}(\mathcal{K})^\ell = W^{1,2}_{-\beta,0}(\mathcal{K})^\ell$ but not a subspace of $V^{1,2}_{-\beta,0}(\mathcal{K})^\ell$. Therefore, the assertion of Lemma 7.7.1 is not true for $p = 2$ and $\delta = 0$. In this case, we use the representation

$$F(v) = \int_{\mathcal{K}} \Big(f^{(0)} \cdot v + \sum_{j=1}^{3} f^{(j)} \cdot \partial_{x_j} v \Big)\, dx \quad \text{for all } v \in \mathcal{H}_{-\beta} \tag{7.7.14}$$

for the functional $F \in \mathcal{H}^*_{-\beta}$, where $f^{(0)} \in V^{0,2}_{\beta+1}(\mathcal{K})^\ell$ and $f^{(j)} \in V^{0,2}_{\beta}(\mathcal{K})^\ell$. The representation (7.7.14) holds analogously to Lemma 7.7.1.

THEOREM 7.7.4. *Suppose that β and κ are such that the closed strip between the lines $\operatorname{Re}\lambda = -\beta - 1/2$ and $\operatorname{Re}\lambda = -\kappa - 1/2$ is free of eigenvalues of the pencil $\mathfrak{A}(\lambda)$. Then the uniquely determined solution $u \in \mathcal{H}_\beta$ of the problem (7.7.13) has the representation*

$$u(x) = \int_{\mathcal{K}} G(x, \xi)\, F(\xi)\, d\xi, \tag{7.7.15}$$

where $G(x, \xi)$ is the Green's matrix introduced in Theorem 7.5.1.

P r o o f. By the representation (7.7.14), the vector-valued function (7.7.15) can be written as

$$u(x) = \int_{\mathcal{K}} G(x, \xi)\, f^{(0)}(\xi)\, d\xi + \sum_{j=1}^{3} \int_{\mathcal{K}} \partial_{\xi_j} G(x, \xi)\, f^{(j)}(\xi)\, d\xi, \tag{7.7.16}$$

where $f^{(0)} \in V^{0,2}_{\beta+1}(\mathcal{K})^\ell$ and $f^{(j)} \in V^{0,2}_{\beta}(\mathcal{K})^\ell$ for $j = 1, 2, 3$. Let ζ_ν be infinitely differentiable functions depending only on $\rho = |x|$ and satisfying the conditions (7.2.3). We consider the vector functions

$$v(x) = \int_{\mathcal{K}} G(x, \xi)\, \zeta_\nu(\xi)\, f^{(0)}(\xi)\, d\xi + \sum_{j=1}^{3} \int_{\mathcal{K}} \partial_{\xi_j} G(x, \xi)\, \zeta_\nu(\xi)\, f^{(j)}(\xi)\, d\xi.$$

Analogously to Lemmas 7.6.1 and 7.7.2, one obtains the estimate

$$\|\zeta_\mu v\|_{V^{1,2}_\beta(\mathcal{K})^\ell} \le c\, 2^{-|\mu-\nu|\varepsilon_0} \Big(\|\zeta_\nu f^{(0)}\|_{V^{0,2}_{\beta+1}(\mathcal{K})^\ell} + \sum_{j=1}^{3} \|\zeta_\nu f^{(j)}\|_{V^{0,2}_{\beta}(\mathcal{K})^\ell} \Big)$$

with positive constants c and ε_0 independent of μ, ν and $f^{(j)}$, $j = 0, 1, 2, 3$. Thus by Lemma 3.5.8, the vector function u defined by (7.7.16) belongs to the space

$V_\beta^{1,2}(\mathcal{K})^\ell$. By Theorem 7.3.7, the solution of the problem (7.7.13) is uniquely determined in this space. □

7.7.4. Regularity results for weak solutions. Our goal is to show that the variational solution $u \in W_{\beta,\delta}^{1,p}(\mathcal{K})^\ell$ belongs to the space $W_{\beta',\delta'}^{l,q}(\mathcal{K})^\ell$ under certain conditions on the data and on the weight parameters β' and δ'. First we consider the case $l = 1$.

THEOREM 7.7.5. *Suppose that there are no eigenvalues of the pencil $\mathfrak{A}$ in the closed strip between the lines $\operatorname{Re}\lambda = 1 - \beta - 3/p$ and $\operatorname{Re}\lambda = 1 - \beta' - 3/q$ and that the components of δ and δ' satisfy the inequalities*

$$\max(0, 1 - \delta_+^{(k)}) < \delta_k + 2/p < 1, \quad \max(0, 1 - \delta_+^{(k)}) < \delta'_k + 2/q < 1 \tag{7.7.17}$$

for $k = 1, \dots, d$. If $u \in W_{\beta,\delta}^{1,p}(\mathcal{K})^\ell$ is a solution of the problem (7.7.1), (7.7.2), where $F \in \mathcal{H}^*_{p',-\beta,-\delta} \cap \mathcal{H}^*_{q',-\beta',-\delta'}$, $p' = p/(p-1)$, $q' = q/(q-1)$, and $g_j \in W_{\beta,\delta}^{1-1/p,p}(\Gamma_j)^\ell \cap W_{\beta',\delta'}^{1-1/q,q}(\Gamma_j)^\ell$ for $j \in I_0$, then $u \in W_{\beta',\delta'}^{1,q}(\mathcal{K})^\ell$. In the case $p = 2$ this result is also true for $\delta = 0$.

P r o o f. By Lemma 7.3.1, we may restrict ourselves to the case of zero Dirichlet data g_j. Then the uniquely determined solution both in $W_{\beta,\delta}^{1,p}(\mathcal{K})^\ell$ and in $W_{\beta',\delta'}^{1,p'}(\mathcal{K})^\ell$ is given by formula (7.7.15), where $G(x, \xi)$ is the Green's matrix introduced in Theorem 7.5.1 for $\kappa = \beta + 3/p - 3/2$. This proves the theorem. □

By the same argument, the following statement holds.

LEMMA 7.7.6. *Let $u \in W_{\beta,\delta}^{1,p}(\mathcal{K})^\ell$ be a solution of the problem (7.7.1), (7.7.2), where $g_j \in W_{\beta,\delta}^{1-1/p,p}(\Gamma_j)^\ell \cap W_{\beta',\delta'}^{2-1/q,q}(\Gamma_j)^\ell$ for $j \in I_0$ and F is a linear and continuous functional on $\mathcal{H}_{p',-\beta,-\delta}$ which has the representation*

$$(F, v)_{\mathcal{K}} = \int_{\mathcal{K}} f \cdot \bar{v}\, dx + \sum_{j \in I_1} \int_{\Gamma_j} g_j \cdot \bar{v}\, dx \tag{7.7.18}$$

with given vector functions $f \in W_{\beta',\delta'}^{0,q}(\mathcal{K})^\ell$, $g_j \in W_{\beta',\delta'}^{1-1/q,q}(\Gamma_j)^\ell$. Suppose that the closed strip between the lines $\operatorname{Re}\lambda = 1 - \beta - 3/p$ and $\operatorname{Re}\lambda = 2 - \beta' - 3/q$ does not contain eigenvalues of the pencil $\mathfrak{A}$ and the components of δ, δ' satisfy the inequalities

$$\max(0, 1 - \delta_+^{(k)}) < \delta_k + 2/p < 1, \qquad \max(0, 2 - \delta_+^{(k)}) < \delta'_k + 2/q < 2$$

for $k = 1, \dots, d$. Then $u \in W_{\beta',\delta'}^{2,q}(\mathcal{K})^\ell$. In the case $p = 2$ this result is also true for $\delta = 0$.

We prove a generalization of Lemma 7.7.6.

THEOREM 7.7.7. *Let $u \in W_{\beta,\delta}^{1,p}(\mathcal{K})^\ell$ be a solution of the problem (7.7.1), (7.7.2), where $g_j \in W_{\beta,\delta}^{1-1/p,p}(\Gamma_j)^\ell \cap W_{\beta',\delta'}^{l-1/q,q}(\Gamma_j)^\ell$ for $j \in I_0$ and F is a linear and continuous functional on $\mathcal{H}_{p',-\beta,-\delta}$ which has the representation (7.7.18), where $f \in W_{\beta',\delta'}^{l-2,q}(\mathcal{K})^\ell$, $g_j \in W_{\beta',\delta'}^{l-1-1/q,q}(\Gamma_j)^\ell$, $l \ge 2$. Suppose that the closed strip between the lines $\operatorname{Re}\lambda = 1 - \beta - 3/p$ and $\operatorname{Re}\lambda = l - \beta' - 3/q$ does not contain eigenvalues of the pencil $\mathfrak{A}(\lambda)$ and that the components of δ, δ' satisfy the inequalities*

$$\max(0, 1 - \delta_+^{(k)}) < \delta_k + 2/p < 1, \qquad \max(0, l - \delta_+^{(k)}) < \delta'_k + 2/q < l.$$

Then $u \in W^{l,q}_{\beta',\delta'}(\mathcal{K})^\ell$. *In the case* $p = 2$ *the result is also valid for* $\delta = 0$.

P r o o f. 1) If $l-\delta'_k-2/q < 2$ for $k = 1,\dots,d$, then $f \in W^{0,q}_{\beta'-l+2,\delta'-l+2}(\mathcal{K})^\ell$, $g_j \in W^{2-d_j-1/q,q}_{\beta'-l+2,\delta'-l+2}(\Gamma_j)^\ell$ and Lemma 7.7.6 implies $u \in W^{2,q}_{\beta'-l+2,\delta'-l+2}(\mathcal{K})^\ell$. Applying Lemma 7.3.4, we conclude that $u \in W^{l,q}_{\beta',\delta'}(\mathcal{K})^\ell$.

2) Suppose that $l-s-1 < \delta'_k + 2/q \le l-s$ for $k = 1,\dots,d$, where s is an integer, $2 \le s \le l-1$. This is only possible if $\delta_+^{(k)} > s$. We prove by induction in s that $u \in W^{l,q}_{\beta',\delta'}(\mathcal{K})^\ell$. Let first $s = 2$. Then $f \in W^{0,q}_{\beta'-l+2,\delta'-l+3-\varepsilon}(\mathcal{K})^\ell$ and $g_j \in W^{2-d_j-1/q,q}_{\beta'-l+2,\delta'-l+3-\varepsilon}(\Gamma_j)^\ell$, where ε is a positive real number such that $l-3 < \delta'_k - \varepsilon + 2/q < l-2$. Using Lemma 7.7.6, we obtain $u \in W^{2,q}_{\beta'-l+2,\delta'-l+3-\varepsilon}(\mathcal{K})^\ell$. This together with Lemma 7.3.4 implies $u \in W^{3,q}_{\beta'-l+3,\delta'-l+4-\varepsilon}(\mathcal{K})^\ell$. Consequently, $\rho\partial_\rho u \in W^{2,q}_{\beta'-l+2,\delta'-l+4-\varepsilon}(\mathcal{K})^\ell$. Moreover,

$$L(D_x)\,\rho\partial_\rho u = \rho\partial_\rho f + 2f \in W^{0,q}_{\beta'-l+2,\delta'-l+3}(\mathcal{K})^\ell,$$

$$\rho\partial_\rho u\big|_{\Gamma_j} = \rho\partial_\rho g_j \in W^{2-1/q,q}_{\beta'-l+2,\delta'-l+3}(\Gamma_j)^\ell \quad \text{for } j \in I_0,$$

$$N(D_x)\,\rho\partial_\rho u\big|_{\Gamma_j} = \rho\partial_\rho g_j + g_j \in W^{1-1/q,q}_{\beta'-l+2,\delta'-l+3}(\Gamma_j)^\ell \quad \text{for } j \in I_1,$$

where $-2/q < \delta'_k - l + 3 \le 1 - 2/q$ for $k = 1,\dots,d$. Thus, it follows from Theorem 7.6.5 that $\rho\partial_\rho u \in W^{2,q}_{\beta'-l+2,\delta'-l+3}(\mathcal{K})^\ell$. Since u belongs to the same space, we derive from Lemma 7.3.5 that $u \in W^{3,q}_{\beta'-l+3,\delta'-l+3}(\mathcal{K})^\ell$. Applying Lemma 7.3.4, we obtain $u \in W^{l,q}_{\beta',\delta'}(\mathcal{K})^\ell$. This proves the theorem for the case $l-3 < \delta'_k + 2/q \le l-2$, $k = 1,\dots,d$.

Suppose that $l-s-1 < \delta'_k + 2/q \le l-s$, $3 \le s \le l-1$, and that the theorem is proved for $l-s < \delta'_k + 2/q \le l-s+1$. Since $f \in W^{l-2,q}_{\beta',\delta'}(\mathcal{K})^\ell \subset W^{l-3,q}_{\beta'-1,\delta'}(\mathcal{K})^\ell$ and $g_j \in W^{l-d_j-1/q,q}_{\beta',\delta'}(\Gamma_j)^\ell \subset W^{l-1-d_j-1/q,q}_{\beta'-1,\delta'}(\Gamma_j)^\ell$, the induction hypothesis implies $u \in W^{l-1,q}_{\beta'-1,\delta'}(\mathcal{K})^\ell$ and consequently $\rho\partial_\rho u \in W^{l-2,q}_{\beta'-2,\delta'}(\mathcal{K})^\ell$. On the other hand,

$$L(D_x)\,\rho\partial_\rho u = \rho\partial_\rho f + 2f \in W^{l-3,q}_{\beta'-1,\delta'}(\mathcal{K})^\ell,$$

$$\rho\partial_\rho u\big|_{\Gamma_j} = \rho\partial_\rho g_j \in W^{l-1-1/q,q}_{\beta'-1,\delta'}(\Gamma_j)^\ell \quad \text{for } j \in I_0,$$

$$N(D_x)\,\rho\partial_\rho u\big|_{\Gamma_j} = \rho\partial_\rho g_j + g_j \in W^{l-2-1/q,q}_{\beta'-1,\delta'}(\Gamma_j)^\ell \quad \text{for } j \in I_1,$$

Hence, it follows from the induction hypothesis that $\rho\partial_\rho u \in W^{l-1,q}_{\beta'-1,\delta'}(\mathcal{K})^\ell$. From this and from Lemma 7.3.5 we conclude that $u \in W^{l,q}_{\beta',\delta'}(\mathcal{K})^\ell$.

3) Finally, we consider the case where $l - s_k - 1 < \delta'_k + 2/q \le l - s_k$ for $k = 1,\dots,d$ with different $s_k \in \{0,\dots,l-1\}$. Then let $\zeta_1,\dots,\zeta_d$ be infinitely differentiable functions on $\overline{\Omega}$ such that $\zeta_k \ge 0$, $\zeta_k = 1$ near $M_k \cap S^2$, and $\sum \zeta_k = 1$. We extend ζ_k to $\mathcal{K}$ by the equality $\zeta_k(x) = \zeta_k(x/|x|)$. Then $\partial_x^\alpha \zeta_k(x) \le c\,|x|^{-|\alpha|}$. Using the first and second parts of the proof, it can be easily shown by induction in l that $\zeta_k u \in W^{l,q}_{\beta',\delta'}(\mathcal{K})^\ell$ for $k = 1,\dots,d$. The proof is complete. □

7.7.5. Existence and uniqueness of solutions in $W^{l,p}_{\beta,\delta}(\mathcal{K})^\ell$. Using the existence and uniqueness result for weak solutions in Theorem 7.7.3 together with the regularity result in Theorem 7.7.7, we obtain the following generalization of Theorem 7.6.4.

THEOREM 7.7.8. *Let* $l \ge 2$, $f \in W^{l-2,p}_{\beta,\delta}(\mathcal{K})^\ell$ *and* $g_j \in W^{l-d_j-1/p,p}_{\beta,\delta}(\Gamma_j)^\ell$ *for* $j = 1, \dots, d$. *We assume that the line* $\operatorname{Re}\lambda = l - \beta - 3/p$ *is free of eigenvalues of the pencil* $\mathfrak{A}(\lambda)$, *the components of* δ *satisfy the inequalities*

$$\max(0, l - \delta_+^{(k)}) < \delta_k + 2/p < l \quad \text{for } k = 1, \dots, d,$$

and the Dirichlet data satisfy the compatibility condition (7.3.5). *Then there exists a uniquely determined solution* $u \in W^{l,p}_{\beta,\delta}(\mathcal{K})^\ell$ *of the boundary value problem* (7.1.1)–(7.1.3).

P r o o f. Let $\delta' = (\delta'_1, \dots, \delta'_d)$ be such that

$$\delta'_k \ge \delta_k - l + 1 \quad \text{and} \quad \max(0, 1 - \delta_+^{(k)}) < \delta'_k + 2/p < 1 \quad \text{for } k = 1, \dots, d.$$

Then $f \in W^{0,p}_{\beta-l+2,\delta'+1}(\mathcal{K})^\ell$, $g_j \in W^{1-1/p,p}_{\beta-l+2,\delta'+1}(\Gamma_j)^\ell = V^{1-1/p,p}_{\beta-l+2,\delta'+1}(\Gamma_j)^\ell$ for $j \in I_1$, and $g_j \in W^{1-1/p,p}_{\beta-l+1,\delta'}(\Gamma_j)^\ell$ for $j \in I_0$. We consider the functional F defined by (7.7.18). Since $\mathcal{H}_{p',l-1-\beta,-\delta'}$ is a subspace of $V^{1,p'}_{l-1-\beta',-\delta'}(\mathcal{K})^\ell$, it follows that $F \in \mathcal{H}^*_{p',l-1-\beta,-\delta'}$. Thus by Theorem 7.7.3, there exists a uniquely determined solution $u \in W^{1,p}_{\beta-l+1,\delta'}(\mathcal{K})^\ell$ of the problem (7.7.1), (7.7.2). Applying Theorem 7.7.7, we conclude that u belongs to the subspace $W^{l,p}_{\beta,\delta}(\mathcal{K})^\ell$ of $W^{1,p}_{\beta-l+1,\delta'}(\mathcal{K})^\ell$. The uniqueness of the solution can be easily deduced from Theorem 7.7.3. □

7.7.6. Asymptotics of weak solutions near the vertex of the cone. In Theorem 7.4.5 we obtained an asymptotic decomposition of weak solutions in weighted L_2 Sobolev spaces. We extend this result to weak solutions in weighted L_p Sobolev spaces.

THEOREM 7.7.9. *Let* $u \in W^{1,p}_{\beta,\delta}(\mathcal{K})^\ell$ *be a solution of the problem* (7.7.1), (7.7.2), *where* $F \in \mathcal{H}^*_{p',-\beta,-\delta} \cap \mathcal{H}^*_{q',-\beta',-\delta'}$, $p' = p/(p-1)$, $q' = q/(q-1)$, *and* $g_j \in W^{1-1/p,p}_{\beta,\delta}(\Gamma_j)^\ell \cap W^{1-1/q,q}_{\beta',\delta'}(\Gamma_j)^\ell$ *for* $j \in I_0$. *Suppose that the lines* $\operatorname{Re}\lambda = 1 - \beta - 3/p$ *and* $\operatorname{Re}\lambda = 1 - \beta' - 3/q$ *do not contain eigenvalues of the pencil* $\mathfrak{A}(\lambda)$ *and that the components of* δ *and* δ' *satisfy the inequalities* (7.7.17). *Then* u *admits the decomposition* (7.4.5), *where* $w \in W^{1,q}_{\beta',\delta'}(\mathcal{K})^\ell$, λ_ν *are the eigenvalues of the pencil* $\mathfrak{A}(\lambda)$ *between the lines* $\operatorname{Re}\lambda = 1 - \beta - 3/p$ *and* $\operatorname{Re}\lambda = 1 - \beta' - 3/q$, *and* $U^{\nu,j,s}$ *are eigenvectors and generalized eigenvectors corresponding to the eigenvalue* λ_ν.

P r o o f. Without loss of generality, we may assume that $g_j = 0$ for $j \in I_0$. Let $\{F_k\} \subset C_0^\infty(\overline{\mathcal{K}}\backslash\{0\})^\ell$ be a sequence converging to F in $\mathcal{H}^*_{p',-\beta,-\delta} \cap \mathcal{H}^*_{q',-\beta',-\delta'}$ Then by Theorems 7.3.7 and 7.7.5, there exist solutions $u_k \in \mathcal{H}_{\beta-3/2+3/p} \cap \mathcal{H}_{p,\beta,\delta}$ and $w_k \in \mathcal{H}_{\beta'-3/2+3/q} \cap W_{q,\beta',\delta'}$ of the problem

$$b_{\mathcal{K}}(u, v) = (F_k, v)_{\mathcal{K}} \quad \text{for all } v \in \mathcal{H}_{p',-\beta,-\delta} \ (v \in \mathcal{H}_{q',-\beta',-\delta'}).$$

By Theorem 7.7.3, the sequence $\{u_k\}$ converges to u in $W^{1,p}_{\beta,\delta}(\mathcal{K})^\ell$, while $\{w_k\}$ converges to the unique solution $w \in W^{1,q}_{\beta',\delta'}(\mathcal{K})^\ell$ of problem (7.7.1), (7.7.2). Let $\mathcal{X}$ be the linear span of the functions

$$\rho^{\lambda_\nu} \sum_{\sigma=0}^{s} \frac{1}{\sigma!} (\log \rho)^\sigma \, U^{\nu,j,s-\sigma}(\omega).$$

Then it follows from Theorem 7.4.5 that $u_k - w_k \in \mathcal{X}$ for all k. Consequently, $u - v \in \mathcal{X}$. This proves the theorem. □

As a consequence of Theorems 7.7.7 and 7.7.9, the following statement holds.

COROLLARY 7.7.10. *Let $u \in W^{1,p}_{\beta,\delta}(\mathcal{K})^\ell$ be a solution of the problem (7.7.1), (7.7.2), where $g_j \in W^{1-1/p,p}_{\beta,\delta}(\Gamma_j)^\ell \cap W^{l-1/q,q}_{\beta',\delta'}(\Gamma_j)^\ell$ for $j \in I_0$ and F is a linear and continuous functional on $\mathcal{H}_{p',-\beta,-\delta}$ which has the representation (7.7.18), where $f \in W^{l-2,q}_{\beta',\delta'}(\mathcal{K})^\ell$, $g_j \in W^{l-1-1/q,q}_{\beta',\delta'}(\Gamma_j)^\ell$, $l \ge 2$. Suppose that the lines $\operatorname{Re}\lambda = 1-\beta-3/p$ and $\operatorname{Re}\lambda = l - \beta' - 3/q$ do not contain eigenvalues of the pencil $\mathfrak{A}(\lambda)$ and that the components of δ, δ' satisfy the inequalities*

$$\max(0, 1-\delta_+^{(k)}) < \delta_k + 2/p < 1, \qquad \max(0, l-\delta_+^{(k)}) < \delta'_k + 2/q < l.$$

Then u admits the decomposition (7.4.5), where $w \in W^{l,q}_{\beta',\delta'}(\mathcal{K})^\ell$, λ_ν are the eigenvalues of the pencil $\mathfrak{A}(\lambda)$ between the lines $\operatorname{Re}\lambda = 1-\beta-3/p$ and $\operatorname{Re}\lambda = l-\beta'-3/q$, and $U^{\nu,j,s}$ are eigenvectors and generalized eigenvectors corresponding to the eigenvalue λ_ν. In the case $p = 2$, the result is also valid for $\delta = 0$.

7.7.7. Regularity results for solutions of the Neumann problem. We consider the Neumann problem

$$L(D_x)\,u = f \ \text{ in } \mathcal{K}, \quad N(D_x)\,u = g_j \ \text{ on } \Gamma_j,\ j = 1,\dots,d, \tag{7.7.19}$$

and assume that the following conditions are satisfied:

- the strip $-1 \le \operatorname{Re}\lambda \le 0$ contains only the eigenvalues $\lambda = 0$ and $\lambda = -1$ of the pencil $\mathfrak{A}(\lambda)$,
- the eigenvectors corresponding to the eigenvalue $\lambda = 0$ are constants, and generalized eigenvectors corresponding to this eigenvalue do not exist (i.e., the eigenvalue $\lambda = 0$ has geometric and algebraic multiplicity ℓ).

These conditions are satisfied e.g. if $A_{j,k} = A^*_{k,j}$ for all j,k,

$$\sum_{i,j=1}^{3} \big(A_{i,j}f_j\,,\,f_i\big)_{\mathbb{C}^\ell} \ge c \sum_{j=1}^{3} \big|f_j\big|^2_{\mathbb{C}^\ell} \quad \text{for all } f_j \in \mathbb{C}^\ell$$

and the cone $\mathcal{K}$ is Lipschitz graph (cf. [**85**, Theorem 12.3.3]).

We denote by Λ'_+ the greatest real number such that the strip

$$-1 < \operatorname{Re}\lambda < \Lambda'_+$$

contains only the eigenvalue $\lambda = 0$ of the pencil $\mathfrak{A}(\lambda)$. Suppose that

$$-1 < 2-\beta-3/p < 0, \quad \max(0, 2-\delta_+^{(k)}) < \delta_k + 2/p \ \text{ for } k = 1,\dots,d$$

and that $u \in W^{2,p}_{\beta,\delta}(\mathcal{K})^\ell$ is a solution of the boundary value problem (7.7.19), where

$$f \in W^{0,p}_{\beta,\delta}(\mathcal{K})^\ell \cap W^{0,p}_{\gamma,\delta}(\mathcal{K})^\ell, \quad g_j \in W^{1-1/p,p}_{\beta,\delta}(\Gamma_j)^\ell \cap W^{1-1/p,p}_{\gamma,\delta}(\Gamma_j)^\ell.$$

If $-1 < 2-\gamma-3/p < \Lambda'_+$ and $2-\gamma+3/p \ne 0$, then we conclude from Theorems 7.7.7 and 7.7.9 that

$$u = c + v\,,$$

where c is a constant vector and $v \in W^{2,p}_{\gamma,\delta}(\mathcal{K})^\ell$. In particular, it follows that $\partial_{x_j} u \in W^{1,p}_{\gamma,\delta}(\mathcal{K})$ for $j = 1,2,3$. We show, that the last result holds also in the case $2-\gamma+3/p = 0$. For this, we employ the estimates of Green's matrix given in Theorem 7.5.8. In the sequel, let $G(x,\xi)$ be the Green's matrix introduced in Theorem 7.5.1 for $\kappa = 0$.

THEOREM 7.7.11. *Let $f \in W^{0,p}_{\beta,\delta}(\mathcal{K})^\ell \cap W^{0,p}_{\gamma,\delta}(\mathcal{K})^\ell$ and $g_j \in W^{1-1/p,p}_{\beta,\delta}(\Gamma_j)^\ell \cap W^{1-1/p,p}_{\gamma,\delta}(\Gamma_j)^\ell$ be given. Suppose that the strip $-1 \le \mathrm{Re}\,\lambda \le 0$ contains only the eigenvalues $\lambda = 0$ and $\lambda = -1$ of the pencil $\mathfrak{A}(\lambda)$, the eigenvalue $\lambda = 0$ has geometric and algebraic multiplicity ℓ, and β, γ, δ_k satisfy the inequalities*

$$-1 < 2 - \beta - 3/p < 0, \quad -1 < 2 - \gamma - 3/p < \Lambda'_+, \quad \max(0, 2 - \delta_+^{(k)}) < \delta_k + 2/p < 2$$

for $k = 1, \dots, d$. Then there exists a unique solution $u \in W^{2,p}_{\beta,\delta}(\mathcal{K})^\ell$ of the boundary value problem (7.7.19) such that $\partial_{x_j} u \in W^{1,p}_{\gamma,\delta}(\mathcal{K})^\ell$ for $j = 1, 2, 3$.

P r o o f. The existence and uniqueness of solutions in $W^{2,p}_{\beta,\delta}(\mathcal{K})^\ell$ was shown in Section 7.6 (cf. Theorem 7.6.4). It remains to show that $\partial_{x_j} u \in W^{1,p}_{\gamma,\delta}(\mathcal{K})^\ell$ for $j = 1, 2, 3$. By Lemma 7.3.1, we may restrict ourselves to the case of zero boundary data g_j. Then the solution u admits the representation

$$u(x) = \int_{\mathcal{K}} G(x,\xi)\, f(\xi)\, d\xi,$$

where $G(x,\xi)$ is the Green's matrix introduced in Theorem 7.5.1 for $\kappa = 0$. We consider the vector function

$$v(x) = \int_{\mathcal{K}} G(x,\xi)\, \zeta_\nu(\xi)\, f(\xi)\, d\xi,$$

where ζ_ν are infinitely differentiable functions depending only on $\rho = |x|$ and satisfying the conditions (7.2.3). As in the proof of Lemma 7.6.1, we get

$$\|\zeta_\mu \partial_x^\alpha v\|_{W^{0,p}_{\gamma-2+|\alpha|,\delta}(\mathcal{K})^\ell} \le c\, 2^{(\mu-\nu)(\gamma-3+\varepsilon+3/p)}\, \|\zeta_\nu f\|_{W^{0,p}_{\gamma,\delta}(\mathcal{K})^\ell}$$

for $\mu > \nu + 2$, $|\alpha| \le 2$, and

$$\|\zeta_\mu v\|_{W^{2,p}_{\gamma,\delta}(\mathcal{K})^\ell} \le c\, \|\zeta_\nu f\|_{W^{0,p}_{\gamma,\delta}(\mathcal{K})^\ell}$$

for $|\mu - \nu| \le 2$. Here ε is an arbitrarily mall positive number. Furthermore, the estimate

$$\|\zeta_\mu \partial_x^\alpha v\|_{W^{0,p}_{\gamma-2+|\alpha|,\delta}(\mathcal{K})^\ell} \le c\, 2^{(\mu-\nu)(\gamma-2+\Lambda'_+-\varepsilon+3/p)}\, \|\zeta_\nu f\|_{W^{0,p}_{\gamma,\delta}(\mathcal{K})^\ell}$$

holds by means of Theorem 7.5.8 for $\mu < \nu - 2$, $1 \le |\alpha| \le 2$. Consequently, the operator

$$W^{0,p}_{\gamma,\delta}(\mathcal{K})^\ell \ni f \to \mathcal{O}f = \partial_{x_j} \int_{\mathcal{K}} G(x,\xi)\, f(\xi)\, d\xi$$

satisfies the conditions of Lemma 3.5.8 for the spaces $\mathcal{X} = W^{0,p}_{\gamma,\delta}(\mathcal{K})^\ell$ and $\mathcal{Y} = W^{1,p}_{\gamma,\delta}(\mathcal{K})^\ell$. Thus, this operator realizes a continuous mapping from $\mathcal{X}$ into $\mathcal{Y}$. This proves the theorem. □

Suppose that $\beta > -3/p$ and $\delta_k > -2/p$ for $k = 1, \dots, d$. Then we define the space $L^{l,p}_{\beta,\delta}(\mathcal{K})$ as the closure of the set $C_0^\infty(\overline{\mathcal{K}})$ with respect to the norm

$$\|u\|_{L^{l,p}_{\beta,\delta}(\mathcal{K})} = \Big(\int_{\mathcal{K}} \sum_{|\alpha|=l} \rho^{p\beta} \prod_{k=1}^{d} \Big(\frac{r_k}{\rho}\Big)^{p\delta_k} |\partial_x^\alpha u(x)|^p\, dx \Big)^{1/p}.$$

The corresponding trace space on Γ_j is denoted by $L^{l-1/p,p}_{\beta,\delta}(\Gamma_j)$. If $\beta > 1-3/p$, then by Hardy's inequality

$$\int_{\mathcal{K}} \rho^{p(\beta-1)} \prod_{k=1}^{d} \Big(\frac{r_k}{\rho}\Big)^{p\delta_k} |u|^p\, dx \le c \int_{\mathcal{K}} \rho^{p\beta} \prod_{k=1}^{d} \Big(\frac{r_k}{\rho}\Big)^{p\delta_k} |\nabla u|^p\, dx$$

for all $u \in C_0^\infty(\overline{\mathcal{K}})$. Therefore, $L^{1,p}_{\beta,\delta}(\mathcal{K}) = W^{1,p}_{\beta,\delta}(\mathcal{K})$ and $L^{1-1/p,p}_{\beta,\delta}(\Gamma_j) = W^{1-1/p,p}_{\beta,\delta}(\Gamma_j)$ for $\beta > 1-3/p$.

As a consequence of Theorem 7.7.11, we obtain the following result.

COROLLARY 7.7.12. *Let the assumptions of Theorem* 7.7.11 *on the spectrum of the pencil* $\mathfrak{A}(\lambda)$ *be satisfied. Furthermore, let* $-1 < 2-\beta-3/p < 0$, $-1 < 2-\gamma-3/p < \min(1,\Lambda'_+)$ *and* $\max(0, 2-\delta_+^{(k)}) < \delta_k + 2/p < 2$ *for* $k = 1,\dots,d$. *Then the operator* $u \to \big(L(D_x)u, N(D_x)u\big)$ *of the Neumann problem* (7.7.19) *is an isomorphism from* $W^{2,p}_{\beta,\delta}(\mathcal{K})^\ell \cap L^{2,p}_{\gamma,\delta}(\mathcal{K})$ *onto the space*

$$W^{0,p}_{\beta,\delta}(\mathcal{K})^\ell \cap W^{0,p}_{\gamma,\delta}(\mathcal{K})^\ell \times \prod_{j=1}^{d} \big(W^{1-1/p,p}_{\beta,\delta}(\Gamma_j)^\ell \cap W^{1-1/p,p}_{\gamma,\delta}(\Gamma_j)^\ell\big).$$

Next, we consider variational solutions of the Neumann problem (7.7.19). By Theorem 7.7.3, there exists a unique solution $u \in W^{1,p}_{\beta,\delta}(\mathcal{K})^\ell$ of the problem

$$b_{\mathcal{K}}(u,v) = (F,v)_{\mathcal{K}} \quad \text{for all } v \in W^{1,p'}_{-\beta,-\delta}(\mathcal{K})^\ell, \tag{7.7.20}$$

for arbitrary $F \in \big(W^{1,p'}_{-\beta,-\delta}(\mathcal{K})^*\big)^\ell$, $p' = p/(p-1)$, if $-1 < 1-\beta-3/p < 0$ and $\max(0, 1-\delta_+^{(k)}) < \delta_k + 2/p < 1$ for $k = 1,\dots,d$.

The following theorem is a corollary of Theorem 7.7.9 if $1-\beta-3/p \ne 0$. Using the estimates of Green's matrix in Theorem 7.5.8, it is possible to prove this assertion without the last restriction on β.

THEOREM 7.7.13. *Suppose that* 0 *and* -1 *are the only eigenvalues of the pencil* $\mathfrak{A}(\lambda)$ *in the strip* $-1 \le \operatorname{Re}\lambda \le 0$ *and that the eigenvalue* $\lambda = 0$ *has geometric and algebraic multiplicity* ℓ. *Furthermore, let* $F \in (W^{1,p'}_{-\beta,-\delta}(\mathcal{K})^*)^\ell \cap (W^{1,p'}_{-\gamma,-\delta}(\mathcal{K})^*)^\ell$, *where*

$$-1 < 1-\beta-3/p < 0, \quad -1 < 1-\gamma-3/p < \Lambda'_+\,, \quad \max(0, 1-\delta_+^{(k)}) < \delta_k + 2/p < 1$$

for $k = 1,\dots,d$. *Then there exists a uniquely determined solution* $u \in W^{1,p}_{\beta,\delta}(\mathcal{K})^\ell$ *of the problem* (7.7.20) *such that* $\partial_{x_j} u \in W^{0,p}_{\gamma,\delta}(\mathcal{K})^\ell$ *for* $j = 1,2,3$.

P r o o f. The unique solution $u \in W^{1,p}_{\beta,\delta}(\mathcal{K})^\ell$ of the problem (7.7.20) has the representation

$$u(x) = \int_{\mathcal{K}} G(x,\xi)\, F(\xi)\, d\xi,$$

where $G(x,\xi)$ is the Green matrix introduced in Theorem 7.5.1 for $\kappa = 0$. By the assumptions of the theorem, the functional F has the form

$$F(v) = \int_{\mathcal{K}} \Big(f^{(0)} \cdot v + \sum_{j=1}^{3} f^{(j)} \cdot \partial_{x_j} v\Big)\, dx \quad \text{for all } v \in W^{1,p'}_{-\gamma,-\delta}(\mathcal{K})^\ell$$

(cf. Lemma 7.7.1), where $f^{(0)} \in V^{0,p}_{\gamma+1,\delta+1}(\mathcal{K})^\ell$ and $f^{(j)} \in V^{0,p}_{\gamma,\delta}(\mathcal{K})^\ell$, $j = 1,2,3$. This means that

$$u(x) = \int_{\mathcal{K}} G(x,\xi)\, f^{(0)}(\xi)\, d\xi + \sum_{j=1}^{3} \int_{\mathcal{K}} \partial_{\xi_j} G(x,\xi)\, f^{(j)}(\xi)\, d\xi.$$

As was shown in the proof of Theorem 7.7.11, the function

$$x \to \partial_{x_j} \int_{\mathcal{K}} G(x,\xi)\, f^{(0)}(\xi)\, d\xi$$

belongs to the space $W^{1,p}_{\gamma+1,\delta+1}(\mathcal{K})^\ell$ which is a subspace of $W^{0,p}_{\gamma,\delta}(\mathcal{K})^\ell$. We consider the operator

$$f^{(j)} \to \mathcal{O}\, f^{(j)} = \partial_{x_j} v, \quad \text{where} \quad v(x) = \int_{\mathcal{K}} \partial_{\xi_j} G(x,\xi)\, f^{(j)}(\xi)\, d\xi.$$

Using Theorem 7.5.8, we obtain the estimate

$$\|\zeta_\mu \mathcal{O}\,(\zeta_\nu f^{(j)})\|_{W^{0,p}_{\gamma,\delta}(\mathcal{K})^\ell} \le c\, 2^{-|\mu-\nu|\varepsilon_0}\, \|\zeta_\nu f^{(j)}\|_{W^{0,p}_{\gamma,\delta}(\mathcal{K})^\ell}$$

analogously to Lemma 7.7.2, where ε_0 is positive. Consequently, the operator $\mathcal{O}$ satisfies the conditions of Lemma 3.5.8 for the spaces $\mathcal{X} = \mathcal{Y} = W^{0,p}_{\gamma,\delta}(\mathcal{K})^\ell$. Thus, this operator realizes a continuous mapping $W^{0,p}_{\gamma,\delta}(\mathcal{K})^\ell \to W^{0,p}_{\gamma,\delta}(\mathcal{K})^\ell$. This proves that $\partial_{x_j} u \in W^{0,p}_{\gamma,\delta}(\mathcal{K})^\ell$ for $j = 1,2,3$. □

Note that $W^{1,p'}_{-\gamma,-\delta}(\mathcal{K}) = L^{1,p'}_{-\gamma,-\delta}(\mathcal{K})$ if $-\gamma > 1 - 3/p'$, i.e., $1 - \gamma - 3/p > -1$. Thus, the following assertion holds as a corollary of Theorem 7.7.13.

COROLLARY 7.7.14. *Let the assumptions of Theorem 7.7.13 on the spectrum of the pencil $\mathfrak{A}(\lambda)$ and on β, γ, δ_k be satisfied. Then the operator $u \to F$ of the problem (7.7.20) is an isomorphism from $W^{1,p}_{\beta,\delta}(\mathcal{K})^\ell \cap L^{1,p}_{\gamma,\delta}(\mathcal{K})^\ell$ onto the space $\big(W^{1,p'}_{-\beta,-\delta}(\mathcal{K})^*\big)^\ell \cap \big(W^{1,p'}_{-\gamma,-\delta}(\mathcal{K})^*\big)^\ell$.*

Finally, we assume that the conditions (i)–(iii) of Subsection 7.4.5 are satisfied. Then we denote by $\mu_+^{(k)}$ the greatest positive real number such that the strip $0 < \operatorname{Re}\lambda < \mu_+^{(k)}$ contains at most the eigenvalue $\lambda = 1$ of the operator pencil $A_k(\lambda)$. Since then the Green's matrix satisfies the estimates given in Theorem 7.5.9, the condition on δ' in Theorem 7.7.7 can be weakened. However, the boundary data must satisfy additional conditions on the edge M_k if $\delta'_k + 2/q \le l - 1$ (cf. Lemma 6.5.7). More precisely, if $\delta'_k + 2/q < l - 1$, then we must assume in addition that

$$w^{j,k} \cdot \big(g_{k_+}|_{M_k}, g_{k_-}|_{M_k}\big) = 0 \quad \text{for } j = 1, \dots, 2\ell - r'_k. \tag{7.7.21}$$

Here g_{k_+}, g_{k_-} are the data on the adjoining faces $\Gamma_{k_+}, \Gamma_{k_-}$ to the edge M_k, r'_k is the rank of the matrices in the condition (iii), and $w^{j,k}$ are certain constant vectors. For example, in the case of the Neumann problem for the Lamé system, the condition (7.7.21) takes the form (cf. Theorem 6.5.10)

$$n^{(k_-)} \cdot g_{k_+} = n^{(k_+)} \cdot g_{k_-} \quad \text{on } M_k\,, \tag{7.7.22}$$

where $n^{(j)}$ denotes the unit outer normal to the face Γ_j. If $\delta'_k + 2/q = l - 1$, then the trace condition (7.7.21) has to be satisfied in the generalized sense, i. e.

$$\int_0^\infty \int_0^{\varepsilon t} t^{q(\beta'-l-1+2/q)}\, r^{-1}\, \big|w^{j,k} \cdot \big(g_{k_+}(r,t),\, g_{k_-}(r,t)\big)\big|^q\, dr\, dt < \infty, \tag{7.7.23}$$

where r, t are Cartesian coordinates on Γ_{k_+} and Γ_{k_-}, $r = \text{dist}(x, M_k)$, and ε is a sufficiently small positive number.

Using Theorem 7.5.9, we obtain a modification of the regularity assertion in Theorem 7.7.7.

THEOREM 7.7.15. *Let $u \in W^{1,p}_{\beta,\delta}(\mathcal{K})^\ell$ be a solution of the problem*

$$b_{\mathcal{K}}(u, v) = (F, v)_{\mathcal{K}} \quad \text{for all } v \in W^{1,p'}_{-\beta',-\delta'}(\mathcal{K})^\ell \tag{7.7.24}$$

where F is a linear and continuous functional on $W^{1,p'}_{-\beta,-\delta}(\mathcal{K})^\ell$ which has the representation (7.7.18) *with vector functions $f \in W^{l-2,q}_{\beta',\delta'}(\mathcal{K})^\ell$ and $g_j \in W^{l-1-1/q,q}_{\beta',\delta'}(\Gamma_j)^\ell$, $l \geq 2$. Suppose that the conditions* (i)–(iii) *of Subsection* 7.4.5 *are satisfied, that the closed strip between the lines* $\text{Re}\,\lambda = 1 - \beta - 3/p$ *and* $\text{Re}\,\lambda = l - \beta' - 3/q$ *is free of eigenvalues of the pencil $\mathfrak{A}(\lambda)$, and that the components of δ and δ' satisfy the inequalities*

$$\max(0, 1 - \delta_+^{(k)}) < \delta_k + 2/p < 1, \qquad \max(0, l - \mu_+^{(k)}) < \delta'_k + 2/q < l.$$

In the case $\delta'_k + 2/q \leq l - 1$, we assume in addition that g_{k_+} and g_{k_-} satisfy the compatibility conditions (7.7.21) *and* (7.7.23), *respectively. Then $u \in W^{l,q}_{\beta',\delta'}(\mathcal{K})^\ell$. In the case $p = 2$, the result holds for $\delta = 0$ as well.*

7.8. Solvability in weighted Hölder spaces

The last section of this chapter deals with the solvability of the boundary value problem (7.1.1)–(7.1.3) in the class of the weighted Hölder spaces $C^{l,\sigma}_{\beta,\delta}(\mathcal{K})$ introduced below. Using the estimates for Green's matrix obtained in Section 7.5, we are able to prove the existence and uniqueness of solutions in the space $C^{2,\sigma}_{\beta,\delta}(\mathcal{K})$ provided that the line $\text{Re}\,\lambda = 2 + \sigma - \beta$ is free of eigenvalues of the pencil $\mathfrak{A}(\lambda)$ and the components of δ satisfy the inequalities

$$2 - \delta_+^{(k)} < \delta_k - \sigma < 2, \quad \delta_k \geq 0, \quad \delta_k - \sigma \text{ is not integer} \tag{7.8.1}$$

for $k = 1, \dots, d$. Furthermore, we are interested in regularity results for the solutions of the boundary value problem (7.1.1)–(7.1.3) in weighted Hölder spaces. We prove that the solution $u \in C^{2,\sigma'}_{\beta',\delta'}(\mathcal{K})^\ell$ belongs to the space $C^{l,\sigma}_{\beta,\delta}(\mathcal{K})$ under certain conditions on the data f, g_j and on the weight parameters β, δ. An analogous result holds for the weak solution $u \in W^{1,p}_{\beta,\delta}(\mathcal{K})$ considered in the preceding section. Finally, we improve the result for the Neumann problem for a special class of elliptic systems which includes the Lamé system.

7.8.1. Weighted Hölder spaces in a polyhedral cone. Let $\mathcal{K}$ be the polyhedral cone introduced in Section 7.1. We denote again by $r_k(x)$ the distance of the point x from the edge M_k and by $r(x)$ the distance from the set $\mathcal{S} = M_1 \cup \dots \cup M_d \cup \{0\}$. The subset $\{x \in \mathcal{K} : r_k(x) < 3r(x)/2\}$ is denoted by $\mathcal{K}_k$. Furthermore, let l be a nonnegative integer, β an arbitrary real number, $\delta = (\delta_1, \dots, \delta_d) \in \mathbb{R}^d$, $\delta_k \geq 0$ for $k = 1, \dots, d$, and $0 < \sigma < 1$. Then the space $C^{l,\sigma}_{\beta,\delta}(\mathcal{K})$ is defined as the set of all l times continuously differentiable functions on

$\overline{\mathcal{K}}\backslash\mathcal{S}$ with finite norm

$$\|u\|_{C^{l,\sigma}_{\beta,\delta}(\mathcal{K})} = \sum_{|\alpha|\le l} \sup_{x\in\mathcal{K}} |x|^{\beta-l-\sigma+|\alpha|} \prod_{k=1}^d \Big(\frac{r_k(x)}{|x|}\Big)^{\max(0,\delta_k-l-\sigma+|\alpha|)} \big|\partial_x^\alpha u(x)\big|$$
$$+ \sum_{k:\, m_k\le l} \sum_{|\alpha|=l-m_k} \sup_{\substack{x,y\in\mathcal{K}_k\\ |x-y|<|x|/2}} |x|^{\beta-\delta_k} \frac{\big|\partial_x^\alpha u(x)-\partial_y^\alpha u(y)\big|}{|x-y|^{m_k+\sigma-\delta_k}}$$
$$+ \sum_{|\alpha|=l} \sup_{\substack{x,y\in\mathcal{K}\\ |x-y|<r(x)/2}} |x|^\beta \prod_{k=1}^d \Big(\frac{r_k(x)}{|x|}\Big)^{\delta_k} \frac{\big|\partial_x^\alpha u(x)-\partial_y^\alpha u(y)\big|}{|x-y|^\sigma},$$

where $m_k = [\delta_k-\sigma]+1$, $[s]$ denotes integral part of s. Replacing the expression

$$\sum_{|\alpha|=l-m_k} \sup_{\substack{x,y\in\mathcal{K}_k\\ |x-y|<|x|/2}} |x|^{\beta-\delta_k} \frac{\big|\partial^\alpha u(x)-\partial^\alpha u(y)\big|}{|x-y|^{m_k+\sigma-\delta_k}}$$

in the definition of the last norm by

$$\sum_{|\alpha|=l-m_k} \Big(\sup_{\substack{x,y\in\mathcal{K}_k\\ |x-y|<r(x)/2}} |x|^{\beta-\delta_k} \frac{\big|\partial^\alpha u(x)-\partial^\alpha u(y)\big|}{|x-y|^{m_k+\sigma-\delta_k}}$$
$$+ \sup_{x\in\mathcal{K}_k} \sup_{1/2<t<3/2} |x|^{\beta-\delta_k} \frac{\big|\partial^\alpha u(x)-\partial^\alpha u(tx)\big|}{|x-tx|^{m_k+\sigma-\delta_k}}\Big),$$

we obtain an equivalent norm in $C^{l,\sigma}_{\beta,\delta}(\mathcal{K})$ (cf. Lemma 6.7.7). Obviously, $N^{l,\sigma}_{\beta,\delta}(\mathcal{K})$ is a subset of $C^{l,\sigma}_{\beta,\delta}(\mathcal{K})$. If $\delta_k \ge l+\sigma$ for $k=1,\dots,d$, then both spaces coincide. The trace space for $C^{l,\sigma}_{\beta,\delta}(\mathcal{K})$ on Γ_j is denoted by $C^{l,\sigma}_{\beta,\delta}(\Gamma_j)$. Finally, we introduce the following notation. If $\delta\in\mathbb{R}^d$ and $s\in\mathbb{R}$, then by $C^{l,\sigma}_{\beta,\delta+s}(\mathcal{K})$ we mean the space $C^{l,\sigma}_{\beta,\delta'}(\mathcal{K})$ with $\delta'=(\delta_1+s,\dots,\delta_d+s)$.

LEMMA 7.8.1. *Let $l+\sigma\ge l'+\sigma'$, $\beta-l-\sigma=\beta'-l'-\sigma'$, $\delta_k\ge 0$, $\delta'_k\ge 0$ and $\delta_k-l-\sigma\le\delta'_k-l'-\sigma'$ for $k=1,\dots,d$. Then there are the topological imbeddings*

$$C^{l,\sigma}_{\beta,\delta}(\mathcal{K}) \subset C^{l',\sigma'}_{\beta',\delta'}(\mathcal{K}) \quad \textit{and} \quad C^{l,\sigma}_{\beta,\delta}(\Gamma_j) \subset C^{l',\sigma'}_{\beta',\delta'}(\Gamma_j).$$

P r o o f. Let $u\in C^{l,\sigma}_{\beta,\delta}(\mathcal{K})$, and let ζ_ν be infinitely differentiable functions depending only on $\rho=|x|$ and satisfying the conditions (7.2.3). Furthermore, let $\tilde u_\nu(x)=u(2^\nu x)$ and $\tilde\zeta_\nu(x)=\zeta_\nu(2^\nu x)$. Then

$$\|u\|_{C^{l',\sigma'}_{\beta',\delta'}(\mathcal{K})} \le c\sup_\nu \|\zeta_\nu u\|_{C^{l',\sigma'}_{\beta',\delta'}(\mathcal{K})} = c\sup_\nu 2^{\nu(\beta'-l'-\sigma')}\,\|\tilde\zeta_\nu\tilde u_\nu\|_{C^{l',\sigma'}_{\beta',\delta'}(\mathcal{K})}.$$

Since the support of $\tilde\zeta_\nu\tilde u_\nu$ is contained in the set $\{x: 1/2\le|x|\le 2\}$, it follows from Lemma 6.7.8 that

$$\|\tilde\zeta_\nu\tilde u_\nu\|_{C^{l',\sigma'}_{\beta',\delta'}(\mathcal{K})} \le c\,\|\tilde\zeta_\nu\tilde u_\nu\|_{C^{l,\sigma}_{\beta,\delta}(\mathcal{K})} = c\;2^{\nu(l+\sigma-\beta)}\,\|\zeta_\nu u\|_{C^{l,\sigma}_{\beta,\delta}(\mathcal{K})},$$

where c is a constant independent of u and ν. Consequently,

$$\|u\|_{C^{l',\sigma'}_{\beta',\delta'}(\mathcal{K})} \le c\sup_\nu \|\zeta_\nu u\|_{C^{l,\sigma}_{\beta,\delta}(\mathcal{K})} \le c'\,\|u\|_{C^{l,\sigma}_{\beta,\delta}(\mathcal{K})}.$$

This proves the lemma. □

Analogously, the next lemma can be proved by means of Theorem 6.7.11.

LEMMA 7.8.2. *Let $u \in C^{l,\sigma}_{\beta,\delta}(\mathcal{K})$, and let k be an integer, $k \ge 0$. Then there exist functions $v \in N^{l,\sigma}_{\beta,\delta}(\mathcal{K})$ and $w \in C^{l+k,\sigma}_{\beta+k,\delta+k}(\mathcal{K})$ such that $u = v + w$ and*

$$\|v\|_{N^{l,\sigma}_{\beta,\delta}(\mathcal{K})} + \|w\|_{C^{l+k,\sigma}_{\beta+k,\delta+k}(\mathcal{K})} \le c\, \|u\|_{C^{l,\sigma}_{\beta,\delta}(\mathcal{K})},$$

where c is a constant independent of u.

P r o o f. Let ζ_ν, $\tilde{\zeta}_\nu$ and $\tilde{u}_\nu$ be the same functions as in the proof of Lemma 7.8.1. Then $\tilde{\zeta}_\nu(x)$ vanishes for $|x| < 1/2$ and $|x| > 2$. Furthermore,

$$2^{\nu(\beta-l-\sigma)}\, \|\tilde{\zeta}_\nu\, \tilde{u}_\nu\|_{C^{l,\sigma}_{\beta,\delta}(\mathcal{K})} = \|\zeta_\nu u\|_{C^{l,\sigma}_{\beta,\delta}(\mathcal{K})} \le c\, \|u\|_{C^{l,\sigma}_{\beta,\delta}(\mathcal{K})}$$

for every ν, where c is a constant independent of u and ν. By Theorem 6.7.11, there exist functions $\tilde{v}_\nu \in N^{l,\sigma}_{\beta,\delta}(\mathcal{K})$ and $\tilde{w}_\nu \in C^{l+k,\sigma}_{\beta+k,\delta+k}(\mathcal{K})$ such that $\tilde{\zeta}_\nu\, \tilde{u}_\nu = \tilde{v}_\nu + \tilde{w}_\nu$ and

$$\|\tilde{v}_\nu\|_{N^{l,\sigma}_{\beta,\delta}(\mathcal{K})} + \|\tilde{w}_\nu\|_{C^{l+k,\sigma}_{\beta+k,\delta+k}(\mathcal{K})} \le c\, \|\tilde{\zeta}_\nu\, \tilde{u}_\nu\|_{C^{l,\sigma}_{\beta,\delta}(\mathcal{K})}.$$

We define $\eta_\nu = \zeta_{\nu-1} + \zeta_\nu + \zeta_{\nu+1}$, $\tilde{\eta}_\nu(x) = \eta_\nu(2^\nu x)$, $v_\nu(x) = \tilde{v}_\nu(2^{-\nu}x)$, and $w_\nu(x) = \tilde{w}_\nu(2^{-\nu}x)$. From the equation $\tilde{\zeta}_\nu\, \tilde{u}_\nu = \tilde{\eta}_\nu\, \tilde{v}_\nu + \tilde{\eta}_\nu\, \tilde{w}_\nu$ it follows that

$$\zeta_\nu u_\nu = \eta_\nu v_\nu + \eta_\nu w_\nu$$

and, consequently,

$$u = v + w, \quad \text{where } v = \sum_{\nu=-\infty}^{+\infty} \eta_\nu v_\nu \quad \text{and } w = \sum_{\nu=-\infty}^{+\infty} \eta_\nu w_\nu .$$

Here,

$$\begin{aligned}
\|v\|_{N^{l,\sigma}_{\beta,\delta}(\mathcal{K})} &\le c \sup_\nu \|\zeta_\nu v\|_{N^{l,\sigma}_{\beta,\delta}(\mathcal{K})} = c \sup_\nu \Big\| \zeta_\nu \sum_{\mu=\nu-2}^{\nu+2} \eta_\mu v_\mu \Big\|_{N^{l,\sigma}_{\beta,\delta}(\mathcal{K})} \\
&\le c_1 \sup_\nu \|v_\nu\|_{N^{l,\sigma}_{\beta,\delta}(\mathcal{K})} = c_1 \sup_\nu 2^{\nu(\beta-l-\sigma)}\, \|\tilde{v}_\nu\|_{N^{l,\sigma}_{\beta,\delta}(\mathcal{K})} \\
&\le c_2 \sup_\nu 2^{\nu(\beta-l-\sigma)}\, \|\tilde{\zeta}_\nu\, \tilde{u}_\nu\|_{C^{l,\sigma}_{\beta,\delta}(\mathcal{K})} \le c_3\, \|u\|_{C^{l,\sigma}_{\beta,\delta}(\mathcal{K})}
\end{aligned}$$

and, analogously,

$$\|w\|_{C^{l+k,\sigma}_{\beta+k,\delta+k}(\mathcal{K})} \le c\, \|u\|_{C^{l,\sigma}_{\beta,\delta}(\mathcal{K})}.$$

This proves the lemma. □

7.8.2. Reduction to zero boundary data. We consider the boundary value problem (7.1.1)–(7.1.3) in the polyhedral cone $\mathcal{K}$. The subsequent two lemmas allow us to restrict ourselves to the case of zero boundary data.

LEMMA 7.8.3. *Let $g_j \in C^{l-d_j,\sigma}_{\beta,\delta}(\Gamma_j)^\ell$ for $j = 1, \dots, d$, where $l \ge 1$ and the components of δ satisfy the inequalities $\delta_k \ge 0$, $\delta_k > l + \sigma - 2$ and $\delta_k \ne l + \sigma - 1$ for $k = 1, \dots, d$. We suppose that $\lambda = 1$ is not an eigenvalue of the pencil $A_k(\lambda)$ if $\delta_k < l + \sigma - 1$. If $\delta_k < l + \sigma$, then we assume in addition that the boundary* *data satisfy the compatibility condition* (7.3.5). *Then there exists a vector function* $u \in C^{l,\sigma}_{\beta,\delta}(\mathcal{K})^\ell$ *satisfying the boundary conditions* (7.1.2), (7.1.3) *and the estimate*

$$\|u\|_{C^{l,\sigma}_{\beta,\delta}(\mathcal{K})^\ell} \le c \sum_{j=1}^{d} \|g_j\|_{C^{l-d_j,\sigma}_{\beta,\delta}(\Gamma_j)^\ell}, \tag{7.8.2}$$

where the constant c is independent of g_j.

P r o o f. The proof is essentially the same as for Lemma 7.3.1. Let ζ_ν be smooth functions depending only on $\rho = |x|$ and satisfying the conditions (7.2.3). We set

$$h_{j,\nu}(x) = 2^{d_j \nu}\, \zeta_\nu(2^\nu x)\, g_j(2^\nu x).$$

Obviously, $h_{j,\nu}(x) = 0$ for $|x| < \frac{1}{2}$ and $|x| > 2$. Consequently by Lemmas 6.8.1 and 6.8.2, there exists a function $v_\nu \in C^{l,\sigma}_{\beta,\delta}(\mathcal{K})$ vanishing outside the set $\{x : 1/4 < |x| < 4\}$ such that

$$v_\nu = h_{j,\nu} \text{ on } \Gamma_j \text{ for } j \in I_0, \quad N(D_x)\, v_\nu = h_{j,\nu} \text{ on } \Gamma_j \text{ for } j \in I_1.$$

and

$$\begin{aligned} \|v_\nu\|_{C^{l,\sigma}_{\beta,\delta}(\mathcal{K})} &\le c \sum_{j=1}^{d} \|\tilde{\zeta}_\nu\, h_{j,\nu}\|_{C^{l-d_j,\sigma}_{\beta,\delta}(\Gamma_j)} \\ &\le c'\, 2^{\nu(l+\sigma-\beta)} \sum_{j=1}^{d} \|\zeta_\nu\, g_j\|_{C^{l-d_j,\sigma}_{\beta,\delta}(\Gamma_j)} \end{aligned}$$

with constants c and c' independent of the functions g_j and ν. We define

$$u_\nu(x) = v_\nu(2^{-\nu} x) \quad \text{and} \quad u = \sum_{\nu=-\infty}^{+\infty} u_\nu\,.$$

Then u satisfies the boundary conditions (7.1.2), (7.1.3) and the estimate (7.8.2). □

Analogously, we deduce the next lemma from the results of Section 6.8. For the proof, we have to employ Lemma 6.8.4 instead of Lemmas 6.8.1 and 6.8.2.

LEMMA 7.8.4. *Let* $f \in C^{l-2,\sigma}_{\beta,\delta}(\mathcal{K})^\ell$ *and* $g_j \in C^{l-d_j,\sigma}_{\beta,\delta}(\Gamma_j)^\ell$ *for* $j = 1,\dots,d$, *where* $l \ge 2$ *and the components of* δ *satisfy the inequalities* $\delta_k \ge 0$, $l+\sigma-3 < \delta_k < l+\sigma-2$ *for* $k = 1,\dots,d$. *We suppose that the boundary data satisfy the compatibility condition* (7.3.5) *and that* $\lambda = 1$ *and* $\lambda = 2$ *are not eigenvalues of the pencil* $A_k(\lambda)$. *Then there exists a vector function* $u \in C^{l,\sigma}_{\beta,\delta}(\mathcal{K})^\ell$ *satisfying the boundary conditions* (7.1.2), (7.1.3) *such that*

$$L(D_x)\, u - f \in N^{l-2,\sigma}_{\beta,\delta}(\mathcal{K})^\ell.$$

Furthermore, the estimate

$$\|u\|_{C^{l,\sigma}_{\beta,\delta}(\mathcal{K})^\ell} \le c \Big(\|f\|_{C^{l-2,\sigma}_{\beta,\delta}(\mathcal{K})^\ell} + \sum_{j=1}^{d} \|g_j\|_{C^{l-d_j,\sigma}_{\beta,\delta}(\Gamma_j)^\ell} \Big)$$

holds with a constant c independent of f and g_j.

7.8.3. A priori estimates for the solution. The results of this subsection are based on analogous results for the problem in a dihedron. For the Dirichlet problem, the following assertion was proved in Section 3.6. Here we prove it for the mixed problem (7.1.1)–(7.1.3).

LEMMA 7.8.5. *Let $u \in W^{2,p}_{loc}(\mathcal{K})^\ell \cap L^\infty_{\beta-l-\sigma,\delta-l-\sigma}(\mathcal{K})^\ell$ be a solution of problem (7.1.1)–(7.1.3). If $f \in N^{l-2,\sigma}_{\beta,\delta}(\mathcal{K})^\ell$ and $g_j \in N^{l-d_j,\sigma}_{\beta,\delta}(\Gamma_j)^\ell$, $l \geq 2$, for $j = 1,\dots,d$, then $u \in N^{l,\sigma}_{\beta,\delta}(\mathcal{K})^\ell$ and*

$$\|u\|_{N^{l,\sigma}_{\beta,\delta}(\mathcal{K})^\ell} \leq c\Big(\|f\|_{N^{l-2,\sigma}_{\beta,\delta}(\mathcal{K})^\ell} + \sum_{j=1}^{d} \|g_j\|_{N^{l-d_j,\sigma}_{\beta,\delta}(\Gamma_j)^\ell} + \|u\|_{L^\infty_{\beta-l-\sigma,\delta-l-\sigma}(\mathcal{K})^\ell}\Big).$$

P r o o f. By Lemma 3.6.4, we may assume without loss of generality $g_j = 0$ for $j = 1,\dots,d$. From Lemma 6.8.6 it follows that $\zeta u \in N^{l,\sigma}_{\beta,\delta}(\mathcal{K})^\ell$ for every smooth function ζ with compact support vanishing in a neighborhood of the origin.

Let ρ be a positive integer, $\mathcal{K}_\rho = \{x \in \mathcal{K} : \rho/2 < |x| < 2\rho\}$, and $\mathcal{K}'_\rho = \{x \in \mathcal{K} : \rho/4 < |x| < 4\rho\}$. Furthermore, let $\tilde{u}(x) = u(\rho x)$ and $\tilde{f}(x) = \rho^2 f(\rho x)$. Obviously, the function $\tilde{u}$ satisfies the differential equation $L(D_x)\tilde{u} = \tilde{f}$ in $\mathcal{K}$ and the homogeneous boundary conditions (7.1.2), (7.1.3). Consequently by Lemma 6.8.6,

$$\|\tilde{u}\|_{N^{l,\sigma}_{\beta,\delta}(\mathcal{K}_1)^\ell} \leq c\Big(\|\tilde{f}\|_{N^{l-2,\sigma}_{\beta,\delta}(\mathcal{K}'_1)^\ell} + \|\tilde{u}\|_{L^\infty_{\beta-l-\sigma,\delta-l-\sigma}(\mathcal{K}'_1)^3}\Big)$$

with a constant c independent of u and ρ. Here the $N^{l,\sigma}_{\beta,\delta}(\mathcal{K}_\rho)$-norm is defined by the right-hand side of (3.6.2), where $\mathcal{K}$ has to be replaced by $\mathcal{K}_\rho$. Since

$$\|\tilde{u}\|_{N^{l,\sigma}_{\beta,\delta}(\mathcal{K}_1)^3} = \rho^{l+\sigma-\beta}\|u\|_{N^{l,\sigma}_{\beta,\delta}(\mathcal{K}_\rho)^3},$$

we arrive at the same estimate for the norm of u in $N^{l,\sigma}_{\beta,\delta}(\mathcal{K}_\rho)^\ell$. The result follows. □

In the same way, the two subsequent lemmas can be proved. Here one has to apply Lemmas 6.8.8 and 6.8.9 instead of Lemma 6.8.6.

LEMMA 7.8.6. *Let $u \in C^{l-1,\sigma}_{\beta-1,\delta-1}(\mathcal{K})^\ell$ be a solution of the boundary value problem (7.1.1)–(7.1.3), where $f \in C^{l-2,\sigma}_{\beta,\delta}(\mathcal{K})^\ell$, $g_j \in C^{l-d_j,\sigma}_{\beta,\delta}(\Gamma_j)^\ell$, $l \geq 2$, $0 < \sigma < 1$, and $\delta_k \geq 1$ for $k = 1,\dots,d$. Then $u \in C^{l,\sigma}_{\beta,\delta}(\mathcal{K})^\ell$ and*

$$\|u\|_{C^{l,\sigma}_{\beta,\delta}(\mathcal{K})^\ell} \leq c\Big(\|f\|_{C^{l-2,\sigma}_{\beta,\delta}(\mathcal{K})^\ell} + \sum_{j=1}^{d} \|g_j\|_{C^{l-d_j,\sigma}_{\beta,\delta}(\Gamma_j)^\ell} + \|u\|_{C^{l-1,\sigma}_{\beta-1,\delta-1}(\mathcal{K})^\ell}\Big)$$

with a constant c independent of f and g_j.

LEMMA 7.8.7. *Let u be a solution of the problem (7.1.1)–(7.1.3) such that $(\rho\partial_\rho)^j u \in C^{l,\sigma}_{\beta,\delta}(\mathcal{K})^\ell$ for $j = 0, 1$, where $l \geq 2$, $0 < \sigma < 1$. If $f \in C^{l-1,\sigma}_{\beta+1,\delta}(\mathcal{K})^\ell$, $g_j \in C^{l+1-d_j,\sigma}_{\beta+1,\delta}(\Gamma_j)^\ell$, $j = 1,\dots,d$, and the strip $l+\sigma-\delta_k \leq \operatorname{Re}\lambda \leq l+1+\sigma-\delta_k$ does not contain eigenvalues of the pencil $A_k(\lambda)$ for $k = 1,\dots,d$, then $u \in C^{l+1,\sigma}_{\beta+1,\delta}(\mathcal{K})^\ell$.*

7.8.4. Weighted L_∞ estimate for the solution and its derivatives. We suppose that the line $\text{Re}\,\lambda = 2+\sigma-\beta$ is free of eigenvalues of the pencil $\mathfrak{A}(\lambda)$ and denote by $\Lambda_- < \text{Re}\,\lambda < \Lambda_+$ the widest strip in the complex plane containing this line which is free of eigenvalues of the pencil $\mathfrak{A}(\lambda)$. Furthermore, let $G(x,\xi)$ be the Green matrix introduced in Theorem 7.5.1, where κ is an arbitrary real number in the interval

$$-\Lambda_+ - 1/2 < \kappa < -\Lambda_- - 1/2\,.$$

Then the vector function

$$u(x) = \int_{\mathcal{K}} G(x,\xi)\, f(\xi)\, d\xi \tag{7.8.3}$$

is a solution of the problem (7.1.1)–(7.1.3) with zero boundary data. Let χ be an arbitrary smooth cut-off function on $[0,\infty)$, $\chi(t) = 1$ for $t < 1/4$, $\chi(t) = 0$ for $t > 1/2$. We put

$$\chi^+(x,\xi) = \chi\Big(\frac{|x-\xi|}{r(x)}\Big), \quad \chi^-(x,\xi) = 1 - \chi^+(x,\xi).$$

Then $\chi^+(x,\xi) = 0$ for $|x-\xi| > r(x)/2$, $\chi^-(x,\xi) = 0$ for $|x-\xi| < r(x)/4$, and

$$\big|\partial_x^\alpha \chi^\pm(x,\xi)\big| \le c\, r(x)^{-|\alpha|}$$

with a constant c independent of x and ξ. We write u in the form

$$u = u^+ + u^-, \quad \text{where } u^\pm(x) = \int_{\mathcal{K}} \chi^\pm(x,\xi)\, G(x,\xi)\, f(\xi)\, d\xi. \tag{7.8.4}$$

LEMMA 7.8.8. *Suppose that* $f \in N^{0,\sigma}_{\beta,\delta}(\mathcal{K})^\ell$. *Then*

$$\sup_{x\in\mathcal{K}} |x|^{\beta-2-\sigma+|\alpha|} \prod_{k=1}^d \Big(\frac{r_k(x)}{|x|}\Big)^{\delta_k-2-\sigma+|\alpha|} |\partial_x^\alpha u^+(x)| \le c\, \|f\|_{L^\infty_{\beta-\sigma,\delta-\sigma}(\mathcal{K})^\ell}\,. \tag{7.8.5}$$

for $|\alpha| \le 1$.

P r o o f. On the support of χ^+, we have $|x| \le 2|\xi| \le 3|x|$ and $r_k(x) \le 2r_k(\xi) \le 3r_k(x)$. These inequalities together with Theorem 7.5.3 imply

$$\begin{aligned}\big|\partial_x^\alpha u^+(x)\big| &\le c\,|x|^{\sigma-\beta} \prod_{k=1}^d \Big(\frac{r_k(x)}{|x|}\Big)^{\sigma-\delta_k} \|f\|_{L^\infty_{\beta-\sigma,\delta-\sigma}(\mathcal{K})^\ell} \int\limits_{|x-\xi|<r(x)/2} |x-\xi|^{-1-|\alpha|}\, d\xi \\ &\le c\, r(x)^{2-|\alpha|}\, |x|^{\sigma-\beta} \prod_{k=1}^d \Big(\frac{r_k(x)}{|x|}\Big)^{\sigma-\delta_k} \|f\|_{L^\infty_{\beta-\sigma,\delta-\sigma}(\mathcal{K})^\ell}\,.\end{aligned}$$

The result follows. □

We prove a similar estimate for u^-.

LEMMA 7.8.9. *Suppose that the line* $\text{Re}\,\lambda = 2+\sigma-\beta$ *is free of eigenvalues of the pencil* $\mathfrak{A}(\lambda)$ *and that the components of* δ *satisfy the conditions* (7.8.1) *for* $k = 1,\dots,d$. *If* $f \in N^{0,\sigma}_{\beta,\delta}(\mathcal{K})^\ell$, *then*

$$\begin{aligned}\sup_{x\in\mathcal{K}} |x|^{\beta-2-\sigma+|\alpha|} \prod_{k=1}^d \Big(\frac{r_k(x)}{|x|}\Big)^{\max(0,\delta_k-2-\sigma+|\alpha|)} |\partial_x^\alpha u^-(x)| \\ \le c_\alpha\, \|f\|_{L^\infty_{\beta-\sigma,\delta-\sigma}(\mathcal{K})^\ell}\end{aligned} \tag{7.8.6}$$

for every multi-index α.

P r o o f. We introduce the vector functions

$$v^{(j)}(x) = \int_{\mathcal{K}_x^{(j)}} \chi^-(x,\xi)\, G(x,\xi)\, f(\xi)\, d\xi$$

for $j = 1, 2, 3$, where

$$\mathcal{K}_x^{(1)} = \{\xi \in \mathcal{K} :\ |x-\xi| > r(x)/4,\ |x|/2 < |\xi| < 2|x|\},$$
$$\mathcal{K}_x^{(2)} = \{\xi \in \mathcal{K} :\ |x-\xi| > r(x)/4,\ |\xi| < |x|/2\},$$
$$\mathcal{K}_x^{(3)} = \{\xi \in \mathcal{K} :\ |x-\xi| > r(x)/4,\ |\xi| > 2|x|\}.$$

By Theorems 7.5.2 and 7.5.3, the estimate

$$\big|\partial_x^\alpha \big(\chi^-(x,\xi)\, G(x,\xi)\big)\big| \le c\, |x-\xi|^{-1-|\alpha|} \Big(\frac{r(x)}{|x-\xi|}\Big)^{\delta_\alpha^+(x)}$$

is valid for $\xi \in \mathcal{K}_x^{(1)}$, where $\delta_\alpha^+(x) = \min(0, \delta_+^{(k(x))} - |\alpha| - \varepsilon)$, ε is an arbitrarily small positive number. Consequently,

$$\begin{aligned} \big|\partial_x^\alpha v^{(1)}(x)\big| \ &\le\ c\, \|f\|_{L^\infty_{\beta-\sigma,\delta-\sigma}(\mathcal{K})^\ell}\, |x|^{\sigma-\beta} \\ &\quad \times \int_{\mathcal{K}_x^{(1)}} |x-\xi|^{-1-|\alpha|} \Big(\frac{r(x)}{|x-\xi|}\Big)^{\delta_\alpha^+(x)} \prod_{k=1}^d \Big(\frac{r_k(\xi)}{|\xi|}\Big)^{\sigma-\delta_k} d\xi. \end{aligned}$$

Applying Lemma 3.5.2 to the integral on the right-hand side, we obtain

$$\big|\partial_x^\alpha v^{(1)}(x)\big| \le c\, \|f\|_{L^\infty_{\beta-\sigma,\delta-\sigma}(\mathcal{K})^\ell}\, |x|^{\sigma-\beta+2-|\alpha|} \prod_{k=1}^d \Big(\frac{r_k(x)}{|x|}\Big)^{\min(\delta_{k,\alpha}^+, 2+\sigma-\delta_k-|\alpha|)},$$

where $\delta_{k,\alpha}^+ = \min(0, \delta_+^{(k)} - |\alpha| - \varepsilon)$. Since $\delta_+^{(k)} > 2+\sigma-\delta_k$, it follows that

$$\big|\partial_x^\alpha v^{(1)}(x)\big| \le c\, \|f\|_{L^\infty_{\beta-\sigma,\delta-\sigma}(\mathcal{K})^\ell}\, |x|^{\sigma-\beta+2-|\alpha|} \prod_{k=1}^d \Big(\frac{r_k(x)}{|x|}\Big)^{\min(0, 2+\sigma-\delta_k-|\alpha|)}.$$

Using the estimate

$$\big|\partial_x^\alpha \big(\chi^-(x,\xi)\, G(x,\xi)\big)\big| \le c\, |x|^{\Lambda_- - |\alpha| + \varepsilon}\, |\xi|^{-1-\Lambda_- -\varepsilon} \prod_{k=1}^d \Big(\frac{r_k(x)}{|x|}\Big)^{\delta_{k,\alpha}^+}$$

for $\xi \in \mathcal{K}_x^{(2)}$ (see Theorem 7.5.5), we obtain

$$\begin{aligned} |\partial_x^\alpha v^{(2)}(x)| \ &\le\ c\, |x|^{\Lambda_- - |\alpha| + \varepsilon} \prod_{k=1}^d \Big(\frac{r_k(x)}{|x|}\Big)^{\delta_{k,\alpha}^+} \|f\|_{L^\infty_{\beta-\sigma,\delta-\sigma}(\mathcal{K})^\ell} \\ &\quad \times \int_{\mathcal{K}_x^{(2)}} |\xi|^{\sigma-\beta-1-\Lambda_- -\varepsilon} \prod_{k=1}^d \Big(\frac{r_k(\xi)}{|\xi|}\Big)^{\sigma-\delta_k} d\xi. \end{aligned}$$

The integral on the right-hand side can be estimated by $c\, |x|^{\sigma-\beta+2-\Lambda_- -\varepsilon}$. Furthermore, $\delta_{k,\alpha}^+ \ge \min(0, 2+\sigma-\delta_k-|\alpha|)$. Consequently,

$$\big|\partial_x^\alpha v^{(2)}(x)\big| \le c\, \|f\|_{L^\infty_{\beta-\sigma,\delta-\sigma}(\mathcal{K})^\ell}\, |x|^{\sigma-\beta+2-|\alpha|} \prod_{k=1}^d \Big(\frac{r_k(x)}{|x|}\Big)^{\min(0, 2+\sigma-\delta_k-|\alpha|)}.$$

Analogously, this estimates holds for $v^{(3)}$. This proves the lemma. □

7.8.5. Weighted Hölder estimates for u^-. We introduce the subsets

$$\mathcal{K}_k = \{x \in \mathcal{K} : r_k(x) < 3r(x)/2\}, \quad k = 1, \ldots, d,$$

of the cone $\mathcal{K}$.

LEMMA 7.8.10. *Let $f \in N^{0,\sigma}_{\beta,\delta}(\mathcal{K})$ and*

$$v(x) = \int_{\mathcal{K}} K(x,\xi)\, f(\xi)\, d\xi.$$

We assume that $\Lambda_- < 2 + \sigma - \beta < \Lambda_+$ and that the components of δ satisfy the condition (7.8.1). Furthermore, we suppose that $K(x,\xi)$ vanishes for $|x-\xi| < r(x)/4$ and that the following estimates are satisfied:

$$|\partial_\rho^s K(x,\xi)| \le c\, |x-\xi|^{-3+m_k-s} \quad \text{for } x \in \mathcal{K}_k,\ |x|/4 < |\xi| < 2|x|,$$

where $s = 0$ or $s = 1$, $m_k = [\delta_k - \sigma] + 1$, and $\rho = |x|$,

$$\begin{aligned} |\partial_\rho K(x,\xi)| &\le c\, |x|^{\Lambda_- - 3 + m_k + \varepsilon}\, |\xi|^{-\Lambda_- - 1 - \varepsilon} \quad \text{for } x \in \mathcal{K}_k,\ |\xi| < 3|x|/4, \\ |\partial_\rho K(x,\xi)| &\le c\, |x|^{\Lambda_+ - 3 + m_k - \varepsilon}\, |\xi|^{-\Lambda_+ - 1 + \varepsilon} \quad \text{for } x \in \mathcal{K}_k,\ |\xi| > 3|x|/2. \end{aligned}$$

Then the estimate

$$|x|^{\beta-\delta_k} \frac{|v(x) - v(y)|}{|x-y|^{m_k+\sigma-\delta_k}} \le c\, \|f\|_{L^\infty_{\beta-\sigma,\delta-\sigma}(\mathcal{K})} \tag{7.8.7}$$

holds for $x \in \mathcal{K}_k$, $y = \tau x$, $4/5 < \tau < 5/4$, $|x-y| > r(x)/4$. Here the constant c is independent of f, x, τ.

P r o o f. Let $x \in \mathcal{K}_k$ and $y = \tau x$, $4/5 < \tau < 5/4$, $|x-y| > r(x)/4$. Then $4|x-y| < \min(|x|,|y|)$. Obviously, $|v(x) - v(y)| \le A_1 + A_2 + A_3$, where

$$A_1 = \int\limits_{\substack{\mathcal{K} \\ |\xi-x|<2|x-y|}} |K(x,\xi)\, f(\xi)|\, d\xi, \qquad A_2 = \int\limits_{\substack{\mathcal{K} \\ |\xi-x|<2|x-y|}} |K(y,\xi)\, f(\xi)|\, d\xi,$$

$$A_3 = \int\limits_{\substack{\mathcal{K} \\ |\xi-x|>2|x-y|}} \big|(K(x,\xi) - K(y,\xi))\, f(\xi)\big|\, d\xi.$$

The inequalities $|x-\xi| < 2|x-y| < |x|/2$ imply $|x|/2 < |\xi| < 3|x|/2$. Therefore,

$$A_1 \le c\, \|f\|_{L^\infty_{\beta-\sigma,\delta-\sigma}(\mathcal{K})}\, |x|^{\sigma-\beta} \int |x-\xi|^{-3+m_k} \prod_{j=1}^d \Big(\frac{r_j(\xi)}{|\xi|}\Big)^{\sigma-\delta_j} d\xi,$$

where the domain of integration is contained in the set of all $\xi \in \mathcal{K}$ satisfying the inequalities $|x|/2 < |\xi| < 2|x|$, $r(x)/4 < |x-\xi| < 2|x-y|$. By virtue of (2.6.6), we obtain

$$A_1 \le c\, \|f\|_{L^\infty_{\beta-\sigma,\delta-\sigma}(\mathcal{K})}\, |x|^{\delta_k-\beta}\, |x-y|^{\sigma-\delta_k+m_k}.$$

Analogously, this estimate holds for A_2. For the proof, one can use the fact that $|\xi - x| < 2|x-y|$ implies $|\xi - y| < 3|x-y|$ and $|y|/4 < |\xi| < 2|y|$.

We consider the expression A_3. By the mean value theorem, the inequality

$$\big|K(x,\xi) - K(y,\xi)\big| \le \big|\partial_\rho K(\tilde{x},\xi)\big| \cdot |x-y|$$

is valid, where $\tilde{x}$ is a certain point on the line between x and y, i.e. $4|x|/5 < |\tilde{x}| < 5|x|/4$. Hence,

$$A_3 \le |x-y| \int\limits_{\substack{\mathcal{K}\\ |\xi-x|>2|x-y|}} \big|\partial_\rho K(\tilde{x},\xi)\, f(\xi)\big|\, d\xi.$$

Here, for the integral over the subset of all $\xi \in \mathcal{K}$ such that $|\xi| < |x|/2$, we obtain

$$\begin{aligned}
&\int \big|\partial_\rho K(\tilde{x},\xi)\, f(\xi)\big|\, d\xi \\
&\le c\, \|f\|_{L^\infty_{\beta-\sigma,\delta-\sigma}(\mathcal{K})}\, |x|^{\Lambda_- -3+m_k+\varepsilon} \int |\xi|^{\sigma-\beta-\Lambda_- -1-\varepsilon} \prod_{j=1}^d \Big(\frac{r_j(\xi)}{|\xi|}\Big)^{\sigma-\delta_j} d\xi \\
&\le c\, \|f\|_{L^\infty_{\beta-\sigma,\delta-\sigma}(\mathcal{K})}\, |x|^{\sigma-\beta+m_k-1}.
\end{aligned}$$

The same estimate holds for the integral over the set of all $\xi \in \mathcal{K}$ such that $|\xi| > 2|x|$, while for the integral over the set of all $\xi \in \mathcal{K}$ satisfying $|x|/2 < |\xi| < 2|x|$ and $|\xi - x| > 2|x-y|$ the estimate

$$\begin{aligned}
&\int \big|\partial_\rho K(\tilde{x},\xi)\, f(\xi)\big|\, d\xi \\
&\le c\, \|f\|_{L^\infty_{\beta-\sigma,\delta-\sigma}(\mathcal{K})}\, |x|^{\sigma-\beta} \int |x-\xi|^{-4+m_k} \prod_{j=1}^d \Big(\frac{r_j(\xi)}{|\xi|}\Big)^{\sigma-\delta_j} d\xi \\
&\le c\, \|f\|_{L^\infty_{\beta-\sigma,\delta-\sigma}(\mathcal{K})}\, |x|^{\delta_k-\beta}\, |x-y|^{\sigma-\delta_k+m_k-1}
\end{aligned}$$

holds. Here we used the inequality $|x-\xi|/2 < |\tilde{x}-\xi| < 3|x-\xi|/2$ and the estimate (2.6.5) with $R = 2|x-y|$. Thus, we obtain

$$\begin{aligned}
A_3 &\le c\, \|f\|_{L^\infty_{\beta-\sigma,\delta-\sigma}(\mathcal{K})}\, |x-y| \left(|x|^{\sigma-\beta+m_k-1} + |x|^{\delta_k-\beta}\, |x-y|^{\sigma-\delta_k+m_k-1}\right) \\
&\le c\, \|f\|_{L^\infty_{\beta-\sigma,\delta-\sigma}(\mathcal{K})}\, |x|^{\delta_k-\beta}\, |x-y|^{\sigma-\delta_k+m_k}.
\end{aligned}$$

This proves the lemma. □

Now we are able, to estimate the $C^{2,\sigma}_{\beta,\delta}(\mathcal{K})$-norm of u^-.

LEMMA 7.8.11. *Let the condition of Lemma 7.8.9 be satisfied. Then* $u^- \in C^{2,\sigma}_{\beta,\delta}(\mathcal{K})^\ell$, *and the inequality*

$$\|u^-\|_{C^{2,\sigma}_{\beta,\delta}(\mathcal{K})^\ell} \le c\, \|f\|_{N^{0,\sigma}_{\beta,\delta}(\mathcal{K})^\ell} \tag{7.8.8}$$

holds with a constant c independent of f.

P r o o f. According to Lemma 7.8.9, the vector function u^- satisfies the estimate (7.8.6). We show that

$$|x|^\beta \prod_{k=1}^d \Big(\frac{r_k(x)}{|x|}\Big)^{\delta_k} \frac{\big|\partial_x^\alpha u^-(x) - \partial_y^\alpha u^-(y)\big|}{|x-y|^\sigma} \le c\, \|f\|_{N^{0,\sigma}_{\beta,\delta}(\mathcal{K})^\ell} \tag{7.8.9}$$

for $|\alpha| = 2$, $|x-y| < r(x)/2$. By the mean value theorem,

$$\partial^\alpha u(x) - \partial^\alpha u(y) = (x-y)\cdot \nabla\partial^\alpha u(\tilde{x}),$$

where $\tilde{x} = x + t(y-x)$, $t \in (0,1)$. Furthermore, the inequalities

$$|x|/2 < |\tilde{x}| < 3|x|/2, \qquad r_k(x)/2 < r_k(\tilde{x}) < 3r_k(x)/2,$$

are satisfied for $|x-y| < r(x)/2$, $k = 1, \dots, d$. From this and from (3.1.2) it follows that

$$|x|^\beta \prod_{k=1}^{d} \Big(\frac{r_k(x)}{|x|}\Big)^{\delta_k} \frac{|\partial_x^\alpha u^-(x) - \partial_y^\alpha u^-(y)|}{|x-y|^\sigma}$$
$$\le |x|^\beta \prod_{k=1}^{d} \Big(\frac{r_k(x)}{|x|}\Big)^{\delta_k} r(x)^{1-\sigma} \, |\nabla \partial^\alpha u(\tilde{x})|$$
$$\le c\, |\tilde{x}|^{\beta+1-\sigma} \prod_{k=1}^{d} \Big(\frac{r_k(\tilde{x})}{|\tilde{x}|}\Big)^{\delta_k - 1+\sigma} |\nabla \partial^\alpha u(\tilde{x})|$$

for $|\alpha| = 2$. This together with (7.8.6) implies (7.8.9). Analogously, the estimate

$$(7.8.10) \qquad |x|^{\beta-\delta_k} \frac{|\partial_x^\alpha u^-(x) - \partial_y^\alpha u^-(y)|}{|x-y|^{m_k+\sigma-\delta_k}} \le c\, \|f\|_{N^{0,\sigma}_{\beta,\delta}(\mathcal{K})^\ell}$$

holds for $|\alpha| = 2 - m_k = 1 - [\delta_k - \sigma]$, $x, y \in \mathcal{K}_k$, $|x-y| < r(x)/2$, $j = 1, \dots, d$.

It remains to prove that (7.8.10) is valid for $|\alpha| = 2 - m_k$, $x \in \mathcal{K}_k$, $y = \tau|x|$, $1/2 < \tau < 3/2$. For $|x-y| < r(x)/2$, this is already shown. For $1/2 < \tau < 4/5$ and $5/4 < \tau < 3/2$, the left-hand side of (7.8.10) does not exceed

$$c\Big(|x|^{\beta-\sigma-m_k} \big|\partial_x^\alpha u^-(x)\big| + |y|^{\beta-\sigma-m_k} \big|\partial_y^\alpha u^-(y)\big|\Big)$$

and can estimated by means of (7.8.6). By Remark 7.5.6, the function

$$K(x,\xi) = \partial_x^\alpha \big(\chi^-(x,\xi)\, G(x,\xi)\big)$$

satisfies the conditions of Lemma 7.8.10 for $|\alpha| = 2 - m_k$ since $\delta_+^{(k)} - 2 + m_k > \sigma - \delta_k + m_k > 0$. Consequently, the estimate (7.8.10) is also satisfied for $|x-y| > r(x)/2$, $\tau \in (4/5, 5/4)$. This proves the estimate (7.8.8). □

7.8.6. Existence of solutions in $C^{2,\sigma}_{\beta,\delta}(\mathcal{K})$. Using the estimates for u^+ and u^- obtained in the last subsection and the estimate of Lemma 7.8.5, we can prove an existence and uniqueness theorem for the boundary value problem (7.1.1)–(7.1.3) in weighted Hölder spaces.

THEOREM 7.8.12. *Let $f \in C^{0,\sigma}_{\beta,\delta}(\mathcal{K})^\ell$ and $g_j \in C^{2-d_j,\sigma}_{\beta,\delta}(\Gamma_j)^\ell$, for $j = 1, \dots, d$, where β is such that the line* $\operatorname{Re}\lambda = 2+\sigma-\beta$ *is free of eigenvalues of the pencil $\mathfrak{A}(\lambda)$ and δ satisfies the condition* (7.8.1). *Furthermore, we assume that the boundary data satisfy the compatibility condition* (7.3.5). *Then the problem* (7.1.1)–(7.1.3) *has a unique solution $u \in C^{2,\sigma}_{\beta,\delta}(\mathcal{K})^\ell$.*

P r o o f. If $\delta_k < \sigma + 1$, then the condition (7.8.1) implies $\delta_+^{(k)} > 1$. In the case $\delta_k < \sigma$ we conclude that $\delta_+^{(k)} > 2$. This means that the number $\lambda = 1$ does not belong to the spectrum of the pencil $A_k(\lambda)$ if $\delta_k < \sigma + 1$ and that $\lambda = 2$ is not an eigenvalue of this pencil if $\delta_k < \sigma$. Consequently, Lemmas 7.8.3 and 7.8.4 allow us to restrict ourselves to the case where $f \in N^{0,\sigma}_{\beta,\delta}(\mathcal{K})^\ell$ and $g_j = 0$ for $j = 1, \dots, d$.

Let κ be an arbitrary real number such that the closed strip between the lines $\operatorname{Re}\lambda = 2 + \sigma - \beta$ and $\operatorname{Re}\lambda = -\kappa - 1/2$ is free of eigenvalues of the pencil $\mathfrak{A}(\lambda)$, and let $G(x,\xi)$ be the Green's matrix introduced in Subsection 7.5.4. We consider the vector functions u^+ and u^- defined by (7.8.4). By Lemma 7.8.11, the vector function u^- belongs to the space $C^{2,\sigma}_{\beta,\delta}(\mathcal{K})^\ell$, while u^+ satisfies (7.8.5). From the definition of u^- and from the equation $L(D_x)\,G(x,\xi) = \delta(x-\xi)\,I_\ell$ it follows that

$$\begin{aligned} L(D_x)\,u^- &= \int_{\mathcal{K}} \chi^-(x,\xi)\,L(D_x)\,G(x,\xi)\,f(\xi)\,d\xi + \int_{\mathcal{K}} K(x,\xi)\,f(\xi)\,d\xi \\ &= \int_{\mathcal{K}} K(x,\xi)\,f(\xi)\,d\xi, \end{aligned}$$

where $K(x,\xi) = [L(D_x),\chi^-(x,\xi)]\,G(x,\xi)$ and $[L,\chi^-] = L\,\chi^- - \chi^-\,L$ denotes the commutator of L and χ. The function $K(x,\xi)$ vanish for $|x-\xi| < r(x)/4$ and for $|x-\xi| > r(x)/2$ and satisfies the estimate

$$\big|\partial_x^\alpha K(x,\xi)\big| \le c_\alpha\, r(x)^{-3-|\alpha|}$$

with a constant c_α independent of x and ξ. Consequently,

$$\begin{aligned} &\Big|\partial_x^\alpha \int_{\mathcal{K}} K(x,\xi)\,f(\xi)\,d\xi\Big| \\ &\le c\,\|f\|_{L^\infty_{\beta-\sigma,\delta-\sigma}(\mathcal{K})}\ |x|^{\sigma-\beta} \prod_{k=1}^d \Big(\frac{r_k(x)}{|x|}\Big)^{\sigma-\delta_k} \int_{\mathcal{K}} \big|\partial_x^\alpha K(x,\xi)\big|\,d\xi \\ &\le c\,\|f\|_{L^\infty_{\beta-\sigma,\delta-\sigma}(\mathcal{K})}\ |x|^{\sigma-\beta-|\alpha|} \prod_{k=1}^d \Big(\frac{r_k(x)}{|x|}\Big)^{\sigma-\delta_k-|\alpha|} \end{aligned}$$

for arbitrary multi-indices α. Using the inequality (3.6.3), we obtain

$$\|L(D_x)\,u^-\|_{N^{0,\sigma}_{\beta,\delta}(\mathcal{K})^\ell} \le c\,\|f\|_{L^\infty_{\beta-\sigma,\delta-\sigma}(\mathcal{K})^\ell}\,.$$

Analogously, we obtain $u^- = 0$ on Γ_j for $j \in I_0$ and

$$\|N(D_x)\,u^-\|_{N^{1,\sigma}_{\beta,\delta}(\mathcal{K})^\ell} \le c\,\|f\|_{L^\infty_{\beta-\sigma,\delta-\sigma}(\mathcal{K})^\ell}\,.$$

for $j \in I_1$. Since $u = u^+ + u^-$ is a solution of the boundary value problem (7.1.1)–(7.1.3), it follows that

$$\begin{aligned} &L(D_x)\,u^+ = f - L(D_x)\,u^- \in N^{0,\sigma}_{\beta,\delta}(\mathcal{K})^\ell, \\ &u^+|_{\Gamma_j} = 0 \text{ for } j \in I_0, \quad N_j(D_x)u^+|_{\Gamma_j} = -N_j(D_x)u^-|_{\Gamma_j} \in N^{1,\sigma}_{\beta,\delta}(\Gamma_j) \text{ for } j \in I_1. \end{aligned}$$

Applying Lemma 7.8.5, we conclude that $u^+ \in N^{2,\sigma}_{\beta,\delta}(\mathcal{K})^\ell$. Therefore, $u \in C^{2,\sigma}_{\beta,\delta}(\mathcal{K})^\ell$. This proves the existence of a solution.

We prove the uniqueness. Let $u \in C^{2,\sigma}_{\beta,\delta}(\mathcal{K})^\ell$ be a solution of the problem

$$L(D_x)\,u = 0 \ \text{ in } \mathcal{K}, \quad (1-d_j)u + d_j\,N(D_x)\,u = 0 \ \text{ on } \Gamma_j,\ j = 1,\dots,d,$$

and let χ be a smooth cut-off function on $\overline{\mathcal{K}}$ equal to one for $|x| < 1$ and to zero for $|x| > 2$. Furthermore, let $\beta' = \beta - \sigma - 3/2$ and δ'_k be real numbers such that $\max(0, \delta_k - \sigma, 2 - \delta_+^{(k)}) - 1 < \delta'_k < 1$. Then $\chi u \in W^{2,2}_{\beta'+\varepsilon,\delta'}(\mathcal{K})^\ell$ and $(1-\chi)u \in W^{2,2}_{\beta'-\varepsilon,\delta'}(\mathcal{K})^\ell$, where ε is an arbitrary positive number. Consequently,

$$L(D_x)\,(\chi u) - L(D_x)\,(u - \chi u) \in W^{0,2}_{\beta'-\varepsilon,\delta'}(\mathcal{K})^3$$

and analogously $N(D_x)\,(\chi u) \in W^{1/2,2}_{\beta'-\varepsilon,\delta'}(\Gamma_j)^\ell$ for $j \in I_1$. Applying Theorem 7.6.5, we obtain $\chi u \in W^{2,2}_{\beta'-\varepsilon,\delta'}(\mathcal{K})^\ell$ if ε is sufficiently small. Hence, $u \in W^{2,2}_{\beta'-\varepsilon,\delta'}(\mathcal{K})^\ell$ and Theorem 7.6.4 implies $u = 0$. The proof of the theorem is complete. □

7.8.7. A regularity assertion for the solution. Our goal is to show that any solution $u \in C^{2,\sigma'}_{\beta',\delta'}(\mathcal{K})^\ell$ of the boundary value problem (7.1.1)–(7.1.3) belongs to the space $C^{l,\sigma}_{\beta,\delta}(\mathcal{K})^\ell$ under certain conditions on f, g_j, β and δ. First we consider the case $l = 2$.

THEOREM 7.8.13. *Let $u \in C^{2,\sigma'}_{\beta',\delta'}(\mathcal{K})^\ell$ be a solution of the boundary value problem (7.1.1)–(7.1.3), where*

$$f \in C^{0,\sigma}_{\beta,\delta}(\mathcal{K})^\ell \cap C^{0,\sigma'}_{\beta',\delta'}(\mathcal{K})^\ell, \quad g_j \in C^{2-d_j,\sigma}_{\beta,\delta}(\Gamma_j)^\ell \cap C^{2-d_j,\sigma'}_{\beta',\delta'}(\Gamma_j)^\ell.$$

We suppose that δ_k, δ'_k are nonnegative numbers such that $\delta_k - \sigma$ and $\delta'_k - \sigma'$ are not integer and

$$2 - \delta_+^{(k)} < \delta_k - \sigma < 2, \quad 2 - \delta_+^{(k)} < \delta'_k - \sigma' < 2$$

for $k = 1, \dots, d$. Furthermore, we assume that the closed strip between the lines $\operatorname{Re}\lambda = 2 + \sigma - \beta$ and $\operatorname{Re}\lambda = 2 + \sigma' - \beta'$ does not contain eigenvalues of the pencil $\mathfrak{A}(\lambda)$. Then $u \in C^{2,\sigma}_{\beta,\delta}(\mathcal{K})^\ell$.

P r o o f. 1) Let first $\delta = \delta'$ and $\sigma = \sigma'$. Then, analogously to Lemmas 7.8.3 and 7.8.4, there exists a vector function $v \in C^{2,\sigma}_{\beta,\delta}(\mathcal{K})^\ell \cap C^{2,\sigma}_{\beta',\delta}(\mathcal{K})^\ell$ such that

$$L(D_x)\, v - f \in N^{0,\sigma}_{\beta,\delta}(\mathcal{K})^3 \cap N^{0,\sigma}_{\beta',\delta}(\mathcal{K})^\ell,$$

$v|_{\Gamma_j} = g_j$ for $j \in I_0$, and $N(D_x)v|_{\Gamma_j} = g_j$ for $j \in I_1$. Therefore, we may assume without loss of generality that $f \in N^{0,\sigma}_{\beta,\delta}(\mathcal{K})^\ell \cap N^{0,\sigma}_{\beta',\delta}(\mathcal{K})^\ell$ and $g_j = 0$. Then, as was shown in the proof Theorem 7.8.12, the solution $u \in C^{2,\sigma}_{\beta',\delta}(\mathcal{K})^\ell$ is given by (7.8.3), where $G(x,\xi)$ is the Green's matrix introduced in Section 7.5 with an arbitrary κ in the interval $(-\Lambda_+ - 1/2\,, -\Lambda_- - 1/2)$. Here $\Lambda_- < \operatorname{Re}\lambda < \Lambda_+$ is the widest strip in the complex plane containing the line $\operatorname{Re}\lambda = 2+\sigma' - \beta'$ which is free of eigenvalues of the pencil $\mathfrak{A}(\lambda)$. However, the uniquely determined solution in $C^{2,\sigma}_{\beta,\delta}(\mathcal{K})^\ell$ has also the representation (7.8.3) with the same Green matrix $G(x,\xi)$. This proves the theorem in the case $\delta = \delta'$, $\sigma = \sigma'$.

2) Let $\delta \neq \delta'$ and/or $\sigma \neq \sigma'$. By Theorem 7.8.12, there exists a unique solution $v \in C^{2,\sigma}_{\beta,\delta}(\mathcal{K})^3$ of problem (7.1.1)–(7.1.3). Obviously,

$$u \in C^{2,\sigma''}_{\beta-\sigma+\sigma'',\delta''}(\mathcal{K})^\ell, \quad v \in C^{2,\sigma''}_{\beta'-\sigma'+\sigma'',\delta''}(\mathcal{K})^\ell,$$

where $\sigma'' = \min(\sigma, \sigma')$ and $\delta'_k = \max(\delta_k - \sigma + \sigma'', \delta'_k - \sigma' + \sigma'', 0)$ for $k = 1, \dots, d$. Thus it follows from the first part of the proof and from Theorem 7.8.12 that $u = v$. □

7.8.8. Solvability in $C^{l,\sigma}_{\beta,\delta}(\mathcal{K})^\ell$. The following statement generalizes the assertion of Theorem 7.8.13.

THEOREM 7.8.14. *Let $u \in C^{2,\sigma'}_{\beta',\delta'}(\mathcal{K})^\ell$ be a solution of the boundary value problem (7.1.1)–(7.1.3), where*

$$f \in C^{l-2,\sigma}_{\beta,\delta}(\mathcal{K})^\ell \cap C^{0,\sigma'}_{\beta',\delta'}(\mathcal{K})^\ell, \quad g_j \in C^{l-d_j,\sigma}_{\beta,\delta}(\Gamma_j)^\ell \cap C^{2-d_j,\sigma'}_{\beta',\delta'}(\Gamma_j)^\ell.$$

We suppose that δ_k, δ'_k are nonnegative numbers such that $\delta_k - \sigma$ and $\delta'_k - \sigma'$ are not integer and

$$l - \delta_+^{(k)} < \delta_k - \sigma < l, \quad 2 - \delta_+^{(k)} < \delta'_k - \sigma' < 2.$$

Furthermore, we assume that the closed strip between the lines $\mathrm{Re}\,\lambda = l+\sigma-\beta$ *and* $\mathrm{Re}\,\lambda = 2+\sigma'-\beta'$ *is free of eigenvalues of the pencil* $\mathfrak{A}(\lambda)$. *Then* $u \in C^{l,\sigma}_{\beta,\delta}(\mathcal{K})^\ell$.

P r o o f. 1) Suppose that $l-2 \le \delta_k < l+\sigma$ for $k=1,\dots,d$. Then

$$f \in C^{0,\sigma}_{\beta-l+2,\delta-l+2}(\mathcal{D})^\ell, \quad g_j \in C^{2-d_j,\sigma}_{\beta-l+2,\delta-l+2}(\Gamma_j)^\ell$$

and Theorem 7.8.13 implies $u \in C^{2,\sigma}_{\beta-l+2,\delta-l+2}(\mathcal{K})^\ell$. Using Lemma 7.8.6, we obtain $u \in C^{l,\sigma}_{\beta,\delta}(\mathcal{K})^\ell$.

2) Let $l-m \le \delta_k < l-m+1$ for $k=1,\dots,d$, where m is an integer, $2\le m\le l$. Then we prove the assertion of the theorem by induction in m. For $m=2$ we may refer to the first part of the proof. Suppose that $m\ge 3$ and the assertion of the theorem is true for $\delta_k \ge l-m+1$. Since

$$f \in C^{l-2,\sigma}_{\beta,\delta}(\mathcal{K})^\ell \subset C^{l-3,\sigma}_{\beta-1,\delta}(\mathcal{K})^\ell, \quad \text{and} \quad g_j \in C^{l-1,\sigma}_{\beta-1,\delta}(\Gamma_j)^\ell,$$

it follows from the induction hypothesis that $u \in C^{l-1,\sigma}_{\beta-1,\delta}(\mathcal{K})^\ell$. Furthermore, Lemma 7.8.6 implies $u \in C^{l,\sigma}_{\beta,\delta+1}(\mathcal{K})^\ell$. Consequently, $\rho\partial_\rho u \in C^{l-1,\sigma}_{\beta-1,\delta+1}(\mathcal{K})^\ell$, where $0<\varepsilon<\min(2, l-\delta_k-2)$. Using the equalities

$$L(D_x)\,(\rho\partial_\rho u) = (\rho\partial_\rho+2)f \in C^{l-3,\sigma}_{\beta-1,\delta}(\mathcal{K})^\ell,$$

$$\rho\partial_\rho u|_{\Gamma_j} = \rho\partial_\rho g_j \in C^{l-1,\sigma}_{\beta-1,\delta}(\Gamma_j)^\ell \ \text{ for } j\in I_0,$$

$$N_j(D_x)\,(\rho\partial_\rho u)|_{\Gamma_j} = (\rho\partial_\rho+1)g_j \in C^{l-2,\sigma}_{\beta-1,\delta}(\Gamma_j)^\ell \ \text{ for } j\in I_1$$

and the induction hypothesis, we conclude that $\rho\partial_\rho u \in C^{l-1,\sigma}_{\beta-1,\delta}(\mathcal{K})^\ell$. Applying Lemma 7.8.7, we obtain $u\in C^{l,\sigma}_{\beta,\delta}(\mathcal{K})^\ell$.

3) Finally, we assume that $l-\delta_k \in (s_k-1, s_k]$ for $k=1,\dots,d$ with different $s_k\in\{0,1,\dots,l\}$. Then let $\psi_1,\dots,\psi_d$ be smooth functions on $\bar\Omega$ such that $\psi_k\ge 0$, $\psi_k=1$ near $M_k\cap S^2$, and $\sum\psi_k=1$. We extend ψ_k to $\mathcal{K}$ by the equality $\psi_k(x)=\psi_k(x/|x|)$. Then $\partial_x^\alpha\psi_k(x)\le c\,|x|^{-|\alpha|}$. Using the first two parts of the proof, one can show by induction in l that $\psi_k u\in C^{l,\sigma}_{\beta,\delta}(\mathcal{K})^l$ for $k=1,\dots,d$. This completes the proof. □

7.8.9. A regularity result for variational solutions. Now we consider the solution $u\in W^{1,p}_{\beta,\delta}(\mathcal{K})^\ell$ of the problem (7.7.1), (7.7.2), where $F\in\mathcal{H}^*_{p',-\beta,-\delta}$, $p'=p/(p-1)$, and $g_j\in W^{1-1/p,p}_{\beta,\delta}(\Gamma_j)^\ell$ for $j\in I_0$. By Theorem 7.7.3, this solution exists under certain conditions on β and δ

THEOREM 7.8.15. *Let F be a linear and continuous functional on $\mathcal{H}_{p',-\beta,-\delta}$ which has the form*

$$(F,v)_{\mathcal{K}} = \int_{\mathcal{K}} f\cdot\bar v\,dx + \sum_{j\in I_1}\int_{\Gamma_j} g_j\cdot\bar v\,dx,$$

where $f\in C^{l-2,\sigma}_{\beta',\delta'}(\mathcal{K})^\ell$, $l\ge 2$, and $g_j\in C^{l-1,\sigma}_{\beta',\delta'}(\Gamma_j)^\ell$ for $j\in I_1$. Furthermore, let $g_j\in V^{1-1/p,p}_{\beta,\delta}(\Gamma_j)^\ell\cap C^{l,\sigma}_{\beta',\delta'}(\Gamma_j)^\ell$ for $j\in I_0$. We assume that the lines $\mathrm{Re}\,\lambda = 1-\beta-3/p$ *and* $\mathrm{Re}\,\lambda = l+\sigma-\beta'$ *are free of eigenvalues of the pencil* $\mathfrak{A}(\lambda)$*, the components of δ and δ' satisfy the conditions*

$$\max(0,1-\delta_+^{(k)}) < \delta_k+2/p < 1,$$

$$l-\delta_+^{(k)} < \delta'_k-\sigma < l, \quad \delta'_k\ge 0, \quad \delta'_k-\sigma \ \textit{not integer}$$

for $k = 1, \dots, d$ *and that the Dirichlet data satisfy the compatibility condition* (7.3.5). *Then the solution* $u \in W^{1,p}_{\beta,\delta}(\mathcal{K})^\ell$ *of the problem* (7.7.1), (7.7.2) *admits the decomposition* (7.4.5), *where* $w \in C^{l,\sigma}_{\beta',\delta'}(\mathcal{K})^\ell$, λ_ν *are the eigenvalues of the pencil* $\mathfrak{A}(\lambda)$ *between the lines* $\operatorname{Re}\lambda = 1 - \beta - 3/p$ *and* $\operatorname{Re}\lambda = l + \sigma - \beta'$, *and* $U^{\nu,j,s}$ *are eigenvectors and generalized eigenvectors corresponding to the eigenvalue* λ_ν. *In particular,* $u \in C^{l,\sigma}_{\beta',\delta'}(\mathcal{K})^\ell$ *if the strip between the lines* $\operatorname{Re}\lambda = 1 - \beta - 3/p$ *and* $\operatorname{Re}\lambda = l + \sigma - \beta'$ *is free of eigenvalues of the pencil* $\mathfrak{A}(\lambda)$. *In the case* $p = 2$, *the result is also valid for* $\delta = 0$.

P r o o f. Since $C^{l,\sigma}_{\beta,\delta}(\mathcal{K}) \subset C^{l,\sigma}_{\beta,\delta''}(\mathcal{K})$ if $\delta_k \le \delta''_k$ for $k = 1, \dots, d$, we may assume without loss of generality that the condition $\delta_k + 2/p > \delta'_k - l - \sigma + 1$ is satisfied in addition to the assumptions of the theorem.

By Theorems 7.8.12 and 7.8.14, there exists a solution $w \in C^{l,\sigma}_{\beta',\delta'}(\mathcal{K})^\ell$ of the boundary value problem (7.1.1)–(7.1.3). Let ζ, η be smooth real-valued functions on $\overline{\mathcal{K}}$ which are equal to one near the vertex, vanish outside the unit ball and satisfy the equality $\zeta\eta = \zeta$. By the inequality $\delta'_k < l + \sigma$, we have

$$\zeta w \in W^{1,2}_{\gamma+\varepsilon,\delta}(\mathcal{K})^\ell \quad \text{and} \quad (1-\zeta)\, w \in W^{1,2}_{\gamma-\varepsilon,\delta}(\mathcal{K})^\ell,$$

where $\gamma = \beta' - l - \sigma + 1 - 3/p$ and ε is an arbitrarily small positive number. From the equation $b(w, \eta v) = (F, \eta v)_{\mathcal{K}}$ it follows that

$$b(\zeta w, v) = (\Phi, v)_{\mathcal{K}} \stackrel{def}{=} (F, \eta v)_{\mathcal{K}} - b\big((1-\zeta)w, \eta v\big)$$

for all $v \in C_0^\infty(\overline{\mathcal{K}}\backslash\{0\})^\ell$, $v|_{\Gamma_j} = 0$ for $j \in I_0$. Obviously, $\Phi \in \mathcal{H}^*_{p',-\beta,-\delta}$. Consequently by Theorems 7.7.5 and 7.7.9, the decomposition

$$\zeta w = \Sigma_1 + V$$

holds, where Σ_1 is a sum of the same form as in (7.4.5) and $V \in W^{1,p}_{\beta,\delta}(\mathcal{K})^\ell$. An analogous decomposition holds for $(1-\zeta)w$. Thus,

$$w = \Sigma_1 + \Sigma_2 + W,$$

where $W \in W^{1,p}_{\beta,\delta}(\mathcal{K})^\ell$. Since Σ_1 and Σ_2 are solutions of the homogeneous problem (7.1.1)-(7.1.3), the vector function W solves the problem (7.7.1), (7.7.2). By the uniqueness of the solution in $W^{1,2}_{\beta,\delta}(\mathcal{K})^\ell$ (cf. Theorems 7.3.7 and 7.7.3), we conclude that $W = u$. This proves the theorem. □

7.8.10. The case of the Neumann problem. We consider the Neumann problem (7.4.9) and assume that the conditions (i)–(iii) of Subsection 7.4.5 are satisfied. Then the Green's matrix satisfies the estimates in Theorems 7.5.2 and 7.5.5 with

$$\delta^+_\alpha(x) = \min(0, \mu_+^{(k(x))} - |\alpha| - \varepsilon), \quad \delta^-_\gamma(\xi) = \min(0, \delta_-^{(k(\xi))} - |\gamma| - \varepsilon),$$
$$\delta^+_{k,\alpha} = \min(0, \mu_+^{(k)} - |\alpha| - \varepsilon), \quad \delta^-_{k,\alpha} = \min(0, \delta_-^{(k)} - |\alpha| - \varepsilon)$$

(cf. Theorem 7.5.9), where $\mu_+^{(k)}$ is the greatest positive real numbers such that the strip

$$0 < \operatorname{Re}\lambda < \mu_+^{(k)}$$

contains at most the eigenvalue $\lambda = \pm 1$ of the operator pencil $A_k(\lambda)$ and $k(x)$ is the smallest integer k such that $r_k(x) = r(x)$. Consequently, the condition (7.8.1) in Lemmas 7.8.9–7.8.11 and Theorem 7.8.12 can be replaced by

$$2 - \mu_+^{(k)} < \delta_k - \sigma < 2, \quad \delta_k - \sigma \text{ is not an integer.}$$

However, if $\delta_k - \sigma < 1$, then for the validity of the assertion of Theorem 7.8.12 it is necessary that the boundary data satisfy the additional compatibility condition

$$w^{(j,k)} \cdot \big(g_{k_+}|_{M_k},\, g_{k_-}|_{M_k} \big) = 0 \quad \text{for } j = 1, \dots, 2\ell - r'_k \tag{7.8.11}$$

(cf. Lemma 6.8.3). Here g_{k_+} and g_{k_-} are the Neumann data on the adjoining faces $\Gamma_{k_+}, \Gamma_{k_-}$ to the edge M_k, $w^{j,k}$ are certain constant vectors, and r'_k is the rank of the matrices in the condition (iii). For example, in the case of the Neumann problem for the Lamé system, the Neumann data must satisfy the condition (7.7.22).

Moreover, the numbers $\delta_+^{(k)}$ in the conditions of Theorems 7.8.13–7.8.15 can be replaced by $\mu_+^{(k)}$. In particular, the following result holds.

THEOREM 7.8.16. *Let $u \in W^{1,p}_{\beta,\delta}(\mathcal{K})^\ell$ be a solution of the problem* (7.7.24), *where F is a linear and continuous functional on $W^{1,p'}_{-\beta,-\delta}(\mathcal{K})^\ell$ which has the representation*

$$(F, v)_{\mathcal{K}} = \int_{\mathcal{K}} f \cdot \bar{v}\, dx + \sum_{j=1}^{d} \int_{\Gamma_j} g_j \cdot \bar{v}\, dx$$

with vector functions $f \in C^{l-2,\sigma}_{\beta',\delta'}(\mathcal{K})^\ell$ and $g_j \in C^{l-1,\sigma}_{\beta',\delta'}(\Gamma_j)^\ell$, $l \ge 2$. Suppose that the conditions (i)–(iii) *of Subsection* 7.4.5 *are satisfied, that the closed strip between the lines* $\operatorname{Re}\lambda = 1 - \beta - 3/p$ *and* $\operatorname{Re}\lambda = l + \sigma - \beta'$ *does not contain eigenvalues of the pencil $\mathfrak{A}(\lambda)$, and that the components of δ, δ' satisfy the inequalities*

$$\max(0, 1 - \delta_+^{(k)}) < \delta_k + 2/p < 1,$$

$$l - \mu_+^{(k)} < \delta'_k - \sigma < l, \quad \delta'_k \ge 0, \quad \delta'_k - \sigma \textit{ is not integer.}$$

If $\delta'_k - \sigma < l - 1$, then we assume in addition that the Neumann data satisfy the compatibility condition (7.8.11) *on the edge M_k. Then u admits the decomposition* (7.4.5), *where $w \in C^{l,\sigma}_{\beta',\delta'}(\mathcal{K})^\ell$, λ_ν are the eigenvalues of the pencil $\mathfrak{A}(\lambda)$ between the lines* $\operatorname{Re}\lambda = 1 - \beta - 3/p$ *and* $\operatorname{Re}\lambda = l + \sigma - \beta'$, *and $U^{\nu,j,s}$ are eigenvectors and generalized eigenvectors corresponding to the eigenvalue λ_ν. In particular, $u \in C^{l,\sigma}_{\beta',\delta'}(\mathcal{K})^\ell$ if the strip between the lines* $\operatorname{Re}\lambda = 1 - \beta - 3/p$ *and* $\operatorname{Re}\lambda = l + \sigma - \beta'$ *is free of eigenvalues of the pencil $\mathfrak{A}(\lambda)$. In the case $p = 2$, the result is also valid for $\delta = 0$.*

Note that this theorem is applicable, in particular, to the Neumann problem (7.4.10) for the Lamé system. Then the numbers $\mu_+^{(k)}$ are given by (7.4.11).

CHAPTER 8

Boundary value problems for second order systems in a bounded polyhedral domain

Here we study the Dirichlet, Neumann and mixed boundary value problems for elliptic second order systems with variable coefficients in a bounded domain $\mathcal{G}$ of polyhedral type. We assume that the corresponding sesquilinear form is coercive and prove that the operator of the boundary value problem is Fredholm both in weighted Sobolev and Hölder spaces. The conditions on the weight parameters β and δ are the same as for the problem in the cone $\mathcal{K}$ which was studied in the last chapter. In Sections 8.1 and 8.2, we prove the Fredholm property by constructing a regularizer for the operator of the boundary value problem. To this end, we apply the existence and uniqueness results obtained for the model problem in a cone. Furthermore, we prove regularity assertions for variational solutions in weighted and nonweighted Sobolev and Hölder spaces.

In Section 8.3, the general results obtained are illustrated by applications to the Laplace equation and the Lamé system in domains with smooth nonintersecting edges and in polyhedral domains. To give an impression of these applications, we state a couple of regularity results for a variational solution $u \in W^{1,2}(\mathcal{G})$ of the Neumann problem

$$-\Delta u = f \ \text{ in } \mathcal{G}, \quad \frac{\partial u}{\partial n} = 0 \ \text{ on } \partial\mathcal{G}\backslash\mathcal{S}$$

in an arbitrary polyhedron $\mathcal{G}$: in Subsection 8.3.4, it is shown that

- $u \in W^{1,p}(\mathcal{G})$ if $f \in (W^{1,p'}(\mathcal{G}))^*$ and $p < 3+\varepsilon$,
- $u \in W^{2,p}(\mathcal{G})$ if $f \in L_p(\mathcal{G})$ and $p < 4/3+\varepsilon$.

Other sharp facts of the same nature proved in Subsection 8.3.5 and valid for convex polyhedra were formulated in the introduction to this book.

8.1. Solvability of the boundary value problem in weighted Sobolev spaces

Let $\mathcal{G}$ be a bounded domain of polyhedral type in $\mathbb{R}^3$ with the faces $\Gamma_1, \dots, \Gamma_N$. Furthermore, let

$$L(x, D_x)\, u = \sum_{i,j=1}^{3} D_{x_j}\big(A_{i,j}(x)\, D_{x_i} u\big) + \sum_{i=1}^{3} A_i(x)\, \partial_{x_i} u + A_0(x)\, u$$

and

$$N(x, D_x)\, u = \sum_{i,j=1}^{3} A_{i,j}(x)\, n_j\, \partial_{x_i} u.$$

We consider the boundary value problem

(8.1.1) $$L(x,D_x)\,u = f \text{ in } \mathcal{G},$$

(8.1.2) $$u = g_j \ \text{ on } \Gamma_j \text{ for } j \in I_0, \quad N(x,D_x)\,u = g_j \ \text{ on } \Gamma_j \text{ for } j \in I_1\,,$$

where I_0 and I_1 are sets of integer numbers such that $I_0 \cup I_1 = \{1,2,\dots,N\}$, $I_0 \cap I_1 = \emptyset$. For the sake of simplicity, we suppose that the coefficients $A_{i,j}$ and A_i are $\ell \times \ell$-matrices of infinitely differentiable functions on $\overline{\mathcal{G}}$. We set $d_j = 0$ for $j \in I_0$, $d_j = 1$ for $j \in I_1$ and define

$$B_j(D_x)\,u = (1-d_j)\,u + d_j\,N(D_x)\,u \quad \text{for } j = 1,\dots,N.$$

Then the boundary conditions (8.1.2) can be written as

(8.1.3) $$B_j(D_x)\,u = g_j \ \text{ on } \Gamma_j \ \text{ for } j = 1,\dots,N.$$

Let b be the sequilinear form

$$b(u,v) = \int_{\mathcal{G}} \Big(\sum_{i,j=1}^{3} A_{i,j}\partial_{x_i}u \cdot \partial_{x_j}\bar{v} + \sum_{i=1}^{3} A_i\partial_{x_i}u \cdot \bar{v} + A_0 u \cdot \bar{v} \Big)\, dx,$$

and let $\mathcal{H}$ be the subspace

$$\mathcal{H} = \{u \in W^{1,2}(\mathcal{G})^\ell : u|_{\Gamma_j} = 0 \text{ for } j \in I_0\}.$$

Throughout this chapter, it is assumed that the form $b(u,v)$ is $\mathcal{H}$-coercive, i.e. there exist positive constants c_1 and c_2 such that

(8.1.4) $$\operatorname{Re} b(u,u) \ge c_1\, \|u\|^2_{W^{1,2}(\mathcal{G})^\ell} - c_2\, \|u\|^2_{L^2(\mathcal{G})^\ell} \quad \text{ for all } u \in \mathcal{H}.$$

One goal of this section is to show that the operator of the boundary value problem (8.1.1), (8.1.2) is Fredholm in weighted Sobolev spaces under certain conditions on the weight parameters. Here, we include variational solutions.

8.1.1. The domain. Throughout this chapter, we assume that $\mathcal{G}$ is a bounded domain in $\mathbb{R}^3$ satisfying the following conditions.

(i) The boundary $\partial\mathcal{G}$ consists of smooth (of class C^∞) open two-dimensional manifolds Γ_j (the faces of $\mathcal{G}$), $j = 1,\dots,N$, smooth curves M_k (the edges), $k = 1,\dots,d$, and vertices $x^{(1)},\dots,x^{(d')}$.

(ii) For every $\xi \in M_k$ there exist a neighborhood $\mathcal{U}_\xi$ and a diffeomorphism (a C^∞ mapping) κ_ξ which maps $\mathcal{G}\cap\mathcal{U}_\xi$ onto $\mathcal{D}_\xi \cap B_1$, where $\mathcal{D}_\xi$ is a dihedron and B_1 is the unit ball.

(iii) For every vertex $x^{(i)}$ there exist a neighborhood $\mathcal{U}_i$ and a diffeomorphism κ_i mapping $\mathcal{G}\cap\mathcal{U}_i$ onto $\mathcal{K}_i \cap B_1$, where $\mathcal{K}_i$ is a polyhedral cone with vertex at the origin.

The set $M_1 \cup \cdots \cup M_d \cup \{x^{(1)},\dots,x^{(d')}\}$ of the singular boundary points is denoted by $\mathcal{S}$. We do not exclude the case $d' = 0$. In this case, the set $\mathcal{S}$ consists only of smooth non-intersecting edges.

8.1.2. Weighted Sobolev spaces. We denote the distance of x from the edge M_k by $r_k(x)$, the distance from the vertex $x^{(j)}$ by $\rho_j(x)$, and the distance from $\mathcal{S}$ by $r(x)$. Let $\mathcal{U}_1,\dots,\mathcal{U}_{d'}$ be domains in $\mathbb{R}^3$ such that

$$\mathcal{U}_1 \cup \dots \cup \mathcal{U}_{d'} \supset \overline{\mathcal{G}} \quad \text{and} \quad \overline{\mathcal{U}}_j \cap \overline{M}_k = \emptyset \ \text{ if } k \notin X_j.$$

Here again X_j denotes the set of the indices k such that $x^{(j)}$ is an end point of the edge M_k. If l is a nonnegative integer, p is a real number, $p > 1$, and $\beta = (\beta_1,\dots,\beta_{d'})$, $\delta = (\delta_1,\dots,\delta_d)$ are tuples of real numbers, $\delta_k > -2/p$ for $k = 1,\dots,d$, then the weighted Sobolev space $W^{l,p}_{\beta,\delta}(\mathcal{G})$ is defined as the closure of the set $C_0^\infty(\overline{\mathcal{G}}\backslash\{x^{(1)},\dots,x^{(d')}\})$ with respect to the norm

$$\|u\|_{W^{l,p}_{\beta,\delta}(\mathcal{G})} = \Big(\sum_{j=1}^{d'} \int_{\mathcal{G}\cap\mathcal{U}_j} \sum_{|\alpha|\le l} \rho_j^{p(\beta_j-l+|\alpha|)} \prod_{k\in X_j} \Big(\frac{r_k}{\rho_j}\Big)^{p\delta_k} |\partial_x^\alpha u|^p\, dx\Big)^{1/p}.$$

Obviously, the space $W^{l,p}_{\beta,\delta}(\mathcal{G})$ does not depend on the choice of the domains $\mathcal{U}_j$. The trace space for $W^{l,p}_{\beta,\delta}(\mathcal{G})$, $l \ge 1$, on the face Γ_j is denoted by $W^{l-1/p,p}_{\beta,\delta}(\Gamma_j)$.

8.1.3. Imbeddings. The following lemma can be easily proved by means of Hölder's inequality.

LEMMA 8.1.1. *Let $1 < q < p < \infty$, $\beta_j + 3/p < \beta'_j + 3/q$ for $j = 1,\dots,d'$, and $0 < \delta_k + 2/p < \delta'_k + 2/q$ for $k = 1,\dots,d$. Then $W^{l,p}_{\beta,\delta}(\mathcal{G})$ is continuously imbedded in $W^{l,q}_{\beta',\delta'}(\mathcal{G})$.*

P r o o f. Let $s = pq/(p-q)$. By Hölder's inequality,

$$\Big\|\rho_j^{\beta'_j-l+|\alpha|} \prod_{k\in X_j} \Big(\frac{r_k}{\rho}\Big)^{\delta'_k} \partial_x^\alpha u\Big\|_{L_q(\mathcal{G}\cap\mathcal{U}_j)} \le c\, \Big\|\rho_j^{\beta_j-l+|\alpha|} \prod_{k\in X_j} \Big(\frac{r_k}{\rho}\Big)^{\delta_k} \partial_x^\alpha u\Big\|_{L_p(\mathcal{G}\cap\mathcal{U}_j)},$$

where

$$c = \Big\|\rho_j^{\beta'_j-\beta_j} \prod_{k\in X_j} \Big(\frac{r_k}{\rho}\Big)^{\delta'_k-\delta_k}\Big\|_{L_s(\mathcal{G}\cap\mathcal{U}_j)} < \infty$$

if $\beta'_j - \beta_j > -3/s$ and $\delta'_k - \delta_k > -2/s$ for $k \in X_j$. This proves the lemma. □

Next we prove that the imbedding $W^{l+1,p}_{\beta',\delta'}(\mathcal{G}) \subset W^{l,p}_{\beta,\delta}(\mathcal{G})$ is compact under certain conditions on the weight parameters.

LEMMA 8.1.2. *Suppose that $\beta'_j - \beta_j \le 1$ for $j = 1,\dots,d'$ and $\delta'_k - \delta_k \le 1$, $\delta_k > -2/p$, $\delta'_k > -2/p$ for $k = 1,\dots,d$. Then $W^{l+1,p}_{\beta',\delta'}(\mathcal{G})$ is continuously imbedded in $W^{l,p}_{\beta,\delta}(\mathcal{G})$. If $\beta'_j - \beta_j < 1$ for $j = 1,\dots,d'$ and $\delta'_k - \delta_k < 1$ for $k = 1,\dots,d$, then the imbedding $W^{l+1,p}_{\beta',\delta'}(\mathcal{G}) \subset W^{l,p}_{\beta,\delta}(\mathcal{G})$ is even compact.*

P r o o f. The imbedding $W^{l+1,p}_{\beta',\delta'}(\mathcal{G}) \subset W^{l,p}_{\beta,\delta}(\mathcal{G})$ follows immediately from Lemma 7.2.2. Suppose that $\beta'_j - \beta_j < 1$ for $j = 1,\dots,d'$ and $\delta'_k - \delta_k < 1$, $\delta_k > -2/p$, $\delta'_k > -2/p$ for $k = 1,\dots,d$. Then there exists a tuple $\delta'' = (\delta''_1,\dots,\delta''_d)$ such that $\delta''_k \ge \delta'_k - 1$ and $-2/p < \delta''_k < \delta_k$ for $k = 1,\dots,d$. We denote by $\mathfrak{M}$ the set of all $u \in W^{l+1,p}_{\beta',\delta'}(\mathcal{G})$ with norm $\le c_0$ and show that $\mathfrak{M}$ is precompact in $W^{l,p}_{\beta,\delta}(\mathcal{G})$, i.e. that for arbitrary positive ε there exists a finite ε-net $\{u_1,\dots,u_m\}$ in $\mathfrak{M}$ such that

$$\min_{1\le j\le m} \|u - u_j\|_{W^{l,p}_{\beta,\delta}(\mathcal{G})} \le \varepsilon \tag{8.1.5}$$

for all $u \in \mathfrak{M}$. Let $\mathcal{G}_\sigma = \{x \in \mathcal{G} : r(x) < \sigma\}$, $\sigma > 0$, and let the $W^{l,p}_{\beta,\delta}(\mathcal{G}_\sigma)$-norm be defined by the same integral as the $W^{l,p}_{\beta,\delta}(\mathcal{G})$-norm, where the domain of integration is replaced by $\mathcal{G}_\sigma$. Obviously,

$$\|u\|_{W^{l,p}_{\beta,\delta}(\mathcal{G}_\sigma)} \le c(\sigma)\, \|u\|_{W^{l,p}_{\beta'-1,\delta''}(\mathcal{G})}$$

for all $u \in W^{l+1,p}_{\beta',\delta'}(\mathcal{G})$, where $c(\sigma) \to 0$ as $\sigma \to 0$. By the continuity of the imbedding $W^{l+1,p}_{\beta',\delta'}(\mathcal{G}) \subset W^{l,p}_{\beta'-1,\delta''}(\mathcal{G})$, the number σ can be chosen such that

$$\|u\|_{W^{l,p}_{\beta,\delta}(\mathcal{G}_\sigma)} \le \varepsilon/3 \tag{8.1.6}$$

for all $u \in \mathfrak{M}$. Since the imbedding $W^{l+1,p}(\mathcal{G}\backslash\overline{\mathcal{G}}_\sigma) \subset W^{l,p}(\mathcal{G}\backslash\overline{\mathcal{G}}_\sigma)$ is compact and the $W^{l,p}(\mathcal{G}\backslash\overline{\mathcal{G}}_\sigma)$-norm is equivalent to the $W^{l,p}_{\beta,\delta}(\mathcal{G}\backslash\overline{\mathcal{G}}_\sigma)$-norm, there exist finitely many functions $u_1, \dots, u_m \in \mathfrak{M}$ such that

$$\min_{1\le j\le m} \|u - u_j\|_{W^{l,p}_{\beta,\delta}(\mathcal{G}\backslash\overline{\mathcal{G}}_\sigma)} \le \varepsilon/3\,.$$

The last inequality together with (8.1.6) implies (8.1.5). This proves the lemma. □

8.1.4. Model problems and corresponding operator pencils. We introduce the operator pencils generated by the boundary value problem (8.1.1), (8.1.2) for the singular boundary points.

1) Let ξ be a point on the edge M_k, and let $\Gamma_{k_+}, \Gamma_{k_-}$ be the faces of $\mathcal{G}$ adjacent to M_k. Then by $\mathcal{D}_\xi$ we denote the dihedron which is bounded by the half-planes $\Gamma^\circ_{k_\pm}$ tangent to $\Gamma_{k_\pm}$ at ξ and consider the model problem

$$L^\circ(\xi, D_x)\, u = f \ \text{ in } \mathcal{D}_\xi, \quad B_{k_\pm}(\xi, D_x) = g_{k_\pm} \ \text{ on } \Gamma^\circ_{k_\pm} \tag{8.1.7}$$

where $L^\circ(\xi, D_x)$ denotes the principal part of $L(x, D_x)$ with coefficients frozen at ξ and

$$B_{k_\pm}(\xi, D_x)u = (1 - d_{k_\pm})u + d_{k_\pm} \sum_{i,j=1}^{3} A_{i,j}(\xi)\, n_j\, \partial_{x_i} u.$$

The operator pencil corresponding to this model problem (see Section 6.1.3) is denoted by $A_\xi(\lambda)$. Furthermore, we denote by $\delta^{(k)}_+$ and $\delta^{(k)}_-$ the greatest positive numbers such that the strip

$$-\delta^{(k)}_- < \operatorname{Re}\lambda < \delta^{(k)}_+$$

contains at most the eigenvalue $\lambda = 0$ of the pencil $A_\xi(\lambda)$ for every $\xi \in M_k$.

2) Let $x^{(i)}$ be a vertex of $\mathcal{G}$ and let J_i be the set of all indices j such that $x^{(i)} \in \overline{\Gamma}_j$. By our assumptions, there exist a neighborhood $\mathcal{U}$ of $x^{(i)}$ and a diffeomorphism κ mapping $\mathcal{G} \cap \mathcal{U}$ onto $\mathcal{K}_i \cap B_1$ and $\Gamma_j \cap \mathcal{U}$ onto $\Gamma^\circ_j \cap B_1$ for $j \in J_i$, where $\mathcal{K}_i$ is a polyhedral cone with vertex at the origin and Γ°_j are the faces of this cone. Without loss of generality, we may assume that the Jacobian matrix $\kappa'(x)$ coincides with the identity matrix I at $x^{(i)}$. We consider the model problem

$$L^\circ(x^{(i)}, D_x)\, u = f \ \text{ in } \mathcal{K}_i, \quad B^\circ_j(x^{(i)}, D_x) = g_j \ \text{ on } \Gamma^\circ_j \text{ for } j \in J_i, \tag{8.1.8}$$

where $L^\circ(x^{(i)}, D_x)$, $B^\circ_j(x^{(i)}, D_x)$ are the principal parts of $L(x, D_x)$ and $B_j(x, D_x)$ with coefficients frozen at $x^{(i)}$. The operator pencil generated by this model problem (see Subsection 7.1.3) is denoted by $\mathfrak{A}_i(\lambda)$.

8.1.5. Normal solvability of the boundary value problem. Let $b_{\mathcal{D}_\xi}(\cdot,\cdot)$ be the sesquilinear form corresponding to the model problem (8.1.7), i.e.

$$b_{\mathcal{D}_\xi}(\cdot,\cdot) = \int_{\mathcal{D}_\xi} \sum_{j,k=1}^{3} A_{j,k}(\xi)\, \partial_{x_j} u \cdot \partial_{x_k} \overline{v}\, dx,$$

and let $\mathcal{H}_{\mathcal{D}_\xi}$ be the subspace

$$\mathcal{H}_{\mathcal{D}_\xi} = \{u \in W^{1,2}(\mathcal{D}_\xi)^\ell : \ (1 - d_{k\pm})\, u = 0 \text{ on } \Gamma^\circ_{k\pm}\}.$$

Analogously, we define the sesquilinear form $b_{\mathcal{K}_i}$ and the subspace $\mathcal{H}_{\mathcal{K}_i}$ for the model problem (8.1.8).

LEMMA 8.1.3. *Suppose that the form $b(\cdot,\cdot)$ is $\mathcal{H}$-coercive. Then the forms $b_{\mathcal{D}_\xi}(\cdot,\cdot)$ and $b_{\mathcal{K}_i}(\cdot,\cdot)$ of the problems* (8.1.7), (8.1.8) *are $\mathcal{H}_{\mathcal{D}_\xi}$- and $\mathcal{H}_{\mathcal{K}_i}$-elliptic, respectively.*

P r o o f. Let $\mathcal{U}$ be a neighborhood of the vertex $x^{(i)}$, and let κ be a diffeomorphism mapping $\mathcal{G} \cap \mathcal{U}$ onto $\mathcal{K}_i \cap B_1$ such that $\kappa'(x^{(i)}) = I$. Here $\mathcal{K}_i$ is a cone with vertex at the origin and B_1 denotes the unit ball. For functions with support in $\mathcal{K}_i \cap B_1$ we introduce the sesquilinear form

$$(8.1.9)\quad \tilde{b}(\tilde{u},\tilde{v}) = \int_{\mathcal{K}_i} \Big(\sum_{j,k=1}^{3} \tilde{A}_{j,k}(x)\, \partial_{x_j}\tilde{u} \cdot \overline{\partial_{x_k}\tilde{v}} + \sum_{j=1}^{3} \tilde{A}_j(x)\, \partial_{x_j}\tilde{u} \cdot \overline{\tilde{v}} + \tilde{A}_0(x) \tilde{u} \cdot \overline{\tilde{v}} \Big)\, dx$$

which is defined by

$$(8.1.10)\qquad \tilde{b}(\tilde{u},\tilde{v}) = b(u,v), \quad \text{where } u(x) = \tilde{u}\big(\kappa(x)\big),\ v(x) = \tilde{v}\big(\kappa(x)\big)$$

for $x \in \mathcal{G} \cap \mathcal{U}$. Then

$$b_{\mathcal{K}_i}(\tilde{u},\tilde{v}) = \int_{\mathcal{K}_i} \sum_{j,k=1}^{3} \tilde{A}_{j,k}(0)\, \partial_{x_j}\tilde{u} \cdot \overline{\partial_{x_k}\tilde{v}}\, dx.$$

Note that $\tilde{A}_{j,k}(0) = A_{j,k}(x^{(i)})$ since $\kappa'(x^{(k)}) = I$.

Let $\tilde{u} \in C_0^\infty(\overline{\mathcal{K}}_i)^\ell$, $\tilde{u}|_{\Gamma_j^\circ} = 0$ for $j \in J_i \cap I_0$ and $\operatorname{supp} \tilde{u} \subset B_\varepsilon$, where B_ε is the ball with radius ε and center at the origin and ε is sufficiently small. Using the inequality (8.1.4) for the function $u(x) = \tilde{u}(\kappa(x))$, we get

$$\operatorname{Re} \tilde{b}(\tilde{u},\tilde{u}) \ge c_0' \, \|\tilde{u}\|^2_{W^{1,2}(\mathcal{K}_i)^\ell} - c_1' \, \|\tilde{u}\|^2_{L_2(\mathcal{K}_i)^\ell}$$

with certain positive constants c_0', c_1'. By Hardy's inequality,

$$\|\tilde{u}\|^2_{L_2(\mathcal{K}_i)^\ell} \le \varepsilon^2 \int_{\mathcal{K}_i} |x|^{-2}\, |\tilde{u}(x)|^2\, dx \le c\, \varepsilon^2 \int_{\mathcal{K}} \sum_{j=1}^{3} \big|\partial_{x_j}\tilde{u}(x)\big|^2\, dx.$$

Furthermore,

$$\begin{aligned} \big|\tilde{b}(\tilde{u},\tilde{u}) - b_{\mathcal{K}_i}(\tilde{u},\tilde{u})\big| \ &\le\ c\,\varepsilon \sum_{j=1}^{3} \|\partial_{x_j}\tilde{u}\|^2_{L_2(\mathcal{K}_i)^\ell} + c\, \|\tilde{u}\|_{L_2(\mathcal{K}_i)^\ell}\, \|\tilde{u}\|_{W^{1,2}(\mathcal{K}_i)^\ell} \\ &\le\ c'\,\varepsilon \sum_{j=1}^{3} \|\partial_{x_j}\tilde{u}\|^2_{L_2(\mathcal{K}_i)^\ell}\, . \end{aligned}$$

Thus for small ε,

$$\operatorname{Re} b_{\mathcal{K}_i}(\tilde{u},\tilde{u}) \ge \frac{c_0'}{2} \sum_{j=1}^{3} \|\partial_{x_j}\tilde{u}\|^2_{L_2(\mathcal{K}_i)^\ell}. \tag{8.1.11}$$

Applying the dilation $x = y/N$, we obtain this estimate (with the same constant $c_0'/2$) for the function $\tilde{v}(x) = \tilde{u}(x/N)$. Consequently, the estimate (8.1.11) is valid for all $\tilde{u} \in C_0^\infty(\overline{\mathcal{K}}_i)^\ell$, $\tilde{u}|_{\Gamma_j^\circ} = 0$ for $j \in J_i \cap I_0$. This proves the $\mathcal{H}_{\mathcal{K}_i}$-ellipticity of the form $b_{\mathcal{K}_i}(\cdot,\cdot)$. Analogously, the $\mathcal{H}_{\mathcal{D}_\xi}$-ellipticity of the form $b_{\mathcal{D}_\xi}(\cdot,\cdot)$ holds. □

The last lemmas allows us to apply the results of Chapter 7 to the model problems (8.1.7) and (8.1.8). For every $k = 1,\dots,d$, let again Γ_{k_+} and Γ_{k_-} be the faces adjacent to the edge M_k. Suppose that $l \ge 2$ and $0 < \delta_k + 2/p < l$ for $k = 1,\dots,d$. Then we define $\mathcal{W}_{l,p,\beta,\delta}$ as the space of all

$$g = (g_1,\dots,g_N) \in \prod_{j=1}^{N} W_{\beta,\delta}^{l-d_j-1/p,p}(\Gamma_j)^\ell$$

satisfying the compatibility condition

$$g_{k_+}|_{M_k} = g_{k_-}|_{M_k} \quad \text{if } d_{k_+} = d_{k_-} = 0. \tag{8.1.12}$$

Furthermore, we denote the operator

$$W_{\beta,\delta}^{l,p}(\mathcal{G})^\ell \ni u \to \big(Lu, B_1u|_{\Gamma_1},\dots,B_Nu|_{\Gamma_N}\big) \in W_{\beta,\delta}^{l-2,p}(\mathcal{G})^\ell \times \mathcal{W}_{l,p,\beta,\delta} \tag{8.1.13}$$

of the boundary value problem (8.1.1), (8.1.2) by $\mathcal{A}$. Our goal is to prove that this operator is Fredholm if the line $\operatorname{Re}\lambda = l - \beta_k - 3/p$ is free of eigenvalues of the pencil $\mathfrak{A}_k(\lambda)$ for $k = 1,\dots,d$ and the components of δ satisfy the inequalities

$$\max(0, l - \delta_+^{(k)}) < \delta_k + 2/p < l \tag{8.1.14}$$

for $k = 1,\dots,d$. To this end, we construct a regularizer for $\mathcal{A}$. First we prove the existence of local regularizers.

LEMMA 8.1.4. *Let $\mathcal{U}$ be a sufficiently small subdomain of $\mathcal{G}$ and let φ be a smooth function with support in $\overline{\mathcal{U}}$. Suppose that there are no eigenvalues of the pencil $\mathfrak{A}_j(\lambda)$ on the line $\operatorname{Re}\lambda = l-\beta_j-3/p$ for $j = 1,\dots,d'$ and that the components of δ satisfy the condition (8.1.14). Then there exists an operator $\mathcal{R}$ continuously mapping the space of all $(f,g) \in W_{\beta,\delta}^{l-2,p}(\mathcal{G})^\ell \times \mathcal{W}_{l,p,\beta,\delta}$, $l \ge 2$, with support in $\overline{\mathcal{U}}$ into the space $W_{\beta,\delta}^{l,p}(\mathcal{G})^\ell$ such that*

$$\varphi\mathcal{A}\mathcal{R}(f,g) = \varphi(f,g)$$

for all $(f,g) \in W_{\beta,\delta}^{l-2,p}(\mathcal{G})^\ell \times \mathcal{W}_{l,p,\beta,\delta}$, $\operatorname{supp}(f,g) \in \overline{\mathcal{U}}$, and

$$\varphi\mathcal{R}\mathcal{A}u = \varphi u$$

for all $u \in W_{\beta,\delta}^{l,p}(\mathcal{G})^\ell$, $\operatorname{supp} u \subset \overline{\mathcal{U}}$.

Suppose that $\beta_j' < \beta_j + 1$ for $j = 1,\dots,d'$, the strip $l - \beta_j - 3/p \le \operatorname{Re}\lambda \le l+1-\beta_j'-3/p$ does not contain eigenvalues of the pencil $\mathfrak{A}_j(\lambda)$ for $j = 1,\dots,d'$, and that the components of δ' satisfy the inequalities $\max(0, l+1-\delta_+^{(k)}) < \delta_k' + 2/p < \delta_k + 1 + 2/p$. Then the operator $\mathcal{R}$ realizes also a continuous mapping from the subspace of all $(f,g) \in W_{\beta',\delta'}^{l-1,p}(\mathcal{G})^\ell \times \mathcal{W}_{l+1,p,\beta'\delta'}$ with support in $\overline{\mathcal{U}}$ into the space $W_{\beta',\delta'}^{l+1,p}(\mathcal{G})^\ell$.

P r o o f. There exists a diffeomorphism κ mapping $\mathcal{U}$ onto a subset $\mathcal{V}$ of a cone $\mathcal{K}$ with vertex at the origin. Applying the coordinate transformation $y = \kappa(x)$ to the problem (8.1.1), (8.1.2), we obtain the equations

$$\tilde{L}\tilde{u} = \tilde{f} \text{ in } \mathcal{V}, \quad \tilde{B}_j\tilde{u} = \tilde{g}_j \text{ on } \kappa(\Gamma_j) \cap \overline{\mathcal{V}}, \ j = 1, \dots, N.$$

where $\tilde{u} = u \circ \kappa^{-1}$.

Suppose first that $\overline{\mathcal{U}}$ contains the vertex $x^{(1)}$ of $\mathcal{G}$, that $\kappa(x^{(1)})$ coincides with the origin and that the Jacobian matrix κ' coincides with the identity matrix at $x^{(1)}$. We denote the principal parts of $\tilde{L}$ and $\tilde{B}_j$ with coefficients frozen at $x = 0$ by $\tilde{L}^\circ$ and $\tilde{B}_j^\circ$, respectively. By Theorem 7.6.4, the operator

$$\tilde{\mathcal{A}}_0 = \big(\tilde{L}^\circ, \{\tilde{B}_j^\circ\}_{j \in J_1}\big)$$

realizes an isomorphism from $W^{l,p}_{\beta_1,\delta}(\mathcal{K})^\ell$ onto the space of all

$$(f, g) \in W^{l-2,p}_{\beta_1,\delta}(\mathcal{K})^\ell \times \prod_{j \in J_1} W^{l-d_j-1/p,p}_{\beta_1,\delta}(\Gamma_j^\circ)^\ell \tag{8.1.15}$$

satisfying the compatibility condition (7.3.5). Let ζ be an infinitely differentiable cut-off function on $[0, \infty)$ equal to 1 in $[0, 1)$ and to zero in $(2, \infty)$. Furthermore, let $\zeta_\varepsilon(x) = \zeta(|x|/\varepsilon)$ for arbitrary positive ε. We introduce the differential operators

$$\tilde{L}_\varepsilon = \zeta_\varepsilon \tilde{L} + (1 - \zeta_\varepsilon)\tilde{L}^\circ \tag{8.1.16}$$

and

$$\tilde{B}_{j,\varepsilon} = \zeta_\varepsilon \tilde{B}_j + (1 - \zeta_\varepsilon)\tilde{B}_j^\circ, \tag{8.1.17}$$

If ε is small, then the difference of the operators $\tilde{\mathcal{A}}_0$ and

$$\tilde{\mathcal{A}}_\varepsilon = \big(\tilde{L}_\varepsilon, \{\tilde{B}_{j,\varepsilon}\}_{j \in J_1}\big)$$

in the operator norm

$$W^{l,p}_{\beta_1,\delta}(\mathcal{K})^\ell \to W^{l-2,p}_{\beta_1,\delta}(\mathcal{K})^\ell \times \prod_{j \in J_1} W^{l-d_j-1/p,p}_{\beta_1,\delta}(\Gamma_j^\circ)^\ell$$

is also small. Thus, the operator $\tilde{\mathcal{A}}_\varepsilon$ realizes an isomorphism from $W^{l,p}_{\beta_1,\delta}(\mathcal{K})^\ell$ onto the space of all vector functions (8.1.15) satisfying the compatibility condition of Theorem 7.3.2 if ε is sufficiently small. We assume that $\mathcal{V}$ is contained in the ball $|x| < \varepsilon$. Then the coefficients of $\tilde{L}_\varepsilon$, $\tilde{B}_\varepsilon$ coincide with those of $\tilde{L}$ and $\tilde{B}$ on the support of $\tilde{f}$ and $\tilde{g}$, respectively. Let

$$u(x) = \tilde{u}(\kappa(x)) \text{ for } x \in \mathcal{U}, \quad \text{where } \tilde{u} = \tilde{\mathcal{A}}_\varepsilon^{-1}(\tilde{f}, \tilde{g}). \tag{8.1.18}$$

Here $\tilde{f}$ and $\tilde{g}$ are extended by zero outside $\mathcal{U}$. The function (8.1.18) can be continuously extended to a function $u \in W^{l,p}_{\beta,\delta}(\mathcal{G})^\ell$. The so defined mapping $(f, g) \to u$ is denoted by $\mathcal{R}$. It can be easily verified that $\mathcal{R}$ has the desired properties. In particular, the continuity of the mapping $\mathcal{R}$ from the subspace of all $(f, g) \in W^{l-1,p}_{\beta',\delta'}(\mathcal{G})^\ell \times \mathcal{W}_{l+1,p,\beta',\delta'}$ with support in $\overline{\mathcal{U}}$ into the space $W^{l+1,p}_{\beta',\delta'}(\mathcal{G})^\ell$ follows from Theorem 7.7.7.

Suppose now that $\overline{\mathcal{U}}$ contains an edge point $\xi \in M_k$ but no points of other edges and no vertices of $\mathcal{G}$. Let Γ_{k_+} and Γ_{k_-} be the faces adjacent to the edge M_k and $d_\pm = d_{k_\pm}$. In this case, we may assume that $\mathcal{V} = \kappa(\mathcal{U})$ is a subset of a dihedron $\mathcal{D}_\xi$ with the faces $\Gamma_{k_\pm}^\circ$, that the edge M of this dihedron coincides with the x_3-axis, and that $\mathcal{V}$ has a positive distance from the origin. We can see $\mathcal{D}_\xi$ also as a polyhedral cone with vertex at the origin. Let $\tilde{L}^\circ$, $\tilde{B}^\circ_{k\pm}$ be the principal parts of $\tilde{L}$ and $\tilde{B}_{k\pm}$, respectively, with coefficients frozen at $\eta = \kappa(\xi)$. Furthermore, let the operators $\tilde{L}_\varepsilon$ and $\tilde{B}_{j,\varepsilon}$ be defined by (8.1.16) and (8.1.17), respectively, where $\zeta_\varepsilon(x) = \zeta(|x-\eta|/\varepsilon)$ and ζ is the above introduced cut-off function. By Theorem 7.6.4, there exists a number β_0 such that the operator

$$\tilde{\mathcal{A}}_0 = \big(\tilde{L}^\circ, \tilde{B}^\circ_{k_+}, \tilde{B}^\circ_{k_-}\big)$$

and, for sufficiently small ε, also the operator

$$\tilde{\mathcal{A}}_\varepsilon = \big(\tilde{L}_\varepsilon, \tilde{B}_{k_+,\varepsilon}, \tilde{B}_{k_-,\varepsilon}\big)$$

realize isomorphisms from $W^{l,p}_{\beta_0,\delta_k}(\mathcal{D}_\xi)^\ell$ onto the space of all

$$(f, g_+, g_-) \in W^{l-2,p}_{\beta_0,\delta_k}(\mathcal{D}_\xi)^\ell \times \prod_{\pm} W^{l-d_\pm-1/p,p}_{\beta_0,\delta_k}(\Gamma^\circ_{k_\pm})^\ell,$$

satisfying the compatibility condition $g_+|_M = g_-|_M$ if $d_+ = d_- = 0$. Since the origin lies outside $\overline{\mathcal{V}}$, the $W^{l,p}_{\beta_0,\delta_k}(\mathcal{D}_\xi)$- and $W^{l,p}_{\delta_k}(\mathcal{D}_\xi)$-norms are equivalent on the subset of functions with support in $\overline{\mathcal{V}}$. We suppose that $\mathcal{U}$ is sufficiently small such that $\mathcal{V}$ is contained in the ball $|x-\eta| < \varepsilon$. Again let $u(x)$ be defined by (8.1.18) for $x \in \mathcal{U}$. This function can be continuously extended to a vector function $u \in W^{l,p}_{\beta,\delta}(\mathcal{G})^\ell$. The so defined mapping $(f,g) \to u$ is denoted by $\mathcal{R}$ and has the desired properties. Analogously, we can construct the local regularizer $\mathcal{R}$ in the case where $\mathcal{U}$ contains no edge points. □

Using local regularizers, one can construct a global regularizer for the boundary value problem (8.1.1), (8.1.2). This is done in the proof of the next theorem.

THEOREM 8.1.5. *Suppose that the line* $\operatorname{Re}\lambda = l - \beta_j - 3/p$ *does not contain eigenvalues of the pencil* $\mathfrak{A}_j(\lambda)$ *for* $j = 1, \dots, d'$ *and that the components of* δ *satisfy the condition* (8.1.14). *Then the operator* (8.1.13) *is Fredholm for* $l \ge 2$.

P r o o f. Let $\{\mathcal{U}_\nu\}$ be a sufficiently fine open covering of $\mathcal{G}$ and let ϕ_ν, ψ_ν be infinitely differentiable functions such that $\operatorname{supp}\phi_\nu \subset \operatorname{supp}\psi_\nu \subset \mathcal{U}_\nu$, $\phi_\nu\psi_\nu = \phi_\nu$, and $\sum \phi_\nu = 1$. For every ν, there exists an operator $\mathcal{R}_\nu$ which has the properties of the operator $\mathcal{R}$ in Lemma 8.1.4 for $\mathcal{U} = \mathcal{U}_\nu \cap \mathcal{G}$. We prove that the operator $\mathcal{R}$ defined by

$$\mathcal{R}(f,g) = \sum_\nu \phi_\nu \mathcal{R}_\nu\, \psi_\nu(f,g)$$

is a left and right regularizer for the operator (8.1.13). First we show that $\mathcal{R}\mathcal{A} - I$ is a compact operator on $W^{l,p}_{\beta,\delta}(\mathcal{G})^\ell$. Here I denotes the identity operator. Obviously,

$$\mathcal{R}\mathcal{A}u = \sum_\nu \phi_\nu \mathcal{R}_\nu \big(\mathcal{A}\psi_\nu u - [\mathcal{A},\psi_\nu]u\big) = u - \sum_\nu \phi_\nu \mathcal{R}_\nu [\mathcal{A},\psi_\nu]u,$$

where $[\mathcal{A},\psi_\nu] = \mathcal{A}\psi_\nu - \psi_\nu\mathcal{A}$ is the commutator of $\mathcal{A}$ and ψ_ν. Since the commutator $[L(x,D_x), \varphi_\nu]$ is a first order differential operator, the mapping $u \to [\mathcal{A},\psi_\nu]u$ is continuous from $W^{l,p}_{\beta,\delta}(\mathcal{G})^\ell$ into $W^{l-1,p}_{\beta',\delta'}(\mathcal{G})^\ell \times \mathcal{W}_{l+1,p,\beta',\delta'}$, where β' and δ' are arbitrary tuples satisfying the inequalities $\beta'_j \ge \beta_j$ for $j = 1, \dots, d'$ and $\delta'_k \ge \delta_k$ for $k = 1, \dots, d$. We can choose β' such that $\beta'_j < \beta_j + 1$ and the strip $l - \beta_j - 3/p \le \operatorname{Re}\lambda \le l + 1 - \beta'_j - 3/p$ is free of eigenvalues of the pencil $\mathfrak{A}_j(\lambda)$ for $j = 1, \dots, d'$. Furthermore, we can assume that the components of δ' satisfy the inequality $\max(0, l+1-\delta_+^{(k)}) < \delta'_k + 2/p < \delta_k + 1 + 2/p$ for $k = 1, \dots, d$. Then by Lemma 8.1.4, the operator $\mathcal{R}_\nu$ realizes a continuous mapping

$$W^{l-1,p}_{\beta',\delta'}(\mathcal{G})^\ell \times \mathcal{W}_{l+1,p,\beta',\delta'} \to W^{l+1,p}_{\beta',\delta'}(\mathcal{G})^\ell.$$

The last space is compactly imbedded in $W^{l,p}_{\beta,\delta}(\mathcal{G})^\ell$ (cf. Lemma 8.1.4). Consequently, the operator $\mathcal{R}\mathcal{A} - I$ is compact on $W^{l,p}_{\beta,\delta}(\mathcal{G})^\ell$. Furthermore,

$$\begin{aligned} \mathcal{A}\mathcal{R}(f,g) &= \sum_\nu \varphi_\nu \mathcal{A}\mathcal{R}_\nu \psi_\nu(f,g) + \sum_\nu [\mathcal{A}, \varphi_\nu]\, \mathcal{R}_\nu \psi_\nu(f,g) \\ &= (f,g) + \sum_\nu [\mathcal{A}, \varphi_\nu]\, \mathcal{R}_\nu \psi_\nu(f,g). \end{aligned}$$

Since the commutator $[L(x, D_x), \varphi_\nu]$ is a first order differential operator, the mapping

$$(f,g) \to \sum_\nu [\mathcal{A}, \varphi_\nu]\, \mathcal{R}_\nu \psi_\nu(f,g)$$

is continuous from $W^{l-2,p}_{\beta,\delta}(\mathcal{G})^\ell \times \mathcal{W}_{l,p,\beta,\delta}$ into $W^{l-1,p}_{\beta',\delta'}(\mathcal{G})^\ell \times \mathcal{W}_{l+1,p,\beta,\delta}$. Since the imbeddings $W^{l-1,p}_{\beta',\delta'}(\mathcal{G}) \subset W^{l-2,p}_{\beta,\delta}(\mathcal{G})$ and $\mathcal{W}_{l+1,p,\beta,\delta} \subset \mathcal{W}_{l,p,\beta,\delta}$ are compact, we conclude that the operator $\mathcal{A}\mathcal{R} - I$ is compact on $W^{l-2,p}_{\beta,\delta}(\mathcal{G})^\ell \times \mathcal{W}_{l,p,\beta,\delta}$. This means that $\mathcal{R}$ is a regularizer for the operator $\mathcal{A}$. Thus, the operator (8.1.13) is Fredholm. □

8.1.6. Regularity results for weak solutions. Suppose that $\delta_k + 2/p < 1$ for $k = 1, \dots, d$. We introduce the space

$$\mathcal{H}_{p,\beta,\delta} = \{u \in W^{1,p}_{\beta,\delta}(\mathcal{G})^\ell :\ u|_{\Gamma_j} = 0 \text{ for } j \in I_0\}.$$

Furthermore, let $\mathcal{W}_{p,\beta,\delta}$ denote the set of all

$$g = \{g_j\}_{j \in I_0} \in \prod_{j \in I_0} W^{1-1/p,p}_{\beta,\delta}(\Gamma_j)^\ell$$

satisfying the compatibility condition (8.1.12). Note that the space $\mathcal{H}_{p,\beta,\delta}$ coincides with $\mathcal{H}$ in the case $p = 2$, $\beta = 0$, $\delta = 0$. Analogously to Theorem 8.1.5, one can show that the operator

$$\mathcal{H}_{p,\beta,\delta} \ni u \to (F, g) \in \mathcal{H}^*_{p',-\beta,-\delta} \times \mathcal{W}_{p,\beta,\delta}$$

($p' = p/(p-1)$) of the problem

$$b(u,v) = (F,v)_{\mathcal{G}} \quad \text{for all } v \in \mathcal{H}_{p',-\beta,-\delta} \tag{8.1.19}$$

$$u|_{\Gamma_j} = g_j \quad \text{for } j \in I_0 \tag{8.1.20}$$

is Fredholm if the line $\operatorname{Re}\lambda = l - \beta_k - 3/p$ does not contain eigenvalues of the pencil $\mathfrak{A}_k(\lambda)$ for $k = 1, \dots, d$ and the components of δ satisfy the condition

$$\max(0, 1 - \delta_+^{(k)}) < \delta_k + 2/p < 1. \tag{8.1.21}$$

In the case $p = 2$, this assertion is also true for $\delta_k = 0$ (then of course the trace condition (8.1.12) has to be understood in the generalized sense, cf. Lemma 6.4.3). Local regularizers can be constructed in the same way as in the proof of Lemma 8.1.4). Here one has to apply Theorems 7.7.3 (Theorem 7.3.7 if $p = 2$ and $\delta = 0$) instead of Theorem 7.6.4.

THEOREM 8.1.6. *Let* $u \in W^{1,p}_{\beta,\delta}(\mathcal{G})^\ell$ *be a solution of the problem* (8.1.19), (8.1.20). *Suppose that* $F \in \mathcal{H}^*_{p',-\beta,-\delta} \cap \mathcal{H}^*_{q',-\beta',-\delta'}$, $g \in \mathcal{W}_{p,\beta,\delta} \cap \mathcal{W}_{q,\beta',\delta'}$, *the closed strip between the lines* $\operatorname{Re}\lambda = 1-\beta_j - 3/p$ *and* $\operatorname{Re}\lambda = 1 - \beta_j - 3/p$ *is free of eigenvalues of the pencil* $\mathfrak{A}_j(\lambda)$ *for* $j = 1, \dots, d'$, *and the components of* δ *and* δ' *satisfy the inequalities*

$$\max(0, 1-\delta_+^{(k)}) < \delta_k + 2/p < 1, \quad \max(0, 1-\delta_+^{(k)}) < \delta'_k + 2/q < 1$$

for $k = 1, \dots, d$. *Then* $u \in W^{1,q}_{\beta',\delta'}(\mathcal{G})^\ell$. *In the case* $p = 2$, *the assertion is also true for* $\delta = 0$.

P r o o f. Suppose that the support of u is contained in a (sufficiently small) neighborhood $\mathcal{U}$ of the vertex $x^{(i)}$. Then there exists a diffeomorphism κ mapping $\mathcal{G} \cap \mathcal{U}$ onto the intersection of a cone $\mathcal{K}_i$ with a neighborhood $\mathcal{V}$ of the origin. The faces of the cone $\mathcal{K}_i$ are denote by Γ_j°, $j \in J_i$. We assume that $\kappa'(x^{(i)}) = I$. Let $\tilde{A}_{j,k}$ and $\tilde{A}_j$ be the coefficients of the sesquilinear form (8.1.10). Furthermore, let $\zeta_\varepsilon(x) = \zeta(|x|/\varepsilon)$, where ζ is an infinitely differentiable function on $[0,\infty)$ such that $\zeta = 1$ in $[0,1]$ and $\zeta = 0$ in $[2,\infty)$. We introduce the sesquilinear forms

$$\tilde{b}_0(u,v) = \int_{\mathcal{K}_i} \sum_{j,k=1}^{3} \tilde{A}_{j,k}(0)\, \partial_{x_j} u \cdot \partial_{x_k} \bar{v}\, dx$$

and

$$\tilde{b}_\varepsilon(u,v) = \int_{\mathcal{K}_i} \Big(\sum_{j,k=1}^{3} \tilde{A}^{(\varepsilon)}_{j,k} \partial_{x_j} u \cdot \partial_{x_k} \bar{v} + \sum_{j=1}^{3} \tilde{A}^{(\varepsilon)}_j \partial_{x_j} u \cdot \bar{v} + \tilde{A}^{(\varepsilon)}_0 u \cdot \bar{v} \Big)\, dx,$$

where

$$\tilde{A}^{(\varepsilon)}_{j,k}(x) = \zeta_\varepsilon\, \tilde{A}_{j,k}(x) + (1-\zeta_\varepsilon)\, \tilde{A}_{j,k}(0) \quad \text{for } j,k = 1,2,3,$$
$$\tilde{A}^{(\varepsilon)}_{j}(x) = \zeta_\varepsilon\, \tilde{A}_{j}(x) \text{ for } j = 0,1,2,3.$$

Obviously, $\tilde{b}_\varepsilon(u,v) = \tilde{b}(u,v)$ if $u(x) = 0$ for $|x| > \varepsilon$. Let $\mathcal{H}^{(i)}_{p,\beta_i,\delta}$ be the space of all $u \in W^{1,p}_{\beta_i,\delta}(\mathcal{K}_i)^\ell$ such that $u = 0$ on Γ_j°, $j \in I_0 \cap J_i$. Furthermore, let $\mathcal{W}^{(i)}_{p,\beta_i,\delta}$ be the set of all

$$g = \{g_j\}_{j \in I_0 \cap J_i} \in W^{1-1/p}_{\beta_i,\delta}(\Gamma_j^\circ)^\ell$$

satisfying the compatibility condition (8.1.12). The form $\tilde{b}_\varepsilon$ generates the linear and continuous operator

$$W^{1,p}_{\beta_i,\delta}(\mathcal{K}_i)^\ell \cap W^{1,q}_{\beta'_i,\delta'}(\mathcal{K}_i)^\ell \ni u$$
$$\to \tilde{A}_\varepsilon u = (F,g) \in \big((\mathcal{H}^{(i)}_{p',-\beta_i,-\delta})^* \cap (\mathcal{H}^{(i)}_{q',-\beta'_i,-\delta'})^* \big) \times \big(\mathcal{W}^{(i)}_{p,\beta_i,\delta} \cap \mathcal{W}^{(i)}_{q,\beta'_i,\delta'} \big)$$

by

$$(F,v)_{\mathcal{K}_i} = \tilde{b}_\varepsilon(u,v) \quad \text{for all } v \in \mathcal{H}^{(i)}_{p',-\beta_i,-\delta} \cap \mathcal{H}^{(i)}_{q',-\beta'_i,-\delta'},$$
$$g_j = u\big|_{\Gamma_j^\circ} \quad \text{for } j \in I_0 \cap J_i.$$

If the support of u is sufficiently small, then the vector function $\tilde{u} = u \circ \kappa^{-1}$ is a solution of the equation

$$\tilde{A}_\varepsilon \tilde{u} = (\tilde{F}, \tilde{g}),$$

where $\tilde{F} \in (\mathcal{H}^{(i)}_{p',-\beta,-\delta})^* \cap (\mathcal{H}^{(i)}_{q',-\beta',-\delta'})^*$ and $\tilde{g} \in \mathcal{W}^{(i)}_{p,\beta,\delta} \cap \mathcal{W}^{(i)}_{q,\beta',\delta'}$. It follows from Theorems 7.7.3 and 7.7.5 that $\tilde{\mathcal{A}}_0$ is an isomorphism. Since the difference of $\tilde{\mathcal{A}}_\varepsilon$ and $\tilde{\mathcal{A}}_0$ is small for small ε, we conclude that $\tilde{\mathcal{A}}_\varepsilon$ is an isomorphism if ε is sufficiently small. Consequently, $\tilde{u} \in W^{1,q}_{\beta'_i,\delta'}(\mathcal{K}_i)^\ell$ and $u \in W^{1,q}_{\beta',\delta'}(\mathcal{G})^\ell$. Analogously, this results holds if the support of u is contained in a sufficiently small neighborhood of an edge point. Using a partition of unity on $\overline{\mathcal{G}}$, we obtain the assertion of the theorem for arbitrary $u \in W^{1,p}_{\beta,\delta}(\mathcal{G})^\ell$. □

Analogously, the following result can be proved by means of Theorem 7.7.7.

THEOREM 8.1.7. *Let $u \in W^{1,p}_{\beta,\delta}(\mathcal{G})^\ell$ be a solution of the problem* (8.1.19), (8.1.20). *Suppose that $g_j \in W^{1-1/p,p}_{\beta,\delta}(\Gamma_j)^\ell \cap W^{l-1/q,q}_{\beta',\delta'}(\Gamma_j)^\ell$ for $j \in I_0$ and that F is a functional on $\mathcal{H}_{p',-\beta,-\delta}$ which has the form*

$$(F, v)_{\mathcal{G}} = \int_{\mathcal{G}} f \cdot \bar{v}\, dx + \sum_{j\in I_1} \int_{\Gamma_j} g_j \cdot \bar{v}\, dx$$

for all $v \in C_0^\infty(\overline{\mathcal{G}}\backslash\{x^{(1)}, \dots, x^{(d')}\})^\ell$, $v|_{\Gamma_j} = 0$ for $j \in I_0$, where $g_j \in W^{l-1-1/q,q}_{\beta',\delta'}(\Gamma_j)^\ell$ for $j \in I_1$, $f \in W^{l-2,q}_{\beta',\delta'}(\mathcal{G})^\ell$. Furthermore, we assume that the closed strip between the lines $\mathrm{Re}\,\lambda = 1 - \beta_j - 3/p$ and $\mathrm{Re}\,\lambda = l - \beta'_j - 3/q$ is free of eigenvalues of the pencil $\mathfrak{A}_j(\lambda)$ for $j = 1, \dots, d'$ and that the components of δ and δ' satisfy the inequalities

$$\max(0, 1 - \delta_+^{(k)}) < \delta_k + 2/p < 1, \quad \max(0, l - \delta_+^{(k)}) < \delta'_k + 2/q < l$$

for $k = 1, \dots, d$. Then $u \in W^{l,q}_{\beta',\delta'}(\mathcal{G})^\ell$. The assertion of the theorem is also true if $p = 2$ and $\delta = 0$.

8.1.7. Regularity results for solutions of the Neumann problem. We consider the Neumann problem

$$(8.1.22) \qquad L(x, D_x)\, u = f \;\text{ in } \mathcal{G}, \quad N(x, D_x)\, u = g_j \;\text{ on } \Gamma_j\,,\; j = 1, \dots, N.$$

Suppose that the conditions of Subsection 7.7.7 are satisfied for the spectra of the pencils $\mathfrak{A}_j(\lambda)$. This means that the strip $-1 \le \mathrm{Re}\,\lambda \le 0$ contains only the eigenvalues $\lambda = 0$ and $\lambda = -1$ and the eigenvalue $\lambda = 0$ has geometric and algebraic multiplicity ℓ (the corresponding eigenvectors are the constant vectors in $\mathbb{C}^\ell$). In this case, we denote by Λ_j the greatest real number such that the strip

$$-1 < \mathrm{Re}\lambda < \Lambda_j$$

contains only the eigenvalue $\lambda = 0$ of the pencil $\mathfrak{A}_j(\lambda)$. Then we can deduce the following regularity results from Theorems 7.7.11 and 7.7.13.

LEMMA 8.1.8. *Let $u \in W^{2,p}_{\beta,\delta}(\mathcal{G})^\ell$ be a solution of the Neumann problem* (8.1.22), *where*

$$f \in W^{0,p}_{\beta,\delta}(\mathcal{G})^\ell \cap W^{0,p}_{\gamma,\delta}(\mathcal{G})^\ell, \quad g_j \in W^{1-1/p,p}_{\beta,\delta}(\Gamma_j)^\ell \cap W^{1-1/p,p}_{\gamma,\delta}(\Gamma_j)^\ell$$

for $j = 1, \dots, N$. We assume that $\max(0, 2 - \delta_+^{(k)}) < \delta_k + 2/p < 2$ for $k = 1, \dots, d$, $-1 < 2 - \beta_j - 3/p < 0$, $-1 < 2 - \gamma_j - 3/p < \Lambda_j$ and $\gamma_j \ge \beta_j - 1$ for $j = 1, \dots, d'$. Then $\partial_{x_j} u \in W^{1,p}_{\gamma,\delta}(\mathcal{G})^\ell$ for $j = 1, 2, 3$.

Proof. Let χ be a smooth cut-off function equal to one in a neighborhood of the vertex $x^{(1)}$ with sufficiently small support. Obviously, $\chi u \in W^{2,p}_{\beta,\delta}(\mathcal{G})^\ell$. We show that $\partial_{x_j}(\chi u) \in W^{1,p}_{\gamma,\delta}(\mathcal{G})^\ell$ for $j = 1, 2, 3$. Since $\gamma_j \ge \beta_j - 1$, the vector function χu satisfies the equations

$$L(x, D_x)\,(\chi u) = F \quad \text{in } \mathcal{G}, \qquad N(x, D_x)\,(\chi u) = G_j \quad \text{on } \Gamma_j\,, j = 1, \dots, N,$$

where

$$F = \chi f + [L(x, D_x), \chi]\, u \in W^{0,p}_{\beta,\delta}(\mathcal{G})^\ell \cap W^{0,p}_{\gamma,\delta}(\mathcal{G})^\ell$$

and $G_j \in W^{1-1/p,p}_{\beta,\delta}(\Gamma_j)^\ell \cap W^{1-1/p,p}_{\gamma,\delta}(\Gamma_j)^\ell$. Here $[L, \chi] = L\chi - \chi L$ denotes the commutator of L and χ. Without loss of generality, we may assume that $x^{(1)}$ is the origin and that $\mathcal{G}$ coincides with the cone $\mathcal{K}$ in a neighborhood of the vertex. We introduce the operators

$$L_\varepsilon = \zeta_\varepsilon\, L(x, D_x) + (1 - \zeta_\varepsilon)\, L^\circ(0, D_x), \quad N_\varepsilon = \zeta_\varepsilon\, N(x, D_x) + (1 - \zeta_\varepsilon)\, N^\circ(0, D_x),$$

where ζ_ε is the same cut-off function as in the proof of Theorem 8.1.6. Since the coefficients of L_ε are constant outside the support of ζ_ε, we can see L_ε as a differential operator on $\mathcal{K}$ if ε is sufficiently small. Furthermore

$$L(x, D_x)\,(\chi u) = L_\varepsilon\,(\chi u) \quad \text{and} \quad N(x, D_x)\,(\chi u) = N_\varepsilon\,(\chi u)$$

if the support of the cut-off function χ is sufficiently small. By Corollary 7.7.12, the operator $\big(L^\circ(0, D_x), N^\circ(0, D_x)\big)$ is an isomorphism from $W^{2,p}_{\beta_1,\delta}(\mathcal{K})^\ell \cap L^{2,p}_{\gamma_1,\delta}(\mathcal{K})^\ell$ onto

$$W^{0,p}_{\beta_1,\delta}(\mathcal{K})^\ell \cap W^{0,p}_{\gamma_1,\delta}(\mathcal{K})^\ell \times \prod_j \big(W^{1-1/p,p}_{\beta,\delta}(\Gamma_j) \cap W^{1-1/p,p}_{\gamma,\delta}(\Gamma_j)\big)^\ell.$$

The same is true for the operator $\big(L_\varepsilon, N_\varepsilon\big)$ if ε is small, because then the norm of the operator $\big(L^\circ(0, D_x) - L_\varepsilon, N^\circ(0, D_x) - N_\varepsilon\big)$ is small. Consequently, $\chi u \in L^{2,p}_{\gamma_1,\delta}(\mathcal{K})^\ell$. Since $\gamma_1 \ge \beta_1 - 1 > 1 - 3/p$, we conclude that $\partial_{x_j} u \in L^{1,p}_{\gamma_1,\delta}(\mathcal{K})^\ell \subset W^{1,p}_{\gamma_1,\delta}(\mathcal{K})^\ell$. Analogously, we obtain $\partial_{x_j} u \in W^{1,p}_{\gamma,\delta}(\mathcal{G})^\ell$ for $j = 1, 2, 3$ if χ is a cut-off function with support in a neighborhood of one of the vertices $x^{(2)}, \dots, x^{(d')}$. This proves the lemma. □

In the same way, an analogous regularity result for the variational solution holds.

LEMMA 8.1.9. *Let $u \in W^{1,p}_{\beta,\delta}(\mathcal{G})^\ell$ be a solution of the problem*

$$b(u, v) = (F, v)_{\mathcal{G}} \quad \textit{for all } v \in W^{1,p'}_{-\beta,-\delta}(\mathcal{G})^\ell,$$

where $F \in \big(W^{1,p'}_{-\beta,-\delta}(\mathcal{G})^* \cap W^{1,p'}_{-\gamma,-\delta}(\mathcal{G})^*\big)^\ell$. We assume $\max(0, 1-\delta_+^{(k)}) < \delta_k + 2/p < 1$ for $k = 1, \dots, d$, $-1 < 1 - \beta_j - 3/p < 0$, $-1 < 1 - \gamma_j - 3/p < \Lambda_j$ and $\gamma_j \ge \beta_j - 1$ for $j = 1, \dots, d'$. Then $\partial_{x_j} u \in W^{0,p}_{\gamma,\delta}(\mathcal{G})^\ell$ for $j = 1, 2, 3$.

Proof. Let χ be a smooth cut-off function equal to one in a neighborhood of the vertex $x^{(1)}$ with sufficiently small support. We may assume again without loss of generality that $\mathcal{G}$ coincides with the cone $\mathcal{K}$ in a neighborhood of the vertex. In order to show that $\partial_{x_j}(\chi u) \in W^{0,p}_{\gamma,\delta}(\mathcal{G})^\ell$ for $j = 1, 2, 3$, we consider the sesquilinear form

$$b_\varepsilon(u,v) = \int_{\mathcal{K}} \Big(\sum_{j,k=1}^{3} A^{(\varepsilon)}_{j,k}(x)\, \partial_{x_j} u \cdot \partial_{x_k} \bar{v} + \sum_{j=1}^{3} A^{(\varepsilon)}_{j}(x)\, \partial_{x_j} u \cdot \bar{v} + A^{(\varepsilon)}_0(x)\, u \cdot \bar{v} \Big)\, dx,$$

where

$$A^{(\varepsilon)}_{j,k}(x) = \zeta_\varepsilon\, A_{j,k}(x) + (1-\zeta_\varepsilon)\, A_{j,k}(0) \quad \text{for } j,k = 1,2,3,$$
$$A^{(\varepsilon)}_{j}(x) = \zeta_\varepsilon\, A_j(x) \text{ for } j = 0,1,2,3,$$

and ζ_ε is the same cut-off function as in the proof of Theorem 8.1.6. Obviously,

$$b_\varepsilon(\chi u, v) = b(\chi u, v)$$

if the support of χ is sufficiently small. By Corollary 7.7.14, the operator $u \to \Phi$ of the problem

$$b_0(u,v) = \int_{\mathcal{K}} \sum_{j,k=1}^{3} A_{j,k}(0)\, \partial_{x_j} u \cdot \partial_{x_k} \bar{v}\, dx = (\Phi, v)_{\mathcal{K}} \quad \text{for all } v \in W^{1,p'}_{-\beta_1,-\delta}(\mathcal{K})^\ell$$

is an isomorphism

$$W^{1,p}_{\beta_1,\delta}(\mathcal{K})^\ell \cap L^{1,p}_{\gamma_1,\delta}(\mathcal{K})^\ell \to \big(W^{1,p'}_{-\beta_1,-\delta}(\mathcal{K})^*\big)^\ell \cap \big(W^{1,p'}_{-\gamma_1,-\delta}(\mathcal{K})^*\big)^\ell.$$

For sufficiently small ε, the same is true for the operator of the problem

$$b_\varepsilon(u,v) = (\Phi, v)_{\mathcal{K}} \quad \text{for all } v \in W^{1,p'}_{-\beta_1,-\delta}(\mathcal{K})^\ell.$$

Thus, $\chi u \in W^{1,p}_{\beta_1,\delta}(\mathcal{K})^\ell \cap L^{1,p}_{\gamma_1,\delta}(\mathcal{K})^\ell$. From this we conclude that $\partial_{x_j}(\chi u) \in W^{0,p}_{\gamma,\delta}(\mathcal{G})^\ell$ for $j = 1,2,3$. This is also true if χ is a smooth cut-off function equal to one in a neighborhood of one of the vertices $x^{(2)}, \dots, x^{(d')}$. The lemma is proved. □

Combining the last two lemmas with Theorems 8.1.6 and 8.1.7, we get the following regularity results for the variational solution $u \in W^{1,2}(\mathcal{G})^\ell$.

THEOREM 8.1.10. *Let $u \in W^{1,2}(\mathcal{G})^\ell$ be a solution of the problem*

$$b(u,v) = (F,v) \quad \text{for all } v \in W^{1,2}(\mathcal{G})^\ell.$$

We assume that the strip $-1 \le \operatorname{Re}\lambda \le 0$ contains only the eigenvalues $\lambda = 0$ and $\lambda = -1$ of the pencils $\mathfrak{A}_j(\lambda)$ and that the eigenvalue $\lambda = 0$ has geometric and algebraic multiplicity ℓ.

1) *If $F \in \big(W^{1,2}(\mathcal{G})^* \cap W^{1,p'}_{-\beta,-\delta}(\mathcal{G})^*\big)^\ell$, where $\max(0, 1-\delta_+^{(k)}) < \delta_k + 2/p < 1$ and $-1 < 1 - \beta_j - 3/p < \min(1, \Lambda_j)$, then $\partial_{x_j} u \in W^{0,p}_{\beta,\delta}(\mathcal{G})^\ell$ for $j = 1,2,3$.*

2) *Suppose that the functional $F \in \big(W^{1,2}(\mathcal{G})^*\big)^\ell$ has the form*

$$(F,v)_{\mathcal{G}} = \int_{\mathcal{G}} f \cdot \bar{v}\, dx + \sum_{j=1}^{N} \int_{\Gamma_j} g_j \cdot \bar{v}\, dx,$$

where $f \in W^{0,p}_{\beta,\delta}(\mathcal{G})^\ell$, $g_j \in W^{1-1/p,p}_{\beta,\delta}(\Gamma_j)^\ell$. If $\max(0, 2-\delta_+^{(k)}) < \delta_k + 2/p < 2$ and $-1 < 2 - \beta_j - 3/p < \min(1, \Lambda_j)$, then $\partial_{x_j} u \in W^{1,p}_{\beta,\delta}(\mathcal{G})^\ell$ for $j = 1,2,3$.

P r o o f. We prove the assertion 1). Let $\gamma = (\gamma_1, \dots, \gamma_{d'})$ be such that

$$-1 < 1 - \gamma_j - 3/p < 0 \quad \text{and} \quad \gamma_j - 1 \le \beta_j \le \gamma_j$$

for $j = 1, \dots, d'$. Then $\big(W^{1,p'}_{-\beta,-\delta}(\mathcal{G})^*\big)^\ell \subset \big(W^{1,p'}_{-\gamma,-\delta}(\mathcal{G})^*\big)^\ell$ and from Theorem 8.1.6 it follows that $u \in W^{1,p}_{\gamma,\delta}(\mathcal{G})^\ell$. Applying Lemma 8.1.9, we conclude that $\partial_{x_j} u \in W^{0,p}_{\beta,\delta}(\mathcal{G})^\ell$ for $j = 1, 2, 3$. Analogously, we obtain the second assertion by means of Theorem 8.1.7 and Lemma 8.1.8. □

REMARK 8.1.11. The condition on δ_k in Theorem 8.1.7 can be weakened for the Neumann problem if the assumptions of Theorem 6.5.9 are valid for the model problem (8.1.7), $\xi \in M_k$. Then it is sufficient that δ'_k satisfies the inequality

$$\max(0, l - \mu_+^{(k)}) < \delta'_k + 2/q < l,$$

where $\mu_+^{(k)}$ is the greatest real number such that the strip $0 < \operatorname{Re}\lambda < \mu_+^{(k)}$ contains at most the eigenvalue $\lambda = 1$ of the pencil $A_\xi(\lambda)$ for $\xi \in M_k$. However, the boundary data must satisfy additional compatibility conditions on the edge M_k if $\delta_k + 2/p \le l - 1$ (cf. Theorem 7.7.15). Analogously, the assertion of Theorem 8.1.10, item 2), holds if

$$\max(0, 2 - \mu_+^{(k)}) < \delta_k + 2/p < 2, \quad -1 < 2 - \beta_j - 3/p < \min(1, \Lambda_j)$$

and the assumptions of Theorem 6.5.9 are valid for the model problem (8.1.7), $\xi \in M_k$.

8.2. Solvability of the boundary value problem in weighted Hölder spaces

Let $\mathcal{G}$ be the same domain of polyhedral type as in foregoing section. We consider again the boundary value problem (8.1.1), (8.1.2). Using the results of Section 7.8, we prove now that the operator of this boundary value problem is Fredholm in weighted Hölder spaces under certain conditions on the weight parameters β and δ. As in the preceding section, we construct a regularizer for the operator of the boundary value problem. Furthermore, we obtain regularity assertions for the solutions in weighted Hölder spaces.

8.2.1. Weighted Hölder spaces. In the following, l is an nonnegative integer, σ is a real number, $0 < \sigma < 1$, and $\beta = (\beta_1, \dots, \beta_{d'})$, $\delta = (\delta_1, \dots, \delta_d)$ are tuples of real numbers, $\delta_k \ge 0$ for $k = 1, \dots, d$. We define the weighted Hölder space $C^{l,\sigma}_{\beta,\delta}(\mathcal{G})$ as follows. Let $\mathcal{U}_1, \dots, \mathcal{U}_{d'}$ be domains in $\mathbb{R}^3$ such that

$$\mathcal{U}_1 \cup \dots \cup \mathcal{U}_{d'} \supset \overline{\mathcal{G}} \quad \text{and} \quad \overline{\mathcal{U}}_j \cap \overline{M}_k = \emptyset \ \text{ if } k \notin X_j.$$

Here again X_j denotes the set of the indices k such that $x^{(j)}$ is an end point of the edge M_k. Furthermore, let

$$\mathcal{G}_j = \mathcal{G} \cap \mathcal{U}_j\,, \quad \mathcal{G}_{j,k} = \{x \in \mathcal{G}_j : \, r_k(x) < 3r(x)/2\},$$

and $m_k = [\delta_k - \sigma] + 1$. Then $C^{l,\sigma}_{\beta,\delta}(\mathcal{G})$ is defined as the set of all l times continuously differentiable functions on $\overline{\mathcal{G}}\backslash\mathcal{S}$ with finite norm

$$\|u\|_{C^{l,\sigma}_{\beta,\delta}(\mathcal{G})} = \sum_{j=1}^{d'} \sum_{|\alpha|\le l} \sup_{x\in\mathcal{G}_j} \rho_j(x)^{\beta_j - l - \sigma + |\alpha|} \prod_{k\in X_j} \Big(\frac{r_k(x)}{\rho_j(x)}\Big)^{\max(0,\delta_k - l - \sigma + |\alpha|)} \big|\partial_x^\alpha u(x)\big|$$
$$+ \sum_{j=1}^{d'} \sum_{\substack{k\in X_j\\ m_k\le l}} \sum_{|\alpha|=l-m_k} \sup_{\substack{x,y\in\mathcal{G}_{j,k}\\ |x-y|<\rho_j(x)/2}} \rho_j(x)^{\beta_j-\delta_k} \frac{\big|\partial_x^\alpha u(x) - \partial_y^\alpha u(y)\big|}{|x-y|^{m_k+\sigma-\delta_k}}$$
$$+ \sum_{j=1}^{d'} \sum_{|\alpha|=l} \sup_{\substack{x,y\in\mathcal{G}_j\\ |x-y|<r(x)/2}} \rho_j(x)^{\beta_j} \prod_{k\in X_j} \Big(\frac{r_k(x)}{\rho_j(x)}\Big)^{\delta_k} \frac{\big|\partial_x^\alpha u(x) - \partial_y^\alpha u(y)\big|}{|x-y|^\sigma}.$$

Analogously, we define the Hölder spaces $C^{l,\sigma}_{\beta,\delta}(\Gamma_j)$ on the faces of the domain $\mathcal{G}$.

LEMMA 8.2.1. *Let $l+\sigma \ge l'+\sigma'$, $\beta_j - l - \sigma \le \beta'_j - l' - \sigma'$ for $j = 1,\dots,d'$, $\delta_k \ge 0$, $\delta'_k \ge 0$ and $\delta_k - l - \sigma \le \delta'_k - l' - \sigma'$ for $k = 1,\dots,d$. Then $C^{l,\sigma}_{\beta,\delta}(\mathcal{G})$ is continuously imbedded in $C^{l',\sigma'}_{\beta',\delta'}(\mathcal{G})$. If $l+\sigma > l'+\sigma'$, $\beta_j - l - \sigma < \beta'_j - l' - \sigma'$ for $j = 1,\dots,d'$ and $\delta_k - l - \sigma < \delta'_k - l' - \sigma'$ for $k = 1,\dots,d$, then the imbedding is even compact. The same assertions are true for the spaces $C^{l,\sigma}_{\beta,\delta}(\Gamma_j)$.*

P r o o f. The continuity of the imbedding $C^{l,\sigma}_{\beta,\delta}(\mathcal{G}) \subset C^{l,\sigma}_{\beta',\delta'}(\mathcal{G})$ can be easily deduced from Lemmas 6.7.8 and 7.8.1. The proof of the compactness is essentially the same as for Lemma 8.1.2. □

8.2.2. Normal Solvability of the boundary value problem. For every $k = 1,\dots,d$ let again Γ_{k_+} and Γ_{k_-} be the faces adjacent to the edge M_k. Suppose that $l \ge 2$,

$$\delta_k \ge 0, \quad l - \delta_+^{(k)} < \delta_k - \sigma < l, \quad \text{and} \quad \delta_k - \sigma \text{ is not integer} \tag{8.2.1}$$

for $k = 1,\dots,d$. Then we define $\mathcal{C}_{l,\sigma,\beta,\delta}$ as the space of all

$$g = (g_1,\dots,g_N) \in \prod_{j=1}^N C^{l-d_j,\sigma}_{\beta,\delta}(\Gamma_j)^\ell$$

satisfying the compatibility condition (8.1.12). Furthermore, we denote the operator

$$C^{l,\sigma}_{\beta,\delta}(\mathcal{G})^\ell \ni u \to \big(Lu, B_1u|_{\Gamma_1},\dots,B_Nu|_{\Gamma_N}\big) \in C^{l-2,\sigma}_{\beta,\delta}(\mathcal{G})^\ell \times \mathcal{C}_{l,\sigma,\beta,\delta} \tag{8.2.2}$$

of the boundary value problem (8.1.1), (8.1.2) by $\mathcal{A}$. We prove that this operator is Fredholm if the line $\mathrm{Re}\,\lambda = l+\sigma-\beta_j$ is free of eigenvalues of the pencil $\mathfrak{A}_j(\lambda)$ for $j = 1,\dots,d'$ and the components of δ satisfy the inequalities (8.2.1). As in the foregoing section, we construct a regularizer for $\mathcal{A}$. We start with the construction of a local regularizer.

LEMMA 8.2.2. *Let $\mathcal{U}$ be a sufficiently small subdomain of $\mathcal{G}$ and let φ be a smooth function with support in $\overline{\mathcal{U}}$. Suppose that there are no eigenvalues of the pencil $\mathfrak{A}_j(\lambda)$ on the line $\mathrm{Re}\,\lambda = l+\sigma-\beta_j$ for $j = 1,\dots,d'$ and that the components of δ satisfy the condition (8.2.1). Then there exists an operator $\mathcal{R}$ continuously* *mapping the space of all* $(f,g) \in C^{l-2,\sigma}_{\beta,\delta}(\mathcal{G})^\ell \times \mathcal{C}_{l,\sigma,\beta,\delta}$ *with support in* $\overline{\mathcal{U}}$ *into the space* $W^{l,p}_{\beta,\delta}(\mathcal{G})^\ell$ *such that*

$$\varphi \mathcal{A}\mathcal{R}(f,g) = \varphi(f,g)$$

for all $(f,g) \in C^{l-2,\sigma}_{\beta,\delta}(\mathcal{G})^\ell \times \mathcal{C}_{l,\sigma,\beta,\delta}$, $\operatorname{supp}(f,g) \in \overline{\mathcal{U}}$, *and*

$$\varphi \mathcal{R}\mathcal{A} u = \varphi u$$

for all $u \in C^{l,\sigma}_{\beta,\delta}(\mathcal{G})^\ell$, $\operatorname{supp} u \subset \overline{\mathcal{U}}$.

Suppose that $\beta'_j < \beta_j + 1$ *for* $j = 1, \dots, d'$, *the strip* $l + \sigma - \beta_j \le \operatorname{Re}\lambda \le l + 1 + \sigma - \beta'_j$ *does not contain eigenvalues of the pencil* $\mathfrak{A}_j(\lambda)$ *for* $j = 1, \dots, d'$, *and that the components of* δ' *are such that* $\delta'_k \ge 0$, $\delta'_k - \sigma$ *is not an integer and* $l + 1 - \delta^{(k)}_+ < \delta'_k - \sigma < \delta_k + 1 - \sigma$ *for* $k = 1, \dots, d$. *Then the operator* $\mathcal{R}$ *realizes also a continuous mapping from the subspace of all* $(f,g) \in C^{l-1,\sigma}_{\beta',\delta'}(\mathcal{G})^\ell \times \mathcal{C}_{l+1,\sigma,\beta'\delta'}$ *with support in* $\overline{\mathcal{U}}$ *into the space* $C^{l+1,\sigma}_{\beta',\delta'}(\mathcal{G})^\ell$.

P r o o f. The proof is in essence the same as for Lemma 8.1.4. We restrict ourselves here to the case where $\overline{\mathcal{U}}$ contains the vertex $x^{(1)}$ of $\mathcal{G}$. Then there exists a diffeomorphism κ mapping $x^{(1)}$ onto the origin and $\mathcal{U}$ onto a subset $\mathcal{V}$ of a cone $\mathcal{K}$ with vertex at the origin and faces Γ°_j, $j \in J_1$. Applying the coordinate transformation $y = \kappa(x)$ to the problem (8.1.1), (8.1.2), we obtain the equations

$$\tilde{L}\tilde{u} = \tilde{f} \ \text{ in } \mathcal{V}, \quad \tilde{B}_j \tilde{u} = \tilde{g}_j \ \text{ on } \kappa(\Gamma_j) \cap \overline{\mathcal{V}}, \quad j = 1, \dots, N,$$

where $\tilde{u}(x) = u(\kappa^{-1}x)$ for $x \in \mathcal{V}$. Let $\tilde{L}^\circ$ and $\tilde{B}^\circ_j$ denote the principal parts of $\tilde{L}$ and $\tilde{B}_j$, respectively, with coefficients frozen at $x = 0$. By Theorems 7.8.12 and 7.8.14, the operator

$$\tilde{\mathcal{A}}_0 = \big(\tilde{L}^\circ, \{\tilde{B}^\circ_j\}_{j \in J_1}\big)$$

realizes an isomorphism from $C^{l,\sigma}_{\beta_1,\delta}(\mathcal{K})^\ell$ onto the space of all

$$(f,g) \in C^{l-2,\sigma}_{\beta_1,\delta}(\mathcal{K})^\ell \times \prod_{j \in J_1} C^{l-d_j,\sigma}_{\beta_1,\delta}(\Gamma^\circ_j)^\ell \tag{8.2.3}$$

satisfying the compatibility condition (7.3.5) of Theorem 7.8.12. Let ε be a sufficiently small positive real number, and let $\tilde{L}_\varepsilon$ and $\tilde{B}_{j,\varepsilon}$ be the same differential operators as in the proof of Lemma 8.1.4. Since the difference of the operators $\tilde{\mathcal{A}}_0$ and

$$\tilde{\mathcal{A}}_\varepsilon = \big(\tilde{L}_\varepsilon, \{\tilde{B}_{j,\varepsilon}\}_{j \in J_1}\big)$$

is small in the operator norm

$$C^{l,\sigma}_{\beta_1,\delta}(\mathcal{K})^\ell \to C^{l-2,\sigma}_{\beta_1,\delta}(\mathcal{K})^\ell \times \prod_{j \in J_1} C^{l-d_j,\sigma}_{\beta_1,\delta}(\Gamma^\circ_j)^\ell,$$

the operator $\tilde{\mathcal{A}}_\varepsilon$ realizes also an isomorphism from $C^{l,\sigma}_{\beta_1,\delta}(\mathcal{K})^\ell$ onto the space of all vector functions (8.2.3) satisfying the compatibility condition of Theorem 7.8.12. We assume that $\mathcal{V}$ is contained in the ball $|x| < \varepsilon$. Then the coefficients of $\tilde{L}_\varepsilon$, $\tilde{B}_\varepsilon$ coincide with those of $\tilde{L}$ and $\tilde{B}$ on the support of $\tilde{f}$ and $\tilde{g}$, respectively. Let $\tilde{f}$ and $\tilde{g}$ be extended by zero outside $\mathcal{U}$. Then we define

$$u(x) = \tilde{u}(\kappa(x)) \ \text{ for } x \in \mathcal{U}, \quad \text{where } \tilde{u} = \tilde{\mathcal{A}}_\varepsilon^{-1}(\tilde{f}, \tilde{g}).$$

This function can be continuously extended to a function in $u \in C^{l,\sigma}_{\beta,\delta}(\mathcal{G})^\ell$. The so defined mapping $(f,g) \to u$ is denoted by $\mathcal{R}$ and has the desired properties. $\square$

THEOREM 8.2.3. *Suppose that the line* $\operatorname{Re}\lambda = l + \sigma - \beta_j$ *does not contain eigenvalues of the pencil* $\mathfrak{A}_j(\lambda)$ *for* $j = 1, \dots, d'$ *and that the components of* δ *satisfy the condition* (8.2.1). *Then the operator* (8.2.2) *is Fredholm for* $l \ge 2$.

Proof. Let the operator $\mathcal{R}$ be defined by

$$\mathcal{R}(f,g) = \sum_\nu \phi_\nu \mathcal{R}_\nu \psi_\nu(f,g),$$

where ϕ_ν, ψ_ν are the same cut-off functions as in the proof of Theorem 8.1.5 and $\mathcal{R}_\nu$ is the local regularizer of Lemma 8.2.2 for the subdomain $\mathcal{U} = \mathcal{U}_\nu \cap \mathcal{G}$. Repeating the proof of Theorem 8.1.5, we conclude from Lemma 8.2.2 that $\mathcal{R}\mathcal{A} - I$ is compact on $C^{l,\sigma}_{\beta,\delta}(\mathcal{G})^\ell$ and $\mathcal{A}\mathcal{R} - I$ is compact on $C^{l-2,\sigma}_{\beta,\delta}(\mathcal{G})^\ell \times \mathcal{C}_{l,\sigma,\beta,\delta}$. This means that $\mathcal{R}$ is a regularizer for the operator $\mathcal{A}$. The result follows. □

8.2.3. Regularity assertions for weak solutions. The following theorem can be proved in the same way as Theorems 8.1.6 and 8.1.7 if one applies Theorem 7.8.15 instead of Theorems 7.7.5 and 7.7.7, respectively.

THEOREM 8.2.4. *Let* $u \in W^{1,p}_{\beta,\delta}(\mathcal{G})^\ell$ *be a solution of the problem* (8.1.19), (8.1.20). *Suppose that* $g_j \in W^{1-1/p,p}_{\beta,\delta}(\Gamma_j)^\ell \cap C^{l,\sigma}_{\beta',\delta'}(\Gamma_j)^\ell$ *for* $j \in I_0$ *and that* F *is a functional on* $\mathcal{H}_{p',-\beta,-\delta}$ *which has the form*

$$(F,v)_{\mathcal{G}} = \int_{\mathcal{G}} f \cdot \bar{v}\, dx + \sum_{j \in I_1} \int_{\Gamma_j} g_j \cdot \bar{v}\, dx$$

for all $v \in C_0^\infty(\overline{\mathcal{G}} \backslash \{x^{(1)}, \dots, x^{(d')}\})^\ell$, $v|_{\Gamma_j} = 0$ *for* $j \in I_0$, *where* $f \in C^{l-2,\sigma}_{\beta',\delta'}(\mathcal{G})^\ell$ *and* $g_j \in C^{l-1,\sigma}_{\beta',\delta'}(\Gamma_j)^\ell$. *Furthermore, we assume that the closed strip between the lines* $\operatorname{Re}\lambda \le 1 - \beta_j - 3/p$ *and* $\operatorname{Re}\lambda \le l + \sigma - \beta'_j$ *is free of eigenvalues of the pencil* $\mathfrak{A}_j(\lambda)$ *for* $j = 1, \dots, d'$, *that the components of* δ *and* δ' *satisfy the inequalities*

$$\max(0, 1 - \delta_+^{(k)}) < \delta_k + 2/p < 1, \quad \delta'_k \ge 0, \ l - \delta_+^{(k)} < \delta'_k - \sigma < l,$$

and that $\delta'_k - \sigma$ *is not integer for* $k = 1, \dots, d$. *Then* $u \in C^{l,\sigma}_{\beta',\delta'}(\mathcal{G})^\ell$. *The assertion of the theorem is also true if* $p = 2$ *and* $\delta = 0$.

REMARK 8.2.5. The condition on δ'_k in Theorem 8.2.4 can be weakened if the assumptions of Theorem 6.5.9 are valid for the model problem (8.1.7), $\xi \in M_k$. Then it is sufficient that δ'_k satisfies the inequalities

$$\delta'_k \ge 0, \quad l - \mu_+^{(k)} < \delta'_k - \sigma < l,$$

where $\mu_+^{(k)}$ is the greatest real number such that the strip $0 < \operatorname{Re}\lambda < \mu_+^{(k)}$ contains at most the eigenvalue $\lambda = 1$ of the pencil $A_\xi(\lambda)$ for $\xi \in M_k$. However, if $\delta'_k - \sigma < l - 1$, then it is necessary that the Neumann data satisfy the additional compatibility condition of Theorem 7.8.16 on the edge M_k.

8.3. Examples

In the last section of this chapter, we discuss some boundary value problems for the Laplace equation and for the Lamé system in domains with smooth nonintersecting edges and in polyhedral domains. We establish regularity results for the variational solutions in nonweighted Sobolev and Hölder spaces. Examples for the Dirichlet problem can be also found at the end of Chapter 4.

8.3.1. Boundary value problems for the Laplace equation in domains with nonintersecting edges. Let $\mathcal{G}$ be a bounded domain in $\mathbb{R}^3$ with boundary $\partial\mathcal{G}$ consisting of smooth (of class C^∞) 2-dimensional manifolds Γ_j, $j = 1,\dots,d+1$, and smooth closed curves M_k (the edges), $k = 1,\dots,d$. We assume that for every $\xi \in M_k$ there exist a neighborhood $\mathcal{U}_\xi$ and a diffeomorphism (a C^∞ mapping) κ_ξ which maps $\mathcal{G} \cap \mathcal{U}_\xi$ onto $\mathcal{D}_\xi \cap B_1$, where $\mathcal{D}_\xi$ is a dihedron and B_1 is the unit ball.

Let I_0, I_1 be disjoint sets of natural numbers such that $I_0 \cup I_1 = \{1,\dots,d+1\}$. We set $d_j = 0$ for $j \in I_0$, $d_j = 1$ for $j \in I_1$, and consider the boundary value problem

$$-\Delta u = f \quad \text{in } \mathcal{G}, \tag{8.3.1}$$

$$(1-d_j)\, u + d_j \frac{\partial u}{\partial n} = 0 \quad \text{on } \Gamma_j\,,\ j = 1,\dots,d+1. \tag{8.3.2}$$

Let $\xi \in M_k$ be an arbitrary edge point, and let Γ_{k_+}, Γ_{k_-} be the adjoining faces to the edge M_k. The eigenvalues of the operator pencil $A_\xi(\lambda)$ (cf. Subsection 8.1.4) are the numbers

$$j\pi/\theta_\xi \ \text{ if } d_{k_+} = d_{k_-}\,, \qquad (j+1/2)\pi/\theta_\xi \ \text{ if } d_{k_+} \neq d_{k_-}\,.$$

Here θ_ξ denotes the inner angle at the edge point ξ, and j is an arbitrary integer, $j \neq 0$ if $d_{k_+} = d_{k_-} = 0$. Consequently, the numbers $\delta_\pm^{(k)}$ introduced in Subsection 8.1.4 are

$$\delta_\pm^{(k)} = \frac{\pi}{\theta_k} \ \text{ if } d_{k_+} = d_{k_-}\,, \qquad \delta_\pm^{(k)} = \frac{\pi}{2\theta_k} \ \text{ if } d_{k_+} \neq d_{k_-}\,,$$

where

$$\theta_k = \max_{\xi \in M_k} \theta_\xi\,.$$

For arbitrary integer $l \geq 0$, real $p \in (1,\infty)$ and $\delta = (\delta_1,\dots,\delta_d) \in \mathbb{R}^d$, $\delta_k > -2/p$, we denote by $W_\delta^{l,p}(\mathcal{G})$ the weighted Sobolev space with the norm

$$\|u\|_{W_\delta^{l,p}(\mathcal{G})} = \Big(\int_{\mathcal{G}} \prod_{k=1}^d r_k(x)^{p\delta_k} \sum_{|\alpha|\leq l} |\partial_x^\alpha u(x)|^p \, dx \Big)^{1/p}.$$

Let $m_k = [\delta_k - \sigma] + 1$, where $[s]$ denotes the integral part of s. Furthermore, let $\mathcal{U}_k$ be a neighborhood of the edge M_k having a positive distance from the edges M_j, $j \neq k$. Then $C_\delta^{l,\sigma}(\mathcal{G})$ is the weighted Hölder space with the norm

$$\begin{aligned} \|u\|_{C_\delta^{l,\sigma}(\mathcal{G})} &= \sum_{|\alpha|\leq l} \sup_{x\in\mathcal{G}} \prod_{k=1}^d r_k(x)^{\max(0,\delta_k-l-\sigma+|\alpha|)} \big|\partial_x^\alpha u(x)\big| \\ &\quad + \sum_{k:\, m_k\leq l} \ \sum_{|\alpha|=l-m_k} \ \sup_{x,y\in\mathcal{G}\cap\mathcal{U}_k} \frac{\big|\partial_x^\alpha u(x) - \partial_y^\alpha u(y)\big|}{|x-y|^{m_k+\sigma-\delta_k}} \\ &\quad + \sum_{|\alpha|=l} \ \sup_{\substack{x,y\in\mathcal{G} \\ |x-y|<r(x)/2}} \ \prod_{k=1}^d r_k(x)^{\delta_k} \frac{\big|\partial_x^\alpha u(x) - \partial_y^\alpha u(y)\big|}{|x-y|^\sigma}\,. \end{aligned}$$

It is evident that the space $C_\delta^{l,\sigma}(\mathcal{G})$ does not depend on the choice of the neighborhoods $\mathcal{U}_k$.

Theorems 8.1.6, 8.1.7 and 8.2.4 imply the following result.

THEOREM 8.3.1. *Let $u \in W^{1,2}(\mathcal{G})$ be a solution of the problem* (8.3.1), (8.3.2).

1) *Suppose that $f \in W_\delta^{l-2,p}(\mathcal{G})$ for $l \ge 2$ and $f \in W_{-\delta}^{1,p'}(\mathcal{G})^*$ for $l = 1$, where $p' = p/(p-1)$ and the components δ_k of δ satisfy the inequalities*

$$\max(0, l - \delta_+^{(k)}) < \delta_k + 2/p < l$$

for $k = 1, \dots, d$. Then $u \in W_\delta^{l,p}(\mathcal{G})$.

2) *If $f \in C_\delta^{l-2,\sigma}(\mathcal{G})$, $l \ge 2$, and the components of δ satisfy the conditions*

$$\delta_k \ge 0, \quad l - \delta_+^{(k)} < \delta_k - \sigma < l, \quad \delta_k - \sigma \text{ is not integer}$$

for $k = 1, \dots, d$, then $u \in C_\delta^{l,\sigma}(\mathcal{G})$.

For $\delta = 0$, we obtain regularity results in nonweighted Sobolev and Hölder spaces. We denote by

$$\theta = \max(\theta_1, \dots, \theta_d)$$

the maximal angle at the edges of the domain $\mathcal{G}$. Then the following assertions hold.

COROLLARY 8.3.2. *Let $u \in W^{1,2}(\mathcal{G})$ be a variational solution of the mixed boundary value problem* (8.3.1), (8.3.2).

1) *If $f \in W^{1,p'}(\mathcal{G})^*$, $p' = p/(p-1) < 2$ and $2/p > 1 - \pi/(2\theta)$, then $u \in W^{1,p}(\mathcal{G})$.*

2) *If $f \in W^{l-2,p}(\mathcal{G})$, $l \ge 2$, $p > 1$ and $2/p > l - \pi/(2\theta)$, then $u \in W^{l,p}(\mathcal{G})$.*

3) *If $f \in C^{l-2,\sigma}(\mathcal{G})$, where $2 < l + \sigma < \pi/(2\theta)$, then $u \in C^{l,\sigma}(\mathcal{G})$.*

The last results can be improved for the Dirichlet and Neumann problems. Then the next corollary holds as a consequence of Theorem 8.3.1.

COROLLARY 8.3.3. *Let $u \in W^{1,2}(\mathcal{G})$ be a solution of the boundary value problem* (8.3.1), (8.3.2), *where $d_j = 0$ for all j or $d_j = 1$ for all j.*

1) *If $f \in W^{1,p'}(\mathcal{G})^*$, $p' = p/(p-1)$, $p > 2$ and $2/p > 1 - \pi/\theta$, then $u \in W^{1,p}(\mathcal{G})$.*

2) *If $f \in W^{l-2,p}(\mathcal{G})$, $l \ge 2$, $p > 1$ and $2/p > l - \pi/\theta$, then $u \in W^{l,p}(\mathcal{G})$.*

3) *If $f \in C^{l-2,\sigma}(\mathcal{G})$, where $2 < l + \sigma < \pi/\theta$, then $u \in C^{l,\sigma}(\mathcal{G})$.*

8.3.2. The Neumann problem for the Lamé system in domains with nonintersecting edges. Let $\mathcal{G}$ be the same domain as in Subsection 8.3.1. Now we consider the boundary value problem

$$(8.3.3) \quad -\mu\Big(\Delta u + \frac{1}{1-2\nu}\nabla\nabla\cdot u\Big) = f \text{ in } \mathcal{G}, \quad \sigma(u)\,n = 0 \text{ on } \Gamma_j,\ j = 1, \dots, d+1.$$

We assume that $\theta_\xi \ne \pi$ for all $\xi \in \mathcal{S} = M_1 \cup \cdots \cup M_d$ and define $\mu_+^{(k)}$ as the greatest real number such that the strip $0 < \operatorname{Re}\lambda < \mu_+^{(k)}$ is free of eigenvalues of the pencil $A_\xi(\lambda)$ (cf. Subsection 8.1.4) for each $\xi \in M_k$. The numbers $\mu_+^{(k)}$ are given by the equalities (7.4.11), i.e.

$$(8.3.4) \qquad \mu_+^{(k)} = \frac{\pi}{\theta_k} \text{ if } \theta_k < \pi, \qquad \mu_+^{(k)} = \frac{\xi_+(\theta_k)}{\theta_k} \text{ if } \theta_k > \pi,$$

where $\xi_+(\theta)$ is the smallest positive solution of the equation

$$\frac{\sin\xi}{\xi} + \frac{\sin\theta}{\theta} = 0.$$

Note that $1/2 < \mu_+^{(k)} < 1$ for $\theta_k > \pi$ (cf. [**85**, Theorem 4.2.2]).

Theorems 8.1.6, 8.1.7 and 8.2.4 (see also Remarks 8.1.11 and 8.2.5) imply the following theorem.

THEOREM 8.3.4. *Let $u \in W^{1,2}(\mathcal{G})^3$ be a solution of the problem* (8.3.3).

1) *Suppose that $f \in W_\delta^{l-2,p}(\mathcal{G})^3$ for $l \ge 2$ and $f \in (W_{-\delta}^{1,p'}(\mathcal{G})^*)^3$ for $l = 1$, where $p' = p/(p-1)$ and the components δ_k of δ satisfy the inequalities*

$$\max(0, l - \mu_+^{(k)}) < \delta_k + 2/p < l$$

for $k = 1, \dots, d$. Then $u \in W_\delta^{l,p}(\mathcal{G})^3$.

2) *If $f \in C_\delta^{l-2,\sigma}(\mathcal{G})^3$, $l \ge 2$, and the components of δ satisfy the conditions*

$$\delta_k \ge 0, \quad l - \mu_+^{(k)} < \delta_k - \sigma < l, \quad \delta_k - \sigma \text{ is not integer}$$

for $k = 1, \dots, d$, then $u \in C_\delta^{l,\sigma}(\mathcal{G})^3$.

In the particular case $\delta_k = 0$, we obtain analogous regularity assertions in nonweighted Sobolev and Hölder spaces. We state here only two results for the case $\theta < \pi$, where $\theta = \max(\theta_1, \dots, \theta_d)$.

COROLLARY 8.3.5. *Let $u \in W^{1,2}(\mathcal{G})^3$ be a variational solution of the problem* (8.3.3), *and let $\theta < \pi$.*

1) *If $f \in (W^{1,p'}(\mathcal{G})^*)^3$, $p' = p/(p-1)$, $2 < p < \infty$, then $u \in W^{1,p}(\mathcal{G})^3$.*

2) *If $f \in L_p(\mathcal{G})$, $1 < p < \infty$ for $\theta \le \pi/2$ and $1 < p < 2\theta/(2\theta - \pi)$ for $\theta > \pi/2$, then $u \in W^{2,p}(\mathcal{G})$.*

Suppose that the angle at every edge point is less than $\pi/2$ as in the figure below.

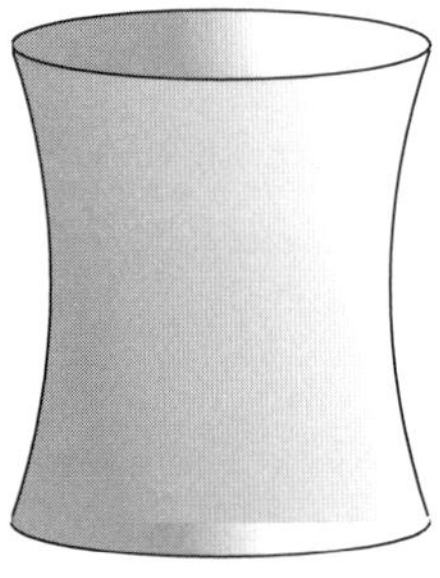

FIGURE 7. A domain for which $u \in C^{2,\sigma}(\mathcal{G})^3$

Then also the following $C^{2,\sigma}$-regularity result for the variational solution $u \in W^{1,2}(\mathcal{G})^3$ holds:

If $f \in C^{0,\sigma}(\mathcal{G})^3$, $0 < \sigma < \min(1, -2 + \pi/\theta)$, then $u \in C^{2,\sigma}(\mathcal{G})^3$.

8.3.3. The Neumann problem for the Lamé system in a polyhedron. Let $\mathcal{G}$ be a polyhedron with the faces $\Gamma_1, \dots, \Gamma_N$, the edges $M_1, \dots, M_d$ and the vertices $x^{(1)}, \dots, x^{(d')}$. The inner angle at the edge M_k is denoted by θ_k. We consider the Neumann problem for the Lamé system

$$(8.3.5) \qquad -\mu\Big(\Delta u + \frac{1}{1-2\nu}\nabla\nabla\cdot u\Big) = f \ \text{ in } \mathcal{G}, \quad \sigma(u)\, n = 0 \ \text{ on } \partial\mathcal{G}\backslash\mathcal{S},$$

where $\mathcal{S}$ denotes the set of all edge points and vertices. Let the numbers $\mu_+^{(k)}$ be defined by (8.3.4). According to [**85**, Theorem 4.3.1], the strip $-1 \le \operatorname{Re}\lambda \le 0$ contains only the eigenvalues $\lambda = 0$ and $\lambda = -1$ of the pencils $\mathfrak{A}_j(\lambda)$ generated by the boundary value problem (8.3.5) at the vertices $x^{(j)}$ (cf. Subsection 8.1.4). The eigenvectors corresponding to the eigenvalue $\lambda = 0$ are constant, generalized eigenvectors corresponding to this eigenvalue do not exist. We denote by Λ_j the greatest real number such that the strip

$$0 < \operatorname{Re}\lambda < \Lambda_j$$

is free of eigenvalues of the pencil $\mathfrak{A}_j(\lambda)$. Then the following lemma holds.

LEMMA 8.3.6. *Let $u \in W^{1,2}(\mathcal{G})^3$ be a variational solution of the problem* (8.3.5), *and let ζ_j be an infinitely differentiable cut-off function in $\overline{\mathcal{G}}$ equal to one near the vertex $x^{(j)}$ and to zero near all other vertices of $\mathcal{G}$.*

1) *Suppose that $f \in (W^{1,2}(\mathcal{G})^*)^3 \cap (W^{1,p'}_{-\beta,-\delta}(\mathcal{G})^*)^3$ if $l = 1$ and $f \in (W^{1,2}(\mathcal{G})^*)^3 \cap W^{l-2,p}_{\beta,\delta}(\mathcal{G})^3$ if $l \ge 2$. Furthermore, we assume that*

$$-1 < l - \beta_j - 3/p < \Lambda_j\,, \quad l - \beta_j - 3/p \ne 0,$$
$$\max(0, l - \mu_+^{(k)}) < \delta_k + 2/p < l \quad \text{for } k = 1, \dots, d.$$

Then there exists a constant vector c_j such that $\zeta_j(u - c_j) \in W^{l,p}_{\beta,\delta}(\mathcal{G})^3$.

2) *If $f \in (W^{1,2}(\mathcal{G})^*)^3 \cap C^{l-2,\sigma}_{\beta,\delta}(\mathcal{G})^3$, $l \ge 2$,*

$$-1 < l + \sigma - \beta_j < \Lambda_j\,, \quad l + \sigma - \beta_j \ne 0$$
$$l - \mu_+^{(k)} < \delta_k - \sigma < l, \quad \delta_k \ge 0, \quad \delta_k - \sigma \text{ is not integer} \quad \text{for } k = 1, \dots, d,$$

then there exists a constant vector c_j such that $\zeta_j(u - c_j) \in C^{l,\sigma}_{\beta,\delta}(\mathcal{G})^3$.

P r o o f. By Hardy's inequality, the Sobolev space $W^{1,2}(\mathcal{G})$ coincides with $W^{1,2}_{0,0}(\mathcal{G})$. Consequently, the assertion 1) follows immediately from Theorems 7.7.9, 7.7.15 and Corollary 7.7.10. For the second assertion, one has to apply Theorem 7.8.15. □

Next, we prove regularity assertions for the derivatives of the variational solution. Here, in contrast to Lemma 8.3.6, the assumption $l - \beta_j - 3/p \ne 0$ is not necessary.

LEMMA 8.3.7. *Let $u \in W^{1,2}(\mathcal{G})^3$ be a variational solution of the problem* (8.3.5).

1) *If $f \in (W^{1,2}(\mathcal{G})^*)^3 \cap (W^{1,p'}_{-\beta,-\delta}(\mathcal{G})^*)^3$, $-1 < 1 - \beta_j - 3/p < \Lambda_j$ for $j = 1, \dots, d'$ and $\max(0, 1 - \mu_+^{(k)}) < \delta_k + 2/p < 1$ for $k = 1, \dots, d$ then $\partial_{x_j} u \in W^{0,p}_{\beta,\delta}(\mathcal{G})^3$ for $j = 1, 2, 3$.*

2) *If $f \in (W^{1,2}(\mathcal{G})^*)^3 \cap W^{0,p}_{\beta,\delta}(\mathcal{G})^3$, $-1 < 2 - \beta_j - 3/p < \Lambda_j$ for $j = 1, \dots, d'$ and $\max(0, 2 - \mu_+^{(k)}) < \delta_k + 2/p < 2$ for $k = 1, \dots, d$ then $\partial_{x_j} u \in W^{1,p}_{\beta,\delta}(\mathcal{G})$ for $j = 1, 2, 3$.*

P r o o f. If $1 - \beta_j - 3/p \ne 0$, then the assertion 1) is a corollary of Lemma 8.3.6. For $1 - \beta_j - 3/p = 0$, this assertion can be deduced from the first part of Theorem 8.1.10 and Remark 8.1.11. For the second assertion, we refer to the second part of Theorem 8.1.10. □

In the special case $\beta = 0$ and $\delta = 0$, we obtain regularity results in weighted Sobolev spaces withought weight.

THEOREM 8.3.8. *Let $u \in W^{1,2}(\mathcal{G})^3$ be a variational solution of the boundary value problem* (8.3.5).

1) *Suppose that $f \in (W^{1,p'}(\mathcal{G})^*)^3$, $p \geq 2$, $3/p > 1 - \Lambda_j$ for $j = 1, \dots, d'$ and $2/p > 1 - \mu_+^{(k)}$ for $k = 1, \dots, d$. Then $u \in W^{1,p}(\mathcal{G})^3$.*

2) *Suppose that $f \in (L_p(\mathcal{G}))^3 \cap (W^{1,2}(\mathcal{G})^*)^3$, $3/p > 2 - \Lambda_j$ for $j = 1, \dots, d'$ and $2/p > 2 - \mu_+^{(k)}$ for $k = 1, \dots, d$. Then $u \in W^{2,p}(\mathcal{G})^3$.*

3) *If $f \in (W^{1,2}(\mathcal{G})^*)^3 \cap C_{2,2}^{0,\sigma}(\mathcal{G})^3$, $0 < \sigma < 1$, $\sigma < \Lambda_j$ for all j and $\sigma < \mu_+^{(k)}$ for all k, then $u \in C^{0,\sigma}(\mathcal{G})^3$.*

P r o o f. 1) Obviously $W_{0,0}^{1,p'}(\mathcal{G}) \subset W^{1,p'}(\mathcal{G})$ and, consequently, $(W^{1,p'}(\mathcal{G}))^* \subset (W_{0,0}^{1,p'}(\mathcal{G}))^*$. Thus, we conclude from Lemma 8.3.7 that $\partial_{x_j} u \in L_p(\mathcal{G})^3$ for $j = 1, 2, 3$. This implies $u \in W^{1,p}(\mathcal{G})^3$.

2) By the second assertion of Lemma 8.3.7, we have $\partial_{x_j} u \in W_{0,0}^{1,p}(\mathcal{G})^3 \subset W^{1,p}(\mathcal{G})^3$ for $j = 1, 2, 3$. Thus, $u \in W^{2,p}(\mathcal{G})^3$.

3) Let ζ_j be infinitely differentiable cut-off function in $\overline{\mathcal{G}}$ equal to one near the vertex $x^{(j)}$ and to zero near all other vertices of $\mathcal{G}$. By Lemma 8.3.6, there exist constant vectors c_j such that $\zeta_j(u - c_j) \in C_{2,2}^{2,\sigma}(\mathcal{G})^3$. The last space is a subspace of $C^{0,\sigma}(\mathcal{G})^3$. This proves the lemma. □

8.3.4. The Neumann problem for the Laplace equation in a polyhedron. Let $\mathcal{G}$ be the same polyhedron as in the preceding subsection. We consider the Neumann problem

$$(8.3.6) \qquad -\Delta u = f \ \text{ in } \mathcal{G}, \quad \frac{\partial u}{\partial n} = 0 \ \text{ on } \partial\mathcal{G}\backslash\mathcal{S}.$$

For this problem, we have

$$\delta_+^{(k)} = \delta_-^{(k)} = \pi/\theta_k \,.$$

Furthermore, $\mathfrak{A}_j(\lambda)$ is the operator of the Neumann problem for the operator $-\delta - \lambda(\lambda+1)$ on Ω_j (cf. [**85**, Section 2.3]). The eigenvalues of the pencils $\mathfrak{A}_j(\lambda)$ are real, the smallest eigenvalue $> -1/2$ is the eigenvalue $\lambda = 0$ with the eigenfunction $U = c = const$. Let Λ_j denote the smallest positive eigenvalue of the operator pencil $\mathfrak{A}_j(\lambda)$.

THEOREM 8.3.9. *Let $u \in W^{1,2}(\mathcal{G})$ be a variational solution of the boundary value problem* (8.3.6).

1) *Suppose that $f \in (W^{1,p'}(\mathcal{G}))^*$, $p \geq 2$, $3/p > 1 - \Lambda_j$ for $j = 1, \dots, d'$ and $2/p > 1 - \pi/\theta_k$ for $k = 1, \dots, d$. Then $u \in W^{1,p}(\mathcal{G})$.*

2) *Suppose that $f \in L_p(\mathcal{G}) \cap (W^{1,2}(\mathcal{G}))^*$, $3/p > 2 - \Lambda_j$ for $j = 1, \dots, d'$ and $2/p > 2 - \pi/\theta_k$ for $k = 1, \dots, d$. Then $u \in W^{2,p}(\mathcal{G})$.*

P r o o f. Forst note that the assertions of Lemmas 8.3.6 and 8.3.7 with $\delta_+^{(k)} = \pi/\theta_k$ instead of $\mu_+^{(k)}$ are valid for the solution of the problem (8.3.6). Repeating the proof of Theorem 8.3.8, we obtain the assertions 1) and 2). □

8.3.5. The Neumann problem for the Laplace equation in a convex polyhedron. Now let $\mathcal{G}$ be a convex polyhedron. Then $\pi/\theta_k > 1$ for all k. Furthermore, it follows from a result of ESCOBAR [**49**] that, in this case, the first positive eigenvalue of the Neumann Beltrami operator $-\delta$ on $\Omega_j = \mathcal{K}_j \cap S^2$ is greater than 2 for $j = 1, \dots, d'$. Another proof for the nonexistence of positive eigenvalues less than 2 was given by MAZ'YA [**107**]. Consequently, $\Lambda_j > 1$ if the cone $\mathcal{K}_j$ is convex. Thus, the following results hold.

THEOREM 8.3.10. *Let* $u \in W^{1,2}(\mathcal{G})$ *be a variational solution of the problem* (8.3.6) *in a convex polyhedron* $\mathcal{G}$.

1) *If* $f \in (W^{1,p'}(\mathcal{G}))^*$, $2 \le p < \infty$, *then* $u \in W^{1,p}(\mathcal{G})$.

2) *If* $f \in L_p(\mathcal{G}) \cap (W^{1,2}(\mathcal{G}))^*$, $1 < p \le 3$, $2/p > 2 - \pi/\theta_k$ *for all* k, *then* $u \in W^{2,p}(\mathcal{G})$.

3) *If* $f \in (W^{1,2}(\mathcal{G}))^* \cap C^{0,\sigma}_{1,1}(\mathcal{G})$, *where* σ *is a sufficiently small positive number (such that* $\sigma + 1 < \Lambda_j$ *and* $\sigma + 1 < \pi/\theta_k$ *for all* j *and* k*), then* $u \in C^{1,\sigma}(\mathcal{G})$.

P r o o f. The first two assertions are direct consequences of Theorem 8.3.9. If $f \in (W^{1,2}(\mathcal{G}))^* \cap C^{0,\sigma}_{1,1}(\mathcal{G})$, then analogously to item 2) of Lemma 8.3.6, we have $\zeta\,(u - c_j) \in C^{2,\sigma}_{1,1}(\mathcal{G}) \subset C^{1,\sigma}_{0,0}(\mathcal{G})$ for an arbitrary smooth cut-off function ζ equal to one near the vertex $x^{(j)}$ and to zero near the other vertices. Consequently, $u \in C^{1,\sigma}(\mathcal{G})$. □

Part 3

Mixed boundary value problems for stationary Stokes and Navier-Stokes systems in polyhedral domains

CHAPTER 9

Boundary value problem for the Stokes system in a dihedron

Let $\mathcal{D}$ be the dihedron $\{x \in \mathbb{R}^3 : x' = (x_1, x_2) \in K,\ x_3 \in \mathbb{R}\}$ with K standing for the angle $\{x' \in \mathbb{R}^2,\ r > 0,\ |\varphi| < \theta/2\}$. We deal with a mixed boundary value problem for the linear Stokes system

$$-\Delta u + \nabla p = f, \qquad -\nabla \cdot u = g$$

in $\mathcal{D}$, where one of the boundary conditions

(i) $u = h$,

(ii) $u_\tau = h, \quad -p + 2\varepsilon_{n,n}(u) = \phi$,

(iii) $u_n = h, \quad 2\varepsilon_{n,\tau}(u) = \phi$,

(iv) $-pn + 2\varepsilon_n(u) = \phi$

is given on each of the faces $\Gamma^\pm = \{x : \varphi = \pm\theta/2\}$. Here n is the outward normal, $u_n = u \cdot n$ is the normal and $u_\tau = u - u_n n$ the tangent component of the velocity u. Furthermore, $\varepsilon(u)$ denotes the matrix with the elements

$$\varepsilon_{i,j}(u) = \frac{1}{2}\Big(\frac{\partial u_j}{\partial x_i} + \frac{\partial u_i}{\partial x_j}\Big),$$

$\varepsilon_n(u)$ is the vector $\varepsilon(u)n$, $\varepsilon_{n,n} = \varepsilon_n(u) \cdot n$ its normal component and $\varepsilon_{n,\tau}(u)$ its tangent component.

The conditions (i)–(iv) are frequently used in the study of steady-state flows of incompressible viscous Newtonian fluids. For example, on solid walls, the Dirichlet condition $u = 0$ is prescribed. A no-friction condition (Neumann condition) $2\varepsilon_n(u) - pn = 0$ may be useful on an artificial boundary such as the exit of a canal or a free surface. Note that the Neumann problem naturally appears in the theory of hydrodynamic potentials (see [**89**]). The slip condition for uncovered fluid surfaces has the form (iii), and conditions for in/out-stream surfaces can be written in the form (ii).

One of the goals of this chapter is to develop a solvability and regularity theory for the boundary value problem in weighted Sobolev spaces. Here, as in Chapter 6, we use the class of the weighted spaces $W_\delta^{l,2}(\mathcal{D})$ with nonhomogeneous norms. In the case, where the Dirichlet condition (i) is prescribed on at least one of the faces Γ^+ and Γ^-, we obtain also solvability and regularity results in the spaces $V_\delta^{l,2}(\mathcal{D})$. However, the general problem with the boundary conditions (i)–(iv) requires the use of weighted spaces with nonhomogenous norms. This makes the proofs more difficult. On the other hand, in some cases (e.g. the Dirichlet problem in a convex dihedron), the results can be improved when considering solutions in weighted spaces with nonhomogeneous norms.

In Section 9.1, we prove the existence and uniqueness of solutions in the space $L^{1,2}(\mathcal{D})^3 \times L_2(\mathcal{D})$, where $L^{1,2}(\mathcal{D})$ is the closure of $C_0^\infty(\overline{\mathcal{D}})$ with respect to the norm (6.1.4). Then we assume that the functions on the right-hand sides of the Stokes system and of the boundary conditions belong to weighted Sobolev spaces. It is necessary that these functions satisfy certain compatibility conditions on the edge of the dihedron. A description of these conditions is given in Section 9.2. The following Section 9.3 deals with the corresponding boundary value problem in a plane angle which arises when considering solutions independent of the variable x_3. The main result of Section 9.4 is the following regularity assertion for the solution $(u,p) \in L^{1,2}(\mathcal{D})^3 \times L_2(\mathcal{D})$ of the boundary value in the dihedron $\mathcal{D}$:

> If $f \in W_\delta^{l-2,2}(\mathcal{D})^3$, $g \in W_\delta^{l-1,2}(\mathcal{D})$, the boundary data belong to corresponding trace spaces, and $\max(0, l-1-\delta_+) < \delta < l-1$, then $\zeta\,(u,p) \in W_\delta^{l,2}(\mathcal{D})^3 \times W_\delta^{l-1,2}(\mathcal{D})$ for an arbitrary smooth function ζ with compact support.

Here, δ_+ is the greatest real number such that the strip $0 < \operatorname{Re}\lambda < \delta_+$ is free of eigenvalues of a certain operator pencil $A(\lambda)$. For the sake of simplicity, we prove this result only for noninteger δ. However, using estimates of Green's matrix, this result can be also proved for arbitrary real δ satisfying the inequalities given above. This is done in the next chapter for the boundary value problem in a polyhedral cone. For some special boundary value problems for the Stokes system (the Dirichlet problem, the mixed problem with the boundary conditions (i) and (iii), and others), the number δ_+ in the condition on δ can be replaced by μ_+ which is the greatest real number such that the strip $0 < \operatorname{Re}\lambda < \mu_+$ contains at most the eigenvalue $\lambda = 1$ of the pencil $\mathfrak{A}(\lambda)$.

Another subject of this chapter is the Green's matrix of boundary value problems for the Stokes system. In Section 9.5, we consider the Green's matrix for the Stokes system in the half-space $x_3 > 0$, where in each case one of the boundary conditions (i)-(iv) is prescribed on the plane $x_3 = 0$. We obtain explicit representations for the elements of this matrix. The Green's matrix for the boundary value problem in the dihedron is studied in Section 9.6. Here, we prove point estimates for its elements $G_{i,j}(x,\xi)$. For example, we obtain the estimate

$$\big|G_{i,j}(x,\xi)\big| \le c\,|x-\xi|^{-1-\delta_{i,4}-\delta_{j,4}} \Big(\frac{|x'|}{|x-\xi|}\Big)^{\min(0,\delta_+-\delta_{i,4}-\varepsilon)} \Big(\frac{|\xi'|}{|x-\xi|}\Big)^{\min(0,\delta_+-\delta_{j,4}-\varepsilon)}$$

if $|x-\xi| > \min(|x'|,|\xi'|)$, where ε is an arbitrarily small positive number, and analogous estimates for the derivatives of $G_{i,j}$. For some special problems, the number δ_+ can be replaced again by μ_+.

The last section is dedicated to solutions of the boundary value problem in weighted Hölder spaces. Here, we prove some auxiliary assertions which are needed in the next chapter.

9.1. Existence of weak solutions of the boundary value problem

We consider the boundary value problem

$$-\Delta u + \nabla p = f, \quad -\nabla\cdot u = g \quad \text{in } \mathcal{D}, \tag{9.1.1}$$

$$S^\pm u = h^\pm, \quad N^\pm(u,p) = \phi^\pm \quad \text{on } \Gamma^\pm. \tag{9.1.2}$$

Here (9.1.2) is one of the boundary conditions (i)–(iv) on every of the faces Γ^+, Γ^-. To be more precise, we introduce integer numbers $d^\pm \in \{0,1,2,3\}$ characterizing the boundary conditions on the faces Γ^+ and Γ^-, respectively, and define

- $S^\pm u = u \quad$ if $d^\pm = 0$,
- $S^\pm u = u - (u \cdot n^\pm)n^\pm, \quad N^\pm(u,p) = -p + 2\varepsilon^\pm_{nn}(u) \quad$ if $d^\pm = 1$,
- $S^\pm u = u \cdot n^\pm, \quad N^\pm(u,p) = 2\varepsilon^\pm_{n,\tau}(u) \quad$ if $d^\pm = 2$,
- $N^\pm(u,p) = -pn^\pm + 2\varepsilon^\pm_n(u) \quad$ if $d^\pm = 3$.

Here $n^\pm = (n_1^\pm, n_2^\pm, 0)$ denotes the outward normal to the face $\Gamma^\pm$. Furthermore,

$$\varepsilon^\pm_n(u) = \varepsilon(u)\, n^\pm, \quad \varepsilon^\pm_{n,n}(u) = \varepsilon^\pm_n(u) \cdot n^\pm, \quad \text{and} \quad \varepsilon^\pm_{n,\tau}(u) = \varepsilon^\pm_n(u) - \varepsilon^\pm_{n,n}(u)\, n^\pm.$$

The condition $N^\pm(u,p) = \phi^\pm$ is absent in (9.1.2) if $d^\pm = 0$, while the condition $S^\pm u = h^\pm$ is absent in the case $d^\pm = 3$. If $d^\pm = 1$, then $h^\pm$ is a vector function tangential to $\Gamma^\pm$ and $\phi^\pm$ is a scalar function. Conversely, $\phi^\pm$ is a vector function tangential to $\Gamma^\pm$ and $h^\pm$ is a scalar function if $d^\pm = 2$. Using Green's formula, we can define variational solutions of the boundary value problem (9.1.1), (9.1.2). The goal of this section is to prove the existence and uniqueness of variational solutions.

9.1.1. Green's formula. We introduce the bilinear form

$$b_{\mathcal{D}}(u,v) = 2\int_{\mathcal{D}} \sum_{i,j=1}^{3} \varepsilon_{i,j}(u)\, \varepsilon_{i,j}(v)\, dx. \tag{9.1.3}$$

Integration by parts yields

$$\begin{aligned} &b_{\mathcal{D}}(u,v) - \int_{\mathcal{D}} p\, \nabla \cdot v\, dx \\ &\quad = \int_{\mathcal{D}} (-\Delta u - \nabla\nabla \cdot u + \nabla p) \cdot v\, dx + \sum_{\pm} \int_{\Gamma^\pm} (-pn^\pm + 2\varepsilon(u)n^\pm) \cdot v\, dx \end{aligned} \tag{9.1.4}$$

for arbitrary $u, v \in C_0^\infty(\overline{\mathcal{D}})^3$, $p \in C_0^\infty(\overline{\mathcal{D}})$. As a consequence of (9.1.4), the following *Green's formula* holds for $u, v \in C_0^\infty(\overline{\mathcal{D}})^3$ and $p, q \in C_0^\infty(\overline{\mathcal{D}})$.

$$\begin{aligned} &\int_{\mathcal{D}} (-\Delta u - \nabla\nabla \cdot u + \nabla p) \cdot v\, dx - \int_{\mathcal{D}} (\nabla \cdot u)\, q\, dx \\ &\quad + \sum_{\pm} \int_{\Gamma^\pm} \big(-pn^\pm + 2\varepsilon(u)n^\pm \big) \cdot v\, dx = \int_{\mathcal{D}} u \cdot (-\Delta v - \nabla\nabla \cdot v + \nabla q)\, dx \\ &\quad - \int_{\mathcal{D}} p\, \nabla \cdot v\, dx + \sum_{\pm} \int_{\Gamma^\pm} u \cdot \big(-qn^\pm + 2\varepsilon(v)n^\pm \big)\, dx. \end{aligned} \tag{9.1.5}$$

9.1.2. Variational solutions. Let $L^{1,2}(\mathcal{D})$ be the closure of the set $C_0^\infty(\overline{\mathcal{D}})$ with respect to the norm (6.1.4). The closure of the set $C_0^\infty(\mathcal{D})$ with respect to this norm is denoted by $\overset{\circ}{L}{}^{1,2}(\mathcal{D})$. Furthermore, we define

$$\mathcal{H}_{\mathcal{D}} = \{u \in L^{1,2}(\mathcal{D})^3 :\ S^\pm u = 0 \text{ on } \Gamma^\pm\}.$$

In the case of the Dirichlet problem ($d^+ = d^- = 0$), the space $\mathcal{H}_{\mathcal{D}}$ coincides with $\overset{\circ}{L}{}^{1,2}(\mathcal{D})^3$. By Hardy's inequality,

$$\int_{\mathcal{D}} |x|^{-2}|u|^2\, dx \le 4\, \|u\|^2_{L^{1,2}(\mathcal{D})} \quad \text{for } u \in C_0^\infty(\overline{\mathcal{D}}).$$

Therefore, the norm in $L^{1,2}(\mathcal{D})$ is equivalent to

$$(9.1.6) \qquad \Big(\int_{\mathcal{D}} \big(|x|^{-2}\, |u|^2 + \sum_{j=1}^{3} |\partial_{x_j} u|^2 \, dx \Big)^{1/2},$$

and $L^{1,2}(\mathcal{D})$ can be also defined as the closure of $C_0^\infty(\overline{\mathcal{D}})$ with respect to the norm (9.1.6). Obviously, every $u \in L^{1,2}(\mathcal{D})$ is quadratically integrable on each compact subset of $\mathcal{D}$. Furthermore, it follows from Hardy's inequality that

$$\|u\|^2_{L_2(\mathcal{D})} \le \int_{\mathcal{D}} (r^{2\delta} + r^{2\delta-2})\, |u|^2 \, dx \le c \int_{\mathcal{D}} r^{2\delta} \big(|u|^2 + |\nabla u|^2\big)\, dx$$

for $u \in C_0^\infty(\overline{\mathcal{D}})$ and $0 < \delta < 1$, where c depends only on δ. Consequently, the imbeddings

$$W_\delta^{1,2}(\mathcal{D}) \subset L_2(\mathcal{D}) \quad \text{and} \quad W_\delta^{2,2}(\mathcal{D}) \subset L^{1,2}(\mathcal{D})$$

are valid for $0 < \delta < 1$.

By (9.1.4), every solution $(u,p) \in W_\delta^{2,2}(\mathcal{D})^3 \times W_\delta^{1,2}(\mathcal{D})$ of the boundary value problem (9.1.1), (9.1.2) satisfies the equality

$$b_{\mathcal{D}}(u,v) - \int_{\mathcal{D}} p\, \nabla \cdot v \, dx = \int_{\mathcal{D}} (f + \nabla g) \cdot v \, dx + \sum_{\pm} \int_{\Gamma^\pm} \phi^\pm \cdot v \, dx$$

for all $v \in C_0^\infty(\overline{\mathcal{D}})^3$, $S^\pm v = 0$ on $\Gamma^\pm$. Here in the case $d^\pm = 1$, when $\phi^\pm$ is a scalar function, we identify $\phi^\pm$ with the vector $\phi^\pm n^\pm$. This means that $\phi^\pm \cdot v$ has to be understood as the product $\phi^\pm n^\pm \cdot v$ in this case.

By a *variational solution* of the problem (9.1.1), (9.1.2) we mean a pair $(u,p) \in L^{1,2}(\mathcal{D})^3 \times L_2(\mathcal{D})$ satisfying the equations

$$(9.1.7) \qquad b_{\mathcal{D}}(u,v) - \int_{\mathcal{D}} p\, \nabla \cdot v \, dx = F(v) \quad \text{for all } v \in \mathcal{H}_{\mathcal{D}},$$

$$(9.1.8) \qquad -\nabla \cdot u = g \ \text{ in } \mathcal{D}, \quad S^\pm u = h^\pm \ \text{ on } \Gamma^\pm,$$

where

$$(9.1.9) \qquad F(v) = \int_{\mathcal{D}} (f + \nabla g) \cdot v \, dx + \sum_{\pm} \int_{\Gamma^\pm} \phi^\pm \cdot v \, dx,$$

provided the functional (9.1.9) belongs to the dual space $\mathcal{H}^*_{\mathcal{D}}$ of $\mathcal{H}_{\mathcal{D}}$. In the case $d^\pm = 1$ we identified again $\phi^\pm$ with $\phi^\pm n^\pm$. For example, $F \in \mathcal{H}^*_{\mathcal{D}}$ if $f \in W_\delta^{0,2}(\mathcal{D})^3$, $g \in W_\delta^{1,2}(\mathcal{D})$, $\phi^\pm \in W_\delta^{1/2,2}(\Gamma^\pm)^{d^\pm}$, $\delta < 1$, and the supports of f, g and $\phi^\pm$ are compact (cf. Lemma 6.4.5).

9.1.3. A property of the operator div. The goal of this subsection is to prove that the operator $\operatorname{div} : \overset{\circ}{L}{}^{1,2}(\mathcal{D})^3 \to L_2(\mathcal{D})$ is surjective. To this end, we show that its dual operator is injective and has closed range.

Let again K be the two-dimensional angle

$$K = \{x' = (x_1, x_2) :\ 0 < r < \infty,\ -\theta/2 < \varphi < \theta/2\},$$

where r, φ denote the polar coordinates of the point x'. Furthermore, let $\overset{\circ}{W}{}^{1,2}(K)$ denote the closure of $C_0^\infty(K)$ with respect to the norm

$$\|u\|_{W^{1,2}(K)} = \Big(\int_K \big(|u|^2 + |\partial_{x_1} u|^2 + |\partial_{x_2} u|^2\big)\, dx \Big)^{1/2}$$

and let $W^{-1,2}(K)$ be its dual space (with respect to the L_2 scalar product in K).

LEMMA 9.1.1. *There exists a constant c such that*

$$\|f\|_{L_2(K)} \le c\left(\|\partial_{x_1} f\|_{W^{-1,2}(K)} + \|\partial_{x_2} f\|_{W^{-1,2}(K)} + \|f\|_{W^{-1,2}(K)}\right) \tag{9.1.10}$$

for all $f \in L_2(K)$.

P r o o f. For bounded Lipschitz domains, the assertion of the lemma can be found e.g. in [**55**, Chapter 2, §2]. Let $\mathcal{U}_j$, $j = 1, 2, \dots$, be pairwise disjoint congruent parallelograms such that $\bar{K} = \bar{\mathcal{U}}_1 \cup \bar{\mathcal{U}}_2 \cup \cdots$. Then

$$\begin{aligned} \|f\|^2_{L_2(K)} &= \sum_{j=1}^{\infty} \|f\|^2_{L_2(\mathcal{U}_j)} \\ &\le c \sum_{j=1}^{\infty} \left(\|\partial_{x_1} f\|^2_{W^{-1,2}(\mathcal{U}_j)} + \|\partial_{x_2} f\|^2_{W^{-1,2}(\mathcal{U}_j)} + \|f\|^2_{W^{-1,2}(\mathcal{U}_j)}\right). \end{aligned} \tag{9.1.11}$$

By Riesz' representation theorem, there exists a function $g \in \overset{\circ}{W}{}^{1,2}(K)$ such that

$$\|f\|_{W^{-1,2}(K)} = \|g\|_{W^{1,2}(K)}, \qquad \int_K f\,\bar{v}\,dx = (g, v)_{W^{1,2}(K)}$$

for all $v \in \overset{\circ}{W}{}^{1,2}(K)$. Furthermore, there exist functions $g_j \in \overset{\circ}{W}{}^{1,2}(\mathcal{U}_j)$, $j = 1, 2, \dots$, such that

$$\|f\|_{W^{-1,2}(\mathcal{U}_j)} = \|g_j\|_{W^{1,2}(\mathcal{U}_j)}, \qquad \int_{\mathcal{U}_j} f\,\bar{v}\,dx = (g_j, v)_{W^{1,2}(\mathcal{U}_j)}$$

for all $v \in \overset{\circ}{W}{}^{1,2}(\mathcal{U}_j)$. Let $g_{j,0}$ be the extension of g_j by zero. Then

$$\|g_j\|^2_{W^{1,2}(\mathcal{U}_j)} = \int_{\mathcal{U}_j} f\,\bar{g}_j\,dx = \int_K f\,\bar{g}_{j,0}\,dx = (g, g_{j,0})_{W^{1,2}(K)} = (g, g_j)_{W^{1,2}(\mathcal{U}_j)}$$

and, therefore

$$\|g_j\|_{W^{1,2}(\mathcal{U}_j)} \le \|g\|_{W^{1,2}(\mathcal{U}_j)}.$$

Consequently,

$$\begin{aligned} \sum_{j=1}^{\infty} \|f\|^2_{W^{-1,2}(\mathcal{U}_j)} &= \sum_{j=1}^{\infty} \|g_j\|^2_{W^{1,2}(\mathcal{U}_j)} \le \sum_{j=1}^{\infty} \|g\|^2_{W^{1,2}(\mathcal{U}_j)} = \|g\|^2_{W^{1,2}(K)} \\ &= \|f\|^2_{W^{-1,2}(K)}. \end{aligned}$$

The same inequality holds for $\partial_{x_j} f$. This together with (9.1.11) implies (9.1.10). □

Next, we give equivalent norms in $L_2(\mathcal{D})$ and in the dual space $L^{-1,2}(\mathcal{D})$ of $\overset{\circ}{L}{}^{1,2}(\mathcal{D})$.

LEMMA 9.1.2. *Let* $F(y,\xi) = \tilde{f}(|\xi|^{-1}y, \xi)$, *where* $\tilde{f}(x',\xi)$ *denotes the Fourier transform of* $f(x',x_3)$ *with respect to the last variable. Then*

$$\|f\|^2_{L_2(\mathcal{D})} = \int_{\mathbb{R}} \xi^{-2} \|F(\cdot,\xi)\|^2_{L_2(K)}\,d\xi \quad \textit{for } f \in L_2(\mathcal{D}),$$

$$\|f\|^2_{L^{-1,2}(\mathcal{D})} = \int_{\mathbb{R}} \xi^{-4} \|F(\cdot,\xi)\|^2_{W^{-1,2}(K)}\,d\xi \quad \textit{for } f \in L^{-1,2}(\mathcal{D}).$$

P r o o f. The first equality follows directly from Parseval's theorem. We prove the second one. By Riesz' representation theorem, there exists a function $u \in \overset{\circ}{L}{}^{1,2}(\mathcal{D})$ such that

(9.1.12)

$$\|f\|_{L^{-1,2}(\mathcal{D})} = \|u\|_{L^{1,2}(\mathcal{D})}, \qquad \int_{\mathcal{D}} f\,\bar{v}\,dx = \int_{\mathcal{D}} \nabla u \cdot \nabla \bar{v}\,dx \quad \text{for all } v \in \overset{\circ}{L}{}^{1,2}(\mathcal{D}).$$

Furthermore, for arbitrary $\xi \in \mathbb{R}$ there exists a function $W(\cdot,\xi) \in \overset{\circ}{W}{}^{1,2}(K)$ such that

$$\|F(\cdot,\xi)\|_{W^{-1,2}(K)} = \|W(\cdot,\xi)\|_{W^{1,2}(K)},$$

$$\int_K F(y,\xi)\,\overline{V(y,\xi)}\,dy = \int_K \big(\nabla_y W(y,\xi)\cdot \nabla_y \overline{V(y,\xi)} + W(y,\xi)\,\overline{V(y,\xi)}\big)\,dy,$$

where $V(y,\xi) = \tilde{v}(|\xi|^{-1}y,\xi)$ and $\tilde{v}$ is the Fourier transform with respect to x_3 of an arbitrary function $v \in \overset{\circ}{L}{}^{1,2}(\mathcal{D})$. It follows from the last equality that

$$\int_K \tilde{f}(x',\xi)\,\overline{\tilde{v}(x',\xi)}\,dx' \;=\; \xi^{-2}\int_K \Big(\nabla_{x'} W(|\xi|x',\xi)\cdot \nabla_{x'}\overline{\tilde{v}(x',\xi)} + \xi^2 W(|\xi|x',\xi)\,\overline{\tilde{v}(x',\xi)}\Big)\,dx'.$$

Comparing this with (9.1.12), we conclude that $W(|\xi|x',\xi) = \xi^2\tilde{u}(x',\xi)$. Consequently,

$$\int_{\mathbb{R}} \|F(\cdot,\xi)\|^2_{W^{-1,2}(K)}\,d\xi = \int_{\mathbb{R}} \xi^{-4}\,\|W(\cdot,\xi)\|^2_{W^{1,2}(K)}d\xi$$
$$= \int_{\mathbb{R}}\int_K \big(|\nabla_{x'}\tilde{u}(x',\xi)|^2 + \xi^2\,|\tilde{u}(x',\xi)|^2\big)\,dx'\,d\xi = \|u\|^2_{L^{1,2}(\mathcal{D})} = \|f\|^2_{L^{-1,2}(\mathcal{D})}.$$

The proof is complete. □

By means of the last lemma, we prove the following properties of the operators grad and div.

THEOREM 9.1.3. 1) *There exists a constant c such that*

$$\|f\|_{L_2(\mathcal{D})} \le c\sum_{j=1}^{3}\|\partial_{x_j} f\|_{L^{-1,2}(\mathcal{D})} \quad \textit{for all } f \in L_2(\mathcal{D}).$$

2) *For arbitrary $f \in L_2(\mathcal{D})$ there exists a vector function $u \in \overset{\circ}{L}{}^{1,2}(\mathcal{D})^3$ such that $\nabla\cdot u = f$ and*

$$\|u\|_{L^{1,2}(\mathcal{D})^3} \le c\,\|f\|_{L_2(\mathcal{D})},$$

where the constant c is independent of f.

P r o o f. Let f be an arbitrary function in $L_2(\mathcal{D})$ and $F(y,\xi) = \tilde{f}(|\xi|^{-1}y,\xi)$, where $\tilde{f}$ denotes the Fourier transform of f with respect to x_3. Then by Lemmas

9.1.1 and 9.1.2,

$$\begin{aligned}\|f\|^2_{L_2(\mathcal{D})} &= \int_{\mathbb{R}} \xi^{-2} \, \|F(\cdot,\xi)\|^2_{L_2(K)} \, d\xi \\ &\leq c \int_{\mathbb{R}} \xi^{-2} \Big(\sum_{j=1}^{2} \|\partial_{y_j} F(\cdot,\xi)\|^2_{W^{-1,2}(K)} + \|F(\cdot,\xi)\|^2_{W^{-1,2}(K)} \Big) \, d\xi \\ &= c \sum_{j=1}^{3} \|\partial_{x_j} f\|^2_{L^{-1,2}(\mathcal{D})}.\end{aligned}$$

In particular, the last inequality implies that the range of the mapping

$$L_2(\mathcal{D}) \ni f \to \nabla f \in L^{-1,2}(\mathcal{D})^3$$

is closed. Moreover, the kernel of this operator is obviously trivial. Consequently, by the closed range theorem, its dual operator $u \to -\nabla \cdot u$ maps $\overset{\circ}{L}{}^{1,2}(\mathcal{D})^3$ onto $L_2(\mathcal{D})$. This proves the theorem. □

9.1.4. Korn's inequality. The classical *Korn's inequality*

$$\sum_{i,j=1}^{N} \|\varepsilon_{i,j}(u)\|^2_{L_2(\Omega)} + \|u\|^2_{L_2(\Omega)^N} \geq c \, \|u\|^2_{W^{1,2}(\Omega)^N}$$

holds for an arbitrary bounded Lipschitz domain in $\mathbb{R}^N$. For the proof, we refer to [**43**] (see also [**55**, Remark 5.5]). Using this result, we get the following inequality.

LEMMA 9.1.4. *There exists a positive constant $c > 0$ such that*

$$b_{\mathcal{D}}(u, \bar{u}) \geq c \, \|u\|^2_{L^{1,2}(\mathcal{D})^3}$$

for all $u \in L^{1,2}(\mathcal{D})^3$.

Proof. There exists a constant c such that

$$b_{\mathcal{D}}(u, \bar{u}) + \|u\|^2_{L_2(\mathcal{D})^3} \geq c \, \|u\|^2_{L^{1,2}(\mathcal{D})^3}$$

for all $u \in C_0^\infty(\overline{\mathcal{D}})^3$, $u(x) = 0$ for $|x| > 1$. We consider the set of all $u \in C_0^\infty(\overline{\mathcal{D}})^3$ with support in the ball $|x| \leq \varepsilon$. For such u, Hardy's inequality implies

$$\int_{\mathcal{D}} |u(x)|^2 \, dx \leq c_1 \int_{\mathcal{D}} |x|^2 \sum_{j=1}^{3} |\partial_{x_j} u(x)|^2 \, dx \leq c_1 \, \varepsilon^2 \, \|u\|^2_{L^{1,2}(\mathcal{D})^3}$$

and therefore

$$b_{\mathcal{D}}(u, \bar{u}) \geq \frac{c}{2} \, \|u\|^2_{L^{1,2}(\mathcal{D})^3}$$

if ε is sufficiently small. Applying the similarity transformation $x = \alpha y$, we obtain the same inequality for arbitrary $u \in C_0^\infty(\overline{\mathcal{D}})^3$. The result follows. □

9.1.5. Existence and uniqueness of variational solutions. The next assertion is a consequence of Lax-Milgram's lemma.

THEOREM 9.1.5. *Let $F \in \mathcal{H}^*_{\mathcal{D}}$, $g \in L_2(\mathcal{D})$, and let $h^\pm$ be such that there exists a vector function $w \in L^{1,2}(\mathcal{D})^3$ satisfying the equalities $S^+ w = h^+$ on Γ^+. Then there exists a unique solution $(u,p) \in L^{1,2}(\mathcal{D})^3 \times L_2(\mathcal{D})$ of the problem* (9.1.7), (9.1.8). *Furthermore, (u,p) satisfies the estimate*

$$\|u\|_{L^{1,2}(\mathcal{D})^3} + \|p\|_{L_2(\mathcal{D})} \leq c \Big(\|F\|_{\mathcal{H}^*_{\mathcal{D}}} + \|g\|_{L_2(\mathcal{D})} + \|w\|_{L^{1,2}(\mathcal{D})^3} \Big)$$

with a constant c independent of F, g and w.

P r o o f. 1) We prove the existence of a solution. By our assumption on $h^\pm$ and by Theorem 9.1.3, we may restrict ourselves to the case $g = 0$, $h^\pm = 0$. Let

$$V = \{u \in \mathcal{H}_\mathcal{D} : \nabla \cdot u = 0\},$$

and let $V^\perp$ be the orthogonal complement of V in $\mathcal{H}_\mathcal{D}$. Then by Lax-Milgram's lemma, there exists a vector function $u \in V$ such that

$$b_\mathcal{D}(u, v) = F(v) \quad \text{for all } v \in V, \quad \|u\|_{L^{1,2}(\mathcal{D})^3} \le c\,\|F\|_{V^*} \le c\,\|F\|_{\mathcal{H}^*_\mathcal{D}}.$$

By Theorem 9.1.3, the operator $B : u \to -\nabla \cdot u$ is an isomorphism from $V^\perp$ onto $L_2(\mathcal{D})$. Hence, the mapping

$$L_2(\mathcal{D}) \ni q \to \ell(q) \stackrel{def}{=} F(B^{-1}q) - b_\mathcal{D}(u, B^{-1}q)$$

defines a linear and continuous functional on $L_2(\mathcal{D})$. By the Riesz representation theorem, there exists a function $p \in L_2(\mathcal{D})^3$ such that

$$\int_\mathcal{D} p\, q\, dx = \ell(q) \quad \text{for all } q \in L_2(\mathcal{D}), \quad \|p\|_{L_2(\mathcal{D})} \le c\Big(\|F\|_{\mathcal{H}^*_\mathcal{D}} + \|u\|_{L^{1,2}(\mathcal{D})^3}\Big).$$

If we set $q = -\nabla \cdot v$, where v is an arbitrary element of $V^\perp$, we obtain

$$-\int_\mathcal{D} p\, \nabla \cdot v\, dx = F(v) - b_\mathcal{D}(u, v) \quad \text{for all } v \in V_0^\perp.$$

Since both sides of the last equality vanish for $v \in V$, we get (9.1.7). Furthermore, $-\nabla \cdot u = 0$ and $S^\pm u|_{\Gamma^\pm} = 0$ for $u \in V$.

2) We prove the uniqueness. Suppose that $(u, p) \in L^{1,2}(\mathcal{D})^3 \times L_2(\mathcal{D})$ is a solution of problem (9.1.7), (9.1.8) with $F = 0$, $g = 0$, $h^\pm = 0$. Then, in particular, $u \in V$ and $b_\mathcal{D}(u, \bar{u}) = 0$. By Lemma 9.1.4, this implies $u = 0$. Consequently,

$$\int_\mathcal{D} p\, \nabla \cdot v\, dx = 0 \quad \text{for all } v \in \mathcal{H}_\mathcal{D}\,.$$

Since v can be chosen such that $\nabla \cdot v = \bar{p}$, we obtain $p = 0$. The proof is complete. $\square$

9.2. Compatibility conditions on the edge

In this section, we consider the boundary value problem (9.1.1), (9.1.2), where the right-hand sides $f, g, h^\pm, \phi^\pm$ are elements of weighted Sobolev spaces. In general, we are concerned with functions $f \in W_\delta^{l-2,s}(\mathcal{G})^3$, $g \in W_\delta^{l-1,s}(\mathcal{G})$ and boundary data $h^\pm \in W_\delta^{l-1/s,s}(\Gamma^\pm)^{3-d^\pm}$, $\phi^\pm \in W_\delta^{l-1-1/s,s}(\Gamma^\pm)^{d^\pm}$. Note that in the case $d^\pm = 1$, the vector function $h^\pm$ is tangential to $\Gamma^\pm$. Therefore, the function space for $h^\pm$ can be identified with $W_\delta^{l-1/s,s}(\Gamma^\pm)^2$ in this case. Conversely, $\phi^\pm$ is a vector function tangential to $\Gamma^\pm$ if $d^\pm = 2$. Then the function space for $\phi^\pm$ can be identified with $W_\delta^{l-1-1/s,s}(\Gamma^\pm)^2$. From the equations (9.1.1) and (9.1.2) it follows that the traces of the functions $f, g, h^\pm, \phi^\pm$ satisfy certain conditions on the edge of the dihedron $\mathcal{D}$. In this section, we establish these compatibility conditions for the case $l - 2 < \delta + 2/s < l$.

9.2.1. The case $l-1<\delta+2/s<l$. Let $f\in W_\delta^{l-2,s}(\mathcal{G})^3$, $g\in W_\delta^{l-1,s}(\mathcal{G})$, $h^\pm\in W_\delta^{l-1/s,s}(\Gamma^\pm)^{3-d^\pm}$, and $\phi^\pm\in W_\delta^{l-1-1/s,s}(\Gamma^\pm)^{d^\pm}$ be given functions. We are interested in conditions on these functions which ensure the existence of a vector function $(u,p)\in W_\delta^{l,s}(\mathcal{D})^3\times W_\delta^{l-1,s}(\mathcal{D})$ such that

$$S^\pm u=h^\pm,\quad N^\pm(u,p)=\phi^\pm \text{ on } \Gamma^\pm, \tag{9.2.1}$$

$$\Delta u-\nabla p+f\in V_\delta^{l-2,s}(\mathcal{D})^3 \text{ and } \nabla\cdot u+g\in V_\delta^{l-1,s}(\mathcal{D}). \tag{9.2.2}$$

If $\delta+2/s>l-1$, then $W_\delta^{l-2,s}(\mathcal{D})\subset V_\delta^{l-2,s}(\mathcal{D})$, $W_\delta^{l-1,s}(\mathcal{D})\subset V_\delta^{l-1,s}(\mathcal{D})$, and it suffices to consider the boundary condition (9.2.1).

The following result for the class of the spaces $V_\delta^{l,p}(\mathcal{D})$ is valid without restrictions on δ.

LEMMA 9.2.1. 1) *For arbitrary* $h^\pm\in V_\delta^{1-1/s,s}(\Gamma^\pm)^{3-d^\pm}$, *there exists a vector function* $u\in V_\delta^{1,s}(\mathcal{D})^3$ *such that* $S^\pm u=h^\pm$ *on* $\Gamma^\pm$ *and*

$$\|u\|_{V_\delta^{1,s}(\mathcal{D})^3}\le c\sum_\pm \|h^\pm\|_{V_\delta^{1-1/s,s}(\Gamma^\pm)^{3-d^\pm}}$$

with a constant c *independent of* h^+ *and* h^-.

2) *For arbitrary* $h^\pm\in V_\delta^{l-1/s,s}(\Gamma^\pm)^{3-d^\pm}$ *and* $\phi^\pm\in V_\delta^{l-1-1/s,s}(\Gamma^\pm)^{d^\pm}$, $l\ge 2$, *there exists a vector function* $u\in V_\delta^{l,s}(\mathcal{D})^3$ *satisfying the boundary conditions*

$$S^\pm u=h^\pm,\quad N^\pm(u,0)=\phi^\pm \quad \text{on } \Gamma^\pm \tag{9.2.3}$$

and the estimate

$$\|u\|_{V_\delta^{l,s}(\mathcal{D})^3}\le c\sum_\pm\Big(\|h^\pm\|_{V_\delta^{l-1/s,s}(\Gamma^\pm)^{3-d^\pm}}+\|\phi^\pm\|_{V_\delta^{l-1-1/s,s}(\Gamma^\pm)^{d^\pm}}\Big). \tag{9.2.4}$$

Moreover, if If $h^\pm$ *and* $\phi^\pm$ *vanish for* $|x'|>1$, *then also* u *can be chosen such that* $u(x)=0$ *for* $|x'|>1$.

P r o o f. The first part of the lemma is a corollary of Lemma 2.2.1. For the proof of the second part, we assume for simplicity that Γ^- coincides with the half-plane $\varphi=0$ (i.e., $x_1>0$, $x_2=0$) and Γ^+ with the half-plane $\varphi=\theta$. Then the boundary conditions

$$S^-u=h^-,\quad N^-(u,0)=\phi^- \quad \text{on } \Gamma^- \tag{9.2.5}$$

have the form $u=h^-$ on Γ^- if $d^-=0$,

$$u_1=h_1^-,\quad u_3=h_3^-,\quad 2\partial_{x_2}u_2=\phi^- \quad \text{on } \Gamma^- \quad \text{if } d^-=1,$$
$$u_2=h^-,\quad -\varepsilon_{1,2}(u)=\phi_1^-,\quad -\varepsilon_{3,2}(u)=\phi_3^- \quad \text{on } \Gamma^- \quad \text{if } d^-=2,$$
$$-2\varepsilon_{j,2}(u)=\phi_j^- \quad \text{on } \Gamma^- \quad \text{for } j=1,2,3 \text{ if } d^-=3.$$

In all cases, the existence of a vector function $u\in V_\delta^{l,s}(\mathcal{D})^3$ satisfying (9.2.5) can be easily deduced from Lemma 2.2.3.

Furthermore, there exists a function $v\in V_\delta^{l,s}(\mathcal{D})^3$ satisfying the boundary conditions

$$S^+v=h^+,\quad N^+(v,0)=\phi^+ \quad \text{on } \Gamma^+.$$

Let $\zeta=\zeta(\varphi)$ be a smooth function on $[0,\theta]$ equal to 1 for $\varphi<\theta/2$ and to zero for $\varphi>3\theta/4$. Then the function $w(x)=\zeta(\varphi)\,u(x)+(1-\zeta(\varphi))\,v(x)$ belongs to the space $V_\delta^{l,s}(\mathcal{D})^3$ and satisfies (9.2.3). □

The analogous result in the class of the spaces $W_\delta^{l,s}(\mathcal{D})$ is not valid without additional conditions on the traces of $h^\pm$ and $\phi^\pm$ on the edge M. If $u \in W_\delta^{l,s}(\mathcal{D})^3$, $\delta < l - 2/s$, then the trace of u on M exists,

$$u|_M \in W^{l-\delta-2/s,s}(M)^3$$

(cf. Lemma 6.2.2). Thus, the boundary conditions $S^\pm u|_{\Gamma^\pm} = h^\pm$ imply in particular that

$$S^\pm u|_M = h^\pm|_M .$$

Here S^+ and S^- are considered as operators on $W^{l-\delta-2/s,s}(M)^3$. Consequently, the boundary data h^+ and h^- satisfy the compatibility condition

$$(h^+|_M \, , h^-|_M) \in R(T), \tag{9.2.6}$$

where $R(T)$ is the range of the operator $T = (S^+, S^-)$. The last condition can be also written in the form

$$A^+ h^+|_M = A^- h^-|_M \, ,$$

where $A^\pm = 1$ if $d^\pm = 2$, and $A^\pm$ are certain constant matrices if $d^\pm \le 1$. For example, the compatibility condition (9.2.6) takes the form

$$\begin{aligned} &h^-|_M = h^+|_M \quad \text{if } d^- = d^+ = 0, \\ &h^-|_M - \big(h^-|_M \cdot n^+\big)\, n^+ = h^+|_M \quad \text{if } d^- = 0,\ d^+ = 1, \\ &h^-|_M \cdot n^+ = h^+|_M \quad \text{if } d^- = 0,\ d^+ = 2, \\ &h_3^+|_M = h_3^-|_M \quad \text{if } d^- = d^+ = 1,\ \sin\theta \ne 0. \end{aligned}$$

No compatibility conditions of the form (9.2.6) appear if

- $d^+ = 3$ or $d^- = 3$,
- $d^+ = d^- = 2$ and $\sin\theta \ne 0$,
- $d^+ = 1,\ d^- = 2$ (or $d^+ = 1,\ d^- = 2$) and $\cos\theta \ne 0$.

LEMMA 9.2.2. *Let $l \ge 1$, $l - 1 < \delta + 2/s < l$, and let $h^\pm \in W_\delta^{l-1/s,s}(\Gamma^\pm)^{3-d^\pm}$ be given vector functions such that $h^\pm(x) = 0$ for $|x'| > 1$. Suppose that h^+ and h^- satisfy the compatibility condition (9.2.6). Then there exists a vector function $u \in W_\delta^{l,s}(\mathcal{D})^3$ such that $u(x) = 0$ for $|x'| > 1$, $S^\pm u = h^\pm$ on $\Gamma^\pm$, and*

$$\|u\|_{W_\delta^{l,s}(\mathcal{D})^3} \le c \sum_\pm \|h^\pm\|_{W_\delta^{l-1/s,s}(\Gamma^\pm)^{3-d^\pm}}$$

with a constant c independent of h^+ and h^-.

P r o o f. By (9.2.6), there exists a vector function $\psi \in W^{l-\delta-2/s,s}(M)^3$ such that $S^\pm\psi = h^\pm|_M$. Let $v \in W_\delta^{l,s}(\mathcal{D})^3$ be an extension of ψ vanishing for $|x'| > 1$. Then the traces of $S^\pm v|_{\Gamma^\pm} - h^\pm$ are zero on M and, consequently, $S^\pm v|_{\Gamma^\pm} - h^\pm \in V_\delta^{l-1/s,s}(\Gamma^\pm)^{3-d^\pm}$ (cf. Theorem 6.2.15). Applying Lemma 9.2.1, we obtain the assertion of the lemma. □

Since $W_\delta^{l-1-1/s,s}(\mathcal{D}) \subset V_\delta^{l-1-1/s,s}(\mathcal{D})$ for $\delta > l - 1 - 2/s$, we deduce the following result from the last two lemmas.

COROLLARY 9.2.3. *Let $l \geq 2$, $l-1 < \delta + 2/s < l$, $h^\pm \in W_\delta^{l-1/s,s}(\Gamma^\pm)^{3-d^\pm}$, and $\phi^\pm \in W_\delta^{l-1-1/s,s}(\Gamma^\pm)^{d^\pm}$. Suppose that $h^\pm$ and $\phi^\pm$ vanish for $|x'| > 1$ and that h^+ and h^- satisfy the compatibility (9.2.6). Then there exists a vector function $u \in W_\delta^{l,s}(\mathcal{D})^3$ satisfying the boundary conditions (9.2.3) and the estimate*

$$\|u\|_{W_\delta^{l,s}(\mathcal{D})^3} \leq c \sum_{\pm} \Big(\|h^\pm\|_{W_\delta^{l-1/s,s}(\Gamma^\pm)^{3-d^\pm}} + \|\phi^\pm\|_{W_\delta^{l-1-1/s,s}(\Gamma^\pm)^{d^\pm}} \Big).$$

Moreover, u can be chosen such that $u(x) = 0$ for $|x'| > 1$.

P r o o f. By Lemma 9.2.2, there exists a vector function $v \in W_\delta^{l,s}(\mathcal{D})^3$ such that $v(x) = 0$ for $|x'| > 1$ and $S^\pm v = h^\pm$ on $\Gamma^\pm$. Furthermore by Lemma 9.2.1, there exists a vector function $w \in V_\delta^{l,s}(\mathcal{D})^3$ such that

$$S^\pm w = 0, \quad N^\pm(w,0) = \phi^\pm - N^\pm(v,0) \quad \text{on } \Gamma^\pm,$$

and $w(x) = 0$ for $|x'| > 1$. Then $u = v + w \in W_\delta^{l,s}(\mathcal{D})^3$, and u satisfies (9.2.4). □

9.2.2. The case where $l-2 < \delta + 2/s \leq l-1$ and $d^+ + d^-$ is odd. Again, our goal is to find conditions on $f, g, h^\pm, \phi^\pm$ which ensure the existence of $u \in W_\delta^{l,s}(\mathcal{D})^3$ and $p \in W_\delta^{l-1,s}(\mathcal{D})$ satisfying the conditions (9.2.1), (9.2.2). If $\delta + 2/s > l-2$, then $W_\delta^{l-2,s}(\mathcal{D}) \subset V_\delta^{l-2,s}(\mathcal{D})$. Consequently, the condition $\Delta u - \nabla p + f \in V_\delta^{l-2,s}(\mathcal{D})^3$ is satisfied for arbitrary $u \in W_\delta^{l,s}(\mathcal{D})^3$, $p \in W_\delta^{l-1,s}(\mathcal{D})^3$ and $f \in W_\delta^{l-2,s}(\mathcal{D})$.

Suppose that $l-2 < \delta + 2/s < l-1$ and $(u,p) \in W_\delta^{2,s}(\mathcal{D})^3 \times W_\delta^{1,s}(\mathcal{D})$ is a vector function satisfying (9.2.1) and (9.2.2). Then the traces of u, $\partial_{x_j} u$ and p on M exist. We put

$$b = u|_M\,, \quad c = (\partial_{x_1} u)|_M\,, \quad d = (\partial_{x_2} u)|_M \quad \text{and} \quad q = p|_M\,.$$

It follows from the equations $S^\pm u = h^\pm$ on $\Gamma^\pm$ that $S^\pm \partial_r u = \partial_r h^\pm$ on $\Gamma^\pm$, and therefore,

$$S^\pm b = h^\pm|_M\,, \tag{9.2.7}$$

$$S^\pm\big(c\cos\tfrac{\theta}{2} \pm d\sin\tfrac{\theta}{2}\big) = (\partial_r h^\pm)|_M\,. \tag{9.2.8}$$

Moreover, $\nabla \cdot u + g \in V_\delta^{l-1,s}(\mathcal{D})$ if and only if the trace of $\nabla \cdot u + g$ on M vanishes, i.e.,

$$c_1 + d_2 + \partial_{x_3} b_3 = -g|_M\,. \tag{9.2.9}$$

Obviously, the trace of $N^\pm(u,p)$ on the edge M can be written as a linear form $M^\pm(c,d,\partial_{x_3} b, q)$. Thus, the equation $N^\pm(u,p) = \phi^\pm$ on $\Gamma^\pm$ implies

$$M^\pm(c,d,\partial_{x_3} b, q) = \phi^\pm|_M\,. \tag{9.2.10}$$

The equations (9.2.8)–(9.2.10) form a linear system of 7 equations with 7 unknowns $c_1, c_2, c_3, d_1, d_2, d_3, q$.

LEMMA 9.2.4. *Suppose that $d^+ + d^-$ is odd. In the case $d^+ + d^- = 3$, we assume that $\sin 2\theta \neq 0$, while the condition $\cos\theta\, \cos 2\theta \neq 0$ is imposed for $d^+ + d^- \in \{1,5\}$. Then the linear system (9.2.8)–(9.2.10) has a unique solution (c,d,q) for arbitrary $h^\pm, \phi^\pm, g$, and b.*

P r o o f. It can be checked directly that, under the given assumptions on $d^\pm$ and θ, there does not exist a pair $(u,p) \neq (0,0)$ of a linear vector function $u = cx_1 + dx_2$ and a constant $p = q$ satisfying

$$-\nabla \cdot u = 0 \ \text{ in } \mathcal{D}, \quad S^\pm u = 0, \quad N^\pm(u,p) = 0 \ \text{ on } \Gamma^\pm. \tag{9.2.11}$$

This means that the system

$$c_1 + d_2 = 0, \quad S^\pm\big(c\cos\tfrac{\theta}{2} \pm d\sin\tfrac{\theta}{2}\big) = 0, \quad \text{and} \quad M^\pm(c,d,0,q) = 0 \tag{9.2.12}$$

has only the trivial solution $c = d = 0$, $q = 0$. Consequently, the inhomogeneous system (9.2.8)–(9.2.10) is uniquely solvable. □

Using Lemmas 9.2.1 and 9.2.4, we prove the following assertion.

LEMMA 9.2.5. *Let the assumptions of Lemma 9.2.4 on d^+, d^-, θ be satisfied, and let $l-2 < \delta + 2/s < l$. Furthermore, let $h^\pm \in W_\delta^{l-1/s,s}(\Gamma^\pm)^{3-d^\pm}$, $\phi^\pm \in W_\delta^{l-1-1/s,s}(\Gamma^\pm)^{d^\pm}$ and $g \in W_\delta^{l-1,s}(\mathcal{D})$ be given functions vanishing for $|x'| > 1$. We suppose that h^+ and h^- satisfy the compatibility condition (9.2.6) on M. Then there exist a vector function $u \in W_\delta^{l,s}(\mathcal{D})^3$ and a function $p \in W_\delta^{l-1,s}(\mathcal{D})$ vanishing for $|x'| > 1$ and satisfying (9.2.1), (9.2.2). Furthermore, the estimate*

$$\begin{aligned} &\|u\|_{W_\delta^{l,s}(\mathcal{D})^3} + \|p\|_{W_\delta^{l-1,s}(\mathcal{D})} \\ &\le c\Big(\sum_\pm \big(\|h^\pm\|_{W_\delta^{l-1/s,s}(\Gamma^\pm)^{3-d^\pm}} + \|\phi^\pm\|_{W_\delta^{l-1-1/s,s}(\Gamma^\pm)^{d^\pm}}\big) + \|g\|_{W_\delta^{l-1,s}(\mathcal{D})}\Big) \end{aligned}$$

is satisfied for u and p.

P r o o f. For $\delta > l-1-2/s$ the assertion of the lemma follows immediately from Corollary 9.2.3. Let $l-2-2/s < \delta < l-1-2/s$. Then there exist functions $b \in W^{l-\delta-2/s,s}(M)^3$, $c, d \in W^{l-1-\delta-2/s,s}(M)^3$ and $q \in W^{l-1-\delta-2/s,s}(M)$ satisfying (9.2.7)–(9.2.10). We set

$$v = Eb + x_1\, Ec + x_2\, Ed, \quad p = Eq,$$

where E is the extension operator (6.2.4). Then

$$S^\pm v|_M = h^\pm|_M\,, \quad (\partial_r S^\pm v)|_M = (\partial_r h^\pm)|_M\,, \quad -(\nabla\cdot v)|_M = g|_M\,.$$

and

$$N^\pm(v,p)|_M = M^\pm(c,d,\partial_{x_3}b,q) = \phi^\pm|_M\,.$$

Consequently according to Theorem 6.2.15,

$$S^\pm v - h^\pm \in V_\delta^{l-1/s,s}(\Gamma^\pm)^{3-d^\pm}, \quad \nabla\cdot v + g \in V_\delta^{l-1,s}(\mathcal{D}),$$

and

$$N^\pm(v,p) - \phi^\pm \in V_\delta^{l-1-1/s,s}(\Gamma^\pm)^{d^\pm}.$$

By Lemma 9.2.1, there exists a vector function $w \in V_\delta^{l,s}(\mathcal{D})^3$, $w(x) = 0$ for $|x'| > 1$ such that

$$S^\pm w = h^\pm - S^\pm v, \quad N^\pm(w,0) = \phi^\pm - N^\pm(v,p) \quad \text{on } \Gamma^\pm.$$

Then the pair $(u,p) = (v+w,p)$ has the desired properties.

In the case $\delta = l-1-2/s$, the lemma can be proved analogously using the relations between the spaces $V_\delta^{l,s}(\mathcal{D})$ and $W_\delta^{l,s}(\mathcal{D})$ in Subsection 6.2.6. □

9.2.3. The case where $l-2<\delta+2/s\le l-1$ and $d^+ + d^-$ is even. If $d^+ + d^-$ is an even number, then the assertion of Lemma 9.2.4 fails for all θ. The reason is that then the homogeneous system (9.2.12) has nontrivial solutions. If d^+ and d^- are both even, then the vector function $(u,p)=(0,1)$ satisfies (9.2.11), i.e. the homogeneous system (9.2.12) has the nontrivial solution $c=d=0$, $q=1$. If d^+ and d^- are odd, then $(u,p)=(x_2,-x_1,0,0)$ satisfies (9.2.11). This means that the system (9.2.12) has the solution $c=(0,-1,0)$, $d=(1,0,0)$, $q=0$. In these cases, the assertion of Lemma 9.2.5 holds only under additional compatibility conditions on the functions $h^\pm$, $\phi^\pm$ and g. These compatibility conditions have the following forms for $d^+=d^-$, $\theta\ne\pi$, $\theta\ne 2\pi$.

$$(9.2.13)\qquad n^-\cdot\partial_r h^+|_M + n^+\cdot\partial_r h^-|_M = \sin\theta\,(g|_M+\partial_{x_3}h_3^+|_M)\ \text{ if } d^+=d^-=0,$$

$$(9.2.14)\qquad 2\big(\partial_r h_2^+|_M+\partial_r h_2^-|_M\big)=\sin\frac{\theta}{2}\,\big(\phi^-|_M-\phi^+|_M\big)\quad\text{if } d^+=d^-=1,$$

$$(9.2.15)\qquad \sin\frac{\theta}{2}\,\big(\partial_r h^+|_M+\partial_r h^-|_M\big)=\phi_2^+|_M-\phi_2^-|_M\quad\text{if } d^+=d^-=2,$$

$$(9.2.16)\qquad \phi^+\cdot n^-=\phi^-\cdot n^+\quad\text{on } M\quad\text{if } d^+=d^-=3.$$

In the case $d^+=2$, $d^-=0$, $\theta\ne\pi/2$, $\theta\ne 3\pi/2$, the compatibility condition

$$(9.2.17)\qquad \partial_r h^+|_M\cos 2\theta-(2n^+\cos\theta+n^-)\cdot\partial_r h^-|_M+\frac{\sin 2\theta}{2}\,(g|_M+\partial_{x_3}h_3^-|_M)$$
$$+2\sin^2\theta\,(\phi_1^+\cos\theta/2+\phi_2^+\sin\theta/2)\big|_M=0$$

holds, while the condition

$$(9.2.18)\qquad 8\cos\frac{\theta}{2}\,\sin\theta\,\partial_r h_2^+|_M+2\sin^2\theta\,(1-2\cos\theta)\,\big(g|_M+\partial_{x_3}h_3^+\big)$$
$$+\sin\frac{\theta}{2}\,(1+2\cos\theta)\,\phi_1^-|_M+\cos\frac{\theta}{2}\,(1-2\cos\theta)\,\phi_2^-|_M=0$$

holds for $d^+=1$, $d^-=3$, $\theta\ne\pi/2$, $\theta\ne 3\pi/2$.

If $h^\pm$, $\phi^\pm$ and g satisfy the conditions (9.2.13)–(9.2.18), then the linear system (9.2.8)–(9.2.10) with the unknowns $c=\partial_{x_1}u|_M$, $d=\partial_{x_2}u|_M$ and $q=p|_M$ is solvable for an arbitrary vector function b satisfying (9.2.7). Thus, the following result holds.

LEMMA 9.2.6. *Let $l-2<\delta+2/s<l$, $\delta\ne l-1-2/s$, and let $h^\pm\in W_\delta^{l-1/s,s}(\Gamma^\pm)^{3-d^\pm}$, $\phi^\pm\in W_\delta^{l-1-1/s,s}(\Gamma^\pm)^{d^\pm}$, $g\in W_\delta^{l-1,s}(\mathcal{D})$ be given functions vanishing for $|x'|>1$. We suppose that h^+ and h^- satisfy the compatibility condition (9.2.6). If $d^+ + d^-$ is even and $\delta<l-1-2/s$, we assume in addition that the compatibility conditions (9.2.13)–(9.2.18) are satisfied. Then there exists a vector function $(u,p)\in W_\delta^{l,s}(\mathcal{D})^3\times W_\delta^{l-1,s}(\mathcal{D})$ vanishing for $|x'|>1$ and satisfying (9.2.1), (9.2.2).*

P r o o f. We restrict ourselves in the proof to the Dirichlet problem. The proof for the other boundary conditions proceeds analogously. If $\delta>l-1-2/s$, then $V_\delta^{l-1,s}(\mathcal{D})\subset W_\delta^{l-1,s}(\mathcal{D})$, and we can refer to Lemma 9.2.2. Suppose that $l-2-2/s<\delta<l-1-2/s$. Then the traces of $h^\pm$, $\partial_r h^\pm$, and g on M exist and one can find vector functions $c,d\in W^{1-\delta-2/s,s}(M)^3$ satisfying

$$c\cos\tfrac{\alpha}{2}\pm d\sin\tfrac{\alpha}{2}=(\partial_r h^\pm)|_M\quad\text{and}\quad c_1+d_2=-\partial_{x_3}h^+|_M-g|_M\,.$$

Analogously to the proof of Lemma 9.2.5, it can be shown that

$$v=Eh^+|_M+x_1\,Ec+x_2\,Ed$$

satisfies the conditions

$$v|_{\Gamma^\pm} - h^\pm \in V_\delta^{l-1/s,s}(\Gamma^\pm)^3 \quad \text{and} \quad \nabla\cdot v + g \in V_\delta^{l-1,s}(\mathcal{D}).$$

Applying Lemma 9.2.1, we obtain the assertion of the lemma. □

REMARK 9.2.7. If $\delta = l-1-2/s$, then the compatibility conditions (9.2.13)–(9.2.18) have to be understood in a generalized sense (cf. Lemma 6.4.3). For example in the case of the Dirichlet problem ($d^+ = d^- = 0$), the condition $h^+|_M = h^-|_M$ and the "generalized trace condition"

$$\int_0^\infty \int_{\mathbb{R}} r^{-1} \Big| n^- \cdot \partial_r h^+(r,x_3) + n^+ \cdot \partial_r h^-(r,x_3) \\ - \big(\overset{\circ}{g}(r,x_3) - \partial_{x_3} h_3^+(r,x_3)\big) \sin\theta \Big|^s \, dx_3\, dr < \infty$$

guarantee the existence of a function $u \in W_\delta^{l,s}(\mathcal{D})^3$ satisfying (9.2.1) and (9.2.2). Here

$$\overset{\circ}{g}(r,x_3) = \frac{1}{\theta}\int_{-\theta/2}^{\theta/2} g(r\cos\varphi, r\sin\varphi, x_3)\, d\varphi$$

denotes the average of g with respect to the variable φ.

9.3. The model problem in an angle

One of the goals of this chapter is the proof of regularity assertions for the variational solution $(u,p) \in L^{1,2}(\mathcal{D})^3 \times L_2(\mathcal{D})$ of the boundary value problem (9.1.1), (9.1.2). For this we need regularity results for the solutions of the corresponding boundary value problem in the two-dimensional angle K. We deal here with a boundary value problem for the system

$$(9.3.1) \qquad -\Delta_{x'}(u_1,u_2) + \nabla_{x'} p = (f_1,f_2), \quad -\nabla_{x'}\cdot(u_1,u_2) = g,$$

$$(9.3.2) \qquad -\Delta_{x'} u_3 = f_3$$

with the unknown vector function (u_1,u_2,u_3,p). Here $\Delta_{x'}$ and $\nabla_{x'}$ denote the Laplace and Nabla operators in the coordinates $x' = (x_1,x_2)$. This boundary value problem arises if one considers the problem (9.1.1), (9.1.2) for vector functions (u,p) not depending on the variable x_3.

9.3.1. The operator pencil $A(\lambda)$. Suppose that (u,p) is a solution of the boundary value problem (9.1.1), (9.1.2) which is independent of x_3. Then (u,p) satisfies the equations (9.3.1), (9.3.2) and the boundary conditions

$$(9.3.3) \qquad \tilde{S}^\pm(u_1,u_2) = (h_1^\pm, h_2^\pm), \quad \tilde{N}^\pm(u_1,u_2,p) = (\phi_1^\pm,\phi_2^\pm) \text{ on } \gamma^\pm,$$

$$(9.3.4) \qquad u_3 = h_3^\pm \text{ on } \gamma^\pm \text{ for } d^\pm \le 1, \quad \frac{\partial u_3}{\partial n^\pm} = \phi_3^\pm \text{ on } \gamma^\pm \text{ for } d^\pm \ge 2.$$

Here the operators $\tilde{S}^\pm$ and $\tilde{N}^\pm$ are defined as follows:

$$\tilde{S}^\pm u' = u' \text{ if } d^\pm = 0, \quad \tilde{S}^\pm u' = u'\cdot\tau^\pm \text{ if } d^\pm = 1, \quad \tilde{S}^\pm u' = u'\cdot n^\pm \text{ if } d^\pm = 2,$$

$$\tilde{N}^\pm(u',p) = -p + 2(\varepsilon(u')n^\pm)\cdot n^\pm \text{ if } d^\pm = 1,$$

$$\tilde{N}^\pm(u',p) = 2\,(\varepsilon(u')n^\pm)\cdot\tau^\pm \text{ if } d^\pm = 2,$$

$$\tilde{N}^\pm(u',p) = -pn^\pm + 2\varepsilon(u')n^\pm \text{ if } d^\pm = 3,$$

where $u' = (u_1, u_2)$, $\varepsilon(u')$ denotes the matrix with the elements $\varepsilon_{i,j}(u)$, $i, j = 1, 2$, $n^\pm$ is the outward normal to $\gamma^\pm$, and $\tau^\pm$ is the unit vector on the side $\gamma^\pm$.

Setting $u = r^\lambda U(\varphi)$ and $p = r^{\lambda-1}P(\varphi)$, we obtain a boundary value problem for the vector function (U, P) on the interval $(-\theta/2, +\theta/2)$ quadratically depending on the parameter $\lambda \in \mathbb{C}$. The operator $A(\lambda)$ of this problem is a continuous mapping

$$W^{2,2}(-\tfrac{\theta}{2}, +\tfrac{\theta}{2})^3 \times W^{1,2}(-\tfrac{\theta}{2}, +\tfrac{\theta}{2}) \to L_2(-\tfrac{\theta}{2}, +\tfrac{\theta}{2})^3 \times W^{1,2}(-\tfrac{\theta}{2}, +\tfrac{\theta}{2}) \times \mathbb{C}^3$$

for arbitrary λ. The spectrum of the pencil $A(\lambda)$ consists only of eigenvalues with finite geometric and algebraic multiplicities.

9.3.2. Eigenvalues of the pencil $A(\lambda)$. We give a description of the spectrum of the pencil $A(\lambda)$ for different d^- and d^+. Without loss of generality, we may assume that $d^+ \geq d^-$. In order to determine the eigenvalues and eigenfunctions of the pencil $A(\lambda)$, one has to find solutions of the homogeneous problem (9.3.1)–(9.3.4) having the form

$$(u, p) = \big(r^\lambda U(\varphi), r^{\lambda-1}P(\varphi)\big).$$

Problem (9.3.1)–(9.3.4) consists of two boundary value problems which can be handled separately. The homogeneous system (9.3.2), (9.3.4) has the nontrivial solutions

$$u_3 = r^{j\pi/\theta} \sin\frac{j\pi\psi}{\theta}, \quad j = \pm 1, \pm 2, \dots, \quad \text{if } d^\pm \leq 1,$$

$$u_3 = r^{j\pi/\theta} \cos\frac{j\pi\psi}{\theta}, \quad j = 0, \pm 1, \pm 2, \dots, \quad \text{if } d^\pm \geq 2,$$

$$u_3 = r^{(j+1/2)\pi/\theta} \sin\frac{(j+1/2)\pi\psi}{\theta}, \quad j = 0, \pm 1, \pm 2, \dots, \quad \text{if } d^- \leq 1 < d^+,$$

where $\psi = \varphi + \theta/2$.

We consider the boundary value problem (9.3.1), (9.3.3) with the unknown vector function (u_1, u_2, p). Let u_r, u_φ be the polar components of the vector function $u = (u_1, u_2)$, i.e.

$$\begin{pmatrix} u_r \\ u_\varphi \end{pmatrix} = \begin{pmatrix} \cos\varphi & \sin\varphi \\ -\sin\varphi & \cos\varphi \end{pmatrix} \cdot \begin{pmatrix} u_1 \\ u_2 \end{pmatrix}.$$

Then the homogeneous system (9.3.1) can be written as

$$\begin{aligned} -\frac{1}{r^2}\Big((r\partial_r)^2 u_r + \partial_\varphi^2 u_r - 2\partial_\varphi u_\varphi - u_r\Big) + \partial_r p &= 0, \\ -\frac{1}{r^2}\Big((r\partial_r)^2 u_\varphi + \partial_\varphi^2 u_\varphi + 2\partial_\varphi u_r - u_\varphi\Big) + \frac{1}{r}\partial_\varphi p &= 0, \\ \partial_r u_r + \frac{1}{r}(u_r + \partial_\varphi u_\varphi) &= 0, \end{aligned}$$

where $0 < r < \infty$, $-\theta/2 < \varphi < \theta/2$. We are interested in solutions (u_r, u_φ, p) which have the form

$$u_r(r, \varphi) = r^\lambda U_r(\varphi), \quad u_\varphi(r, \varphi) = r^\lambda U_\varphi(\varphi), \quad p(r, \varphi) = r^{\lambda-1} P(\varphi).$$

For the vector function (U_r, U_φ, P), we obtain the following system of ordinary differential equations on the interval $-\theta/2 < \varphi < \theta/2$:

$$\text{(9.3.5)} \qquad \left\{ \begin{array}{l} -U_r'' + (1-\lambda^2)\, U_r + 2\, U_\varphi' + (\lambda - 1)\, P = 0, \\ -U_\varphi'' + (1-\lambda^2)\, U_\varphi - 2\, U_r' + P' = 0, \\ U_\varphi' + (\lambda + 1)\, U_r = 0. \end{array} \right.$$

If $\lambda \neq 0$, then

$$\begin{pmatrix} U_r^{(1)} \\ U_\varphi^{(1)} \\ P^{(1)} \end{pmatrix} = \begin{pmatrix} \cos(1+\lambda)\varphi \\ -\sin(1+\lambda)\varphi \\ 0 \end{pmatrix}, \qquad \begin{pmatrix} U_r^{(2)} \\ U_\varphi^{(2)} \\ P^{(2)} \end{pmatrix} = \begin{pmatrix} (1-\lambda)\cos(1-\lambda)\varphi \\ -(1+\lambda)\sin(1-\lambda)\varphi \\ -4\lambda\cos(1-\lambda)\varphi \end{pmatrix},$$

$$\begin{pmatrix} U_r^{(3)} \\ U_\varphi^{(3)} \\ P^{(3)} \end{pmatrix} = \begin{pmatrix} \sin(1+\lambda)\varphi \\ \cos(1+\lambda)\varphi \\ 0 \end{pmatrix}, \qquad \begin{pmatrix} U_r^{(4)} \\ U_\varphi^{(4)} \\ P^{(4)} \end{pmatrix} = \begin{pmatrix} (1-\lambda)\sin(1-\lambda)\varphi \\ (1+\lambda)\cos(1-\lambda)\varphi \\ -4\lambda\sin(1-\lambda)\varphi \end{pmatrix}.$$

are linearly independent solutions of this system. In the case $\lambda = 0$, the second and the fourth vector have to be replaced by

$$\begin{pmatrix} -\cos\varphi + 2\varphi\sin\varphi \\ -\sin\varphi + 2\varphi\cos\varphi \\ -4\cos\varphi \end{pmatrix} \quad \text{and} \quad \begin{pmatrix} -\sin\varphi - 2\varphi\cos\varphi \\ \cos\varphi + 2\varphi\sin\varphi \\ -4\sin\varphi \end{pmatrix},$$

respectively. Thus, the general solution of the system (9.3.5) has the form

$$\begin{pmatrix} U_r \\ U_\varphi \\ P \end{pmatrix} = \sum_{k=1}^{4} c_k \begin{pmatrix} U_r^{(k)} \\ U_\varphi^{(k)} \\ P^{(k)} \end{pmatrix}. \tag{9.3.6}$$

Inserting

$$\begin{pmatrix} u_1 \\ u_2 \end{pmatrix} = r^\lambda \begin{pmatrix} U_r(\varphi)\cos\varphi - U_\varphi(\varphi)\sin\varphi \\ U_r(\varphi)\sin\varphi + U_\varphi(\varphi)\cos\varphi \end{pmatrix} \quad \text{and} \quad p = r^{\lambda-1}\, P(\varphi)$$

into the boundary conditions (9.3.3), we obtain a linear algebraic system for the coefficients $c_1, \dots, c_4$ in (9.3.6). The coefficient determinant of this system is a function of λ.

The eigenvalues of the pencil $A(\lambda)$ are the zeros of the just mentioned function and the exponents λ in the solutions

$$u_3 = r^\lambda\, U_3(\varphi)$$

of the homogeneous problem (9.3.2), (9.3.4). Here we list the eigenvalues of the operator pencil $A(\lambda)$ for the different combinations of boundary conditions on γ^+ and γ^-.

(1) In the case of the Dirichlet problem ($d^- = d^+ = 0$), the spectrum of the pencil $A(\lambda)$ consists of the numbers $j\pi/\theta$, where j is an arbitrary nonzero integer, and of the nonzero solutions of the equation

$$\lambda \sin\theta \pm \sin(\lambda\theta) = 0.$$

(2) $d^- = 0$, $d^+ = 1$: Then the spectrum consists of the numbers $j\pi/\theta$, where $j = \pm 1, \pm 2, \dots$, and of the nonzero solutions of the equation

$$\lambda \sin(2\theta) + \sin(2\lambda\theta) = 0.$$

(3) $d^- = 0$, $d^+ = 2$: Then the spectrum consists of the numbers $(j+1/2)\pi/\theta$, where j is an arbitrary integer, and of the nonzero solutions of the equation

$$\lambda \sin(2\theta) - \sin(2\lambda\theta) = 0.$$

(4) $d^- = 0,\ d^+ = 3$: Then the numbers $(j+1/2)\pi/\theta$, $j = 0, \pm1, \pm2, \ldots$, and the solutions of the equation

$$\lambda \sin\theta \pm \cos(\lambda\theta) = 0$$

are eigenvalues of the pencil $A(\lambda)$.

(5) $d^- = d^+ = 1$: Then the spectrum consists of the numbers $j\pi/\theta$ and $\pm1 + k\pi/\theta$, where j, k are arbitrary integers, $j \neq 0$.

(6) $d^- = 1,\ d^+ = 2$: Then the spectrum consists of the numbers $j\pi/(2\theta)$ and $\pm1 + k\pi/(2\theta)$, where j and k are arbitrary odd integers.

(7) $d^- = 1,\ d^+ = 3$: Then the numbers $(j+1/2)\pi/\theta$, $j = 0, \pm1, \pm2, \ldots$, and all solutions of the equation

$$\lambda \sin(2\theta) - \sin(2\lambda\theta) = 0.$$

belong to the spectrum.

(8) $d^- = d^+ = 2$: Then the spectrum consists of the numbers $j\pi/\theta$ and $\pm1 + k\pi/\theta$, where j, k are arbitrary integers.

(9) $d^- = 2,\ d^+ = 3$: Then the numbers $j\pi/\theta$, $j = 0, \pm1, \ldots$, and the solutions of the equation

$$\lambda \sin(2\theta) + \sin(2\lambda\theta) = 0$$

belong to the spectrum.

(10) In the case of the Neumann problem ($d^- = d^+ = 3$), the spectrum of the pencil $A(\lambda)$ consists of the numbers $j\pi/\theta$, where j is an arbitrary integer, and of all solutions of the equation

$$\lambda \sin\theta \pm \sin(\lambda\theta) = 0.$$

We refer to the papers by KALEX [**69, 70**], ORLT and SÄNDIG [**167**] and for the cases (1) and (3) also to the paper by MAZ'YA, PLAMENEVSKIĬ, STUPELIS [**125**] and to the book [**85**]. Figure 8 below shows the dependence of the eigenvalues of the pencil $A(\lambda)$ on the angle θ in the cases (1) and (10). The dashed lines represent the curves $\lambda = j\pi/\theta$, the dotted lines represent the solutions of the equation $\lambda \sin\theta + \sin(\lambda\theta) = 0$, and the solid lines correspond to the solutions of the equation $\lambda \sin\theta - \sin(\lambda\theta) = 0$.

Note that the line $\operatorname{Re}\lambda = 0$ may contain only the eigenvalue $\lambda = 0$ and that $\lambda = 0$ is an eigenvalue only if one of the following conditions is satisfied:

- $d^+ + d^- \geq 4$,
- $d^+ = d^- = 1$ and $\theta \in \{\pi, 2\pi\}$,
- $d^+ = 1,\ d^- = 2$ (or $d^- = 1,\ d^+ = 2$) and $\theta \in \{\frac{\pi}{2}, \frac{3\pi}{2}\}$.

The eigenvectors corresponding to the eigenvalue $\lambda = 0$ are of the form $(U, P) = (C, 0)$, where C is constant, and have rank 2 (i.e., have a generalized eigenvector).

In what follows, let δ_+ be the greatest real number such that the strip

$$0 < \operatorname{Re}\lambda < \delta_+$$

does not contain eigenvalues of the pencil $A(\lambda)$.

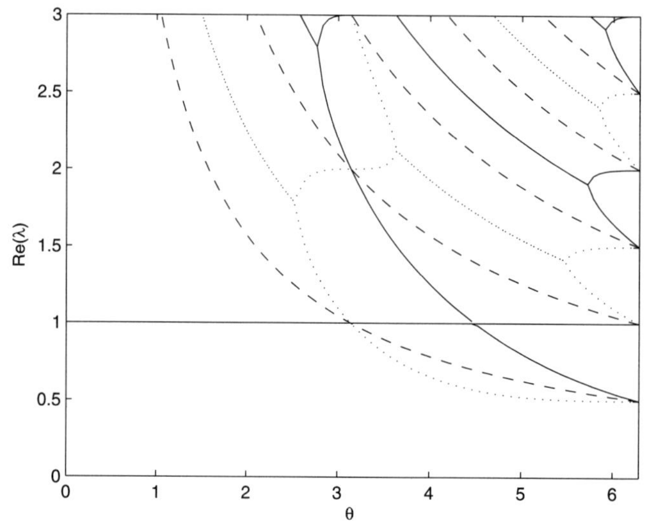

FIGURE 8. Real parts of the eigenvalues for the cases (1) and (10) in the strip $0 < \operatorname{Re}\lambda < 3$

9.3.3. Regularity results for the solutions of the problem (9.3.1)–(9.3.4). First, we prove a $W_\delta^{2,2}$-regularity assertion.

LEMMA 9.3.1. *Let* $(u,p) \in W^{1,2}(K)^3 \times L_2(K)$ *be a solution of the boundary value problem* (9.3.1)–(9.3.4) *vanishing outside the unit ball. If* $f \in W_\delta^{0,2}(K)^3$, $g \in W_\delta^{1,2}(K)$, $h^\pm = 0$, $\phi^\pm = 0$, *and* $\max(0, 1-\delta_+) < \delta < 1$, *then* $u \in W_\delta^{2,2}(K)^3$ *and* $p \in W_\delta^{1,2}(K)$. *Furthermore*

$$\|u\|_{W_\delta^{2,2}(K)^3} + \|p\|_{W_\delta^{1,2}(K)} \le c\Big(\|f\|_{W_\delta^{0,2}(K)^3} + \|g\|_{W_\delta^{1,2}(K)}\Big)$$

with a constant c *independent of* u *and* p.

P r o o f. Since the support of (u,p) is compact, the vector function (u,p) belongs to the weighted space $V_\varepsilon^{1,2}(K)^3 \times V_\varepsilon^{0,2}(K)$ for arbitrary $\varepsilon > 0$. According to Theorem 1.2.6, the vector function u admits the representation

$$u = c + d\,\log r + v$$

with constant vectors c, d and $v \in V_\delta^{2,2}(K)^3$. Furthermore, $p \in V_\delta^{1,2}(K)$ and

$$\|v\|_{V_\delta^{2,2}(K)^3} + \|p\|_{V_\delta^{1,2}(K)} \le c\Big(\|f\|_{V_\delta^{0,2}(K)^3} + \|g\|_{V_\delta^{1,2}(K)}\Big).$$

Since $u \in W_0^{1,2}(K)^3$, we conclude that $d = 0$. The result follows. □

For the proof of higher order regularity assertions, we need the following lemma.

LEMMA 9.3.2. *Let* l *be an integer,* $l \ge 1$. *Furthermore, let* f *be a homogenous vector polynomial of degree* $l-2$ ($f = 0$ *if* $l = 1$), *let* g *be a homogeneous polynomial of degree* $l-1$, *and let* $h^\pm = c^\pm r^l$, $\phi^\pm = d^\pm r^{l-1}$, *where* $c^\pm \in \mathbb{C}^{3-d^\pm}$ *and* $d^\pm \in \mathbb{C}^{d^\pm}$.

We suppose that $\lambda = l$ is not an eigenvalue of the pencil $A(\lambda)$. Then there exist uniquely determined homogeneous polynomials

$$(9.3.7) \qquad u = \sum_{i+j=l} c_{i,j}\, x_1^i\, x_2^j, \qquad p = \sum_{i+j=l-1} d_{i,j}\, x_1^i\, x_2^j,$$

where $c_{i,j} \in \mathbb{C}^3$, $d_{i,j} \in \mathbb{C}$, such that (u,p) is a solution of (9.3.1)–(9.3.4).

P r o o f. Inserting (9.3.7) into (9.3.1)–(9.3.4), we obtain a linear system of $4l+3$ equations with $4l+3$ unknowns $c_{i,j}, d_{i,j}$. Since $\lambda = l$ is not an eigenvalue of $A(\lambda)$, the corresponding homogenous system has only the trivial solution. Therefore, the inhomogeneous system is uniquely solvable. This proves the lemma. □

We employ the last lemma and Theorem 1.2.6 to prove the following regularity assertion in the class of the spaces $W_\delta^{l,s}(K)$.

LEMMA 9.3.3. *Let $(u,p) \in W_\delta^{l,s}(K)^3 \times W_\delta^{l-1,s}(K)$ be a solution of the problem* (9.3.1)–(9.3.4) *with data*

$$f \in W_{\delta'}^{l-1,s}(K)^3, \ g \in W_{\delta'}^{l,s}(K), \ h^\pm \in W_{\delta'}^{l+1-1/s,s}(\gamma^\pm)^{3-d^\pm}, \ \phi^\pm \in W_{\delta'}^{l-1/s}(\gamma^\pm)^{d^\pm},$$

where $l \ge 2$, $-2/s < \delta \le \delta' \le \delta+1$ and $\delta + 2/s$ is not integer. Furthermore, we assume that the strip $l - \delta - 2/s \le \operatorname{Re}\lambda \le l+1-\delta' - 2/s$ does not contain eigenvalues of the pencil $A(\lambda)$. Then $u \in W_{\delta'}^{l+1,s}(K)^3$ and $p \in W_{\delta'}^{l,s}(K)$.

P r o o f. Let $k = [l-\delta-2/s]$ denote the integral part of $l-\delta-2/s$. Then

$$k < l-\delta-2/s < k+1 \quad \text{and} \quad k < l+1-\delta'-2/s < k+2.$$

The derivatives of u and p up to orders k and $k-1$, respectively, are continuous at the origin (cf. [**84**, Lemma 7.1.3]). We denote by $u^{(1)}$ the Taylor polynomial of degree k of u and by $p^{(1)}$ the Taylor polynomial of degree $k-1$ of p, $u^{(1)} = 0$ if $k < 0$, $p^{(1)} = 0$ if $k < 1$. Furthermore, let ζ be a smooth cut-off function in $\mathbb{R}^2$ equal to one for $|x'| < 1$ and to zero for $|x'| > 2$. By Lemma 6.2.12, the functions u and p admit the representations

$$u = u^{(0)} + \zeta u^{(1)}, \quad p = p^{(0)} + \zeta p^{(1)},$$

where $u^{(0)} \in V_\delta^{l,s}(K)^3$ and $p^{(0)} \in V_\delta^{l-1,s}(K)$. From (9.3.1)–(9.3.4) it follows that

$$-\Delta_{x'}\big(u_1^{(0)}, u_2^{(0)}\big) + \nabla_{x'} p^{(0)} = (F_1, F_2), \quad -\nabla_{x'} \cdot \big(u_1^{(0)}, u_2^{(0)}\big) = G \text{ in } K$$

and

$$-\Delta_{x'} u_3^{(0)} = F_3 \text{ in } K,$$

where $F = (F_1,F_2,F_3) \in V_\delta^{l-2,s}(K)^3 \cap W_{\delta'}^{l-1,s}(K)^3$, $G \in V_\delta^{l-1,s}(K) \cap W_{\delta'}^{l,s}(K)$. Furthermore,

$$\tilde{S}^\pm(u_1^{(0)}, u_2^{(0)}) = H^\pm \in V_\delta^{l-1/s,s}(\gamma^\pm) \cap W_{\delta'}^{l+1-1/s}(\gamma^\pm)$$

and

$$\tilde{N}^\pm(u_1^{(0)}, u_2^{(0)}, p^{(0)}) = \Phi^\pm \in V_\delta^{l-1-1/s,s}(\gamma^\pm) \cap W_{\delta'}^{l-1/s,s}(\gamma^\pm).$$

Analogous inclusions hold for the traces of $u_3^{(0)}$ and $\partial u_3^{(0)}/\partial n^\pm$, respectively. Since $F \in V_\delta^{l-2,s}(K)^3$, the derivatives of F up to order $k-2$ are zero at the origin (cf. Lemma 6.2.12).

Suppose first that $l+1-\delta'-2/s < k+1$. Then we conclude from Lemma 6.2.12 that $F \in V_{\delta'}^{l-1,s}(K)^3$ and, analogously, $G \in V_{\delta'}^{l,s}(K)$, $H^\pm \in V_{\delta'}^{l+1-1/s}(\gamma^\pm)$,

$\Phi^\pm \in V_{\delta'}^{l-1/s,s}(\gamma^\pm)$. Applying Corollary 1.2.7, we obtain $u^{(0)} \in V_{\delta'}^{l+1,s}(K)^3$ and $p^{(0)} \in V_{\delta'}^{l,s}(K)$. This implies $u \in W_{\delta'}^{l+1,s}(K)^3$ and $p \in W_{\delta'}^{l,s}(K)$.

We consider the case $l+1-\delta'-2/s > k+1$. Then the representations

$$F = \zeta \sum_{i+j=k-1} F_{i,j}\, x_1^i\, x_2^j + \tilde{F}, \quad G = \zeta \sum_{i+j=k} G_{i,j}\, x_1^i\, x_2^j + \tilde{G}$$

follow from Lemma 6.2.12, where $\tilde{F} \in V_{\delta'}^{l-1,s}(K)^3$ and $\tilde{G} \in V_{\delta'}^{l,s}(K)$, $F_{i,j}$ and $G_{i,j}$ are constants. Analogously,

$$H^\pm = \zeta\, c^\pm\, r^{k+1} + \tilde{H}^\pm, \qquad \Phi^\pm = \zeta\, d^\pm\, r^k + \tilde{\Phi}^\pm,$$

where $\tilde{H}^\pm \in V_{\delta'}^{l+1-1/s,s}(\gamma^\pm)$ and $\tilde{\Phi}^\pm \in V_{\delta'}^{l-1/s,s}(\gamma^\pm)$. Since $\lambda = k+1$ is not an eigenvalue of the pencil $A(\lambda)$, there exist a homogeneous polynomial $u^{(2)}$ of degree $k+1$ and a homogeneous polynomial $p^{(2)}$ of degree k satisfying the equations

$$\begin{pmatrix} -\Delta_{x'}\big(u_1^{(2)}, u_2^{(2)}\big) + \nabla_{x'} p^{(2)} \\ -\Delta_{x'} u_3^{(0)} \end{pmatrix} = \sum_{i+j=k-1} F_{i,j}\, x_1^i\, x_2^j,$$

and

$$-\nabla_{x'} \cdot \big(u_1^{(2)}, u_2^{(2)}\big) = \sum_{i+j=k} G_{i,j}\, x_1^i\, x_2^j$$

in K and the corresponding boundary conditions on the sides $\gamma^\pm$. Then $v = u^{(0)} - \zeta u^{(2)} \in V_\delta^{l,s}(K)^3$, $q = p^{(0)} - \zeta p^{(2)} \in V_\delta^{l-1,s}(K)$,

$$-\Delta_{x'}(v_1, v_2) + \nabla_{x'} q \in V_{\delta'}^{l-1,s}(K)^2, \quad -\nabla_{x'} \cdot (v_1, v_2) \in V_{\delta'}^{l,s}(K)^2,$$

and

$$-\Delta_{x'} v_3 \in V_{\delta'}^{l-1,s}(K)^2$$

Furthermore,

$$\tilde{S}^\pm(v_1, v_2) \in V_{\delta'}^{l+1-1/s,s}(\gamma^\pm), \quad \tilde{N}^\pm(v_1, v_2) \in V_{\delta'}^{l-1/s,s}(\gamma^\pm),$$

$$v_3 \in V_{\delta'}^{l+1-1/s,s}(\gamma^\pm) \ \text{ for } d^\pm \le 1, \qquad \frac{\partial v_3}{\partial n^\pm} \in V_{\delta'}^{l-1/s,s}(\gamma^\pm) \ \text{ for } d^\pm \ge 2.$$

By Corollary 1.2.7, we obtain $(v, q) \in V_{\delta'}^{l+1,s}(K)^3 \times V_{\delta'}^{l,s}(K)$. Therefore, $(u, p) \in W_{\delta'}^{l+1,s}(K)^3 \times W_{\delta'}^{l,s}(K)$.

The case $l+1-\delta'-2/s = k+1$ can be handled analogously using the relation between the space $V_\delta^{l,s}(K)$ and $W_\delta^{l,s}(K)$ for integer $\delta + 2/s > 0$ (see the proof of Lemma 6.3.7). □

9.4. Solvability in weighted L_2 Sobolev spaces

We return to the boundary value problem (9.1.1), (9.1.2). In Section 9.1, we proved the existence and uniqueness of a solution (u, p) in the space $L^{1,2}(\mathcal{D})^3 \times L_2(\mathcal{D})$. Now our goal is to show that $\zeta(u, p)$ belongs to $W_\delta^{l,2}(\mathcal{D})^3 \times W_\delta^{l-1,2}(\mathcal{D})$ for an arbitrary smooth function ζ with compact support if the data $f, g, h^\pm, \phi^\pm$ are from the corresponding weighted spaces. For this, we must suppose that δ satisfies the inequalities

(9.4.1) $$\max(0, l-1-\delta_+) < \delta < l-1.$$

Furthermore, we obtain a modification of the regularity result for the case $d^+ + d^- \le 3$ and for the case where $d^+ + d^-$ is even.

9.4.1. Some estimates for the solutions. The proofs of the following two lemmas are based on local estimates for solutions of elliptic boundary value problems in domains with smooth boundaries. In the sequel, let $W^{l,s}_{loc}(\overline{\mathcal{D}}\backslash M)$ denote the set of all functions u on $\mathcal{D}$ such that $\zeta u \in W^{l,s}(\mathcal{D})$ for all $\zeta \in C_0^\infty(\overline{\mathcal{D}}\backslash M)$.

We start with an estimate in the spaces $V_\delta^{l,s}(\mathcal{D})$ which is analogous to that in Theorem 2.2.9.

LEMMA 9.4.1. *Let (u,p) be a solution of the boundary value problem* (9.1.1), (9.1.2) *such that* $u \in V_{\delta-1}^{l-1,s}(\mathcal{D})^3 \cap W^{2,s}_{loc}(\overline{\mathcal{D}}\backslash M)^3$, $p \in V_{\delta-1}^{l-2,s}(\mathcal{D}) \cap W^{1,s}_{loc}(\overline{\mathcal{D}}\backslash M)$, $l \ge 2$. *If* $f \in V_\delta^{l-2,s}(\mathcal{D})^3$, $g \in V_\delta^{l-1,s}(\mathcal{D})$, $h^\pm \in V_\delta^{l-1/s,s}(\Gamma^\pm)^{3-d^\pm}$ *and* $\phi^\pm \in V_\delta^{l-1-1/s,s}(\Gamma^\pm)^{d^\pm}$, *then* $u \in V_\delta^{l,s}(\mathcal{D})^3$, $p \in V_\delta^{l-1,s}(\mathcal{D})$ *and*

$$\|u\|_{V_\delta^{l,s}(\mathcal{D})^3} + \|p\|_{V_\delta^{l-1,s}(\mathcal{D})} \le c\Big(\|u\|_{V_{\delta-1}^{l-1,s}(\mathcal{D})^3} + \|p\|_{V_{\delta-1}^{l-2,s}(\mathcal{D})} + \|f\|_{V_\delta^{l-2,s}(\mathcal{D})^3}$$
$$+\|g\|_{V_\delta^{l-1,s}(\mathcal{D})} + \sum_\pm \|h^\pm\|_{V_\delta^{l-1/s,s}(\Gamma^\pm)^{3-d^\pm}} + \sum_\pm \|\phi^\pm\|_{V_\delta^{l-1-1/s,s}(\Gamma^\pm)^{d^\pm}}\Big).$$

P r o o f. By Lemma 9.2.1, we may restrict ourselves to the case of zero boundary data $h^\pm$ and $\phi^\pm$. Let $\zeta_{j,k}$ be infinitely differentiable functions such that

$$\operatorname{supp}\zeta_{j,k} \subset \{x:\ 2^{k-1} < |x'| < 2^{k+1},\ j-1 < 2^{-k}x_3 < j+1\}$$

and

$$\sum_{j,k=-\infty}^{+\infty} \zeta_{j,k} = 1, \quad \big|\partial_x^\alpha \zeta_{j,k}(x)\big| \le c\, 2^{-k|\alpha|},$$

where c is independent of j,k and x. Furthermore, let

$$\eta_{j,k} = \sum_{i=j-1}^{j+1} \sum_{l=k-1}^{k+1} \zeta_{i,l}.$$

Obviously, $\eta_{j,k} = 1$ on $\operatorname{supp}\zeta_{j,k}$. We introduce the functions $\tilde\zeta_{j,k}(x) = \zeta_{j,k}(2^k x)$, $\tilde\eta_{j,k}(x) = \eta_{j,k}(2^k x)$, $\tilde u(x) = 2^{-k}u(2^k x)$, $\tilde p(x) = p(2^k x)$, $\tilde f(x) = 2^k f(2^k x)$, and $\tilde g(x) = g(2^k x)$. Then the support of $\zeta_{j,k}$ is contained in the set

$$\{x:\ 1/2 < r(x) < 2,\ j-1 < x_3 < j+1\},$$

and the derivatives of $\zeta_{j,k}$ are bounded by constants independent of j and k. Furthermore, the vector function $(\tilde u, \tilde p)$ satisfies the equations

$$-\Delta\tilde u + \nabla\tilde p = \tilde f, \quad -\nabla\cdot\tilde u = \tilde g \ \text{ in } \mathcal{D}$$

and the homogeneous boundary conditions (9.1.2) on $\Gamma^\pm$. Using local estimates of solutions of elliptic boundary value problems in smooth domains (cf. Theorem 1.1.7), we obtain

$$\|\tilde\zeta_{j,k}\tilde u\|^2_{V_\delta^{l,s}(\mathcal{D})^3} + \|\tilde\zeta_{j,k}\tilde p\|^2_{V_\delta^{l-1,s}(\mathcal{D})} \le c\Big(\|\tilde\eta_{j,k}\tilde f\|^2_{V_\delta^{l-2,s}(\mathcal{D})^3} + \|\tilde\eta_{j,k}\tilde g\|^2_{V_\delta^{l-1,s}(\mathcal{D})}$$
$$+ \|\tilde\eta_{j,k}\tilde u\|^2_{V_{\delta-1}^{l-1,s}(\mathcal{D})^3} + \|\tilde\eta_{j,k}\tilde p\|^2_{V_{\delta-1}^{l-2,s}(\mathcal{D})}\Big)$$

with a constant c independent of j and k. This implies

$$\|\zeta_{j,k}u\|^2_{V_\delta^{l,s}(\mathcal{D})^3} + \|\zeta_{j,k}p\|^2_{V_\delta^{l-1,s}(\mathcal{D})} \le c\Big(\|\eta_{j,k}f\|^2_{V_\delta^{l-2,s}(\mathcal{D})^3} + \|\eta_{j,k}g\|^2_{V_\delta^{l-1,s}(\mathcal{D})}$$
$$+ \|\eta_{j,k}u\|^2_{V_{\delta-1}^{l-1,s}(\mathcal{D})^3} + \|\eta_{j,k}p\|^2_{V_{\delta-1}^{l-2,s}(\mathcal{D})}\Big)$$

with the same constant c. It can be proved analogously to Lemma 2.1.4 that the norm in $V_\delta^{l,s}(\mathcal{D})$ is equivalent to

$$\Big(\sum_{j,k=-\infty}^{+\infty} \|\zeta_{j,k} u\|^2_{V_\delta^{l,s}(\mathcal{D})} \Big)^{1/2}.$$

Using this and the last inequality, we conclude that $u \in V_\delta^{l,s}(\mathcal{D})^3$, $p \in V_\delta^{l-1,s}(\mathcal{D})$. Moreover, the desired estimate for u and p holds. $\square$

An analogous estimate for the case $l = 1$ is given in the next lemma. Here, $\mathcal{H}^*_{s',-\delta;\mathcal{D}}$ denotes the dual space of

$$\mathcal{H}_{s',-\delta;\mathcal{D}} = \{ u \in V_{-\delta}^{1,s'}(\mathcal{D})^3 : \ S^\pm u = 0 \text{ on } \Gamma^\pm \}.$$

LEMMA 9.4.2. *Suppose that $u \in W^{1,s}_{loc}(\overline{\mathcal{D}}\backslash M)^3 \cap V^{0,s}_{\delta-1}(\mathcal{D})^3$ and $p \in W^{0,s}_{loc}(\overline{\mathcal{D}}\backslash M) \cap V^{-1,s}_{\delta-1}(\mathcal{D})$. If (u,p) satisfies the equations*

$$(9.4.2) \quad b_{\mathcal{D}}(u,v) - \int_{\mathcal{D}} p\, \nabla \cdot v\, dx = F(v) \quad \textit{for all } v \in C_0^\infty(\overline{\mathcal{D}}\backslash M)^3, \ S^\pm v|_{\Gamma^\pm} = 0,$$

$-\nabla u = g$ in $\mathcal{D}$ and the boundary conditions $S^\pm u = h^\pm$ on $\Gamma^\pm$, where

$$F \in \mathcal{H}^*_{s',-\delta;\mathcal{D}}, \quad g \in V_\delta^{0,s}(\mathcal{D}), \quad h^\pm \in V_\delta^{1-1/s,s}(\Gamma^\pm)^{3-d^\pm},$$

$s' = s/(s-1)$, $0 < \delta < 1$, then $u \in V_\delta^{1,s}(\mathcal{D})^3$, $p \in V_\delta^{0,s}(\mathcal{D})$, and

$$\|u\|_{V_\delta^{1,s}(\mathcal{D})^3} + \|p\|_{V_\delta^{0,s}(\mathcal{D})} \le c \Big(\|F\|_{\mathcal{H}^*_{s',-\delta;\mathcal{D}}} + \|g\|_{V_\delta^{0,s}(\mathcal{D})} + \sum_\pm \|h^\pm\|_{V_\delta^{1-1/s,s}(\Gamma^\pm)} + \|u\|_{V^{0,s}_{\delta-1}(\mathcal{D})^3} + \|p\|_{V^{-1,s}_{\delta-1}(\mathcal{D})} \Big)$$

with a constant independent of (u,p).

P r o o f. By Lemma 9.2.1, we may assume without loss of generality that $h^\pm = 0$. Let $\zeta_{j,k}$, $\eta_{j,k}$ be the cut-off functions introduced in the proof of Lemma 9.4.1. Then, by the same arguments as in the proof of Lemma 9.4.1, we obtain the estimate

$$\|\zeta_{j,k}u\|^2_{V_\delta^{1,s}(\mathcal{D})^3} + \|\zeta_{j,k}p\|^2_{V_\delta^{0,s}(\mathcal{D})} \le c \Big(\|\eta_{j,k}F\|^2_{\mathcal{H}^*_{s',-\delta;\mathcal{D}}} + \|\eta_{j,k}g\|^2_{V_\delta^{0,s}(\mathcal{D})} + \|\eta_{j,k}u\|^2_{V^{0,s}_{\delta-1}(\mathcal{D})^3} + \|\eta_{j,k}p\|^2_{V^{-1,s}_{\delta-1}(\mathcal{D})} \Big)$$

with a constant c independent of j and k. Summing up over all j and k, we arrive at the desired estimate for u and p. $\square$

We prove a similar assertion for the weighted spaces $W_\delta^{l,s}$ (cf. Theorem 6.4.1).

LEMMA 9.4.3. *Let (u,p) be a solution of the boundary value problem (9.1.1), (9.1.2), and let ζ, η be infinitely differentiable functions on $\overline{\mathcal{D}}$ with compact supports such that $\eta = 1$ in a neighborhood of* supp ζ. *Suppose that $\eta u \in W^{2,s}_{loc}(\overline{\mathcal{D}}\backslash M)^3 \cap W^{l-1,s}_{\delta-1}(\mathcal{D})^3$, $\eta p \in W^{1,s}_{loc}(\overline{\mathcal{D}}\backslash M) \cap W^{l-2,s}_{\delta-1}(\mathcal{D})$, $l \ge 2$, $\delta > 1 - 2/s$, $\eta f \in W_\delta^{l-2,s}(\mathcal{D})^3$, $\eta g \in W_\delta^{l-1,s}(\mathcal{D})$, $\eta h^\pm \in W_\delta^{l-1/s,s}(\Gamma^\pm)^{3-d^\pm}$ and $\eta \phi^\pm \in W_\delta^{l-1-1/s}(\Gamma^\pm)^{d^\pm}$. Then $\zeta u \in W_\delta^{l,s}(\mathcal{D})^3$ and $\zeta p \in W_\delta^{l-1,s}(\mathcal{D})$.*

P r o o f. If $\delta > l-2/s$, then $W_\delta^{l,s}(\mathcal{D})^\ell \subset V_\delta^{l,s}(\mathcal{D})^\ell$, $W_\delta^{l-1,s}(\mathcal{D})^\ell \subset V_\delta^{l-1,s}(\mathcal{D})^\ell$, and the assertion follows immediately from Theorem 9.4.1. Let $1-2/s < \delta \le l-2/s$. Then by Theorems 6.2.7 and 6.2.11, ζu and ζp admit the representations

$$\zeta u = v + u^{(0)}, \quad \zeta p = q + p^{(0)},$$

where $v, p, u^{(0)}, p^{(0)}$ are functions with compact supports, $v \in W_\delta^{l,s}(\mathcal{D})^3$, $q \in W_\delta^{l-1,s}(\mathcal{D})$, $u^{(0)} \in V_{\delta-1}^{l-1,s}(\mathcal{D})^3$, and $p^{(0)} \in V_{\delta-1}^{l-2,s}(\mathcal{D})$. Let $[\Delta,\zeta] = \Delta\zeta - \zeta\Delta$ denote the commutator of Δ and ζ. Then

$$\begin{aligned} -\Delta u^{(0)} + \nabla p^{(0)} &= -\Delta(\zeta u - v) + \nabla(\zeta p - q) \\ &= \zeta f - [\Delta,\zeta]\, u + p\nabla\zeta + \Delta v - \nabla q \in W_\delta^{l-2,s}(\mathcal{D})^3. \end{aligned}$$

Furthermore, $-\Delta u^{(0)} + \nabla p^{(0)} \in V_{\delta-1}^{l-3,s}(\mathcal{D})^3$ if $l \ge 3$ and, consequently,

$$-\Delta u^{(0)} + \nabla p^{(0)} \in V_\delta^{l-2,s}(\mathcal{D})^3.$$

Analogously, we obtain $\nabla \cdot u^{(0)} \in V_\delta^{l-1,s}(\mathcal{D})$, $S^\pm u^{(0)}|_{\Gamma^\pm} \in V_\delta^{l-1/s,s}(\Gamma^\pm)^{3-d^\pm}$, and $N^\pm(u^{(0)},p^{(0)})|_{\Gamma^\pm} \in V_\delta^{l-1-1/s,s}(\Gamma^\pm)^{d^\pm}$. Using Lemma 9.4.1, we conclude that $u^{(0)} \in V_\delta^{l,s}(\mathcal{D})^3$ and $p^{(0)} \in V_\delta^{l-1,s}(\mathcal{D})$. The result follows. □

9.4.2. Smoothness of x_3-derivatives. Our goal is to prove that the variational solution (u,p) of the problem (9.1.1), (9.1.2) belongs to $W_\delta^{2,2}(\mathcal{D})^3 \times W_\delta^{1,2}(\mathcal{D})$ if $f \in W_\delta^{0,2}(\mathcal{D})^3$, $g \in W_\delta^{1,2}(\mathcal{D})$, $h^\pm \in W_\delta^{3/2,2}(\Gamma^\pm)^{3-d^\pm}$ and $\phi^\pm \in W_\delta^{1/2,2}(\Gamma^\pm)^{d^\pm}$, $0<\delta<1$. For this purpose, we show in this subsection that $\partial_{x_3} u \in V_\delta^{1,2}(\mathcal{D})^3$ and $\partial_{x_3} p \in V_\delta^{0,2}(\mathcal{D})$ under the above assumptions on the data. Lemma 9.2.1 allows us to restrict ourselves to zero boundary data $h^\pm$ and $\phi^\pm$.

LEMMA 9.4.4. *Let $(u,p) \in L^{1,2}(\mathcal{D})^3 \times L_2(\mathcal{D})$ be a solution of the problem* (9.1.7), (9.1.8). *Suppose that $g \in V_\delta^{1,2}(\mathcal{D})$, $0<\delta<1$, $h^\pm = 0$, and that the functional F has the form*

$$F(v) = \int_{\mathcal{D}} f \cdot v\, dx,$$

where $f \in V_\delta^{0,2}(\mathcal{D})^3$. We assume furthermore that f and g have compact supports. Then $\partial_{x_3} u \in V_\delta^{1,2}(\mathcal{D})^3$, $\partial_{x_3} p \in V_\delta^{0,2}(\mathcal{D})$ and

$$\|\partial_{x_3} u\|_{V_\delta^{1,2}(\mathcal{D})^3} + \|\partial_{x_3} p\|_{V_\delta^{0,2}(\mathcal{D})} \le c\Big(\|f\|_{V_\delta^{0,2}(\mathcal{D})^3} + \|g\|_{V_\delta^{1,2}(\mathcal{D})}\Big).$$

P r o o f. First note that $F \in \mathcal{H}_{\mathcal{D}}^*$ and $g \in L_2(\mathcal{D})$ under the assumptions of the lemma (cf. Lemma 6.4.5). For arbitrary real h let

$$u_h(x) = h^{-1}\big(u(x', x_3+h) - u(x', x_3)\big).$$

Obviously,

$$b_{\mathcal{D}}(u_h, v) - \int_{\mathcal{D}} p_h\, \nabla\cdot v\, dx = b_{\mathcal{D}}(u, v_{-h}) - \int_{\mathcal{D}} p\, \nabla \cdot v_{-h}\, dx = F(v_{-h}) = \int_{\mathcal{D}} f_h \cdot v\, dx$$

for all $v \in \mathcal{H}_{\mathcal{D}}$, $-\nabla\cdot u_h = g_h$ in $\mathcal{D}$, and $S^\pm u_h = 0$ on $\Gamma^\pm$. Consequently by Theorem 9.1.5, there exists a constant c independent of u, p and h such that

$$\|u_h\|^2_{L^{1,2}(\mathcal{D})^3} + \|p_h\|^2_{L_2(\mathcal{D})} \le c\Big(\|f_h\|^2_{\mathcal{H}_{\mathcal{D}}^*} + \|g_h\|^2_{L_2(\mathcal{D})}\Big). \tag{9.4.3}$$

Let $\tilde{g}(x',\xi)$ be the Fourier transform of $g(x',x_3)$ with respect to the variable x_3. Then

$$\int_0^\infty h^{2\delta-1}\|g_h\|^2_{L_2(\mathcal{D})}\,dh = \int_0^\infty \int_{\mathbb{R}}\int_K h^{2\delta-3}\,|e^{i\xi h}-1|^2\,|\tilde{g}(x',\xi)|^2\,dx'\,d\xi\,dh$$

$$= c\int_{\mathbb{R}}\int_K |\xi|^{2-2\delta}\,|\tilde{g}(x',\xi)|^2\,dx'\,d\xi \le c\int_{\mathbb{R}}\int_K r^{2\delta-2}\,(1+r^2\xi^2)\,|\tilde{g}(x',\xi)|^2\,dx'\,d\xi$$

and consequently

$$\int_0^\infty h^{2\delta-1}\|g_h\|^2_{L_2(\mathcal{D})}\,dh \le c\,\|g\|^2_{V_\delta^{1,2}(\mathcal{D})}\,. \tag{9.4.4}$$

Furthermore,

$$\int_0^\infty h^{2\delta-1}\|f_h\|^2_{\mathcal{H}^*_{\mathcal{D}}}\,dh \le c\,\|f\|^2_{V_\delta^{0,2}(\mathcal{D})^3} \tag{9.4.5}$$

and

$$\|\partial_{x_3}u\|^2_{V^{0,2}_{\delta-1}(\mathcal{D})^3} \le c\int_0^\infty h^{2\delta-1}\,\|u_h\|^2_{\mathcal{H}_{\mathcal{D}}}\,dh. \tag{9.4.6}$$

(see (6.5.2) and (6.5.3)). Since

$$\Big|\int_{\mathcal{D}}\partial_{x_3}p\;v\,dx\Big|^2 \le \int_{\mathbb{R}}\int_K |\xi|^{2\delta}\,|\tilde{v}(x',\xi)|^2\,dx'\,d\xi\;\int_{\mathbb{R}}\int_K |\xi|^{2-2\delta}\,|\tilde{p}(x',\xi)|^2\,dx'\,d\xi$$

$$\le \int_{\mathbb{R}}\int_K r^{-2\delta}\,(1+r^2|\xi|^2)\,|\tilde{v}(x',\xi)|^2\,dx'\,d\xi\;\int_{\mathbb{R}}\int_K |\xi|^{2-2\delta}\,|\tilde{p}(x',\xi)|^2\,dx'\,d\xi$$

$$\le \|v\|^2_{V^{1,2}_{1-\delta}(\mathcal{D})}\int_{\mathbb{R}}\int_K |\xi|^{2-2\delta}\,|\tilde{p}(x',\xi)|^2\,dx'\,d\xi$$

and

$$\int_{\mathbb{R}}\int_K |\xi|^{2-2\delta}\,|\tilde{p}(x',\xi)|^2\,dx'\,d\xi = c\int_0^\infty\int_{\mathbb{R}}\int_K h^{2\delta-3}|e^{i\xi h}-1|^2\,|\tilde{p}(x',\xi)|^2\,dx'\,d\xi\,dh$$

$$= c\int_0^\infty h^{2\delta-1}\|p_h\|^2_{L_2(\mathcal{D})}\,dh,$$

we obtain

$$\|\partial_{x_3}p\|^2_{V^{-1,2}_{\delta-1}(\mathcal{D})} \le \int_{\mathbb{R}}\int_K |\xi|^{2-2\delta}\,|\tilde{p}(x',\xi)|^2\,dx'\,d\xi \le c\int_0^\infty h^{2\delta-1}\|p_h\|^2_{L_2(\mathcal{D})}\,dh.$$

The last inequality along with (9.4.3)–(9.4.6) implies

$$\|\partial_{x_3}u\|_{V^{0,2}_{\delta-1}(\mathcal{D})^3} + \|\partial_{x_3}p\|_{V^{-1,2}_{\delta-1}(\mathcal{D})^3} \le c\left(\|f\|_{V_\delta^{0,2}(\mathcal{D})^3} + \|g\|_{V_\delta^{1,2}(\mathcal{D})}\right).$$

Using Lemma 9.4.2, we conclude that $\partial_{x_3}u \in V_\delta^{1,2}(\mathcal{D})^3$ and $\partial_{x_3}p \in V_\delta^{0,2}(\mathcal{D})$. Furthermore, the desired estimate for $\partial_{x_3}u$ and $\partial_{x_3}p$ holds. □

9.4.3. Existence of solutions in $L^{2,2}_\delta(\mathcal{D})^3 \times L^{1,2}_\delta(\mathcal{D})$. In Section 9.3, we proved a regularity result for the solutions of the model problem in the angle K. Combining this result with the regularity assertion for the x_3-derivative of the solution given in Lemma 9.4.4, we obtain the following statement.

THEOREM 9.4.5. *Suppose that* $\max(0, 1-\delta_+) < \delta < 1$ *and that*

$$f \in W^{0,2}_\delta(\mathcal{D})^3, \quad g \in W^{1,2}_\delta(\mathcal{D}), \quad h^\pm \in W^{3/2,2}_\delta(\Gamma^\pm)^{3-d^\pm}, \quad \phi^\pm \in W^{1/2,2}_\delta(\Gamma^\pm)^{d^\pm}$$

are functions with compact supports. We assume that h^+ *and* h^- *satisfy the compatibility condition* (9.2.6). *Furthermore, let* $(u,p) \in L^{1,2}(\mathcal{D})^3 \times L_2(\mathcal{D})$ *be the uniquely determined solution of the problem* (9.1.7), (9.1.8), *where the functional* F *is given by* (9.1.9). *Then* $(u,p) \in L^{2,2}_\delta(\mathcal{D})^3 \times L^{1,2}_\delta(\mathcal{D})$ *and*

$$\|u\|_{L^{2,2}_\delta(\mathcal{D})^3} + \|p\|_{L^{1,2}_\delta(\mathcal{D})} \le c\Big(\|f\|_{W^{0,2}_\delta(\mathcal{D})^3} + \|g\|_{W^{1,2}_\delta(\mathcal{D})} + \sum_\pm \|h^\pm\|_{W^{3/2,2}_\delta(\Gamma^\pm)^{3-d^\pm}} + \sum_\pm \|\phi^\pm\|_{W^{1/2,2}_\delta(\Gamma^\pm)^{d^\pm}}\Big) \tag{9.4.7}$$

with a constant c *independent of* f, g, $h^\pm$, *and* $\phi^\pm$.

P r o o f. First note that the conditions of the theorem ensure the existence of a uniquely determined solution in $L^{1,2}(\mathcal{D})^3 \times L_2(\mathcal{D})$. By Lemma 9.2.1, we may assume without loss of generality that $h^\pm = 0$ and $\phi^\pm = 0$. Let ζ be an infinitely differentiable function depending only on $r = |x'|$ such that $\zeta = 1$ for $r < 1/2$ and $\zeta = 0$ for $r > 1$. Then $\zeta u \in W^{1,2}(\mathcal{D})^3$. Furthermore by Lemma 9.4.4, $\partial_{x_3}(\zeta u) \in V^{1,2}_\delta(\mathcal{D})^3$ and $\partial_{x_3}(\zeta p) \in V^{0,2}_\delta(\mathcal{D})$. Consequently, the vector function $\zeta\,(u,p)$ satisfies the equations

$$-\Delta_{x'}\,(\zeta u_1, \zeta u_2) + \nabla_{x'}(\zeta p) = (F_1, F_2), \quad -\nabla_{x'} \cdot (\zeta u_1, \zeta u_2) = G, \tag{9.4.8}$$

$$-\Delta_{x'}(\zeta u_3) = F_3\,, \tag{9.4.9}$$

where $F(\cdot, x_3) \in W^{0,2}_\delta(K)^3$ and $G(\cdot, x_3) \in W^{1,2}_\delta(K)$ for almost all x_3. Furthermore, $(\zeta u)(\cdot, x_3)$ and $(\zeta p)(\cdot, x_3)$ satisfy the boundary conditions (9.3.3), (9.3.4) with boundary data in $W^{3/2,2}_\delta(\gamma^\pm)$ and $W^{1/2,2}_\delta(\gamma^\pm)$, respectively. Applying Lemma 9.3.1, we get $(\zeta u)(\cdot, x_3) \in W^{2,2}_\delta(K)^3$, $(\zeta p)(\cdot, x_3) \in W^{1,2}_\delta(K)$, and

$$\|(\zeta u)(\cdot, x_3)\|^2_{W^{2,2}_\delta(K)^3} + \|(\zeta p)(\cdot, x_3)\|^2_{W^{1,2}_\delta(K)} \le c\Big(\|f(\cdot, x_3)\|^2_{W^{0,2}_\delta(K)^3} + \|g(\cdot, x_3)\|^2_{W^{1,2}_\delta(K)} + \|\partial_{x_3}(\zeta u)(\cdot, x_3)\|^2_{W^{1,2}_\delta(K)^3} + \|\partial_{x_3}(\zeta p)(\cdot, x_3)\|^2_{W^{0,2}_\delta(K)}\Big)$$

with a constant c independent of x_3. Integrating this inequality with respect to x_3 and using Lemma 9.4.4, we arrive at the inequality

$$\|\zeta u\|^2_{W^{2,2}_\delta(\mathcal{D})^3} + \|\zeta p\|^2_{W^{1,2}_\delta(\mathcal{D})} \le c\Big(\|f(\cdot, x_3)\|^2_{W^{0,2}_\delta(\mathcal{D})^3} + \|g\|^2_{W^{1,2}_\delta(\mathcal{D})}\Big).$$

We consider the vector function $(1-\zeta)\,(u,p)$. Obviously, $(1-\zeta)p \in V^{0,2}_{\delta-1}(\mathcal{D})$ and $(1-\zeta)u \in V^{1,2}_{\delta-1}(\mathcal{D})^3$. The last follows from Hardy's inequality. Furthermore,

$$-\Delta(u - \zeta u) + \nabla(p - \zeta p) \in V^{0,2}_\delta(\mathcal{D})^3, \quad \nabla \cdot (u - \zeta u) \in W^{1,2}_\delta(\mathcal{D}) \subset V^{1,2}_\delta(\mathcal{D}),$$

$S^\pm(u - \zeta u) = 0$ on $\Gamma^\pm$ and $N^\pm(u - \zeta u, p - \zeta p) \in V^{1/2,2}_\delta(\Gamma^\pm)^{d^\pm}$. Thus, we conclude from Lemma 9.4.1 that $(1-\zeta)u \in V^{2,2}_\delta(\mathcal{D})^3$ and $(1-\zeta)p \in V^{1,2}_\delta(\mathcal{D})$. Moreover, the estimate

$$\|(1-\zeta)u\|^2_{V^{2,2}_\delta(\mathcal{D})^3} + \|(1-\zeta)p\|^2_{V^{1,2}_\delta(\mathcal{D})} \le c\Big(\|f(\cdot,x_3)\|^2_{W^{0,2}_\delta(\mathcal{D})^3} + \|g\|^2_{W^{1,2}_\delta(\mathcal{D})}\Big)$$

holds. This completes the proof. □

Next we prove the analogous regularity assertion for the x_3-derivatives of the solution.

THEOREM 9.4.6. *Let ζ, η be smooth functions with compact supports such that $\eta = 1$ in a neighborhood of $\operatorname{supp}\zeta$, and let $(u,p) \in L^{1,2}(\mathcal{D})^3 \times L_2(\mathcal{D})$ be a solution of (9.1.7), (9.1.8), where F is a functional on $\mathcal{H}_\mathcal{D}$ which has the form (9.1.9). Here it is assumed that*

$$\eta\partial^j_{x_3} f \in W^{0,2}_\delta(\mathcal{D})^3,\quad \eta\partial^j_{x_3} g \in W^{1,2}_\delta(\mathcal{D}),\quad \eta\partial^j_{x_3}\phi^\pm \in W^{1/2,2}_\delta(\Gamma^\pm)^{d^\pm}$$

for $j = 0,1,\dots,k$, where $\max(0,1-\delta_+) < \delta < 1$. Furthermore, we suppose that $\eta\partial^j_{x_3} h^\pm \in W^{3/2,2}_\delta(\Gamma^\pm)^{3-d^\pm}$ for $j=0,1,\dots,k$ and that $h^\pm$ satisfy the compatibility condition (9.2.6). Then $\zeta\partial^j_{x_3}(u,p) \in W^{2,2}_\delta(\mathcal{D})^3 \times W^{1,2}_\delta(\mathcal{D})$ for $j=0,1,\dots,k$ and

$$(9.4.10)\quad \sum_{j=0}^k \Big(\|\zeta\partial^j_{x_3}u\|_{W^{2,2}_\delta(\mathcal{D})^3} + \|\zeta\partial^j_{x_3}p\|_{W^{1,2}_\delta(\mathcal{D})}\Big) \le c\Big(\|\eta u\|_{W^{1,2}_\delta(\mathcal{D})^3}$$

$$+ \|\eta p\|_{W^{0,2}_\delta(\mathcal{D})} + \sum_{j=0}^k \Big(\|\eta\partial^j_{x_3}f\|_{W^{0,2}_\delta(\mathcal{D})^3} + \|\eta\partial^j_{x_3}g\|_{W^{1,2}_\delta(\mathcal{D})}$$

$$+ \sum_\pm \|\eta\partial^j_{x_3}h^\pm\|_{W^{3/2,2}_\delta(\Gamma^\pm)^{3-d^\pm}} + \sum_\pm \|\eta\partial^j_{x_3}\phi^\pm\|_{W^{1/2,2}_\delta(\Gamma^\pm)^{d^\pm}}\Big)\Big).$$

P r o o f. For $k=0$ the assertion follows immediately from Theorem 9.4.5. Let the conditions of the theorem on F, g and $h^\pm$ with $k=1$ be satisfied. Moreover, we suppose that $\eta\partial^j_{x_3}h^\pm \in V^{3/2,2}_\delta(\Gamma^\pm)^{3-d^\pm}$ for $j=0$ and $j=1$. By χ and χ_1 we denote smooth functions such that $\chi=1$ in a neighborhood of $\operatorname{supp}\zeta$, $\chi_1 = 1$ in a neighborhood of $\operatorname{supp}\chi$, and $\eta=1$ in a neighborhood of $\operatorname{supp}\chi_1$. Furthermore, for an arbitrary function v on $\mathcal{D}$ or $\Gamma^\pm$, we set

$$v_h(x',x_3) = h^{-1}\big(v(x',x_3+h) - v(x',x_3)\big).$$

Obviously, $(u_h,p_h) \in \mathcal{H}\times L_2(\mathcal{D})$ for arbitrary real h. Consequently by Theorem 9.4.5,

$$(9.4.11)\quad \|\zeta u_h\|_{W^{2,2}_\delta(\mathcal{D})^3} + \|\zeta p_h\|_{W^{1,2}_\delta(\mathcal{D})} \le c\Big(\|\chi f_h\|_{W^{0,2}_\delta(\mathcal{D})^3} + \|\chi g_h\|_{W^{1,2}_\delta(\mathcal{D})}$$

$$+\sum_\pm \|\chi h^\pm_h\|_{V^{3/2,2}_\delta(\Gamma^\pm)^{3-d^\pm}} + \sum_\pm \|\chi\phi^\pm_h\|_{V^{1/2,2}_\delta(\Gamma^\pm)^{3-d^\pm}}$$

$$+\|\chi u_h\|_{W^{1,2}_\delta(\mathcal{D})^3} + \|\chi p_h\|_{W^{0,2}_\delta(\mathcal{D})}\Big)$$

with a constant c independent of h, where χ is a smooth function, $\chi=1$ in a neighborhood of $\operatorname{supp}\zeta$, $\eta=1$ in a neighborhood of $\operatorname{supp}\chi$. Here $\chi f_h = (\chi f)_h - \chi_h f$ and, for sufficiently small $|h|$,

$$\|(\chi f)_h\|^2_{W^{0,2}_\delta(\mathcal{D})^3} = \int_\mathcal{D} r^{2\delta}\Big|\int_0^1 \frac{\partial(\chi f)}{\partial x_3}(x',x_3+th)\,dt\Big|^2 dx \le \|\partial_{x_3}(\chi f)\|^2_{W^{0,2}_\delta(\mathcal{D})},$$

$$\|\chi_h f\|^2_{W^{0,2}_\delta(\mathcal{D})^3} \le c\,\|\eta f\|_{W^{0,2}_\delta(\mathcal{D})^3}.$$

Analogous estimates hold for the norms of χg_h, $\chi h_h^\pm$ and $\chi\phi_h^\pm$, χu_h, and χp_h on the right-hand side of (9.4.11). Here one can use the equivalence of the norm in $V_\delta^{l-1/2,2}(\Gamma^\pm)$ to the second norm in Lemma 2.1.10. Hence, the right-hand side and therefore also the limit (as $h \to 0$) of the left-hand of (9.4.11) are dominated by

$$c\sum_{j=0}^{1}\Big(\|\eta\partial_{x_3}^j f\|_{W_\delta^{0,2}(\mathcal{D})^3} + \|\eta\partial_{x_3}^j g\|_{W_\delta^{1,2}(\mathcal{D})} + \sum_\pm \|\eta\partial_{x_3}^j h^\pm\|_{V_\delta^{3/2,2}(\Gamma^\pm)^{3-d^\pm}}$$
$$+\sum_\pm \|\eta\partial_{x_3}^j \phi^\pm\|_{V_\delta^{1/2,2}(\Gamma^\pm)^{3-d^\pm}} + \|\chi_1\partial_{x_3}^j u\|_{W_\delta^{1,2}(\mathcal{D})^3} + \|\chi_1\partial_{x_3}^j p\|_{W_\delta^{0,2}(\mathcal{D})}\Big).$$

By Theorem 9.4.5, the norm of $\chi_1\partial_{x_3}(u,p)$ in $W_\delta^{1,2}(\mathcal{D})^3 \times W_\delta^{0,2}$ is dominated by the right-hand side of (9.4.10) with $k=0$. This proves (9.4.10) for $k=1$.

Suppose now that $\eta\partial_{x_3}^j h^\pm \in W_\delta^{3/2,2}(\Gamma^\pm)^{3-d^\pm}$ for $j=0,1$ and that the compatibility condition (9.2.6) is satisfied. Then there exists a vector function $\psi \in W^{k+1-\delta,2}(M)^3$ such that $S^\pm\psi = (\eta h^\pm)|_M$. Let $v \in W_\delta^{k+2,2}(\mathcal{D})^3$ be an extension of ψ. Then $\chi\partial_{x_3}^j(h^\pm - S^\pm v)|_M = 0$ for $j=0,\dots,k$ and consequently

$$\chi\partial_{x_3}^j(h^\pm - S^\pm v|_{\Gamma^\pm}) \in V_\delta^{3/2,2}(\Gamma^\pm)^{3-d^\pm}.$$

Now we can apply the above proved result to the vector function $(u-v,p)$ and obtain $\zeta\partial_{x_3}(u-v,p) \in W_\delta^{2,2}(\mathcal{D})^3 \times W_\delta^{1,2}(\mathcal{D})$.

3) For $k>1$ the assertion of the theorem holds by induction. □

9.4.4. Regularity results for the solutions. The next lemma is based on the analogous result for the model problem in an angle (cf. Lemma 9.3.3).

LEMMA 9.4.7. *Let ζ, η be smooth functions on $\overline{\mathcal{D}}$ with compact supports such that $\eta=1$ in a neighborhood of* $\operatorname{supp}\zeta$, *and let (u,p) be a solution of the boundary value problem* (9.1.1), (9.1.2) *such that*

$$\eta u \in W_\delta^{l,s}(\mathcal{D})^3, \quad \eta p \in W_\delta^{l-1,s}(\mathcal{D}), \quad \eta\partial_{x_3}u \in W_{\delta'}^{l,s}(\mathcal{D})^3, \quad \eta\partial_{x_3}p \in W_{\delta'}^{l-1,s}(\mathcal{D})$$

where $l\ge 2$, $-2/s < \delta \le \delta' \le \delta+1$. Furthermore, we assume that

$$\eta f \in W_{\delta'}^{l-1,s}(\mathcal{D})^3, \quad \eta g \in W_{\delta'}^{l,s}(\mathcal{D}),$$
$$\eta h^\pm \in W_{\delta'}^{l+1-1/s,s}(\Gamma^\pm)^{3-d^\pm}, \quad \eta\phi^\pm \in W_{\delta'}^{l-1/s,s}(\Gamma^\pm)^{d^\pm}.$$

If the strip $l-\delta-2/s \le \operatorname{Re}\lambda \le l+1-\delta'-2/s$ is free of eigenvalues of the pencil $A(\lambda)$, then $\zeta u \in W_{\delta'}^{l+1,s}(\mathcal{D})^3$ and $\zeta p \in W_{\delta'}^{l,s}(\mathcal{D})$.

P r o o f. By the assumptions on (u,p) and $\partial_{x_3}(u,p)$, the vector function $\zeta(u,p)$ satisfies the equations (9.4.8) and (9.4.9), where $F(\cdot,x_3) \in W_{\delta'}^{l-1,s}(K)^3$ and $G(\cdot,x_3) \in W_{\delta'}^{l,s}(K)^3$ for almost all x_3. Furthermore, $(\zeta u)(\cdot,x_3)$ and $(\zeta p)(\cdot,x_3)$ satisfy the boundary conditions (9.3.3), (9.3.4), where the right-hand sides are functions in $W_{\delta'}^{l+1-1/s,s}(\gamma^\pm)$ and $W_{\delta'}^{l-1/s,s}(\gamma^\pm)$, respectively. From this and from Lemma 9.3.3 we conclude that $(\zeta u)(\cdot,x_3) \in W_{\delta'}^{l+1,s}(K)^3$ and $(\zeta p)(\cdot,x_3) \in W_{\delta'}^{l,s}(\mathcal{D})$.

Furthermore, the estimate

$$\begin{aligned}
&\|(\zeta u)(\cdot,x_3)\|^s_{W^{l+1,s}_{\delta'}(K)^3} + \|(\zeta p)(\cdot,x_3)\|^s_{W^{l,s}_{\delta'}(K)} \\
&\quad \le c\Big(\|(\eta f)(\cdot,x_3\|^s_{W^{l-1,s}_{\delta'}(K)^3} + \|(\eta g)(\cdot,x_3)\|^s_{W^{l,s}_{\delta'}(K)} \\
&\qquad + \|(\eta h^\pm)(\cdot,x_3)\|^s_{W^{l+1-1/s,s}_{\delta'}(\gamma^\pm)^{3-d^\pm}} + \|(\eta\phi^\pm)(\cdot,x_3)\|^s_{W^{l-1/s,s}_{\delta'}(\gamma^\pm)^{d^\pm}} \\
&\qquad + \|(\eta u)(\cdot,x_3)\|^s_{W^{l,s}_{\delta}(K)^3} + \|(\eta p)(\cdot,x_3)\|^s_{W^{l-1,s}_{\delta}(K)} \\
&\qquad + \|\eta(\cdot,x_3)\partial_{x_3}u(\cdot,x_3)\|^s_{W^{l,s}_{\delta'}(K)^3} + \|\eta(\cdot,x_3)\,\partial_{x_3}p(\cdot,x_3)\|^s_{W^{l-1,s}_{\delta'}(K)}\Big)
\end{aligned}$$

with a constant c independent of u, p and x_3. Integrating this inequality with respect to x_3, we obtain an analogous estimate for the norms of ζu and ζp in $W^{l+1,s}_{\delta'}(\mathcal{D})^3$ and $W^{l,s}_{\delta'}(\mathcal{D})$, respectively. □

REMARK 9.4.8. The assertion of Lemma 9.4.7 is also valid in the class of the spaces $V^{l,s}_\delta$. Then one has to employ Corollary 1.2.7 instead of Lemma 9.3.3 in the proof.

The last lemma allows us to generalize the regularity assertion of Theorem 9.4.6.

THEOREM 9.4.9. *Let ζ, η be the same cut-off functions as in Theorem 9.4.6, and let $(u,p) \in L^{1,2}(\mathcal{D})^3 \times L_2(\mathcal{D})$ be a solutions of (9.1.7), (9.1.8). Suppose that F is a functional on $\mathcal{H}_\mathcal{D}$ which has the form (9.1.9), where*

$$\eta\partial^j_{x_3} f \in W^{l-2,2}_\delta(\mathcal{D})^3, \quad \eta\partial^j_{x_3} g \in W^{l-1,2}_\delta(\mathcal{D}), \quad \eta\partial^j_{x_3}\phi^\pm \in W^{l-3/2,2}_\delta(\Gamma^\pm)^{d^\pm}$$

for $j = 0,1,\dots,k$. Here, $l \ge 2$, δ is a noninteger number satisfying the inequalities (9.4.1). Furthermore, we assume that $\eta\partial^j_{x_3} h^\pm \in W^{l-1/2,2}_\delta(\Gamma^\pm)^{3-d^\pm}$ for $j = 0,1,\dots,k$ and that $h^\pm$ satisfy the compatibility condition (9.2.6). Then $\zeta\partial^j_{x_3}(u,p) \in W^{l,2}_\delta(\mathcal{D})^3 \times W^{l-1,2}_\delta(\mathcal{D})$ for $j = 0,1,\dots,k$. Furthermore, an estimate analogous to (9.4.10) holds.

P r o o f. 1) First, we consider the case $k = 0$, $\max(0, l-1-\delta_+) < \delta < 1$. In this case, we prove the assertion by induction in l. For $l = 2$ we refer to Theorem 9.4.6. Suppose the assertion is proved for a certain integer $l = s \ge 2$ and that the conditions of the theorem are satisfied for $l = s+1$. We denote by χ and χ_1 the same cut-off functions as in the proof of Theorem 9.4.6. Then by the induction hypothesis,

$$\chi_1(u,p) \in W^{s,2}_\delta(\mathcal{D})^3 \times W^{s-1,2}_\delta(\mathcal{D}), \quad \chi_1\partial_{x_3}(u,p) \in W^{s-1,2}_\delta(\mathcal{D})^3 \times W^{s-2,2}_\delta(\mathcal{D}).$$

If $s \ge 3$, this implies that $\chi_1\partial_{x_3}(u,p) \in L^{1,2}(\mathcal{D})^3 \times L_2(\mathcal{D})$. By the induction hypothesis, we obtain $\chi\partial_{x_3}(u,p) \in W^{s,2}_\delta(\mathcal{D})^3 \times W^{s-1,2}_\delta(\mathcal{D})$. For $s = 2$ the last inclusion follows from Theorem 9.4.6. Consequently,

$$\chi\partial^j_{x_3}(u,p) \in W^{s,2}_\delta(\mathcal{D})^3 \times W^{s-1,2}_\delta(\mathcal{D}) \quad \text{for } j = 0 \text{ and } j = 1.$$

Using this result and Lemma 9.4.7, we conclude that $\zeta(u,p) \in W^{s+1,2}_\delta(\mathcal{D})^3 \times W^{s,2}_\delta(\mathcal{D})$.

2) Now let $k = 0$ and $\max(\sigma, l-1-\delta_+) < \delta < \sigma+1$, where σ is an integer, $1 \le \sigma < l-2-\delta_+$. Then $\max(0, l-1-\sigma-\delta_+) < \delta-\sigma < 1$, $\eta f \in W^{l-\sigma-2,2}_{\delta-\sigma}(\mathcal{D})^3$, $\eta g \in$

$W^{l-\sigma-1,2}_{\delta-\sigma}(\mathcal{D})$, $\eta h^\pm \in W^{l-\sigma-1/2,2}_{\delta-\sigma}(\Gamma^\pm)^{3-d^\pm}$, and $\eta\phi^\pm \in W^{l-\sigma-3/2,2}_{\delta-\sigma}(\Gamma^\pm)^{d^\pm}$. Therefore by the first part of the proof, we obtain $\chi(u,p) \in W^{l-\sigma,2}_{\delta-\sigma}(\mathcal{D})^3 \times W^{l-\sigma-1,2}_{\delta-\sigma}(\mathcal{D})$. From this and from Lemma 9.4.3 we conclude that $\zeta(u,p) \in W^{l,2}_\delta(\mathcal{D})^3 \times W^{l-1,2}_\delta(\mathcal{D})$. Thus, the theorem is proved for $k = 0$.

3) Next, we prove the assertion for $k > 0$, $\max(l-2, l-1-\delta_+) < \delta < l-1$. If $\max(l-2, l-1-\delta_+) < \delta < l-1$, then it follows from Theorem 9.4.6 that $\chi\partial^j_{x_3}(u,p) \in W^{2,2}_{\delta-l+2}(\mathcal{D})^3 \times W^{1,2}_{\delta-l+2}(\mathcal{D})$ for $j = 0, 1, \dots, k$. Using Lemma 9.4.3, we obtain $\zeta\partial^j_{x_3}(u,p) \in W^{l,2}_\delta(\mathcal{D})^3 \times W^{l-1,2}_\delta(\mathcal{D})$ for $j = 0, 1, \dots, k$.

4) It remains to consider the case $k > 0$, $\max(0, l-1-\delta_+) < \delta < l-2$. In this case, we prove the assertion by induction in k. Suppose the theorem is proved for $k-1$. From parts 1) and 2) of the proof we conclude that $\chi(u,p) \in W^{l,2}_\delta(\mathcal{D})^3 \times W^{l-1,2}_\delta(\mathcal{D})$ and $\chi\partial_{x_3}(u,p) \in W^{l-1,2}_\delta(\mathcal{D})^3 \times W^{l-2,2}_\delta(\mathcal{D})$. In particular, this means that $\chi\partial_{x_3}(u,p) \in L^{1,2}(\mathcal{D})^3 \times L_2(\mathcal{D})$. Thus by the induction hypothesis, $\zeta\partial^j_{x_3}\partial_{x_3}(u,p) \in W^{l,2}_\delta(\mathcal{D})^3 \times W^{l-1,2}_\delta(\mathcal{D})$ for $j = 0, \dots, k-1$. The proof of the theorem is complete. □

9.4.5. A modification of the regularity result in the case $d^- + d^+ \le 3$. We consider the case $d^- + d^+ \le 3$ and suppose that

(9.4.12) $\qquad \theta \notin \{\pi, 2\pi\}$ if $d^- = d^+ = 1$, $\quad \theta \notin \{\pi/2, 3\pi/2\}$ if $d^- \cdot d^+ = 2$.

Then the number $\lambda = 0$ does not belong to the spectrum of the pencil $A(\lambda)$. Furthermore in this case, the conditions

$$S^+u|_M = 0, \quad S^-u|_M = 0$$

for a vector function $u \in W^{2,2}_\delta(\mathcal{D})^3$, $0 < \delta < 1$, imply

$$u|_M = 0.$$

Using this fact, we can easily deduce the following result from Theorem 9.4.6.

THEOREM 9.4.10. *Let ζ, η be the same cut-off functions as in Theorem 9.4.6, and let $(u,p) \in L^{1,2}(\mathcal{D})^3 \times L_2(\mathcal{D})$ be a solution of problem (9.1.7), (9.1.8). We suppose that $d^- + d^+ \le 3$ and that the angle θ satisfies the condition (9.4.12). Furthermore, we assume that*

$$\eta\partial^j_{x_3} f \in V^{0,2}_\delta(\mathcal{D})^3, \quad \eta\partial^j_{x_3} g \in V^{1,2}_\delta(\mathcal{D}),$$
$$\eta\partial^j_{x_3} h^\pm \in V^{3/2,2}_\delta(\Gamma^\pm)^{3-d^\pm}, \quad \eta\partial^j_{x_3}\phi^\pm \in V^{1/2,2}_\delta(\Gamma^\pm)^{d^\pm},$$

for $j = 0, 1, \dots, k$, where $\max(0, 1-\delta_+) < \delta < 1$, and that the functional F on $\mathcal{H}_\mathcal{D}$ has the form (9.1.9). Then $\zeta\partial^j_{x_3}(u,p) \in V^{2,2}_\delta(\mathcal{D})^3 \times V^{1,2}_\delta(\mathcal{D})$ for $j = 1, \dots, k$ and an estimate analogous to (9.4.10) holds.

Proof. According to Theorem 9.4.6, we have $\zeta\partial^j_{x_3}(u,p) \in W^{2,2}_\delta(\mathcal{D})^3 \times V^{1,2}_\delta(\mathcal{D})$ for $j = 0, \dots, k$. Furthermore,

$$S^\pm\zeta\partial^j_{x_3}u|_M = \zeta\partial^j_{x_3}h^\pm|_M = 0 \quad \text{for } j = 0, 1, \dots, k$$

and therefore $\zeta\partial^j_{x_3}u|_M = 0$. From this and from Lemma 6.2.5 we conclude that $\zeta\partial^j_{x_3}u \in V^{2,2}_\delta(\mathcal{D})^3$. This proves the lemma. □

Furthermore, the following generalization of the last theorem can be proved in the same manner as Theorem 9.4.9 (see Remark 9.4.8).

THEOREM 9.4.11. *Let ζ, η be the same cut-off functions as in Theorem* 9.4.6, *and let* $(u,p) \in L^{1,2}(\mathcal{D})^3 \times L_2(\mathcal{D})$ *be a solution of the problem* (9.1.7), (9.1.8). *Suppose that*

$$\eta\partial_{x_3}^j f \in V_\delta^{l-2,2}(\mathcal{D})^3, \quad \eta\partial_{x_3}^j g \in V_\delta^{l-1,2}(\mathcal{D}),$$

$$\eta\partial_{x_3}^j h^\pm \in V_\delta^{l-1/2,2}(\Gamma^\pm)^{3-d^\pm}, \quad \eta\partial_{x_3}^j \phi^\pm \in V_\delta^{l-3/2,2}(\Gamma^\pm)^{d^\pm}$$

for $j = 0, 1, \dots, k$, *and that the functional* F *has the form* (9.1.9). *If* δ *is not an integer and satisfies the condition* (9.4.1), *then* $\zeta\partial_{x_3}^j(u,p) \in V_\delta^{l,2}(\mathcal{D})^3 \times V_\delta^{l-1,2}(\mathcal{D})$ *for* $j = 0, 1, \dots, k$.

REMARK 9.4.12. In the case, where $\lambda = 0$ is not an eigenvalue of the pencil $A(\lambda)$, one can establish a solvability and regularity theory in the class of the spaces $V_\delta^{l,2}(\mathcal{D})$ which is analogous to the theory for the Dirichlet problem in Chapter 2. Then it is possible to show that the results of Theorems 9.4.10 and 9.4.11 are valid for arbitrary δ satisfying the inequalities

$$l - 1 - \delta_+ < \delta < l - 1 + \delta_+ .$$

For the Dirichlet problem and for the case $d^- = 0$, $d^+ = 2$, we refer to [**125**]. However, we will use in this book only the result under the sharper condition (9.4.1).

9.4.6. The case where $d^+ + d^-$ is even. The number $\lambda = 1$ is an eigenvalue of the pencil $A(\lambda)$ for all angles θ if $d^+ + d^-$ is even. Moreover, $\lambda = 1$ is the eigenvalue with smallest positive real part if

$$d^+ + d^- \text{ is even and } \theta < \frac{\pi}{m}, \tag{9.4.13}$$

where $m = 1$ for $d^+ + d^- \in \{0, 6\}$, $m = 2$ for $d^+ + d^- \in \{2, 4\}$. Then the eigenvalue $\lambda = 1$ has geometric and algebraic multiplicity 1. This means in particular, that there do not exist generalized eigenvectors to the eigenvalue $\lambda = 1$. For even d^+ and d^-, the eigenvector $(U, P) = (0, 1)$ corresponds to this eigenvalue. If d^+ and d^- are both odd, then $(U, P) = (\sin\varphi, -\cos\varphi, 0, 0)$ is an eigenvector corresponding to this eigenvalue. In the case where $d^+ + d^-$ is even and $\theta < \pi/m$, we denote by μ_+ the greatest real number such that the strip

$$0 < \text{Re}\,\lambda < \mu_+ \tag{9.4.14}$$

contains only the eigenvalue $\lambda = 1$. If $\theta \ge \pi/m$ or $d^+ + d^-$ is odd, then we put $\mu_+ = \delta_+$. The result of Theorem 9.4.9 can be improved in the case (9.4.13). However in this case, the boundary data and the function g must satisfy additional compatibility conditions on the edge if $\delta < l - 2$ (see Subsection 9.2.3).

THEOREM 9.4.13. *Let* (u,p) *be a solution of the boundary value problem* (9.1.1), (9.1.2), *and let* ζ, η *be the same cut-off functions as in Theorem* 9.4.6. *Suppose that* $d^+ + d^-$ *is even,* $\eta(u,p) \in L^{1,2}(\mathcal{D})^3 \times L_2(\mathcal{D})$, $\eta f \in W_\delta^{l-2,2}(\mathcal{D})^2$, $\eta g \in W_\delta^{l-1,2}(\mathcal{D})$, $\eta h^\pm \in W_\delta^{l-1/2,2}(\Gamma^\pm)^3$, $l \ge 3$, $\max(0, l - 1 - \mu_+) < \delta < l - 2$, δ *is not integer, and that* $g, h^\pm, \phi^\pm$ *satisfy the compatibility conditions in Subsection* 9.2.3. *Then* $\zeta u \in W_\delta^{l,2}(\mathcal{D})^3$ *and* $\zeta p \in W_\delta^{l-1,2}(\mathcal{D})$.

P r o o f. We restrict ourselves to the Dirichlet problem. The proof for the other boundary value problems proceeds analogously. First, we prove the result for $l = 3$. Let χ, χ_1 be smooth functions on $\overline{\mathcal{D}}$ such that $\chi = 1$ in a neighborhood of

$\operatorname{supp}\zeta$, $\chi_1 = 1$ in a neighborhood of $\operatorname{supp}\chi$, and $\eta = 1$ in a neighborhood of $\operatorname{supp}\chi_1$. Then by Theorem 9.4.6, $\chi_1\partial_{x_3}u \in W_\delta^{2,2}(\mathcal{D})^3$ and $\chi_1\partial_{x_3}p \in W_\delta^{1,2}(\mathcal{D})$. Suppose that g, h^+, h^- satisfy the compatibility conditions $h^+|_M = h^-|_M$ and (9.2.13). Then there exist vectors $c(x_3)$ and $d(x_3)$ such that

$$-c_1(x_3) - d_2(x_3) = g(0,x_3) + (\partial_{x_3}h_3^+)(0,x_3),$$
$$\cos\frac{\theta}{2}\,c(x_3) \pm \sin\frac{\theta}{2}\,d(x_3) = (\partial_r h^\pm)(0,x_3)$$

for all $x_3 \in M \cap \operatorname{supp}\zeta$ and

$$|c(x_3)| + |d(x_3)| \le C\Big(|g(0,x_3)| + |(\partial_{x_3}h_3^+)(0,x_3)| + \sum_\pm |(\partial_r h^\pm)(0,x_3)|\Big)$$

with a constant independent of x_3. We set

$$v(x',x_3) = u(x',x_3) - h^+(0,x_3) - c(x_3)x_1 - d(x_3)x_2.$$

Since $\chi_1(\cdot,x_3)u(\cdot,x_3) \in W_\delta^{2,2}(K)^3$ and $v(0,x_3) = 0$, it follows that $\chi_1(\cdot,x_3)u(\cdot,x_3) \in V_\delta^{2,2}(K)^3$ (see Lemma 6.2.12). The functions $v(\cdot,x_3)$ and $p(\cdot,x_3)$ satisfy the equations

$$-\Delta_{x'}\big(v_1(\cdot,x_3), v_2(\cdot,x_3)\big) + \nabla_{x'}p(\cdot,x_3) = \big(F_1(\cdot,x_3), F_2(\cdot,x_3)\big),$$
$$-\nabla_{x'}\cdot\big(v_1(\cdot,x_3), v_2(\cdot,x_3)\big) = G(\cdot,x_3),$$
$$-\Delta_{x'}v_3(\cdot,x_3) = F_3(\cdot,x_3)$$

in K and the boundary condition $v(\cdot,x_3) = H^\pm(\cdot,x_3)$ on $\gamma^\pm$, where

$$F_j(x) = f_j(x) + \partial_{x_3}^2 u_j(x) \text{ for } j=1, j=2, \quad F_3(x) = f_3(x) + \partial_{x_3}^2 u_3 - \partial_{x_3}p,$$
$$G(x',x_3) = g(x',x_3) - g(0,x_3) + \partial_{x_3}\big(u_3(x',x_3) - u_3(0,x_3)\big),$$
$$H^\pm(r,x_3) = h^\pm(r,x_3) - h^\pm(0,x_3) - (\partial_r h^\pm)(0,x_3)\,r.$$

Obviously, $\chi(\cdot,x_3)\,F_j(\cdot,x_3) \in W_\delta^{1,2}(K) \subset V_\delta^{1,2}(K)$ for $j=1,2,3$ and for almost all $x_3 \in M \cap \operatorname{supp}\chi$. Furthermore, since

$$G(0,x_3) = 0 \quad\text{and}\quad H^\pm(0,x_3) = (\partial_r H^\pm)(0,x_3) = 0,$$

it follows that $\chi(\cdot,x_3)\,G(\cdot,x_3) \in V_\delta^{2,2}(K)$ and $\chi(\cdot,x_3)\,H^\pm(\cdot,x_3) \in V_\delta^{5/2,2}(\Gamma^\pm)^3$. Under the assumptions of the theorem, the strip $0 < \operatorname{Re}\lambda \le 2-\delta$ contains only the eigenvalue $\lambda = 1$ of the pencil $A(\lambda)$. The corresponding eigenvector is the constant vector $(U,P) = (0,1)$, while generalized eigenvectors corresponding to this eigenvalue do not exist. Thus by Theorem 1.2.6,

$$\zeta(\cdot,x_3)\big(v(\cdot,x_3), p(\cdot,x_3) - p(0,x_3)\big) \in V_\delta^{3,2}(K)^3 \times V_\delta^{1,2}(K).$$

This implies $\zeta(\cdot,x_3)\,u(\cdot,x_3) \in W_\delta^{3,2}(K)^3$ and $\zeta(\cdot,x_3)\,p(\cdot,x_3) \in W_\delta^{2,2}(K)$. Moreover, the norms of $\zeta(\cdot,x_3)$ and $\zeta(\cdot,x_3)\,p(\cdot,x_3)$ can be estimated by the corresponding norms of $\eta(\cdot,x_3)f(\cdot,x_3)$, $\eta(\cdot,x_3)g(\cdot,x_3)$, $\eta(\cdot,x_3)h^\pm(\cdot,x_3)$, $\eta(\cdot,x_3)\partial_{x_3}u(\cdot,x_3)$, and $\eta(\cdot,x_3)\partial_{x_3}p(\cdot,x_3)$. Therefore as in the proof of Theorem 9.4.5, we obtain $\zeta\,(u,p) \in W_\delta^{3,2}(\mathcal{D})^3 \times W_\delta^{2,2}(\mathcal{D})$. This proves the theorem for $l=3$. For $l>3$, we obtain the result analogously to Theorem 9.4.9 by means of Lemma 9.4.7. □

Furthermore, the following assertion holds (see the fourth part of the proof of Theorem 9.4.9).

COROLLARY 9.4.14. *Let (u,p) be a solution of problem* (9.1.1), (9.1.2), *and let ζ, η be the same cut-off functions as in Theorem* 9.4.6. *Suppose that $d^+ + d^-$ is even, $\eta(u,p) \in L^{1,2}(\mathcal{D})^3 \times L_2(\mathcal{D})$, $\eta\partial_{x_3}^j f \in W_\delta^{l-2,2}(\mathcal{D})^2$, $\eta\partial_{x_3}^j g \in W_\delta^{l-1,2}(\mathcal{D})$, $\eta\partial_{x_3}^j h^\pm \in W_\delta^{l-1/2,2}(\Gamma^\pm)^3$ for $j = 0,1,\dots,k$, $l \ge 3$, $\max(0, l-1-\mu_+) < \delta < l-2$, δ is not integer, and that $g, h^\pm, \phi^\pm$ satisfy the compatibility conditions in Subsection* 9.2.3. *Then $\zeta\partial_{x_3}^j(u,p) \in W_\delta^{l,2}(\mathcal{D})^3 \times W_\delta^{l-1,2}(\mathcal{D})$ for $j = 0,1,\dots,k$.*

9.5. Green's matrix of the problems in a half-space

In this section, we consider the Stokes system in the half-space

$$\mathbb{R}^3_+ = \{x = (x_1,x_2,x_3) \in \mathbb{R}^3 : x_3 > 0\}$$

with one of the boundary conditions (i)–(iv) on the plane $x_3 = 0$. We establish explicit formulas for the elements of the Green's matrix of this problem. The formulas for the Dirichlet problem and for the Stokes system with the boundary condition (iii) were derived by MAZ'YA, PLAMENEVSKIĬ, STUPELIS [**125**].

9.5.1. Definition of Green's matrix. On the plane $x_3 = 0$, the boundary conditions (i)–(iv) take the following form:

(i) $u(x) = 0$,

(ii) $u_1(x) = u_2(x) = -p + 2\partial_{x_3}u_3(x) = 0$,

(iii) $u_3(x) = \partial_{x_3}u_1(x) = \partial_{x_3}u_2(x) = 0$,

(iv) $\partial_{x_1}u_3(x) + \partial_{x_3}u_1(x) = \partial_{x_2}u_3(x) + \partial_{x_3}u_2(x) = -p + 2\partial_{x_3}u_3(x) = 0$.

Let

$$\mathcal{G}^+(x,\xi) = \big(\mathcal{G}^+_{i,j}(x,\xi)\big)_{i,j=1}^4$$

be a 4×4 matrix with the elements $\mathcal{G}^+_{i,j}(x,\xi)$ and let

$$\vec{\mathcal{G}}^+_j = (\mathcal{G}^+_{1,j}, \mathcal{G}^+_{2,j}, \mathcal{G}^+_{3,j})^t$$

denote the column vector with the components $\mathcal{G}^+_{1,j}, \mathcal{G}^+_{2,j}, \mathcal{G}^+_{3,j}$. The matrix $\mathcal{G}^+(x,\xi)$ is called *Green's matrix* for the boundary value problem to the Stokes system

$$-\Delta u + \nabla p = f, \quad -\nabla\cdot u = g \quad \text{in } \mathbb{R}^3_+ \tag{9.5.1}$$

with one of the boundary conditions (i)–(iv) if the vectors $\vec{\mathcal{G}}^+_j = (\mathcal{G}^+_{1,j}, \mathcal{G}^+_{2,j}, \mathcal{G}^+_{3,j})^t$ and the function $\mathcal{G}^+_{4,j}$ satisfy the equations

$$-\Delta_x \vec{\mathcal{G}}^+_j(x,\xi) + \nabla_x \mathcal{G}^+_{4,j}(x,\xi) = \delta(x-\xi)\,\big(\delta_{1,j}, \delta_{2,j}, \delta_{3,j}\big)^t, \tag{9.5.2}$$

$$-\nabla_x \cdot \vec{\mathcal{G}}^+_j(x,\xi) = \delta_{4,j}\delta(x-\xi) \tag{9.5.3}$$

for $x, \xi \in \mathbb{R}^3_+$, $j = 1,2,3,4$, and the corresponding boundary condition for $x_3 = 0$, $\xi \in \mathbb{R}^3_+$. Here $\delta_{i,j}$ denotes the Kronecker symbol and $(a_1,a_2,a_3)^t$ the column vector with the components a_1, a_2, a_3. Then the vectors $\vec{\mathcal{H}}^+_i = (\mathcal{G}^+_{i,1}, \mathcal{G}^+_{i,2}, \mathcal{G}^+_{i,3})^t$ and the function $\mathcal{G}^+_{i,4}$ satisfy the equations

$$-\Delta_\xi \vec{\mathcal{H}}^+_i(x,\xi) + \nabla_\xi \mathcal{G}^+_{i,4}(x,\xi) = \delta(x-\xi)\,\big(\delta_{i,1}, \delta_{i,2}, \delta_{i,3}\big)^t, \tag{9.5.4}$$

$$-\nabla_\xi \cdot \vec{\mathcal{H}}^+_i(x,\xi) = \delta_{i,4}\delta(x-\xi) \tag{9.5.5}$$

for $x, \xi \in \mathbb{R}^3_+$, $i = 1,2,3,4$, and the corresponding boundary condition for $\xi_3 = 0$. By (9.5.4), (9.5.5) and by Green's formula in the half-space (cf. (9.1.5)), the solution

(u, p) of the Stokes system (9.5.1) with the boundary conditions (i), (ii), (iii) or (iv) is given by the formulas

$$\begin{aligned} u_i(x) &= \int_{\mathbb{R}^3_+} \big(f(\xi) + \nabla_\xi g(\xi)\big) \cdot \vec{\mathcal{H}}_i^+(x,\xi)\, d\xi + \int_{\mathbb{R}^3_+} g(\xi)\, \mathcal{G}^+_{i,4}(x,\xi)\, d\xi, \ i = 1,2,3, \\ p(x) &= -g(x) + \int_{\mathbb{R}^3_+} \big(f(\xi) + \nabla_\xi g(\xi)\big) \cdot \vec{\mathcal{H}}_4^+(x,\xi)\, d\xi + \int_{\mathbb{R}^3_+} g(\xi)\, \mathcal{G}^+_{4,4}(x,\xi)\, d\xi. \end{aligned}$$

Note that the simpler formulas

$$\begin{aligned} u_i(x) &= \int_{\mathbb{R}^3_+} f(\xi) \cdot \vec{\mathcal{H}}_i^+(x,\xi)\, d\xi + \int_{\mathbb{R}^3_+} g(\xi)\, \mathcal{G}^+_{i,4}(x,\xi)\, d\xi, \\ p(x) &= \int_{\mathbb{R}^3_+} f(\xi) \cdot \vec{\mathcal{H}}_4^+(x,\xi)\, d\xi + \int_{\mathbb{R}^3_+} g(\xi)\, \mathcal{G}^+_{4,4}(x,\xi)\, d\xi \end{aligned}$$

are valid in the cases of the boundary conditions (i) and (iii).

9.5.2. Green's matrix for the Dirichlet problem in the half-space. We denote by $\mathcal{G}(x,\xi) = \big(\mathcal{G}_{i,j}(x,\xi)\big)_{i,j=1}^4$ the Green's matrix of Stokes system in $\mathbb{R}^3$, i.e., the matrix satisfying (9.5.2) and (9.5.3) for $x, \xi \in \mathbb{R}^3$, $j = 1,2,3,4$. The components of $\mathcal{G}(x,\xi)$ are (cf. [**89**])

(9.5.6) $$\mathcal{G}_{i,j}(x,\xi) = \frac{1}{8\pi}\Big(\frac{\delta_{i,j}}{|x-\xi|} + \frac{(x_i-\xi_i)(x_j-\xi_j)}{|x-\xi|^3}\Big) \quad \text{for } i,j = 1,2,3,$$

(9.5.7) $$\mathcal{G}_{4,j}(x,\xi) = -\mathcal{G}_{j,4}(x,\xi) = \frac{1}{4\pi}\,\frac{x_j-\xi_j}{|x-\xi|^3} \quad \text{for } j = 1,2,3,$$

(9.5.8) $$\mathcal{G}_{4,4}(x,\xi) = -\delta(x-\xi).$$

Note that $\mathcal{G}_{i,j}(x,\xi) = \mathcal{G}_{j,i}(\xi,x)$ for $i,j = 1,2,3,4$.

Using the above given Green's matrix $\mathcal{G}(x,\xi)$, we construct the Green's matrix $\mathcal{G}^+(x,\xi)$ of the Dirichlet problem for the Stokes system in the half-space $\mathbb{R}^3_+$. For this, we need the following lemma.

LEMMA 9.5.1. *Let $x = (x_1, x_2, x_3)$ and $\xi = (\xi_1, \xi_2, \xi_3)$ be arbitrary points in $\mathbb{R}^3_+$. Then*

$$\int_{\mathbb{R}^2} \frac{x_3\, dy'}{|x-(y',0)|^3\, |\xi-(y',0)|} = \frac{2\pi}{|x-\xi^*|},$$

where $\xi^* = (\xi_1, \xi_2, -\xi_3)$.

P r o o f. Let

$$(Ff)(\eta') = \frac{1}{2\pi}\int_{\mathbb{R}^2} f(z')\, e^{-iz'\cdot\eta'}\, dz'$$

denote the Fourier transform of f. Then

$$F\left(|z'|^2 + (x_3+\xi_3)^2\right)^{-1/2} = \frac{1}{2\pi}\int_{\mathbb{R}^2} \frac{e^{-iz'\cdot\eta'}\, dz'}{(|z'|^2 + (x_3+\xi_3)^2)^{1/2}} = \frac{e^{-(x_3+\xi_3)\,|\eta'|}}{|\eta'|}$$

and consequently

$$\frac{1}{\big(|z'|^2+(x_3+\xi_3)^2\big)^{1/2}} = F^{-1}\Big(e^{-x_3|\eta'|}\,\frac{e^{-\xi_3|\eta'|}}{|\eta'|}\Big)$$
$$= \frac{1}{2\pi}\Big(F^{-1}\,e^{-x_3|\eta'|}\Big) * \Big(F^{-1}\,\frac{e^{-\xi_3|\eta'|}}{|\eta'|}\Big)$$
$$= \frac{1}{2\pi}\int_{\mathbb{R}^2} \frac{x_3\,dy'}{(|z'-y'|^2+x_3^2)^{3/2}\,(|y'|^2+\xi_3^2)^{1/2}}\,.$$

The result follows. □

In the next theorem, let $\mathcal{G}^+(x,\xi)$ be the Green's matrix of the Dirichlet problem for the Stokes system, i.e., the elements of $\mathcal{G}^+(x,\xi)$ satisfy the equations (9.5.2), (9.5.3) and the boundary condition

$$\mathcal{G}^+_{i,j}(x,\xi) = 0 \quad \text{for } x_3=0,\ i=1,2,3,\ j=1,2,3,4.$$

THEOREM 9.5.2. *The elements of the matrix* $\mathcal{G}^+(x,\xi)$ *are given for* $i=1,2$ *by*

$$\mathcal{G}^+_{i,j}(x,\xi) = \mathcal{G}_{i,j}(x,\xi) - \mathcal{G}_{i,j}(x,\xi^*) - \frac{1}{4\pi}\,\frac{\partial^2}{\partial x_i\partial\xi_j}\,\frac{x_3\xi_3}{|x-\xi^*|}, \quad j=1,2,$$
$$\mathcal{G}^+_{i,3}(x,\xi) = \mathcal{G}_{i,3}(x,\xi) + \mathcal{G}_{i,3}(x,\xi^*) + \frac{1}{4\pi}\,\frac{\partial}{\partial x_i}\Big(\frac{x_3}{|x-\xi^*|} + \frac{x_3\xi_3(x_3+\xi_3)}{|x-\xi^*|^3}\Big),$$
$$\mathcal{G}^+_{i,4}(x,\xi) = \mathcal{G}_{i,4}(x,\xi) - \mathcal{G}_{i,4}(x,\xi^*) - \frac{1}{2\pi}\,\frac{\partial^2}{\partial x_i\partial\xi_3}\,\frac{x_3}{|x-\xi^*|},$$

where $\xi^*=(\xi_1,\xi_2,-\xi_3)$, *and* $\mathcal{G}_{i,j}$ *is defined by* (9.5.6)–(9.5.8). *The other elements of Green's matrix are given by the formulas*

$$\mathcal{G}^+_{3,3}(x,\xi) = \mathcal{G}_{3,3}(x,\xi) + \mathcal{G}_{3,3}(x,\xi^*) - \frac{1}{4\pi}\Big(1-\xi_3\frac{\partial}{\partial\xi_3}\Big)\Big(1-x_3\frac{\partial}{\partial x_3}\Big)\frac{1}{|x-\xi|^*},$$
$$\mathcal{G}^+_{3,4}(x,\xi) = \mathcal{G}_{3,4}(x,\xi) - \mathcal{G}_{3,4}(x,\xi^*) + \frac{1}{2\pi}\,\frac{\partial}{\partial\xi_3}\Big(\frac{1}{|x-\xi^*|} + \frac{x_3(x_3+\xi_3)}{|x-\xi^*|^3}\Big),$$
$$\mathcal{G}^+_{4,4}(x,\xi) = -\delta(x-\xi) - \frac{1}{\pi}\,\frac{\partial^2}{\partial\xi_3^2}\,\frac{1}{|x-\xi^*|},$$
$$\mathcal{G}^+_{i,j}(x,\xi) = \mathcal{G}^+_{j,i}(\xi,x) \quad \textit{for } i,j=1,2,3,4.$$

P r o o f. We extend f and g to functions $\tilde f$ and $\tilde g$ on $\mathbb{R}^3$ as follows:

$$\tilde f(\xi) = f(\xi),\quad \tilde g(\xi) = g(\xi) \quad \text{for } \xi_3>0,$$
$$\tilde f_j(\xi) = -(-1)^{\delta_{j,3}} f_j(\xi^*),\quad \tilde g(\xi) = -g(\xi^*) \quad \text{for } \xi_3<0.$$

Then the vector function (u,p) with the components

$$u_i(x) = \sum_{j=1}^{3} \int_{\mathbb{R}^3} \tilde{f}_j(\xi)\, \mathcal{G}_{i,j}(x,\xi)\, d\xi + \int_{\mathbb{R}^3} \tilde{g}(\xi)\, \mathcal{G}_{i,4}(x,\xi)$$

$$= \sum_{j=1}^{3} \int_{\mathbb{R}^3_+} f_j(\xi) \left(\mathcal{G}_{i,j}(x,\xi) - (-1)^{\delta_{j,3}} \mathcal{G}_{i,j}(x,\xi^*) \right) d\xi$$

$$+ \int_{\mathbb{R}^3} g(\xi) \left(\mathcal{G}_{i,4}(x,\xi) + \mathcal{G}_{i,4}(x,\xi^*) \right) d\xi,$$

$$p(x) = -g(x) + \sum_{j=1}^{3} \int_{\mathbb{R}^3} \tilde{f}_j(\xi)\, \mathcal{G}_{4,j}(x,\xi)\, d\xi$$

$$= -g(x) + \sum_{j=1}^{3} \int_{\mathbb{R}^3_+} f_j(\xi) \left(\mathcal{G}_{4,j}(x,\xi) - (-1)^{\delta_{j,3}} \mathcal{G}_{4,j}(x,\xi^*) \right) d\xi$$

satisfies (9.5.1) and the boundary conditions

$$u_1 = u_2 = 0$$

on the plane $x_3 = 0$. Furthermore,

$$u_3(x',0) = -\frac{1}{4\pi} \sum_{j=1}^{2} \int_{\mathbb{R}^3_+} f_j(\xi) \Big(\frac{(x_j - \xi_j)\xi_3}{|x' - \xi|^3}$$

$$+ \frac{1}{4\pi} \int_{\mathbb{R}^3_+} f_3(\xi) \left(\frac{1}{|x' - \xi|} + \frac{\xi_3^2}{|x' - \xi|^3} \right) d\xi + \frac{1}{2\pi} \int_{\mathbb{R}^3_+} g(\xi)\, \frac{\xi_3}{|x' - \xi|}\, d\xi.$$

Here for the sake of brevity, we wrote $|x' - \xi|$ instead of $|(x',0) - \xi|$. We denote the vector function on the right-hand side of the last equation by $\Phi(x')$ and construct a solution (v,q) of the problem

$$-\Delta v + \nabla q = 0, \quad \nabla \cdot v = 0 \ \text{ in } \mathbb{R}^3_+, \tag{9.5.9}$$

$$v_1(x',0) = v_2(x',0) = 0, \quad v_3(x',0) = \Phi(x') \ \text{ for } x' \in \mathbb{R}^2. \tag{9.5.10}$$

Let v and q be defined as follows:

$$v_i(x) = \frac{3}{2\pi} \int_{\mathbb{R}^2} \frac{(x_i - y_i) x_3^2}{|x - y'|^5}\, \Phi(y')\, dy', \quad i = 1,2,$$

$$v_3(x) = \frac{3}{2\pi} \int_{\mathbb{R}^2} \frac{x_3^3}{|x - y'|^5}\, \Phi(y')\, dy',$$

$$q(x) = -\frac{1}{\pi} \frac{\partial}{\partial x_3} \int_{\mathbb{R}^2} \frac{x_3}{|x - y'|}\, \Phi(y')\, dy'.$$

Since v_i and q are double layer potentials for the Stokes system, the vector function (v,q) satisfies (9.5.9). Using the equalities

$$\int_{\mathbb{R}^2} \frac{(x_i - y_i) x_3^2}{|x - y'|^5}\, dy' = \lim_{r\to\infty} \frac{x_3^2}{3} \int\limits_{|x'-y'|\le r} \frac{\partial}{\partial y_i} \frac{1}{|x - y'|^3}\, dy' \tag{9.5.11}$$

$$= \frac{x_3^2}{3} \lim_{r\to\infty} \frac{1}{(r^2 + x_3^2)^{3/2}} \int\limits_{|x'-y'|=r} \cos(y' - x', y_i)\, ds = 0 \ \text{ for } i = 1,2,$$

and

$$\frac{3}{2\pi}\int_{\mathbb{R}^2}\frac{x_3^3}{|x-y'|^5}\,dy' = 3\int_0^\infty \frac{x_3^3\,\rho}{(\rho^2+x_3^2)^{5/2}}\,d\rho = 1, \tag{9.5.12}$$

we obtain

$$\begin{aligned} v_i(x',0) &= \frac{3}{2\pi}\lim_{x_3\to 0}\int_{\mathbb{R}^2}\frac{(x_i-y_i)x_3^2}{|x-y'|^5}\big(\Phi(y')-\Phi(x')\big)\,dy' \\ &= \frac{3}{2\pi}\lim_{x_3\to 0}\int_{\mathbb{R}^2}\frac{z_i}{(1+|z'|^2)^{5/2}}\big(\Phi(x'+x_3z')-\Phi(x')\big)\,dz' = 0 \end{aligned}$$

for $i=1,2$ and

$$\begin{aligned} v_3(x',0) &= \Phi(x') + \frac{3}{2\pi}\lim_{x_3\to 0}\int_{\mathbb{R}^2}\frac{x_3^3}{|x-y'|^5}\big(\Phi(y')-\Phi(x')\big)\,dy' \\ &= \Phi(x') + \frac{3}{2\pi}\lim_{x_3\to 0}\int_{\mathbb{R}^2}\frac{1}{(1+|z'|^2)^{5/2}}\big(\Phi(x'+x_3z')-\Phi(x')\big)\,dz' = \Phi(x'). \end{aligned}$$

Consequently, the vector function $(U,P)=(u-v,p-q)$ satisfies the Stokes system (9.5.1) and the homogeneous Dirichlet condition on the plane $x_3=0$. It remains to represent v and q in terms of f and g. For $i=1,2$ we get

$$\begin{aligned} v_i(x) &= -\frac{3}{8\pi^2}\sum_{j=1}^2\int_{\mathbb{R}^3_+} f_j(\xi)\int_{\mathbb{R}^2}\frac{(x_i-y_i)(y_j-\xi_j)x_3^2\xi_3}{|x-y'|^5\,|y'-\xi|^3}\,dy'\,d\xi \\ &\quad +\frac{3}{8\pi^2}\int_{\mathbb{R}^3_+} f_3(\xi)\int_{\mathbb{R}^2}\frac{(x_i-y_i)x_3^2}{|x-y'|^5}\Big(\frac{1}{|y'-\xi|}+\frac{\xi_3^2}{|y'-\xi|^3}\Big)\,dy'\,d\xi \\ &\quad +\frac{3}{4\pi^2}\int_{\mathbb{R}^3_+} g(\xi)\int_{\mathbb{R}^2}\frac{(x_i-y_i)x_3^2\xi_3}{|x-y'|^5\,|y'-\xi|^3}\,dy'\,d\xi. \end{aligned}$$

Using the formulas

$$\begin{aligned} \frac{(x_i-y_i)(y_j-\xi_j)x_3^2\xi_3}{|x-y'|^5\,|y'-\xi|^3} &= -\frac{x_3\xi_3}{3}\,\frac{\partial^2}{\partial x_i\partial\xi_j}\,\frac{x_3}{|x-y'|^3\,|y'-\xi|}, \\ \frac{(x_i-y_i)x_3^2}{|x-y'|^5\,|y'-\xi|} &= -\frac{x_3}{3}\,\frac{\partial}{\partial x_i}\,\frac{x_3}{|x-y'|^3\,|y'-\xi|}, \\ \frac{(x_i-y_i)x_3^2\xi_3}{|x-y'|^5\,|y'-\xi|^3} &= \frac{x_3}{3}\,\frac{\partial^2}{\partial x_i\partial\xi_3}\,\frac{x_3}{|x-y'|^3\,|y'-\xi|} \end{aligned}$$

for $i,j=1,2$ and Lemma 9.5.1, we obtain

$$\begin{aligned} v_i(x) &= \frac{1}{4\pi}\sum_{j=1}^2\int_{\mathbb{R}^3_+} f_j(\xi)\,\frac{\partial^2}{\partial x_i\partial\xi_j}\,\frac{x_3\xi_3}{|x-\xi^*|}\,d\xi \\ &\quad -\frac{1}{4\pi}\int_{\mathbb{R}^3_+} f_3(\xi)\Big(\frac{\partial}{\partial x_i}\,\frac{x_3}{|x-\xi^*|}-\xi_3\,\frac{\partial^2}{\partial x_i\partial\xi_3}\,\frac{x_3}{|x-\xi^*|}\Big)d\xi \\ &\quad +\frac{1}{2\pi}\int_{\mathbb{R}^3_+} g(\xi)\,\frac{\partial^2}{\partial x_i\partial\xi_3}\,\frac{x_3}{|x-\xi^*|}\,d\xi \end{aligned}$$

for $i = 1, 2$. Analogously,

$$\begin{aligned} v_3(x) &= -\frac{3}{8\pi^2} \sum_{j=1}^{2} \int_{\mathbb{R}^3_+} f_j(\xi) \int_{\mathbb{R}^2} \frac{(y_j - \xi_j) x_3^3 \xi_3}{|x - y'|^5 \, |y' - \xi|^3} \, dy' \, d\xi \\ &+ \frac{3}{8\pi^2} \int_{\mathbb{R}^3_+} f_3(\xi) \int_{\mathbb{R}^2} \frac{x_3^3}{|x - y'|^5} \Big(\frac{1}{|y' - \xi|} + \frac{\xi_3^2}{|y' - \xi|^3} \Big) \, dy' \, d\xi \\ &+ \frac{3}{4\pi^2} \int_{\mathbb{R}^3_+} g(\xi) \int_{\mathbb{R}^2} \frac{x_3^3 \xi_3}{|x - y'|^5 \, |y' - \xi|^3} \, dy' \, d\xi. \end{aligned}$$

Here the formulas

$$\begin{aligned} \frac{(y_j - \xi_j) x_3^3 \xi_3}{|x - y'|^5 \, |y' - \xi|^3} &= \Big(\frac{\xi_3}{3} \frac{\partial}{\partial \xi_j} - \frac{x_3 \xi_3}{3} \frac{\partial^2}{\partial x_3 \, \partial \xi_j} \Big) \frac{x_3}{|x - y'|^3 \, |y' - \xi|}, \\ \frac{x_3^3}{|x - y'|^5 \, |y' - \xi|} &= \frac{1}{3} \Big(1 - x_3 \frac{\partial}{\partial x_3} \Big) \frac{x_3}{|x - y'|^3 \, |y' - \xi|}, \\ \frac{x_3^3 \xi_3}{|x - y'|^5 \, |y' - \xi|^3} &= -\frac{1}{3} \Big(\frac{\partial}{\partial \xi_3} - x_3 \frac{\partial^2}{\partial x_3 \partial \xi_3} \Big) \frac{x_3}{|x - y'|^3 \, |y' - \xi|} \end{aligned}$$

and Lemma 9.5.1 yield

$$\begin{aligned} v_3(x) &= -\frac{1}{4\pi} \sum_{j=1}^{2} \int_{\mathbb{R}^3_+} f_j(\xi) \Big(\frac{\partial}{\partial \xi_j} \frac{\xi_3}{|x - \xi^*|} - \frac{\partial^2}{\partial x_3 \, \partial \xi_j} \frac{x_3 \xi_3}{|x - \xi^*|} \Big) \, d\xi \\ &+ \frac{1}{4\pi} \int_{\mathbb{R}^3_+} f_3(\xi) \Big(1 - \xi_3 \frac{\partial}{\partial \xi_3} \Big) \Big(1 - x_3 \frac{\partial}{\partial x_3} \Big) \frac{1}{|x - \xi|^*} \, d\xi \\ &- \frac{1}{2\pi} \int_{\mathbb{R}^3_+} g(\xi) \Big(\frac{\partial}{\partial \xi_3} - x_3 \frac{\partial^2}{\partial x_3 \, \partial \xi_3} \Big) \frac{1}{|x - \xi^*|} \, d\xi. \end{aligned}$$

Finally, from the representation

$$\begin{aligned} q(x) &= \frac{1}{4\pi^2} \sum_{j=1}^{2} \int_{\mathbb{R}^3_+} f_j(\xi) \frac{\partial}{\partial x_3} \int_{\mathbb{R}^2} \frac{(y_j - \xi_j) x_3 \xi_3}{|x - y'|^3 \, |y' - \xi|^3} \, dy' \, d\xi \\ &- \frac{1}{4\pi^2} \int_{\mathbb{R}^3_+} f_3(\xi) \frac{\partial}{\partial x_3} \int_{\mathbb{R}^2} \frac{x_3}{|x - y'|^3} \Big(\frac{1}{|y' - \xi|} + \frac{\xi_3^2}{|y' - \xi|^3} \Big) \, dy' \, d\xi \\ &- \frac{1}{2\pi^2} \int_{\mathbb{R}^3_+} g(\xi) \frac{\partial}{\partial x_3} \int_{\mathbb{R}^2} \frac{x_3 \xi_3}{|x - y'|^3 \, |y' - \xi|^3} \, dy' \, d\xi \end{aligned}$$

and from the formulas

$$\begin{aligned} \frac{(y_j - \xi_j) x_3}{|x - y'|^3 \, |y' - \xi|^3} &= \frac{\partial}{\partial \xi_j} \frac{x_3}{|x - y'|^3 \, |y' - \xi|} \qquad \text{for } j = 1, 2, \\ \frac{x_3 \xi_3}{|x - y'|^3 \, |y' - \xi|^3} &= -\frac{\partial}{\partial \xi_3} \frac{x_3}{|x - y'|^3 \, |y' - \xi|} \end{aligned}$$

we conclude that

$$\begin{aligned} q(x) &= \frac{1}{\pi} \int_{\mathbb{R}^3_+} g(\xi) \frac{\partial^2}{\partial x_3 \partial \xi_3} \frac{1}{|x - \xi^*|} \, d\xi + \frac{1}{2\pi} \sum_{j=1}^{2} \int_{\mathbb{R}^3_+} f_j(\xi) \frac{\partial^2}{\partial x_3 \partial \xi_j} \frac{\xi_3}{|x - \xi^*|} \, d\xi \\ &- \frac{1}{2\pi} \int_{\mathbb{R}^3_+} f_3(\xi) \Big(\frac{\partial}{\partial x_3} - \xi_3 \frac{\partial^2}{\partial x_3 \, \partial \xi_3} \Big) \frac{1}{|x - \xi^*|} \, d\xi. \end{aligned}$$

Now the assertion of the lemma follows directly from the above obtained representation of the solution $(U, P) = (u - v, p - q)$. □

9.5.3. Green's matrix for the Stokes system with the boundary condition (ii). Next we consider the Stokes system in $\mathbb{R}^3_+$ with the boundary condition

$$u_1(x) = u_2(x) = -p + 2\partial_{x_3} u_3(x) = 0$$

on the plane $x_3 = 0$.

THEOREM 9.5.3. *The elements of the Green's matrix for the problem* (9.5.1) *with the boundary condition* (ii) *are*

$$\mathcal{G}^+_{i,j}(x,\xi) = \mathcal{G}_{i,j}(x,\xi) - (-1)^{\delta_{j,3}} \mathcal{G}_{i,j}(x,\xi^*) \quad \textit{for } i+j \le 7,$$
$$\mathcal{G}^+_{4,4}(x,\xi) = -\delta(x-\xi),$$

where $\xi^* = (\xi_1, \xi_2, -\xi_3)$, *and* $\mathcal{G}_{i,j}$ *is defined by* (9.5.6)–(9.5.8).

P r o o f. Let $F(\xi) = f(\xi) + \nabla_\xi g(\xi)$. Furthermore, we define $\tilde{F}(\xi)$ and $\tilde{g}(\xi)$ as

$$\tilde{F}(\xi) = F(\xi), \quad \tilde{g}(\xi) = g(\xi) \quad \text{for } \xi_3 > 0,$$
$$\tilde{F}_j(\xi) = -(-1)^{\delta_{j,3}} F_j(\xi^*), \quad \tilde{g}(\xi) = -g(\xi^*) \quad \text{for } \xi_3 < 0.$$

Then the vector function with the components

$$\begin{aligned} u_i(x) &= \sum_{j=1}^{3} \int_{\mathbb{R}^3} \tilde{F}_j(\xi)\, \mathcal{G}_{i,j}(x,\xi)\, d\xi + \int_{\mathbb{R}^3} \tilde{g}(\xi)\, \mathcal{G}_{i,4}(x,\xi)\, d\xi \\ &= \sum_{j=1}^{3} \int_{\mathbb{R}^3_+} F_j(\xi) \left(\mathcal{G}_{i,j}(x,\xi) - (-1)^{\delta_{j,3}} \mathcal{G}_{i,j}(x,\xi^*) \right) d\xi \\ &\quad + \int_{\mathbb{R}^3} g(\xi) \left(\mathcal{G}_{i,4}(x,\xi) - \mathcal{G}_{i,4}(x,\xi^*) \right) d\xi, \\ p(x) &= -2g(x) + \sum_{j=1}^{3} \int_{\mathbb{R}^3} \tilde{F}_j(\xi)\, \mathcal{G}_{4,j}(x,\xi)\, d\xi \\ &= -2g(x) + \sum_{j=1}^{3} \int_{\mathbb{R}^3_+} F_j(\xi) \left(\mathcal{G}_{4,j}(x,\xi) - (-1)^{\delta_{j,3}} \mathcal{G}_{4,j}(x,\xi^*) \right) d\xi \end{aligned}$$

satisfies (9.5.1). Furthermore, since $\mathcal{G}_{i,j}(x',0,\xi) = (-1)^{\delta_{j,3}} \mathcal{G}_{i,j}(x',0,\xi^*)$ for $i = 1, 2$, it follows that

$$u_1(x',0) = u_2(x',0) = 0.$$

We show that $\Phi(x) = -p(x) + 2\partial_{x_3} u_3(x)$ vanishes for $x_3 = 0$. Using (9.5.6)–(9.5.8), we get

$$\Phi(x) = 2g(x) - \frac{3}{4\pi} \Psi(x) - \frac{1}{2\pi} \frac{\partial}{\partial x_3} \int_{\mathbb{R}^3_+} g(\xi) \frac{\partial}{\partial \xi_3} \Big(\frac{1}{|x-\xi|} + \frac{1}{|x-\xi^*|} \Big)\, d\xi,$$

where

$$\Psi(x) = \sum_{j=1}^{2} \int\limits_{\mathbb{R}^3_+} F_j(\xi)\Big(\frac{(x_j-\xi_j)(x_3-\xi_3)^2}{|x-\xi|^5} - \frac{(x_j-\xi_j)(x_3+\xi_3)^2}{|x-\xi^*|^5}\Big)\, d\xi$$
$$+ \int\limits_{\mathbb{R}^3_+} F_3(\xi)\Big(\frac{(x_3-\xi_3)^3}{|x-\xi|^5} + \frac{(x_3+\xi_3)^3}{|x-\xi^*|^5}\Big)\, d\xi.$$

Integrating by parts, we obtain

$$\Phi(x) = 2g(x) - \frac{3}{4\pi}\Psi(x) + \frac{1}{2\pi}\frac{\partial}{\partial x_3}\int_{\mathbb{R}^3_+} \frac{\partial g(\xi)}{\partial \xi_3}\Big(\frac{1}{|x-\xi|} + \frac{1}{|x-\xi^*|}\Big)\, d\xi$$
$$+\frac{1}{\pi}\frac{\partial}{\partial x_3}\int_{\mathbb{R}^2} g(\xi',0)\,\frac{1}{|x-(\xi',0)|}\, d\xi'.$$

Here $\Psi(x) \to 0$ as $x_3 \to 0$. Furthermore, the expression

$$\frac{\partial}{\partial x_3}\int_{\mathbb{R}^3_+} \frac{\partial g(\xi)}{\partial \xi_3}\Big(\frac{1}{|x-\xi|} + \frac{1}{|x-\xi^*|}\Big)\, d\xi = -\int_{\mathbb{R}^3_+} \frac{\partial g(\xi)}{\partial \xi_3}\Big(\frac{x_3-\xi_3}{|x-\xi|^3} + \frac{x_3+\xi_3}{|x-\xi^*|^3}\Big)\, d\xi$$

tends to zero as $x_3 \to 0$, and from the equality

$$\frac{1}{2\pi}\int_{\mathbb{R}^2} \frac{x_3}{|x-(\xi',0)|^3}\, d\xi' = x_3\int_0^\infty \frac{\rho\, d\rho}{(\rho^2+x_3^2)^{3/2}} = 1$$

it follows that

$$\frac{1}{\pi}\frac{\partial}{\partial x_3}\int_{\mathbb{R}^2} g(\xi',0)\,\frac{1}{|x-(\xi',0)|}\, d\xi' = -\frac{1}{\pi}\int_{\mathbb{R}^2} g(\xi',0)\,\frac{x_3}{|x-(\xi',0)|}\, d\xi'$$
$$= \frac{1}{\pi}\int_{\mathbb{R}^2} \frac{(g(x',0)-g(\xi',0))x_3}{|x-(\xi',0)|^3}\, d\xi' - 2g(x',0)$$
$$= \frac{1}{\pi}\int_{\mathbb{R}^2} \frac{g(x',0)-g(x'+x_3z',0)}{(1+|z'|^2)^{3/2}}\, dz' - 2g(x',0) \to -2g(x',0) \quad \text{as } x_3 \to 0.$$

Consequently, $\Phi(x',0) = 0$. This means, the vector function (u,p) introduced above is a solution of the system (9.5.1) with the boundary condition (ii). The result follows. □

9.5.4. Green's matrix for the Stokes system with the boundary condition (iii). In the next theorem, we obtain representations for the elements of the Green's matrix for the Stokes system in $\mathbb{R}^3_+$ with the boundary condition

(9.5.13) $$u_3(x) = \partial_{x_3}u_1(x) = \partial_{x_3}u_2(x) = 0$$

on the plane $x_3 = 0$.

THEOREM 9.5.4. *The elements of the Green's matrix for the Stokes system in $\mathbb{R}^3_+$ with the boundary condition* (9.5.13) *are*

$$\mathcal{G}^+_{i,j}(x,\xi) = \mathcal{G}_{i,j}(x,\xi) + (-1)^{\delta_{j,3}}\mathcal{G}_{i,j}(x,\xi^*) \quad \textit{for } i+j \le 7,$$
$$\mathcal{G}^+_{4,4}(x,\xi) = -\delta(x-\xi),$$

where $\xi^* = (\xi_1,\xi_2,-\xi_3)$, *and* $\mathcal{G}_{i,j}$ *is defined by* (9.5.6)–(9.5.8).

P r o o f. Let the extensions $\tilde{f}$ and $\tilde{g}$ of f and g be defined as

$$\tilde{f}(\xi) = f(\xi), \quad \tilde{g}(\xi) = g(\xi) \quad \text{for } \xi_3 > 0,$$
$$\tilde{f}_j(\xi) = (-1)^{\delta_{j,3}} f_j(\xi^*), \quad \tilde{g}(\xi) = g(\xi^*) \quad \text{for } \xi_3 < 0.$$

Then the functions

$$\begin{aligned} u_i(x) &= \sum_{j=1}^{3} \int_{\mathbb{R}^3} \tilde{f}_j(\xi)\, \mathcal{G}_{i,j}(x,\xi)\, d\xi + \int_{\mathbb{R}^3} \tilde{g}(\xi)\, \mathcal{G}_{i,4}(x,\xi)\, d\xi \\ &= \sum_{j=1}^{3} \int_{\mathbb{R}^3_+} f_j(\xi) \big(\mathcal{G}_{i,j}(x,\xi) + (-1)^{\delta_{j,3}} \mathcal{G}_{i,j}(x,\xi^*)\big)\, d\xi \\ &\quad + \int_{\mathbb{R}^3} g(\xi) \big(\mathcal{G}_{i,4}(x,\xi) + \mathcal{G}_{i,4}(x,\xi^*)\big)\, d\xi \end{aligned}$$

and

$$\begin{aligned} p(x) &= -g(x) + \sum_{j=1}^{3} \int_{\mathbb{R}^3} \tilde{f}_j(\xi)\, \mathcal{G}_{4,j}(x,\xi)\, d\xi \\ &= -g(x) + \sum_{j=1}^{3} \int_{\mathbb{R}^3_+} f_j(\xi) \big(\mathcal{G}_{4,j}(x,\xi) + (-1)^{\delta_{j,3}} \mathcal{G}_{4,j}(x,\xi^*)\big)\, d\xi \end{aligned}$$

satisfy (9.5.1). Furthermore, one can easily check that u satisfies the boundary conditions (9.5.13). This proves the theorem. □

9.5.5. Green's matrix for the Neumann problem for the Stokes system. Finally, we consider the Stokes system in $\mathbb{R}^3_+$ with the the Neumann boundary condition

$$\partial_{x_1} u_3(x) + \partial_{x_3} u_1(x) = \partial_{x_2} u_3(x) + \partial_{x_3} u_2(x) = -p + 2\partial_{x_3} u_3(x) = 0$$

on the plane $x_3 = 0$.

THEOREM 9.5.5. *The elements $\mathcal{G}^+_{i,j}(x,\xi)$ of the Green's matrix of the Neumann problem for the Stokes system in the half-space $\mathbb{R}^3_+$ are given for $i = 1,2$ by*

$$\mathcal{G}^+_{i,j}(x,\xi) = \mathcal{G}_{i,j}(x,\xi) + \mathcal{G}_{i,j}(x,\xi^*) + \frac{1}{4\pi} \frac{\partial^2}{\partial x_i \partial \xi_j} \frac{x_3 \xi_3}{|x-\xi^*|}, \qquad j = 1,2,$$

$$\mathcal{G}^+_{i,3}(x,\xi) = \mathcal{G}_{i,3}(x,\xi) - \mathcal{G}_{i,3}(x,\xi^*) - \frac{1}{4\pi} \frac{\partial}{\partial x_i} \Big(\frac{x_3}{|x-\xi^*|} + \frac{x_3 \xi_3 (x_3+\xi_3)}{|x-\xi^*|^3} \Big),$$

$$\mathcal{G}^+_{i,4}(x,\xi) = \mathcal{G}_{i,4}(x,\xi) + \mathcal{G}_{i,4}(x,\xi^*) + \frac{1}{2\pi} \frac{\partial^2}{\partial x_i \partial \xi_3} \frac{x_3}{|x-\xi^*|}.$$

The other elements of the Green's matrix are given by the formulas

$$\mathcal{G}^+_{3,3}(x,\xi) = \mathcal{G}_{3,3}(x,\xi) - \mathcal{G}_{3,3}(x,\xi^*) + \frac{1}{4\pi} \Big(1 - x_3 \frac{\partial}{\partial x_3}\Big) \Big(1 - \xi_3 \frac{\partial}{\partial \xi_3}\Big) \frac{1}{|x-\xi|^*},$$

$$\mathcal{G}^+_{3,4}(x,\xi) = \mathcal{G}_{3,4}(x,\xi) + \mathcal{G}_{3,4}(x,\xi^*) - \frac{1}{2\pi} \frac{\partial}{\partial \xi_3} \Big(\frac{1}{|x-\xi^*|} + \frac{x_3(x_3+\xi_3)}{|x-\xi^*|^3} \Big),$$

$$\mathcal{G}^+_{4,4}(x,\xi) = -\delta(x-\xi) + \frac{1}{\pi} \frac{\partial^2}{\partial \xi_3^2} \frac{1}{|x-\xi^*|},$$

$$\mathcal{G}^+_{i,j}(x,\xi) = \mathcal{G}^+_{j,i}(\xi,x) \quad \textit{for } i,j = 1,2,3,4.$$

P r o o f. Let $F(\xi) = f(\xi) + \nabla_\xi g(\xi)$. We define the functions $\tilde{F}$ and $\tilde{g}$ by

$$\tilde{F}(\xi) = F(\xi), \quad \tilde{g}(\xi) = g(\xi) \quad \text{for } \xi_3 > 0,$$
$$\tilde{F}_j(\xi) = (-1)^{\delta_{j,3}} F_j(\xi^*), \quad \tilde{g}(\xi) = g(\xi^*) \quad \text{for } \xi_3 < 0.$$

Then the functions

$$\begin{aligned} u_i(x) &= \sum_{j=1}^3 \int_{\mathbb{R}^3} \tilde{F}_j(\xi)\, \mathcal{G}_{i,j}(x,\xi)\, d\xi + \int_{\mathbb{R}^3} \tilde{g}(\xi)\, \mathcal{G}_{i,4}(x,\xi) \\ &= \sum_{j=1}^3 \int_{\mathbb{R}^3_+} F_j(\xi)\, \big(\mathcal{G}_{i,j}(x,\xi) + (-1)^{\delta_{j,3}} \mathcal{G}_{i,j}(x,\xi^*)\big)\, d\xi \\ &\quad + \int_{\mathbb{R}^3_+} g(\xi)\, \big(\mathcal{G}_{i,4}(x,\xi) + \mathcal{G}_{i,4}(x,\xi^*)\big)\, d\xi, \end{aligned}$$

$$\begin{aligned} p(x) &= -2g(x) + \sum_{j=1}^3 \int_{\mathbb{R}^3} \tilde{F}_j(\xi)\, \mathcal{G}_{4,j}(x,\xi)\, d\xi \\ &= -2g(x) + \sum_{j=1}^3 \int_{\mathbb{R}^3_+} F_j(\xi)\, \big(\mathcal{G}_{4,j}(x,\xi) + (-1)^{\delta_{j,3}} \mathcal{G}_{4,j}(x,\xi^*)\big)\, d\xi \end{aligned}$$

satisfy (9.5.1) and the boundary conditions

$$\partial_{x_1} u_3(x) + \partial_{x_3} u_1(x) = \partial_{x_2} u_3(x) + \partial_{x_3} u_2(x) = 0$$

on the plane $x_3 = 0$. We consider the function

$$\begin{aligned} &-p(x) + 2\partial_{x_3} u_3(x) \\ &= -\frac{3}{4\pi} \sum_{j=1}^2 \int_{\mathbb{R}^3_+} F_j(\xi) \Big(\frac{(x_j - \xi_j)(x_3 - \xi_3)^2}{|x - \xi|^5} + \frac{(x_j - \xi_j)(x_3 + \xi_3)^2}{|x - \xi^*|^5} \Big)\, d\xi \\ &\quad - \frac{3}{4\pi} \int_{\mathbb{R}^3_+} F_3(\xi) \Big(\frac{(x_3 - \xi_3)^3}{|x - \xi|^5} - \frac{(x_3 + \xi_3)^3}{|x - \xi^*|^5} \Big)\, d\xi + 2g(x) \\ &\quad - \frac{1}{2\pi} \frac{\partial}{\partial x_3} \int_{\mathbb{R}^3_+} g(\xi)\, \frac{\partial}{\partial \xi_3} \Big(\frac{1}{|x - \xi|} - \frac{1}{|x - \xi^*|} \Big)\, d\xi. \end{aligned}$$

The restriction to the plane $x_3 = 0$ is

$$\begin{aligned} \Phi(x') \;=\; & -\frac{3}{2\pi} \sum_{j=1}^2 \int_{\mathbb{R}^3_+} F_j(\xi)\, \frac{(x_j - \xi_j)\xi_3^2}{|(x',0) - \xi|^5}\, d\xi + \frac{3}{2\pi} \int_{\mathbb{R}^3_+} F_3(\xi)\, \frac{\xi_3^3}{|(x',0) - \xi|^5}\, d\xi \\ & + 2g(x',0) + \frac{1}{\pi} \int_{\mathbb{R}^3_+} \frac{\partial g(\xi)}{\partial \xi_3}\, \frac{\xi_3}{|(x',0) - \xi|^3}\, d\xi. \end{aligned}$$

The vector (v, q) with the components

$$\begin{aligned} v_i(x) &= -\frac{1}{4\pi} \int_{\mathbb{R}^2} \frac{(x_i - y_i)x_3}{|x - y'|^3}\, \Phi(y')\, dy', \quad i = 1, 2, \\ v_3(x) &= -\frac{1}{4\pi} \int_{\mathbb{R}^2} \Big(\frac{1}{|x - y'|} + \frac{x_3^2}{|x - y'|^3} \Big)\, \Phi(y')\, dy', \\ q(x) &= \frac{1}{2\pi} \frac{\partial}{\partial x_3} \int_{\mathbb{R}^2} \frac{1}{|x - y'|}\, \Phi(y')\, dy' \end{aligned}$$

(here, for the sake of brevity, we wrote y' instead of $(y',0)$) satisfies the equations $-\Delta v + \nabla q = 0$, $\nabla \cdot v = 0$. Furthermore, it follows from the equalities (9.5.11) and (9.5.12) that

$$\begin{aligned} -q + 2\frac{\partial v_3}{\partial x_3} &= \frac{3}{2\pi}\int_{\mathbb{R}^2} \frac{x_3^3\,\Phi(y')}{|x-y'|^5}\,dy' = \Phi(x') + \frac{3}{2\pi}\int_{\mathbb{R}^2} \frac{x_3^3\,\big(\Phi(y')-\Phi(x')\big)}{|x-y'|^5}\,dy' \\ &= \Phi(x') + \frac{3}{2\pi}\int_{\mathbb{R}^2} \frac{\Phi(x'+x_3z')-\Phi(x')\big)}{(1+|z'|^2)^{5/2}}\,dz' \to \Phi(x') \quad \text{as } x_3 \to 0 \end{aligned}$$

and

$$\begin{aligned} \frac{\partial v_i}{\partial x_3} + \frac{\partial v_3}{\partial x_i} &= \frac{3}{2\pi}\int_{\mathbb{R}^2} \frac{(x_i-y_i)x_3^2}{|x-y'|^5}\,\Phi(y')\,dy' \\ &= -\frac{3}{2\pi}\int_{\mathbb{R}^2} \frac{z_i\big(\Phi(x'+x_3z')-\Phi(x')\big)}{(1+|z'|^2)^{5/2}}\,dz' \to 0 \quad \text{as } x_3\to 0,\ i=1,2. \end{aligned}$$

Consequently, the vector function $(u-v, p-q)$ is a solution of the system (9.5.1) satisfying the Neumann condition (iv). It remains to represent v and q in terms of F and g. For $i=1,2$ we get

$$\begin{aligned} v_i(x) = &-\frac{1}{8\pi^2}\sum_{j=1}^{2}\int_{\mathbb{R}^3_+} F_j(\xi)\,\frac{\partial^2}{\partial x_i\partial \xi_j}\int_{\mathbb{R}^2}\frac{x_3\xi_3^2}{|x-y'|\,|y'-\xi|^3}\,dy'\,d\xi \\ &+\frac{1}{8\pi^2}\int_{\mathbb{R}^3_+} F_3(\xi)\,\frac{\partial}{\partial x_i}\int_{\mathbb{R}^2}\Big(\frac{2x_3\xi_3}{|x-y'|\,|y'-\xi|^3} - \frac{\partial}{\partial\xi_3}\frac{x_3\xi_3^2}{|x-y'|\,|y'-\xi|^3}\Big)\,dy'\,d\xi \\ &-\frac{1}{2\pi}\int_{\mathbb{R}^2}\frac{(x_i-y_i)x_3}{|x-y'|^3}\,g(y',0)\,dy' \\ &+\frac{1}{4\pi^2}\int_{\mathbb{R}^3_+}\frac{\partial g(\xi)}{\partial \xi_3}\,\frac{\partial}{\partial x_i}\int_{\mathbb{R}^2}\frac{x_3\xi_3}{|x-y'|\,|y'-\xi|^3}\,dy'\,d\xi. \end{aligned}$$

Using the equality

$$\int_{\mathbb{R}^2}\frac{\xi_3\,dy'}{|x-y'|\,|y'-\xi|^3} = \frac{2\pi}{|x-\xi^*|} \tag{9.5.14}$$

(cf. Lemma 9.5.1), we obtain

$$\begin{aligned} v_i(x) = &\frac{1}{4\pi}\int_{\mathbb{R}^3_+} F_3(\xi)\,\frac{\partial}{\partial x_i}\Big(\frac{x_3}{|x-\xi^*|} + \frac{x_3\xi_3(x_3+\xi_3)}{|x-\xi^*|^3}\Big)\,d\xi \\ &-\frac{1}{4\pi}\sum_{j=1}^{2}\int_{\mathbb{R}^3_+} F_j(\xi)\,\frac{\partial^2}{\partial x_i\partial\xi_j}\frac{x_3\xi_3}{|x-\xi^*|}\,d\xi - \frac{1}{2\pi}\int_{\mathbb{R}^3_+} g(\xi)\,\frac{\partial^2}{\partial x_i\partial\xi_3}\frac{x_3}{|x-\xi^*|}\,d\xi \end{aligned}$$

for $i = 1, 2$. Analogously, $v_3(x)$ is equal to

$$\frac{1}{8\pi^2}\sum_{j=1}^{2}\int_{\mathbb{R}^3_+} F_j(\xi)\,\frac{\partial}{\partial \xi_j}\int_{\mathbb{R}^2}\Big(\frac{2\xi_3^2}{|x-y'|\,|y'-\xi|^3} - \frac{\partial}{\partial x_3}\frac{x_3\xi_3^2}{|x-y'|\,|y'-\xi|^3}\Big)\,dy'\,d\xi$$
$$-\frac{1}{8\pi^2}\int_{\mathbb{R}^3_+} F_3(\xi)\int_{\mathbb{R}^2}\Big(\frac{1}{|x-y'|} - x_3\,\frac{\partial}{\partial x_3}\frac{1}{|x-y'|}\Big)\Big(1-\xi_3\,\frac{\partial}{\partial \xi_3}\Big)\,\frac{\xi_3}{|y'-\xi|^3}\,dy'\,d\xi$$
$$-\frac{1}{2\pi}\int_{\mathbb{R}^2}\Big(\frac{1}{|x-y'|} + \frac{x_3^2}{|x-y'|^3}\Big)\,g(y',0)\,dy'$$
$$-\frac{1}{4\pi^2}\int_{\mathbb{R}^3_+}\frac{\partial g(\xi)}{\partial \xi_3}\int_{\mathbb{R}^2}\Big(\frac{\xi_3}{|x-y'|\,|y'-\xi|^3} - \frac{\partial}{\partial \xi_3}\frac{x_3^2}{|x-y'|^3\,|y'-\xi|}\Big)\,dy'\,d\xi.$$

This and (9.5.14) imply

$$\begin{aligned} v_3(x) &= \frac{1}{4\pi}\sum_{j=1}^{2}\int_{\mathbb{R}^3_+} F_j(\xi)\,\frac{\partial}{\partial \xi_j}\Big(\frac{\xi_3}{|x-\xi^*|} + \frac{x_3\xi_3(x_3+\xi_3)}{|x-\xi^*|^3}\Big)\,d\xi \\ &\quad - \frac{1}{4\pi}\int_{\mathbb{R}^3_+} F_3(\xi)\,\Big(1 - x_3\frac{\partial}{\partial x_3}\Big)\Big(1-\xi_3\,\frac{\partial}{\partial \xi_3}\Big)\,\frac{1}{|x-\xi|^*}\,d\xi \\ &\quad + \frac{1}{2\pi}\int_{\mathbb{R}^3_+} g(\xi)\,\frac{\partial}{\partial \xi_3}\Big(\frac{1}{|x-\xi^*|} + \frac{x_3(x_3+\xi_3)}{|x-\xi^*|^3}\Big)\,d\xi. \end{aligned}$$

Finally, since $q(x)$ is equal to

$$-\frac{1}{4\pi^2}\sum_{j=1}^{2}\frac{\partial}{\partial x_3}\int_{\mathbb{R}^3_+} F_j(\xi)\,\frac{\partial}{\partial \xi_j}\int_{\mathbb{R}^2}\frac{\xi_3^2}{|x-y'|\,|y'-\xi|^3}\,dy'\,d\xi$$
$$+\frac{1}{4\pi^2}\frac{\partial}{\partial x_3}\int_{\mathbb{R}^3_+} F_3(\xi)\int_{\mathbb{R}^2}\Big(\frac{2\xi_3}{|x-y'|\,|y'-\xi|^3} - \frac{\partial}{\partial \xi_3}\frac{\xi_3^2}{|x-y'|\,|y'-\xi|^3}\Big)\,d\xi$$
$$+\frac{1}{\pi}\frac{\partial}{\partial x_3}\int_{\mathbb{R}^2}\frac{g(y')\,dy'}{|x-y'|} + \frac{1}{2\pi^2}\frac{\partial}{\partial x_3}\int_{\mathbb{R}^3_+}\frac{\partial g(\xi)}{\partial \xi_3}\int_{\mathbb{R}^2}\frac{\xi_3}{|x-y'|\,|y'-\xi|^3}\,dy'\,d\xi,$$

we arrive at the equality

$$\begin{aligned} q(x) &= -\frac{1}{2\pi}\sum_{j=1}^{2}\int_{\mathbb{R}^3_+} F_j(\xi)\,\frac{\partial^2}{\partial x_3\partial \xi_j}\frac{\xi_3}{|x-\xi^*|}\,d\xi \\ &\quad + \frac{1}{2\pi}\int_{\mathbb{R}^3_+} F_3(\xi)\,\frac{\partial}{\partial x_3}\Big(\frac{1}{|x-\xi^*|} + \frac{\xi_3(x_3+\xi_3)}{|x-\xi^*|^3}\Big)\,d\xi \\ &\quad - \frac{1}{\pi}\int_{\mathbb{R}^3_+} g(\xi)\,\frac{\partial^2}{\partial x_3\partial \xi_3}\frac{1}{|x-\xi^*|}\,d\xi. \end{aligned}$$

Now the assertion of the lemma follows directly from the representation of the solution $(u-v, p-q)$ just obtained. □

9.5.6. Common properties of the above Green's matrices. Let $\mathcal{G}^+(x,\xi)$ be the Green's matrix for the Stokes system in the half-space $\mathbb{R}^3_+$ with one of the boundary conditions (i)–(iv) on the plane $x_3 = 0$. Then the following theorem holds.

THEOREM 9.5.6. 1) $\mathcal{G}^+_{i,j}(x,\xi) = \mathcal{G}^+_{j,i}(\xi,x)$ *for* $i,j = 1,2,3,4$.

2) *The elements of the matrix* $G(x,\xi)$ *satisfy the estimate*

$$|\partial_x^\alpha \partial_\xi^\gamma \mathcal{G}^+_{i,j}(x,\xi)| \le c_{\alpha,\gamma}\, |x-\xi|^{-1-\delta_{i,4}-\delta_{j,4}-|\alpha|-|\gamma|},$$

where the constants $c_{\alpha,\gamma}$ *are independent of* x *and* ξ.

3) *There exist vector functions* $\vec{P}_i(x,\xi) = (P_{i,1}(x,\xi), P_{i,2}(x,\xi), P_{i,3}(x,\xi))$, $i = 1,\dots,4$, *such that*

$$\mathcal{G}^+_{i,4}(x,\xi) = -\nabla_\xi \cdot \vec{P}_i(x,\xi), \qquad P_{i,3}(x,\xi)|_{\xi_3=0} = 0,$$

and

$$|\partial_x^\alpha \partial_\xi^\gamma \vec{P}_i(x,\xi)| \le c_{\alpha,\gamma}\, |x-\xi|^{-1-\delta_{i,4}-|\alpha|-|\gamma|}. \tag{9.5.15}$$

P r o o f. The assertions 1) and 2) are obvious. The functions $\mathcal{G}^+_{i,4}$, $i = 1,2$, are of the form $\partial_{\xi_j} P_i(x,\xi)$ with a function P_i satisfying (9.5.15). Thus, assertion 3) is true for $i = 1$ and $i = 2$. In the case of the Dirichlet condition (i), the function $\mathcal{G}^+_{3,4}(x,\xi)$ has the form

$$\mathcal{G}^+_{3,4}(x,\xi) = -\frac{1}{4\pi}\frac{\partial}{\partial\xi_3}\Big(\frac{1}{|x-\xi|} - \frac{1}{|x-\xi^*|} - 2\,\frac{x_3(x_3+\xi_3)}{|x-\xi^*|^3}\Big).$$

Using the equality

$$\frac{\partial}{\partial\xi_3}\,\frac{x_3(x_3+\xi_3)}{|x-\xi^*|^3} = -x_3\frac{\partial^2}{\partial\xi_3^2}\,\frac{1}{|x-\xi^*|} = x_3\sum_{j=1}^{2}\frac{\partial^2}{\partial\xi_j^2}\,\frac{1}{|x-\xi^*|},$$

we obtain

$$\mathcal{G}^+_{3,4}(x,\xi) = \frac{1}{2\pi}\sum_{j=1}^{2}\frac{\partial^2}{\partial\xi_j^2}\,\frac{x_3}{|x-\xi^*|} - \frac{1}{4\pi}\frac{\partial}{\partial\xi_3}\Big(\frac{1}{|x-\xi|} - \frac{1}{|x-\xi^*|}\Big).$$

Furthermore, the representation

$$\begin{aligned}\mathcal{G}^+_{3,4}(x,\xi) &= \frac{1}{4\pi}\sum_{j=1}^{2}\frac{\partial}{\partial\xi_j}\Big(\frac{(x_j-\xi_j)(x_3-\xi_3)}{|x-\xi|^3} + \frac{(x_j-\xi_j)(x_3-\xi_3)}{|x-\xi^*|^3}\Big)\\ &\quad + \frac{1}{4\pi}\frac{\partial}{\partial\xi_3}\Big(\frac{(x_3-\xi_3)^2}{|x-\xi|^3} - \frac{x_3^2-\xi_3^2}{|x-\xi^*|^3}\Big)\end{aligned}$$

holds for the boundary condition (ii), while

$$\mathcal{G}^+_{3,4}(x,\xi) = -\frac{1}{4\pi}\frac{\partial}{\partial\xi_3}\Big(\frac{1}{|x-\xi|} - \frac{1}{|x-\xi^*|}\Big)$$

in the case of the boundary condition (iii) and

$$\begin{aligned}\mathcal{G}^+_{3,4}(x,\xi) &= -\frac{1}{2\pi}\sum_{j=1}^{2}\frac{\partial}{\partial\xi_j}\,\frac{(x_j-\xi_j)(x_3+\xi_3)}{|x-\xi^*|^3}\\ &\quad - \frac{1}{4\pi}\frac{\partial}{\partial\xi_3}\Big(\frac{1}{|x-\xi|} - \frac{1}{|x-\xi^*|} - 2\frac{\xi_3(x_3+\xi_3)}{|x-\xi^*|^3}\Big)\end{aligned}$$

in the case of the Neumann condition (iv). This proves the assertion 3) for $i = 3$. Finally using the equalities

$$-\delta(x-\xi) = \frac{1}{4\pi}\,\Delta_\xi\Big(\frac{1}{|x-\xi|}+\frac{1}{|x-\xi^*|}\Big)$$
$$= \frac{1}{4\pi}\sum_{j=1}^{2}\frac{\partial^2}{\partial\xi_j^2}\Big(\frac{1}{|x-\xi|}+\frac{1}{|x-\xi^*|}\Big)+\frac{1}{4\pi}\frac{\partial}{\partial\xi_3}\Big(\frac{x_3-\xi_3}{|x-\xi|^3}-\frac{x_3+\xi_3}{|x-\xi^*|^3}\Big)$$

and

$$\frac{\partial^2}{\partial\xi_3^2}\frac{1}{|x-\xi^*|} = -\sum_{j=1}^{2}\frac{\partial^2}{\partial\xi_j^2}\frac{1}{|x-\xi^*|},$$

we obtain the assertion 3) for $i = 4$. The proof of the theorem is complete. □

9.6. Green's matrix for the boundary value problem in a dihedron

This section is dedicated to the *Green's matrix* for the boundary value problem (9.1.1), (9.1.2) in the dihedron $\mathcal{D}$. In particular, we are interested in point estimates for the elements of the matrix $G(x,\xi)$. We consider the cases $|x-\xi| < \min(|x'|,|\xi'|)$ and $|x-\xi| > \min(|x'|,|\xi'|)$ separately.

9.6.1. Existence of Green's matrix. Let

$$G(x,\xi) = \big(G_{i,j}(x,\xi)\big)_{i,j=1}^{4}$$

be the Green's matrix of the boundary value problem (9.1.1), (9.1.2). This means that the vector functions $\vec{G}_j = (G_{1,j}, G_{2,j}, G_{3,j})^t$ and the functions $G_{4,j}$, $j = 1,2,3,4$, are solutions of the problem

$$-\Delta_x\vec{G}_j(x,\xi)+\nabla_x G_{4,j}(x,\xi) = \delta(x-\xi)\,(\delta_{1,j},\delta_{2,j},\delta_{3,j})^t,$$
$$-\nabla_x\cdot\vec{G}_j(x,\xi) = \delta(x-\xi)\,\delta_{4,j}\quad\text{for } x,\xi\in\mathcal{D},$$
$$S^\pm\vec{G}_j(x,\xi)=0,\quad N^\pm\big(\vec{G}_j(x,\xi),G_{4,j}(x,\xi)\big)=0\quad\text{for } x\in\Gamma^\pm,\ \xi\in\mathcal{D}.$$

We prove the existence, uniqueness and some basic properties of the Green's matrix.

THEOREM 9.6.1. 1) *There exists a unique Green's matrix $G(x,\xi)$ such that the function $x \to \zeta(x,\xi)\,G_{i,j}(x,\xi)$ belongs to $L^{1,2}(\mathcal{D})$ for $i = 1,2,3$, $\xi\in\mathcal{D}$, and to $L_2(\mathcal{D})$ for $i=4$, $\xi\in\mathcal{D}$. Here $\zeta(\cdot,\xi)$ is an arbitrary infinitely differentiable function which is bounded together with all derivatives and equal to zero in a neighborhood of the point $x=\xi$.*

2) *The functions $G_{i,j}(x,\xi)$ are infinitely differentiable with respect to $x,\xi\in\overline{\mathcal{D}}\backslash M$, $x\neq\xi$.*

3) *The elements of the Green's matrix $G(x,\xi)$ satisfy the equalities*

$$G_{i,j}(tx,t\xi) = t^{-1-\delta_{i,4}-\delta_{j,4}}\,G_{i,j}(x,\xi)\quad \textit{for arbitrary } t>0,\ x,\xi\in\mathcal{D} \tag{9.6.1}$$

and

$$G_{i,j}(x,\xi) = G_{j,i}(\xi,x) \tag{9.6.2}$$

for $i,j = 1,2,3,4$.

4) *Every solution $(u,p)\in C_0^\infty(\overline{\mathcal{D}})^4$ of the Stokes system (9.1.1) satisfying the homogeneous boundary condition (9.1.2) is given by the formulas*

$$u_i(x) = \int_{\mathcal{D}}\big(f(\xi)+\nabla_\xi g(\xi)\big)\cdot\vec{H}_i(x,\xi)\,d\xi+\int_{\mathcal{D}} g(\xi)\,G_{i,4}(x,\xi)\,d\xi \tag{9.6.3}$$

for $i = 1, 2, 3$ *and*

$$p(x) = -g(x) + \int_{\mathcal{D}} \big(f(\xi) + \nabla_\xi g(\xi)\big) \cdot \vec{H}_4(x,\xi)\, d\xi + \int_{\mathcal{D}} g(\xi)\, G_{4,4}(x,\xi)\, d\xi, \tag{9.6.4}$$

where $\vec{H}_i$ *denotes the vector function* $(G_{i,1}, G_{i,2}, G_{i,3})^t$.

P r o o f. 1) Let $\mathcal{G}(x,\xi)$ be the Green's matrix of the Stokes system in $\mathbb{R}^3$ (see (9.5.6)–(9.5.8)). We define the matrix $G(x,\xi) = (G_{i,j}(x,\xi))_{i,j=1}^4$ of the boundary value problem (9.1.1), (9.1.2) by the formula

$$G(x,\xi) = \psi(x,\xi)\, \mathcal{G}(x,\xi) + \mathcal{R}(x,\xi). \tag{9.6.5}$$

Here $\psi(x,\xi)$ is a smooth function equal to 1 for small $|x-\xi|$ and to zero for large $|x-\xi|$ and for x near the edge M. The vector function $\big(\vec{\mathcal{R}}_j, \mathcal{R}_{4,j}\big) = \big(\mathcal{R}_{1,j}, \mathcal{R}_{2,j}, \mathcal{R}_{3,j}, \mathcal{R}_{4,j}\big)$ is the uniquely determined variational solution of the Stokes system

$$\begin{aligned} &-\Delta_x \vec{\mathcal{R}}_j(x,\xi) + \nabla_x \mathcal{R}_{4,j}(x,\xi) \\ &\qquad = \delta(x-\xi)\,(\delta_{1,j}, \delta_{2,j}, \delta_{3,j}) + \Delta_x\big(\psi(x,\xi)\vec{\mathcal{G}}_j(x,\xi)\big) - \nabla_x\big(\psi(x,\xi)\mathcal{G}_{4,j}(x,\xi)\big), \\ &-\nabla_x \vec{\mathcal{R}}_j(x,\xi) = \delta_{4,j}\delta(x-\xi) + \nabla_x\big(\psi(x,\xi)\vec{\mathcal{G}}_j(x,\xi)\big) \quad \text{for } x \in \mathcal{D} \end{aligned}$$

with the boundary conditions

$$\begin{aligned} &S^\pm \vec{\mathcal{R}}_j(x,\xi) = -S^\pm\big(\psi(x,\xi)\vec{\mathcal{G}}_j(x,\xi)\big), \\ &N^\pm\big(\vec{\mathcal{R}}_j(x,\xi), \mathcal{R}_{4,j}(x,\xi)\big) = -N^\pm\big(\psi(x,\xi)\vec{\mathcal{G}}_j(x,\xi), \psi(x,\xi)\mathcal{G}_{4,j}(x,\xi)\big) \quad \text{on } \Gamma^\pm \end{aligned}$$

($j = 1, 2, 3, 4$, $\xi \in \mathcal{D}$). Since the right-hand sides of the above equations are smooth, have compact supports and vanish near the edge, this problem has a unique solution $\big(\vec{\mathcal{R}}_j(\cdot,\xi), \mathcal{R}_{4,j}(\cdot,\xi)\big)$ in the space $L^{1,2}(\mathcal{D})^3 \times L_2(\mathcal{D})$ for arbitrary $\xi \in \mathcal{D}$ (cf. Theorem 9.1.5).

2) is obvious.

3) The equality (9.6.1) follows immediately from the definition of $G(x,\xi)$. We show that Green's formula (9.1.5) is valid for $u(x) = \vec{G}_i(x,y)$, $p(x) = G_{i,4}(x,y)$, $v(x) = \vec{G}_j(x,z)$, $q(x) = G_{4,j}(x,z)$, where y and z are arbitrary points in $\mathcal{D}$. Let ζ be a smooth function on the interval $(0,\infty)$, $\zeta(t) = 1$ for $t < 1/2$, $\zeta(t) = 0$ for $t > 1$, and let ε be a sufficiently small positive number. By Theorem 9.4.9, the vector functions $\eta\big(\vec{G}_j(\cdot,\xi), G_{4,j}(\cdot,\xi)\big)$ belong to $W_\delta^{l,2}(\mathcal{D})^3 \times W_\delta^{l-1,2}(\mathcal{D})$ with a certain $\delta < l-1$ for an arbitrary smooth function η with compact support equal to zero in a neighborhood of ξ. Consequently (9.1.5) is valid for the following vector functions (u,p) and (v,q).

$$\begin{aligned} \text{(i)}\ \ & u(x) = \zeta\Big(\frac{x-y}{\varepsilon}\Big)\, \vec{G}_i(x,y), \quad p(x) = \zeta\Big(\frac{x-y}{\varepsilon}\Big)\, G_{4,i}(x,y), \\ & v(x) = \Big(1 - \zeta\Big(\frac{x-z}{\varepsilon}\Big)\Big)\, \vec{G}_j(x,z), \quad q(x) = \Big(1 - \zeta\Big(\frac{x-z}{\varepsilon}\Big)\Big)\, G_{4,j}(x,z) \\ \text{(ii)}\ \ & u(x) = \Big(1 - \zeta\Big(\frac{x-y}{\varepsilon}\Big)\Big)\, \vec{G}_i(x,y), \quad p(x) = \Big(1 - \zeta\Big(\frac{x-y}{\varepsilon}\Big)\Big)\, G_{4,i}(x,y), \\ & v(x) = \zeta\Big(\frac{x-z}{\varepsilon}\Big)\, \vec{G}_j(x,z), \quad q(x) = \zeta\Big(\frac{x-z}{\varepsilon}\Big)\, G_{4,j}(x,z), \end{aligned}$$

$$\text{(iii)}\ u(x) = \big(1 - \zeta(\frac{x-y}{\varepsilon})\big)\,\vec{G}_i(x,y), \quad p(x) = \big(1 - \zeta(\frac{x-y}{\varepsilon})\big)\,G_{4,i}(x,y),$$
$$v(x) = \big(\zeta(\frac{x-z}{R}) - \zeta(\frac{x-z}{\varepsilon})\big)\,\vec{G}_j(x,z),$$
$$q(x) = \big(\zeta(\frac{x-z}{R}) - \zeta(\frac{x-z}{\varepsilon})\big)\,G_{4,j}(x,z).$$

Here R is an arbitrary positive number. From this it follows that (9.1.5) is also valid for

$$\text{(iv)}\ u(x) = \big(1 - \zeta(\frac{x-y}{\varepsilon})\big)\,\vec{G}_i(x,y), \quad p(x) = \big(1 - \zeta(\frac{x-y}{\varepsilon})\big)\,G_{4,i}(x,y),$$
$$v(x) = \big(1 - \zeta(\frac{x-z}{\varepsilon})\big)\,\vec{G}_j(x,z), \quad q(x) = \big(1 - \zeta(\frac{x-z}{\varepsilon})\big)\,G_{4,j}(x,z)$$

To see this, one has to show that all integrals in (9.1.5) tend to zero as $R \to \infty$ if

$$u(x) = \big(1 - \zeta(\frac{x-y}{\varepsilon})\big)\,\vec{G}_i(x,y), \quad p(x) = \big(1 - \zeta(\frac{x-y}{\varepsilon})\big)\,G_{4,i}(x,y), \tag{9.6.6}$$
$$v(x) = \big(1 - \zeta(\frac{x-z}{R})\big)\,\vec{G}_j(x,z), \quad q(x) = \big(1 - \zeta(\frac{x-z}{R})\big)\,G_{4,j}(x,z). \tag{9.6.7}$$

For the integral

$$\int_{\mathcal{D}} (-\Delta u - \nabla\nabla\cdot u + \nabla p)\cdot v\,dx$$

this is evident. It vanishes for large R, since (u,p) and v have disjoint supports. Furthermore for these (u,p) and (v,q) and for sufficiently large R, the inequality

$$\Big|\int_{\mathcal{D}} u\cdot(-\Delta v - \nabla\nabla\cdot v + \nabla q)\,dx\Big|^2$$
$$\le c\Big(\int\limits_{\mathcal{D}_R(z)} R^{-1}|\vec{G}_i(x,y)|\,\big(R^{-1}|\vec{G}_j(x,z)| + |\nabla\vec{G}_j(x,z)| + |G_{4,j}(x,z)|\big)\,dx\Big)^2$$
$$\le c \int\limits_{\mathcal{D}_R(z)} \frac{|\vec{G}_i(x,y)|^2}{|x|^2}\,dx \int\limits_{\mathcal{D}_R(z)} \Big(\frac{|\vec{G}_j(x,z)|^2}{|x|^2} + |\nabla\vec{G}_j(x,z)|^2 + |G_{4,j}(x,z)|^2\Big)\,dx$$

holds, where $\mathcal{D}_R(z) = \{x : R/2 < |x-z| < R\}$. The right-hand side of the last inequality tends to zero as $R \to \infty$, since $G(x,\xi)$ can be written in the form (9.6.5) and the norm in $L^{1,2}(\mathcal{D})$ is equivalent to (9.1.6). Analogously, it can be shown that the other integrals in (9.1.5) tend to zero as $R \to \infty$ if u, v, p, q are given by (9.6.6), (9.6.7).

Finally, all integrals in (9.1.5) vanish if

$$\text{(v)}\ u(x) = \zeta(\frac{x-y}{\varepsilon})\,\vec{G}_i(x,y),\ p(x) = \zeta(\frac{x-y}{\varepsilon})\,G_{4,i}(x,y),$$
$$v(x) = \zeta(\frac{x-z}{\varepsilon})\,\vec{G}_j(x,z),\ q(x) = \zeta(\frac{x-z}{\varepsilon})\,G_{4,j}(x,z),$$

and ε is sufficiently small. Then the supports of (u,p) and (v,q) are disjoint. From the validity of (9.1.5) for the vector functions (i), (ii), (iv), (v) it follows that this formula is applicable to $u(x) = \vec{G}_i(x,y)$, $p(x) = G_{4,i}(x,y)$, $v(x) = \vec{G}_j(x,z)$, and $q(x) = G_{4,j}(x,z)$. This implies $G_{i,j}(y,z) = G_{j,i}(z,y)$.

4) follows immediately from (9.1.5) and (9.6.2). □

9.6.2. Estimates for Green's matrix. The case $|x-\xi| < \min(|x'|, |\xi'|)$. Our goal is to obtain point estimates for the elements of the matrix $G(x, \xi)$. We start with the case $|x - \xi| < \min(|x'|, |\xi'|)$.

THEOREM 9.6.2. *If $|x - \xi| < \min(|x'|, |\xi'|)$, then the estimate*

$$\big|\partial_x^\alpha \partial_\xi^\beta G_{i,j}(x,\xi)\big| \le c\,|x-\xi|^{-1-\delta_{i,4}-\delta_{j,4}-|\alpha|-|\beta|} \tag{9.6.8}$$

is valid with a constant c independent of x and ξ. For $j = 1, \dots, 4$ the representation

$$G_{4,j}(x,\xi) = -\nabla_x \cdot \vec{P}_j(x,\xi) + Q_j(x,\xi)$$

holds, where $\vec{P}_j(x,\xi) \cdot n^\pm = 0$ for $x \in \Gamma^\pm$, $\xi \in \mathcal{D}$, and $\vec{P}_j$, Q_j satisfy the estimates

$$|\partial_x^\alpha \partial_\xi^\gamma \vec{P}_j(x,\xi)| \le c_{\alpha,\gamma}\,|x-\xi|^{-1-\delta_{j,4}-|\alpha|-|\gamma|}, \tag{9.6.9}$$

$$|\partial_x^\alpha \partial_\xi^\gamma Q_j(x,\xi)| \le c_{\alpha,\gamma}\,|\xi'|^{-2-\delta_{j,4}-|\alpha|-|\gamma|} \tag{9.6.10}$$

for $|x - \xi| < \min(|x'|, |\xi'|)$.

P r o o f. Let $\mathcal{D}_+$ be the set of all $x \in \mathcal{D}$ such that $\text{dist}(x, \Gamma^+) < 2\,\text{dist}(x, \Gamma^-)$, and let t_0 denote the distance of the set $\{x \in \mathcal{D}_+ : |x'| = 1\}$ from Γ^-. Furthermore, let ζ be a smooth function on $(0, \infty)$, $\zeta(t) = 1$ for $t < t_0/2$, $\zeta(t) = 0$ for $t > t_0$.

Suppose that $\xi \in \mathcal{D}_+$. Then the function $x \to \zeta(|\xi'|^{-1}|x - \xi|)$ vanishes on Γ^-. We write the Green's matrix $G(x, \xi)$ in the form

$$G(x,\xi) = \zeta\Big(\frac{|x-\xi|}{|\xi'|}\Big)\,\mathcal{G}^+(x,\xi) + R^+(x,\xi), \tag{9.6.11}$$

where $\mathcal{G}^+(x,\xi)$ is the Green's matrix of the problem in the half-space with the boundary conditions $S^+u = 0$, $N^+(u,p) = 0$ on the plane containing Γ^+. Let $R^+_{i,j}(x,\xi)$ denote the elements of the matrix $R^+(x,\xi)$, and let $\vec{R}^+_j$ be the vector with the components $R^+_{1,j}, R^+_{2,j}, R^+_{3,j}$. Then the vector functions $\partial_\xi^\gamma\big(\vec{R}^+_j(\cdot,\xi), R^+_{4,j}(\cdot,\xi)\big)$ are the unique solutions in $L^{1,2}(\mathcal{D})^3 \times L_2(\mathcal{D})$ of the problems

$$\begin{aligned}
&-\Delta_x \partial_\xi^\gamma \vec{R}^+_j(x,\xi) + \nabla_x R^+_{4,j}(x,\xi) = \vec{\Phi}_j(x,\xi) \ \text{ for } x \in \mathcal{D},\\
&-\nabla_x \cdot \partial_\xi^\gamma \vec{R}^+_j(x,\xi) = \Psi_j(x,\xi) \ \text{ for } x \in \mathcal{D},\\
&S^+\partial_\xi^\gamma \vec{R}^+_j(x,\xi) = 0,\ N^+\partial_\xi^\gamma\big(\vec{R}^+_j(x,\xi), R^+_{4,j}(x,\xi)\big) = \vec{\Upsilon}_j \ \text{ for } x \in \Gamma^+,\\
&S^-\partial_\xi^\gamma \vec{R}^+_j(x,\xi) = 0,\ N^-\partial_\xi^\gamma\big(\vec{R}^+_j(x,\xi), R^+_{4,j}(x,\xi)\big) = 0 \ \text{ for } x \in \Gamma^-
\end{aligned}$$

for $j = 1, 2, 3, 4$, where

$$\begin{aligned}
\vec{\Phi}_j &= -\Delta_x \partial_\xi^\gamma\Big(\vec{G}_j - \zeta\Big(\frac{|x-\xi|}{|\xi'|}\Big)\vec{\mathcal{G}}^+_j\Big) + \nabla_x \partial_\xi^\gamma\Big(G_{4,j} - \zeta\Big(\frac{|x-\xi|}{|\xi'|}\Big)\mathcal{G}^+_{4,j}\Big),\\
\Psi_j &= -\nabla_x \cdot \partial_\xi^\gamma\Big(\vec{G}_j - \zeta\Big(\frac{|x-\xi|}{|\xi'|}\Big)\vec{\mathcal{G}}^+_j\Big).
\end{aligned}$$

The functions $\vec{\Phi}_j$, Ψ_j and $\vec{\Upsilon}_j$ are infinitely differentiable with respect to x and vanish for $|x - \xi| < t_0|\xi'|/2$ and $|x - \xi| > t_0|\xi'|$. Furthermore, all derivatives $\partial_x^\alpha \vec{\Phi}_j(x,\xi)$, $\partial_x^\alpha \Psi_j(x,\xi)$ and $\partial_x^\alpha \vec{\Upsilon}_j(x,\xi)$ are bounded by constants independent of x and ξ if $\xi \in \mathcal{D}_+$ and $|\xi'| = 1$. Consequently, there exist constants $c_{\alpha,\gamma}$ such that

$$\big|\partial_x^\alpha \partial_\xi^\gamma R^+_{i,j}(x,\xi)\big| \le c_{\alpha,\gamma} \quad \text{for } \xi \in \mathcal{D}_+,\ |\xi'| = 1,\ |x - \xi| < \min(|x'|, 1).$$

(Note that $|\xi'|/2 < |x'| < 2|\xi'|$ if $|x-\xi| < \min(|x'|,|\xi'|)$.) Since the functions $R^+_{i,j}(x,\xi)$ (as well as $G_{i,j}(x,\xi)$ and $\mathcal{G}^+_{i,j}(x,\xi)$) are positively homogeneous of degree $-1-\delta_{i,4}-\delta_{j,4}$, we conclude that

$$|\partial_x^\alpha \partial_\xi^\gamma R^+_{i,j}(x,\xi)| \le c_{\alpha,\gamma}\, |\xi'|^{-1-\delta_{i,4}-\delta_{j,4}-|\alpha|-|\gamma|} \tag{9.6.12}$$

for $\xi \in \mathcal{D}_+$, $|x-\xi| < \min(|x'|,|\xi'|)$. Analogously, this estimate holds for $\xi \in \mathcal{D}_- = \{x \in \mathcal{D} : \operatorname{dist}(x,\Gamma^-) < 2\operatorname{dist}(x,\Gamma^+)\}$ and $|x-\xi| < \min(|x'|,|\xi'|)$. This proves the estimate (9.6.8).

By Theorem 9.5.6, there exists a vector function $\vec{P}^+_j$ satisfying the estimate

$$|\partial_x^\alpha \partial_\xi^\gamma \vec{P}^+_j(x,\xi)| \le c_{\alpha,\gamma}\, |x-\xi|^{-1-\delta_{j,4}-|\alpha|-|\gamma|}.$$

such that $-\nabla_x \cdot \vec{P}^+_j(x,\xi) = \mathcal{G}^+_{4,j}(x,\xi)$ and $\vec{P}^+_j(x,\xi)\cdot n^+ = 0$ for $x \in \Gamma^+$. Thus, it follows from (9.6.11) that

$$G_{4,j}(x,\xi) = -\zeta\Big(\frac{|x-\xi|}{|\xi'|}\Big)\, \nabla_x \cdot \vec{P}^+_j(x,\xi) + R^+_{4,j}(x,\xi)$$

for $\xi \in \mathcal{D}_+$, where $R^+_{4,j}$ satisfies (9.6.12). Analogously, we obtain the representation

$$G_{4,j}(x,\xi) = -\zeta\Big(\frac{|x-\xi|}{|\xi'|}\Big)\, \nabla_x \cdot \vec{P}^-_j(x,\xi) + R^-_{4,j}(x,\xi)$$

for $\xi \in \mathcal{D}_-$, where $\vec{P}^-_j(x,\xi)\cdot n^- = 0$ for $x \in \Gamma^-$,

$$|\partial_x^\alpha \partial_\xi^\gamma \vec{P}^-_j(x,\xi)| \le c_{\alpha,\gamma}\, |x-\xi|^{-1-\delta_{j,4}-|\alpha|-|\gamma|}$$

and $R^-_{4,j}$ satisfies the estimate (9.6.12). Let $\eta^+(\xi)$ be a function depending only on $\xi'/|\xi'|$, infinitely differentiable for $\xi' \neq 0$ such that $\eta^+(\xi) = 1$ for $\xi \in \mathcal{D}\backslash\mathcal{D}_-$, $\eta^+(\xi) = 0$ for $\xi \in \mathcal{D}\backslash\mathcal{D}_+$. Furthermore, let $\eta^-(\xi) = 1-\eta^+(\xi)$. Then

$$\begin{aligned} G_{4,j}(x,\xi) &= -\zeta\Big(\frac{|x-\xi|}{|\xi'|}\Big) \sum_{\pm} \eta^\pm(\xi)\, \nabla_x \cdot \vec{P}^\pm_j(x,\xi) + \sum_{\pm} \eta^\pm(\xi)\, R^\pm_{4,j}(x,\xi) \\ &= -\nabla_x \cdot \vec{P}_j(x,\xi) + Q_j(x,\xi), \end{aligned}$$

where

$$\vec{P}_j(x,\xi) = \sum_{\pm} \eta^\pm(\xi)\, \zeta\Big(\frac{|x-\xi|}{|\xi'|}\Big)\, \vec{P}^\pm_j(x,\xi),$$

$$Q_j(x,\xi) = \Big(\nabla_x \zeta\Big(\frac{|x-\xi|}{|\xi'|}\Big)\Big) \cdot \sum_{\pm} \eta^\pm(\xi)\, \vec{P}^\pm_j(x,\xi) + \sum_{\pm} \eta^\pm(\xi)\, R^\pm_{4,j}(x,\xi).$$

Obviously $\vec{P}_j$ and Q_j satisfy (9.6.9) and (9.6.10), respectively. If $x \in \Gamma^+$, then $\vec{P}^+_j(x,\xi)\cdot n = 0$. If moreover $|x-\xi| < t_0|\xi'|$, then $\xi \in \mathcal{D}\backslash\mathcal{D}_-$ and, consequently, $\eta^-(\xi) = 0$. Thus, $\vec{P}_j(x,\xi)\cdot n = 0$ for $x \in \Gamma^+$. Analogously, this equality holds for $x \in \Gamma^-$. The proof is complete. □

9.6.3. Estimates for Green's matrix. The case $|x-\xi| > \min(|x'|,|\xi'|)$. Let again δ_+ be the greatest real number such that the strip $0 < \operatorname{Re}\lambda < \delta_+$ is free of eigenvalues of the pencil $A(\lambda)$. Furthermore, we define the number μ_+ as in Subsection 9.4.6. If $d^+ + d^-$ is even and $\theta < \pi/m$, where $m = 1$ for $d^+ = d^-$, $m = 2$ for $d^+ \neq d^-$, then μ_+ is the greatest real number such that the strip

$$0 < \operatorname{Re}\lambda < \mu_+$$

contains only the eigenvalue $\lambda = 1$. In all other cases, we put $\mu_+ = \delta_+$. Using the regularity results in Section 9.4, we obtain the following assertion analogously to Lemma 6.6.4.

LEMMA 9.6.3. *Let $x_0 \in \mathcal{D}$, $\operatorname{dist}(x_0, M) \le 4$, and let $\mathcal{B}$ be a ball with radius 1 and center at x_0. Furthermore, let ζ, η be smooth functions with support in $\mathcal{B}$ such that $\eta = 1$ in a neighborhood of $\operatorname{supp}\zeta$. If $\eta(u,p) \in L^{1,2}(\mathcal{D})^3 \times L_2(\mathcal{D})$,*

$$-\Delta u + \nabla p = 0, \quad \nabla \cdot u = 0 \quad in\ \mathcal{D} \cap \mathcal{B}, \tag{9.6.13}$$

$$S^\pm u = 0, \quad N^\pm(u,p) = 0 \quad on\ \Gamma^\pm \cap \mathcal{B}, \tag{9.6.14}$$

then

$$\sup_{x\in\mathcal{D}} |x'|^{\max(0,|\alpha|-\mu_+ +\varepsilon)} \big|\zeta(x)\partial_{x'}^\alpha \partial_{x_3}^j u(x)\big| \le c\,\big(\|\eta u\|_{L^{1,2}(\mathcal{D})^3} + \|\eta p\|_{L_2(\mathcal{D})}\big)$$

and

$$\sup_{x\in\mathcal{D}} |x'|^{\max(0,|\alpha|+1-\mu_+ +\varepsilon)} \big|\zeta(x)\partial_{x'}^\alpha \partial_{x_3}^j p(x)\big| \le c\,\big(\|\eta u\|_{L^{1,2}(\mathcal{D})^3} + \|\eta p\|_{L_2(\mathcal{D})}\big)$$

for arbitrary α and j, where ε is an arbitrary positive number and c is independent of u and p.

P r o o f. Suppose that $k < \mu_+ \le k+1$, where k is a nonnegative integer. Then the number $\delta = k+1-\mu_+ +\varepsilon$ lies in the interval $(0,1)$ for sufficiently small $\varepsilon > 0$. Let χ be a function from $C_0^\infty(\mathcal{B})$ such that $\zeta\chi = \zeta$ and $\eta\chi = \chi$. Since the strip $0 < \operatorname{Re}\lambda \le k+1-\delta = \mu_+ - \varepsilon$ does not contain eigenvalues of the pencil $A(\lambda)$, it follows from Theorems 9.4.9 and 9.4.13 that

$$\partial_{x_3}^j(\chi u) \in W_\delta^{k+2,2}(\mathcal{D})^3 \quad \text{and} \quad \partial_{x_3}^j(\chi p) \in W_\delta^{k+1,2}(\mathcal{D})$$

for $j = 0, 1, \dots$. Using Lemma 9.4.3, we also get $\partial_{x_3}^j(\chi u) \in W_{\delta+\nu}^{k+\nu+2,2}(\mathcal{D})^3$, $\partial_{x_3}^j(\chi p) \in W_{\delta+\nu}^{k+\nu+1,2}(\mathcal{D})$ for $j, \nu = 0, 1, \dots$ and

$$\begin{aligned} &\|\partial_{x_3}^j(\chi u)\|_{W_{\delta+\nu}^{k+\nu+2,2}(\mathcal{D})^3} + \|\partial_{x_3}^j(\chi p)\|_{W_{\delta+\nu}^{k+\nu+1,2}(\mathcal{D})} \\ &\quad \le c\,\big(\|\eta u\|_{L^{1,2}(\mathcal{D})^3} + \|\eta p\|_{L_2(\mathcal{D})}\big). \end{aligned} \tag{9.6.15}$$

In particular, $\partial_{x'}^\alpha \partial_{x_3}^j(\chi p) \in W_\delta^{2,2}(\mathcal{D})$ for $0 \le |\alpha| \le k-1$. Since $W_\delta^{2,2}(K)$ is continuously imbedded in $C(\bar{K})$, we conclude that

$$\sup_{x'\in K, x_3\in\mathbb{R}} \big|\partial_{x'}^\alpha \partial_{x_3}^j(\chi p)(x', x_3)\big| \le c \sup_{x_3\in\mathbb{R}} \|\partial_{x'}^\alpha \partial_{x_3}^j(\chi p)(\cdot, x_3)\|_{W_\delta^{2,2}(K)}$$

Furthermore, using the imbedding $W^{1,2}(M) \subset C(M)$, we obtain

$$\begin{aligned} &\sup_{x_3\in\mathbb{R}} \|\partial_{x'}^\alpha \partial_{x_3}^j(\chi p)(\cdot, x_3)\|_{W_\delta^{2,2}(K)} \\ &\quad \le c\Big(\|\partial_{x'}^\alpha \partial_{x_3}^j(\chi p)\|_{W_\delta^{2,2}(\mathcal{D})} + \|\partial_{x'}^\alpha \partial_{x_3}^{j+1}(\chi p)\|_{W_\delta^{2,2}(\mathcal{D})}\Big). \end{aligned}$$

This implies

$$\sup_{x\in\mathcal{D}} \big|\partial_{x'}^\alpha \partial_{x_3}^j(\chi p)(x)\big| \le c\big(\|\eta u\|_{L^{1,2}(\mathcal{D})^3} + \|\eta p\|_{L_2(\mathcal{D})}\big).$$

If $|\alpha| \ge k$, then we conclude from (9.6.15) that

$$\partial_{x'}^\alpha \partial_{x_3}^j(\chi p) \in W_{\delta-k+1+|\alpha|}^{2,2}(\mathcal{D}) \subset V_{\delta-k+1+|\alpha|}^{2,2}(\mathcal{D}).$$

Applying Lemma 1.2.3, the imbedding $W_2^1(M) \subset C(M)$ and (9.6.15), we arrive at

$$\sup_{x'\in K, x_3\in\mathbb{R}} |x'|^{\delta-k+|\alpha|} \big|\partial_{x'}^\alpha \partial_{x_3}^j (\chi p)(x',x_3)\big|$$

$$\le c \sup_{x_3\in\mathbb{R}} \|\partial_{x'}^\alpha \partial_{x_3}^j (\chi p)(\cdot,x_3)\|_{V^{2,2}_{\delta-k+1+|\alpha|}(K)}$$

$$\le c \Big(\|\partial_{x'}^\alpha \partial_{x_3}^j (\chi p)\|_{V^{2,2}_{\delta-k+1+|\alpha|}(\mathcal{D})} + \|\partial_{x'}^\alpha \partial_{x_3}^{j+1} (\chi p)\|_{V^{2,2}_{\delta-k+1+|\alpha|}(\mathcal{D})} \Big)$$

$$\le c \big(\|\eta u\|_{L^{1,2}(\mathcal{D})^3} + \|\eta p\|_{L_2(\mathcal{D})} \big)$$

for $|\alpha| \ge k$. Consequently,

$$\sup_{x\in\mathcal{D}} |x'|^{\max(0,|\alpha|+1-\mu_+ +\varepsilon)} \big|\zeta(x)\partial_{x'}^\alpha \partial_{x_3}^j p(x)\big| \le c \big(\|\eta u\|_{L^{1,2}(\mathcal{D})^3} + \|\eta p\|_{L_2(\mathcal{D})} \big).$$

Analogously, it can be shown (see the proof of Lemma 6.6.4) that

$$\sup_{x\in\mathcal{D}} |x'|^{\max(0,|\alpha|-\mu_+ +\varepsilon)} \big|\zeta(x)\partial_{x'}^\alpha \partial_{x_3}^j u(x)\big| \le c \big(\|\eta u\|_{L^{1,2}(\mathcal{D})^3} + \|\eta p\|_{L_2(\mathcal{D})} \big).$$

The proof is complete. □

If $d^+ + d^- \le 3$ and θ satisfies the condition (9.4.12) (i.e., if $\lambda = 0$ is not an eigenvalue of the pencil $A(\lambda)$), then the estimate in the last lemma can be improved for $|\alpha| < \delta_+$ by means of Theorem 9.4.11.

LEMMA 9.6.4. *Let $\mathcal{B}, \zeta, \eta$ be as in Lemma 9.6.3, and let $\eta(u,p) \in L^{1,2}(\mathcal{D})^3 \times L_2(\mathcal{D})$. Furthermore, we assume that (u,p) satisfies the equations (9.6.13) and (9.6.14). If $d^+ + d^- \le 3$ and θ satisfies the condition (9.4.12), then*

$$\sup_{x\in\mathcal{D}} |x'|^{|\alpha|-\delta_+ +\varepsilon} \Big(\big|\zeta(x)\partial_{x'}^\alpha \partial_{x_3}^j u(x)\big| + |x'|\, \big|\zeta(x)\partial_{x'}^\alpha \partial_{x_3}^j p(x)\big| \Big)$$

$$\le c \big(\|\eta u\|_{L^{1,2}(\mathcal{D})^3} + \|\eta p\|_{L_2(\mathcal{D})} \big).$$

P r o o f. Let ε be a small positive number such that $\delta = 1-\delta_+ +\varepsilon$ is not integer. Furthermore, let $\chi \in C_0^\infty(\mathcal{B})$ be such that $\zeta\chi = \zeta$ and $\eta\chi = \chi$. By Theorem 9.4.11,

$$\|\partial_{x_3}^j (\chi u)\|_{V^{k+2,2}_{\delta+k}(\mathcal{D})} + \|\partial_{x_3}^j (\chi p)\|_{V^{k+1,2}_{\delta+k}(\mathcal{D})} \le c \big(\|\eta u\|_{L^{1,2}(\mathcal{D})^3} + \|\eta p\|_{L_2(\mathcal{D})} \big)$$

for arbitrary integer k. Using this inequality, Lemma 1.2.3 and the continuity of the imbedding $W_2^1(M) \subset C(M)$, we obtain

$$\sup_{(x',x_3)\in\mathcal{D}} |x'|^{\delta-1+|\alpha|} \big|\partial_{x'}^\alpha \partial_{x_3}^j (\chi u)(x',x_3)\big| \le c \sup_{x_3\in\mathbb{R}} \|\partial_{x'}^\alpha \partial_{x_3}^j (\chi u)(\cdot,x_3)\|_{V^{2,2}_{\delta+|\alpha|}(\mathcal{D})^3}$$

$$\le c \Big(\|\partial_{x'}^\alpha \partial_{x_3}^j (\chi u)\|_{V^{2,2}_{\delta+|\alpha|}(\mathcal{D})^3} + \|\partial_{x'}^\alpha \partial_{x_3}^{j+1} (\chi u)\|_{V^{2,2}_{\delta+|\alpha|}(\mathcal{D})^3} \Big)$$

$$\le c \big(\|\eta u\|_{L^{1,2}(\mathcal{D})^3} + \|\eta p\|_{L_2(\mathcal{D})} \big).$$

Analogously,

$$\sup_{(x',x_3)\in\mathcal{D}} |x'|^{\delta+|\alpha|} \big|\partial_{x'}^\alpha \partial_{x_3}^j (\chi p)(x',x_3)\big| \le c \big(\|\eta u\|_{L^{1,2}(\mathcal{D})^3} + \|\eta p\|_{L_2(\mathcal{D})} \big).$$

The result follows. □

The last two lemmas allow us to estimate the elements of Green's matrix and their derivatives in the case $|x-\xi| \ge \min(|x'|,|\xi'|)$.

THEOREM 9.6.5. *The elements of the Green's matrix* $G(x,\xi)$ *satisfy the estimate*

$$(9.6.16)\qquad \begin{aligned} &\big|\partial_{x'}^{\alpha}\partial_{x_3}^{\sigma}\partial_{\xi'}^{\beta}\partial_{\xi_3}^{\tau}G_{i,j}(x,\xi)\big| \\ &\le c\,|x-\xi|^{-T-|\alpha|-|\beta|-\sigma-\tau}\Big(\frac{|x'|}{|x-\xi|}\Big)^{\sigma_{i,\alpha}}\Big(\frac{|\xi'|}{|x-\xi|}\Big)^{\sigma_{j,\beta}}, \end{aligned}$$

for $|x-\xi|\ge\min(|x'|,|\xi'|)$, *where* $T=1+\delta_{i,4}+\delta_{j,4}$, ε *is an arbitrarily small positive number, and* $\sigma_{i,\alpha}=\min(0,\mu_+-|\alpha|-\delta_{i,4}-\varepsilon)$.

If $d^++d^-\le 3$ *and the angle* θ *satisfies the condition* (9.4.12), *then the estimate* (9.6.16) *holds also with* $\sigma_{i,\alpha}=\delta_+-|\alpha|-\delta_{i,4}-\varepsilon$ *for* $|\alpha|<\delta_+-\delta_{i,4}$.

P r o o f. By (9.6.1), it suffices to prove the estimate for $|x-\xi|=2$. Then, under the assumption $\min(|x'|,|\xi'|)\le|x-\xi|$, we have $\max(|x'|,|\xi'|)\le 4$. Let $\mathcal{B}_x$, $\mathcal{B}_\xi$ be balls with centers x and ξ, respectively, and radius 1. Furthermore, let ζ and η be infinitely differentiable functions with supports in $\mathcal{B}_x$ and $\mathcal{B}_\xi$ equal to one in neighborhoods of x and ξ, respectively. By Lemmas 9.6.3 and 9.6.4,

$$(9.6.17)\qquad \begin{aligned} &\sum_{j=1}^{4}|\xi'|^{-\sigma_{j,\beta}}\big|\partial_{x'}^{\alpha}\partial_{x_3}^{\sigma}\partial_{\xi'}^{\beta}\partial_{\xi_3}^{\tau}G_{i,j}(x,\xi)\big| \\ &\le c\Big(\sum_{j=1}^{3}\|\eta(\cdot)\partial_{x'}^{\alpha}\partial_{x_3}^{\sigma}G_{i,j}(x,\cdot)\|_{\mathcal{H}}+\|\eta(\cdot)\partial_{x'}^{\alpha}\partial_{x_3}^{\sigma}G_{i,4}(x,\cdot)\|_{L_2(\mathcal{D})}\Big) \end{aligned}$$

for $i=1,2,3,4$. Let F and g be smooth functions, and let

$$\begin{aligned} u_i(y) &= \int_{\mathcal{D}}\eta(z)\,F(z)\cdot\vec{H}_i(y,z)\,dz+\int_{\mathcal{D}}\eta(z)\,g(z)\,G_{i,4}(y,z)\,dz,\quad i=1,2,3,\\ p(y) &= -\eta(y)\,g(y)+\int_{\mathcal{D}}\eta(z)\,F(z)\cdot\vec{H}_4(y,z)\,dz+\int_{\mathcal{D}}\eta(z)\,g(z)\,G_{4,4}(y,z)\,dz. \end{aligned}$$

By (9.1.4) and Theorem 9.6.1, the vector (u,p) is a solution of the problem

$$\begin{aligned} &b_{\mathcal{D}}(u,v)-\int_{\mathcal{D}}p\nabla\cdot v\,dx=\int_{\mathcal{D}}\eta(y)\,F(y)\cdot v(y)\,dy\ \text{ for all } v\in\mathcal{H}_{\mathcal{D}},\\ &-\nabla\cdot u=\eta g\ \text{ in } \mathcal{D},\quad S^{\pm}u=0\ \text{ on } \Gamma^{\pm}. \end{aligned}$$

Since ηF vanishes in $\mathcal{B}_x$, we conclude from Lemmas 9.6.3 and 9.6.4 that

$$\begin{aligned} &|x'|^{-\sigma_{1,\alpha}}\big|\partial_{x'}^{\alpha}\partial_{x_3}^{\sigma}u(x)\big|+|x'|^{-\sigma_{4,\alpha}}\big|\partial_{x'}^{\alpha}\partial_{x_3}^{\sigma}p(x)\big| \\ &\qquad\le c\,\big(\|\zeta u\|_{L^{1,2}(\mathcal{D})^3}+\|\zeta p\|_{L_2(\mathcal{D})}\big)\le c\,\big(\|F\|_{\mathcal{H}_{\mathcal{D}}^*}+\|g\|_{L_2(\mathcal{D})}\big), \end{aligned}$$

where c is independent of x and ξ. Consequently (for every fixed $x\in\mathcal{D}$), the functionals

$$\begin{aligned} &(F,g)\to|x'|^{-\sigma_{i,\alpha}}\partial_{x'}^{\alpha}\partial_{x_3}^{\sigma}u_i(x)\\ &\quad=|x'|^{-\sigma_{i,\alpha}}\int_{\mathcal{D}}\eta(z)\Big(\sum_{j=1}^{3}F_j(z)\,\partial_{x'}^{\alpha}\partial_{x_3}^{\sigma}G_{i,j}(x,z)+g(z)\,\partial_{x'}^{\alpha}\partial_{x_3}^{\sigma}G_{i,4}(x,z)\Big)\,dz, \end{aligned}$$

$i=1,2,3$, and

$$\begin{aligned} &(F,g)\to|x'|^{-\sigma_{4,\alpha}}\partial_{x'}^{\alpha}\partial_{x_3}^{\sigma}p(x)\\ &\quad=|x'|^{-\sigma_{4,\alpha}}\int_{\mathcal{D}}\eta(z)\Big(\sum_{j=1}^{3}F_j(z)\,\partial_{x'}^{\alpha}\partial_{x_3}^{\sigma}G_{4,j}(x,z)+g(z)\,\partial_{x'}^{\alpha}\partial_{x_3}^{\sigma}G_{4,4}(x,z)\Big)\,dz \end{aligned}$$

can be extended to linear and continuous functionals on $\mathcal{H}_{\mathcal{D}}^* \times L_2(\mathcal{D})$. The norms of these functionals are bounded by constants independent of x. Therefore,

$$\sum_{j=1}^{3} \|\eta(\cdot)\partial_{x'}^\alpha \partial_{x_3}^\sigma G_{i,j}(x,\cdot)\|_{\mathcal{H}} + \|\eta(\cdot)\partial_{x'}^\alpha \partial_{x_3}^\sigma G_{i,4}(x,\cdot)\|_{L_2(\mathcal{D})} \le c\,|x'|^{\sigma_{i,\alpha}}$$

for $i = 1, 2, 3, 4$. From this and from (9.6.17) we obtain the assertion of the theorem. □

9.7. Some estimates of solutions in weighted Hölder spaces

In this section, we prove some auxiliary assertions which are needed in the next chapter, where we consider a mixed boundary value problem for the Stokes system in a polyhedral cone with the boundary conditions (i)–(iv) on its faces. First, we deal with the compatibility conditions on the edge M for the data g, $h^\pm$ and $\phi^\pm$. Then we prove some local regularity results for the solutions in the classes of the spaces $N_\delta^{l,\sigma}$ and $C_\delta^{l,\sigma}$.

9.7.1. Reduction to zero boundary data. First, we formulate a direct corollary of Lemma 2.8.2 (cf. Lemma 9.2.1).

LEMMA 9.7.1. *Let $h^\pm \in N_\delta^{l,\sigma}(\Gamma^\pm)^{3-d^\pm}$, $\phi^\pm \in N_\delta^{l-1,\sigma}(\Gamma^\pm)^{d^\pm}$, $l \ge 1$, be given. Then there exists a vector function $u \in N_\delta^{l,\sigma}(\mathcal{D})^3$ such that*

$$S^\pm u = h^\pm, \quad N^\pm(u,0) = \phi^\pm \quad \textit{on } \Gamma^\pm.$$

The norm of u can be estimated by the norms of $h^\pm$ and $\phi^\pm$.

The following analogous result in the space $C_\delta^{l,\sigma}$ holds only under additional assumptions on the boundary data. If $u \in C_\delta^{l,\sigma}(\mathcal{D})^3$, $\delta < l+\sigma$, then there exists the trace $u|_M \in C^{l-k,k-\delta+\sigma}(M)^3$, $k = [\delta - \sigma] + 1$, and from the boundary conditions (9.1.2) it follows that $S^\pm u|_M = h^\pm|_M$. Consequently, the boundary data h^+ and h^- must satisfy the compatibility condition (9.2.6).

LEMMA 9.7.2. *Let $h^\pm \in C_\delta^{l,\sigma}(\Gamma^\pm)^{3-d^\pm}$ and $\phi^\pm \in C_\delta^{l-1,\sigma}(\Gamma^\pm)^{d^\pm}$, where $0 \le l-1 < \delta-\sigma < l$. Suppose that h^+ and h^- satisfy the compatibility condition* (9.2.6) *on M. Then there exists a vector function $u \in C_\delta^{l,\sigma}(\mathcal{D})^3$ such that $S^\pm u = h^\pm$, $N^\pm(u,0) = \phi^\pm$ on $\Gamma^\pm$ and*

$$\|u\|_{C_\delta^{l,\sigma}(\mathcal{D})^3} \le c \sum_\pm \Big(\|h^\pm\|_{C_\delta^{l,\sigma}(\Gamma^\pm)^{3-d^\pm}} + \|\phi^\pm\|_{C_\delta^{l-1,\sigma}(\Gamma^\pm)^{d^\pm}} \Big),$$

where c is a constant independent of $h^\pm$ and $\phi^\pm$.

P r o o f. By (9.2.6), there exists a vector function $\psi \in C^{0,l+\sigma-\delta}(M)^3$ such that $S^\pm \psi = h^\pm|_M$. Let $v \in C_\delta^{l,\sigma}(\mathcal{D})^3$ be an extension of ψ. Then the trace of the vector function $h^\pm - S^\pm v|_{\Gamma^\pm}$ on M is equal to zero and consequently

$$h^\pm - S^\pm v|_{\Gamma^\pm} \in N_\delta^{l,\sigma}(\Gamma^\pm)^{3-d^\pm}$$

(cf. Lemma 6.7.10). Furthermore,

$$\phi^\pm - N^\pm(v,0)|_{\Gamma^\pm} \in C_\delta^{l-1,\sigma}(\Gamma^\pm)^{d^\pm} \subset N_\delta^{l-1,\sigma}(\Gamma^\pm)^{d^\pm}.$$

Thus according to Lemma 9.7.1, there exists a vector function $w \in N_\delta^{l,\sigma}(\mathcal{D})^3$ such that $S^\pm w = h^\pm - S^\pm v$ and $N^\pm(w,0) = \phi^\pm - N^\pm(v,0)$ on $\Gamma^\pm$. Then $u = v + w$ has the desired properties. □

Now let $g \in C_\delta^{l-1,\sigma}(\mathcal{D})$, $h^\pm \in C_\delta^{l,\sigma}(\Gamma^\pm)^{3-d^\pm}$, and $\phi \in C_\delta^{l,\sigma}(\Gamma^\pm)^{d^\pm}$, where

$$l-2 < \delta - \sigma < l-1.$$

Then the traces of $g, h^\pm, \partial_r h^\pm$ and $\phi^\pm$ on M exist. We suppose that there exists a vector function $(u,p) \in C_\delta^{l,\sigma}(\mathcal{D})^3 \times C_\delta^{l-1,\sigma}(\mathcal{D})$ such that

$$(9.7.1) \qquad S^\pm u = h^\pm, \quad N^\pm(u,p) = \phi^\pm \ \text{ on } \Gamma^\pm \quad \text{and} \quad -\nabla\cdot u = g \ \text{ on } M.$$

For this, it is necessary that the traces

$$b = u|_M\,, \quad c = (\partial_{x_1}u)|_M\,, \quad d = (\partial_{x_2}u)|_M \quad \text{and} \quad q = p|_M$$

satisfy the equations (9.2.7)–(9.2.10). As in Section 9.2, we consider the cases of odd $d^+ + d^-$ and even $d^+ + d^-$ separately. If $d^+ + d^-$ is odd and moreover

$$(9.7.2) \quad \sin 2\theta \neq 0 \ \text{ for } d^+ + d^- = 3, \qquad \cos\theta\,\cos 2\theta \neq 0 \ \text{ for } d^+ + d^- \in \{1,5\},$$

then the system of the equations (9.2.8)–(9.2.10) with the unknowns c, d and q is uniquely solvable for arbitrary $h^\pm, \phi^\pm, g, b$ (cf. Lemma 9.2.4). Thus, the following lemma holds analogously to Lemma 9.2.5.

LEMMA 9.7.3. *Let $d^+ + d^-$ be odd, and let the angle θ satisfy the condition* (9.7.2). *We suppose that $g \in C_\delta^{l-1,\sigma}(\mathcal{D})$, $h^\pm \in C_\delta^{l,\sigma}(\Gamma^\pm)^{3-d^\pm}$, $\phi^\pm \in C_\delta^{l-1,\sigma}(\Gamma^\pm)^{d^\pm}$, where $l \geq 1$ and $l-2 < \delta-\sigma < l-1$. Furthermore, we assume that $h^\pm(x)$, $\phi^\pm(x)$ and $g(x)$ vanish for $|x'| > 1$ and that the vector functions $h^\pm$ satisfy the compatibility condition* (9.2.6) *on M. Then there exists a vector function $(u,p) \in C_\delta^{l,\sigma}(\mathcal{D})^3 \times C_\delta^{l-1,\sigma}(\mathcal{D})$ satisfying* (9.7.1).

P r o o f. Let $b \in C^{1,l-1-\delta+\sigma}(M)^3$, $c, d \in C^{0,l-1-\delta+\sigma}(M)^3$, $q \in C^{0,l-1-\delta+\sigma}(M)$ solve the system (9.2.7)–(9.2.10). We set

$$v = Eb + x_1\, Ec + x_2\, Ed, \quad p = Eq,$$

where E is the extension operator (6.2.4). Then $v \in C_\delta^{l,\sigma}(\mathcal{D})^3$, $p \in C_\delta^{l-1,\sigma}(\mathcal{D})$,

$$v|_M = b, \quad \partial_{x_1}v|_M = c, \quad \partial_{x_2}v|_M = d \quad \text{and} \quad p|_M = q.$$

Consequently,

$$S^\pm v|_M = h^\pm|_M, \quad \partial_r S^\pm v|_M = \partial_r h^\pm|_M, \quad -\nabla\cdot v|_M = g|_M, \quad N^\pm(v,p)|_M = \phi^\pm|_M.$$

Thus by Lemma 6.7.10,

$$S^\pm v - h^\pm \in N_\delta^{l,\sigma}(\Gamma^\pm)^{3-d^\pm} \quad \text{and} \quad N^\pm(v,p) - \phi^\pm \in N_\delta^{l-1,\sigma}(\Gamma^\pm)^{d^\pm}.$$

By Lemma 9.7.1, there exists a vector function $w \in N_\delta^{l,\sigma}(\mathcal{D})^3$ such that

$$S^\pm w = h^\pm - S^\pm v, \quad N^\pm(w,0) = \phi^\pm \ \text{ on } \Gamma^\pm.$$

Then $(u,p) = (v+w,p)$ satisfies (9.7.1). □

If $d^+ + d^-$ is even, then the assertion of Lemma 9.7.3 is only valid under additional compatibility conditions on the functions $g, h^\pm, \phi^\pm$. In this case, the following assertion can be proved in the exact same manner as Lemma 9.2.6.

LEMMA 9.7.4. *Let* $l-2<\delta-\sigma<l-1$, *and let* $h^\pm \in C_\delta^{l,\sigma}(\Gamma^\pm)^{3-d^\pm}$, $\phi^\pm \in C_\delta^{l-1,\sigma}(\Gamma^\pm)^{d^\pm}$, $g \in C_\delta^{l-1,\sigma}(\mathcal{D})$ *be given functions vanishing for* $|x'|>1$. *We suppose that* $d^+ + d^-$ *is even and that* g, $h^\pm$ *and* $\phi^\pm$ *satisfy the compatibility conditions* (9.2.6) *and* (9.2.13)–(9.2.18) *on the edge* M. *Then there exists a vector function* $u \in C_\delta^{l,\sigma}(\mathcal{D})^3$ *vanishing for* $|x'|>1$ *and satisfying* (9.7.1).

Furthermore, we will employ the following result.

LEMMA 9.7.5. *Let* $l \ge 2$, $\delta \ge 0$, $l-3<\delta-\sigma<l-2$, *and let* $f \in C_\delta^{l-2,\sigma}(\mathcal{D})^3$, $g \in C_\delta^{l-1,\sigma}(\mathcal{D})$, $h^\pm \in C_\delta^{l,\sigma}(\Gamma^\pm)^{3-d^\pm}$, *and* $\phi \in C_\delta^{l-1,\sigma}(\Gamma^\pm)^{3-d^\pm}$ *be given functions vanishing for* $|x'|>1$. *We suppose that* h^+ *and* h^- *satisfy the compatibility condition* (9.2.6) *and that* $\lambda=2$ *is not an eigenvalue of the pencil* $A(\lambda)$. *If* $d^+ + d^-$ *is odd, then we assume in addition that the angle* θ *satisfies the condition* (9.7.2). *In the case of even* $d^+ + d^-$, *we assume that the compatibility conditions* (9.2.13)–(9.2.18) *are satisfied. Then there exists a vector function* $(u,p) \in C_\delta^{l,\sigma}(\mathcal{D})^3 \times C_\delta^{l-1,\sigma}(\mathcal{D})$ *such that*

$$\Delta u - \nabla p + f \in N_{\beta,\delta}^{l-2,\sigma}(\mathcal{D})^3, \quad \nabla\cdot u + g \in N_{\beta,\delta}^{l-1,\sigma}(\mathcal{D}),$$
$$S^\pm u|_{\Gamma^\pm} = h^\pm, \quad N^\pm(u,p)|_{\Gamma^\pm} = \phi^\pm .$$

P r o o f. Under the conditions of the lemma, the system (9.2.7)–(9.2.10) with the unknowns b, c, d, q is solvable. Let $b \in C^{2,l-2-\delta+\sigma}(M)^3$, $c, d \in C^{1,l-2-\delta+\sigma}(M)^3$, $q \in C^{1,l-2-\delta+\sigma}(M)$ be a solution of this system. We set

$$V = Eb + x_1\, Ec + x_2\, Ed, \quad P = Eq,$$

where E is the extension operator (6.2.4). Then $V \in C_\delta^{l,\sigma}(\mathcal{D})^3$ and $P \in C_\delta^{l-1,\sigma}(\mathcal{D})$. Furthermore,

$$-(\nabla\cdot V)|_M = g|_M\,, \quad N^\pm(V,P)|_M = \phi^\pm|_M \quad \text{and} \quad \partial_r^j S^\pm V|_M = \partial_r^j h^\pm|_M$$

for $j=0$ and $j=1$. Since $\lambda=2$ is not an eigenvalue of the pencil $A(\lambda)$, there exist vector functions

$$W(x) = \sum_{i+j=2} w^{(i,j)}(x_3)\, \frac{x_1^i\, x_2^j}{i!\, j!}\,, \qquad Q(x) = q_1(x_3)\, x_1 + q_2(x_3)\, x_2$$

with coefficients $w^{(i,j)}, q_1, q_2 \in C^{0,l-2-\delta+\sigma}(M)$ such that

$$-\partial_{x_1}^2 W - \partial_{x_2}^2 W + \nabla Q = f + \Delta V - \nabla P \quad \text{on } M,$$
$$-\partial_{x_j}\big(\partial_{x_1} W_1 + \partial_{x_2} W_2\big) = \partial_{x_j}\big(g - \nabla\cdot V\big) \quad \text{on } M \text{ for } j=1,2,$$
$$\partial_r^2 S^\pm W = \partial_r^2\big(h^\pm - S^\pm V\big) \quad \text{on } M,$$
$$\partial_r N^\pm(D_{x'},0)\,(W,Q) = \partial_r\big(\phi^\pm - N^\pm(D_x)\,(V,P)\big) \quad \text{on } M$$

(cf. Lemma 9.3.2). We set

$$w(x) = \sum_{i+j=2} (Ew^{(i,j)})(x)\, \frac{x_1^i\, x_2^j}{i!\, j!} \quad \text{and} \quad p_0(x) = (Eq_1)(x)\, x_1 + (Eq_2)(x)\, x_2$$

Then obviously,

$$
\begin{aligned}
&-\Delta w + \nabla p_0 = f + \Delta V - \nabla P \quad \text{on } M,\\
&-\partial_{x_j} \nabla\cdot w = \partial_{x_j}\left(g - \nabla\cdot V\right) \quad \text{on } M \text{ for } j = 1,2,\\
&\partial_r^2 S^\pm w = \partial_r^2\left(h^\pm - S^\pm V\right) \quad \text{on } M,\\
&\partial_r N^\pm(D_x)\,(w,p_0) = \partial_r\left(\phi^\pm - N^\pm(D_x)\,(V,P)\right) \quad \text{on } M.
\end{aligned}
$$

Moreover,

$$(\nabla\cdot w)|_M = 0, \quad N^\pm(w,p_1)|_M = 0 \quad \text{and} \quad \partial_r^j S^\pm w|_M = 0 \text{ for } j \le 1.$$

Thus, it follows from Lemma 6.7.10 that

$$-\Delta(V+w) + \nabla(P+p_0) - f \in N_\delta^{l-2,\sigma}(\mathcal{D})^3, \quad \nabla\cdot(V+w) + g \in N_\delta^{l-2,\sigma}(\mathcal{D}),$$

Since moreover

$$S^\pm(V+w) - h^\pm \in N_\delta^{l,\sigma}(\Gamma^\pm)^{3-d^\pm}$$

and

$$N^\pm(V+w, P+p_0) - \phi^\pm \in N_\delta^{l-1,\sigma}(\Gamma^\pm)^{d^\pm},$$

there exists a vector function $v \in N_\delta^{l,\sigma}(\mathcal{D})^3$ such that

$$S^\pm v = h^\pm - S^\pm(V+w) \quad \text{and} \quad N^\pm(v,0) = \phi^\pm - N^\pm(V+w, P+p_0)$$

on $\Gamma^\pm$ (cf. Lemma 9.7.1). Consequently, the vector function

$$(u,p) = (v + V + w, P + p_0)$$

has the desired properties. □

9.7.2. Local regularity results in the spaces $N_\delta^{l,\sigma}$. For the next lemma, we refer to [**152**, Theorem 6.3.7] and [**7**, Theorem 9.3].

LEMMA 9.7.6. *Let G_1, G_2 be bounded subdomains of $\mathbb{R}^3$ such that $\overline{G}_1 \subset G_2$, $G_1 \cap \mathcal{D} \ne \emptyset$ and $\overline{G}_1 \cap M = \emptyset$. If (u,p) is a solution of* (9.1.1), (9.1.2), *$u \in W^{2,s}(\mathcal{D}\cap G_2)^3$, $p \in W^{1,s}(\mathcal{D}\cap G_2)$, $f \in C^{l-2,\sigma}(\mathcal{D}\cap G_2)^3$, $g \in C^{l-2,\sigma}(\mathcal{D}\cap G_2)$, $h^\pm \in C^{l,\sigma}(\Gamma^\pm\cap G_2)^{3-d^\pm}$, $\phi^\pm \in C^{l-1,\sigma}(\Gamma^\pm\cap G_2)^{d^\pm}$, $l \ge 2$, $0 < \sigma < 1$, then $u \in C^{l,\sigma}(\mathcal{D}\cap G_1)^3$, $p \in C^{l-1,\sigma}(\mathcal{D}\cap G_2)$, and*

$$
\begin{aligned}
\|u\|_{C^{l,\sigma}(\mathcal{D}\cap G_1)^\ell} + \|p\|_{C^{l-1,\sigma}(\mathcal{D}\cap G_2)} \le c\Big(&\|f\|_{C^{l-2,\sigma}(\mathcal{D}\cap G_2)^\ell} + \|g\|_{C^{l-2,\sigma}(\mathcal{D}\cap G_2)}\\
&+ \sum_\pm \|h^\pm\|_{C^{l,\sigma}(\Gamma^\pm\cap G_2)} + \sum_\pm \|\phi^\pm\|_{C^{l-1,\sigma}(\Gamma^\pm\cap G_2)} + \|(u,p)\|_{L_\infty(\mathcal{D}\cap G_2)^4}\Big)
\end{aligned}
$$

with a constant c independent of u and p.

Using the last lemma, we prove an estimate for the norm of (u,p) in the space $N_\delta^{l,\sigma}(\mathcal{D})^3 \times N_\delta^{l-1,\sigma}(\mathcal{D})$.

LEMMA 9.7.7. *Let $(u,p) \in L^\infty_{\delta-l-\sigma}(\mathcal{D})^3 \times L^\infty_{\delta-l+1-\sigma}(\mathcal{D})$ be a solution of the boundary value problem* (9.1.1), (9.1.2) *such that $\chi(u,p) \in W^{2,s}(\mathcal{D})^3 \times W^{1,s}(\mathcal{D})$*

for every $\chi \in C_0^\infty(\bar{\mathcal{D}}\backslash M)$. *If* $f \in N_\delta^{l-2,\sigma}(\mathcal{D})^3$, $l \ge 2$, $g \in N_\delta^{l-1,\sigma}(\mathcal{D})$, $h^\pm \in N_\delta^{l,\sigma}(\Gamma^\pm)^{3-d^\pm}$, $\phi^\pm \in N_\delta^{l-1,\sigma}(\Gamma^\pm)^{d^\pm}$, *then* $u \in N_\delta^{l,\sigma}(\mathcal{D})^3$, $p \in N_\delta^{l,\sigma}(\mathcal{D})$, *and*

$$\|u\|_{N_\delta^{l,\sigma}(\mathcal{D})^3} + \|p\|_{N_\delta^{l-1,\sigma}(\mathcal{D})} \le c\Big(\|f\|_{N_\delta^{l-2,\sigma}(\mathcal{D})^3} + \|g\|_{N_\delta^{l-1,\sigma}(\mathcal{D})} + \sum_\pm \|h^\pm\|_{N_\delta^{l,\sigma}(\Gamma^\pm)} + \sum_\pm \|\phi^\pm\|_{N_\delta^{l-1,\sigma}(\Gamma^\pm)} + \|u\|_{L^\infty_{\delta-l-\sigma}(\mathcal{D})^3} + \|p\|_{L^\infty_{\delta-l+1-\sigma}(\mathcal{D})}\Big). \tag{9.7.3}$$

P r o o f. By Lemma 9.7.1, we may restrict ourselves to the case $h^\pm = 0$, $\phi^\pm = 0$. For an arbitrary point $y \in \mathcal{D}$, we denote by B_y the ball with center y and radius $|y'|/2$ and by B'_y the ball with center y and radius $3|y'|/4$. For an arbitrary subdomain $\mathcal{U} \subset \mathcal{D}$, let the norms in $N_{\beta,\delta}^{l,\sigma}(\mathcal{U})$ and $L_\delta^\infty(\mathcal{U})$ be defined by (2.7.1) and (2.8.4), respectively, where $\mathcal{D}$ is replaced by $\mathcal{U}$. If $|y'| = 1$, then these norms are equivalent to the $C^{l,\sigma}$- and L_∞-norms, respectively, and Lemma 9.7.6 implies

$$\|u\|_{N_\delta^{l,\sigma}(B_y\cap\mathcal{D})^3} + \|p\|_{N_\delta^{l-1,\sigma}(B_y\cap\mathcal{D})} \le c\Big(\|f\|_{N_\delta^{l-2,\sigma}(B'_y\cap\mathcal{D})^3} + \|g\|_{N_\delta^{l-1,\sigma}(B_y\cap\mathcal{D})} + \|u\|_{L^\infty_{\delta-l-\sigma}(B'_y\cap\mathcal{D})^3} + \|p\|_{L^\infty_{\delta-l+1-\sigma}(B'_y\cap\mathcal{D})}\Big) \tag{9.7.4}$$

with a constant c independent of y. Let $|y'| \ne 1$ and $z = |y'|^{-1}y$. We introduce the functions $\tilde{u}(\xi) = u(|y'|\xi)$, $\tilde{p}(\xi) = |y'|\,p(|y'|\xi)$, $\tilde{f}(\xi) = |y'|^2 f(|y'|\xi)$, and $\tilde{g}(\xi) = |y'|\,g(|y'|\xi)$. Then

$$-\Delta\tilde{u} + \nabla\tilde{p} = \tilde{f}, \quad -\nabla\cdot\tilde{u} = \tilde{g} \ \text{ in } \mathcal{D}, \quad S^\pm\tilde{u} = 0, \quad N^\pm(\tilde{u},\tilde{p}) = 0 \ \text{ on } \Gamma^\pm.$$

Therefore, by (9.7.4),

$$\|\tilde{u}\|_{N_\delta^{l,\sigma}(B_z\cap\mathcal{D})^3} + \|\tilde{p}\|_{N_\delta^{l-1,\sigma}(B_z\cap\mathcal{D})} \le c\Big(\|\tilde{f}\|_{N_\delta^{l-2,\sigma}(B'_z\cap\mathcal{D})^3} + \|\tilde{g}\|_{N_\delta^{l-1,\sigma}(B_z\cap\mathcal{D})} + \|\tilde{u}\|_{N^0_{\delta-l-\sigma}(B'_z\cap\mathcal{D})^3} + \|\tilde{p}\|_{N^0_{\delta-l+1-\sigma}(B'_z\cap\mathcal{D})}\Big).$$

Using the inequalities

$$c_1\,|y'|^{l+\sigma-\delta}\|u\|_{N_\delta^{l,\sigma}(B_y\cap\mathcal{D})^3} \le \|\tilde{u}\|_{N_\delta^{l,\sigma}(B_z\cap\mathcal{D})^3} \le c_2\,|y'|^{l+\sigma-\delta}\|u\|_{N_\delta^{l,\sigma}(B_y\cap\mathcal{D})^3}$$

and the analogous inequalities for the norms of $\tilde{p}$, $\tilde{f}$ and $\tilde{g}$, we obtain the estimate (9.7.4) for arbitrary $y \in \mathcal{D}$. This proves the lemma. □

We will also use the following modification of Lemma 9.7.7.

LEMMA 9.7.8. *Let* ζ, η *be smooth functions with compact supports,* $\eta = 1$ *in a neighborhood of* supp ζ. *Furthermore, let* $(u,p) \in W^{2,s}_{loc}(\bar{\mathcal{D}}\backslash M)^3 \times W^{1,s}_{loc}(\bar{\mathcal{D}}\backslash M)$ *be a solution of the boundary value problem* (9.1.1), (9.1.2) *such that* $\eta(u,p) \in L^\infty_{\delta-l-\sigma}(\mathcal{D})^3 \times L^\infty_{\delta-l+1-\sigma}(\mathcal{D})$. *If moreover* $\eta f \in N_\delta^{l-2,\sigma}(\mathcal{D})^3$, $\eta g \in N_\delta^{l-1,\sigma}(\mathcal{D})$, $\eta h^\pm \in N_\delta^{l,\sigma}(\Gamma^\pm)^{3-d^\pm}$, $\eta\phi^\pm \in N_\delta^{l-1,\sigma}(\Gamma^\pm)^{d^\pm}$, $l \ge 2$, *then* $\zeta u \in N_\delta^{l,\sigma}(\mathcal{D})^3$, $\zeta p \in N_\delta^{l-1,\sigma}(\mathcal{K})$, *and*

$$\|\zeta u\|_{N_\delta^{l,\sigma}(\mathcal{D})^3} + \|\zeta p\|_{N_\delta^{l-1,\sigma}(\mathcal{D})} \le c\Big(\|\eta f\|_{N_\delta^{l-2,\sigma}(\mathcal{D})^3} + \|\eta g\|_{N_\delta^{l-1,\sigma}(\mathcal{D})} + \sum_\pm \|\eta h^\pm\|_{N_\delta^{l,\sigma}(\Gamma^\pm)^{3-d^\pm}} + \sum_\pm \|\eta\phi^\pm\|_{N_\delta^{l-1,\sigma}(\Gamma^\pm)^{d^\pm}} + \|\eta u\|_{L^\infty_{\delta-l-\sigma}(\mathcal{D})^3} + \|\eta p\|_{L^\infty_{\delta-l+1-\sigma}(\mathcal{D})}\Big).$$

P r o o f. We may restrict ourselves to the case $h^\pm = 0$, $\phi^\pm = 0$. Let $\mathcal{U}$ be a neighborhood of $\operatorname{supp}\zeta$ such that $\eta = 1$ in a neighborhood of $\mathcal{U}$. Obviously, we obtain an equivalent norm in $N_\delta^{l,\sigma}(\mathcal{D})$ if we replace the expression $\langle u\rangle_{l,\sigma,\delta;\mathcal{D}}$ in (2.7.1) by

$$\langle u\rangle'_{l,\sigma,\delta;\mathcal{D}} = \sum_{|\alpha|=l} \sup_{\substack{x,y\in\mathcal{D}\\ |x-y|<\varepsilon|x'|}} |x'|^\delta \frac{|\partial_x^\alpha u(x) - \partial_y^\alpha u(y)|}{|x-y|^\sigma},$$

where ε is an arbitrarily small positive number. Using this norm with sufficiently small ε, we get

$$\|\zeta u\|_{N_\delta^{l,\sigma}(\mathcal{D})^3} + \|\zeta p\|_{N_\delta^{l-1,\sigma}(\mathcal{D})} \le c\Big(\|u\|_{N_\delta^{l,\sigma}(\mathcal{U}\cap\mathcal{D})^3} + \|p\|_{N_\delta^{l-1,\sigma}(\mathcal{U}\cap\mathcal{D})}\Big).$$

Here, we used the same notation as in the proof of Lemma 9.7.7. Furthermore, the estimate (9.7.4) is also valid if we denote by B_y and B'_y the balls centered at y with radii $\varepsilon|y'|$ and $2\varepsilon|y'|$, respectively. If ε is sufficiently small, then it follows that

$$\|u\|_{N_\delta^{l,\sigma}(\mathcal{U}\cap\mathcal{D})^3} + \|p\|_{N_\delta^{l-1,\sigma}(\mathcal{U}\cap\mathcal{D})} \le c\Big(\|f\|_{N_\delta^{l-2,\sigma}(\mathcal{U}'\cap\mathcal{D})^3} + \|g\|_{N_\delta^{l-1,\sigma}(\mathcal{U}'\cap\mathcal{D})}$$
$$+ \|u\|_{L^\infty_{\delta-l-\sigma}(\mathcal{U}'\cap\mathcal{D})^3} + \|p\|_{L^\infty_{\delta-l+1-\sigma}(\mathcal{U}'\cap\mathcal{D})}\Big),$$

where $\mathcal{U}' = \{x\in\mathcal{D} : \eta(x) = 1\}$. This proves the lemma. □

Next, we prove an assertion which is analogous to the regularity result in weighted Sobolev spaces in Lemma 9.4.7.

LEMMA 9.7.9. *Let ζ, η be smooth functions with compact supports such that $\eta = 1$ in a neighborhood of $\operatorname{supp}\zeta$, and let (u,p) be a solution of problem* (9.1.1), (9.1.2). *We assume that*

$$\eta\partial_{x_3}^j(u,p) \in N_\delta^{l,\sigma}(\mathcal{D})^3 \times N_\delta^{l-1,\sigma}(\mathcal{D})$$

for $j = 0$ and $j = 1$ and that

$$\eta f \in N_\delta^{l-1,\sigma}(\mathcal{D})^3, \quad \eta g \in N_\delta^{l,\sigma}(\mathcal{D}), \quad \eta h^\pm \in N_\delta^{l+1,\sigma}(\Gamma^\pm)^{3-d^\pm}, \quad \eta\phi^\pm \in N_\delta^{l,\sigma}(\Gamma^\pm)^{d^\pm},$$

where $l\ge 2$, $0<\sigma<1$. If the strip $l+\sigma-\delta \le \operatorname{Re}\lambda \le l+1+\sigma-\delta$ does not contain eigenvalues of the pencil $A(\lambda)$, then $\zeta(u,p) \in N_\delta^{l+1,\sigma}(\mathcal{D})^3 \times N_\delta^{l,\sigma}(\mathcal{D})$.

P r o o f. Let χ be a smooth cut-off function such that $\chi = 1$ in a neighborhood of $\operatorname{supp}\zeta$ and $\eta = 1$ in a neighborhood of $\operatorname{supp}\chi$. We denote by $\Delta_{x'}$, $\nabla_{x'}$ the Laplace and Nabla operators in the coordinates $x' = (x_1,x_2)$. Then $-\Delta_{x'}(\chi u_3) = F_3$, where

$$F_3 = \chi\big(f_3 + \partial_{x_3}^2 u_3 - \partial_{x_3}p\big) - 2\nabla_{x'}\chi\cdot\nabla_{x'}u_3 - u_3\,\Delta_{x'}\chi \in N_\delta^{l-1,\sigma}(\mathcal{D}).$$

Furthermore, χu_3 satisfies the boundary conditions

$$\chi u_3\big|_{\Gamma^\pm} = H_3^\pm \in N_\delta^{l+1,\sigma}(\Gamma^\pm) \text{ for } d^\pm \le 1,$$
$$\frac{\partial(\chi u_3)}{\partial n^\pm}\Big|_{\Gamma^\pm} = \Phi_3^\pm \in N_\delta^{l,\sigma}(\Gamma^\pm) \text{ for } d^\pm \ge 2$$

on $\Gamma^\pm$, where

$$H_3^\pm = \chi h_3^\pm \text{ and } \Phi_3^\pm = \chi\left(\phi_3^\pm - n^\pm\cdot\partial_{x_3}u\right) + u_3\,\partial\chi/\partial n^\pm.$$

Analogously, we obtain the equations

$$-\Delta_{x'}(\chi u') + \nabla_{x'}(\chi p) = F', \quad -\nabla_{x'} \cdot (\chi u') = G,$$
$$\tilde{S}^{\pm}(\chi u')|_{\Gamma^\pm} = H'^{\pm}, \quad \tilde{N}^{\pm}(\chi u', \chi p)|_{\Gamma^\pm} = \Phi'^{\pm}$$

for the vector function $(u',p) = (u_1,u_2,p)$, where $F' \in N_\delta^{l-1,\sigma}(\mathcal{D})^2$, $G \in N_\delta^{l,\sigma}(\mathcal{D})$, $H'^{\pm} \in N_\delta^{l+1,\sigma}(\Gamma^\pm)$, $\Phi'^{\pm} \in N_\delta^{l,\sigma}(\Gamma^\pm)$. Here, $\tilde{S}^{\pm}u' = S^{\pm}(u',0)$ and $\tilde{N}^{\pm}(u',p) = N^{\pm}(u',0,p)$. Using [**117**, Theorem 8.4], we obtain $(\chi u)(\cdot,x_3) \in N_\delta^{l+1,\sigma}(K)^3$, $(\chi p)(\cdot,x_3) \in N_\delta^{l,\sigma}(K)$, and

$$\|(\chi u)(\cdot,x_3)\|_{N_\delta^{l+1,\sigma}(K)^3} + \|(\chi p)(\cdot,x_3)\|_{N_\delta^{l,\sigma}(K)}$$
$$\le c\Big(\|F(\cdot,x_3)\|_{N_\delta^{l-1,\sigma}(K)^3} + \|G(\cdot,x_3)\|_{N_\delta^{l,\sigma}(K)} + \sum_\pm \|H^\pm\|_{N_\delta^{l+1,\sigma}(\gamma^\pm)}$$
$$+ \sum_\pm \|\Phi^\pm\|_{N_\delta^{l,\sigma}(\gamma^\pm)}\Big)$$

with a constant c independent of x_3. In particular, $\chi(u,p) \in L^\infty_{\delta-l-1-\sigma}(\mathcal{D})^3 \times L^\infty_{\delta-l-\sigma}(\mathcal{D})$, and the assertion of the lemma follows from Lemma 9.7.8. □

9.7.3. Local regularity results in the spaces $C_\delta^{l,\sigma}$.

First, we prove the analog to Lemma 9.7.8 in the class of the spaces $C_\delta^{l,\sigma}$.

LEMMA 9.7.10. *Let $(u,p) \in W^{2,s}_{loc}(\bar{\mathcal{D}}\backslash M)^3 \times W^{1,s}_{loc}(\bar{\mathcal{D}}\backslash M)$ be a solution of the boundary value problem* (9.1.1), (9.1.2), *and let ζ, η be smooth functions with compact supports such that $\eta = 1$ in a neighborhood of supp ζ. Suppose that $\eta u \in C_{\delta-1}^{l-1,\sigma}(\mathcal{D})^3$, $\eta p \in C_{\delta-1}^{l-2,\sigma}(\mathcal{D})$, $\eta f \in C_\delta^{l-2,\sigma}(\mathcal{D})^3$, $\eta g \in C_\delta^{l-1,\sigma}(\mathcal{D})$, $\eta h^\pm \in C_\delta^{l,\sigma}(\Gamma^\pm)^{3-d^\pm}$, $\eta\phi^\pm \in C_\delta^{l-1,\sigma}(\Gamma^\pm)^{d^\pm}$, where $l \ge 2$, $\delta \ge 1$, $0 < \sigma < 1$. Then $\zeta(u,p) \in C_\delta^{l,\sigma}(\mathcal{D})^3 \times C_\delta^{l-1,\sigma}(\mathcal{D})$.*

P r o o f. By Theorem 6.7.11, the functions ηu and ηp admit the decompositions

$$\eta u = v + V, \quad \eta p = q + Q,$$

where $v \in N_{\delta-1}^{l-1,\sigma}(\mathcal{D})^3$, $q \in N_{\delta-1}^{l-2,\sigma}(\mathcal{D})$, $V \in C_\delta^{l,\sigma}(\mathcal{D})^3$, and $Q \in C_\delta^{l-1,\sigma}(\mathcal{D})$. Let χ be a smooth cut-off function equal to one in a neighborhood of supp ζ such that $\eta = 1$ in a neighborhood of supp χ. Then

$$\chi(-\Delta v + \nabla q) = \chi f + \chi(\Delta V - \nabla Q) \in C_\delta^{l-2,\sigma}(\mathcal{D})^3$$

and

$$-\chi\nabla\cdot v = \chi g + \chi\nabla\cdot V \in C_\delta^{l-1,\sigma}(\mathcal{D}).$$

In the case $\delta \ge l-2+\sigma$, the space $C_\delta^{l-2,\sigma}(\mathcal{D})$ coincides with $N_\delta^{l-2,\sigma}(\mathcal{D})$. If $1 \le \delta < l-2+\sigma$, then $l \ge 3$ and $\chi(-\Delta v + \nabla q) \in N_{\delta-1}^{l-3,\sigma}(\mathcal{D})^3$. From Lemma 6.7.10 it follows that $N_{\delta-1}^{l-3,\sigma}(\mathcal{D}) \cap C_\delta^{l-2,\sigma}(\mathcal{D}) \subset N_\delta^{l-2,\sigma}(\mathcal{D})$. Therefore in both cases, we obtain $\chi(-\Delta v + \nabla q) \in N_\delta^{l-2,\sigma}(\mathcal{D})^3$. Analogously, $\chi\nabla\cdot v \in N_\delta^{l-1,\sigma}(\mathcal{D})$, $\chi S^\pm v \in N_\delta^{l,\sigma}(\Gamma^\pm)^{3-d^\pm}$, and $\chi N^\pm(v,q) \in N_\delta^{l-1,\sigma}(\Gamma^\pm)^{d^\pm}$. This together with Lemma 9.7.8 implies $\zeta v \in N_\delta^{l,\sigma}(\mathcal{D})^3$, $\zeta q \in N_\delta^{l-1,\sigma}(\mathcal{D})$. The result follows. □

Finally, we extend the result of Lemma 9.7.9 to the class of the spaces $C_\delta^{l,\sigma}$.

LEMMA 9.7.11. *Let* ζ, η *be the same functions as in Lemma* 9.7.10, *and let* (u,p) *be a solution of problem* (9.1.1), (9.1.2) *such that*

$$\eta \partial_{x_3}^j (u,p) \in C_\delta^{l,\sigma}(\mathcal{D})^3 \times C_\delta^{l-1,\sigma}(\mathcal{D})$$

for $j=0$ *and* $j=1$, *where* $l \ge 2$, $0<\sigma<1$, $\delta-\sigma$ *is not integer. If*

$$\eta f \in C_\delta^{l-1,\sigma}(\mathcal{D})^3, \quad \eta g \in C_\delta^{l,\sigma}(\mathcal{D}), \quad \eta h^\pm \in C_\delta^{l+1,\sigma}(\Gamma^\pm)^{3-d^\pm}, \quad \eta\phi^\pm \in C_\delta^{l,\sigma}(\Gamma^\pm)^{d^\pm},$$

and the strip $l+\sigma-\delta \le \operatorname{Re}\lambda \le l+1+\sigma-\delta$ *does not contain eigenvalues of the pencil* $A(\lambda)$, *then* $\zeta(u,p) \in C_\delta^{l+1,\sigma}(\mathcal{D})^3 \times C_\delta^{l,\sigma}(\mathcal{D})$.

P r o o f. Suppose that $k-1<\delta-\sigma<k$, where k is an integer, $k \le l$. Then both ηu and $\partial_{x_3}(\eta u)$ belong to $C^{l-k,k-\delta+\sigma}(\mathcal{D})^3$. Consequently, the traces $u^{(i,j)}$ of $\partial_{x_1}^i \partial_{x_2}^j (\eta u)$ on M are elements of the space $C^{l-k-i-j+1,k-\delta+\sigma}(M)^3$ for $i+j \le l-k$. Analogously, the traces $p^{(i,j)}$ of $\partial_{x_1}^i \partial_{x_2}^j(\eta p)$ on M belong to $C^{l-k-i-j,k-\delta+\sigma}(M)$ for $i+j \le l-k-1$. By Theorem 6.7.11, the functions ζu and ζp admit the decompositions

$$\zeta u = \zeta \sum_{i+j\le l-k} \frac{Eu^{(i,j)}}{i!j!}\, x_1^i x_2^j + v, \qquad \zeta p = \zeta \sum_{i+j \le l-k-1} \frac{Ep^{(i,j)}}{i!j!}\, x_1^i x_2^j + q,$$

where E is the extension operator (6.2.4), $\partial_{x_3}^j v \in N_\delta^{l,\sigma}(\mathcal{D})^3$, and $\partial_{x_3}^j q \in N_\delta^{l-1,\sigma}(\mathcal{D})$ for $j=0,1$. From the properties of the extension operator E it follows that

$$\zeta u - v \in C_\delta^{l+1,\sigma}(\mathcal{D})^3, \quad \zeta p - q \in C_\delta^{l,\sigma}(\mathcal{D}).$$

Therefore,

$$-\Delta v + \nabla q \in C_\delta^{l-1,\sigma}(\mathcal{D})^3, \quad -\nabla\cdot v \in C_\delta^{l-1,\sigma}(\mathcal{D}),$$
$$S^\pm v \in C_\delta^{l+1,\sigma}(\Gamma^\pm)^{3-d^\pm}, \quad N^\pm(v,q) \in C_\delta^{l+1,\sigma}(\Gamma^\pm)^{d^\pm}.$$

Since $v \in N_\delta^{l,\sigma}(\mathcal{D})^3$ and $q \in N_\delta^{l-1,\sigma}(\mathcal{D})$, the traces of $\partial_x^\alpha(\nabla q - \Delta v)$ on M vanish for $|\alpha| \le l-k-2$, while the traces of $\partial_x^\alpha \nabla\cdot v$ on M vanish for $|\alpha| \le l-k-1$. Furthermore, $\partial_r^j S^\pm v = 0$ on M for $j \le l-k$ and $\partial_r^j N^\pm(v,q)=0$ on M for $j \le l-k-1$. We set

$$f^{(i,j)} = \partial_{x_1}^i \partial_{x_2}^j (\nabla q - \Delta v)\big|_M \quad \text{for } i+j = l-k-1,$$
$$g^{(i,j)} = -\partial_{x_1}^i \partial_{x_2}^j \nabla\cdot v\big|_M \quad \text{for } i+j=l-k,$$
$$H^\pm = \partial_r^{l-k+1} S^\pm v|_M \quad \text{and} \quad \Phi^\pm = \partial_r^{l-k} N^\pm(v,q)|_M\,.$$

Obviously $f^{(i,j)}, g^{(i,j)}, H^\pm$, and $\Phi^\pm$ belong to the space $C^{0,k+\sigma-\delta}(M)$. Then by Theorem 6.7.11,

$$\eta\Big(\nabla q - \Delta v - \sum_{i+j=l-k-1} \frac{Ef^{(i,j)}}{i!\,j!}\, x_1^i x_2^j\Big) \in N_\delta^{l-1,\sigma}(\mathcal{D})^3,$$
$$\eta\Big(\nabla\cdot v + \sum_{i+j=l-k} \frac{Eg^{(i,j)}}{i!\,j!}\, x_1^i x_2^j\Big) \in N_\delta^{l,\sigma}(\mathcal{D}).$$

Analogously,

$$\eta\Big(S^\pm v - \frac{EH^\pm}{(l-k+1)!}\, r^{l-k+1}\Big) \in N_\delta^{l+1,\sigma}(\Gamma^\pm)^{3-d^\pm},$$
$$\eta\Big(N^\pm(v,q) - \frac{E\Phi^\pm}{(l-k)!}\, r^{l-k}\Big) \in N_\delta^{l,\sigma}(\Gamma^\pm)^{d^\pm}.$$

Since $\lambda = l-k+1$ is not an eigenvalue of the pencil $A(\lambda)$, there exist homogeneous vector-valued polynomials $U^{(i,j,\mu)}(x_1,x_2)$ of degree $l-k+1$ and homogeneous polynomials $P^{(i,j,\mu)}(x_1,x_2)$ of degree $l-k$, $\mu = 1,2,3,4$, $i+j = l-k-1-\delta_{\mu,4}$, satisfying the equations

$$-\Delta U_\nu^{(i,j,\mu)} + \partial_{x_\nu} P^{(i,j,\mu)} = \delta_{\mu,\nu} \frac{x_1^i\, x_2^j}{i!\, j!} \quad \text{for } \nu = 1,2,3,$$

$$-\nabla \cdot U^{(i,j,\mu)} = \delta_{\mu,4} \frac{x_1^i\, x_2^j}{i!\, j!}$$

and the boundary conditions $S^\pm U^{(i,j,\mu)} = 0$, $N^\pm(U^{(i,j,\mu)}, P^{(i,j,\mu)}) = 0$ on $\Gamma^\pm$ (cf. Lemma 9.3.2). We define

$$U = \sum_{\mu=1}^{3} \sum_{i+j=l-k-1} U^{(i,j,\mu)}\, Ef_\mu^{(i,j)} + \sum_{i+j=l-k} U^{(i,j,4)}\, Eg^{(i,j)},$$

$$P = \sum_{\mu=1}^{3} \sum_{i+j=l-k-1} P^{(i,j,\mu)}\, Ef_\mu^{(i,j)} + \sum_{i+j=l-k} P^{(i,j,4)}\, Eg^{(i,j)}.$$

Then $\eta(U,P) \in C_\delta^{l+1,\sigma}(\mathcal{D})^3 \times C_\delta^{l,\sigma}(\mathcal{D})$ and $\eta\, \partial_{x_3}^j(U,P) \in N_\delta^{l,\sigma}(\mathcal{D})^3 \times N_\delta^{l-1,\sigma}(\mathcal{D})$ for $j = 0,1$. Using the fact that $\eta\partial_{x_\nu} Ef^{(i,j)} \in N_{\delta+l-k+1}^{l,\sigma}(\mathcal{D})^3$ and $\eta\partial_{x_\nu} Eg^{(i,j)} \in N_{\delta+l-k+1}^{l,\sigma}(\mathcal{D})$ for $\nu = 1,2,3$, we conclude that

$$\eta\,\big(\Delta(U-v) - \nabla(P-q)\big) \in N_\delta^{l-1,\sigma}(\mathcal{D})^3, \quad \eta\,\nabla\cdot(U-v) \in N_\delta^{l,\sigma}(\mathcal{D}),$$

$$S^\pm U|_{\Gamma^\pm} = 0, \quad \eta\, N^\pm(U,P)|_{\Gamma^\pm} \in N_\delta^{l,\sigma}(\Gamma^\pm)^{d^\pm}.$$

Analogously there exists a vector function (V,Q) such that $\eta(V,Q) \in C_\delta^{l+1,\sigma}(\mathcal{D})^3 \times C_\delta^{l,\sigma}(\mathcal{D})$, $\eta\partial_{x_3}^j(V,Q) \in N_\delta^{l,\sigma}(\mathcal{D})^3 \times N_\delta^{l-1,\sigma}(\mathcal{D})$ for $j = 0,1$,

$$\eta\,(\Delta V - \nabla Q) \in N_\delta^{l-1,\sigma}(\mathcal{D})^3, \quad \eta\,\nabla\cdot V \in N_\delta^{l,\sigma}(\mathcal{D})$$

$$\eta\, S^\pm(V-v) \in N_\delta^{l+1,\sigma}(\Gamma^\pm)^{3-d^\pm}, \quad \eta\, N^\pm(V-v,Q-q) \in N_\delta^{l,\sigma}(\Gamma^\pm)^{d^\pm}.$$

Applying Lemma 9.7.9 to the vector-function $(U+V-v, P+Q-q)$, we obtain $\chi(U+V-v, P+Q-q) \in N_\delta^{l+1,\sigma}(\mathcal{D})^3 \times N_\delta^{l,\sigma}(\mathcal{D})$, where χ is the same cut-off function as in the proof of Lemma 9.7.9. This proves the lemma. □

CHAPTER 10

Mixed boundary value problems for the Stokes system in a polyhedral cone

We consider mixed boundary value problems for the Stokes system in a polyhedral cone $\mathcal{K}$, where the boundary conditions (i)–(iv) of the preceding chapter are arbitrarily combined on the the faces $\Gamma_1, \dots, \Gamma_d$ of the cone. In some special case, e.g. if the Dirichlet condition is prescribed on at least one of the adjoining faces of every edge, it is possible to establish a solvability and regularity theory in the weighted spaces $V^{l,p}_{\beta,\delta}(\mathcal{K})$ and $N^{l,\sigma}_{\beta,\delta}(\mathcal{K})$ with homogeneous norms. However, the general case requires the use of weighted spaces with nonhomogeneous norms.

The main results of this chapter (Theorems 10.5.6, 10.6.5, 10.6.9, 10.7.13, 10.7.16, 10.8.8) concern the existence and uniqueness of solutions in the weighted spaces $W^{l,s}_{\beta,\delta}(\mathcal{K})$ and $C^{l,\sigma}_{\beta,\delta}(\mathcal{K})$ and regularity assertions for the solutions. For example, we prove that the boundary value problem has a uniquely determined solution (u,p) in the space $W^{l,s}_{\beta,\delta}(\mathcal{K})^3 \times W^{l-1,s}_{\beta,\delta}(\mathcal{K})$ (for data in appropriate weighted spaces) if the line $\operatorname{Re}\lambda = l - \beta - 3/s$ is free of eigenvalues of a certain operator pencil $\mathfrak{A}(\lambda)$ introduced in Section 10.1 and the components of δ satisfy the inequalities

$$\max(0, l - \mu_+^{(k)}) < \delta_k + 2/p < l,$$

where $\mu_+^{(k)}$ are positive numbers determined by the spectrum of the operator pencils $A_k(\lambda)$ (see Sections 9.3 and 10.1). Here variational solutions in the space $W^{1,s}_{\beta,\delta}(\mathcal{K})^3 \times W^{0,s}_{\beta,\delta}(\mathcal{K})$ are also included.

As for the mixed boundary value problem in Chapter 6, we start with the case $s = 2$ (Sections 10.1–10.3). In Section 10.4, we study the Green's matrix for the boundary value problem. In particular, we prove point estimates for the elements of this matrix. For this, we apply the a priori estimates of the solutions in the weighted spaces $W^{l,2}_{\beta,\delta}(\mathcal{K})$ obtained in the preceding section. In the case $|x| > 2|\xi|$, the estimate for the elements of Green's matrix has the form

$$\begin{aligned} \big|\partial_x^\alpha \partial_\xi^\gamma G_{i,j}(x,\xi)\big| \;&\le\; c\,|x|^{\Lambda_- - \delta_{i,4} - |\alpha| + \varepsilon} \prod_{k=1}^d \Big(\frac{r_k(x)}{|x|}\Big)^{\min(0,\mu_+^{(k)} - |\alpha| - \delta_{i,4} - \varepsilon)} \\ &\quad\times\, |\xi|^{-\Lambda_- - 1 - \delta_{j,4} - |\gamma| - \varepsilon} \prod_{k=1}^d \Big(\frac{r_k(\xi)}{|\xi|}\Big)^{\min(0,\mu_+^{(k)} - |\gamma| - \delta_{j,4} - \varepsilon)}. \end{aligned}$$

This estimate and similar estimates for the cases $|\xi| > 2|x|$ and $|x|/2 < |\xi| < 2|x|$ enable us to extend the solvability and regularity theorems obtained in Sections 10.2 and 10.3 to weighted L_s Sobolev spaces $W^{l,s}_{\beta,\delta}(\mathcal{K})$ (Sections 10.5 and 10.6) and weighted Hölder spaces $C^{l,\sigma}_{\beta,\delta}(\mathcal{K})$ (Section 10.7 and 10.8). For the special case, where the Dirichlet condition is prescribed on at least one of the two adjoining faces of every edge, we obtain analogous results in the spaces $V^{l,s}_{\beta,\delta}(\mathcal{K})$ and $N^{l,\sigma}_{\beta,\delta}(\mathcal{K})$.

10.1. The boundary value problem and corresponding operator pencils

Let $\mathcal{K} = \{x \in \mathbb{R}^3 : \ x/|x| = \omega \in \Omega\}$ be the same polyhedral cone as in Chapter 7. The boundary $\partial\mathcal{K}$ consists of the vertex $x = 0$, the edges (half-lines) $M_1, \dots, M_d$, and plane faces $\Gamma_1, \dots, \Gamma_d$. For every $j = 1, \dots, d$, let d_j be one of the natural numbers 0,1,2,3. We consider the boundary value problem

$$-\Delta u + \nabla p = f, \quad -\nabla \cdot u = g \ \text{ in } \mathcal{K}, \tag{10.1.1}$$

$$S_j u = h_j\,, \quad N_j(u,p) = \phi_j \quad \text{on } \Gamma_j,\ j = 1, \dots, d. \tag{10.1.2}$$

Here S_j is defined as

$$S_j u = u \text{ if } d_j = 0, \quad S_j u = u_n = u \cdot n \text{ if } d_j = 2, \quad S_j u = u_\tau = u - u_n n \text{ if } d_j = 1.$$

The operators N_j are defined as

$$N_j(u,p) = -p + 2\varepsilon_{n,n}(u) \text{ if } d_j = 1, \qquad N_j(u,p) = 2\varepsilon_{n,\tau}(u) \text{ if } d_j = 2,$$
$$N_j(u,p) = -pn + 2\varepsilon_n(u) \text{ if } d_j = 3.$$

In the case $d_j = 0$, the condition $N_j(u,p) = \phi_j$ does not appear in (10.1.2), whereas the condition $S_j u = h_j$ does not appear for $d_j = 3$. If $d_j = 1$, then h_j is a vector function tangential to Γ_j and ϕ_j is a scalar function. Conversely, ϕ_j is a vector function tangential to Γ_j and h_j is a scalar function if $d_j = 2$.

10.1.1. Variational solutions. Let $V^{l,2}_\beta(\mathcal{K}) = W^{l,2}_{\beta,0}(\mathcal{K})$ be the closure of the set $C_0^\infty(\overline{\mathcal{K}}\backslash\{0\})$ with respect to the norm

$$\|u\|_{V^{l,2}_\beta(\mathcal{K})} = \Big(\int_{\mathcal{K}} \sum_{|\alpha|\le l} |x|^{2(\beta-l+|\alpha|)} \, \big|\partial_x^\alpha u(x)\big|^2 \, dx \Big)^{1/2},$$

and let $\overset{\circ}{V}{}^{l,2}_\beta(\mathcal{K})$ be the closure of the set $C_0^\infty(\mathcal{K})$ in $V^{l,2}_\beta(\mathcal{K})$. By Hardy's inequality, the norm in the space $V_0^{1,2}(\mathcal{K})$ is equivalent to the norm

$$\|u\|_{L^{1,2}(\mathcal{K})} = \|\nabla u\|_{L_2(\mathcal{K})^3}\,.$$

We introduce the subspace

$$\mathcal{H}_{\mathcal{K}} = \{u \in V_0^{1,2}(\mathcal{K})^3 : \ S_j u = 0 \text{ on } \Gamma_j,\ j = 1, \dots, d\}$$

and the bilinear form

$$b_{\mathcal{K}}(u,v) = 2 \int_{\mathcal{K}} \sum_{i,j=1}^{3} \varepsilon_{i,j}(u)\, \varepsilon_{i,j}(v)\, dx. \tag{10.1.3}$$

A vector function $(u,p) \in V_0^{1,2}(\mathcal{K})^3 \times L_2(\mathcal{K})$ is called *variational solution* of the problem (10.1.1), (10.1.2) if

$$b_{\mathcal{K}}(u,v) - \int_{\mathcal{K}} p\, \nabla \cdot v\, dx = F(v) \ \text{ for all } v \in \mathcal{H}_{\mathcal{K}}, \tag{10.1.4}$$

$$-\nabla \cdot u = g \ \text{ in } \mathcal{K}, \quad S_j u = h_j \ \text{ on } \Gamma_j,\ j = 1, \dots, d, \tag{10.1.5}$$

where

$$(10.1.6) \qquad F(v) = \int_{\mathcal{K}} (f + \nabla g) \cdot v \, dx + \sum_{j=1}^{d} \int_{\Gamma_j} \phi_j \cdot v \, dx,$$

provided the functional (10.1.6) belongs to the dual space $\mathcal{H}_{\mathcal{K}}^*$ of $\mathcal{H}_{\mathcal{K}}$. Here in the case $d_j = 1$, we identified the scalar function ϕ_j with the vector function $\phi_j n$. This means that $\phi_j \cdot v$ has to be understood as the scalar product $\phi_j n \cdot v$ in this case.

By the *Green's formula*

$$(10.1.7) \qquad b_{\mathcal{K}}(u,v) - \int_{\mathcal{K}} p \nabla \cdot v \, dx$$
$$= \int_{\mathcal{K}} (-\Delta u - \nabla \nabla \cdot u + \nabla p) \cdot v \, dx + \sum_{j=1}^{d} \int_{\Gamma_j} (-pn + 2\varepsilon(u)\, n) \cdot v \, dx,$$

every solution $(u,p) \in W_{1,\delta}^{2,2}(\mathcal{K})^3 \times W_{1,\delta}^{1,2}(\mathcal{K})$ of the boundary value problem (10.1.1), (10.1.2), where $0 < \delta_k < 1$ for $k = 1, \dots, d$, is also a variational solution of this problem.

The existence and uniqueness of variational solutions can be proved in the same way as for the problem in the dihedron $\mathcal{D}$. For this, we need the following lemma.

LEMMA 10.1.1. *For every $g \in L_2(\mathcal{K})$ there exists a vector function $u \in \overset{\circ}{V}{}_0^{1,2}(\mathcal{K})^3$ such that $\nabla \cdot u = g$ in $\mathcal{K}$ and*

$$\|u\|_{V_0^{1,2}(\mathcal{K})^3} \le c \, \|g\|_{L_2(\mathcal{K})} ,$$

where c is a constant independent of g.

P r o o f. For $k = 1, \dots, d$ let ψ_k and χ_k be a smooth functions on Ω with support in a small neighborhood of the corner $M_k \cap S^2$ which are equal to one near this corner. Furthermore, let $\psi_k \chi_k = \psi_k$ for $k = 1, \dots, d$ and $\psi_1 + \cdots + \psi_d = 1$. We extend ψ_k and χ_k by the equalities $\psi_k(x) = \psi_k(x/|x|)$ and $\chi_k(x) = \chi_k(x/|x|)$ to functions on $\mathcal{K}$. Let $\mathcal{D}_k$ be the dihedron bounded by the faces $\Gamma_k^{\pm} \supset \Gamma_{k_\pm}$, where Γ_{k_+} and Γ_{k_-} are the faces of the cone $\mathcal{K}$ adjacent to the edge M_k. By Theorem 9.1.3, there exist vector functions $v_k \in \overset{\circ}{L}{}^{1,2}(\mathcal{D}_k)^3$ such that

$$\nabla \cdot v_k = \psi_k g \quad \text{and} \quad \|v_k\|_{L^{1,2}(\mathcal{D}_k)^3} \le c \, \|\psi_k g\|_{L_2(\mathcal{K})} , \quad k = 1, \dots, d.$$

We define $u_k(x) = \chi_k(x)\, v_k(x)$ for $x \in \mathcal{K} \cap \operatorname{supp} \chi_k$ and extend this function by zero outside the support of η_k. By Hardy's inequality, $u_k \in \overset{\circ}{V}{}_0^{1,2}(\mathcal{K})^3$ and

$$\|u_k\|_{V_0^{1,2}(\mathcal{K})^3} \le c \, \|\psi_k g\|_{L_2(\mathcal{K})}$$

for $k = 1, \dots, d$. Let Ω_0 be a smooth subdomain of Ω such that $\chi_1 + \cdots + \chi_d = 1$ on $\Omega \backslash \Omega_0$, and let $\mathcal{K}_0$ be the cone $\{x \in \mathbb{R}^3 : x/|x| \in \Omega_0\}$. Then there exists a vector function $u_0 \in \overset{\circ}{V}{}_0^{1,2}(\mathcal{K}_0)^3$ such that

$$\nabla \cdot u_0 = -\sum_{k=1}^{d} (\nabla \chi_k) \cdot v_k \quad \text{and} \quad \|u_0\|_{V_0^{1,2}(\mathcal{K})^3} \le c \sum_{k=1}^{d} \|\psi_k g\|_{L_2(\mathcal{K})} .$$

This stems from the existence and uniqueness of weak solutions of the Dirichlet problem for the Stokes system in the cone $\mathcal{K}_0$ (see e.g. [**84**, Sections 6.1 and 8.2]). We extend u_0 by zero outside $\mathcal{K}_0$ and define $u = u_0 + u_1 + \cdots + u_d$. Then $\nabla \cdot u = g$.

This proves the lemma. □

Furthermore analogously to Lemma 9.1.4, the inequality

$$b_{\mathcal{K}}(u,\bar{u}) \ge c\,\|u\|^2_{L^{1,2}(\mathcal{K})^3}$$

holds for all $u \in L^{1,2}(\mathcal{K})^3$ with a positive constant c independent of u. Using this inequality and Lemma 10.1.1, we arrive at the following statement. Its proof is a word-by-word repetition of the proof of Theorem 9.1.5.

THEOREM 10.1.2. *Let $F \in \mathcal{H}^*_{\mathcal{K}}$, $g \in L_2(\mathcal{K})$, and let the functions h_j be such that there exists a vector function $w \in V_0^{1,2}(\mathcal{K})^3$ satisfying the equalities $S_j w = h_j$ on Γ_j, $j = 1,\dots,d$. Then there exists a unique solution $(u,p) \in V_0^{1,2}(\mathcal{K})^3 \times L_2(\mathcal{K})$ of the problem* (10.1.4), (10.1.5). *Furthermore, (u,p) satisfies the estimate*

$$\|u\|_{V_0^{1,2}(\mathcal{K})^3} + \|p\|_{L_2(\mathcal{K})} \le c\Big(\|F\|_{\mathcal{H}^*_{\mathcal{K}}} + \|g\|_{L_2(\mathcal{K})} + \|w\|_{V_0^{1,2}(\mathcal{K})^3}\Big)$$

with a constant c independent of F, g and w.

10.1.2. The operator pencil $A_k(\lambda)$. Let $\Gamma_{k_\pm}$ be the faces of $\mathcal{K}$ adjacent to the edge M_k, and let θ_k be the angle at the edge M_k. By $A_k(\lambda)$ we denote the operator pencil introduced in Section 9.3, where $S^\pm = S_{k_\pm}$ and $N^\pm = N_{k_\pm}$. Furthermore, let $\delta_+^{(k)}$ be the greatest positive number such that the strip

$$0 < \operatorname{Re}\lambda < \delta_+^{(k)}$$

is free of eigenvalues of the pencil $A_k(\lambda)$. If $d_{k_+} + d_{k_-}$ is even, then $\lambda = 1$ is always an eigenvalue of the pencil $A_k(\lambda)$. It is the eigenvalue with smallest positive real part of the pencil $A_k(\lambda)$ if in addition $\theta_k < \pi/m_k$, where

$$m_k = 1 \text{ for } d_{k_+} + d_{k_-} \in \{0,6\}, \quad m_k = 2 \text{ for } d_{k_+} + d_{k_-} \in \{2,4\}.$$

We define the numbers $\mu_+^{(k)}$ as follows. If $d_{k_+} + d_{k_-}$ is odd or $d_{k_+} + d_{k_-}$ is even and $\theta_k \ge \pi/m_k$, then we put $\mu_+^{(k)} = \delta_+^{(k)}$. For even $d_{k_+} + d_{k_-}$ and $\theta_k < \pi/m_k$, let $\mu_+^{(k)}$ be the greatest real number such that the strip

$$0 < \operatorname{Re}\lambda < \mu_+^{(k)}$$

contains only the eigenvalue $\lambda = 1$ of the pencil $A_k(\lambda)$. In the last case, we have $\mu_+^{(k)} > \delta_+^{(k)} = 1$.

10.1.3. The operator pencil $\mathfrak{A}(\lambda)$. Let $\rho = |x|$ and $\omega = x/|x|$. Furthermore, let Ω be the intersection of the cone $\mathcal{K}$ with the unit sphere S^2, and let $\gamma_j = \Gamma_j \cap S^2$ be the sides of Ω. We introduce the subspace

$$\mathcal{H}_\Omega = \{u \in W^{1,2}(\Omega)^3 : S_j u = 0 \text{ on } \gamma_j \text{ for } j = 1,\dots,d\}$$

and the bilinear form

$$a\Big(\Big(\begin{smallmatrix} u \\ p \end{smallmatrix}\Big),\Big(\begin{smallmatrix} v \\ q \end{smallmatrix}\Big);\lambda\Big) = \frac{1}{\log 2}\int\limits_{\substack{\mathcal{K} \\ 1<|x|<2}} \Big(2\sum_{i,j=1}^{3} \varepsilon_{i,j}(U)\cdot\varepsilon_{i,j}(V) - P\nabla\cdot V - (\nabla\cdot U)\,Q\Big)\,dx,$$

where $U = \rho^\lambda u(\omega)$, $V = \rho^{-1-\lambda}v(\omega)$, $P = \rho^{\lambda-1}p(\omega)$, $Q = \rho^{-2-\lambda}q(\omega)$, $u,v \in V_\Omega$, $p,q \in L_2(\Omega)$, and $\lambda \in \mathbb{C}$. The bilinear form $a(\cdot,\cdot;\lambda)$ generates the linear and continuous operator

$$\mathfrak{A}(\lambda) : \mathcal{H}_\Omega \times L_2(\Omega) \to \mathcal{H}^*_\Omega \times L_2(\Omega)$$

by

$$\int_\Omega \mathfrak{A}(\lambda)\begin{pmatrix} u \\ p \end{pmatrix} \cdot \begin{pmatrix} v \\ q \end{pmatrix} d\omega = a\Big(\begin{pmatrix} u \\ p \end{pmatrix}, \begin{pmatrix} v \\ q \end{pmatrix}; \lambda\Big), \quad u, v \in V_\Omega,\ p, q \in L_2(\Omega).$$

The spectrum of the operator pencil $\mathfrak{A}(\lambda)$ consists of isolated points, the eigenvalues of this pencil. Detailed information on the spectrum can be found in [**85**, Chapters 5,6]. In particular, the line $\operatorname{Re}\lambda = -1/2$ is free of eigenvalues of the pencil $\mathfrak{A}(\lambda)$. Furthermore, the number λ_0 is an eigenvalue if and only if $-1-\bar\lambda_0$ is an eigenvalue.

We consider the restriction $\mathfrak{A}_\delta(\lambda)$ of the operator $\mathfrak{A}(\lambda)$ to the subspace $(W_\delta^{2,2}(\Omega)^3 \cap \mathcal{H}_\Omega) \times W_\delta^{1,2}(\Omega)$, $0 < \delta < 1$. For the definition of the weighted space $W_\delta^{l,2}(\Omega)$, we refer to Subsection 7.1.4. The operator $\mathfrak{A}_\delta(\lambda)$ is defined as the mapping

$$\begin{aligned}&(W_\delta^{2,2}(\Omega)^3 \cap \mathcal{H}_\Omega) \times W_\delta^{1,2}(\Omega) \ni (u,p)\\ &\qquad\qquad \to \big(f, g, \{\phi_j\}\big) \in W_\delta^{0,2}(\Omega)^3 \times W_\delta^{1,2}(\Omega) \times \prod_{j=1}^d W_\delta^{1/2,2}(\gamma_j)^{d_j}\end{aligned}$$

where

$$\begin{aligned}&f(\omega) = \rho^{2-\lambda}(-\Delta U + \nabla P), \quad g(\omega) = -\rho^{1-\lambda}\nabla\cdot U, \quad \phi_j = \rho^{1-\lambda}N_j(U,P),\\ &U(x) = \rho^\lambda u(\omega), \quad P(x) = \rho^{\lambda-1}p(\omega).\end{aligned}$$

THEOREM 10.1.3. *Let* $\max(0, 1-\delta_+^{(k)}) < \delta_k < 1$ *for* $k = 1,\dots,d$. *Then the following assertions are true.*

1) *The spectra of the pencils* $\mathfrak{A}(\lambda)$ *and* $\mathfrak{A}_\delta(\lambda)$ *coincide.*

2) *There exist positive constants* N *and* ε *such that the operator* $\mathfrak{A}_\delta(\lambda)$ *is an isomorphism for all* λ *in the set* (7.1.18). *Furthermore, every solution* $(u,p) \in (W_\delta^{2,2}(\Omega)^3 \cap \mathcal{H}_\Omega) \times W_\delta^{1,2}(\Omega)$ *of the equation* $\mathfrak{A}_\delta(\lambda)\,(u,p) = \big(f, g, \{\phi_j\}\big)$ *satisfies the estimate*

$$\begin{aligned}(10.1.8)\qquad &\sum_{j=0}^2 |\lambda|^{2-j}\,\|u\|_{W_\delta^{j,2}(\Omega)^3} + \sum_{j=0}^1 |\lambda|^{1-j}\,\|p\|_{W_\delta^{j,2}(\Omega)}\\ &\quad \le c\Big(\|f\|_{W_\delta^{0,2}(\Omega)^3} + \sum_{j=0}^1 |\lambda|^{1-j}\,\|g\|_{W_\delta^{j,2}(\Omega)}\\ &\qquad + \sum_{j=1}^d \Big(\|\phi_j\|_{W_\delta^{1/2,2}(\gamma_j)^{d_j}} + |\lambda|^{1/2}\,\|\phi_j\|_{W_\delta^{0,2}(\gamma_j)^{d_j}}\Big)\Big)\end{aligned}$$

if λ *lies in the set* (7.1.18). *Here the constant* c *is independent of* (u,p) *and* λ.

P r o o f. By the same arguments as in the proof of Theorem 7.1.2, we conclude that all eigenvectors of the pencil $\mathfrak{A}(\lambda)$ belong to the space $W_\delta^{2,2}(\Omega)^3 \times W_\delta^{1,2}(\Omega)$ and, consequently, are also eigenvectors of the pencil $\mathfrak{A}_\delta(\lambda)$. This proves the first part.

In order to prove the estimate in the second part, we introduce smooth functions $\psi_0,\dots,\psi_d$ on Ω such that $\psi_0+\dots+\psi_d = 1$ and $\psi_j = 0$ near $M_k \cap S^2$ for $j \ne k$. Furthermore, we assume that the supports of $\psi_1,\dots,\psi_d$ are sufficiently small. Then by Theorem 1.1.10, the estimate (10.1.8) is valid for $\psi_0 u$. The estimate for $\psi_1 u,\dots,\psi_d u$ can be proved in the exact same manner as in the proof of Theorem 7.1.2. The only difference is the appearance of an additional function p. $\square$

10.2. Solvability of the boundary value problem in weighted L_2 Sobolev spaces

The goal of this section ist to show that the boundary value problem (10.1.1), (10.1.2) has a uniquely determined solution in the space $W^{2,2}_{\beta,\delta}(\mathcal{K})^3 \times W^{1,2}_{\beta,\delta}(\mathcal{K})$ if the line $\text{Re}\,\lambda = -\beta + 1/2$ is free of eigenvalues of the pencil $\mathfrak{A}(\lambda)$ and the components of δ satisfy the inequalities

$$\max(0, 1 - \delta_+^{(k)}) < \delta_k < 1 \quad \text{for } k = 1, \dots, d. \tag{10.2.1}$$

In fact, it is sufficient that $\max(-1, 1 - \mu_+^{(k)}) < \delta_k < 1$ for $k = 1, \dots, d$, as will be shown in Section 10.5 by means of estimates for the Green's matrix.

Furthermore, we prove in this section that the boundary value problem is uniquely solvable in the weighted space $V^{1,2}_\beta(\mathcal{K})^3 \times V^{0,2}_\beta(\mathcal{K})$ provided that no eigenvalues of the pencil $\mathfrak{A}(\lambda)$ are situated on the line $\text{Re}\,\lambda = -\beta - 1/2$.

10.2.1. Reduction to zero boundary data.

Let g, h_j, ϕ_j be given functions, $g \in W^{l-1,s}_{\beta,\delta}(\mathcal{K})$, $h_j \in W^{l-1/s,s}_{\beta,\delta}(\Gamma_j)^{3-d_j}$, $\phi_j \in W^{l-1-1/s,s}_{\beta,\delta}(\Gamma_j)^{d_j}$. We are interested in conditions which ensure the existence of a vector function $u \in W^{l,s}_{\beta,\delta}(\mathcal{K})$ satisfying the equations

$$S_j u = h_j, \quad N_j(u,p) = \phi_j \text{ on } \Gamma_j,\ j = 1, \dots, d, \quad \nabla \cdot u + g \in V^{l-1,s}_{\beta,\delta}(\mathcal{K}). \tag{10.2.2}$$

For functions in the corresponding $V^{l,s}_{\beta,\delta}$-spaces, the following assertion can be easily deduced from Lemma 9.2.1.

LEMMA 10.2.1. *Let $h_j \in V^{l-1/s,s}_{\beta,\delta}(\Gamma_j)^{3-d_j}$, $\phi_j \in V^{l-1-1/s,s}_{\beta,\delta}(\Gamma_j)^{d_j}$, $l \ge 2$. Then there exists a vector function $u \in V^{l,s}_{\beta,\delta}(\mathcal{K})^3$ such that $S_j u = h_j$ and $N_j(u, 0) = \phi_j$ on Γ_j and*

$$\|u\|_{V^{l,s}_{\beta,\delta}(\mathcal{K})^3} \le c \sum_{j=1}^{n} \Big(\|h_j\|_{V^{l-1/s,s}_{\beta,\delta}(\Gamma_j)^{3-d_j}} + \|\phi_j\|_{V^{l-1-1/s,s}_{\beta,\delta}(\Gamma_j)^{d_j}} \Big) \tag{10.2.3}$$

with a constant c independent of h_j and ϕ_j.

P r o o f. Let ζ_k be infinitely differentiable functions depending only on $\rho = |x|$ such that

$$\text{supp}\,\zeta_k \subset (2^{k-1}, 2^{k+1}), \qquad \sum_{k=-\infty}^{+\infty} \zeta_k = 1, \quad |(\rho\partial_\rho)^j \zeta_k(\rho)| \le c_j\,, \tag{10.2.4}$$

where c_j are constants independent of k. We set

$$h_{j,k}(x) = \zeta_k(2^k x)\, h_j(2^k x), \quad \phi_{j,k}(x) = 2^k\, \zeta_k(2^k x)\, \phi_j(2^k x).$$

Obviously, these functions vanish for $|x| < 1/2$ and $|x| > 2$. Consequently by Lemma 9.2.1, there exist vector functions $v_k \in V^{l,s}_{\beta,\delta}(\mathcal{K})^3$ such that

$$S_j v_k = h_{j,k}\,, \quad N_j(v_k, 0) = \phi_{j,k} \text{ on } \Gamma_j \text{ for } j = 1, \dots, d,$$

$v_k(x) = 0$ for $|x| < 1/4$ and $|x| > 4$, and

$$\|v_k\|_{V^{l,s}_{\beta,\delta}(\mathcal{K})^3} \le c \sum_{j=1}^{n} \Big(\|h_{j,k}\|_{V^{l-1/s,s}_{\beta,\delta}(\Gamma_j)^{3-d_j}} + \|\phi_{j,k}\|_{V^{l-1-1/s,s}_{\beta,\delta}(\Gamma_j)^{d_j}} \Big). \tag{10.2.5}$$

Hence, the functions $u_k(x) = v_k(2^{-k}x)$ satisfy the equations

$$S_j u_k = \zeta_k h_j\,, \quad N_j(u_k,0) = \zeta_k \phi_j \quad \text{on } \Gamma_j,$$

Furthermore, $u_k(x) = 0$ for $|x| < 2^{k-2}$ and $|x| > 2^{k+2}$. Moreover, u_k satisfies (10.2.5) with $\zeta_k h_j$ and $\zeta_k \phi_j$ instead of $h_{j,k}$ and $\phi_{j,k}$ on the right-hand side. Here the constant c is independent of k, h_j and ϕ_j. Consequently, the vector function

$$u = \sum_{k=-\infty}^{+\infty} u_k$$

satisfies the equations $S_j u = h_j$ on and $N_j(u,0) = \phi_j$ on Γ_j for $j = 1,\dots,d$. The inequality (10.2.3) follows from the equivalence of the norms in $V^{l,s}_{\beta,\delta}(\mathcal{K})$ and $V^{l-1/s,s}_{\beta,\delta}(\Gamma_j)$ to the norms

$$\|u\| = \Big(\sum_{k=-\infty}^{+\infty} \|\zeta_k u\|^2_{V^{l,s}_{\beta,\delta}(\mathcal{K})} \Big)^{1/s} \quad \text{and} \quad \|h\| = \Big(\sum_{k=-\infty}^{+\infty} \|\zeta_k h\|^2_{V^{l-1/s,s}_{\beta,\delta}(\Gamma_j)} \Big)^{1/s},$$

respectively (see Lemmas 3.1.2 and 3.1.5). □

An analogous result in $W^{l,s}_{\beta,\delta}(\mathcal{K})$ is only valid under additional compatibility conditions on the boundary data. We start with the simple case where $\delta_k > l-1-2/s$ for $k = 1,\dots,d$.

Let Γ_{k_+} and Γ_{k_-} denote the faces of the cone $\mathcal{K}$ adjacent to the edge M_k, and let θ_k be the inner angle at M_k. If $u \in W^{l,s}_{\beta,\delta}(\mathcal{K})^3$ and $\delta_k < l - 2/s$, then the trace of u on M_k exists, and from the equations $S_j u = h_j$ on Γ_j it follows that the pair $\big(h_{k_+}|_{M_j}, h_{k_-}|_{M_j}\big)$ belongs to the range of the matrix operator (S_{k_+}, S_{k_-}). This condition can be also written in the form

$$A_k^+ h_{k_+}|_{M_k} = A_k^- h_{k_-}|_{M_k}\,, \tag{10.2.6}$$

where A_k^+, A_k^- are certain constant matrices (cf. Section 9.2). Using Corollary 9.2.3, one can prove the following result analogously to Lemma 10.2.1.

LEMMA 10.2.2. *Let $h_j \in W^{l-1/s,s}_{\beta,\delta}(\Gamma_j)^{3-d_j}$ and $\phi_j \in W^{l-1-1/s,s}_{\beta,\delta}(\Gamma_j)^{d_j}$, $l \ge 2$, $\delta_k > l-1-2/s$, and $\delta_k \ne l-2/s$ for $k = 1,\dots,d$. In the case where $\delta_k < l-2/s$, we assume that h_{k_+} and h_{k_-} satisfy the compatibility condition* (10.2.6). *Then there exists a vector function $u \in W^{l,s}_{\beta,\delta}(\mathcal{K})^3$ satisfying the equations $S_j u = h_j$ and $N_j(u,0) = \phi_j$ on Γ_j and an estimate analogous to* (10.2.3).

If $u \in W^{l,s}_{\beta,\delta}(\mathcal{K})^3$ and $\delta_k < l-1-2/s$, then the traces of the first order derivatives of u exist on the edge M_k. Therefore in general, the traces of g, $\phi^\pm$ and of the derivatives of $h^\pm$ must satisfy certain compatibility conditions on the edge M_k. We restrict ourselves here to the case where $\delta_k > l-2-2/s$. Then the following lemma can be easily deduced from Lemmas 9.2.5 and 9.2.6.

LEMMA 10.2.3. *Let $h_j \in W^{l-1/s,s}_{\beta,\delta}(\Gamma_j)^{3-d_j}$, $\phi_j \in W^{l-1-1/s,s}_{\beta,\delta}(\Gamma_j)^{d_j}$ and $g \in W^{l-1,s}_{\beta,\delta}(\mathcal{K})$, where $l \ge 2$, $\delta_k > l-2-2/s$, $\delta_k \ne l-1-2/s$ and $\delta_k \ne l-2/s$ for $k = 1,\dots,d$. We suppose that h_j, ϕ_j and g satisfy the compatibility conditions of Section 9.2 on the edge M_k if $\delta_k < l-2/s$ (cf. Lemmas 9.2.5 and 9.2.6). Then* *there exist a vector function* $u \in W^{l,s}_{\beta,\delta}(\mathcal{K})^3$ *and a function* $p \in W^{l-1,s}_{\beta,\delta}(\mathcal{K})$ *satisfying the conditions* (10.2.2) *and the estimate*

$$\|u\|_{W^{l,s}_{\beta,\delta}(\mathcal{K})^3} + \|p\|_{W^{l-1,s}_{\beta,\delta}(\mathcal{K})} \leq c\Big(\|g\|_{W^{l-1,s}_{\beta,\delta}(\mathcal{K})} + \sum_{j=1}^{d} \|h_j\|_{W^{l-1/s,s}_{\beta,\delta}(\Gamma_j)^{3-d_j}} + \sum_{j=1}^{d} \|\phi_j\|_{W^{l-1-1/s,s}_{\beta,\delta}(\Gamma_j)^{d_j}}\Big).$$

REMARK 10.2.4. If $\delta_k = l-1-2/s$, then for the existence of u and p satisfying (10.2.2) it is necessary that the compatibility conditions on the edge M_k are satisfied in the generalized sense (cf. Remark 9.2.7). For example in the case $d_{k_+} = d_{k_-} = 0$ (Dirichlet conditions on the faces adjacent to M_k), the functions h_{k_+}, h_{k_-} and g must satisfy the conditions $h_{k_+}|_{M_k} = h_{k_-}|_{M_k}$ and

$$\int_0^\infty \int_0^{\varepsilon t} t^{s(\beta-l-1+2/s)}\, r^{-1} \Big| n_{k_-} \cdot \partial_r h_{k_+}(r,t) + n_{k_+} \cdot \partial_r h_{k_-}(r,t) - \big(\mathring{g}(r,t) - \partial_t (h_{k_+}(r,t) \cdot e_k)\big) \sin\theta_k \Big|^s \, dr\, dt < \infty.$$

Here ε is a small positive number, r and t are Cartesian coordinates on Γ_{k_+} and Γ_{k_-}, $r = \text{dist}(x, M_k)$, e_k is the unit vector on M_k and $\mathring{g}(r,t)$ denotes the average of g with respect to the angle φ in the plane perpendicular to M_k.

10.2.2. Some a priori estimates for the solutions. In the next lemma, let ζ_k be infinitely differentiable functions depending only on $\rho = |x|$ and satisfying the conditions (10.2.4). Furthermore, let $\eta_k = \zeta_{k-1} + \zeta_k + \zeta_{k+1}$. By $W^{l,s}_{loc}(\overline{\mathcal{K}}\backslash\mathcal{S})$, we denote the set of all functions u on $\mathcal{K}$ such that $\zeta u \in W^{l,s}(\mathcal{K})$ for all $\zeta \in C_0^\infty(\overline{\mathcal{K}}\backslash\mathcal{S})$. Then the following assertion can be easily deduced from Lemma 9.4.1.

LEMMA 10.2.5. *Let* $(u,p) \in W^{l,s}_{loc}(\bar{\mathcal{K}}\backslash\mathcal{S})^3 \times W^{l-1,s}_{loc}(\bar{\mathcal{K}}\backslash\mathcal{S})$, $l \geq 2$, *be a solution of the boundary value problem* (10.1.1), (10.1.2). *If* $\eta_k u \in V^{l-1,s}_{\beta-1,\delta-1}(\mathcal{K})^3$, $\eta_k p \in V^{l-2,s}_{\beta-1,\delta-1}(\mathcal{K})$, $\eta_k f \in V^{l-2,s}_{\beta,\delta}(\mathcal{K})^3$, $\eta_k g \in V^{l-1,s}_{\beta,\delta}(\mathcal{K})$, $\eta_k h_j \in V^{l-1/s,s}_{\beta,\delta}(\Gamma_j)^{3-d_j}$, *and* $\eta_k \phi_j \in V^{l-1-1/s,s}_{\delta}(\Gamma_j)^{d_j}$, *then* $\zeta_k u \in V^{l,s}_{\beta,\delta}(\mathcal{K})^3$, $\zeta_k p \in V^{l-1,s}_{\beta,\delta}(\mathcal{K})$ *and*

$$(10.2.7)\quad \|\zeta_k u\|_{V^{l,s}_{\beta,\delta}(\mathcal{K})^3} + \|\zeta_k p\|_{V^{l-1,s}_{\beta,\delta}(\mathcal{K})}$$
$$\leq c\Big(\|\eta_k f\|_{V^{l-2,s}_{\beta,\delta}(\mathcal{K})^3} + \|\eta_k g\|_{V^{l-1,s}_{\beta,\delta}(\mathcal{K})} \sum_{j=1}^{d} \|\eta_k h_j\|_{V^{l-1/s,s}_{\beta,\delta}(\Gamma_j)^{3-d_j}} + \sum_{j=1}^{d} \|\eta_k \phi_j\|_{V^{l-1-1/s,s}_{\beta,\delta}(\Gamma_j)^{d_j}} + \|\eta_k u\|_{V^{l-1,s}_{\beta-1,\delta-1}(\mathcal{K})^3} + \|\eta_k p\|_{V^{l-2,s}_{\beta-1,\delta-1}(\mathcal{K})}\Big).$$

Here the constant c *is independent of* (u,p) *and* k.

P r o o f. By Lemma 10.2.1, we may assume without loss of generality that $h_j = 0$ and $\phi_j = 0$ for $j = 1, \ldots, d$. Let $\tilde{\zeta}_k(x) = \zeta_k(2^k x)$, $\tilde{\eta}_k(x) = \eta_k(2^k x)$, $\tilde{u}(x) = u(2^k x)$, and $\tilde{p}(x) = 2^k p(2^k x)$. By (10.2.4), the support of $\tilde{\zeta}_k$ is contained in the set $\{x : 1/2 < |x| < 2\}$ and the derivatives of $\tilde{\zeta}_k$ are bounded by constants independent of k. Obviously,

$$-\Delta\tilde{u} + \nabla\tilde{p} = \tilde{f}, \quad -\nabla\tilde{u} = \tilde{g} \ \text{ in } \mathcal{K},$$
$$S_j\tilde{u} = 0, \ N_j(\tilde{u},\tilde{p}) = 0 \ \text{ on } \Gamma_j, \ j = 1,\dots,d,$$

where $\tilde{f}(x) = 2^{2k}f(2^kx)$ and $\tilde{g}(x) = 2^kg(2^kx)$. Consequently by Lemma 9.4.1, $\tilde{\zeta}_k\tilde{u} \in V^{l,s}_{\beta,\delta}(\mathcal{K})^3$, $\tilde{\zeta}_k\tilde{p} \in V^{l-1,s}_{\beta,\delta}(\mathcal{K})$, and

$$\|\tilde{\zeta}_k\tilde{u}\|_{V^{l,s}_{\beta,\delta}(\mathcal{K})^3} + \|\tilde{\zeta}_k\tilde{p}\|_{V^{l-1,s}_{\beta,\delta}(\mathcal{K})} \le c\Big(\|\tilde{\eta}_k\tilde{u}\|_{V^{l-1,s}_{\beta-1,\delta-1}(\mathcal{K})^3} + \|\tilde{\eta}_k\tilde{p}\|_{V^{l-2,s}_{\beta-1,\delta-1}(\mathcal{K})} + \|\tilde{\eta}_k\tilde{f}\|_{V^{l-2,s}_{\beta,\delta}(\mathcal{K})^3} + \|\tilde{\eta}_k\tilde{g}\|_{V^{l-1,s}_{\beta,\delta}(\mathcal{K})}\Big).$$

where c is independent of u, p, and k. Applying the coordinate change $2^kx = y$, we obtain (10.2.7). □

Using the last lemma and Lemma 3.1.2, we obtain an estimate of the norm of the solution (u,p) in the space $V^{l,s}_{\beta,\delta}(\mathcal{K})^3 \times V^{l-1,s}_{\beta,\delta}(\mathcal{K})$.

LEMMA 10.2.6. *Let $(u,p) \in W^{l,s}_{loc}(\bar{\mathcal{K}}\backslash\mathcal{S})^3 \times W^{l-1,s}_{loc}(\bar{\mathcal{K}}\backslash\mathcal{S})$, $l \ge 2$, be a solution of the boundary value problem* (10.1.1), (10.1.2). *If $u \in V^{l-1,s}_{\beta-1,\delta-1}(\mathcal{K})^3$, $p \in V^{l-2,s}_{\beta-1,\delta-1}(\mathcal{K})$, $f \in V^{l-2,s}_{\beta,\delta}(\mathcal{K})^3$, $g \in V^{l-1,s}_{\beta,\delta}(\mathcal{K})$, $h_j \in V^{l-1/s,s}_{\beta,\delta}(\Gamma_j)^{3-d_j}$, and $\phi_j \in V^{l-1-1/s,s}_{\beta,\delta}(\Gamma_j)^{d_j}$, then $u \in V^{l,s}_{\beta,\delta}(\mathcal{K})^3$, $p \in V^{l-1,s}_{\beta,\delta}(\mathcal{K})$ and*

(10.2.8)
$$\|u\|_{V^{l,s}_{\beta,\delta}(\mathcal{K})^3} + \|p\|_{V^{l-1,s}_{\beta,\delta}(\mathcal{K})} \le c\Big(\|f\|_{V^{l-2,s}_{\beta,\delta}(\mathcal{K})^3} + \|g\|_{V^{l-1,s}_{\beta,\delta}(\mathcal{K})}\sum_{j=1}^{d}\|h_j\|_{V^{l-1/s,s}_{\beta,\delta}(\Gamma_j)^{3-d_j}} + \sum_{j=1}^{d}\|\phi_j\|_{V^{l-1-1/s,s}_{\beta,\delta}(\Gamma_j)^{d_j}} + \|u\|_{V^{l-1,s}_{\beta-1,\delta-1}(\mathcal{K})^3} + \|p\|_{V^{l-2,s}_{\beta-1,\delta-1}(\mathcal{K})}\Big).$$

Analogously, one can prove the next lemma by means of Lemmas 7.2.1, 9.4.3.

LEMMA 10.2.7. *Let $(u,p) \in W^{l,s}_{loc}(\bar{\mathcal{K}}\backslash\mathcal{S})^3 \times W^{l-1,s}_{loc}(\bar{\mathcal{K}}\backslash\mathcal{S})$, $l \ge 2$, be a solution of the boundary value problem* (10.1.1), (10.1.2). *If $u \in W^{l-1,s}_{\beta-1,\delta-1}(\mathcal{K})^3$, $p \in W^{l-2,s}_{\beta-1,\delta-1}(\mathcal{K})$, $f \in W^{l-2,s}_{\beta,\delta}(\mathcal{K})^3$, $g \in W^{l-1,s}_{\beta,\delta}(\mathcal{K})$, $h_j \in W^{l-1/s,s}_{\beta,\delta}(\Gamma_j)^{3-d_j}$, and $\phi_j \in W^{l-1-1/s,s}_{\beta,\delta}(\Gamma_j)^{d_j}$, $\delta_k > 1-2/s$, then $u \in W^{l,s}_{\beta,\delta}(\mathcal{K})^3$ and $p \in W^{l-1,s}_{\beta,\delta}(\mathcal{K})$ Furthermore, an estimate analogous to* (10.2.8) *is valid for the norm of (u,p) in $W^{l,s}_{\beta,\delta}(\mathcal{K})^3 \times W^{l-1,s}_{\beta,\delta}(\mathcal{K})$.*

Finally, Lemma 9.4.7 leads to the following regularity result.

LEMMA 10.2.8. *Let $(u,p) \in W^{l,s}_{\beta,\delta}(\mathcal{K})^3 \times W^{l-1,s}_{\beta,\delta}(\mathcal{K})$ be a solution of the problem* (10.1.1), (10.1.2) *such that*

$$\rho\partial_\rho u \in W^{l,s}_{\beta,\delta'}(\mathcal{K})^3, \quad \rho\partial_\rho p \in W^{l-1,s}_{\beta,\delta'}(\mathcal{K}),$$

where $l \ge 2$, $-2/s < \delta_k \le \delta'_k \le \delta_k + 1$ for $k = 1,\dots,d$. Furthermore, we assume that

$$f \in W^{l-1,s}_{\beta+1,\delta'}(\mathcal{K})^3, \quad g \in W^{l,s}_{\beta+1,\delta'}(\mathcal{K}),$$
$$h_j \in W^{l+1-1/s,s}_{\beta+1,\delta'}(\Gamma_j)^{3-d_j} \quad \text{and} \quad \phi_j \in W^{l-1/s,s}_{\beta+1,\delta'}(\Gamma_j)^{d_j}.$$

If the strip $l-\delta_k-2/s \le \operatorname{Re}\lambda \le l+1-\delta'_k-2/s$ *does not contain eigenvalues of the pencils* $A_k(\lambda)$ *for* $k=1,\dots,d$, *then* $u\in W^{l+1,s}_{\beta+1,\delta'}(\mathcal{K})^3$ *and* $p\in W^{l,s}_{\beta+1,\delta'}(\mathcal{K})$.

P r o o f. By Lemma 10.2.7, the solution (u,p) belongs to $W^{l+1,s}_{\beta+1,\delta+1}(\mathcal{K})^3 \times W^{l,s}_{\beta+1,\delta+1}(\mathcal{K})$. Let ζ_k, η_k, $\tilde\zeta_k$, $\tilde\eta_k$, $\tilde u$ and $\tilde p$ be the same functions as in the proof of Lemma 10.2.5. Furthermore, let $\tilde f(x)=2^{2k}f(2^k x)$ and $\tilde g(x)=2^k g(2^k x)$, $\tilde h_j(x)=h_j(2^k x)$ and $\tilde\phi_j(x)=2^k\phi_j(2^k x)$. Since $\tilde\zeta_k(x)=0$ for $|x|<1/2$ and $|x|>2$, we conclude from Lemma 9.4.7 that $\tilde\zeta_k\tilde u\in W^{l+1,s}_{\beta+1,\delta'}(\mathcal{K})^3$, $\tilde\zeta_k\tilde p\in W^{l,s}_{\beta+1,\delta'}(\mathcal{K})$ and

$$\|\tilde\zeta_k\tilde u\|^s_{W^{l+1,s}_{\beta+1,\delta'}(\mathcal{K})^3}+\|\tilde\zeta_k\tilde p\|^s_{W^{l,s}_{\beta+1,\delta'}(\mathcal{K})^3}\le c\Big(\|\tilde\eta_k\tilde u\|^s_{W^{l,s}_{\beta,\delta}(\mathcal{K})^3}+\|\tilde\eta_k\rho\partial_\rho\tilde u\|^s_{W^{l,s}_{\beta,\delta'}(\mathcal{K})^3}$$
$$+\|\tilde\eta_k\tilde p\|^s_{W^{l-1,s}_{\beta,\delta}(\mathcal{K})}+\|\tilde\eta_k\rho\partial_\rho\tilde p\|^s_{W^{l-1,s}_{\beta,\delta'}(\mathcal{K})}+\|\tilde\eta_k\tilde f\|^s_{W^{l-1,s}_{\beta+1,\delta'}(\mathcal{K})^3}+\|\tilde\eta_k\tilde g\|^s_{W^{l,s}_{\beta+1,\delta'}(\mathcal{K})}$$
$$+\sum_{j=1}^d\|\tilde\eta_k\tilde h_j\|^s_{W^{l+1-1/s,s}_{\beta+1,\delta'}(\Gamma_j)}+\sum_{j=1}^d\|\tilde\eta_k\tilde\phi_j\|^s_{W^{l-1/s,s}_{\beta+1,\delta'}(\Gamma_j)^\ell}\Big),$$

where c is independent of k. Multiplying the last inequality by $2^{ks(\beta-l)+3k}$ and substituting $2^k=y$, we obtain the same inequality with $\zeta_k,\eta_k,u,p,f,g,h_j,\phi_j$ instead of $\tilde\zeta_k,\tilde\eta_k,\tilde u,\tilde p,\tilde f,\tilde g,\tilde h_j,\tilde\phi_j$. Now the assertion of the lemma results from Lemma 7.2.1. □

10.2.3. Existence and uniqueness of solutions. Using the Mellin transform (3.3.12) and Theorem 10.1.3, we obtain and existence and uniqueness theorem for the problem (10.1.1), (10.1.2) in the weighted space $W^{2,2}_{\beta,\delta}(\mathcal{K})^3\times W^{1,2}_{\beta,\delta}(\mathcal{K})$.

THEOREM 10.2.9. *Suppose that the line* $\operatorname{Re}\lambda=-\beta+1/2$ *is free of eigenvalues of the pencil* $\mathfrak{A}(\lambda)$ *and that the components of* δ *satisfy the inequalities* (10.2.1). *Then the boundary value problem* (10.1.1), (10.1.2) *is uniquely solvable in* $W^{2,2}_{\beta,\delta}(\mathcal{K})^3\times W^{1,2}_{\beta,\delta}(\mathcal{K})$ *for arbitrary* $f\in W^{0,2}_{\beta,\delta}(\mathcal{K})^3$, $g\in W^{1,2}_{\beta,\delta}(\mathcal{K})$, $h_j\in W^{3/2,2}_{\beta,\delta}(\Gamma_j)^{3-d_j}$ *satisfying* (10.2.6), *and* $\phi_j\in W^{1/2,2}_{\beta,\delta}(\Gamma_j)^{d_j}$.

P r o o f. By Lemma 10.2.2, we may assume without loss of generality that $h_j=0$ and $\phi_j=0$ for $j=1,\dots,d$. Applying the Mellin transform $\mathcal{M}=\mathcal{M}_{\rho\to\lambda}$ with respect to $\rho=|x|$ to the equations (10.1.1), (10.1.2), we obtain the parameter-dependent problem

$$\mathfrak{A}_\delta(\lambda)\big(\mathcal{M}u,\mathcal{M}\rho p\big)=\big(\mathcal{M}\rho^2 f,\mathcal{M}\rho g,0\big) \tag{10.2.9}$$

where $\mathfrak{A}_\delta(\lambda)$ is the operator introduced in Section 10.1, $(\mathcal{M}\rho^2 f)(\cdot,\lambda)\in W^{0,2}_\delta(\Omega)^3$ and $(\mathcal{M}\rho g)(\cdot,\lambda)\in W^{1,2}_\delta(\Omega)$ for almost all λ. By Theorem 10.1.3, the problem (10.2.9) is uniquely solvable in $W^{2,2}_\delta(\Omega)^3\times W^{1,2}_\delta(\Omega)$ for every λ on the line $\operatorname{Re}\lambda=-\beta+1/2$, and the solution satisfies the estimate

$$\sum_{j=0}^2|\lambda|^{4-2j}\,\|(\mathcal{M}u)(\lambda,\cdot)\|^2_{W^{j,2}_\delta(\Omega)^3}+\sum_{j=0}^1|\lambda|^{2-2j}\,\|(\mathcal{M}\rho p)(\lambda,\cdot)\|^2_{W^{j,1}_\delta(\Omega)}$$
$$\le c\Big(\|(\mathcal{M}\rho^2 f)(\lambda,\cdot)\|^2_{W^{0,2}_\delta(\Omega)^3}+\sum_{j=0}^1|\lambda|^{2-2j}\,\|(\mathcal{M}\rho g)(\lambda,\cdot)\|^2_{W^{j,1}_\delta(\Omega)}\Big)$$

or, what is the same,

$$(10.2.10)\sum_{j=0}^{2} |\lambda|^{4-2j} \, \|(\mathcal{M}u)(\lambda,\cdot)\|^2_{W^{j,2}_\delta(\Omega)^3} + \sum_{j=0}^{1} |\lambda|^{2-2j} \, \|(\mathcal{M}p)(\lambda-1,\cdot)\|^2_{W^{j,1}_\delta(\Omega)}$$
$$\le c\Big(\|(\mathcal{M}f)(\lambda-2,\cdot)\|^2_{W^{0,2}_\delta(\Omega)^3} + \sum_{j=0}^{1} |\lambda|^{2-2j} \, \|(\mathcal{M}g)(\lambda-1,\cdot)\|^2_{W^{j,1}_\delta(\Omega)}\Big).$$

Consequently, the vector function (u,p) with the components

$$u_j(\rho,\omega) = \frac{1}{2\pi i} \int\limits_{-\beta+1/2-i\infty}^{-\beta+1/2+i\infty} \rho^\lambda \, (\mathcal{M}u_j)(\lambda,\omega)\, d\lambda, \quad j=1,2,3,$$

$$p(\rho,\omega) = \frac{1}{2\pi i\rho} \int\limits_{-\beta+1/2-i\infty}^{-\beta+1/2+i\infty} \rho^\lambda \, (\mathcal{M}p)(\lambda-1,\omega)\, d\lambda$$

solves the problem (10.1.1), (10.1.2). From (10.2.10) and from the equivalence of the $W^{l,s}_{\beta,\delta}(\mathcal{K})$-norm to (7.3.7) it follows that $u \in W^{2,2}_{\beta,\delta}(\mathcal{K})^3$, $p \in W^{1,2}_{\beta,\delta}(\mathcal{K})$ and

$$\|u\|_{W^{2,2}_{\beta,\delta}(\mathcal{K})^3} + \|p\|_{W^{1,2}_{\beta,\delta}(\mathcal{K})} \le c\Big(\|f\|_{W^{0,2}_{\beta,\delta}(\mathcal{K})^3} + \|g\|_{W^{1,2}_{\beta,\delta}(\mathcal{K})}\Big).$$

This proves the theorem. □

10.2.4. Variational solutions in weighted L_2 Sobolev spaces. Let again $V^{l,2}_\beta(\mathcal{K}) = W^{l,2}_{\beta,0}(\mathcal{K})$ for any integer $k \ge 0$. Variational solutions of the boundary value problem (10.1.1), (10.1.2) in the space $V^{1,2}_\beta(\mathcal{K})^3 \times V^{0,2}_\beta(\mathcal{K})$ are defined in the same way as for the case $\beta = 0$ in Section 10.1. Let

$$\mathcal{H}_\beta = \{u \in V^{1,2}_\beta(\mathcal{K})^3 : \; S_j u = 0 \text{ on } \Gamma_j, \; j=1,\dots,d\}.$$

The vector function $(u,p) \in V^{1,2}_\beta(\mathcal{K})^3 \times V^{0,2}_\beta(\mathcal{K})$ is called *variational solution* of problem (10.1.1), (10.1.2) if

$$(10.2.11) \qquad b_{\mathcal{K}}(u,v) - \int_{\mathcal{K}} p \nabla\cdot v\, dx = F(v) \quad \text{for all } v \in \mathcal{H}_{-\beta},$$

$$(10.2.12) \qquad -\nabla\cdot u = g \;\text{ in } \mathcal{K}, \quad S_j u = h_j \;\text{ on } \Gamma_j, \; j=1,\dots,d.$$

We restrict ourselves to the case of zero Dirichlet data h_j and define the operator $\mathcal{A}_\beta$ as the mapping

$$\mathcal{H}_\beta \times V^{0,2}_\beta(\mathcal{K}) \ni (u,p) \to (F,g) \in \mathcal{H}^*_{-\beta} \times V^{0,2}_\beta(\mathcal{K}),$$

where

$$F(v) = b_{\mathcal{K}}(u,v) - \int_{\mathcal{K}} p \nabla\cdot v\, dx \;\text{ for all } v \in \mathcal{H}^*_{-\beta} \quad\text{and}\quad g = -\nabla\cdot u.$$

Our goal is to prove that $\mathcal{A}_\beta$ is an isomorphism if the line $\operatorname{Re}\lambda = -\beta - 1/2$ is free of eigenvalues of the pencil $\mathfrak{A}(\lambda)$. For this, we need the following lemma.

LEMMA 10.2.10. *Let $u \in \mathcal{H}_\beta$, $p \in V^{0,2}_\beta(\mathcal{K})$ and $(F,g) = \mathcal{A}_\beta(u,p)$. Then the estimate*

$$\|u\|_{V^{1,2}_\beta(\mathcal{K})^3} + \|p\|_{V^{0,2}_\beta(\mathcal{K})} \le c\Big(\|F\|_{\mathcal{H}^*_{-\beta}} + \|g\|_{V^{0,2}_\beta(\mathcal{K})} + \|u\|_{V^{0,2}_{\beta-1}(\mathcal{K})^3} + \|p\|_{(V^{1,2}_{1-\beta}(\mathcal{K}))^*}\Big)$$

is valid with a constant c independent of u and p.

P r o o f. The proof proceeds analogously to Lemma 7.3.6. Let $\zeta_k = \zeta_k(\rho)$ be smooth real-valued functions satisfying the conditions (10.2.4), and let $\eta_k = \zeta_{k-1} + \zeta_k + \zeta_{k+1}$. By Lemma 7.2.1, the norm in $V_\beta^{l,2}(\mathcal{K})$ is equivalent to

$$\Big(\sum_{k=-\infty}^{+\infty} \|\zeta_k u\|^2_{V_\beta^{l,2}(\mathcal{K})} \Big)^{1/2}.$$

Analogously, the norms in $(V_{1-\beta}^{1,2}(\mathcal{K}))^*$ and $\mathcal{H}^*_{-\beta}$ are equivalent to

$$\Big(\sum_{k=-\infty}^{+\infty} \|\zeta_k p\|^2_{(V_{1-\beta}^{1,2}(\mathcal{K}))^*} \Big)^{1/2}$$

and

$$\Big(\sum_{k=-\infty}^{+\infty} \|\zeta_k F\|^2_{\mathcal{H}^*_{-\beta}} \Big)^{1/2},$$

respectively. We show that

$$\begin{aligned} &\|\zeta_k u\|^2_{V_\beta^{1,2}(\mathcal{K})^3} + \|\zeta_k p\|^2_{V_\beta^{0,2}(\mathcal{K})} \\ &\quad \le c\Big(\|\zeta_k F\|^2_{\mathcal{H}^*_{-\beta}} + \|\zeta_k g\|^2_{V_\beta^{0,2}(\mathcal{K})} + \|\eta_k p\|^2_{(V_{1-\beta}^{1,2}(\mathcal{K}))^*} \Big) \\ &\quad\quad + \varepsilon\, \|\eta_k u\|^2_{V_\beta^{1,2}(\mathcal{K})^3} + C(\varepsilon)\, \|\eta_k u\|^2_{V_{\beta-1}^{0,2}(\mathcal{K})^3}, \end{aligned} \tag{10.2.13}$$

where c, ε and $C(\varepsilon)$ are constants independent of k and ε can be chosen arbitrarily small. First let $k = 0$. As in the proof of Lemma 7.3.6, we obtain

$$b_{\mathcal{K}}(\zeta_0 u, v) - b_{\mathcal{K}}(u, \zeta_0 v) = c(u, v),$$

where

$$\big|c(u,v)\big| \le \Big(\varepsilon\, \|\eta_0 u\|_{V_0^{1,2}(\mathcal{K})^3} + C(\varepsilon)\, \|\eta_0 u\|_{L_2(\mathcal{K})^3} \Big)\, \|v\|_{V_0^{1,2}(\mathcal{K})^3}.$$

Consequently,

$$b_{\mathcal{K}}(\zeta_0 u, v) - \int_{\mathcal{K}} \zeta_0 p\, \nabla \cdot v\, dx = F(\zeta_0 v) + c(u,v) - \int_{\mathcal{K}} p\, (\nabla \zeta_0) \cdot v\, dx.$$

Furthermore, $-\nabla \cdot (\zeta_0 u) = \zeta_0 g - (\nabla \zeta_0) \cdot u$ in $\mathcal{K}$ and $S_j \zeta_0 u = 0$ on Γ_j. Applying Theorem 10.1.2, we arrive at the inequality

$$\begin{aligned} \|\zeta_0 u\|^2_{V_0^{1,2}(\mathcal{K})^3} + \|\zeta_0 p\|^2_{L_2(\mathcal{K})} &\le c\Big(\|\zeta_0 F\|^2_{\mathcal{H}^*_{\mathcal{K}}} + \|\zeta_0 g\|^2_{L_2(\mathcal{K})} + \|\eta_0 p\|^2_{(V_0^{1,2}(\mathcal{K}))^*} \Big) \\ &\quad + \varepsilon\, \|\eta_0 u\|^2_{V_0^{1,2}(\mathcal{K})^3} + C(\varepsilon)\, \|\eta_0 u\|^2_{L_2(\mathcal{K})^3}. \end{aligned}$$

Since ζ_0 and η_0 are zero outside the set $\{x : 1/4 < |x| < 4\}$, the last inequality implies (10.2.13) for $k = 0$. By means of the transformation $x = 2^k y$, we obtain (10.2.13) with the same constants c, ε and $C(\varepsilon)$ for $k \ne 0$. Summing up in (10.2.13), we get

$$\begin{aligned} \|u\|^2_{V_\beta^{1,2}(\mathcal{K})^3} + \|p\|^2_{V_\beta^{0,2}(\mathcal{K})} &\le c\Big(\|F\|^2_{\mathcal{H}^*_{-\beta}} + \|g\|^2_{V_\beta^{0,2}(\mathcal{K})} + \|p\|^2_{(V_{1-\beta}^{1,2}(\mathcal{K}))^*} \Big) \\ &\quad + \varepsilon\, \|u\|^2_{V_\beta^{1,2}(\mathcal{K})^3} + C(\varepsilon)\, \|u\|^2_{V_{\beta-1}^{0,2}(\mathcal{K})^3}, \end{aligned}$$

where ε can be chosen arbitrarily small. This proves the lemma. □

THEOREM 10.2.11. *If the line* $\operatorname{Re}\lambda = -\beta - 1/2$ *is free of eigenvalues of the pencil* $\mathfrak{A}(\lambda)$, *then the operator* $\mathcal{A}_\beta$ *is an isomorphism.*

P r o o f. We show first that

$$\|u\|_{V_\beta^{1,2}(\mathcal{K})^3} + \|p\|_{V_\beta^{0,2}(\mathcal{K})} \le c\Big(\|F\|_{\mathcal{H}^*_{-\beta}} + \|g\|_{V_\beta^{0,2}(\mathcal{K})}\Big) \tag{10.2.14}$$

for all $u \in \mathcal{H}_\beta$, $p \in V_\beta^{0,2}(\mathcal{K})$, $(F,g) = \mathcal{A}_\beta(u,p)$. Suppose that $u \in \mathcal{H}_\beta \subset W^{0,2}_{\beta-1,\varepsilon-1}(\mathcal{K})^3$, $p \in V_\beta^{0,2}(\mathcal{K})$, $w \in W^{0,2}_{1-\beta,1-\varepsilon}(\mathcal{K})^3$, and $\psi \in W^{1,2}_{1-\beta,1-\varepsilon}(\mathcal{K})$, where ε is a sufficiently small positive number. By Theorem 10.2.9, there exists a solution $(v,q) \in W^{2,2}_{1-\beta,1-\varepsilon}(\mathcal{K})^3 \times W^{1,2}_{1-\beta,1-\varepsilon}(\mathcal{K})$ of the problem

$$-\Delta v - \nabla\nabla\cdot v + \nabla q = w, \quad -\nabla\cdot v = \psi \ \text{ in } \mathcal{K},$$
$$S_j v = 0, \ N_j(v,q) = 0 \ \text{ on } \Gamma_j,\ j = 1,\dots,d,$$

satisfying the estimate

$$\|v\|_{W^{2,2}_{1-\beta,1-\varepsilon}(\mathcal{K})^3} + \|q\|_{W^{1,2}_{1-\beta,1-\varepsilon}(\mathcal{K})} \le c\Big(\|w\|_{W^{0,2}_{1-\beta,1-\varepsilon}(\mathcal{K})^3} + \|\psi\|_{W^{1,2}_{1-\beta,1-\varepsilon}(\mathcal{K})}\Big).$$

According to Green's formula (cf. formula (9.1.4)), the equality

$$\begin{aligned}\int_{\mathcal{K}} u\cdot w\,dx + \int_{\mathcal{K}} p\,\psi\,dx &= b_{\mathcal{K}}(u,v) - \int_{\mathcal{K}} p\,\nabla\cdot v\,dx - \int_{\mathcal{K}} (\nabla\cdot u)\,q\,dx \\ &= F(v) + \int_{\mathcal{K}} g\,q\,dx\end{aligned}$$

holds. Hence,

$$\begin{aligned}\Big|\int_{\mathcal{K}} u\cdot w\,dx + \int_{\mathcal{K}} p\,\psi\,dx\Big| &\le \|F\|_{\mathcal{H}^*_{-\beta}}\,\|v\|_{V^{1,2}_{-\beta}(\mathcal{K})^3} + \|g\|_{V_\beta^{0,2}(\mathcal{K})}\,\|q\|_{V^{0,2}_{-\beta}(\mathcal{K})} \\ &\le c\,\|\mathcal{A}_\beta(u,p)\|_{\mathcal{H}^*_{-\beta}\times V_\beta^{0,2}(\mathcal{K})}\Big(\|v\|_{W^{2,2}_{1-\beta,1-\varepsilon}(\mathcal{K})^3} + \|q\|_{W^{1,2}_{1-\beta,1-\varepsilon}(\mathcal{K})}\Big) \\ &\le c\,\|\mathcal{A}_\beta(u,p)\|_{\mathcal{H}^*_{-\beta}\times V_\beta^{0,2}(\mathcal{K})}\Big(\|w\|_{W^{0,2}_{1-\beta,1-\varepsilon}(\mathcal{K})^3} + \|\psi\|_{W^{1,2}_{1-\beta,1-\varepsilon}(\mathcal{K})}\Big).\end{aligned}$$

Setting $w = \rho^{2(\beta-1)}\prod(r_j/\rho)^{2(\varepsilon-1)}u$ and $\psi = 0$, we obtain

$$\|u\|^2_{W^{0,2}_{\beta-1,\varepsilon-1}(\mathcal{K})^3} \le c\,\|\mathcal{A}_\beta(u,p)\|_{\mathcal{H}^*_{-\beta}\times V_\beta^{0,2}(\mathcal{K})}\,\|u\|_{W^{0,2}_{\beta-1,\varepsilon-1}(\mathcal{K})^3}$$

and therefore,

$$\|u\|_{V^{0,2}_{\beta-1}(\mathcal{K})^3} \le c\,\|u\|_{W^{0,2}_{\beta-1,\varepsilon-1}(\mathcal{K})^3} \le c\,\|\mathcal{A}_\beta(u,p)\|_{\mathcal{H}^*_{-\beta}\times V_\beta^{0,2}(\mathcal{K})}.$$

Analogously for $w = 0$ and arbitrary $\psi \in V^{1,2}_{1-\beta}(\mathcal{K})$, we get

$$\begin{aligned}\Big|\int_{\mathcal{K}} p\,\psi\,dx\Big| &\le c\,\|\mathcal{A}_\beta(u,p)\|_{\mathcal{H}^*_{-\beta}\times V_\beta^{0,2}(\mathcal{K})}\,\|\psi\|_{W^{1,2}_{1-\beta,1-\varepsilon}(\mathcal{K})} \\ &\le c\,\|\mathcal{A}_\beta(u,p)\|_{\mathcal{H}^*_{-\beta}\times V_\beta^{0,2}(\mathcal{K})}\,\|\psi\|_{V^{1,2}_{1-\beta}(\mathcal{K})}\end{aligned}$$

which implies the inequality

$$\|p\|_{(V^{1,2}_{1-\beta}(\mathcal{K}))^*} \le c\,\|\mathcal{A}_\beta(u,p)\|_{\mathcal{H}^*_{-\beta}\times V_\beta^{0,2}(\mathcal{K})}\,.$$

Using Lemma 10.2.10, we arrive at (10.2.14). By (10.2.14), the operator $\mathcal{A}_\beta$ is injective and has closed range. Furthermore by Theorem 10.2.9, the range of $\mathcal{A}_\beta$ contains the set $W^{0,2}_{\beta-1,\varepsilon-1}(\mathcal{K})^3 \times W^{1,2}_{\beta-1,\varepsilon-1}(\mathcal{K})$ which is dense in $\mathcal{H}^*_{-\beta} \times V_\beta^{0,2}(\mathcal{K})$.

This proves the theorem. □

10.3. Regularity results for variational solutions

We consider solutions of the variational problem (10.2.11), (10.2.12). The goal of this section is to establish regularity results for these solutions (and for their derivatives in ρ) in the spaces $W^{l,2}_{\beta,\delta}(\mathcal{K})$. In this way, we also obtain an existence and uniqueness theorem for solutions of the boundary value problem (10.1.1), (10.1.2) in the space $W^{l,2}_{\beta,\delta}(\mathcal{K})^3 \times W^{l-1,2}_{\beta,\delta}(\mathcal{K})$. Here β must be such that the line $\operatorname{Re}\lambda = l-\beta-3/2$ is free of eigenvalues of the pencil $\mathfrak{A}(\lambda)$, while the components of δ satisfy the inequalities

$$\max(0, l-1-\delta_+^{(k)}) < \delta_k < l-1 \quad \text{for } k = 1,\dots,d. \tag{10.3.1}$$

In some cases, the last condition can be weakened. Furthermore, for some other cases (e.g. if the Dirichlet condition is prescribed on at least one of the adjoining faces of every edge), we get analogous existence and regularity results in the class of the space $V^{l,2}_{\beta,\delta}(\mathcal{K})$.

10.3.1. A $W^{l,2}_{\beta,\delta}$-regularity result for the solution.

The proof of the next theorem is based on Theorem 9.4.9.

THEOREM 10.3.1. *Let $(u,p) \in V^{1,2}_{\beta-l+1}(\mathcal{K})^3 \times V^{0,2}_{\beta-l+1}(\mathcal{K})$ be a solution of problem*

$$b_{\mathcal{K}}(u,v) - \int_{\mathcal{K}} p\nabla\cdot v\,dx = F(v) \quad \textit{for all } v \in \mathcal{H}_{l-1-\beta}, \tag{10.3.2}$$

$$-\nabla\cdot u = g \ \textit{ in } \mathcal{K}, \quad S_j u = h_j \ \textit{ on } \Gamma_j,\ j = 1,\dots,d. \tag{10.3.3}$$

Suppose that the functional $F \in \mathcal{H}^*_{l-1-\beta}$ has the form

$$F(v) = \int_{\mathcal{K}} (f+\nabla g)\cdot v\,dx + \sum_{j=1}^{d}\int_{\Gamma_j} \phi_j\cdot v\,dx, \tag{10.3.4}$$

where

$$f \in W^{l-2,2}_{\beta,\delta}(\mathcal{K})^3, \quad g \in W^{l-1,2}_{\beta,\delta}(\mathcal{K}), \quad \phi_j \in W^{l-3/2,2}_{\beta,\delta}(\Gamma_j)^{d_j},$$

$l \ge 2$, the components of δ are not integer and satisfy (10.3.1) *Furthermore, we assume that the vector functions $h_j \in W^{l-1/2,2}_{\beta,\delta}(\Gamma_j)^{3-d_j}$ satisfy the compatibility condition* (10.2.6) *on the edges M_k. Then $u \in W^{l,2}_{\beta,\delta}(\mathcal{K})^3$, $p \in W^{l-1,2}_{\beta,\delta}(\mathcal{K})$, and (u,p) solves the boundary value problem* (10.1.1), (10.1.2).

P r o o f. Let ε be a real number such that $0<\varepsilon<1$ and $\delta_k - l + 2 \le 1-\varepsilon$ for all k. Then $f+\nabla g \in V^{0,2}_{\beta-l+2,1-\varepsilon}(\mathcal{K})^3$ and $\phi_j \in W^{l-3/2,2}_{\beta,l-1-\varepsilon}(\Gamma_j)^{d_j} = V^{l-3/2,2}_{\beta,l-1-\varepsilon}(\Gamma_j)^{d_j}$. Since $\mathcal{H}_{l-1-\beta} \subset V^{1,2}_{l-1-\beta,\varepsilon}(\mathcal{K})^3 \subset V^{0,2}_{l-2-\beta,\varepsilon-1}(\mathcal{K})^3$, it follows that the functional F defined by (10.3.4) is linear and continuous on $\mathcal{H}_{l-1-\beta}$.

Let $\zeta_k, \eta_k, \tilde\zeta_k, \tilde\eta_k$ be the same cut-off functions as in the proof of Theorem 7.4.1. Furthermore, let $v(x) = u(2^k x)$ and $q(x) = 2^k p(2^k x)$. By Theorem 9.4.9, the vector functions $\zeta_k(u,p)$ and $\tilde\zeta_k(v,q)$ belong to $W^{l,2}_{\beta,\delta}(\mathcal{K})^3 \times W^{l-1,2}_{\beta,\delta}(\mathcal{K})$ for every integer k.

Furthermore,

$$\|\tilde\zeta_k v\|^2_{W^{l,2}_{\beta,\delta}(\mathcal{K})^3} + \|\tilde\zeta_k q\|^2_{W^{l-1,2}_{\beta,\delta}(\mathcal{K})} \le c\Big(\|\tilde\eta_k v\|^2_{V^{1,2}_{\beta-l+1}(\mathcal{K})^3} + \|\tilde\eta_k q\|^2_{V^{0,2}_{\beta-l+1}(\mathcal{K})}$$
$$+\|\tilde\eta_k(-\Delta v+\nabla q)\|^2_{W^{l-2,2}_{\beta,\delta}(\mathcal{K})^3} + \|\tilde\eta_k \nabla\cdot v\|^2_{W^{l-1,2}_{\beta,\delta}(\mathcal{K})}$$
$$+\sum_{j=1}^d \|\tilde\eta_k S_j v\|^2_{W^{l-3/2,2}_{\beta,\delta}(\Gamma_j)^{3-d_j}} + \sum_{j=1}^d \|\tilde\eta_k N_j(v,q)\|^2_{W^{l-1/2,2}_{\beta,\delta}(\Gamma_j)^{d_j}}\Big).$$

Here the constant c is independent of k. Multiplying the last estimate by $2^{2k(\beta-l)+3k}$ and substituting $2^k x = y$, we arrive at the inequality

$$\|\zeta_k u\|^2_{W^{l,2}_{\beta,\delta}(\mathcal{K})^3} + \|\zeta_k p\|^2_{W^{l-1,2}_{\beta,\delta}(\mathcal{K})}$$
$$\le c\Big(\|\eta_k u\|^2_{V^{1,2}_{\beta-l+1}(\mathcal{K})^3} + \|\eta_k p\|^2_{V^{0,2}_{\beta-l+1}(\mathcal{K})} + \|\eta_k f\|^2_{W^{l-2,2}_{\beta,\delta}(\mathcal{K})^3} + \|\eta_k g\|^2_{W^{l-1,2}_{\beta,\delta}(\mathcal{K})}$$
$$+\sum_{j=1}^d \|\eta_k h_j\|^2_{W^{l-3/2,2}_{\beta,\delta}(\Gamma_j)^{3-d_j}} + \sum_{j=1}^d \|\eta_k \phi_j\|^2_{W^{l-1/2,2}_{\beta,\delta}(\Gamma_j)^{d_j}}\Big).$$

Adding over all k and using Lemma 7.2.1, we obtain the assertion of the theorem. □

10.3.2. An existence and uniqueness theorem. As a consequence of Theorems 10.2.11 and 10.3.1, we obtain an existence and uniqueness result in the space $W^{l,2}_{\beta,\delta}(\mathcal{K})^3 \times W^{l-1,2}_{\beta,\delta}(\mathcal{K})$.

THEOREM 10.3.2. *Let $f \in W^{l-2,2}_{\beta,\delta}(\mathcal{K})^3$, $g \in W^{l-1,2}_{\beta,\delta}(\mathcal{K})$, $h_j \in W^{l-1/2,2}_{\beta,\delta}(\Gamma_j)^{3-d_j}$ and $\phi_j \in W^{l-3/2,2}_{\beta,\delta}(\Gamma_j)^{d_j}$, $l \ge 2$. Suppose that the line $\operatorname{Re}\lambda = l-\beta-3/2$ does not contain eigenvalues of the pencil $\mathfrak{A}(\lambda)$ and that the components of δ are not integer and satisfy the inequalities* (10.3.1). *Furthermore, we assume that the boundary data h_j satisfy the compatibility condition* (10.2.6). *Then the boundary value problem* (10.1.1), (10.1.2) *has a uniquely determined solution $(u,p) \in W^{l,2}_{\beta,\delta}(\mathcal{K})^3 \times W^{l-1,2}_{\beta,\delta}(\mathcal{K})$.*

P r o o f. By the same arguments as in the proof of Theorem 10.3.1, the expression on the right-hand side of (10.3.4) defines a continuous functional on $\mathcal{H}_{l-1-\beta}$. Furthermore, $g \in V^{0,2}_{\beta-l+1}(\mathcal{K})$ and $h_j \in W^{1/2,2}_{\beta-l+1,0}(\Gamma_j)^{3-d_j}$. Consequently by Theorem 10.2.11, there exists a uniquely determined solution $(u,p) \in V^{1,2}_{\beta-l+1}(\mathcal{K})^3 \times V^{0,2}_{\beta-l+1}(\mathcal{K})$ of the problem (10.3.2), (10.3.3). From Theorem 10.3.1 it follows that $(u,p) \in W^{l,2}_{\beta,\delta}(\mathcal{K})^3 \times W^{l-1,2}_{\beta,\delta}(\mathcal{K})$. The last space is a subspace of $V^{1,2}_{\beta-l+1}(\mathcal{K})^3 \times V^{0,2}_{\beta-l+1}(\mathcal{K})$. This proves the theorem. □

REMARK 10.3.3. If $d_{k_+} + d_{k_-}$ is even, then the condition (10.3.1) in Theorems 10.3.1 and 10.3.2 can be replaced by the weaker condition

$$\max(0, l-1-\mu_+^{(k)}) < \delta_k < l-1. \tag{10.3.5}$$

However, if $\delta_k < l-2$, then one must assume in addition that the data g, h_{k_+}, h_{k_-}, ϕ_{k_+} and ϕ_{k_-} satisfy the compatibility conditions of Section 9.2.3 (cf. Lemma 9.2.6).

10.3.3. Asymptotics of solutions near the vertex of the cone. First we prove a theorem on the asymptotics of solutions in the weighted space $W^{2,2}_{\beta,\delta}(\mathcal{K})^3 \times W^{1,2}_{\beta,\delta}(\mathcal{K})$.

THEOREM 10.3.4. *Let $(u,p) \in W^{2,2}_{\beta,\delta}(\mathcal{K})^3 \times W^{1,2}_{\beta,\delta}(\mathcal{K})$ be a solution of the boundary value problem* (10.1.1), (10.1.2), *where*

$$f \in W^{0,2}_{\beta,\delta}(\mathcal{K})^3 \cap W^{0,2}_{\beta',\delta'}(\mathcal{K})^3, \quad g \in W^{1,2}_{\beta,\delta}(\mathcal{K}) \cap W^{1,2}_{\beta',\delta'}(\mathcal{K}),$$

$$h_j \in W^{3/2,2}_{\beta,\delta}(\Gamma_j)^{3-d_j} \cap W^{3/2,2}_{\beta',\delta'}(\Gamma_j)^{3-d_j}, \quad \phi_j \in W^{1/2,2}_{\beta,\delta}(\Gamma_j)^{d_j} \cap W^{1/2,2}_{\beta',\delta'}(\Gamma_j)^{d_j}.$$

Suppose that the lines $\operatorname{Re}\lambda = -\beta + 1/2$ *and* $\operatorname{Re}\lambda = -\beta' + 1/2$ *do not contain eigenvalues of the pencil $\mathfrak{A}(\lambda)$ and that the components of δ and δ' satisfy the inequalities* (10.2.1). *Then (u,p) admits the decomposition*

(10.3.6)
$$\begin{pmatrix} u \\ p \end{pmatrix} = \sum_{\nu=1}^{N} \sum_{j=1}^{I_\nu} \sum_{s=0}^{\kappa_{\nu,j}-1} c_{\nu,j,s} \sum_{\sigma=0}^{s} \frac{1}{\sigma!} (\log\rho)^\sigma \begin{pmatrix} \rho^{\lambda_\nu} u^{(\nu,j,s-\sigma)}(\omega) \\ \rho^{\lambda_\nu - 1} p^{(\nu,j,s-\sigma)}(\omega) \end{pmatrix} + \begin{pmatrix} w \\ q \end{pmatrix}$$

where $(w,q) \in W^{2,2}_{\beta',\delta'}(\mathcal{K})^3 \times W^{1,2}_{\beta',\delta'}(\mathcal{K})$ is a solution of the problem (10.1.1), (10.1.2), *λ_ν are the eigenvalues of the pencil $\mathfrak{A}(\lambda)$ between the lines* $\operatorname{Re}\lambda = -\beta + 1/2$ *and* $\operatorname{Re}\lambda = -\beta' + 1/2$, *and $\big(u^{(\nu,j,s)}, p^{(\nu,j,s)}\big)$ are eigenvectors and generalized eigenvectors corresponding to the eigenvalue λ_ν.*

P r o o f. By Lemma 10.2.2, we may assume without loss of generality that $h_j = 0$ and $\phi_j = 0$ for $j = 1, \dots, d$. Then the uniquely determined solution $(u,p) \in W^{2,2}_{\beta,\delta}(\mathcal{K})^3 \times W^{1,2}_{\beta,\delta}(\mathcal{K})$ is given by the formula

$$\big(u(\rho,\omega),\, \rho p(\rho,\omega)\big) = \frac{1}{2\pi i} \int_{-\beta+1/2-i\infty}^{-\beta+1/2+i\infty} \rho^\lambda \left(\mathfrak{A}_\delta(\lambda)\right)^{-1} \tilde{F}(\lambda,\omega)\, d\lambda,$$

where

$$\tilde{F}(\lambda,\omega) = \big(\tilde{f}(\lambda-2,\omega)\,,\, \tilde{g}(\lambda-1,\omega)\,,\, 0\big),$$

$\tilde{f}(\lambda,\omega)$ and $\tilde{g}(\lambda,\omega)$ are the Mellin transforms of f and g, respectively (see the proof Theorem 10.2.9). It follows from the assumptions on f and g that $\tilde{f}(\lambda,\cdot) \in W^{0,2}_\delta(\Omega)^3 \cap W^{0,2}_{\delta'}(\Omega)^3$ and $\tilde{g}(\lambda,\cdot) \in W^{1,2}_\delta(\Omega) \cap W^{1,2}_{\delta'}(\Omega)$. Since the components of δ and δ' satisfy the inequalities (10.2.1), we get

$$\left(\mathfrak{A}_\delta(\lambda)\right)^{-1} \tilde{F}(\lambda,\cdot) = \left(\mathfrak{A}_{\delta'}(\lambda)\right)^{-1} \tilde{F}(\lambda,\omega)$$

Furthermore, the function $\tilde{F}$ is holomorphic in the strip between the lines $\operatorname{Re}\lambda = -\beta+1/2$ and $\operatorname{Re}\lambda = -\beta'+1/2$. Hence, the only singularities of $\rho^\lambda \left(\mathfrak{A}_\delta(\lambda)\right)^{-1}\tilde{F}(\lambda,\omega)$ between the lines $\operatorname{Re}\lambda = -\beta+1/2$ and $\operatorname{Re}\lambda = -\beta'+1/2$ are the poles of $\left(\mathfrak{A}_\delta(\lambda)\right)^{-1}$, i.e. the eigenvalues $\lambda_1, \dots, \lambda_N$ of the pencil $\mathfrak{A}(\lambda)$ (see Theorem 10.1.3). Since the integral

$$\int_{-\beta+1/2}^{-\beta'+1/2} \rho^{t+iR} \left(\mathfrak{A}_\delta(t+iR)\right)^{-1} \tilde{F}(t+iR,\omega)\, dt$$

tends to zero as $R \to \pm\infty$ (cf. [**84**, Lemma 5.4.1]), Cauchy's formula yields

$$\begin{pmatrix} u \\ \rho p \end{pmatrix} = \sum_{\nu=1}^{N} \operatorname{Res} \rho^{\lambda} \left(\mathfrak{A}_\delta(\lambda)\right)^{-1} \tilde{F}(\lambda,\omega)\big|_{\lambda=\lambda_\nu} + \begin{pmatrix} w \\ \rho q \end{pmatrix},$$

where

$$\begin{pmatrix} w(\rho,\omega) \\ \rho\, q(\rho,\omega) \end{pmatrix} = \frac{1}{2\pi i} \int\limits_{-\beta'+1/2-i\infty}^{-\beta'+1/2+i\infty} \rho^{\lambda} \left(\mathfrak{A}_\delta(\lambda)\right)^{-1} \tilde{F}(\lambda,\omega)\, d\lambda,$$

i.e., (w,q) is the solution of the problem (10.1.1), (10.1.2) in the space $W^{2,2}_{\beta',\delta'}(\mathcal{K})^3 \times W^{1,2}_{\beta',\delta'}(\mathcal{K})$. Using the representation

$$\left(\mathfrak{A}_\delta(\lambda)\right)^{-1} = \sum_{j=1}^{I_\nu} \sum_{s=0}^{\kappa_{\nu,j}-1} \frac{P^{(\nu)}_{j,s}}{(\lambda-\lambda_\nu)^{\kappa_{\nu,j}-s}} + \Gamma_\nu(\lambda)$$

for the resolvent in a neighborhood of λ_ν, where Γ_ν is holomorphic in a neighborhood of λ_ν and $P^{(\nu)}_{j,s}$ are mappings into the space of eigenvectors and generalized eigenvectors (see [**81**, Theorem 1.2.1]), we obtain the desired decomposition for (u,p). □

We prove the analogous result for the variational solution $(u,p) \in V^{1,2}_\beta(\mathcal{K})^3 \times V^{0,2}_\beta(\mathcal{K})$ of the boundary value problem (10.1.1), (10.1.2).

THEOREM 10.3.5. *Let $(u,p) \in \mathcal{H}_\beta \times V^{0,2}_\beta(\mathcal{K})$ be a solution of the problem* (10.2.11), (10.2.12), *where*

$$F \in \mathcal{H}^*_{-\beta} \cap \mathcal{H}^*_{-\beta'}, \quad g \in V^{0,2}_\beta(\mathcal{K}) \cap V^{0,2}_{\beta'}(\mathcal{K})$$

and $h_j = 0$ for $j = 1,\dots,d$. If the lines $\operatorname{Re}\lambda = -\beta-1/2$ and $\operatorname{Re}\lambda = -\beta'-1/2$ are free of eigenvalues of the pencil $\mathfrak{A}(\lambda)$, then (u,p) has the representation (10.3.6), *where $(w,q) \in \mathcal{H}_{\beta'} \times V^{0,2}_{\beta'}(\mathcal{K})$ is the uniquely determined solution of the equation*

$$\mathcal{A}_{\beta'}(w,q) = (F,g),$$

(cf. Theorem 10.2.11*) and λ_ν are the eigenvalues of the pencil $\mathfrak{A}(\lambda)$ between the lines $\operatorname{Re}\lambda = -\beta-1/2$ and $\operatorname{Re}\lambda = -\beta'-1/2$.*

P r o o f. We may assume without loss of generality that $\beta' < \beta$. Let (w,q) be the uniquely determined solution of the problem (10.2.11), (10.2.12) in the space $\mathcal{H}_{\beta'} \times V^{0,2}_{\beta'}(\mathcal{K})$, and let χ be an infinitely differentiable function with compact support equal to one near the origin. Then $\chi(u-w) \in V^{1,2}_\beta(\mathcal{K})^3$ and $\chi(p-q) \in V^{0,2}_\beta(\mathcal{K})$. Since $\chi v \in V^{1,2}_{-\beta}(\mathcal{K})^3 \cap V^{1,2}_{-\beta'}(\mathcal{K})^3$ for arbitrary $v \in V^{1,2}_{-\beta}(\mathcal{K})^3$, we obtain

$$\begin{aligned} b_{\mathcal{K}}\big(\chi(u-w),v\big) &- \int_{\mathcal{K}} \chi(p-q)\, \nabla\cdot v\, dx \\ &= b_{\mathcal{K}}\big(u-w,\chi v\big) - \int_{\mathcal{K}} (p-q)\, \nabla\cdot(\chi v)\, dx + \Phi(v) = \Phi(v) \end{aligned}$$

for arbitrary $v \in \mathcal{H}_{-\beta}$, where

$$\Phi(v) = 2\int\limits_{\mathcal{K}} \sum_{i,j=1}^{3} \Big(\frac{\partial\chi}{\partial x_i}\,(u_j-w_j)\,\varepsilon_{i,j}(v) - \varepsilon_{i,j}(u-w)\,\frac{\partial\chi}{\partial x_i}\, v_j\Big)\, dx + \int\limits_{\mathcal{K}} (p-q)\,(\nabla\chi)\cdot v\, dx.$$

Integration by parts yields

$$\Phi(v) = \int_{\mathcal{K}} f \cdot v\, dx + \sum_{k=1}^{d} \int_{\Gamma_k} \phi^{(k)} \cdot v\, dx,$$

where f and $\phi^{(k)}$ are vector functions with the components

$$\begin{aligned} f_j &= -\sum_{i=1}^{3} \frac{\partial}{\partial x_i}\Big(\frac{\partial \chi}{\partial x_j}(u_i - w_i) + \frac{\partial \chi}{\partial x_i}(u_j - w_j)\Big) \\ &\quad -2\sum_{i=1}^{3} \varepsilon_{i,j}(u-w)\frac{\partial \chi}{\partial x_i} + \frac{\partial \chi}{\partial x_j}(p-q) \end{aligned}$$

and

$$\phi_j^{(k)} = \sum_{i=1}^{3}\Big(\frac{\partial \chi}{\partial x_j}(u_i - w_i) + \frac{\partial \chi}{\partial x_i}(u_j - w_j)\Big) n_i = \frac{\partial \chi}{\partial x_j}(u-w)\cdot n + \frac{\partial \chi}{\partial n}(u_j - w_j),$$

respectively. Since $v \in \mathcal{H}_{-\beta}$, the vector $\phi^{(k)}$ can be replaced by its normal component if $d_k = 1$ and its tangent component on Γ_k if $d_k = 2$. Obviously,

$$f \in W^{0,2}_{\beta+1,0}(\mathcal{K})^\ell \cap W^{0,2}_{\beta'+1,0}(\mathcal{K})^\ell \quad \text{and} \quad \phi^{(k)} \in W^{1/2,2}_{\beta+1,0}(\Gamma_j)^{d_k} \cap W^{1/2,2}_{\beta'+1,0}(\Gamma_j)^{d_k}.$$

Applying Theorem 10.3.1, we conclude that $\chi(u-w) \in W^{2,2}_{\beta+1,\delta}(\mathcal{K})^3$ and $\chi(p-q) \in W^{1,2}_{\beta+1,\delta}(\mathcal{K})$, where the components of δ are arbitrary real numbers satisfying the inequalities (10.3.1). Furthermore, it follows from Theorem 10.3.4 that

$$\chi \begin{pmatrix} u-w \\ p-q \end{pmatrix} = \Sigma + \begin{pmatrix} V \\ Q \end{pmatrix}, \tag{10.3.7}$$

where Σ is the sum in the decomposition (10.3.6) and

$$(V,Q) \in W^{2,2}_{\beta'+1,\delta}(\mathcal{K})^3 \times W^{1,2}_{\beta'+1,\delta}(\mathcal{K}) \subset V^{1,2}_{\beta'}(\mathcal{K})^3 \times V^{0,2}_{\beta'}(\mathcal{K})$$

is a solution of the boundary value problem

$$-\Delta V + \nabla Q = -\Delta(\chi u - \chi w) + \nabla(\chi p - \chi q), \quad \nabla \cdot V = \nabla \cdot (\chi u - \chi w) \ \text{in } \mathcal{K},$$

$$S_j V = 0, \quad N_j(D_x)(V,Q) = N_j(D_x)(\chi u - \chi w) \ \text{on } \Gamma_j,\ j = 1,\dots,d.$$

By (10.3.7),

$$\begin{pmatrix} u \\ p \end{pmatrix} = \Sigma + \begin{pmatrix} W \\ R \end{pmatrix}, \quad \text{where } \begin{pmatrix} W \\ R \end{pmatrix} = \begin{pmatrix} V \\ q \end{pmatrix} + \chi \begin{pmatrix} w \\ q \end{pmatrix} + (1-\chi)\begin{pmatrix} u \\ p \end{pmatrix}.$$

Obviously, $(W,R) \in V^{1,2}_{\beta'}(\mathcal{K})^3 \times V^{0,2}_{\beta'}(\mathcal{K})$ and $\mathcal{A}_{\beta'}(W,R) = (F,g)$. Consequently, $(W,R) = (w,q)$. The proof is complete. □

10.3.4. A regularity result for the ρ-derivatives of the solution. We prove the following regularity theorem for the ρ-derivatives of the solutions of the problem (10.2.11), (10.2.12) analogously to Theorem 7.4.6.

THEOREM 10.3.6. *Let $u \in V^{1,2}_{\beta-l+1}(\mathcal{K})^3 \times V^{0,2}_{\beta-l+1}(\mathcal{K})$ be a solution of the problem*

$$b_{\mathcal{K}}(u,v) - \int_{\mathcal{K}} p\nabla \cdot v\, dx = \int_{\mathcal{K}} (f + \nabla g)\cdot v\, dx \quad \textit{for all } v \in \mathcal{H}_{l-1-\beta},$$

$$-\nabla \cdot u = g \ \textit{in } \mathcal{K}, \quad S_j u = 0 \ \textit{on } \Gamma_j,\ j = 1,\dots,d,$$

where

$$(\rho\partial_\rho)^\nu f \in W^{l-2,2}_{\beta,\delta}(\mathcal{K})^3, \quad (\rho\partial_\rho)^\nu g \in W^{l-1,2}_{\beta,\delta}(\mathcal{K}) \ \ \textit{for } \nu = 0,1,\dots,N.$$

Suppose that $l \ge 2$, *the line* $\operatorname{Re}\lambda = l-\beta-3/2$ *is free of eigenvalues of the pencil* $\mathfrak{A}(\lambda)$ *and that the components of* δ *are not integer and satisfy the inequalities* (10.3.5). *In the case where* (d_{k_+}, d_{k_-}) *is one of the pairs* $(0,0)$, $(0,2)$, $(2,0)$, $(1,3)$, $(3,1)$, *we assume in addition that* $g|_{M_k} = 0$ *if* $\delta_k < l-2$. *Then* $(\rho\partial_\rho)^\nu (u,p) \in W^{l,2}_{\beta,\delta}(\mathcal{K})^3 \times W^{l-1,2}_{\beta,\delta}(\mathcal{K})$ *for* $\nu = 0,1,\dots,N$ *and*

$$\sum_{\nu=0}^{N} \Big(\|(\rho\partial_\rho)^\nu u\|_{W^{l,2}_{\beta,\delta}(\mathcal{K})^3} + \|(\rho\partial_\rho)^\nu p\|_{W^{l-1,2}_{\beta,\delta}(\mathcal{K})} \Big) \le c \sum_{\nu=0}^{N} \Big(\|(\rho\partial_\rho)^\nu f\|_{W^{l-2,2}_{\beta,\delta}(\mathcal{K})^3} + \|(\rho\partial_\rho)^\nu g\|_{W^{l-1,2}_{\beta,\delta}(\mathcal{K})^3} \Big).$$

P r o o f. For $N = 0$ we refer to Theorem 10.3.1 and Remark 10.3.3. We assume that $N > 0$. Let t be an arbitrary real number, $1/2 < t < 1$. As in the proof of Theorem 7.4.6, we define $\phi_t(x) = \dfrac{\phi(x) - \phi(tx)}{1-t}$ for an arbitrary function ϕ. Then

$$-\Delta u_t(x) + \nabla p_t(x) = f_t(x) + (1+t)\, f(tx) - t\, (\nabla p)(tx),$$
$$-\nabla \cdot u_t(x) = g_t(x) + g(tx)$$

for $x \in \mathcal{K}$. Furthermore, $S_j u_t = 0$ on Γ_j for $j = 1,\dots,d$. If $d_j = 2$, then also $N_j(u_t,p_t) = \varepsilon_{n,\tau}(u_t) = 0$ on Γ_j, while $N_j(D_x)\,(u_t(x),p_t(x)) = p(tx)$ for $x \in \Gamma_j$, $d_j = 1$, and $N_j(D_x)\,(u_t(x),p_t(x)) = p(tx)\, n$ for $x \in \Gamma_j$, $d_j = 3$. Consequently by Theorems 10.3.1 and 10.3.2,

$$\|u_t\|_{W^{l,2}_{\beta,\delta}(\mathcal{K})^3} + \|p_t\|_{W^{l-1,2}_{\beta,\delta}(\mathcal{K})} \le c \Big(\|f_t\|_{W^{l-2,2}_{\beta,\delta}(\mathcal{K})^3} + \|g_t\|_{W^{l-1,2}_{\beta,\delta}(\mathcal{K})} + \|f(t\cdot)\|_{W^{l-2,2}_{\beta,\delta}(\mathcal{K})^3} + \|g(t\cdot)\|_{W^{l-1,2}_{\beta,\delta}(\mathcal{K})} + \|p(t\cdot)\|_{W^{l-1,2}_{\beta,\delta}(\mathcal{K})} \Big)$$

with a constant c independent of u,p and t. Applying the equality (7.4.8), we obtain

$$\|f_t\|_{W^{l-2,2}_{\beta,\delta}(\mathcal{K})^3} + \|g_t\|_{W^{l-1,2}_{\beta,\delta}(\mathcal{K})} \le c\, \|\rho\partial_\rho f\|_{W^{l-2,2}_{\beta,\delta}(\mathcal{K})^3} + \|\rho\partial_\rho g\|_{W^{l-1,2}_{\beta,\delta}(\mathcal{K})} \Big)$$

with a constant c independent of t. Furthermore,

$$\|f(t\cdot)\|_{W^{l-2,2}_{\beta,\delta}(\mathcal{K})^3} + \|g(t\cdot)\|_{W^{l-1,2}_{\beta,\delta}(\mathcal{K})} \le c \Big(\|f\|_{W^{l-2,2}_{\beta,\delta}(\mathcal{K})^3} + \|g\|_{W^{l-1,2}_{\beta,\delta}(\mathcal{K})} \Big)$$

and

$$\|p(t\cdot)\|_{W^{l-1,2}_{\beta,\delta}(\mathcal{K})} \le c\, \|p\|_{W^{l-1,2}_{\beta,\delta}(\mathcal{K})} \le c \Big(\|f\|_{W^{l-2,2}_{\beta,\delta}(\mathcal{K})^3} + \|g\|_{W^{l-1,2}_{\beta,\delta}(\mathcal{K})} \Big).$$

Using the last inequalities and the fact that $(u_t,p_t) \to \rho\partial_\rho(u,p)$ as $t \to 1$, we conclude that $\rho\partial_\rho(u,p) \in W^{l,2}_{\beta,\delta}(\mathcal{K})^3 \times W^{l-1,2}_{\beta,\delta}(\mathcal{K})$. Repeating this argument, we obtain $(\rho\partial_\rho)^\nu(u,p) \in W^{l,2}_{\beta,\delta}(\mathcal{K})^3 \times W^{l-1,2}_{\beta,\delta}(\mathcal{K})$ for $\nu = 2,\dots,N$. Furthermore, the desired estimate holds. □

10.3.5. A modification of the regularity result in the case $d_{k_+}+d_{k_-}\le 3$. Let M_k be an edge of the cone $\mathcal{K}$. We denote the angle at the edge M_k by θ_k and the faces adjacent to M_k by Γ_{k_+} and Γ_{k_-}. Suppose that $d_{k_+}+d_{k_-}\le 3$. In addition, we assume that

$$(10.3.8)\quad \theta_k \not\in \{\pi, 2\pi\} \text{ if } d_{k_+}=d_{k_-}=1 \quad \text{and} \quad \theta_k \not\in \{\frac{\pi}{2}, \frac{3\pi}{2}\} \text{ if } d_{k_+}\cdot d_{k_-}=2.$$

Then the number $\lambda = 0$ does not belong to the spectrum of the pencil $A_k(\lambda)$ and it is possible to prove regularity results analogous to Theorems 10.3.1 and 10.3.6 in the class of the weighted Sobolev spaces $V^{l,s}_{\beta,\delta}(\mathcal{K})$ (cf. Theorem 9.4.10).

In the following theorem, let ψ_k, χ_k be smooth cut-off functions on $\overline{\Omega}$ equal to one near the corner $\overline{\Omega}\cap M_k$ such that $\chi_k = 1$ in a neighborhood of $\operatorname{supp}\psi_k$ and $\chi_k = 0$ in a neighborhood of the corners $\overline{\Omega}\cap M_j$, $j\ne k$. We extend ψ_k and χ_k by

$$\psi_k(x) = \psi_k(x/|x|), \quad \chi_k(x) = \chi_k(x/|x|)$$

to functions on $\mathcal{K}$.

THEOREM 10.3.7. *Suppose that $d_{k_+}+d_{k_-}\le 3$, the angle θ_k satisfies the condition* (10.3.8) *for $d_{k_+}\cdot d_{k_-}\ne 0$, and that the functions f, g, h_j, ϕ_j are such that*

$$\chi_k(\rho\partial_\rho)^\nu f \in V^{l-2,2}_{\beta,\delta}(\mathcal{K})^3, \quad \chi_k(\rho\partial_\rho)^\nu g \in V^{l-1,2}_{\beta,\delta}(\mathcal{K}),$$

$$\chi_k(\rho\partial_\rho)^\nu h_j \in V^{l-1/2,2}_{\beta,\delta}(\Gamma_j)^{3-d_j}, \quad \chi_k(\rho\partial_\rho)^\nu \phi_j \in V^{l-3/2,2}_{\beta,\delta}(\Gamma_j)^{d_j},$$

for $\nu = 0,1,\dots,N$, where $l\ge 2$ and δ_k is a noninteger number satisfying (10.3.1). *If $(u,p)\in V^{1,2}_{\beta-l+1}(\mathcal{K})^3\times V^{0,2}_{\beta-l+1}(\mathcal{K})$ is a solution of the problem* (10.3.2), (10.3.3) *with a functional $F\in \mathcal{H}^*_{l-1-\beta}$ of the form* (10.3.4), *then $\psi_k(\rho\partial_\rho)^\nu u\in V^{l,2}_{\beta,\delta}(\mathcal{K})^3$ and $\psi_k(\rho\partial_\rho)^\nu p \in V^{l-1,2}_{\beta,\delta}(\mathcal{K})$ for $\nu = 0,1,\dots,N$.*

P r o o f. Using Theorem 10.3.6, we get $\psi_k(\rho\partial_\rho)^\nu u\in W^{l,2}_{\beta,\delta}(\mathcal{K})^3$ and $\psi_k(\rho\partial_\rho)^\nu p\in W^{l-1,2}_{\beta,\delta}(\mathcal{K})$ for $\nu=1,\dots,N$. Let ζ be an arbitrary smooth function with compact support vanishing near the vertex of the cone. By Theorem 9.4.11, $\zeta\psi_k(\rho\partial_\rho)^\nu u\in V^{l,2}_{\beta,\delta}(\mathcal{K})^3$ and $\zeta\psi_k(\rho\partial_\rho)^\nu p\in V^{l-1,2}_{\beta,\delta}(\mathcal{K})$ for $\nu=1,\dots,N$. Therefore,

$$\partial_x^\alpha \psi_k(\rho\partial_\rho)^\nu u = 0 \text{ on } M_k \text{ for } |\alpha| < l-1-\delta_k\,,$$

$$\partial_x^\alpha \psi_k(\rho\partial_\rho)^\nu p = 0 \text{ on } M_k \text{ for } |\alpha| < l-2-\delta_k\,.$$

This implies $\psi_k(\rho\partial_\rho)^\nu u\in V^{l,2}_{\beta,\delta}(\mathcal{K})^3$ and $\psi_k(\rho\partial_\rho)^\nu p\in V^{l-1,2}_{\beta,\delta}(\mathcal{K})$ for $\nu=1,\dots,N$ (cf. Lemma 6.2.5). □

10.4. Green's matrix of the boundary value problem in a cone

By a *Green's matrix* for the boundary value problem (10.1.1), (10.1.2), we mean a 4×4-matrix

$$G(x,\xi) = \big(G_{i,j}(x,\xi)\big)_{i,j=1}^4$$

for which the vector functions $\vec{G}_j = (G_{1,j}, G_{2,j}, G_{3,j})^t$ and the function $G_{4,j}$ are solutions of the problem

$$(10.4.1)\quad -\Delta_x \vec{G}_j(x,\xi) + \nabla_x G_{4,j}(x,\xi) = \delta(x-\xi)\,(\delta_{1,j},\delta_{2,j},\delta_{3,j})^t \quad \text{for } x,\xi\in\mathcal{K},$$

$$(10.4.2)\quad -\nabla_x\cdot \vec{G}_j(x,\xi) = \delta_{4,j}\,\delta(x-\xi) \quad \text{for } x,\xi\in\mathcal{K},$$

$$(10.4.3)\quad S_k\vec{G}_j(x,\xi)=0,\quad N_k(\partial_x)\big(\vec{G}_j(x,\xi), G_{4,j}(x,\xi)\big)=0 \quad \text{for } x\in\Gamma_k,\ \xi\in\mathcal{K},$$

$k=1,\dots,d$, $j=1,\dots,4$. In Section 10.2 we proved the existence and uniqueness of variational solutions in the weighted space $V_\kappa^{1,2}(\mathcal{K})\times V_\kappa^{0,2}(\mathcal{K})$. Using this result, one can construct Green's matrices for different values of κ. We fix here a number κ such that the line $\operatorname{Re}\lambda = -\kappa-1/2$ does not contain eigenvalues of the pencil $\mathfrak{A}(\lambda)$ and prove estimates for the elements of the corresponding Green's matrix.

10.4.1. Existence of Green's matrix. First, we prove the existence, uniqueness and some basic properties of the Green's matrix.

THEOREM 10.4.1. *Suppose that the line* $\operatorname{Re}\lambda=-\kappa-1/2$ *is free of eigenvalues of the pencil* $\mathfrak{A}(\lambda)$. *Then the following assertions are true.*

1) *There exists a unique Green's matrix* $G(x,\xi)$ *such that the function* $x\to \zeta(|x-\xi|/r(\xi))\,G_{i,j}(x,\xi)$ *belongs to* $V_\kappa^{1,2}(\mathcal{K})$ *for* $i=1,2,3$ *and to* $V_\kappa^{0,2}(\mathcal{K})$ *for* $i=4$, *where* ζ *is an arbitrary smooth function on* $(0,\infty)$ *equal to one in* $(1,\infty)$ *and to zero in* $(0,1/2)$. *The functions* $G_{i,j}(x,\xi)$ *are infinitely differentiable with respect to* $x,\xi\in\bar{\mathcal{K}}\backslash\mathcal{S}$, $x\neq\xi$.

2) *The elements of the matrix* $G(x,\xi)$ *satisfy the equality*

$$(10.4.4)\qquad G_{i,j}(tx,t\xi) = t^{-1-\delta_{i,4}-\delta_{j,4}}\,G_{i,j}(x,\xi)$$

for $x,\xi\in\mathcal{K}$, $t>0$, $i,j=1,\dots,4$.

3) *The functions* $\xi\to\zeta(|x-\xi|/r(x))\,G_{i,j}(x,\xi)$ *belong to* $V_{-\kappa}^{1,2}(\mathcal{K})$ *for* $j=1,2,3$ *and to* $V_{-\kappa}^{0,2}(\mathcal{K})$ *for* $j=4$. *The vector functions* $\vec{H}_i=(G_{i,1},G_{i,2},G_{i,3})^t$ *and the functions* $G_{i,4}$, $i=1,2,3,4$, *are solutions of the problem*

$$-\Delta_\xi\vec{H}_i(x,\xi)+\nabla_\xi G_{i,4}(x,\xi)=\delta(x-\xi)\,(\delta_{i,1},\delta_{i,2},\delta_{i,3})^t \quad \textit{for } x,\xi\in\mathcal{K},$$

$$-\nabla_\xi\cdot\vec{H}_i(x,\xi)=\delta_{i,4}\,\delta(x-\xi)\quad \textit{for } x,\xi\in\mathcal{K},$$

$$S_k\vec{H}_i(x,\xi)=0,\quad N_k(\partial_\xi)\big(\vec{H}_i(x,\xi),G_{i,4}(x,\xi)\big)=0\quad \textit{for } x\in\mathcal{K},\ \xi\in\Gamma_k,$$

$k=1,\dots,d$. *This means that every solution* $(u,p)\in C_0^\infty(\overline{\mathcal{K}})^4$ *of the equation* (10.1.1) *satisfying the homogeneous boundary condition* (10.1.2) *is given by the formulas*

$$(10.4.5)\qquad u_i(x)=\int_{\mathcal{K}}\big(f(\xi)+\nabla_\xi g(\xi)\big)\cdot\vec{H}_i(x,\xi)\,d\xi+\int_{\mathcal{K}} g(\xi)\,G_{i,4}(x,\xi)\,d\xi,$$

$i=1,2,3$, *and*

$$(10.4.6)\quad p(x)=-g(x)+\int_{\mathcal{K}}\big(f(\xi)+\nabla_\xi g(\xi)\big)\cdot\vec{H}_4(x,\xi)\,d\xi+\int_{\mathcal{K}} g(\xi)\,G_{4,4}(x,\xi)\,d\xi.$$

P r o o f. 1) We define Green's matrix by the formula

$$G(x,\xi)=\big(1-\zeta(\frac{|x-\xi|}{r(\xi)})\big)\,\mathcal{G}(x,\xi)+R(x,\xi),$$

where $\mathcal{G}(x,\xi)$ is the Green's matrix of Stokes system in $\mathbb{R}^3$. The existence of a matrix $R(x,\xi)$, where $R_{i,j}(\cdot,\xi) \in V_\kappa^{1,2}(\mathcal{K})$, $i = 1,2,3$, and $R_{4,j}(\cdot,\xi) \in V_\kappa^{0,2}(\mathcal{K})$ are such that $G(x,\xi)$ satisfies (10.4.1)–(10.4.3) follows from Theorem 10.2.11. The smoothness of $G(x,\xi)$ for $x \neq \xi$, $x,\xi \in \bar{\mathcal{K}}\backslash\mathcal{S}$ follows from the manner of its construction.

2) The equation (10.4.4) results directly from the definition of $G(x,\xi)$.

3) If the line $\operatorname{Re}\lambda = -\kappa - 1/2$ is free of eigenvalues of the pencil $\mathfrak{A}(\lambda)$, then the same is true for the line $\operatorname{Re}\lambda = \kappa - 1/2$. By the first part of the proof, there exists a matrix

$$H(x,\xi) = \big(H_{i,j}(x,\xi)\big)_{i,j=1}^4$$

such that the vectors $\vec{H}_j = (H_{1,j}, H_{2,j}, H_{3,j})$ and the function $H_{4,j}$ solve the problem

$$-\Delta_x \vec{H}_j(\xi,x) + \nabla_x H_{4,j}(\xi,x) = \delta(x-\xi)\,(\delta_{j,1},\delta_{j,2},\delta_{j,3})^t \quad \text{for } x,\xi \in \mathcal{K},$$
$$-\nabla_x \cdot \vec{H}_j(\xi,x) = \delta_{j,4}\,\delta(x-\xi) \quad \text{for } x,\xi \in \mathcal{K},$$
$$S_k \vec{H}_j(\xi,x) = 0, \quad N_k(\partial_x)\big(\vec{H}_j(\xi,x), H_{4,j}(\xi,x)\big) = 0 \quad \text{for } x \in \Gamma_k,\ \xi \in \mathcal{K},$$

$k = 1,\dots,d$, and the functions $\xi \to \zeta(|x-\xi|/r(x))\,H_{i,j}(x,\xi)$ belong to $V_{-\kappa}^{1,2}(\mathcal{K})$ for $i = 1,2,3$ and to $V_{-\kappa}^{0,2}(\mathcal{K})$ for $i = 4$. By the same arguments as in the proof of Theorem 9.6.1, one can apply the *Green's formula*

$$\int_{\mathcal{K}} (-\Delta u - \nabla\nabla\cdot u + \nabla p)\cdot v\,dx - \int_{\mathcal{K}} (\nabla\cdot u)\,q\,dx$$
$$+ \sum_{j=1}^d \int_{\Gamma_j} \big(-pn + 2\varepsilon(u)\,n\big)\cdot v\,dx = \int_{\mathcal{K}} u\cdot(-\Delta v - \nabla\nabla\cdot v + \nabla q)\,dx$$
$$- \int_{\mathcal{K}} p\,\nabla\cdot v\,dx + \sum_{j=1}^d \int_{\Gamma_j} u\cdot\big(-qn + 2\varepsilon(v)\,n\big)\,dx$$

to the functions $u(x) = \vec{G}_i(x,y)$, $p(x) = G_{4,i}(x,y)$, $v(x) = \vec{H}_j(z,x)$ and $q(x) = H_{4,j}(z,x)$. Then we obtain $H_{i,j}(z,y) = G_{j,i}(z,y)$ for $i,j = 1,2,3,4$. This proves the third assertion. □

10.4.2. Estimates of Green's matrix. The case $|x|/2 < |\xi| < 2|x|$. First, we estimate the elements of Green's matrix and their derivatives for $|x-\xi| < \min(r(x), r(\xi))$. Obviously, the last inequality implies

$$|x|/2 < |\xi| < 2|x| \quad \text{and } r(x)/2 < r(\xi) < 2r(x).$$

THEOREM 10.4.2. *Let $G_{i,j}(x,\xi)$ be the elements of the Green's matrix introduced in Theorem* 10.4.1. *If $|x-\xi| < \min(r(x), r(\xi))$, then*

$$\big|\partial_x^\alpha \partial_\xi^\gamma G_{i,j}(x,\xi)\big| \le c\,|x-\xi|^{-1-\delta_{i,4}-\delta_{j,4}-|\alpha|-|\gamma|}. \tag{10.4.7}$$

Furthermore, the representation

$$G_{4,j}(x,\xi) = -\nabla_x \cdot \vec{P}_j(x,\xi) + Q_j(x,\xi)$$

is valid for $j = 1,\dots,4$ where $\vec{P}_j(x,\xi)\cdot n$ for $x \in \Gamma_k$, $k = 1,\dots,d$, $\xi \in \mathcal{K}$, and

$$|\partial_x^\alpha \partial_\xi^\gamma \vec{P}_j(x,\xi)| \le c_{\alpha,\gamma}\,|x-\xi|^{-1-\delta_{j,4}-|\alpha|-|\gamma|}, \tag{10.4.8}$$

$$|\partial_x^\alpha \partial_\xi^\gamma Q_j(x,\xi)| \le c_{\alpha,\gamma}\,r(\xi)^{-2-\delta_{j,4}-|\alpha|-|\gamma|} \tag{10.4.9}$$

for $|x-\xi| < \min(r(x), r(\xi))$.

P r o o f. Let M_ν be a fixed edge of the cone $\mathcal{K}$, and let $\Gamma_{\nu_+}, \Gamma_{\nu_-}$ be the faces adjoining to M_ν. By $\mathcal{D}_\nu$, we denote the dihedron whose faces are supersets of Γ_{ν_+} and Γ_{ν_-}. Furthermore, let $\mathcal{K}_\nu = \{x \in \mathcal{K} : r_\nu(x) < 2r(x)\}$.

Suppose that $\xi \in \mathcal{K}_\nu$. We write the Green's matrix $G(x,\xi)$ in the form

$$G(x,\xi) = \zeta\Big(\frac{|x-\xi|}{r(\xi)}\Big)\, G^{(\nu)}(x,\xi) + R^{(\nu)}(x,\xi), \tag{10.4.10}$$

where $G^{(\nu)}(x,\xi)$ is the Green's matrix of the problem in the dihedron $\mathcal{D}_\nu$ and ζ is an infinitely differentiable function on $[0,\infty)$ such that $\zeta(t) = 1$ for $t < \delta$ and $\zeta(t) = 0$ for $t > 2\delta$. Here δ is a sufficiently small positive number depending on the cone $\mathcal{K}$, $\delta < 1/4$. Then the function $x \to \zeta\big(|x-\xi|/r(\xi)\big)$ vanishes on the edges of the cone $\mathcal{K}$. Let $\vec{R}_j^{(\nu)}$ denote the vector with the components $R_{1,j}^{(\nu)}, R_{2,j}^{(\nu)}, R_{3,j}^{(\nu)}$. The vector functions $\partial_\xi^\gamma\big(\vec{R}_j^{(\nu)}(\cdot,\xi), R_{4,j}^{(\nu)}(\cdot,\xi)\big)$ are the unique solutions in $V_\kappa^{1,2}(\mathcal{K})^3 \times V_\kappa^{0,2}(\mathcal{K})$ of the problems

$$-\Delta_x \partial_\xi^\gamma \vec{R}_j^{(\nu)}(x,\xi) + \nabla_x R_{4,j}^{(\nu)}(x,\xi) = \vec{\Phi}_j(x,\xi) \quad \text{for } x \in \mathcal{K},$$

$$-\nabla_x \cdot \partial_\xi^\gamma \vec{R}_j^{(\nu)}(x,\xi) = \Psi_j(x,\xi) \quad \text{for } x \in \mathcal{K},$$

$$S_k \partial_\xi^\gamma \vec{R}_j^{(\nu)}(x,\xi) = 0, \;\; N_k \partial_\xi^\gamma\big(\vec{R}_j^{(\nu)}(x,\xi), R_{4,j}^{(\nu)}(x,\xi)\big) = \vec{\Upsilon}_j \quad \text{for } x \in \Gamma_k,$$

$j = 1,2,3,4$, where

$$\vec{\Phi}_j = -\Delta_x \partial_\xi^\gamma\Big(\vec{G}_j - \zeta\Big(\frac{|x-\xi|}{r(\xi)}\Big)\vec{G}_j^{(\nu)}\Big) + \nabla_x \partial_\xi^\gamma\Big(G_{4,j} - \zeta\Big(\frac{|x-\xi|}{r(\xi)}\Big)G_{4,j}^{(\nu)}\Big),$$

$$\Psi_j = -\nabla_x \cdot \partial_\xi^\gamma\Big(\vec{G}_j - \zeta\Big(\frac{|x-\xi|}{|\xi'|}\Big)\vec{G}_j^{(\nu)}\Big).$$

The functions $\vec{\Phi}_j$, Ψ_j and $\vec{\Upsilon}_j$ are infinitely differentiable with respect to x and vanish for $|x-\xi| < r(\xi)/4$ and $|x-\xi| > r(\xi)/2$. Furthermore, all derivatives $\partial_x^\alpha \vec{\Phi}_j(x,\xi)$, $\partial_x^\alpha \Psi_j(x,\xi)$ and $\partial_x^\alpha \vec{\Upsilon}_j(x,\xi)$ are bounded by constants independent of x and ξ if $\xi \in \mathcal{K}_\nu$ and $r(\xi) = 1$. Consequently, there exist constants $c_{\alpha,\gamma}$ such that

$$\big|\partial_x^\alpha \partial_\xi^\gamma R_{i,j}^{(\nu)}(x,\xi)\big| \le c_{\alpha,\gamma} \quad \text{for } \xi \in \mathcal{K}_\nu,\; r(\xi) = 1,\; |x-\xi| < \min(r(\xi), 1).$$

(Note that $1/2 < r(x) < 2$ if $r(\xi) = 1$ and $|x-\xi| < \min(r(x), 1)$.) Since the functions $R_{i,j}^{(\nu)}(x,\xi)$ (as well as $G_{i,j}(x,\xi)$ and $G_{i,j}^{(\nu)}(x,\xi)$) are positively homogeneous of degree $-1-\delta_{i,4}-\delta_{j,4}$, we conclude that

$$\big|\partial_x^\alpha \partial_\xi^\gamma R_{i,j}^{+}(x,\xi)\big| \le c_{\alpha,\gamma}\, r(\xi)^{-1-\delta_{i,4}-\delta_{j,4}-|\alpha|-|\gamma|} \tag{10.4.11}$$

for $\xi \in \mathcal{K}_\nu$, $|x-\xi| < \min(r(x), r(\xi))$. This proves the estimate (10.4.7).

By Theorem 9.6.2, the functions $G_{4,j}^{(\nu)}(x,\xi)$ admit the representation

$$G_{4,j}^{(\nu)}(x,\xi) = -\nabla_x \cdot \vec{P}_j^{(\nu)}(x,\xi) + Q_j^{(\nu)}(x,\xi) \quad \text{for } x,\xi \in \mathcal{D}_\nu,$$

where $\vec{P}_i^{(\nu)}(x,\xi) \cdot n = 0$ for $x \in \Gamma_{\nu+}$,

$$\big|\partial_x^\alpha \partial_\xi^\gamma \vec{P}_j^{(\nu)}(x,\xi)\big| \le c_{\alpha,\gamma}\, |x-\xi|^{-1-\delta_{j,4}-|\alpha|-|\gamma|}$$

and

$$\big|\partial_x^\alpha \partial_\xi^\gamma Q_j^{(\nu)}(x,\xi)\big| \le c_{\alpha,\gamma}\, |x-\xi|^{-2-\delta_{j,4}-|\alpha|-|\gamma|}.$$

Let η_ν, $\nu = 1,\dots,d$, be infinitely differentiable functions on the sphere S^2, $\eta_\nu = 1$ in a neighborhood of $M_\nu \cap S^2$. These functions can be extended by the equality $\eta_\nu(x) = \eta_\nu\big(x/|x|\big)$ to the cone $\mathcal{K}$. We assume that

$$\operatorname{supp}\eta_\nu \cap \mathcal{K} \subset \mathcal{D}_\nu, \quad \operatorname{supp}\eta_\nu \cap \partial\mathcal{K} \subset \overline{\Gamma}_{\nu_+} \cup \overline{\Gamma}_{\nu_-}, \quad \text{and} \quad \sum_{\nu=1}^d \eta_\nu = 1 \text{ in } \mathcal{K}$$

Using (10.4.10), we obtain

$$\begin{aligned} G_{4,j}(x,\xi) &= \sum_{\nu=1}^d \eta_\nu(\xi)\,\Big(\zeta\Big(\frac{|x-\xi|}{r(\xi)}\Big)\,G^{(\nu)}_{4,j}(x,\xi) + R^{(\nu)}_{4,j}(x,\xi)\Big) \\ &= \sum_{\nu=1}^d \eta_\nu(\xi)\,\Big(\zeta\Big(\frac{|x-\xi|}{r(\xi)}\Big)\,\Big(-\nabla_x\cdot \vec{P}^{(\nu)}_j(x,\xi) + Q^{(\nu)}_j(x,\xi)\Big) + R^{(\nu)}_{4,j}(x,\xi)\Big) \\ &= -\nabla_x\cdot \vec{P}^{(\nu)}_j(x,\xi) + Q^{(\nu)}_j(x,\xi), \end{aligned}$$

where

$$\begin{aligned} \vec{P}^{(\nu)}_j(x,\xi) &= \sum_{\nu=1}^d \eta_\nu(\xi)\,\zeta\Big(\frac{|x-\xi|}{r(\xi)}\Big)\,\vec{P}^{(\nu)}_j(x,\xi) \\ Q^{(\nu)}_j(x,\xi) &= \Big(\nabla_x\,\zeta\Big(\frac{|x-\xi|}{r(\xi)}\Big)\Big)\cdot \sum_{\nu=1}^d \eta_\nu(\xi)\,\vec{P}^{(\nu)}_j(x,\xi) \\ &\qquad + \sum_{\nu=1}^d \eta_\nu(\xi)\,\Big(\zeta\Big(\frac{|x-\xi|}{r(\xi)}\Big)\,Q^{(\nu)}_j(x,\xi)\Big) + R^{(\nu)}_{4,j}(x,\xi)\Big). \end{aligned}$$

Obviously, $\vec{P}^{(\nu)}_j(x,\xi)$ and $Q^{(\nu)}_j(x,\xi)$ satisfy the inequalities (10.4.8) and (10.4.9). It remains to show that $\vec{P}_j(x,\xi)\cdot n = 0$ on the faces of $\mathcal{K}$. Let $x \in \Gamma_k$ and let n be the outer unit normal to Γ_k. If $M_\nu \subset \overline{\Gamma}_k$, then $\vec{P}_j(x,\xi)\cdot n = 0$. If $M_\nu \cap \overline{\Gamma}_k = \emptyset$, then

$$\eta_\nu(\xi)\,\zeta\Big(\frac{|x-\xi|}{r(\xi)}\Big) = 0$$

for $\xi \in \mathcal{K}$. This follows from the fact that $\eta_\nu(\xi) = 0$ if $M_\nu \cap \overline{\Gamma}_k = \emptyset$ and $|x-\xi| < 2\delta r(\xi)$ with sufficiently small δ. Consequently, $\vec{P}_j(x,\xi)\cdot n = 0$. □

As an immediate consequence of the last two theorems, the following assertion concerning the elements $G_{i,4}(x,\xi)$ holds.

COROLLARY 10.4.3. *For $i = 1,\dots,4$ the representation*

$$G_{i,4}(x,\xi) = -\nabla_\xi\cdot \vec{\mathcal{P}}_i(x,\xi) + \mathcal{Q}_i(x,\xi)$$

holds, where $\vec{\mathcal{P}}_i(x,\xi)\cdot n$ for $\xi \in \Gamma_k$, $x \in \mathcal{D}$, and $\vec{\mathcal{P}}_i$ and $\mathcal{Q}_i$ satisfy the estimates

$$\begin{aligned} |\partial_x^\alpha \partial_\xi^\gamma \vec{\mathcal{P}}_i(x,\xi)| &\le c_{\alpha,\gamma}\, |x-\xi|^{-1-\delta_{i,4}-|\alpha|-|\gamma|}, \\ |\partial_x^\alpha \partial_\xi^\gamma \mathcal{Q}_i(x,\xi)| &\le c_{\alpha,\gamma}\, r(x)^{-2-\delta_{i,4}-|\alpha|-|\gamma|} \end{aligned}$$

for $|x-\xi| < \min(r(x), r(\xi))$.

Next, we consider the case where $|x|/2 < |\xi| < 2|x|$ and $|x-\xi| > \min(r(x), r(\xi))$. Let $\delta_+^{(k)}$ and $\mu_+^{(k)}$ be the same numbers as in Section 10.1. For an arbitrary point $x \in \mathcal{K}$, we denote by $k(x)$ the smallest integer k such that $r(x) = r_k(x)$. Furthermore, we set

$$\delta_x = \delta_+^{(k(x))} \quad \text{and} \quad \mu_x = \mu_+^{(k(x))}.$$

THEOREM 10.4.4. *Let $G_{i,j}(x,\xi)$ be the elements of the Green's matrix introduced in Theorem 10.4.1. If $|x|/2 < |\xi| < 2|x|$, $|x-\xi| > \min(r(x), r(\xi))$, then*

$$\big|\partial_x^\alpha \partial_\xi^\gamma G_{i,j}(x,\xi)\big| \le c\,|x-\xi|^{-1-\delta_{i,4}-\delta_{j,4}-|\alpha|-|\gamma|} \Big(\frac{r(x)}{|x-\xi|}\Big)^{\sigma_{i,\alpha}(x)} \Big(\frac{r(\xi)}{|x-\xi|}\Big)^{\sigma_{j,\gamma}(\xi)},$$

where $\sigma_{i,\alpha}(x) = \min(0, \mu_x - |\alpha| - \delta_{i,4} - \varepsilon)$. In the case where $\lambda = 0$ is not an eigenvalue of the pencil $A_{k(x)}(\lambda)$ (i.e. if $d_{k_+} + d_{k_-} \le 3$ and the edge angle θ_k satisfies the condition (10.3.8) for $d_{k_+} \cdot d_{k_-} \ne 0$), the last estimate is also valid with $\sigma_{i,\alpha}(x) = \delta_x - \delta_{i,4} - |\alpha| - \varepsilon$ for $|\alpha| < \delta_x - \delta_{i,4}$.

P r o o f. Suppose that $|x|/2 < |\xi| < 2|x|$ and $2 = |x-\xi| > \min(r(x), r(\xi))$. Then $\max(r(x), r(\xi)) < 4$, $|x| > 2/3$ and $|\xi| > 2/3$. We denote by $\mathcal{B}(x)$ and $\mathcal{B}(\xi)$ the balls with radius $1/2$ centered at x and ξ, respectively. Furthermore, let ζ and η be smooth cut-off functions with supports in $\mathcal{B}(x)$ and $\mathcal{B}(\xi)$, respectively, which are equal to one in balls with radius $1/4$ concentric to $\mathcal{B}(x)$ and $\mathcal{B}(\xi)$, respectively. Then

$$-\Delta_y \partial_x^\alpha \vec{H}_i(x,y) + \nabla_y \partial_x^\alpha G_{i,4}(x,y) = 0, \quad -\nabla_y \cdot \partial_x^\alpha \vec{H}_i(x,y) = 0 \ \text{ for } y \in \mathcal{K} \cap \mathcal{B}(\xi)$$

and

$$S_k \partial_x^\alpha \vec{H}_i(x,y) = 0, \quad N_k(\partial_y)\, \partial_x^\alpha \big(\vec{H}_i(x,y), G_{i,4}(x,y)\big) = 0$$

for $y \in \Gamma_j \cap \mathcal{B}(\xi)$, $j = 1, \dots, d$. Therefore, Lemma 9.6.3 yields

$$\begin{aligned} &r(\xi)^{-\sigma_{j,\gamma}(\xi)} \big|\partial_x^\alpha \partial_\xi^\gamma G_{i,j}(x,\xi)\big| \\ &\quad \le c \Big(\|\eta(\cdot)\, \partial_x^\alpha \vec{H}_i(x,\cdot)\|_{V_0^{1,2}(\mathcal{K})^3} + \|\eta(\cdot)\, \partial_x^\alpha G_{i,4}(x,\cdot)\|_{L_2(\mathcal{K})} \Big). \end{aligned} \tag{10.4.12}$$

Let $F \in \mathcal{H}_{-\kappa}^*$ and $g \in V_\kappa^{0,2}(\mathcal{K})$. By Theorem 10.2.11, the problem

$$b_{\mathcal{K}}(u,v) - \int_{\mathcal{K}} p \nabla \cdot v = \int_{\mathcal{K}} \eta(y)\, F(y) \cdot v(y)\, dy \quad \text{for all } v \in \mathcal{H}_{-\kappa},$$
$$-\nabla \cdot u = \eta g \ \text{ in } \mathcal{K}, \quad S_j u = 0 \ \text{ on } \Gamma_j,\ j = 1, \dots, d,$$

has a unique solution $(u,p) \in \mathcal{H}_\kappa \times V_\kappa^{0,2}(\mathcal{K})$. The components of this solution are given by the formulas

$$u_i(y) = \int_{\mathcal{K}} \eta(z) \big(F(z) \cdot \vec{H}_i(y,z) + g(z)\, G_{i,4}(y,z)\big)\, dz,\ i = 1,2,3, \tag{10.4.13}$$

$$p(y) = -\eta(y)\, g(y) + \int_{\mathcal{K}} \eta(z) \big(F(z) \cdot \vec{H}_4(y,z) + g(z)\, G_{4,4}(y,z)\big)\, dz. \tag{10.4.14}$$

Since $\eta F = 0$ and $\eta g = 0$ in $\mathcal{B}(x)$, we conclude from Lemma 9.6.3 that

$$\begin{aligned} &r(x)^{\max(0,|\alpha|-\mu_x+\varepsilon)} \big|\partial_x^\alpha u(x)\big| + r(x)^{\max(0,|\alpha|+1-\mu_x+\varepsilon)} \big|\partial_x^\alpha p(x)\big| \\ &\quad \le c \Big(\|\zeta u\|_{V_0^{1,2}(\mathcal{K})^3} + \|\zeta p\|_{L_2(\mathcal{K})} \Big) \le c\,|x|^{-\kappa} \Big(\|\zeta u\|_{V_\kappa^{1,2}(\mathcal{K})^3} + \|\zeta p\|_{V_\kappa^{0,2}(\mathcal{K})} \Big). \end{aligned}$$

This means that the functionals

$$\mathcal{H}^*_{-\kappa} \times V^{0,2}_{-\kappa}(\mathcal{K}) \ni (F,g) \to |x|^\kappa\, r(x)^{\max(0,|\alpha|-\mu_x+\varepsilon)}\, \partial_x^\alpha u_i(x)$$
$$= |x|^\kappa\, r(x)^{\max(0,|\alpha|-\mu_x+\varepsilon)} \int_{\mathcal{K}} \eta(z)\, \big(F(z)\cdot \partial_x^\alpha \vec{H}_i(x,z) + g(z)\, \partial_x^\alpha G_{i,4}(x,z)\big)\, dz,$$

$i = 1,2,3$, and

$$\mathcal{H}^*_{-\kappa} \times V^{0,2}_{-\kappa}(\mathcal{K}) \ni (F,g) \to |x|^\kappa\, r(x)^{\max(0,|\alpha|+1-\mu_x+\varepsilon)} \partial_x^\alpha p(x)$$
$$= |x|^\kappa r(x)^{\max(0,|\alpha|+1-\mu_x+\varepsilon)} \int_{\mathcal{K}} \eta(z)\, \big(F(z)\cdot \partial_x^\alpha \vec{H}_4(x,z) + g(z)\, \partial_x^\alpha G_{4,4}(x,z)\big)\, dz,$$

are continuous and their norms are bounded by constants independent of x. Consequently,

$$|x|^\kappa\, r(x)^{\max(0,|\alpha|+\delta_{i,4}-\mu_x+\varepsilon)}\, \|\eta(\cdot)\, \partial_x^\alpha\, \vec{H}_i(x,\cdot)\|_{V^{1,2}_{-\kappa}(\mathcal{K})^3} \le c.$$

and

$$|x|^\kappa\, r(x)^{\max(0,|\alpha|+\delta_{i,4}-\mu_x+\varepsilon)}\, \|\eta(\cdot)\, \partial_x^\alpha\, G_{i,4}(x,\cdot)\|_{V^{0,2}_{-\kappa}(\mathcal{K})} \le c.$$

This together with (10.4.12) proves the first part of the theorem for the case $|x-\xi| = 2$. If $|x-\xi| \neq 2$, then it suffices to apply (10.4.4). The assertion for the case $d_{k_+} + d_{k_-} \le 3$ can be proved analogously by means of Lemma 9.6.4. □

10.4.3. Estimates of Green's matrix. The cases $|x| > 2|\xi|$, $|\xi| > 2|x|$. Let κ be a fixed number such that no eigenvalues of the pencil $\mathfrak{A}(\lambda)$ lie on the line $\operatorname{Re}\lambda = -\kappa - 1/2$. Furthermore let

$$\Lambda_- < \operatorname{Re}\lambda < \Lambda_+$$

be the widest strip in the complex plane which is free of eigenvalues of this pencil and contains the line $\operatorname{Re}\lambda = -\kappa - 1/2$. Then the following estimates of the Green's matrix hold analogously to Theorem 7.5.5.

THEOREM 10.4.5. *The elements of the Green's matrix introduced in Theorem* 10.4.1 *satisfy the estimate*

$$\big|\partial_x^\alpha \partial_\xi^\gamma G_{i,j}(x,\xi)\big| \le c\, \frac{|x|^{\Lambda_- -\delta_{i,4}-|\alpha|+\varepsilon}}{|\xi|^{\Lambda_- +1+\delta_{j,4}+|\gamma|+\varepsilon}} \prod_{k=1}^d \Big(\frac{r_k(x)}{|x|}\Big)^{\sigma_{k,i,\alpha}} \prod_{k=1}^d \Big(\frac{r_k(\xi)}{|\xi|}\Big)^{\sigma_{k,j,\gamma}}$$

for $|x| > 2|\xi|$, where ε is an arbitrarily small positive number and

$$\sigma_{k,i,\alpha} = \min(0, \mu_+^{(k)} - |\alpha| - \delta_{i,4} - \varepsilon).$$

In the case where $\lambda = 0$ is not an eigenvalue of the pencil $A_k(\lambda)$ (i.e. if $d_{k_+} + d_{k_-} \le 3$ and the edge angle θ_k satisfies the condition (10.3.8) *for $d_{k_+}\cdot d_{k_-} \neq 0$), the above estimate is even valid with $\sigma_{k,i,\alpha} = \delta_+^{(k)} - |\alpha| - \delta_{i,4} - \varepsilon$ for $|\alpha| < \delta_+^{(k)} - \delta_{i,4}$. Analogously,*

$$\big|\partial_x^\alpha \partial_\xi^\gamma G_{i,j}(x,\xi)\big| \le c\, \frac{|x|^{\Lambda_+ -\delta_{i,4}-|\alpha|-\varepsilon}}{|\xi|^{\Lambda_+ +1+\delta_{j,4}+|\gamma|-\varepsilon}} \prod_{k=1}^d \Big(\frac{r_k(x)}{|x|}\Big)^{\sigma_{k,i,\alpha}} \prod_{k=1}^d \Big(\frac{r_k(\xi)}{|\xi|}\Big)^{\sigma_{k,j,\gamma}}$$

for $|\xi| > 2|x|$.

P r o o f. Suppose that $|x| = 1$. We denote by ζ and η smooth functions on $\overline{\mathcal{K}}$ such that $\zeta(\xi) = 1$ for $|\xi| < 1/2$, $\eta = 1$ in a neighborhood of $\operatorname{supp}\zeta$, and $\eta(\xi) = 0$ for $|\xi| > 3/4$. Furthermore, let l be an integer, $l > \max \mu_+^{(k)} + 1$, $l \ge 3$. By Theorem 10.4.1,

$$\eta(\xi)\big(- \Delta_\xi \partial_x^\alpha \vec{H}_i(x,\xi) + \nabla_\xi \partial_x^\alpha G_{i,4}(x,\xi)\big) = 0, \quad \eta(\xi)\, \nabla_\xi \cdot \vec{H}_i(x,\xi) = 0$$

for $\xi \in \mathcal{K}$, $i = 1,2,3,4$, and

$$\eta(\xi) S_k \partial_x^\alpha \vec{H}_i(x,\xi) = 0, \quad \eta(\xi) N_k(\partial_\xi)\, \big(\partial_x^\alpha \vec{H}_i(x,\xi), \partial_x^\alpha G_{i,4}(x,\xi)\big) = 0$$

for $\xi \in \Gamma_k$, $k = 1,\dots,d$, $i = 1,2,3,4$. Since

$$\eta(\cdot)\partial_x^\alpha \vec{H}_i \in V_{-\kappa}^{1,2}(\mathcal{K})^3 \quad \text{and} \quad \eta(\cdot)\partial_x^\alpha G_{i,4}(x,\cdot) \in V_{-\kappa}^{0,2}(\mathcal{K}),$$

we conclude from Theorems 10.3.5 and 10.3.6 that

$$\zeta(\cdot)\, (|\xi|\partial_{|\xi|})^j \partial_\xi^\gamma \partial_x^\alpha \vec{H}_i(x,\cdot) \in W_{\beta+|\gamma|,\delta+|\gamma|}^{l,2}(\mathcal{K})^3,$$
$$\zeta(\cdot)\, (|\xi|\partial_{|\xi|})^j \partial_\xi^\gamma \partial_x^\alpha \vec{G}_{i,4}(x,\cdot) \in W_{\beta+|\gamma|,\delta+|\gamma|}^{l-1,2}(\mathcal{K})$$

for $j = 0,1,\dots$, where $\beta = l + \Lambda_- + \varepsilon - 1/2$, $\delta_k = l - 1 - \mu_+^{(k)} + \varepsilon$. If $\lambda = 0$ is not an eigenvalue of the pencil $A_k(\lambda)$, then also

$$\zeta(\cdot)\, \psi_k(\cdot)\, (|\xi|\partial_{|\xi|})^j \partial_\xi^\gamma \partial_x^\alpha \vec{H}_i(x,\cdot) \in V_{\beta+|\gamma|,\delta'+|\gamma|}^{l,2}(\mathcal{K})^3,$$
$$\zeta(\cdot)\, \psi_k(\cdot)\, (|\xi|\partial_{|\xi|})^j \partial_\xi^\gamma \partial_x^\alpha \vec{G}_{i,4}(x,\cdot) \in V_{\beta+|\gamma|,\delta'+|\gamma|}^{l-1,2}(\mathcal{K}),$$

where $\delta_k' = l - 1 - \delta_+^{(k)} + \varepsilon$ (here ψ_k is the same cut-off function as in Theorem 10.3.7). Using Lemma 7.5.4, we obtain

$$(10.4.15) \qquad |\xi|^{\Lambda_- + 1 + \delta_{j,4} + |\gamma| + \varepsilon} \prod_{k=1}^d \Big(\frac{r_k(\xi)}{|\xi|}\Big)^{-\sigma_{k,j,\gamma}} \big|\partial_x^\alpha \partial_\xi^\gamma G_{i,j}(x,\xi)\big|$$
$$\le c\, \Big(\|\eta(\cdot)\partial_x^\alpha \vec{H}_i(x,\cdot)\|_{V_{-\kappa}^{1,2}(\mathcal{K})^3} + \|\eta(\cdot)\partial_x^\alpha G_{i,4}(x,\cdot)\|_{V_{-\kappa}^{0,2}(\mathcal{K})} \Big)$$

for $i,j = 1,2,3,4$, and $|\xi| < 1/2$, where c is independent of x and ξ. By Theorem 10.2.11, the problem

$$b_{\mathcal{K}}(u,v) - \int_{\mathcal{K}} p\, \nabla\cdot v = \int_{\mathcal{K}} \eta(y)\, F(y)\cdot v(y)\, dy \quad \text{for all } v \in \mathcal{H}_{-\kappa}\,,$$
$$-\nabla\cdot u = \eta g \ \text{ in } \mathcal{K}, \quad S_j u = 0 \ \text{ on } \Gamma_j,\ j = 1,\dots,d,$$

has a unique solution $(u,p) \in \mathcal{H}_\kappa \times V_\kappa^{0,2}(\mathcal{K})$ whose components are given by the formulas (10.4.13) and (10.4.14). Let χ_1 and χ_2 be smooth cut-off function, $\chi_2 = 1$ in a neighborhood of x, $\chi_1 = 1$ in a neighborhood of χ_2, $\chi_1(y) = 0$ for $|x-y| > 1/4$. Since χ_1 and η have disjunct supports, we have

$$\chi_1(-\Delta u + \nabla p) = 0,\ \chi_1 \nabla\cdot u = 0,\ \chi_1 S_k u = 0,\ \chi_1 N_k(u,p) = 0.$$

Consequently, it follows from Theorems 10.3.5 and 10.3.6 that

$$\chi_2 (\rho\partial_\rho)^j \partial_x^\alpha u \in W_{\beta',\delta+|\alpha|}^{l,2}(\mathcal{K})^3, \quad \chi_2(\rho\partial_\rho)^j \partial_x^\alpha p \in W_{\beta',\delta+|\alpha|}^{l-1,2}(\mathcal{K})$$

for arbitrary integer j and for arbitrary β'. In the case where $\lambda = 0$ is not an eigenvalue of the pencil $A_k(\lambda)$, Theorem 10.3.7 implies

$$\chi_2\psi_k(\rho\partial_\rho)^j \partial_x^\alpha u \in V_{\beta',\delta'+|\alpha|}^{l,2}(\mathcal{K})^3, \quad \chi_2\psi_k(\rho\partial_\rho)^j \partial_x^\alpha p \in V_{\beta',\delta'+|\alpha|}^{l-1,2}(\mathcal{K}).$$

Thus by Lemma 7.5.4,

$$\prod_{k=1}^{d} r_k(x)^{-\sigma_{k,1,\alpha}} \, |\partial_x^\alpha u(x)| + \prod_{k=1}^{d} r_k(x)^{-\sigma_{k,4,\alpha}} \, |\partial_x^\alpha p(x)| \le c \left(\|F\|_{\mathcal{H}^*_{-\kappa}} + \|g\|_{V_\kappa^{0,2}(\mathcal{K})} \right).$$

This means that the functionals

$$\mathcal{H}^*_{-\kappa} \times V^{0,2}_{-\kappa}(\mathcal{K}) \ni (F,g) \to \prod_{k=1}^{d} r_k(x)^{-\sigma_{k,i,\alpha}} \partial_x^\alpha u_i(x)$$
$$= \prod_{k=1}^{d} r_k(x)^{-\sigma_{k,i,\alpha}} \int_{\mathcal{K}} \eta(z) \left(F(z) \cdot \partial_x^\alpha \vec{H}_i(x,z) + g(z) \, \partial_x^\alpha G_{i,4}(x,z) \right) dz,$$

$i = 1, 2, 3$, and

$$\mathcal{H}^*_{-\kappa} \times V^{0,2}_{-\kappa}(\mathcal{K}) \ni (F,g) \to \prod_{k=1}^{d} r_k(x)^{-\sigma_{k,4,\alpha}} \partial_x^\alpha p(x)$$
$$= \prod_{k=1}^{d} r_k(x)^{-\sigma_{k,4,\alpha}} \int_{\mathcal{K}} \eta(z) \left(F(z) \cdot \partial_x^\alpha \vec{H}_4(x,z) + g(z) \, \partial_x^\alpha G_{4,4}(x,z) \right) dz,$$

are continuous and their norms are bounded by constants independent of x. Consequently,

$$\|\eta(\cdot) \partial_x^\alpha \vec{H}_i(x,\cdot)\|_{\mathcal{H}_{-\kappa}} + \|\eta(\cdot) \partial_x^\alpha G_{i,4}(x,\cdot)\|_{V^{0,2}_{-\kappa}(\mathcal{K})} \le c \prod_{k=1}^{d} r_k(x)^{\sigma_{k,i,\alpha}}$$

for $i, 1, 2, 3$ and

$$\|\eta(\cdot) \partial_x^\alpha \vec{H}_4(x,\cdot)\|_{\mathcal{H}_{-\kappa}} + \|\eta(\cdot) \partial_x^\alpha G_{4,4}(x,\cdot)\|_{V^{0,2}_{-\beta}(\mathcal{K})} \le c \prod_{k=1}^{d} r_k(x)^{\sigma_{k,4,\alpha}}.$$

Combining the last two inequalities with (10.4.15), we obtain the assertion of the theorem for $|x| = 1$, $|\xi| < 1/2$. Since $G_{i,j}(x,\xi)$ is positively homogeneous of degree $-1 - \delta_{i,4} - \delta_{j,4}$, the estimate holds also for arbitrary x, ξ, $|\xi| < |x|/2$. The proof for the case $|\xi| > 2|x|$ proceeds analogously. □

REMARK 10.4.6. The estimates in Theorems 10.4.4, 10.4.5 can be improved for the ρ-derivatives by means of Theorems 9.6.5 and 10.3.6. In particular,

$$|\partial_\rho G_{i,j}(x,\xi)| \le c |x-\xi|^{-2-\delta_{i,4}-\delta_{j,4}} \Big(\frac{r(x)}{|x-\xi|} \Big)^{\min(0,\mu_x - \delta_{i,4} - \varepsilon)} \Big(\frac{r(\xi)}{|x-\xi|} \Big)^{\min(0,\mu_\xi - \delta_{j,4} - \varepsilon)}$$

for $|x|/2 < |\xi| < 2|x|$, $|x-\xi| > \min(r(x), r(\xi))$,

$$|\partial_\rho G_{i,j}(x,\xi)| \le c \, \frac{|x|^{\Lambda_- - 1 - \delta_{i,4} + \varepsilon}}{|\xi|^{\Lambda_- + 1 + \delta_{j,4} + \varepsilon}} \prod_{k=1}^{d} \Big(\frac{r_k(x)}{|x|} \Big)^{\sigma_{k,i,0}} \prod_{k=1}^{d} \Big(\frac{r_k(\xi)}{|\xi|} \Big)^{\sigma_{k,j,0}}$$

for $|x| > 2|\xi|$, and

$$|\partial_\rho G_{i,j}(x,\xi)| \le c \, \frac{|x|^{\Lambda_+ - 1 - \delta_{i,4} - \varepsilon}}{|\xi|^{\Lambda_+ + 1 + \delta_{j,4} - \varepsilon}} \prod_{k=1}^{d} \Big(\frac{r_k(x)}{|x|} \Big)^{\sigma_{k,i,0}} \prod_{k=1}^{d} \Big(\frac{r_k(\xi)}{|\xi|} \Big)^{\sigma_{k,j,0}}$$

for $|\xi| > 2|x|$, where $\sigma_{k,i,0} = \min(0, \mu_+^{(k)} - \delta_{i,4} - \varepsilon)$.

10.5. Solvability of the boundary value problem in weighted L_p Sobolev spaces

In this section, we prove the existence and uniqueness of solutions of the boundary value problem (10.1.1), (10.1.2) in the space $W^{2,s}_{\beta,\delta}(\mathcal{K})^3 \times W^{1,s}_{\beta,\delta}(\mathcal{K})$ with arbitrary real s, $1 < s < \infty$. By Lemma 10.2.3, we can assume without loss of generality, that the boundary data are zero and the right-hand sides f, g are from the spaces $V^{0,s}_{\beta,\delta}(\mathcal{K})^3$ and $V^{1,s}_{\beta,\delta}(\mathcal{K})$, respectively. Then the components of the solution (u,p) are given by the formulas (10.4.5) and (10.4.6). Using the estimates of Green's matrix obtained in the preceding section, we can show that $u \in W^{2,s}_{\beta,\delta}(\mathcal{K})^3$ and $p \in W^{1,s}_{\beta,\delta}(\mathcal{K})$ if the line $\mathrm{Re}\,\lambda = 2 - \beta - 3/s$ is free of eigenvalues of the operator pencil $\mathfrak{A}(\lambda)$ and the components δ_k satisfy the inequalities

$$\max(0, 2 - \mu_+^{(k)}) < \delta_k + 2/s < 2 \quad \text{for } k = 1, \dots, d. \tag{10.5.1}$$

10.5.1. Auxiliary estimates for the solution.

Let $f \in V^{0,s}_{\beta,\delta}(\mathcal{K})^3$ and $g \in V^{1,s}_{\beta,\delta}(\mathcal{K})$ be given functions, and let the boundary data h_j and ϕ_j be zero. We suppose that the line $\mathrm{Re}\,\lambda = 2 - \beta - 3/s$ does not contain eigenvalues of the pencil $\mathfrak{A}(\lambda)$ and that the components of δ satisfy the inequalities (10.5.1). By

$$\Lambda_- < \mathrm{Re}\,\lambda < \Lambda_+$$

we denote the widest strip in the complex plane which contains the line $\mathrm{Re}\,\lambda = 2 - \beta - 3/s$ and which is free of eigenvalues of the pencil $\mathfrak{A}(\lambda)$. Furthermore, let $G(x,\xi)$ be the Green's matrix introduced in Theorem 10.4.1 with an arbitrary κ in the interval

$$-\Lambda_+ - 1/2 < \kappa < -\Lambda_- - 1/2\,.$$

Then the vector function (u,p) with the components (10.4.5), (10.4.6) is a solution of the boundary value problem (10.1.1), (10.1.2). Our goal is to show that the formulas (10.4.5) and (10.4.6) define a continuous mapping

$$W^{0,s}_{\beta,\delta}(\mathcal{K})^3 \times V^{1,s}_{\beta,\delta}(\mathcal{K}) \ni (f,g) \to (u,p) \in W^{2,s}_{\beta,\delta}(\mathcal{K})^3 \times W^{1,s}_{\beta,\delta}(\mathcal{K}).$$

To this end, we consider the vector function (v,q) with the components

$$\begin{aligned} v_i(x) = &\int_{\mathcal{K}} \Big(\zeta_\nu(\xi)\, f(\xi) + \nabla_\xi\big(\zeta_\nu(\xi)\, g(\xi)\big)\Big) \cdot \vec{H}_i(x,\xi)\, d\xi \\ &+ \int_{\mathcal{K}} \zeta_\nu(\xi)\, g(\xi)\, G_{i,4}(x,\xi)\, d\xi, \end{aligned} \tag{10.5.2}$$

$i = 1, 2, 3$, and

$$\begin{aligned} q(x) = -\zeta_\nu(x)\, g(x) + &\int_{\mathcal{K}} \Big(\zeta_\nu(\xi)\, f(\xi) + \nabla_\xi\big(\zeta_\nu(\xi)\, g(\xi)\big)\Big) \cdot \vec{H}_4(x,\xi)\, d\xi \\ &+ \int_{\mathcal{K}} \zeta_\nu(\xi)\, g(\xi)\, G_{4,4}(x,\xi)\, d\xi, \end{aligned} \tag{10.5.3}$$

where ζ_ν are infinitely differentiable functions depending only on $\rho = |x|$ and satisfying the conditions (10.2.4).

LEMMA 10.5.1. *Suppose that the line $\mathrm{Re}\,\lambda = 2 - \beta - 3/s$ does not contain eigenvalues of the pencil $\mathfrak{A}(\lambda)$ and that the components of δ satisfy the inequalities*

(10.5.1). *Then the vector function* (v, q) *with the components* (10.5.2), (10.5.3) *satisfies the estimate*

$$\|\zeta_\mu v\|_{W^{2,s}_{\beta,\delta}(\mathcal{K})^3} + \|\zeta_\mu q\|_{W^{1,s}_{\beta,\delta}(\mathcal{K})} \le c\, 2^{-|\mu-\nu|\varepsilon_0} \Big(\|\zeta_\nu f\|_{W^{0,s}_{\beta,\delta}(\mathcal{K})^3} + \|\zeta_\nu g\|_{V^{1,s}_{\beta,\delta}(\mathcal{K})} \Big)$$

for arbitrary $f \in W^{0,s}_{\beta,\delta}(\mathcal{K})^3$, $g \in V^{1,s}_{\beta,\delta}(\mathcal{K})$, $|\mu-\nu| > 2$. *Here* c *and* ε_0 *are positive constants independent of* f, g, μ, ν.

P r o o f. Let $\chi_\mu(x) = 1$ for $2^{\mu-1} \le |x| \le 2^{\mu+1}$ and $\chi_\mu(x) = 0$ else. Then

$$\|\zeta_\mu v\|_{W^{2,s}_{\beta,\delta}(\mathcal{K})^3} \le c \sum_{|\alpha|\le 2} \|\chi_\mu \partial_x^\alpha v\|_{V^{0,s}_{\beta-2+|\alpha|,\delta}(\mathcal{K})^3}$$

and

$$\|\zeta_\mu q\|_{W^{1,s}_{\beta,\delta}(\mathcal{K})} \le c \Big(\|\chi_\mu q\|_{V^{0,s}_{\beta-1,\delta}(\mathcal{K})} + \|\chi_\mu \nabla q\|_{V^{0,s}_{\beta,\delta}(\mathcal{K})^3} \Big)$$

with a constant c independent of v, q and μ. Suppose first that $\mu > \nu + 2$. Then by Theorem 10.4.5,

$$\big|\partial_x^\alpha \vec{H}_i(x,\xi)\big| \le c \frac{|x|^{\Lambda_- - \delta_{i,4} - |\alpha| + \varepsilon}}{|\xi|^{\Lambda_- + 1 + \varepsilon}} \prod_{k=1}^d \Big(\frac{r_k(x)}{|x|}\Big)^{\sigma_{k,i,\alpha}}$$

and

$$\big|\partial_x^\alpha G_{i,4}(x,\xi)\big| \le c \frac{|x|^{\Lambda_- - \delta_{i,4} - |\alpha| + \varepsilon}}{|\xi|^{\Lambda_- + 2 + \varepsilon}} \prod_{k=1}^d \Big(\frac{r_k(x)}{|x|}\Big)^{\sigma_{k,i,\alpha}} \prod_{k=1}^d \Big(\frac{r_k(\xi)}{|\xi|}\Big)^{\min(0,\mu_+^{(k)}-1-\varepsilon)}$$

for $2^{\mu-1} \le |x| \le 2^{\mu+1}$, $2^{\nu-1} \le |\xi| \le 2^{\nu+1}$, where ε is an arbitrarily small positive number and $\sigma_{k,i,\alpha} = \min(0, \mu_+^{(k)} - |\alpha| - \delta_{i,4} - \varepsilon)$. Applying Lemma 3.5.1, we obtain

$$\begin{aligned} &\|\chi_\mu \partial_x^\alpha v\|_{V^{0,s}_{\beta-2+|\alpha|,\delta}(\mathcal{K})^3} \\ &\le c\, 2^{(\mu-\nu)(\beta-2+\Lambda_-+\varepsilon+3/s)} \Big(\|\zeta_\nu f + \nabla(\zeta_\nu g)\|_{V^{0,s}_{\beta,\delta}(\mathcal{K})^3} + \|\zeta_\nu g\|_{V^{0,s}_{\beta-1,\delta-1}(\mathcal{K})} \Big) \end{aligned}$$

for $|\alpha| \le 2$ and

$$\begin{aligned} &\|\chi_\mu q\|_{V^{0,s}_{\beta-1,\delta}(\mathcal{K})} + \|\chi_\mu \nabla q\|_{V^{0,s}_{\beta,\delta}(\mathcal{K})} \\ &\le c\, 2^{(\mu-\nu)(\beta-2+\Lambda_-+\varepsilon+3/s)} \Big(\|\zeta_\nu f + \nabla(\zeta_\nu g)\|_{V^{0,s}_{\beta,\delta}(\mathcal{K})^3} + \|\zeta_\nu g\|_{V^{0,s}_{\beta-1,\delta-1}(\mathcal{K})} \Big). \end{aligned}$$

Here the constant ε can be chosen such that $\beta - 2 + \Lambda_- + \varepsilon + 3/s < 0$. This proves the lemma in the case $\mu > \nu + 2$. Analogously, the assertion of the lemma holds in the case $\mu < \nu - 2$. □

Next, we prove the estimate in Lemma 10.5.1 for the case $|\mu + \nu| \le 2$. To this end, we introduce the functions

$$\chi^+(x,\xi) = \chi\Big(\frac{|x-\xi|}{r(x)}\Big), \qquad \chi^-(x,\xi) = 1 - \chi^+(x,\xi). \tag{10.5.4}$$

Here χ is a smooth function on $[0,\infty)$ such that $\chi(t) = 1$ for $0 \le t \le 1/2$ and $\chi(t) = 0$ for $t \ge 3/4$. Instead of v_i and q we consider the functions

$$\begin{aligned} v_i^\pm(x) = &\int_{\mathcal{K}} \chi^\pm(x,\xi) \Big(\zeta_\nu(\xi)\, f(\xi) + \nabla_\xi\big(\zeta_\nu(\xi)\, g(\xi)\big) \Big) \cdot \vec{H}_i(x,\xi)\, d\xi \\ &+ \int_{\mathcal{K}} \chi^\pm(x,\xi)\, \zeta_\nu(\xi)\, g(\xi)\, G_{i,4}(x,\xi)\, d\xi, \end{aligned} \tag{10.5.5}$$

$i = 1, 2, 3$, and

$$q^\pm(x) = \int_{\mathcal{K}} \chi^\pm(x,\xi)\Big(\zeta_\nu(\xi)\, f(\xi) + \nabla_\xi\big(\zeta_\nu(\xi)\, g(\xi)\big)\Big)\cdot \vec{H}_4(x,\xi)\, d\xi \\ - \zeta_\nu(x)\, g(x)/2 + \int_{\mathcal{K}} \chi^\pm(x,\xi)\, \zeta_\nu(\xi)\, g(\xi)\, G_{4,4}(x,\xi)\, d\xi. \tag{10.5.6}$$

Then $v = v^+ + v^-$ and $q = q^+ + q^-$.

LEMMA 10.5.2. *Let v^- and q^- be given by* (10.5.5) *and* (10.5.6), *respectively. If the components of δ satisfy the inequalities* (10.5.1), *then*

$$\|\zeta_\mu v^-\|_{W^{2,s}_{\beta,\delta}(\mathcal{K})^3} + \|\zeta_\mu q^-\|_{W^{1,s}_{\beta,\delta}(\mathcal{K})} \le c\Big(\|\zeta_\nu f\|_{W^{0,s}_{\beta,\delta}(\mathcal{K})^3} + \|\zeta_\nu g\|_{V^{1,s}_{\beta,\delta}(\mathcal{K})}\Big)$$

for arbitrary $f \in W^{0,s}_{\beta,\delta}(\mathcal{K})^3$, $g \in V^{1,s}_{\beta,\delta}(\mathcal{K})$, $|\mu - \nu| \le 2$. Here c is a constant independent of f, g, μ, ν.

P r o o f. We set $\Phi_\nu = \zeta_\nu f + \nabla(\zeta_\nu g)$. Furthermore, let χ_μ be the same function as in the proof of Lemma 10.5.1. Then

$$\|\zeta_\mu v^-\|_{W^{2,s}_{\beta,\delta}(\mathcal{K})^3} \le c \sum_{|\alpha|\le 2} \|\chi_\mu \partial_x^\alpha v^-\|_{V^{0,s}_{\beta-2+|\alpha|,\delta}(\mathcal{K})^3}$$

and

$$\|\zeta_\mu q^-\|_{W^{1,s}_{\beta,\delta}(\mathcal{K})} \le c\Big(\|\chi_\mu q^-\|_{V^{0,s}_{\beta-1,\delta}(\mathcal{K})} + \|\chi_\mu \nabla q^-\|_{V^{0,s}_{\beta,\delta}(\mathcal{K})^3}\Big).$$

Thus, it suffices to show that

$$\|\chi_\mu \partial_x^\alpha v^-\|_{V^{0,s}_{\beta-2+|\alpha|,\delta}(\mathcal{K})^3} \le c\Big(\|\Phi_\nu\|_{V^{0,s}_{\beta,\delta}(\mathcal{K})^3} + \|\zeta_\nu g\|_{V^{0,s}_{\beta-1,\delta-1}(\mathcal{K})}\Big) \tag{10.5.7}$$

for $|\alpha| \le 2$ and

$$\|\chi_\mu \partial_x^\alpha q^-\|_{V^{0,s}_{\beta-1+|\alpha|,\delta}(\mathcal{K})^3} \le c\Big(\|\Phi_\nu\|_{V^{0,s}_{\beta,\delta}(\mathcal{K})^3} + \|\zeta_\nu g\|_{V^{0,s}_{\beta-1,\delta-1}(\mathcal{K})}\Big) \tag{10.5.8}$$

for $|\alpha| \le 1$. By Theorem 10.4.4,

$$\big|\partial_x^\alpha\big(\chi^-(x,\xi)\, G_{i,j}(x,\xi)\big)\big| \le c\, |x-\xi|^{-T-|\alpha|} \Big(\frac{r(x)}{|x-\xi|}\Big)^{\sigma_{i,\alpha}(x)} \Big(\frac{r(\xi)}{|x-\xi|}\Big)^{\sigma_{j,0}(\xi)},$$

where $T = 1 + \delta_{i,4} + \delta_{j,4}$, $\sigma_{i,\alpha}(x) = \min(0, \mu_x - |\alpha| - \delta_{i,4} - \varepsilon)$, and $\sigma_{j,0}(\xi) = \min(0, \mu_\xi - \delta_{j,4} - \varepsilon)$. Applying Lemma 3.5.3, we obtain (10.5.7) for $1 \le |\alpha| \le 2$ and (10.5.8) for $|\alpha| \le 1$. In the case $|\alpha| = 0$, the estimate (10.5.7) can be proved in the same way as the inequality (7.6.5) in the proof of Lemma 7.6.1. □

We prove the same estimate for the vector function (v^+, q^+).

LEMMA 10.5.3. *Let (v^+, q^+) be defined by* (10.5.5), (10.5.6). *Then*

$$\|\zeta_\mu v^+\|_{W^{2,s}_{\beta,\delta}(\mathcal{K})^3} + \|\zeta_\mu q^+\|_{W^{1,s}_{\beta,\delta}(\mathcal{K})} \le c\Big(\|\zeta_\nu f\|_{W^{0,s}_{\beta,\delta}(\mathcal{K})^3} + \|\zeta_\nu g\|_{V^{1,s}_{\beta,\delta}(\mathcal{K})}\Big)$$

for arbitrary $f \in W^{0,s}_{\beta,\delta}(\mathcal{K})^3$, $g \in V^{1,s}_{\beta,\delta}(\mathcal{K})$, $|\mu - \nu| \le 2$, where c is a constant independent of f, g, μ, ν.

P r o o f. We show first that $v^+ \in V^{1,s}_{\beta-1,\delta-1}(\mathcal{K})^3$, $q^+ \in V^{0,s}_{\beta-1,\delta-1}(\mathcal{K})$ and

$$\|v^+\|_{V^{1,s}_{\beta-1,\delta-1}(\mathcal{K})^3} + \|q^+\|_{V^{0,s}_{\beta-1,\delta-1}(\mathcal{K})} \le c\Big(\|\Phi_\nu\|_{V^{0,s}_{\beta,\delta}(\mathcal{K})} + \|\zeta_\nu g\|_{V^{1,s}_{\beta,\delta}(\mathcal{K})}\Big), \tag{10.5.9}$$

where $\Phi_\nu = \zeta_\nu f + \nabla(\zeta_\nu g)$. By the representation

$$G_{i,4}(x,\xi) = -\nabla_\xi \cdot \vec{\mathcal{P}}_i(x,\xi) + \mathcal{Q}_i(x,\xi)$$

(cf. Corollary 10.4.3), the components of the vector function v^+ have the representation

$$\begin{aligned} v_i^+(x) &= \int_{\mathcal{K}} \chi^+(x,\xi)\, \vec{H}_i(x,\xi)\cdot\Phi_\nu(\xi)\, d\xi + \int_{\mathcal{K}} \chi^+(x,\xi)\, G_{i,4}(x,\xi)\, \zeta_\nu(\xi)\, g(\xi)\, d\xi \\ &= \int_{\mathcal{K}} \chi^+(x,\xi)\, \vec{H}_i(x,\xi)\cdot\Phi_\nu(\xi)\, d\xi + \int_{\mathcal{K}} \chi^+(x,\xi)\, \vec{\mathcal{P}}_i(x,\xi)\cdot\nabla_\xi\big(\zeta_\nu(\xi)\, g(\xi)\big)\, d\xi \\ &\quad + \int_{\mathcal{K}} \Big(\chi^+(x,\xi)\, \mathcal{Q}_i(x,\xi) + \vec{\mathcal{P}}_i(x,\xi)\cdot\nabla_\xi\chi^+(x,\xi)\Big)\, \zeta_\nu(\xi)\, g(\xi)\, d\xi, \end{aligned}$$

where

$$\big|\partial_x^\alpha\big(\chi^+(x,\xi)\, \vec{H}_i(x,\xi)\big)\big| + \big|\partial_x^\alpha\big(\chi^+(x,\xi)\, \vec{\mathcal{P}}_i(x,\xi)\big)\big| \le c\, |x-\xi|^{-1-|\alpha|}$$

and

$$\Big|\partial_x^\alpha\Big(\chi^+(x,\xi)\, \mathcal{Q}_i(x,\xi) + \vec{\mathcal{P}}_i(x,\xi)\cdot\nabla_\xi\chi^+(x,\xi)\Big)\Big| \le c\, r(x)^{-2-|\alpha|}.$$

Using Lemma 3.5.4, we obtain

$$\|\partial_x^\alpha v_i^+\|_{V^{0,s}_{\beta-2+|\alpha|,\delta-2+|\alpha}(\mathcal{K})} \le c\,\Big(\|\Phi_\nu\|_{V^{0,s}_{\beta,\delta}(\mathcal{K})} + \|\zeta_\nu g\|_{V^{1,s}_{\beta,\delta}(\mathcal{K})}\Big)$$

for $|\alpha| \le 1$ and analogously

$$\|q^+\|_{V^{0,s}_{\beta-1,\delta-1}(\mathcal{K})} \le c\,\Big(\|\Phi_\nu\|_{V^{0,s}_{\beta,\delta}(\mathcal{K})} + \|\zeta_\nu g\|_{V^{1,s}_{\beta,\delta}(\mathcal{K})}\Big).$$

This proves (10.5.9). Furthermore by Lemma 3.5.5,

$$\text{(10.5.10)}\quad \begin{aligned} &\int_{\Gamma_j} |x|^{s(\beta-2+|\alpha|)+1} \prod_{k=1}^d \Big(\frac{r_k(x)}{|x|}\Big)^{s(\delta_k-2+|\alpha|)+1} \big|\partial_x^\alpha v^+(x)\big|^s\, dx \\ &\qquad \le c\,\big(\|\Phi_\nu\|^s_{W^{0,s}_{\beta,\delta}(\mathcal{K})^3} + \|\zeta_\nu g\|^s_{V^{1,s}_{\beta,\delta}(\mathcal{K})}\big) \end{aligned}$$

for $j = 1,\dots,d$, $|\alpha| \le 1$, and

$$\text{(10.5.11)}\quad \begin{aligned} &\int_{\Gamma_j} |x|^{s(\beta-1)+1} \prod_{k=1}^d \Big(\frac{r_k(x)}{|x|}\Big)^{s(\delta_k-1)+1} |q^+(x)|^s\, dx \\ &\qquad \le c\,\big(\|\Phi_\nu\|^s_{W^{0,s}_{\beta,\delta}(\mathcal{K})^3} + \|\zeta_\nu g\|^s_{V^{1,s}_{\beta,\delta}(\mathcal{K})}\big). \end{aligned}$$

By the definition of $v^\pm$ and $q^\pm$, the equations

$$-\Delta v^+ + \nabla q^+ = \zeta_\nu f + \Delta v^- - \nabla q^-, \quad -\nabla\cdot v^+ = \zeta_\nu g + \nabla\cdot v^-$$

are valid in $\mathcal{K}$. Let $\eta_\mu = \zeta_{\mu-1} + \zeta_\mu + \zeta_{\mu+1}$. From Lemma 10.5.2 it follows that

$$\eta_\mu(\zeta_\nu f + \Delta v^- - \nabla g^-) \in V^{0,s}_{\beta,\delta}(\mathcal{K})^3, \quad \eta_\mu(\zeta_\nu g + \nabla\cdot v^-) \in W^{1,s}_{\beta,\delta}(\mathcal{K}).$$

and

$$\begin{aligned} &\|\eta_\mu(\zeta_\nu f + \Delta v^- - \nabla g^-)\|_{V^{0,s}_{\beta,\delta}(\mathcal{K})^3} + \|\eta_\mu(\zeta_\nu g + \nabla\cdot v^-)\|_{W^{1,s}_{\beta,\delta}(\mathcal{K})} \\ &\qquad \le c\,\big(\|\zeta_\nu f\|_{W^{0,s}_{\beta,\delta}(\mathcal{K})^3} + \|\zeta_\nu g\|_{W^{1,s}_{\beta,\delta}(\mathcal{K})}\big). \end{aligned}$$

Since $v^+ \in V^{1,s}_{\beta-1,\delta-1}(\mathcal{G})$, we conclude that

$$\eta_\mu\big(\zeta_\nu g + \nabla\cdot v^-\big) = -\eta_\mu \nabla\cdot v^+ \in W^{1,s}_{\beta,\delta}(\mathcal{K}) \cap V^{0,s}_{\beta-1,\delta-1}(\mathcal{K}) = V^{1,s}_{\beta,\delta}(\mathcal{K})$$

and

$$\|\eta_\mu(\zeta_\nu g + \nabla\cdot v^-)\|_{V^{1,s}_{\beta,\delta}(\mathcal{K})} \le c\,\big(\|\zeta_\nu f\|_{W^{0,s}_{\beta,\delta}(\mathcal{K})^3} + \|\zeta_\nu g\|_{W^{1,s}_{\beta,\delta}(\mathcal{K})}\big)\,.$$

Furthermore by Lemma 10.5.2,

$$\begin{aligned}\|\eta_\mu S_j v^+\|_{W^{2-1/s,s}_{\beta,\delta}(\Gamma_j)^{3-d_j}} &= \|\eta_\mu S_j v^-\|_{W^{2-1/s,s}_{\beta,\delta}(\Gamma_j)^{3-d_j}} \\ &\le c\,\big(\|\zeta_\nu f\|_{W^{0,s}_{\beta,\delta}(\mathcal{K})^3} + \|\zeta_\nu g\|_{W^{1,s}_{\beta,\delta}(\mathcal{K})}\big)\end{aligned}$$

and, analogously,

$$\|\eta_\mu N_j(v^+,q^+)\|_{W^{1-1/s,s}_{\beta,\delta}(\Gamma_j)^{d_j}} \le c\,\big(\|\zeta_\nu f\|_{W^{0,s}_{\beta,\delta}(\mathcal{K})^3} + \|\zeta_\nu g\|_{W^{1,s}_{\beta,\delta}(\mathcal{K})}\big).$$

This together with (10.5.10) and (10.5.11) implies

$$\begin{aligned}\|\eta_\mu S_j v^+\|_{V^{2-1/s,s}_{\beta,\delta}(\Gamma_j)^{3-d_j}} &+ \|\eta_\mu N_j(v^+,q^+)\|_{V^{1-1/s,s}_{\beta,\delta}(\Gamma_j)^{d_j}} \\ &\le c\,\big(\|\zeta_\nu f\|_{W^{0,s}_{\beta,\delta}(\mathcal{K})^3} + \|\zeta_\nu g\|_{W^{1,s}_{\beta,\delta}(\mathcal{K})}\big)\,.\end{aligned}$$

Applying Lemma 10.2.5 to the vector function (v^+, q^+), we obtain

$$(10.5.12)\qquad \|\zeta_\mu v^+\|_{V^{2,s}_{\beta,\delta}(\mathcal{K})^3} + \|\zeta_\mu q^+\|_{V^{1,s}_{\beta,\delta}(\mathcal{K})} \le c\,\Big(\|\zeta_\nu f\|_{V^{0,s}_{\beta,\delta}(\mathcal{K})^3} + \|\zeta_\nu g\|_{V^{1,s}_{\beta,\delta}(\mathcal{K})}\Big).$$

The result follows. □

10.5.2. Existence of solutions in $W^{2,s}_{\beta,\delta}(\mathcal{K})^3 \times W^{1,s}_{\beta,\delta}(\mathcal{K})$. Lemmas 10.5.1–10.5.3 together with Lemma 3.5.8 enable us to prove the following existence theorem.

THEOREM 10.5.4. *Let $f \in W^{0,s}_{\beta,\delta}(\mathcal{K})^3$, $g \in W^{1,s}_{\beta,\delta}(\mathcal{K})$, $h_j \in W^{2-1/s,s}_{\beta,\delta}(\Gamma_j)^{3-d_j}$ and $\phi_j \in W^{1-1/s,s}_{\beta,\delta}(\Gamma_j)^{d_j}$ for $j=1,\dots,d$. We assume that the line $\operatorname{Re}\lambda = 2-\beta-3/s$ is free of eigenvalues of the pencil $\mathfrak{A}(\lambda)$, the components of δ satisfy the inequalities* (10.5.1) *and that g, h_j, ϕ_j satisfy the compatibility conditions of Lemma* 10.2.3 *and Remark* 10.2.4 *on the edges M_k. Then there exists a solution $(u,p) \in W^{2,s}_{\beta,\delta}(\mathcal{K})^3 \times W^{1,s}_{\beta,\delta}(\mathcal{K})$ of the boundary value problem* (10.1.1), (10.1.2) *satisfying the estimate*

$$\begin{aligned}\|u\|_{W^{2,s}_{\beta,\delta}(\mathcal{K})^3} + \|p\|_{W^{1,s}_{\beta,\delta}(\mathcal{K})} \le c\,\Big(&\|f\|_{W^{0,s}_{\beta,\delta}(\mathcal{K})^3} + \|g\|_{W^{1,s}_{\beta,\delta}(\mathcal{K})} \\ &+ \sum_{j=1}^d \|h_j\|_{W^{2-1/s,s}_{\beta,\delta}(\Gamma_j)^{3-d_j}} + \sum_{j=1}^d \|\phi_j\|_{W^{1-1/s,s}_{\beta,\delta}(\Gamma_j)^{d_j}}\Big)\end{aligned}$$

with a constant c independent of f, g, h_j, ϕ_j.

P r o o f. By Lemma 10.2.3 (see also Remark 10.2.4), there exists a vector function $(v,q) \in W^{2,s}_{\beta,\delta}(\mathcal{K})^3 \times W^{1,s}_{\beta,\delta}(\mathcal{K})$ satisfying the conditions

$$S_j v = h_j,\quad N_j(v,q) = \phi_j \text{ on } \Gamma_j,\ j=1,\dots,d,\quad \nabla\cdot v + g \in V^{1,s}_{\beta,\delta}(\mathcal{K}).$$

Thus, we may assume without loss of generality that $h_j = 0$, $\phi_j = 0$ for $j=1,\dots,d$ and $g \in V^{1,s}_{\beta,\delta}(\mathcal{K})$. Let $G(x,\xi)$ be the Green's matrix introduced in Theorem 10.4.1 for an arbitrary κ in the interval $-\Lambda_+ - 1/2 < \kappa < -\Lambda_- - 1/2$, where $\Lambda_- < \operatorname{Re}\lambda < \Lambda_+$ is the widest strip in the complex plane which is free of eigenvalues of the pencil $\mathfrak{A}(\lambda)$ and which contains the line $\operatorname{Re}\lambda = 2-\beta-3/s$. Then the vector function

(u,p) with the components (10.4.5) and (10.4.6) is a solution of the boundary value problem (10.1.1), (10.1.2). According to Lemmas 10.5.1–10.5.3, the operator

$$W^{0,s}_{\beta,\delta}(\mathcal{K})^3 \times W^{1,s}_{\beta,\delta}(\mathcal{K}) \ni (f,g) \to \mathcal{O}(f,g) = (u,p)$$

satisfies the conditions of Lemma 3.5.8 for the spaces $\mathcal{X} = W^{0,s}_{\beta,\delta}(\mathcal{K})^3 \times W^{1,s}_{\beta,\delta}(\mathcal{K})$ and $\mathcal{Y} = W^{2,s}_{\beta,\delta}(\mathcal{K})^3 \times W^{1,s}_{\beta,\delta}(\mathcal{K})$. Consequently, this operator realizes a continuous mapping from $\mathcal{X}$ into $\mathcal{Y}$. This proves the theorem. □

10.5.3. Uniqueness of the solution. The uniqueness of the solution in Theorem 10.5.4 can be proved in same way as for the mixed boundary value problem considered in Section 7.6. In the case $s < 2$ we pass to the coordinates t, ω, where $t = \log \rho = \log |x|$ and $\omega = x/|x|$. Let $W^{l,s}_\delta(\mathbb{R} \times \Omega)$ be the weighted Sobolev space with the norm (7.6.9). For an arbitrary function $v \in W^{l,s}_\delta(\mathbb{R} \times \Omega)$ we define by v_ε the mollification with respect to the variable t of v, i.e.,

$$v_\varepsilon(t,\omega) = \int_{\mathbb{R}} v(\tau,\omega)\, h_\varepsilon(t-\tau)\, d\tau,$$

where $h_\varepsilon(t) = \varepsilon^{-1} h(t/\varepsilon)$ and h is a smooth function with compact support such that $\int h(t)\, dt = 1$. Since

$$\partial_\omega^\alpha \partial_t^{j+k} v_\varepsilon(\omega,t) = \int_{\mathbb{R}} (\partial_\omega^\alpha \partial_t^k v)(\omega,\tau)\, h_\varepsilon^{(j)}(t-\tau)\, d\tau,$$

it follows that $\partial_t^j v_\varepsilon \in W^{l,s}_\delta(\mathbb{R} \times \Omega)$ for $v \in W^{l,s}_\delta(\mathbb{R} \times \Omega)$, $\varepsilon > 0$, $j = 0, 1, \ldots$.

LEMMA 10.5.5. *Suppose that the line* $\operatorname{Re}\lambda = 2 - \beta - 3/s$ *is free of eigenvalues of the pencil* $\mathfrak{A}(\lambda)$ *and that the components of* δ *satisfy the inequalities* (10.5.1). *Then the homogeneous boundary value problem*

$$-\Delta u + \nabla p = 0, \quad -\nabla \cdot u = 0 \quad \textit{in } \mathcal{K}, \tag{10.5.13}$$

$$S_j u = 0, \quad N_j(u,p) = 0 \quad \textit{on } \Gamma_j,\ j = 1, \ldots, d, \tag{10.5.14}$$

has only the trivial solution $(u,p) = (0,0)$ *in* $W^{2,s}_{\beta,\delta}(\mathcal{K})^3 \times W^{1,s}_{\beta,\delta}(\mathcal{K})$.

P r o o f. Let $(u,p) \in W^{2,s}_{\beta,\delta}(\mathcal{K})^3 \times W^{2,s}_{\beta,\delta}(\mathcal{K})$ be a solution of the problem (10.5.13), (10.5.14). We suppose first that $s \le 2$. Since $W^{l,s}_{\beta,\delta}(\mathcal{K}) \subset W^{l,s}_{\beta,\delta'}(\mathcal{K})$ if $\delta_k \le \delta'_k$ for $k = 1, \ldots, d$, we may assume without loss of generality that the components of δ satisfy the inequalities $\max(1, 2 - \delta_+^{(k)}) < \delta_k + 2/s < 2$. Using Lemma 10.2.7, we obtain $u \in W^{3,s}_{\beta+1,\delta+1}(\mathcal{K})^3$ and $p \in W^{2,s}_{\beta+1,\delta+1}(\mathcal{K})$. We set $v = \rho^{\beta-2+3/s} u$ and $q = \rho^{\beta-1+3/s} p$. Then, in the coordinates $t = \log|x|$ and $\omega = x/|x|$, we have $v \in W^{3,s}_{\delta+1}(\mathbb{R} \times \Omega)^3$ and $q \in W^{2,s}_{\delta+1}(\mathbb{R} \times \Omega) = V^{2,s}_{\delta+1}(\mathbb{R} \times \Omega)$. Consequently,

$$\partial_t^j v_\varepsilon \in W^{3,s}_{\delta+1}(\mathbb{R} \times \Omega)^3, \quad \partial_t^j q_\varepsilon \in W^{2,s}_{\delta+1}(\mathbb{R} \times \Omega)$$

for $j = 0, 1, 2, \ldots$. From Lemma 2.1.2 we conclude that

$$v_\varepsilon \in W^{1,2}_{\delta-2+2/s}(\mathbb{R} \times \Omega)^3 \subset W^{1,2}_0(\mathbb{R} \times \Omega)^3, \quad q_\varepsilon \in W^{0,2}_{\delta-2+2/s}(\mathbb{R} \times \Omega)^3 \subset L_2(\mathbb{R} \times \Omega).$$

Thus, the functions $u_\varepsilon = \rho^{-\beta+2-3/s} v_\varepsilon$ and $p_\varepsilon = \rho^{-\beta+1-3/s} q_\varepsilon$ (as functions in x) belong to the spaces $W^{1,2}_{\beta+3/s-5/2,0}(\mathcal{K})^3$ and $W^{0,2}_{\beta+3/s-5/2,0}(\mathcal{K})$, respectively. It can be easily seen that $(u_\varepsilon, p_\varepsilon)$ is also a solution of the homogeneous problem (10.5.13), (10.5.14). By Theorem 10.2.11, this problem has no nonzero solutions in the space

$W^{1,2}_{\beta+3/s-5/2,0}(\mathcal{K})^3 \times W^{0,2}_{\beta+3/s-5/2,0}(\mathcal{K})$. Therefore, $u_\varepsilon = 0$, $p_\varepsilon = 0$ which implies $u = 0$ and $p = 0$.

Now, let $s \ge 2$ and let ζ be a smooth cut-off function on $\overline{\mathcal{K}}$ equal to one for $|x| < 1$ and to zero for $|x| > 2$. Furthermore, we set $\beta' = \beta - (3s-6)/(2s)$ and $\delta'_k = \delta_k - 1 + 2/s$ for $k = 1, \dots, d$. Then by Hölder's inequality,

$$\int_{\mathcal{K}} \rho^{2(\beta'+\varepsilon-2+|\alpha|)} \prod_k \Big(\frac{r_k}{\rho}\Big)^{2(\delta'_k+\varepsilon)} |\partial_x^\alpha(\zeta u)|^2\, dx$$

$$\le \Big(\int_{\mathcal{K}} \rho^{s(\beta-2+|\alpha|)} \prod_k \Big(\frac{r_k}{\rho}\Big)^{s\delta_k} |\partial_x^\alpha(\zeta u)|^s\, dx\Big)^{2/s} \Big(\int\limits_{\substack{\mathcal{K}\\ |x|\le 2}} \prod \Big(\frac{r_k}{\rho}\Big)^{s'\varepsilon-2} \frac{dx}{\rho^{3-s'\varepsilon}}\Big)^{2/s'},$$

where $s' = 2s/(s-2)$. The second integral on the right is finite if $\varepsilon > 0$. Consequently, $\zeta u \in W^{2,2}_{\beta'+\varepsilon,\delta'+\varepsilon}(\mathcal{K})^3$ and $\zeta p \in W^{1,2}_{\beta'+\varepsilon,\delta'+\varepsilon}(\mathcal{K})$. Analogously, we obtain $(1-\zeta)u \in W^{2,2}_{\beta'-\varepsilon,\delta'-\varepsilon}(\mathcal{K})^3$ and $(1-\zeta)p \in W^{1,2}_{\beta'-\varepsilon,\delta'-\varepsilon}(\mathcal{K})$. This implies

$$-\Delta(\zeta u) + \nabla(\zeta p) = \Delta\big((1-\zeta)u\big) - \nabla\big((1-\zeta)p\big) \in W^{0,2}_{\beta'-\varepsilon,\delta'-\varepsilon}(\mathcal{K})^3$$

and analogously $\nabla\cdot(\zeta u) \in W^{0,2}_{\beta'-\varepsilon,\delta'-\varepsilon}(\mathcal{K})$, $S_j(\zeta u) \in W^{3/2,2}_{\beta'-\varepsilon,\delta'-\varepsilon}(\Gamma_j)^{3-d_j}$, $N_j(\zeta u, \zeta p) \in W^{1/2,2}_{\beta'-\varepsilon,\delta'-\varepsilon}(\Gamma_j)^{d_j}$. This together with Theorem 10.3.4 yields $\zeta u \in W^{2,2}_{\beta'-\varepsilon,\delta'-\varepsilon}(\mathcal{K})^3$ and $\zeta p \in W^{1,2}_{\beta'-\varepsilon,\delta'-\varepsilon}(\mathcal{K})$. Obviously, the same is true for u and p. Hence $u = 0$ and $p = 0$ (cf. Theorem 10.2.9). The proof of the lemma is complete. □

The last lemma and the existence result of Theorem 10.5.4 imply the following statement.

THEOREM 10.5.6. *Let $f \in W^{0,s}_{\beta,\delta}(\mathcal{K})^3$, $g \in W^{1,s}_{\beta,\delta}(\mathcal{K})$, $h_j \in W^{2-1/s,s}_{\beta,\delta}(\Gamma_j)^{3-d_j}$ and $\phi_j \in W^{1-1/s,s}_{\beta,\delta}(\Gamma_j)^{d_j}$ for $j = 1, \dots, d$. If the conditions of Theorem* 10.5.4 *are satisfied, then the boundary value problem* (10.1.1), (10.1.2) *is uniquely solvable in $W^{2,s}_{\beta,\delta}(\mathcal{K})^3 \times W^{1,s}_{\beta,\delta}(\mathcal{K})$.*

The solution in Theorem 10.5.4 was constructed by means of the Green's matrix introduced in Theorem 10.4.1. Here the number κ in Theorem 10.4.1 was chosen such that the closed strip between the lines $\operatorname{Re}\lambda = 2-\beta-3/p$ and $\operatorname{Re}\lambda = -\kappa-1/2$ is free of eigenvalues of the pencil of the pencil $\mathfrak{A}(\lambda)$. This means in particular that solutions both in the spaces $W^{2,s}_{\beta,\delta}(\mathcal{K})^3 \times W^{1,s}_{\beta,\delta}(\mathcal{K})$ and $W^{2,s'}_{\beta',\delta'}(\mathcal{K})^3 \times W^{1,s'}_{\beta',\delta'}(\mathcal{K})$ can be represented by the same Green's matrix if the closed strip between the lines $\operatorname{Re}\lambda = 2-\beta-3/s$ and $\operatorname{Re}\lambda = 2-\beta'-3/s'$ does not contain eigenvalues of the pencil $\mathfrak{A}(\lambda)$ and the components of δ and δ' satisfy the condition (10.5.1). Thus, the following regularity assertion holds.

THEOREM 10.5.7. *Let $(u,p) \in W^{2,s'}_{\beta',\delta'}(\mathcal{K})^3 \times W^{1,s'}_{\beta',\delta'}(\mathcal{K})$ be a solution of problem* (10.1.1), (10.1.2), *where*

$$f \in W^{0,s}_{\beta,\delta}(\mathcal{K})^3 \cap W^{0,s'}_{\beta',\delta'}(\mathcal{K})^3, \quad g \in W^{1,s}_{\beta,\delta}(\mathcal{K}) \cap W^{1,s'}_{\beta',\delta'}(\mathcal{K}),$$

$$h_j \in W^{2-1/s,s}_{\beta,\delta}(\Gamma_j)^{3-d_j} \cap W^{2-1/s',s'}_{\beta',\delta'}(\Gamma_j)^{3-d_j},$$

$$\phi_j \in W^{1-1/s,s}_{\beta,\delta}(\Gamma_j)^{d_j} \cap W^{1-1/s',s'}_{\beta',\delta'}(\Gamma_j)^{d_j}.$$

Suppose that the closed strip between the lines $\mathrm{Re}\,\lambda = 2-\beta-3/s$ *and* $\mathrm{Re}\,\lambda = 2-\beta'-3/s'$ *does not contain eigenvalues of the pencil* $\mathfrak{A}(\lambda)$, *that* δ *and* δ' *satisfy the inequalities*

$$\max(0, 2-\mu_+^{(k)}) < \delta_k + 2/s < 2, \quad \max(0, 2-\mu_+^{(k)}) < \delta'_k + 2/s' < 2$$

and that g, h_j, ϕ_j *satisfy the compatibility conditions of Lemma* 10.2.3 *and Remark* 10.2.4 *on the edges* M_k. *Then* $u \in W^{2,s}_{\beta,\delta}(\mathcal{K})^3$ *and* $p \in W^{1,s}_{\beta,\delta}(\mathcal{K})$.

10.5.4. The case $d_{k_+} + d_{k_-} \le 3$. Let again Γ_{k_+} and Γ_{k_-} denote the adjoining faces to the edge M_k. If $\lambda = 0$ is not an eigenvalue of the pencil $A_k(\lambda)$ (i.e. $d_{k_+}+d_{k_-} \le 3$ and the edge angle θ_k satisfies the condition (10.3.8) for $d_{k_+}\cdot d_{k_-} \ne 0$), then we can use the sharper estimates of the Green's matrix in Theorems 10.4.4 and 10.4.5. This leads to the following result.

THEOREM 10.5.8. *Let* $f \in V^{0,s}_{\beta,\delta}(\mathcal{K})^3$, $g \in V^{1,s}_{\beta,\delta}(\mathcal{K})$, $h_j \in V^{2-1/s,s}_{\beta,\delta}(\Gamma_j)^{3-d_j}$ *and* $\phi_j \in V^{1-1/s,s}_{\beta,\delta}(\Gamma_j)^{d_j}$ *for* $j = 1,\dots,d$. *We assume that* $d_{k_+} + d_{k_-} \le 3$ *for all* k *and the edge angles* θ_k *satisfy the condition* (10.3.8) *for* $d_{k_+}\cdot d_{k_-} \ne 0$. *Furthermore, we suppose that the line* $\mathrm{Re}\,\lambda = 2-\beta-3/s$ *is free of eigenvalues of the pencil* $\mathfrak{A}(\lambda)$ *and that the components of* δ *satisfy the inequalities*

$$2-\delta_+^{(k)} < \delta_k + 2/s < 2 + \delta_+^{(k)} \quad \textit{for } k = 1,\dots,d.$$

Then the problem (10.1.1), (10.1.2) *is uniquely solvable in* $V^{2,s}_{\beta,\delta}(\mathcal{K})^3 \times V^{1,s}_{\beta,\delta}(\mathcal{K})$.

P r o o f. Let (v,q) and $(v^\pm, q^\pm)$ be the same vector functions as in Subsection 10.5.1. Under the conditions of our theorem, the elements of the Green's matrix satisfy the estimates in Theorems 10.4.2, 10.4.4 and 10.4.5 with the exponents $\sigma_{i,\alpha}(x) = \delta_x - \delta_{i,4} - |\alpha| - \varepsilon$ and $\sigma_{k,i,\alpha} = \delta_+^{(k)} - \delta_{i,4} - |\alpha| - \varepsilon$, respectively. Thus analogously to the proofs of Lemmas 10.5.1 and 10.5.2, we obtain

$$\|\zeta_\mu v\|_{V^{2,s}_{\beta,\delta}(\mathcal{K})^3} + \|\zeta_\mu q\|_{V^{1,s}_{\beta,\delta}(\mathcal{K})} \le c\, 2^{-|\mu-\nu|\varepsilon_0}\Big(\|\zeta_\nu f\|_{V^{0,s}_{\beta,\delta}(\mathcal{K})^3} + \|\zeta_\nu g\|_{V^{1,s}_{\beta,\delta}(\mathcal{K})}\Big)$$

for $|\mu-\nu| > 2$ and

$$\|\zeta_\mu v^-\|_{V^{2,s}_{\beta,\delta}(\mathcal{K})^3} + \|\zeta_\mu q^-\|_{V^{1,s}_{\beta,\delta}(\mathcal{K})} \le c\Big(\|\zeta_\nu f\|_{V^{0,s}_{\beta,\delta}(\mathcal{K})^3} + \|\zeta_\nu g\|_{V^{1,s}_{\beta,\delta}(\mathcal{K})}\Big).$$

for $|\mu-\nu| \le 2$. As was shown in the proof of Lemma 10.5.3, the same estimate (cf. (10.5.12)) is valid for the vector function $(\zeta_\mu v^+, \zeta_\mu q^+)$ if $|\mu-\nu| \le 2$. Consequently, the operator

$$V^{0,s}_{\beta,\delta}(\mathcal{K})^3 \times V^{1,s}_{\beta,\delta}(\mathcal{K}) \ni (f,g) \to \mathcal{O}(f,g) = (u,p),$$

where (u,p) is the vector function with the components (10.4.5) and (10.4.6), satisfies the conditions of Lemma 3.5.8 for the spaces $\mathcal{X} = V^{0,s}_{\beta,\delta}(\mathcal{K})^3 \times V^{1,s}_{\beta,\delta}(\mathcal{K})$ and $\mathcal{Y} = V^{2,s}_{\beta,\delta}(\mathcal{K})^3 \times V^{1,s}_{\beta,\delta}(\mathcal{K})$. This proves the solvability of problem (10.1.1), (10.1.2) in $\mathcal{Y}$. The uniqueness of the solution can be proved in same way as in Theorem 10.5.6. □

10.6. Variational solutions of the boundary value problem in weighted L_p Sobolev spaces

Now we are interested in variational solutions of the boundary value problem (10.1.1), (10.1.2) in weighted L_s Sobolev spaces with arbitrary s, $1 < s < \infty$. These solutions are defined as in the case $s = 2$ by means of the bilinear form

$b_{\mathcal{K}}(u,v)$. In particular, we prove that there exists a uniquely determined solution $(u,p) \in W^{1,s}_{\beta,\delta}(\mathcal{K})^3 \times W^{0,s}_{\beta,\delta}(\mathcal{K})$ if the line $\text{Re}\,\lambda = 1-\beta-3/s$ is free of eigenvalues of the pencil $\mathfrak{A}(\lambda)$ and the components of δ satisfy the inequalities

$$\max(0, 1-\delta_+^{(k)}) < \delta_k + 2/s < 1 \quad \text{for } k = 1,\dots,d. \tag{10.6.1}$$

For this, we employ again the estimates of the Green's matrix obtained in Section 10.4. Furthermore, we prove regularity assertions for the variational solution in the class of the weighted Sobolev spaces $W^{l,s}_{\beta,\delta}$.

10.6.1. Definition of weak solutions. Let $b_{\mathcal{K}}$ be the bilinear form (10.1.3), and let

$$\mathcal{H}_{s,\beta,\delta} = \{u \in W^{1,s}_{\beta,\delta}(\mathcal{K})^3 : \ S_j u|_{\Gamma_j} = 0, \ j = 1,\dots,d\}.$$

Obviously, the bilinear form $b_{\mathcal{K}}$ is continuous on $W^{1,s}_{\beta,\delta}(\mathcal{K})^3 \times W^{1,s'}_{-\beta,-\delta}(\mathcal{K})^3$, where $s' = s/(s-1)$. As in the case $s=2$, $\delta = 0$ (cf. Subsection 10.2.4), a vector function $(u,p) \in W^{1,s}_{\beta,\delta}(\mathcal{K})^3 \times W^{0,s}_{\beta,\delta}(\mathcal{K})$ is called *variational solution* of the boundary value problem (10.1.1), (10.1.2) if

$$b_{\mathcal{K}}(u,v) - \int_{\mathcal{K}} p\,\nabla\cdot v\,dx = F(v) \quad \text{for all } v \in \mathcal{H}_{s',-\beta,-\delta}, \tag{10.6.2}$$

$$-\nabla\cdot u = g \ \text{ in } \mathcal{K}, \quad S_j u = h_j \ \text{ on } \Gamma_j,\ j = 1,\dots,d. \tag{10.6.3}$$

Here F is a linear and continuous functional on $\mathcal{H}_{s',-\beta,-\delta}$. By Green's formula (10.1.7), every solution $(u,p) \in W^{2,s}_{\beta+1,\delta}(\mathcal{K})^3 \times W^{1,s}_{\beta+1,\delta}(\mathcal{K})$ of the boundary value problem (10.1.1), (10.1.2) is also a solution of the problem (10.6.2), (10.6.3), where F is defined by (10.1.6).

We suppose in this section that the line $\text{Re}\,\lambda = 1-\beta-3/s$ is free of eigenvalues of the pencil $\mathfrak{A}(\lambda)$ and that the components of δ satisfy the inequalities (10.6.1). Then $-\delta_k > 1 - 2/s'$ and therefore

$$\mathcal{H}_{s',-\beta,-\delta} \subset W^{1,s'}_{-\beta,-\delta}(\mathcal{K}) = V^{1,s'}_{-\beta,-\delta}(\mathcal{K})$$

(cf. Lemma 7.2.2).

The assertion of the next lemma was proved in Section 7.7 for the special case where $\mathcal{H}_{s,\beta,\delta} = \{u \in W^{1,s}_{\beta,\delta}(\mathcal{K})^\ell : u|_{\Gamma_j} = 0 \text{ for } j \in I_0\}$ (cf. Lemma 7.7.1). For the space $\mathcal{H}_{s,\beta,\delta}$ just defined, the result can be proved by the same arguments.

LEMMA 10.6.1. *Let $F \in \mathcal{H}^*_{s',-\beta,-\delta}$, where $0 < \delta_k + 2/s < 1$ for $k = 1,\dots,d$. Then there exist vector functions $f^{(0)} \in W^{0,s}_{\beta+1,\delta+1}(\mathcal{K})^3$ and $f^{(k)} \in W^{0,s}_{\beta,\delta}(\mathcal{K})^3$, $k = 1,2,3$, such that*

$$F(v) = \int_{\mathcal{K}} \Big(f^{(0)}\cdot v + \sum_{k=1}^{3} f^{(k)}\cdot \partial_{x_k} v\Big)\,dx \quad \textit{for all } v \in \mathcal{H}_{s',-\beta,-\delta} \tag{10.6.4}$$

and

$$\|f^{(0)}\|_{W^{0,s}_{\beta+1,\delta+1}(\mathcal{K})^\ell} + \sum_{k=1}^{3} \|f^{(k)}\|_{W^{0,s}_{\beta,\delta}(\mathcal{K})^\ell} \le c\,\|F\|_{\mathcal{H}^*_{s',-\beta,-\delta}}.$$

Here c is a constant independent of F.

10.6.2. Representation of weak solutions by Green's matrix. We suppose that the line $\mathrm{Re}\,\lambda = 1-\beta-3/s$ is free of eigenvalues of the pencil $\mathfrak{A}(\lambda)$ and denote by

$$\Lambda_- < \mathrm{Re}\,\lambda < \Lambda_+$$

the widest strip in the complex plane which contains the line $\mathrm{Re}\,\lambda = 1-\beta-3/s$ and which is free of eigenvalues of the pencil $\mathfrak{A}(\lambda)$. Furthermore, let κ be an arbitrary real number in the interval

$$-\Lambda_+ - 1/2 < \kappa < -\Lambda_- - 1/2\,.$$

We restrict ourselves to the case where $h_j = 0$ for $j = 1,\dots,d$ and suppose that F is a linear and continuous functional on $\mathcal{H}_{s',-\beta,-\delta}$ of the form (10.6.4). Then analogously to (10.4.5), (10.4.6), the following representations for the components of the solution (u,p) of the problem (10.6.2), (10.6.3) hold.

$$(10.6.5)\quad u_i(x) = \int_{\mathcal{K}} \Big(f^{(0)}(\xi)\cdot \vec{H}_i(x,\xi) + \sum_{k=1}^{3} f^{(k)}(\xi)\cdot \partial_{\xi_k}\vec{H}_i(x,\xi)\Big)\, d\xi$$
$$+ \int_{\mathcal{K}} g(\xi)\, G_{i,4}(x,\xi)\, d\xi, \quad i=1,2,3,$$

$$(10.6.6)\quad p(x) = -g(x) + \int_{\mathcal{K}} \Big(f^{(0)}(\xi)\cdot \vec{H}_4(x,\xi) + \sum_{k=1}^{3} f^{(k)}(\xi)\cdot \partial_{\xi_k}\vec{H}_4(x,\xi)\Big)\, d\xi$$
$$+ \int_{\mathcal{K}} g(\xi)\, G_{4,4}(x,\xi)\, d\xi.$$

Our goal is to prove that $(u,p) \in W^{1,s}_{\beta,\delta}(\mathcal{K})^3 \times W^{0,s}_{\beta,\delta}(\mathcal{K})$ for arbitrary functions $f^{(0)} \in W^{0,s}_{\beta+1,\delta+1}(\mathcal{K})^3$, $f^{(k)} \in W^{0,s}_{\beta,\delta}(\mathcal{K})^3$, $k=1,2,3$, and $g \in W^{0,s}_{\beta,\delta}(\mathcal{K})$. By what has been shown in the previous section, the mapping $f^{(0)} \to (U,P)$ defined by

$$(10.6.7)\qquad U_i(x) = \int_{\mathcal{K}} f^{(0)}(\xi)\cdot \vec{H}_i(x,\xi)\, d\xi, \; i=1,2,3,$$

$$(10.6.8)\qquad P(x) = \int_{\mathcal{K}} f^{(0)}(\xi)\cdot \vec{H}_4(x,\xi)\, d\xi$$

is continuous from $V^{0,s}_{\beta+1,\delta+1}(\mathcal{K})^3$ into the subspace $W^{2,s}_{\beta+1,\delta+1}(\mathcal{K})^3 \times W^{1,s}_{\beta+1,\delta+1}(\mathcal{K})$ of $W^{1,s}_{\beta,\delta}(\mathcal{K})^3 \times W^{0,s}_{\beta,\delta}(\mathcal{K})$. Therefore, we may restrict ourselves to the case $f^{(0)} = 0$.

10.6.3. Auxiliary estimates. Let ζ_ν be infinitely differentiable functions depending only on $\rho = |x|$ and satisfying the conditions (10.2.4). We consider the vector function (v,q) with the components

$$(10.6.9)\quad v_i(x) = \sum_{k=1}^{3} \int_{\mathcal{K}} \zeta_\nu(\xi)\, f^{(k)}(\xi)\cdot \partial_{\xi_k}\vec{H}_i(x,\xi)\, d\xi + \int_{\mathcal{K}} \zeta_\nu(\xi)\, g(\xi)\, G_{i,4}(x,\xi)\, d\xi,$$

$i = 1,2,3$, and

$$(10.6.10)\qquad q(x) = -\zeta_\nu(x)\, g(x) + \sum_{k=1}^{3} \int_{\mathcal{K}} \zeta_\nu(\xi)\, f^{(k)}(\xi)\cdot \partial_{\xi_k}\vec{H}_4(x,\xi)\, d\xi$$
$$+ \int_{\mathcal{K}} \zeta_\nu(\xi)\, g(\xi)\, G_{4,4}(x,\xi)\, d\xi.$$

LEMMA 10.6.2. *Suppose that the line* $\mathrm{Re}\,\lambda = 1 - \beta - 3/s$ *does not contain eigenvalues of the pencil* $\mathfrak{A}(\lambda)$ *and that the components of* δ *satisfy the inequalities* (10.6.1). *Then the vector function* (v, q) *with the components* (10.6.9), (10.6.10) *satisfies the estimate*

$$\|\zeta_\mu v\|_{W^{1,s}_{\beta,\delta}(\mathcal{K})^3} + \|\zeta_\mu q\|_{W^{0,s}_{\beta,\delta}(\mathcal{K})} \le c\, 2^{-|\mu-\nu|\varepsilon_0} \Big(\sum_{k=1}^{3} \|\zeta_\nu f^{(k)}\|_{V^{0,s}_{\beta,\delta}(\mathcal{K})^3} + \|\zeta_\nu g\|_{V^{0,s}_{\beta,\delta}(\mathcal{K})} \Big)$$

for arbitrary $f^{(k)} \in W^{0,s}_{\beta,\delta}(\mathcal{K})^3$, $k = 1,2,3$, $g \in V^{0,s}_{\beta,\delta}(\mathcal{K})$, $|\mu - \nu| > 2$, *where* c *and* ε_0 *are positive constants independent of* $f^{(k)}, g, \mu, \nu$.

P r o o f. Let $\chi_\mu(x) = 1$ for $2^{\mu-1} \le |x| \le 2^{\mu+1}$ and $\chi_\mu(x) = 0$ else. Then

$$\|\zeta_\mu v\|_{W^{1,s}_{\beta,\delta}(\mathcal{K})^3} \le c \sum_{|\alpha|\le 1} \|\chi_\mu \partial_x^\alpha v\|_{V^{0,s}_{\beta-1+|\alpha|,\delta}(\mathcal{K})^3}$$

with a constant c independent of v and μ. If $\mu > \nu + 2$, then by Theorem 10.4.5,

$$\begin{aligned} &\big|\partial_x^\alpha \partial_{\xi_k} \vec{H}_i(x,\xi)\big| + \big|\partial_x^\alpha G_{i,4}(x,\xi)\big| \\ &\quad \le c\, \frac{|x|^{\Lambda_- - \delta_{i,4} - |\alpha| + \varepsilon}}{|\xi|^{\Lambda_- + 2 + \varepsilon}} \prod_{k=1}^{d} \Big(\frac{r_k(x)}{|x|}\Big)^{\sigma_{k,i,\alpha}} \prod_{k=1}^{d} \Big(\frac{r_k(\xi)}{|\xi|}\Big)^{\min(0,\mu_+^{(k)}-1-\varepsilon)} \end{aligned}$$

for $2^{\mu-1} \le |x| \le 2^{\mu+1}$, $2^{\nu-1} \le |\xi| \le 2^{\nu+1}$, where ε is an arbitrarily small positive number and $\sigma_{k,i,\alpha} = \min(0, \mu_+^{(k)} - |\alpha| - \delta_{i,4} - \varepsilon)$. Applying Lemma 3.5.1, we obtain

$$\begin{aligned} &\|\chi_\mu \partial_x^\alpha v\|_{V^{0,s}_{\beta-1+|\alpha|,\delta}(\mathcal{K})^3} \\ &\quad \le c\, 2^{(\mu-\nu)(\beta-1+\Lambda_-+\varepsilon+3/s)} \Big(\sum_{k=1}^{3} \|\zeta_\nu f^{(k)}\|_{V^{0,s}_{\beta,\delta}(\mathcal{K})^3} + \|\zeta_\nu g\|_{V^{0,s}_{\beta,\delta}(\mathcal{K})} \Big) \end{aligned}$$

for $|\alpha| \le 1$ and

$$\begin{aligned} &\|\zeta_\mu q\|_{V^{0,s}_{\beta,\delta}(\mathcal{K})} \le c\, \|\chi_\mu q\|_{V^{0,s}_{\beta,\delta}(\mathcal{K})} \\ &\quad \le c\, 2^{(\mu-\nu)(\beta-1+\Lambda_-+\varepsilon+3/s)} \Big(\sum_{k=1}^{3} \|\zeta_\nu f^{(k)}\|_{V^{0,s}_{\beta,\delta}(\mathcal{K})^3} + \|\zeta_\nu g\|_{V^{0,s}_{\beta,\delta}(\mathcal{K})} \Big). \end{aligned}$$

Here the constant ε can be chosen such that $\beta - 1 + \Lambda_- + \varepsilon + 3/s < 0$. Thus, the assertion of the lemma is true for $\mu > \nu + 2$. In the same way, the assertion of the lemma holds in the case $\mu < \nu - 2$. □

In order to prove the estimate of Lemma 10.6.2 for the case $|\mu + \nu| \le 2$, we consider the functions

$$\begin{aligned} v_i^\pm(x) = &\sum_{k=1}^{3} \int_{\mathcal{K}} \chi^\pm(x,\xi)\, \zeta_\nu(\xi)\, f^{(k)}(\xi) \cdot \partial_{\xi_k} \vec{H}_i(x,\xi)\, d\xi \\ &+ \int_{\mathcal{K}} \chi^\pm(x,\xi)\, \zeta_\nu(\xi)\, g(\xi)\, G_{i,4}(x,\xi)\, d\xi, \end{aligned} \tag{10.6.11}$$

$i = 1, 2, 3$, and

$$(10.6.12)\qquad q^\pm(x) = \sum_{k=1}^{3} \int_{\mathcal{K}} \chi^\pm(x,\xi)\, \zeta_\nu(\xi)\, f^{(k)}(\xi)\cdot \partial_{\xi_k} \vec{H}_4(x,\xi)\, d\xi + \int_{\mathcal{K}} \chi^\pm(x,\xi)\, \zeta_\nu(\xi)\, g(\xi)\, G_{4,4}(x,\xi)\, d\xi - \zeta_\nu(x)\, g(x)/2\,,$$

where $\chi^\pm$ are the same functions (10.5.4) as in the previous section. Then $v = v^+ + v^-$ and $q = q^+ + q^-$.

LEMMA 10.6.3. *Suppose that the inequalities* (10.6.1) *are satisfied for the components of* δ. *Then the vector function* (v^-, q^-) *with the components* (10.6.11), (10.6.12) *satisfies the estimate*

$$\|\zeta_\mu v^-\|_{W^{1,s}_{\beta,\delta}(\mathcal{K})^3} + \|\zeta_\mu q^-\|_{W^{0,s}_{\beta,\delta}(\mathcal{K})} \le c\Big(\sum_{k=1}^{3} \|\zeta_\nu f^{(k)}\|_{V^{0,s}_{\beta,\delta}(\mathcal{K})^3} + \|\zeta_\nu g\|_{V^{0,s}_{\beta,\delta}(\mathcal{K})}\Big)$$

for arbitrary $f^{(k)} \in V^{0,s}_{\beta,\delta}(\mathcal{K})^3$, $k = 1, 2, 3$, $g \in V^{0,s}_{\beta,\delta}(\mathcal{K})$, $|\mu - \nu| \le 2$, *where* c *is a constant independent of* $f^{(k)}, g, \mu, \nu$.

P r o o f. Let χ_μ be the same function as in the proof of Lemma 10.6.2. Then

$$\|\zeta_\mu v^-\|_{W^{1,s}_{\beta,\delta}(\mathcal{K})^3} \le c \sum_{|\alpha|\le 1} \|\chi_\mu \partial_x^\alpha v^-\|_{V^{0,s}_{\beta-1+|\alpha|,\delta}(\mathcal{K})^3}\,.$$

By Theorem 10.4.4,

$$\big|\partial_x^\alpha\big(\chi^-(x,\xi)\,\partial_{\xi_k} G_{i,j}(x,\xi)\big)\big| \le \frac{c}{|x-\xi|^{2+\delta_{i,4}+|\alpha|}} \Big(\frac{r(x)}{|x-\xi|}\Big)^{\sigma_{i,\alpha}(x)} \Big(\frac{r(\xi)}{|x-\xi|}\Big)^{\sigma_{4,0}(\xi)}$$

for $j = 1, 2, 3$ and

$$\big|\partial_x^\alpha\big(\chi^-(x,\xi)\, G_{i,4}(x,\xi)\big)\big| \le \frac{c}{|x-\xi|^{2+\delta_{i,4}+|\alpha|}} \Big(\frac{r(x)}{|x-\xi|}\Big)^{\sigma_{i,\alpha}(x)} \Big(\frac{r(\xi)}{|x-\xi|}\Big)^{\sigma_{4,0}(\xi)},$$

where $\sigma_{i,\alpha}(x) = \min(0, \mu_x - |\alpha| - \delta_{i,4} - \varepsilon)$, and $\sigma_{4,0}(\xi) = \min(0, \mu_\xi - 1 - \varepsilon)$. Applying Lemma 3.5.3, we obtain

$$\|\chi_\mu \partial_x^\alpha v^-\|_{V^{0,s}_{\beta-1+|\alpha|,\delta}(\mathcal{K})^3} \le c\Big(\sum_{k=1}^{3} \|\zeta_\nu f^{(k)}\|_{V^{0,s}_{\beta,\delta}(\mathcal{K})^3} + \|\zeta_\nu g\|_{V^{0,s}_{\beta,\delta}(\mathcal{K})}\Big)$$

for $|\alpha| \le 1$ and

$$\|\zeta_\mu q^-\|_{W^{0,s}_{\beta,\delta}(\mathcal{K})} \le c\,\|\chi_\mu q^-\|_{V^{0,s}_{\beta,\delta}(\mathcal{K})} \le c\Big(\sum_{k=1}^{3} \|\zeta_\nu f^{(k)}\|_{V^{0,s}_{\beta,\delta}(\mathcal{K})^3} + \|\zeta_\nu g\|_{V^{0,s}_{\beta,\delta}(\mathcal{K})}\Big).$$

This proves the lemma. □

We prove an analogous estimate for the vector function $\zeta_\mu(v^+, q^+)$.

LEMMA 10.6.4. *Let the components of the vector function* (v^+, q^+) *be given by* (10.6.11) *and* (10.6.12). *Then*

$$(10.6.13)\qquad \|\zeta_\mu v^+\|_{V^{1,s}_{\beta,\delta}(\mathcal{K})^3} + \|\zeta_\mu q^+\|_{V^{0,s}_{\beta,\delta}(\mathcal{K})} \le c\Big(\sum_{k=1}^{3} \|\zeta_\nu f^{(k)}\|_{V^{0,s}_{\beta,\delta}(\mathcal{K})^3} + \|\zeta_\nu g\|_{V^{0,s}_{\beta,\delta}(\mathcal{K})}\Big)$$

for arbitrary $f^{(k)} \in V^{0,s}_{\beta,\delta}(\mathcal{K})^3$, $k = 1,2,3$, $g \in V^{0,s}_{\beta,\delta}(\mathcal{K})$, $|\mu - \nu| \le 2$, *where* c *is a constant independent of* f, g, μ, ν.

P r o o f. By Theorem 10.4.2,

$$\big|\chi^+(x,\xi)\,\partial_{\xi_k}\vec{H}_i(x,\xi)\big| + \big|\chi^+(x,\xi)\,G_{i,4}(x,\xi)\big| \le c\,|x-\xi|^{-2}$$

for $i = 1,2,3$. Consequently, the estimate

$$\|v^+\|_{V^{0,s}_{\beta-1,\delta-1}(\mathcal{K})^3} \le c\Big(\sum_{k=1}^{3} \|\zeta_\nu f^{(k)}\|_{V^{0,s}_{\beta,\delta}(\mathcal{K})^3} + \|\zeta_\nu g\|_{V^{0,s}_{\beta,\delta}(\mathcal{K})}\Big) \tag{10.6.14}$$

can be deduced from Lemma 3.5.4. Furthermore by Theorem 10.4.2,

$$G_{4,j}(x,\xi) = -\nabla_x \cdot \vec{P}_j(x,\xi) + Q_j(x,\xi) \quad \text{for } j = 1,\dots,4,$$

where $\vec{P}_j(x,\xi)\cdot n$ for $x \in \Gamma_k$, $k = 1,\dots,d$, $\xi \in \mathcal{K}$, and

$$|\partial_\xi^\gamma \vec{P}_j(x,\xi)| \le c_{\alpha,\gamma}\,|x-\xi|^{-1-\delta_{j,4}-|\gamma|}, \tag{10.6.15}$$

$$|\partial_\xi^\gamma Q_j(x,\xi)| \le c_{\alpha,\gamma}\,r(x)^{-2-\delta_{j,4}-|\gamma|} \tag{10.6.16}$$

for $|x-\xi| < \min(r(x), r(\xi))$. Using this representation, we obtain

$$q^+ = -\nabla \cdot W + R\,,$$

where

$$W(x) = \sum_{k=1}^{3}\sum_{j=1}^{3} \int_{\mathcal{K}} \zeta_\nu(\xi)\, f_j^{(k)}(\xi)\,\chi^+(x,\xi)\,\partial_{\xi_k}\vec{P}_j(x,\xi)\,d\xi + \int_{\mathcal{K}} \zeta_\nu(\xi)\,g(\xi)\,\chi^+(x,\xi)\,\vec{P}_4(x,\xi)\,d\xi$$

and

$$R(x) = \sum_{k=1}^{3}\sum_{j=1}^{3} \int_{\mathcal{K}} \zeta_\nu(\xi)\, f_j^{(k)}(\xi)\Big(\chi^+(x,\xi)\,\partial_{\xi_k}Q_j(x,\xi) + \partial_{\xi_k}\vec{P}_j(x,\xi)\cdot\nabla_x\chi^+(x,\xi)\Big)d\xi$$
$$+ \int_{\mathcal{K}} \zeta_\nu(\xi)\,g(\xi)\Big(\chi^+(x,\xi)\,Q_4(x,\xi) + \vec{P}_4(x,\xi)\cdot\nabla_x\chi^+(x,\xi)\Big)d\xi - \frac{\zeta_\nu(x)g(x)}{2}.$$

The inequalities (10.6.15), (10.6.16) imply

$$\big|\chi^+(x,\xi)\,\partial_{\xi_k}Q_j(x,\xi) + \partial_{\xi_k}\vec{P}_j(x,\xi)\cdot\nabla\chi^+(x,\xi)\big| \le c\,r(x)^{-1}\,|x-\xi|^{-2}$$

for $j = 1,2,3$ and

$$\big|\chi^+(x,\xi)\,Q_4(x,\xi) + \vec{P}_4(x,\xi)\cdot\nabla\chi^+(x,\xi)\big| \le c\,r(x)^{-1}\,|x-\xi|^{-2}.$$

Thus, we obtain the estimate

$$\|r\,R\|_{V^{0,s}_{\beta-1,\delta-1}(\mathcal{K})^3} \le c\Big(\|r\,g\|_{V^{0,s}_{\beta-1,\delta-1}(\mathcal{K})^3} + \sum_{k=1}^{3} \|\zeta_\nu f^{(k)}\|_{V^{0,s}_{\beta,\delta}(\mathcal{K})^3} + \|\zeta_\nu g\|_{V^{0,s}_{\beta,\delta}(\mathcal{K})}\Big)$$

analogously to (10.6.14). This implies

$$\|R\|_{V^{0,s}_{\beta,\delta}(\mathcal{K})^3} \le c\Big(\sum_{k=1}^{3} \|\zeta_\nu f^{(k)}\|_{V^{0,s}_{\beta,\delta}(\mathcal{K})^3} + \|\zeta_\nu g\|_{V^{0,s}_{\beta,\delta}(\mathcal{K})}\Big).$$

Furthermore, it follows from (10.6.15) and Lemma 3.5.4 that

$$\|W\|_{V^{0,s}_{\beta-1,\delta-1}(\mathcal{K})^3} \le c\Big(\sum_{k=1}^3 \|\zeta_\nu f^{(k)}\|_{V^{0,s}_{\beta,\delta}(\mathcal{K})^3} + \|\zeta_\nu g\|_{V^{0,s}_{\beta,\delta}(\mathcal{K})}\Big).$$

Since $W \cdot n = 0$ on Γ_j, we get

$$\Big|\int_{\mathcal{K}} (\nabla \cdot W)\, w\, dx\Big| = \Big|\int_{\mathcal{K}} W \cdot \nabla w\, dx\Big| \le \|W\|_{V^{0,s}_{\beta-1,\delta-1}(\mathcal{K})^3}\, \|w\|_{V^{1,s'}_{1-\beta,1-\delta}(\mathcal{K})}$$
$$\le c\Big(\sum_{k=1}^3 \|\zeta_\nu f^{(k)}\|_{V^{0,s}_{\beta,\delta}(\mathcal{K})^3} + \|\zeta_\nu g\|_{V^{0,s}_{\beta,\delta}(\mathcal{K})}\Big)\, \|w\|_{V^{1,s'}_{1-\beta,1-\delta}(\mathcal{K})}$$

for all $w \in V^{1,s'}_{1-\beta,1-\delta}(\mathcal{K})$, $s' = s/(s-1)$. Thus,

$$\|q^+\|_{(V^{1,s'}_{1-\beta,1-\delta}(\mathcal{K}))^*} \le c\Big(\sum_{k=1}^3 \|\zeta_\nu f^{(k)}\|_{V^{0,s}_{\beta,\delta}(\mathcal{K})^3} + \|\zeta_\nu g\|_{V^{0,s}_{\beta,\delta}(\mathcal{K})}\Big). \tag{10.6.17}$$

The vector function (v^+, q^+) satisfies the equations

$$b_{\mathcal{K}}(v^+, w) - \int_{\mathcal{K}} q^+\, \nabla \cdot w\, dx = \Phi(w) \quad \text{for all } w \in \mathcal{H}_{s',-\beta,-\delta}\,,$$
$$-\nabla \cdot v^+ = \zeta_\nu g + \nabla \cdot v^- \text{ in } \mathcal{K}, \quad S_j v^+\big|_{\Gamma_j} = -S_j v^-\big|_{\Gamma_j},\ j = 1, \dots, d,$$

where

$$\Phi(w) = -b_{\mathcal{K}}(v^-, w) + \int_{\mathcal{K}} q^-\, \nabla \cdot w\, dx + \int_{\mathcal{K}} \zeta_\nu f^{(k)} \cdot \partial_{x_k} w\, dx.$$

Let $\eta_\mu = \zeta_{\mu-1} + \zeta_\mu + \zeta_{\mu+1}$. By Lemma 10.6.3, the functional $\eta_\mu \Phi$ is continuous on $\mathcal{H}_{s',-\beta,-\delta}$ and

$$\|\eta_\mu \Phi\|_{\mathcal{H}^*_{s',-\beta,-\delta}} \le c\Big(\sum_{k=1}^3 \|\zeta_\nu f^{(k)}\|_{V^{0,s}_{\beta,\delta}(\mathcal{K})^3} + \|\zeta_\nu g\|_{V^{0,s}_{\beta,\delta}(\mathcal{K})}\Big) \tag{10.6.18}$$

Furthermore,

$$\|\eta_\mu \nabla \cdot v^+\|_{V^{0,s}_{\beta,\delta}(\mathcal{K})} + \sum_{j=1}^d \|\eta_\mu S_j v^+\|_{V^{1-1/s,s}_{\beta,\delta}(\Gamma_j)^{3-d_j}}$$
$$\le c\Big(\sum_{k=1}^3 \|\zeta_\nu f^{(k)}\|_{V^{0,s}_{\beta,\delta}(\mathcal{K})^3} + \|\zeta_\nu g\|_{V^{0,s}_{\beta,\delta}(\mathcal{K})}\Big). \tag{10.6.19}$$

Using (10.6.14), (10.6.17)–(10.6.19) and the estimate

$$\|\zeta_\mu v^+\|_{V^{1,s}_{\beta,\delta}(\mathcal{K})^3} + \|\zeta_\mu q^+\|_{V^{0,s}_{\beta,\delta}(\mathcal{K})} \le c\Big(\|\eta_\mu \Phi\|_{\mathcal{H}^*_{s',-\beta,-\delta}} + \|\eta_\mu \nabla \cdot v^+\|_{V^{0,s}_{\beta,\delta}(\mathcal{K})}$$
$$+ \sum_{j=1}^d \|\eta_\mu S_j v^+\|_{V^{1-1/s,s}_{\beta,\delta}(\Gamma_j)^{3-d_j}} + \|\eta_\mu v^+\|_{V^{0,s}_{\beta-1,\delta-1}(\mathcal{K})^3} + \|\eta_\mu q^-\|_{(V^{1,s'}_{1-\beta,1-\delta}(\mathcal{K}))^*}\Big)$$

(cf. Lemma 10.2.5), we obtain (10.6.13). □

10.6.4. Existence and uniqueness of solutions in $W^{1,s}_{\beta,\delta}(\mathcal{K})^3 \times W^{0,s}_{\beta,\delta}(\mathcal{K})$. Lemmas 10.5.1–10.5.3 together with Lemma 3.5.8 enable us to prove the following statement.

THEOREM 10.6.5. *Let $F \in \mathcal{H}^*_{s',-\beta,-\delta}$, $g \in W^{0,s}_{\beta,\delta}(\mathcal{K})$ and $h_j \in W^{1-1/s,s}_{\beta,\delta}(\Gamma_j)^{3-d_j}$ for $j = 1,\dots,d$. We assume that the line $\mathrm{Re}\,\lambda = 1-\beta-3/s$ is free of eigenvalues of the pencil $\mathfrak{A}(\lambda)$, the components of δ satisfy the inequalities* (10.6.1) *and that the boundary data h_j satisfy the compatibility condition* (10.2.6) *on the edges M_k. Then there exists a unique solution $(u,p) \in W^{1,s}_{\beta,\delta}(\mathcal{K})^3 \times W^{0,s}_{\beta,\delta}(\mathcal{K})$ of the problem* (10.6.2), (10.6.3).

P r o o f. By Lemma 10.2.2, there exists a vector function $v \in W^{1,s}_{\beta,\delta}(\mathcal{K})^3$ satisfying the condition $S_j v = h_j$ on Γ_j, $j = 1,\dots,d$ and the estimate

$$\|v\|_{W^{1,s}_{\beta,\delta}(\mathcal{K})^3} \le c \sum_{j=1}^{d} \|h_j\|_{W^{1-1/s,s}_{\beta,\delta}(\Gamma_j)^{3-d_j}}\,.$$

Thus, we may assume without loss of generality that $h_j = 0$ for $j = 1,\dots,d$. By Lemma 10.6.1, the functional F has the form (10.6.4), where $f^{(0)} \in W^{0,s}_{\beta+1,\delta+1}(\mathcal{K})^3$ and $f^{(k)} \in W^{0,s}_{\beta,\delta}(\mathcal{K})^3$ for $k = 1,2,3$. Since the problem

$$b_{\mathcal{K}}(u,v) - \int_{\mathcal{K}} p\,\nabla\cdot v\,dx = \int_{\mathcal{K}} f^{(0)}(x)\cdot v(x)\,dx \quad \text{for all } v \in \mathcal{H}_{s',-\beta,-\delta},$$

$$-\nabla\cdot u = 0 \ \text{ in } \mathcal{K}, \quad S_j u = 0 \ \text{ on } \Gamma_j,\ j = 1,\dots,d,$$

has a uniquely determined solution in the subspace $W^{2,s}_{\beta+1,\delta+1}(\mathcal{K})^3 \times W^{1,s}_{\beta+1,\delta+1}(\mathcal{K})$ of $W^{1,s}_{\beta,\delta}(\mathcal{K})^3 \times W^{0,s}_{\beta,\delta}(\mathcal{K})$ (see Theorem 10.5.6), we may assume without loss of generality that $f^{(0)} = 0$. Let $G(x,\xi)$ be the Green's matrix introduced in Theorem 10.4.1 for an arbitrary κ in the interval $-\Lambda_+ - 1/2 < \kappa < -\Lambda_- - 1/2$, where $\Lambda_- < \mathrm{Re}\,\lambda < \Lambda_+$ is the widest strip in the complex plane which is free of eigenvalues of the pencil $\mathfrak{A}(\lambda)$ and which contains the line $\mathrm{Re}\,\lambda = 1-\beta-3/s$. Then the vector function (u,p) with the components (10.6.5) and (10.6.6) is a solution of the problem (10.6.2), (10.6.3). According to Lemmas 10.6.2–10.6.4, the operator

$$W^{0,s}_{\beta,\delta}(\mathcal{K})^4 \ni \left(f^{(1)}, f^{(2)}, f^{(3)}, g\right) \to \mathcal{O}\left(f^{(1)}, f^{(2)}, f^{(3)}, g\right) = (u,p)$$

satisfies the conditions of Lemma 3.5.8 for the spaces $\mathcal{X} = W^{0,s}_{\beta,\delta}(\mathcal{K})^4$ and $\mathcal{Y} = W^{1,s}_{\beta,\delta}(\mathcal{K})^3 \times W^{0,s}_{\beta,\delta}(\mathcal{K})$. Consequently, this operator realizes a continuous mapping from $\mathcal{X}$ into $\mathcal{Y}$, i.e., problem (10.6.2), (10.6.3) is solvable in $\mathcal{Y}$.

We prove the uniqueness. Let $(u,p) \in \mathcal{Y}$ be a solution of problem (10.6.2), (10.6.3) with zero data F, g, h_j. Then (u,p) is also a solution of the homogeneous boundary value problem (10.5.13), (10.5.14), and from Lemma 10.2.7 it follows that $(u,p) \in W^{2,s}_{\beta+1,\delta+1}(\mathcal{K})^3 \times W^{1,s}_{\beta+1,\delta+1}(\mathcal{K})$. Thus, Theorem 10.5.6 implies $(u,p) = (0,0)$. The proof is complete. □

REMARK 10.6.6. The condition (10.6.1) in Theorem 10.6.5 excludes the case $s = 2$, $\delta_k = 0$. However by Theorem 10.2.11, the problem

$$b_{\mathcal{K}}(u,v) - \int_{\mathcal{K}} p\,\nabla\cdot v\,dx = F(v) \quad \text{for all } v \in \mathcal{H}_{2,-\beta,0},$$

$$-\nabla\cdot u = g \ \text{ in } \mathcal{K}, \quad S_j u = 0 \ \text{ on } \Gamma_j,\ j = 1,\dots,d,$$

has a unique solution $(u,p) \in W^{1,2}_{\beta,0}(\mathcal{K})^3 \times W^{0,2}_{\beta,0}(\mathcal{K})^3$ for arbitrary $F \in \mathcal{H}^*_{2,-\beta,0}$ and $g \in W^{0,2}_{\beta,0}(\mathcal{K})$ if the line $\text{Re}\,\lambda = -\beta - 1/2$ is free of eigenvalues of the pencil $\mathfrak{A}(\lambda)$. It can be shown analogously to Lemma 7.7.1 that an arbitrary functional $F \in \mathcal{H}^*_{2,-\beta,0}$ has the form

$$F(v) = \int_{\mathcal{K}} \Big(f^{(0)} \cdot v + \sum_{k=1}^{3} f^{(k)} \cdot \partial_{x_k} v \Big)\, dx \quad \text{for all } v \in \mathcal{H}_{2,-\beta,0}\,,$$

where $f^{(0)} \in W^{0,s}_{\beta+1,0}(\mathcal{K})^3$ and $f^{(k)} \in W^{0,s}_{\beta,0}(\mathcal{K})^3$, $k = 1,2,3$. Then the components of the solution $(u,p) \in W^{1,2}_{\beta,0}(\mathcal{K})^3 \times W^{0,2}_{\beta,0}(\mathcal{K})^3$ are also given by the formulas (10.6.5) and (10.6.6), where $G(x,\xi)$ is the Green's matrix introduced in Theorem 10.4.1 and κ is such that the closed strip between the lines $\text{Re}\,\lambda = -\beta - 1/2$ and $\text{Re}\,\lambda = -\kappa - 1/2$ does not contain eigenvalues of the pencil $\mathfrak{A}(\lambda)$.

10.6.5. Regularity results for variational solutions. Our goal is to show that the solution $(u,p) \in W^{1,s}_{\beta,\delta}(\mathcal{K})^3 \times W^{0,s}_{\beta,\delta}(\mathcal{K})$ of the problem (10.6.2), (10.6.3) belongs to the space $W^{l,q}_{\beta',\delta'}(\mathcal{K})^3 \times W^{l-1,q}_{\beta',\delta'}(\mathcal{K})$ under certain natural conditions on the data and on the weight parameters. We start with the case $l = 1$.

THEOREM 10.6.7. *Suppose that the closed strip between the lines* $\text{Re}\,\lambda = 1 - \beta - 3/s$ *and* $\text{Re}\,\lambda = 1 - \beta' - 3/q$ *is free of eigenvalues of the pencil* $\mathfrak{A}$ *and that the components of* δ *and* δ' *satisfy the inequalities*

$$\max(0, 1 - \delta_+^{(k)}) < \delta_k + 2/s < 1, \quad \max(0, 1 - \delta_+^{(k)}) < \delta'_k + 2/q < 1$$

for $k = 1, \ldots, d$. *If* $(u,p) \in W^{1,s}_{\beta,\delta}(\mathcal{K})^3 \times W^{0,s}_{\beta,\delta}(\mathcal{K})$ *is a variational solution of the problem* (10.6.2), (10.6.3), *where*

$$F \in \mathcal{H}^*_{s,-\beta,-\delta} \cap \mathcal{H}^*_{q',-\beta',-\delta'}, \quad s' = s/(s-1), \quad q' = q/(q-1),$$

$$g \in W^{0,s}_{\beta,\delta}(\mathcal{K}) \cap W^{0,q}_{\beta',\delta'}(\mathcal{K}), \quad h_j \in W^{1-1/s,s}_{\beta,\delta}(\Gamma_j)^{3-d_j} \cap W^{1-1/q,q}_{\beta',\delta'}(\Gamma_j)^{3-d_j},$$

then $(u,p) \in W^{1,q}_{\beta',\delta'}(\mathcal{K})^3 \times W^{0,q}_{\beta',\delta'}(\mathcal{K})$. *In the case* $s = 2$ *this result is also true for* $\delta = 0$.

P r o o f. By Lemma 10.2.2, we may restrict ourselves to the case of zero boundary data h_j. Then the uniquely determined solution both in $W^{1,s}_{\beta,\delta}(\mathcal{K})^3 \times W^{0,s}_{\beta,\delta}(\mathcal{K})$ and in $W^{1,q}_{\beta',\delta'}(\mathcal{K})^3 \times W^{0,q}_{\beta',\delta'}(\mathcal{K})$ is given by the formulas (10.6.5) and (10.6.6), where $G(x,\xi)$ is the Green's matrix introduced in Theorem 10.4.1 for $\kappa = \beta + 3/s - 3/2$. This proves the theorem. □

By the same argument, the subsequent statement is true.

LEMMA 10.6.8. *Let* $(u,p) \in W^{1,s}_{\beta,\delta}(\mathcal{K})^3 \times W^{0,s}_{\beta,\delta}(\mathcal{K})$ *be a solution of the problem* (10.6.2), (10.6.3), *where* $g \in W^{0,s}_{\beta,\delta}(\mathcal{K}) \cap W^{1,q}_{\beta',\delta'}(\mathcal{K})$, $h_j \in W^{1-1/s,s}_{\beta,\delta}(\Gamma_j)^{3-d_j} \cap W^{2-1/q,q}_{\beta',\delta'}(\Gamma_j)^{3-d_j}$ *and* F *is a linear and continuous functional on* $\mathcal{H}_{s',-\beta,-\delta}$ *which has the representation*

$$F(v) = \int_{\mathcal{K}} (f + \nabla g) \cdot v\, dx + \sum_{j=1}^{d} \int_{\Gamma_j} \phi_j \cdot v\, dx \tag{10.6.20}$$

with given vector functions $f \in W^{0,q}_{\beta',\delta'}(\mathcal{K})^3$, $\phi_j \in W^{1-1/q,q}_{\beta',\delta'}(\Gamma_j)^{d_j}$. *Suppose that the closed strip between the lines* $\operatorname{Re}\lambda = 1-\beta-3/s$ *and* $\operatorname{Re}\lambda = 2-\beta'-3/q$ *does not contain eigenvalues of the pencil* $\mathfrak{A}(\lambda)$ *and the components of* δ, δ' *satisfy the inequalities*

$$\max(0, 1-\delta_+^{(k)}) < \delta_k + 2/s < 1, \qquad \max(0, 2-\delta_+^{(k)}) < \delta'_k + 2/q < 2$$

for $k = 1, \dots, d$. *Then* $(u,p) \in W^{2,q}_{\beta',\delta'}(\mathcal{K})^3 \times W^{1,q}_{\beta',\delta'}(\mathcal{K})$. *In the case* $s = 2$ *this result is also true for* $\delta = 0$.

We prove a generalization of the last lemma.

THEOREM 10.6.9. *Let* $(u,p) \in W^{1,s}_{\beta,\delta}(\mathcal{K})^3 \times W^{0,s}_{\beta,\delta}(\mathcal{K})$ *be a solution of the problem* (10.6.2), (10.6.3), *where* $g \in W^{0,s}_{\beta,\delta}(\mathcal{K}) \cap W^{l-1,q}_{\beta',\delta'}(\mathcal{K})$, $h_j \in W^{1-1/s,s}_{\beta,\delta}(\Gamma_j)^{3-d_j} \cap W^{l-1/q,q}_{\beta',\delta'}(\Gamma_j)^{3-d_j}$ *and* F *is a linear and continuous functional on* $\mathcal{H}_{s',-\beta,-\delta}$ *which has the representation* (10.6.20) *with given vector functions* $f \in W^{l-2,q}_{\beta',\delta'}(\mathcal{K})^3$, $\phi_j \in W^{l-1-1/q,q}_{\beta',\delta'}(\Gamma_j)^{d_j}$. *Suppose that the closed strip between the lines* $\operatorname{Re}\lambda = 1-\beta-3/s$ *and* $\operatorname{Re}\lambda = l-\beta'-3/q$ *does not contain eigenvalues of the pencil* $\mathfrak{A}(\lambda)$ *and the components of* δ, δ' *satisfy the inequalities*

$$\max(0, 1-\delta_+^{(k)}) < \delta_k + 2/s < 1, \qquad \max(0, l-\delta_+^{(k)}) < \delta'_k + 2/q < l$$

for $k = 1, \dots, d$. *Then* $(u,p) \in W^{l,q}_{\beta',\delta'}(\mathcal{K})^3 \times W^{l-1,q}_{\beta',\delta'}(\mathcal{K})$. *In the case* $s = 2$ *this result is also true for* $\delta = 0$.

P r o o f. 1) First we consider the case where

$$\delta'_k + 2/q > l-2 \quad \text{for } k = 1, \dots, d.$$

Then $f \in W^{0,q}_{\beta'-l+2,\delta'-l+2}(\mathcal{K})^3$, $g \in W^{1,q}_{\beta'-l+2,\delta'-l+2}(\mathcal{K})$, $h_j \in W^{2-1/q,q}_{\beta'-l+2,\delta'-l+2}(\Gamma_j)^{3-d_j}$ and $\phi_j \in W^{1-1/q,q}_{\beta'-l+2,\delta'-l+2}(\Gamma_j)^{d_j}$. Hence, it follows from Lemma 10.6.8 that

$$(u,p) \in W^{2,q}_{\beta'-l+2,\delta'-l+2}(\mathcal{K})^3 \times W^{1,q}_{\beta'-l+2,\delta'-l+2}(\mathcal{K}).$$

Applying Lemma 10.2.7, we obtain $(u,p) \in W^{l,q}_{\beta',\delta'}(\mathcal{K})^3 \times W^{l-1,q}_{\beta',\delta'}(\mathcal{K})$.

2) Suppose that

$$l-3 < \delta'_k + 2/q \le l-2 \quad \text{for } k = 1, \dots, d.$$

This is only possible if $l \ge 3$ and $\delta_+^{(k)} > 2$. Let the components of γ satisfy the inequalities

$$\delta'_k - l + 2 + 2/q \le 0 < \gamma_k + 2/q < \delta'_k - l + 3 + 2/q \le 1.$$

Then $f \in W^{0,q}_{\beta'-l+2,\gamma}(\mathcal{K})^3$, $g \in W^{1,q}_{\beta'-l+2,\gamma}(\mathcal{K})$, $h_j \in W^{2-1/q,q}_{\beta'-l+2,\gamma}(\Gamma_j)^{3-d_j}$ and $\phi_j \in W^{1-1/q,q}_{\beta'-l+2,\gamma}(\Gamma_j)^{d_j}$. Thus, we conclude from Lemma 10.6.8 that

$$(u,p) \in W^{2,q}_{\beta'-l+2,\gamma}(\mathcal{K})^3 \times W^{1,q}_{\beta'-l+2,\gamma}(\mathcal{K}).$$

This together with Lemma 10.2.7 yields $(u,p) \in W^{3,q}_{\beta'-l+3,\gamma+1}(\mathcal{K})^3 \times W^{2,q}_{\beta'-l+3,\gamma+1}(\mathcal{K})$. Consequently, $\rho\partial_\rho(u,p) \in W^{2,q}_{\beta'-l+2,\gamma+1}(\mathcal{K})^3 \times W^{1,q}_{\beta'-l+2,\gamma+1}(\mathcal{K})$. The vector function $(\rho\partial_\rho u, \rho\partial_\rho p + p)$ is a solution of the boundary value problem

$$(10.6.21) \qquad -\Delta(\rho\partial_\rho u) + \nabla(\rho\partial_\rho p + p) = (\rho\partial_\rho + 2)f, \quad -\nabla\cdot(\rho\partial_\rho u) = (\rho\partial_\rho + 1)g,$$

$$(10.6.22) \qquad S_j\rho\partial_\rho u = \rho\partial_\rho h_j\,, \quad N_j(\rho\partial_\rho u, \rho\partial_\rho p + p) = (\rho\partial_\rho + 1)\phi_j \ \text{ on } \Gamma_j,$$

where

$$(\rho\partial_\rho+2)f\in W^{0,q}_{\beta'-l+2,\delta'-l+3}(\mathcal{K})^3,\quad (\rho\partial_\rho+1)g\in W^{1,q}_{\beta'-l+2,\delta'-l+3}(\mathcal{K}),$$

$$\rho\partial_\rho h_j\in W^{2-1/q,q}_{\beta'-l+2,\delta'-l+3}(\Gamma_j)^{3-d_j},\quad (\rho\partial_\rho+1)\phi_j\in W^{1-1/q,q}_{\beta'-l+2,\delta'-l+3}(\Gamma_j)^{d_j}.$$

Since $0<\delta'_k-l+3+2/q\le 1$, it follows from Theorem 10.5.7 that

$$(\rho\partial_\rho u,\rho\partial_\rho p+p)\in W^{2,q}_{\beta'-l+2,\delta'-l+3}(\mathcal{K})^3\times W^{2,q}_{\beta'-l+2,\delta'-l+3}(\mathcal{K}).$$

Obviously, the vector function (u,p) belongs to the same space. Therefore by Lemma 10.2.8, $(u,p)\in W^{3,q}_{\beta'-l+3,\delta'-l+3}(\mathcal{K})^3\times W^{2,q}_{\beta'-l+3,\delta'-l+3}(\mathcal{K})$. Using Lemma 10.2.7, we obtain $(u,p)\in W^{l,q}_{\beta',\delta'}(\mathcal{K})^3\times W^{l-1,q}_{\beta',\delta'}(\mathcal{K})$. This proves the theorem for the case $l-3<\delta'_k+2/q\le l-2$, $k=1,\dots,d$.

3) We prove by induction in N that the assertion of the theorem is true if

$$l-N-1<\delta'_k+2/q\le l-N\quad\text{for } k=1,\dots,d, \tag{10.6.23}$$

where N is an integer, $2\le N\le l-1$. For $N=2$, we refer to the second part of the proof. Suppose that the components of δ' satisfy the inequalities (10.6.23) with a certain $N\in\{3,\dots,l-1\}$ and that the theorem is proved for $l-N<\delta'_k+2/q\le l-N+1$. Since $f\in W^{l-3,q}_{\beta'-1,\delta'}(\mathcal{K})^3$, $g\in W^{l-2,q}_{\beta'-1,\delta'}(\mathcal{K})$, $h_j\in W^{l-1-1/q,q}_{\beta'-1,\delta'}(\Gamma_j)^{3-d_j}$ and $\phi_j\in W^{l-2-1/q,q}_{\beta'-1,\delta'}(\Gamma_j)^{d_j}$, it follows from the induction hypothesis that

$$(u,p)\in W^{l-1,q}_{\beta'-1,\delta'}(\mathcal{K})^3\times W^{l-2,q}_{\beta'-1,\delta'}(\mathcal{K}).$$

Consequently, $\rho\partial_\rho(u,p)\in W^{l-2,q}_{\beta'-2,\delta'}(\mathcal{K})^3\times W^{l-2,q}_{\beta'-2,\delta'}(\mathcal{K})$. Moreover, the vector function $(\rho\partial_\rho u,\rho\partial_\rho p+p)$ is a solution of the boundary value problem (10.6.21), (10.6.22), where

$$(\rho\partial_\rho+2)f\in W^{l-3,q}_{\beta'-1,\delta'}(\mathcal{K})^3,\quad (\rho\partial_\rho+1)g\in W^{l-2,q}_{\beta'-1,\delta'}(\mathcal{K}),$$

$$\rho\partial_\rho h_j\in W^{l-1-1/q,q}_{\beta'-1,\delta'}(\Gamma_j)^{3-d_j},\quad (\rho\partial_\rho+1)\phi_j\in W^{1-1/q,q}_{\beta'-l+2,\delta'-l+3}(\Gamma_j)^{d_j}.$$

Hence, we conclude from the induction hypothesis that

$$(\rho\partial_\rho u,\rho\partial_\rho p+p)\in W^{l-1,q}_{\beta'-1,\delta'}(\mathcal{K})^3\times W^{l-2,q}_{\beta'-1,\delta'}(\mathcal{K}).$$

Applying Lemma 10.2.8, we obtain $(u,p)\in W^{l,q}_{\beta',\delta'}(\mathcal{K})^3\times W^{l-1,q}_{\beta',\delta'}(\mathcal{K})$.

4) Finally, we consider the case where $l-N_k-1<\delta'_k+2/q\le l-N_k$ for $k=1,\dots,d$ with different $N_k\in\{0,\dots,l-1\}$. Then let $\psi_1,\dots,\psi_d$ be infinitely differentiable functions on $\overline{\Omega}$ such that $\psi_k\ge 0$, $\psi_k=1$ near $M_k\cap S^2$, and $\sum\psi_k=1$. We extend ψ_k to $\mathcal{K}$ by the equality $\psi_k(x)=\psi_k(x/|x|)$. Then by the first three parts of the proof, $\psi_k(u,p)\in W^{l,q}_{\beta',\delta'}(\mathcal{K})^3\times W^{l-1,q}_{\beta',\delta'}(\mathcal{K})$ for $k=1,\dots,d$. This proves the theorem. □

REMARK 10.6.10. If $d_{k_+}+d_{k_-}$ is even, then the condition $\max(0,l-\delta_+^{(k)})<\delta'_k+2/q<l$ in Theorem 10.6.9 can be replaced by the weaker condition

$$\max(0,l-\mu_+^{(k)})<\delta'_k+2/q<l.$$

However, then one must assume in addition that $g,h_{k_+},h_{k_-},\phi_{k_+},\phi_{k_-}$ satisfy the compatibility conditions of Lemma 10.2.3 on the edge M_k if $\delta'_k+2/q<l-1$ or the same conditions in the generalized sense (cf. Remark 10.2.4) if $\delta'_k+2/q=l-1$.

10.6.6. Existence of solutions in $W^{l,s}_{\beta,\delta}(\mathcal{K})^3 \times W^{l-1,s}_{\beta,\delta}(\mathcal{K})$. The last theorem together with the existence and uniqueness result for weak solutions imply the following theorem.

THEOREM 10.6.11. *Suppose that the line* $\mathrm{Re}\,\lambda = l-\beta-3/s$ *is free of eigenvalues of the pencil* $\mathfrak{A}(\lambda)$ *and that the components of* δ *satisfy the inequalities*

$$\max(0, l-\delta_+^{(k)}) < \delta_k + 2/s < l. \tag{10.6.24}$$

Then the problem (10.1.1), (10.1.2) *is uniquely solvable in* $W^{l,s}_{\beta,\delta}(\mathcal{K})^3 \times W^{l-1,s}_{\beta,\delta}(\mathcal{K})$ *for arbitrary* $f \in W^{l-2,s}_{\beta,\delta}(\mathcal{K})^3$, $g \in W^{l-1,s}_{\beta,\delta}(\mathcal{K})$, $h_j \in W^{l-1/s,s}_{\beta,\delta}(\Gamma_j)^{3-d_j}$ *satisfying* (10.2.6), *and* $\phi_j \in W^{l-1-1/s,s}_{\beta,\delta}(\Gamma_j)^{d_j}$.

P r o o f. Let $\delta' = (\delta'_1, \dots, \delta'_d)$ be such that

$$\delta'_k \ge \delta_k - l + 1 \quad \text{and} \quad \max(0, 1-\delta_+^{(k)}) < \delta'_k + 2/s < 1 \quad \text{for } k = 1, \dots, d.$$

Then $f + \nabla g \in W^{0,s}_{\beta-l+2,\delta'+1}(\mathcal{K})^3$ and $\phi_j \in W^{1-1/s,s}_{\beta-l+2,\delta'+1}(\Gamma_j)^{d_j} = V^{1-1/s,s}_{\beta-l+2,\delta'+1}(\Gamma_j)^{d_j}$. Since moreover $\mathcal{H}_{s',l-1-\beta,-\delta'} \subset V^{1,s'}_{l-1-\beta,-\delta'}(\mathcal{K})^3$ for $s' = s/(s-1)$, it follows that the functional F defined by (10.6.20) is linear and continuous on $\mathcal{H}_{s',l-1-\beta,-\delta'}$. Furthermore, $g \in W^{0,s}_{\beta-l+1,\delta'}(\mathcal{K})$ and $h_j \in W^{1-1/s,s}_{\beta-l+1,\delta'}(\Gamma_j)^{3-d_j}$. Thus by Theorem 10.6.5, there exists a uniquely determined solution $(u,p) \in W^{1,s}_{\beta-l+1,\delta'}(\mathcal{K})^3 \times W^{0,s}_{\beta-l+1,\delta'}(\mathcal{K})$ of the problem (10.6.2), (10.6.3). Applying Theorem 10.6.9, we conclude that $(u,p) \in W^{l,s}_{\beta,\delta}(\mathcal{K})^3 \times W^{l,s}_{\beta,\delta}(\mathcal{K})$. The last space is a subspace of $W^{1,s}_{\beta-l+1,\delta'}(\mathcal{K})^3 \times W^{0,s}_{\beta-l+1,\delta'}(\mathcal{K})$. This proves the theorem. □

REMARK 10.6.12. If $d_{k_+} + d_{k_-}$ is even, then the condition (10.6.24) in Theorem 10.6.11 can be replaced by the weaker condition

$$\max(0, l-\mu_+^{(k)}) < \delta_k + 2/s < l$$

(cf. Theorem 10.5.4 for $l=2$). However, then one must assume in addition that $g, h_{k_+}, h_{k_-}, \phi_{k_+}, \phi_{k_-}$ satisfy the compatibility conditions of Lemma 10.2.3 on the edge M_k if $\delta_k + 2/s < l-1$ or the same conditions in the generalized sense (cf. Remark 10.2.4) if $\delta_k + 2/s = l-1$.

10.6.7. The case $d_{k_+} + d_{k_-} \le 3$. In the case where $\lambda = 0$ is not an eigenvalue of the pencils $A_k(\lambda)$, $k = 1, \dots, d$ (this means, if $d_{k_+} + d_{k_-} \le 3$ for all k and the edge angles θ_k satisfy the condition (10.3.8) for $d_{k_+} \cdot d_{k_-} \ne 0$), we obtain an existence and uniqueness theorem for the problem (10.6.2), (10.6.3) in the weighted space $V^{1,s}_{\beta,\delta}(\mathcal{K})^3 \times V^{0,s}_{\beta,\delta}(\mathcal{K})$.

THEOREM 10.6.13. *Let* $F \in \mathcal{H}^*_{s',-\beta,-\delta}$, $g \in V^{0,s}_{\beta,\delta}(\mathcal{K})$ *and* $h_j \in V^{1-1/s,s}_{\beta,\delta}(\Gamma_j)^{3-d_j}$ *for* $j = 1, \dots, d$. *We assume that* $d_{k_+} + d_{k_-} \le 3$ *for all* k, *the edge angles* θ_k *satisfy the condition* (10.3.8) *for* $d_{k_+} \cdot d_{k_-} \ne 0$, *the line* $\mathrm{Re}\,\lambda = 1-\beta-3/s$ *is free of eigenvalues of the pencil* $\mathfrak{A}(\lambda)$, *and the components of* δ *satisfy the inequalities*

$$1-\delta_+^{(k)} < \delta_k + 2/s < 1 + \delta_+^{(k)}$$

for $k = 1, \dots, d$. *Then there exists a unique solution* $(u,p) \in V^{1,s}_{\beta,\delta}(\mathcal{K})^3 \times V^{0,s}_{\beta,\delta}(\mathcal{K})$ *of the problem* (10.6.2), (10.6.3).

P r o o f. Under the assumptions of the theorem, we can use the estimates of Green's matrix in Theorems 10.4.2, 10.4.4 and 10.4.5 with the exponents $\sigma_{i,\alpha}(x) = \delta_x - \delta_{i,4} - |\alpha| - \varepsilon$ and $\sigma_{k,i,\alpha} = \delta_+^{(k)} - |\alpha| - \delta_{i,4} - \varepsilon$, respectively. Then we obtain the same estimates as in Lemmas 10.6.2 and 10.6.3 for the $V_{\beta,\delta}^{1,s}(\mathcal{K})^3 \times V_{\beta,\delta}^{0,s}(\mathcal{K})$-norms of $\zeta_\mu(v,q)$ and $\zeta_\mu(v^-,q^-)$. Thus, the statement of Theorem 10.6.13 holds analogously to Theorem 10.6.5. □

Furthermore, the following regularity result for variational solutions in the weighted Sobolev spaces $V_{\beta,\delta}^{l,s}(\mathcal{K})$ holds.

THEOREM 10.6.14. *Let $(u,p) \in V_{\beta,\delta}^{1,s}(\mathcal{K})^3 \times V_{\beta,\delta}^{0,s}(\mathcal{K})$ be a solution of the problem* (10.6.2), (10.6.3). *Suppose that $d_{k_+} + d_{k_-} \le 3$ for all k and that the edge angles θ_k satisfy the condition* (10.3.8) *for $d_{k_+} \cdot d_{k_-} \ne 0$. Furthermore, we assume that $g \in V_{\beta,\delta}^{0,s}(\mathcal{K}) \cap V_{\beta',\delta'}^{l-1,q}(\mathcal{K})$, $h_j \in V_{\beta,\delta}^{1-1/s,s}(\Gamma_j)^{3-d_j} \cap V_{\beta',\delta'}^{l-1/q,q}(\Gamma_j)^{3-d_j}$ and that F is a linear and continuous functional on $\mathcal{H}_{s',-\beta,-\delta}$ which has the representation* (10.6.20) *with given vector functions $f \in V_{\beta',\delta'}^{l-2,q}(\mathcal{K})^3$, $\phi_j \in V_{\beta',\delta'}^{l-1-1/q,q}(\Gamma_j)^{d_j}$. If the closed strip between the lines $\operatorname{Re}\lambda = 1-\beta-3/s$ and $\operatorname{Re}\lambda = l-\beta'-3/q$ is free of eigenvalues of the pencil $\mathfrak{A}(\lambda)$ and the components of δ, δ' satisfy the inequalities*

$$1-\delta_+^{(k)} < \delta_k + 2/s < 1+\delta_+^{(k)}, \qquad l-\delta_+^{(k)} < \delta'_k + 2/q < l+\delta_+^{(k)}$$

for $k = 1,\dots,d$, then $(u,p) \in V_{\beta',\delta'}^{l,q}(\mathcal{K})^3 \times V_{\beta',\delta'}^{l-1,q}(\mathcal{K})$.

P r o o f. For $l = 1$ the assertion is true by the same arguments as in the proof of Theorem 10.6.7. Applying Lemma 10.2.6, we obtain the statement for arbitrary $l \ge 1$. □

10.7. Solvability in weighted Hölder spaces

Now we are interested in solutions of the boundary value problem (10.1.1), (10.1.2) in the weighted Hölder spaces $C_{\beta,\delta}^{l,\sigma}(\mathcal{K})$. Using the estimates for Green's matrix obtained in Section 10.4, we are able to prove the existence and uniqueness of solutions in the space $C_{\beta,\delta}^{2,\sigma}(\mathcal{K})^3 \times C_{\beta,\delta}^{1,\sigma}(\mathcal{K})$ provided that the line $\operatorname{Re}\lambda = 2+\sigma-\beta$ is free of eigenvalues of the pencil $\mathfrak{A}(\lambda)$ and the components of δ satisfy the inequalities

$$\text{(10.7.1)} \qquad \delta_k \ge 0, \quad 2-\mu_+^{(k)} < \delta_k - \sigma < 2, \quad \delta_k - \sigma \text{ is not integer}$$

for $k = 1,\dots,d$. Furthermore, we are interested in regularity results for the solutions of the boundary value problem (10.1.1), (10.1.2) in weighted Hölder spaces. We prove that the solution $u \in C_{\beta',\delta'}^{2,\sigma'}(\mathcal{K})^3 \times C_{\beta',\delta'}^{1,\sigma'}(\mathcal{K})$ belongs to the space $C_{\beta,\delta}^{l,\sigma}(\mathcal{K}) \times C_{\beta,\delta}^{l-1,\sigma}(\mathcal{K})$ under certain conditions on the data f, g, h_j, ϕ_j and on the weight parameters β, δ. An analogous result holds for the weak solution $u \in V_\kappa^{1,2}(\mathcal{K})^3 \times V_\kappa^{0,2}(\mathcal{K})$ considered in Section 10.2. For some special cases (for example, if the Dirichlet condition is prescribed on at least one of the adjoining faces of every edge), we obtain analogous results in the weighted spaces $N_{\beta,\delta}^{l,\sigma}$.

10.7.1. Reduction to zero boundary data. The next lemma is an analog to Lemma 10.2.1.

LEMMA 10.7.1. *Let $h_j \in N^{l,\sigma}_{\beta,\delta}(\Gamma_j)^{3-d_j}$, $\phi_j \in N^{l-1,\sigma}_{\beta,\delta}(\Gamma_j)^{d_j}$, $j = 1,\dots,d$, $l \ge 1$, be given. Then there exists a vector function $u \in N^{l,\sigma}_{\beta,\delta}(\mathcal{K})^3$ such that*

$$S_j u = h_j\,, \quad N_j(u,0) = \phi_j \quad \text{on } \Gamma_j\,,\ j = 1,\dots,d,$$

and

$$\|u\|_{N^{l,\sigma}_{\beta,\delta}(\mathcal{K})^3} \le c \sum_{j=1}^{d} \Big(\|h_j\|_{N^{l,\sigma}_{\beta,\delta}(\Gamma_j)^{3-d_j}} + \|\phi_j\|_{N^{l-1,\sigma}_{\beta,\delta}(\Gamma_j)^{d_j}} \Big),$$

where c is a constant independent of h_j and ϕ_j.

P r o o f. Let ζ_k be infinitely differentiable functions depending only on $\rho = |x|$ and satisfying the conditions (10.2.4). We set

$$h_{k,j}(x) = \zeta_k(2^k x)\, h_j(2^k x), \qquad \phi_{k,j}(x) = 2^k\, \zeta_k(2^k x)\, \phi_j(2^k x).$$

The supports of $h_{k,j}$ and $\phi_{k,j}$ are contained in the set $\{x : 1/2 < |x| < 2\}$. Consequently by Lemma 9.7.1, there exists a vector function $v_k \in N^{l,\sigma}_{\beta,\delta}(\mathcal{K})^3$ such that $v_k(x) = 0$ for $|x| < 1/4$ and $|x| > 4$,

$$S_j v_k = h_{k,j}\,, \quad N_j(v_k,0) = \phi_{k,j} \quad \text{on } \Gamma_j\,,\ j = 1,\dots,d,$$

and

$$\|v_k\|_{N^{l,\sigma}_{\beta,\delta}(\mathcal{K})^3} \le c \sum_{j=1}^{N} \Big(\|h_{k,j}\|_{N^{l,\sigma}_{\beta,\delta}(\Gamma_j)^{3-d_j}} + \|\phi_{k,j}\|_{N^{l-1,\sigma}_{\beta,\delta}(\Gamma_j)^{d_j}} \Big). \tag{10.7.2}$$

Here, c is a constant independent of k. Hence, the functions $u_k(x) = v_k(2^{-k}x)$ satisfy the equations

$$S_j u_k = \zeta_k\, h_j\,, \quad N_j(u_k,0) = \zeta_k\, \phi_j \quad \text{on } \Gamma_j\,,\ j = 1,\dots,d,$$

and the estimate (10.7.2) with $\zeta_k h_j$, $\zeta_k \phi_j$ instead of $h_{k,j}$ and $\phi_{k,j}$, respectively. Thus, the vector function $u = \sum u_k$ has the desired properties. □

An analogous result in $C^{l,\sigma}_{\beta,\delta}(\mathcal{K})$ can be deduced only under additional compatibility conditions on the boundary data. We denote again by Γ_{k_+} and Γ_{k_-} the faces of the cone $\mathcal{K}$ adjacent to the edge M_k and by θ_k the inner angle at M_k. If $u \in C^{l,\sigma}_{\beta,\delta}(\mathcal{K})$ and $\delta_k < l + \sigma$, then the trace of u on M_k exists and from the equations $S_j u = h_j$ on Γ_j it follows that h_{k_+} and h_{k_-} satisfy the compatibility condition (10.2.6) on the edge M_k. Using Lemmas 9.7.2–9.7.5 instead of Lemma 9.7.1, one can prove the following assertion in the same way as Lemma 10.7.1.

LEMMA 10.7.2. *Let $f \in C^{l-2,\sigma}_{\beta,\delta}(\mathcal{K})^3$, $g \in C^{l-1,\sigma}_{\beta,\delta}(\mathcal{K})$, $h_j \in C^{l,\sigma}_{\beta,\delta}(\Gamma_j)^{3-d_j}$, and $\phi_j \in C^{l-1,\sigma}_{\beta,\delta}(\Gamma_j)^{d_j}$, where $l \ge 2$, $\delta_k \ge 0$, $l - 3 < \delta_k - \sigma < l$ and $\delta_k - \sigma$ is not integer for $k = 1,\dots,d$. Suppose that the boundary data h_j satisfy the compatibility condition (10.2.6). Furthermore, we assume that the edge angle θ_k satisfies the condition (9.7.2) if $d_{k_+} + d_{k_-}$ is odd and $\delta_k < l + \sigma - 1$, and that the functions $g, h_{k_\pm}, \phi_{k_\pm}$ satisfy the compatibility conditions (9.2.13)–(9.2.18) if $d_{k_+} + d_{k_-}$ is even and $\delta_k < l + \sigma - 1$. In the case $\delta_k < l + \sigma - 2$, we assume in addition that $\lambda = 2$ is not an eigenvalue of the pencil $A_k(\lambda)$. Then there exists a vector function $(u,p) \in C^{l,\sigma}_{\beta,\delta}(\mathcal{K})^3 \times C^{l-1,\sigma}_{\beta,\delta}(\mathcal{K})$ such that*

$$\Delta u - \nabla p + f \in N^{l-2,\sigma}_{\beta,\delta}(\mathcal{K})^3, \qquad \nabla \cdot u + g \in N^{l-1,\sigma}_{\beta,\delta}(\mathcal{K}), \tag{10.7.3}$$

$$S_j u = h_j, \quad N_j(u,p) = \phi_j \ \text{on } \Gamma_j,\ j = 1,\dots,d. \tag{10.7.4}$$

The norms of u and p can be estimated by the norms of f, g, h_j and ϕ_j. The same assertion is true for $l=1$ without the condition on f.

10.7.2. Some estimates for solutions of the boundary value problem. We prove the following estimate for solutions in the weighted spaces $N^{l,\sigma}_{\beta,\delta}(\mathcal{K})$ by means of an analogous estimate for solutions of the boundary value problem in a dihedron (cf. Lemma 9.7.7).

LEMMA 10.7.3. *Suppose that*

$$u \in W^{2,s}_{loc}(\bar{\mathcal{K}}\backslash\mathcal{S})^3 \cap L^\infty_{\beta-l-\sigma,\delta-l-\sigma}(\mathcal{K})^3, \quad p \in W^{1,s}_{loc}(\bar{\mathcal{K}}\backslash\mathcal{S}) \cap L^\infty_{\beta-l+1-\sigma,\delta-l+1-\sigma}(\mathcal{K}),$$

$l \ge 2$. If (u,p) is a solution of problem (10.1.1), (10.1.2), where $f \in N^{l-2,\sigma}_{\beta,\delta}(\mathcal{K})^3$, $g \in N^{l-1,\sigma}_{\beta,\delta}(\mathcal{K})$, $h_j \in N^{l,\sigma}_{\beta,\delta}(\Gamma_j)^{3-d_j}$ and $\phi_j \in N^{l-1,\sigma}_{\beta,\delta}(\Gamma_j)^{d_j}$ for $j=1,\dots,d$, then $(u,p) \in N^{l,\sigma}_{\beta,\delta}(\mathcal{K})^3 \times N^{l-1,\sigma}_{\beta,\delta}(\mathcal{K})$.

$$\|u\|_{N^{l,\sigma}_{\beta,\delta}(\mathcal{K})^3} + \|p\|_{N^{l-1,\sigma}_{\beta,\delta}(\mathcal{K})} \le c\Big(\|f\|_{N^{l-2,\sigma}_{\beta,\delta}(\mathcal{K})^\ell} + \|g\|_{N^{l-1,\sigma}_{\beta,\delta}(\mathcal{K})} + \sum_{j=1}^d \|h_j\|_{N^{l,\sigma}_{\beta,\delta}(\Gamma_j)^{3-d_j}} + \sum_{j=1}^d \|\phi_j\|_{N^{l-1,\sigma}_{\beta,\delta}(\Gamma_j)^{3-d_j}} + \|u\|_{L^\infty_{\beta-l-\sigma,\delta-l-\sigma}(\mathcal{K})^3} + \|p\|_{L^\infty_{\beta-l+1-\sigma,\delta-l+1-\sigma}(\mathcal{K})}\Big).$$

P r o o f. By Lemma 10.7.1, we may assume without loss of generality that $h_j=0$ and $\phi_j=0$. From Lemma 9.7.8 it follows that $\zeta u \in N^{l,\sigma}_{\beta,\delta}(\mathcal{K})^3$ and $\zeta p \in N^{l-1,\sigma}_{\beta,\delta}(\mathcal{K})$ for every smooth function ζ with compact support vanishing in a neighborhood of the origin.

Let ρ be a positive integer, $\mathcal{K}_\rho = \{x\in\mathcal{K}:\ \rho/2<|x|<2\rho\}$, and $\mathcal{K}'_\rho = \{x\in\mathcal{K}:\ \rho/4<|x|<4\rho\}$. Furthermore, let $\tilde{u}(x)=u(\rho x)$, $\tilde{p}(x)=\rho\, p(\rho x)$, $\tilde{f}(x)=\rho^2 f(\rho x)$, and $\tilde{g}(x)=\rho\, g(\rho x)$. Then

$$-\Delta\tilde{u}+\nabla\tilde{p}=\tilde{f} \quad\text{and}\quad -\nabla\cdot\tilde{u}=\tilde{g}$$

in $\mathcal{K}$. Moreover, $\tilde{u}$ and $\tilde{p}$ satisfy the boundary conditions (10.1.2) with zero boundary data. Consequently by Lemma 9.7.8,

$$\|\tilde{u}\|_{N^{l,\sigma}_{\beta,\delta}(\mathcal{K}_1)^3} + \|\tilde{p}\|_{N^{l-1,\sigma}_{\beta,\delta}(\mathcal{K}_1)} \le c\Big(\|\tilde{f}\|_{N^{l-2,\sigma}_{\beta,\delta}(\mathcal{K}'_1)^3} + \|\tilde{g}\|_{N^{l-1,\sigma}_{\beta,\delta}(\mathcal{K}'_1)} + \|\tilde{u}\|_{N^0_{\beta-l-\sigma,\delta-l-\sigma}(\mathcal{K}'_1)^3} + \|\tilde{p}\|_{N^0_{\beta-l+1-\sigma,\delta-l+1-\sigma}(\mathcal{K}'_1)}\Big) \tag{10.7.5}$$

with a constant c independent of u,p and ρ. Here the norm in $N^{l,\sigma}_{\beta,\delta}(\mathcal{K}_\rho)$ is defined by (3.6.2), where $\mathcal{K}$ has to be replaced by $\mathcal{K}_\rho$. Since

$$\|\tilde{u}\|_{N^{l,\sigma}_{\beta,\delta}(\mathcal{K}_1)^3} = \rho^{l+\sigma-\beta}\|u\|_{N^{l,\sigma}_{\beta,\delta}(\mathcal{K}_\rho)^3},$$

we obtain an analogous estimate for the norms of u and p in $N^{l,\sigma}_{\beta,\delta}(\mathcal{K}_\rho)^3$ and $N^{l-1,\sigma}_{\beta,\delta}(\mathcal{K}_\rho)$, respectively. Here, the constant c is the same as in (10.7.5). The result follows. □

In the same way, the subsequent two lemmas hold by means of Lemmas 9.7.10 and 9.7.11.

LEMMA 10.7.4. *Let* $(u,p) \in W^{2,s}_{loc}(\bar{\mathcal{K}}\backslash\mathcal{S})^3 \times W^{1,s}_{loc}(\bar{\mathcal{K}}\backslash\mathcal{S})$ *be a solution of the problem* (10.1.1), (10.1.2). *Suppose that* $u \in C^{l-1,\sigma}_{\beta-1,\delta-1}(\mathcal{K})^3$, $p \in C^{l-2,\sigma}_{\beta-1,\delta-1}(\mathcal{K})$, $f \in C^{l-2,\sigma}_{\beta,\delta}(\mathcal{K})^3$, $g \in C^{l-1,\sigma}_{\beta,\delta}(\mathcal{K})$, $h_j \in C^{l,\sigma}_{\beta,\delta}(\Gamma_j)^{3-d_j}$ *and* $\phi_j \in C^{l-1,\sigma}_{\beta,\delta}(\Gamma_j)^{d_j}$, *where* $l \ge 2$, $\delta_k \ge 1$ *for* $k = 1,\dots,d$, $0 < \sigma < 1$. *Then* $(u,p) \in C^{l,\sigma}_{\beta,\delta}(\mathcal{K})^3 \times C^{l-1,\sigma}_{\beta,\delta}(\mathcal{K})$.

LEMMA 10.7.5. *Let* (u,p) *be a solution of the boundary value problem* (10.1.1), (10.1.2) *such that*

$$(\rho\partial_\rho)^j(u,p) \in C^{l,\sigma}_{\beta,\delta}(\mathcal{K})^3 \times C^{l-1,\sigma}_{\beta,\delta}(\mathcal{K})$$

for $j = 0,1$, *where* $l \ge 2$ *and* $0 < \sigma < 1$. *If*

$$f \in C^{l-1,\sigma}_{\beta+1,\delta}(\mathcal{K})^3, \quad g \in C^{l,\sigma}_{\beta+1,\delta}(\mathcal{K}), \quad h_j \in C^{l+1,\sigma}_{\beta+1,\delta}(\Gamma_j)^{3-d_j}, \quad \phi_j \in C^{l,\sigma}_{\beta+1,\delta}(\Gamma_j)^{d_j},$$

$j = 1,\dots,d$, *and the strip* $l+\sigma-\delta_k \le \operatorname{Re}\lambda \le l+1+\sigma-\delta_k$ *does not contain eigenvalues of the pencil* $A_k(\lambda)$, $k = 1,\dots,d$, *then* $(u,p) \in C^{l+1,\sigma}_{\beta+1,\delta}(\mathcal{K})^3 \times C^{l,\sigma}_{\beta+1,\delta}(\mathcal{K})$.

10.7.3. Representation of the solution by Green's matrix. Let $f \in N^{0,\sigma}_{\beta,\delta}(\mathcal{K})^3$ and $g \in N^{1,\sigma}_{\beta,\delta}(\mathcal{K})$ be given. We suppose that the line $\operatorname{Re}\lambda = 2+\sigma-\beta$ is free of eigenvalues of the pencil $\mathfrak{A}(\lambda)$ and that the components of δ satisfy the conditions (10.7.1). Then we denote by $\Lambda_- < \operatorname{Re}\lambda < \Lambda_+$ the widest strip in the complex plane containing the line $\operatorname{Re}\lambda = 2+\sigma-\beta$ which is free of eigenvalues of the pencil $\mathfrak{A}(\lambda)$. Furthermore, let $G(x,\xi)$ be the Green's matrix introduced in Theorem 10.4.1, where κ is an arbitrary real number in the interval

$$-\Lambda_+ - 1/2 < \kappa < -\Lambda_- - 1/2\,.$$

Then the vector function (u,p) with the components

$$u_i(x) = \sum_{j=1}^{3} \int_{\mathcal{K}} \big(f_j(\xi) + \partial_{\xi_j} g(\xi)\big)\, G_{i,j}(x,\xi)\, d\xi + \int_{\mathcal{K}} g(\xi)\, G_{i,4}(x,\xi)\, d\xi, \tag{10.7.6}$$

$i = 1,2,3$, and

$$p(x) = -g(x) + \sum_{j=1}^{3} \int_{\mathcal{K}} \big(f_j(\xi) + \partial_{\xi_j} g(\xi)\big)\, G_{4,j}(x,\xi)\, d\xi + \int_{\mathcal{K}} g(\xi)\, G_{4,4}(x,\xi)\, d\xi \tag{10.7.7}$$

is a solution of the boundary value problem (10.1.1), (10.1.2) with zero boundary data h_j and ϕ_j.

Let χ be an arbitrary smooth cut-off function on $[0,\infty)$, $\chi(t) = 1$ for $t < 1/4$, $\chi(t) = 0$ for $t > 1/2$. We set

$$\chi^+(x,\xi) = \chi\Big(\frac{|x-\xi|}{r(x)}\Big), \quad \chi^-(x,\xi) = 1 - \chi^+(x,\xi). \tag{10.7.8}$$

Then $\chi^+(x,\xi) = 0$ for $|x-\xi| > r(x)/2$, $\chi^-(x,\xi) = 0$ for $|x-\xi| < r(x)/4$, and

$$\big|\partial_x^\alpha \chi^\pm(x,\xi)\big| \le c\, r(x)^{-|\alpha|}$$

with a constant c independent of x and ξ. We write u and p in the form

$$u = u^+ + u^-, \quad p = p^+ + p^-,$$

where $u^\pm$ is the vector with the components

$$(10.7.9)\qquad u_i^\pm(x) = \sum_{j=1}^{3} \int_{\mathcal{K}} \big(f_j(\xi) + \partial_{\xi_j} g(\xi)\big)\, \chi^\pm(x,\xi)\, G_{i,j}(x,\xi)\, d\xi + \int_{\mathcal{K}} g(\xi)\, \chi^\pm(x,\xi)\, G_{i,4}(x,\xi)\, d\xi,$$

$i = 1, 2, 3$, and

$$(10.7.10)\qquad p^\pm(x) = \sum_{j=1}^{3} \int_{\mathcal{K}} \big(f_j(\xi) + \partial_{\xi_j} g(\xi)\big)\, \chi^\pm(x,\xi)\, G_{4,j}(x,\xi)\, d\xi + \int_{\mathcal{K}} g(\xi)\, \chi^\pm(x,\xi)\, G_{4,4}(x,\xi)\, d\xi - \frac{1}{2}\big(g(x) \pm g(x)\big).$$

10.7.4. Weighted L_∞ estimates for u^+ and p^+. Our goal is to show that both (u^+, p^+) and (u^-, p^-) belong to the space $C^{2,\sigma}_{\beta,\delta}(\mathcal{K})^3 \times C^{1,\sigma}_{\beta,\delta}(\mathcal{K})$. We start with two estimate for the vector function (u^+, p^+).

LEMMA 10.7.6. *Suppose that the line* $\operatorname{Re}\lambda = 2 + \sigma - \beta$ *is free of eigenvalues of the pencil* $\mathfrak{A}(\lambda)$ *and that the components of* δ *satisfy the condition conditions* (10.7.1). *Then the estimates*

$$(10.7.11)\qquad |x|^{\beta-2-\sigma} \prod_{k=1}^{d} \Big(\frac{r_k(x)}{|x|}\Big)^{\delta_k-2-\sigma} |u^+(x)| \le c\Big(\|f\|_{N^{0,\sigma}_{\beta,\delta}(\mathcal{K})^3} + \|g\|_{N^{1,\sigma}_{\beta,\delta}(\mathcal{K})}\Big),$$

$$(10.7.12)\qquad |x|^{\beta-1-\sigma} \prod_{k=1}^{d} \Big(\frac{r_k(x)}{|x|}\Big)^{\delta_k-1-\sigma} \Big(\sum_{j=1}^{3} |\partial_{x_j} u^+(x)| + |p^+(x)|\Big) \le c\Big(\|f\|_{N^{0,\sigma}_{\beta,\delta}(\mathcal{K})^3} + \|g\|_{N^{1,\sigma}_{\beta,\delta}(\mathcal{K})}\Big)$$

are satisfied for arbitrary $f \in N^{0,\sigma}_{\beta,\delta}(\mathcal{K})^3$ *and* $g \in N^{1,\sigma}_{\beta,\delta}(\mathcal{K})$.

P r o o f. We consider the expressions

$$A = \sum_{j=1}^{3} \Big| \int_{\mathcal{K}} \big(f_j(\xi) + \partial_{\xi_j} g(\xi)\big)\, \partial_{x_\nu}\big(\chi^+(x,\xi)\, G_{i,j}(x,\xi)\big)\, d\xi \Big| \quad \text{and}$$

$$B = \Big| \int_{\mathcal{K}} g(\xi)\, \partial_{x_\nu}\big(\chi^+(x,\xi)\, G_{i,4}(x,\xi)\big)\, d\xi \Big|$$

for $\nu = 1, 2, 3$. Obviously, $\big|\partial_{x_\nu} u_i^+(x)\big| \le A + B$. On the support of χ^+, the inequalities

$$|x|/2 \le |\xi| \le 3|x|/2, \qquad r_k(x)/2 \le r_k(\xi) \le 3r_k(x)/2$$

are satisfied for $k = 1, \dots, d$. This together with Theorem 10.4.2 implies

$$\begin{aligned} A &\le c\,|x|^{\sigma-\beta} \prod_{k=1}^{d} \Big(\frac{r_k(x)}{|x|}\Big)^{\sigma-\delta_k} \Big(\|f\|_{L^\infty_{\beta-\sigma,\delta-\sigma}(\mathcal{K})^3} + \|\nabla g\|_{L^\infty_{\beta-\sigma,\delta-\sigma}(\mathcal{K})}\Big) \\ &\quad \times \int\limits_{|x-\xi|<r(x)/2} |x-\xi|^{-2}\, d\xi \\ &\le c\,|x|^{\sigma-\beta+1} \prod_{k=1}^{d} \Big(\frac{r_k(x)}{|x|}\Big)^{\sigma-\delta_k+1} \Big(\|f\|_{L^\infty_{\beta-\sigma,\delta-\sigma}(\mathcal{K})^3} + \|\nabla g\|_{L^\infty_{\beta-\sigma,\delta-\sigma}(\mathcal{K})}\Big) \end{aligned}$$

By the representation

$$G_{i,4}(x,\xi) = -\nabla_\xi \cdot \vec{\mathcal{P}}_i(x,\xi) + \mathcal{Q}_i(x,\xi)$$

in Corollary 10.4.3, the term B can be written as

$$\begin{aligned} B &= \Big|\partial_{x_\nu} \int_{\mathcal{K}} \Big(g(\xi)\,\big(\nabla_\xi \chi^+(x,\xi)\cdot \vec{\mathcal{P}}_i(x,\xi) + \mathcal{Q}_i(x,\xi)\big) \\ &\qquad + \chi^+(x,\xi)\,\nabla_\xi g(\xi)\cdot \vec{\mathcal{P}}_i(x,\xi)\Big)\, d\xi\Big|. \end{aligned}$$

Using the properties of $\mathcal{P}_i$, $\mathcal{Q}_i$ in Corollary 10.4.3, we conclude that

$$\begin{aligned} B &\le c\,|x|^{\sigma-\beta+1} \prod_{k=1}^{d} \Big(\frac{r_k(x)}{|x|}\Big)^{\sigma-\delta_k+1} \|g\|_{L^\infty_{\beta-1-\sigma,\delta-1-\sigma}(\mathcal{K})} \int\limits_{|x-\xi|<r(x)/2} r(x)^{-3}\, d\xi \\ &\quad + c\,|x|^{\sigma-\beta} \prod_{k=1}^{d} \Big(\frac{r_k(x)}{|x|}\Big)^{\sigma-\delta_k} \|\nabla g\|_{L^\infty_{\beta-\sigma,\delta-\sigma}(\mathcal{K})} \int\limits_{|x-\xi|<r(x)/2} |x-\xi|^{-2}\, d\xi \\ &\le c\,|x|^{\sigma-\beta+1} \prod_{k=1}^{d} \Big(\frac{r_k(x)}{|x|}\Big)^{\sigma-\delta_k+1} \Big(\|g\|_{L^\infty_{\beta-1-\sigma,\delta-1-\sigma}(\mathcal{K})^3} + \|\nabla g\|_{L^\infty_{\beta-\sigma,\delta-\sigma}(\mathcal{K})}\Big). \end{aligned}$$

Thus,

$$|x|^{\beta-1-\sigma} \prod_{k=1}^{d} \Big(\frac{r_k(x)}{|x|}\Big)^{\delta_k-1-\sigma} |\partial_{x_\nu} u^+(x)| \le c\,\Big(\|f\|_{N^{0,\sigma}_{\beta,\delta}(\mathcal{K})^3} + \|g\|_{N^{1,\sigma}_{\beta,\delta}(\mathcal{K})}\Big).$$

In the same way, we obtain the desired weighted L_∞-estimates for u and p. □

10.7.5. Weighted L_∞ estimates for the derivatives of u^-, p^-. Next, we show that

$$(10.7.13) \qquad \begin{aligned} &|x|^{\beta-2-\sigma+|\alpha|} \prod_{k=1}^{d} \Big(\frac{r_k(x)}{|x|}\Big)^{\max(0,\delta_k-2-\sigma+|\alpha|)} |\partial_x^\alpha u^-(x)| \\ &\qquad \le c\,\Big(\|f\|_{N^{0,\sigma}_{\beta,\delta}(\mathcal{K})^3} + \|g\|_{N^{1,\sigma}_{\beta,\delta}(\mathcal{K})}\Big) \end{aligned}$$

and

$$(10.7.14) \qquad \begin{aligned} &|x|^{\beta-1-\sigma+|\alpha|} \prod_{k=1}^{d} \Big(\frac{r_k(x)}{|x|}\Big)^{\max(0,\delta_k-1-\sigma+|\alpha|)} |\partial_x^\alpha p^-(x)| \\ &\qquad \le c\,\Big(\|f\|_{N^{0,\sigma}_{\beta,\delta}(\mathcal{K})^3} + \|g\|_{N^{1,\sigma}_{\beta,\delta}(\mathcal{K})}\Big) \end{aligned}$$

for an arbitrary multi-index α. To this end, we prove the subsequent two lemmas.

LEMMA 10.7.7. *Let $f \in N^{t,\sigma}_{\beta,\delta}(\mathcal{K})$ and*

$$v(x) = \int\limits_{\substack{\mathcal{K} \\ |\xi|<|x|/2}} K(x,\xi)\, f(\xi)\, d\xi,$$

where the kernel K satisfies the estimate

$$|K(x,\xi)| \le c\, \frac{|x|^{\Lambda_- - s+\varepsilon}}{|\xi|^{\Lambda_- +1+t+\varepsilon}} \prod_{k=1}^{d} \Big(\frac{r_k(x)}{|x|}\Big)^{\min(0,\mu_+^{(k)}-s-\varepsilon)} \prod_{k=1}^{d} \Big(\frac{r_k(\xi)}{|\xi|}\Big)^{\min(0,\mu_+^{(k)}-t-\varepsilon)}.$$

with nonnegative integers s,t. Suppose that $\Lambda_- < 2+\sigma-\beta$ and that the components of δ satisfy the conditions (10.7.1). *Then*

$$(10.7.15)\qquad |x|^{\beta-2+s-\sigma} \prod_{k=1}^{d} \Big(\frac{r_k(x)}{|x|}\Big)^{\max(0,\delta_k-2+s-\sigma)} |v(x)| \le c\, \|f\|_{L^\infty_{\beta-t-\sigma,\delta-t-\sigma}(\mathcal{K})}$$

with a constant c independent of f and x.

P r o o f. Obviously,

$$(10.7.16)\ |v(x)| \le c\, |x|^{\Lambda_- -s+\varepsilon} \prod_{k=1}^{d} \Big(\frac{r_k(x)}{|x|}\Big)^{\min(0,\mu_+^{(k)}-s-\varepsilon)} \|f\|_{L^\infty_{\beta-t-\sigma,\delta-t-\sigma}(\mathcal{K})}$$
$$\times \int\limits_{\substack{\mathcal{K} \\ |\xi|<|x|/2}} |\xi|^{-\Lambda_- -1-\beta+\sigma-\varepsilon} \prod_{k=1}^{d} \Big(\frac{r_k(\xi)}{|\xi|}\Big)^{\min(0,\mu_+^{(k)}-t-\varepsilon)-\delta_k+t+\sigma} d\xi.$$

By the conditions on β and δ, we have

$$-\Lambda_- - 1 - \beta + \sigma > -3 \quad \text{and} \quad \min(0,\mu_+^{(k)} - t - \varepsilon) - \delta_k + t + \sigma > -2.$$

Hence, the integral on the right-hand side of (10.7.16) is equal to $c\,|x|^{-\Lambda_- +2-\beta+\sigma-\varepsilon}$. Therefore,

$$|v(x)| \le c\, |x|^{-\beta+2-s+\sigma} \prod_{k=1}^{N} \Big(\frac{r_k(x)}{|x|}\Big)^{\min(0,\mu_+^{(k)}-s-\varepsilon)} \|f\|_{L^\infty_{\beta-t-\sigma,\delta-t-\sigma}(\mathcal{K})}\,.$$

Using the inequality

$$\min(0,\mu_+^{(k)} - s - \varepsilon) \ge \min(0, 2 - \delta_k + \sigma - s) = -\max(0, \delta_k - 2 + s - \sigma),$$

we obtain the desired estimate for v. □

In the same manner, we can prove the next lemma.

LEMMA 10.7.8. *Let $f \in N^{t,\sigma}_{\beta,\delta}(\mathcal{K})$ and*

$$v(x) = \int\limits_{\substack{\mathcal{K} \\ |\xi|>2|x|}} K(x,\xi)\, f(\xi)\, d\xi,$$

where the kernel K satisfies the estimate

$$|K(x,\xi)| \le c\, \frac{|x|^{\Lambda_+ - s-\varepsilon}}{|\xi|^{\Lambda_+ +1+t-\varepsilon}} \prod_{k=1}^{d} \Big(\frac{r_k(x)}{|x|}\Big)^{\min(0,\mu_+^{(k)}-s-\varepsilon)} \prod_{k=1}^{d} \Big(\frac{r_k(\xi)}{|\xi|}\Big)^{\min(0,\mu_+^{(k)}-t-\varepsilon)}.$$

with nonnegative integers s,t. Suppose that $\Lambda_+ > 2+\sigma-\beta$ and that the components of δ satisfy the conditions (10.7.1). *Then* (10.7.15) *is valid with a constant c independent of f and x.*

P r o o f. As in the proof of Lemma 10.7.7, we get

$$|v(x)| \le c\,|x|^{\Lambda_+-s-\varepsilon} \prod_{k=1}^{d} \Big(\frac{r_k(x)}{|x|}\Big)^{\min(0,\mu_+^{(k)}-s-\varepsilon)} \|f\|_{L^\infty_{\beta-t-\sigma,\delta-t-\sigma}(\mathcal{K})}$$

$$\times \int\limits_{\substack{\mathcal{K}\\ |\xi|>2|x|}} |\xi|^{-\Lambda_+-1-\beta+\sigma+\varepsilon} \prod_{k=1}^{d} \Big(\frac{r_k(\xi)}{|\xi|}\Big)^{\min(0,\mu_+^{(k)}-t-\varepsilon)-\delta_k+t+\sigma} d\xi$$

$$\le c\,|x|^{-\beta+2-s+\sigma} \prod_{k=1}^{d} \Big(\frac{r_k(x)}{|x|}\Big)^{\min(0,\mu_+^{(k)}-s-\varepsilon)} \|f\|_{L^\infty_{\beta-t-\sigma,\delta-t-\sigma}(\mathcal{K})}$$

$$\le c\,|x|^{-\beta+2-s+\sigma} \prod_{k=1}^{d} \Big(\frac{r_k(x)}{|x|}\Big)^{-\max(0,\delta_k-2+s-\sigma)} \|f\|_{L^\infty_{\beta-t-\sigma,\delta-t-\sigma}(\mathcal{K})}\,.$$

This proves the lemma. □

Note that the functions $\partial_x^\alpha G_{i,j}(x,\xi)$ satisfy the conditions on the kernel K in Lemmas 10.7.7 and 10.7.8 with $s=|\alpha|+\delta_{i,4}$, $t=\delta_{j,4}$ for $|\xi|<|x|/2$ and $|\xi|>2|x|$, respectively.

Next, we prove an estimate for the integral over the zone $|x|/2<|\xi|<2|x|$.

LEMMA 10.7.9. *Let $f\in N^{t,\sigma}_{\beta,\delta}(\mathcal{K})$ and*

$$v(x) = \int\limits_{\substack{\mathcal{K}\\ |x|/2<|\xi|<2|x|}} K(x,\xi)\, f(\xi)\, d\xi,$$

where $K(x,\xi)$ vanishes for $|x-\xi|<r(x)/4$ and satisfies the estimate

$$|K(x,\xi)| \le c\,|x-\xi|^{-1-s-t} \Big(\frac{r(x)}{|x-\xi|}\Big)^{\min(0,\mu_x-s-\varepsilon)} \Big(\frac{r(\xi)}{|x-\xi|}\Big)^{\min(0,\mu_\xi-t-\varepsilon)}.$$

with nonnegative integers s,t, where μ_x and μ_ξ are the same constants as in Theorem 10.4.4. Suppose that the components of δ satisfy the condition (10.7.1). *Then* (10.7.15) *is valid with a constant c independent of f and x.*

P r o o f. Obviously,

$$|v(x)| \le c\,|x|^{\sigma+t-\beta}\, \|f\|_{L^\infty_{\beta-t-\sigma,\delta-t-\sigma}(\mathcal{K})} \int\limits_{\substack{\mathcal{K}\\ |x|/2<|\xi|<2|x|}} |K(x,\xi)| \prod_{k=1}^{d} \Big(\frac{r_k(\xi)}{|\xi|}\Big)^{\sigma+t-\delta_k} d\xi.$$

Applying Lemma 3.5.2, we obtain (10.7.15). □

Now it is easy to prove the estimates (10.7.13) and (10.7.14) for the functions u^- and p^-.

THEOREM 10.7.10. *Suppose that $\Lambda_- < 2+\sigma-\beta < \Lambda_+$ and that the components of δ satisfy the conditions* (10.7.1). *Furthermore, let (u^-,p^-) be the vector function with the components* (10.7.9), (10.7.10), *where $f\in N^{0,\sigma}_{\beta,\delta}(\mathcal{K})^3$, $g\in N^{1,\sigma}_{\beta,\delta}(\mathcal{K})$ and*

$G(x,\xi)$ is the Green's matrix introduced in Theorem 10.4.1. *Then u^- and p^- satisfy* (10.7.13) and (10.7.14).

P r o o f. By Theorems 10.4.4 and 10.4.5, the function

$$K(x,\xi) = \partial_x^\alpha \big(\chi^-(x,\xi) G_{i,j}(x,\xi)\big)$$

satisfies the conditions of Lemmas 10.7.7–10.7.9 with $s = |\alpha| + \delta_{i,4}$, $t = \delta_{j,4}$ for $|\xi| < |x|/2$, $|\xi| > 2|x|$ and $|x|/2 < |\xi| < 2|x|$, respectively. Applying these lemmas, we obtain the desired estimates for u^- and p^-. □

10.7.6. Weighted Hölder estimates for u^-, p^-. In the next lemma, let

$$\mathcal{K}_k = \{x \in \mathcal{K} : r_k(x) < 3r(x)/2\}$$

for $k = 1, \dots, d$. Furthermore, let $m_k = [\delta_k - \sigma] + 1$, where $[s]$ denotes the integral part of s.

LEMMA 10.7.11. *Let $f \in N^{t,\sigma}_{\beta,\delta}(\mathcal{K})$ and*

$$v(x) = \int_{\mathcal{K}} K(x,\xi)\, f(\xi)\, d\xi.$$

We assume that $\Lambda_- < 2 + \sigma - \beta < \Lambda_+$ and that the components of δ satisfy the conditions (10.7.1). *Furthermore, we assume that $K(x,\xi)$ vanishes for $|x-\xi| < r(x)/4$ and that the following estimates are satisfied for the derivatives with respect to $\rho = |x|$ are valid:*

$$|\partial_\rho^s K(x,\xi)| \le c\, |x-\xi|^{-3-s-t+m_k} \Big(\frac{r(\xi)}{|x-\xi|}\Big)^{\min(0,\mu_\xi - t-\varepsilon)}$$
$$\text{for } x \in \mathcal{K}_k,\ |x|/4 < |\xi| < 2|x|,\ s = 0,1,$$

$$|\partial_\rho K(x,\xi)| \le c\, \frac{|x|^{\Lambda_- -3+m_k+\varepsilon}}{|\xi|^{\Lambda_- +1+t+\varepsilon}} \prod_{j=1}^d \Big(\frac{r_j(\xi)}{|\xi|}\Big)^{\min(0,\mu_+^{(j)}-t-\varepsilon)} \quad \text{if } x \in \mathcal{K}_k,\ |\xi| < \frac{3|x|}{4},$$

$$|\partial_\rho K(x,\xi)| \le c\, \frac{|x|^{\Lambda_+ -3+m_k-\varepsilon}}{|\xi|^{\Lambda_+ +1+t-\varepsilon}} \prod_{j=1}^d \Big(\frac{r_j(\xi)}{|\xi|}\Big)^{\min(0,\mu_+^{(j)}-t-\varepsilon)} \quad \text{if } x \in \mathcal{K}_k,\ |\xi| > \frac{3|x|}{2}.$$

Then

$$|x|^{\beta-\delta_k} \frac{|v(x)-v(y)|}{|x-y|^{\sigma-\delta_k+m_k}} \le c\, \|f\|_{L^\infty_{\beta-t-\sigma,\delta-t-\sigma}(\mathcal{K})} \tag{10.7.17}$$

for $x \in \mathcal{K}_k$, $y = \tau x$, $4/5 < \tau < 5/4$, $|x-y| > r(x)/4$. Here c is a constant independent of f, x and τ.

P r o o f. Let $x \in \mathcal{K}_k$ and $y = \tau x$, $4/5 < \tau < 5/4$, $|x-y| > r(x)/4$. Then $4|x-y| < \min(|x|,|y|)$. We introduce the integrals

$$A_1 = \int\limits_{\substack{\mathcal{K}\\ |\xi-x|<2|x-y|}} |K(x,\xi)\, f(\xi)|\, d\xi, \quad A_2 = \int\limits_{\substack{\mathcal{K}\\ |\xi-x|<2|x-y|}} |K(y,\xi)\, f(\xi)|\, d\xi,$$

$$A_3 = \int\limits_{\substack{\mathcal{K}\\ |\xi-x|>2|x-y|}} \big|(K(x,\xi) - K(y,\xi))\, f(\xi)\big|\, d\xi.$$

Then $|v(x)-v(y)| \le A_1+A_2+A_3$. From the inequalities $|x-\xi| < 2|x-y| < |x|/2$ it follows that $|x|/2 < |\xi| < 3|x|/2$. Therefore,

$$A_1 \le c\,\|f\|_{L^\infty_{\beta-t-\sigma,\delta-t-\sigma}(\mathcal{K})}\, |x|^{t+\sigma-\beta} \times \int |x-\xi|^{-3-t+m_k} \Big(\frac{r(\xi)}{|x-\xi|}\Big)^{\min(0,\mu_\xi-t-\varepsilon)} \prod_{j=1}^d \Big(\frac{r_j(\xi)}{|\xi|}\Big)^{t+\sigma-\delta_j} d\xi,$$

where the domain of integration is contained in the set of all $\xi \in \mathcal{K}$ satisfying the inequalities $|x|/2 < |\xi| < 2|x|$, $r(x)/4 < |x-\xi| < 2|x-y|$. By virtue of (2.6.6), we obtain

$$A_1 \le c\,\|f\|_{L^\infty_{\beta-t-\sigma,\delta-t-\sigma}(\mathcal{K})}\, |x|^{\delta_k-\beta}\, |x-y|^{\sigma-\delta_k+m_k}$$

since $m_k+\sigma-\delta_k > 0$ and $\min(0,\mu_+^{(j)}-t-\varepsilon)+t+\sigma-\delta_j > -2$. Analogously, this estimate holds for A_2. For the proof, one can use the fact that the inequality $|\xi-x| < 2|x-y|$ implies $|\xi-y| < 3|x-y|$ and $|y|/4 < |\xi| < 2|y|$.

We consider A_3. Using the inequality

$$\big|K(x,\xi)-K(y,\xi)\big| \le \big|\partial_\rho K(\tilde{x},\xi)\big|\cdot|x-y|,$$

where $\tilde{x}$ is a certain point on the line between x and y, i.e. $4|x|/5 < |\tilde{x}| < 5|x|/4$, we obtain

$$A_3 \le |x-y| \int\limits_{\substack{\mathcal{K}\\ |\xi-x|>2|x-y|}} \big|\partial_\rho K(\tilde{x},\xi)\, f(\xi)\big|\, d\xi.$$

For the integral over the set $\{\xi\in\mathcal{K} : |\xi| < |x|/2\}$, we obtain

$$\begin{aligned}\int \big|\partial_\rho K(\tilde{x},\xi)\, f(\xi)\big|\, d\xi &\le c\,\|f\|_{L^\infty_{\beta-t-\sigma,\delta-t-\sigma}(\mathcal{K})}\, |x|^{\Lambda_- -3+m_k+\varepsilon} \\ &\quad\times \int |\xi|^{\sigma-\beta-\Lambda_- -1-\varepsilon} \prod_{j=1}^d \Big(\frac{r_j(\xi)}{|\xi|}\Big)^{t+\sigma-\delta_j+\min(0,\mu_+^{(j)}-t-\varepsilon)} d\xi \\ &\le c\,\|f\|_{L^\infty_{\beta-t-\sigma,\delta-t-\sigma}(\mathcal{K})}\, |x|^{\sigma-\beta+m_k-1}.\end{aligned}$$

The same estimate is valid for the integral over the set $\{\xi\in\mathcal{K} : |\xi| > 2|x|\}$, while for the integral over the set $\{\xi \in \mathcal{K} : |x|/2 < |\xi| < 2|x|,\ |\xi-x| > 2|x-y|\}$ the estimate

$$\begin{aligned}\int \big|\partial_\rho K(\tilde{x},\xi)\, f(\xi)\big|\, d\xi &\le c\,\|f\|_{L^\infty_{\beta-t-\sigma,\delta-t-\sigma}(\mathcal{K})}\, |x|^{t+\sigma-\beta} \\ &\quad\times \int |x-\xi|^{-4-t+k_\nu} \Big(\frac{r(\xi)}{|x-\xi|}\Big)^{\min(0,\mu_\xi-t-\varepsilon)} \prod_{j=1}^d \Big(\frac{r_j(\xi)}{|\xi|}\Big)^{t+\sigma-\delta_j} d\xi \\ &\le c\,\|f\|_{L^\infty_{\beta-t-\sigma,\delta-t-\sigma}(\mathcal{K})}\, |x|^{\delta_k-\beta}\, |x-y|^{\sigma-\delta_k+m_k-1}\end{aligned}$$

holds. Here, we employed the inequality $|x-\xi|/2 < |\tilde{x}-\xi| < 3|x-\xi|/2$ and the estimate (2.6.5) with $R = 2|x-y|$. Thus,

$$\begin{aligned} A_3 &\le c\,\|f\|_{L^\infty_{\beta-t-\sigma,\delta-t-\sigma}(\mathcal{K})}\, |x-y|\, \big(|x|^{\sigma-\beta+m_k-1} + |x|^{\delta_k-\beta}\, |x-y|^{\sigma-\delta_k+m_k-1}\big) \\ &\le c\,\|f\|_{L^\infty_{\beta-t-\sigma,\delta-t-\sigma}(\mathcal{K})}\, |x|^{\delta_k-\beta}\, |x-y|^{\sigma-\delta_k+m_k}.\end{aligned}$$

This proves the lemma. □

Using the last lemma and Theorem 10.7.10, we can estimate all terms in the $C^{2,\sigma}_{\beta,\delta}(\mathcal{K})$-norm of u^- and in the $C^{1,\sigma}_{\beta,\delta}(\mathcal{K})$-norm of p^-. This leads to the following theorem.

THEOREM 10.7.12. *Let the conditions of Theorem* 10.7.10 *be satisfied. Then* $u^- \in C^{2,\sigma}_{\beta,\delta}(\mathcal{K})^3$, $p^- \in C^{1,\sigma}_{\beta,\delta}(\mathcal{K})$ *and*

$$\|u^-\|_{C^{2,\sigma}_{\beta,\delta}(\mathcal{K})^3} + \|p^-\|_{C^{1,\sigma}_{\beta,\delta}(\mathcal{K})} \le c\Big(\|f\|_{N^{0,\sigma}_{\beta,\delta}(\mathcal{K})^3} + \|g\|_{N^{1,\sigma}_{\beta,\delta}(\mathcal{K})}\Big) \tag{10.7.18}$$

with a constant c independent of f and g.

P r o o f. According to Theorem 10.7.10, the vector function u^- satisfies (10.7.13). We show that

$$|x|^\beta \prod_{k=1}^d \Big(\frac{r_k(x)}{|x|}\Big)^{\delta_k} \frac{\big|\partial_x^\alpha u^-(x) - \partial_y^\alpha u^-(y)\big|}{|x-y|^\sigma} \le c\Big(\|f\|_{N^{0,\sigma}_{\beta,\delta}(\mathcal{K})^3} + \|g\|_{N^{1,\sigma}_{\beta,\delta}(\mathcal{K})}\Big) \tag{10.7.19}$$

for $|\alpha| = 2$, $|x-y| < r(x)/2$. By the mean value theorem,

$$\partial^\alpha u(x) - \partial^\alpha u(y) = (x-y)\cdot \nabla\partial^\alpha u(\tilde{x}),$$

where $\tilde{x} = x + t(y-x)$, $t \in (0,1)$. Furthermore, the inequalities

$$|x|/2 < |\tilde{x}| < 3|x|/2, \quad r_k(x)/2 < r_k(\tilde{x}) < 3r_k(x)/2$$

are valid for $|x-y| < r(x)/2$, $k = 1,\dots,d$. From this and from (3.1.2) it follows that

$$\begin{aligned} &|x|^\beta \prod_{k=1}^d \Big(\frac{r_k(x)}{|x|}\Big)^{\delta_k} \frac{\big|\partial_x^\alpha u^-(x) - \partial_y^\alpha u^-(y)\big|}{|x-y|^\sigma} \\ &\quad\le |x|^\beta\, r(x)^{1-\sigma} \prod_{k=1}^d \Big(\frac{r_k(x)}{|x|}\Big)^{\delta_k} |\nabla\partial^\alpha u(\tilde{x})| \\ &\quad\le c\,|\tilde{x}|^{\beta+1-\sigma} \prod_{k=1}^N \Big(\frac{r_k(\tilde{x})}{|\tilde{x}|}\Big)^{\delta_k-1+\sigma} |\nabla\partial^\alpha u(\tilde{x})| \end{aligned}$$

for $|\alpha| = 2$, $|x-y| < r(x)/2$. This together with (10.7.13) implies (10.7.19). Analogously, the estimate

$$|x|^{\beta-\delta_k} \frac{\big|\partial_x^\alpha u^-(x) - \partial_y^\alpha u^-(y)\big|}{|x-y|^{m_k+\sigma-\delta_k}} \le c\Big(\|f\|_{N^{0,\sigma}_{\beta,\delta}(\mathcal{K})^3} + \|g\|_{N^{1,\sigma}_{\beta,\delta}(\mathcal{K})}\Big) \tag{10.7.20}$$

holds for $|\alpha| = 2 - m_k = 1 - [\delta_k - \sigma]$, $x, y \in \mathcal{K}_k$, $|x-y| < r(x)/2$, $k = 1,\dots,d$.

Next we prove that (10.7.20) is valid for $|\alpha| = 2 - m_k$, $x \in \mathcal{K}_k$, $y = \tau|x|$, $1/2 < \tau < 3/2$. For $|x-y| < r(x)/2$ this is already shown. For $1/2 < \tau < 4/5$ or $5/4 < \tau < 3/2$, the left-hand side of (10.7.20) does not exceed

$$c\Big(|x|^{\beta-\sigma-m_k}\big|\partial_x^\alpha u^-(x)\big| + |y|^{\beta-\sigma-m_k}\big|\partial_y^\alpha u^-(y)\big|\Big)$$

and can estimated by means of (10.7.13). For $|x-y| > r(x)/2$, $\tau \in (4/5, 5/4)$, we may refer to Lemma 10.7.11, since $K(x,\xi) = \partial_x^\alpha\big(\chi^-(x,\xi)\, G_{i,j}(x,\xi)\big)$ satisfies the conditions of this lemma for $|\alpha| = 2 - \delta_{i,4} - m_k$, $t = \delta_{j,4}$, $m_k \le 2 - \delta_{i,4}$ (cf. Theorems 10.4.4, 10.4.5 and Remark 10.4.6).

Hence, the norm of u^- in $C^{2,\sigma}_{\beta,\delta}(\mathcal{K})^3$ is majorized by the right-hand side of (10.7.18). Analogously, the desired estimate for p^- can be proved by means of (10.7.14) and Lemma 10.7.11. The proof is complete. □

10.7.7. Solvability in $C^{2,\sigma}_{\beta,\delta}(\mathcal{K})^3 \times C^{1,\sigma}_{\beta,\delta}(\mathcal{K})$. The estimates for the vector functions (u^+,p^+) and (u^-,p^-) in the preceding three subsections are used in the proof of the following existence and uniqueness theorem for solutions of the boundary value problem (10.1.1), (10.1.2).

THEOREM 10.7.13. *Let $f \in C^{0,\sigma}_{\beta,\delta}(\mathcal{K})^3$, $g \in C^{1,\sigma}_{\beta,\delta}(\mathcal{K})$, $h_j \in C^{2,\sigma}_{\beta,\delta}(\Gamma_j)^{3-d_j}$, $\phi_j \in C^{1,\sigma}_{\beta,\delta}(\Gamma_j)^{d_j}$, $j = 1,\dots,d$, We suppose that the line* $\operatorname{Re}\lambda = 2+\sigma-\beta$ *is free of eigenvalues of the pencil $\mathfrak{A}(\lambda)$, the components of δ satisfy the conditions* (10.7.1), *and that the compatibility conditions of Lemma* 10.7.2 *are satisfied on the edges M_k. Then the problem* (10.1.1), (10.1.2) *has a unique solution $(u,p) \in C^{2,\sigma}_{\beta,\delta}(\mathcal{K})^3 \times C^{1,\sigma}_{\beta,\delta}(\mathcal{K})$.*

P r o o f. If $\delta_k < \sigma$, then it follows from (10.7.1) that $\mu_+^{(k)} > 2$. This means that then the number $\lambda = 2$ does not belong to the spectrum of the pencil $A_k(\lambda)$. Hence, Lemma 10.7.2 allows us to restrict ourselves in the proof to the case where $f \in N^{0,\sigma}_{\beta,\delta}(\mathcal{K})^3$, $g \in N^{1,\sigma}_{\beta,\delta}(\mathcal{K})$, $h_j = 0$ and $\phi_j = 0$ for $j = 1,\dots,d$.

Let κ be an arbitrary real number such that the closed strip between the lines $\operatorname{Re}\lambda = 2+\sigma-\beta$ and $\operatorname{Re}\lambda = -\kappa - 1/2$ is free of eigenvalues of the pencil $\mathfrak{A}(\lambda)$, and let $G(x,\xi)$ be the Green's matrix introduced in Theorem 10.4.1. We consider the vector-functions $(u^\pm,p^\pm)$ with the components (10.7.9) and (10.7.10).

By Theorem 10.7.12, we have $(u^-,p^-) \in C^{2,\sigma}_{\beta,\delta}(\mathcal{K})^3 \times C^{1,\sigma}_{\beta,\delta}(\mathcal{K})$, while (u^+,p^+) satisfies the estimates (10.7.11), (10.7.12). From the definition of u^-,p^- and from the equation

$$-\Delta_x G_{i,j}(x,\xi) + \partial_{x_i} G_{4,j}(x,\xi) = \delta_{i,j}\,\delta(x-\xi), \quad i = 1,2,3,$$

we deduce

$$-\Delta u_i^-(x) + \partial_{x_i} p^-(x) = -\sum_{j=1}^{3} \int_{\mathcal{K}} \big(f_j(\xi) + \partial_{\xi_j} g(\xi)\big)\, K_{i,j}(x,\xi)\, d\xi - \int_{\mathcal{K}} g(\xi)\, K_{i,4}(x,\xi)\, d\xi,$$

where

$$\begin{aligned} K_{i,j}(x,\xi) &= G_{i,j}(x,\xi)\,\Delta_x \chi^-(x,\xi) + 2\nabla_x G_{i,j}(x,\xi)\cdot \nabla_x \chi^-(x,\xi) \\ &\quad - G_{4,j}(x,\xi)\,\partial_{x_i}\chi^-(x,\xi). \end{aligned}$$

The functions $K_{i,j}(x,\xi)$ vanish for $|x-\xi| < r(x)/4$ and for $|x-\xi| > r(x)/2$ and satisfy the estimate

$$\big|\partial_x^\alpha K_{i,j}(x,\xi)\big| \le c_\alpha\, r(x)^{-3-|\alpha|-\delta_{4,j}}$$

with constants c_α independent of x and ξ. Consequently,

$$\begin{aligned}
&\Big|\partial_x^\alpha \int_{\mathcal{K}} \big(f_j(\xi) + \partial_{\xi_j} g(\xi)\big)\, K_{i,j}(x,\xi)\, d\xi\Big| \\
&\le c\, \|f_j + \partial_{x_j} g\|_{L^\infty_{\beta-\sigma,\delta-\sigma}(\mathcal{K})}\ |x|^{\sigma-\beta} \prod_{k=1}^d \Big(\frac{r_k(x)}{|x|}\Big)^{\sigma-\delta_k} \int_{\mathcal{K}} \big|\partial_x^\alpha K_{i,j}(x,\xi)\big|\, d\xi \\
&\le c\, \|f_j + \partial_{x_j} g\|_{L^\infty_{\beta-\sigma,\delta-\sigma}(\mathcal{K})}\ |x|^{\sigma-\beta-|\alpha|} \prod_{k=1}^d \Big(\frac{r_k(x)}{|x|}\Big)^{\sigma-\delta_k-|\alpha|}
\end{aligned}$$

for arbitrary multi-indices α, $j = 1,2,3$, and analogously

$$\Big|\partial_x^\alpha \int_{\mathcal{K}} g(\xi)\, K_{i,4}(x,\xi)\, d\xi\Big| \le c\, \|g\|_{L^\infty_{\beta-1-\sigma,\delta-1-\sigma}(\mathcal{K})}\ |x|^{\sigma-\beta-|\alpha|} \prod_{k=1}^d \Big(\frac{r_k(x)}{|x|}\Big)^{\sigma-\delta_k-|\alpha|}.$$

In particular,

$$\begin{aligned}
&\|-\Delta u^- + \nabla p^-\|_{L^\infty_{\beta-\sigma,\delta-\sigma}(\mathcal{K})^3} + \sum_{j=1}^3 \|\partial_{x_j}(-\Delta u^- + \nabla p^-)\|_{L^\infty_{\beta+1-\sigma,\delta+1-\sigma}(\mathcal{K})^3} \\
&\le c\Big(\sum_{j=1}^3 \|f_j + \partial_{x_j} g\|_{L^\infty_{\beta-\sigma,\delta-\sigma}(\mathcal{K})} + \|g\|_{L^\infty_{\beta-1-\sigma,\delta-1-\sigma}(\mathcal{K})}\Big).
\end{aligned}$$

Using the inequality (3.6.3), we get

$$\|-\Delta u^- + \nabla p^-\|_{N^{0,\sigma}_{\beta,\delta}(\mathcal{K})^3} \le c\big(\|f\|_{N^{0,\sigma}_{\beta,\delta}(\mathcal{K})^3} + \|g\|_{N^{1,\sigma}_{\beta,\delta}(\mathcal{K})}\big).$$

The same estimate holds for the norms of $\nabla\cdot u^-$ in $N^{1,\sigma}_{\beta,\delta}(\mathcal{K})$ and and $N_j(u^-,p^-)$ in $N^{1,\sigma}_{\beta,\delta}(\Gamma_j)^{d_j}$. Furthermore, $S_j u^- = 0$ on Γ_j for $j = 1,\dots,d$. Since $(u,p) = (u^+ + u^-, p^+ + p^-)$ is a solution of problem (10.1.1), (10.1.2), we arrive at

$$\begin{aligned}
&-\Delta u^+ + \nabla p^+ = f + \Delta u^- - \nabla p^- \in N^{0,\sigma}_{\beta,\delta}(\mathcal{K})^3, \\
&-\nabla\cdot u^+ = g + \nabla\cdot u^- \in N^{1,\sigma}_{\beta,\delta}(\mathcal{K}), \\
&S_j u^+|_{\Gamma_j} = 0, \quad N_j(u^+,p^+)|_{\Gamma_j} = -N_j(u^-,p^-)|_{\Gamma_j} \in N^{1,\sigma}_{\beta,\delta}(\Gamma_j), \quad j = 1,\dots,d.
\end{aligned}$$

Applying Lemma 10.7.3, we conclude that $(u^+,p^+) \in N^{2,\sigma}_{\beta,\delta}(\mathcal{K})^3 \times N^{1,\sigma}_{\beta,\delta}(\mathcal{K})$. Therefore, $(u,p) \in C^{2,\sigma}_{\beta,\delta}(\mathcal{K})^3 \times C^{1,\sigma}_{\beta,\delta}(\mathcal{K})$. This proves the existence of a solution.

We verify the uniqueness. Let $(u,p) \in C^{2,\sigma}_{\beta,\delta}(\mathcal{K})^3 \times C^{1,\sigma}_{\beta,\delta}(\mathcal{K})$ be a solution of problem (10.1.1), (10.1.2) with zero data f, g, h_j, ϕ_j, and let χ be a smooth cut-off function on $\overline{\mathcal{K}}$ equal to one for $|x| < 1$ and to zero for $|x| > 2$. Furthermore, let $\beta' = \beta - \sigma - 3/2$ and let δ'_k be real numbers such that

$$\max(0, \delta_k - \sigma, 2 - \mu_+^{(k)}) - 1 < \delta'_k < 1.$$

Then $\chi(u,p) \in W^{2,2}_{\beta'+\varepsilon,\delta'}(\mathcal{K})^3 \times W^{1,2}_{\beta'+\varepsilon,\delta'}(\mathcal{K})$ and $(1-\chi)(u,p) \in W^{2,2}_{\beta'-\varepsilon,\delta'}(\mathcal{K})^3 \times W^{1,2}_{\beta'-\varepsilon,\delta'}(\mathcal{K})$, where ε is an arbitrary positive number. Consequently,

$$-\Delta(\chi u) + \nabla(\chi p) = \Delta(u - \chi u) - \nabla(p - \chi p) \in W^{0,2}_{\beta'-\varepsilon,\delta'}(\mathcal{K})^3$$

and analogously

$$\nabla\cdot(\chi u) \in W^{1,2}_{\beta'-\varepsilon,\delta'}(\mathcal{K}), \quad S_j(\chi u) = 0, \quad N_j(\chi u, \chi p) \in W^{1/2,2}_{\beta'-\varepsilon,\delta'}(\Gamma_j)^{d_j}.$$

Applying Theorem 10.5.7, we obtain $\chi(u,p) \in W^{2,2}_{\beta'-\varepsilon,\delta'}(\mathcal{K})^3 \times W^{1,2}_{\beta'-\varepsilon,\delta'}(\mathcal{K})$ if ε is sufficiently small. Hence, $(u,p) \in W^{2,2}_{\beta'-\varepsilon,\delta'}(\mathcal{K})^3 \times W^{1,2}_{\beta'-\varepsilon,\delta'}(\mathcal{K})$ and Theorem 10.5.6 implies $u = 0$, $p = 0$. The proof of the theorem is complete. □

10.7.8. Regularity results for the solution. At the end of the section, we deal with regularity assertions for solutions of the boundary value problem (10.1.1), (10.1.2) in weighted Hölder spaces. We start with a result for the solution $(u,p) \in C^{2,\sigma}_{\beta,\delta}(\mathcal{K}) \times C^{1,\sigma}_{\beta,\delta}(\mathcal{K})$ which was considered in the preceding subsection.

THEOREM 10.7.14. *Let $(u,p) \in C^{2,\sigma'}_{\beta',\delta'}(\mathcal{K}) \times C^{1,\sigma'}_{\beta',\delta'}(\mathcal{K})$ be a solution of problem* (10.1.1), (10.1.2), *where*

$$f \in C^{0,\sigma}_{\beta,\delta}(\mathcal{K})^3 \cap C^{0,\sigma'}_{\beta',\delta'}(\mathcal{K})^3, \quad g \in C^{1,\sigma}_{\beta,\delta}(\mathcal{K}) \cap C^{1,\sigma'}_{\beta',\delta'}(\mathcal{K}),$$

$$h_j \in C^{2,\sigma}_{\beta,\delta}(\Gamma_j)^{3-d_j} \cap C^{2,\sigma'}_{\beta',\delta'}(\Gamma_j)^{3-d_j}, \quad \phi_j \in C^{1,\sigma}_{\beta,\delta}(\Gamma_j)^{d_j} \cap C^{1,\sigma'}_{\beta',\delta'}(\Gamma_j)^{d_j}.$$

We suppose that the closed strip between the lines $\operatorname{Re}\lambda = 2+\sigma-\beta$ *and* $\operatorname{Re}\lambda = 2+\sigma'-\beta'$ *does not contain eigenvalues of the pencil* $\mathfrak{A}(\lambda)$ *and that* δ_k, δ'_k *are nonnegative numbers such that* $\delta_k - \sigma$ *and* $\delta'_k - \sigma'$ *are not integer and*

$$2 - \mu_+^{(k)} < \delta_k - \sigma < 2, \quad 2 - \mu_+^{(k)} < \delta'_k - \sigma' < 2$$

for $k = 1,\dots,d$. *Furthermore, we assume that* g, h_j, ϕ_j *satisfy the same compatibility conditions as in Theorem* 10.7.13. *Then* $(u,p) \in C^{2,\sigma}_{\beta,\delta}(\mathcal{K})^3 \times C^{1,\sigma}_{\beta,\delta}(\mathcal{K})$.

P r o o f. 1) First let $\delta = \delta'$ and $\sigma = \sigma'$. Then analogously to Lemma 10.7.2, there exist $v \in C^{2,\sigma}_{\beta,\delta}(\mathcal{K})^3 \cap C^{2,\sigma}_{\beta',\delta}(\mathcal{K})^3$ and $q \in C^{1,\sigma}_{\beta,\delta}(\mathcal{K}) \cap C^{1,\sigma}_{\beta',\delta}(\mathcal{K})$ such that

$$\Delta v - \nabla q + f \in N^{0,\sigma}_{\beta,\delta}(\mathcal{K})^3 \cap N^{0,\sigma}_{\beta',\delta}(\mathcal{K})^3, \quad \nabla\cdot v + g \in N^{1,\sigma}_{\beta,\delta}(\mathcal{K}) \cap N^{1,\sigma}_{\beta',\delta}(\mathcal{K})$$

$S_j v = h_j$ and $N_j(v,q) = \phi_j$ on Γ_j for $j = 1,\dots,d$. Therefore, we may assume without loss of generality that $f \in N^{0,\sigma}_{\beta,\delta}(\mathcal{K})^3 \cap N^{0,\sigma}_{\beta',\delta}(\mathcal{K})^3$, $g \in N^{1,\sigma}_{\beta,\delta}(\mathcal{K}) \cap N^{1,\sigma}_{\beta',\delta}(\mathcal{K})$, $h_j = 0$, and $\phi_j = 0$. Then, as was shown in the proof Theorem 10.7.13, the uniquely determined solution $(u,p) \in C^{2,\sigma}_{\beta',\delta}(\mathcal{K}) \times C^{1,\sigma}_{\beta',\delta}(\mathcal{K})$ is given by (10.7.6), (10.7.7), where $G(x,\xi)$ is the Green's matrix introduced in Theorem 10.4.1 with $\kappa = \beta - \sigma - 5/2$. However, the uniquely determined solution in $C^{2,\sigma}_{\beta,\delta}(\mathcal{K})^3 \times C^{1,\sigma}_{\beta,\delta}(\mathcal{K})$ has also the representation (10.7.6), (10.7.7) with the same Green's matrix $G(x,\xi)$. This proves the theorem in the case $\delta = \delta'$, $\sigma = \sigma'$.

2) By Theorem 10.7.13, there exists a unique solution $(v,q) \in C^{2,\sigma}_{\beta,\delta}(\mathcal{K})^3 \times C^{1,\sigma}_{\beta,\delta}(\mathcal{K})$ of problem (10.1.1), (10.1.2). We set

$$\sigma'' = \min(\sigma,\sigma'), \quad \delta'_k = \max(\delta_k - \sigma + \sigma'', \delta'_k - \sigma' + \sigma'', 0).$$

Then $(u,p) \in C^{2,\sigma''}_{\beta-\sigma+\sigma'',\delta''}(\mathcal{K})^3 \times C^{1,\sigma''}_{\beta-\sigma+\sigma'',\delta''}(\mathcal{K})$ and $(v,q) \in C^{2,\sigma''}_{\beta'-\sigma'+\sigma'',\delta''}(\mathcal{K})^3 \times C^{1,\sigma''}_{\beta'-\sigma'+\sigma'',\delta''}(\mathcal{K})$. Consequently, it follows from the first part of the proof that $(u,p) = (v,q)$. □

Next, we prove a generalization of Theorem 10.7.14 involving weighted Hölder spaces of higher order.

THEOREM 10.7.15. *Let* $(u,p) \in C^{2,\sigma'}_{\beta',\delta'}(\mathcal{K}) \times C^{1,\sigma'}_{\beta',\delta'}(\mathcal{K})$ *be a solution of problem* (10.1.1), (10.1.2), *where*

$$f \in C^{l-2,\sigma}_{\beta,\delta}(\mathcal{K})^3 \cap C^{0,\sigma'}_{\beta',\delta'}(\mathcal{K})^3, \quad g \in C^{l-1,\sigma}_{\beta,\delta}(\mathcal{K}) \cap C^{1,\sigma'}_{\beta',\delta'}(\mathcal{K}),$$

$$h_j \in C^{l,\sigma}_{\beta,\delta}(\Gamma_j)^{3-d_j} \cap C^{2,\sigma'}_{\beta',\delta'}(\Gamma_j)^{3-d_j}, \quad \phi_j \in C^{l-1,\sigma}_{\beta,\delta}(\Gamma_j)^{d_j} \cap C^{1,\sigma'}_{\beta',\delta'}(\Gamma_j)^{d_j}.$$

We suppose that the closed strip between the lines $\operatorname{Re}\lambda = l+\sigma-\beta$ *and* $\operatorname{Re}\lambda = 2+\sigma'-\beta'$ *does not contain eigenvalues of the pencil* $\mathfrak{A}(\lambda)$ *and that* δ_k, δ'_k *are nonnegative numbers such that* $\delta_k - \sigma$ *and* $\delta'_k - \sigma'$ *are not integer and*

$$l - \mu_+^{(k)} < \delta_k - \sigma < l, \quad 2 - \mu_+^{(k)} < \delta'_k - \sigma' < 2.$$

Furthermore, we assume that the data g, h_j, ϕ_j *satisfy the same compatibility conditions as in Lemma* 10.7.2 *if* $\delta_k < l+\sigma-1$. *Then* $(u,p) \in C^{l,\sigma}_{\beta,\delta}(\mathcal{K}) \times C^{l-1,\sigma}_{\beta,\delta}(\mathcal{K})$.

P r o o f. 1) Suppose that $l-2 \le \delta_k < l+\sigma$ for $k=1,\dots,d$. Then

$$f \in C^{0,\sigma}_{\beta-l+2,\delta-l+2}(\mathcal{K})^3, \quad g \in C^{1,\sigma}_{\beta-l+2,\delta-l+2}(\mathcal{K}),$$

$$h_j \in C^{2,\sigma}_{\beta-l+2,\delta-l+2}(\Gamma_j)^{3-d_j}, \quad \phi_j \in C^{1,\sigma}_{\beta-l+2,\delta-l+2}(\Gamma_j)^{d_j},$$

and Theorem 10.7.14 implies $(u,p) \in C^{2,\sigma}_{\beta-l+2,\delta-l+2}(\mathcal{K})^3 \times C^{1,\sigma}_{\beta-l+2,\delta-l+2}(\mathcal{K})$. Using Lemma 10.7.4, we obtain $(u,p) \in C^{l,\sigma}_{\beta,\delta}(\mathcal{K})^3 \times C^{l-1,\sigma}_{\beta,\delta}(\mathcal{K})$.

2) Let $l-N \le \delta_k < l-N+1$ for $k=1,\dots,d$, where N is an integer, $2 \le m \le l$. Then we prove the assertion of the theorem by induction in N. For $N=2$ we may refer to the first part of the proof. Suppose that $N \ge 3$ and the assertion of the theorem is true for $\delta_k \ge l-N+1$. Since

$$f \in C^{l-2,\sigma}_{\beta,\delta}(\mathcal{K})^3 \subset C^{l-3,\sigma}_{\beta-1,\delta}(\mathcal{K})^3, \quad g \in C^{l-2,\sigma}_{\beta-1,\delta}(\mathcal{K}), \quad h_j \in C^{l-1,\sigma}_{\beta-1,\delta}(\Gamma_j)^{3-d_j},$$

and $\phi_j \in C^{l-2,\sigma}_{\beta-1,\delta}(\Gamma_j)^{d_j}$, the induction hypothesis implies

$$(u,p) \in C^{l-1,\sigma}_{\beta-1,\delta}(\mathcal{K})^3 \times C^{l-2,\sigma}_{\beta-1,\delta}(\mathcal{K}).$$

Furthermore, it follows from Lemma 10.7.4 that $(u,p) \in C^{l,\sigma}_{\beta,\delta+1}(\mathcal{K})^3 \times C^{l-1,\sigma}_{\beta,\delta+1}(\mathcal{K})$. Thus, $\rho\partial_\rho u \in C^{l-1,\sigma}_{\beta-1,\delta+1}(\mathcal{K})^3 \subset C^{2,\sigma}_{\beta-l+2,2-\varepsilon}(\mathcal{K})^3$, where $0<\varepsilon<\min(2, l-\delta_k-2)$. Analogously, $\rho\partial_\rho p \in C^{1,\sigma}_{\beta-l+2,2-\varepsilon}(\mathcal{K})$. Since the vector function $(\rho\partial_\rho u, \rho\partial_\rho p + p)$ is a solution of the boundary value problem (10.6.21), (10.6.22), we conclude from the induction hypothesis that $\rho\partial_\rho(u,p) \in C^{l-1,\sigma}_{\beta-1,\delta}(\mathcal{K})^3 \times C^{l-2,\sigma}_{\beta-1,\delta}(\mathcal{K})$. Applying Lemma 10.7.5, we obtain $(u,p) \in C^{l,\sigma}_{\beta,\delta}(\mathcal{K})^3 \times C^{l-1,\sigma}_{\beta,\delta}(\mathcal{K})$.

3) Finally, we assume that $l-\delta_k \in (N_k-1, N_k]$ for $k=1,\dots,d$ with different $N_k \in \{0,1,\dots,l\}$. Then let $\psi_1,\dots,\psi_d$ be smooth functions on $\bar{\Omega}$ such that $\psi_k \ge 0$, $\psi_k = 1$ near $M_k \cap S^2$, and $\sum \psi_k = 1$. We extend ψ_k to $\mathcal{K}$ by the equality $\psi_k(x) = \psi_k(x/|x|)$. Then by the first two parts of the proof, $\psi_k(u,p) \in C^{l,\sigma}_{\beta,\delta}(\mathcal{K})^3 \times C^{l-1,\sigma}_{\beta,\delta}(\mathcal{K})$ for $k=1,\dots,d$. The result follows. □

10.7.9. A regularity result for the variational solution. In Subsection 10.2.4 we considered variational solutions of the boundary value problem (10.1.1), (10.1.2) in the weighted Sobolev space $V^{1,2}_\kappa(\mathcal{K})^3 \times V^{0,2}_\kappa(\mathcal{K})$. We prove a regularity assertion in weighted Hölder spaces for these solutions.

THEOREM 10.7.16. *Let* $(u,p) \in V_\kappa^{1,2}(\mathcal{K})^3 \times V_\kappa^{0,2}(\mathcal{K})$ *be a solution of the problem* (10.2.11), (10.2.12), *where* $g \in V_\kappa^{0,2}(\mathcal{K}) \cap C_{\beta,\delta}^{l-1,\sigma}(\mathcal{K})$, $h_j \in W_{\kappa,0}^{1/2,2}(\Gamma_j)^{3-d_j} \cap C_{\beta,\delta}^{l,\sigma}(\Gamma_j)^{3-d_j}$ *and* F *is a linear and continuous functional on* $\mathcal{H}_{-\kappa}$ *which has the form*

$$F(v) = \int_{\mathcal{K}} (f + \nabla g)\cdot v\,dx + \sum_{j=1}^{d} \int_{\Gamma_j} \phi_j \cdot v\,dx$$

with given vector functions $f \in C_{\beta,\delta}^{l-2,\sigma}(\mathcal{K})^3$ *and* $\phi_j \in C_{\beta,\delta}^{l-1,\sigma}(\Gamma_j)^{d_j}$. *We suppose that the lines* $\operatorname{Re}\lambda = -\kappa - 1/2$ *and* $\operatorname{Re}\lambda = l+\sigma-\beta$ *do not contain eigenvalues of the pencil* $\mathfrak{A}(\lambda)$ *and that the components of* δ *satisfy the condition*

$$(10.7.21) \qquad l - \mu_+^{(k)} < \delta_k - \sigma < l, \quad \delta_k \ge 0, \quad \delta_k - \sigma \text{ is not integer}$$

for $k = 1,\dots,d$. *Furthermore, we assume that the data* g, h_j, ϕ_j *satisfy the same compatibility conditions as in Lemma* 10.7.2. *Then* (u,p) *admits the decomposition* (10.3.6), *where* (w,q) *is the solution of the boundary value problem* (10.1.1), (10.1.2) *in the space* $C_{\beta,\delta}^{l,\sigma}(\mathcal{K})^3 \times C_{\beta,\delta}^{l-1,\sigma}(\mathcal{K})$, λ_ν *are the eigenvalues of the pencil* $\mathfrak{A}(\lambda)$ *between the lines* $\operatorname{Re}\lambda = -\kappa-1/2$ *and* $\operatorname{Re}\lambda = l+\sigma-\beta$, *and* $\big(u^{(\nu,j,s)}, p^{(\nu,j,s)}\big)$ *are eigenvectors and generalized eigenvectors corresponding to the eigenvalue* λ_ν.

P r o o f. By Theorems 10.7.13 and 10.7.15, there exists a solution $(w,q) \in C_{\beta,\delta}^{l,\sigma}(\mathcal{K})^3 \times C_{\beta,\delta}^{l-1,\sigma}(\mathcal{K})$ of the problem (10.1.1), (10.1.2). Let ζ, η be smooth functions on $\bar{\mathcal{K}}$ equal to one near the origin which vanish outside the unit ball and satisfy the equality $\zeta\eta = \zeta$. Then

$$\zeta(w,q) \in V_{\beta'+\varepsilon}^{1,2}(\mathcal{K})^3 \times V_{\beta'+\varepsilon}^{0,2}(\mathcal{K}), \quad (1-\zeta)(w,q) \in V_{\beta'-\varepsilon}^{1,2}(\mathcal{K})^3 \times V_{\beta'-\varepsilon}^{0,2}(\mathcal{K}),$$

where $\beta' = \beta - l - \sigma - 1/2$ and ε is an arbitrarily small positive number. Since (w,q) satisfies (10.2.11) for all $v \in C_0^\infty(\bar{\mathcal{K}}\backslash\{0\})^3$, $S_j v = 0$ on Γ_j, we obtain

$$b_{\mathcal{K}}(\zeta w, v) - \int_{\mathcal{K}} \zeta q\, \nabla\cdot v\,dx = \Phi(v),$$

where

$$\Phi(v) = F(\eta v) - b_{\mathcal{K}}\big((1-\zeta)w, \eta v\big) + \int_{\mathcal{K}} (1-\zeta) q\, \nabla\cdot(\eta v)\,dx.$$

Obviously, $\Phi \in \mathcal{H}_{-\kappa}^*$. Furthermore,

$$-\nabla\cdot(\zeta w) = \zeta g - w\cdot\nabla\zeta \in V_\kappa^{0,2}(\mathcal{K}) \quad\text{and}\quad S_j(\zeta w) = \zeta h_j \in W_{\kappa,0}^{1/2,2}(\Gamma_j)^{3-d_j}.$$

By Theorem 10.3.5, there exists a sum Σ_1 of the same form as in (10.3.6) such that

$$\zeta(w,q) - \Sigma_1 \in V_\kappa^{1,2}(\mathcal{K})^3 \times V_\kappa^{0,2}(\mathcal{K}).$$

Analogously, we obtain $(1-\zeta)\,(w,q) - \Sigma_2 \in V_\kappa^{1,2}(\mathcal{K})^3 \times V_\kappa^{0,2}(\mathcal{K})$, where Σ_2 is also a sum of the same form as in (10.3.6). Consequently, the vector function (w,q) admits the decomposition

$$(w,q) = \Sigma_1 + \Sigma_2 + (w',q'),$$

where $(w',q') \in V_\kappa^{1,2}(\mathcal{K})^3 \times V_\kappa^{0,2}(\mathcal{K})$. Since Σ_1 and Σ_2 are solutions of the homogeneous problem (10.1.1), (10.1.2), the vector function (w',q') solves the problem (10.2.11), (10.2.12). By virtue of the uniqueness of the solution in $V_\kappa^{1,2}(\mathcal{K})^3 \times V_\kappa^{0,2}(\mathcal{K})$ (see Theorem 10.2.11), we conclude that $(w',q') = (u,p)$. This proves the theorem. $\square$

10.7.10. The case $d_{k_+} + d_{k_-} \le 3$. If $\lambda = 0$ is not an eigenvalue of the pencils $A_k(\lambda)$ (i.e., if $d_{k_+} + d_{k_-} \le 3$ and the edge angle θ_k satisfies the condition (10.3.8) for $d_{k_+} \cdot d_{k_-} \ne 0$), then we can use the sharper estimates of the Green's matrix in Theorems 10.4.4 and 10.4.5. This leads to the following result.

THEOREM 10.7.17. *Let $f \in N^{l-2,\sigma}_{\beta,\delta}(\mathcal{K})^3$, $g \in N^{l-1,\sigma}_{\beta,\delta}(\mathcal{K})$, $h_j \in N^{l,\sigma}_{\beta,\delta}(\Gamma_j)^{3-d_j}$ and $\phi_j \in N^{l-1,\sigma}_{\beta,\delta}(\Gamma_j)^{d_j}$ for $j = 1, \dots, d$. We assume that $d_{k_+} + d_{k_-} \le 3$ for all k and that the edge angles θ_k satisfy the condition (10.3.8) for $d_{k_+} \cdot d_{k_-} \ne 0$. Furthermore, we suppose that the line $\operatorname{Re}\lambda = l + \sigma - \beta$ is free of eigenvalues of the pencil $\mathfrak{A}(\lambda)$ and that the components of δ satisfy the inequalities*

$$l - \delta_+^{(k)} < \delta_k - \sigma < l + \delta_+^{(k)} \quad \textit{for } k = 1, \dots, d.$$

Then problem (10.1.1), (10.1.2) is uniquely solvable in $N^{l,\sigma}_{\beta,\delta}(\mathcal{K})^3 \times N^{l-1,\sigma}_{\beta,\delta}(\mathcal{K})$.

P r o o f. By Lemma 10.7.1, we may assume without loss of generality that $h_j = 0$ and $\phi_j = 0$ for all j. Let κ be an arbitrary real number such that the closed strip between the lines $\operatorname{Re}\lambda = l + \sigma - \beta$ and $\operatorname{Re}\lambda = -\kappa - 1/2$ is free of eigenvalues of the pencil $\mathfrak{A}(\lambda)$. Furthermore, let $G(x,\xi)$ be the Green's matrix introduced in Theorem 10.4.1 (with the above κ), and let $(u^\pm, p^\pm)$ be the same vector functions as in Subsection 10.7.3. Under the conditions of our theorem, the elements of the Green's matrix satisfy the estimates in Theorems 10.4.4 and 10.4.5 with the exponents $\sigma_{i,\alpha}(x) = \delta_x - \delta_{i,4} - |\alpha| - \varepsilon$ and $\sigma_{k,i,\alpha} = \delta_+^{(k)} - |\alpha| - \delta_{i,4} - \varepsilon$, respectively. Analogously to the proofs of Lemma 10.7.6 and Theorem 10.7.10, we obtain

$$\begin{aligned} &\|u^\pm\|_{L^\infty_{\beta-l-\sigma,\delta-l-\sigma}(\mathcal{K})^3} + \|p^\pm\|_{L^\infty_{\beta-l+1-\sigma,\delta-l+1-\sigma}(\mathcal{K})} \\ &\quad \le c\left(\|f\|_{N^{l,\sigma}_{\beta,\delta}(\mathcal{K})^3} + \|g\|_{N^{l-1,\sigma}_{\beta,\delta}(\mathcal{K})}\right). \end{aligned}$$

Hence, the same estimate is satisfied for the solution $(u,p) = (u^+,p^+) + (u^-,p^-)$ of the boundary value problem (10.1.1), (10.1.2). Using Lemma 10.7.3, we obtain $(u,p) \in N^{l,\sigma}_{\beta,\delta}(\mathcal{K})^3 \times N^{l-1,\sigma}_{\beta,\delta}(\mathcal{K})$. The uniqueness of the solution holds analogously to the proof of Theorem 10.7.13. □

Finally, the following regularity results in the weighted Hölder spaces $N^{l,\sigma}_{\beta,\delta}(\mathcal{K})$ hold.

THEOREM 10.7.18. *Suppose that $d_{k_+} + d_{k_-} \le 3$ for all k, the edge angles θ_k satisfy the condition (10.3.8) for $d_{k_+} \cdot d_{k_-} \ne 0$ and that $(u,p) \in N^{l,\sigma}_{\beta,\delta}(\mathcal{K})^3 \times N^{l-1,\sigma}_{\beta,\delta}(\mathcal{K})$ is a solution of the boundary value problem (10.1.1), (10.1.2).*

1) *If $f \in N^{l'-2,\sigma'}_{\beta',\delta'}(\mathcal{K})^3$, $l, l' \ge 2$, $g \in N^{l'-1,\sigma'}_{\beta',\delta'}(\mathcal{K})$, $h_j \in N^{l',\sigma'}_{\beta',\delta'}(\Gamma_j)^{3-d_j}$, $\phi_j \in N^{l'-1,\sigma'}_{\beta',\delta'}(\Gamma_j)^{d_j}$, the closed strip between the lines $\operatorname{Re}\lambda = l + \sigma - \beta$ and $\operatorname{Re}\lambda = l' + \sigma' - \beta'$ is free of eigenvalues of the pencil $\mathfrak{A}(\lambda)$, and the components of δ, δ' satisfy the inequalities*

$$l - \delta_+^{(k)} < \delta_k - \sigma < l + \delta_+^{(k)}, \quad l' - \delta_+^{(k)} < \delta'_k - \sigma' < l' + \delta_+^{(k)}$$

for $k = 1, \dots, d$, then $(u,p) \in N^{l',\sigma'}_{\beta',\delta'}(\mathcal{K})^3 \times N^{l'-1,\sigma'}_{\beta',\delta'}(\mathcal{K})$.

2) *If $f \in V^{l'-2,s}_{\beta',\delta'}(\mathcal{K})^3$, $l, l' \ge 2$, $g \in V^{l'-1,s}_{\beta',\delta'}(\mathcal{K})$, $h_j \in V^{l'-1/s,s}_{\beta',\delta'}(\Gamma_j)^{3-d_j}$, $\phi_j \in V^{l'-1-1/s,s}_{\beta',\delta'}(\Gamma_j)^{d_j}$, the closed strip between the lines $\operatorname{Re}\lambda = l + \sigma - \beta$ and $\operatorname{Re}\lambda =$*

$l' - \beta' - 2/s$ is free of eigenvalues of the pencil $\mathfrak{A}(\lambda)$, and the components of δ, δ' satisfy the inequalities

$$l - \delta_+^{(k)} < \delta_k - \sigma < l + \delta_+^{(k)}, \quad l' - \delta_+^{(k)} < \delta'_k + 2/s < l' + \delta_+^{(k)}$$

for $k = 1, \dots, d$, then $(u,p) \in V_{\beta',\delta'}^{l',s}(\mathcal{K})^3 \times V_{\beta',\delta'}^{l'-1,s}(\mathcal{K})$.

For the proof, it suffices to note that the (uniquely determined) solutions of the boundary value problem (10.1.1), (10.1.2) in the spaces $N_{\beta,\delta}^{l,\sigma}(\mathcal{K})^3 \times N_{\beta,\delta}^{l-1,\sigma}(\mathcal{K})$, $N_{\beta',\delta'}^{l',\sigma'}(\mathcal{K})^3 \times N_{\beta',\delta'}^{l'-1,\sigma'}(\mathcal{K})$ and $V_{\beta',\delta'}^{l',s}(\mathcal{K})^3 \times V_{\beta',\delta'}^{l'-1,s}(\mathcal{K})$ have the representation (10.7.6), (10.7.7) with the same Green's matrix $G(x,\xi)$.

10.8. Weak solutions in weighted Hölder spaces

In this section, we are interested in solutions of the boundary value problem (10.1.1), (10.1.2) in the weighted space $C_{\beta,\delta}^{1,\sigma}(\mathcal{K})^3 \times C_{\beta,\delta}^{0,\sigma}(\mathcal{K})$ for the data

$$f \in C_{\beta,\delta}^{-1,\sigma}(\mathcal{K})^3, \quad g \in C_{\beta,\delta}^{0,\sigma}(\mathcal{K}), \quad h_j \in C_{\beta,\delta}^{1,\sigma}(\Gamma_j)^{3-d_j}, \quad \phi_j \in C_{\beta,\delta}^{0,\sigma}(\Gamma_j)^{d_j}.$$

Here, $C_{\beta,\delta}^{-1,\sigma}(\mathcal{K})$ denotes the set of all distributions represented in the form

$$f = f^{(0)} + \sum_{k=1}^{3} \partial_{x_k} f^{(k)}, \quad \text{where } f^{(0)} \in C_{\beta+1,\delta+1}^{0,\sigma}(\mathcal{K}), \ f^{(k)} \in C_{\beta,\delta}^{0,\sigma}(\mathcal{K}), \ k = 1,2,3.$$

The existence of solutions is proved again by means of the estimates for the Green's matrix given in Section 10.4.

10.8.1. Representation of weak solutions by Green's matrix. Let κ be an arbitrary real number such that the line $\operatorname{Re}\lambda = -\kappa - 1/2$ does not contain eigenvalues of the pencil $\mathfrak{A}(\lambda)$, and let $G(x,\xi)$ be the Green's matrix introduced in Theorem 10.4.1. Furthermore, let

$$\Lambda_- < \operatorname{Re}\lambda < \Lambda_+$$

be the widest strip in the complex plane which contains the line $\operatorname{Re}\lambda = -\kappa - 1/2$ and which is free of eigenvalues of the pencil $\mathfrak{A}(\lambda)$. We consider the functions (10.7.6), (10.7.7) for $f \in C_{\beta,\delta}^{-1,\sigma}(\mathcal{K})^3$ and $g \in N_{\beta,\delta}^{0,\sigma}(\mathcal{K})$. The components of f can be written as

$$f_j = F_j + \nabla \cdot \Phi^{(j)}, \quad F_j \in C_{\beta+1,\delta+1}^{0,\sigma}(\mathcal{K}), \ \Phi^{(j)} \in C_{\beta,\delta}^{0,\sigma}(\mathcal{K})^3. \tag{10.8.1}$$

We assume that β and δ are subject to the inequalities

$$\Lambda_- < 1 + \sigma - \beta < \Lambda_+, \tag{10.8.2}$$

$$1 - \mu_+^{(k)} < \delta_k - \sigma < 1, \quad \delta_k \ge 0, \quad \delta_k \ne \sigma \quad \text{for } k = 1, \dots, d. \tag{10.8.3}$$

If the components of δ satisfy (10.8.3), then $C_{\beta+1,\delta+1}^{0,\sigma}(\mathcal{K}) = N_{\beta+1,\delta+1}^{0,\sigma}(\mathcal{K})$. Furthermore, every $\Phi^{(j)} \in C_{\beta,\delta}^{0,\sigma}(\mathcal{K})^3$ admits the decomposition

$$\Phi^{(j)} = \Psi^{(j)} + f^{(j)}, \quad \text{where } \Psi^{(j)} \in C_{\beta+1,\delta+1}^{1,\sigma}(\mathcal{K})^3, \ f^{(j)} \in N_{\beta,\delta}^{0,\sigma}(\mathcal{K})^3$$

(cf. Lemma 7.8.2). Consequently, the functions f_j in (10.8.1) have also the representation

$$f_j = f_j^{(0)} + \nabla \cdot f^{(j)}, \quad f_j^{(0)} \in N_{\beta+1,\delta+1}^{0,\sigma}(\mathcal{K}), \ f^{(j)} \in N_{\beta,\delta}^{0,\sigma}(\mathcal{K})^3. \tag{10.8.4}$$

Here, $f_j^{(0)} = F_j + \nabla \cdot \Psi^{(j)}$. If the vector function $f^{(j)}$ belongs to the subspace $N^{1,\sigma}_{\beta+1,\delta+1}(\mathcal{K})^3$ of $N^{0,\sigma}_{\beta,\delta}(\mathcal{K})^3$, then integration by parts yields

$$\int_{\mathcal{K}} (\nabla \cdot f^{(j)})\, v\, dx = -\int_{\mathcal{K}} f^{(j)} \cdot \nabla v\, dx + \sum_{\nu=1}^{d} \int_{\Gamma_\nu} f^{(j)} v \cdot n\, dx$$

for arbitrary $v \in C^\infty_{\sigma,\beta,\delta}$. Here, $C^\infty_{\sigma,\beta,\delta}$ denotes the set of all $v \in C^\infty(\bar{\mathcal{K}} \backslash \mathcal{S})$ such that

$$\int_{\mathcal{K}} |x|^{\sigma-\beta} \prod_{k=1}^{d} \Big(\frac{r_k(x)}{|x|}\Big)^{\sigma-\delta_k} \big(r(x)^{-1}\, |v(x)| + |\nabla v(x)|\big)\, dx < \infty.$$

Therefore, it is natural to extend the distributions f_j by

$$f_j(v) = \int_{\mathcal{K}} \big(f_j^{(0)}\, v - f^{(j)} \cdot \nabla v\big)\, dx + \sum_{\nu=1}^{d} \int_{\Gamma_\nu} f^{(j)} v \cdot n\, dx$$

to arbitrary $v \in C^\infty_{\sigma,\beta,\delta}$. Then we obtain the representations

$$u_i = u_i^+ + u_i^-, \quad p = p^+ + p^-$$

for the functions (10.7.6), (10.7.7), where

$$\begin{aligned}(10.8.5)\quad u_i^\pm(x) = &\sum_{j=1}^{3} \int_{\mathcal{K}} f_j^{(0)}(\xi)\, \chi^\pm(x,\xi)\, G_{i,j}(x,\xi)\, d\xi \\ &- \sum_{j=1}^{3} \int_{\mathcal{K}} f^{(j)}(\xi) \cdot \nabla_\xi \big(\chi^\pm(x,\xi)\, G_{i,j}(x,\xi)\big)\, d\xi \\ &+ \int_{\mathcal{K}} g(\xi) \Big(\chi^\pm(x,\xi)\, G_{i,4}(x,\xi) - \sum_{j=1}^{3} \partial_{\xi_j} \big(\chi^\pm(x,\xi)\, G_{i,j}(x,\xi)\big)\Big)\, d\xi \\ &+ \sum_{j=1}^{3} \sum_{\nu=1}^{d} \int_{\Gamma_\nu} \big(f^{(j)}(\xi) \cdot n + g(\xi)\, n_j\big)\, \chi^\pm(x,\xi)\, G_{i,j}(x,\xi)\, d\xi\end{aligned}$$

and

$$\begin{aligned}(10.8.6)\quad p^\pm(x) = &-\frac{1}{2}\big(g(x) \pm g(x)\big) + \sum_{j=1}^{3} \int_{\mathcal{K}} f_j^{(0)}(\xi)\, \chi^\pm(x,\xi)\, G_{4,j}(x,\xi)\, d\xi \\ &- \sum_{j=1}^{3} \int_{\mathcal{K}} f^{(j)}(\xi) \cdot \nabla_\xi \big(\chi^\pm(x,\xi)\, G_{4,j}(x,\xi)\big)\, d\xi \\ &+ \int_{\mathcal{K}} g(\xi) \Big(\chi^\pm(x,\xi)\, G_{4,4}(x,\xi) - \sum_{j=1}^{3} \partial_{\xi_j} \big(\chi^\pm(x,\xi)\, G_{4,j}(x,\xi)\big)\Big)\, d\xi \\ &+ \sum_{j=1}^{3} \sum_{\nu=1}^{d} \int_{\Gamma_\nu} \big(f^{(j)}(\xi) \cdot n + g(\xi)\, n_j\big)\, \chi^\pm(x,\xi)\, G_{4,j}(x,\xi)\, d\xi.\end{aligned}$$

Here, χ^+ and χ^- are the same functions (10.7.8) as in the foregoing section.

10.8.2. Weighted L_∞ estimates for u^+ and p^+. First we prove an analog to Lemma 10.7.6.

LEMMA 10.8.1. *Suppose that the conditions* (10.8.2) *and* (10.8.3) *are satisfied for β and the components of δ. Then the estimate*

$$(10.8.7)\quad \|u^+\|_{L^\infty_{\beta-1-\sigma,\delta-1-\sigma}(\mathcal{K})^3} + \|p^+\|_{L^\infty_{\beta-\sigma,\delta-\sigma}(\mathcal{K})} \le c\Big(\sum_{j=1}^3 \|f_j^{(0)}\|_{N^{0,\sigma}_{\beta+1,\delta+1}(\mathcal{K})} + \sum_{j=1}^3 \|f^{(j)}\|_{N^{0,\sigma}_{\beta,\delta}(\mathcal{K})^3} + \|g\|_{N^{0,\sigma}_{\beta,\delta}(\mathcal{K})}\Big)$$

is valid with a constant c independent of $f_j^{(0)}$, $f^{(j)}$, and g.

P r o o f. Since

$$(10.8.8)\quad |x-\xi| \le r_k(x)/2, \quad |x|/2 \le |\xi| \le 3|x|/2, \quad r_k(x)/2 \le r_k(\xi) \le 3r_k(x)/2$$

on the support of χ^+, we can easily prove the estimate for u^+ by means of the inequality

$$(10.8.9)\quad \big|\partial_\xi^\alpha\big(\chi^+(x,\xi)\,G_{i,j}(x,\xi)\big)\big| \le c\,|x-\xi|^{-1-|\alpha|-\delta_{i,4}-\delta_{j,4}}$$

(cf. Theorem 10.4.2). In the same way, the integrals

$$\int_{\mathcal{K}} f_j^{(0)}(\xi)\,\chi^+(x,\xi)\,G_{4,j}(x,\xi)\,d\xi$$

in the representation of p^+ can be estimated. We consider the expressions

$$A_j = -\int_{\mathcal{K}} f^{(j)}(\xi)\cdot\nabla_\xi\big(\chi^+(x,\xi)\,G_{4,j}(x,\xi)\big)\,d\xi + \sum_{\nu=1}^d \int_{\Gamma_\nu} f^{(j)}(\xi)\cdot\chi^+(x,\xi)\,G_{4,j}(x,\xi)\,n\,d\xi,$$

$$B_j = -\int_{\mathcal{K}} g(\xi)\,\partial_{\xi_j}\big(\chi^+(x,\xi)\,G_{4,j}(x,\xi)\big)\,d\xi + \sum_{\nu=1}^d \int_{\Gamma_\nu} g(\xi)\,\chi^+(x,\xi)\,G_{4,j}(x,\xi)\,n_j\,d\xi,$$

and

$$C = \int_{\mathcal{K}} g(\xi)\,\chi^+(x,\xi)\,G_{4,4}(x,\xi)\,d\xi$$

in the representation of (10.8.6) of p^+. Obviously,

$$A_j = -\int_{\mathcal{K}} \big(f^{(j)}(\xi)-f^{(j)}(x)\big)\cdot\nabla_\xi\big(\chi^+(x,\xi)\,G_{4,j}(x,\xi)\big)\,d\xi + \sum_{\nu=1}^d \int_{\Gamma_\nu} \big(f^{(j)}(\xi)-f^{(j)}(x)\big)\cdot\chi^+(x,\xi)\,G_{4,j}(x,\xi)\,n\,d\xi.$$

Hence by (10.8.8) and (10.8.9),

$$|A_j| \le c|x|^{-\beta} \prod_{k=1}^{d} \Big(\frac{r_k(x)}{|x|}\Big)^{-\delta_k} \|f^{(j)}\|_{N^{0,\sigma}_{\beta,\delta}(\mathcal{K})^3}$$
$$\times\Big(\int_{\mathcal{K}} |x-\xi|^\sigma \,\big|\nabla_\xi(\chi^+ G_{4,j})(x,\xi)\big|\,d\xi + \sum_{\nu=1}^{d}\int_{\Gamma_\nu} |x-\xi|^\sigma\,\big|(\chi^+ G_{4,j})(x,\xi)\big|\,d\xi\Big)$$
$$\le c|x|^{-\beta} \prod_{k=1}^{d} \Big(\frac{r_k(x)}{|x|}\Big)^{-\delta_k} r(x)^\sigma\,\|f^{(j)}\|_{N^{0,\sigma}_{\beta,\delta}(\mathcal{K})^3}$$
$$\le c|x|^{-\beta+\sigma} \prod_{k=1}^{d} \Big(\frac{r_k(x)}{|x|}\Big)^{-\delta_k+\sigma} \|f^{(j)}\|_{N^{0,\sigma}_{\beta,\delta}(\mathcal{K})^3}.$$

Analogously, this estimates holds for B_j. For the estimation of C, we employ the representation

$$G_{4,4}(x,\xi) = -\nabla_\xi\cdot\vec{\mathcal{P}}(x,\xi) + \mathcal{Q}(x,\xi),$$

where $\vec{\mathcal{P}}(x,\xi)\cdot n = 0$ for $\xi\in\Gamma_\nu$,

$$|\partial_\xi^\alpha \vec{\mathcal{P}}(x,\xi)| \le c\,|x-\xi|^{-2-|\alpha|}, \quad |\mathcal{Q}(x,\xi)| \le c\,r(\xi)^{-3} \ \text{ for } |x-\xi| < r(x)/2$$

(cf. Corollary 10.4.3). Then

$$C = \int_{\mathcal{K}} (g(\xi)-g(x))\,\chi^+(x,\xi)\,\nabla_\xi\cdot\vec{\mathcal{P}}(x,\xi)\,d\xi - \int_{\mathcal{K}} g(x)\,\vec{\mathcal{P}}(x,\xi)\cdot\nabla_\xi\chi^+(x,\xi)\,d\xi$$
$$+\int_{\mathcal{K}} g(\xi)\,\chi^+(x,\xi)\mathcal{Q}(x,\xi)\,d\xi.$$

Using (10.8.8) and (10.8.9), we arrive at the inequalities

$$|x|^\beta \prod_{k=1}^{d}\Big(\frac{r_k(x)}{|x|}\Big)^{\delta_k}\Big|\int_{\mathcal{K}}\Big((g(\xi)-g(x))\,\chi^+(x,\xi)\,\nabla_\xi\cdot\mathcal{P}(x,\xi)\,d\xi\Big|$$
$$\le c\,\|g\|_{N^{0,\sigma}_{\beta,\delta}(\mathcal{K})} \int\limits_{|x-\xi|<r(x)/2} |x-\xi|^{-3+\sigma}\,d\xi \le c\,r(x)^\sigma\,\|g\|_{N^{0,\sigma}_{\beta,\delta}(\mathcal{K})}$$

and

$$|x|^{\beta-\sigma}\prod_{k=1}^{d}\Big(\frac{r_k(x)}{|x|}\Big)^{\delta_k-\sigma}\Big|\int_{\mathcal{K}}\Big(g(x)\,\vec{\mathcal{P}}(x,\xi)\cdot\nabla_\xi\chi^+(x,\xi) - g(\xi)\,\chi^+(x,\xi)\mathcal{Q}(x,\xi)\Big)\,d\xi\Big|$$
$$\le c\,\|g\|_{L^\infty_{\beta-\sigma,\delta-\sigma}(\mathcal{K})}.$$

This proves the desired estimate for p^+. □

10.8.3. Weighted L_∞ estimates for u^-, p^-. We show that the functions u^- and p^- defined by (10.8.5), (10.8.6) satisfy the estimates

$$|x|^{\beta-1-\sigma+|\alpha|} \prod_{k=1}^{d} \Big(\frac{r_k(x)}{|x|}\Big)^{\max(0,\delta_k-1-\sigma+|\alpha|)} \big|\partial_x^\alpha u^-(x)\big| + |x|^{\beta-\sigma+|\alpha|} \prod_{k=1}^{d} \Big(\frac{r_k(x)}{|x|}\Big)^{\max(0,\delta_k-\sigma+|\alpha|)} \big|\partial_x^\alpha p^-(x)\big| \le c\Big(\sum_{j=1}^{3} \|f_j^{(0)}\|_{N^{0,\sigma}_{\beta+1,\delta+1}(\mathcal{K})} + \sum_{j=1}^{3} \|f^{(j)}\|_{N^{0,\sigma}_{\beta,\delta}(\mathcal{K})^3} + \|g\|_{N^{0,\sigma}_{\beta,\delta}(\mathcal{K})}\Big) \tag{10.8.10}$$

for arbitrary $x \in \mathcal{K}$ and for arbitrary multi-indices α. The integrals over $\mathcal{K}$ in the representations for u^- and p^- can be estimated by means of Lemmas 10.7.7–10.7.9. In order to estimate the integrals over the faces Γ_ν, we prove the following lemma.

LEMMA 10.8.2. *Let $g \in N^{0,\sigma}_{\beta,\delta}(\mathcal{K})$ and let*

$$v(x) = \int_{\Gamma_\nu} K(x,\xi)\, g(\xi)\, d\xi,$$

where $K(x,\xi)$ vanishes for $|x-\xi| < r(x)/4$ and satisfies the inequalities

$$\big|K(x,\xi)\big| \le c\, \frac{|x|^{\Lambda_- -s+\varepsilon}}{|\xi|^{\Lambda_- +1+\varepsilon}} \prod_{k=1}^{d} \Big(\frac{r_k(x)}{|x|}\Big)^{\min(0,\mu_+^{(k)}-s-\varepsilon)} \quad \text{for } |\xi| < |x|/2,$$

$$\big|K(x,\xi)\big| \le c\, \frac{|x|^{\Lambda_+ -s-\varepsilon}}{|\xi|^{\Lambda_+ +1-\varepsilon}} \prod_{k=1}^{d} \Big(\frac{r_k(x)}{|x|}\Big)^{\min(0,\mu_+^{(k)}-s-\varepsilon)} \quad \text{for } |\xi| > 2|x|,$$

$$\big|K(x,\xi)\big| \le c\, |x-\xi|^{-1-s} \Big(\frac{r(x)}{|x-\xi|}\Big)^{\min(0,\mu_x-s-\varepsilon)} \quad \text{for } |x|/2 < |\xi| < 2|x|$$

with an arbitrarily small positive ε. Here μ_x is the same constant as in Theorem 10.4.4. Suppose that β and δ satisfy the conditions (10.8.2) and (10.8.3), respectively. Then

$$|x|^{\beta-1-\sigma+s} \prod_{k=1}^{d} \Big(\frac{r_k(x)}{|x|}\Big)^{\max(0,\delta_k-1-\sigma+s)} \big|v(x)\big| \le c\, \|g\|_{L^\infty_{\beta-\sigma,\delta-\sigma}(\mathcal{K})}\,. \tag{10.8.11}$$

P r o o f. By the assumptions on $K(x,\xi)$, we have

$$|x|^{\beta-1-\sigma+s} \prod_{k=1}^{d} \Big(\frac{r_k(x)}{|x|}\Big)^{\max(0,\delta_k-1-\sigma+s)} \Big| \int\limits_{\substack{\Gamma_\nu \\ |\xi|<|x|/2}} K(x,\xi)\, g(\xi)\, d\xi\Big|$$

$$\le c\, |x|^{\beta-1-\sigma+\Lambda_-+\varepsilon} \prod_{k=1}^{d} \Big(\frac{r_k(x)}{|x|}\Big)^{\max(0,\delta_k-1-\sigma+s)+\min(0,\mu_+^{(k)}-s-\varepsilon)} \|g\|_{L^\infty_{\beta-\sigma,\delta-\sigma}(\mathcal{K})}$$

$$\times \int\limits_{|\xi|<|x|/2} |\xi|^{\sigma-\beta-\Lambda_--1-\varepsilon} \prod_{k=1}^{d} \Big(\frac{r_k(\xi)}{|\xi|}\Big)^{\sigma-\delta_k} d\xi.$$

Since $\sigma-\beta-\Lambda_--1 > -2$, $\sigma-\delta_k > -1$ and $\max(0,\delta_k-1-\sigma+s)+\min(0,\mu_+^{(k)}-s-\varepsilon) \ge 0$, the right-hand side of the last inequality does not exceed the right hand side of

(10.8.11). Analogously,

$$|x|^{\beta-1-\sigma+s}\prod_{k=1}^{d}\Big(\frac{r_k(x)}{|x|}\Big)^{\max(0,\delta_k-1-\sigma+s)}\Big|\int\limits_{\substack{\Gamma_\nu\\ |\xi|>2|x|}} K(x,\xi)\,g(\xi)\,d\xi\Big| \le c\,\|g\|_{L^\infty_{\beta-\sigma,\delta-\sigma}(\mathcal{K})}\,.$$

We consider the expression

$$A=|x|^{\beta-1-\sigma+s}\prod_{k=1}^{d}\Big(\frac{r_k(x)}{|x|}\Big)^{\max(0,\delta_k-1-\sigma+s)}\Big|\int\limits_{\substack{\Gamma_\nu\\ |x|/2<|\xi|<2|x|}} K(x,\xi)\,g(\xi)\,d\xi\Big|$$

and assume that M_1 is the nearest edge to x. If $\overline{M_1}\cap\overline{\Gamma}_\nu=\{0\}$, then there exists a positive constant c such that $|x-\xi|>c|x|$ for $\xi\in\Gamma_\nu$, and we obtain

$$\begin{aligned} A &\le c\,|x|^{\beta-2-\sigma}\Big(\frac{r_1(x)}{|x|}\Big)^{\max(0,\delta_1-1-\sigma+s)+\min(0,\mu_+^{(1)}-s-\varepsilon)}\|g\|_{L^\infty_{\beta-\sigma,\delta-\sigma}(\mathcal{K})}\\ &\quad\times\int\limits_{\substack{\Gamma_\nu\\ |x|/2<|\xi|<2|x|}}|\xi|^{\sigma-\beta}\prod_{k=1}^{d}\Big(\frac{r_k(\xi)}{|\xi|}\Big)^{\sigma-\delta_k}d\xi. \end{aligned}$$

Since the integral on the right-hand side of the last inequality does not exceed $c\,|x|^{\sigma-\beta+2}$, we conclude that

$$A\le c\,\|g\|_{L^\infty_{\beta-\sigma,\delta-\sigma}(\mathcal{K})}\,. \tag{10.8.12}$$

If $M_1\subset\overline{\Gamma}_\nu$ and $g(\xi)=0$ for $r_1(\xi)<2\,r(\xi)$, then $|x-\xi|>c|x|$ for $\xi\in\Gamma_\nu\cap\operatorname{supp}g$. Therefore, the inequality (10.8.12) holds as in the case $M_1\cap\overline{\Gamma}_\nu=\{0\}$.

Suppose finally that $M_1\subset\overline{\Gamma}_\nu$ and $g(\xi)=0$ for $r_1(\xi)>3\,r(\xi)$. Then

$$|g(\xi)|\le c\,|\xi|^{\sigma-\beta}\Big(\frac{r_1(\xi)}{|\xi|}\Big)^{\sigma-\delta_1}\|g\|_{L^\infty_{\beta-\sigma,\delta-\sigma}(\mathcal{K})}$$

and

$$\begin{aligned} A &\le c\,|x|^{\delta_1-1-\sigma+s}\Big(\frac{r_1(x)}{|x|}\Big)^{\max(0,\delta_1-1-\sigma+s)} r_1(x)^{\min(0,\mu_+^{(1)}-s-\varepsilon)}\,\|g\|_{L^\infty_{\beta-\sigma,\delta-\sigma}(\mathcal{K})}\\ &\quad\times\int|x-\xi|^{-1-s-\min(0,\mu_+^{(1)}-s-\varepsilon)}\,r_1(\xi)^{\sigma-\delta_1}\,d\xi, \end{aligned}$$

where the domain of integration is the set of all $\xi\in\Gamma_\nu$ satisfying the inequalities $|x|/2<|\xi|<2|x|$ and $|x-\xi|>r(x)/4$. Analogously to Lemma 2.6.1, the following two inequalities hold with a constant c independent of x and $R\ge r_1(x)/4$.

$$\int\limits_{\substack{\Gamma_\nu\\ r_1(x)/4<|x-\xi|<R}}|x-\xi|^\alpha\,r_1(\xi)^\gamma\,d\xi\le c\,R^{2+\alpha+\gamma}\quad\text{if }\alpha+\gamma>-2,\ \gamma>-1, \tag{10.8.13}$$

$$\int\limits_{\substack{\Gamma_\nu\\ |x-\xi|>R}}|x-\xi|^\alpha\,r_1(\xi)^\gamma\,d\xi\le c\,R^{2+\alpha+\gamma}\quad\text{if }\alpha+\gamma<-2,\ \gamma>-1. \tag{10.8.14}$$

If $\delta_1-1-\sigma+s<0$, then $\mu_1>s$ and (10.8.13) with $R=2|x|$ implies

$$A\le c\,|x|^{\delta_1-1-\sigma+s}\,\|g\|_{L^\infty_{\beta-\sigma,\delta-\sigma}(\mathcal{K})}\int|x-\xi|^{-1-s}\,r_1(\xi)^{\sigma-\delta_1}\,d\xi\le c\,\|g\|_{L^\infty_{\beta-\sigma,\delta-\sigma}(\mathcal{K})}\,.$$

If $\delta_1 - 1 - \sigma + s > 0$, then $-1 - s - \min(0, \mu_+^{(1)} - s - \varepsilon) + \sigma - \delta_1 < -2$, and we derive the same estimate using (10.8.14) with $R = r_1(x)/4$.

Thus, the inequality (10.8.12) is valid for an arbitrary function g vanishing for $r_1(\xi) < 2r(\xi)$ or for $r_1(\xi) > 3r(\xi)$. It remains to note that every $g \in N^{1,\sigma}_{\beta,\delta}(\mathcal{K})$ can be written as a sum $g = g_1 + g_2$, where $g_1(\xi) = 0$ for $r_1(\xi) < 2r(\xi)$, $g_2(\xi) = 0$ for $r_1(\xi) > 3r(\xi)$, and

$$\|g_1\|_{L^\infty_{\beta-\sigma,\delta-\sigma}(\mathcal{K})} + \|g_2\|_{L^\infty_{\beta-\sigma,\delta-\sigma}(\mathcal{K})} \le c\, \|g\|_{L^\infty_{\beta-\sigma,\delta-\sigma}(\mathcal{K})}.$$

This completes the proof. □

Utilizing Lemmas 10.7.7–10.7.9 and 10.8.2, we get a weighted L_∞ estimate for the vector function (u^-, p^-).

LEMMA 10.8.3. *Suppose that β and δ satisfy* (10.8.2) *and* (10.8.3), *respectively. Then the vector function (u^-,p^-) with the components* (10.8.5), (10.8.6) *satisfies the estimate* (10.8.10) *for all $x \in \mathcal{K}$, $f_j^{(0)} \in N^{0,\sigma}_{\beta+1,\delta+1}(\mathcal{K})$, $f^{(j)} \in N^{0,\sigma}_{\beta,\delta}(\mathcal{K})^3$, $j = 1,2,3$, $g \in N^{0,\sigma}_{\beta,\delta}(\mathcal{K})$, and every multi-index α.*

P r o o f. If β and δ_k satisfy the conditions (10.8.2) and (10.8.3), then $\beta' = \beta + 1$ and $\delta'_k = \delta_k + 1$ satisfy the conditions of Lemmas 10.7.7–10.7.9, respectively. Furthermore, the functions

$$K(x,\xi) = \partial_x^\alpha\big(\chi^-(x,\xi)\, G_{i,j}(x,\xi)\big), \quad j = 1,2,3,$$

satisfy the assumptions of these lemmas with $s = |\alpha| + \delta_{i,4}$ and $t = 0$ (see Theorems 10.4.4 and 10.4.5). Consequently,

$$|x|^{\beta'-2-\sigma+|\alpha|+\delta_{i,4}} \prod_{k=1}^d \Big(\frac{r_k(x)}{|x|}\Big)^{\max(0,\delta'_k-2-\sigma+|\alpha|+\delta_{i,4})}$$
$$\times \Big|\partial_x^\alpha \int_{\mathcal{K}} f_j^{(0)}(\xi)\, \chi^-(x,\xi)\, G_{i,j}(x,\xi)\, d\xi\Big|$$
$$\le c\, \|f_j^{(0)}\|_{L^\infty_{\beta'-\sigma,\delta'-\sigma}(\mathcal{K})} \le c\, \|f_j^{(0)}\|_{N^{0,\sigma}_{\beta+1,\delta+1}(\mathcal{K})}\,.$$

Using the fact that $K(x,\xi) = \partial_x^\alpha \partial_{\xi_\nu}\big(\chi^-(x,\xi)\, G_{i,j}(x,\xi)\big)$, $j = 1,2,3$, satisfies the conditions of Lemmas 10.7.7–10.7.9 with $s = |\alpha| + \delta_{i,4}$ and $t = 1$, we obtain

$$|x|^{\beta'-2-\sigma+|\alpha|+\delta_{i,4}} \prod_{k=1}^d \Big(\frac{r_k(x)}{|x|}\Big)^{\max(0,\delta'_k-2-\sigma+|\alpha|+\delta_{i,4})}$$
$$\times \Big|\partial_x^\alpha \int_{\mathcal{K}} f^{(j)}(\xi) \cdot \nabla_\xi\big(\chi^-(x,\xi)\, G_{i,j}(x,\xi)\big)\, d\xi\Big|$$
$$\le c\, \|f^{(j)}\|_{L^\infty_{\beta'-1-\sigma,\delta'-1-\sigma}(\mathcal{K})^3} = c\, \|f^{(j)}\|_{N^{0,\sigma}_{\beta,\delta}(\mathcal{K})^3}\,.$$

Since the functions $\partial_x^\alpha(\chi^-\, G_{i,4})$ and $\partial_x^\alpha \partial_{\xi_j}(\chi^-\, G_{i,j})$, $j = 1,2,3$, satisfy the conditions of Lemmas 10.7.7–10.7.9 with $s = |\alpha| + \delta_{i,4}$ and $t = 1$, the same inequality with g instead of $f^{(j)}$ holds for the integral

$$\int_{\mathcal{K}} g(\xi)\, \Big(\chi^-(x,\xi)\, G_{i,4}(x,\xi) - \sum_{j=1}^3 \partial_{\xi_j}\big(\chi^-(x,\xi)\, G_{i,j}(x,\xi)\big)\Big)\, d\xi.$$

Finally, using Lemma 10.8.2 with $K(x,\xi) = \partial_x^\alpha\big(\chi^-(x,\xi)\,G_{i,j}(x,\xi)\big)$, $s = |\alpha| + \delta_{i,4}$, we obtain

$$|x|^{\beta-1-\sigma+|\alpha|+\delta_{i,4}} \prod_{k=1}^d \Big(\frac{r_k(x)}{|x|}\Big)^{\max(0,\delta_k-1-\sigma+|\alpha|+\delta_{i,4})} \times \Big| \int_{\Gamma_\nu} \big(f^{(j)}(\xi)\cdot n + g(\xi)\,n_j\big)\,\chi^-(x,\xi)\,G_{i,j}(x,\xi)\,d\xi\Big| \le c\Big(\|f^{(j)}\|_{L^\infty_{\beta-\sigma,\delta-\sigma}(\mathcal{K})^3} + \|g\|_{L^\infty_{\beta-\sigma,\delta-\sigma}(\mathcal{K})}\Big).$$

The proof of the lemma is complete. □

10.8.4. Hölder estimates for u^-, p^-. For the proof of weighted Hölder estimates, we employ the following lemma which holds analogously to Lemma 10.7.11.

LEMMA 10.8.4. *Let $x \in \mathcal{K}_k = \{\xi \in \mathcal{K} : r_k(\xi) < 3r(\xi)/2\}$, $f \in N^{0,\sigma}_{\beta,\delta}(\mathcal{K})$, and*

$$v(x) = \int_{\Gamma_j} K(x,\xi)\,f(\xi)\,d\xi.$$

We assume that β,δ satisfy the conditions (10.8.2), (10.8.3), $K(x,\xi)$ vanishes for $|x-\xi| > r(x)/4$ and that there exists a constant c independent of x and ξ such that

$$\begin{aligned} |\partial_\rho^s K(x,\xi)| &\le c\,|x-\xi|^{-2-s+m_k} && \text{for } |x|/4 < |\xi| < 2|x|,\ s = 0,1,\\ |\partial_\rho K(x,\xi)| &\le c\,|x|^{\Lambda_- -2+m_k+\varepsilon}\,|\xi|^{-\Lambda_- -1-\varepsilon} && \text{for } |\xi| < 3|x|/4,\\ |\partial_\rho K(x,\xi)| &\le c\,|x|^{\Lambda_+ -2+m_k-\varepsilon}\,|\xi|^{-\Lambda_+ -1-t+\varepsilon} && \text{for } |\xi| > 3|x|/2, \end{aligned}$$

where $m_k = 1 + [\delta_k - \sigma]$ and $\rho = |x|$. Then

$$|x|^{\beta-\delta_k}\,\frac{|v(x)-v(y)|}{|x-y|^{\sigma-\delta_k+m_k}} \le c\,\|f\|_{L^\infty_{\beta-\sigma,\delta-\sigma}(\mathcal{K})}$$

for $y = \tau x$, $4/5 < \tau < 5/4$, $|x-y| > r(x)/4$. Here, c is a constant independent of f, x, τ.

The last lemma enables us to estimate the $C^{1,\sigma}_{\beta,\delta}(\mathcal{K})$-norm of u^- and the $C^{0,\sigma}_{\beta,\delta}(\mathcal{K})$-norm of p^-.

THEOREM 10.8.5. *Suppose that β and δ satisfy the conditions (10.8.2) and (10.8.3). Then the vector function (u^-,p^-) with the components (10.8.5), (10.8.6) satisfies the estimate*

$$(10.8.15)\quad \|u^-\|_{C^{1,\sigma}_{\beta,\delta}(\mathcal{K})^3} + \|p^-\|_{C^{0,\sigma}_{\beta,\delta}(\mathcal{K})} \le c\Big(\sum_{j=1}^3 \|f_j^{(0)}\|_{N^{0,\sigma}_{\beta+1,\delta+1}(\mathcal{K})} + \sum_{j=1}^3 \|f^{(j)}\|_{N^{0,\sigma}_{\beta,\delta}(\mathcal{K})^3} + \|g\|_{N^{0,\sigma}_{\beta,\delta}(\mathcal{K})}\Big)$$

for arbitrary $f_j^{(0)} \in N^{0,\sigma}_{\beta+1,\delta+1}\mathcal{K})$, $f^{(j)} \in N^{0,\sigma}_{\beta,\delta}(\mathcal{K})^3$, $j = 1,2,3$, $g \in N^{0,\sigma}_{\beta,\delta}(\mathcal{K})$.

P r o o f. By Lemma 10.8.3, the vector function (u^-,p^-) satisfies (10.8.10). From this it follows analogously to the proof of Theorem 10.7.12 that

$$|x|^\beta \prod_{k=1}^{d}\Big(\frac{r_k(x)}{|x|}\Big)^{\delta_k} \frac{|\partial_x^\alpha u^-(x)-\partial_y^\alpha u^-(y)|}{|x-y|^\sigma}$$
$$\le c\Big(\sum_{j=1}^{3}\big(\|f_j^{(0)}\|_{N^{0,\sigma}_{\beta+1,\delta+1}(\mathcal{K})}+\|f^{(j)}\|_{N^{0,\sigma}_{\beta,\delta}(\mathcal{K})^3}\big)+\|g\|_{N^{0,\sigma}_{\beta,\delta}(\mathcal{K})}\Big)$$

for $|\alpha|=1$, $x,y\in\mathcal{K}$, $|x-y|<r(x)/2$, and

$$|x|^{\beta-\delta_k}\frac{|\partial_x^\alpha u^-(x)-\partial_y^\alpha u^-(y)|}{|x-y|^{m_k+\sigma-\delta_k}}$$
$$\le c\Big(\sum_{j=1}^{3}\big(\|f_j^{(0)}\|_{N^{0,\sigma}_{\beta+1,\delta+1}(\mathcal{K})}+\|f^{(j)}\|_{N^{0,\sigma}_{\beta,\delta}(\mathcal{K})^3}\big)+\|g\|_{N^{0,\sigma}_{\beta,\delta}(\mathcal{K})}\Big)$$

for $|\alpha|=1-m_k=-[\delta_k-\sigma]$, $x,y\in\mathcal{K}_k$, $|x-y|<r(x)/2$, $k=1,\dots,d$. Here $\mathcal{K}_k$ is the same set as in Lemma 10.8.4. The function u_i^- is the sum of

$$v_i^-(x)=\sum_{j=1}^{3}\int_{\mathcal{K}}\Big(f_j^{(0)}(\xi)\,\chi^-(x,\xi)\,G_{i,j}(x,\xi)\;-f^{(j)}(\xi)\cdot\nabla_\xi\big(\chi^-(x,\xi)\,G_{i,j}(x,\xi)\big)\Big)\,d\xi$$
$$+\int_{\mathcal{K}}g(\xi)\,\Big(\chi^-(x,\xi)\,G_{i,4}(x,\xi)-\sum_{j=1}^{3}\partial_{\xi_j}\big(\chi^-(x,\xi)\,G_{i,j}(x,\xi)\big)\Big)\,d\xi$$

and

$$w_i^-(x)=\sum_{j=1}^{3}\sum_{\nu=1}^{d}\int_{\Gamma_\nu}\big(f^{(j)}(\xi)\cdot n+g(\xi)\,n_j\big)\,\chi^-(x,\xi)\,G_{i,j}(x,\xi)\,d\xi.$$

If β and δ_k satisfy (10.8.2) and (10.8.3), then $\beta'=\beta+1$ and $\delta'_k=\delta_k+1$ satisfy the conditions of Lemma 10.7.11. Consequently,

$$|x|^{\beta'-\delta'_k}\frac{|\partial_x^\alpha v_i^-(x)-\partial_y^\alpha v_i^-(y)|}{|x-y|^{m'_k+\sigma-\delta'_k}}$$
$$\le c\Big(\sum_{j=1}^{3}\big(\|f_j^{(0)}\|_{L^\infty_{\beta'-\sigma,\delta'-\sigma}(\mathcal{K})}+\|f^{(j)}\|_{L^\infty_{\beta-\sigma,\delta-\sigma}(\mathcal{K})^3}\big)+\|g\|_{L^\infty_{\beta-\sigma,\delta-\sigma}(\mathcal{K})}\Big)$$

for $x\in\mathcal{K}_k$, $y=\tau x$, $1/2<\tau<3/2$, $|\alpha|=2-m'_k$, where $m'_k=m_k+1=[\delta_k-\sigma]+2$. By Lemma 10.8.4, the same estimate holds for w_i^-. Thus, the norm of u^- in $C^{1,\sigma}_{\beta,\delta}(\mathcal{K})^3$ can be estimated by the right-hand side of (10.8.15). Analogously, this is true for the norm of p^- in $C^{0,\sigma}_{\beta,\delta}(\mathcal{K})$. □

As a consequence of Lemma 10.8.3 and Theorem 10.8.5, the following sharper estimate for (u^-,p^-) holds.

COROLLARY 10.8.6. *Let the conditions of Theorem* 10.8.5 *be satisfied. Then*

$$\|u\|_{C^{2,\sigma}_{\beta+1,\delta+1}(\mathcal{K})^3}+\|p\|_{C^{1,\sigma}_{\beta+1,\delta+1}(\mathcal{K})}$$
$$\le c\Big(\sum_{j=1}^{3}\|f_j^{(0)}\|_{N^{0,\sigma}_{\beta+1,\delta+1}(\mathcal{K})}+\sum_{j=1}^{3}\|f^{(j)}\|_{N^{0,\sigma}_{\beta,\delta}(\mathcal{K})^3}+\|g\|_{N^{0,\sigma}_{\beta,\delta}(\mathcal{K})}\Big)$$

with a constant c independent of $f_j^{(0)}$, $f^{(j)}$ and g.

P r o o f. All terms in the norm of (u^-,p^-) in $C^{2,\sigma}_{\beta+1,\delta+1}(\mathcal{K})^3 \times C^{1,\sigma}_{\beta,\delta}(\mathcal{K})$ with the exception of

$$\langle u^- \rangle_{2,\sigma,\beta+1,\delta+1;\mathcal{K}}$$

$$= \sum_{|\alpha|=2} \sup_{|x-y|<r(x)/2} |x|^{\beta+1} \prod_{k=1}^{d} \Big(\frac{r_k(x)}{|x|}\Big)^{\delta_k+1} \frac{|\partial_x^\alpha u^-(x) - \partial_y^\alpha u^-(y)|}{|x-y|^\sigma}$$

and

$$\langle p^- \rangle_{1,\sigma,\beta+1,\delta+1;\mathcal{K}}$$

$$= \sum_{|\alpha|=1} \sup_{|x-y|<r(x)/2} |x|^{\beta+1} \prod_{k=1}^{d} \Big(\frac{r_k(x)}{|x|}\Big)^{\delta_k+1} \frac{|\partial_x^\alpha p^-(x) - \partial_y^\alpha p^-(y)|}{|x-y|^\sigma}$$

appear in the left-hand sides of (10.8.10) and (10.8.15). However by (3.6.3), the last two terms are dominated by the left-hand side of (10.8.10). This proves the lemma. □

10.8.5. Existence of weak solutions. By [**152**, Theorem 6.4.8], the local regularity result of Lemma 9.7.6 is also valid if $l = 1$. Therefore, the following analog to Lemma 10.7.3 for the case $l = 1$ holds.

LEMMA 10.8.7. *Suppose that*

$$u \in W^{1,s}_{loc}(\bar{\mathcal{K}}\backslash\mathcal{S})^3 \cap L^\infty_{\beta-1-\sigma,\delta-1-\sigma}(\mathcal{K})^3, \quad p \in W^{0,s}_{loc}(\bar{\mathcal{K}}\backslash\mathcal{S}) \cap L^\infty_{\beta-\sigma,\delta-\sigma}(\mathcal{K}).$$

If (u,p) is a solution of the boundary value problem (10.1.1), (10.1.2), *where $f \in N^{-1,\sigma}_{\beta,\delta}(\mathcal{K})^3$, $g \in N^{0,\sigma}_{\beta,\delta}(\mathcal{K})$, $h_j \in N^{1,\sigma}_{\beta,\delta}(\Gamma_j)^{3-d_j}$ and $\phi_j \in N^{0,\sigma}_{\beta,\delta}(\Gamma_j)^{d_j}$ for $j = 1,\dots,d$, then $u \in N^{1,\sigma}_{\beta,\delta}(\mathcal{K})^3$ and $p \in N^{0,\sigma}_{\beta,\delta}(\mathcal{K})$.*

Using the last lemma and the estimates for $(u^\pm,p^\pm)$ in the preceding subsections, we prove the following existence and uniqueness theorem.

THEOREM 10.8.8. *Let $f \in C^{-1,\sigma}_{\beta,\delta}(\mathcal{K})^3$, $g \in C^{0,\sigma}_{\beta,\delta}(\mathcal{K})$, $h_j \in C^{1,\sigma}_{\beta,\delta}(\Gamma_j)^{3-d_j}$, $\phi_j \in C^{0,\sigma}_{\beta,\delta}(\Gamma_j)^{d_j}$, $j = 1,\dots,d$. We suppose that the line $\operatorname{Re}\lambda = 1+\sigma-\beta$ is free of eigenvalues of the pencil $\mathfrak{A}(\lambda)$ and that the components of δ satisfy the condition* (10.8.3). *Furthermore, we assume that the compatibility conditions of Lemma* 10.7.2 *are satisfied on the edges M_k. Then the problem* (10.1.1), (10.1.2) *has a unique solution $(u,p) \in C^{1,\sigma}_{\beta,\delta}(\mathcal{K})^3 \times C^{0,\sigma}_{\beta,\delta}(\mathcal{K})$.*

P r o o f. Lemma 10.7.2 allows us to restrict ourselves in the proof to the case where $g \in N^{0,\sigma}_{\beta,\delta}(\mathcal{K})$, $h_j = 0$ and $\phi_j = 0$ for $j = 1,\dots,d$. Let κ be an arbitrary real number such that the closed strip between the lines $\operatorname{Re}\lambda = 1+\sigma-\beta$ and $\operatorname{Re}\lambda = -\kappa - 1/2$ is free of eigenvalues of the pencil $\mathfrak{A}(\lambda)$, and let $G(x,\xi)$ be the Green's matrix introduced in Theorem 10.4.1. We consider the vector-functions $(u^\pm,p^\pm)$ with the components (10.8.5) and (10.8.6). By Corollary 10.8.6, $(u^-,p^-) \in C^{2,\sigma}_{\beta+1,\delta+1}(\mathcal{K})^3 \times C^{1,\sigma}_{\beta+1,\delta+1}(\mathcal{K})$. Consequently, $\Delta u^- - \nabla p^- \in N^{0,\sigma}_{\beta+1,\delta+1}(\mathcal{K})^3$ and

$$-\Delta u^+ + \nabla p^+ = f + \Delta u^- - \nabla p^- \in N^{-1,\sigma}_{\beta,\delta}(\mathcal{K})^3.$$

Furthermore, we obtain

$$-\nabla\cdot u^+ \in N^{0,\sigma}_{\beta,\delta}(\mathcal{K}),\quad S_j u^+|_{\Gamma_j}=0,\quad N_j(u^+,p^+)|_{\Gamma_j}\in N^{0,\sigma}_{\beta,\delta}(\Gamma_j)^{d_j}$$

analogously to the proof of Theorem 10.7.13. From this and from Lemmas 10.8.1 and 10.8.7 we conclude that $(u^+,p^+)\in N^{1,\sigma}_{\beta,\delta}(\mathcal{K})^3\times N^{0,\sigma}_{\beta,\delta}(\mathcal{K})$. This together with Theorem 10.8.5 implies that

$$(u,p)=(u^+,p^+)+(u^-,p^-)\in C^{1,\sigma}_{\beta,\delta}(\mathcal{K})^3\times C^{0,\sigma}_{\beta,\delta}(\mathcal{K}).$$

The uniqueness of the solution (u,p) holds analogously to the proof of Theorem 10.7.13. □

10.8.6. A regularity result for the variational solution. We consider the variational solutions $(u,p)\in V^{1,2}_\kappa(\mathcal{K})^3\times V^{0,2}_\kappa(\mathcal{K})$ of the boundary value problem (10.1.1), (10.1.2). The following regularity assertion for this solution holds by the same arguments as in the proof of Theorem 10.7.16, where we obtained the analogous result for $l>1$.

THEOREM 10.8.9. *Let $(u,p)\in V^{1,2}_\kappa(\mathcal{K})^3\times V^{0,2}_\kappa(\mathcal{K})$ be a solution of the problem* (10.2.11), (10.2.12), *where $g\in V^{0,2}_\kappa(\mathcal{K})\cap C^{0,\sigma}_{\beta,\delta}(\mathcal{K})$, $h_j\in W^{1/2,2}_{\kappa,0}(\Gamma_j)^{3-d_j}\cap C^{1,\sigma}_{\beta,\delta}(\Gamma_j)^{3-d_j}$ and F is a linear and continuous functional on $\mathcal{H}_{-\kappa}$ which has the form*

$$F(v)=\int_{\mathcal{K}}(f+\nabla g)\cdot v\,dx+\sum_{j=1}^d\int_{\Gamma_j}\phi_j\cdot v\,dx$$

with given vector functions $f\in C^{-1,\sigma}_{\beta,\delta}(\mathcal{K})^3$ and $\phi_j\in C^{0,\sigma}_{\beta,\delta}(\Gamma_j)^{d_j}$. We suppose that the lines $\operatorname{Re}\lambda=-\kappa-1/2$ *and* $\operatorname{Re}\lambda=l+\sigma-\beta$ *do not contain eigenvalues of the pencil $\mathfrak{A}(\lambda)$, that the components of δ satisfy the condition* (10.8.3) *and that the data g,h_j,ϕ_j satisfy the same compatibility conditions as in Lemma* 10.7.2. *Then (u,p) admits the decomposition* (10.3.6), *where (w,q) is the solution of the problem* (10.1.1), (10.1.2) *in the space $C^{1,\sigma}_{\beta,\delta}(\mathcal{K})^3\times C^{0,\sigma}_{\beta,\delta}(\mathcal{K})$, λ_ν are the eigenvalues of the pencil $\mathfrak{A}(\lambda)$ between the lines* $\operatorname{Re}\lambda=-\kappa-1/2$ *and* $\operatorname{Re}\lambda=l+\sigma-\beta$, *and $\big(u^{(\nu,j,s)},p^{(\nu,j,s)}\big)$ are eigenvectors and generalized eigenvectors corresponding to the eigenvalue λ_ν.*

10.8.7. Example. Let $(u,p)\in V^{1,2}_0(\mathcal{K})^3\times L_2(\mathcal{K})$ be the variational solution of the Dirichlet problem

(10.8.16) $$b_{\mathcal{K}}(u,v)-\int_{\mathcal{K}}p\,\nabla\cdot v\,dx=F(v)\quad\text{for all } v\in\mathcal{H}_{\mathcal{K}},$$

(10.8.17) $$-\nabla\cdot u=g\ \text{ in }\mathcal{K},\quad u=0\ \text{ on }\Gamma_j,\ j=1,\dots,d,$$

where

$$\mathcal{H}_{\mathcal{K}}=\{u\in V^{1,2}_0(\mathcal{K})^3:\ u=0\text{ on }\Gamma_j,\ j=1,\dots,d\}\ =\overset{\circ}{V}{}^{1,2}_{0,0}(\mathcal{K})^3.$$

We assume that $g\in C^{0,\sigma}(\mathcal{K})$, $g=0$ on M_k for $k=1,\dots,d$, and $F\in C^{-1,\sigma}(\mathcal{K})^3$, i.e. F is a distribution of the form

(10.8.18) $$F=F^{(0)}+\sum_{j=1}^3\partial_{x_j}F^{(j)},\quad\text{where } F^{(j)}\in C^{0,\sigma}(\mathcal{K})^3,\ j=0,1,2,3.$$

Every distribution (10.8.18) can be written in the form

$$F = \Phi^{(0)} + \sum_{j=1}^{3} \partial_{x_j} \Phi^{(j)}, \quad \text{where } \Phi^{(j)} \in C^{0,\sigma}_{0,0}(\mathcal{K})^3,\ j = 0, 1, 2, 3.$$

For example, we can set

$$\Phi^{(0)} = F^{(0)} - \Big(\chi(x) + \frac{1}{3}\, x \cdot \nabla\chi\Big)\, F^{(0)}(0)$$

and

$$\Phi^{(j)} = F^{(j)} - F^{(j)}(0) + \frac{1}{3}\,\chi(x)\, x_j F^{(0)}(0) \ \text{ for } j = 1, 2, 3,$$

where χ is a differentiable function on $\overline{\mathcal{K}}$, $\chi(x) = 1$ for $|x| \le 1$, $\chi(x) = 0$ for $|x| \ge 2$. Thus, $F \in C^{-1,\sigma}_{0,0}(\mathcal{K})^3 = N^{-1,\sigma}_{0,0}(\mathcal{K})^3$. Furthermore, every $g \in C^{0,\sigma}(\mathcal{K})$ satisfying $g = 0$ on M_k for $k = 1, \dots, d$ belongs to the space $N^{0,\sigma}_{0,0}(\mathcal{K})$.

According to [**85**, Theorems 5.5.5 and 5.5.6], the strip $-1 \le \operatorname{Re}\lambda \le 0$ is free of eigenvalues of the pencil $\mathfrak{A}(\lambda)$. If the cone is convex, then the eigenvalue with smallest positive real part is $\Lambda_1 = 1$. This eigenvalue is simple and has the eigenvector $(0, 0, 0, 1)$. Moreover, $\mu_+^{(k)} = \pi/\theta_k > 1$ if the angle θ_k at the edge M_k is less than π. Thus, we obtain the following regularity result.

THEOREM 10.8.10. *Let $(u, p) \in V_0^{1,2}(\mathcal{K})^3 \times L_2(\mathcal{K})$ be a solution of the problem* (10.8.16), (10.8.17), *where $F \in V_{0,0}^{-1,2}(\mathcal{K})^3 \cap C^{-1,\sigma}(\mathcal{K})^3$, $g \in L_2(\mathcal{K}) \cap C^{0,\sigma}(\mathcal{K})$, $g = 0$ on M_k for $k = 1, \dots, d$. If $\mathcal{K}$ is convex and σ is such that $1 + \sigma < \pi/\theta_k$ and the strip $1 < \operatorname{Re}\lambda < 1 + \sigma$ contains only the eigenvalue $\Lambda_1 = 1$ of the pencil $\mathfrak{A}(\lambda)$, then $\zeta(u, p) \in C^{1,\sigma}(\mathcal{K})^3 \times C^{0,\sigma}(\mathcal{K})$ for every smooth function ζ with compact support.*

P r o o f. Under the above assumptions, the decomposition (10.3.6) for (u, p) in Theorem 10.8.9 has the simple form

$$(u, p) = (0, c) + (w, q), \quad \text{where } (w, q) \in C^{1,\sigma}_{0,0}(\mathcal{K})^3 \times C^{0,\sigma}_{0,0}(\mathcal{K})$$

and c is a constant. Consequently, $\zeta(u, p) \in C^{1,\sigma}(\mathcal{K})^3 \times C^{0,\sigma}(\mathcal{K})$ for every smooth function ζ with compact support.

CHAPTER 11

Mixed boundary value problems for the Stokes and Navier-Stokes systems in a bounded domain of polyhedral type

In this chapter, we turn to mixed boundary value problems for the stationary Stokes and Navier-Stokes systems in a three-dimensional bounded domain $\mathcal{G}$ of polyhedral type with components of the velocity and/or the friction prescribed on the boundary. In other words, one of the boundary conditions (i)–(iv) in the introduction to Chapter 9 is given on each face Γ_j of the domain $\mathcal{G}$. The boundary value problem for the Stokes system is considered in Section 11.1, where we prove the existence of variational solutions of the boundary value problem and obtain regularity results for this solution both in weighted Sobolev and Hölder spaces.

In Section 11.2, we extend some of the regularity results to the nonlinear *Navier-Stokes system*

$$-\nu\,\Delta u + (u\cdot\nabla)\,u + \nabla p = f, \qquad \nabla\cdot u = g.$$

Various applications of the general theorems of Section 11.2 are discussed in Section 11.3. Here we formulate particular regularity results from Section 11.3. Suppose that the data belong to the corresponding Sobolev or Hölder spaces and satisfy certain compatibility conditions on the edges. Then the following smoothness of the weak solution is guaranteed and is the best possible.

- If (u,p) is a solution of the Dirichlet problem in an arbitrary polyhedron or a solution of the Neumann problem in an arbitrary Lipschitz graph polyhedron, then
$$(u,p)\in W^{1,3+\varepsilon}(\mathcal{G})^3\times L_{3+\varepsilon}(\mathcal{G}),$$
$$(u,p)\in W^{2,4/3+\varepsilon}(\mathcal{G})^3\times W^{1,4/3+\varepsilon}(\mathcal{G}),$$
$$u\in C^{0,\varepsilon}(\mathcal{G})^3.$$
Here ε is a positive number depending on the domain $\mathcal{G}$.
- If (u,p) is a solution of the Dirichlet problem in a convex polyhedron, then
$$(u,p)\in W^{1,s}(\mathcal{G})^3\times L_s(\mathcal{G}) \quad \text{for all } s,\ 1<s<\infty,$$
$$(u,p)\in W^{2,2+\varepsilon}(\mathcal{G})^3\times W^{1,2+\varepsilon}(\mathcal{G}),$$
$$(u,p)\in C^{1,\varepsilon}(\mathcal{G})^3\times C^{0,\varepsilon}(\mathcal{G}).$$
- If (u,p) is a solution of the mixed problem in an arbitrary polyhedron with the Dirichlet and Neumann boundary conditions prescribed arbitrarily on different faces, then
$$(u,p)\in W^{2,8/7+\varepsilon}(\mathcal{G})^3\times W^{1,8/7+\varepsilon}(\mathcal{G}).$$

- Let (u,p) be a solution of the mixed boundary value problem with the slip condition (iii) on one face Γ_1 and Dirichlet condition on the other faces. Then

$$(u,p) \in W^{1,3+\varepsilon}(\mathcal{G})^3 \times L_{3+\varepsilon}(\mathcal{G}) \quad \text{if } \theta < 3\pi/2,$$
$$(u,p) \in W^{2,2+\varepsilon}(\mathcal{G})^3 \times W^{2,2+\varepsilon}(\mathcal{G}) \quad \text{if } \mathcal{G} \text{ is convex and } \theta < \pi/2,$$
$$(u,p) \in C^{1,\varepsilon}(\mathcal{G})^3 \times C^{0,\varepsilon}(\mathcal{G}) \quad \text{if } \mathcal{G} \text{ is convex and } \theta < \pi/2,$$

where θ is the maximal angle between Γ_1 and the adjoining faces.

Sections 11.4 and 11.5 deal with the Green's matrix of the Dirichlet problem for the Stokes system. We prove point estimates of its elements for general domains of polyhedral type. In the case of a convex polyhedron, these estimates are improved in Section 11.5. The last section of the chapter (Section 11.6) is dedicated to L_∞ estimates for solutions of the Dirichlet problem for the Stokes and Navier-Stokes systems. First, we prove a weighted L_∞-estimate for the solution of the Stokes system by using point estimates for Green's matrix. As a particular case of the just mentioned weighted L_∞-estimate, we obtain the inequality

$$\|u\|_{L_\infty(\mathcal{G})} \le c\, \|h\|_{L_\infty(\partial\mathcal{G})}$$

for solutions of the Stokes system $\Delta u + \nabla p = 0$ with the Dirichlet condition $u = h$ on $\partial\mathcal{G}$. For solutions of the nonlinear Navier-Stokes system, the estimate

$$\|u\|_{L_\infty(\mathcal{G})} \le F\big(\|h\|_{L_\infty(\partial\mathcal{G})}\big)$$

with an explicitly written function F holds.

11.1. Mixed boundary value problems for the Stokes system

Let $\mathcal{G}$ be a bounded domain of polyhedral type with the faces $\Gamma_1,\dots,\Gamma_N$, the edges $M_1,\dots,M_d$ and the vertices $x^{(1)},\dots,x^{(d')}$. We assume that the conditions of Chapter 8 are satisfied for the domain $\mathcal{G}$ and consider the boundary value problem

$$-\Delta u + \nabla p = f, \quad -\nabla\cdot u = g \ \text{ in } \mathcal{G}, \tag{11.1.1}$$
$$S_j u = h_j, \quad N_j(u,p) = \phi_j \ \text{ on } \Gamma_j,\ j = 1,\dots,N. \tag{11.1.2}$$

Here S_j and N_j as well as the numbers $d_j \in \{0,1,2,3\}$ are defined as in Section 10.1, i.e.

$$S_j u = u \text{ if } d_j = 0,$$
$$S_j u = u_\tau\,, \quad N_j(u,p) = -p + 2\varepsilon_{n,n}(u) \text{ if } d_j = 1,$$
$$S_j u = u_n\,, \quad N_j(u,p) = 2\varepsilon_{n,\tau}(u) \text{ if } d_j = 2,$$
$$N_j(u,p) = -pn + 2\varepsilon_n(u) \ \text{ if } d_j = 3.$$

In the case $d_j = 0$ the condition $N_j(u,p) = \phi_j$ does not appear in (11.1.2), whereas the condition $S_j u = h_j$ does not appear for $d_j = 3$.

11.1.1. Existence of variational solutions. We are interested in this section in variational solutions of the boundary value problem (11.1.1), (11.1.2). We introduce the spaces

$$\mathcal{H} = \{u \in W^{1,2}(\mathcal{G})^3 : \ S_j u = 0 \text{ on } \Gamma_j, \ j = 1, \dots, N\}$$

and

$$\mathcal{H}_0 = \{u \in \mathcal{H} : \ \nabla \cdot u = 0\}.$$

Furthermore, we denote by $L_{\mathcal{H}}$ the set of all $u \in \mathcal{H}$ such that $\varepsilon_{i,j}(u) = 0$ for $i,j = 1,2,3$. It can be easily seen that $L_{\mathcal{H}}$ is contained in the span of all all constant vectors and of the vectors $(x_2, -x_1, 0)$, $(0, x_3, -x_2)$, $(-x_3, 0, x_1)$. In particular, $L_{\mathcal{H}} \subset \mathcal{H}_0$.

Let the bilinear form b be defined as

$$b(u,v) = 2 \int_{\mathcal{G}} \sum_{i,j=1}^{3} \varepsilon_{i,j}(u)\, \varepsilon_{i,j}(v)\, dx. \tag{11.1.3}$$

We consider the problem

$$b(u,v) - \int_{\mathcal{G}} p \nabla \cdot v\, dx = F(v) \ \text{ for all } v \in \mathcal{H}, \tag{11.1.4}$$

$$-\nabla \cdot u = g \ \text{ in } \mathcal{G}, \quad S_j u = h_j \ \text{ on } \Gamma_j, \ j = 1, \dots, N, \tag{11.1.5}$$

where F is a given linear and continuous functional on $\mathcal{H}$, $g \in L_2(\mathcal{G})$ and $h_j \in W^{1/2,2}(\Gamma_j)^{3-d_j}$. Note again that, in the case $d_j = 1$, the vector function h_j is tangent to Γ_j. For this reason, the function space for h_j can be identified with $W^{1/2,2}(\Gamma_j)^2$ then.

We assume that the vector functions h_j are such that there exists a vector function $w \in W^{1,2}(\mathcal{G})^3$ satisfying the boundary conditions $S_j w = h_j$ on Γ_j for $j = 1, \dots, N$. This means, the boundary data h_j must satisfy certain compatibility conditions on the edges of the domain (cf. Section 10.2). By Green's formula (cf. (10.1.7)), every solution $(u,p) \in W^{2,2}(\mathcal{G})^3 \times W^{1,2}(\mathcal{G})$ satisfies the equations (11.1.4), (11.1.5) with the functional

$$F(v) = \int_{\mathcal{G}} (f + \nabla g) \cdot v\, dx + \sum_{j=1}^{N} \int_{\Gamma_j} \phi_j \cdot v\, dx. \tag{11.1.6}$$

Here in the case $d_j = 1$ we identified the scalar function ϕ_j with the vector function $\phi_j n$.

LEMMA 11.1.1. *Let $g \in L_2(\mathcal{G})$, and let $h_j \in W^{1/2,2}(\Gamma_j)^{3-d_j}$ be such that there exists a vector function $w \in W^{1,2}(\mathcal{G})^3$, $S_j w = h_j$ on Γ_j for $j = 1, \dots, N$. In the case where $d_j \in \{0, 2\}$ for all j, we assume in addition that*

$$\int_{\mathcal{G}} g\, dx + \sum_{j:\, d_j = 0} \int_{\Gamma_j} h_j \cdot n\, dx + \sum_{j:\, d_j = 2} \int_{\Gamma_j} h_j\, dx = 0. \tag{11.1.7}$$

Then there exists a vector function $u \in W^{1,2}(\mathcal{G})^3$ such that $\nabla \cdot u = -g$ in $\mathcal{G}$ and $S_j u = h_j$ on Γ_j for $j = 1, \dots, N$.

P r o o f. Let $w \in W^{1,2}(\mathcal{G})^3$, $S_j w = h_j$ on Γ_j for $j = 1, \dots, N$. We have to show that there exists a vector function $v \in \mathcal{H}$ such that $\nabla \cdot v = -g - \nabla \cdot w$. Then $u = v + w$ is the desired vector function.

Let first $d_j \in \{0,2\}$ for all j. By [**55**, Chapter 1,Corollary 2.4], for the existence of a vector function $w \in \overset{\circ}{W}{}^{1,2}(\mathcal{G})^3 \subset \mathcal{H}$ satisfying the equation $\nabla \cdot v = -g - \nabla \cdot w$ it is necessary and sufficient that

$$\int_{\mathcal{G}} (g + \nabla \cdot w)\, dx = 0.$$

The last condition is equivalent to (11.1.7).

We consider the case where $d_j \in \{1,3\}$ for at least one $j = j_0$. Let $\phi \in C_0^\infty(\Gamma_{j_0})$ be such that

$$\int_{\Gamma_{j_0}} \phi(x)\, dx = 1.$$

Then there exists a vector function $\psi \in W^{1,2}(\mathcal{G})^3$ such that

$$\psi = 0 \quad \text{on } \Gamma_j \text{ for } j \neq j_0, \quad \psi_n = \phi, \ \psi_\tau = 0 \quad \text{on } \Gamma_{j_0}$$

(cf. Lemma 10.2.1). Since $d_{j_0} \in \{1,3\}$, the vector function ψ belongs to $\mathcal{H}$. We introduce the function

$$G = g + \nabla \cdot w - c\, \nabla \cdot \psi, \quad \text{where } c = \int_{\mathcal{G}} (g + \nabla \cdot w)\, dx.$$

Since

$$\int_{\mathcal{G}} G\, dx = c\Big(1 - \int_{\mathcal{G}} \nabla \cdot \psi\, dx\Big) = c\Big(1 - \int_{\Gamma_{j_0}} \phi\, dx\Big) = 0,$$

there exists a vector function $W \in \overset{\circ}{W}{}^{1,2}(\mathcal{G})^3 \subset \mathcal{H}$ such that $-\nabla \cdot W = G$. Consequently, $v = W - c\psi$ satisfies the equation $\nabla \cdot v = -g - \nabla \cdot w$. The result follows. $\square$

The necessity of the condition (11.1.7) in Lemma 11.1.1 is obvious. Moreover, since $b(u,v) = 0$ and $\nabla \cdot v = 0$ for $v \in L_{\mathcal{H}}$, for the solvability of the problem (11.1.4), (11.1.5) it is necessary that

$$F(v) = 0 \quad \text{for all } v \in L_{\mathcal{H}}\,. \tag{11.1.8}$$

Thus, we obtain the following existence and uniqueness theorem for the variational problem (11.1.4), (11.1.5).

THEOREM 11.1.2. *Let g and h_j be as in Lemma 11.1.1, and let $F \in \mathcal{H}^*$ be a functional satisfying the condition (11.1.8). Then there exists a solution $(u,p) \in W^{1,2}(\mathcal{G})^3 \times L_2(\mathcal{G})$ of the problem (11.1.4), (11.1.5). Here p is uniquely determined if $d_j \in \{1,3\}$ for at least one j and unique up to constants if $d_j \in \{0,2\}$ for all j. The vector function u is unique up to elements of the set $L_{\mathcal{H}}$.*

P r o o f. 1) First let $g = 0$ and $h_j = 0$ for $j = 1,\dots,N$. We denote by $L_{\mathcal{H}}^\perp$ the orthogonal complement of $L_{\mathcal{H}}$ in $\mathcal{H}_0$. By Korn's inequality,

$$b(u,\bar{u}) \geq c\, \|u\|^2_{W^{1,2}(\mathcal{G})^3} \quad \text{for all } v \in L_{\mathcal{H}}^\perp. \tag{11.1.9}$$

Here c is a positive constant. Consequently, there exists a unique vector function $u \in L_{\mathcal{H}}^\perp$ such that $b(u,v) = F(v)$ for all $v \in L_{\mathcal{H}}^\perp$. Since both $b(u,v)$ and $F(v)$ vanish for $v \in L_{\mathcal{H}}$, it follows that

$$b(u,v) = F(v) \quad \text{for all } v \in \mathcal{H}_0\,. \tag{11.1.10}$$

Let $\mathcal{H}_0^\perp$ denote the orthogonal complement of $\mathcal{H}_0$ in $\mathcal{H}$. By Lemma 11.1.1, the operator $B = -\mathrm{div}$ is an isomorphism from $\mathcal{H}_0^\perp$ onto $L_2(\mathcal{G})$ if $d_j \in \{1,3\}$ for at least one j and onto the space

$$\overset{\circ}{L}_2(\mathcal{G}) = \{q \in L_2(\mathcal{G}) : \int_{\mathcal{G}} q(x)\, dx = 0\}$$

if $d_j \in \{0,2\}$ for all j. Suppose that $d_j \in \{0,2\}$ for all j. Then we consider the mapping

$$L_2(\mathcal{G}) \ni q \to \ell(q) = F(B^{-1} \overset{\circ}{q}) - b(u, B^{-1} \overset{\circ}{q}),$$

where

$$\overset{\circ}{q} = q - \frac{1}{|\mathcal{G}|} \int_{\mathcal{G}} q(x)\, dx \in \overset{\circ}{L}_2(\mathcal{G}).$$

Obviously, ℓ defines a linear and continuous functional on $L_2(\mathcal{G})$. Consequently, there exists a function $p \in L_2(\mathcal{G})$ such that

$$\int_{\mathcal{G}} p\, q\, dx = \ell(q) \quad \text{for all } q \in L_2(\mathcal{G}).$$

Therefore,

$$(11.1.11) \qquad -\int_{\mathcal{G}} p\, \nabla \cdot v\, dx = \ell(-\nabla \cdot v) = F(v) - b(u,v) \quad \text{for all } v \in \mathcal{H}_0^\perp.$$

In the case where $d_j \in \{1,3\}$ for at least one j, the existence of a function $p \in L_2(\mathcal{G})$ satisfying (11.1.10) follows analogously from the continuity of the mapping

$$L_2(\mathcal{G}) \ni q \to \ell(q) = F(B^{-1}q) - b(u, B^{-1}q) \in \mathbb{C}.$$

Combining (11.1.10) and (11.1.11), we conclude that u and p satisfy (11.1.4). This proves the existence of a solution.

2) Now let g and h_j be arbitrary functions satisfying the conditions of Lemma 11.1.1. Then there exists a vector function $w \in W^{1,2}(\mathcal{G})^3$ such that $\nabla \cdot w = -g$ in $\mathcal{G}$ and $S_j w = h_j$ on Γ_j, $j = 1, \dots, N$. By the first part of the proof, there exists a solution (u,p) of the problem

$$b(u,v) - \int_{\mathcal{G}} p\, \nabla \cdot v\, dx = F(v) - b(w,v) \quad \text{for all } v \in \mathcal{H},$$

$$-\nabla \cdot u = 0 \ \text{ in } \mathcal{G}, \quad S_j u = 0 \ \text{ on } \Gamma_j,\ j = 1, \dots, N.$$

Then $(u + w, p)$ is a solution of the problem (11.1.4), (11.1.5).

3) We prove the uniqueness. Let $u \in \mathcal{H}_0$ and $p \in L_2(\mathcal{G})$ satisfy (11.1.4) with $F = 0$. Then in particular, $b(u, \bar{u}) = 0$. Obviously, $b(u,\bar{u}) = b(u - w, \bar{u} - \bar{w})$, where w is the orthogonal projection of $u \in \mathcal{H}_0$ onto $L_{\mathcal{H}}$. Using (11.1.9), we obtain $u - w = 0$, i.e., $u \in L_{\mathcal{H}}$. However, then $b(u,v) = 0$ for all $v \in \mathcal{H}$ and therefore

$$\int_{\mathcal{G}} p\, \nabla \cdot v = 0 \ \text{ for all } v \in \mathcal{H}.$$

If $d_j \in \{1,3\}$ for at least one j, then v can be chosen such that $\nabla v = \bar{p}$, and we obtain $p = 0$. If $d_j \in \{0,2\}$ for all j, then we obtain

$$\int_{\mathcal{G}} p\, q\, dx = 0 \quad \text{for all } q \in L_2(\mathcal{G}),\ \int_{\mathcal{G}} q\, dx = 0.$$

This means that p is constant. The proof is complete. □

11.1.2. Model problems and corresponding operator pencils. Suppose that the following conditions on the domain $\mathcal{G}$ are satisfied.

(i) The boundary $\partial\mathcal{G}$ consists of smooth (of class C^∞) open two-dimensional manifolds Γ_j (the faces of $\mathcal{G}$), $j = 1, \dots, N$, smooth curves M_k (the edges), $k = 1, \dots, d$, and vertices $x^{(1)}, \dots, x^{(d')}$.

(ii) For every $\xi \in M_k$ there exist a neighborhood $\mathcal{U}_\xi$ and a diffeomorphism (a C^∞ mapping) κ_ξ which maps $\mathcal{G} \cap \mathcal{U}_\xi$ onto $\mathcal{D}_\xi \cap B_1$, where $\mathcal{D}_\xi$ is a dihedron and B_1 is the unit ball.

(iii) For every vertex $x^{(j)}$ there exist a neighborhood $\mathcal{U}_j$ and a diffeomorphism κ_j mapping $\mathcal{G} \cap \mathcal{U}_j$ onto $\mathcal{K}_j \cap B_1$, where $\mathcal{K}_j$ is a polyhedral cone with vertex at the origin.

We introduce the operator pencils generated by the boundary value problem (11.1.1), (11.1.2) for the edge points and vertices of the domain $\mathcal{G}$.

1) Let ξ be a point on an edge M_k, and let $\Gamma_{k_+}, \Gamma_{k_-}$ be the faces of $\mathcal{G}$ adjacent to ξ. Then by $\mathcal{D}_\xi$ we denote the dihedron which is bounded by the half-planes $\Gamma^\circ_{k_\pm}$ tangent to $\Gamma_{k_\pm}$ at ξ. The angle between the half-planes $\Gamma^\circ_{k_+}$ and $\Gamma^\circ_{k_-}$ is denoted by $\theta(\xi)$. We consider the model problem

$$-\Delta u + \nabla p = f, \quad -\nabla \cdot u = g \quad \text{in } \mathcal{D}_\xi,$$

$$S_{k_\pm} u = h_{k_\pm}, \quad N_{k_\pm}(u,p) = \phi_{k_\pm} \quad \text{on } \Gamma^\circ_{k_\pm}.$$

The operator pencil corresponding to this model problem (see Section 2.2) is denoted by $A_\xi(\lambda)$. Furthermore, let $\delta_+(\xi)$ be the greatest positive real number such that the strip

$$0 < \operatorname{Re}\lambda < \delta_+(\xi)$$

is free of eigenvalues of this pencil, and let

$$\delta_+^{(k)} = \inf_{\xi \in M_k} \delta_+(\xi).$$

If $d_{k_+} + d_{k_-}$ is even and $\theta(\xi) < \pi/m_k$, where $m_k = 1$ for $d_{k_+} + d_{k_-} \in \{0, 6\}$, $m_k = 2$ for $d_{k_+} + d_{k_-} \in \{2, 4\}$, then $\lambda = 1$ is the smallest positive eigenvalue of the pencil $A_\xi(\lambda)$. In this case, we denote by $\mu_+(\xi)$ the greatest positive real number such that the strip

$$0 < \operatorname{Re}\lambda < \mu_+(\xi)$$

contains only the eigenvalue $\lambda = 1$. In all other cases, we put $\mu_+(\xi) = \delta_+(\xi)$. Furthermore, we define

$$\mu_+^{(k)} = \inf_{\xi \in M_k} \mu_+(\xi).$$

2) Let $x^{(j)}$ be a corner of $\mathcal{G}$ and let I_j be the set of all indices k such that $x^{(j)} \in \overline{\Gamma}_k$. By our assumptions, there exist a neighborhood $\mathcal{U}$ of $x^{(j)}$ and a diffeomorphism κ mapping $\mathcal{G} \cap \mathcal{U}$ onto $\mathcal{K}_j \cap B_1$ and $\Gamma_k \cap \mathcal{U}$ onto $\Gamma^\circ_k \cap B_1$ for $k \in I_j$, where $\mathcal{K}_j$ is a polyhedral cone with vertex 0 and Γ°_k are the faces of this cone. Without loss of generality, we may assume that the Jacobian matrix $\kappa'(x)$ is equal to the identity matrix I at the point $x^{(j)}$. We consider the model problem

$$-\Delta u + \nabla p = f, \quad -\nabla \cdot u = g \quad \text{in } \mathcal{K}_j,$$

$$S_k u = h_k, \quad N_k(u,p) = \phi_k \quad \text{on } \Gamma^\circ_k \text{ for } k \in I_j.$$

The operator pencil generated by this model problem (see Subsection 10.1.3) is denoted by $\mathfrak{A}_j(\lambda)$.

11.1.3. Solvability in weighted Sobolev spaces. Let $s' = s/(s-1)$ and

$$\mathcal{H}_{s',-\beta,-\delta;\mathcal{G}} = \{u \in V^{1,s'}_{-\beta,-\delta}(\mathcal{G})^3 \ : \ S_j u = 0 \text{ on } \Gamma_j, \ j = 1, \dots, N\}.$$

Note that $V^{1,s'}_{-\beta,-\delta}(\mathcal{G}) = W^{1,s'}_{-\beta,-\delta}(\mathcal{G})$ if the components of δ satisfy the inequality $\delta_k + 2/s < 1$ for $k = 1, \dots, d$. Our goal is to show that the solution $(u,p) \in W^{1,2}(\mathcal{G})^3 \times L_2(\mathcal{G})$ of the problem (11.1.4), (11.1.5) belongs to $W^{1,s}_{\beta,\delta}(\mathcal{G})^3 \times W^{0,s}_{\beta,\delta}(\mathcal{G})$ if $F \in \mathcal{H}^*_{s',-\beta,-\delta;\mathcal{G}}$, $g \in W^{0,s}_{\beta,\delta}(\mathcal{G})$, $h_j \in W^{1-1/s,s}_{\beta,\delta}(\Gamma_j)^{3-d_j}$ and β, δ satisfy the same conditions as in in the corresponding regularity results of Chapter 10. To this end, we consider the following perturbed Stokes problem in the cone $\mathcal{K}_j$.

$$b_1(u,v) + \int_{\mathcal{K}_j} p\, L_1 v\, dx = F(v) \ \text{ for all } v \in \mathcal{H}_{\mathcal{K}_j}\,, \tag{11.1.12}$$

$$L_1 u = g \ \text{ in } \mathcal{K}_j, \quad S_k u = h_k \ \text{ on } \Gamma_k^\circ, \ k \in I_j. \tag{11.1.13}$$

Here

$$\mathcal{H}_{\mathcal{K}_j} = \{v \in W^{1,2}(\mathcal{K}_j)^3 \ : \ S_k v = 0 \text{ on } \Gamma_k^\circ, \ k \in I_j\},$$

$$b_1(u,v) = 2 \int_{\mathcal{K}_j} \sum_{k,l=1}^{3} \varepsilon_{k,l}(u)\, \varepsilon_{k,l}(v)\, dx + \sum_{\mu,\nu,k,l=1}^{3} \int_{\mathcal{K}_j} b_{\mu,\nu,k,l}(x)\, \frac{\partial u_k}{\partial x_\mu}\, \frac{\partial u_l}{\partial x_\nu}\, dx,$$

$$L_1 v = -\nabla \cdot v + \sum_{k,l=1}^{3} c_{k,l}(x) \frac{\partial v_k}{\partial x_l}\,.$$

We assume that

$$\sum_{\mu,\nu,k,l} |b_{\mu,\nu,k,l}(x)| + \sum_{k,l} |c_{k,l}(x)| < \varepsilon \tag{11.1.14}$$

with sufficiently small ε and that

$$F \in \mathcal{H}^*_{\mathcal{K}_j} \cap \mathcal{H}^*_{s',-\beta,-\delta;\mathcal{K}_j}, \quad g \in L_2(\mathcal{K}_j) \cap W^{0,s}_{\beta,\delta}(\mathcal{K}_j), \tag{11.1.15}$$

$$h_k \in W^{1/2,2}(\Gamma_k^\circ)^{3-d_k} \cap W^{1-1/s,s}_{\beta,\delta}(\Gamma_k^\circ)^{3-d_k} \ \text{ for } k \in I_j\,, \tag{11.1.16}$$

where

$$\mathcal{H}_{s,\beta,\delta;\mathcal{K}_j} = \{u \in W^{1,2}_{\beta,\delta}(\mathcal{K}_j)^3, \ S_k u = 0 \text{ on } \Gamma_k^\circ, \ k \in I_j\}.$$

Using Theorems 10.6.5 and 10.6.7, we obtain the following assertion.

LEMMA 11.1.3. *Let $(u,p) \in W^{1,2}(\mathcal{K}_j)^3 \times L_2(\mathcal{K}_j)$ be a solution of the problem (11.1.12), (11.1.13) with the data (11.1.15) and (11.1.16). Suppose that the closed strip between the lines $\operatorname{Re}\lambda = -1/2$ and $\operatorname{Re}\lambda = 1 - \beta_j - 3/s$ is free of eigenvalues of the pencil $\mathfrak{A}_j(\lambda)$ for $j = 1, \dots, d'$, that the components of δ satisfy the inequalities*

$$\max(0, 1 - \mu_+^{(k)}) < \delta_k + 2/s < 1, \tag{11.1.17}$$

and that the number ε in (11.1.14) is sufficiently small. Then $u \in W^{1,s}_{\beta,\delta}(\mathcal{K}_j)^3$ and $p \in W^{0,s}_{\beta,\delta}(\mathcal{K}_j)$.

P r o o f. Let $\mathcal{W}_{s,\beta,\delta}$ be the space of all

$$\{h_k\}_{k \in I_j} \in \prod_{k \in I_j} W^{1-1/s,s}_{\beta,\delta}(\Gamma_k^\circ)^{3-d_k}$$

such that there exists a vector function $u \in W^{1,s}_{\beta,\delta}(\mathcal{K}_j)$ satisfying $S_k u = h_k$ on Γ_k° for $k \in I_j$. We define A as the operator

$$W^{1,2}(\mathcal{K}_j)^3 \times L_2(\mathcal{K}_j) \ni (u,p) \to (F,g,h_j) \in \mathcal{H}^*_{\mathcal{K}_j} \times L_2(\mathcal{K}_j) \times \mathcal{W}_{2,0,0} \tag{11.1.18}$$

where F, g and h_k are given by (11.1.12) and (11.1.13). Furthermore, let A_0 be the operator (11.1.18), where

$$F(v) = 2\int_{\mathcal{K}_j} \sum_{i,j=1}^{3} \varepsilon_{i,j}(u)\,\varepsilon_{i,j}(v)\,dx - \int_{\mathcal{K}_j} p\nabla\cdot v\,dx, \quad g = -\nabla\cdot u, \quad h_j = S_j u.$$

By Theorem 10.1.2, the operator A_0 is an isomorphism. Furthermore, it follows from Theorems 10.6.5 and 10.6.7 that A_0 is an isomorphism

$$\Big(W^{1,2}(\mathcal{K}_j)^3 \times L_2(\mathcal{K}_j)\Big) \cap \Big(W^{1,s}_{\beta,\delta}(\mathcal{K}_j)^3 \times W^{0,s}_{\beta,\delta}(\mathcal{K}_j)\Big) \to \Big(\mathcal{H}^*_{\mathcal{K}_j} \times L_2(\mathcal{K}_j) \times \mathcal{W}_{2,0,0}\Big) \cap \Big(\mathcal{H}^*_{s',-\beta,-\delta;\mathcal{K}_j} \times W^{0,s}_{\beta,\delta}(\mathcal{K}_j) \times \mathcal{W}_{s,\beta,\delta}\Big). \tag{11.1.19}$$

By (11.1.14), the operator norm (11.1.19) of $A - A_0$ is less than $c\varepsilon$. Hence for sufficiently small ε, the operator $A - A_0$ is also an isomorphism (11.1.19). The result follows. □

We prove the analogous assertion for the problem (11.1.4), (11.1.5).

THEOREM 11.1.4. *Let $(u,p) \in W^{1,2}(\mathcal{G})^3 \times L_2(\mathcal{G})$ be a solution of the problem (11.1.4), (11.1.5), where $F \in \mathcal{H}^* \cap \mathcal{H}^*_{s',-\beta,-\delta;\mathcal{G}}$, $g \in L_2(\mathcal{G}) \cap W^{0,s}_{\beta,\delta}(\mathcal{G})$ and $h_j \in W^{1/2,2}(\Gamma_j)^{3-d_j} \cap W^{1-1/s,s}_{\beta,\delta}(\Gamma_j)^{3-d_j}$. Suppose that the closed strip between the lines $\mathrm{Re}\,\lambda = -1/2$ and $\mathrm{Re}\,\lambda = 1 - \beta - 3/s$ does not contain eigenvalues of the pencils $\mathfrak{A}_j(\lambda)$, $j = 1,\dots,d'$, and that the components of δ satisfy the condition (11.1.17). Then $u \in W^{1,s}_{\beta,\delta}(\mathcal{G})^3$ and $p \in W^{0,s}_{\beta,\delta}(\mathcal{G})$.*

P r o o f. It suffices to prove the theorem for vector functions (u,p) with small supports. For solutions with arbitrary support the assertion then can be easily proved by means of a partition of unity on $\mathcal{G}$. Let the support of (u,p) be contained in a sufficiently small neighborhood $\mathcal{U}$ of the vertex $x^{(j)}$, and let κ be a diffeomorphism mapping $\mathcal{G} \cap \mathcal{U}$ onto $\mathcal{K}_j \cap B$, where $\mathcal{K}_j$ is a cone with vertex at the origin and B is a ball centered at the origin. We assume that $\kappa'(x^{(j)}) = I$. Then the vector function $(w(x), q(x)) = \big(u(\kappa^{-1}(x)), p(\kappa^{-1}(x))\big)$ is a solution of a perturbed Stokes problem (11.1.12), (11.1.13), where the coefficients $b_{\mu,\nu,k,l}$ and $c_{k,l}$ are zero at the origin and bounded by small constants on the support of (w,q). Applying Lemma 11.1.3, we obtain $(w,q) \in W^{1,s}_{\beta_j,\delta}(\mathcal{K}_j)^3 \times W^{0,s}_{\beta_j,\delta}(\mathcal{K}_j)$ and, therefore $(u,p) \in W^{1,s}_{\beta,\delta}(\mathcal{G})^3 \times W^{0,s}_{\beta,\delta}(\mathcal{G})$. For vector functions (u,p) with support in a neighborhood of an edge point, the assertion of the theorem can be proved analogously. □

Analogously, the next theorem can be proved by means of Theorem 10.6.9 (see also Remark 10.6.10).

THEOREM 11.1.5. *Let $(u,p) \in W^{1,2}(\mathcal{G})^3 \times L_2(\mathcal{G})$ be a solution of the problem (11.1.12), (11.1.13), where $g \in W^{l-1,s}_{\beta,\delta}(\mathcal{G})$, $h_j \in W^{l-1/s,s}_{\beta,\delta}(\Gamma_j)^{3-d_j}$, and $F \in \mathcal{H}^*$ has the representation (11.1.6) with given vector functions $f \in W^{l-2,s}_{\beta,\delta}(\mathcal{G})^3$, $\phi_j \in$*

$W_{\beta,\delta}^{l-1-1/s,s}(\Gamma_j)^{d_j}$. *We suppose that the closed strip between the lines* $\mathrm{Re}\,\lambda = -1/2$ *and* $\mathrm{Re}\,\lambda = l - \beta_j - 3/s$ *does not contain eigenvalues of the pencils* $\mathfrak{A}_j(\lambda)$, $j = 1, \dots, d'$, *and that*

$$\max(0, l - \mu_+^{(k)}) < \delta_k + 2/s < l$$

for $k = 1, \dots, d$. *Furthermore, we assume that* g, h_j *and* ϕ_j *satisfy compatibility conditions on the edges* M_k *which guarantee that there exists a vector function* $(w, q) \in W_{\beta,\delta'}^{l,s}(\mathcal{G})^3 \times W_{\beta,\delta'}^{l-1,s}(\mathcal{G})$ *such that*

$$S_j w = h_j, \quad N_j(w, q) = \phi_j \quad \text{on } \Gamma_j, \; j = 1, \dots, N, \quad \nabla \cdot w + g \in V_{\beta,\delta'}^{l-1,s}(\mathcal{G})$$

for arbitrary δ' *with components* $\delta'_k \ge \delta_k$, $\delta'_k > l - 2 - 2/s$, $k = 1, \dots, d$. *Then* $u \in W_{\beta,\delta}^{l,s}(\mathcal{G})^3$ *and* $p \in W_{\beta,\delta}^{l-1,s}(\mathcal{G})$.

REMARK 11.1.6. The compatibility conditions on g, h_j, ϕ_j in Theorem 11.1.5 where described in Chapters 9 and 10 (cf. Lemma 10.2.3 and Remark 10.2.4).

11.1.4. Solvability in weighted Hölder spaces. The following theorem can be deduced in the same way as Theorem 11.1.4. Here one has to apply Theorems 10.7.15, 10.7.16 for $l \ge 2$ and Theorems 10.8.8, 10.8.9 for $l = 1$ instead of Theorems 10.6.5, 10.6.7.

THEOREM 11.1.7. *Let* $(u, p) \in W^{1,2}(\mathcal{G})^3 \times L_2(\mathcal{G})$ *be a variational solution of the boundary value problem* (11.1.1), (11.1.2), *where* $f \in C_{\beta,\delta}^{l-2,\sigma}(\mathcal{G})^3$, $g \in C_{\beta,\delta}^{l-1,\sigma}(\mathcal{G})$, $h_j \in C_{\beta,\delta}^{l,\sigma}(\Gamma_j)^{3-d_j}$ *and* $\phi_j \in C_{\beta,\delta}^{l-1,\sigma}(\Gamma_j)^{d_j}$, $l \ge 1$, $0 < \sigma < 1$. *We suppose that the strip* $-1/2 < \mathrm{Re}\,\lambda < l + \sigma - \beta_j$ *does not contain eigenvalues of the pencil* $\mathfrak{A}_j(\lambda)$, $j = 1, \dots, d'$, *that*

$$l - \mu_+^{(k)} < \delta_k - \sigma < l,$$

and that $\delta_k - \sigma$ *is not integer for* $k = 1, \dots, d$. *Furthermore, we assume that* g, h_j, ϕ_j *satisfy compatibility conditions on the edges* M_k *which guarantee the existence of a vector function* $(w, q) \in C_{\beta,\delta'}^{l,s}(\mathcal{G})^3 \times C_{\beta,\delta'}^{l-1,s}(\mathcal{G})$ *such that*

$$S_j w = h_j, \quad N_j(w, q) = \phi_j \quad \text{on } \Gamma_j, \; j = 1, \dots, N, \quad \nabla \cdot w + g \in N_{\beta,\delta'}^{l-1,s}(\mathcal{G})$$

for arbitrary δ' *with components* $\delta'_k \ge \delta_k$, $\delta'_k > l - 2 + \sigma$, $k = 1, \dots, d$ *(cf. Lemma* 10.7.2*). Then* $(u, p) \in C_{\beta,\delta}^{l,\sigma}(\mathcal{G})^3 \times C_{\beta,\delta}^{l-1,\sigma}(\mathcal{G})$.

11.1.5. Regularity assertions in weighted spaces with homogeneous norms. We suppose that $d_{k_+} \cdot d_{k_-} = 0$ for each $k = 1, \dots, d$, i.e. the Dirichlet condition is prescribed on at least one of the adjoining faces of every edge M_k. Then we can apply the regularity results of Theorems 10.6.14 and 10.7.18 in the spaces $V_{\beta,\delta}^{l,s}$ and $N_{\beta,\delta}^{l,\sigma}$, respectively. Analogously to Theorems 11.1.5 and 11.1.7, we obtain the subsequent statements.

THEOREM 11.1.8. *Let* $(u, p) \in W^{1,2}(\mathcal{G})^3 \times L_2(\mathcal{G})$ *be a solution of the problem* (11.1.12), (11.1.13), *where* $g \in V_{\beta,\delta}^{l-1,s}(\mathcal{G})$, $h_j \in V_{\beta,\delta}^{l-1/s,s}(\Gamma_j)^{3-d_j}$, *and* F *is a linear and continuous functional on* $\mathcal{H}$ *which has the form* (11.1.6) *with given vector functions* $f \in V_{\beta,\delta}^{l-2,s}(\mathcal{G})^3$, $\phi_j \in V_{\beta,\delta}^{l-1-1/s,s}(\Gamma_j)^{d_j}$ *if* $l \ge 2$. *In the case* $l = 1$, *we suppose that* $F \in \mathcal{H}^* \cap \mathcal{H}^*_{s',-\beta,-\delta;\mathcal{G}}$. *Furthermore, we assume that the closed strip between the lines* $\mathrm{Re}\,\lambda = -1/2$ *and* $\mathrm{Re}\,\lambda = l - \beta_j - 3/s$ *does not contain eigenvalues of the pencils* $\mathfrak{A}_j(\lambda)$, $j = 1, \dots, d'$, *and that*

$$l - \delta_+^{(k)} < \delta_k + 2/s < l + \delta_+^{(k)}$$

for $k = 1, \dots, d$. *Then* $u \in V^{l,s}_{\beta,\delta}(\mathcal{G})^3$ *and* $p \in V^{l-1,s}_{\beta,\delta}(\mathcal{G})$.

THEOREM 11.1.9. *Let* $(u,p) \in W^{1,2}(\mathcal{G})^3 \times L_2(\mathcal{G})$ *be a solution of problem* (11.1.12), (11.1.13), *where* $g \in N^{l-1,\sigma}_{\beta,\delta}(\mathcal{G})$, $h_j \in N^{l,\sigma}_{\beta,\delta}(\Gamma_j)^{3-d_j}$ *and* F *is a linear and continuous functional on* $\mathcal{H}$ *which has the form* (11.1.6) *with given vector functions* $f \in N^{l-2,\sigma}_{\beta,\delta}(\mathcal{G})^3$, $\phi_j \in N^{l-1,\sigma}_{\beta,\delta}(\Gamma_j)^{d_j}$, $l \geq 2$, $0 < \sigma < 1$. *Suppose that the strip* $-1/2 < \operatorname{Re}\lambda < l + \sigma - \beta_j$ *does not contain eigenvalues of the pencil* $\mathfrak{A}_j(\lambda)$, $j = 1, \dots, d'$, *and that*

$$l - \delta_+^{(k)} < \delta_k - \sigma < l + \delta_+^{(k)}$$

for $k = 1, \dots, d$. *Then* $(u,p) \in N^{l,\sigma}_{\beta,\delta}(\mathcal{G})^3 \times N^{l-1,\sigma}_{\beta,\delta}(\mathcal{G})$.

11.2. Regularity results for solutions of the Navier-Stokes system

Let $\mathcal{G}$ be the same domain as in Section 11.1, and let S_j and N_j be the same operators as in the boundary value problem (11.1.1), (11.1.2). We consider the problem

$$(11.2.1) \qquad -\nu\Delta u + \sum_{j=1}^{3} u_j\, \partial_{x_j} u + \nabla p = f, \quad -\nabla \cdot u = g \quad \text{in } \mathcal{G},$$

$$(11.2.2) \qquad S_j u = h_j, \quad N_j(u,p) = \phi_j \quad \text{on } \Gamma_j,\ j = 1, \dots, N.$$

Here S_j are the same operators as in Section 11.1 and the operators N_j are defined as

$$N_j(u,p) = -p + 2\nu\varepsilon_{n,n}(u) \text{ if } d_j = 1, \qquad N_j(u,p) = 2\nu\varepsilon_{n,\tau}(u) \text{ if } d_j = 2,$$
$$N_j(u,p) = -pn + 2\nu\varepsilon_n(u) \text{ if } d_j = 3$$

(in the case $d_j = 0$, the condition $N_j(u,p) = \phi_j$ does not appear). We are interested again in variational solutions $W^{1,2}(\mathcal{G})^3 \times L_2(\mathcal{G})$. In the case of the Dirichlet problem (and even in the case of the mixed boundary value problem with the boundary condition (i) and (iii)), such solutions exist for arbitrary f (see the books by LADYZHENSKAYA [**89**], TEMAM [**194**], GIRAULT and RAVIART [**55**]). Using a fixed point argument, we prove the existence of solutions of the more general problem (11.2.1), (11.2.2) for small data. Furthermore, we obtain regularity results for the solutions both in weighted Sobolev and Hölder spaces.

11.2.1. Existence of variational solutions.

We introduce the bilinear form

$$(11.2.3) \qquad b(u,v) = 2\nu \int_{\mathcal{G}} \sum_{i,j=1}^{3} \varepsilon_{i,j}(u)\, \varepsilon_{i,j}(v)\, dx.$$

By a *variational solution* of the problem (11.2.1), (11.2.2), we mean a vector function $(u,p) \in W^{1,2}(\mathcal{G})^3 \times L_2(\mathcal{G})$ satisfying the equations

$$(11.2.4) \quad b(u,v) + \int_{\mathcal{G}} \sum_{j=1}^{3} u_j \frac{\partial u}{\partial x_j} \cdot v\, dx - \int_{\mathcal{G}} p\, \nabla \cdot v\, dx = F(v) \quad \text{for all } v \in \mathcal{H},$$

$$(11.2.5) \quad -\nabla \cdot u = g \ \text{ in } \mathcal{G}, \quad S_j u = h_j \ \text{ on } \Gamma_j,\ j = 1, \dots, N.$$

Here again $\mathcal{H}$ denotes the subspace $\{u \in W^{1,2}(\mathcal{G})^3 : S_j u|_{\Gamma_j} = 0,\ j = 1, \dots, N\}$ and F is the functional (11.1.6). Note that the functional

$$v \to \int_{\mathcal{G}} u_j \frac{\partial u}{\partial x_j} \cdot v\, dx$$

is continuous on $W^{1,2}(\mathcal{G})^3$ for arbitrary $u \in W^{1,2}(\mathcal{G})^3$. This follows from the inequality

$$\Big| \int_{\mathcal{G}} u_j \frac{\partial u}{\partial x_j} \cdot v\, dx \Big| \le \|u_j\|_{L_4(\mathcal{G})}\, \|\partial_{x_j} u\|_{L_2(\mathcal{G})^3}\, \|v\|_{L_4(\mathcal{G})^3}$$

and from the imbedding $W^{1,2}(\mathcal{G}) \subset L_4(\mathcal{G})$.

Let the operator Q be defined by

$$Qu = (u \cdot \nabla)\, u.$$

Obviously, Q realizes a mapping $W^{1,2}(\mathcal{G}) \to \mathcal{H}^*$. Furthermore, there exist constants c_1, c_2 such that

$$\|Qu\|_{\mathcal{H}^*} \le c\, \|u\|^2_{W^{1,2}(\mathcal{G})^3} \quad \text{for all } u \in W^{1,2}(\mathcal{G}) \tag{11.2.6}$$

and

$$\|Qu - Qv\|_{\mathcal{H}^*} \le c \left(\|u\|_{W^{1,2}(\mathcal{G})^3} + \|v\|_{W^{1,2}(\mathcal{G})^3} \right) \|u - v\|_{W^{1,2}(\mathcal{G})^3} \tag{11.2.7}$$

for all $u, v \in W^{1,2}(\mathcal{G})^3$. Using the last two estimates together with Theorem 11.1.2, we prove the following statement.

THEOREM 11.2.1. *Let g and h_j be as in Theorem* 11.1.2. *Furthermore, we suppose that $L_{\mathcal{H}} = \{0\}$ and that*

$$\|F\|_{\mathcal{H}^*} + \|g\|_{L_2(\mathcal{G})} + \sum_{j=1}^{N} \|h_j\|_{W^{1/2,2}(\Gamma_j)^{3-d_j}}$$

is sufficiently small. Then there exists a solution $(u,p) \in W^{1,2}(\mathcal{G})^3 \times L_2(\mathcal{G})$ of the problem (11.2.4), (11.2.5). *Here u is unique on the set of all functions with norm less than a certain positive ε, p is unique if $d_j \in \{1,3\}$ for at least one j, otherwise p is unique up to a constant.*

P r o o f. Let $(u^{(0)}, p^{(0)}) \in W^{1,2}(\mathcal{G})^3 \times L_2(\mathcal{G})$ be the variational solution of the linear problem

$$-\nu \Delta u + \nabla p = f, \quad -\nabla \cdot u = g \quad \text{in } \mathcal{G}, \tag{11.2.8}$$

$$S_j u = h_j, \quad N_j(u,p) = \phi_j \quad \text{on } \Gamma_j,\ j = 1, \dots, N, \tag{11.2.9}$$

By our assumptions on F, g and h_j, we may assume that

$$\|u^{(0)}\|_{W^{1,2}(\mathcal{G})^3} \le \varepsilon_1,$$

where ε_1 is a small positive number. Let $\mathcal{H}_0$ denote the set of all $v \in \mathcal{H}$ such that $\nabla \cdot v = 0$. We set $w = u - u^{(0)}$ and $q = p - p^{(0)}$. Then (u,p) is a solution of the problem (11.2.4), (11.2.5) if and only if $(w,q) \in \mathcal{H}_0 \times L_2(\mathcal{G})$ and

$$b(w,v) - \int_{\mathcal{G}} q \nabla \cdot v\, dx = -\int_{\mathcal{G}} Q(w + u^{(0)}) \cdot v\, dx \quad \text{for all } v \in \mathcal{H}. \tag{11.2.10}$$

By Theorem 11.1.2, there exists a linear and continuous mapping

$$\mathcal{H}^* \ni \Phi \to A\Phi = (w,q) \in \mathcal{H}_0 \times L_2(\mathcal{G})$$

defined by

$$b(w,v) - \int_{\mathcal{G}} q\,\nabla\cdot v\,dx = \Phi(v) \text{ for all } v\in\mathcal{H}, \qquad \int_{\mathcal{G}} q\,dx = 0 \text{ if } d_j\in\{0,2\} \text{ for all } j.$$

We write (11.2.10) as

$$(w,q) = T(w,q), \quad \text{where } T(w,q) = -AQ(w+u^{(0)}).$$

By (11.2.7), the operator T is contractive on the set of all $(w,q)\in\mathcal{H}_0\times L_2(\mathcal{G})$ with norm $\le \varepsilon_2$ if ε_1 and ε_2 are sufficiently small. Hence there exist $w\in W^{1,2}(\mathcal{G})^3$ and $q\in L_2(\mathcal{G})$ satisfying (11.2.10). The result follows. □

REMARK 11.2.2. If $d_j\in\{0,2\}$ for all j, then

$$(11.2.11) \qquad \int_{\mathcal{G}} \sum_{j=1}^{3} u_j \frac{\partial v}{\partial x_j}\cdot v\,dx = 0 \text{ for all } v\in W^{1,2}(\mathcal{G})^3,\ u\in\mathcal{H},\ \nabla\cdot u = 0$$

(cf. [**55**, Lemma IV.2.2]). Thus, analogously to [**55**, Theorem IV.2.3], the problem (11.2.4), (11.2.5) hast at least one solution for arbitrary $F\in\mathcal{H}^*$, $g=0$, and for all $h_j\in W^{1/2,2}(\Gamma_j)^{3-d_j}$ satisfying the condition

$$\sum_{j:\,d_j=0}\int_{\Gamma_j} h_j\cdot n\,dx + \sum_{j:\,d_j=2}\int_{\Gamma_j} h_j\,dx = 0.$$

11.2.2. Regularity assertions for variational solutions in weighted Sobolev spaces. Our goal is to extend the results of Theorems 11.1.4–11.1.7 to the nonlinear problem (11.2.1), (11.2.2). We start with regularity results in weighted Sobolev spaces.

LEMMA 11.2.3. *Let $1<q\le s<\infty$, $3/q\le 1+3/s$, $\beta_j+3/q\le\beta'_j+1+3/s$ for $j=1,\dots,d'$, $\delta_k+3/q\le\delta'_k+1+3/s$ for $k=1,\dots,d$, and $s'=s/(s-1)$. Then $V^{0,q}_{\beta,\delta}(\mathcal{G})$ is continuously imbedded in $\big(V^{1,s'}_{-\beta',-\delta'}(\mathcal{G})\big)^*$.*

P r o o f. The assertion of the lemma follows immediately from the continuity of the imbedding $V^{1,s'}_{-\beta',-\delta'}(\mathcal{G})\subset V^{0,q'}_{-\beta,-\delta}(\mathcal{G})$, where $q'=q/(q-1)$ (cf. Lemma 4.1.3). □

COROLLARY 11.2.4. *Let $u\in L_6(\mathcal{G})$ and $v\in W^{1,s}_{\beta,\delta}(\mathcal{G})$, $s>6/5$, $\beta'_j\ge\beta_j-1/2$ for $j=1,\dots,d'$, and $\delta'_k\ge\delta_k-1/2$ for $k=1,\dots,d$. Then $u\,\partial_{x_i}v\in\big(V^{1,s'}_{-\beta',-\delta'}(\mathcal{G})\big)^*$ for $i=1,2,3$.*

P r o o f. Let $q=6s/(s+6)$. By Hölders inequality,

$$\|u\,\partial_{x_i}v\|_{V^{0,q}_{\beta,\delta}(\mathcal{G})} \le \|u\|_{L_6(\mathcal{G})}\,\|\partial_{x_i}v\|_{V^{0,s}_{\beta,\delta}(\mathcal{G})}.$$

Furthermore, the space $V^{0,q}_{\beta,\delta}(\mathcal{G})$ is continuously imbedded in $\big(V^{1,s'}_{-\beta',-\delta'}(\mathcal{G})\big)^*$ if $\beta'_j\ge\beta_j-1/2$ and $\delta'_k\ge\delta_k-1/2$. This proves the corollary. □

Using the last lemma and Theorem 11.1.4, we obtain a $W^{1,s}_{\beta,\delta}$-regularity result for solutions of the boundary value problem (11.2.1), (11.2.2).

THEOREM 11.2.5. *Let $(u,p)\in W^{1,2}(\mathcal{G})^3\times L_2(\mathcal{G})$ be a solution of the problem (11.2.4), (11.2.5), where $F\in\mathcal{H}^*\cap\mathcal{H}^*_{s',-\beta,-\delta;\mathcal{G}}$, $g\in L_2(\mathcal{G})\cap W^{0,s}_{\beta,\delta}(\mathcal{G})$ and $h_j\in W^{1/2,2}(\Gamma_j)^{3-d_j}\cap W^{1-1/s,s}_{\beta,\delta}(\Gamma_j)^{3-d_j}$, $s>6/5$. Suppose that the closed strip between* *the lines* $\mathrm{Re}\,\lambda = -1/2$ *and* $\mathrm{Re}\,\lambda = 1 - \beta - 3/s$ *does not contain eigenvalues of the pencils* $\mathfrak{A}_j(\lambda)$, $j = 1, \dots, d'$, *and that the components of* δ *satisfy the condition* (11.1.17). *Then* $u \in W^{1,s}_{\beta,\delta}(\mathcal{G})^3$ *and* $p \in W^{0,s}_{\beta,\delta}(\mathcal{G})$.

P r o o f. 1) First, let $3/2 \le s \le 3$. Since $u_j \in W^{1,2}(\mathcal{G}) \subset L_6(\mathcal{G})$ and $\partial_{x_j} u \in L_2(\mathcal{G})^3$, it follows that $u_j\, \partial_{x_j} u \in L_{3/2}(\mathcal{G})^3$. From this and from Lemma 11.2.3 we conclude that $(u \cdot \nabla)\, u \in \big(V^{1,s'}_{-1+3/s,-1+3/s}(\mathcal{G})^*\big)^3$. Hence, (u,p) satisfies the equations

$$b(u,v) - \int_{\mathcal{G}} p\, \nabla \cdot v\, dx = \Phi(v) \quad \text{for all } v \in \mathcal{H}, \tag{11.2.12}$$

$$-\nabla \cdot u = g \quad \text{in } \mathcal{G}, \quad S_j u = h_j \quad \text{on } \Gamma_j,\ j = 1, \dots, N, \tag{11.2.13}$$

where $\Phi = F - (u \cdot \nabla) u \in \mathcal{H}^*_{s',-\beta',-\delta';\mathcal{G}}$ and the components of β', δ' are

$$\beta'_j = \max(\beta_j, 1 - 3/s), \quad \delta'_k = \max(\delta_k, 1 - 3/s).$$

Applying Theorem 11.1.4, we obtain $(u,p) \in W^{1,s}_{\beta',\delta'}(\mathcal{G})^3 \times W^{0,s}_{\beta',\delta'}(\mathcal{G})$. Corollary 11.2.4 yields $(u \cdot \nabla)\, u \in \big(V^{1,s'}_{-\beta'+1/2,-\delta'+1/2}(\mathcal{G})^3\big)^*$. Therefore, $\Phi \in \mathcal{H}^*_{s',-\beta'',-\delta'';\mathcal{G}}$, where

$$\beta''_j = \max(\beta_j, 1/2 - 3/s), \quad \delta''_k = \max(\delta_k, 1/2 - 3/s).$$

Using again Theorem 11.1.4, we conclude that $(u,p) \in W^{1,s}_{\beta'',\delta''}(\mathcal{G})^3 \times W^{0,s}_{\beta'',\delta''}(\mathcal{G})$. Repeating the last consideration, we obtain $(u,p) \in W^{1,s}_{\beta,\delta}(\mathcal{G})^3 \times W^{0,s}_{\beta,\delta}(\mathcal{G})$.

2) Next, let $6/5 < s < 3/2$. Applying Lemma 4.1.2, we obtain

$$(u \cdot \nabla) u \in L_{3/2}(\mathcal{G})^3 = V^{0,3/2}_{0,0}(\mathcal{G})^3 \subset V^{0,s}_{\beta'+1,\delta'+1}(\mathcal{G})^3 \subset \big(V^{1,s'}_{-\beta',-\delta'}(\mathcal{G})^*\big)^3,$$

where $\beta'_j = \max(\beta_j, 1 + \varepsilon - 3/s)$ and $\delta'_k = \max(\delta_k, 1/3 - 2/s + \varepsilon)$ with an arbitrarily small positive ε. Hence, the functional Φ in (11.2.12) belongs to the space $\mathcal{H}^*_{s',-\beta',-\delta';\mathcal{G}}$ and Theorem 11.1.4 implies $(u,p) \in W^{1,s}_{\beta',\delta'}(\mathcal{G})^3 \times W^{0,s}_{\beta',\delta'}(\mathcal{G})$. From this and from Corollary 11.2.4 it follows that $\Phi \in \mathcal{H}^*_{s',-\beta'',-\delta;\mathcal{G}}$, where $\beta''_j = \max(\beta_j, \beta'_j - 1/2)$. Therefore by Theorem 11.1.4, $(u,p) \in W^{1,s}_{\beta'',\delta}(\mathcal{G})^3 \times W^{0,s}_{\beta'',\delta}(\mathcal{G})$. Repeating the last argument, we obtain $(u,p) \in W^{1,s}_{\beta,\delta}(\mathcal{G})^3 \times W^{0,s}_{\beta,\delta}(\mathcal{G})$.

3) Now let $3 < s \le 6$. Then by Lemma 4.1.2, the imbeddings

$$W^{0,s}_{\beta,\delta}(\mathcal{G}) \subset W^{0,3}_{\beta',\delta'}(\mathcal{G}), \quad W^{1-1/s,s}_{\beta,\delta}(\Gamma_j) \subset W^{2/3,3}_{\beta',\delta'}(\Gamma_j)$$

and $\mathcal{H}^*_{s',-\beta,-\delta;\mathcal{G}} \subset \mathcal{H}^*_{3/2,-\beta',-\delta';\mathcal{G}}$ are valid, where $\beta'_j = \beta + 3/s - 1 + \varepsilon$, $\delta'_k = \delta_k + 2/s - 2/3 + \varepsilon$, ε is an arbitrarily small positive number. For sufficiently small ε, we conclude from part 1) that $u \in W^{1,3}_{\beta',\delta'}(\mathcal{G})^3$. Furthermore by Hölders inequality,

$$\|u_j\, \partial_{x_j} u\|_{W^{0,2}_{\beta',\delta'}(\mathcal{G})^3} \le \|u_j\|_{L_6(\mathcal{G})}\, \|\partial_{x_j} u\|_{W^{0,3}_{\beta',\delta'}(\mathcal{G})^3}\,.$$

Since $W^{0,2}_{\beta',\delta'}(\mathcal{G}) \subset \mathcal{H}^*_{s',-\beta,-\delta;\mathcal{G}}$ if $\varepsilon < 1/3$ (see Lemma 11.2.3), we get $F - (u \cdot \nabla)\, u \in \mathcal{H}^*_{s',-\beta,-\delta;\mathcal{G}}$. Using Theorem 11.1.4, we obtain $(u,p) \in W^{1,s}_{\beta,\delta}(\mathcal{G})^3 \times W^{0,s}_{\beta,\delta}(\mathcal{G})$.

4) Finally, let $s > 6$. Then by Lemma 4.1.2, the imbeddings

$$W^{0,s}_{\beta,\delta}(\mathcal{G}) \subset W^{0,6}_{\beta',\delta'}(\mathcal{G}), \quad W^{1-1/s,s}_{\beta,\delta}(\Gamma_j) \subset W^{5/6,6}_{\beta',\delta'}(\Gamma_j)$$

and $\mathcal{H}^*_{s',-\beta,-\delta;\mathcal{G}} \subset \mathcal{H}^*_{6/5,-\beta',-\delta';\mathcal{G}}$, are valid, where $\beta'_j = \beta + 3/s - 1/2 + \varepsilon$, $\delta'_k = \delta_k + 2/s - 1/3 + \varepsilon$, ε is an arbitrarily small positive number. For sufficiently small ε, we conclude from part 4) that $u \in W^{1,6}_{\beta',\delta'}(\mathcal{G})^3$. Since $u \in L_6(\mathcal{G})^3$, it follows that $(u \cdot \nabla)\, u \in W^{0,3}_{\beta',\delta'}(\mathcal{G})^3$. The last space is imbedded in $\mathcal{H}^*_{s',-\beta,-\delta;\mathcal{G}}$ if $\varepsilon \le 1/3$ (see Lemma 11.2.3). Therefore, (u,p) is a solution of the problem (11.2.12), (11.2.13), where $\Phi \in \mathcal{H}^*_{s',-\beta,-\delta;\mathcal{G}}$. Applying Theorem 11.1.4, we obtain $(u,p) \in W^{1,s}_{\beta,\delta}(\mathcal{G})^3 \times W^{0,s}_{\beta,\delta}(\mathcal{G})$. The proof of the theorem is complete. □

REMARK 11.2.6. The assertion of Theorem 11.2.5 is also true for $s \le 6/5$ if the components of β are not too small.

Indeed, if $s < 6/5$ then we obtain first

$$(u,p) \in W^{1,s}_{\beta',\delta'}(\mathcal{G})^3 \times W^{0,s}_{\beta',\delta'}(\mathcal{G}),$$

where $\beta'_j = \max(\beta_j, 1+\varepsilon-3/s)$ and $\delta'_k = \max(\delta_k, 1/3-2/s+\varepsilon)$ (see the proof of the last theorem for the case $s = 6/5$). Using this, we conclude in the second step that

$$(u,p) \in W^{1,s}_{\beta'',\delta''}(\mathcal{G})^3 \times W^{0,s}_{\beta'',\delta''}(\mathcal{G}),$$

where $\beta''_j = \max(\beta_j, \theta\beta'_j + 3/t - 3/s - 1/2)$ and $\delta''_k = \max(\delta, \theta\delta'_k + 3/t - 3/s - 1/2)$, $t = \varepsilon + 6/5$, $\theta = s(2-t)/(t(2-s))$. Here, ε is an arbitrarily small positive number. In the next step, we obtain the weight parameters $\max(\beta_j, \theta\beta''_j + 3/t - 3/s - 1/2)$ and $\max(\delta, \theta\delta''_k + 3/t - 3/s - 1/2)$. If

$$\beta_j + 3/s > -\frac{6}{5}\left((6-5s)^{-1} - 1\right),$$

then, after a finite number of steps, we arrive at the inclusion $(u,p) \in W^{1,s}_{\beta,\delta}(\mathcal{G})^3 \times W^{0,s}_{\beta,\delta}(\mathcal{G})$.

For the proof of an analogous $W^{2,s}_{\beta,\delta}$-regularity result, we need the following lemma.

LEMMA 11.2.7. *Let $u \in W^{2,s}_{\beta,\delta}(\mathcal{G}) \cap W^{1,2}(\mathcal{G})$, $1 < s < 6$, $\beta_j + 3/s \le 5/2$ for all j, $\delta_k + 2/s > 0$, $\delta'_k + 2/s > 0$ and $\delta'_k \ge \delta_k - 1/2$ for all k. Then $u\,\nabla u \in V^{0,s}_{\beta-1/2,\delta'}(\mathcal{G})^3$.*

P r o o f. 1) Suppose that $\delta_k + 2/s > 1$ for $k = 1, \dots, d$. Then

$$\nabla u \in W^{1,s}_{\beta,\delta}(\mathcal{G})^3 = V^{1,s}_{\beta,\delta}(\mathcal{G})^3 \subset V^{0,q}_{\beta-1/2,\delta-1/2}(\mathcal{G})^3,$$

where $q = 6s/(6-s)$ (cf. Lemma 4.1.3). From this, from the assumption $u \in W^{1,2}(\mathcal{G}) \subset L_6(\mathcal{G})$ and from Hölder's inequality we get $u\,\nabla u \in V^{0,s}_{\beta-1/2,\delta-1/2}(\mathcal{G})^3$ and

$$\|u\,\nabla u\|_{V^{0,s}_{\beta-1/2,\delta-1/2}(\mathcal{G})^3} \le c\,\|u\|_{L_6(\mathcal{G})}\,\|\nabla u\|_{V^{0,q}_{\beta-1/2,\delta-1/2}(\mathcal{G})^3}\,.$$

2) We consider the case where $0 < \delta_k + 2/s \le 1$ for all k. Then by Lemma 7.2.5, the function u admits the decomposition (cf. Lemma 7.2.5)

$$u = v + w, \quad v \in V^{2,s}_{\beta,\delta}(\mathcal{G}), \quad w \in W^{4,s}_{\beta+2,\delta+2}(\mathcal{G}).$$

Let $\mathcal{U}_1, \dots, \mathcal{U}_{d'}$ be domains in $\mathbb{R}^3$ such that

$$\mathcal{U}_1 \cup \dots \cup \mathcal{U}_{d'} \supset \overline{\mathcal{G}} \quad \text{and} \quad \overline{\mathcal{U}}_j \cap \overline{M}_k = \emptyset \text{ if } k \notin X_j\,.$$

Here again X_j denotes the set of all indices k such that the vertex $x^{(j)}$ is an end point of the edge M_k. Using Lemma 7.2.3, we obtain

$$\rho_j^{\beta_j-5/2+3/s} w \in L_6(\mathcal{G} \cap \mathcal{U}_j), \quad \rho_j^{\beta_j-2+3/s} w \in L_\infty(\mathcal{G} \cap \mathcal{U}_j)$$

and

$$\rho_j^{\beta_j-1+2/s}\,\nabla w \in L_{3s}(\mathcal{G}\cap\mathcal{U}_j)^3$$

for $j = 1,\dots,d'$. Since $\beta_j - 5/2 + 3/2 \le 0$, it follows in particular that $w \in L_6(\mathcal{G})$ and therefore also $v \in L_6(\mathcal{G})$. We estimate the norms of $v\nabla u$ and $w\nabla u$ in $V^{0,s}_{\beta-1/2,\delta-1/2}(\mathcal{G})^3$. Let $q = 6s/(6-s)$. Using Hölder's inequality and Lemma 4.1.3, we obtain

$$\|v\nabla v\|_{V^{0,s}_{\beta-1/2,\delta-1/2}(\mathcal{G})^3} \le \|v\|_{L_6(\mathcal{G})}\, \|\nabla v\|_{V^{0,q}_{\beta-1/2,\delta-1/2}(\mathcal{G})^3} \le c\,\|v\|_{L_6(\mathcal{G})}\, \|\nabla v\|_{V^{1,s}_{\beta,\delta}(\mathcal{G})^3}\,.$$

By Lemma 4.1.3, $V^{2,s}_{\beta,\delta}(\mathcal{G})$ is continuously imbedded in $V^{0,3s/2}_{\beta-2+1/s,\delta-2+1/s}(\mathcal{G})$. Consequently,

$$\begin{aligned}
&\|v\,\nabla w\|_{V^{0,s}_{\beta-1/2,\delta-1/2}(\mathcal{G})^3}\\
&\quad\le c\,\|v\|_{V^{0,3s/2}_{\beta-2+1/s,\delta-2+1/s}(\mathcal{G})} \sum_{j=1}^{d'} \Big\|\rho_j^{3/2-1/s} \prod_{k\in X_j}\Big(\frac{r_k}{\rho_j}\Big)^{3/2-1/s}\nabla w\Big\|_{L_{3s}(\mathcal{G}\cap\mathcal{U}_j)^3}\\
&\quad\le c\,\|v\|_{V^{2,s}_{\beta,\delta}(\mathcal{G})} \sum_{j=1}^{d'} \Big\|\rho_j^{\beta_j-1+2/s}\,\nabla w\Big\|_{L_{3s}(\mathcal{G}\cap\mathcal{U}_j)^3}\,.
\end{aligned}$$

Thus, $v\nabla u \in V^{0,s}_{\beta-1/2,\delta-1/2}(\mathcal{G})^3 \subset V^{0,s}_{\beta-1/2,\delta'}(\mathcal{G})^3$. Using the continuity of the imbedding $W^{2,s}_{\beta,\delta}(\mathcal{G}) \subset W^{1,s}_{3/2-3/s,\delta'}(\mathcal{G})$, we obtain

$$\begin{aligned}
&\|w\,\nabla u\|_{V^{0,s}_{\beta-1/2,\delta'}(\mathcal{G})^3}\\
&\quad\le c\sum_{j=1}^{d'} \Big\|\rho_j^{\beta_j-2+3/s} w\Big\|_{L_\infty(\mathcal{G}\cap\mathcal{U}_j)} \Big\|\rho_j^{3/2-3/s}\prod_{k\in X_j}\Big(\frac{r_k}{\rho_j}\Big)^{\delta'_k}\nabla u\Big\|_{L_s(\mathcal{G}\cap\mathcal{U}_j)^3}\\
&\quad\le c\sum_{j=1}^{d'} \Big\|\rho_j^{\beta_j-2+3/s} w\Big\|_{L_\infty(\mathcal{G}\cap\mathcal{U}_j)} \|u\|_{W^{2,s}_{\beta,\delta}(\mathcal{G})^3}\,.
\end{aligned}$$

This proves the lemma for the case $\delta_k + 2/s < 1$, $k = 1,\dots,d$.

3) The case where $\delta_k + 2/s \le 1$ for some but not all k can be reduced to the cases 1) and 2) using suitable cut-off functions. □

We prove the analog to Theorem 11.1.5 for the nonlinear problem (11.2.1), (11.2.2) in the case $l = 2$.

THEOREM 11.2.8. *Let $(u,p) \in W^{1,2}(\mathcal{G})^3 \times L_2(\mathcal{G})$ be a solution of the problem (11.2.4), (11.2.5), where $g \in W^{1,s}_{\beta,\delta}(\mathcal{G})$, $h_j \in W^{2-1/s,s}_{\beta,\delta}(\Gamma_j)^{3-d_j}$, and $F \in \mathcal{H}^*$ has the representation (11.1.6) with given vector functions $f \in W^{0,s}_{\beta,\delta}(\mathcal{G})^3$, $\phi_j \in W^{1-1/s,s}_{\beta,\delta}(\Gamma_j)^{d_j}$. We suppose that the closed strip between the lines $\operatorname{Re}\lambda = -1/2$ and $\operatorname{Re}\lambda = 2-\beta-3/s$ does not contain eigenvalues of the pencils $\mathfrak{A}_j(\lambda)$, $j = 1,\dots,d'$, and that*

$$\max(0, 2-\mu_+^{(k)}) < \delta_k + 2/s < 2$$

for $k = 1,\dots,d$. Furthermore, we assume that g, h_j and ϕ_j satisfy compatibility conditions on the edges M_k which guarantee that there exists a vector function

$(w,q) \in W^{2,s}_{\beta,\delta}(\mathcal{G})^3 \times W^{1,s}_{\beta,\delta}(\mathcal{G})$ such that

$$S_j w = h_j, \quad N_j(w,q) = \phi_j \ \text{ on } \Gamma_j,\ j=1,\dots,N, \quad \nabla\cdot w + g \in V^{1,s}_{\beta,\delta}(\mathcal{G})$$

(cf. Lemma 10.2.3 and Remark 10.2.4). Then $u \in W^{2,s}_{\beta,\delta}(\mathcal{G})^3$ and $p \in W^{1,s}_{\beta,\delta}(\mathcal{G})$.

P r o o f. 1) First we consider the case $1 < s \le 3/2$. Let $\mathcal{U}_j$ and X_j be the same sets as in the proof of Lemma 11.2.7. We set $q = 3s/(3-2s)$ if $s < 3/2$, $q = \infty$ if $s = 3/2$. Since

$$\|u_i \partial_{x_i} u\|_{W^{0,s}_{\beta',\delta'}(\mathcal{G})^3} \le c\, \|u_i\|_{L_6(\mathcal{G})}\ \|\partial_{x_i} u\|_{L_2(\mathcal{G})^3} \sum_{j=1}^{d'} \Big\| \rho_j^{\beta'_j} \prod_{k\in X_j} \Big(\frac{r_k}{\rho_j}\Big)^{\delta'_k} \Big\|_{L_q(\mathcal{G}\cap\mathcal{U}_j)},$$

we obtain $(u\cdot\nabla)\, u \in W^{0,s}_{\beta',\delta'}(\mathcal{G})^3$ if $\beta'_j > -3/q$ and $\delta'_k > -2/q$ or, what is the same, if $\beta'_j + 3/s > 2$ for $j = 1,\dots,d'$ and $\delta'_k + 2/s > 4/3$ for $k = 1,\dots,d$. Let $\beta'_j = \max(\beta_j, 2 - 3/s + \varepsilon)$, $\delta'_k = \max(\delta_k, 4/3 - 2/s + \varepsilon)$, where ε is a sufficiently small positive number. Then (u,p) satisfies the equations

$$-\nu\Delta u + \nabla p = f - (u\cdot\nabla)\, u, \quad -\nabla\cdot u = g \ \text{ in } \mathcal{G}, \tag{11.2.14}$$

$$S_j u = h_j, \quad N_j(u,p) = \phi_j \ \text{ on } \Gamma_j,\ j = 1,\dots,N, \tag{11.2.15}$$

where $f - (u\cdot\nabla)\, u \in W^{0,s}_{\beta',\delta'}(\mathcal{G})^3$, $g \in W^{1,s}_{\beta',\delta'}(\mathcal{G})$, $h_j \in W^{2-1/s,s}_{\beta',\delta'}(\Gamma_j)^{3-d_j}$ and $\phi_j \in W^{1-1/s,s}_{\beta',\delta'}(\Gamma_j)^{d_j}$. Theorem 11.1.5 yields $(u,p) \in W^{2,s}_{\beta',\delta'}(\mathcal{G})^3 \times W^{1,s}_{\beta',\delta'}(\mathcal{G})$. Applying Lemma 11.2.7, we obtain $f - (u\cdot\nabla)\, u \in W^{0,s}_{\beta'',\delta''}(\mathcal{G})^3$, where

$$\beta''_j = \max(\beta_j, 3/2 - 3/s + \varepsilon), \quad \delta''_k = \max(\delta_k, 5/6 - 2/s + \varepsilon).$$

Hence, it follows from Theorem 11.1.5 that $(u,p) \in W^{2,s}_{\beta'',\delta''}(\mathcal{G})^3 \times W^{1,s}_{\beta'',\delta''}(\mathcal{G})$. Repeating this procedure, we obtain $(u,p) \in W^{2,s}_{\beta,\delta}(\mathcal{G})^3 \times W^{1,s}_{\beta,\delta}(\mathcal{G})$.

2) Next, we consider the case $3/2 < s \le 2$. Let ε be a positive number less than $1/2$ such that $\delta_k + 2/s < 2 - \varepsilon$ for all k. Then by Lemma 4.1.2, $W^{l,s}_{\beta,\delta}(\mathcal{G}) \subset W^{l,3/2}_{\beta+3/s-2+\varepsilon,2/3-\varepsilon}(\mathcal{G})$, and from part 1) it follows that

$$(u,p) \in W^{2,3/2}_{\beta+3/s-2+\varepsilon,2/3-\varepsilon}(\mathcal{G})^3 \times W^{1,3/2}_{\beta+3/s-2+\varepsilon,2/3-\varepsilon}(\mathcal{G})$$

if ε is sufficiently small. In particular,

$$\partial_{x_i} u \in W^{1,3/2}_{\beta+3/s-2+\varepsilon,2/3-\varepsilon}(\mathcal{G})^3 = V^{1,3/2}_{\beta+3/s-2+\varepsilon,2/3-\varepsilon}(\mathcal{G})^3 \subset V^{0,3}_{\beta+3/s-2+\varepsilon,2/3-\varepsilon}(\mathcal{G})^3$$

(cf. Lemma 4.1.3). Let $\delta'_k = \max(\delta_k, 5/3 - 2/s)$ for $k = 1,\dots,d$. Then

$$\|u_i\, \partial_{x_i} u\|_{W^{0,s}_{\beta,\delta'}(\mathcal{G})^3} \le c\, \|u_i\|_{L_6(\mathcal{G})} \sum_{j=1}^{d'} \Big\| \rho_j^{\beta_j+3/s-2+\varepsilon} \prod_{k\in X_j} \Big(\frac{r_k}{\rho_j}\Big)^{2/3-\varepsilon} \partial_{x_i} u \Big\|_{L_3(\mathcal{G}\cap\mathcal{U}_j)^3},$$

where

$$c = \sum_{j=1}^{d'} \Big\| \rho_j^{2-3/s-\varepsilon} \prod_{k\in X_j} \Big(\frac{r_k}{\rho_j}\Big)^{\delta'_k - 2/3+\varepsilon} \Big\|_{L_{2s/(2-s)}(\mathcal{G})} < \infty.$$

Consequently, $f - (u\cdot\nabla)\, u \in W^{0,s}_{\beta,\delta'}(\mathcal{G})^3$, and Theorem 11.1.5 implies $(u,p) \in W^{2,s}_{\beta,\delta'}(\mathcal{G})^3 \times W^{1,s}_{\beta,\delta'}(\mathcal{G})$. Using Lemma 11.2.7, we obtain $f - (u\cdot\nabla)\, u \in W^{0,s}_{\beta,\delta''}(\mathcal{G})^3$, where $\delta''_k = \max(\delta_k, \delta'_k - 1/2)$. Again Theorem 11.1.5 yields $(u,p) \in W^{2,s}_{\beta,\delta''}(\mathcal{G})^3 \times W^{1,s}_{\beta,\delta''}(\mathcal{G})$. Repeating this argument, we obtain $(u,p) \in W^{2,s}_{\beta,\delta}(\mathcal{G})^3 \times W^{1,s}_{\beta,\delta}(\mathcal{G})$.

3) Let $2 < s \le 3$, and let ε be a positive number less than $1/2$ such that $\delta_k + 2/s < 2 - \varepsilon$ for all k. Then $W^{l,s}_{\beta,\delta}(\mathcal{G}) \subset W^{l,2}_{\beta+3/s-3/2+\varepsilon,1-\varepsilon}(\mathcal{G})$ (cf. Lemma 4.1.2). Therefore by part 2), $(u,p) \in W^{2,2}_{\beta+3/s-3/2+\varepsilon,1-\varepsilon}(\mathcal{G})^3 \times W^{1,2}_{\beta+3/s-3/2+\varepsilon,1-\varepsilon}(\mathcal{G})$ provided ε is sufficiently small. Consequently,

$$\partial_{x_i} u \in W^{1,2}_{\beta+3/s-3/2+\varepsilon,1-\varepsilon}(\mathcal{G})^3 = V^{1,2}_{\beta+3/s-3/2+\varepsilon,1-\varepsilon}(\mathcal{G})^3 \subset V^{0,6}_{\beta+3/s-3/2+\varepsilon,1-\varepsilon}(\mathcal{G})^3.$$

Let $\delta'_k = \max(\delta_k, 5/3 - 2/s)$ for $k = 1, \dots, d$. Then

$$\|u_i\, \partial_{x_i} u\|_{W^{0,s}_{\beta,\delta'}(\mathcal{G})^3} \le c\, \|u_i\|_{L_6(\mathcal{G})}\, \|\partial_{x_i} u\|_{V^{0,6}_{\beta+3/s-3/2+\varepsilon,1-\varepsilon}(\mathcal{G})^3}\,,$$

where

$$c = \sum_{j=1}^{d'} \Big\| \rho_j^{3/2-3/s-\varepsilon} \prod_{k\in X_j} \Big(\frac{r_k}{\rho_j}\Big)^{\delta'_k-1+\varepsilon} \Big\|_{L_{3s/(3-s)}(\mathcal{G}\cap\mathcal{U}_j)} < \infty.$$

Thus, $f - (u\cdot\nabla)\, u \in W^{0,s}_{\beta,\delta'}(\mathcal{G})^3$ and $(u,p) \in W^{2,s}_{\beta,\delta'}(\mathcal{G})^3 \times W^{1,s}_{\beta,\delta'}(\mathcal{G})$ (by Theorem 11.1.5). Hence by Lemma 11.2.7, $f - (u\cdot\nabla)\, u \in W^{0,s}_{\beta,\delta''}(\mathcal{G})^3$, where $\delta''_k = \max(\delta_k, 7/6 - 2/s)$, and Theorem 11.1.5 implies $(u,p) \in W^{2,s}_{\beta,\delta''}(\mathcal{G})^3 \times W^{1,s}_{\beta,\delta''}(\mathcal{G})$. Repeating the last argument, we obtain $(u,p) \in W^{2,s}_{\beta,\delta}(\mathcal{G})^3 \times W^{1,s}_{\beta,\delta}(\mathcal{G})$.

4) Finally, let $s > 3$. We define $\delta'_k = \max(\delta_k, 1-2/s+\varepsilon)$, where ε is a sufficiently small positive number. Then

$$g \in W^{1,s}_{\beta,\delta'}(\mathcal{G}) \subset W^{0,s}_{\beta-1,\delta'-1}(\mathcal{G}), \quad h_j \in W^{2-1/s,s}_{\beta,\delta'}(\Gamma_j)^{3-d_j} \subset W^{1-1/s,s}_{\beta-1,\delta'-1}(\Gamma_j)^{3-d_j}.$$

Furthermore, the functional (11.1.6) belongs to the space $\mathcal{H}^*_{s',1-\beta,1-\delta';\mathcal{G}}$. Since $\max(1-\mu_+^{(k)}, 0) < \delta'_k - 1 + 2/s < 1$, it follows from Theorem 11.2.5 that $u \in W^{1,s}_{\beta-1,\delta'-1}(\mathcal{G})^3$. Then by Lemma 7.2.4,

$$\rho_j^{\beta_j-2+3/s} \prod_{k\in X_j} \Big(\frac{r_k}{\rho_j}\Big)^{\sigma_k} u \in L_\infty(\mathcal{G}\cap\mathcal{U}_j)^3,$$

where $\sigma_k = 0$ for $\delta_k < 2-3/s$, $\sigma_k = 1/s+\varepsilon$ for $2-3/s \le \delta_k < 2-2/s$. By Hölder's inequality, the estimate

$$\begin{aligned} & \|u_i\, \partial_{x_i} u\|_{W^{0,s}_{\beta,\delta}(\mathcal{G})} \\ & \le c \sum_{j=1}^{d'} \Big\|\rho_j^{\beta_j-2+3/s} \prod_{k\in X_j} \Big(\frac{r_k}{\rho_j}\Big)^{\sigma_k} u_i\Big\|_{L_\infty(\mathcal{G}\cap\mathcal{U}_j)} \Big\|\rho_j^{2-3/s} \prod_{k\in X_j} \Big(\frac{r_k}{\rho_j}\Big)^{\delta_k-\sigma_k} \partial_{x_i} u\Big\|_{L_s(\mathcal{G}\cap\mathcal{U}_j)^3} \\ & \le c \sum_{j=1}^{d'} \Big\|\rho_j^{\beta_j-2+3/s} \prod_{k\in X_j} \Big(\frac{r_k}{\rho_j}\Big)^{\sigma_k} u_i\Big\|_{L_\infty(\mathcal{G}\cap\mathcal{U}_j)} \|\partial_{x_i} u\|_{W^{0,s}_{\beta-1,\delta'-1}(\mathcal{G})^3} \end{aligned}$$

holds. For the last inequality, we used the fact that $\beta - 1 \le 2 - 3/s$ and $\delta'_k - 1 \le \delta_k - \sigma_k$. Consequently, (u,p) is a solution of the problem (11.2.12), (11.2.13), where $\Phi = F - (u\cdot\nabla)\, u$ is a functional of the form (11.1.6) with $f \in W^{0,s}_{\beta,\delta}(\mathcal{G})^3$, $\phi_j \in W^{1-1/s,s}_{\beta,\delta}(\mathcal{G})^{d_j}$. Applying Theorem 11.1.5, we obtain $u \in W^{2,s}_{\beta,\delta}(\mathcal{G})^3$ and $p \in W^{1,s}_{\beta,\delta}(\mathcal{G})$. □

11.2.3. Regularity results in weighted Hölder spaces. In order to extend the result of Theorem 11.1.7 to the boundary value problem (11.2.1), (11.2.2), we consider first the nonlinear term in the Navier-Stokes system.

LEMMA 11.2.9. *Let $u \in C^{2,\sigma}_{\beta,\delta}(\mathcal{G})$, where $\beta_j \le 3+\sigma$ for $j=1,\dots,d'$, $0 \le \delta_k < 2+\sigma$, $\delta_k - \sigma$ is not an integer for $k=1,\dots,d$. Then $u\,\partial_{x_i} u \in C^{0,\sigma}_{\beta,\delta'}(\mathcal{G})$, where δ' is an arbitrary tuple with the components $\delta'_k \ge \max(0,\delta_k-1)$, $k=1,\dots,d$.*

P r o o f. Let $\mathcal{U}_j$ and X_j be the same sets as in the proof of Lemma 11.2.7. We have to show that there exists a finite constant C such that

$$\rho_j(x)^{\beta_j-\sigma} \prod_{k\in X_j} \Big(\frac{r_k}{\rho_j}\Big)^{\max(0,\delta'_k-\sigma)} \big|u(x)\partial_{x_i}u(x)\big| \le C \tag{11.2.16}$$

for $x \in \mathcal{G}\cap\mathcal{U}_j$,

$$\rho_j(x)^{\beta_j} \prod_{k\in X_j} \Big(\frac{r_k(x)}{\rho_j(x)}\Big)^{\delta'_k} \frac{|u(x)\partial_{x_i}u(x) - u(y)\partial_{y_i}u(y)|}{|x-y|^\sigma} \le C \tag{11.2.17}$$

for $x,y \in \mathcal{G}\cap\mathcal{U}_j$, $|x-y| < r(x)/2$ and

$$\rho_j(x)^{\beta_j-\delta'_k} \frac{|u(x)\partial_{x_i}u(x) - u(y)\partial_{y_i}u(y)|}{|x-y|^{\sigma-\delta'_k}} \le C \tag{11.2.18}$$

for $x,y \in \mathcal{G}_{j,k}$, $|x-y| < \rho_j(x)/2$, $\delta'_k < \sigma$. Here by $\mathcal{G}_{j,k}$ we denoted the set $\{x \in \mathcal{G}\cap\mathcal{U}_j : r_k(x) < 3r(x)/2\}$. The inequality (11.2.16) follows immediately from the estimate

$$\rho_j(x)^{\beta_j-2-\sigma+|\alpha|} \prod_{k\in X_j} \Big(\frac{r_k}{\rho_j}\Big)^{\max(0,\delta_k-2-\sigma+|\alpha|)} |\partial_x^\alpha u(x)| \le \|u\|_{C^{2,\sigma}_{\beta,\delta}(\mathcal{G})} \tag{11.2.19}$$

($|\alpha| \le 2$) and from the inequalities $\beta_j \le 3+\sigma$, $\delta'_k \ge \delta_k - 1$. Furthermore,

$$\frac{|u(x)-u(y)|}{|x-y|^\sigma} \le c\,|x-y|^{1-\sigma} \sup_z \big|\nabla u(z)\big|$$

for $x,y \in \mathcal{G}\cap\mathcal{U}_j$, $|x-y| < r(x)/2$, where the supremum is taken over all $z \in \mathcal{G}\cap\mathcal{U}_j$ such that $|x-z| < r(x)/2$. Using the inequalities $\rho_j(x)/2 < \rho_j(z) < 3\rho_j(x)/2$, $r_k(x)/2 < r_k(z) < 3r_k(x)/2$ and (11.2.19), we obtain

$$\frac{|u(x)-u(y)|}{|x-y|^\sigma} \le c\,r(x)^{1-\sigma}\,\rho_j(x)^{1+\sigma-\beta_j} \prod_{k\in X_j} \Big(\frac{r_k(x)}{\rho_j(x)}\Big)^{-\max(0,\delta_k-1-\sigma)}$$

for $x,y \in \mathcal{G}\cap\mathcal{U}_j$, $|x-y| < r(x)/2$. Analogously,

$$\frac{\big|\partial_{x_i}u(x) - \partial_{y_i}u(y)\big|}{|x-y|^\sigma} \le c\,r(x)^{1-\sigma}\,\rho_j(x)^{\sigma-\beta_j} \prod_{k\in X_j} \Big(\frac{r_k(x)}{\rho_j(x)}\Big)^{-\max(0,\delta_k-\sigma)}.$$

for $x, y \in \mathcal{G} \cap \mathcal{U}_j$, $|x-y| < r(x)/2$. Hence,

$$\begin{aligned}
&\frac{|u(x)\partial_{x_i}u(x) - u(y)\partial_{y_i}u(y)|}{|x-y|^\sigma} \\
&\quad \le \frac{|u(x)-u(y)|}{|x-y|^\sigma}\,\big|\partial_{x_i}u(x)\big| + \frac{\big|\partial_{x_i}u(x) - \partial_{y_i}u(y)\big|}{|x-y|^\sigma}\,|u(y)| \\
&\quad \le c\,r(x)^{1-\sigma}\,\rho_j(x)^{2+2\sigma-2\beta_j} \\
&\qquad \times \Big(\prod_{k\in X_j} \Big(\frac{r_k(x)}{\rho_j(x)}\Big)^{-2\max(0,\delta_k-1-\sigma)} + \prod_{k\in X_j} \Big(\frac{r_k(x)}{\rho_j(x)}\Big)^{-\max(0,\delta_k-\sigma)} \Big)
\end{aligned}$$

for $x, y \in \mathcal{G} \cap \mathcal{U}_j$, $|x-y| < r(x)/2$. From this estimate and from the inequalities

$$c_1\, r(x) \le \rho_j(x) \prod_{k\in X_j} \frac{r_k(x)}{\rho_j(x)} \le c_2\, r(x),$$

$\delta'_k + 1 - \sigma \ge 2\max(0, \delta_k - \sigma - 1)$, $\delta'_k + 1 - \sigma \ge \max(0, \delta_k - \sigma)$, and $\beta_j \le 3 + \sigma$, we obtain (11.2.17). Analogously, we obtain the estimate

$$\rho_j(x)^{\beta_j - \delta'_k}\, \frac{|u(x)-u(y)|}{|x-y|^{\sigma-\delta'_k}}\, |\partial_{x_i}u(x)| \le C$$

for $\delta'_k < \sigma$, $x, y \in \mathcal{G}_{j,k}$, $|x-y| < \rho_j(x)/2$. Since $\partial_{x_i}u(x) \in C^{1,\sigma}_{\beta,\delta}(\mathcal{G}) \subset C^{1,\sigma}_{\beta,\delta'+1}(\mathcal{G})$, there exists a constant C such that

$$\rho_j(x)^{\beta_j - 1 - \delta'_k}\, \frac{|\partial_{x_i}u(x) - \partial_{y_i}u(y)|}{|x-y|^{\sigma-\delta'_k}} \le C$$

for $\delta'_k < \sigma$, $x, y \in \mathcal{G}_{j,k}$, $|x-y| < \rho_j(x)/2$. This together with (11.2.19) implies

$$\rho_j(x)^{\beta_j - \delta'_k}\, \frac{|\partial_{x_i}u(x) - \partial_{y_i}u(y)|}{|x-y|^{\sigma-\delta'_k}}\, |u(x)| \le C$$

for $\delta'_k < \sigma$, $x, y \in \mathcal{G}_{j,k}$, $|x-y| < \rho_j(x)/2$. Thus, the estimate (11.2.18) holds. The proof is complete. □

The next lemma can be deduced directly from the definition of the space $N^{l,\sigma}_{\beta,\delta}(\mathcal{G})$.

LEMMA 11.2.10. *Let f and g be arbitrary functions from $N^{l,\sigma}_{\beta,\delta}(\mathcal{G})$ and $N^{l,\sigma}_{\beta',\delta'}(\mathcal{G})$, respectively. Then $fg \in N^{l,\sigma}_{\beta+\beta'-l-\sigma,\delta+\delta'-l-\sigma}(\mathcal{G})$.*

The last two lemmas together Theorems 11.1.7 and 11.2.8 enable us to prove the following $C^{2,\sigma}_{\beta,\delta}$-regularity result for the solution of the boundary value problem (11.2.1), (11.2.2).

THEOREM 11.2.11. *Let $(u,p) \in W^{1,2}(\mathcal{G})^3 \times L_2(\mathcal{G})$ be a variational solution of the problem (11.2.1), (11.2.2), where $f \in C^{0,\sigma}_{\beta,\delta}(\mathcal{G})^3$, $g \in C^{1,\sigma}_{\beta,\delta}(\mathcal{G})$, $h_j \in C^{2,\sigma}_{\beta,\delta}(\Gamma_j)^{3-d_j}$ and $\phi_j \in C^{1,\sigma}_{\beta,\delta}(\Gamma_j)^{d_j}$. We assume that g, h_j and ϕ_j are such that there exist $w \in C^{2,\sigma}_{\beta,\delta}(\mathcal{G})^3$ and $q \in C^{1,\sigma}_{\beta,\delta}(\mathcal{G})$ satisfying the conditions*

$$S_j w = h_j, \quad N_j(w,q) = \phi_j \quad \textit{on } \Gamma_j,\ j = 1,\dots,N, \quad \nabla\cdot w + g \in N^{1,\sigma}_{\beta,\delta}(\mathcal{G}).$$

Furthermore, we suppose that $\beta_j - \sigma < 5/2$ for $j = 1,\dots,d$, the strip $-1/2 < \operatorname{Re}\lambda \le 2 + \sigma - \beta_j$ is free of eigenvalues the pencil $\mathfrak{A}_j(\lambda)$ for $j = 1,\dots,d'$, the *components of δ are nonnegative and satisfy the inequalities $2-\mu_+^{(k)} < \delta_k - \sigma < 2$, $\delta_k \neq \sigma$, $\delta_k \neq 1+\sigma$. Then $u \in C^{2,\sigma}_{\beta,\delta}(\mathcal{G})^3$ and $p \in C^{1,\sigma}_{\beta,\delta}(\mathcal{G})$.*

P r o o f. Suppose first that $\delta_k > \sigma$ for $k = 1, \dots, d$. Let ε be an arbitrarily small positive number and s an arbitrary real number greater than 1. We set $\beta'_j = \beta_j - \sigma - 3/s + \varepsilon$ for $j = 1, \dots, d'$ and $\delta'_k = \delta_k - \sigma - 2/s + \varepsilon$ for $k = 1, \dots, d$. From our assumptions on f, g, h_j and ϕ_j it follows that

$$f \in W^{0,s}_{\beta',\delta'}(\mathcal{G})^3, \quad g \in W^{1,s}_{\beta',\delta'}(\mathcal{G}), \quad h_j \in W^{2-1/s,s}_{\beta',\delta'}(\Gamma_j)^{3-d_j}, \quad \phi_j \in W^{1-1/s,s}_{\beta',\delta'}(\Gamma_j)^{d_j}.$$

Using Theorem 11.2.8, we obtain $u \in W^{2,s}_{\beta',\delta'}(\mathcal{G})^3$. Consequently,

$$u \in W^{0,s}_{\beta'-2,-2/s+\varepsilon}(\mathcal{G})^3, \quad \partial_{x_i} u \in W^{0,s}_{\beta'-1,\delta''}(\mathcal{G})^3, \quad \partial_{x_i}\partial_{x_j} u \in W^{0,s}_{\beta',\delta'}(\mathcal{G})^3$$

for $i, j = 1, 2, 3$, where $\delta''_k = \max(\delta_k - \sigma - 1, 0) - 2/s + \varepsilon$ for $k = 1, \dots, d$. This together with Hölder's inequality implies

$$u_j\, \partial_{x_j} u \in W^{0,s/2}_{2\beta'-3,\delta''-2/s+\varepsilon}(\mathcal{G})^3, \quad u_j\, \partial_{x_i}\partial_{x_j} u \in W^{0,s/2}_{2\beta'-2,\delta'-2/s+\varepsilon}(\mathcal{G})^3,$$

and

$$(\partial_{x_i} u_j)\, \partial_{x_j} u \in W^{0,s/2}_{2\beta'-2,2\delta''}(\mathcal{G})^3.$$

Therefore, $u_j\, \partial_{x_j} u \in W^{1,s/2}_{2\beta'-2,\delta'-2/s+\varepsilon}(\mathcal{G})^3$. The numbers ε and s can be chosen such that $\beta_j - \sigma \le 3 - 2\varepsilon$ for $j = 1, \dots, d'$, $\delta_k - \sigma > 2/s$ for $k = 1, \dots, d$, $\varepsilon + 1/s \le 1/2$ and $1 - 6/s > \sigma$. Then $2\beta'_j - 2 \le \beta_j - \sigma + 1 - 6/s$, $\delta'_k - 2/s + \varepsilon \le \delta_k - \sigma + 1 - 6/s$ and $\delta_k - \sigma + 1 - 6/s > 1 - 4/s$. Consequently,

$$u_j\, \partial_{x_j} u \in W^{1,s/2}_{\beta-\sigma+1-6/s,\delta-\sigma+1-6/s}(\mathcal{G})^3 = V^{1,s/2}_{\beta-\sigma+1-6/s,\delta-\sigma+1-6/s}(\mathcal{G})^3 \subset N^{0,\sigma}_{\beta,\delta}(\mathcal{G})^3$$

(cf. Lemma 3.6.2). Hence, (u,p) is a solution of the problem (11.2.14), (11.2.15), where $f - (u \cdot \nabla)\, u \in C^{0,\sigma}_{\beta,\delta}(\mathcal{G})^3$. Applying Theorem 11.1.7, we obtain $(u,p) \in C^{2,\sigma}_{\beta,\delta}(\mathcal{G})^3 \times C^{1,\sigma}_{\beta,\delta}(\mathcal{G})$.

Suppose now that $\delta'_k < \sigma$ for at least one k. By the first part of the proof, we obtain $u \in C^{2,\sigma}_{\beta,\gamma}(\mathcal{G})^3$, where $\gamma_k = \max(\delta_k, \sigma + \varepsilon)$, ε is an arbitrarily small positive real number. Then Lemma 11.2.9 implies $f - (u \cdot \nabla)\, u \in C^{0,\sigma}_{\beta,\delta}(\mathcal{G})^3$. Thus, it follows from Theorem 11.1.7 that $(u,p) \in C^{2,\sigma}_{\beta,\delta}(\mathcal{G})^3 \times C^{1,\sigma}_{\beta,\delta}(\mathcal{G})$. The proof is complete. □

Finally, we prove the analogous $C^{1,\sigma}_{\beta,\delta}$-regularity result.

THEOREM 11.2.12. *Let $(u,p) \in W^{1,2}(\mathcal{G})^3 \times L_2(\mathcal{G})$ be a variational solution of the problem (11.2.1), (11.2.2), where $f \in C^{-1,\sigma}_{\beta,\delta}(\mathcal{G})^3$, $g \in C^{0,\sigma}_{\beta,\delta}(\mathcal{G})$, $h_j \in C^{1,\sigma}_{\beta,\delta}(\Gamma_j)^{3-d_j}$ and $\phi_j \in C^{0,\sigma}_{\beta,\delta}(\Gamma_j)^{d_j}$. We suppose that g, h_j and ϕ_j are such that there exist $w \in C^{1,\sigma}_{\beta,\delta}(\mathcal{G})^3$ and $q \in C^{1,\sigma}_{\beta,\delta}(\mathcal{G})$ satisfying the conditions*

$$S_j w = h_j, \quad N_j(w,q) = \phi_j \quad \text{on } \Gamma_j,\ j = 1, \dots, n, \qquad \nabla \cdot w + g \in N^{0,\sigma}_{\beta,\delta}(\mathcal{G}).$$

Furthermore, we assume that $\beta_j - \sigma < 3/2$ for $j = 1, \dots, d'$, the strip $-1/2 < \operatorname{Re}\lambda \le 1 + \sigma - \beta_j$ is free of eigenvalues of the pencils $\mathfrak{A}_j(\lambda)$ for $j = 1, \dots, d'$, and that the components of δ are nonnegative and satisfy the inequalities $1 - \mu_+^{(k)} < \delta_k - \sigma < 1$, $\delta_k \neq \sigma$. Then $u \in C^{1,\sigma}_{\beta,\delta}(\mathcal{G})^3$ and $p \in C^{0,\sigma}_{\beta,\delta}(\mathcal{G})$.

P r o o f.) First let $\delta_k > \sigma$ for $k = 1, \dots, d$. Then $g \in W^{0,s}_{\beta',\delta'}(\mathcal{G})$ and $h_j \in W^{1-1/s,s}_{\beta',\delta'}(\Gamma_j)^{3-d_j}$, where $\beta'_j = \beta_j - \sigma - 3/s + \varepsilon$, $\delta'_k = \delta_k - \sigma - 2/s + \varepsilon$, ε is an arbitrarily mall positive number, and $s > 1$. Furthermore, the functional (11.1.6) belongs to the space $\mathcal{H}^*_{s',-\beta',-\delta';\mathcal{G}}$. Using Theorem 11.2.5, we obtain $(u,p) \in W^{1,s}_{\beta',\delta'}(\mathcal{G})^3 \times W^{0,s}_{\beta',\delta'}(\mathcal{G})$. We consider the term

$$(u \cdot \nabla)\, u = \sum_{j=1}^{3} \partial_{x_j}(u_j u) - \sum_{j=1}^{3} u\, \partial_{x_j} u_j = \sum_{j=1}^{3} \partial_{x_j}(u_j u) + gu.$$

Since $u_i \in W^{0,s}_{\beta'-1,-2/s+\varepsilon}(\mathcal{G})$ and $\partial_{x_k} u_i \in W^{0,s}_{\beta',\delta'}(\mathcal{G})$, we get

$$\begin{aligned} u_j u \in W^{1,s/2}_{2\beta'-1,\delta'-2/s+\varepsilon}(\mathcal{G})^3 &\subset W^{1,s/2}_{\beta-\sigma+1-6/s,\delta-\sigma+1-6/s}(\mathcal{G})^3 \\ &= V^{1,s/2}_{\beta-\sigma+1-6/s,\delta-\sigma+1-6/s}(\mathcal{G})^3 \subset N^{0,\sigma}_{\beta,\delta}(\mathcal{G})^3 \end{aligned}$$

if ε is sufficiently small and s is sufficiently large (cf. Lemma 3.6.2). Furthermore, it follows from the inclusions $g \in N^{0,\sigma}_{\beta,\delta}(\mathcal{G})$,

$$u \in W^{1,s}_{\beta',1-2/s+\varepsilon}(\mathcal{G})^3 = V^{1,s}_{\beta',1-2/s+\varepsilon}(\mathcal{G})^3 \subset N^{0,\sigma}_{\beta-1+\varepsilon,\sigma+\varepsilon+1/s}(\mathcal{G})^3$$

and Lemma 11.2.10 that

$$gu \in N^{0,\sigma}_{2\beta-1+\varepsilon-\sigma,\delta+\varepsilon+1/s}(\mathcal{G}) \subset N^{0,\sigma}_{\beta+1,\delta+1}(\mathcal{G}).$$

Consequently, $(u\cdot\nabla)\, u \in C^{-1,\sigma}_{\beta,\delta}(\mathcal{G})^3$ and Theorem 11.1.7 implies $(u,p) \in C^{1,\sigma}_{\beta,\delta}(\mathcal{G})^3 \times C^{0,\sigma}_{\beta,\delta}(\mathcal{G})$.

2) Suppose that $\delta_k < \sigma$ for some k. By the first part of the proof, the vector function (u,p) belongs to the space $C^{1,\sigma}_{\beta,\gamma}(\mathcal{G})^3 \times C^{0,\sigma}_{\beta,\gamma}(\mathcal{G})$, where $\gamma_k = \max(\delta_k, \sigma+\varepsilon)$ for $k = 1,\dots,d$ and ε is an arbitrarily small positive number. In particular, $u_j \in N^{0,\sigma}_{\beta-1,\gamma}(\mathcal{G})$, $\partial_{x_j} u \in N^{0,\sigma}_{\beta,\gamma}(\mathcal{G})^3$. From Lemma 11.2.10 we conclude that

$$u_j \partial_{x_j} u \in N^{0,\sigma}_{2\beta-1-\sigma,2\gamma-\sigma}(\mathcal{G})^3.$$

The last space is contained in $N^{0,\sigma}_{\beta+1,\delta+1}(\mathcal{G})^3$ for sufficiently small ε. Applying Theorem 11.1.7, we obtain $(u,p) \in C^{1,\sigma}_{\beta,\delta}(\mathcal{G})^3 \times C^{0,\sigma}_{\beta,\delta}(\mathcal{G})$. □

11.3. Regularity results for particular boundary value problems

Here we establish some regularity assertions for weak solutions of special boundary value problems for the Navier-Stokes system in the class of the nonweighted spaces $W^{l,s}(\mathcal{G})$. We assume that $\mathcal{G}$ is a polyhedron with the faces Γ_j and edges M_k. The angle at the edge M_k is denoted by θ_k. For the sake of simplicity, we restrict ourselves to homogeneous boundary conditions

$$S_j u = 0, \quad N_j(u,p) = 0 \quad \text{on } \Gamma_j,\ j = 1,\dots,N. \tag{11.3.1}$$

Analogous results are valid for inhomogeneous boundary conditions provided the boundary data satisfy certain compatibility conditions on the edges.

11.3.1. The Dirichlet problem for the Navier-Stokes system. For arbitrary $f \in W^{-1,2}(\mathcal{G})^3$, there exists a solution $(u,p) \in W^{1,2}(\mathcal{G})^3 \times L_2(\mathcal{G})$ of the Dirichlet problem

$$(11.3.2) \quad -\nu\Delta u + (u\cdot\nabla)\,u + \nabla p = f, \quad -\nabla u = 0 \text{ in } \mathcal{G}, \quad u = 0 \text{ on } \Gamma_j,\ j = 1,\dots,N$$

(see e.g. [**55**, Theorem IV.2.1]). The regularity results established below are based on the following properties of the operator pencils $\mathfrak{A}_j(\lambda)$ (cf. [**85**, Chapter 5]). The eigenvalues of these pencils satisfy the inequalities

$$(11.3.3) \qquad \Big|\operatorname{Re}\lambda + \frac{1}{2}\Big| \ge \frac{1}{2} + \frac{\mu}{\mu+4}\,.$$

Here $\mu(\mu+1)$ is the first eigenvalue of the Beltrami operator on a subdomain of the sphere, $\mu > 0$. If the cone $\mathcal{K}_j$ is contained in a half-space, then the strip

$$-1/2 \le \operatorname{Re}\lambda \le 1$$

contains only the eigenvalue $\lambda = 1$ of the pencil $\mathfrak{A}_j(\lambda)$. This eigenvalue has only the eigenvector $(0,0,0,c)$, $c = const.$, and no generalized eigenvectors. The eigenvalues of the pencil $\mathfrak{A}_j(\lambda)$ in the strip $-1/2 \le \operatorname{Re}\lambda \le 1$ are real and monotonous with respect to the cone $\mathcal{K}_j$. Moreover, the eigenvalues for a circular cone are solutions of a certain transcendental equation (cf. [**86**] or [**85**, Section 5.6]).

The numbers $\mu_+^{(k)}$ can be easily calculated. In the case $\theta_k < \pi$, we have $\mu_+^{(k)} = \pi/\theta_k$. If $\theta_k > \pi$, then $\mu_+^{(k)}$ is the smallest positive solution of the equation

$$(11.3.4) \qquad \sin(\mu\theta_k) + \mu\sin\theta_k = 0.$$

Note that $\mu_+^{(k)} > 1/2$ for every $\theta_k < 2\pi$, $\mu_+^{(k)} > 2/3$ if $\theta_k < 3\arccos(1/4) \approx 1.2587\pi$, $\mu_+^{(k)} > 1$ if $\theta_k < \pi$, and $\mu_+^{(k)} > 4/3$ if $\theta_k < 3\pi/4$. Using these facts together with Theorems 11.2.5 and 11.2.8, we obtain the following assertions.

THEOREM 11.3.1. *Let $(u,p) \in W^{1,2}(\mathcal{G})^3 \times L_2(\mathcal{G})$ be a solution of the boundary value problem* (11.3.2).

1) *If $f \in (W^{-1,s}(\mathcal{G}))^3$, $2 < s \le 3$, $s' = s/(s-1)$, then $(u,p) \in W^{1,s}(\mathcal{G})^3 \times L_s(\mathcal{G})$. If the polyhedron $\mathcal{G}$ is convex, then this assertion is true for all $s > 2$.*

2) *If $f \in W^{-1,2}(\mathcal{G})^3 \cap L_s(\mathcal{G})^3$, $1 < s \le 4/3$, then $(u,p) \in W^{2,s}(\mathcal{G})^3 \times W^{1,s}(\mathcal{G})$. If $\theta_k < 3\arccos(1/4) \approx 1.2587\pi$ for $k = 1,\dots,d$, then this result is true for $1 < s \le 3/2$. If $\mathcal{G}$ is convex, then this result is valid for $1 < s \le 2$. If, moreover, the angles at the edges are less than $3\pi/4$, then the result holds even for $1 < s < 3$.*

P r o o f. 1) If $s' < 2$, then $\overset{\circ}{W}{}^{1,s'}(\mathcal{G}) = \overset{\circ}{V}{}^{1,s'}_{0,0}(\mathcal{G})$ and therefore $W^{-1,s}(\mathcal{G}) = (\overset{\circ}{V}{}^{1,s'}_{0,0}(\mathcal{G}))^*$. Applying Theorem 11.2.5 with $\beta = 0$, $\delta = 0$, we obtain the first assertion.

2) The second assertion follows immediately from Theorem 11.2.8 and from the equality $W^{1,s}(\mathcal{G}) = W^{1,s}_{0,0}(\mathcal{G})$ for $s < 3$. □

Furthermore, the subsequent regularity assertion in Hölder spaces holds.

THEOREM 11.3.2. *Let $(u,p) \in W^{1,2}(\mathcal{G})^3 \times L_2(\mathcal{G})$ be a solution of the boundary value problem* (11.3.2). *If $\mathcal{G}$ is convex, $f \in C^{-1,\sigma}(\mathcal{G})$, and σ is sufficiently small (such that $1+\sigma < \pi/\theta_k$ and there are no eigenvalues of the pencils $\mathfrak{A}_j$ in the strip $1 < \operatorname{Re}\lambda \le 1+\sigma$), then $(u,p) \in C^{1,\sigma}(\mathcal{G})^3 \times C^{0,\sigma}(\mathcal{G})$.*

P r o o f. Let ε be a positive number, $\varepsilon < 1-\sigma$. Since $C^{-1,\sigma}(\mathcal{G}) \subset C^{-1,\sigma}_{\sigma+\varepsilon,0}(\mathcal{G})$, it follows from Theorem 11.2.12 that $(u,p) \in C^{1,\sigma}_{\sigma+\varepsilon,0}(\mathcal{G})^3 \times C^{0,\sigma}_{\sigma+\varepsilon,0}(\mathcal{G})$ if $\sigma < \mu_+^{(k)} - 1$. In particular, we have $u \in C^{1,\sigma}_{\sigma+\varepsilon,0}(\mathcal{G})^3 \subset C^{0,\sigma}(\mathcal{G})^3$. This implies $u_j u \in C^{0,\sigma}(\mathcal{G})^3$. Since $\nabla \cdot u = 0$, it follows that $(u\cdot\nabla)\,u = \sum_j \partial_{x_j}(u_j u) \in C^{-1,\sigma}(\mathcal{G})^3$. Thus, (u,p) satisfies the Stokes system

$$-\nu\,\Delta u + \nabla p = f - (u\cdot\nabla)u, \quad -\nabla\cdot u = 0,$$

where $f-(u\cdot\nabla)u \in C^{-1,\sigma}(\mathcal{G}) \subset C^{-1,\sigma}_{0,0}(\mathcal{G})$. Hence by Theorem 10.8.9, the solution (u,p) admits the decomposition

$$\begin{pmatrix} u(x) \\ p(x) \end{pmatrix} = \begin{pmatrix} 0 \\ p(x^{(j)}) \end{pmatrix} + \begin{pmatrix} w(x) \\ q(x) \end{pmatrix}$$

in a neighborhood of the vertex $x^{(j)}$, where $(w,q) \in C^{1,\sigma}_{0,0}(\mathcal{G})^3 \times C^{0,\sigma}_{0,0}(\mathcal{G})$. This is true for every vertex $x^{(j)}$, $j=1,\dots,d'$. Consequently, $(u,p) \in C^{1,\sigma}(\mathcal{G})^3 \times C^{0,\sigma}(\mathcal{G})$. $\square$

For special domains, it is possible to obtain precise regularity results. Let for example $\mathcal{G}$ be a step-shaped domain as in the first two pictures below with angles $\pi/2$ or $3\pi/2$ at every edge or have the form of two beams, where one lies on the other, as in the third picture. Note that the third polyhedron is not a Lipschitz graph domain.

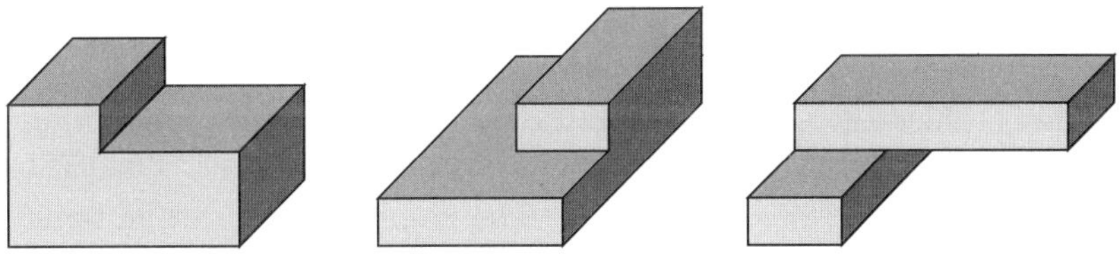

FIGURE 9. Polyhedral domains considered in Subsection 11.3.1

The greatest edge angle is $3\pi/2$, and we obtain $\min \mu_+^{(k)} = 0.54448373....$ Moreover for every vertex, there exists a circular cone with the same vertex and aperture $3\pi/2$ which contains the polyhedron. For every vertex of the left polyhedron, there exists even a cone with aperture π (a half-space) containing the polyhedron. Consequently, the smallest positive eigenvalue of the pencils $\mathfrak{A}_j(\lambda)$ does not exceed the first eigenvalue for a circular cone with vertex $3\pi/2$. A numerical calculation (see [**85**, Section 5.6]) shows that this eigenvalue is greater than $(3\min\mu_+^{(k)}-1)/2$. Using Theorems 11.2.5 and 11.2.8 for $\beta=0$ and $\delta=0$, we obtain the following results:

$$(u,p) \in W^{1,s}(\mathcal{G})^3 \times L_s(\mathcal{G}) \quad \text{if } f \in W^{-1,s}(\mathcal{G}),\ s < 2/(1-\min\mu_+^{(k)}) = 4.3905...,$$
$$(u,p) \in W^{2,s}(\mathcal{G})^3 \times W^{1,s}(\mathcal{G}) \quad \text{if } f \in L_s(\mathcal{G}),\ s < 2/(2-\min\mu_+^{(k)}) = 1.3740....$$

Here the condition on s is sharp.

We give some explanations concerning the examples in the introduction of this monograph (the flow outside a regular polyhedron G). By Theorem 11.2.8, the regularity result $(u,p) \in W^{2,s} \times W^{1,s}$ in an arbitrary bounded subdomain of the complement of G holds if $s < 2/(2-\mu_+^{(k)})$ and there are no eigenvalues of the pencils $\mathfrak{A}_j(\lambda)$ in the strip $-1/2 < \operatorname{Re}\lambda < 2-3/s$. Here, $\mu_+^{(k)}$ is the smallest positive solution of the equation (11.3.4), where $\theta_k = \theta$ is the edge angle in the exterior of G, $\sin\theta$ is equal to $-\frac{2}{3}\sqrt{2}$ if G is a regular tetrahedron or octahedron, -1 if G is a cube, $-\frac{2}{5}\sqrt{5}$ if G is a regular dodecahedron, and $-2/3$ if G is a regular icosahedron. The smallest positive solutions of (11.3.4) are $\mu_+^{(k)} = 0.52033360...$ for the regular tetrahedron, $\mu_+^{(k)} = 0.54448373...$ for a cube, $\mu_+^{(k)} = 0.58489758...$ for the regular octahedron, $\mu_+^{(k)} = 0.60487306...$ for the regular dodecahedron, and $\mu_+^{(k)} = 0.68835272...$ for the regular icosahedron. In the case of a regular tetrahedron, cube, octahedron or dodecahedron, the inequality $s < 2/(2-\mu_+^{(k)})$ implies $2 - 3/s < 3\mu_+^{(k)}/2 - 1 < 0$. Then the absence of eigenvalues of the pencil $\mathfrak{A}_j(\lambda)$ in the strip $-1/2 < \operatorname{Re}\lambda < 2 - 3/s$ follows from [**85**, Theorem 5.5.6]. The exterior of a regular icosahedron is contained in a right circular cone with aperture less than 255°. Numerical results for right circular cones together with the monotonicity of the eigenvalues of the pencils $\mathfrak{A}_j(\lambda)$ in the interval $[-1/2, 1)$ show that also for this polyhedron, the strip $-1/2 < \operatorname{Re}\lambda < 3\mu_+^{(k)}/2 - 1$ is free of eigenvalues of the pencil $\mathfrak{A}_j(\lambda)$. Thus, the above mentioned regularity result holds for all $s < 2/(2-\mu_+^{(k)})$. This inequality cannot be weakened.

11.3.2. The Neumann problem for the Navier-Stokes system. We consider a weak solution $u \in W^{1,2}(\mathcal{G})^3 \times L_2(\mathcal{G})$ of the Neumann problem

$$-\nu\,\Delta u + (u\cdot\nabla)\,u + \nabla p = f, \quad -\nabla u = 0 \text{ in } \mathcal{G}, \tag{11.3.5}$$

$$-pn + 2\nu\varepsilon_n(u) = 0 \text{ on } \Gamma_j,\ j = 1,\dots,N. \tag{11.3.6}$$

For this problem it is known that the strip $-1 \le \operatorname{Re}\lambda \le 0$ contains only the eigenvalues $\lambda = 0$ and $\lambda = 1$ of the operator pencils $\mathfrak{A}_j(\lambda)$ (see [**85**, Theorem 6.3.2]) if $\mathcal{G}$ is a Lipschitz graph polyhedron. The numbers $\mu_+^{(k)}$ are the same as for the Dirichlet problem. Therefore, the following assertions are valid.

THEOREM 11.3.3. *Let $(u,p) \in W^{1,2}(\mathcal{G})^3 \times L_2(\mathcal{G})$ be a solution of the boundary value problem* (11.3.5), (11.3.6).

1) *If $f \in (W^{1,s'}(\mathcal{G})^*)^3$, $2 < s < 3$, then $(u,p) \in W^{1,s}(\mathcal{G})^3 \times L_s(\mathcal{G})$.*

2) *If $f \in (W^{1,2}(\mathcal{G})^*)^3 \cap L_s(\mathcal{G})^3$, $1 < s \le 4/3$, then $(u,p) \in W^{2,s}(\mathcal{G})^3 \times W^{1,s}(\mathcal{G})$. If the angles θ_k are less than $3\arccos(1/4)$, then the last result is true for $1 < s < 3/2$.*

11.3.3. The mixed problem with Dirichlet and Neumann boundary conditions. We assume that on each face Γ_j either the Dirichlet condition $u = 0$ or the Neumann condition $-pn + 2\varepsilon_n(u) = 0$ is prescribed. If on the adjoining faces of the edge M_k the same boundary conditions are given, then $\mu_+^{(k)} > 1/2$. If on one of the adjoining faces the Dirichlet condition and on the other face the Neumann condition is given, then $\mu_+^{(k)} > 1/4$. This implies the following result.

THEOREM 11.3.4. *Let $(u,p) \in W^{1,2}(\mathcal{G})^3 \times L_2(\mathcal{G})$ be a variational solution of the boundary value problem* (11.2.1), (11.2.2), *where $f \in \mathcal{H}^* \cap L_s(\mathcal{G})^3$, $1 < s \le 8/7$, $g = 0$, $h_j = 0$ and $\phi_j = 0$. We assume that on each face Γ_j either the Dirichlet or the Neumann condition is prescribed. Then $(u,p) \in W^{2,s}(\mathcal{G})^3 \times W^{1,s}(\mathcal{G})$.*

11.3.4. The mixed problem with boundary conditions (i)–(iii). We consider the boundary value problem (11.2.1), (11.2.2) in the case where $d_j \le 2$ for all j (i.e., the Neumann condition does not appear in the boundary conditions). Here we assume that the Dirichlet condition is given on at least one of the adjoining faces of every edge. Then by [**85**, Theorem 6.1.5], the strip $-1 \le \operatorname{Re}\lambda \le 0$ is free of eigenvalues of the pencils $\mathfrak{A}_j(\lambda)$. Furthermore, $\mu_+^{(k)} > 1/2$ if the Dirichlet condition is given on both adjoining faces of the edge M_k. For the other indices k, the inequality $\mu_+^{(k)} > 1/4$ holds. If $\theta_k < 3\pi/2$, then $\mu_+^{(k)} > 1/3$. Using this information and Theorems 11.1.4, 11.1.5, we obtain the following statement analogously to Theorem 11.3.1.

THEOREM 11.3.5. *Let* $(u,p) \in W^{1,2}(\mathcal{G})^3 \times L_2(\mathcal{G})$ *be a variational solution of the problem* (11.2.1), (11.2.2), *where* $d_j \le 2$ *for all* j *and* $d_{k_+}\, d_{k_-} = 0$ *for all* k.

1) *Suppose that* $2 < s \le 8/3$, $f \in (W^{1,s'}(\mathcal{G})^*)^3$, $g = 0$, $h_j = 0$ *and* $\phi_j = 0$. *Then* $(u,p) \in W^{1,s}(\mathcal{G})^3 \times L_s(\mathcal{G})$. *If* $\theta_k < 3\pi/2$ *for* $d_{k_+} + d_{k_-} \ne 0$, *then this result is even true for* $2 < s \le 3$.

2) *Suppose that* $1 < s \le 8/7$, $f \in \mathcal{H}^* \cap L_s(\mathcal{G})^3$, $g = 0$, $h_j = 0$ *and* $\phi_j = 0$. *Then* $(u,p) \in W^{2,s}(\mathcal{G})^3 \times W^{1,s}(\mathcal{G})$. *If* $\theta_k < 3\arccos(1/4)$ *for* $d_{k_+} + d_{k_-} = 0$, $\theta_k < \frac{3}{2}\arccos(1/4)$ *for* $d_{k_+} + d_{k_-} = 1$, *and* $\theta_k < 3\pi/4$ *for* $d_{k_+} + d_{k_-} = 2$, *then the last result is true for* $1 < s \le 3/2$.

Note that in the last case, the inequality $\mu_+^{(k)} > 2/3$ is satisfied for $k = 1, \dots, d$.

Finally, we consider the problem (11.2.1), (11.2.2) in the case $d_1 = \cdots = d_{N-1} = 0$, $d_N = 2$. This means that the Dirichlet condition $u = 0$ is given on the faces $\Gamma_1, \dots, \Gamma_{N-1}$, while the boundary condition $u_n = 0$, $2\varepsilon_{n,\tau}(u) = 0$ is given on Γ_N.

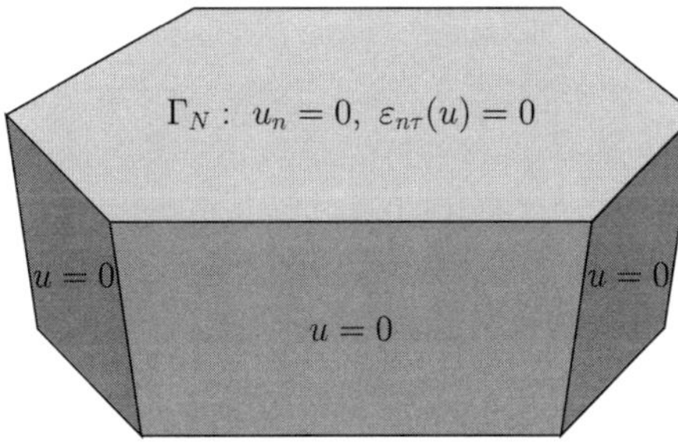

FIGURE 10. A special mixed boundary value problem for the Navier-Sokes system in a polyhedron

Let I be the set of all k such that $M_k \subset \bar{\Gamma}_N$ and $I' = \{1, \dots, d\} \backslash I$. We suppose that the polyhedron $\mathcal{G}$ is convex and that $\theta_k < \pi/2$ for $k \in I$. Then $\mu_+^{(k)} > 1$ for all k, and the strip $-1/2 \le \operatorname{Re}\lambda < 1$ is free of eigenvalues of the pencils $\mathfrak{A}_j(\lambda)$ (see [**85**, Th.6.2.7]). If $\theta_k < 3\pi/8$ for $k \in I$ and $\theta_k < 3\pi/4$ for $k \in I'$, then even $\mu_+^{(k)} > 4/3$. This implies the following result.

THEOREM 11.3.6. *Let* $(u,p) \in W^{1,2}(\mathcal{G})^3 \times L_2(\mathcal{G})$ *be a variational solution of the problem* (11.2.1), (11.2.2), *where* $d_1 = \cdots = d_{N-1} = 0$, $d_N = 2$. *We suppose that the polyhedron* $\mathcal{G}$ *is convex and that* $\theta_k < \pi/2$ *for* $k \in I$.

1) *If* $f \in \mathcal{H}^* \cap L_s(\mathcal{G})^3$, $1 < s \le 2$, $g = 0$, $h_j = 0$ *and* $\phi_j = 0$, *then* $(u,p) \in W^{2,s}(\mathcal{G})^3 \times W^{1,s}(\mathcal{G})$. *If moreover* $\theta_k < 3\pi/8$ *for* $k \in I$ *and* $\theta_k < 3\pi/4$ *for* $k \in I'$, *then this assertion is even true for* $1 < s < 3$.

2) *If* $f \in (W^{1,2}(\mathcal{G})^*)^3 \cap C^{-1,\sigma}(\mathcal{G})$, $g = 0$, $h_j = 0$, $\phi_j = 0$ *for all* j, *and* σ *is sufficiently small, then* $(u,p) \in C^{1,\sigma}(\mathcal{G}) \times C^{0,\sigma}(\mathcal{G})$.

11.4. Green's matrix of the Dirichlet problem for the Stokes system

In this and in the following sections, we restrict ourselves to the Dirichlet problem

$$(11.4.1) \qquad -\Delta u + \nabla p = f, \quad -\nabla \cdot u = g \ \text{ in } \mathcal{G},$$

$$(11.4.2) \qquad u = h_j \quad \text{on } \Gamma_j\,, \ j = 1, \dots, N.$$

We are interested in the properties of the Green's matrix for this problem. In particular, we obtain point estimates for the elements of this matrix and their derivatives.

11.4.1. Existence of Green's matrix. Let $h_j = 0$ for $j = 1, \dots, N$. Then by Theorem 11.1.2, the boundary value problem (11.4.1), (11.4.2) is solvable in $W^{1,2}(\mathcal{G})^3 \times L_2(\mathcal{G})$ for arbitrary $f \in W^{-1,2}(\mathcal{G})^3$ and $g \in L_2(\mathcal{G})$ satisfying the condition

$$\int_{\mathcal{G}} g(x)\, dx = 0.$$

The solution (u,p) is unique up to vectors $(0,c)$, where c is a constant.

Let $G(x,\xi)$ be the matrix with the elements $G_{i,j}(x,\xi)$, $i,j = 1,2,3,4$. The matrix $G(x,\xi)$ is called *Green's matrix* for the problem (11.4.1), (11.4.2) if the vector functions $\vec{G}_j = (G_{1,j}, G_{2,j}, G_{3,j})^t$ and the function $G_{4,j}$ are solutions of the problems

$$-\Delta_x \vec{G}_j(x,\xi) + \nabla_x G_{4,j}(x,\xi) = \delta(x-\xi)\,(\delta_{1,j}, \delta_{2,j}, \delta_{3,j})^t \quad \text{for } x,\xi \in \mathcal{G},$$

$$-\nabla_x \cdot \vec{G}_j(x,\xi) = \big(\delta(x-\xi) - (\text{meas}(\mathcal{G}))^{-1}\big)\,\delta_{4,j} \quad \text{for } x,\xi \in \mathcal{G},$$

$$\vec{G}_j(x,\xi) = 0 \ \text{ for } x \in \partial\mathcal{G}, \ \xi \in \mathcal{G}$$

and $G_{4,j}$ satisfies the condition

$$\int_{\mathcal{G}} G_{4,j}(x,\xi)\, dx = 0 \text{ for } \xi \in \mathcal{G}, \ j = 1,2,3,4.$$

THEOREM 11.4.1. *There exists a uniquely determined Green's matrix* $G(x,\xi)$ *such that the vector functions* $x \to \zeta(x,\xi)\,\big(\vec{G}_j(x,\xi), G_{4,j}(x,\xi)\big)$ *belong to the space* $\overset{\circ}{W}{}^{1,2}(\mathcal{G})^3 \times L_2(\mathcal{G})$ *for each* $\xi \in \mathcal{G}$ *and for every infinitely differentiable function* $\zeta(\cdot,\xi)$ *on* $\overline{\mathcal{G}}$ *equal to zero in a neighborhood of the point* $x = \xi$. *Furthermore,*

$$G_{i,j}(x,\xi) = G_{j,i}(\xi,x)$$

for $x,\xi \in \mathcal{G}$ *and* $i,j = 1,2,3,4$.

P r o o f. The first assertion can be easily deduced from Theorem 11.1.2. By the same arguments as in the proof of Theorem 9.6.1, the *Green's formula*

$$\int_{\mathcal{G}} (-\Delta u + \nabla p) \cdot v\, dx - \int_{\mathcal{G}} (\nabla \cdot u)\, q\, dx = \int_{\mathcal{G}} u \cdot (-\Delta v + \nabla q)\, dx - \int_{\mathcal{G}} p\, \nabla \cdot v\, dx$$

is valid for the functions $u(x) = \vec{G}_i(x,y)$, $p(x) = G_{4,i}(x,y)$, $v(x) = \vec{G}_j(x,z)$ and $q(x) = G_{4,j}(x,z)$. Thus, $G_{i,j}(y,z) = G_{j,i}(z,y)$ for $i,j = 1,2,3,4$. This proves the theorem. □

As a consequence of the last theorem, the vector functions $(\vec{H}_i, G_{i,4})$, where $\vec{H}_i = (G_{i,1}, G_{i,2}, G_{i,3})^t$, are solutions of the problems

$$-\Delta_\xi \vec{H}_i(x,\xi) + \nabla_\xi G_{i,4}(x,\xi) = \delta(x-\xi)\,(\delta_{i,1},\delta_{i,2},\delta_{i,3})^t \quad \text{for } x,\xi \in \mathcal{G},$$
$$-\nabla_\xi \cdot \vec{H}_i(x,\xi) = \big(\delta(x-\xi) - (\text{meas}(\mathcal{G}))^{-1}\big)\,\delta_{i,4} \quad \text{for } x,\xi \in \mathcal{G},$$
$$\vec{H}_i(x,\xi) = 0 \ \text{ for } x \in \mathcal{G},\ \xi \in \partial\mathcal{G}$$

for $i = 1,2,3,4$. Furthermore, $\int_{\mathcal{G}} G_{i,4}(x,\xi)\,d\xi = 0$ for $x \in \mathcal{G}$, $i = 1,2,3,4$.

If $f \in C_0^\infty(\overline{\mathcal{G}}\backslash\mathcal{S})^3$, $g \in C_0^\infty(\overline{\mathcal{G}}\backslash\mathcal{S})$ and $h \in C_0^\infty(\partial\mathcal{G}\backslash\mathcal{S})^3$, then the vector-function (u,p) with the components

$$\begin{aligned} u_i(x) &= \int_{\mathcal{G}} f(\xi)\cdot \vec{H}_i(x,\xi)\,d\xi + \int_{\mathcal{G}} g(\xi)\,G_{i,4}(x,\xi)\,d\xi \\ &\quad + \int_{\partial\mathcal{G}\backslash\mathcal{S}} h(\xi)\cdot\Big(-\frac{\partial \vec{H}_i(x,\xi)}{\partial n_\xi} + G_{i,4}(x,\xi)\,n_\xi\Big)\,d\xi, \end{aligned}$$

$i = 1,2,3$, and

$$\begin{aligned} p(x) &= \int_{\mathcal{G}} f(\xi)\cdot \vec{H}_4(x,\xi)\,d\xi + \int_{\mathcal{G}} g(\xi)\,G_{4,4}(x,\xi)\,d\xi \\ &\quad + \int_{\partial\mathcal{G}\backslash\mathcal{S}} h(\xi)\cdot\Big(-\frac{\partial \vec{H}_4(x,\xi)}{\partial n_\xi} + G_{4,4}(x,\xi)\,n_\xi\Big)\,d\xi, \end{aligned}$$

is the uniquely determined solution of the problem

$$-\Delta u + \nabla p = f, \quad -\nabla\cdot u = g - C \ \text{ in } \mathcal{G},$$
$$u = h \ \text{ on } \partial\mathcal{G}\backslash\mathcal{S}, \quad \int_{\mathcal{G}} p(x)\,dx = 0,$$

where

$$C = (\text{meas}(\mathcal{G}))^{-1}\Big(\int_{\mathcal{G}} g(x)\,dx + \int_{\partial\mathcal{G}\backslash\mathcal{S}} h(x)\cdot n\,dx\Big).$$

The formulas for u and p result directly from *Green's formula*

$$\begin{aligned} &\int_{\mathcal{G}} (-\Delta u + \nabla p)\cdot v\,dx - \int_{\mathcal{G}} (\nabla\cdot u)\,q\,dx + \int_{\partial\mathcal{G}\backslash\mathcal{S}} \Big(\frac{\partial u}{\partial n} - pn\Big)\cdot v\,dx \\ &= \int_{\mathcal{G}} u\cdot(-\Delta v + \nabla q)\,dx - \int_{\mathcal{G}} p\,\nabla\cdot v\,dx + \int_{\partial\mathcal{G}\backslash\mathcal{S}} u\cdot\Big(\frac{\partial v}{\partial n} - qn\Big)\,dx. \end{aligned}$$

11.4.2. Estimates for Green's matrix: the case where x and ξ lie in neighborhoods of different vertices. For $j = 1,\dots,d'$, let Λ_j^+ be the greatest real number such that the strip

$$-1/2 < \text{Re}\,\lambda < \Lambda_j^+$$

is free of eigenvalues of the pencil $\mathfrak{A}_j(\lambda)$. According to [**85**, Theorem 5.5.6], the numbers Λ_j^+ are positive. Since $\lambda = 1$ is always an eigenvalue of the pencils $\mathfrak{A}_j(\lambda)$ and $A_k(\lambda)$, we have $\Lambda_j^+ \le 1$ and $\delta_+^{(k)} \le 1$.

In the subsequent theorems, we give point estimates for the elements $G_{i,j}(x,\xi)$ of the Green's matrix. First, we consider the case where x and ξ lie in neighborhoods of different vertices.

THEOREM 11.4.2. *Let $\mathcal{U}_\mu$ and $\mathcal{U}_\nu$ be sufficiently small neighborhoods of the vertices $x^{(\mu)}$ and $x^{(\nu)}$, respectively. If $x \in \mathcal{G} \cap \mathcal{U}_\mu$ and $\xi \in \mathcal{G} \cap \mathcal{U}_\nu$, $\mu \neq \nu$, then*

$$\begin{aligned} \big|D_x^\alpha D_\xi^\gamma G_{i,j}(x,\xi)\big| \ &\le \ c_{\alpha,\gamma}\, \rho_\mu(x)^{\Lambda_\mu^+ - \delta_{i,4} - |\alpha| - \varepsilon}\, \rho_\nu(\xi)^{\Lambda_\nu^+ - \delta_{j,4} - |\gamma| - \varepsilon} \\ &\quad \times \prod_{k\in X_\mu} \Big(\frac{r_k(x)}{\rho_\mu(x)}\Big)^{\delta_+^{(k)} - \delta_{i,4} - |\alpha| - \varepsilon} \prod_{k\in X_\nu} \Big(\frac{r_k(\xi)}{\rho_\nu(\xi)}\Big)^{\delta_+^{(k)} - \delta_{j,4} - |\gamma| - \varepsilon}, \end{aligned}$$

where ε is an arbitrarily small positive number. The constant $c_{\alpha,\gamma}$ is independent of x and ξ.

P r o o f. Let ζ, η be smooth cut-off functions with sufficiently small supports such that $\zeta = 1$ on $\mathcal{U}_\nu$ and $\eta = 1$ in a neighborhood of $\operatorname{supp}\zeta$. Since

$$\eta(\xi)\,\big(-\Delta_\xi \vec{H}_i(x,\xi) + \nabla_\xi G_{i,4}(x,\xi)\big) = 0$$

and

$$\eta(\xi)\,\nabla_\xi \cdot \vec{H}_i(x,\xi) = \eta(\xi)\,(\operatorname{meas}(\mathcal{G}))^{-1}\,\delta_{i,4}$$

for $x \in \mathcal{G} \cap \mathcal{U}_\mu$ and $\xi \in \mathcal{G}$, we conclude from Theorem 11.1.8 that

$$\zeta(\cdot)\, D_x^\alpha \vec{H}_i(x,\cdot) \in V_{\beta,\delta}^{l,s}(\mathcal{G})^3, \quad \zeta(\cdot)\, D_x^\alpha G_{i,4}(x,\cdot) \in V_{\beta,\delta}^{l-1,s}(\mathcal{G}).$$

Furthermore, the inequality

$$\begin{aligned} &\|\zeta(\cdot)\, D_x^\alpha \vec{H}_i(x,\cdot)\|_{V_{\beta,\delta}^{l,s}(\mathcal{G})^3} + \|\zeta(\cdot)\, D_x^\alpha G_{i,4}(x,\cdot)\|_{V_{\beta,\delta}^{l-1,s}(\mathcal{G})} \\ &\qquad \le c\,\Big(\|\eta(\cdot) D_x^\alpha \vec{H}_i(x,\cdot)\|_{W^{1,2}(\mathcal{G})^3} + \|\eta(\cdot) D_x^\alpha G_{i,4}(x,\cdot)\|_{L_2(\mathcal{G})}\Big) \end{aligned}$$

holds with a constant independent of $x \in \mathcal{G}\cap\mathcal{U}_\mu$, where $-1-\Lambda_\nu^+ < l-\beta_\nu-3/s < \Lambda_\nu^+$ and $|\delta_k - l + 2/s| < \delta_+^{(k)}$. (Note that $\eta \in V_{\beta,\delta}^{l-1,s}(\mathcal{G})$ if $l - \beta_\nu - 3/s < 1$ and $l - \delta_k - 2/s < 1$.) Hence, Lemma 3.1.4 yields

$$\begin{aligned} (11.4.3)\quad & \rho_\nu(\xi)^{\beta_\nu - l + |\gamma| + 3/s} \prod_{k\in X_\nu} \Big(\frac{r_k(\xi)}{\rho_\nu(\xi)}\Big)^{\delta_k - l + |\gamma| + 3/s} \big|D_x^\alpha D_\xi^\gamma \vec{H}_i(x,\xi)\big| \\ &\quad + \rho_\nu(\xi)^{\beta_\nu - l + 1 + |\gamma| + 3/s} \prod_{k\in X_\nu} \Big(\frac{r_k(\xi)}{\rho_\nu(\xi)}\Big)^{\delta_k - l + 1 + |\gamma| + 3/s} \big|D_x^\alpha D_\xi^\gamma G_{i,4}(x,\xi)\big| \\ &\quad \le c\,\Big(\|\eta(\cdot) D_x^\alpha \vec{H}_i(x,\cdot)\|_{W^{1,2}(\mathcal{G})^3} + \|\eta(\cdot) D_x^\alpha G_{i,4}(x,\cdot)\|_{L_2(\mathcal{G})}\Big) \end{aligned}$$

for $\xi \in \mathcal{U}_\nu$. Let $f \in W^{-1,2}(\mathcal{G})^3$, $g \in L_2(\mathcal{G})$, $C = (\operatorname{meas}(\mathcal{G}))^{-1} \int_{\mathcal{G}} \eta g\, dx$, and let

$$\begin{aligned} u_i(y) &= \int_{\mathcal{G}} \vec{H}_i(y,z)\,\eta(z)\, f(z)\, dz + \int_{\mathcal{G}} G_{i,4}(y,z)\,\eta(z)\, g(z)\, dz, \quad i = 1,2,3, \\ p(y) &= \int_{\mathcal{G}} \vec{H}_4(y,z)\,\eta(z)\, f(z)\, dz + \int_{\mathcal{G}} G_{4,4}(y,z)\,\eta(z)\, g(z)\, dz \end{aligned}$$

be the components of the unique solution of the system

$$-\Delta u + \nabla p = \eta f, \quad -\nabla \cdot u = \eta g - C$$

in $\overset{\circ}{W}{}^{1,2}(\mathcal{G})^3 \times L_2(\mathcal{G})$ satisfying the condition

$$\int_{\mathcal{G}} p(y)\, dy = 0.$$

We denote by χ and ψ smooth cut-off functions such that $\chi = 1$ in $\mathcal{U}_\mu$ and $\psi = 1$ in a neighborhood of $\operatorname{supp}\chi$. Since

$$\psi\big(-\Delta u + \nabla p\big) = \psi\eta f = 0 \quad \text{and} \quad \psi\,\nabla\cdot u = C\psi,$$

we obtain (again by Theorem 11.1.8) the estimate

$$\begin{aligned} \|\chi u\|_{V^{l,s}_{\beta,\delta}(\mathcal{G})^3} + \|\chi p\|_{V^{l-1,s}_{\beta,\delta}(\mathcal{G})} &\le c\Big(\|\psi u\|_{W^{1,2}(\mathcal{G})^3} + \|\psi p\|_{L_2(\mathcal{G})}\Big) \\ &\le c\Big(\|f\|_{W^{-1,2}(\mathcal{G})^3} + \|g\|_{L_2(\mathcal{G})}\Big), \end{aligned}$$

where $-1-\Lambda_\mu^+ < l - \beta_\mu - 3/s < \Lambda_\mu^+$ and $|\delta_k - l + 2/s| < \delta_+^{(k)}$. Applying Lemma 3.1.4, we get

$$\begin{aligned} &\rho_\mu(x)^{\beta_\mu - l + |\alpha| + 3/s} \prod_{k\in X_\mu} \Big(\frac{r_k(x)}{\rho_\mu(x)}\Big)^{\delta_k - l + |\alpha| + 3/s} \big|D_x^\alpha u(x)\big| \\ &\quad + \rho_\mu(x)^{\beta_\mu - l + 1 + |\alpha| + 3/s} \prod_{k\in X_\mu} \Big(\frac{r_k(x)}{\rho_\mu(x)}\Big)^{\delta_k - l + 1 + |\alpha| + 3/s} \big|D_x^\alpha p(x)\big| \\ &\quad \le c\Big(\|f\|_{W^{-1,2}(\mathcal{G})^3} + \|g\|_{L_2(\mathcal{G})}\Big). \end{aligned}$$

Thus, the mapping

$$\begin{aligned} (f,g) \to{}& \rho_\mu(x)^{\beta_\mu - l + |\alpha| + 3/s} \prod_{k\in X_\mu} \Big(\frac{r_k(x)}{\rho_\mu(x)}\Big)^{\delta_k - l + |\alpha| + 3/s} D_x^\alpha u_i(x) \\ ={}& \rho_\mu(x)^{\beta_\mu - l + |\alpha| + 3/s} \prod_{k\in X_\mu} \Big(\frac{r_k(x)}{\rho_\mu(x)}\Big)^{\delta_k - l + |\alpha| + 3/s} \\ &\times \Big(\int_{\mathcal{G}} D_x^\alpha \vec{H}_i(x,z)\,\eta(z)\, f(z)\, dz + \int_{\mathcal{G}} D_x^\alpha G_{i,4}(x,z)\,\eta(z)\, g(z)\, dz\Big) \end{aligned}$$

is continuous from $W^{-1,2}(\mathcal{G})^3 \times L_2(\mathcal{G})$ into $\mathbb{C}$ for every $x \in \mathcal{G}\cap\mathcal{U}_\mu$, $i = 1,2,3$, and its norm can be estimated by a constant independent of x. Therefore,

$$\begin{aligned} &\|\eta(\cdot)\, D_x^\alpha \vec{H}_i(x,\cdot)\|_{W^{1,2}(\mathcal{G})^3} + \|\eta(\cdot)\, D_x^\alpha G_{i,4}(x,\cdot)\|_{L_2(\mathcal{G})} \\ &\quad \le c\,\rho_\mu(x)^{l - |\alpha| - \beta_\mu - 3/s} \prod_{k\in X_\mu} \Big(\frac{r_k(x)}{\rho_\mu(x)}\Big)^{l - |\alpha| - \delta_k - 3/s} \end{aligned}$$

for $i = 1,2,3$. Analogously,

$$\begin{aligned} &\|\eta(\cdot)\, D_x^\alpha \vec{H}_4(x,\cdot)\|_{W^{1,2}(\mathcal{G})^3} + \|\eta(\cdot)\, D_x^\alpha G_{4,4}(x,\cdot)\|_{L_2(\mathcal{G})} \\ &\quad \le c\,\rho_\mu(x)^{l - 1 - |\alpha| - \beta_\mu - 3/s} \prod_{k\in X_\mu} \Big(\frac{r_k(x)}{\rho_\mu(x)}\Big)^{l - 1 - |\alpha| - \delta_k - 3/s}. \end{aligned}$$

This together with (11.4.3) implies

$$\big|D_x^\alpha D_\xi^\gamma G_{i,j}(x,\xi)\big| \le c\,\rho_\mu(x)^{l-|\alpha|-\beta_\mu-\delta_{i,4}-3/s}\;\rho_\nu(\xi)^{l-|\gamma|-\beta_\nu-\delta_{j,4}-3/s}$$
$$\times \prod_{k\in X_\mu}\Big(\frac{r_k(x)}{\rho_\mu(x)}\Big)^{l-|\alpha|-\delta_k-\delta_{i,4}-3/s} \prod_{k\in X_\nu}\Big(\frac{r_k(\xi)}{\rho_\nu(\xi)}\Big)^{l-|\gamma|-\delta_k-\delta_{j,4}-3/s}$$

for $x\in\mathcal{G}\cap\mathcal{U}_\mu$, $\xi\in\mathcal{G}\cap\mathcal{U}_\nu$. We can choose $\beta_\mu,\beta_\nu,\delta_k$ and s such that

$$l-\beta_\mu-3/s=\Lambda_\mu^+-\varepsilon,\quad l-\beta_\nu-3/s=\Lambda_\nu^+-\varepsilon,\quad \delta_k-l+3/s=-\delta_+^{(k)}+\varepsilon.$$

This proves the theorem. □

11.4.3. Two auxiliary estimates for solutions of the Stokes system. We assume now that x and ξ lie in a neighborhood $\mathcal{U}_\nu$ of the same vertex $x^{(\nu)}$. In order to estimate $G(x,\xi)$ for this case, we prove the subsequent two lemmas. The first one is an analog to Lemma 5.1.5.

LEMMA 11.4.3. *Let η_ε be an infinitely differentiable function such that $\eta_\varepsilon(x)=1$ for $c_1\varepsilon<\rho_\nu(x)<c_2\varepsilon$, $\eta_\varepsilon(x)=0$ for $\rho_\nu(x)>c_3\varepsilon$, where ε is a sufficiently small positive number. Suppose that $-1-\Lambda_\nu^+<l-\beta_\nu-3/s<\Lambda_\nu^+$ and that the components of δ satisfy the inequalities $|\delta_k-l+2/s|<\delta_+^{(k)}$. If $\eta_\varepsilon(u,p)\in \overset{\circ}{W}{}^{1,2}(\mathcal{G})^3\times L_2(\mathcal{G})$, $\eta_\varepsilon(-\Delta u+\nabla p)=0$, $\eta_\varepsilon\nabla\cdot u=C\,\eta_\varepsilon$ with a constant C, then*

$$\|\zeta_\varepsilon u\|_{V^{l,s}_{\beta,\delta}(\mathcal{G})^3}+\|\zeta_\varepsilon p\|_{V^{l-1,s}_{\beta,\delta}(\mathcal{G})}\le c\,\varepsilon^{\beta_\nu-l+3/s-1/2}\Big(\|\eta_\varepsilon u\|_{W^{1,2}(\mathcal{G})^3}+\|\eta_\varepsilon p\|_{L_2(\mathcal{G})}\Big).$$

Here c is a constant independent of u,p,ε and ζ_ε is an infinitely differentiable function such that $\zeta_\varepsilon(x)=1$ for $c_1'\varepsilon<\rho_j(x)<c_2'\varepsilon$, $0\le c_1\le c_1'<c_2'<c_2$, $\zeta_\varepsilon(x)=0$ for $\rho_j(x)<c_1\varepsilon$ and $\rho_j(x)>c_2\varepsilon$. (The case $c_1=c_1'=0$ is included in this lemma.)

P r o o f. By our conditions on $\mathcal{G}$, there exists a diffeomorphism κ mapping $\mathcal{U}_\nu\cap\mathcal{G}$ onto $\mathcal{K}\cap\mathcal{V}_\nu$, where $\mathcal{K}$ is a cone with vertex at the origin and $\mathcal{V}_\nu$ is a neighborhood of the origin. In the new coordinates $y=\kappa(x)$, the equations $\eta_\varepsilon(-\Delta u+\nabla p)=0$ and $\eta_\varepsilon\nabla\cdot u=C\,\eta_\varepsilon$ take the form

$$\eta_\varepsilon(y)\,\big(L(y,D_y)\,u(y)+A(y)\,\nabla_y p(y)\big)=0,\quad \eta_\varepsilon(y)\,A(y)\nabla_y\cdot u(y)=C\eta_\varepsilon,$$

where $L(y,D_y)$ is a linear second order differential operator and $A(y)$ is a matrix such that $L^\circ(0,D_y)=-\Delta_y$ and $A(0)$ is the identity matrix. Here, we wrote $u(y),p(y),\eta_\varepsilon(y)$ instead of $u(\kappa^{-1}(y)),p(\kappa^{-1}(y))$ and $\eta_\varepsilon(\kappa^{-1}(y))$, respectively. Consequently, the functions

$$v(y)=u\big(\kappa^{-1}(\varepsilon y)\big)\quad\text{and}\quad q(y)=\varepsilon\,p\big(\kappa^{-1}(\varepsilon y)\big)$$

satisfy the equations

$$\tilde\eta_\varepsilon\,\Big(\varepsilon^2\,L(\varepsilon y,\varepsilon^{-1}D_y)\,v(y)+A(\varepsilon y)\,\nabla_y q(y)\Big)=0,\quad \tilde\eta_\varepsilon\,A(\varepsilon y)\nabla_y\cdot v(y)=C\varepsilon\tilde\eta_\varepsilon\,,$$

where the function $\tilde\eta_\varepsilon(y)=\eta_\varepsilon\big(\kappa^{-1}(\varepsilon y)\big)$ is equal to one for $c_1\varepsilon<\big|\kappa^{-1}(\varepsilon y)\big|<c_2\varepsilon$. Let $\tilde\zeta$ be an infinitely differentiable function such that $\tilde\eta_\varepsilon=1$ in a neighborhood of $\operatorname{supp}\tilde\zeta$. Applying Theorem 11.1.8, we obtain

$$\|\tilde\zeta v\|_{V^{l,s}_{\beta_\nu,\delta}(\mathcal{K})^3}+\|\tilde\zeta q\|_{V^{l-1,s}_{\beta_\nu,\delta}(\mathcal{K})}\le c\,\Big(\|\tilde\eta_\varepsilon v\|_{W^{1,2}(\mathcal{K})^3}+\|\tilde\eta_\varepsilon q\|_{L_2(\mathcal{K})}\Big).\tag{11.4.4}$$

Since the operator $\tilde{\eta}_\varepsilon\big(\varepsilon^2 L(\varepsilon y,\varepsilon^{-1}D_y) + \Delta_y\big)$ is small in the operator norm $V^{l,s}_{\beta_\nu,\delta}(\mathcal{K}) \to V^{l-2,s}_{\beta_\nu,\delta}(\mathcal{K})$ and the operator $\tilde{\eta}_\varepsilon\big(A(\varepsilon y) - I\big)\nabla_y$ is small in the operator norm $V^{l,s}_{\beta_\nu,\delta}(\mathcal{K}) \to V^{l-1,s}_{\beta_\nu,\delta}(\mathcal{K})^3$ for small ε, the constant c in the last inequality can be chosen independent of ε. Inequality (11.4.4) together with the equalities

$$\|\tilde{\zeta} v\|_{V^{l,s}_{\beta_\nu,\delta}(\mathcal{K})^3} + \|\tilde{\zeta} q\|_{V^{l-1,s}_{\beta_\nu,\delta}(\mathcal{K})} = \varepsilon^{l-\beta_\nu-3/s}\Big(\|\zeta_\varepsilon u\|_{V^{l,s}_{\beta_\nu,\delta}(\mathcal{K})^3} + \|\zeta_\varepsilon p\|_{V^{l-1,s}_{\beta_\nu,\delta}(\mathcal{K})}\Big),$$

where $\zeta_\varepsilon(x) = \tilde{\zeta}(x/\varepsilon)$, and

$$\|\tilde{\eta}_\varepsilon v\|_{W^{1,2}(\mathcal{K})^3} + \|\tilde{\eta}_\varepsilon q\|_{L_2(\mathcal{K})} = \varepsilon^{-1/2}\Big(\|\eta_\varepsilon u\|_{W^{1,2}(\mathcal{K})^3} + \|\eta_\varepsilon p\|_{L_2(\mathcal{K})}\Big)$$

yields

$$\begin{aligned}\|\zeta_\varepsilon u\|_{V^{l,s}_{\beta_\nu,\delta}(\mathcal{K})^3} &+ \|\zeta_\varepsilon p\|_{V^{l-1,s}_{\beta_\nu,\delta}(\mathcal{K})} \\ &\le c\,\varepsilon^{-1/2-l+\beta_\nu+3/s}\Big(\|\eta_\varepsilon u\|_{W^{1,2}(\mathcal{K})^3} + \|\eta_\varepsilon q\|_{L_2(\mathcal{K})}\Big).\end{aligned}$$

Returning to the coordinates $x = \kappa^{-1}(y)$, we get the desired inequality in the domain $\mathcal{G}$. □

Furthermore, we need the following analog to Lemma 2.9.11.

LEMMA 11.4.4. *Let $\mathcal{B}_R(x_0)$ be the ball with radius R centered at the point $x_0 \in \mathcal{G}$. We assume that $x^{(\nu)}$ is the nearest vertex to x_0, M_k is the nearest edge to x_0, that $R \le \varepsilon_0\rho_\nu(x_0)$ and $r_k(x_0) \le 6R/\varepsilon_0$ with a sufficiently small constant ε_0. Furthermore, let η be an infinitely differentiable cut-off function with support in $\mathcal{B}_R(x_0)$ which is equal to one in the ball $\mathcal{B}_{R/2}(x_0)$. If $\eta(u,p) \in \overset{\circ}{W}{}^{1,2}(\mathcal{G})^3 \times L_2(\mathcal{G})$, $\eta(-\Delta u + \nabla p) = 0$ and $\eta\nabla\cdot u = C\eta$ with a certain constant C, then*

$$\begin{aligned}(11.4.5)\qquad r_k(x_0)^{|\alpha|-\delta_+^{(k)}+\varepsilon}&\Big(\big|(D_x^\alpha u)(x_0)\big| + r(x_0)\,\big|(D_x^\alpha p)(x_0)\big|\Big) \\ &\le c\,R^{\varepsilon-\delta_+^{(k)}-1/2}\Big(\|\eta u\|_{W^{1,2}(\mathcal{G})^3} + \|\eta p\|_{L_2(\mathcal{G})}\Big).\end{aligned}$$

Here ε is an arbitrarily small positive real number and c is independent of u, p, x_0 and R.

P r o o f. Suppose first that $r_k(x_0) \le 2R$. If $R \le \varepsilon_0\rho_\nu(x_0)$ with a sufficiently small ε_0, then there exists a diffeomorphism κ mapping $\mathcal{G}\cap\mathcal{B}_R(x_0)$ onto the intersection of a dihedron $\mathcal{D}$ with a neighborhood $\mathcal{U}$ of the point $y_0 = \kappa(x_0)$. We may assume without loss of generality that $\kappa(\mathcal{B}_R(x_0)) \subset \mathcal{B}_{2R}(y_0)$, $\kappa(\mathcal{B}_{R/2}(x_0)) \supset \mathcal{B}_{R/4}(y_0)$, that the nearest edge point to y_0 is the origin, the Jacobian matrix κ' is the identity matrix at $\kappa^{-1}(0)$, and that $r(y_0) \le 2c_0R$. In the new coordinates $y = \kappa(x)$, the equations $\eta(-\Delta u + \nabla p) = 0$ and $\eta\nabla\cdot u = C\eta$ take the form

$$\eta\,\big(L(y,D_y)\,u + A(y)\,\nabla_y p\big) = 0, \qquad \eta\,A(y)\,\nabla_y\cdot u = C\eta,$$

where, as in the proof of Lemma 11.4.3, $L(y,D_y)$ is a linear second order differential operator and $A(y)$ is a matrix such that $L^\circ(0,D_y) = -\Delta_y$ and $A(0)$ is the identity matrix. Here, we wrote $u(y), p(y), \eta(y)$ instead of $u(\kappa^{-1}(y)), p(\kappa^{-1}(y))$ and $\eta(\kappa^{-1}(y))$, respectively. We set

$$z_0 = \frac{y_0}{R},\quad v(z) = u(\kappa^{-1}(Rz)),\quad q(z) = R\,p(\kappa^{-1}(Rz)),\quad \zeta(z) = \eta(\kappa^{-1}(Rz)).$$

Then $r(z_0) \le 2c_0$, $\zeta(z) = 1$ for $|z - z_0| < 1/4$ and $\zeta(z) = 0$ for $|z - z_0| > 2$. Furthermore, the vector function (v, q) satisfies the equations

$$\zeta(z)\,\big(R^2\,L(Rz, R^{-1}D_z)\,v(z) + A(Rz)\,\nabla_z q(z)\big) = 0,$$
$$\zeta(z)\,A(Rz)\,\nabla_z \cdot v(z) = C\,R\,\zeta(z).$$

Let χ be a smooth cut-off function equal to one near z_0 and to zero outside the ball $|z - z_0| < 1/4$. Then it follows from Theorem 11.1.8 that $\chi v \in V_\delta^{l,s}(\mathcal{D})^3$ and $\chi q \in V_\delta^{l-1,s}(\mathcal{D})$, where $l - \delta_+^{(k)} < \delta + 2/s < l + \delta_+^{(k)}$. Moreover,

$$\|\chi v\|_{V_\delta^{l,s}(\mathcal{D})^3} + \|\chi q\|_{V_\delta^{l-1,s}(\mathcal{D})} \le c\,\big(\|\zeta v\|_{V_0^{1,2}(\mathcal{D})^3} + \|\zeta q\|_{L_2(\mathcal{D})}\big).$$

Since the operator $\zeta\,\big(R^2\,L(Rz, R^{-1}D_z) + \Delta_z\big)$ is small in the operator norm $V_\delta^{l,s}(\mathcal{D}) \to V_\delta^{l-2,s}(\mathcal{D})$ and the operator $\zeta\,\big(A(Rz) - I\big)\nabla_z$ is small in the operator norm $V_\delta^{l,s}(\mathcal{D}) \to V_\delta^{l-1,s}(\mathcal{D})^3$ for small R, the constant c in the last inequality can be assumed to be independent of R (cf. Lemma 2.9.10). Applying Lemma 2.1.3, we get

$$\begin{aligned} r(z_0)^{\delta-l+|\alpha|+3/s}\,\big|(D_z^\alpha v)(z_0)\big| + r(z_0)^{\delta-l+1+|\alpha|+3/s}\,\big|(D_z^\alpha q)(z_0)\big| \\ \le c\,\big(\|\zeta v\|_{V_0^{1,2}(\mathcal{D})^3} + \|\zeta q\|_{L_2(\mathcal{D})}\big) \end{aligned} \tag{11.4.6}$$

for $|\alpha| < l - 1 - 3/s$. We can choose δ, l and s sucht that $\delta - l + 3/s = \varepsilon - \delta_+^{(k)}$ with an arbitrary fixed $\varepsilon > 0$. Using the inequalities

$$\|\zeta v\|_{V_0^{1,2}(\mathcal{D})^3} + \|\zeta q\|_{L_2(\mathcal{D})} \le c\,R^{-1/2}\,\big(\|\eta u\|_{W^{1,2}(\mathcal{G})^3} + \|\eta p\|_{L_2(\mathcal{G})}\big),$$
$$\big|(D_x^\alpha u)(x_0)\big| \le c\,R^{-|\alpha|}\,\big|(D_z^\alpha v)(z_0)\big|, \quad \big|(D_x^\alpha p)(x_0)\big| \le c\,R^{-|\alpha|-1}\,\big|(D_z^\alpha q)(z_0)\big|,$$

we obtain

$$\begin{aligned} r(y_0)^{|\alpha|-\delta_+^{(k)}+\varepsilon}\,\big|(D_x^\alpha u)(x_0)\big| + r(y_0)^{|\alpha|+1-\delta_+^{(k)}+\varepsilon}\,\big|(D_x^\alpha p)(x_0)\big| \\ \le c\,R^{\varepsilon-\delta_+^{(k)}-1/2}\,\big(\|\eta u\|_{W^{1,2}(\mathcal{G})^3} + \|\eta p\|_{L_2(\mathcal{G})}\big). \end{aligned}$$

This proves (11.4.5) for the case $r_k(x_0) \le 2R$.

Suppose now that $2R < r_k(x_0) < 6R/\varepsilon_0$. Then there exists a diffeomorphism κ mapping $\mathcal{G} \cap \mathcal{B}_R(x_0)$ onto the intersection of a half-space with a neighborhood $\mathcal{U}$ of the point $y_0 = \kappa(x_0)$. Thus, we obtain the inequality

$$\big|(D_z^\alpha v)(z_0)\big| + \big|(D_z^\alpha q)(z_0)\big| \le c\,\big(\|\zeta v\|_{V_0^{1,2}(\mathcal{D})^3} + \|\zeta q\|_{L_2(\mathcal{D})}\big)$$

instead of (11.4.6). Consequently,

$$R^{|\alpha|}\,\big|(D_x^\alpha u)(x_0)\big| + R^{|\alpha|+1}\,\big|(D_x^\alpha p)(x_0)\big| \le c\,R^{-1/2}\,\big(\|\eta u\|_{W^{1,2}(\mathcal{G})^3} + \|\eta p\|_{L_2(\mathcal{G})}\big).$$

Since $2R < r_k(x_0) < 6R/\varepsilon_0$, this implies (11.4.5). The proof of the lemma is complete. □

11.4.4. Estimates for Green's matrix: the case where x and ξ lie in a neighborhood of the same vertex. In this subsection, we assume that x and ξ lie in a neighborhood $\mathcal{U}_\nu$ of the same vertex $x^{(\nu)}$. First, we consider the cases $\rho_\nu(\xi) < \rho_\nu(x)/2$ and $\rho_\nu(\xi) > 2\rho_\nu(x)$.

THEOREM 11.4.5. *Let $\mathcal{U}_\nu$ be a sufficiently small neighborhood of the vertex $x^{(\nu)}$. If $x, \xi \in \mathcal{G} \cap \mathcal{U}_\nu$ and $\rho_\nu(\xi) < \rho_\nu(x)/2$, then*

$$\begin{aligned} \big|D_x^\alpha D_\xi^\gamma G_{i,j}(x,\xi)\big| \ & \le c\, \rho_\nu(x)^{-1-\Lambda_\nu^+ -\delta_{i,4}-|\alpha|+\varepsilon}\, \rho_\nu(\xi)^{\Lambda_\nu^+ -\delta_{j,4}-|\gamma|-\varepsilon} \\ & \quad \times \prod_{k\in X_\nu} \Big(\frac{r_k(x)}{\rho_\nu(x)}\Big)^{\delta_+^{(k)}-\delta_{i,4}-|\alpha|-\varepsilon} \prod_{k\in X_\nu} \Big(\frac{r_k(\xi)}{\rho_\nu(\xi)}\Big)^{\delta_+^{(k)}-\delta_{j,4}-|\gamma|-\varepsilon}, \end{aligned}$$

where X_ν is the set of all indices k such that $x^{(\nu)}$ is an end point of the edge M_k and ε is an arbitrarily small positive number. Analogously, the estimate

$$\begin{aligned} \big|D_x^\alpha D_\xi^\gamma G_{i,j}(x,\xi)\big| \ & \le c\, \rho_\nu(x)^{\Lambda_\nu^+ -\delta_{i,4}-|\alpha|-\varepsilon}\, \rho_\nu(\xi)^{-1-\Lambda_\nu^+ -\delta_{j,4}-|\gamma|+\varepsilon} \\ & \quad \times \prod_{k\in X_\nu} \Big(\frac{r_k(x)}{\rho_\nu(x)}\Big)^{\delta_+^{(k)}-\delta_{i,4}-|\alpha|-\varepsilon} \prod_{k\in X_\nu} \Big(\frac{r_k(\xi)}{\rho_\nu(\xi)}\Big)^{\delta_+^{(k)}-\delta_{j,4}-|\gamma|-\varepsilon} \end{aligned}$$

holds for $x, \xi \in \mathcal{U}_\nu$, $\rho_\nu(\xi) > 2\rho_\nu(x)$.

Proof. Since $G_{i,j}(x,\xi) = G_{j,i}(\xi,x)$, it suffices to prove the first assertion. Let $x \in \mathcal{G}\cap\mathcal{U}_\nu$ and let ζ, η be smooth functions such that $\zeta(\xi) = 1$ for $\rho_\nu(\xi) < \rho_\nu(x)/2$, $\eta = 1$ in a neighborhood of $\operatorname{supp}\zeta$, and $\eta(\xi) = 0$ for $\rho_\nu(\xi) > 3\rho_\nu(x)/4$. Since

$$\eta(\xi)\,\big(-\Delta_\xi \vec{H}_i(x,\xi) + \nabla_\xi G_{i,4}(x,\xi)\big) = 0$$

and

$$\eta(\xi)\,\nabla_\xi \cdot \vec{H}_i(x,\xi) = \eta(\xi)\,(\text{meas}(\mathcal{G}))^{-1}\,\delta_{i,4}\,,$$

it follows from Lemma 11.4.3 that

$$\begin{aligned} & \|\zeta(\cdot)\, D_x^\alpha \vec{H}_i(x,\cdot)\|_{V_{\beta,\delta}^{l,s}(\mathcal{G})^3} + \|\zeta(\cdot)\, D_x^\alpha G_{i,4}(x,\cdot)\|_{V_{\beta,\delta}^{l-1,s}(\mathcal{G})} \\ & \quad \le c\, \rho_\nu(x)^{\beta_\nu - l + 3/s - 1/2} \Big(\|\eta(\cdot) D_x^\alpha \vec{H}_i(x,\cdot)\|_{W^{1,2}(\mathcal{G})^3} + \|\eta(\cdot) D_x^\alpha G_{i,4}(x,\cdot)\|_{L_2(\mathcal{G})}\Big), \end{aligned}$$

where $-1-\Lambda_\nu^+ < l - \beta_\nu - 3/s < \Lambda_\nu^+$ and $|\delta_k - l + 2/s| < \delta_+^{(k)}$. Applying Lemma 3.1.4, we obtain

$$\begin{aligned} & \rho_\nu(\xi)^{\beta_\nu - l + |\gamma| + 3/s} \prod_{k\in X_\nu} \Big(\frac{r_k(\xi)}{\rho_\nu(\xi)}\Big)^{\delta_k - l + |\gamma| + 3/s} \big|D_x^\alpha D_\xi^\gamma \vec{H}_i(x,\xi)\big| \\ & \quad + \rho_\nu(\xi)^{\beta_\nu - l + 1 + |\gamma| + 3/s} \prod_{k\in X_\nu} \Big(\frac{r_k(\xi)}{\rho_\nu(\xi)}\Big)^{\delta_k - l + 1 + |\gamma| + 3/s} \big|D_x^\alpha D_\xi^\gamma G_{i,4}(x,\xi)\big| \\ & \quad \le c\, \rho_\nu(x)^{\beta_\nu - l + 3/s - 1/2} \Big(\|\eta(\cdot) D_x^\alpha \vec{H}_i(x,\cdot)\|_{W^{1,2}(\mathcal{G})^3} + \|\eta(\cdot) D_x^\alpha G_{i,4}(x,\cdot)\|_{L_2(\mathcal{G})}\Big) \end{aligned}$$

for $\rho_\nu(\xi) < \rho_\nu(x)/2$. Let f, g, u and p be the same functions as in the proof of Theorem 11.4.2. Furthermore, we denote by ψ a smooth cut-off function such that $\psi(y) = 1$ for $|\rho_\nu(x) - \rho_\nu(y)| < \rho_\nu(x)/8$ and $\psi(y) = 0$ for $|\rho_\nu(x) - \rho_\nu(y)| > \rho_\nu(x)/4$. Since

$$\psi\,\big(-\Delta u + \nabla p\big) = \psi \eta f = 0 \quad \text{and} \quad \psi\,\nabla\cdot u = C\psi,$$

we obtain (again by Lemmas 11.4.3 and 3.1.4) the estimate

$$\rho_\nu(x)^{|\alpha|+1/2} \prod_{k\in X_\nu} \Big(\frac{r_k(x)}{\rho_\nu(x)}\Big)^{\delta_k-l+|\alpha|+3/s} \big|D_x^\alpha u(x)\big|$$
$$+\rho_\nu(x)^{|\alpha|+3/2} \prod_{k\in X_\nu} \Big(\frac{r_k(x)}{\rho_\nu(x)}\Big)^{\delta_k-l+1+|\alpha|+3/s} \big|D_x^\alpha p(x)\big|$$
$$\le c\Big(\|\psi u\|_{W^{1,2}(\mathcal{G})^3} + \|\psi p\|_{L_2(\mathcal{G})}\Big) \le c\Big(\|f\|_{W^{-1,2}(\mathcal{G})^3} + \|g\|_{L_2(\mathcal{G})}\Big).$$

Arguing as in the proof of Theorem 11.4.2, we conclude that

$$\|\eta(\cdot)\, D_x^\alpha \vec{H}_i(x,\cdot)\|_{W^{1,2}(\mathcal{G})^3} + \|\eta(\cdot)\, D_x^\alpha G_{i,4}(x,\cdot)\|_{L_2(\mathcal{G})}$$
$$\le c\,\rho_\nu(x)^{-|\alpha|-\delta_{i,4}-1/2} \prod_{k\in X_\nu} \Big(\frac{r_k(x)}{\rho_\nu(x)}\Big)^{l-|\alpha|-\delta_k-\delta_{i,4}-3/s}.$$

This implies

$$\big|D_x^\alpha D_\xi^\gamma G_{i,j}(x,\xi)\big| \le c\,\rho_\nu(x)^{\beta_\nu-l-1-|\alpha|-\delta_{i,4}+3/s}\; \rho_\nu(\xi)^{l-|\gamma|-\beta_\nu-\delta_{j,4}-3/s}$$
$$\times \prod_{k\in X_\nu} \Big(\frac{r_k(x)}{\rho_\nu(x)}\Big)^{l-|\alpha|-\delta_k-\delta_{i,4}-3/s} \prod_{k\in X_\nu} \Big(\frac{r_k(\xi)}{\rho_\nu(\xi)}\Big)^{l-|\gamma|-\delta_k-\delta_{j,4}-3/s}.$$

Here we can choose β_ν, δ_k and s such that $l-\beta_\nu-3/s = \Lambda_\nu^+ - \varepsilon$ and $l-\delta_k-3/s = \delta_+^{(k)} - \varepsilon$. This proves the theorem. □

Next, we consider the case where x and ξ lie in a neighborhood $\mathcal{U}_\nu$ of the same vertex $x^{(\nu)}$ and $\rho_\nu(x)/2 < \rho_\nu(\xi) < 2\rho_\nu(x)$. Then the following theorem can be proved similarly to Theorems 11.4.2 and 11.4.5.

THEOREM 11.4.6. *Let $\mathcal{U}_\nu$ be a sufficiently small neighborhood of the vertex $x^{(\nu)}$. If $x,\xi \in \mathcal{G}\cap\mathcal{U}_\nu$, $\rho_\nu(x)/2 < \rho_\nu(\xi) < 2\rho_\nu(x)$ and $|x-\xi| > \min(r(x), r(\xi))$, then*

$$\big|D_x^\alpha D_\xi^\gamma G_{i,j}(x,\xi)\big| \le c_{\alpha,\gamma} |x-\xi|^{-1-\delta_{i,4}-\delta_{j,4}-|\alpha|-|\gamma|} \Big(\frac{r(x)}{|x-\xi|}\Big)^{\delta_+^{(k(x))}-\delta_{i,4}-|\alpha|-\varepsilon}$$
$$\times \Big(\frac{r(\xi)}{|x-\xi|}\Big)^{\delta_+^{(k(\xi))}-\delta_{j,4}-|\gamma|-\varepsilon},$$

where $k(x)$ denotes the smallest integer k such that $r_k(x) = r(x)$ and ε is an arbitrarily small positive number.

P r o o f. Let $\mathcal{B}(x)$ and $\mathcal{B}(\xi)$ be balls with the radius $R = c_0|x-\xi|$ centered at x and ξ, respectively, where c_0 is a sufficiently small constant, $c_0 < 1/2$. Furthermore, let ζ and η be smooth cut-off functions with supports in $\mathcal{B}(x)$ and $\mathcal{B}(\xi)$, respectively, which are equal to one in balls with radius $c_0|x-\xi|/2$ centered at x and ξ, respectively. We assume that $|D^\alpha\zeta| + |D^\alpha\eta| \le c_\alpha\, |x-\xi|^{-|\alpha|}$ for every multi-index α, where c_α is a constant depending only on α. Obviously,

$$-\Delta_y \vec{H}_i(x,y) + \nabla_y G_{i,4}(x,y) = 0 \quad\text{and}\quad \nabla_y\cdot \vec{H}_i(x,y) = \eta(y)\,(\text{meas}(\mathcal{G}))^{-1}\,\delta_{i,4},$$

for $y \in \mathcal{G} \cap \mathcal{B}(\xi)$. Suppose that M_k is the nearest edge to ξ. Since $R \le 3c_0\rho_\nu(\xi)$ and $r_k(\xi) \le 2|x-\xi| = 2R/c_0$, we can apply Lemma 11.4.4 and obtain

$$(11.4.7)\quad r_k(\xi)^{|\gamma|-\delta_+^{(k)}+\varepsilon}\Big(\big|D_x^\alpha D_\xi^\gamma \vec{H}_i(x,\xi)\big| + r_k(\xi)\,\big|D_x^\alpha D_\xi^\gamma G_{i,4}(x,\xi)\big|\Big)$$
$$\le c\,|x-\xi|^{\varepsilon-\delta_+^{(k)}-1/2}\Big(\|\eta(\cdot)D_x^\alpha \vec{H}_i(x,\cdot)\|_{W^{1,2}(\mathcal{G})^3} + \|\eta(\cdot)D_x^\alpha G_{i,4}(x,\cdot)\|_{L_2(\mathcal{G})}\Big).$$

Let $f \in W^{-1,2}(\mathcal{K})^\ell$, $g \in L_2(\mathcal{G})$, and let u,p be the same functions as in the proof of Theorem 11.4.2. Since

$$\zeta\big(-\Delta u + \nabla p\big) = \zeta\eta f = 0 \quad\text{and}\quad \zeta\,\nabla\cdot u = C\zeta,$$

where $C = (\text{meas}(\mathcal{G}))^{-1}\displaystyle\int_{\mathcal{G}} \eta g\,dx$, Lemma 11.4.4 yields

$$r(x)^{|\alpha|-\delta_+^{(k(x))}+\varepsilon}\Big(\big|\partial_x^\alpha u(x)\big| + r(x)\,\big|\partial_x^\alpha p(x)\big|\Big)$$
$$\le c\,|x-\xi|^{\varepsilon-\delta_+^{(k(x))}-1/2}\big(\|\zeta u\|_{W^{1,2}(\mathcal{G})^3} + \|\zeta p\|_{L_2(\mathcal{G})}\big)$$
$$\le c\,|x-\xi|^{\varepsilon-\delta_+^{(k(x))}-1/2}\big(\|f\|_{W^{-1,2}(\mathcal{G})^3} + \|g\|_{L_2(\mathcal{G})}\big).$$

Arguing as in the proof of Theorem 11.4.2, we obtain the estimate

$$\|\eta(\cdot)D_x^\alpha \vec{H}_i(x,\cdot)\|_{W^{1,2}(\mathcal{G})^3} + \|\eta(\cdot)D_x^\alpha G_{i,4}(x,\cdot)\|_{L_2(\mathcal{G})}$$
$$\le c\,r(x)^{\delta_+^{(k(x))}-\delta_{i,4}-|\alpha|-\varepsilon}\,|x-\xi|^{\varepsilon-\delta_+^{(k(x))}-1/2}$$

for $i = 1,2,3,4$. This along with (11.4.7) proves the theorem. □

It remains to consider the case where x and ξ lie in a neighborhood $\mathcal{U}_\nu$ of the same vertex $x^{(\nu)}$ and $|x-\xi| < \min(r(x), r(\xi))$. Note that the last inequality implies $\rho_\nu(x)/2 < \rho_\nu(\xi) < 2\rho_\nu(x)$.

THEOREM 11.4.7. *If $x,\xi \in \mathcal{G}\cap\mathcal{U}_\nu$ and $|x-\xi| < \min(r(x), r(\xi))$, then*

$$\big|D_x^\alpha D_\xi^\gamma G_{i,j}(x,\xi)\big| \le c_{\alpha,\gamma}\,|x-\xi|^{-1-\delta_{i,4}-\delta_{j,4}-|\alpha|-|\gamma|}.$$

The constant $c_{\alpha,\gamma}$ is independent of x and ξ.

P r o o f. Suppose that $|x-\xi| < \min(r(x), r(\xi)) = 2\delta$ and that δ is small. Then there exists a ball B_δ with radius δ and distance $> 3\delta/4$ from the set $\mathcal{S}$ of the edge points containing both x and ξ. Let B'_δ be a ball concentric to B_δ with radius $3\delta/2$. We put $\mathcal{G}_\delta = \mathcal{G}\cap B'_\delta$, $\Gamma_\delta = \partial\mathcal{G}\cap B'_\delta$, $\tilde{\mathcal{G}}_\delta = \delta^{-1}\mathcal{G}_\delta = \{y = x/\delta : x \in \mathcal{G}_\delta\}$, and $\tilde{\Gamma}_\delta = \delta^{-1}\Gamma_\delta$. Since the distance of the sets $\tilde{\mathcal{G}}_\delta$ and $\tilde{\Gamma}_\delta$ from the edges of $\delta^{-1}\mathcal{G}$ is greater than $1/4$, there exists a Green's matrix $g(y,\eta)$ satisfying the equations

$$-\Delta_y \vec{g}_j(y,\eta) + \nabla_y g_{4,j}(y,\eta) = \delta(y-\eta)\,(\delta_{1,j},\delta_{2,j},\delta_{3,j})^t \quad \text{for } y,\eta \in \tilde{\mathcal{G}}_\delta\,,$$
$$-\nabla_y\cdot\vec{g}_j(y,\eta) = \big(\delta(y-\eta) - C\big)\,\delta_{4,j} \quad \text{for } y,\eta \in \tilde{\mathcal{G}}_\delta\,,$$
$$\vec{g}_j(y,\eta) = 0 \quad \text{for } y \in \tilde{\Gamma}_\delta\,,\ \eta \in \tilde{\mathcal{G}}_\delta$$

with a certain constant C and the estimates

$$\big|D_y^\alpha D_\eta^\gamma g_{i,j}(y,\eta)\big| \le c_{\alpha,\gamma}\,|y-\eta|^{-1-\delta_{i,4}-\delta_{j,4}-|\alpha|-|\gamma|}$$

with constants $c_{\alpha,\gamma}$ independent of y, η and δ. Here $\vec{g}_j$ denotes the vector with the components $g_{1,j}, g_{2,j}, g_{3,j}$. Obviously, the matrix $G^{(0)}(x,\xi)$ with the elements

$$G^{(0)}_{i,j}(x,\xi) = \delta^{-1-\delta_{i,4}-\delta_{j,4}}\, g_{i,j}(\delta^{-1}x, \delta^{-1}\xi)$$

satisfies the equations

$$-\Delta_x \vec{G}^{(0)}_j(x,\xi) + \nabla_x G^{(0)}_{4,j}(x,\xi) = \delta(x-\xi)\,(\delta_{1,j}, \delta_{2,j}, \delta_{3,j})^t \quad \text{for } x, \xi \in \mathcal{G}_\delta\,,$$
$$-\nabla_x \cdot \vec{G}^{(0)}_j(x,\xi) = \big(\delta(x-\xi) - C\delta^{-3}\big)\,\delta_{4,j} \quad \text{for } x, \xi \in \mathcal{G}_\delta\,,$$
$$\vec{G}^{(0)}_j(x,\xi) = 0 \quad \text{for } x \in \Gamma_\delta\,,\ \xi \in \mathcal{G}_\delta$$

and the inequalities

$$\big|D^\alpha_x D^\gamma_\xi G^{(0)}_{i,j}(x,\xi)\big| \le c_{\alpha,\gamma}\, |x-\xi|^{-1-\delta_{i,4}-\delta_{j,4}-|\alpha|-|\gamma|}. \tag{11.4.8}$$

Let $\xi \in \mathcal{G} \cap B_\delta$, and let $\zeta \in C_0^\infty(B'_\delta)$ be a cut-off function equal to one in the ball concentric to B_δ with radius $5\delta/4$. Then the matrix-valued function

$$U(x,\xi) = G(x,\xi) - \zeta(x)\, G^{(0)}(x,\xi)$$

satisfies the equations

$$-\Delta_x \vec{U}_j(x,\xi) + \nabla_x U_{4,j}(x,\xi) = \vec{F}_j(x,\xi) \quad \text{for } x, \xi \in \mathcal{G}\,,$$
$$-\nabla_x \cdot \vec{U}_j(x,\xi) = \Phi_j \quad \text{for } x, \xi \in \mathcal{G}\,,$$
$$\vec{U}_j(x,\xi) = 0 \quad \text{for } x \in \partial\mathcal{G},\ \xi \in \mathcal{G}$$

(here $\vec{U}_j$ denotes the vector $(U_{1,j}, U_{2,j}, U_{3,j})^t$), where

$$\big|D^\alpha_x D^\gamma_\xi \vec{F}_j(x,\xi)\big| \le c\,\delta^{-3-\delta_{j,4}-|\alpha|-|\gamma|}$$

and

$$\big|D^\alpha_x D^\gamma_\xi \Phi_j(x,\xi)\big| \le c\,\delta^{-2-\delta_{j,4}-|\alpha|-|\gamma|}.$$

Moreover, $\vec{F}_j(x,\xi) = 0$ and $\Phi_j(x,\xi) = -(\text{meas}(\mathcal{G}))^{-1}$ for $x \in \mathcal{G}\backslash B'_\delta$. From the uniqueness of solutions of the Dirichlet problem for the Stokes system in the space $V^{l,2}_{l-1,l-1}(\mathcal{G})^3 \times V^{l-1,2}_{l-1,l-1}(\mathcal{G})$ (cf. Theorems 11.1.2 and 11.1.8) it follows that

$$\|D^\gamma_\xi \vec{U}_j(x,\xi)\|_{V^{l,2}_{l-1,l-1}(\mathcal{G})^3} + \|D^\gamma_\xi U_{4,j}(x,\xi)\|_{V^{l-1,2}_{l-1,l-1}(\mathcal{G})}$$
$$\le c\Big(\|D^\gamma_\xi \vec{F}_j(x,\xi)\|_{V^{l-2,2}_{l-1,l-1}(\mathcal{G})^3} + \|D^\gamma_\xi \Phi_j(x,\xi)\|_{V^{l-1,2}_{l-1,l-1}(\mathcal{G})}\Big) \le c\,\delta^{-\delta_{j,4}-|\gamma|-1/2}$$

for arbitrary integer $l \ge 2$. As in the proof of Theorem 5.1.3, the last inequality yields

$$\|D^\alpha_x D^\gamma_\xi U_{i,j}(x,\xi)\|_{L_\infty(\mathcal{G}\cap B_\delta)} \le c\,\delta^{-1-\delta_{i,4}-\delta_{j,4}-|\alpha|-|\gamma|}\,.$$

This together with (11.4.8) implies

$$\big|D^\alpha_x D^\gamma_\xi G_{i,j}(x,\xi)\big| \le c_{\alpha,\gamma}\, |x-\xi|^{-1-\delta_{i,4}-\delta_{j,4}-|\alpha|-|\gamma|}\,.$$

The theorem is proved. □

11.5. Green's matrix in a convex polyhedral domain

If the polyhedral domain $\mathcal{G}$ is convex, then $\lambda = 1$ is the only eigenvalue of the pencils $\mathfrak{A}_j(\lambda)$ in the strip $-1/2 < \operatorname{Re}\lambda \le 1$ and the only eigenvalue of the pencils $A_\xi(\lambda)$ in the strip $0 < \operatorname{Re}\lambda \le 1$ (cf. [**85**, Theorems 5.1.2, 5.5.5]). This means that the numbers Λ_j^+ and $\delta_+^{(k)}$ in the estimates of the Green's matrix in the last theorems are equal to one. However, both for the pencils $\mathfrak{A}_j(\lambda)$ and $A_\xi(\lambda)$, the eigenvalue $\lambda = 1$ is simple and the corresponding eigenvectors have the form $(0, c)$, where c is a constant. This allows us to improve the estimates for Green's matrix given in the preceding section.

In this section, we use another definition of Green's matrix. Let ϕ be an infinitely differentiable function on $\overline{\mathcal{G}}$ which vanishes in a neighborhood of the edges such that

$$\int_{\mathcal{G}} \phi(x)\, dx = 1.$$

The matrix

$$G(x,\xi) = \big(G_{i,j}(x,\xi)\big)_{i,j=1,2,3,4}$$

is called *Green's matrix* for the problem (11.4.1), (11.4.2) if the vector functions $\vec{G}_j = (G_{1,j}, G_{2,j}, G_{3,j})^t$ and the function $G_{4,j}$ are solutions of the problems

$$-\Delta_x \vec{G}_j(x,\xi) + \nabla_x G_{4,j}(x,\xi) = \delta(x-\xi)\,(\delta_{1,j},\delta_{2,j},\delta_{3,j})^t \quad \text{for } x,\xi \in \mathcal{G},$$

$$-\nabla_x \cdot \vec{G}_j(x,\xi) = \big(\delta(x-\xi) - \phi(x)\big)\,\delta_{4,j} \quad \text{for } x,\xi \in \mathcal{G},$$

$$\vec{G}_j(x,\xi) = 0 \ \text{ for } x \in \partial\mathcal{G},\ \xi \in \mathcal{G}$$

and $G_{4,j}$ satisfies the condition

$$\int_{\mathcal{G}} G_{4,j}(x,\xi)\,\phi(x)\,dx = 0 \text{ for } \xi \in \mathcal{G},\ j = 1,2,3,4.$$

The existence and uniqueness of this Green's matrix holds analogously to Theorem 11.4.1. Note that

$$G_{i,j}(x,\xi) = G_{j,i}(\xi,x) \quad \text{for } x,\xi \in \mathcal{G},\ i,j = 1,2,3,4.$$

Furthermore, if $f \in W^{-1,2}(\mathcal{G})^3$, $g \in L_2(\mathcal{G})$, $g \perp 1$, and $(u,p) \in \overset{\circ}{W}{}^{1,2}(\mathcal{G})^3 \times L_2(\mathcal{G})$ is the uniquely determined solution of the Stokes system (11.4.1) satisfying the condition

$$\int_{\mathcal{G}} p(x)\,\phi(x)\,dx = 0,$$

then the components of (u,p) admit the representations

$$u_i(x) = \int_{\mathcal{G}} f(\xi)\cdot \vec{G}_i(\xi,x)\,d\xi + \int_{\mathcal{G}} g(\xi)\,G_{4,i}(\xi,x)\,d\xi,\ i = 1,2,3,$$

$$p(x) = \int_{\mathcal{G}} f(\xi)\cdot \vec{G}_4(\xi,x)\,d\xi + \int_{\mathcal{G}} g(\xi)\,G_{4,4}(x,\xi)\,d\xi.$$

The goal of this section is to obtain global estimates for the elements of Green's matrix in the case of a convex polyhedral type domain. In particular, we prove that

$$\big|\partial_x^\alpha \partial_\xi^\gamma G_{i,j}(x,\xi)\big| \le c\,|x-\xi|^{-1-\delta_{i,4}-\delta_{j,4}-|\alpha|-|\gamma|}$$

for $|\alpha| \le 1 - \delta_{i,4}$, $|\gamma| \le 1 - \delta_{j,4}$ if the polyhedron $\mathcal{G}$ is convex. Furthermore, we obtain Hölder estimates for the derivatives of $G_{i,j}(x,\xi)$.

11.5.1. Estimates for the case where x and ξ lie in a neighborhood of the same vertex. We are interested again in point estimates for the elements of $G(x,\xi)$. First, we consider the case where x and ξ lie in a sufficiently small neighborhood $\mathcal{U}_\nu$ of the same vertex. For this case, we need a modification of Lemma 11.4.3. Let again $\mu_+^{(k)}$ be the greatest real number such that the strip $0 < \operatorname{Re}\lambda < \mu_+^{(k)}$ contains at most the eigenvalue $\lambda = 1$ of the pencils $A_\xi(\lambda)$ for $\xi \in M_k$. By Λ'_ν, we denote the greatest real number such that the strip $-1/2 < \operatorname{Re}\lambda < \Lambda'_\nu$ contains at most the eigenvalue $\lambda = 1$ of the pencil $\mathfrak{A}_\nu(\lambda)$. Furthermore, we define

$$\tilde{\Lambda}_\nu = \min(2, \Lambda'_\nu) \quad \text{and} \quad \tilde{\mu}_k = \min(2, \mu_+^{(k)}).$$

In the next lemma, let $\zeta_\varepsilon, \eta_\varepsilon$ be the same infinitely differentiable functions as in Lemma 11.4.3. In particular, $\eta_\varepsilon(x) = 1$ for $c_1\varepsilon < \rho_\nu(x) < c_2\varepsilon$ and $\zeta_\varepsilon(x) = 1$ for $c'_1\varepsilon < \rho_\nu(x) < c'_2\varepsilon$, where $0 \le c_1 \le c'_1 < c'_2 < c_2$.

LEMMA 11.5.1. *Suppose that $l \ge 2$, $0 < l - \beta_\nu - 3/s < \tilde{\Lambda}_\nu$ and that the components of δ satisfy the inequalities $0 < l - \delta_k - 2/s < \tilde{\mu}_k$ for $k \in X_\nu$. If $\eta_\varepsilon(u,p) \in \overset{\circ}{W}{}^{1,2}(\mathcal{G})^3 \times L_2(\mathcal{G})$, $\eta_\varepsilon(-\Delta u + \nabla p) = 0$, $\eta_\varepsilon \nabla\cdot u = 0$, then*

$$(11.5.1) \qquad \|\zeta_\varepsilon u\|_{V^{l,s}_{\beta,\delta}(\mathcal{G})^3} + \|\nabla(\zeta_\varepsilon p)\|_{V^{l-2,s}_{\beta,\delta}(\mathcal{G})} \le c\,\varepsilon^{\beta_\nu - l + (6-s)/(2s)} \Big(\|\eta_\varepsilon u\|_{W^{1,2}(\mathcal{G})^3} + \|\eta_\varepsilon p\|_{L_2(\mathcal{G})} \Big).$$

Moreover, if $\tilde{\Lambda}_\nu > 1$ and $\tilde{\mu}_k > 1$ for $k \in X_\nu$, then

$$(11.5.2) \qquad \big|\zeta_\varepsilon(x)\, p(x)\big| \le c\,\varepsilon^{-3/2} \Big(\|\eta_\varepsilon u\|_{W^{1,2}(\mathcal{G})^3} + \|\eta_\varepsilon p\|_{L_2(\mathcal{G})} \Big).$$

Here, the constant c is independent of u, p and ε.

P r o o f. We use the same notation as in the proof of Lemma 11.4.3. Then we obtain the equations

$$\tilde{\eta}_\varepsilon \Big(\varepsilon^2 L(\varepsilon y, \varepsilon^{-1} D_y)\, v(y) + A(\varepsilon y)\, \nabla_y q(y) \Big) = 0, \quad \tilde{\eta}_\varepsilon\, A(\varepsilon y) \nabla_y \cdot v(y) = 0$$

for the functions $v(y) = u\big(\kappa^{-1}(\varepsilon y)\big)$ and $q(y) = \varepsilon\, p\big(\kappa^{-1}(\varepsilon y)\big)$. Here, $L(y, D_y)$ is a linear second order differential operator and $A(y)$ is a matrix such that $L^\circ(0, D_y) = -\Delta_y$ and $A(0)$ is the identity matrix. Let $\gamma = \beta_\nu - l + (6-s)/(2s)$. Since the strip $-1/2 < \operatorname{Re}\lambda < -\gamma - 1/2$ contains at most the eigenvalue $\lambda = 1$ of the pencil $\mathfrak{A}_\nu(\lambda)$, we conclude from Theorem 10.3.5 that

$$(v, q) = (0, c_0) + (V, Q)$$

in a neighborhood of the origin, where $(V, Q) \in W^{1,2}_{\gamma,0}(\mathcal{K})^3 \times W^{0,2}_{\gamma,0}(\mathcal{K})$ and c_0 is a constant. Applying Theorem 10.6.9, we obtain $(V, Q) \in W^{l,s}_{\beta,\delta}(\mathcal{K})^3 \times W^{l-1,s}_{\beta,\delta}(\mathcal{K})$. Consequently, $\tilde{\zeta} v \in W^{l,s}_{\beta,\delta}(\mathcal{K})^3$ and $\tilde{\zeta}\nabla q \in W^{l-2,s}_{\beta,\delta}(\mathcal{K})^3$. Furthermore, the estimate

$$(11.5.3) \qquad \|\tilde{\zeta} v\|_{W^{l,s}_{\beta_\nu,\delta}(\mathcal{K})^3} + \|\tilde{\zeta}(q - c_0)\|_{W^{l-1,s}_{\beta_\nu,\delta}(\mathcal{K})} + |c_0| \le c \Big(\|\tilde{\eta}_\varepsilon v\|_{W^{1,2}(\mathcal{K})^3} + \|\tilde{\eta}_\varepsilon q\|_{L_2(\mathcal{K})} \Big)$$

holds. By the same arguments as in the proof of Lemma 11.4.3, the constant c in the last estimate is independent of ε. The Dirichlet condition $\tilde{\zeta} v = 0$ on $\partial\mathcal{K}$ implies

$\tilde{\zeta}v = \partial_{x_j}(\tilde{\zeta}v) = 0$ on the edges of $\mathcal{K}$ for $j = 1, 2, 3$. Since $\delta_k + 2/s > l - 2$, we get $\tilde{\zeta}v \in V^{l,s}_{\beta_\nu,\delta}(\mathcal{K})^3$ and

$$\|\tilde{\zeta}v\|_{V^{l,s}_{\beta_\nu,\delta}(\mathcal{K})^3} \le c\, \|\tilde{\zeta}v\|_{W^{l,s}_{\beta_\nu,\delta}(\mathcal{K})^3}$$

(cf. Lemma 6.2.5). Furthermore, $W^{l-2,s}_{\beta_\nu,\delta}(\mathcal{K}) = V^{l-2,s}_{\beta_\nu,\delta}(\mathcal{K})$. Thus, it follows from (11.5.3) that

$$\|\tilde{\zeta}v\|_{V^{l,s}_{\beta_\nu,\delta}(\mathcal{K})^3} + \|\nabla(\tilde{\zeta}q)\|_{V^{l-2,s}_{\beta_\nu,\delta}(\mathcal{K})^3} \le c\left(\|\tilde{\eta}_\varepsilon v\|_{W^{1,2}(\mathcal{K})^3} + \|\tilde{\eta}_\varepsilon q\|_{L_2(\mathcal{K})}\right)$$

which implies

$$\begin{aligned} &\|\zeta_\varepsilon u\|_{V^{l,s}_{\beta_\nu,\delta}(\mathcal{K})^3} + \|\nabla(\zeta_\varepsilon p)\|_{V^{l-2,s}_{\beta_\nu,\delta}(\mathcal{K})^3} \\ &\qquad \le c\, \varepsilon^{\beta_\nu - l + (6-s)/(2s)} \left(\|\eta_\varepsilon u\|_{W^{1,2}(\mathcal{K})^3} + \|\eta_\varepsilon q\|_{L_2(\mathcal{K})}\right). \end{aligned}$$

Returning to the coordinates $x = \kappa^{-1}(y)$, we obtain (11.5.1). If $\tilde{\Lambda}_\nu > 1$ and $\tilde{\mu}_k > 1$ for $k \in X_\nu$, then we can set $\beta_\nu = l - 1 - 3/s$ and choose l, s and δ_k such that $\delta_k < l - 1 - 3/s$. In this case, $W^{l-1,s}_{\beta_\nu,\delta}(\mathcal{K}) \subset L_\infty(\mathcal{K})$ (cf. Lemma 7.2.3), and (11.5.3) yields

$$\left|\tilde{\zeta}(y)\, q(y)\right| \le c\left(\|\tilde{\eta}_\varepsilon v\|_{W^{1,2}(\mathcal{K})^3} + \|\tilde{\eta}_\varepsilon q\|_{L_2(\mathcal{K})}\right)$$

for $c_1' < |y| < c_2'$. This implies (11.5.2). □

The last lemma enables us to improve the results of Theorem 11.4.5 in the case of a convex polyhedral domain.

THEOREM 11.5.2. *Suppose that $\mathcal{G}$ is convex, $x, \xi \in \mathcal{G} \cap \mathcal{U}_\nu$ and $\rho_\nu(\xi) < \rho_\nu(x)/2$. Then*

$$\begin{aligned} \left|\partial_x^\alpha \partial_\xi^\gamma G_{i,j}(x,\xi)\right| \;&\le\; c\,\rho_\nu(x)^{-1-\tilde{\Lambda}_\nu-\delta_{i,4}-|\alpha|+\varepsilon}\, \rho_\nu(\xi)^{\tilde{\Lambda}_\nu-\delta_{j,4}-|\gamma|-\varepsilon} \\ &\quad \times \prod_{k\in X_\nu} \left(\frac{r_k(x)}{\rho_\nu(x)}\right)^{\tilde{\mu}_k-\delta_{i,4}-|\alpha|-\varepsilon} \prod_{k\in X_\nu} \left(\frac{r_k(\xi)}{\rho_\nu(\xi)}\right)^{\tilde{\mu}_k-\delta_{j,4}-|\gamma|-\varepsilon} \end{aligned}$$

for $|\alpha| \ge \delta_{i,4}$, $|\gamma| \ge \delta_{j,4}$,

$$\left|\partial_x^\alpha G_{i,4}(x,\xi)\right| \;\le\; c\,\rho_\nu(x)^{-2-\delta_{i,4}-|\alpha|} \prod_{k\in X_\nu} \left(\frac{r_k(x)}{\rho_\nu(x)}\right)^{\tilde{\mu}_k-\delta_{i,4}-|\alpha|-\varepsilon}$$

for $|\alpha| \ge \delta_{i,4}$,

$$\left|\partial_\xi^\gamma G_{4,j}(x,\xi)\right| \;\le\; c\,\rho_\nu(x)^{-2-\tilde{\Lambda}_\nu+\varepsilon}\, \rho_\nu(\xi)^{\tilde{\Lambda}_\nu-\delta_{j,4}-|\gamma|-\varepsilon} \prod_{k\in X_\nu} \left(\frac{r_k(\xi)}{\rho_\nu(\xi)}\right)^{\tilde{\mu}_k-\delta_{j,4}-|\gamma|-\varepsilon}$$

for $|\gamma| \ge \delta_{j,4}$, and

$$\left|G_{4,4}(x,\xi)\right| \le c\,\rho_\nu(x)^{-3}.$$

Here, ε is an arbitrarily small positive number.

P r o o f. Let $x \in \mathcal{G} \cap \mathcal{U}_\nu$. Furthermore, let η be a smooth cut-off function such that $\eta(\xi) = 1$ for $\rho_\nu(\xi) \le 5\rho_\nu(x)/8$ and $\eta(\xi) = 0$ for $\rho_\nu(\xi) \ge 3\rho_\nu(x)/4$. We consider the vector functions $\vec{H}_i = (G_{i,1}, G_{i,2}, G_{i,3})^t$ and the functions $G_{i,4}$. Since

$$-\Delta_\xi \vec{H}_i(x,\xi) + \nabla_\xi G_{i,4}(x,\xi) = \delta(x-\xi)\,(\delta_{i,1}, \delta_{i,2}, \delta_{i,3})^t \quad \text{for } x,\xi \in \mathcal{G},$$
$$-\nabla_\xi \cdot \vec{H}_i(x,\xi) = \big(\delta(x-\xi) - \phi(\xi)\big)\,\delta_{i,4} \quad \text{for } x,\xi \in \mathcal{G},$$
$$\vec{H}_i(x,\xi) = 0 \quad \text{for } x \in \mathcal{G},\ \xi \in \partial\mathcal{G}$$

and $\phi = 0$ near the set $\mathcal{S}$ of the edge points and vertices, it follows that

$$\eta(\xi)\,\big(-\Delta_\xi \partial_x^\alpha \vec{H}_i(x,\xi) + \nabla_\xi \partial_x^\alpha G_{i,4}(x,\xi)\big) = 0 \quad \text{for } \xi \in \mathcal{G}$$

and

$$\eta(\xi)\,\nabla_\xi \cdot \partial_x^\alpha \vec{H}_i(x,\xi) = 0$$

for $|\alpha| \ge \delta_{i,4}$. Let l be sufficiently large, and let β and δ_k be real numbers satisfying the inequalities

$$1 < l - \beta - 3/s < \tilde{\Lambda}_\nu, \quad l - \tilde{\mu}_k < \delta_k + 2/s < l - 1 - 1/s. \tag{11.5.4}$$

Then by Lemmas 3.1.4 and 11.5.1,

$$\rho_\nu(\xi)^{\beta-l+|\gamma|+3/s} \prod_{k\in X_\nu} \Big(\frac{r_k(\xi)}{\rho_\nu(\xi)}\Big)^{\delta_k-l+|\gamma|+3/s} \big|\partial_x^\alpha \partial_\xi^\gamma \vec{H}_i(x,\xi)\big| \le c\,A_i \tag{11.5.5}$$

for $\rho_\nu(\xi) < \rho_\nu(x)/2$, where

$$A_i = \rho_\nu(x)^{\beta-l+(6-s)/(2s)} \Big(\|\eta(\cdot)\partial_x^\alpha \vec{H}_i(x,\cdot)\|_{\overset{\circ}{W}{}^{1,2}(\mathcal{G})^3} + \|\eta(\cdot)\partial_x^\alpha G_{i,4}(x,\cdot)\|_{L_2(\mathcal{G})}\Big).$$

Furthermore,

$$\rho_\nu(\xi)^{\beta-l+1+|\gamma|+3/s} \prod_{k\in X_\nu} \Big(\frac{r_k(\xi)}{\rho_\nu(\xi)}\Big)^{\delta_k-l+1+|\gamma|+3/s} \big|\partial_x^\alpha \partial_\xi^\gamma G_{i,4}(x,\xi)\big| \le c\,A_i \tag{11.5.6}$$

for $\rho_\nu(\xi) < \rho_\nu(x)/2$, $|\gamma| \ge 1$, and

$$\begin{aligned} \big|\partial_x^\alpha G_{i,4}(x,\xi)\big| \ \le\ & c\,\rho_\nu(x)^{-3/2} \Big(\|\eta(\cdot)\,\partial_x^\alpha \vec{H}_i(x,\cdot)\|_{\overset{\circ}{W}{}^{1,2}(\mathcal{K}_\nu)^3} \\ & + \|\eta(\cdot)\,\partial_x^\alpha G_{i,4}(x,\cdot)\|_{L_2(\mathcal{K}_\nu)}\Big). \end{aligned} \tag{11.5.7}$$

Let $f \in W^{-1,2}(\mathcal{G})^3$, $g \in L_2(\mathcal{G})$, and let ψ be an infinitely differentiable function such that $\psi(y) = 1$ for $|\rho_\nu(x) - \rho_\nu(y)| < \rho_\nu(x)/8$ and $\psi(y) = 0$ for $|\rho_\nu(x) - \rho_\nu(y)| > \rho_\nu(x)/4$. We consider the vector function (u,p) with the components

$$u_i(x) = \int_{\mathcal{G}} \eta(y)\,f(y)\cdot \vec{H}_i(x,y)\,dy + \int_{\mathcal{G}} \eta(y)\,g(y)\,G_{i,4}(x,y)\,dy, \tag{11.5.8}$$

$$p(x) = \int_{\mathcal{G}} \eta(y)\,f(y)\cdot \vec{H}_4(x,y)\,dy + \int_{\mathcal{G}} \eta(y)\,g(y)\,G_{4,4}(x,y)\,dy, \tag{11.5.9}$$

which satisfies the equations

$$-\Delta u + \nabla p = \eta f, \qquad -\nabla\cdot u = \eta g - C\phi, \quad \text{where } C = \int_{\mathcal{G}} \eta(y)\,g(y)\,dy.$$

Since $\psi(-\Delta u + \nabla p) = 0$ and $\psi\nabla\cdot u = 0$, it follows from Lemmas 3.1.4 and 11.5.1 that

$$\begin{aligned} & \rho_\nu(x)^{\beta-l+|\alpha|+3/s} \prod_{k\in X_\nu} \Big(\frac{r_k(x)}{\rho_\nu(x)}\Big)^{\delta_k-l+|\alpha|+3/s} \big|\partial_x^\alpha u(x)\big| \\ & \quad \le c\,\rho_\nu(x)^{\beta-l+(6-s)/(2s)} \Big(\|\psi u\|_{\overset{\circ}{W}{}^{1,2}(\mathcal{G})^3} + \|\psi p\|_{L_2(\mathcal{G})}\Big). \end{aligned}$$

Consequently,

$$\left|\partial_x^\alpha u(x)\right| \le c\,\rho_\nu(x)^{-|\alpha|-1/2} \prod_{k\in X_\nu} \Big(\frac{r_k(x)}{\rho_\nu(x)}\Big)^{l-\delta_k-|\alpha|-3/s} \|(f,g)\|.$$

Here by $\|(f,g)\|$, we mean the norm of (f,g) in $W^{-1,2}(\mathcal{G})^3 \times L_2(\mathcal{G})$. Furthermore,

$$\left|\partial_x^\alpha p(x)\right| \le c\,\rho_\nu(x)^{-|\alpha|-3/2} \prod_{k\in X_\nu} \Big(\frac{r_k(x)}{\rho_\nu(x)}\Big)^{l-1-\delta_k-|\alpha|-3/s} \|(f,g)\|$$

for $|\alpha| \ge 1$ and

$$\left|p(x)\right| \le c\,\rho_\nu(x)^{-3/2}\,\|(f,g)\|.$$

Thus,

$$\left\|\eta(\cdot)\,\partial_x^\alpha \vec{H}_i(x,\cdot)\right\|_{\overset{\circ}{W}{}^{1,2}(\mathcal{G})^3} \le c\,\rho_\nu(x)^{-|\alpha|-\delta_{i,4}-1/2} \prod_{k\in X_\nu} \Big(\frac{r_k(x)}{\rho_\nu(x)}\Big)^{l-\delta_{i,4}-\delta_k-|\alpha|-3/s}$$

for $|\alpha| \ge \delta_{i,4}$, while

$$\left\|\eta(\cdot)\,\vec{H}_4(x,\cdot)\right\|_{\overset{\circ}{W}{}^{1,2}(\mathcal{G})^3} \le c\,\rho_\nu(x)^{-3/2}.$$

Analogously,

$$\left\|\eta(\cdot)\,\partial_x^\alpha G_{i,4}(x,\cdot)\right\|_{L_2(\mathcal{G})} \le c\,\rho_\nu(x)^{-|\alpha|-\delta_{i,4}-1/2} \prod_{k\in X_\nu} \Big(\frac{r_k(x)}{\rho_\nu(x)}\Big)^{l-\delta_{i,4}-\delta_k-|\alpha|-3/s}$$

for $|\alpha| \ge \delta_{i,4}$ and

$$\left\|\eta(\cdot)\,G_{4,4}(x,\cdot)\right\|_{L_2(\mathcal{G})^3} \le c\,\rho_\nu(x)^{-3/2}.$$

We can choose l, s, β and δ_k such that

$$l-\beta-3/s = \Lambda_\nu - \varepsilon, \quad l-\delta_k-3/s = \mu_k' - \varepsilon, \quad 1/s < \varepsilon.$$

Combining then the last four inequalities with (11.5.5)–(11.5.7), we obtain the desired estimates for the elements of the matrix $G(x,\xi)$. □

Using the equality $G_{i,j}(x,\xi) = G_{j,i}(\xi,x)$, one can deduce analogous estimates for the elements of Green's matrix in the case $\rho_\nu(\xi) > 2\rho_\nu(x)$.

Next, we consider the case where x and ξ lie in a neighborhood $\mathcal{U}_\nu$ of the vertex $x^{(\nu)}$ and $\rho_\nu(x)/2 < \rho_\nu(\xi) < 2\rho_\nu(x)$. In the following theorem, let $\tilde\mu$ denote the minimum of the numbers $\tilde\mu_1, \dots, \tilde\mu_d$.

THEOREM 11.5.3. 1) *If $x,\xi \in \mathcal{G}\cap\mathcal{U}_\nu$ and $|x-\xi| < \min(r(x), r(\xi))$, then*

$$\left|\partial_x^\alpha \partial_\xi^\gamma G_{i,j}(x,\xi)\right| \le c_{\alpha,\gamma}\,|x-\xi|^{-1-\delta_{i,4}-\delta_{j,4}-|\alpha|-|\gamma|}.$$

The constant $c_{\alpha,\gamma}$ is independent of x and ξ.

2) *Suppose that $x,\xi \in \mathcal{G}\cap\mathcal{U}_\nu$, $\rho_\nu(x)/2 < \rho_\nu(\xi) < 2\rho_\nu(x)$ and $|x-\xi| > \min(r(x), r(\xi))$. Then*

$$(11.5.10)\qquad \left|\partial_x^\alpha \partial_\xi^\gamma G_{i,j}(x,\xi)\right| \le c_{\alpha,\gamma}|x-\xi|^{-1-\delta_{i,4}-\delta_{j,4}-|\alpha|-|\gamma|} \times\Big(\frac{r(x)}{|x-\xi|}\Big)^{\tilde\mu-\delta_{i,4}-|\alpha|-\varepsilon} \Big(\frac{r(\xi)}{|x-\xi|}\Big)^{\tilde\mu-\delta_{j,4}-|\gamma|-\varepsilon}$$

for $|\alpha| \ge \delta_{i,4}$, $|\gamma| \ge \delta_{j,4}$,

$$(11.5.11)\qquad \left|\partial_x^\alpha G_{i,4}(x,\xi)\right| \le c_{\alpha,\gamma}|x-\xi|^{-2-\delta_{i,4}-|\alpha|}\Big(\frac{r(x)}{|x-\xi|}\Big)^{\tilde\mu-\delta_{i,4}-|\alpha|-\varepsilon}$$

for $|\alpha| \ge \delta_{i,4}$,

$$(11.5.12)\qquad \big|\partial_\xi^\gamma G_{4,j}(x,\xi)\big| \le c_{\alpha,\gamma}|x-\xi|^{-2-\delta_{j,4}-|\gamma|}\Big(\frac{r(\xi)}{|x-\xi|}\Big)^{\tilde\mu-\delta_{j,4}-|\gamma|-\varepsilon}$$

for $|\gamma| \ge \delta_{j,4}$, *and* $|G_{4,4}(x,\xi)| \le c\,|x-\xi|^{-3}$. *Here,* ε *is an arbitrarily small positive number.*

P r o o f. For the first estimate, we refer to Theorem 11.4.7. Furthermore, the estimate

$$(11.5.13)\quad \big|\partial_x^\alpha\partial_\xi^\gamma G_{i,j}(x,\xi)\big| \le c\,|x-\xi|^{-1-\delta_{i,4}-\delta_{j,4}-|\alpha|-|\gamma|} \times\Big(\frac{r(x)}{|x-\xi|}\Big)^{\min(0,\tilde\mu-\delta_{i,4}-|\alpha|-\varepsilon)}\Big(\frac{r(\xi)}{|x-\xi|}\Big)^{\min(0,\tilde\mu-\delta_{j,4}-|\gamma|-\varepsilon)}$$

holds for $\rho_\nu(\xi)/8 < \rho_\nu(x) < 8\rho_\nu(\xi)$ and $4|x-\xi| > 3\min(r(x),r(\xi))$ (cf. Theorem 10.4.4). Thus in particular, the estimate (11.5.10) is valid for $|\alpha|+\delta_{i,4} \ge 2$ and $|\gamma|+\delta_{j,4} \ge 2$, the estimate (11.5.11) is valid for $|\alpha|+\delta_{i,4} \ge 2$, while (11.5.12) holds for $|\gamma|+\delta_{j,4} \ge 2$.

We prove (11.5.10) for $i,j=1,2,3$, $|\alpha|=1$ and $|\gamma| \ge 2$. If $|x-\xi| < 4r(x)$, then the estimate

$$\big|\partial_x^\alpha\partial_\xi^\gamma G_{i,j}(x,\xi)\big| \le c\,|x-\xi|^{-2-|\gamma|}\Big(\frac{r(x)}{|x-\xi|}\Big)^{\tilde\mu-1-\varepsilon}\Big(\frac{r(\xi)}{|x-\xi|}\Big)^{\tilde\mu-|\gamma|-\varepsilon}$$

follows immediately from (11.5.13). Let $|x-\xi| > 4r(x)$. Then we denote by y the nearest point to x on the set

$$\mathcal{S}_\nu = \bigcup_{k\in X_\nu} \overline{M}_k\,.$$

It can be easily verified that the inequalities

$$\rho_\nu(\xi)/8 < \rho_\nu(z) < 8\rho_\nu(\xi) \quad\text{and}\quad 3|x-\xi| < 4|z-\xi| < 5|x-\xi|$$

are satisfied for the point $z = y + t(x-y)$, where $t\in(0,1)$. Moreover, $r(z) = t|x-y| = t\,r(x) > 3\min(r(z),r(\xi))/4$. Consequently by (11.5.13),

$$\begin{aligned}\big|(\nabla_x\partial_x^\alpha\partial_\xi^\gamma G_{i,j})(z,\xi)\big| &\le c\,|z-\xi|^{-3-|\gamma|}\Big(\frac{r(z)}{|z-\xi|}\Big)^{\tilde\mu-2-\varepsilon}\Big(\frac{r(\xi)}{|z-\xi|}\Big)^{\tilde\mu-|\gamma|-\varepsilon}\\ &\le c\,|x-\xi|^{-3-|\gamma|}\Big(\frac{t\,r(x)}{|x-\xi|}\Big)^{\tilde\mu-2-\varepsilon}\Big(\frac{r(\xi)}{|x-\xi|}\Big)^{\tilde\mu-|\gamma|-\varepsilon}.\end{aligned}$$

It follows from Theorem 11.3.2 that $G_{i,j}(\cdot,\xi)\in C^{1,\sigma}$ in a neighborhood of $\mathcal{S}$ for arbitrary $\xi\in\mathcal{G}$, $i\ne 4$. Since $G_{i,j}(x,\xi)=0$ for $x\in\partial\mathcal{G}$ and $i\ne 4$, we conclude that

$$(11.5.14)\qquad G_{i,j}(x,\xi)=0 \quad\text{and}\quad \nabla_x G_{i,j}(x,\xi)=0 \quad \text{for } x\in\mathcal{S},\ i\ne 4.$$

This implies

$$\begin{aligned}\big|\partial_x^\alpha\partial_\xi^\gamma G_{i,j}(x,\xi)\big| &= \big|(\partial_x^\alpha\partial_\xi^\gamma G_{i,j})(x,\xi) - (\partial_x^\alpha\partial_\xi^\gamma G_{i,j})(y,\xi)\big|\\ &= \Big|\int_0^1 \frac{d}{dt}(\partial_x^\alpha\partial_\xi^\gamma G_{i,j})(y+t(x-y),\xi)\,dt\Big|\\ &\le |x-y|\int_0^1 \big|(\nabla_x\partial_x^\alpha\partial_\xi^\gamma G_{i,j})(y+t(x-y),\xi)\big|\,dt\\ &\le c\,|x-\xi|^{-2-|\gamma|}\Big(\frac{r(x)}{|x-\xi|}\Big)^{\tilde\mu-1-\varepsilon}\Big(\frac{r(\xi)}{|x-\xi|}\Big)^{\tilde\mu-|\gamma|-\varepsilon}.\end{aligned}$$

Thus, (11.5.10) is proved for $i, j = 1, 2, 3$, $|\alpha| = 1$, $|\gamma| \geq 2$. Repeating this argument, we get

$$\left|\partial_\xi^\gamma G_{i,j}(x,\xi)\right| \leq |x-\xi|^{-1-|\gamma|} \Big(\frac{r(x)}{|x-\xi|}\Big)^{\tilde{\mu}-\varepsilon} \Big(\frac{r(\xi)}{|x-\xi|}\Big)^{\tilde{\mu}-|\gamma|-\varepsilon}$$

for $i, j = 1, 2, 3$, $|\gamma| \geq 2$. This proves (11.5.10) for $i, j = 1, 2, 3$, $|\gamma| \geq 2$ and arbitrary α. Analogously, the inequality (11.5.10) for $i = 1, 2, 3$, $j = 4$, $|\gamma| \geq 1$ and the inequality (11.5.11) for $i = 1, 2, 3$ hold.

Using the fact that

$$G_{i,j}(x,\xi) = 0 \quad \text{and} \quad \nabla_\xi G_{i,j}(x,\xi) = 0 \quad \text{for } \xi \in \mathcal{S},\ j \neq 4$$

(cf. (11.5.14)), we obtain the inequalities (11.5.10) and (11.5.12) for $j = 1, 2, 3$ and $|\gamma| \leq 1$ by the same arguments as above. This completes the proof. □

11.5.2. Estimates for the case where x and ξ lie in neighborhoods of different vertices. Now we consider the case where x and ξ lie in neighborhoods $\mathcal{U}_\mu$ and $\mathcal{U}_\nu$ of different vertices $x^{(\mu)}$ and $x^{(\nu)}$, respectively. Then the following modification of Theorem 11.4.2 holds.

THEOREM 11.5.4. *Suppose that $\mathcal{G}$ is convex, $x \in \mathcal{U}_\mu$ and $\xi \in \mathcal{U}_\nu$, where $\mathcal{U}_\mu$ and $\mathcal{U}_\nu$ are neighborhoods of the vertices $x^{(\mu)}$ and $x^{(\nu)}$, respectively, which have a positive distance. Then*

$$\begin{aligned}\left|\partial_x^\alpha \partial_\xi^\gamma G_{i,j}(x,\xi)\right| \;&\leq\; c\, \rho_\mu(x)^{\tilde{\Lambda}_\mu-\delta_{i,4}-|\alpha|-\varepsilon}\, \rho_\nu(\xi)^{\tilde{\Lambda}_\nu-\delta_{j,4}-|\gamma|-\varepsilon} \\ &\times \prod_{k\in X_\mu} \Big(\frac{r_k(x)}{\rho_\mu(x)}\Big)^{\tilde{\mu}_k-\delta_{i,4}-|\alpha|-\varepsilon} \prod_{k\in X_\nu} \Big(\frac{r_k(\xi)}{\rho_\nu(\xi)}\Big)^{\tilde{\mu}_k-\delta_{j,4}-|\gamma|-\varepsilon}.\end{aligned}$$

for $|\alpha| \geq \delta_{i,4}$, $|\gamma| \geq \delta_{j,4}$,

$$\left|\partial_x^\alpha G_{i,4}(x,\xi)\right| \leq c\, \rho_\mu(x)^{\tilde{\Lambda}_\mu-\delta_{i,4}-|\alpha|-\varepsilon} \prod_{k\in X_\mu} \Big(\frac{r_k(x)}{\rho_\mu(x)}\Big)^{\tilde{\mu}_k-\delta_{i,4}-|\alpha|-\varepsilon}$$

for $|\alpha| \geq \delta_{i,4}$,

$$\left|\partial_\xi^\gamma G_{4,j}(x,\xi)\right| \leq c\, \rho_\nu(\xi)^{\tilde{\Lambda}_\nu-\delta_{j,4}-|\gamma|-\varepsilon} \prod_{k\in X_\nu} \Big(\frac{r_k(\xi)}{\rho_\nu(\xi)}\Big)^{\tilde{\mu}_k-\delta_{j,4}-|\gamma|-\varepsilon}$$

for $|\gamma| \geq \delta_{j,4}$, and $\left|G_{4,4}(x,\xi)\right| \leq c$. Here, c is a constant independent of x and ξ, and ε is an arbitrarily small positive number.

P r o o f. Let ζ and η be smooth cut-off functions, $\zeta = 1$ on $\mathcal{U}_\mu$, $\eta = 1$ on $\mathcal{U}_\nu$, $\zeta = 0$ near all edges M_k, $k \notin X_\mu$, $\eta = 0$ near all edges M_k, $k \notin X_\nu$. We assume that the distance of the supports of ζ and η is positive. Let $A_{i,\alpha}$ denote the expression

$$A_{i,\alpha} = \left\|\eta(\cdot)\, \partial_x^\alpha \vec{H}_i(x,\cdot)\right\|_{\overset{\circ}{W}{}^{1,2}(\mathcal{G})^3} + \left\|\eta(\cdot)\, \partial_x^\alpha G_{i,4}(x,\cdot)\right\|_{L_2(\mathcal{G})}.$$

Here again $\vec{H}_i$ is the vector with the components $G_{i,1}, G_{i,2}, G_{i,3}$. Since

$$\eta(z) \big(-\Delta_z \vec{H}_i(x,z) + \nabla_z G_{i,4}(x,z)\big) = 0$$

and

$$\eta(z)\, \nabla_z \cdot \vec{H}_i(x,z) = \eta(z)\, \varphi(z)\, \delta_{i,4}$$

for $z \in \mathcal{G}$, it follows from Lemmas 3.1.4 and 11.5.1 that

$$\big|\partial_x^\alpha \partial_\xi^\gamma \vec{H}_i(x,\xi)\big| \le c\,\rho_\nu(\xi)^{l-\beta-|\gamma|-3/s} \prod_{k\in X_\nu} \Big(\frac{r_k(\xi)}{\rho_\nu(\xi)}\Big)^{l-\delta_k-|\gamma|-3/s} A_{i,\alpha}$$

for all γ,

$$\big|\partial_x^\alpha \partial_\xi^\gamma G_{i,4}(x,\xi)\big| \le c\,\rho_\nu(\xi)^{l-1-\beta-|\gamma|-3/s} \prod_{k\in X_\nu} \Big(\frac{r_k(\xi)}{\rho_\nu(\xi)}\Big)^{l-1-\delta_k-|\gamma|-3/s} A_{i,\alpha}$$

for $|\gamma| \ge 1$ and

$$\big|\partial_x^\alpha G_{i,4}(x,\xi)\big| \le c\,A_{i,\alpha}\,,$$

where c is a constant independent of x and ξ, l is a sufficiently large integer, β and δ_k are real numbers satisfying the inequalities (11.5.4).

Let $f \in W^{-1,2}(\mathcal{G})$ and $g \in L_2(\mathcal{G})$. We consider the vector function with the components (11.5.8) and (11.5.9). Since

$$\zeta(-\Delta u + \nabla p) = 0 \quad \text{and} \quad \zeta\nabla\cdot u = C\zeta\varphi, \quad \text{where } C = \int_{\mathcal{G}} \eta(y)\,g(y)\,dy,$$

Lemmas 3.1.4 and 11.5.1 and yield

$$|\partial_x^\alpha u(x)| \le c\,\rho_\mu(x)^{l-\beta'-|\alpha|-3/s} \prod_{k\in X_\mu} \Big(\frac{r_k(x)}{\rho_\mu(x)}\Big)^{l-\delta_k-|\alpha|-3/s} \|(f,g)\|$$

for all α,

$$|\partial_x^\alpha p(x)| \le c\,\rho_\mu(x)^{l-1-\beta'-|\alpha|-3/s} \prod_{k\in X_\mu} \Big(\frac{r_k(x)}{\rho_\mu(x)}\Big)^{l-1-\delta_k-|\alpha|-3/s} \|(f,g)\|$$

for $|\alpha| \ge 1$, and

$$|p(x)| \le c\,\|(f,g)\|.$$

Here by $\|(f,g)\|$, we mean the norm of (f,g) in the space $W^{-1,2}(\mathcal{G})^3 \times L_2(\mathcal{G})$, and β' is an arbitrary real number such that $1 < l - \beta' - 3/s < \tilde{\Lambda}_\mu$. Thus, we obtain the following estimates for the expressions $A_{i,\alpha}$:

$$A_{i,\alpha} \le c\,\rho_\mu(x)^{l-\beta'-\delta_{i,4}-|\alpha|-3/s} \prod_{k\in X_\mu} \Big(\frac{r_k(x)}{\rho_\mu(x)}\Big)^{l-\delta_k-\delta_{i,4}-|\alpha|-3/s} \quad \text{for } |\alpha| \ge \delta_{i,4}$$

and $A_{4,0} \le c$. We can choose l, s, β, β' and δ_k such that

$$l-\beta-3/s = \Lambda_\nu - \varepsilon, \quad l-\beta'-3/s = \Lambda_\mu - \varepsilon, \quad l-\delta_k-3/s = \mu_k' - \varepsilon,$$

where $\varepsilon > 1/s$. Combining then the last two inequalities for $A_{i,\alpha}$ with the estimates for $\partial_x^\alpha \partial_\xi^\gamma \vec{H}_i(x,\xi)$ and $\partial_x^\alpha \partial_\xi^\gamma G_{i,4}(x,\xi)$ in the first part of the proof, we obtain the assertion of the theorem. □

11.5.3. Hölder estimates for the elements of Green's matrix. As a consequence of the estimates in Theorems 11.5.2–11.5.4, we obtain the following result. Here, we use the inequalities

$$\tilde{\mu}_k > 1 \quad \text{and} \quad \tilde{\Lambda}_\nu > 1$$

which are valid in the case of a convex polyhedral type domain.

THEOREM 11.5.5. *Let $\mathcal{G}$ be a convex polyhedral type domain. Then the elements of the matrix $G(x,\xi)$ satisfy the estimate*

$$\big|\partial_x^\alpha \partial_\xi^\gamma G_{i,j}(x,\xi)\big| \le c\,|x-\xi|^{-1-\delta_{i,4}-\delta_{j,4}-|\alpha|-|\gamma|} \tag{11.5.15}$$

for $|\alpha| \le 1-\delta_{i,4}$, $|\gamma| \le 1-\delta_{j,4}$.

P r o o f. If x and ξ lie in (sufficiently small) neighborhoods of different vertices, then the assertion of the lemma follows immediately from the estimates in Theorem 11.5.4. We assume that x and ξ lie in a neighborhood $\mathcal{U}_\nu$ of the same vertex $x^{(\nu)}$. Then by Theorem 11.5.2,

$$\big|\partial_x^\alpha \partial_\xi^\gamma G_{i,j}(x,\xi)\big| \le c\,\rho_\nu(x)^{-1-\delta_{i,4}-\delta_{j,4}-|\alpha|-|\gamma|}$$

for $\rho_\nu(\xi) < \rho_\nu(x)/2$, $|\alpha|+\delta_{i,4} \le 1$, $|\gamma|+\delta_{j,4} \le 1$. Using the inequality $|x-\xi| < 3\rho_\nu(x)/2$ for $\rho_\nu(\xi) < \rho_\nu(x)/2$, we obtain (11.5.15). For $\rho_\nu(x) < \rho_\nu(\xi)/2$, we obtain the same estimate by means of the equality $G_{i,j}(x,\xi) = G_{j,i}(\xi,x)$ and Theorem 11.5.2. In the case $\rho_\nu(x)/2 < \rho_\nu(\xi) < 2\rho_\nu(x)$, the inequality (11.5.15) follows from Theorem 11.5.3. □

The following Hölder estimate for the elements $\partial_\xi^\gamma G_{i,j}(x,\xi)$, $i=1,2,3$, can be easily deduced from the last theorem.

COROLLARY 11.5.6. *Let $\mathcal{G}$ be a convex polyhedral type domain. Then the estimate*

$$\begin{aligned}&\frac{\big|\partial_\xi^\gamma G_{i,j}(x,\xi) - \partial_\xi^\gamma G_{i,j}(y,\xi)\big|}{|x-y|^\sigma} \\ &\qquad \le c\,\big(|x-\xi|^{-1-\sigma-\delta_{j,4}-|\gamma|} + |y-\xi|^{-1-\sigma-\delta_{j,4}-|\gamma|}\big)\end{aligned} \tag{11.5.16}$$

with arbitrary $\sigma \in (0,1)$ holds for $x,y,\xi \in \mathcal{G}$, $x \ne y$, $i=1,2,3$, $|\gamma| \le 1-\delta_{j,4}$. Analogously, the estimate

$$\begin{aligned}&\frac{\big|\partial_x^\alpha G_{i,j}(x,\xi) - \partial_x^\alpha G_{i,j}(x,\eta)\big|}{|\xi-\eta|^\sigma} \\ &\qquad \le c\,\big(|x-\xi|^{-1-\sigma-\delta_{i,4}-|\alpha|} + |x-\eta|^{-1-\sigma-\delta_{i,4}-|\alpha|}\big)\end{aligned} \tag{11.5.17}$$

is satisfied for $x,y,\xi \in \mathcal{G}$, $x \ne y$, $j=1,2,3$, $|\alpha| \le 1-\delta_{i,4}$.

P r o o f. If $|x-\xi| < 2|x-y|$, then $|y-\xi| < 3|x-y|$ and, consequently,

$$\begin{aligned}\big|\partial_\xi^\gamma G_{i,j}(x,\xi) - \partial_\xi^\gamma G_{i,j}(y,\xi)\big| &\le c\,\big(|x-\xi|^{-1-\delta_{j,4}-|\gamma|} + |y-\xi|^{-1-\delta_{j,4}-|\gamma|}\big)\\ &\le c\,3^\sigma\,\big(|x-\xi|^{-1-\sigma-\delta_{j,4}-|\gamma|} + |y-\xi|^{-1-\sigma-\delta_{j,4}-|\gamma|}\big)\,|x-y|^\sigma.\end{aligned}$$

Suppose that $|x-\xi| > 2|x-y|$. Then by the mean value theorem,

$$\big|\partial_\xi^\gamma G_{i,j}(x,\xi) - \partial_\xi^\gamma G_{i,j}(y,\xi)\big| \le |x-y|\,\big|\nabla_z \partial_\xi^\gamma G_{i,j}(z,\xi)\big|,$$

where $z = x+t(y-x)$, $0<t<1$. Since $|z-\xi| > |x-\xi|/2 > |x-y|$, Theorem 11.5.5 yields

$$\begin{aligned}\big|\partial_\xi^\gamma G_{i,j}(x,\xi) - \partial_\xi^\gamma G_{i,j}(y,\xi)\big| &\le c\,|x-y|\,|z-\xi|^{-2-\delta_{j,4}-|\gamma|}\\ &\le c\,|x-y|^\sigma\,|x-\xi|^{-1-\sigma-\delta_{j,4}-|\gamma|}.\end{aligned}$$

This proves (11.5.16). The estimate (11.5.17) follows from (11.5.16) and from the equality $G_{i,j}(x,\xi) = G_{j,i}(\xi,x)$. □

Now we are interested in Hölder estimates for the derivatives $\partial_x^\alpha \partial_\xi^\gamma G_{i,j}(x,\xi)$ of Green's matrix. First, we assume that $|x-\xi| < m|x-y|$ with a certain positive m.

LEMMA 11.5.7. *Let $\mathcal{G}$ be a convex polyhedral type domain, and let m be an arbitrary positive number. Then the estimate*

$$\frac{\big|\partial_x^\alpha \partial_\xi^\gamma G_{i,j}(x,\xi) - \partial_y^\alpha \partial_\xi^\gamma G_{i,j}(y,\xi)\big|}{|x-y|^\sigma} \le c\left(|x-\xi|^{-2-\sigma-\delta_{j,4}-|\gamma|} + |y-\xi|^{-2-\sigma-\delta_{j,4}-|\gamma|}\right) \tag{11.5.18}$$

is satisfied for $i=1,2,3$, $|\alpha|=1$, $|\gamma| \le 1-\delta_{j,4}$, $x,y,\xi \in \mathcal{G}$, $x \ne y$, $|x-\xi| < m|x-y|$.

P r o o f. Suppose that $i \ne 4$, $|\alpha| = 1$ and $|\gamma| \le 1-\delta_{j,4}$. By Theorem 11.5.5,

$$\frac{\big|\partial_x^\alpha \partial_\xi^\gamma G_{i,j}(x,\xi)\big|}{|x-y|^\sigma} \le c\,\frac{|x-\xi|^{-2-\delta_{j,4}-|\gamma|}}{|x-y|^\sigma} \le c\,m^\sigma\,|x-\xi|^{-2-\sigma-\delta_{j,4}-|\gamma|}$$

for $|x-\xi| < m|x-y|$. Analogously,

$$\frac{\big|\partial_x^\alpha \partial_\xi^\gamma G_{i,j}(x,\xi)\big|}{|x-y|^\sigma} \le c\,(m+1)^\sigma\,|x-\xi|^{-2-\sigma-\delta_{j,4}-|\gamma|}$$

since $|y-\xi| < (m+1)|x-y|$. This proves the lemma. □

Next, we consider the case $|x-\xi| > m|x-y|$. In the following lemma, we assume that x and ξ lie in neighborhoods $\mathcal{U}_\mu$ and $\mathcal{U}_\nu$ of different vertices.

LEMMA 11.5.8. *Let $\mathcal{G}$ be convex, and let σ be a positive number, $\sigma < \tilde\Lambda_\nu - 1$ for all ν, and $\sigma < \tilde\mu_k - 1$ for all k. Furthermore, let δ be an arbitrary fixed positive number, and let m be sufficiently large. Then* (11.5.18) *is satisfied for $i=1,2,3$, $|\alpha|=1$, $|\gamma| \le 1-\delta_{j,4}$, $x,y \in \mathcal{G}\cap\mathcal{U}_\mu$, $\xi \in \mathcal{G}\cap\mathcal{U}_\nu$, $x \ne y$, $|x-\xi| > m|x-y|$, $|x-\xi| > \delta$.*

P r o o f. First note that the inequalities $|x-\xi| > m|x-y|$ and $|x-\xi| > \delta$ imply $|y-\xi| > (m-1)|x-y|$ and $|y-\xi| > (m-1)\delta/m$. Suppose first that $r(x) < m|x-y|$. Then $r(y) < (m+1)|x-y|$, and Theorem 11.5.4 together with (3.1.2) yield

$$\begin{aligned}\big|\partial_x^\alpha \partial_\xi^\gamma G_{i,j}(x,\xi)\big| &\le c\,\rho_\mu(x)^{\Lambda_\mu-1-\varepsilon}\prod_{k\in I_\mu}\Big(\frac{r_k(x)}{\rho_\mu(x)}\Big)^{\mu_k'-1-\varepsilon}\\ &\le c\,r(x)^\sigma \le c\,m^\sigma|x-y|^\sigma\end{aligned}$$

and, analogously,

$$\big|\partial_y^\alpha \partial_\xi^\gamma G_{i,j}(y,\xi)\big| \le c\,(m+1)^\sigma\,|x-y|^\sigma$$

for $i \ne 4$, $|\alpha|=1$, $|\gamma| \le 1-\delta_{j,4}$. This implies (11.5.18).

If $r(x) > m|x-y|$, then there exists a point $z = x+t(y-x)$, $0<t<1$, such that

$$\begin{aligned}&\big|\partial_x^\alpha \partial_\xi^\gamma G_{i,j}(x,\xi) - \partial_y^\alpha \partial_\xi^\gamma G_{i,j}(y,\xi)\big| \le |x-y|\,\big|\nabla_z \partial_z^\alpha \partial_\xi^\gamma G_{i,j}(z,\xi)\big|\\ &\le c\,|x-y|\,\rho_\mu(z)^{\Lambda_\mu-2-\varepsilon}\prod_{k\in I_\mu}\Big(\frac{r_k(z)}{\rho_\mu(z)}\Big)^{\mu_k'-2-\varepsilon}\\ &\le c\,|x-y|\,r(z)^{\sigma-1} \le c\,(m-1)^{\sigma-1}|x-y|^\sigma.\end{aligned}$$

This proves the lemma. □

The last lemma allows us to restrict ourselves in the proof of the following two lemmas to the case when x and ξ lie in a neighborhood of the same vertex $x^{(\nu)}$.

LEMMA 11.5.9. *Let $\mathcal{G}$ be a convex polyhedral type domain, and let σ be a positive number, $\sigma < \tilde{\Lambda}_\nu - 1$ for all ν, and $\sigma < \tilde{\mu}_k - 1$ for all k. Then the estimate (11.5.18) is satisfied for $i = 1,2,3$, $|\alpha| = 1$, $|\gamma| \le 1 - \delta_{j,4}$, $x, y, \xi \in \mathcal{G}$, $x \ne y$, $|x-\xi| > m|x-y| > r(x)$.*

Proof. Since $|x-\xi| > m|x-y|$, where m is sufficiently large, we may suppose that x and y lie in a neighborhood $\mathcal{U}_\nu$ of the same vertex $x^{(\nu)}$. We assume that ξ lies in the same neighborhood and consider the following cases:

(1) $\xi \in \mathcal{U}_\nu$, $\rho_\nu(\xi) < \rho_\nu(x)/2$,

(2) $\xi \in \mathcal{U}_\nu$, $\rho_\nu(\xi) > 2\rho_\nu(x)$,

(3) $\xi \in \mathcal{U}_\nu$, $\rho_\nu(x)/2 < \rho_\nu(\xi) < 2\rho_\nu(x)$.

We start with *Case 1.* Then obviously $|x-\xi| < \rho_\nu(x) + \rho_\nu(\xi) < 3\rho_\nu(x)/2$, and Theorem 11.5.2 together with (3.1.2) yields

$$\begin{aligned} \big|\partial_x^\alpha \partial_\xi^\gamma G_{i,j}(x,\xi)\big| &\le c\,\rho_\nu(x)^{-2-\delta_{j,4}-|\gamma|} \Big(\frac{r(x)}{\rho_\nu(x)}\Big)^\sigma \\ &\le c\,|x-\xi|^{-2-\sigma-\delta_{j,4}-|\gamma|}\,|x-y|^\sigma \end{aligned}$$

for $i = 1,2,3$, $|\alpha| = 1$, $|\gamma| + \delta_{j,4} \le 1$. Since

$$\rho_\nu(y) > \rho_\nu(x) - |x-y| > \rho_\nu(x) - \frac{|x-\xi|}{m} > \Big(1 - \frac{3}{2m}\Big)\rho_\nu(x) > \Big(2 - \frac{3}{m}\Big)\rho_\nu(\xi),$$

we obtain analogously

$$\big|\partial_y^\alpha \partial_\xi^\gamma G_{i,j}(y,\xi)\big| \le c\,|y-\xi|^{-2-\sigma-\delta_{j,4}-|\gamma|}\,|x-y|^\sigma.$$

This proves (11.5.18) for Case 1.

Case 2. In this case, $|x-\xi| < 3\rho_\nu(\xi)/2$ and

$$\rho_\nu(y) < \rho_\nu(x) + |x-y| < \rho_\nu(x) + \frac{|x-\xi|}{m} < \Big(\frac{1}{2} + \frac{3}{2m}\Big)\,\rho_\nu(\xi).$$

Using Theorem 11.5.2 and the inequality $G_{i,j}(x,\xi) = G_{j,i}(\xi,x)$, we obtain

$$\begin{aligned} \big|\partial_x^\alpha \partial_\xi^\gamma G_{i,j}(x,\xi)\big| &\le c\,\rho_\nu(x)^{\Lambda_\nu-1-\varepsilon}\,\rho_\nu(\xi)^{-1-\Lambda_\nu-\delta_{j,4}-|\gamma|+\varepsilon}\Big(\frac{r(x)}{\rho_\nu(x)}\Big)^\sigma \\ &\le c\,\rho_\nu(\xi)^{-2-\sigma-\delta_{j,4}-|\gamma|}\,r(x)^\sigma \le c\,|x-\xi|^{-2-\sigma-\delta_{j,4}-|\gamma|}\,|x-y|^\sigma \end{aligned}$$

and, analogously,

$$\big|\partial_y^\alpha \partial_\xi^\gamma G_{i,j}(y,\xi)\big| \le c\,|y-\xi|^{-2-\sigma-\delta_{j,4}-|\gamma|}\,|x-y|^\sigma$$

for $i = 1,2,3$, $|\alpha| = 1$, $|\gamma| + \delta_{j,4} \le 1$. Thus, (11.5.18) holds for Case 2.

Case 3. Then $|x-\xi| < 3\rho_\nu(\xi)$, and from the conditions of the lemma it follows that

$$\Big(\frac{1}{2} - \frac{3}{m}\Big)\,\rho_\nu(\xi) < \rho_\nu(y) < \Big(2 + \frac{3}{m}\Big)\,\rho_\nu(\xi).$$

Furthermore, $r(x) < |x-\xi|$ and

$$r(y) < (m+1)\,|x-y| < \frac{m+1}{m}\,|x-\xi| < \frac{m+1}{m-1}\,|y-\xi|.$$

Applying Theorem 11.5.3, we obtain

$$\begin{aligned}\big|\partial_x^\alpha\partial_\xi^\gamma G_{i,j}(x,\xi)\big| &\le c\,|x-\xi|^{-2-\delta_{j,4}-|\gamma|}\Big(\frac{r(x)}{|x-\xi|}\Big)^\sigma\\ &\le c\,|x-\xi|^{-2-\sigma-\delta_{j,4}-|\gamma|}\,|x-y|^\sigma\end{aligned}$$

and analogously

$$\big|\partial_y^\alpha\partial_\xi^\gamma G_{i,j}(y,\xi)\big| \le c\,|y-\xi|^{-2-\sigma-\delta_{j,4}-|\gamma|}\,|x-y|^\sigma$$

for $|\alpha|=1$, $|\gamma|+\delta_{j,4}\le 1$. Consequently, the inequality (11.5.18) is also valid in Case 3. This completes the proof. □

It remains to prove (11.5.18) for the case $m|x-y|<\min(|x-\xi|, r(x))$.

LEMMA 11.5.10. *Let $\mathcal{G}$ be a convex polyhedral type domain, and let σ be a positive number, $\sigma<\tilde\Lambda_\nu-1$ for all ν, and $\sigma<\tilde\mu_k-1$ for all k. Then the estimate (11.5.18) is satisfied for $i=1,2,3$, $|\alpha|=1$, $|\gamma|\le 1-\delta_{j,4}$, $x,y,\xi\in\mathcal{G}$, $x\ne y$, $|x-\xi|>m|x-y|$, $r(x)>m|x-y|$.*

P r o o f. By the mean value theorem,

$$\big|\partial_x^\alpha\partial_\xi^\gamma G_{i,j}(x,\xi)-\partial_y^\alpha\partial_\xi^\gamma G_{i,j}(y,\xi)\big|\le |x-y|\,\big|\nabla_z\partial_z^\alpha\partial_\xi^\gamma G_{i,j}(z,\xi)\big|, \tag{11.5.19}$$

where $z=x+t(y-x)$, $0<t<1$. Since $m|x-y|<|x-\xi|$ and m is assumed to be sufficiently large, we may suppose that x,ξ and y lie in a neighborhood $\mathcal{U}_\nu$ of the same vertex $x^{(\nu)}$. We estimate the right-hand side of (11.5.19) for $|\alpha|=1$ and $|\gamma|\le 1-\delta_{j,4}$. To this end, we consider the same cases as in the proof of Lemma 11.5.9.

Case 1: $\xi\in\mathcal{U}_\nu$, $\rho_\nu(\xi)<\rho_\nu(x)/2$. Since $|x-\xi|<3\rho_\nu(x)/2$ and

$$\begin{aligned}\rho_\nu(z) &> \rho_\nu(x)-|x-y|>\rho_\nu(x)-\frac{|x-\xi|}{m}>\Big(1-\frac{3}{2m}\Big)\rho_\nu(x)\\ &> \Big(2-\frac{3}{m}\Big)\rho_\nu(\xi),\end{aligned}$$

Theorem 11.5.2 yields

$$\big|\nabla_z\partial_z^\alpha\partial_\xi^\gamma G_{i,j}(z,\xi)\big|\le c\,\rho_\nu(z)^{-3-\delta_{j,4}-|\gamma|}\Big(\frac{r(z)}{\rho_\nu(z)}\Big)^{\sigma-1}$$

for $i=1,2,3$, $|\alpha|=1$, $|\gamma|+\delta_{j,4}\le 1$. Using the inequalities $r(z)>(m-1)|x-y|$ and

$$\rho_\nu(z)>\rho_\nu(x)-|x-y|>\Big(\frac{2}{3}-\frac{1}{m}\Big)\,|x-\xi|,$$

we obtain

$$\big|\nabla_z\partial_z^\alpha\partial_\xi^\gamma G_{i,j}(z,\xi)\big|\le c\,|x-\xi|^{-2-\sigma-\delta_{j,4}-|\gamma|}\,|x-y|^{\sigma-1}.$$

This together with (11.5.19) implies (11.5.18).

Case 2: $\xi\in\mathcal{U}_\nu$, $\rho_\nu(x)<\rho_\nu(\xi)/2$. Then $|x-\xi|<3\rho_\nu(\xi)/2$ and

$$\rho_\nu(z)<\rho_\nu(x)+|x-y|<\rho_\nu(x)+\frac{|x-\xi|}{m}<\Big(\frac{1}{2}+\frac{3}{2m}\Big)\,\rho_\nu(\xi).$$

Theorem 11.5.2 and the equality $G_{i,j}(x,\xi)=G_{j,i}(\xi,x)$ imply

$$\big|\nabla_z\partial_z^\alpha\partial_\xi^\gamma G_{i,j}(z,\xi)\big| \le c\,\rho_\nu(z)^{\Lambda_\nu-2-\varepsilon}\,\rho_\nu(\xi)^{-1-\Lambda_\nu-\delta_{j,4}-|\gamma|+\varepsilon}\,\Big(\frac{r(z)}{\rho_\nu(z)}\Big)^{\sigma-1}$$

for $i = 1,2,3$, $|\alpha| = 1$, $|\gamma| + \delta_{j,4} \le 1$. The number ε can be chosen such that $\Lambda_\nu - 1 - \sigma - \varepsilon \ge 0$. Then we get

$$\begin{aligned} \big|\nabla_z \partial_z^\alpha \partial_\xi^\gamma G_{i,j}(z,\xi)\big| &\le c\,\rho_\nu(\xi)^{-2-\sigma-\delta_{j,4}-|\gamma|}\, r(z)^{\sigma-1} \\ &\le c\,|x-\xi|^{-2-\sigma-\delta_{j,4}-|\gamma|}\,|x-y|^{\sigma-1}. \end{aligned}$$

This proves (11.5.18) for the case $\xi \in \mathcal{U}_\nu$, $\rho_\nu(x) < \rho_\nu(\xi)/2$.

Case 3: $\xi \in \mathcal{U}_\nu$, $\rho_\nu(x)/2 < \rho_\nu(\xi) < 2\rho_\nu(x)$. In this case,

$$\Big(\frac{1}{2} - \frac{3}{m}\Big)\,\rho_\nu(\xi) < \rho_\nu(z) < \Big(2 + \frac{3}{m}\Big)\,\rho_\nu(\xi).$$

Furthermore, by the conditions of the lemma,

$$\Big(1 - \frac{1}{m}\Big)\,|x-\xi| < |z-\xi| < \Big(1 + \frac{1}{m}\Big)\,|x-\xi|$$

and

$$\Big(1 - \frac{1}{m}\Big)\,r(x) < r(z) < \Big(1 + \frac{1}{m}\Big)\,r(x).$$

Let $i \ne 4$, $|\alpha| = 1$ and $|\gamma| + \delta_{j,4} \le 1$. Using Theorem 11.5.3, we get

$$\begin{aligned} \big|\nabla_z \partial_z^\alpha \partial_\xi^\gamma G_{i,j}(z,\xi)\big| &\le c\,|z-\xi|^{-3-\delta_{j,4}-|\gamma|}\,\Big(\frac{r(z)}{|z-\xi|}\Big)^{\sigma-1} \\ &\le c\,|x-\xi|^{-2-\sigma-\delta_{j,4}-|\gamma|}\,r(x)^{\sigma-1} \\ &\le c\,|x-\xi|^{-2-\sigma-\delta_{j,4}-|\gamma|}\,|x-y|^{\sigma-1} \end{aligned}$$

for $|z-\xi| > \min(r(z), r(\xi))$ and

$$\begin{aligned} \big|\nabla_z \partial_z^\alpha \partial_\xi^\gamma G_{i,j}(z,\xi)\big| &\le c\,|z-\xi|^{-3-\delta_{j,4}-|\gamma|} \\ &\le c\,|x-\xi|^{-2-\sigma-\delta_{j,4}-|\gamma|}\,|x-y|^{\sigma-1} \end{aligned}$$

for $|z-\xi| < \min(r(z), r(\xi))$. This together with (11.5.19) implies (11.5.18). The proof is complete. □

Lemmas 11.5.7–11.5.10 imply the following theorem.

THEOREM 11.5.11. *Suppose that $\mathcal{G}$ is a convex domain of polyhedral type and that σ is a positive number, $\sigma < \tilde{\mu}_k - 1$ for all k, and $\sigma < \tilde{\Lambda}_\nu - 1$ for all ν. Then the estimate*

$$\begin{aligned} &\frac{\big|\partial_x^\alpha \partial_\xi^\gamma G_{i,j}(x,\xi) - \partial_y^\alpha \partial_\xi^\gamma G_{i,j}(y,\xi)\big|}{|x-y|^\sigma} \\ &\quad \le c\,\big(|x-\xi|^{-2-\sigma-\delta_{j,4}-|\gamma|} + |y-\xi|^{-2-\sigma-\delta_{j,4}-|\gamma|}\big) \end{aligned} \tag{11.5.20}$$

holds for $x \ne y$, $i = 1,2,3$, $|\alpha| = 1$, $|\gamma| \le 1 - \delta_{j,4}$. Analogously,

$$\begin{aligned} &\frac{\big|\partial_x^\alpha \partial_\xi^\gamma G_{i,j}(x,\xi) - \partial_x^\alpha \partial_\xi^\gamma G_{i,j}(x,\eta)\big|}{|\xi-\eta|^\sigma} \\ &\quad \le c\,\big(|x-\xi|^{-2-\sigma-\delta_{i,4}-|\alpha|} + |x-\eta|^{-2-\sigma-\delta_{i,4}-|\alpha|}\big) \end{aligned}$$

for $j = 1,2,3$, $|\alpha| \le 1 - \delta_{i,4}$, $|\gamma| = 1$.

Note that the second inequality of Theorem 11.5.11 results directly from the estimate (11.5.20) and from the equality $G_{i,j}(x,\xi) = G_{j,i}(\xi,x)$.

11.6. Maximum modulus estimates for solutions of the Stokes and Navier-Stokes system

We consider the boundary value problem

$$(11.6.1)\qquad -\Delta u + \nabla p = 0, \quad \nabla \cdot u = 0 \ \text{ in } \mathcal{G},$$

$$(11.6.2)\qquad u = h \ \text{ on } \partial\mathcal{G}$$

in a bounded domain of polyhedral type. Using the estimates of the Green's matrix in Section 11.4, we obtain weighted L_∞ estimates for the solution (u,p) and for the derivatives of u. In particular, it follows from these results that

$$\|u\|_{L_\infty(\mathcal{G})^3} \le c\, \|h\|_{L_\infty(\partial\mathcal{G})^3}$$

with a constant c independent of h.

Furthermore, we are interested in L_∞-estimates for solutions of the nonlinear Navier-Stokes system

$$(11.6.3)\qquad -\nu\, \Delta u + (u\cdot\nabla)\, u + \nabla p = 0, \quad \nabla \cdot u = 0 \ \text{ in } \mathcal{G},$$

$$(11.6.4)\qquad u = h \ \text{ on } \partial\mathcal{G}.$$

We obtain an inequality of the form

$$\|u\|_{L_\infty(\mathcal{G})^3} \le F\left(\|h\|_{L_\infty(\partial\mathcal{G})^3}\right),$$

with a certain function F. The function F constructed here has the form

$$F(t) = c_0 t\, e^{c_1 t/\nu},$$

where c_0 and c_1 are positive constants independent of ν.

11.6.1. Generalized solutions of the boundary value problem. We assume that $h \in \mathcal{V}^{0,\infty}_{\beta,\delta}(\partial\mathcal{G})^3$ (cf. Subsection 5.2.2), i.e. there exists an *approximating sequence* $\{h^{(\nu)}\}_{\nu=1}^\infty \subset C_0^\infty(\partial\mathcal{G}\backslash\mathcal{S})^3$ such that

$$\|h^{(\nu)}\|_{L^\infty_{\beta,\delta}(\partial\mathcal{G})^3} \le c < \infty \ \text{ for } j = 1,\dots,N,$$

$$h^{(\nu)} \to h \quad \text{a.e. on } \partial\mathcal{G}.$$

Here the components of β and δ satisfy the inequalities

$$(11.6.5)\qquad |\delta_k| < \delta_+^{(k)} \quad \text{for } k = 1,\dots,d,$$

and

$$(11.6.6)\qquad |\beta_j - 1/2| < \Lambda_j^+ + 1/2 \quad \text{for } j = 1,\dots,d',$$

where Λ_j^+ denotes the greatest real number such that the strip $-1/2 < \text{Re}\,\lambda < \Lambda_j^+$ is free of eigenvalues of the pencil $\mathfrak{A}_j(\lambda)$. Since the spectra of the pencils $\mathfrak{A}_j(\lambda)$ and $A_\xi(\lambda)$ contain the eigenvalue $\lambda = 1$, we have $\delta_+^{(k)} \le 1$ and $\Lambda_j^+ \le 1$. In particular, the conditions (11.6.5) and (11.6.6) imply that $\beta_j < 2$ and $\delta_k < 1$. Consequently,

$$\Big|\int_{\mathcal{G}} h(x)\cdot n\, dx\Big| \le c\, \|h\|_{L^\infty_{\beta,\delta}(\partial\mathcal{G})^3}$$

for arbitrary $h \in L^\infty_{\beta,\delta}(\mathcal{G})^3$. We suppose that

$$(11.6.7)\qquad \int_{\mathcal{G}} h(x)\cdot n\, dx = 0.$$

Let $G(x,\xi)$ be the Green's matrix introduced in Theorem 11.4.1. Furthermore, let $\vec{H}_i(x,\xi)$ denote the vector $\big(G_{i,1}(x,\xi), G_{i,2}(x,\xi), G_{i,3}(x,\xi)\big)^t$. Then the vector function $(u^{(\nu)}, p^{(\nu)})$ with the components

$$u_i^{(\nu)}(x) = \int\limits_{\partial\mathcal{G}\backslash\mathcal{S}} h^{(\nu)}(\xi)\cdot\Big(-\frac{\partial \vec{H}_i(x,\xi)}{\partial n_\xi} + G_{i,4}(x,\xi)\, n_\xi\Big)\, d\xi,\ i=1,2,3,$$

$$p^{(\nu)}(x) = \int\limits_{\partial\mathcal{G}\backslash\mathcal{S}} h^{(\nu)}(\xi)\cdot\Big(-\frac{\partial \vec{H}_4(x,\xi)}{\partial n_\xi} + G_{4,4}(x,\xi)\, n_\xi\Big)\, d\xi$$

is the uniquely determined solution of the problem

$$-\Delta u + \nabla p = 0, \quad \nabla\cdot u = C_\nu \ \text{ in } \mathcal{G},$$
$$u = h^{(\nu)} \ \text{ on } \partial\mathcal{G}\backslash\mathcal{S}, \qquad \int_{\mathcal{G}} p(x)\, dx = 0$$

(see Subsection 11.4.1), where

$$C_\nu = (\text{meas}(\mathcal{G}))^{-1}\int_{\partial\mathcal{G}\backslash\mathcal{S}} h^{(\nu)}(x)\cdot n\, dx \to 0 \ \text{ as } \nu\to\infty.$$

The sequence $\big(u^{(\nu)}(x), p^{(\nu)}(x)\big)$ converges for every $x\in\mathcal{G}$. Obviously, the limit $\big(u(x), p(x)\big)$ does not depend on the choice of the approximating sequence $\{h^{(\nu)}\}$. The components of (u,p) admit the representation

$$u_i(x) = \int\limits_{\partial\mathcal{G}\backslash\mathcal{S}} h(\xi)\cdot\Big(-\frac{\partial \vec{H}_i(x,\xi)}{\partial n_\xi} + G_{i,4}(x,\xi)\, n_\xi\Big)\, d\xi,\ i=1,2,3,$$

$$p(x) = \int\limits_{\partial\mathcal{G}\backslash\mathcal{S}} h(\xi)\cdot\Big(-\frac{\partial \vec{H}_4(x,\xi)}{\partial n_\xi} + G_{4,4}(x,\xi)\, n_\xi\Big)\, d\xi.$$

The vector function (u,p) is called *generalized solution* of the problem (11.6.1), (11.6.2).

11.6.2. Weighted L_∞estimates of generalized solutions. Let (u,p) be a generalized solution of the boundary value problem (11.6.1), (11.6.2). We prove a weighted L_∞-estimate for the vector function u.

THEOREM 11.6.1. *Suppose that $h\in\mathcal{V}^{0,\infty}_{\beta,\delta}(\partial\mathcal{G})^3$, where β and δ satisfy the conditions* (11.6.5) *and* (11.6.6), *respectively, and that h satisfies the condition* (11.6.7). *Then*

$$\|u\|_{L^\infty_{\beta,\delta}(\mathcal{G})^3} \le c\sum_{\nu=1}^{N}\|h\|_{L^\infty_{\beta,\delta}(\Gamma_\nu)^3} \tag{11.6.8}$$

with a constant c independent of h_j.

P r o o f. Suppose that x lies in a neighborhood $\mathcal{U}_1$ of the vertcx $x^{(1)}$. We have to estimate the integral

$$\int\limits_{\Gamma_\nu} K(x,\xi)\, h_j(\xi)\, d\xi,$$

where $K(x,\xi)$ is one of the functions

$$K(x,\xi) = \frac{\partial G_{i,j}(x,\xi)}{\partial n_\xi} \quad \text{and} \quad K(x,\xi) = G_{i,4}(x,\xi)\, n_j,$$

$i,j = 1,2,3$. First, we consider the integral

$$\int\limits_{\Gamma_\nu \cap \mathcal{U}_l} K(x,\xi)\, h_j(\xi)\, d\xi \tag{11.6.9}$$

for $x \in \mathcal{G} \cap \mathcal{U}_1$, where $\mathcal{U}_l$ is a (sufficiently small) neighborhood of the vertex $x^{(l)}$, $l \neq 1$. Using Theorem 11.4.2, we obtain

$$\Big| \int\limits_{\Gamma_\nu \cap \mathcal{U}_l} K(x,\xi)\, h_j(\xi)\, d\xi \Big| \le c\, \|h_j\|_{L^\infty_{\beta,\delta}(\Gamma_\nu)}\, \rho_1(x)^{\Lambda_1^+ - \varepsilon} \prod_{k \in X_1} \Big(\frac{r_k(x)}{\rho_1(x)}\Big)^{\delta_+^{(k)} - \varepsilon}$$

$$\times \int\limits_{\Gamma_\nu \cap \mathcal{U}_l} \rho_l(\xi)^{\Lambda_l^+ - 1 - \beta_l - \varepsilon} \prod_{k \in X_l} \Big(\frac{r_k(\xi)}{\rho_l(\xi)}\Big)^{\delta_+^{(k)} - 1 - \delta_k - \varepsilon} d\xi$$

$$\le c\, \|h_j\|_{L^\infty_{\beta,\delta}(\Gamma_\nu)}\, \rho_1(x)^{-\beta_1} \prod_{k \in X_1} \Big(\frac{r_k(x)}{\rho_1(x)}\Big)^{-\delta_k}.$$

Next, we show that the integral

$$I(x) = \int\limits_{\Gamma_\nu \cap \mathcal{U}_1} K(x,\xi)\, h_j(\xi)\, d\xi$$

satisfies the estimate

$$\rho_1(x)^{\beta_1} \prod_{k \in X_1} \Big(\frac{r_k(x)}{\rho_1(x)}\Big)^{\delta_k} |I(x)| \le c\, \|h_j\|_{L^\infty_{\beta,\delta}(\Gamma_\nu)} \tag{11.6.10}$$

with a constant c independent of $x \in \mathcal{G} \cap \mathcal{U}_1$. Let A_1, A_2, A_3, A_4 be the following subsets of $\mathcal{U}_1 \cap \Gamma_\nu$:

$$A_1 = \{\xi \in \mathcal{U}_1 \cap \Gamma_\nu :\ \rho_1(\xi) > 2\rho_1(x)\},$$
$$A_2 = \{\xi \in \mathcal{U}_1 \cap \Gamma_\nu :\ \rho_1(\xi) < \rho_1(x)/2\},$$
$$A_3 = \{\xi \in \mathcal{U}_1 \cap \Gamma_\nu :\ \rho_1(x)/2 < \rho_1(\xi) < 2\rho_1(x),\ |x-\xi| > \min(r(x), r(\xi))\},$$
$$A_4 = \{\xi \in \mathcal{U}_1 \cap \Gamma_\nu :\ \rho_1(x)/2 < \rho_1(\xi) < 2\rho_1(x),\ |x-\xi| < \min(r(x), r(\xi))\}.$$

By Theorems 11.4.5 and 11.4.6, the following estimates are valid for $x \in \mathcal{U}_1$ and $\xi \in \Gamma_\nu \cap \mathcal{U}_1$:

$$|K(x,\xi)| \le c\, \frac{\rho_1(x)^{\Lambda_1^+ - \varepsilon}}{\rho_1(\xi)^{\Lambda_1^+ + 2 - \varepsilon}} \prod_{k \in X_1} \Big(\frac{r_k(x)}{\rho_1(x)}\Big)^{\delta_+^{(k)} - \varepsilon} \prod_{k \in X_1} \Big(\frac{r_k(\xi)}{\rho_1(\xi)}\Big)^{\delta_+^{(k)} - 1 - \varepsilon} \quad \text{for } \xi \in A_1,$$

$$|K(x,\xi)| \le c\, \frac{\rho_1(\xi)^{\Lambda_1^+ - 1 - \varepsilon}}{\rho_1(x)^{\Lambda_1^+ + 1 - \varepsilon}} \prod_{k \in X_1} \Big(\frac{r_k(x)}{\rho_1(x)}\Big)^{\delta_+^{(k)} - \varepsilon} \prod_{k \in X_1} \Big(\frac{r_k(\xi)}{\rho_1(\xi)}\Big)^{\delta_+^{(k)} - 1 - \varepsilon} \quad \text{for } \xi \in A_2,$$

$$|K(x,\xi)| \le c\, |x-\xi|^{-2} \Big(\frac{r(x)}{|x-\xi|}\Big)^{\delta_+^{(k(x))} - \varepsilon} \Big(\frac{r(\xi)}{|x-\xi|}\Big)^{\delta_+^{(k(\xi))} - 1 - \varepsilon} \quad \text{for } \xi \in A_3.$$

Furthermore, it follows from Theorem 11.4.7 that

$$|K(x,\xi)| \le c\, d(x)\, |x-\xi|^{-3} \quad \text{for } \xi \in A_4,$$

where $d(x)$ denotes the distance of x from the boundary $\partial\mathcal{G}$ (see the proof of Corollary 5.1.4). We write $I(x)$ as the sum

$$I(x) = I_1 + I_2 + I_3 + I_4,$$

where I_μ is the integral of $K(x,\xi)\, h_j(\xi)$ over the set A_μ, $\mu = 1,2,3,4$. Then

$$\begin{aligned} |I_1| &\le c\, \|h_j\|_{L^\infty_{\beta,\delta}(\Gamma_\nu)}\, \rho_1(x)^{\Lambda_1^+ - \varepsilon} \prod_{k\in X_1} \Big(\frac{r_k(x)}{\rho_1(x)}\Big)^{\delta_+^{(k)}-\varepsilon} \\ &\qquad \times \int_{A_1} \rho_1(\xi)^{-\beta_1-\Lambda_1^+-2+\varepsilon} \prod_{k\in X_1} \Big(\frac{r_k(\xi)}{\rho_1(\xi)}\Big)^{-\delta_k+\delta_+^{(k)}-1-\varepsilon} d\xi \\ &\le c\, \|h_j\|_{L^\infty_{\beta,\delta}(\Gamma_\nu)}\, \rho_1(x)^{-\beta_1} \prod_{k\in X_1} \Big(\frac{r_k(x)}{\rho_1(x)}\Big)^{\delta_+^{(k)}-\varepsilon} \\ &\le c\, \|h_j\|_{L^\infty_{\beta,\delta}(\Gamma_\nu)}\, \rho_1(x)^{-\beta_1} \prod_{k\in X_1} \Big(\frac{r_k(x)}{\rho_1(x)}\Big)^{-\delta_k}. \end{aligned}$$

Analogously, the inequality

$$|I_\mu| \le c\, \|h_j\|_{L^\infty_{\beta,\delta}(\Gamma_\nu)}\, \rho_1(x)^{-\beta_1} \prod_{k\in X_1} \Big(\frac{r_k(x)}{\rho_1(x)}\Big)^{-\delta_k} \tag{11.6.11}$$

holds for $\mu = 2$. We may suppose without loss of generality that M_1 is the nearest edge to x. Then we denote by A_3', A_3'' the following subsets of A_3:

$$A_3' = \{\xi \in A_3 :\ r_1(\xi) < \min_{2\le k\le d} r_k(\xi)\}, \qquad A_3'' = A_3 \backslash A_3'.$$

Since $|x-\xi| > c\,\rho_1(x)$ for $\xi \in A_3''$, we obtain

$$\begin{aligned} \Big| \int_{A_3''} K(x,\xi)\, h_j(\xi)\, d\xi \Big| &\le c\, \|h_j\|_{L^\infty_{\beta,\delta}(\Gamma_\nu)}\, \rho_1(x)^{-\beta_1-1-\delta_+^{(1)}+2\varepsilon}\, r_1(x)^{\delta_+^{(1)}-\varepsilon} \\ &\qquad \times \int_{A_3''} \rho_1(\xi)^{-\delta_+^{(k(\xi))}}\, r(\xi)^{\delta_-^{(k(\xi))}-1-\varepsilon} \prod_{k\in X_1} \Big(\frac{r_k(\xi)}{\rho_1(\xi)}\Big)^{-\delta_k} d\xi. \end{aligned}$$

The integral on the right-hand side is dominated by $c\,\rho_1(x)^{1-\varepsilon}$. Consequently,

$$\begin{aligned} \Big| \int_{A_3''} K(x,\xi)\, h_j(\xi)\, d\xi \Big| &\le c\, \|h_j\|_{L^\infty_{\beta,\delta}(\Gamma_\nu)}\, \rho_1(x)^{-\beta_1} \Big(\frac{r(x)}{\rho_1(x)}\Big)^{\delta_+^{(1)}-\varepsilon} \\ &\le c\, \|h_j\|_{L^\infty_{\beta,\delta}(\Gamma_\nu)}\, \rho_1(x)^{-\beta_1} \Big(\frac{r(x)}{\rho_1(x)}\Big)^{-\delta_1}. \end{aligned}$$

Furthermore,

$$\begin{aligned} \Big| \int_{A_3'} K(x,\xi)\, h_j(\xi)\, d\xi \Big| &\le c\, \|h_j\|_{L^\infty_{\beta,\delta}(\Gamma_\nu)}\, \rho_1(x)^{\delta_1-\beta_1}\, r_1(x)^{\delta_+^{(1)}-\varepsilon} \\ &\qquad \times \int_{A_3'} |x-\xi|^{-1-2\delta_+^{(1)}+2\varepsilon}\, r_1(\xi)^{\delta_+^{(1)}-1-\delta_1-\varepsilon}\, d\xi. \end{aligned}$$

As was shown in the proof of Lemma 5.2.1, the integral on the right-hand side satisfies the estimate

$$\int_{A_3'} |x-\xi|^{-1-2\delta_+^{(1)}+2\varepsilon}\, r_1(\xi)^{\delta_+^{(1)}-1-\delta_1-\varepsilon}\, d\xi \le c\, r_1(x)^{-\delta_1-\delta_+^{(1)}+\varepsilon}.$$

This implies

$$\Big|\int_{A_3'} K(x,\xi)\, h_j(\xi)\, d\xi\Big| \le c\, \|h_j\|_{L^\infty_{\beta,\delta}(\Gamma_\nu)}\, \rho_1(x)^{-\beta_1} \Big(\frac{r_1(x)}{\rho_1(x)}\Big)^{-\delta_1}.$$

Thus, the estimate (11.6.11) is valid for $\mu = 3$. Finally, using the estimate for $K(x,\xi)$ in A_4, we obtain

$$|I_4| \le c\, d(x)\, \|h_j\|_{L^\infty_{\beta,\delta}(\Gamma_\nu)}\, \rho_1(x)^{-\beta_1} \prod_{k\in X_1} \Big(\frac{r_k(x)}{\rho_1(x)}\Big)^{-\delta_k} \int_{A_4} |x-\xi|^{-3}\, d\xi.$$

Since

$$\int_{\Gamma_\nu} |x-\xi|^{-3}\, dx \le c\, d(x)^{-1},$$

it follows that (11.6.11) is true for $\mu = 4$. This proves the inequality (11.6.10). The proof of the theorem is complete. □

Let again $d(x)$ denote the distance of the point x from the boundary $\partial\mathcal{G}$. We prove the following weighted L_∞ estimates for the function p and the derivatives of u.

LEMMA 11.6.2. *Suppose that the components of β and δ satisfy the conditions* (11.6.5) *and* (11.6.6), *respectively. Then the generalized solution of the problem* (11.6.1), (11.6.2) *satisfies the estimate*

$$\sum_{k=1}^{3} \|d\,\partial_{x_k} u\|_{L^\infty_{\beta,\delta}(\mathcal{G})^3} + \|d\,p\|_{L^\infty_{\beta,\delta}(\mathcal{G})} \le c \sum_{\nu=1}^{N} \|h\|_{L^\infty_{\beta,\delta}(\Gamma_\nu)^3}.$$

P r o o f. Suppose that x lies in a sufficiently small neighborhood $\mathcal{U}_1$ of the vertex $x^{(1)}$. We have to estimate the integral

$$\int_{\Gamma_\nu} K(x,\xi)\, h_j(\xi)\, d\xi\,,$$

where $K(x,\xi)$ is one of the following functions:

$$\frac{\partial}{\partial x_k}\frac{\partial}{\partial n_\xi} G_{i,j}(x,\xi), \quad \frac{\partial G_{i,4}(x,\xi)}{\partial x_k}\, n_j(\xi), \quad \frac{\partial}{\partial n_\xi} G_{4,j}(x,\xi), \quad G_{4,4}(x,\xi)\, n_j(\xi),$$

$i,j = 1,2,3$. Analogously to the proof of Theorem 11.6.1, we obtain

$$\Big|\int\limits_{\Gamma_\nu\cap\mathcal{U}_l} K(x,\xi)\, h_j(\xi)\, d\xi\Big| \le c\, \|h_j\|_{L^\infty_{\beta,\delta}(\Gamma_\nu)}\, \rho_1(x)^{\Lambda_1^+-1-\varepsilon} \prod_{k\in X_1} \Big(\frac{r_k(x)}{\rho_1(x)}\Big)^{\delta_+^{(k)}-1-\varepsilon}$$

$$\times \int\limits_{\Gamma_\nu\cap\mathcal{U}_l} \rho_1(\xi)^{\Lambda_l^+-1-\beta_l-\varepsilon} \prod_{k\in X_l} \Big(\frac{r_k(\xi)}{\rho_l(\xi)}\Big)^{\delta_+^{(k)}-1-\delta_k-\varepsilon} d\xi$$

$$\le c\, \|h_j\|_{L^\infty_{\beta,\delta}(\Gamma_\nu)}\, r(x)^{-1}\, \rho_1(x)^{-\beta_1} \prod_{k\in X_1} \Big(\frac{r_k(x)}{\rho_1(x)}\Big)^{-\delta_k}$$

if $\mathcal{U}_l$ is a (sufficiently small) neighborhood of the vertex $x^{(l)}$, $l \ne 1$. We consider the integrals

$$I_\mu = \int_{A_\mu} K(x,\xi)\, h_j(\xi)\, d\xi, \quad \mu = 1,2,3,4,$$

where A_1, A_2, A_3, A_4 are the same subsets of Γ_ν as in the proof of Theorem 11.6.1. According to Theorems 11.4.5–11.4.7, the function $K(x,\xi)$ satisfies the following estimates for $x \in \mathcal{U}_1$ and $\xi \in \Gamma_\nu \cap \mathcal{U}_1$:

$$\big|K(x,\xi)\big| \le c\, \frac{\rho_1(x)^{\Lambda_1^+ -1-\varepsilon}}{\rho_1(\xi)^{\Lambda_1^+ +2-\varepsilon}} \prod_{k\in X_1} \Big(\frac{r_k(x)}{\rho_1(x)}\,\frac{r_k(\xi)}{\rho_1(\xi)}\Big)^{\delta_+^{(k)}-1-\varepsilon} \quad \text{for } \xi \in A_1,$$

$$\big|K(x,\xi)\big| \le c\, \frac{\rho_1(\xi)^{\Lambda_1^+ -1-\varepsilon}}{\rho_1(x)^{\Lambda_1^+ +2-\varepsilon}} \prod_{k\in X_1} \Big(\frac{r_k(x)}{\rho_1(x)}\,\frac{r_k(\xi)}{\rho_1(\xi)}\Big)^{\delta_+^{(k)}-1-\varepsilon} \quad \text{for } \xi \in A_2,$$

$$\big|K(x,\xi)\big| \le c\, |x-\xi|^{-3} \Big(\frac{r(x)}{|x-\xi|}\Big)^{\delta_+^{(k(x))}-1-\varepsilon} \Big(\frac{r(\xi)}{|x-\xi|}\Big)^{\delta_+^{(k(\xi))}-1-\varepsilon} \quad \text{for } \xi \in A_3,$$

$$\big|K(x,\xi)\big| \le c\, |x-\xi|^{-3} \quad \text{for } \xi \in A_4.$$

Thus, we obtain

$$\begin{aligned} |I_1| \;&\le\; c\, \|h_j\|_{L^\infty_{\beta,\delta}(\Gamma_\nu)}\, \rho_1(x)^{\Lambda_1^+-1-\varepsilon} \prod_{k\in X_1} \Big(\frac{r_k(x)}{\rho_1(x)}\Big)^{\delta_+^{(k)}-1-\varepsilon} \\ &\qquad \times \int_{A_1} \rho_1(\xi)^{-\beta_1-\Lambda_1^+-2+\varepsilon} \prod_{k\in X_1} \Big(\frac{r_k(\xi)}{\rho_1(\xi)}\Big)^{-\delta_k+\delta_+^{(k)}-1-\varepsilon} d\xi \\ &\le\; c\, \|h_j\|_{L^\infty_{\beta,\delta}(\Gamma_\nu)}\, \rho_1(x)^{-\beta_1-1} \prod_{k\in X_1} \Big(\frac{r_k(x)}{\rho_1(x)}\Big)^{\delta_+^{(k)}-1-\varepsilon} \\ &\le\; c\, \|h_j\|_{L^\infty_{\beta,\delta}(\Gamma_\nu)}\, r(x)^{-1}\, \rho_1(x)^{-\beta_1} \prod_{k\in X_1} \Big(\frac{r_k(x)}{\rho_1(x)}\Big)^{-\delta_k}. \end{aligned}$$

In the same way, this estimate holds for I_2. Furthermore, by the same arguments as in the proof of Theorem 11.6.1, we get

$$|I_3| \le c\, \|h_j\|_{L^\infty_{\beta,\delta}(\Gamma_\nu)}\, r(x)^{-1}\, \rho_1(x)^{-\beta_1} \prod_{k\in X_1} \Big(\frac{r_k(x)}{\rho_1(x)}\Big)^{-\delta_k}$$

and

$$\begin{aligned} |I_4| \;&\le\; c\, \|h_j\|_{L^\infty_{\beta,\delta}(\Gamma_\nu)}\, \rho_1(x)^{-\beta_1} \prod_{k\in X_1} \Big(\frac{r_k(x)}{\rho_1(x)}\Big)^{-\delta_k} \int_{A_4} |x-\xi|^{-3}\, d\xi \\ &\le\; c\, \|h_j\|_{L^\infty_{\beta,\delta}(\Gamma_\nu)}\, d(x)^{-1} \rho_1(x)^{-\beta_1} \prod_{k\in X_1} \Big(\frac{r_k(x)}{\rho_1(x)}\Big)^{-\delta_k}. \end{aligned}$$

Thus,

$$|I_\mu| \le c\, \|h_j\|_{L_\infty(\Gamma_\nu)}\, d(x)^{-1}\, \rho_1(x)^{-\beta_1} \prod_{k\in X_1} \Big(\frac{r_k(x)}{\rho_1(x)}\Big)^{-\delta_k} \quad \text{for } \mu = 1,2,3,4.$$

The result follows. □

11.6.3. The maximum principle for solutions of the Stokes system. It is known that $\Lambda_j^+ > 0$ (cf. [**85**, Theorem 5.5.6]). Thus, we can set $\beta_j = 0$ and $\delta_k = 0$ in Theorem 11.6.1 and Lemma 11.6.2. This leads to the following result.

THEOREM 11.6.3. *Let* (u,p) *be the generalized solution of the problem* (11.6.1), (11.6.2), *where* $h \in \mathcal{V}_{0,0}^{0,\infty}(\partial\mathcal{G})^3$ *and* h *satisfies the condition* (11.6.7). *Then the estimates*

$$\|u\|_{L_\infty(\mathcal{G})^3} \le c\,\|h\|_{L_\infty(\partial\mathcal{G})^3}$$

and

$$\sum_{k=1}^{3} \|d\,\partial_{x_k} u\|_{L_\infty(\mathcal{G})^3} + \|d\,p\|_{L_\infty(\mathcal{G})} \le c\,\|h\|_{L_\infty(\partial\mathcal{G})^3}$$

hold with a constant c *independent of* h.

11.6.4. A L_∞ estimate for the solution of the inhomogeneous Stokes system. By Theorems 11.1.2 and 11.1.8, the boundary value problem

$$-\Delta u + \nabla p = f, \quad \nabla\cdot u = 0 \ \text{ in } \mathcal{G}, \quad u = 0 \ \text{ on } \partial\mathcal{G} \tag{11.6.12}$$

is uniquely solvable (up to vector functions of the form $(0,c)$, where c is a constant) in $V_{0,0}^{1,s}(\mathcal{G})^3 \times V_{0,0}^{0,s}(\mathcal{G})$ for arbitrary $f \in V_{0,0}^{-1,s}(\mathcal{G})^3$, $s \ge 2$, if $2/s > 1 - \delta_+^{(k)}$ for all k and $3/s > 1 - \Lambda_j^+$ for all j. The solution satisfies the estimate

$$\|u\|_{V_{0,0}^{1,s}(\mathcal{G})^3} \le c\,\|f\|_{V_{0,0}^{-1,s}(\mathcal{G})^3} \tag{11.6.13}$$

for arbitrary $f \in V_{0,0}^{-1,s}(\mathcal{G})^3$. By Theorem 11.4.1, the components of the vector function u are given by the formula

$$u_i(x) = \int_{\mathcal{G}} \sum_{j=1}^{3} G_{i,j}(x,\xi)\, f_j(\xi)\, d\xi. \tag{11.6.14}$$

LEMMA 11.6.4. *Suppose that* $f = \partial_{x_j} g$, *where* $j \in \{1,2,3\}$. *If* $g \in L_s(\mathcal{G})^3$, $s > 3$, *then*

$$\|u\|_{L_\infty(\mathcal{G})^3} \le c\,\|g\|_{L_s(\mathcal{G})^3}\,. \tag{11.6.15}$$

If $g \in L_3(\mathcal{G})^3$, *then*

$$\|u\|_{L_s(\mathcal{G})^3} \le c\,\|g\|_{L_3(\mathcal{G})^3} \tag{11.6.16}$$

for arbitrary s, $1 < s < \infty$.

P r o o f. Let $g \in L_s(\Omega)$, $s > 3$. Since $\delta_+^{(k)} > 1/2$ and $\Lambda_j^+ > 0$ (cf. [**85**, Theorems 5.1.2 and 5.5.6], there exists a positive number ε such that

$$3 + \varepsilon < s, \quad \frac{2}{3+\varepsilon} > 1 - \delta_+^{(k)}, \quad \frac{3}{3+\varepsilon} > 1 - \Lambda_j^+$$

for all j and k. Thus, it follows from (11.6.13) and from the continuity of the imbeddings

$$V_{0,0}^{1,3+\varepsilon}(\mathcal{G}) \subset W^{1,3+\varepsilon}(\mathcal{G}) \subset L_\infty(\mathcal{G})$$

that

$$\|u\|_{L_\infty(\mathcal{G})} \le c_1\,\|u\|_{W^{1,3+\varepsilon}(\mathcal{G})} \le c_2\,\|u\|_{V_{0,0}^{1,3+\varepsilon}(\mathcal{G})} \le c_3\,\|g\|_{L_{3+\varepsilon}(\mathcal{G})} \le c\,\|g\|_{L_s(\mathcal{G})}\,.$$

Analogously, we obtain

$$\|u\|_{L_s(\mathcal{G})} \le c_1\,\|u\|_{W^{1,3}(\mathcal{G})} \le c_2\,\|u\|_{V_{0,0}^{1,3}(\mathcal{G})} \le c\,\|g\|_{L_3(\mathcal{G})}$$

if $g \in L_3(\mathcal{G})$. The lemma is proved. □

11.6.5. Representation of the solution of problem (11.6.1), (11.6.2). The following lemma concerns again the solution (u, p) of the linear boundary value problem (11.6.1), (11.6.2). We show that u is the rotation of a vector function $b \in W^{1,6}(\mathcal{G}))^3$. This fact is used in the next subsection, where we obtain a L_∞ estimate for the solution of the nonlinear Navier-Stokes system.

LEMMA 11.6.5. *Let (u,p) be a solution of the boundary value problem* (11.6.1), (11.6.2), *where $h \in L_\infty(\partial\mathcal{G})^3$. Then there exists a vector function $b \in W^{1,6}(\mathcal{G}))^3$ such that $u = \operatorname{rot} b$ and*

$$\|b\|_{W^{1,6}(\mathcal{G})^3} \le c\, \|h\|_{L_\infty(\partial\mathcal{G})^3} \tag{11.6.17}$$

with a constant c independent of h.

P r o o f. Let B_ρ be a ball with radius ρ centered at the origin and such that $\overline{\mathcal{G}} \subset B_\rho$. Furthermore, let $(u^{(1)}, q)$ be a solution of the problem

$$-\Delta u^{(1)} + \nabla q = 0, \quad \nabla \cdot u^{(1)} = 0 \text{ in } B_\rho \backslash \overline{\mathcal{G}}, \quad u^{(1)}|_{\partial\mathcal{G}} = h, \quad u^{(1)}|_{\partial B_\rho} = 0.$$

Obviously, the vector function

$$w(x) = \begin{cases} u(x) & \text{for } x \in \mathcal{G}, \\ u^{(1)}(x) & \text{for } x \in B_\rho \backslash \mathcal{G} \end{cases}$$

satisfies the equality $\nabla \cdot w = 0$ in the sense of distributions in B_ρ. By Theorem 11.6.3, the L_∞ norms of u and $u^{(1)}$ can be estimated by the L_∞ norm of h. Hence,

$$\|w\|_{L_6(B_\rho)} \le c\, \|h\|_{L_\infty(\partial\mathcal{G})} ,$$

where c is a constant independent of h. Suppose that there exists a vector function $v \in W^{2,6}(B_\rho)^3$ satisfying the equations

$$-\Delta v = w \;\text{ in } B_\rho, \quad \nabla \cdot v = 0 \;\text{ on } \partial B_\rho \tag{11.6.18}$$

and the inequality

$$\|v\|_{W^{2,6}(B_\rho)^3} \le c\, \|w\|_{L_6(B_\rho)^3} . \tag{11.6.19}$$

Since $\Delta(\nabla \cdot v) = -\nabla \cdot w = 0$ in B_ρ it follows that $\nabla \cdot v = 0$ in B_ρ. Consequently for the vector function $b = \operatorname{rot} v$, we obtain

$$\operatorname{rot} b = \operatorname{rot} \operatorname{rot} v = -\Delta v + \operatorname{grad} \operatorname{div} v = w \;\text{ in } B_\rho$$

and

$$\|b\|_{W^{1,6}(B_\rho)^3} \le c_1\, \|v\|_{W^{2,6}(B_\rho)^3} \le c\, c_1\, \|w\|_{L_6(B_\rho)^3} \le c_2\, \|h\|_{L_\infty(\partial\mathcal{G})} .$$

It remains to show that the problem (11.6.18) has a solution v subject to (11.6.19). To this end, we consider the boundary value problem

$$-\Delta v = w \;\text{ in } B_\rho, \quad \frac{\partial v_r}{\partial r} + \frac{2}{r}\, v_r = v_\theta = v_\varphi = 0 \;\text{ on } \partial B_\rho, \tag{11.6.20}$$

where r, φ, θ are the spherical coordinates ($r = |x|$, $\tan\varphi = x_2/x_1$, $\cos\theta = x_3/r$) and v_r, v_θ, v_φ are the spherical components of the vector function v, i.e.

$$\begin{pmatrix} v_r \\ v_\theta \\ v_\varphi \end{pmatrix} = \begin{pmatrix} \sin\theta\, \cos\varphi & \sin\theta\, \sin\varphi & \cos\theta \\ \cos\theta\, \cos\varphi & \cos\theta\, \sin\varphi & -\sin\theta \\ -\sin\varphi & \cos\varphi & 0 \end{pmatrix} \begin{pmatrix} v_1 \\ v_2 \\ v_3 \end{pmatrix} .$$

On the set of all v satisfying the boundary conditions in (11.6.20), we have

$$\begin{aligned}
-\int_{B_\rho} \Delta v \cdot \bar{v}\, dx &= \sum_{j=1}^{3} \int_{B_\rho} \left|\partial_{x_j} v\right|^2 dx - \rho^{-1} \int_{\partial B_\rho} \frac{\partial v}{\partial r} \cdot \bar{v}\, d\sigma \\
&= \sum_{j=1}^{3} \int_{B_\rho} \left|\partial_{x_j} v\right|^2 dx - \rho^{-1} \int_{\partial B_\rho} \frac{\partial v_r}{\partial r} \cdot \bar{v}_r\, d\sigma \\
&= \sum_{j=1}^{3} \int_{B_\rho} \left|\partial_{x_j} v\right|^2 dx + 2\rho^{-2} \int_{\partial B_\rho} |v_r|^2\, d\sigma.
\end{aligned}$$

Since the quadratic form on the right-hand side is coercive, the problem (11.6.20) is uniquely solvable in $W^{1,2}(B_\rho)^3$. By Theorem 1.1.7, the solution belongs to the space $W^{2,6}(B_\rho)^3$ and satisfies (11.6.19) if $w \in L_6(B_\rho)^3$. From (11.6.20) and from the equality

$$\nabla \cdot v = \frac{\partial v_r}{\partial r} + \frac{2}{r} v_r + \frac{1}{r} \frac{\partial v_\theta}{\partial \theta} + \frac{\cot \theta}{r} v_\theta + \frac{1}{r \sin \theta} \frac{\partial v_\varphi}{\partial \varphi}$$

it follows that $\nabla \cdot v = 0$ on ∂B_ρ. The proof of the lemma is complete. □

11.6.6. A maximum modulus estimate for solutions of the Navier-Stokes system. Now we consider the solution of the boundary value problem (11.6.3), (11.6.4), where $h \in L_\infty(\partial \mathcal{G})^3$.

THEOREM 11.6.6. *Let (u, p) be a solution of the problem* (11.6.3), (11.6.4), *where $\mathcal{G}$ is a domain of polyhedral type. Then u satisfies the estimate*

$$\|u\|_{L_\infty(\mathcal{G})^3} \le F\left(\|h\|_{L_\infty(\partial\mathcal{G})^3}\right), \tag{11.6.21}$$

with the function

$$F(t) = c_0 t\, e^{c_1 t/\nu},$$

where c_0 and c_1 are positive constants independent of ν.

P r o o f. Suppose first that $\nu = 1$. Let (w, q) be the solution of the problem

$$-\Delta w + \nabla q = 0, \quad \nabla \cdot w = 0, \quad w|_{\partial \mathcal{G}} = h, \qquad \int_{\mathcal{G}} q(x)\, dx = 0.$$

According to Theorem 11.6.3, the vector function w satisfies the estimate

$$\|w\|_{L_\infty(\mathcal{G})^3} \le c\, \|h\|_{L_\infty(\partial\mathcal{G})^3}\,. \tag{11.6.22}$$

Furthermore, the vector function $(u - w, p - q)$ satisfies the equations

$$-\Delta(u - w) + \nabla(p - q) = -(u \cdot \nabla)\, u, \quad \nabla \cdot (u - w) = 0 \ \text{ in } \mathcal{G}$$

and the boundary condition $u - w = 0$ on $\partial\mathcal{G}$. By (11.6.14), the function u admits the decomposition $u = w + W$, where W is the vector function with the components

$$\begin{aligned}
W_i(x) &= -\int_{\mathcal{G}} \sum_{j=1}^{3} G_{i,j}(x,\xi) \left(u(\xi) \cdot \nabla_\xi\right) u_j(\xi)\, d\xi \\
&= -\int_{\mathcal{G}} \sum_{j=1}^{3} G_{i,j}(x,\xi)\, \nabla_\xi \cdot \left(u_j(\xi)\, u(\xi)\right) d\xi, \quad i = 1, 2, 3.
\end{aligned}$$

Using (11.6.15), we get

$$\begin{aligned}(11.6.23)\quad \|u\|_{L_\infty(\mathcal{G})^3} &\le \|w\|_{L_\infty(\mathcal{G})^3} + \|W\|_{L_\infty(\mathcal{G})^3} \\ &\le \|w\|_{L_\infty(\mathcal{G})^3} + c\sum_{i,j=1}^{3} \|u_i\, u_j\|_{L_{s/2}(\mathcal{G})} \le \|w\|_{L_\infty(\mathcal{G})^3} + c\,\|u\|^2_{L_s(\mathcal{G})^3}\end{aligned}$$

for arbitrary $s > 6$. From (11.6.16) it follows that

$$\begin{aligned}(11.6.24)\quad \|u\|_{L_s(\mathcal{G})^3} &\le \|w\|_{L_s(\mathcal{G})^3} + \|W\|_{L_s(\mathcal{G})^3} \\ &\le \|w\|_{L_s(\mathcal{G})^3} + c\sum_{i,j=1}^{3} \|u_i\, u_j\|_{L_3(\mathcal{G})} \le c_1\,\|w\|_{L_\infty(\mathcal{G})^3} + c_2\,\|u\|^2_{L_6(\mathcal{G})^3}\,.\end{aligned}$$

Combining (11.6.22), (11.6.23) and (11.6.24), we obtain

$$(11.6.25)\qquad \|u\|_{L_\infty(\mathcal{G})^3} \le c_3\Big(\|h\|_{L_\infty(\partial\mathcal{G})^3} + \|h\|^2_{L_\infty(\partial\mathcal{G})^3} + \|u\|^4_{L_6(\mathcal{G})^3}\Big)$$

with a certain constant c_3 independent of h.

It remains to estimate the norm of u in $L_6(\mathcal{G})^3$. Let $\delta(x)$ be the *regularized distance* of x from the boundary $\partial\mathcal{G}$ (cf. [**192**, Chapter 6, §2]), i.e. δ is an infinitely differentiable function on $\mathcal{G}$ satisfying the inequalities

$$c_1\, d(x) \le \delta(x) \le c_2\, d(x), \qquad \big|\partial_x^\alpha \delta(x)\big| \le c_\alpha\, d(x)^{1-|\alpha|}$$

with certain positive constants c_1, c_2, c_α, where $d(x)$ denotes the distance of x from $\partial\mathcal{G}$. Furthermore, let ρ and κ be positive numbers, and let χ be an infinitely differentiable function such that $0 \le \chi \le 1$, $\chi(t) = 0$ for $t \le 0$, and $\chi(t) = 1$ for $t \ge 1$. We define the cut-off function ζ on $\mathcal{G}$ by

$$\zeta(x) = \chi\Big(\kappa\, \log\frac{\rho}{\delta(x)}\Big).$$

This function has the following properties:

$$\begin{aligned}&0 \le \zeta(x) \le 1, \qquad \zeta(x) = 0 \ \text{ for } \delta(x) \ge \rho, \\ &\zeta(x) = 1 \ \text{ for } \delta(x) \le \varepsilon\rho, \ \text{ where } \varepsilon = e^{-1/\kappa}, \\ &|\nabla\zeta(x)| \le c\,\frac{\kappa}{d(x)}, \quad \big|\partial_{x_i}\partial_{x_j}\zeta(x)\big| \le c\,\frac{\kappa}{d(x)^2} \ \text{ for } i,j = 1,2,3.\end{aligned}$$

By Lemma 11.6.5, the vector function w admits the representation $w = \operatorname{rot} b$ with a vector function $b \in W^{1,6}(\mathcal{G})^3$ satisfying (11.6.17). We put

$$u = v + V, \quad \text{where} \quad V = \operatorname{rot}(\zeta b) = \zeta w + \nabla\zeta \times b.$$

Then v satisfies the equations

$$(11.6.26)\qquad -\Delta v + (u\cdot\nabla)\, v + (v\cdot\nabla)\, V = \Delta V - (V\cdot\nabla)\, V - \nabla p, \quad \nabla\cdot v = 0$$

in $\mathcal{G}$ and the boundary condition $v|_{\partial\mathcal{G}} = 0$. Since

$$\int_{\mathcal{G}} \big((u\cdot\nabla)\, v\big)\cdot v\, dx = 0,$$

it follows from (11.6.26) that

$$(11.6.27)\qquad \sum_{j=1}^{3} \|\nabla v_j\|^2_{L_2(\mathcal{G})^3} - \sum_{j=1}^{3}\int_{\mathcal{G}} v_j\, V\cdot\frac{\partial v}{\partial x_j}\, dx = L(v),$$

where

$$L(v) = \int_{\mathcal{G}} \Big(\Delta V - (V \cdot \nabla)V - \nabla p\Big) \cdot v \, dx = \sum_{j=1}^{3} \int_{\mathcal{G}} \Big(- \nabla V_j \cdot \nabla v_j + V_j \, V \cdot \frac{\partial v}{\partial x_j}\Big) \, dx$$

$$= - \int_{\mathcal{G}} \Big(w \cdot v \, \Delta\zeta + 2w \cdot (\nabla\zeta \cdot \nabla) \, v + q \, v \cdot \nabla\zeta \Big) \, dx$$

$$- \sum_{j=1}^{3} \int_{\mathcal{G}} \nabla(\nabla\zeta \times b)_j \cdot \nabla v_j \, dx + \sum_{j=1}^{3} \int_{\mathcal{G}} V_j \, V \cdot \frac{\partial v}{\partial x_j} \, dx$$

(here $(\nabla\zeta \times b)_j$ denotes the jth component of the vector $\nabla\zeta \times b$). We estimate the functional $L(v)$. Using the inequalities

$$|\nabla\zeta| \le c \, \frac{\kappa}{\varepsilon\rho}, \qquad \big|d \, \Delta\zeta\big| \le c \, \frac{\kappa}{\varepsilon\rho},$$

$$\int_{\mathcal{G}} d(x)^{-2} \, \big|u(x)\big|^2 \, dx \le c \int_{\mathcal{G}} \big|\nabla u(x)\big|^2 \, dx$$

(the last follows from Hardy's inequality) and (11.6.22), we obtain

$$\begin{aligned} \Big| \int_{\mathcal{G}} w \cdot v \, \Delta\zeta \, dx \Big| \;&\le\; \|w\|_{L_\infty(\mathcal{G})^3} \, \|d\Delta\zeta\|_{L_2(\mathcal{G})} \, \|d^{-1} v\|_{L_2(\mathcal{G})^3} \\ &\le\; c \, \frac{\kappa}{\varepsilon\rho} \, \|h\|_{L_\infty(\partial\mathcal{G})^3} \sum_{j=1}^{3} \|\nabla v_j\|_{L_2(\mathcal{G})} \end{aligned}$$

and

$$\begin{aligned} \Big| \int_{\mathcal{G}} w \cdot (\nabla\zeta \cdot \nabla) \, v \, dx \Big| \;&\le\; \|w\|_{L_\infty(\mathcal{G})^3} \, \|\nabla\zeta\|_{L_2(\mathcal{G})^3} \sum_{j=1}^{3} \|\nabla v_j\|_{L_2(\mathcal{G})^3} \\ &\le\; c \, \frac{\kappa}{\varepsilon\rho} \, \|h\|_{L_\infty(\partial\mathcal{G})^3} \sum_{j=1}^{3} \|\nabla v_j\|_{L_2(\mathcal{G})^3} \, . \end{aligned}$$

Analogously by the second inequality of Theorem 11.6.3,

$$\begin{aligned} \Big| \int_{\mathcal{G}} q \, v \cdot \nabla\zeta \, dx \Big| \;&\le\; \|qd\|_{L_\infty(\mathcal{G})} \, \|d^{-1} v\|_{L_2(\mathcal{G})^3} \, \|\nabla\zeta\|_{L_2(\mathcal{G})^3} \\ &\le\; c \, \frac{\kappa}{\varepsilon\rho} \, \|h\|_{L_\infty(\partial\mathcal{G})^3} \sum_{j=1}^{3} \|\nabla v_j\|_{L_2(\mathcal{G})^3} \, . \end{aligned}$$

The inequality (11.6.17) implies

$$\begin{aligned} &\Big| \int_{\mathcal{G}} \nabla(\nabla\zeta \times b)_j \cdot \nabla v_j \, dx \Big| \\ &\le c \Big(\|\nabla\zeta\|_{L_\infty(\mathcal{G})^3} \, \|\nabla b\|_{L_2(\mathcal{G})^3} + \sum_{i,k} \|\frac{\partial^2\zeta}{\partial x_i \partial x_k}\|_{L_\infty(\mathcal{G})} \, \|b\|_{L_2(\mathcal{G})^3} \Big) \, \|\nabla v_j\|_{L_2(\mathcal{G})^3} \\ &\le c \, \frac{\kappa}{\varepsilon^2\rho^2} \, \|h\|_{L_\infty(\partial\mathcal{G})^3} \, \|\nabla v_j\|_{L_2(\mathcal{G})} \end{aligned}$$

and

$$\begin{aligned}\Big|\int_{\mathcal{G}} V_j\, V\cdot\frac{\partial v}{\partial x_j}\,dx\Big| &\le \|V\|^2_{L_4(\mathcal{G})^3}\,\|\partial_{x_j}v\|_{L_2(\mathcal{G})^3}\\ &\le 2\Big(\|\zeta w\|^2_{L_4(\mathcal{G})^3}+\|\nabla\zeta\times b\|^2_{L_4(\mathcal{G})}\Big)\,\|\partial_{x_j}v\|_{L_2(\mathcal{G})^3}\\ &\le c\Big(1+\frac{\kappa^2}{\varepsilon^2\rho^2}\Big)\,\|h\|^2_{L_\infty(\partial\mathcal{G})}\,\|\partial_{x_j}v\|_{L_2(\mathcal{G})^3}\,.\end{aligned}$$

Thus,

$$\begin{aligned}(11.6.28)\quad \big|L(v)\big| &\le C_1\Big(\frac{\kappa}{\varepsilon^2\rho^2}\,\|h\|_{L_\infty(\partial\mathcal{G})^3}+\big(1+\frac{\kappa^2}{\varepsilon^2\rho^2}\big)\,\|h\|^2_{L_\infty(\partial\mathcal{G})^3}\Big)\\ &\qquad\times\sum_{j=1}^3\|\partial_{x_j}v\|_{L_2(\mathcal{G})^3}\,,\end{aligned}$$

where C_1 is a constant independent of ρ and κ. Furthermore,

$$\begin{aligned}\Big|\sum_{j=1}^3\int_{\mathcal{G}} v_j\, V\cdot\frac{\partial v}{\partial x_j}\,dx\Big| &= \Big|\sum_{j=1}^3\int_{\mathcal{G}} v_j(\zeta w+\nabla\zeta\times b)\cdot\frac{\partial v}{\partial x_j}\,dx\Big|\\ &\le\Big(\|\zeta d\|_{L_\infty(\mathcal{G})}\,\|w\|_{L_\infty(\mathcal{G})^3}+\|d\,\nabla\zeta\|_{L_\infty(\mathcal{G})^3}\,\|b\|_{L_\infty(\mathcal{G})^3}\Big)\\ &\qquad\times\sum_{j=1}^3\|d^{-1}v_j\|_{L_2(\mathcal{G})}\,\|\partial_{x_j}v\|_{L_2(\mathcal{G})^3}\\ &\le C_2\,(\rho+\kappa)\,\|h\|_{L_\infty(\partial\mathcal{G})^3}\sum_{j=1}^3\|\partial_{x_j}v\|_{L_2(\mathcal{G})^3}\,,\end{aligned}$$

where C_2 is independent of h,ρ,κ. The numbers ρ and κ can be chosen such that

$$C_2\,(\rho+\kappa)\,\|h\|_{L_\infty(\partial\mathcal{G})^3}\le 1/2.$$

Then (11.6.27) and (11.6.28) yield

$$\sum_{j=1}^3\|\partial_{x_j}v\|_{L_2(\mathcal{G})}\le 2\,C_1\Big(\frac{\kappa}{\varepsilon^2\rho^2}\,\|h\|_{L_\infty(\partial\mathcal{G})^3}+\big(1+\frac{\kappa^2}{\varepsilon^2\rho^2}\big)\,\|h\|^2_{L_\infty(\partial\mathcal{G})^3}\Big).$$

By the continuity of the imbedding $W^{1,2}(\mathcal{G})\subset L_6(\mathcal{G})$, the same estimate (with another constant C_1) holds for the norm of v in $L_6(\mathcal{G})^3$. Since $|\nabla\zeta|\le c\kappa/(\varepsilon\rho)$, we furthermore get

$$\|V\|_{L_6(\mathcal{G})^3}\le\|\zeta w\|_{L_6(\mathcal{G})^3}+\|\nabla\zeta\times b\|_{L_6(\mathcal{G})}\le C_3\,\big(1+\kappa/(\varepsilon\rho)\big)\,\|h\|_{L_\infty(\partial\mathcal{G})^3}$$

(cf. (11.6.17) and (11.6.22)) and consequently

$$\begin{aligned}\|u\|_{L_6(\mathcal{G})^3} &\le \|V\|_{L_6(\mathcal{G})^3}+\|v\|_{L_6(\mathcal{G})^3}\\ &\le C_4\Big(\big(1+\frac{\kappa}{\varepsilon\rho}+\frac{\kappa}{\varepsilon^2\rho^2}\big)\,\|h\|_{L_\infty(\partial\mathcal{G})^3}+\big(1+\frac{\kappa^2}{\varepsilon^2\rho^2}\big)\,\|h\|^2_{L_\infty(\partial\mathcal{G})}\Big).\end{aligned}$$

If we put

$$\kappa=\rho=\frac{1}{4C_2\,\|h\|_{L_\infty(\partial\mathcal{G})^3}}\quad\text{and}\quad\varepsilon=e^{-1/\kappa}=e^{-4C_2\|h\|_{L_\infty(\partial\mathcal{G})^3}}\,,$$

we obtain

$$\|u\|_{L_6(\mathcal{G})^3} \leq C_5 \left(\|h\|_{L_\infty(\partial\mathcal{G})^3}\, e^{4C_2\|h\|_{L_\infty(\partial\mathcal{G})^3}} + \|h\|^2_{L_\infty(\partial\mathcal{G})^3}\, e^{8C_2\|h\|_{L_\infty(\partial\mathcal{G})^3}} \right).$$

This together with (11.6.25) implies (11.6.21) for $\nu = 1$. If $\nu \neq 1$, then we consider the vector function $(\nu^{-1}u, \nu^{-2}p)$ instead of (u, p). □

Historical remarks

1. Bibliographical notes to chapters

Chapter 1 (smooth domains and isolated singularities). In addition to the references given in Sections 1.1 and 1.2, we note that a historical survey on elliptic boundary value problems in domains with smooth boundaries and in domains with isolated singularities on the boundary can be found in the book [**84**], which contains many references related to this topic. Therefore, in what follows, we will refer only to works dealing with boundary singularities of positive dimension.

Chapter 2 (Dirichlet problem, nonintersecting edges). The material in this chapter is an adaption to the Dirichlet problem of a more general framework in the papers [**118, 119, 120**] by MAZ'YA and PLAMENEVSKIĬ. The main difference is that, in contrast to these papers, we allow the right-hand side of the differential equation to belong to a weighted Sobolev or Hölder space of negative order. An earlier exposition of solvability and regularity results in Hilbert-Sobolev spaces of integer order was given by the same authors in [**114**]. A theory of the Dirichlet problem in Hilbert-Sobolev spaces of fractional order was developed by DAUGE [**31**]. In [**118, 119, 120**], arbitrary elliptic equations supplied with different boundary conditions on the faces of a n-dimensional dihedron were considered.

In particular, as shown in [**118**], the boundary value problem is solvable in the weighted Sobolev space $V_\delta^{l,p}$ if the kernel and cokernel of the operator of the corresponding parameter-depending model problem in the plane cross-section angle (cf. Section 2.3) are trivial. This condition can be easily checked for the Dirichlet problem and for a broad class of strongly elliptic problems. In general, the algebraic verification of the triviality of the kernel and cokernel just mentioned is an open problem, but the answer is known in some special cases, see MAZ'YA and PLAMENEVSKIĬ [**112, 114, 115**], MAZ'YA [**103**], KOMECH [**73**], ESKIN [**51**]. It is proved by KOZLOV [**77**] that, under some requirements on the elliptic operator, one can achieve the triviality of the kernel of the model problem in question by prescribing a finite number of complementary conditions on the edge.

Note that the results in [**118**] were derived using an operator multiplier theorem for the Fourier transform. The approach in the present book goes up to the paper [**119**], where estimates of solutions in weighted Hölder spaces were obtained by means of point estimates for Green's functions.

Various aspects of the elliptic theory for manifolds with edges (parametrices, Fredholm property, index) were studied in numerous works by SCHULZE and his collaborators by methods of the theory of pseudo-differential operators (see for example the monograph by NAZAIKINSKIĬ, SAVIN, SCHULZE and STERNIN [**154**]).

Properties of the Dirichlet problem for the Laplacian stated in Subsections 2.6.6 and 2.8.6 are corollaries of the general Theorems 2.6.5 and 2.8.8. However, particular cases of these results were obtained previously by specific methods of the theory of second order elliptic equations with real coefficients. For instance, coercive estimates of solutions of the Dirichlet problem for second order elliptic equations in the weighted spaces $V_\delta^{2,2}$ were obtained by KONDRAT'EV [**75**]. The paper [**12**] of APUSHKINSKAYA and NAZAROV is dedicated to Hölder estimates for solutions to the Dirichlet problem for quasilinear elliptic equations in domains with smooth closed edges of arbitrary dimension.

Chapters 3 and 4 (Dirichlet problem in domains of polyhedral type). Pointwise estimates for Green's matrix of the Dirichlet problem for strongly elliptic equations of higher order were obtained in our paper [**129**]. In the same paper, one can find estimates of solutions in weighted L_p-Sobolev spaces similar to those in Sections 3.5 and 4.1. The Hölder estimates in Sections 3.6 and 4.2 were not published before.

In the paper [**113**] MAZ'YA and PLAMENEVSKIĬ introduced a large class of multi-dimensional manifolds with edges of different dimensions intersecting under nonzero angles. This class of manifolds contains polyhedra in $\mathbb{R}^N$ as a very special case. A solvability theory for general elliptic boundary value problems on such manifolds in weighted L_2-Sobolev spaces was developed in [**116**] by an induction argument in dimensions of singular strata. It is assumed in this paper that kernels and cokernels of all model problems generated by edges of different dimensions are trivial, which is the case, in particular, for the Dirichlet problem. This material is reproduced in the book by NAZAROV and PLAMENEVSKIĬ [**160**].

A L_2-theory for the Dirichlet problem for general elliptic equations in three-dimensional polyhedral domains was also established in the papers by LUBUMA, NICAISE [**92**] and NICAISE [**164**]. Some regularity results related to the Dirichlet problem for the Laplace equation in a polyhedral domain were obtained by HANNA and SMITH [**65**], GRISVARD [**58, 60**], DAUGE [**31**], AMMANN and NISTOR [**11**]. BUFFA, COSTABEL and DAUGE [**18**] stated regularity assertions for the Laplace and Maxwell equations in isotropic and anisotropic weighted Sobolev spaces. The Dirichlet problem for the Lamé system (and for the Laplace equation as a particular case) in a broad class of piecewise smooth domains without cusps was investigated in detail by MAZ'YA and PLAMENEVSKIĬ [**124**].

Chapter 5 (Miranda-Agmon maximum principle). The main results of this chapter were obtained in our papers [**129**] and [**130**], the Hölder estimates for the derivatives of Green's matrix in convex polyhedral type domains presented in Subsection 5.1.5 were proved by GUZMAN, LEYKEKHMAN, ROSSMANN and SCHATZ [**64**].

The history starts with the estimate

$$\|u\|_{C^{m-1}(\overline{\mathcal{G}})} \le c\Big(\sum_{k=1}^{m}\Big\|\frac{\partial^{k-1}u}{\partial n^{k-1}}\Big\|_{C^{m-k}(\partial\mathcal{G})} + \|u\|_{L_1(\mathcal{G})}\Big), \tag{11.6.29}$$

for solutions of strongly elliptic equations $Lu = 0$ of order $2m > 2$ proved in the case of smooth boundaries by MIRANDA [**144, 145**] for two-dimensional and by AGMON

[**5**] for higher-dimensional domains. SCHULZE [**182, 183**] justified analogous C^k-estimates for solutions of strongly elliptic systems and for more general boundary conditions $D_\nu^{m_k} = g_k$ on $\partial\mathcal{G}$, where $m_k \le 2m-1$.

MAZ'YA and PLAMENEVSKIĬ [**122**] proved the estimate (11.6.29) for solutions of the biharmonic equation in a three-dimensional domain with conical vertices. As shown independently in MAZ'YA, ROSSMANN [**130**] and PIPHER, VERCHOTA [**169**], this estimate fails if the dimension is greater than 3. In [**169, 170**] PIPHER and VERCHOTA proved the estimate (11.6.29) for solutions of the biharmonic and polyharmonic equations in Lipschitz domains.

Chapter 6 (systems of second order, nonintersecting edges). The results of this chapter are borrowed from our paper [**133**]. Even when dealing only with the Dirichlet problem, we obtain new results in comparison with Chapter 2. Here the data and the solutions belong to a wider class of spaces with nonhomogeneous norms which include classical nonweighted Sobolev spaces. These spaces were earlier used in the paper [**128**] of MAZ'YA and ROSSMANN, where general elliptic boundary value problems were considered under the assumption of the unique solvability of model problems in a plane cross-section angle.

The first treatment of the Neumann problem for the equation $\Delta u = 0$ in the presence of a smooth edge on the boundary was given as early as 1916 by CARLEMAN [**19**], who used methods of potential theory. For the same problem see the works by MAZ'YA and PLAMENEVSKIĬ [**112, 115**] and SOLONNIKOV and ZAJACKOWSKI [**204**], where solutions in the spaces $W_\delta^{l,2}$ were considered. Analogous results in the weighted Sobolev spaces $W_\delta^{l,p}$ and weighted Hölder spaces $C_\delta^{l,\sigma}$ were obtained in the preprint [**190**] by Solonnikov. Furthermore, the Green's function for the Neumann problem was estimated in [**190**]. An L_2-theory for more general boundary value problems including the Neumann problem was developed in papers by NAZAROV [**155, 156**], ROSSMANN [**177**], NAZAROV, PLAMENEVSKIĬ [**158, 159**] (see also the book of NAZAROV and PLAMENEVSKIĬ [**160**]). The elliptic oblique derivative problem in domains with nonintersecting edges was treated by MAZ'YA and PLAMENEVSKIĬ [**112**].

NAZAROV and SWEERS [**161**] investigated the $W^{2,2}$-solvability of the biharmonic equation with prescribed boundary value of the solution and its Laplacian in a three-dimensional domain with variable opening at the edge, where some interesting effects arise for a critical opening.

If the domain is smooth and the role of an edge is played by a smooth surface of codimension 1 in the boundary separating different boundary conditions, another approach to mixed problems based on the Wiener-Hopf method was used starting in the 1960s (see the monograph by ESKIN [**50**]). A similar approach proved to be effective in the study of boundary value problems for domains with two-dimensional cracks and interior cuspidal edges (see DUDUCHAVA and WENDLAND [**39**], DUDUCHAVA and NATROSHVILI [**38**], CHKADUA [**20**], CHKADUA, DUDUCHAVA [**21**] *et al.*). In particular in [**39**], the Wiener-Hopf method was developed for systems of boundary pseudo-differential equations which allowed to manage without the factorization of corresponding matrix symbols and to investigate the asymptotics of the solution to the crack problem in an anisotropic medium.

Exterior cuspidal edges which require different methods were studied by DAUGE [**35**], SCHULZE, TARKHANOV [**187**], RABINOVICH, SCHULZE, TARKHANOV [**173,**

174], MAZ'YA, NETRUSOV, POBORCHI [**111**] and MAZ'YA, POBORCHI [**126**].

Chapters 7 and 8 (second order systems in domains of polyhedral type). These chapters contain a somewhat extended exposition of the results obtained by the authors in [**133, 134, 135**]. New features in comparison with Chapters 3 and 4 are the use of nonhomogeneous Sobolev and Hölder norms, and the inclusion of the Neumann problem.

The Neumann problem for the Laplace equation in a polyhedral cone was earlier studied in the preprint [**57**] of GRACHEV and MAZ'YA, where the authors obtained estimates for the solutions in weighted Sobolev and Hölder spaces and pointwise estimates of the Green's matrix. DAUGE [**34**] proved regularity assertions for solutions of the Neumann problem for second order elliptic equations with real coefficients in nonweighted L_p-Sobolev spaces. Regularity results in weighted L_2-Sobolev spaces for general self-adjoint systems were proved by NAZAROV and PLAMENEVSKIĬ [**157**]. The behavior of the solution of the Neumann problem for the Lamé system near the vertex of a polyhedron is studied in the book by Grisvard [**62**].

Mixed boundary value problems for the Laplace equation with Dirichlet and Neumann conditions are considered e.g. in the above mentioned works by DAUGE [**34**] and GRISVARD [**62**]. The same problems were studied by EBMEYER [**44**], EBMEYER and FREHSE [**45**] for nonlinear second order equations in N-dimensional domains, $N \geq 3$, with piecewise smooth boundaries. NICAISE [**163**] obtained regularity results for solutions of mixed boundary value problems to the Lamé system in L_2-Sobolev spaces.

MAZ'YA [**102, 103**] and DAUGE [**32**] studied oblique derivative problems in domains of polyhedral type. Transmission problems in polyhedral domains were handled in papers by COSTABEL, DAUGE, NICAISE [**29**], CHIKOUCHE, MERCIER, NICAISE [**22, 23**], KNEES [**72**], ELSCHNER, REHBERG, SCHMIDT [**48**], ELSCHNER, KAISER, REHBERG, SCHMIDT [**47**].

The conditions ensuring the solvability and regularity of solutions near the vertices depend on information about eigenvalues of the operator pencils $A_k(\lambda)$ and $\mathfrak{A}_j(\lambda)$ introduced in Section 8.1. Information of this nature is collected in the book by KOZLOV, MAZ'YA and ROSSMANN [**85**]. The pencil generated by the Neumann problem for elliptic differential operators of arbitrary order was investigated by KOZLOV and MAZ'YA [**80**]. Assuming that the cone is convex, it was shown by ESCOBAR [**49**] and in another way by MAZ'YA [**107**] that the first positive eigenvalue of the pencil $\delta + \lambda(\lambda + N - 2)$ with zero Neumann conditions satisfies the sharp inequality $\lambda_1 \geq 1$. Earlier DAUGE [**34**] found a rougher estimate $\lambda_1 > (\sqrt{5} - 1)/2$ in the three-dimensional case.

For special problems and special domains, eigenvalues of operator pencils generated by the Neumann problem were calculated numerically by LEGUILLON and SANCHEZ-PALENCIA [**90**], DIMITROV [**40**], DIMITROV, ANDRÄ and SCHNACK [**41**] *et al.*

Chapter 9 (Stokes and Navier-Stokes systems, nonintersecting edges). This chapter is an extended version of our paper [**136**] concerning the mixed boundary value problem for the Stokes system in a dihedron. Some related results can be found in the earlier paper of SOLONNIKOV [**189**] and MAZ'YA, PLAMENEVSKIĬ and STUPELIS [**125**], where the Dirichlet problem and a particular mixed boundary value problem were studied in connection with a nonlinear hydrodynamical problem with free boundary. A detailed exposition of the results obtained in [**125**] can be found in STUPELIS [**193**]. In contrast to [**136**], the paper [**125**] deals with solutions in weighted Sobolev and Hölder spaces with homogenous norms.

Chapters 10 and 11 (Stokes and Navier-Stokes systems, domains of polyhedral type). These chapters contain results obtained by the authors in [**136**]–[**140**] and [**179**]. The starting point for the development of this theory was the paper by MAZ'YA and PLAMENEVSKIĬ [**124**] dedicated to the Dirichlet problem.

The inequality (11.3.3) for the eigenvalues of the pencil generated by the Dirichlet problem for the Stokes system obtained by MAZ'YA and PLAMENEVSKIĬ in [**123**] was the first result of this nature. More estimates for the eigenvalues can be found in the paper by DAUGE [**33**]. A detailed analysis of these eigenvalues including a variational principle for real ones was developed by KOZLOV, MAZ'YA and SCHWAB [**86**] (see also the book [**85**]). The only paper, where the eigenvalues corresponding to the Neumann were touched upon, is that of KOZLOV and MAZ'YA [**79**]. Some results corresponding to various mixed type problems were obtained by KOZLOV, MAZ'YA and ROSSMANN [**83**] (see also the book [**85**]).

The results in Section 11.5 have a long history which begins with ODQUIST's inequality

$$\|u\|_{L_\infty(\mathcal{G})} \le c\,\|u\|_{L_\infty(\partial\mathcal{G})}$$

for the solutions of the Stokes system (11.6.1) (see [**166**]). A proof of this inequality for domains with smooth boundaries is given e.g. in the book by LADYZHENSKAYA [**89**]. We refer also to the papers of MAZ'YA and KRESIN [**108**], NAUMANN [**153**], KRATZ [**87**] and MAREMONTI [**93**]. Using point estimates of the Green's matrix, MAZ'YA and PLAMENEVSKIĬ [**123, 124**] proved this inequality for solutions of the Stokes system in three-dimensional domains with conical points and in domains of polyhedral type.

For the nonlinear problem (11.6.1), (11.6.2), SOLONNIKOV [**191**] showed that solutions satisfy the estimate (11.6.21) with a certain unspecified function F if the boundary $\partial\mathcal{G}$ is smooth. An estimate of this form can be also deduced from the results in a paper of MAREMONTI and RUSSO [**94**]. MAZ'YA and PLAMENEVSKIĬ [**124**] proved for domains of polyhedral type that the solution u of (11.6.1), (11.6.2) with finite Dirichlet integral is continuous in $\overline{\mathcal{G}}$ if h is continuous on $\partial\mathcal{G}$. However, the paper [**124**] contains no estimates for the maximum modulus of u. In our paper [**140**], we proved the inequality (11.6.21) for domains of polyhedral type and obtained the representation $F(t) = c_0 t\, e^{c_1 t/\nu}$ for the function c.

2. Bibliographical notes to other related material

The whole theme of elliptic boundary value problems in nonregular domains is so rich that obviously we could touch upon only a small part of it. In order to illustrate the variety of results in this area, we give here some references related to topics outside of this book without aiming at complete satisfaction to a certain extent.

Asymptotics of solutions near edges and vertices. The asymptotic expansions of solutions near boundary singularities are not treated in this book, but this theme was thoroughly studied simultaneously with solvability properties and became a broad area of research. The asymptotics of solutions of the Dirichlet problem for elliptic equations of second order in a neighborhood of an edge was described by KONDRAT'EV [**76**] and NIKISHKIN [**165**] and for the Laplace equation by GRISVARD [**61**]. Asymptotic formulas for solutions to general elliptic boundary value problems were proved by MAZ'YA and PLAMENEVSKIĬ [**114**], MAZ'YA AND ROSSMANN [**127, 128**], DAUGE [**30**], NAZAROV and PLAMENEVSKIĬ [**160**]. It was assumed in the last works that the edges do not contain "critical" points, i.e. that there is no bifurcation in singularities. The case of critical edge points was discussed in the papers by REMPEL and SCHULZE [**175**], and SCHULZE [**184, 185**]. Explicit asymptotic formulas for such cases were derived by COSTABEL and DAUGE [**25**], MAZ'YA and ROSSMANN [**132**]. The asymptotics of solutions near polyhedral vertices was studied by VON PETERSDORFF and STEPHAN [**202**] and DAUGE [**36**] for second order equations. The last paper is a masterful survey of the area. We also mention a comprehensive study of singularities of solutions to the Maxwell equation by COSTABEL and DAUGE [**26, 27, 28**].

Lipschitz graph and other domains. Needless to say, there are other areas in the theory of elliptic boundary value problems differing both by classes of domains and the methods of research. First of all, there exists a rich theory dealing with Lipschitz graph boundaries and based on refined methods of harmonic analysis. We refer only to the survey monograph by KENIG [**71**] and more recent works by ADOLFSSON, PIPHER [**4**], BROWN, PERRY, SHEN [**17**], BROWN [**15**], BROWN, SHEN [**16**], DEURING, VON WAHL [**37**], DINDOŠ, MITREA [**42**], EBMEYER [**44**], EBMEYER, FREHSE [**45, 46**], JAKAB, MITREA, MITREA [**66**], JERISON, KENIG [**67**], MAYBORODA, MITREA [**95, 96**], MITREA [**146**], MITREA, MONNIAUX [**147**], MITREA, TAYLOR [**148**]–[**151**], PIPHER, VERCHOTA [**168, 171**], SHEN [**180, 181**] and VERCHOTA [**198, 199**].

Successful attempts to apply these methods, which are based on the so-called Rellich's identity, to non-Lipschitz graph polyhedral domains in $\mathbb{R}^3$ and $\mathbb{R}^4$ were undertaken by VERCHOTA [**197**], VERCHOTA and VOGEL [**200, 201**], VENOUZIOU and VERCHOTA [**196**].

Asymptotic formulas for solutions of the Dirichlet problem for strongly elliptic equations of arbitrary order near the Lipschitz graph boundary were found by KOZLOV and MAZ'YA [**82**]. The same boundary value problem with data in Besov spaces was treated in MAZ'YA, MITREA and SHAPOSHNIKOVA [**109**] under an assumption on the boundary formulated in terms of the space BMO. Sharp conditions of the $W^{2,2}$-solvability of the Dirichlet problem for the Laplace equation in a domain in C^1 but not in C^2 were derived in MAZ'YA [**101**].

Additional information was derived for boundary value problems in arbitrary convex domains (KADLEC [**68**], ADOLFSSON [**1, 2**], ADOLFSSON, JERISON [**3**], FROMM [**52, 53**], FROMM, JERISON [**54**], KOZLOV, MAZ'YA [**82**], MAZ'YA [**107**], MAYBORODA, MAZ'YA [**97**]).

Introducing classes of Lipschitz graph domains characterized in terms of Sobolev multipliers, MAZ'YA and SHAPOSHNIKOVA obtained sharp results on solutions in $W^{l,p}(\Omega)$ ([**141**], [**142**], [**143**]).

It proved to be possible to obtain substantial information on properties of elliptic boundary value problems without imposing a priori restrictions on the class of domains, such as criteria of solvability and discreteness of spectrum formulated with the help of isoperimetric and isocapacitary inequalities, capacitary inner diameter and other potential theoretic terms (see Maz'ya [**99, 100, 104, 106**], Alvino, Cianchi, Maz'ya, Mercaldo [**10**], Cianchi, Maz'ya [**24**]). Wiener type criteria of regularity of a boundary point and pointwise estimates for solutions and their derivatives in unrestricted domains belong to another direction in the same area (see Maz'ya [**105**], Mayboroda, Maz'ya [**98**]).

In conclusion, we only list as key words some other classes of nonsmooth domains which appear in the studies of elliptic boundary value problems: nontangentially accessible domains, uniform domains, John domains, Jordan domains, Nikodym domains, Sobolev domains, extension domains etc.

Bibliography

[1] ADOLFSSON, V., *L^2-integrability of the second order derivatives for Poisson's equation in nonsmooth domains,* Math. Scand. **70** (1992) 146-160.

[2] ADOLFSSON, V., *L^p-integrability of the second order derivatives of Green potentials in convex domains.* Pacific J. Math. **159** (1993) 2, 201–225.

[3] ADOLFSSON, V., JERISON, D., *L^p-integrability of the second order derivatives for the Neumann problem in convex domains*, Indiana Univ. Math. J. **43** (1994) 1123-1138.

[4] ADOLFSSON, V, PIPHER, J., *The inhomogeneous Dirichlet problem for Δ^2 in Lipschitz domains*, J. Funct. Anal. **159** (1998) 1, 137-190.

[5] AGMON, S., *Maximum theorems for solutions of higher order elliptic equations*, Bull. Amer. Math. Soc. **66** (1960) 77-80.

[6] AGMON, S., DOUGLIS, A., NIRENBERG, L., *Estimates near the boundary for solutions of elliptic partial differential equations satisfying general boundary conditions I,* Comm. Pure Appl. Math. **12** (1959) 623-727,

[7] AGMON, S., DOUGLIS, A., NIRENBERG, L., *Estimates near the boundary for solutions of elliptic partial differential equations satisfying general boundary conditions II,* Comm. Pure Appl. Math. **17** (1964) 35-92.

[8] AGMON, S., NIRENBERG, L., *Properties of solutions of ordinary elliptic equations in Banach space,* Comm. Pure Appl. Math. **16** (1963) 121-163.

[9] AGRANOVICH, M. S., VISHIK, M. I., *Elliptic problems with parameter and parabolic problems of general type,* Uspekhi Mat. Nauk **19** (1964) 3, 53-161 (in Russian).

[10] ALVINO, A., CIANCHI, A., MAZ'YA, V. G., MERCALDO, A., *Well-posed elliptic Neumann problems involving irregular data and domains*, to appear.

[11] AMMANN, B., NISTOR, V., *Weighted Sobolev spaces and regularity for polyhedral domains*, Comput. Methods Appl. Mech. Engrg. **196** (2007) 37-40, 3650–3659.

[12] APUSHKINSKAYA, D. E., NAZAROV, A. I., *The Dirichlet problem for quasilinear elliptic equations in domains with smooth closed edges*, Function theory and partial differential equations. J. Math. Sci. (New York) **105** (2001) 5, 2299–2318.

[13] BERS, L., JOHN, F., SCHECHTER, M., *Partial differential equations*, Lectures in Applied Mathematics, Interscience Publishers, New York - London - Sydney 1964.

[14] BORSUK, M., KONDRATIEV, V., *Elliptic boundary value problems of second order in piecewise smooth domains,* North-Holland Mathematical Library 69, Elsevier, Amsterdam 2006.

[15] BROWN, R. M., *The mixed problem for Laplace's equation in a class of Lipschitz domains*, Comm. Partial Differential Equations 19 (1994), no. 7-8, 1217-1233.

[16] BROWN, R., SHEN, Z., *Estimates for the Stokes operator in Lipschitz domains*, Indiana Univ. J. **44** (1995) 1183-1206.

[17] BROWN, R., PERRY, P., SHEN, Z., *On the dimension of the attractor for the nonhomogeneous Navier-Stokes equations in non-smooth domains*, Indiana Univ. Math. J. **49** (2000) 1, 81-112.

[18] BUFFA, A., COSTABEL, M., DAUGE, M., *Anisotropic regularity results for Laplace and Maxwell operators in a polyhedron*, C. R., Math., Acad. Sci. Paris 336 (2003) 7, 565-570.

[19] CARLEMAN, T, *Über das Neumann-Poincarésche Problem für ein Gebiet mit Ecken*, Inaugural-Dissertation, Uppsala 1916.

[20] CHKADUA, O., *Solvability and asymptotics of solutions of crack-type boundary-contact problems of the couple-stress elasticity.* Dedicated to the 100th birthday anniversary of Professor Victor Kupradze. Georgian Math. J. **10** (2003) 3, 427–465.

[21] CHKADUA, O., DUDUCHAVA, R., *Asymptotics of solutions to some boundary value problems of elasticity for bodies with cuspidal edges*, Mem. Differential Equations Math. Phys. **15** (1998) 29-58.

[22] CHIKOUCHE, W., MERCIER, D., NICAISE, S., *Regularity of the solution of some unilateral boundary value problems in polygonal and polyhedral domains*, Comm. Partial Differential Equations **28** (2003), 11-12, 1975–2001.

[23] CHIKOUCHE, W., MERCIER, D., NICAISE, S., *Regularity of the solution of some unilateral boundary value problems in polygonal and polyhedral domains*, Comm. Partial Differential Equations **29** (2004), 1-2, 43-70.

[24] CIANCHI, A., MAZ'YA, V. G., *Neumann problems and isocapacitary inequalities*, J. Math. Pures Appl. **(9) 89** (2008) 1, 71-105.

[25] Costabel, M., Dauge, M., *General edge asymptotics of solutions of second order elliptic boundary value problems I, II,* Proc. Roy. Soc. Edinburgh **123A** (1993) 109-155, 157-184.

[26] COSTABEL, M., DAUGE, M., *Singularités des équations de Maxwell dans un polyèdre*, C. R. Acad. Sci., Paris, Sér. I 324 (1997) 9, 1005-1010.

[27] COSTABEL, M., DAUGE, M., *Maxwell and Lamé eigenvalues on polyhedra*, Math. Meth. Appl. Sci. 22 (1999), no. 3, 243–258.

[28] COSTABEL, M., DAUGE, M., *Singularities of electromagnetic fields in polyhedral domains*, Arch. Rational Mech. Anal. **151** (2000) 221-276.

[29] COSTABEL, M., DAUGE, M., NICAISE, S., *Singularities of eddy current problems*, Math. Model. Numer. Anal. **37** (2003) 5, 807–831.

[30] DAUGE, M., *Elliptic boundary value problems in corner domains - smoothness and asymptotics of solutions,* Lecture Notes in Mathematics, Vol. 1341, Springer-Verlag Berlin 1988.

[31] DAUGE, M., *Problèmes de Neumann et de Dirichlet sur un polyèdre dans $\mathbb{R}^3$: regularité dans des espaces de Sobolev L_p*, C. R. Acad. Sci. Paris **307** Série I (1988) 27-32.

[32] DAUGE, M., *Higher order oblique derivatives problems on polyhedral domains*, Comm. Partial Differential Equations **14** (1989) 1193-1227.

[33] DAUGE, M., *Stationary Stokes and Navier-Stokes systems on two- and three-dimensional with corners. Part 1: Linearized equations,* SIAM J. Math. Anal. **20** (1989) 74-97.

[34] DAUGE, M., *Neumann and mixed problems on curvilinear polyedra*, Integr. Equat. Oper. Th. **15** (1992) 227-261.

[35] DAUGE, M., *Strongly elliptic problems near cuspidal points and edges*, Partial Differential Equations and Functional Analysis, 93-110. Progr. Nonlinear Differential Equations Appl. **22**, Birkhäuser Boston, MA, 1996.

[36] DAUGE, M., *Singularities of corner problems and problems of corner singularities*, ESAIM, Proc. Vol. 6, 19-40 (1999).

[37] DEURING, P., VON WAHL, W., *Strong solutions of the Navier-Stokes system in Lipschitz bounded domains*, Math. Nachr. **171** (1995) 111-148.

[38] DUDUCHAVA, R., NATROSHVILI, D., *Mixed crack type problem in anisotropic elasticity*, Math. Nachr. bf 191 (1998), 83–107.

[39] DUDUCHAVA, R., WENDLAND, W., *The Wiener-Hopf method for systems of pseudodifferential equations with an application to crack problems*, Integral Equations Operator Theory **23** (1995) 3, 294–335.

[40] DIMITROV, A., *On singularities in the solution of three-dimensional Stokes flow and incompressible elasticity problems with corners*, Int. J. Numer. Methods Eng. **60** (2004) 4, 773-801.

[41] DIMITROV, A., ANDRÄ, H., SCHNACK, E., *Efficient computation of corner singularities in 3d-elasticity*, Int. J. Numer. Meth. Eng. **52** (2001) 8, 805-827.

[42] DINDOŠ, M., MITREA, M., *The stationary Navier-Stokes system in nonsmooth manifolds: The Poisson problem in Lipschitz and C^1 domains*, Arch. Ration. Mech. Anal. **174** (2004) 1, 1-47.

[43] DUVAUT, G, LIONS, J.-L., *Les Inéquations en Mécanique et en Physique*, Dunod, Paris 1972.

[44] EBMEYER, C., *Mixed boundary value problems for nonlinear elliptic systems in n-dimensional Lipschitzian domains*, Z. Anal. Anwendungen **18** (1999) 3, 539-555.

[45] EBMEYER, C., FREHSE, J., *Mixed boundary value problems for nonlinear elliptic equations in multi-dimensional non-smooth domains*, Math. Nachr. **203** (1999) 47-74.

[46] EBMEYER, C., FREHSE, J., *Steady Navier-Stokes equations with mixed boundary value conditions in three-dimensional Lipschitzian domains,* Math. Ann. **319** (2001) 349-381.

[47] ELSCHNER, J., KAISER, H.-C., REHBERG, J., SCHMIDT, G., *$W^{1,q}$ regularity results for elliptic transmission problems on heterogeneous polyhedra*, Math. Models Methods Appl. Sci. **17** 4, 593-615.

[48] ELSCHNER, J., REHBERG, J., SCHMIDT, G., *Optimal regularity for elliptic transmission problems including C^1 interfaces*, Interfaces Free Bound **9**, 2, 233-252.

[49] ESCOBAR, J. F., *Uniqueness theorems on conformal deformation of metrics, Sobolev inequalities, and an eigenvalue estimate*, Comm. Pure Appl. Math. **43** (1990) 7, 857-883.

[50] ESKIN, G. I., *Boundary value problems for elliptic pseudodifferential equations*, Translations of Mathematical Monographs, 52, Amer. Math. Soc., Providence, R.I., 1981.

[51] ESKIN, G., *Boundary value problems for second-order elliptic equations in domains with corners*, Pseudodifferential operators and applications (Notre Dame, Ind., 1984), 105û131, Proc. Sympos. Pure Math., 43, Amer. Math. Soc., Providence, RI, 1985.

[52] FROMM, S. J., *Potential space estimates for Green potentials in convex domains*, Proc. Amer. Math. Soc. **119** (1993) 1, 225-233.

[53] FROMM, S. J., *Regularity of the Dirichlet problem in convex domains in the plane*, Michigan Math. J. **41** (1994) 3, 491–507.

[54] FROMM, S. J., JERISON, D., *Third derivative estimates for Dirichlet's problem in convex domains*. Duke Math. J. **73** (1994), 2, 257–268.

[55] GIRAULT, V., RAVIART, P.-A., *Finite element approximation of the Navier-Stokes equation*, Springer-Verlag 1979.

[56] GOHBERG, I., SIGAL, E. I., *Operator generalization of the theorem on the logarithmic residue and Rouché's theorem*, Mat. Sb.**84** (1971) 4, 607-629; English transl. in: Math. USSR Sb. **13** (1971).

[57] GRACHEV, N. V., MAZ'YA, V. G., *Estimates for the Neumann problem in a polyhedral cone*, Preprint LiTH-MAT-R-91-28, University of Linköping 1991.

[58] GRISVARD, P., *Behavior of the solutions of an elliptic boundary value problem in a polygonal or polyhedral domain*, Proc. 3rd Symp. Numer. Solut. Partial Differ. Equat., College Park 1975, 207-274 (1976).

[59] GRISVARD, P., *Behaviour of the solutions of an elliptic boundary value problem in a polygonal or polyhedral domain*, Synspade 1975, Ed. Hubbard 1975.

[60] GRISVARD, P., *Elliptic problems in nonsmooth domains*, Monographs and Studies in Mathematics **21** Pitman, Boston 1985.

[61] GRISVARD, P., *Edge behavior of the solution of an elliptic problem*, Math. Nachr. **132** (1987) 281-299.

[62] GRISVARD, P., *Singularities in boundary value problems*, Masson, Paris 1992.

[63] GRÜTER, M., WIDMAN, K.-O., *The Green function for uniformly elliptic equations*, Manuscripta Math. **37** (1982) 303-342.

[64] GUZMAN, J., LEYKEKHMAN, D., ROSSMANN, J., SCHATZ, A. H., *Hölder estimates for Green's functions on convex polyhedral domains and their applications to finite element methods*, Numerische Mathematik **112** (2009) 221-243

[65] HANNA, M. S., SMITH, K. T., *Some remarks on the Dirichlet problem in piecewise smooth domains*, Comm. Pure Appl. Math. **20** (1967) 575-593.

[66] JAKAB, T., MITREA, I., MITREA, M., *Sobolev estimates for the Green potential associated with the Robin-Laplacian in Lipschitz domains satisfying a uniform exteriour ball condition*, Sobolev Spaces in mathematics II, Applications in Analysis and Partial Differential Equations, International Mathematical Series, Vol. 9, Springer, 2008.

[67] JERISON, D. S., KENIG, C. E., *The inhomogeneous Dirichlet problem in Lipschitz domains*, J. Funct. Anal. **130** (1995) 161-219.

[68] KADLEC, J., *The regularity of the solution of the Poisson problem in a domain whose boundary is similar to that of a convex domain*, Czechoslovak. Math. J. **14 (89)** (1964) 386-393 (in Russian).

[69] KALEX, H.-U., *Zur Lösungstheorie für die Gleichungen der wärmegekoppelten Strömung zäher, incompressibler Fluide mit temperaturabhängigen Materialkoeffizienten in zweidimensionalen Gebieten mit nichtglatten Randdaten*, Thesis, Techn. Univ. Dresden 1990.

[70] KALEX, H.-U., *On the flow of a temperature dependent Bingham fluid in non-smooth bounded two-dimensional domains*, Z. Anal. Anwendungen **11** (1992) 4, 509-530.

[71] KENIG, C. E., *Harmonic analysis techniques for second order elliptic boundary value problems*, CBMS Regional Conference Series in Mathematics **83**, AMS, Providence, RI, 1994.

[72] KNEES, D., *On the regularity of weak solutions of quasilinear elliptic transmission problems on polygedral domains*, Z. Anal. Anwendungen **23** (2004) 509-546.

[73] KOMECH, A. I., *Elliptic boundary value problems on manifolds with piecewise smooth boundary,* Mat. Sb. **92** (1973) 89-134 (in Russian).

[74] KONDRAT'EV, V. A., *Boundary value problems for elliptic equations in domains with conical or angular points,* Trudy Moskov. Mat. Obshch. **16** (1967) 209-292, English transl. in: Trans. Moscow Math. Soc. **16** (1967) 227-313.

[75] KONDRAT'EV, V. A., *On the smoothness of solutions of the Dirichlet problem for elliptic equations of second order in a piecewise smooth domain,* Differ. Uravn. **6** (1970) 10, 1831-1843 (in Russian).

[76] KONDRAT'EV, V. A., *Singularities of the solutions of the Dirichlet problem for elliptic equations of second order in a neighborhood of an edge,* Differ. Uravn. **13** (1977) 11, 2026-2032 (in Russian).

[77] KOZLOV, V. A., *A theorem on a strong zero for an elliptic boundary value problem in a corner,* Mat. Sbornik **180** (1989), 6, 831–849; English transl. in: Math. USSR Sbornik **67** (1990) 1, 283–302.

[78] KOZLOV, V. A., *Behavior of solutions to the Dirichlet problem for elliptic systems in convex domains,* Comm. Partial Differential Equations **34** (2009) 24-51.

[79] KOZLOV, V. A., MAZ′YA, V. G., *Spectral properties of operator bundles generated by elliptic boundary value problems in a cone,* Funktsional. Anal. i Prilozhen. **22** (1988), 38-46; English transl. in: Funct. Anal. Appl. **22** (1988) 2, 114-121.

[80] KOZLOV, V. A., MAZ'YA, V. G., *On the spectrum of an operator pencil generated by the Neumann problem in a cone,* Algebra i Analiz **3** (1991), 111-131; English transl. in: St. Petersburg Math. J. **3** (1992) 2, 333-353.

[81] KOZLOV, V. A., MAZ'YA, V. G., *Differential equations with operator coefficients,* Springer Monographs in Mathematics, Springer Berlin - Heidelberg - New York 1999.

[82] KOZLOV, V. A., MAZ'YA, V. G., *Asymptotic formula for solutions to elliptic equations near the Lipschitz boundary,* Ann. Mat. Pura ed Appl. **184** (2005), 185-213.

[83] KOZLOV, V. A., MAZ'YA, V. G., ROSSMANN, J., *Conic singularities of solutions to problems in hydrodynamics of a viscous fluid with a free surface,* Math. Scand. **83** (1998) 103-141.

[84] KOZLOV, V. A., MAZ'YA, V. G., ROSSMANN, J., *Elliptic boundary value problems in domains with point singularities,* Mathematical Surveys and Monographs **52**, Amer. Math. Soc., Providence, Rhode Island 1997.

[85] KOZLOV, V. A., MAZ'YA, V. G., ROSSMANN, J., *Spectral problems associated with corner singularities of solutions to elliptic equations,* Mathematical Surveys and Monographs **85**, Amer. Math. Soc., Providence, Rhode Island 2001.

[86] KOZLOV, V. A., MAZ'YA, V. G., SCHWAB, C., *On singularities of solutions to the Dirichlet problem of hydrodynamics near the vertex of a cone,* J. Reine Angew. Math. **456** (1994) 65-97.

[87] KRATZ, W., *On the maximum modulus theorem for Stokes function,* Appl. Anal. **58** (1995) 293-302.

[88] KRESS, R., *Linear integral equations,* Springer-Verlag Berlin - Heidelberg - New York - London - Paris - Tokyo - Hong Kong 1989.

[89] LADYZHENSKAYA, O. A., *The mathematical theory of viscous incompressible flow,* Gordon and Breach, New York 1969.

[90] LEGUILLON, D., SANCHEZ-PALENCIA, E., *Computation of singular sulutions in elliptic problems and elasticity,* Masson, John Wiley, Paris 1987.

[91] LIONS, J.-L., MAGENES, E., *Problèmes aux limites non homogènes et applications,* Dunod, Paris 1968.

[92] LUBUMA, J. M.-S., NICAISE, S., *Dirichlet problems in polyhedral domains. I: Regularity of the solutions,* Math. Nachr. **168** (1994) 243-261.

[93] MAREMONTI, P., *On the Stokes equations: The maximum modulus theorem,* Math. Models Methods Appl. Sci. **10** (2000) 7, 1047-1072.

[94] MAREMONTI, P., RUSSO, R., *On existence and uniqueness of classical solutions to the stationary Navier-Stokes equations and to the traction problem of linear elastostatics,* Russo, R. (ed.) Classical problems in mechanics, Rome: Aracne Editrice. Quad. Mat. **1** (1997) 171-251.

[95] Mayboroda, S., Mitrea, M., *Sharp estimates for Green potentials on non-smooth domains,* Math. Res. Lett. **11** (2004) 4, 481-492.

[96] Mayboroda, S., Mitrea, M., *The Poisson problem for the Lamé system on low-dimensional Lipschitz domains,* Intgral Methods in Science an Engineering (2006) 137-160.

[97] MAYBORODA, S., MAZ'YA, V., *Boundedness of the Hessian of a biharmonic function in a convex domain*, Comm. Partial Diff. Eq. **33** (2008), no 8.

[98] MAYBORODA, S., MAZ'YA, V., *Boundedness of the gradient and Wiener test for the solutions of the biharmonic equation*, to appear in Invent. Math.

[99] MAZ'YA, V. G., *On Neumann's problem in domains with nonregular boundaries*, Sibirsk. Matem. Zh. **9** (1968) 6, 1322-1350.

[100] MAZ'YA, V. G., *On weak solutions of the Dirichlet and Neumann problems*, Trans. Moscow Math. Soc. **20** (1969) 135-172.

[101] MAZ'YA, V. G., *On the coercivity of the Dirichlet problem in a domain with nonregular boundary*, Izv. Vyssh. Uchebn. Zaved. Matem. **4** (1973) 54-76.

[102] MAZ'YA, V. G., *On a problem with oblique derivative for a domain of polyhedral type*, Dokl. Akad. Nauk SSSR **211** (1973) 1; English transl. in: Soviet Math Dokl. **14** (1973) 4.

[103] MAZ'YA, V. G., *On the oblique derivative problemin a domain with edges of different dimensions*, Vestnik Leningrad. Univ. Mat. Mekh. Astronom. **7** (1973) 34-39; English transl. in: Vestnik Leningrad. Univ. Math. **6** (1979) 148-154.

[104] MAZ'YA, V. G., *Sobolev spaces*, Springer 1985.

[105] MAZ'YA, V. G., *The Wiener test for higher order elliptic equations*, Duke Math. J. **115** (2002) 3, 479-512.

[106] MAZ'YA, V. G., *Analytic criteria in the qualitative spectral analysis of the Schrödinger operator, Spectral theory and mathematical physics* a Festschrift in honor of Barry Simon's 60th birthday, Proc. Sympos. Pure Math., Part 1, Amer. Math. Soc., Providence, RI , 76 , 2006,257-288 .

[107] MAZ'YA, V. G., *Boundedness of the gradient of the solution to the Neumann-Laplace problem in a convex domain*, C.R. Acad. Sci. Paris, Ser. I **347** (2009) 517-520.

[108] MAZ'YA, V. G., KRESIN, G. I., *On the maximum principle for strongly elliptic and parabolic second order systems with constant coefficients*, Mat. Sb. **125** (1984), 458-480; English transl. in: Math. USSR-Sb. **53** (1986), 457-479.

[109] MAZ'YA, V. G., MITREA, M., SHAPOSHNIKOVA, T., *The Dirichlet problem in Lipschitz domains for higher order elliptic systems with rough coefficients*, to appear.

[110] MAZ'YA, V. G., NAZAROV, S. A., PLAMENEVSKIĬ, B. A. *Asymptotische Theorie elliptischer Randwertaufgaben in singulär gestörten Gebieten,* Vol. 1, 2, Akademie-Verlag, Berlin 1991.

[111] MAZ'YA, V. G., NETRUSOV, YU. V., POBORCHI, *Boundary values of functions from Sobolev spaces in some non-Lipschitzian domains*, Algebra i Analiz **11** (1999) 1, 141-170; English transl. in: St. Petersburg Math. J. **11** (2000) 1, 107-128.

[112] MAZ'YA, V. G., PLAMENEVSKIĬ, B. A., *Problem with oblique derivative in a domain with piecewise smooth boundary,* Funktsional. Anal. i Prilozh. **5** (1971) 3, 102-103; English transl. in: Functional Anal. Appl. **5** (1971).

[113] MAZ'YA, V. G., PLAMENEVSKIĬ, B. A. *A certain class of manifolds with singularities*, Izv. Vyssh. Uchebn. Zaved. Matematika **11** (1972) 46-52 (in Russian).

[114] MAZ'YA, V. G., PLAMENEVSKIĬ, B. A., *On elliptic boundary value problems in domains with piecewise smooth boundary*, Trudy Simp. Mekh. Sploshnoi Sredy 1971, Metsniereba, Tbilissi 1973, 171-182 (in Russian).

[115] MAZ'YA, V. G., PLAMENEVSKIĬ, B. A., *On boundary value problems for a second order elliptic equation in a domain with edges*, Vestnik Leningrad. Univ. Mat. Mekh. Astron. **1** (1975) 102-108; English transl. in: Vestnik Leningrad Univ. Math. **8** (1980) 99-106.

[116] MAZ'YA, V. G., PLAMENEVSKIĬ, B. A., *Elliptic boundary value problems on manifolds with singularities*, Probl. Mat. Anal. **6** 85-142, Izdat. Leningrad. Univ. 1977.

[117] MAZ'YA, V. G., PLAMENEVSKIĬ, B. A., *Estimates in L_p and Hölder classes and the Miranda-Agmon maximum principle for solutions of elliptic boundary value problems in domains with singular points on the boundary,* Math. Nachr. **81** (1978) 25-82, English transl. in: Amer. Math. Soc. Transl., Vol. 123 (1984) 1-56.

[118] MAZ'YA, V. G., PLAMENEVSKIĬ, B. A., *L_p estimates for solutions of elliptic boundary value problems in domains with edges,* Tr. Moskov. Mat. Obs. **37** (1978) 49-93; English transl. in: Trans. Moscow. Math. Soc. **37** (1980) 49-97.

[119] MAZ'YA, V. G., PLAMENEVSKIĬ, B. A., *Estimates of the Green function and Schauder estimates for solutions of elliptic boundary value problems in a dihedral angle*, Sibirsk. Mat. Zh. **19** (1978) 5, 1065-1082 (in Russian).

[120] MAZ'YA, V. G., PLAMENEVSKIĬ, B. A., *Schauder estimates of solutions of elliptic boundary value problems in domains with edges on the boundary,* Partial Differential Equations (Proc. Sem. S. L. Sobolev, 1978, No. 2) Inst. Mat. Sibirsk. Otdel. Akad. Nauk SSSR, Novosibirsk, 1978, 69-102; English transl. in: Amer. Math. Soc. Transl. **123** (1984) 141-169.

[121] MAZ'YA, V. G., PLAMENEVSKIĬ, B. A., *On the asymptotics of the fundamental solutions of elliptic boundary value problems in regions with conical points,* Probl. Mat. Anal. **7** (1979) 100-145, English transl. in: Sel. Math. Sov. **4** (1985) 4, 363-397.

[122] MAZ'YA, V. G., PLAMENEVSKIĬ, B. A., *On the maximum principle for the biharmonic equation in a domain with conical points*, Izv. Vyssh. Uchebn. Zaved. **2** (1981) 52-59 (in Russian).

[123] MAZ'YA, V. G., PLAMENEVSKIĬ, B. A., *On properties of solutions of three-dimensional problems of elasticity theory and hydrodynamics in domains with isolated singular points* (in Russian), Dinamika Sploshn. Sredy **50** (1981) 99-121; English transl. in: Amer. Math. Soc. Transl. **123** (1984) 109-123.

[124] MAZ'YA, V. G., PLAMENEVSKIĬ, B. A., *The first boundary value problem for classical equations of mathematical physics in domains with piecewise smooth boundaries*, Part 1: Z. Anal. Anwendungen **2** (1983) 4, 335-359, Part 2: Z. Anal. Anwendungen **2** (1983)6, 523-551 (in Russian).

[125] MAZ'YA, V. G., PLAMENEVSKIĬ, B. A., STUPYALIS, L. I., *The three-dimensional problem of steady-state motion of a fluid with a free surface*, Diff. Uravn. i Primen.-Trudy Sem. Protsessy Optimal. Upravl. **23** (1979); English transl. in: Amer. Math. Soc. Transl. **123** (1984) 171-268.

[126] MAZ'YA, V. G, POBORCHI, S., *Differentiable Functions on Bad Domains*, World Scientific, 1997, XVIII.

[127] MAZ'YA, V. G., ROSSMANN, J., *Über die Lösbarkeit und die Asymptotik der Lösungen elliptischer Randwertaufgaben in Gebieten mit Kanten I–III*, Preprint P-MATH 07/84, 30/84, 31/84 Institut für Math., Akademie der Wissenschaften der DDR, Berlin 1984.

[128] MAZ'YA, V. G., ROSSMANN, J., *Über die Asymptotik der Lösungen elliptischer Randwertaufgaben in der Umgebung von Kanten*, Math. Nachr. **138** (1988) 27-53.

[129] MAZ'YA, V. G., ROSSMANN, J., *On the Agmon-Miranda maximum principle for solutions of elliptic equations in polyhedral and polygonal domains*, Ann. Global Anal. Geom. **9** (1991) 3, 253-303.

[130] MAZ'YA, V. G., ROSSMANN, J., *On the Agmon-Miranda maximum principle for solutions of strongly elliptic equations in domains of $\mathbb{R}^n$ with conical ponts*, Ann. Global Anal. Geom. **10** (1992) 125-150.

[131] MAZ'YA, V. G., ROSSMANN, J., *On a problem of Babuška (Stable asymptotics of solutions to the Dirichlet problem for elliptic equations of second order in domains with angular points*, Math. Nachr. **155** (1992) 199-220.

[132] MAZ'YA, V. G., ROSSMANN, J., *On the behaviour of solutions to the Dirichlet problem for second order elliptic equations near edges and polyhedral vertices with critical angles*, Z. Anal. Anwendungen **13** (1994) 1, 19-47.

[133] MAZ'YA, V. G., ROSSMANN, J., *Point estimates for Green's matrix to boundary value problems for second order elliptic systems in a polyhedral cone*, Z. Angew. Math. Mech. **82** (2002) 5, 291-316.

[134] MAZ'YA, V. G., ROSSMANN, J., *Weighted L_p estimates of solutions of boundary value problems for second order elliptic systems in polyhedral domains*, Z. Angew. Math. Mech. **83** (2003) 7, 435-467.

[135] MAZ'YA, V. G., ROSSMANN, J., *Schauder estimates for solutions to boundary value problems for second order elliptic systems in polyhedral domains*, Applicable Analysis **83** (2004) 3, 271-308.

[136] MAZ'YA, V. G., ROSSMANN, J., *Pointwise estimates for Green's kernel of a mixed boundary value problem to the Stokes system in a polyhedral cone*, Math. Nachr. **278** (2005) 15, 1766-1810.

[137] Maz'ya, V. G., Rossmann, J., *Schauder estimates for solutions to a mixed boundary value problem for Stokes system in polyhedral domains*, Math. Meth. Appl. Sci. **29** (2006) 9, 965-1017.

[138] MAZ'YA, V. G., ROSSMANN, J., *L_p estimates of solutions to mixed boundary value problems for the Stokes system in polyhedral domains*, Math. Nachr. **280** (2007) 7, 751-793.

[139] MAZ'YA, V. G., ROSSMANN, J., *Mixed boundary value problems for the Navier-Stokes system in polyhedral domains*, Arch. Rational Mech. Anal. **194** (2009) 2, 669-712.

[140] MAZ'YA, V. G., ROSSMANN, J., *A maximum modulus estimate for solutions of the Navier-Stokes system in domains of polyhedral type*, Math. Nachr. **282** (2009) 3, 459-469.

[141] MAZ'YA, V. G., SHAPOSHNIKOVA, T., *Theory of multipliers in spaces of differentiable functions*, Pitman 1985.

[142] MAZ'YA, V. G., SHAPOSHNIKOVA, T., *Theory of Sobolev Multipliers with Applications to Differential and Integral Operators*, Grundlehren der Mathematischen Wissenschaften, vol. 337, Springer 2008.

[143] MAZ'YA, V. G., SHAPOSHNIKOVA, T., *Higher regularity in the classical layer potential theory for Lipschitz domains*, to appear.

[144] MIRANDA, C., *Formule die maggiorazione e teorema di esistenza per le funzioni biarmoniche in due variabili*, Giorn. Math. Battaglini bf 78 (1949), 97-118.

[145] MIRANDA, C., *Teorema del massimo modulo u teorema di esistenza e di unicita per il problema Dirichlet relative alle equazioni ellitiche in due variabili*, Ann. Mat. Pura Appl. **46** (1958), 265-311.

[146] MITREA, M., *Boundary value problems for Dirac operators and Maxwell's equations in non-smooth domains*, Math. Meth. Appl. Sci. **25** (2002) no. 16-18, 1355-1369.

[147] MITREA, M., MONNIAUX, S., *The regularity of the Stokes operator and the Fujita-Kato approach to the Navier-Stokes initial value prpblem in Lipschitz domains*, J. Funct. Anal. **254** (2008) 6, 1522-1574.

[148] MITREA, M., TAYLOR, M., *Potential theory on Lipschitz domains in Riemannian manifolds: Sobolev-Besov space results and the Poisson problem*, J. Funct. Anal. **176** (2000) 1-79.

[149] MITREA, M., TAYLOR, M., *Navier-Stokes equations on Lipschitz domains in Riemannian manifolds*, Math. Ann.**321** (2001) 4, 955-987.

[150] MITREA, M., TAYLOR, M., *Sobolev and Besov space estimates for solutions to second order PDE on Lipschitz domains in manifolds with Dini or Hölder continuous metric tensors*, Comm. Partial Differential Equations **30** (2005) 1-37.

[151] MITREA, M., TAYLOR, M., *The Poisson problem in weighted Sobolev spaces on Lipschitz domains*, Indiana Univ. Math. J. **55** (2006) 1063-1089.

[152] MORREY, C. B., *Multiple integrals in the calculus of variations*, Springer Berlin - Heidelberg - New York 1966.

[153] NAUMANN, J. *On a maximum principle for weak solutions of the stationary Stokes system*, Ann. Sc. Norm. Super. Pisa, Cl. Sci. IV. Ser. 15, No. 1 (1988) 149-168.

[154] NAZAIKINSKIĬ, V. E., SAVIN, A. YU, SCHULZE, B.-W., STERNIN, B. YU., *Elliptic theory on singular manifolds*, Differential and Integral Equations and Their Applications, 7. Chapman Hall/CRC, Boca Raton, FL, 2006.

[155] NAZAROV, S. A., *Estimates near the edge of the solution of the Neumann problem for an elliptic system*, Vestnik Leningrad. Univ. **1** (1988) 1, 37-42.

[156] NAZAROV, S. A., *The polynomial property of self-adjoint elliptic boundary value problems and an algebraic description of their attributes*, Uspekhi Mat. Nauk **54** (1999) 5, 77-142; English transl. in: Russian Mat. Surveys **54** (1999) 5, 947-1014.

[157] NAZAROV, S. A., PLAMENEVSKIĬ, B. A., *Neumann problem for self-adjoint systems in a domain with piecewise smooth boundary*, Trudy Leningrad. Mat. Obshch. **1** (1991) 174-211.

[158] NAZAROV, S. A., PLAMENEVSKIĬ, B. A., *Elliptic problems with the radiation condition on edges of the boundary*, Mat. Sb. **183** (1992) 13-44; English transl. in: Russian Acad. Sci. Sb. Math. **77** (1994) 1, 149-176.

[159] NAZAROV, S. A., PLAMENEVSKIĬ, B. A., *Self-adjoint elliptic problems with the radiation condition on edges of the boundary*, Algebra i Analiz **4** (1992) 196-225; English transl. in: St. Petersburg Math. J. **4** (1993) 3, 569-594.

[160] NAZAROV, S. A., PLAMENEVSKIĬ, B. A., *Elliptic problems in domains with piecewise smooth boundaries,* De Gruyter Expositions in Mathematics **13**, Berlin-New York 1994.

[161] NAZAROV, S. A., SWEERS, G., *Boundary value problems for the biharmonic equation and the iterated Laplacian in a three-dimensional domain with an edge*, Zap. Nauchn. Sem. S.-Peterburg. Otdel. Mat. Inst. Steklov. (POMI) 336 (2006), Kraev. Zadachi Mat. Fiz. i Smezh. Vopr. Teor. Funkts. 37, 153–198, 276–277; English transl. in: J. Math. Sci. (N. Y.) 143 (2007), no. 2, 2936–2960.

[162] NICAISE, S., *Polygonal interface problems,* Verlag Peter Lang GmbH, Frankfurt/Main 1993.

[163] NICAISE, S., *Regularity of the weak solution of the Lamé system in nonsmooth domains*, Bandle, C. (ed.) et al., Progress in partial differential equations: calculus of variations, applications. Pitman Res. Notes Math. Ser. 267 (1993) 272-284.

[164] NICAISE, S., *Regularity of the solutions of elliptic systems in polyhedral domains*, Bull. Belg. Math. Soc. **4** (1997) 411-429.

[165] NIKISHKIN, V. A., *Singularities of the solutions of the Dirichlet problem for elliptic differential equations of second order in a neighborhood of an edge*, Vestnik Moskov. Univ. Ser. Mat. **34** (1979) 51-62 (in Russian).

[166] ODQUIST, F. K. G., *Über die Randwertaufgaben der Hydrodynamik zäher Flüssigkeiten*, Math. Zeitschrift **32** (1930) 329-375.

[167] ORLT, M., SÄNDIG, A.-M., *Regularity of viscous Navier-Stokes flows in nonsmooth domains*, Costabel, M. (ed.) et al., Boundary value problems and integral equations in nonsmooth domains, Proceedings of the conference at the CIRM, Luminy, France, May 3-7, 1993, Marcel Dekker, New York, Lect. Notes Pure Appl. Math. **167** (1995) 185-201.

[168] PIPHER, J., VERCHOTA, G. C., *The Dirichlet problem in L_p for the biharmonic equation on Lipschitz domains*, Amer. J. Math. **114** (1992) 923-972.

[169] PIPHER, J., VERCHOTA, G. C., *A maximum principle for biharmonic functions in Lipschitz and C^1 domains*, Comment. Math. Helv. **68** (1993) 3, 385-414.

[170] PIPHER, J., VERCHOTA, G. C., *Maximum principles for the polyharmonic equation on Lipschitz domains*, Potential Analysis **4** (1995) 6, 615-636.

[171] PIPHER, J., VERCHOTA, G. C., *Dilation invariant estimates and the boundary Gårding inequality for higher order elliptic operators*, Ann. of Math. **142** (1995) 1, 1-38.

[172] PLAMENEVSKIĬ, B. A., *Ellipic boundary value problems in domains with piecewise smooth boundary*, Partial Differential Equations IX, Encyclopedia of Mathematical Sciences, Vol. 79 (M.S. Agranovich, Yu.V. Egorov, M.A. Shubin Eds.) Springer Berlin - Heidelberg 1997.

[173] RABINOVICH, V., SCHULZE, B.-W., TARKHANOV, N., *Elliptic boundary value problems in domains with intersecting cuspidal edges*, Dokl. Akad. Nauk **375** (2000) 6, 737-741.

[174] RABINOVICH, V., SCHULZE, B.-W., TARKHANOV, N., *Boundary value problems in oscillating cuspidal wedges*, Rocky Mountain J. Math. 34 (2004) 4, 1399–1471.

[175] REMPEL, S., SCHULZE, B.-W., *Branching of asymptotics for elliptic operators on manifolds with edges*, Partial differential equations, Banach Cent. Publ. **19** (1987) 231-236.

[176] ROITBERG, YA., *Elliptic boundary value problems in the spaces of distributions*, Kluwer Academic Publihers Dordrecht-Boston-London 1996.

[177] ROSSMANN, J., *The asymptotics of the solutions of linear elliptic variational problems in domains with edges,* Z. Anal. Anwendungen **9** (1990) 3, 565-575.

[178] ROSSMANN, J., *On two classes of weighted Sobolev-Slobodetskiĭ spaces in a dihedral angle*, Partial Differential Equations, Banach Center Publications **27** (1992) 399-424.

[179] ROSSMANN, J., *Hölder estimates for Green's matrix of the Stokes system in convex polyhedra*, Around the Research of Vladimir Maz'ya II. Partial Differential Equations, pp. 315-336, Springer (2010).

[180] SHEN, Z., *A note on the Dirichlet problem for the Stokes system in Lipschitz domains*, Proc. Amer. Math. Soc. **123** (1995) 801-811.

[181] SHEN, Z., *Necessary and sufficient conditions for the solvability of the L_p Dirichlet problem on Lipschitz domains*, Math. Ann. **336** (2006) 697-725.

[182] SCHULZE, B.-W., *Abschätzungen in Normen gleichmäßiger Konvergenz für elliptische Randwert-Probleme*, Math. Nachr. **67** (1975) 303-315.

[183] SCHULZE, B.-W., *On a priori estimates in uniform norms for strongly elliptic systems*, Sibirsk. Mat. Zh. **16** (1975) 2, 384-393 (in Russian).

[184] SCHULZE, B.-W., *Regularity with continuous and branching asymptotics for elliptic operators on manifolds with edges*, Integral Equations Operator Theory **11** (1988) 557-602.

[185] SCHULZE, B.-W., *Elliptic regularity with continuous and branching edge asymptotics*, Internat. Ser. Numer. Math. **94** (1990) 275-284.

[186] SCHULZE, B.-W., *Pseudo-differential operators on manifolds with singularities,* Studies in Mathematics and Its Applications. 24. North-Holland, Amsterdam etc., 1991.

[187] SCHULZE, B.-W., TARKHANOV, N. *Ellipticity and parametrices on domains with cuspidal edges*, Contemp. Math. **242** Amer. Math. Soc., Providence 1999, 217-256.

[188] SOLONNIKOV, V. A., *On Green's matrices for elliptic boundary. II*, Proc. Steklov Inst. Math. **116** (1971) 187-226.

[189] SOLONNIKOV, V. A., *The solvability of the three-dimensional problem with a free boundary for a system o stationary Navier-Stokes equations*, Zap. Nauchn. Sem. Leningrad. Otd. Math. Inst. **84** (1979) 252-285.

[190] SOLONNIKOV, V. A., *Estimates of solutions of the Neumann problem for a second order elliptic equation in domains with erdges on the boundary*, Preprint P-4-83, Leningrad. Otdel. Mat. Inst. Steklov (LOMI), 1983.

[191] SOLONNIKOV, V. A., *On a maximum modulus estimate of the solution of the stationary problem for the Navier-Stokes equations*, Zap. Nauchn. Sem. S.-Petersburg Otdel. Mat. Inst. Steklov (POMI) **249** (1997) 294-302; English transl. in: J. Math. Sci. (New York) **101** (2000) 5, 3563-3569.

[192] STEIN, E. M., *Singular integrals an differentiability properties of functions*, Princeton University Press, Princeton, New Jersey 1970.

[193] STUPELIS, L., *Navier-Stokes equations in irregular domains*, Mathematics and its Applications, 326. Kluwer Academic Publishers Group, Dordrecht, 1995.

[194] TEMAM, R., *Navier-Stokes equations, theory and numerical analysis*, North-Holland, Amsterdam 1977

[195] TRIEBEL, H., *Interpolation theory, function spaces, differential operators*, Deutscher Verlag der Wissenschaften Berlin 1978.

[196] VENOUZIOU, M., VERCHOTA, G. C., *The mixed problem for harmonic functions in polyhedra of* $\mathbb{R}^3$, Preprint math.AP 0803.0957v1 arXiv.org (2008).

[197] VERCHOTA, G., *The use of Rellich identities on certain nongraph boundaries*, Harmonic analysis and boundary value problems (Fayetteville, AR, 2000), 127–138, Contemp. Math., 277, Amer. Math. Soc., Providence, RI, 2001.

[198] VERCHOTA, G. C., *The biharmonic Neumann problem in Lipschitz domains*, Acta Math. **194** (2005) 217-279.

[199] VERCHOTA, G., *Counterexamples and uniqueness for* $L_p(\partial\Omega)$ *oblique derivative problems*, J. Funct. Anal. **245** (2007) 2, 413-437.

[200] VERCHOTA, G. C., VOGEL, A. L., *A multidirectional Dirichlet problem*, J. Geom. Anal. **13** (2003) 3, 495-520.

[201] VERCHOTA, G. C., VOGEL, A. L., *The multidirectional Neumann problem in* $\mathbb{R}^4$, Math. Ann. **335** (2006) 3, 571-644.

[202] VON PETERSDORFF, T., STEPHAN, E. P., *Decomposition in edge and corner singularities for the solution of of the Dirichlet problem for the Laplacian in a polyhedron*, Math. Nachr. **149** (1990) 71-104.

[203] WLOKA, J. *Partial differential equations*, Cambridge University Press 1987.

[204] ZAJACKOWSKI, W., SOLONNIKOV, V. A., *On the Neumann problem for second order elliptic equations in domains with edges on the boundary*, Zap. Nauchn. Sem. Leningrad. Otdel. Mat. Inst. Steklov. (LOMI) **127** (1983) 7-48 (in Russian).

List of Symbols

Chapter 1

Chapter 2

Chapter 3

$A_k(\lambda)$ operator pencil, 96
$\delta_+^{(k)}, \delta_-^{(k)}$ positive real numbers, 96
$V_\delta^{l,p}(\Omega)$ weighted Sobolev space, 96
$\overset{\circ}{V}{}_\delta^{l,p}(\Omega)$ weighted Sobolev space, 96
$\mathcal{L}(\lambda)$ differential operator, 97
$\mathfrak{A}(\lambda)$ operator pencil, 97
$L^+(D_x)$ formally adjoint operator, 97
$\mathfrak{A}^+(\lambda)$ operator pencil, 97
$\mathfrak{A}_{l,\delta}(\lambda)$ operator pencil, 97
$\mathcal{M}_{\rho\to\lambda}$ Mellin transform, 103
I_ℓ identity matrix, 110
$G(x,\xi)$ Green's matrix, 110
κ a fixed real number, 111
$\delta_+(x)$ function in $\mathcal{K}$, 112
Λ_+, Λ_- real numbers, 115, 123
$N_{\beta,\delta}^{l,\sigma}(\mathcal{K})$ weighted Hölder space, 130
$N_{\beta,\delta}^{l,\sigma}(\Gamma_j)$ weighted Hölder space, 130
$L_{\beta,\delta}^\infty(\mathcal{K})$ weighted L_∞ space, 133

Chapter 4

$L(x,D_x)$ differential operator, 141
$\mathcal{G}$ domain of polyhedral type, 142
$\Gamma_1,\dots,\Gamma_N$ faces of $\mathcal{G}$, 142
$M_1,\dots,M_d$ edges of $\mathcal{G}$, 142
$x^{(1)},\dots,x^{(d')}$ vertices of $\mathcal{G}$, 142
$\mathcal{S}$ set of all vertices and edge points, 142
$r_k(x)$ distance from M_k, 142
$\rho_j(x)$ distance from $x^{(j)}$, 142
$r(x)$ distance from $\mathcal{S}$, 142
X_j set of all indices k such that $x^{(j)}\in\overline{M}_k$, 142
$V_{\beta,\delta}^{l,p}(\mathcal{G})$ weighted Sobolev space, 142
$\overset{\circ}{V}{}_\delta^{l,p}(\mathcal{G})$ weighted Sobolev space, 143
$V_{\beta,\delta}^{-l,p}(\mathcal{G})$ weighted Sobolev space, 143
$(\cdot,\cdot)_\mathcal{G}$ scalar product in $L_2(\mathcal{G})$, 143
$V_{\beta,\delta}^{l-1/p,p}(\Gamma_j)$ trace space, 144
$A_\xi(\lambda)$ operator pencil, 146
$\delta_+^{(k)}, \delta_-^{(k)}$ positive real numbers, 146
$\mathfrak{A}_i(\lambda)$ operator pencil, 146
$\mathcal{A}_{l,p,\beta,\delta}$ operator of the boundary value problem, 146
$N_{\beta,\delta}^{l,\sigma}(\mathcal{G})$ weighted Hölder space, 151
$N_{\beta,\delta}^{l,\sigma}(\Gamma_j)$ weighted Hölder space, 151

Chapter 5

$G(x,\xi)$ Green's matrix, 162
$L^+(x,D_x)$ formally adjoint operator, 162
Λ_j^-, Λ_j^+ real numbers, 163
$L_{\beta,\delta}^\infty(\mathcal{G})$ weighted L_∞ space, 175
$V_{\beta,\delta}^{l,\infty}(\mathcal{G})$ weighted Sobolev space, 175
$\mathcal{V}_{\beta,\delta}^{l,\infty}(\mathcal{G})$ weighted Sobolev space, 175
$\mathcal{V}_{\beta,\delta}^{m-1,\infty}(\partial\mathcal{G})$ weighted Sobolev space, 175
$S_{j,k}(x,D_x)$, $T_{j,k}(x,D_x)$ differential operators on Γ_j, 176
$\mathcal{G}$ domain in $\mathbb{R}^N$ with conical points, 188
$\mathcal{S}$ set of the vertices, 188
$\rho_j(x)$ distance from the vertex $x^{(j)}$, 188
$\mathfrak{A}_j(\lambda)$ operator pencil, 189
Λ_j^+, Λ_j^- real numbers, 189
$V_\beta^{l,\infty}(\mathcal{G})$ weighted Sobolev space, 190
$\mathcal{V}_\beta^{l,\infty}(\mathcal{G})$ weighted Sobolev space, 190
$\mathcal{V}_\beta^{m-1,\infty}(\partial\mathcal{G})$ weighted Sobolev space, 190
$S_k(x,D_x)$, $T_k(x,D_x)$ differential operators on $\partial\mathcal{G}\backslash\mathcal{S}$, 191
$H^{(i)}(x,\xi)$ column of the adjoint Green's matrix, 191

Chapter 6

$\mathcal{D}$ dihedron, 213
$L(D_x)$ differential operator, 214
$A_{j,k}$ coefficients of $L(D_x)$, 214
$b_\mathcal{D}(\cdot,\cdot)$ sesquilinear form, 214
$N^\pm(D_x)$ conormal derivative on $\Gamma^\pm$, 215
$n^\pm$ outer unit normal to $\Gamma^\pm$, 215
$d^\pm$ numbers of the set $\{0,1\}$, 215
$B^\pm(D_x)$ differential operator in the boundary conditions, 215
$L^{1,2}(\mathcal{D})$ function space, 215
$\mathcal{H}_\mathcal{D}$ subspace of $L^{1,2}(\mathcal{D})$, 215
$\mathcal{H}_\mathcal{D}^*$ dual space of $\mathcal{H}_\mathcal{D}$, 215
$(\cdot,\cdot)_\mathcal{D}$ scalar product in $L_2(\mathcal{D})^\ell$, 215
$\mathcal{L}_0(\lambda)$, $\mathcal{B}_0^\pm(\lambda)$ parameter-depending differential operators, 216
$x'=(x_1,x_2)$ 217

Chapter 7

Chapter 8

Chapter 9

Chapter 10

$\mathcal{H}_{\mathcal{K}}$ subspace of $V_0^{1,2}(\mathcal{K})$, 444
$b_{\mathcal{K}}(\cdot,\cdot)$ bilinear form, 444
$A_k(\lambda)$ operator pencil, 446
$\delta_+^{(k)}$ positive real number, 446
$\mu_+^{(k)}$ positive real number, 446
$\mathcal{H}_\Omega$ subspace of $W^{1,2}(\Omega)^3$, 446
$\mathfrak{A}(\lambda)$ operator pencil, 446
$\mathfrak{A}_\delta(\lambda)$ restriction of $\mathfrak{A}(\lambda)$, 446
$\mathcal{H}_\beta$ subspace of $V_\beta^{1,2}(\mathcal{K})^3$, 453
$\mathcal{A}_\beta$ operator of the boundary value problem, 453
$G(x,\xi)$ Green's matrix, 462
δ_x positive real number, 467
μ_x positive real number, 467
κ a fixed real number, 468
Λ_+, Λ_- real numbers, 468, 471, 480, 493, 507
$\mathcal{H}_{s,\beta,\delta}$ subspace of $W_{\beta,\delta}^{1,s}(\mathcal{K})$, 479
$C_{\beta,\delta}^{-1,\sigma}(\mathcal{K})$ function space, 507

Chapter 11

$\mathcal{G}$ domain of polyhedral type, 520
Γ_j faces of $\mathcal{G}$, 520
M_k edges of $\mathcal{G}$, 520
$x^{(i)}$ vertices of $\mathcal{G}$, 520
S_j, N_j operators in the boundary conditions, 520, 528
d_j natural numbers, 520
$\mathcal{H}$ subspace of $W^{1,2}(\mathcal{G})^3$, 521
$\mathcal{H}_0$ subspace of $\mathcal{H}$, 521
$L_{\mathcal{H}}$ subspace of $\mathcal{H}$, 521
$b(\cdot,\cdot)$ bilinear form, 521, 528
$\overset{\circ}{L}_2(\mathcal{G})$ subspace of $L_2(\mathcal{G})$, 523
$\theta(\xi)$ angle at the edge point ξ, 524
$A_\xi(\lambda)$ operator pencil, 524
$\delta_+(\xi)$ positive real number, 524
$\delta_+^{(k)}$ positive real number, 524
$\mu_+(\xi)$ positive real number, 524
$\mu_+^{(k)}$ positive real number, 524
I_j set of indices, 524
$\mathcal{K}_j$ cone, 524
$\mathfrak{A}_j(\lambda)$ operator pencil, 524
$\mathcal{H}_{s,\beta,\delta;\mathcal{G}}$ subspace of $V_{\beta,\delta}^{1,s}(\mathcal{G})^3$, 525
X_j set of indices, 532
θ_k angle at the edge M_k, 539
$G(x,\xi)$ Green's matrix, 544
Λ_j^+ positive real number, 545
Λ'_ν positive real number, 556
$\tilde{\Lambda}_\nu = \min(2,\Lambda'_\nu)$, 556
$\tilde{\mu}_k = \min(2,\mu_+^{(k)})$, 556
$\tilde{\mu} = \min(\tilde{\mu}_1,\dots,\tilde{\mu}_d)$, 559
$d(x) = \mathrm{dist}(x,\partial\mathcal{G})$, 571

List of Examples

Index